Applied Superconductivity 1999

Applied Superconductivity 1999

Proceedings of EUCAS 1999, the Fourth European Conference
on Applied Superconductivity, held in Sitges, Spain
14-17 September 1999

Volume 2. Electronic Applications

Edited by X Obradors, F Sandiumenge and J Fontcuberta

Institute of Physics Conference Series Number 167
Institute of Physics Publishing, Bristol and Philadelphia

0305-2346/00 $30.00+.00

CODEN IPHSAC 167 1-811 (2000)

British Library Cataloguing in Publication Data

A catalogue record for this book is available from the British Library.

ISBN 0 7503 0745 5 Vol. 1
    0 7503 0746 3 Vol. 2
    0 7503 0694 7 (2 vol. set)

Library of Congress Cataloging-in-Publication Data are available

Published by Institute of Physics Publishing, wholly owned by The Institute of Physics, London
Institute of Physics Publishing, Dirac House, Temple Back, Bristol BS1 6BE, UK
US Office: Institute of Physics Publishing, The Public Ledger Building, Suite 1035, 150 South Independence Mall West, Philadelphia, PA 19106, USA

Printed in the UK by J W Arrowsmith Ltd, Bristol

Contents

VOLUME 1. LARGE SCALE APPLICATIONS

Wires, Tapes and Coated Conductors

Flux pinning

Fault current limiters and related materials

Motors, bearings and levitation

Cables, transformers and generators

Systems aspects, high fields and current leads

VOLUME 2. ELECTRONIC APPLICATIONS

Materials related to electronic applications

Josephson junctions

SQUIDS and SQUIDS applications

Mixers and detectors

Oscillators and volt standards

Digital applications

Systems aspects related to electronic applications

Inst. Phys. Conf. Ser. No 167
Paper presented at Applied Superconductivity, Spain, 14-17 September 1999

1

Series production of large area $YBa_2Cu_3O_7$ – films for microwave and electrical power applications

W Prusseit, Furtner S, Nemetschek R

THEVA Dünnschichttechnik GmbH, Hauptstr. 1b, D-85386 Eching-Dietersheim, Germany

ABSTRACT: The series production of YBCO films on a daily basis is making high demands on the reliability and reproducibility of the deposition technique on the one hand, but also flexibility in the film material and substrate choice on the other hand. Beyond that, a strict quality management and customer support are absolute necessities to serve a market wich is getting closer towards real technical applications and where engineers are mainly concerned with device layout and performance instead of material issues. These often overlooked points are even more time consuming and challenging than the deposition process itself.

1.   INTRODUCTION

During the past ten years of thin films development many groups have set up laboratory scale deposition systems to produce high quality YBCO films. For the coating of substrates larger than 2 inches in diameter there are much less facilities worldwide but also for such areas nearly every technique ranging from PLD, sputtering, CVD to evaporation has been demonstrated to yield excellent films. However, when it comes to commercial production on an everyday basis this is quite a different story. Economic efficiency, throughput, and reliability become major issues and a real mature deposition technique is required. On the one hand it should guarantee high reproducibility, uniformity, and quality of the resulting films. On the other hand the numerous applications require flexibility with respect to the substrate material and geometry as well as the film material itself. In some respect reproducibility and flexibility are even contrasting, since the reproducibility is easier achieved, if the same deposition is repeated every day without changes, whereas flexibility implies adaptation of the deposition process to varying requirements.

However, even beyond the deposition itself it takes a lot more until an industrial customer really dares to start a real product development depending on external film supply. Quality management ranging from the substrates and materials used for deposition to the inspection of the outgoing films is an absolute necessity. As clients changing from physicists to engineers the question of how to prepare the material becomes of secondary interest. Instead, service and support covering all material related issues get extremely important. As a consequence, the responsibility of the film material supplier is extended to the selection of the appropriate substrate, realization of photolithographic masks, patterning, and dicing.

2. THERMAL EVAPORATION PROCESS

For many years the YBCO deposition technique developed by Berberich (1993) at the Technical University of Munich has demonstrated its capability to yield excellent large area films. From the point of economy it is rated the best candidate for commercial film production. It has turned out as a very reproducible method and due to its elaborate heater design it is very flexible as well, since it can accommodate even different substrate materials and sizes in the same deposition run and without changing the process parameters.

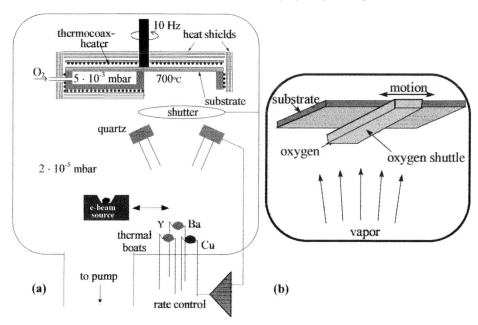

Fig. 1 YBCO - deposition principle with turntable substrate holder (a) and oxygen shuttle (b)

The basic principle of the original turntable substrate holder is depicted in Fig. 1(a). The metals which build up the YBCO lattice are evaporated from resistively heated boat sources in high vacuum ambient. The composition of the film is controlled by quartz crystal monitors individually focussed at the appropriate boat source. The central idea ,however, is the spatial separation between oxidation and metal deposition. In the classic heater scheme this is realized by mounting the substrates on a turntable, whereas oxygen is introduced by an oxygen pocket wich covers a sector of the whole area and is such closely spaced underneath the substrates that a pressure enhancement of 2-3 orders of magnitude can be realized against the surrounding high vacuum ambient. The turntable with its diameter of 22 cm can accommodate a large number of substrates, e.g. twelve 2" or three 4" wafers, which can be simultaneously coated in a single run. Since the substrates are simply laid into machined openings the backside is not deteriorated and deposition also works double sided. The high demand on the reliability of the process becomes directly evident when considering that the value of the substrates alone, i.e. the risk of loss, is in the range of several thousand dollars.

In an advanced version, the frame of reference has been changed to evaporate on even larger rectangular area Kinder (1997). In this arrangement the substrate is held fixed but the oxygen is supplied by an oscillating oxygen shuttle instead (cf. Fig. 1(b)). The rate control has also been modified employing atomic absorption spectroscopy (AAS) and the evaporation sources can be refilled in situ from reservoirs Utz (1997). Consequently, this setup can be operated continuously for a extended deposition time and feeding a long metal tape through the heater it is our approach to coated conductors.

3. APPLICATION TAILORED FILMS

For many promising products involving high temperature superconductors the material is used in form of thin films. However, the requirements on such films are as numerous and manifold as the applications. Current main stream development covers such various products as resistive fault current limiters, microwave filters for mobile or satellite communication, NMR and MRI pickup coils, Squid sensors, and superconducting electronics. While power and high frequency applications usually require highest possible critical current densities, low surface resistance, and good power handling, Squid applications need low noise figures. Superconducting electronics demand really smooth films, since YBCO is only one part in a multilayer structure where surface roughness can result in pinholes and shorts which are not tolerable. All these different and sometimes contradictory requirements cannot be usually met by the same YBCO film. Rather we offer different film types grown under different conditions with properties and specifications optimized to certain applications and the customer can chose from a selection of films tailored for his specific needs.

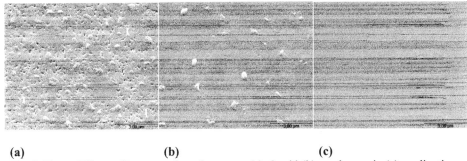

(a)   **(b)**   **(c)**
Fig. 2 Three different film types for microwave (a), Squid (b) or electronic (c) applications

The client structure is still twofold. On the one hand there are scientists in research labs with very specific requirements. In situ contact and cap layers, and even substitution of yttrium by other rare earths elements require a high degree of flexibility. On the other hand in the course of real product development there is a gradual shift from scientists towards engineers. Their main concern, however, is the device layout and performance – not material issues. This tendency renders customer support and service increasingly important. Hence, strong emphasis is laid on technical consulting before an order is placed and films are produced.

4. QUALITY MANAGEMENT

When a customer decides to base his product development on an external film supplier the producer of such films undertakes a strong commitment. For e.g. filters and fault current limiters hundreds of films with very high standards on uniformity and quality are required. It is often neglected that maintaining an appropriate quality management is at least of the same importance and can even cost more time than the film fabrication itself. For industrial users

4

such a quality management which keeps record of every fabrication step is a matter of course. Filters which are part of a platform in a manned space mission have to obey even stricter regulations, covering safety aspects as well as warranted long term performance.

Since the film producer is only a link in long chain his responsibility starts with the inspection of the substrates and approval of all employed materials and chemicals. Records of all handling and fabrication steps complement the processing history up to the final quality inspections and certification of the films. The latter covers visual inspection, electron microscopy, and non destructive measurement of the warranted film specs like transition temperature, critical current density, and uniformity.

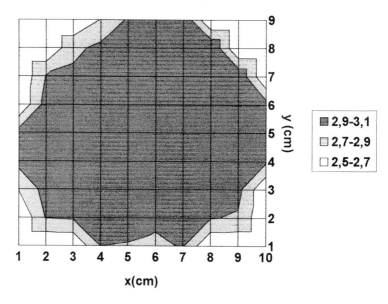

Fig. 3 Inductive j_c-scan of 4" YBCO film on sapphire at 77 K (scale in MA/cm^2)

The considerations above may demonstrate that the time of the pioneering work is over and the fabrication of YBCO films has developed into a real producing business. With the emerging of a market for high temperature superconductors, YBCO film suppliers have to be prepared to meet the standards set to allow a reasonable long term product development.

REFERENCES

Berberich P, Utz B, Prusseit W, and Kinder H 1993 Physica C **219**, 497
Kinder H, Berberich P, Prusseit W, Rieder-Zecha S, Semerad R, and Utz B 1997 Proc. of Int. Workshop on Supercond. Hawaii, June 15-18 1997, 141
Utz B, Rieder-Zecha S, and Kinder H 1997 IEEE Trans. on Appl. Supercond. 7, 1181

Inst. Phys. Conf. Ser. No 167
Paper presented at Applied Superconductivity, Spain, 14-17 September 1999
© 2000 IOP Publishing Ltd

The study of thermal fatigue behavior of YBCO superconductor thin films with oxide buffer layers

F. Yang, K. H. Wu and G. Larkins

College of Engineering and Applied Sciences, Florida International University, Miami, FL 33174, USA

ABSTRACT: Thermal cycling behavior of the $YBa_2Cu_3O_7$ (YBCO) high-temperature superconductor (HTS) thin films on the (100) Si substrates with different buffer layers, including MgO, ZrO_2 and YSZ, was systematically studied in this paper. These HTS assemblies underwent thermal cycling from $-197°C$ to room temperature, and their thermal fatigue reliability was evaluated using the SEM measurement. It was found that YBCO superconductor thin films on the buffered (100) Si substrate eventually cracked and spalled with various mechanisms. The cyclic life of the superconductor assemblies is primarily determined by the extent of CTE mismatch between the YBCO film and the buffer layer and the substrate used. Defects in the buffer layers, inhomogeneity of the film composition and cluster size, and poor interface coherence and adhesion will assist the crack initiation and eventually shorten the cyclic life of the superconductor assemblies.

1.   INTRODUCTION

Recently, greater attention has been given to YBCO superconductor thin films deposited on Si substrates in the field of hybrid microelectronics and microwave applications, primarily because of its advantages, including low price, low RF loss, high thermal conductivity and excellent compatibility, etc. However, the interdiffusion between YBCO films and Si substrates results in poor superconducting properties if the film is directly deposited on the substrate. In order to overcome the problem, a single oxide layer, such as MgO (Witanachi 1989), CeO2 (Inoue 1990) and YSZ (Mogro-Campero 1988), has been introduced as a buffer layer between the YBCO and the Si substrate for the film growth. The thermal expansion mismatch between different layered materials is one of the main factors causing the reliability problem. Other influential factors include the decohesion resistance of the interface and structure mismatch between layers.

This paper focuses on the study of the thermomechanical behavior of the superconductor YBCO thin films on the (100) Si substrate with different oxide buffer layers under temperature cyclic loads. MgO, ZrO_2 and YSZ are selected as buffer layers to determine the effect of buffer layers on the thermal cycling behavior of the assembly. The results of this study indicate that spalling is the dominant failure mechanism of the superconductor YBCO thin films, and the fatigue life of the thin films varies with the type of buffer layer and the uniformity of the composition and surface condition of the film.

2.   EXPERIMENTS

In this study, p-type (100) oriented single-side polished Si wafers with 15mm x 15mm square pieces were used as the substrate. These pieces were degreased in acetone, isopropanol and deionized water and blown dry using N_2. Afterward, these oxide-covered substrates were stored and deglazed in

10:1 DI Water to HF solution just prior to loading into the system. After deglazing, the sample was immediately placed in the deposition chamber and the chamber was evacuated. The oxide buffer layers were sputtered on the Si samples in a custom-built, turbo-pumped system using 2" US-gun magnetron sources and pulsed DC sputtering in an argon-oxygen atmosphere. The buffer layer thickness was approximately 100 nm.

Immediately after the oxide deposition the substrate was mounted onto a stainless-steel heater block using silver paste and the whole assembly was allowed to cure for an hour in room air at 50°C. After curing, the chamber was pumped down to a base pressure of 1 x 10^{-6} Torr and the heater was brought up to deposition temperature. All YBCO film depositions were conducted in a pulsed laser deposition system using a triple Nd-YAG laser. The final total film thickness was approximately 200 nm. The detailed sputtering process was reported in the work of Vlasov (1999).

The samples were sealed in a copper-stuffed copper tube for the thermal cycling test. The temperature change (ΔT) was selected as 217°C (from −197°C to room temperature). The lower-end temperature was controlled using liquid nitrogen. A JEOL JSM-35CF Scanning Electron Microscope was used to characterize the morphology and failure mechanism of the samples, and an EDS analysis was conducted to detect the localized compositions.

3.   RESULTS AND DISCUSSION

Fig. 1a shows the typical microstructure of the as-deposited YBCO superconductor thin film on MgO buffered Si substrate. Figures 1b, c and d show the progressive failure patterns of the thin film during the thermal cycling process. The micrograph in Fig. 1a clearly indicates that there is no crack to be noticed prior to the thermal cycling. After cycling from −197°C to room temperature for approximately 400 cycles, the cracks start to develop (Fig. 1b). Further cycling leads to the spalling of localized areas, and eventually numerous large spalled holes developed in the sample (Figs. 1c and d). For the YBCO/MgO/Si system, the average spalled area size is approximately 4-5 μm. The results of the EDS analysis show that the composition of the spalled area (Fig. 1b) is SiO$_2$, while the composition of the material at the bottom of the hole (Figs. 1c and d) is Si, not MgO. The presence of isolated SiO$_2$ zones in the YBCO film and the fact that only Si was detected at the bottom of the hole (but no MgO) may have indicated that perhaps during the sputtering process, those areas were not completely covered with MgO. With the presence of high temperature (765°C) and an oxygen-rich environment during the subsequent YBCO sputtering process, the exposed Si substrate at these

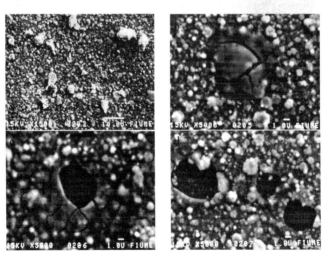

Fig. 1   The microstructure changes during the thermal cycling process of the YBCO/MgO/Si assembly. (a) before thermal cycling, (b) - (d) during the cycling.

MgO-deficient area starts to oxide and continues to grow. As a consequence, a localized SiO_2 zone was formed.

Because of differential CTE among the constituent layers, during the cooling cycle of the thermal cycling process, the SiO_2 area is subjected to compression while the film is subjected to tension. The compression existing inside the SiO_2 film could cause buckling in the debonded SiO_2 area and eventually lead to the cracking and spalling of the SiO_2 zones. Evans et al. (1995, 1997) believe that the non-planar interface between the film and the substrate and the excessive residual stress in the materials could result in a brittle debonding at the interface. The defect in the MgO buffer and the compressive stress existing in the SiO_2 zone may have been the key reasons responsible for the failure of the assembly.

Fig. 2a shows the microstructure of the $YBCO/ZrO_2/Si$ superconductor assembly prior to thermal cycling. Figure 2b shows the failure pattern of the assembly after the thermal cycling for 200 cycles. Here failure is defined as the presence of noticeable spalling of the superconductor films. Similar to the $YBCO/MgO/Si$ system, the original ZrO_2 buffered thin film shown (Fig. 2a) exhibits no noticeable defects or cracks and the YBCO composition is uniform throughout the sample. As thermal cycling proceeded, cracking started to develop (Fig. 2b), and eventually spalling took place. An EDS analysis indicated that the composition of the crack area, which was just prior to spalling, was a mixture of YBCO and ZrO_2. The composition of the area beneath the spalled area was ZrO_2. It was noticed that the morphology of the cracked area was dramatically different from the original sample (Fig. 2b), and these areas were never noticed in the as-sputtered samples. The change of morphology of the cracked zone and the presence of the ZrO_2 composition in the area may have been caused by the debonding between the YBCO film and the ZrO_2 buffer layer. Thus, the existence of the gap at the debonded interface altered the scattering characteristics of the secondary electron of the EDS system and, as a consequence, both ZrO_2 and YBCO appeared in the EDS spectrum. Further thermal cycling enhanced the buckling and cracking in this particular area and eventually caused spalling of the film.

Fig. 2   The microstructure changes during the thermal cycling process of the $YBCO/ZrO_2/Si$ assembly. (a) before thermal cycling, (b) after 200 cycles.

The $YBCO/YSZ/Si$ system demonstrates almost the sample cyclic life as the $YBCO/ZrO_2/Si$ (~175 cycles). Nevertheless, unlike the MgO or ZrO_2 buffered systems, no cracked areas were noticed prior to failure and only spalling took place. Upon carefully examining the surface of the YBCO film, one noticed that relatively large YBCO clusters at a size of 5-10 μm (50-100 times the film thickness) were present on the surface of the film. After spalling, the number of the clusters was substantially reduced. This leads to the speculation by the authors that the YBCO clusters may be responsible for the spalling of the film. A preliminary finite element study carried out in this study indicates that a significant stress concentration was developed at the joint area between the cluster and the film. It is believed that the high stress at the base of the cluster could promote debonding and cracking of the assembly. Eventually, the clusters snapped off from the film under repetitive thermal cycling and spalling took place.

In order to study the influence of buffer layers on the thermal fatigue life of the thin films, samples with the YBCO directly deposited on the $LaAlO_3$ substrate were introduced for comparison. Because of excellent structural coherence between YBCO and $LaAlO_3$, this system demonstrates the greatest cyclic life, as shown in Table 1. Spalling also occurred in the YBCO/ LaAlO$_3$ system and the failure mechanism is the same as that in the YBCO/YSZ/Si system.

Table 1. Number of cycles to failure of various YBCO systems

Materials	YBCO/ LaAlO$_3$	YBCO/MgO/Si	YBCO/ZrO$_2$/Si	YBCO/YSZ/Si
Number of cycles to failure (N_f)	450	400	200	175

The data in Table 1 indicate that the YBCO thin film without a buffer layer has a longer thermal cycling life than those with buffer layers. Yang et al (1999) studied the thermal cycling behavior of YBCO thick films, and had developed a mathematical model to describe the fatigue life of the thin film assembly. The model could accurately predicted the thermal fatigue life of the YBCO/LaAlO$_3$ system, however, it failed to predict the lives of those samples with oxide buffer layers. It is believed that crystal-structure mismatch between the buffer layer and the YBCO could possibly exert a significant influence on the fatigue life of those samples.

4. CONCLUSIONS

The following conclusions can be drawn from the samples studied in this work:

a. The superconductor systems without buffer layers, e.g. YBCO/LaAlO$_3$, demonstrate the longest thermal fatigue life, as compared to those with oxide buffer layers.

b. The system with the least CTE and structural mismatch offers a superior thermal cyclic stability.

c. Spalling is the primary failure mechanism of the oxide-buffered YBCO system. Poor cohesion, defects in the buffer layers, large YBCO clusters, together with the compressive stress in the assembly will promote spalling and failure of the system.

d. A good design and accurate control of processing parameters are the key to better thermal stability of the HTS systems.

5. ACKNOWLEDGMENTS

This work has been supported by the United States Air Force's Office of Scientific Research under grant number F49620-95-1-0519.

REFERENCES

Evans A G and Hutchinson J W 1995 Acta Metall. Mater. **43**, 2507
Evans A G, He M Y and Hutchinson J W 1997 Acta Mater. **45**, 3543
Inoue T, Yamamoto Y, Koyama S, Suzuki S and Ueda Y 1990 Appl. Phys. Lett. **56**, 1332
Mogro-Campero A and Turner A G 1988 Appl. Phys. Lett. **52**, 1185
Witanachchi S, Patel S, Kwok H S and Shaw D T 1989 Appl. Phys. Lett. **54**, 578
Yang F, Liu Y, Wu K H and Larkins G 1999 IEEE Trans. Appl. Supercond. **9**, 1975
Vlasov Yu, Lacambra A, Soto R, Larkins G L Jr, Stampe P and Kennedy R 1999 IEEE Trans. Appl. Supercond. **9**, 1642

Inst. Phys. Conf. Ser. No 167
Paper presented at Applied Superconductivity, Spain, 14-17 September 1999

New method for fabricating ribbon-like thin films of Bi-2212 on Ag substrate and their superconducting properties

S Arisawa[1,2], H Miao[2], H Fujii[1,2], A Ishii[1], Y Takano[1], Y Satoh[3], T Hatano[1], and K Togano[1,2]

[1] National Research Institute for Metals, Sengen, Tsukuba-shi, 305-0047, Japan.

[2] CREST, Japan Science and Technology Corporation, Japan.

[3] Tsukuba University.

ABSTRACT: We have developed simple technique to grow ribbon-like single crystalline films of Bi-2212 on Ag substrates. The process is performed under ambient pressure only by heating the Bi-2212 pellet on the edge of the substrates. The ribbons are extremely thin and stuck firmly to the substrates, resulting in the formation of a new type of composite of Bi-2212 ribbons/Ag. The ribbon-like films showed the superconducting transition with the onset temperature of ~75 K.

1.   INTRODUCTION

Since the discovery of Bi-based superconductors (Maeda 1988), a variety of methods have been proposed for synthesizing the material of the series. For most electronic device applications, these materials must be single crystalline thin films. Common techniques to synthesize Bi-based superconductor films are vapor processes such as sputtering (Karimoto 1997), molecular beam epitaxy (Bodin 1992b), ion beam deposition (Fujita 1991), and pulsed laser deposition (Bedekar 1992a) etc. It is still desirable to develop various technologies to fabricate single crystalline materials for further applications. Recently, we have developed a new method to grow superconducting ribbon-like thin films of Bi-2212 on an Ag substrate under ambient pressure (Arisawa 1999). The process can be extended to other high-temperature superconductors and has potential for some device applications because of the unique texture of the films and the simplicity of the process. The ribbon-like thin films fabricated by this process have very unique textures, but it is not clear how such textures are formed. It is thus very important to clarify the growth mechanism of ribbon-like thin films. In this paper, we report on the fabrication and characterization of the ribbon-like thin films fabricated by this new and very simple technique.

2.   EXPERIMENTAL DETAILS

A study of the structure of ribbon-like thin films was carried out by AFM. The typical textures of the ribbon-like thin films observed by optical microscopy have been presented in

10

previous literature (Arisawa 1999). For the AFM investigation, an Ag(100) single-crystal substrate with a diameter of 10 mm was used, and the heat treatment was carried out in a tube furnace. For X-ray analyses and R-T measurement, a polycrystalline rolled Ag substrate with a thickness of 50 μm was employed. The heat treatment schedule employed for fabricating the ribbon-like thin films is shown in Fig. 1. The substrates with the pellet were then heated up to 905°C and kept at that temperature for 10 minutes. They were then gradually cooled down to 860°C and kept at that temperature for 2 hours, followed by furnace cooling. The atmosphere was 100% oxygen under ambient pressure.

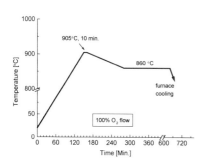

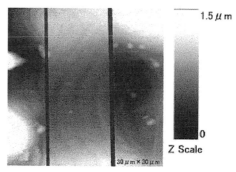

Fig. 1. Heat treatment schedule employed for fabricating ribbon-like thin films.

Fig. 2. A typical image of a ribbon crystal. This sample was prepared on a single crystal Ag(100).

3. RESULTS AND DISCUSSION

Optical microscopy has already revealed that the widths of the ribbons ranged from several microns to a couple of tens of microns. We deduced from the colors of the ribbons that some ribbons had very small thickness of the order of the submicron (Arisawa 1999).

A typical AFM image of a ribbon crystal grown on single crystalline Ag is shown in Fig. 2. The region between the solid lines is the ribbon-like thin film. Small hills with contours on the surface are supposed to be caused by the recrystallization of Ag during the heat treatment. The thickness of the ribbon-like thin films varied from place to place. The ribbon shown in the figure had a thickness of as little as ~0.1μm, which agreed with the deduction mentioned above. Improvement of the experimental conditions to reduce the formation of the hills is under investigation. The composition of the ribbons was characterized by EDX. The ribbons had all the elements of Bi, Sr, Ca, and Cu with a composition ratio of Bi:Sr:Ca:Cu

=2:(1.5-1.6):(0.7-1.0):(0.8-1.2). Therefore, the composition is fairly Cu-poor compared to the stoichiometric Bi-2212 phase, suggesting the intergrowth of the Bi-2201 phase.

The X-ray diffraction pattern of the prepared sample is depicted in Fig. 3. As can be seen from the spectrum, it was confirmed that the c-axis oriented Bi-2212 phase was successfully formed though the ribbons were composed of a couple of phases. The Cu-free phase found by EDX analysis was identified as Bi_2SrO_4 and some other unidentified impurity phases were observed.

The degree of the intergrowth was analyzed based on the lattice constant precision plot from the crystallographic point of view. This analysis proved that the split of the 2212 peaks around 6 degree was originated from the mixture of the n=2.0 and n=1.7, where n's are the averaged numbers of CuO_2 planes per half unit cell.

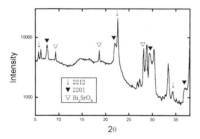

Fig. 3. X-ray diffraction pattern of the ribbon-like thin films on Ag substrate. c-axis oriented Bi-2212 phase and some impurity phases were observed.

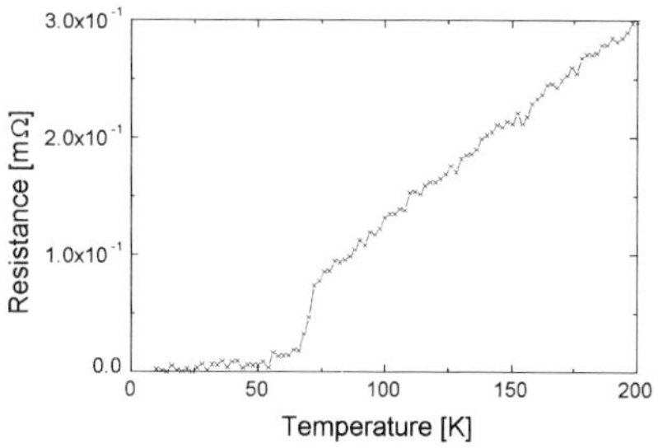

Fig. 4. Temperature dependence of the resistance of the ribbon-like thin films on Ag substrate. The excitation current was 300 μA.

A small rectangular piece of ～1mm x 3mm on which several thin ribbons running in parallel was carefully cut off and its electrical transport property was investigated by the standard four probe method. Figure 4 represents the relationship between resistance and temperature for the probe current of 300μA. To eliminate the error originated from the current flowing in the substrate, the conversion from the resistance into resistivity was deliberately neglected.It was concluded that the ribbons were superconducting with the onset transition temperature around 75 K.

4. SUMMARY

We have developed a novel technique for fabricating superconducting Bi-2212 ribbon-like thin films on Ag substrates, on which the starting material of Bi-2212 powder was placed and melted.

The optimal temperature for growing ribbons under 1 atm O_2 was 905 °C. Although the 2212 ribbon-like films were mixed with smaller amount of 2201 phases, the quality of 2212 phase itself evaluated by the intergrowth analysis was fairly good. We are now investigating the growth mechanism of the ribbons, which could be informative to have further improvement of the quality of the crystals. This novel fabrication form of Bi-2212 films may by useful for fields such as device applications etc.

REFERNCES

Arisawa A, Miao H, Fujii H, Ishii A, Labat S, Hatano T, and Togano K. 1999 Physica C **314,**155.

Bedekar M, Safari A, and Wilber W. 1992a Physica C **202**, 42.

Bodin P, Sakai S, and Kasai Y. 1992b Jpn. J. Appl. Phys. **31** L949.

Fujita J, Yoshitake T, Satoh T, Ichihashi T, and Igarashi H. 1991 IEEE Trans. Mag. **27**, 1205.

Maeda H, Tanaka Y, Fukutomi M, and Asano T. 1988 Jpn. J. Appl. Phys. **27**, L209.

Karimoto S, Kubo S, Tsuru K, and Suzuki M. 1997 Jpn. J. Appl. Phys. **36**, 84.

Inst. Phys. Conf. Ser. No 167
Paper presented at Applied Superconductivity, Spain, 14-17 September 1999
© 2000 IOP Publishing Ltd

SrTiO₃ as a novel substrate for the electrodeposition of HTSC films

M.S. Martín-González, E. Morán, M.Á. Alario-Franco

Laboratorio de Química del Estado Sólido, Departamento Química Inorgánica, Facultad de Ciencias Químicas, Universidad Complutense, 28040-Madrid Spain (UE).

ABSTRACT: In the present work, we show that by pre-reducing $SrTiO_3$ *i.e.* making it conducting, one can obviate the need of a buffer layer for the electrodeposition of cations. Moreover, in the subsequent oxidation of the metallic film the substrate becomes again an insulator and the film ·superconducting, respectively. For this purpose (001) oriented $SrTiO_3$ single crystals were reduced and, afterwards, Y, Ba and Cu were simultaneously co-deposited from a DMSO solution of their salts. The superconducting properties of these films are T_c=92 K and $J_c \approx 3250$ A/cm².

1.   INTRODUCTION

Up to the present time, the research on superconductivity has been focused on three main streams: explanation of the superconducting properties (see, for example, Orlando 1991), search of materials with higher T_c (Owens 1998) and development of techniques suitable for the application of the HTSC materials to commercial devices (Seeber 1998). In this last subject, one of the most promising techniques comes from electrochemical science: the well-known cathodic electrodeposition. Preparation of HTSC thick films via electrodeposition techniques provides a low-cost and fast procedure, which has the added advantage of being widely employed in industry. In summary, the process requires the deposition of the constituent elements of the superconductor from the adequate electrolyte on a *conducting substrate*, followed by an oxidative heating. In this sense, $SrTiO_3$ is one of the most interesting substrates since most, not to say all HTSC materials have a structure that derives from the perovskite one. However, as it is well known, $SrTiO_3$ is insulating, therefore, up to now, for using it in an electrodeposition process a conducting layer (Ag in most of cases) had to be previously deposited by using other methods.

In this work, the obtaining of $YBa_2Cu_3O_{7-x}$ thick films on $SrTiO_3$ without using the buffer layer is reported (Spanish Patent P9802655). The process consists, basically, on first reducing the substrate at high temperature, so that it becomes conducting and then the simultaneous co-deposition of the cations is carried out. The precursor film is, subsequently, heat-treated in air in order to obtain the superconducting phase.

2.   EXPERIMENTAL

Single crystals of $SrTiO_3$ (100) (Crystal-GmbH) of 5X5X1 mm³ were reduced with Zr metal in quartz sealed ampoules at ≈1320 K for 3 days. The reduced crystals were black in the outside and dark blue in the inside; obviously, as in most reactions with single crystals, kinetic barriers seem to be important. Yet, the crystals were conducting enough so as to be used as a substrate for electrodeposition. A full description of the reduced crystals will be published elsewhere (Aguirre to be published).

A typical electrolytic bath for the growing of Y-123 films consisted on a solution of the nitrates in dimethyl sulfoxide with 20 mM Y, 36 mM Ba and 44 mM Cu; this is analogous to that used previously for metallic substrates e.g. silver (Bhattacharya 1991, Martín-González 1999a). The temperature was fixed to 25 °C by means of a thermostatic bath (Clifton NE4-DCE PLUS). The electrodeposited alloy films were heated at 900 °C for 24 h. in air and post-annealed in O_2 at 450 °C for 24 h. In this process the reduced strontium titanate gets back to its stoichiometry and insulating state. The fact that the substrate is insulating is certainly of interest in potential field effect device applications (Xi 1991, Mannhart 1991 and Seeber 1998).

The electrodeposition process used was a chronoamperometry carried out using a conventional three-electrode cell: the reference electrode was Ag/AgCl and platinum was used as a counter electrode the working electrode was a reduced single crystal (5X5X1 mm³), installed in a home made crystal holder, Fig. 1, that allow a silver wire to be pressed against the crystal so as to make a good contact.

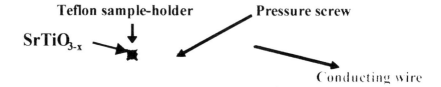

Fig. 1. Home made crystal holder

A VersaStat potentiostat/galvanostat Model 253 fitted with a PC computer interface was used for controlling the electrolytic process. The characterization of the films by X-ray diffraction was done using a Siemens D-5000 powder diffractometer. EDS microanalysis and morphological examination were carried out using a JEOL-JSM 6400 scanning electron microscope (SEM).

Superconducting temperatures and the M(H) loops were recorded using a Quantum Design MPMS XL SQUID. The critical current densities in $A \cdot cm^{-2}$ were obtained, for a slab plane geometry, using the Bean model (Bean 1964, Malozemoff 1989).

3.   RESULTS AND DISCUSSION.

The X-ray diffraction pattern in Fig. 2 shows that "YBCO" can indeed be deposited on SrTiO₃ without a buffer layer. It can be seen that, besides the prominent substrate peak (appearing at d=1.94 Å), the main "reflections" of YBCO are also present. CuO has been detected as an impurity

In order to learn about the morphology of the films, SEM was performed: Fig. 3. Both the (a) top and (b) side views show the films to be formed by a columnar type of microstructure. This does imply the presence of empty spaces between the crystals. In others words, the films are well formed, but not continuous. In considering the lack of contact between the columns, it is worth mentioning that the thickness of the electrodeposited "metallic" precursor layer was of the order of 20 μm (Martín-González to be published) while in the oxidised superconducting material the average height is of 30 μm. Obviously, this suggests a lateral –i.e. 2D- contraction of the film in the oxidation process. It is worth mentioning in here that in our earlier studies of deposition of YBCO on silver substrates (Martín-González 1999b); the lack of full coverage was in much extent improved by repeated sequential electrodeposition (Martín-González 1999c). Perhaps a similar n-fold electrodeposition could be applied here.

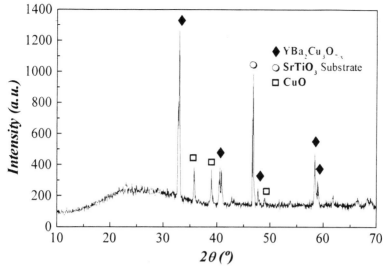

Fig. 2. X-ray diffraction pattern of YBa₂Cu₃O₇₋ₓ/SrTiO₃

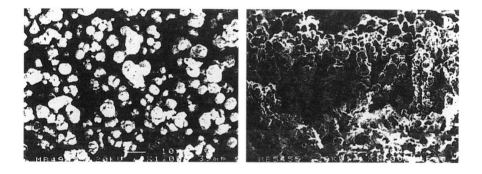

Fig. 3. a) Top and b) side views of YBa₂Cu₃O₇₋ₓ/SrTiO₃ films as observed by scanning electron microscopy

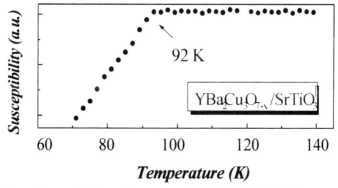

Fig. 4. Magnetic susceptibility vs. temperature of YBa₂Cu₃O₇₋ₓ/SrTiO₃ films.

Magnetic susceptibility measurements, Fig. 4, indicate that the superconducting transition temperature of the obtained films is of the order of 92 K, although the transition is somewhat large. This may be due to a certain degree of heterogeneity in the oxygen stoichiometry of the superconducting grains.

J_c values at zero field were $\approx$ 3250 A/cm^2, which are comparable to some films electrodeposited on silver (e.g. $J_c \approx$ 3000 A/cm^2 (Ondoño-Castillo 1997)). There is, nevertheless a large range for improvement since much higher values of J_c have already been obtained on silver (e.g. $J_c \approx$ 9500 A/cm^2 (Martín-González 1999a)) in a single deposition process; a value that is improved by re-deposition (e.g. $J_c \approx$ 11500 A/cm^2 (Martín-González 1999c)) or even by sequential, ation by cation, deposition (e.g. $J_c \approx$ 6600 A/cm^2 (Régnier 1997)). At high field, however, J_c decreases markedly and becomes zero at 3 Teslas (30 KOe).

4.   CONCLUSIONS

It is possible to obtain superconducting films on a SrTiO$_3$ substrate by electrodeposition, because the pre-reduction of strontium titanate allows it to be used as a cathode for obtaining HTSC films without a buffer layer.

The critical current densities observed in YBa$_2$Cu$_3$O$_{7-x}$/SrTiO$_3$ are $\approx$ 3250 A/cm^2.

5.   REFERENCES

Aguirre, M. Martín-González, M.S. Morán E. and Alario Franco, M. Á. to be published.

Bean, C.P., *Phys. Rev. Lett.* 1964, **8**, 250.

Bhattacharya, R. N. Noufi, R. Roybal, L. L. and Ahrenkiel, R. K., J. Electrochem. Soc., 1991, **138**, 1643.

Malozemoff, A.P., 1989, in *Physical Properties of High Temperature Superconductors I*, de. D.M. Ginsberg, World Scientific, p. 73.

Mannhart, J. Bednorz, J.G. Muller, K.A., Schlom, 1991, Z. Phys. **B83**, 307

Martín González, M. S. García Jaca, J. Morán, E.and Alario Franco, M. Á., 1999, J. Mater. Chem., **9**, 1293.

Martín González, M. S. García Jaca, J. Morán, E.and Alario Franco, M. Á., 1999, J. Mater. Chem, **9**, 137.

Martín-González M.S. "Doctoral Thesis" to be published

Martín-González, M.S. Morán, E. Sáez-Puche, R. Alario-Franco, M.Á., 1999 (Proceedings of the Materials Research Society –MRS-, Boston, Massachusetts, USA 1998, Ed. S.M. Kauzlarich, E.M. McCarron III, A.W. Sleight, H-C. Zur Loye) 547, 287.

Ondoño-Castillo, S.; and Casañ-Pastor, N., 1997, Physica C, **276, 251**.

Orlando T.P. and Delin K.A., 1991, "Foundations of Applied Superconductivity", Addison-Wesley.

Owens F. J. and Poole, C.P.Jr ., 1998 , "The New Superconductors", Plenum.

Régnier, P., Poissonnet, S., Villars, G. and Louchet, C., 1997, Physica C, **282-287**, 2575.

Seeber, B., 1998, "Handbook of Applied Superconductivity", Institute of Physics Publishing Bristol and Philadelphia.

Spanish Patent P9802655

Xi, X.X. Li, Q., Doughty, C., Kwon, C., Bhattacharya, S., Findikolu, A.T., Venkatesan, T., 1991, Appl. Phys. Lett. **59**, 3470.

Inst. Phys. Conf. Ser. No 167
Paper presented at Applied Superconductivity, Spain, 14-17 September 1999
© 2000 IOP Publishing Ltd

Structure and properties of $(Sr,Ca)CuO_2$-$BaCuO_2$ superlattices grown by layer-by-layer PLD

Karen Verbist[1], Gustaav van Tendeloo[1], Gertjan Koster[2], Guus J.H.M. Rijnders[2], Dave H.A. Blank[2] and Horst Rogalla[2]

[1]EMAT, University of Antwerp (RUCA), Antwerp, Belgium

[2]Low Temperature Physics, Applied Physics, University of Twente, Enschede, Netherlands

ABSTRACT: We report on the growth and characterization of $(Sr,Ca)CuO_2$-$BaCuO_2$ superlattices. To impose a true layer-by-layer growth mode, the interval deposition technique has been used, resulting in high quality superlattice structures. To our knowledge, this is the first report on characterization of Ba-containing artificial layered infinite layer compounds by HREM. These materials suffer considerably from radiation damage during ion milling and HREM observations. Nevertheless, with precautions these structures can be observed in HREM. The superlattice period has been deduced from electron diffraction patterns and XRD measurements. Under certain deposition conditions and thickness' of the single Ba containing layers in the superlattice it was observed that the $BaCuO_2$ material is converted to $Ba_2CuO_{4-\delta}$. Image simulations to interpret the HREM contrast are performed.

1   INTRODUCTION

Using thin film deposition, one can control the functionality of different layers by sequential deposition of distinct $ACuO_2$ (A=Ca,Sr,Ba) blocks. In addition, one benefits from the stabilizing effect of the substrate on these highly unstable structures. For superlattices created by sequential deposition containing only $SrCuO_2$ and $CaCuO_2$ (Norton 1994a, Tsukamoto 1997, Koster 1998a), no superconductivity was found. However, Norton *et al.* have demonstrated superconducting artificial layered structures with Ba and Sr_{1-x},Ca_x by Pulsed Laser Deposition (PLD) (Norton 1994b), with a maximum T_c of ~70K. By periodically substituting layers with smaller cations by layers with larger cations, extra oxygen may be incorporated in the latter and act as charge reservoirs. They only observed superconductivity in structures with more than one $BaCuO_2$ block. Later, Balestrino *et al.* (1998a) found also superconductivity ($T_{c,max}$~80K) using the same method, however, only in case of superlattices with Ba and Ca, i.e., Sr containing structures showed no trace of superconductivity. This was attributed to Ba/Sr interdiffusion. Both groups verified the artificial periodicity of the as-grown films with X-ray diffraction (XRD), after calibration of the growth rates of the individual constituents by thickness measurements: No *in situ* growth rate monitoring has been used and, more importantly, the growth mode has not been identified in these studies. Because these structures are very unstable, High Resolution Electron Microscopy (HREM) analysis of these films, using high-energy electrons, is extremely difficult. The relation of superconductivity and the structure is not clear yet, e.g., whether the relation found for the number of different layers with superconductivity (Balestrino 1998b) has a fundamental origin or is determined by the crystalline quality.

The x=1 member of the $Cu_xBa_2Ca_{n-1}Cu_nO_y$ family has been fabricated in thin film through sequential deposition of Sr_{1-x},Ca_x (smaller ions) containing layers and Ba (larger ion) containing layers, in order to obtain layers with different functionalities (e.g., charge reservoir by incorporation of excess oxygen). Superlattices are fabricated by depositing sequentially $ACuO_2$ (A= Ba, Sr or Ca) from different targets, $BaCuO_2$ (BCO), $SrCuO_2$ (SCO), $Sr_{0.7}Ca_{0.3}CuO_2$ (SCCO) and $CaCuO_2$ (CCO), using the interval deposition technique (Koster, 1999a). The structures have been analysed with XRD and HREM and the morphology of the surface is monitored *in situ* with high pressure Reflection High Energy Electron Diffraction (RHEED, Rijnders 1997) and *ex situ* with Atomic Force Microscopy (AFM).

2   EXPERIMENTAL

Superlattice structures have been grown by PLD on treated $SrTiO_3$ substrates ensuring a single terminated surface (Koster 1998b), i.e., TiO_2. The surface termination was subsequently changed by deposition of SrO. Samples with growth sequences of $[(Sr_{1-x},Ca_x)CuO_2)_a-(BaCuO_2)_b]_n$ and n=20; a=b=4, n=20; a=2; b=3 and n=20; a=b=2 have been investigated by HREM. The number of pulses needed for one unit-cell layer is estimated by counting the oscillation period during the deposition of each component at fixed deposition conditions. If possible, this estimation is verified with X-ray reflectivity thickness measurements. θ–2θ scans (X'pert diffractometer, Philips, the Netherlands, Cu-Kα source) show satellite peaks, originating from the artificial superstructures. The positions of these satellite peaks were used to confirm the deposition rate of each constituent.

The HREM analysis was performed on a JEOL 4000EX operating at 400 kV and 376 kV. The point resolution of this instrument is 1.7 Å. The HREM sample preparation involves standard polishing and ionmilling at low angle and low voltage.

The electrical properties of the superlattices were measured with the four-probe technique, inside a He cryostat.

3   RESULTS AND DISCUSSION

Figure   1 a) shows an HREM overview of the hetero-structure as previously studied with XRD (Koster 1999b) a). The top of the hetero-structure contains a protective $SrTiO_3$ layer. The dark layers, 20 in total, are SCO, and the lighter layers, also 20, are BCO. Figure   1 b) shows a HREM image recorded near the top $SrTiO_3$ buffer layer. The BCO layers clearly have a periodicity which differs from the expected 'perovskite' spacing. The zigzag contrast in the closely spaced double rows is typical for the superconductor $(La,Ba)_2CuO_4$, where Ba layers are present and are shifted by (½, ½) in the (a,b)-plane, with respect to each other. The periodicity in the central row is clearly shifted by ½ a_0 after crossing the double row, as indicated by arrows in Fig. 5.1 b). The lattice parameter of the Ba-block along the c-axis measured on the image is 13.2-13.6 Å. All these features combined let us conclude that the "$BaCuO_2$ layers" actually consist of Ba_2CuO_x. Similar structural features have been observed in SCO films by others (Sugii 1993). We observed mostly 2 unit-cells

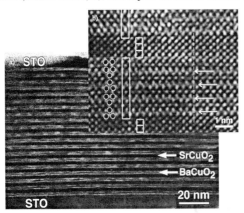

Figure   1: *a) Overview TEM image of a SCO/BCO superlattice based on a 'wrong' deposition rate estimate for the $BaCuO_2$ layers, the white contrast corresponds to BCO, b) HREM image of the same film, the white circles indicate the positions of the Ba ions in the Ba_2CuO_x model.*

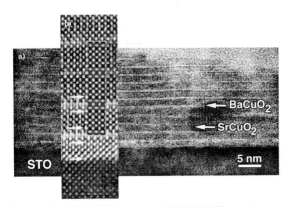

Figure 2: *a) HREM image of a 3SCCO/2BCO structure, b) enlargement, the arrows indicate the copper oxide planes.*

thick layers of Ba_2CuO_x, with a total thickness of 27.2 Å. The observed periodicity in the image, 3 unit-cells of SCO and 2 unit-cells of Ba_2CuO_x, yields a total thickness of about 38 Å. Locally we observe regions with 2 or 4 unit-cells of SCO. It is interesting to note that the incorporation of Ba double layers results in a shift of the Cu sublattice, observed as a ½ a_0 shift. In the case of ½ unit-cell of Ba_2CuO_x this would result in a shift of the Cu sublattice in consecutive SCO layers. Also steps in the Ba_2CuO_x layer with heights of ½ unit-cell would induce this shift. However, such steps are rarely observed; all the SCO layers are aligned, as indicated with the vertical line in Fig. 1 b).

The contrast in the BCO layers could be matched with simulated images of $La_2CuO_{4-\delta}$ (with all La replace by Ba) for a defocus of –20 nm and thickness of 2 nm. The ratio of a/c for $SrCuO_2$ is 1.03 and the ratio for the Ba_2CuO_x is 3.97. Figure 1 b) shows the positions of Ba (white circles) according to a model for the Ba_2CuO_x compound. These observations explain the unexpected superlattice value found with XRD (Koster 1999b), which corresponds to 2 Ba_2CuO_x blocks and 3 SCO blocks, and point to the need to re-consider the deposition rate calibration in the case of BCO.

Figure 2 a) shows a HREM observation of a superlattice structure deposited after recalibration of the deposition rate of BCO, see also Fig. 3 c). The HREM image was taken close to the STO substrate. The superlattice in this region is nearly perfect. Fig. 2 b) is an enlargement of Fig. 2 a) from which the noise due to an amorphous overlayer has been removed taken at defocus of –70 nm with a thickness of about 2 nm, the contrast can be interpreted as Cu represented by white dots. The image shows only contrast typical for perovskite blocks. The lattice spacings for each layer (measured from the images with calibration on STO) is 4.08 to 4.25 Å for BCO and 3.43 to 3.63 Å for SCCO. The super period measured from computer diffractograms obtained from the image of Fig. 2 b) is 19 to 20.4 Å.

Examples of typical θ–2θ scans are given in Fig. 3. The small width and the 'correct' positions of the satellite peaks, indicate high quality of the superstructures obtained using PLD. Superlattice periods found for different compositions are systematically larger than the ones expected from simple summation of the thickness of the individual blocks. By assuming a larger value than 4.2 Å (i.e., 4.4 Å) for the BCO blocks, a better fit to the measured values is obtained, similar to the observations of Balestrino et al. (1998a).

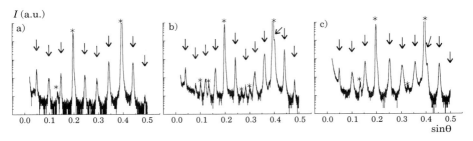

Figure 3: *θ–2θ scans of a) 2BCO/2SCO (Λ~15.7 Å), b) 2BCO/3SCO (Λ~19.1 Å) and c) 2BCO/2CCO (Λ~15.2 Å) superlattices. The (00l) reflections of the films are indicated with arrows, and the peaks corresponding to the substrate are indicated with an asterix.*

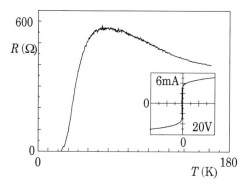

Figure 4: *Resistance versus temperature of a 2CCO/2BCO deposited at 575 °C and 20 Pa; here, the inset is the I-V curve measured at ~8K*

Metal-like behaviour and even superconducting transitions to zero resistance were only observed for films with at least one CCO block, deposited at temperature around 575 °C with an oxygen pressure >20Pa. Note, that the temperature/pressure window for good quality CCO containing films, is small, as determined from the RHEED measurements. The highest $T_{c,onset}$ was ~50K and $T_{c,zero}$~12K, see Fig. 4 c). This film showed an increase in resistance before getting into the superconducting state. Furthermore, the broad transition indicates that this is a multiphase sample. To confirm superconductivity, we performed an IV characterization at ~8K, see the inset of Fig. 4.

4   CONCLUSIONS

We have shown that PLD in combination with an *in situ* analysis technique, the fabrication of superlattice structures of oxide materials down to the atomic level becomes feasible.

We proved that the interval deposition method imposes a layer-by-layer growth mode almost irrespective of the deposition conditions. The structural analysis with XRD and HREM of the superlattices indicates first of all a high quality structure for, both, the overall crystallinity as well as in the imposed periodicity along the c-axis. Furthermore, the correct estimate for the deposition rate per constituent was confirmed by the period of the superlattices. A systematic deviation turned out to be caused by the compressive stress in the film.

This indicates that superconductivity in these materials is related to defect structures responsible for charge-carrier doping of the copper oxide planes. Our most perfect structures, e.g., $SrCuO_2/BaCuO_2$ superlattices, never show superconductivity, whereas for the superconducting films, defects are expected to be present.

REFERENCES

Balestrino G, Martellucci S, Medaglia P G, Paoletti A, Petrocelli G 1998a, *Physica C* 302, 78

Balestrino G, Martellucci S, Medaglia P G, Paoletti A, Petrocelli G and Varlamov A A 1998b, *Phys. Rev. B* 58, R8925

Koster G, Rijnders A J H M, Blank D H A and Rogalla H 1999a Appl. Phys. Lett. 74, 3729

Koster G, Rijnders A J H M, Blank D H A and Rogalla H 1999b, *Mat. Res. Soc. Sym. Proc.* accepted for publication

Koster G, Rijnders A J H M, Blank D H A and Rogalla H 1998a, *Mat. Res. Soc. Symp. Proc.* 502, 255

Koster G, Kropman B L, Rijnders A J H M, Blank D H A and Rogalla H 1998b, Appl. Phys. Lett. 73, 2929

Norton D P, Chakounakos B C, Lowndes D H and Budai J D 1994a *Appl. Phys. Lett.* 65, 2869

Norton D P, Chakounakos B C, Lowndes D H and Budai J D, Sales B C, Thompson J R, Christen D K 1994b, *Science* 265, 2074

Rijnders A J H M, Koster G, Blank D H A and Rogalla H 1997, Appl. Phys. Lett. 70, 1888

Sugii N, Ichikawa M, Hayashi K, Kubo K, Yamamoto K and Yamauchi H 1993 *Physica C* 213, 345

Tsukamoto A, Wen J G, Nakanishi K, Tanabe K 1997 *Physica C* 292, 17

Inst. Phys. Conf. Ser. No 167
Paper presented at Applied Superconductivity, Spain, 14-17 September 1999
© 2000 IOP Publishing Ltd

Bi₂Sr₂CaCu₂O₈₊ₓ epitaxial thin films on silicon substrates

S. Ingebrandt, J. C. Martínez, M. Basset, M. Mauer, M. Maier, H. Adrian

Institut für Physik, Johannes Gutenberg-Universität, 55099 Mainz, Germany

S. Linzen, P. Seidel

Friedrich-Schiller-Universität Jena, Institut für Festkörperphysik, Helmholtzweg 5, D-07743 Jena, Germany

ABSTRACT: Most of high temperature superconductors can be used at temperatures where classical Si-based devices can still operate. This inspired several groups to develop the deposition of $YBa_2Cu_3O_{7-x}$ on silicon. In this work, by the use of YSZ/CeO_2 buffer layer, we succeeded to grow on similar substrates $Bi_2Sr_2CaCu_2O_{8+x}$ epitaxial thin films with $T_C(R=R_n/2)=88.7$ K.

1. INTRODUCTION

Most of the high temperature superconducting cuprates present critical temperatures in T-regions where classical Si-based devices can still operate. The enormous potential of this combination inspired a number of groups to work on the deposition of $YBa_2Cu_3O_{7-x}$ (YBCO) thin films on Si substrates (Ajimine 1991, Haakenaasen 1994, Boikov 1995, Méchin 1996, Seidel 1996). More recently, YBCO-SQUIDs (Superconducting Quantum Interference Devices) deposited on silicon were realised with step-edge (Linzen 1995) and bi-crystal junctions (Linzen 1997). These devices could operate in temperatures up to 77 K.

The deposition of $Bi_2Sr_2CaCu_2O_{8+x}$ (Bi2212) on Si substrates could bring some advantages in comparison to YBCO. Due to the extreme bi-dimensional character of the Bi2212 structure, we would expect this compound to support better the stress induced by the reduced thermal coefficient of silicon (3.8×10^{-6} K^{-1} according to Vasiliev 1995). On the other hand we would expect as well, the degradation of the superconducting properties at the substrate/superconductor interface, to be reduced in comparison to YBCO. This was suggested by the large T_c values obtained for very thin Bi2212 films deposited on $SrTiO_3$ (STO) (Frey 1998a).

2. SAMPLE PREPARATION

In order to get a large number of identical 10 x 10 mm^2 substrates, two kinds of buffer layers have been prepared by pulsed laser deposition on 2" silicon substrates. One set contained a 60 nm thick YSZ (yttrium stabilised zirconia) layer while the second had a YSZ/CeO_2 double layer with thicknesses of 60 and 30 nm respectively. We deposited then ex-situ the Bi2212 layers by dc sputtering from a single target. More details about the sputtering chamber are described in the reference Wagner 1993. The best films were deposited at 830°C and 3 hPa O_2 pressure. In order to improve the superconducting properties the samples were annealed during 20 minutes at 450°C and 10 hPa O_2. Later on the films were patterned by standard photolithography and wet chemical etching with diluted HNO_3.

The Bi2212 thin films grown directly on YSZ showed no superconducting properties. An examination of the surface by scanning electron microscopy revealed a strong interdiffusion

22

between film and substrate. On the other hand the films deposited on Si/YSZ/CeO2 show no foreign phases. From atomic force microscopy (AFM) we determined a surface roughness of only 8.2 nm (RMS). No cracks could be observed for thickness up to 70 nm. AFM measurements done in patterned samples were used for calibrating the sputtering rate.

3. CHARACTERISATION BY X-RAY DIFFRACTION

In Fig. 1 we show a typical θ-2θ scan of a 40 nm thick Bi2212 thin film deposited on Si/YSZ/CeO$_2$. The main reflections correspond to the Bi2212 layer. The peaks due to CeO$_2$ and Si can be observed as well. The full width at half maximum of the rocking curve of the (0010) reflex was typically of the order of 0.9°. This value is considerably larger than the typical values of Bi2212 deposited on SrTiO$_3$ (STO) substrates (about 0.3° according to Wagner 1993).

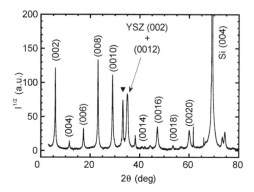

Fig. 1. θ-2θ scan of a Si/YSZ/CeO$_2$/Bi2212 heterostructure. The inverted triangle corresponds to the (002) reflection of the CeO$_2$ layer. The reflections from the Bi2212 film are indicated in the figure. The reflection (0010) of Bi2212 presents a rocking curve of 0.9°.

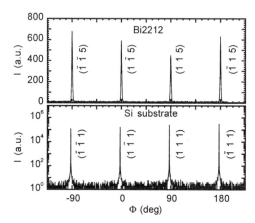

Fig. 2. Φ-scan of a Si/YSZ/CeO$_2$/Bi2212 heterostructure with thicknesses 60, 20 and 40 nm respectively. As shown in the figure Bi2212 grows cube to cube on silicon substrates.

From 4-circle X-ray diffraction we determined a=0.5485(10) nm, b=0.5490(12) nm and c=3.051(8) nm. If we compare these results with the expected values for films on STO (a=0.5414,

b=0.5418 and c=3.089), we see that there is a slight lattice distortion. The ab-plane is 1.3% larger and the c-axis 1.2% shorter.

In Fig. 2 we show Φ-scans for the (115)-family of reflexes of Bi2212 and the (111)-family for silicon. From these measurements we see that indeed Bi2212 grows cube to cube on the buffered silicon substrate.

4. RBS MEASUREMENTS

The film quality was as well investigated by Rutherford Back Scattering. In Fig. 3 we show a typical RBS-spectrum of a Bi2212 film on Si. The open circles represent the channelling experiment while the full circles are the random spectrum. The full line corresponds to a simulated spectrum for our system. For the simulation we used a film thickness of 65 nm and a ratio of 2-2-1-2 for the Bi-Sr-Ca-Cu cations. By comparing the measured and experimental spectra, we see that our films show a small Bi deficit. This could be due to either interdiffusion or, most likely, to a loss of Bi during the sputtering process.

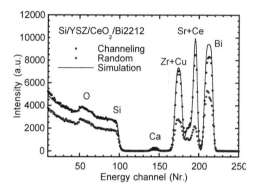

Fig 3. RBS spectrum of a Bi2212 film deposited on Si with a YSZ/CeO$_2$ buffer layer. From these measurements we determined thicknesses of 45, 25 and 65 nm for the YSZ, CeO$_2$ and Bi2212 layers respectively. The Bi2212 film shows a parameter χ_{min} of 57%.

From the ratio between the Bi peaks of the random and channelling spectra, it is possible to check the crystalline quality of the Bi2212 film. For the data above we found a ratio of 57%. This is a very good value when compared to the 23% obtained for a 180 nm thick Bi2212 deposited on STO (Wagner 1994). Due to the larger rocking curve observed by X-rays, we think that this result could be due to the reduced thickness of our films (65 nm).

5. TRANSPORT MEASUREMENTS

The transport measurements have been carried out on films having a thickness of 70 nm. The bridges were patterned with 200 µm width and 1.8 mm length. All measurements have been done in a four-point geometry. In Fig. 4 we show a resistivity measurement of a Bi2212 thin film deposited on a silicon substrate with a YSZ/CeO$_2$ buffer layer. The room temperature resistivity of about 600 µΩ.cm is similar to typical values of Bi2212 films on STO (between 200 and 1000 µΩ.cm). We observe as well a T_{c0} onset of 97 K and a T_c(R=0) of 76 K. This gives a transition width of about 21 K. The reduced quality of the transport properties could be due to the Bi deficit observed by RBS. On the other hand, this could as well be due to the large stress induced by the substrate on the Bi2212 film. A hint of these stresses is given by the deformation of the crystal lattice observed by 4-circle measurements.

24

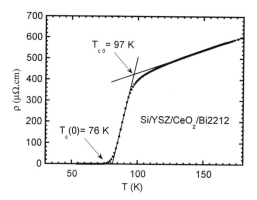

Fig 4. Resistive transition of a 70 nm thick Bi2212 thin film deposited on silicon. From these data we see a $T_c(R=0)$ of 76 K and an onset at T_{c0} = 97 K. The half transition is at 88.7 K. No cracks could be observed by scanning electron microscopy.

6. CONCLUSIONS

We demonstrate that it is possible to deposit superconducting Bi2212 thin films on silicon substrates. In order to reduce interdiffusion between the substrate and the superconductor a double YSZ/CeO$_2$ buffer layer is necessary. From our X-ray measurements we determine a cube to cube c-axis growth of the Bi2212 film. A slight deformation of the lattice, which could be due to the stresses induced by the substrate, has been observed. From RBS measurements we observe a Bi deficiency which could be either due to interdiffusion or to Bi loss during deposition. The Bi2212 films showed superconducting properties with a wide transition going from 76 up to 97 K. So far we could not determine either these reduced quality could be due to the Bi deficiency or internal stresses. Further investigations with targets having different stoichiometries could clarify this point.

ACKNOWLEDGEMENTS

The authors would like to acknowledge the financial support by the TMR-Program (*EG-ERBFMBICT972217*) and the Innovationsstiftung Rheinland-Pfalz (*8036-286261/222*), and the German BMBF (contract N° 13N6808A).

REFERENCES

Ajimine E.M., et al. 1991 Appl.Phys.Lett. **59** 2889
Boikov Y.A. et al. 1995 J.Appl.Phys. **78** 4591
Frey U et al 1998 Appl. Supercond. **6** 429
Haakenaasen R. et al. 1994 Appl.Phys.Lett. **64** 1573
Linzen S, Schmidl F, Dörrer L, Seidel P 1995 Appl.Phys.Lett **67** 2235
Linzen S, et al 1998 Applied Supercond, IOP Conf. Series **158** 699
Méchin L. et al. 1996 Physica C **269** 124
Seidel P. et al. 1996 Journal de Physique IV **C3** 361
Vasilev A L et al 1995 Physica C **244** 373
Wagner P et al 1993 Physica C **215** 123
Wagner P et al 1994 Journ. of Supercond. **7** 217

Inst. Phys. Conf. Ser. No 167
Paper presented at Applied Superconductivity, Spain, 14-17 September 1999
© 2000 IOP Publishing Ltd

Defects in YBCO films on CeO$_2$ buffered sapphire and LaAlO$_3$ and their impact on the microwave properties

J Einfeld, P Lahl, R Kutzner, R Wördenweber and *G Kästner

Forschungszentrum Jülich, ISI, 52425 Jülich, Germany
*MPI für Mikrostrukturphysik, 06120 Halle, Germany

ABSTRACT: In this paper we demonstrate, that defects are important for the improvement of the mechanical stability and hf properties of HTS thin films. Enhanced quasiparticle scattering at the defects results in lower R_S values compared to structurally more perfect films which contain less defects. The experimental observations are explained in terms of the two-fluid model with a d-wave superfluid in combination with thermally excited quasiparticles with a Drude-shaped conductivity spectrum. The reduction of R_S due to quasiparticle scattering at defects is of high interest for applications.

1. INTRODUCTION

The use of high-T_C superconductor (HTS) thin films for microwave applications requires the deposition of HTS material onto large-area, high-quality technical substrates with suitable microwave properties. An excellent candidate is *r*-cut sapphire with extremely low microwave loss tangent (values of tan$\delta \approx 10^{-6}$-10^{-8} are reported for low temperature and 10GHz). For hf applications, film thicknesses d larger than the effective penetration depth λ_{eff} are required. In the past the preparation of high-quality YBCO films on sapphire thicker than 250-300nm ($\approx \lambda_{eff}$ for YBCO at 77K) was found problematic due to the large lattice mismatch and, therefore, the formation of microcracks (Zaitsev 1997). Recently we introduced a method to increase the critical thickness up to $d_C > 700$nm by intentionally introducing structural defects into the YBCO film, which mechanically stabilise the ceramic film (Zaitsev 1997). Here, we examine the impact of these defects on the microwave properties of the film. We demonstrate, that the temperature dependence of the surface resistance R_S depends upon the density of defects in the YBCO films. The microwave surface resistance decreases with decreasing scattering time τ. The experimental observations can be explained in terms of a theoretical model. We demonstrate, that even on structurally better matched substrates defects can be generated via sputtering technique, which leads to a reduction of R_S.

2. DEFECTS IN YBCO FILMS

The YBCO films were prepared by on-axis dc magnetron sputtering on *r*-cut sapphire substrates in-situ buffered with rf-sputtered (001) CeO$_2$. Structural defects have been created in the YBCO film during deposition by adjusting the ion energy. Via this method the growing film is exposed to ion bombardment similar to ion plating processes. Fig. 1 represents the surface morphology of a series of YBCO films prepared with the same set of parameters but at different ion energies (Fig. 1a to 1c: decreasing ion energy). The morphology strongly depends on the ion energy. At high ion energy the surface is relatively rough (Fig. 1a). With decreasing ion energy the surface gets smoother but some outgrowths and holes are still present (Fig. 1b). Finally, at even lower ion energy very smooth and outgrowth free surfaces are achieved with a peak-to-valley roughness of less than

10nm (Fig. 1c). TEM analysis of these differently prepared YBCO films shows different defects. Rough films contain a high density of Y_2O_3 precipitates (Fig. 2). Furthermore, a-axis oriented grains are present which in some cases are associated with Y_2O_3 precipitates. While small precipitates are overgrown by YBCO, large precipitates cause pores (Fig. 1b) or even microcracks (Fig. 1a). With lowering the ion energy number and size of the defects are reduced. For optimised ion energy secondary phases could hardly be observed by TEM analysis, but small precipitates might still be present as deduced from X-ray diffraction experiments. These hardly visible defects or the tension caused by these defects in the YBCO films might be responsible for the modification in the microwave properties of these films.

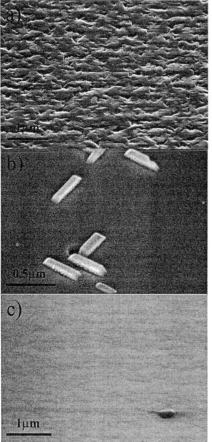

Fig. 1. Surface morphology of YBCO films prepared by dc sputtering with different ion energies (a) to c): decreasing ion energy).

3. IMPACT OF THE DEFECTS ON HF PROPERTIES

Fig. 3 shows the temperature dependence of the surface resistance of our thick YBCO films on sapphire. For comparison, $R_s(T)$ for three different YBCO films, a thermally evaporated YBCO film on $LaAlO_3$, a high-pressure dc sputtered YBCO film on $LaAlO_3$ and an electron beam coevaporated YBCO film on MgO, is shown (Hein 1998). The temperature dependence of R_s can clearly be distinguished for the YBCO films on different substrates. The YBCO films on the structurally well-matched $LaAlO_3$ exhibit a shoulder in $R_s(T)$, whereas this effect is suppressed in the surface resistance of the film on MgO and, even more, on sapphire. These differences in the temperature dependence of R_s can be explained in terms of the two-fluid model with a d-wave superfluid in combination with thermally excited quasiparticles characterised by a Drude-shaped conductivity spectrum, i.e. the defects are considered to be microscopic. Extended defects (e.g. grain boundaries) which in terms of different weak-link models (e.g. Hylton 1988, Hylton 1989) will mainly cause a general increase of R_s are not considered here. For temperatures not too close to T_c (i.e. $\sigma_1 \ll \sigma_2$) the microwave surface resistance is given by (Gorter 1934, London 1935)

$$R_s(T) = 0.5 \, \mu_0^2 \, \omega^2 \, \sigma_1(T) \, \lambda^3(T) \tag{1}$$

with the angular frequency ω, the complex conductivity $\sigma = \sigma_1 - i \, \sigma_2$ and the penetration depth λ. σ_1 can be expressed in terms of the Drude model, which for $\omega\tau \ll 1$ is given by

$$\sigma_1(T) = n_{qp}(T) \, \tau(T) \, e^2 / m \tag{2}$$

Here n_{qp} denotes the quasiparticle density, τ the scattering time of the quasiparticles, e and m the quasiparticle's charge and effective mass, respectively. The temperature dependence of λ is given by

$$\lambda(T) = (\mu_0 \, \omega \, \sigma_2(T))^{-1/2} \tag{3}$$

with

$$\sigma_2(T) = 2 \, n_s(T) \, e^2 / m \, \omega \tag{4}$$

where n_s is the density of the superfluid. In the framework of this model, the temperature dependence of the surface resistance is dominated by the temperature dependence of the density and scattering time of the excited quasiparticles. The density of quasiparticles is expected to increase with temperature according to $n_{qp} \propto (T/T_c)^4$ (Gorter 1934), whereas the scattering time decreases with temperature. In contrast to the exponential decay which is expected for the s-wave superconductor, a power-law behaviour is predicted and experimentally determined for the d-wave superconductor, which can be fitted by the empirical expression (Hosseini 1999)

$$\tau^{-1}(T) = \tau_a^{-1} + \tau_b^{-1}(T/T_c)^\gamma \qquad (5)$$

The relaxation is caused mainly by electron-phonon scattering, electron-electron scattering and scattering by impurities and other defects. The impurity and defect induced scattering mainly influences the low-temperature part of the scattering expressed by τ_a. Scattering rates of the thermally excited quasiparticles derived from experimental data (Bonn 1993, Hosseini 1999) measured on YBCO single crystals are given in Fig. 4. The experimental data can be fitted according to eq. (5) using one set of parameters $\gamma=6$ and $\tau_b=5\times10^{-14}$s. The residual value of the two different sets of crystals naturally depend strongly upon the sample quality. They are given by $\tau_a=2\times10^{-11}$s and $\tau_a=2\times10^{-12}$s for the sets of experimental values, which indicates the large difference in sample quality with respect to the density of scattering centers observed even in single crystals. The defect density (and thus the density of scattering centers) in epitaxial films is expected to be even larger than observed for single crystals. Therefore, an analogue temperature dependence for an even larger residual scattering rate ($\tau_a=2\times10^{-13}$s) has been added to Fig. 4.

Fig. 2. TEM analysis of YBCO/CeO$_2$/sapphire prepared with high ion energy (inset).

The resulting temperature dependences of the surface resistance are sketched in the inset. For increasing density of defects the shape of the $R_S(T)$ curve changes drastically. For low defect densities a peak in $R_S(T)$ occurs at $T/T_c\approx0.4$. With increasing density of scattering centers the position of the peak shifts to higher temperature and the peak gradually transforms to a plateau which finally vanishes. Thus, for high defect density R_S decreases monotonically with temperature, an effect, which might be responsible for the modified temperature dependence of our YBCO film on sapphire (Fig. 3).

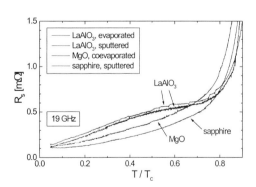

Fig. 3. Temperature dependence of the surface resistance for YBCO films on different substrate materials (Hein 1998).

Finally, we examined, whether differences in lattice mismatch between substrate and YBCO cause the difference in $R_S(T)$. We intentionally grew a series of YBCO films with different structural perfection on LaAlO3 substrates using the sputtering deposition process utilised for YBCO deposition on sapphire. Furthermore, CeO$_2$ buffer layers of different thickness were added. YBCO films on LaAlO3 without buffer layer exhibit a FWHM of the YBCO (005) X-ray rocking curve of $\Delta\omega=0.12°$, whereas a CeO2 buffer layer of about 25nm thickness results in an increase of $\Delta\omega$ to $\Delta\omega=0.57°$. Fig. 5 shows the temperature dependence of the surface resistance for this series of YBCO films. Surprisingly there seems to be no difference between the temperature dependence of R_S for the five different YBCO films. R_S decreases monotonically with temperature in all cases. Even our sputtered YBCO films with highest structural perfection ($\Delta\omega(005)=0.12°$) contain defects, which result in small τ_a values. Therefore, it can be concluded, that the deposition process generates structural imperfections independent of the choice of the substrate material.

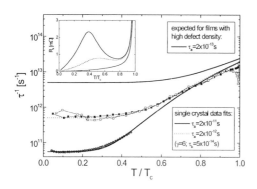

Fig. 4. Scattering rate of the thermally excited quasiparticles of YBCO. The symbols represent experimental values deduced from $R_s(T)$ and $\lambda(T)$ measurements on YBCO single crystals (see text). The lines represent fits of the experimental values according to eq. (5). The resulting temperature dependences of the surface resistance are given in the inset.

4. CONCLUSION

Defects are important for the improvement of the mechanical and hf properties of HTS thin films. They reduce the strain in YBCO films and, therefore, lead to an increased critical film thickness. The films are smooth, crack- and outgrowth-free up to larger thicknesses. Enhanced quasiparticle scattering at the defects results in lower R_S values compared to films containing smaller densities of defects. This experimental observation is described in terms of the two-fluid model with a d-wave superfluid in combination with thermally excited quasiparticles with a Drude-shaped conductivity spectrum. The reduction of R_S due to quasiparticle scattering at defects is of high interest for applications.

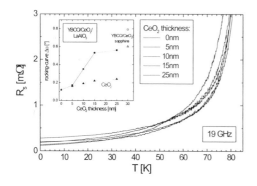

Fig. 5. Temperature dependence of R_s for YBCO films on LaAlO3 with different structural perfection (inset).

REFERENCES

Bonn D A et al 1993 Phys. Rev. B **47**, 11314

Gorter C J et al 1934 Phys. Z. **35**, 963

Hein M A et al 1998 Euroconference "Polarons: Condensation, Pairing, Magnetism", Erice, Sicily, Italy

Hosseini A et al 1999 Phys. Rev. B **60**, 1349

Hylton T L et al 1988 Appl. Phys. Lett. **53**, 1343

Hylton T L et al 1989 Phys. Rev. B **39**, 9042

London F et al 1935 Proc. Roy. Soc. A **149**, 71

Zaitsev A G et al 1997 Applied Superconductivity Inst. Phys. Conf. **158**, 25

Inst. Phys. Conf. Ser. No 167
Paper presented at Applied Superconductivity, Spain, 14-17 September 1999
© 2000 IOP Publishing Ltd

Epitaxial (101) YBa$_2$Cu$_3$O$_7$ thin films on (103) NdGaO$_3$ substrates

Y Y Divin, U Poppe and C L Jia

Institut für Festkörperforschung, Forschungszentrum Jülich GmbH, D-52425 Jülich, Germany

J W Seo

Institute de Physique, Université de Neuchatel, CH-2000, Switzerland and IBM –Zürich Research Laboratory, CH-8803 Rüschlikon, Switzerland

V Glyantsev

Conductus Inc., Sunnyvale, CA 94086, USA

ABSTRACT: We have fabricated epitaxial (101) oriented YBa$_2$Cu$_3$O$_7$ thin-films by high-pressure dc sputtering on about 18$^{\rm O}$ tilted vicinal NdGaO$_3$ substrates. The microstructure and electrical transport of these tilted c-axis films have been studied. Electron microscopy revealed that these films have a single-domain microstructure with a-b planes tilted by about 18$^{\rm O}$ with respect to the normal of the film surface. Additionally, a (101) domain was observed to grow on NdGaO$_3$ substrates with a surface of variable tilt like it is the case for step-edge Josephson junctions near the bottom part of the step. The special behaviour on 18$^{\rm O}$ tilted NdGaO$_3$ substrates is explained by a growth model which minimises the strain of the growing film.

1.   INTRODUCTION

The high-temperature superconductor YBa$_2$Cu$_3$O$_7$ is one of the promising materials for applications in superconducting electronics. Often c-axis oriented YBa$_2$Cu$_3$O$_7$ thin films are not advantageous for the preparation of planar-type multilayered Josephson junctions due to an extremely short coherence length $\xi_c \approx 0.2$ nm along the c-axis. The coherence length in the a-b plane is one order of magnitude larger and the fabrication of Josephson junctions (Divin et al 1995) and other devices like intrinsic Josephson junctions (Kleiner et al 1992) or intrinsic thermopiles (Lengfellner et al 1992), may be simplified using films with a significant c-axis component in the plane of the film.

Films with a-b planes perpendicular to the surface are usually obtained at relatively low deposition temperatures, which result in degraded superconducting properties. One way to avoid this problem or the appearance of grain boundaries is to use films with tilted c-axis on vicinal substrates.

The anisotropy of electrical properties of uncracked YBa$_2$Cu$_3$O$_7$ thin films with tilted c-axis has recently been studied by Divin et al (1994). The YBa$_2$Cu$_3$O$_7$ thin films were deposited by dc sputtering at high oxygen pressures on NdGaO$_3$ substrates with the orientation of substrate normal tilted by an angle α ($0 \leq \alpha \leq 45^{\rm O}$) with respect to the [100] orientation in pseudo cubic notation. As NdGaO$_3$ can be regarded as a substrate material with nearly cubic perovskite structure, we will use this pseudo cubic notation within this paper for simplicity. Usually the tilt of the films followed the tilt of the substrate and with an increase of the tilt angle α, additionally, some narrow columnar-shaped domains with perpendicular orientation of the c-axis were found in high-resolution transmission electron microscopy (HRTEM). A singular exception from this behaviour was observed at a tilt angle of about 18$^{\rm O}$ corresponding to a (103) substrate orientation. The microstructure and transport properties of tilted c-axis films and the impact on step-edge junctions will be presented.

2.   EXPERIMENTAL DETAILS

The YBa$_2$Cu$_3$O$_7$ thin films were deposited by dc sputtering from a stoichiometric target at high oxygen pressure (Poppe et al 1992). The sputtering parameters were optimised to obtain high-quality c-axis YBa$_2$Cu$_3$O$_7$ thin films on single crystal (001) NdGaO$_3$ substrates. The distance between the target and the substrate was around 11mm, the oxygen pressure was about 3.2 mbar, and the current was kept at values of 200 mA. Heater temperatures were held at about 920°C during the sputtering procedure. Cross-sectional specimens for HRTEM were prepared on films grown on vicinal cut or stepped NdGaO$_3$ substrates obtained by argon milling and studied in a JEOL 4000 EX microscope. Patterning for resistivity and critical current measurements of the thin films was made by photolithography and etching with a 0.3% bromine solution in propanol. For the anisotropy measurements two sets of orthogonal bridges, each with a width of 20 or 200 μm and a length of 1 mm were used.

3.   EXPERIMENTAL RESULTS

As already mentioned we observed that the YBa$_2$Cu$_3$O$_7$ thin films grown on vicinal cut sub-strates showed a (101) instead of the (109) orientation at the special tilt angle of α=18.5^o+.5° corresponding to a (103) substrate orientation in pseudo cubic notation. Usually the c-axis orien-tation of YBa$_2$Cu$_3$O$_7$ continuously follows the (001) orientation of the substrate with increasing tilt angle (Divin et al 1994) and, therefore, a (109) film orientation could be expected for this value of α.

In Fig. 1 a cross–sectional HRTEM picture of such a (101) oriented YBa$_2$Cu$_3$O$_7$ film on a (103) NdGaO$_3$ substrate is shown. The unit cells of the stepped substrate with a pseudo cubic lattice constant a$_{sub}$ and the unit cells of the film with c ≈3a ≈ 3a$_{sub}$ are indicated at the sharp interface. The film showed a single domain. The CuO$_2$ -layers in the sputtered material are tilted at a large angle of about 72°, and all of them are reaching the surface of the film. No antidomains with complementary (109) orientation were observed in areas of several μm. For a film thickness below about 100 nm no cracks could be detected within the film. The high quality of the (101) oriented YBa$_2$Cu$_3$O$_7$ thin films is reflected in their transport properties. As expected from the microstructure the films show a high anisotropy of the specific resistance at 100 K and of the critical current density at 77 K, which was about two orders of magnitude. In Fig. 2 and Fig. 3 the temperature dependencies of the resistivity and of the critical current density are shown. The curves measured with a transport current along and perpendicular to the CuO$_2$ -layers of the YBa$_2$Cu$_3$O$_7$ film are denoted by a and b, respectively.

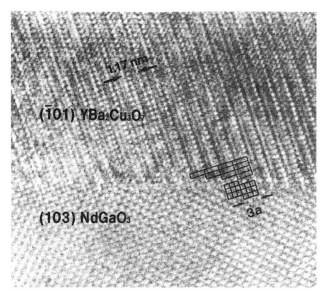

Fig.   1.   Cross–sectional HRTEM picture of a (101) oriented YBa$_2$Cu$_3$O$_7$ film on (103) NdGaO$_3$

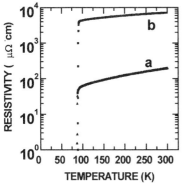

Fig. 2 Anisotropy of the resistivity
a- perpendicular, b - along the tilt

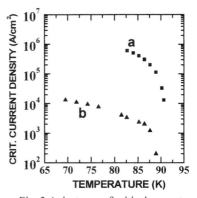

Fig. 3 Anisotropy of critical current
a- perpendicular, b - along the tilt

The microstructure of $YBa_2Cu_3O_7$ films which were dc-sputtered on step-edge $NdGaO_3$ (001) oriented substrates also was investigated by HRTEM. As shown in the overview image on the left part of Fig. 4, the film on the steep-step flank turns its c-axis by 90° in comparison with the film on the horizontal surface of the substrate. As a result, two grain boundaries are formed. This agrees with the observation in $YBa_2Cu_3O_7$ /$SrTiO_3$ or $LaAlO_3$ systems in which the films were prepared using the laser ablation technique (Jia et al 1990). However, an additional special feature was found in the step-edge films prepared by sputtering on the $NdGaO_3$. A small a- or b-axis oriented domain was usually occurring near the lower edge of the step where the surface of the substrate is sloped by a certain angle of about 18°. Such a small domain is denoted by two white arrows at the interface in the right part of Fig. 4, which shows a magnification of the lower part of the step. For this region we locally have the same crystallographic relation between substrate and film as in the case of Fig.1 for a 18° tilted substrate. The grain boundaries within the film and the interface between film and substrate are marked by white and black thin lines, respectively. Due to a series connection of the four grain boundary Josephson junctions at the step-edge additional structures at voltages below 1 mV were observed in the current-voltage characteristics of the junctions.

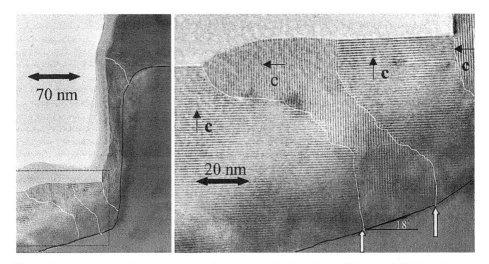

Fig. 4. Cross–sectional electron microscopic overview (left) and magnified detail (right) of a step-edge Josephson junction on a (001) oriented $NdGaO_3$ substrate.

32

4. DISCUSSION

To explain the exceptional growth behaviour at the special tilt of the (103) substrate orientation one possible growth model which minimises the energy of the growing film is suggested. It is known that at monatomic surface steps on slightly vicinal substrates often antiphase boundaries nucleate in c-axis oriented films. This happens as the c-axis length of the film is about three times the step height a_{sub}. Especially for a high surface step density like in the case of the (103) substrate orientation it can be energetically favourable even to grow with another orientation in order to reduce the strain introduced by antiphase boundaries in the film. Fig. 5 and 6 explain possible growth situations for a (103) substrate orientation.

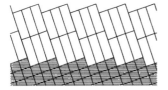

Fig. 5. Growth model of a (109) oriented film $YBa_2Cu_3O_7$ film on a (103) oriented substrate.

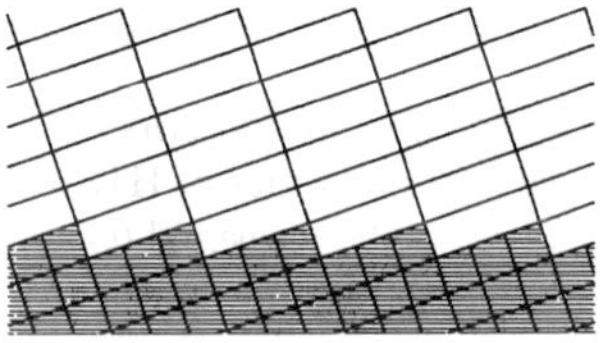

Fig. 6. Growth model of a (101) oriented on a (103) oriented substrate

In Fig. 5 an antiphase boundary appears for every substrate terrace only three unit cells wide. On the other hand for the growth mode with a (101) film orientation described in Fig. 6 no antiphase boundaries are observed. Despite the fact that the growth conditions -namely high substrate temperature and low deposition rate- usually would favour a nearly c-axis oriented (109) growth described in Fig. 5 the energy is lower for the (101) film. Another hypothetical growth mode of a (109) oriented film would be a situation where large regions with non regular atomic positions appear near the interface instead of antiphase boundaries. Also for this case or for relatively flat, reconstructed sustrate surfaces the disturbed regions are much smaller for a (101) oriented film and therefore energetically favourable. It should be noted that this model is only applicable for films where the c-axis length is nearly an integer multiple of the a- and b-axis length like for $YBa_2Cu_3O_7$ with $c \approx 3a \approx 3b \approx 3a_{sub}$. This leads to an avoidance of antiphase boundaries only for special low index substrate orientations like (103) or (106). For other layered compounds like e.g. $HgBa_2Ca_2Cu_3O_{9-x}$ or $TlBa_2Ca_2Cu_3O_{9-x}$ with $c \approx 4a \approx 4a_{sub}$ the growth model predicts a (104) or (108) substrate orientation.

5. CONCLUSIONS

The unique microstructure and electrical transport properties of epitaxial (101) $YBa_2Cu_3O_7$ thin-films on (103) $NdGaO_3$ substrates have been studied and explained by a growth model which minimises the strain of the growing film. For the application of high-T_c thin films on planar-, step edge-, or intrinsic Josephson junctions as well as other devices, like intrinsic thermopiles, the use of this special substrate orientation may be beneficial.

REFERENCES

Divin Y Y, Poppe U, Seo J W, Kabius B and Urban K 1994 Physica C **235-240,** 675
Divin Y Y, Poppe U, Seo J W, Kabius B, Urban K and Shadrin P M 1995 Proceedings ofEUCAS'95, Inst. of Physics Conf. Ser. **148**, IOP Publishing Ltd., Bristol, 1359
Jia C L, Kabius B, Urban K, Schubert J, Zander W, Braginski A I 1990 Physica C **196**, 211
Kleiner R, Steinmeyer F, Kunkel G and Müller P 1992 Phys. Rev. Lett.**68**, 2394
Lengfellner H, Kremb G, Schnellbögl A, Betz J, Renk K F, Prettl W, 1992, Appl.Phys.Lett. **60** 501
Poppe U, Klein N, Dähne U, Soltner H, Jia C L, Kabius B, Urban K, Lübig A, Schmidt K, Hensen S, Orbach S, Müller G and Piel H 1992 J. Appl. Phys. **71**, 5572

Inst. Phys. Conf. Ser. No 167
Paper presented at Applied Superconductivity, Spain, 14-17 September 1999

Spatial variation of the non-linear surface impedance of HTS films.

L Hao and J C Gallop

National Physical Laboratory, Queens Rd., Teddington, TW11 0LW, UK

ABSTRACT: We describe further development of a novel technique for the characterisation of microwave properties of HTS films which allows the spatial variation of this important physical parameter to be measured (Gallop et al. 1997). The method is capable of rapid measurement of the non-linear surface impedance and can in principle distinguish between global heating effects at high incident microwave power and microscopic intrinsic non-linear processes.

1.   INTRODUCTION

Spatially resolved measurements of HTS microwave surface impedance represent an important requirement for the assessment of large area HTS thin films which are of considerable importance to the development of microwave applications of the cuprate superconductors (Hein (1997), Newman & Lyons(1993)). We describe here further developments of a technique (Gallop et al. 1997) which involves scanning a dielectric puck resonator across the surface of a large area HTS film, sampling the surface impedance at a number of discrete frequencies between 5 and 15 GHz. The surface impedance can also be rapidly measured as a function of microwave magnetic field strength. Spatial resolution for the prototype system is as small as 1-2 mm. The key to the method is that the dielectric puck is made of a high permittivity, low-loss single crystal rutile (TiO_2). The relative permittivity ε_r of rutile is anisotropic; along the c-axis ε_r varies from 250 to 160 as the temperature varies from 70K to 300 K whereas the ab plane ε_r varies from 103 to 85 over the same range (Klein et al. 1995). The loss tangent of single crystal rutile is also sufficiently small below 100K that high quality factor (Q) dielectric resonators can be made using it. The high permittivity means that the physical dimensions of the resonator are small for a given resonant frequency and also that the microwave fields are almost totally confined within the volume of the rutile puck.

The surface resistance and spatial variations in surface reactance can be measured by using a loop oscillator which is interrupted by a fast microwave switch. The decay of microwave power in the resonator is then monitored as a function of time to determine the power dependent surface impedance parameters. This process is extremely fast and straightforward and the loop oscillator configuration permits the use of only relatively inexpensive microwave components.

Dielectric resonators have been routinely used to measure the average microwave surface impedance properties of HTS films (see e.g. Klein et al. (1995), Shen et al. (1992)). A high permittivity rutile puck has also been positioned at various points on a large area YBCO film to provide quality control information (Lemaitre et al. (1994)). Here the development is the introduction of a miniature rutile puck which may be continuously moved around over the surface of a HTS film held at 77 K by immersion in liquid nitrogen (LN_2).

2.   EXPERIMENTAL ARRANGEMENT

Fig. 1 shows a schematic of the experimental configuration which we currently use.

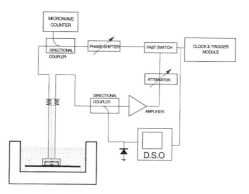

Fig. 1 Schematic of spatially resolved surface impedance measurement system and loop oscillator electronics.

The puck resonator is mounted over a large HTS thin or thick film, cooled to 77 K by immersion in liquid nitrogen so that the lower surface of the puck is held in contact with the HTS by spring loaded pressure. The film may be translated in two dimensions beneath the resonator and at each position the centre frequency and linewidth of a number of high Q modes may be measured. This allows the microwave surface impedance Z_s to be analysed at each point in principle for a range of discrete frequencies representing different resonant modes of the rutile puck. Microwave power dependence of $Z_s(H_{rf})$ (i.e. the non-linear response of the films to the microwave magnetic field H_{rf}) may also be determined as described in section 3 below. The HTS film, which can be up to 60mm in diameter, is mounted on an insulating plate firmly fixed to the inside base of a small glass vacuum flask (130 mm inside diameter and 90 mm deep). The vacuum flask is mounted on a precision x-y translation stage capable of 100 mm travel in either direction. The 3.5 mm rutile puck is fixed in a single crystal quartz housing which in turn is fixed inside a high conductivity copper housing which carries two coupling probes connected to SMA connectors. There is a gap of 3 mm between the copper housing and upper surface of the puck, sufficient to make negligible the Q contribution due to losses in the copper. The copper housing is held vertical in a rigid support which holds the lower surface of the rutile puck in good mechanical contact with the HTS film by means of a compression spring. The SMA connectors are linked to the microwave measurement equipment by low-loss coaxial cables. In the first demonstration of this technique (Gallop et al., 1997) a CW synthesised source (2 GHz to 20 GHz) was coupled into the resonant structure and transmitted power is coupled out, via a broadband low-noise amplifier to a calibrated detector. The synthesiser may be swept, under computer control, through the width of a resonance response (usually that of the TE_{011} mode since this has the highest Q for this configuration), the detector output being recorded so that a Lorentzian lineshape may be fitted to the data. This provides best fit parameters for centre frequency, linewidth and amplitude. The puck is then positioned over a new part of the HTS film before the process is repeated.

3.   PULSED LOOP OSCILLATOR METHOD

More recently we have implemented an alternative, and much faster, method, in which the output coaxial line is fed via a variable phase shifter, attenuator, fast electronic microwave switch and room temperature filter to a low noise microwave amplifier with around 38dB of gain and a saturation output power level of +13dBm. The output from the amplifier is fed back to the input coaxial coupler of the resonator and when the phase shift around the loop has been adjusted to be accurately $2n\pi$ (n an integer) the noise signals in the resonator are amplified and an oscillation at the selected resonant frequency builds up. The attenuator must be adjusted to give stable amplitude of oscillation a little below the gain compression power level of the amplifier. The resonator is first filled up with electromagnetic energy (photons) until a steady state is reached where the input power is equal to the sum of the power dissipated in, and radiated from, the resonator. A trigger pulse opens the fast switch, cutting off the input power to the resonator, sustained oscillation stops and the microwave power emitted by the resonator decays exponentially with time. The stored energy in the resonator U(t) and the radiated power P(t) both decay exponentially with time:

$$P(t) \propto U(t) = U(0)exp(-2\pi\Delta ft) \quad ...(1)$$

where Δf is the 3dB linewidth of the resonance at frequency f_o and is related to the loaded Q_l by the expression:

$$Q_l = f_o/\Delta f \quad ....(2)$$

Equations (1) and (2) show that Q_l, the loaded Q factor can be measured accurately by timing the decay, using a log-scaled detector (such as a spectrum analyser or a calibrated diode detector). Fig. 2 shows the logarithmic output of a spectrum analyser as a function of time following opening of the microwave switch within the loop. Note the very straight portion of the decay (until the noise floor of the analyser is approached) indicating that the non-linearity of the resonator is very low. For non-linear superconductors Q_l is a function of the rf field amplitude H_{rf} and the instantaneous slope $dP(t)/dt$ can be used to calculate $Q(H_{r.})$.

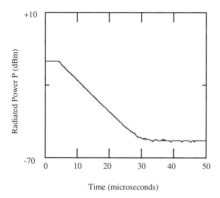
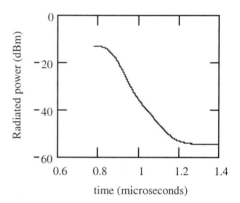

Fig. 2. Plot of a) accurately exponential slow decay (YBCO thin film) b) fast decay with two distinct linear regions (indicating severe non-linear response) at defect in YBCO thick film.

At present measurements have mainly been made using the TE_{011} mode at around 11.5 GHz but a number of other modes in the rutile puck in the frequency range from around 5 GHz to 15 GHz have Q values high enough to make useful measurements. Note that the simple loop oscillator method is only capable of determining the non-linear surface resistance, not the imaginary component $X_s(H_{rf})$ corresponding to the non-linear reactance although, by measuring the loop oscillation frequency using a microwave counter when the switch is closed, the *spatial variation* of the full-power surface reactance may be obtained. A further development has been initiated which allows the latter component to be measured also. This involves reintroducing a spectrally pure tuneable microwave synthesised source. If the loop oscillator output is fed to a mixer which is supplied by a synthesised local oscillator operating at fixed frequency close to that of the loop oscillator the decaying signal from the mixer output following the opening of the fast microwave switch will reflect the changes in the frequency of the loop oscillator as a function of microwave power in the oscillator which, in turn, reflects the non-linear variation of the surface reactance of the superconductor.

36

4. NON-LINEAR PROCESSES AND THEIR MEASUREMENT

An understanding of the non-linear processes which may limit the performance of thin films of HTS for microwave applications is crucial. Many sources of non-linearity may be identified (Hein (1997)), these include intrinsic effects arising from the bulk properties of the HTS film, extrinsic effects which are often related to the presence of defects of various kinds such as grain boundaries and global heating effects relating to the thermal properties of the film, its substrate and its support structure.

4.1 Thermal Modelling of Local and Global Heating in the System

The thermal processes involved in either local or global heating related non-linear response can be reduced in a first approximation to the one dimensional problem of the transfer of microwave energy into Joule heating, either at a localised defect or uniformly throughout the superconducting film, followed by the transfer of heat through the body of the superconductor itself, then through any thermal boundary resistance between the film ad its substrate. The next step in the transfer involves heat flow through the substrate via a possibly significant second thermal boundary resistance to a heat sink maintained at a base temperature. The process for local heating can only be reasonably approximated by a 1D model when the defect in which Joule heating is concentrated has a lateral minimum dimension which is greater than any of the following characteristic lengths: i) the film thickness itself, generally ~200nm for a thin film but anything up to 50micron for a thick film, ii) the substrate (thickness usually in the range 0.2 to 1mm). From an applications viewpoint the time constants associated with various heat transfer processes must be understood. These depend on both the heat capacities of the structures between which heat is being transferred and the thermal conductances of the components through which it is transferred. Of course in reality the process is a distributed problem but it turns out that the 'lumped-circuit' approximation generally used is reasonable and has the added advantage of allowing separation of the above processes into ideally different time scales which may prove useful in identifying the nature and location of the dissipative process. By varying the 'on' cycle time of the microwave loop oscillator between decrement measurements and observing a resulting decay time reduction we have clear evidence of a heating process with a time constant of around 30-50ms for YBCO thick film samples. This implies that the global heating processes is limited by heat transfer from substrate to cooling plate and suggests that improvements in heat transfer at this point may be significant for improved device performance.

5. CONCLUSIONS AND FUTURE WORK

The scanning rutile puck operated in the decrement mode has been demonstrated as a quick and relatively cheap technique for measuring the non-linear spatially resolved surface impedance of films of high temperature superconductor. We have indicated here how the method may be extended to include the determination of the time variation of the surface reactance (and hence its non-linear behaviour). In addition the method should allow detection of non-linear response arising from global heating and a measurement of the characteristic time involved in this process.

REFERENCES

J C Gallop, L Hao & F Abbas (1997) *Physica* C **282-287** pp 1579-80
M A Hein (1997) *Supercond. Sci Technol.* **10** pp. 867-71
N Klein, C Zuccaro, U Daehne, H Schulz, N Tellmann, R Kutzner, A G Zaitsev & R Woerdenweber, J. Appl. Phys. Vol. 78 pp 6683-6 1995
Y Lemaitre, L M Mercandalli, B Dessertenne, D Mansart, B Marcilhac & J C Mage (1994) *Physica C* **235-40** pp 643-4
N. Newman & W.G. Lyons (1993) *J. Supercond.* **6** pp. 119-60
Z Y Shen, C Wilker, P Pang, W L Holstein, D Face & D J Kountz (1992) *IEEE Trans. MTT* **40** pp 2424-32

Inst. Phys. Conf. Ser. No 167
Paper presented at Applied Superconductivity, Spain, 14-17 September 1999
© *2000 IOP Publishing Ltd*

Even and Odd Hall Effects in YBCO Thin Films

V Shapiro, A Verdyan, I Lapsker and J Azoulay

Center for Technological Education, Holon Institute of Technology Arts and Science, P.O. Box 305, Holon 58102, Israel

ABSTRACT: Hall resistivity measurements as a function of temperature in the vicinity of T_c were carried out on thin film YBCO superconductors. On the same films we have observed the anomalous odd Hall effect and also the even Hall effect which was insensitive to the direction of magnetic field along c-axis of the film. The even Hall effect is discussed on the basis of directional motion of vortices along certain net of channels.

1. INTRODUCTION

The anomalous Hall effect has been caused by a vortex motion in high- temperature superconductors has been extensively studied by several groups of researchers since 1988. Hall voltage sign change was detected upon crossing T_c (for example see Hagen et al 1993). In general Hall voltage was calculated as the algebraic half-sum of the cross voltages measured in opposite directions of the magnetic field. Such method is suitable for odd Hall effect measurements (odd in magnetic field). There are only few works, in which observation of even Hall effect in $YBa_2Cu_3O_7$ in the resistive state near T_c was reported (Kopelevich et al 1989, Kholkin et al 1991, Morgoon et al 1996, Villard et al 1996). Even Hall effect (EHE) - is a phenomenon known to exist in type-II superconductors in the mixed state in which a non-reversible component of the transverse voltage does not change sign upon inversion of the magnetic field. Kopelevich et al (1989) studied EHE on ceramics specimens in magnetic fields up to 1 T at 77K. In the work of Holkin et al. (1991) EHE has been studied in films obtained by magnetron spattering on MgO substrates. In the works of Morgoon et al. (1996) and Villard et al (1996) EHE was observed on single crystals with unidirectional twins. EHE was also found in the classical superconductors - cold rolled samples of Pb-In alloys and Nb (Niessen et al 1965). Kopelevich et al. (1989) have explained the EHE effect on the basis of directed motion of vortices along certain channels. In the classical superconductors directional motion is realized by the preference of the flux lines to move in the rolling direction of cold-rolled specimens (Niessen et al 1965). In the YBCO ceramics (Kopelevich et al 1989) and films (Holkin et al 1991) it is conceivable to assume that such channels can be weak links, while in single crystals (Morgoon et al 1996, Villard et al 1996) they are assumed to be grain boundaries or twin boundaries. It was shown by Mawatari (1997) that the off-diagonal components σ_{xy} and ρ_{xy} have large symmetric contributions, which do not change their signs upon reversing the magnetic field direction due to the anisotropy of pinning and of vortex viscosity. According to the model proposed by Meilikhov (1993) the magnetic flux percolates through a random net of quasirectilinear channels of easy motion under Lorenz force. He received the nonzero expression for root-mean square of the Hall voltage $<U_H^2>^{1/2} \propto L_\perp^{1/2}$ ($L_\perp$ - lateral extent of the sample). In this model $<U_H>=0$ which means that the odd Hall effect is absent. The presence of odd Hall effect can be taken into account (in the limits of the considered model) by introducing an angle between Lorenz force direction and the average direction of the magnetic flux flow. This angle determines the existence of the odd Hall effect.

2. EXPERIMENTAL

All the films in this work were synthesized in a vacuum system by evaporation of pulverized stoichiometric mixture of Y, BaF$_2$ and Cu powder from a resistively heated tungsten boat onto a c oriented well polished MgO substrate. The deposited thin film with a typical thickness of 5000Å was subsequently annealed in situ in oxygen atmosphere (Azoulay et al 1991). The films thus obtained were characterized by X-ray diffraction for c-axis orientation and only those with highly c-axis oriented were selected for measurements.

The longitudinal resistivity ρ_{xx} versus temperature shown for different values of magnetic fields **B** in fig. 1 displays the expected transition broadening (about 30 K) as compared to the one with zero magnetic field

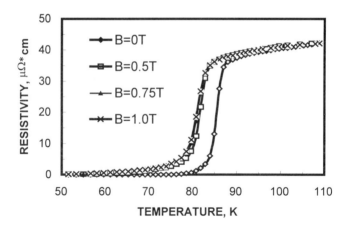

Fig.1. The longitudinal resistivity ρ_{xx} vs. temperature for different values of magnetic fields are along the c-direction.

Odd Hall voltage is received as the half difference of the cross voltage for opposite directions of the magnetic field: $U_H= [U_{xy} (B) -U_{xy} (-B)] /2$. As the signal of the even Hall effect on our samples is many times larger then the signal of the odd effect we have used the following procedure. Spurious signal, caused by the nonequipotentiality of the cross contacts was subtracted from the $U_{xy} (T)$ signal. We have therefore used the correction equation $U_H(T)=U_{xy}(T)-k \cdot U_{xx}(T)$ to eliminate this spurious signal where k assumed to be a temperature independent constant was found from the condition $U_H=0$ at $T=T_c$.

3. RESULTS AND DISCUSSION

The Hall coefficient as a function of temperature was conventionally obtained by subtracting the Hall voltage values U_H for both magnetic field directions. The results summarized in fig. 3 showing the well known sign reversal are similar to those obtained by Hagen et al. (1993).

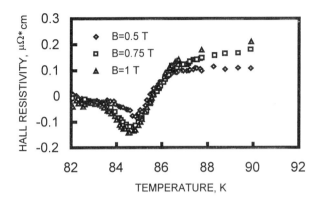

Fig.2. Odd Hall effect as a function of temperature.

The value of even Hall effect (fig. 3) much exceeds the value of odd Hall effect. The temperature range of the EHE is wider, than that of the odd Hall effect as one can see by comparing fig. 3 with fig. 2.

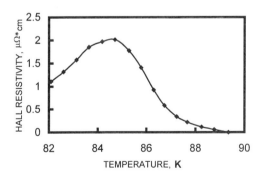

Fig.3. Temperature dependence of even Hall effect, **B**=0.5 T.

It is possible to approximate the dependence of U_H/U_J on $1/J$ by linear line in the range of $J = 100 - 800$ A/cm^2 (fig. 4). The results of our measurements are best explained by the presence of a network of channels along which the magnetic flux can move easily in the studied samples. In accordance with the work of Meilikhov (1993) the average spacing between channels of easy motion of magnetic flux decreases with current.

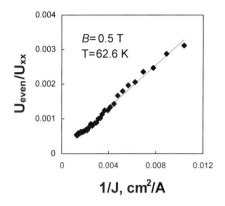

Fig.4 Dependence of U_H/U_J on $1/J$.

4. CONCLUSIONS

The even Hall effect and its insensitivity to the external magnetic field inversion for small fields were discussed in terms of guided motion of vortices along certain channels. It is most probable that they are due to the grain boundaries and weak links.

ACKNOWLEDGMENTS

The authors would like to express their gratitude to L. Burlachkov and K. Kikoin for useful discussion.

The Israel ministry of Science supported this work.

REFERENCE

Azoulay J 1991 Phys. Rev. B, **44**, 7018

Hagen S J, Smith A W, Rajeswari M, Peng J L, Li Z Y, Greene R L, Mao S N, Xi XX, Bhattacharya S, Qi Li and Lobb C J 1993 Phys. Rev. B **97**, 1064

Kholkin A L, Lemanov V V, Kopelevich Ya V 1991 Pisma v JTF (Russian) **17**, 32.

Kopelevich Ya V, Lemanov V V, Sonin E V and Kholkin A L 1989 JETP Lett. **50**, 212

Mawatari Y 1997 Phys. Rev. B **56**, 3433

Meilikhov E Z 1993 Physica C **209**, 566

Morgoon V N, Shklovskij V A, Bindiltti V, Bondarenko A V, Jardim R F, Becerra C C and Satori A F 1996 Czech. J. Phys. **46** suppl. Pt.S3, 1751

Niessen A K, van Suchtelen J, Staas F A and Druyvesteyn W F 1965 Phylips Res. Repts. **20**, 226

Villard C, Koren G, Cohen D, Polturak E, Thrane B and Chateignnier D 1996 Phys. Rev. Lett. **77**, 3913

Inst. Phys. Conf. Ser. No 167
Paper presented at Applied Superconductivity, Spain, 14-17 September 1999
© 2000 IOP Publishing Ltd

Surface resistance of YBCO thin films at THz-frequencies

I Wilke[1], C T Rieck[1], T Kaiser[2], C Jaekel[3], H Kurz[3]

[1]Universität Hamburg, Jungiusstrasse 11, D-20355 Hamburg, Germany
[2]Fachbereich Physik, Universität GH Wuppertal, Gaußstrasse 20, D-42097 Wuppertal, Germany
[3]Institut für Halbleitertechnik II, RWTH Aachen, Sommerfeldstrasse 24, D-52056 Aachen, Germany

ABSTRACT: We report on measurements of the surface resistance of $YBa_2Cu_3O_x$ thin films at frequencies between 0.087THz and 2THz by time-domain Terahertz-transmission spectroscopy and resonant microwave spectroscopy. The temperature and frequency dependence of the surface resistance of $YBa_2Cu_3O_x$ thin films in the THz-range is successfully explained by a weak coupling model of d-wave superconductivity which incorporates inelastic and elastic scattering. The surface resistance of $YBa_2Cu_3O_x$ thin films at THz-frequencies is compared to the surface resistance of gold and niobium. The advantages of $YBa_2Cu_3O_x$ thin films for superconducting THz-electronic devices are discussed.

1. INTRODUCTION

Since the discovery of high temperature superconductors (HTS) the application of these materials to high frequency electronics has received great interest (Likharev et al 1989). Major advances in the implementation of HTS electronic devices at microwave frequencies (Lancaster 1997, Hein 1999) have tremendously increased the interest in the development of active and passive HTS electronic devices operating at THz-frequencies. Promising applications of HTS THz-electronics are e.g. Rapid-Single-Flux-Quantum (RSFQ) circuits (Shokor et al 1995, Ruck et al 1997), HTS transmission lines (Nuss et al 1989 & 1991a, Osbahr et al 1999) or antennas (Chen et al 1996) as well as Josephson junctions (Beuven et al 1997) as emitters and detectors of THz-radiation. An important part of the design of these devices are numerical simulations (Töpfer et al 1996, Kohjiro et al 1997) of their performance. The accuracy of the simulations relies on the input of precisely measured material properties of HTS at THz-frequencies as well as on theoretical models describing the mechanism of HTS superconductivity. One of the key parameters describing high frequency electromagnetic properties of HTS is the surface resistance because it determines the dissipation of ac currents oscillating at THz-frequencies. In this report an experimental and theoretical study of the THz-surface resistance of $YBa_2Cu_3O_x$ film is presented. The surface resistance data are compared to a theory of d-wave superconductivity. The advantages of $YBa_2Cu_3O_x$ thin films for HTS THz-electronic applications in comparison to normal conducting metals and conventional superconductors are discussed.

2. EXPERIMENTS

2.1 Time-domain Terahertz-transmission spectroscopy

For the measurement of the surface resistance of the $YBa_2Cu_3O_x$ thin films two different experimental techniques have been used. At f=0.087THz the surface resistance of the

42

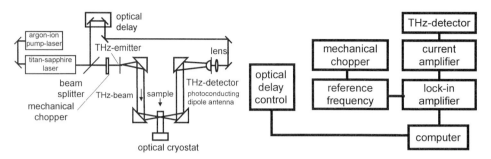

Fig.1a Schematic of the time-domain THz-transmission spectrometer

Fig.1b Data acquisition for time-domain THz-transmission spectroscopy.

$YBa_2Cu_3O_x$ thin film has been measured by resonant microwave spectroscopy. The experimental set-up and the data analysis for this method have been explained in detail by Hensen et al 1997.

Between 0.5THz and 2.0THz the surface resistance of the $YBa_2Cu_3O_x$ thin film has been experimentally determined by time-domain THz-transmission spectroscopy (TDTTS) (Nuss et al 1991b, Jaekel et al 1994 & 1996, Gao et al 1995). Schematics of the experimental setup and data acquisition are displayed in Fig.1a & b. TDTTS measures in the time-domain the transmission of a picosecond electromagnetic transient through the superconducting film $E^F(t)$ and through the bare substrate $E^S(t)$ as a reference. The Fourier components of the transmitted $E^F(\omega,T)$ and reference $E_S(\omega,T)$ THz-electric fields are obtained through fast Fourier transformation. They define the transmission as $t(\omega,T) = E^F(\omega,T)/E^S(\omega,T)$.

$$t(\omega,T) = \frac{E_1^F + iE_2^F}{E_1^S + iE_2^S} = \frac{4n_2}{(1+n_2)(n_2+n_3)\exp(-in_2\omega d/c) + (1-n_2)(n_2-n_3)\exp(in_2\omega d/c)} \quad (1)$$

The complex index of refraction of the superconducting thin film $n_2=n+ik$ is related to the complex transmission $t(\omega,T)$ through eq.(1) (Born and Wolf 1975). In this equation d is the thickness of the superconducting thin film, c is the velocity of light, n_3 is the index of refraction of the substrate and $\omega=2\pi f$ the frequency. In our analysis the complex index of refraction of the superconducting thin film $n_2=n+ik$ is determined by numerically solving eq.(1). Then the dielectric function ε and the conductivity σ are calculated by dielectric conversion as $n_2^2=\varepsilon=\varepsilon_1+i\varepsilon_2$ and as $\sigma=\sigma_1-i\sigma_2=i\omega\varepsilon_0\varepsilon$. The surface resistance R_S is determined from the real and imaginary part of the dynamic conductivity $\sigma=\sigma_1(\omega,T)-i\sigma_2(\omega,T)$ as $R_S(\omega,T)=((\mu_0\omega/2)(|\sigma|(\omega,T)-\sigma_2(\omega,T))/(|\sigma|^2))^{1/2}$ (Jackson 1975).

During the last decade TDTTS has proved to be a unique experimental method for the quick broadband measurement of the surface resistance of HTS thin films at THz-frequencies. As an optical technique it does not require electrical contacts or patterning of the HTS thin film. Most importantly TDTTS is a phase-sensitive method which enables the determination of the real and imaginary part of the complex conductivity without Kramers-Kronig analysis.

2.2 YBa2Cu3Ox thin film sample

We have investigated a 80nm thin c-axis oriented epitaxial $YBa_2Cu_3O_x$ thin film which has been deposited by laser ablation (Heinsohn 1998) on a <001> oriented MgO substrate $(10\times10\times1)mm^3$ in size. The superconducting transition temperature of the sample is $T_C=85.2K$. The $YBa_2Cu_3O_x$ thin film exhibits a high degree of the c-axis orientation $\delta_C=91\%$ and an oxygen content $x=6.95\pm0.05$. The optical examination of the $YBa_2Cu_3O_x$ thin film reveals a smooth surface with a marginal amount of precipitation.

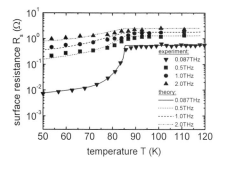
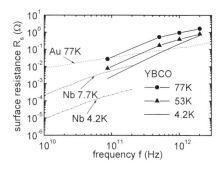

Fig.2 Experimental determined surface resistance of the YBa$_2$Cu$_3$O$_x$ thin film compared to theory. The fit parameters are a=0.11, b$_1$=15, b$_2$=1 λ_L=150.7nm, Γ^{el}_N=1.2meV, Γ^{inel}(T$_C$)=35.8meV, δ_N=0.4π.

Fig.3 Comparison of the surface resistance of YBa$_2$Cu$_3$O$_x$ with the surface resistance of gold (Au) and niobium (Nb).

3. RESULTS AND DISCUSSION

The experimentally determined surface resistance of the YBa$_2$Cu$_3$O$_x$ thin film as a function of temperature is displayed in Fig.2. The temperature dependence of the surface resistance of YBa$_2$Cu$_3$O$_x$ thin films at THz-frequencies is described quantitatively by a weak coupling model of d-wave superconductivity. This theoretical description, which incorporates the d-wave symmetry of the order parameter $\Phi=\Delta_0\cos2\phi$ has been extensively studied for YBa$_2$Cu$_3$O$_x$ thin films at f=0.087THz previously (Hensen 1997). By comparison with our measurements of the surface resistance at frequencies above f=0.087THz the validity of this model at higher THz-frequencies is now tested. The most important parameters in this model are the elastic scattering rate Γ^{el}_N and the inelastic scattering rate Γ^{inel}. It has been demonstrated that inelastic and elastic scattering contribute to the total scattering within our range of temperatures. The elastic scattering rate Γ^{el}_N depends on the density of point defects n$_{imp}$ which are the scattering centers and on the density of states N(0) at the Fermi level. Inelastic scattering is represented by a temperature dependent phenomenological scattering rate. $\Gamma^{inel}(t)=\Gamma^{inel}(T_C)f_{phen}(t)$ with $f_{phen}(t)=at^3+(1-a)\exp\{(b_1(t-1)[1+b_2(t-1)^2]\}$ and t=T/T$_C$. Its temperature dependence is similar to the temperature dependence of inelastic scattering based on spin fluctuation exchange within the Nested-Fermi-Liquid (NFL) model (Ruvalds 1996). While the NFL-model takes the full frequency dependence into account the phenomenological model neglects the frequency dependence involved in inelastic scattering. The solid line in Fig.2 is a fit of this theory to the measured surface resistance at f=0.087THz. The details of the calculations are explained in Hensen (1997). Subsequently, the surface resistance of the YBa$_2$Cu$_3$O$_x$ thin film at f>0.087THz has been calculated with these fit parameters and is indicated in Fig.2 as broken lines. The only fit parameter which had to be adjusted separately for frequencies between 0.5THz and 2THz is the background conductivity σ_{back}. Apart from this, the comparison of experiment and theory exhibits an excellent agreement of the temperature and frequency dependence of the surface resistance of the YBa$_2$Cu$_3$O$_x$ thin film up to the highest measured frequency of f=2.0THz. Thus, our model of the surface resistance of YBa$_2$Cu$_3$O$_x$ thin films has been successfully tested and proved to be valid up to 2THz. This is an improvement of previous work (Nuss 1991a & 1991b, Rueders 1995, Jaekel 1994 &1996, Gao 1995). where the surface resistance is either described by a generalized two-fluid model or by BCS theory and its approximations because the d-wave nature of HTS was not yet known and theories of d-wave superconductivity were not available yet. Therefore, by introducing our weak coupling model of d-wave superconductivity to the theoretical description of the surface resistance at THz-frequencies and demonstrating good agreement with experiments up to f=2THz we provide a powerful tool for the calculation of the surface resistance of $YBa_2Cu_3O_x$ thin films as a function of temperature at THz-frequencies.

With regard to the application of $YBa_2Cu_3O_x$ thin films in HTS THz-electronics the experimental and theoretical results of the surface resistance of $YBa_2Cu_3O_x$ thin films are compared to data of the surface resistance of gold (Au) and niobium (Nb) (Fig.3). The surface resistance data for $YBa_2Cu_3O_x$ at T=77K and T=53K originates from our experiments. Below T=53K the surface resistance of $YBa_2Cu_3O_x$ has been calculated according to our theory. The surface resistance data for Nb is from Orbach-Wettig 1994 and the data for gold from Nuss 1991a.

The comparison reveals that for cooling with liquid nitrogen (T=77K) $YBa_2Cu_3O_x$ has a lower surface resistance than Au for frequencies below 0.1THz and for cooling with liquid helium (T=4.2K) for frequencies below 0.5THz. The surface resistance of $YBa_2Cu_3O_x$ is always higher than the surface resistance of Nb at T=4.2K as long as Nb is superconducting (f≤0.7THz). $YBa_2Cu_3O_x$ is superior than Nb for temperatures above 7.7K and frequencies below 0.2THz.

As a result $YBa_2Cu_3O_x$ is the material of choice for HTS THz-electronic devices, particularly transmission lines or antennas for frequencies below 0.2THz at temperatures between 7.7K and 77K. We remark that this temperature range corresponds to the operating range of state-of-the-art closed cycle coolers.

We would like to thank K.O. Subke, M.Schilling, K.Scharnberg, M.Khazan (Universität Hamburg), P.Kuzl (Czech Academy of Sciences, Prague), M.Hein, G.Müller (Universität GH Wuppertal) and P.Haring, RWTH Aachen for discussion and support of this work..

REFERENCES

Beuven S, Harnack O, Amatuni L, Kohlstedt H, Darula M 1997 IEEE Trans.Appl.Supercond.**7**, 2591

Born M and Wolf E 1975 Principles of Optics 5th edition, Pergamon Press

Chen J, Myoren H, Nakajima K and Yamashita T 1996 Physica C **293**, 288

Gao F, Whitaker J F, Usher C, Hou S Y, Phillips J M 1995 IEEE Trans. Appl. Supercond. **5**

Hensen S, Müller G, Rieck C T, Scharnberg K 1997 Phys. Rev. B **56**, 6237

Hein M A 1999 *High temperature superconducting thin films at microwave frequencies*, Springer Tracts in Modern Physics **155**, (Heidelberg:Springer)

Heinsohn J K 1998 Physica C **299**, 99

Jackson J D 1975 Classical Electrodynamics 2nd edition, John Wiley & Sons

Jaekel C, Waschke C, Roskos H G, Kurz H, Prusseit W, Kinder H 1994 Appl. Phys. Lett. **64**, 3326

Jaekel C, Kyas G, Roskos H G, Kurz H, Kabius B, Meertens D, Prusseit W, Utz B 1996 J. Appl. Phys. **80**, 3488

Kohjiro S, Kikuchi T, Kiryu S, Shoji A 1997 IEEE Trans. Appl. Supercond. **7**, 2343

Lancaster M J 1997 *Passive microwave device application of HTS*, Cambridge University Press

Likharev K L, Semenov V K and Zorin A B 1989 IEEE Trans. Magn. **MAG-25**, 1290

Nuss M C and Goosen K W 1989 IEEE J .Quant. Electron. **25**, 2596

Nuss M C, Goosen K W, Mankiewich P M, O'Malley M O, Marshall J L, Howard R E 1991a IEEE Trans. Magn. **MAG-27**, 863

Nuss M C, Goosen K W, Mankiewich P M, O'Malley M L 1991b Appl. Phys. Lett. **58**, 2561

Orbach-Werbig S 1994 dissertation **WUB-DIS 94-9**, Uni GH Wuppertal

Osbahr C J, Larsen B H, Holst T, Shen Y, Keiding S R 1999 Appl. Phys. Lett. **74**, 1892

Ruck B, Oelzle B and Sodke E 1997 Supercond. Sci. Technol. **10**, 991

Rueders F, Hollrichter C, Copetti C A, Foerster A, Buchal Ch 1995 J.Appl. Phys. **77** 5282

Ruvalds J 1996 Supercond. Sci. Technol.**9**, 905

Shokor S, Nadgorny B, Gurvitch M, Semenov V, Polyakov Yu, Likharev K L1995 Appl. Phys. Lett. **67**,2869

Toepfer H, Harnisch T and Uhlmann F H 1996 J.de Physique **IV(C3)**, 345

Inst. Phys. Conf. Ser. No 167
Paper presented at Applied Superconductivity, Spain, 14-17 September 1999
© 2000 IOP Publishing Ltd

The Influence of Reaction Layers on the Properties of Tl-2212 Thin Films on LaAlO$_3$

D M C Hyland, C J Eastell*, A P Jenkins, C R M Grovenor* and D Dew-Hughes

Dept of Engineering Science, Parks Road, Oxford, OX1 3PJ, UK.
*Dept of Materials, Parks Road, Oxford, OX1 3PH, UK.

ABSTRACT It is widely thought that LaAlO$_3$ (LAO) is a particularly stable and unreactive substrate material suitable for the growth of a wide range of superconducting films. Here we present some results on the reaction of LAO substrates with the HTS film gained as part of the quality control exercise in our Tl-2212 thin film fabrication programme. A number of films exhibiting poorer microwave properties have an amorphous reaction layer forming between the LAO substrate and the superconducting film.

1   INTRODUCTION.

It is widely thought that LaAlO$_3$ (LAO) is a particularly stable and unreactive substrate material (*Bramley et al*, 1999) suitable for the growth of a wide range of superconducting films. However, Tl-based superconducting phases are known to react with substrate materials such as sapphire (*Bramley et al*, 1999), necessitating a CeO$_2$ buffer layer, and Y-ZrO$_2$ giving strong formation of BaZrO$_3$ at 898°C (*Lee et al*, 1989). Most publications indicate no reaction between Tl-HTS films and LAO substrates. No evidence of reaction layers was found for either Tl-2212 grown at 850°C on LAO (*Holstein et al*, 1993), or in HREM studies of Tl-2223 grown on LAO by MOCVD and annealed at 820°C (*Hinds et al*, 1994). HREM studies also find no reaction between YBCO and LAO (*Ramesh et al*, 1991), but for YBCO on LAO processed at a much higher temperature, 955°C, a reaction layer tentatively identified as Ba$_3$Al$_2$O$_6$ was found (*Guo et al*, 1995). A thorough study of (Hg, Tl)-1223 thin films on LAO found an amorphous reaction layer formed in some films processed between 750 and 850°C (*Vasilev et al*, 1996).

Thus LAO seems relatively unreactive with Tl-HTS up to about 850°C. If higher processing temperatures are used then chemical reaction at the interface may become possible. There has been no reported correlation of the effects of these reaction layers on the superconducting properties of the films. Here we present some results on the reaction of LAO substrates with the HTS film gained as part of the quality control exercise in our Tl-2212 thin film fabrication programme. A number of films exhibiting poorer microwave properties have a reaction layer forming between the LAO substrate and the superconducting film.

2   EXPERIMENTAL DETAILS.

Precursor films were r.f. sputtered from a Ba-Ca-Cu-O ceramic target of 2:1:2 stoichiometry onto various sized LAO substrates (1cm^2 to 2 inch diameter). These films were then ex-situ annealed in a specially sealed furnace with a source of Tl_2O vapour. All films described in this paper were nominally annealed in air for 20 minutes at 850°C.

The HTS films were examined by XRD and SEM to determine phase purity and surface morphology before their R_s was measured using the dielectric resonator technique (*Kobayashi*, 1980). This process acts as a quality control before the higher quality films are patterned into microwave devices as part of our microwave communications research programme (*Jenkins et al*, 1999). TEM examinations were carried out on a JEOL 2010 microscope, and details of specimen preparation are published elsewhere (*O'Connor*, 1998).

3 RESULTS.

All data shown here comes from identically processed films. Tl-2212 films where reaction layers had formed could be easily determined visually by the presence of a white layer seen by looking through the reverse side of the LAO substrate. TEM cross-sections were examined (see Figs 1 and 2), and EDAX spectra obtained from some characteristic features (Fig 3).

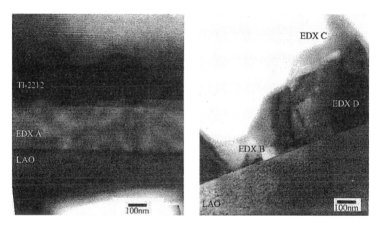

Figs 1 (lhs) and 2 (rhs) showing the reactions between Tl-2212 and LAO, the features of which are described in the text

Substrate atoms would appear to have reacted severely with the Tl-2212 to form a reaction layer containing Tl, Ba, Ca and Cu as well as a significant amount of Al (Fig3a), but also diffused through the film to form La-rich particles on the film surface (Fig 3c). Occasional Al-rich particles were found at the film / substrate interface as shown in Figs 2 & 3b. There also appears to be a substantial concentration of La substituting in the Tl-2212 superconducting phase, Fig 3d. However the most important structural modification found in films with a reaction layer is the formation of polycrystalline Tl-2212 over parts

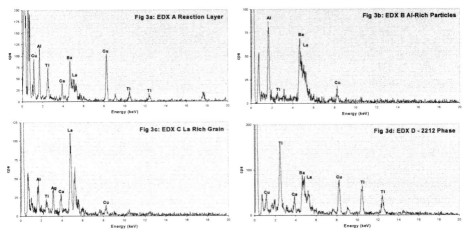

Fig 3a-d EDX data from corresponding features in the TEM micrographs

of the surface of the film identification of unaligned 2212 peaks from these grains are clearly seen in Fig 4.

The R_s data for this film (A) was determined (see fig 5) and is compared to those from (B), a film with a-axis needle-like outgrowths but no reaction layer, and (C) a high quality film with no a-axis grains or reaction layer. While the a-axis needles degrade the microwave properties quite severely, as has been described elsewhere (*O'Connor et al*, 1998), the formation of the reaction layer has a much more damaging effect on the R_s values. This is easy to understand if the microwave currents have to flow through polycrystalline material at the surface.

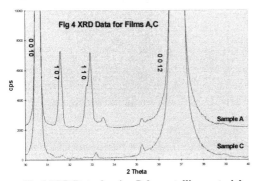

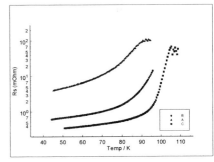

Fig 4 XRD Data showing Polycrystalline material   **Fig 5 R_s Data @ 10GHz for three Tl-2212 films**

4   DISCUSSION AND CONCLUSION

The reaction we observe between some superconducting films and the LAO substrate has a seriously detrimental effect on the microwave properties of the film. The surface resistance of 2212 films has been shown to depend on many factors; phase purity, film thickness and surface morphology, for example. The surface morphology, and more particularly the growth of a-axis oriented needles, will increase R_s values significantly because of the poor superconducting properties of 2212 out of the a-b plane. However we are demonstrating here that any reaction layers can reduce the microwave quality of the Tl-2212 films much more seriously by introducing regions of polycrystalline material presumably as a result of the 2212 film not being able to nucleate epitaxially on the underlying substrate.

From the TEM examination of the samples, not only is there a reaction between the substrate and film resulting in the Al-rich layer, but there would appear to be a "poisoning" of the superconductor by La clearly coming from the substrate. The effect of La on other Tl-HTS phases has been seen in Tl-1212 based superconductors where La substitutions suppress T_c perhaps by the preferential formation of Tl-1201 (*Xin et al,* 1991). We suggest that incorporation of La atoms within the Tl-2212 unit cell will lead to changes in the structure also reducing T_c. The suppression of T_c or at least broadening of the superconducting transition is clear from the R_s data (fig 5), although the material still appears phase pure 2212 in the XRD analysis.

While we have clearly identified the presence of a reaction at the LAO / Tl-2212 interface and the formation of polycrystalline Tl-2212 in some films, we have no explanation of why films processed under essentially identical conditions only occasionally show these features. One possible explanation is that our process temperature of about 850°C is exactly where the substrate / film reaction becomes sufficiently fast to produce a detectable amorphous film. There could also be a corresponding real furnace temperature fluctuation from run to run. We are currently studying this possibility in order to improve our yield of high quality films from about 60%.

REFERENCES

Bramley A P, O'Connor J D & Grovenor C R M, 1999, *Supercond Sci Technol* **12**, R57

Guo L P, Tian Y J, Liu J Z, Xu S F, Li L, Zhao Z X, Chen Z H, Cui D F, Lu H B, Zhou Y L & Yang G Z, 1995, *Appl Phys Lett*, **66**, 3357

Hinds B J, Schultz D L, Neumayer D A, Han B, Schneider J L, Hogan T P & Kannewirf C R, 1994, *Appl Phys Lett*, **65**, 231

Holstein W L, Parisi L A, Flippen R B & Swartzfager D G, 1993, *J Mater Res*, **8**, 962

Jenkins A P, Dew-Hughes D, Edwards E, Hyland D & Grovenor C, *To be presented at EUCAS 1999*

Kobayashi Y & Tanaka S, 1980, *IEEE MTT*, **28**, 1077

Lee W Y, Salem J, Lee V, Deline V, Huang T C, Savoy R, Duran J & Sandstrom R L, 1989, *Physica C*, **160**, 511

O'Connor J D, Jenkins A P, Grovenor C R M, Goringe M J & Dew-Hughes D, 1998, *Supercond Sci Technol* **11**, 207

Ramesh R, Inam A, Hwang, D M, Ravi T S, Sands T, Xi X, Wu X D , Li Q, Venkatesan T & Kilaas R, 1991, *J Mater Res*, **6**, 2264

Vasilev A L, Kvam E P, Foong F & Liou S H, 1996, *Physica C*, **269**, 181

Xin Y, Sheng Z Z, Gu D X & Pederson D O, 1991, *Physica C*, **177**, 183

Inst. Phys. Conf. Ser. No 167
Paper presented at Applied Superconductivity, Spain, 14-17 September 1999
© 2000 IOP Publishing Ltd

Plasma spraying of superconducting YBCO and DYBCO coatings

R Enikov, D Oliver, I Nedkov, T Koutzarova, O Vankov, Ch Ghelev, N Mihailov, V Tsaneva

Institute of Electronics BAS, 72 Tzarigradsko Shaussee, 1784 Sofia, Bulgaria

ABSTRACT: The processes which take place during the spraying of thick YBCO and DYBCO coatings in an air-plasma jet were investigated. The influence of the plasmatron's power parameters, the powder feeding and the subsequent heat-treatment on the coating's superconducting characteristics, were clarified. Coatings with the onset of the superconducting transition at 84K (YBCO) and 89 (DYBCO) were obtained.

1. INTRODUCTION

One of the promising techniques for fabrication of superconducting coatings with a thickness of hundreds of microns using YBCO (or DYBCO) ceramics is the spraying in a low-temperature plasma jet [Kathikeyan (1988), Konaka (1988), Krilov (1990)]. Direct current plasmatrons have been predominantly used with power up to several tens of kW and Ar or $Ar + H_2$ as plasma gas. These studies have been aimed at the quick formation of superconducting coatings on a large area for the purposes of manufacturing magnetic screens or flexible ribbons to be used in superconducting electromagnet coils. There are no data published in the available literature, however, concerning the technological characteristics of the deposition and heat-treatment processes. As our own experience made it evident, the fabrication of superconducting coatings imposes strict requirements on the plasmatron's operating conditions. The aim of the work presented was to establish the plasmatron's operating mode necessary for superconducting coatings formation.

2. EXPERIMENTAL SET-UP

The coatings were sprayed by means of a d.c. plasmatron characterized by gas-vortex plasma-column stabilization. Air blown tangentially into the cathode region was used as plasma-forming gas. The gas vortex thus formed stabilized the arc within the low-pressure zone created along the channel axis. The cathode was made of Hf, the anode, of copper. The anode spot was rotated by an axial magnetic field in order to limit the anode erosion. The plasma jet was discharged through a nozzle with a special design that eliminated the tangential velocity component (Fig. 1). Holes in the nozzle enabled us to feed the powder material used for coating deposition. The plasmatron's operating characteristics were: electric power – up to 40 kW; air consumption rate – $(1 - 1.5)$ g/s; jet enthalpy – up to 18 MJ/kg; jet axis temperature – up to 7 000 K. The plasmatron was mounted on a manipulator allowing a 3-axis motion during operation in order to cover large areas from different distances. The powder was carried by air into the plasma jet using a vibration feeding device ensuring a constant consumption rate during the experiment; the volume consumption rate was 0.5 cm^3/min.

3. POWDER PREPARATION

Fig. 1. Process of plasma spraying.

The powder with composition $YBa_2Cu_3O_7$ and $Dy_{0.4}Y_{0.6}Ba_2Cu_3O_7$ used for spraying was prepared following the classical ceramic technology that involved a solid-phase reaction; the initial batch consisted of Y_2O_3 , CuO, $BaCO_3$ and Dy_2O_3 with 0.999 purity.

The choice of Dy-substituted YBCO in the experiment has to do with the relatively high stability of the superconducting properties of this system as compared with the classical YBCO, as well as with the possibility offered by the magnetic ion for formation of pinning centers. These facts were discussed in another work [Nedkov (1998)].

The preliminary baking took place at 950 ^{0}C. The material obtained was ground and the powder was pressed again into pellets the pellets were ground again and the powder was sieved. The fraction with grain size of 40 μm was only used for plasma spraying.

4. EXPERIMENT

With the aim of establishing the optimal spraying mode, we varied the plasmatron's power parameters: current, voltage, gas consumption rate. The substrates used were 1 mm thick wafers of Al_2O_3 and Cu. To improve the coating adhesion, they were sand-blasted before the experiments. The distance between the nozzle slit and the substrate was 8 – 12 cm. The time needed for depositing a 0.2 mm layer with a 10 cm^2 area was 3 s. After the spraying the samples were carried out in a channel furnace in an oxygen flow. The treatment temperature was 950 ^{0}C with an isothermal delay of 1 hour. The cooling down to room temperature was done without oxygen flow. The superconducting properties were checked by using the four-probe d.c. measuring technique. None of the experimental samples showed a superconducting behavior after the spraying. This is typical for the plasma-spraying technique,

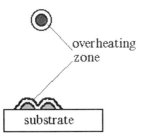

Fig. 2. Process of coating's formation.

whereby the particles lose some of the bound oxygen due to the heating in the jet; annealing is needed to restore their characteristics [Kathikeyan (1988), Konaka (1988)].

When the plasmatron's power is high (the jet enthalpy exceeds 12 MJ/kg), the coatings, although having good adhesion, do not exhibit superconducting properties even after the heat-treatment. The meaning of this is that the particles were melted in the plasma jet, but their chemical structure has been irreversibly changed. After the heat-treatment, the coatings contain oxides of Y, Ba, and Cu, as evidenced by the XRD spectra. These data are similar for both the composition studied. Fig.3 illustrates the XRD pattern for YBCO only. At low values of the enthalpy, the coatings have poor mechanical properties and disintegrate due to the bad adhesion to the substrate. The optimal situation would be when the particles reaching the substrate are with molten outer surface, while their inner part remains with unchanged chemical composition (Fig.2). The high temperature gradient along a particle's radius is the result of the low value of the heat-conduction coefficient, characteristic for ceramic materials, and of the high rate of heating of the particles in the jet (exceeding 10^4 K/s).

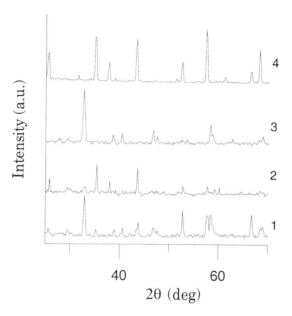

Fig. 3. XRD patterns of YBCO samples submitted to heat treatmments.

In Fig.3 we present XRD patterns of the initial superconducting powder YBCO(1), of the coating deposited (2), of the coating following a heat-treatment in oxygen (3), and of the

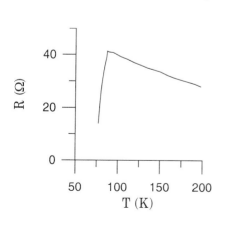

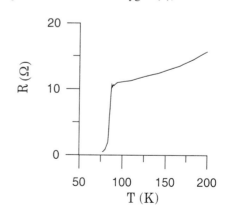

Fig. 4. Temperature dependence of the resistance of YBCO on Al_2O_3 substrate.

Fig. 5. Temperature dependence of the resistance of DYBCO on Cu substrate.

Al_2O_3 substrate (4). The peaks typical for a superconducting structure are negligible in pattern (2), but appear in pattern (3), where one can also see the peaks originating from the substrate. In fig.4 and fig.5 we present the temperature dependence of the electrical resistivity of the YBCO and DYBCO coatings. The resistance begins to drop at 84 K for YBCO and at 89 K for DYBCO.

The results obtained for Dy-substituted YBCO show that this system is more suitable for the deposition technique discussed, which opens up possibilities for further optimization of the method.

5.   CONCLUSION

We have demonstrated that the important factors affecting the superconducting properties of plasma-jet deposited layers are the particles temperature and time of residence in the jet. These can be controlled via the proper choice of the plasmatron's power parameters. The presence of Dy increases the temperature of the superconducting transition. This system does not require a very precise control of the oxygen content in the technological medium. The introduction of magnetic ions improves the electric properties of the layer because of the active rope of the magnetic cation in the formation of pinning centers.

This work was supported by National Foundation for Scientific Research at the Bulgarian Ministry of Education, Science, and Technologies under contract F639.

REFERENCES

Nedkov I, Koutzarova T, Miteva S, Ausloos M, Bougrine H and Cloots R 1998 Proc. XIV [th]
    Int. Conf. on Microwave Ferrites ICMF'98, Eger, Hungary, pp.224
Konaka T, Sankawa I et al.., 1988, Jap.J.Appl.Phys. **27**, 1092
Kathikeyan J, Sreekumar K P et al.,1988, J.Phys.D:Appl.Phys. **21**,1246
Krilov I K, Kudinov V V et al., 1990, Izvestia SO AN USSR Tech. Sci. **1**, 125

Inst. Phys. Conf. Ser. No 167
Paper presented at Applied Superconductivity, Spain, 14-17 September 1999
© 2000 IOP Publishing Ltd

Microwave Properties of Screen-Printed Bi2223 Thick Films on Ba(Sn,Mg,Ta)O$_3$ Dielectric Ceramics

Y Kintaka[1], T Tatekawa[1], N Matsui[1], H Tamura[1], Y Ishikawa[1] and A Oota[2]

[1]Murata Manufacturing Co. Ltd., Tenjin, Nagaokakyo, Kyoto 617-8555, Japan
[2]Toyohashi University of Technology, Tempaku-cho, Toyohashi, Aichi 441-8580, Japan

ABSTRACT: The TM$_{010}$ mode microwave resonator was fabricated with screen-printed (Bi,Pb)$_2$Sr$_2$Ca$_2$Cu$_3$O$_x$ thick films and a Ba(Sn,Mg,Ta)O$_3$ dielectric ceramic disk (3mm thick and 35mm in diameter) of relative dielectric constant εr=24. The typical value of the unloaded quality factor was 30,000 at 2.1GHz, 70K, which corresponds to surface resistance of 0.8mΩ. By addressing this, the 1.8GHz TM$_{110}$ dual-mode resonator was designed and fabricated using a 25mm-cube of Ba(Sn,Mg,Ta)O$_3$ ceramic with (Bi,Pb)$_2$Sr$_2$Ca$_2$Cu$_3$O$_x$ thick films on every 6 surface.

1. INTRODUCTION

Several types of high temperature superconductor (HTS) filter for cellular base station have been investigated and succeeded in demonstrating low insertion loss and sharp skirt characteristics. However, since most of them consist of micro-strip line resonators showing the edge effect, there are difficulties in high power handling capability. To avoid this, we proposed another type of HTS filters using a dielectric resonator composed of planar HTS films and a monoblock dielectric ceramics, and found the crucial facts for its realization at cryogenic temperatures well below 77 K as follows: (1) practical dielectrics Ba(Sn,Mg,Ta)O$_3$ (BSMT) can be served as dielectric ceramics for this purpose because of low dielectric losses and also a monotonic variation with decreasing temperature; (2) screen-printed (Bi,Pb)$_2$Sr$_2$Ca$_2$Cu$_3$O$_x$ (Bi2223) thick films can be fabricated directly on BSMT without significant chemical reactions, so that they show considerably low surface resistance (Tatekawa et al 1999). All these facts encourage us strongly for its realization.

In this paper, we present the results for the surface resistance (Rs) of screen-printed Bi2223 thick films on BSMT disks, as a function of temperature and also as a function of maximum surface magnetic field under microwave operation on a TM$_{010}$ mode. The third-order intermodulation distortion characteristics are also reported. From a practical point of view, we fabricated the 1.8GHz TM dual-mode resonator using a 25mm BSMT cube with Bi2223 thick films as electrodes on every 6 surface and present the results for the unloaded quality factor (Qu).

2. EXPERIMENTAL PROCEDURE

2.1 Sample Preparation

The Bi2223 thick films were fabricated by a screen printing method. Superconducting paste which consists of calcined powder of Bi-Pb-Sr-Ca-Cu-O and organic vehicles was screen-printed directly on both sides of BSMT disks (3mm thick and 35mm in diameter) without any buffer layers. The films were heated at 400℃ to evaporate organic substances and subjected to cold isostatic pressing (CIP) at 0.2GPa. After such mechanical treatments, the sample was sintered at 830℃ for 50h. A combination process for CIP and sintering was repeated until total sintering time reached 200h.

2.2 Measurements

The TM$_{010}$ mode resonator was constructed as Fig.1 and mounted in the helium-filled gas chamber of the cryo-cooler. The value of resonant frequency f$_0$ and unloaded quality factor Qu were measured as a function of temperature between 20 and 130K by Network Analyser (HP8720C). Each temperature was controlled within ±0.05K by a heater during measuring run.

The intermodulation distortion characteristics of each resonator were measured using two tone signals with frequencies f$_0$+25kHz and f$_0$-25kHz.

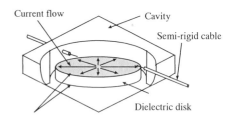

Fig.1   Structure of the TM$_{010}$ mode resonator.

3.   RESULTS AND DISCUSSION

3.1   TM$_{010}$ mode resonator

The temperature dependence of Qu and Rs at 2.1GHz on a TM$_{010}$ mode is shown in Fig.2. The Qu of Ag electrode resonator with same structure and its Rs are also shown for comparison. The value of Qu for the Bi2223 thick film electrode resonator reaches 30,000 at 70K and 80,000 at 20K, which correspond to Rs of 0.8mΩ and 0.3mΩ, respectively. Note that the values of Rs are that of interface side of thick film (not of free surface side of thick film). The Rs value is 1/7 of that of Ag at 70K and 1/10 at 20K.

Fig.3 shows the Rs value at 70 K as a function of maximum surface magnetic field for Bi2223 thick films on BSMT substrate. As can be seen, the Rs value for the film on BSMT increases abruptly at around 10A/m with increasing the maximum field. This field value (~10A/m) is about 2 orders of magnitude lower than the values for YBa$_2$Cu$_3$O$_{7-\delta}$ thin film (Nguyen et al 1993) and Tl$_2$Ba$_2$CaCu$_2$O$_8$

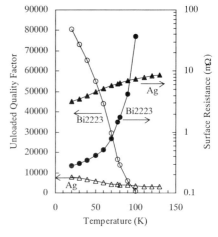

Fig.2   Temperature dependence of Qu (open symbols) and Rs (solid symbols) at 2.1GHz on a TM$_{010}$ mode dielectric resonator: Bi2223, circles; Ag, triangles.

thin film (Shen et al 1997). This insufficient power handling capability for Bi2223 thick film on BSMT is attributed to microstructures of Bi2223 thick film, as will be mentioned later on. To check the feasibility for Bi2223 thick film, we also investigated the Rs value of the thick film on Ag substrate, which has better microstructure. The data at 70K and 3.5GHz on a TE$_{011}$ mode are shown by closed circles in Fig.3. As can be seen, the power handling capability of the film on Ag is at least one order of magnitude higher than that on BSMT.

Fig.4 shows the third-order intermodulation distortion characteristics of Bi2223 thick films on both BSMT and Ag substrates, where the difference in the output power between the fundamental and the third harmonic was plotted against the max surface magnetic field. As can be seen, the distortion characteristics for Bi2223 thick film on BSMT turn out to be –60dBc at 1A/m.

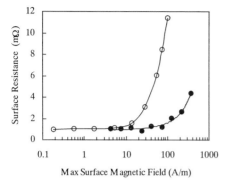

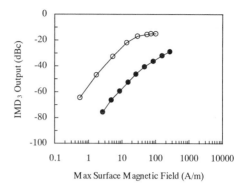

Fig. 3   Maximum surface magnetic field dependence of Rs for Bi2223 thick film on BSMT substrate on a TM_{010} mode at 2.1GHz, 70K (open circle) and the film on Ag substrate on a TE_{011} mode at 3.5 GHz, 70K (closed circle).

Fig. 4   The third-order intermodulation distortion characteristics for Bi2223 thick film on BSMT substrate on a TM_{010} mode at 2.1GHz, 70K (open circle) and the film on Ag substrate on a TE_{011} mode at 3.5 GHz, 70K (closed circle).

The SEM image of a fractured surface of Bi2223 thick film on BSMT substrate is shown in Fig.5. It is evident that Bi2223 film has low density, poor grain growth and low degree of c-axis alignment. It seems that the thick film has too many weak links to make the power handling capability sufficient. Furthermore, X-ray diffraction pattern (Fig.6) shows that the thick film includes large amount of $Bi_2Sr_2CaCu_2O_x$ (Bi2212) phase. In fact, higher temperature sintering enables to obtain the Bi2223 thick film on BSMT substrates with almost single phase, high density and grain alignment. However, such thick films on BSMT show higher Rs value ($\sim$3mΩ at 2.1GHz, 70K) than the film shown in Figs.5 and 6. This degradation is probably attributed to microscopic interface reaction between the Bi2223 and the BSMT ceramics, which is under investigation.

As shown in Figs. 3 and 4, the thick film on Ag has better power handling capability than that on BSMT. The difference in the capability between them is thought to be ascribed to the difference in the density, grain size, phase purity and also a degree of grain alignments in microstructures. Thus, further improvement for the morphology of Bi2223 thick film on BSMT is required to approach the level for the film on Ag.

3.2   TM$_{110}$ dual-mode resonator

Fig.5   SEM image of fractured surface of Bi2223 thick film on BSMT substrate

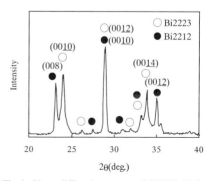

Fig.6   X-ray diffraction pattern of Bi2223 thick film on BSMT substrate

We designed and fabricated the 1.8GHz TM_{110} dual-mode resonator shown in Fig.7, where two resonating modes are coupled with through-holes. The 25mm-cubic BSMT ceramics with 8 through-holes was fabricated and Bi2223 thick films were printed on every 6 surfaces. Then, the composite was subjected to a combination process of CIP and sintering. Moreover, silver paste electrodes were printed on every 12 side-edges to keep good connectivity between planar Bi2223 electrodes. During microwave measurement, coupling probes were inserted into through-holes.

Fig.8 shows the temperature dependence of Qu and Rs for one mode of the dual-mode resonator, where the Rs value for Bi2223 thick films is the apparent value including Rs of the side-edge Ag. Note that the data for another mode are similar to the values shown in Fig.8. For comparison, the values of Qu and Rs for Ag electrode dual-mode resonator with same structure are also included. In comparison, the Qu value of the resonator using Bi2223 thick films reaches 40,000 at 70K, which is 4 times higher than the resonator using Ag electrodes. The Rs value of Bi2223 itself is estimated to be $1m\Omega$ at 70K, comparable to the results of the TM_{010} mode disk resonator, by separating the contribution of side-edge Ag from apparent Rs.

For miniaturization of a microwave filter, the polyhedral-electrode resonator like the present TM dual-mode type is preferable to the TM_{010}-mode disk resonator, because the former has better volume efficiency of Qu than the latter and also enables us to make the dual-mode operation easily.

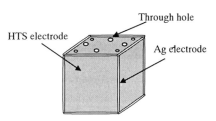

Fig.7 Structure of the TM_{110} dual-mode resonator

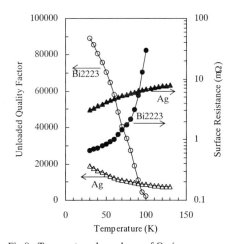

Fig.8 Temperature dependence of Qu (open symbols) and Rs (solid symbols) at 1.8 GHz on a TM dual-mode dielectric resonator: Bi2223, circles; Ag, triangles.

6. CONCLUSIONS

The unloaded quality factor for TM_{010} mode resonator consisting of a BSMT ceramic disk and planar Bi2223 thick film electrodes showed the value of 30,000 at 2.1GHz, 70K, which is 7 times higher than that of the resonator with Ag electrodes. The corresponding value of the surface resistance of Bi2223 thick film was 0.8mΩ. The 1.8GHz TM_{110} dual-mode resonator was fabricated using a 25mm cube of BSMT ceramics and 6 planar Bi2223 thick films as electrodes, and showed 40,000 of Qu at 70K. Further improvement for grain growth, grain orientation and phase purity of Bi2223 thick films on BSMT is our future work toward realization.

REFERENCES

Nguyen P P, Oates D E, Dresselhaus G and Dresselhaus M S 1993 Phys. Rev. B **48**, 6400

Shen Z, Wilker C, Pang P, Face D W, Carter C F III and Harrington C M 1997 IEEE Trans. Appl. Supercond. **7**, 2446

Tatekawa T, Matsui N, Kintaka Y, Ishikawa Y, Fujikawa K, Tanaka M and Oota A 1999 IEEE Trans. Appl. Supercond. **9**, 1940

Inst. Phys. Conf. Ser. No 167
Paper presented at *Applied Superconductivity, Spain, 14-17 September 1999*
© 2000 IOP Publishing Ltd

Ferroelectric field effect in $YBa_2Cu_3O_{7-\delta}$ films

R Aidam, D Fuchs and R Schneider

Forschungszentrum Karlsruhe, Institut für Festkörperphysik, P.O.B. 3640, D-76021 Karlsruhe, Germany

ABSTRACT: Ferroelectric superconducting field effect transistors (FSuFETs) consisting of $YBa_2Cu_3O_{7-\delta}$ (YBCO) channels and $Pb(Zr_{0.54}Ti_{0.46})O_3$ (PZT) gate insulators were fabricated. In order to prevent contamination of the YBCO layers by PZT, 20 nm thick $SrTiO_3$ buffer layers were inserted between YBCO and PZT. Ferroelectric polarization charging effects of the resistivity, the critical temperature and the critical current density of ultrathin YBCO films clearly reflected the ferroelectric hysteresis of the PZT layer. By reversing the polarization state a nonvolatile change of the YBCO properties was measured.

1. INTRODUCTION

For the development of high temperature superconducting electronics three terminal devices are of principal importance. One device, presented by Mannhart et al. (1991), is the superconducting field effect transistor (SuFET). It consists of an ultrathin $YBa_2Cu_3O_{7-\delta}$ (YBCO) channel with a maximum thickness of approximately 10 nm, an insulating gate with a high electrical polarization and a gate electrode, which typically consists of a metal such as gold (Au). The modulation of the electrical properties of the channel depends on the polarization of the gate insulator. Ferroelectric $Pb(Zr_{0.54}Ti_{0.46})O_3$ (PZT) thin films might be attractive as a gate insulator. Remanent polarizations P_r of up to 60 $\mu C/cm^2$ have been achieved by Kanno et al. (1997) in this material. In addition, YBCO suited as an excellent sublayer for the epitaxial growth of PZT films. Such devices are called ferroelectric superconducting field effect transistors (FSuFETs). However, our studies showed, that the properties of ultrathin YBCO films thinner than 20 nm degraded dramatically by the deposition of the PZT top layer, while YBCO films thicker than 100 nm remained unaffected by the PZT deposition.

In this paper, it is shown, that the quality of the YBCO films can be improved by inserting a 20 nm thick $SrTiO_3$ (STO) film between YBCO and PZT in the FSuFET structure. STO thin films can be epitaxially grown on top of YBCO without material interdiffusion and they are suitable sublayers for epitaxial PZT growth. Since the buffer layer changes the properties of the insulating gate of the FSuFET, its influence on the remanent polarization and the coercive voltage of the ferroelectric hysteresis has to be minimized.

In our Au/PZT/STO/YBCO heterostructures with channels thinner than 10 nm modulations of the channel resistance, R_{DS}, the critical temperature, T_C, and the critical current density, j_C, clearly reflected the hysteresis loop of the gate insulator. The modulations amounted to 10 %, 1 K and 16 %, respectively.

2. EXPERIMENTAL

Heterostructures were prepared in situ by the inverted cylindrical magnetron sputtering technique presented by Geerk et al. (1989) on (100) STO single crystals. The deposition parameters of the used materials are listed in Tab. 1.

FSuFET structures were patterned during the single deposition steps by using a set of moveable metal masks over the substrate. A 1.4 mm long and 0.5 mm wide YBCO channel was covered with a STO buffer layer and a PZT gate insulator. The channel linked two electrodes on which Au banks served as contact pads for four probe measurements of the electrical properties of the channel. A Au film on top of the PZT layer was used as a gate electrode. Polarization versus electrical field hysteresis loops were measured with a commercial pulsed testing system using a two probe arrangement.

	YBCO	STO	PZT	Au
target composition	$YBa_2Cu_3O_{7-\delta}$	$SrTiO_3$	$Pb_{1.1}(Zr_{0.54}Ti_{0.46})O_3$	Au
pressure	4×10^{-1} mbar	2.5×10^{-1} mbar	2.5×10^{-1} mbar	3×10^{-2} mbar
O_2 / Ar ratio	1	1	1	pure Ar
deposition temperature	800 °C	760 °C	570 °C	20 °C
power	80 W dc	100 W rf	150 W rf	2 W dc
deposition rate	0.11 nm / s	0.021 nm / s	0.045 nm / s	0.08 nm / s
orientation	001	100	100	polycrystalline

Tab. 1 Deposition parameters of the used materials

3.   RESULTS AND DISCUSSION

3.1   Gate Insulator

The effect of an insulating buffer layer on the potential difference U_{PZT} in the PZT film, the polarization P and the permittivity ε of the insulating bilayer is calculated, using the terms described in Tab. 2. U_{PZT} decreases by the use of an insulating buffer layer, depending on the permittivity ε_{buf} and the thickness d_{buf} of the buffer layer:

$$U_{PZT} = \frac{U}{\dfrac{\varepsilon_{PZT} d_{buf}}{\varepsilon_{buf} d_{PZT}} + 1} \tag{1}$$

Equation (1) shows, that the buffer layer should be of minimum thickness and maximum ε_{buf}. The total polarization P of the bilayer can be calculated as:

$$\vec{P} = \int_V \rho(\vec{r})\vec{r}dV = \int_{V_{buf}} \rho_{buf}(\vec{r})\vec{r}dV + \int_{V_{PZT}} \rho_{PZT}(\vec{r})\vec{r}dV \tag{2a}$$

$$P(U = 0) = \pm P_r \frac{d_{PZT}}{d_{buf} + d_{PZT}} \tag{2b}$$

The use of a nonferroelectric buffer layer leads to a minor reduction of the remanent polarization P(0) of the bilayer, according to equation (2b). This is valid as long as the buffer layer is significantly thinner than the PZT layer, i.e. $d_{buf} \ll d_{PZT}$. The permittivity ε of the insulator is nearly unaffected by the buffer layer as long as the permittivity ε_{buf} of the buffer layer is comparable with or larger than the permittivity ε_{PZT} of PZT:

$$\varepsilon = \varepsilon_{PZT} \frac{\varepsilon_{buf}(d_{buf} + d_{PZT})}{\varepsilon_{PZT} d_{buf} + \varepsilon_{buf} d_{PZT}} \approx \varepsilon_{PZT} . \tag{3}$$

A comparison of hysteresis loops of ferroelectric capacitors with and without buffer layer is shown in Fig. 1. By the use of the buffer layer the loop becomes broader, i.e. the coercive voltage increases, because the electric field in the PZT layer is weakened according to equation (1). The observed increase of 50 % in the coercive voltage is in rather good agreement with equation (1) if the high permittivity $\varepsilon_{PZT} \approx 10000$ at the coercive field is considered. The remanent polarization decreases by 20 % in the experiment whereas equation (2) predicts only a reduction of 4 %. However, this deviation is still within the P_r variation of PZT films deposited under identical conditions. The slope of the saturation tips and therefore the permittivity ε is unaffected by the use of a buffer layer as predicted by equation (3).

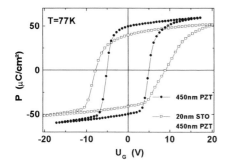

Fig. 1 Comparison of hysteresis loops of ferroelectric capacitors with 450 nm thick PZT layers with and without 20 nm thick STO buffer layers, measured at 77 K.

U, U_{PZT}	potential difference at the bilayer and the PZT layer
$\varepsilon_{PZT}, \varepsilon_{buf}$	permittivity of PZT and the buffer layer
ε_{eff}	effective permittivity of the bilayer
d_{PZT}, d_{buf}	thickness of PZT and the buffer layer
V, V_{PZT}, V_{buf}	volume of the bilayer, PZT and the buffer layer
$\rho, \rho_{pzt}, \rho_{buf}$	charge carrier concentration of bilayer, PZT, buffer layer
P	polarization
P_r	remanent polarization

Tab. 2 Terms used in equations (1) to (3)

3.2 YBCO Channels

In this section the influence of the STO buffer layer on the electrical properties of the YBCO channels in unpoled FSuFET structures is discussed. In Fig. 2 and Fig. 3 T_C and j_C values of YBCO channels versus film thickness d_{YBCO} are shown. 100 nm thick channels reached critical temperatures above 80 K and critical current densities of 8×10^5 A/cm² at 4.2 K independent of the use of a buffer layer. Decreasing YBCO film thickness yielded decreasing values for T_C and j_C. The reduction of T_C and j_C was more distinct for films without a buffer layer. Consequently, superconductivity in bufferless structures was only achieved in films thicker than 20 nm. T_C and j_C had a tendency to higher values for structures with a buffer layer. In this case, superconducting channels were fabricated down to a minimum thickness of 7 nm. These channels had critical temperatures between 30 and 40 K and critical current densities ranging from 1×10^4 A/cm² to 2×10^5 A/cm² at 4.2 K. Films thinner than 7 nm showed no superconductivity at all and a dramatic increase of the resistance R_{DS}.

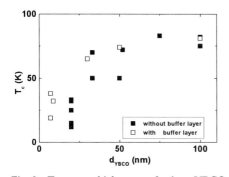

Fig. 2 T_C vs. thickness of the YBCO channels in FSuFET structures with and without a 20 nm thick buffer layer.

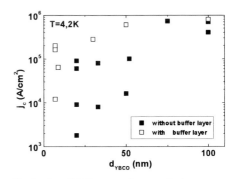

Fig. 3 j_C (4.2K) vs. d_{YBCO} of the same channels as in Fig. 2

3.3 Ferroelectric Polarization Charging of YBCO Films

In field effect transistors the density of mobile charge carriers in the channel and therefore the conductivity is changed due to the charge induced by the polarization of the gate insulator. In Fig. 4(a), the hysteresis loop of the insulating bilayer of a FSuFET consisting of a 8.8 nm thick

60

channel, a 20 nm thick buffer layer and a 1000 nm thick PZT layer is presented. This loop is clearly reflected in the R_{DS} versus gate voltage V_G characteristic shown in Fig. 4(b). A nonvolatile change at $V_G = 0$ of 7.9 % and a maximum change of 10 % of R_{DS} is observed. According to Figs. 4(c) and 4(d), the superconducting transition temperature $T_C (V_G)$ and the critical current density $j_C(V_G)$ characteristics are also hysteretic with nonvolatile changes of 0.65 K and 9 %, respectively. The changes at the extreme V_G values $\pm V_{G,max}$ are calculated as 0.95 K and 15.6 %, respectively. Details of the analysis of the measured modulations are presented by Aidam (1999).

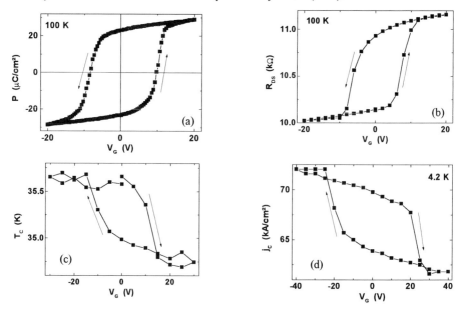

Fig. 4   P vs. V_G at 100 K (a), R_{DS} vs. V_G at 100 K (b), T_C vs. V_G (c) and j_C vs. V_G at 4.2 K characteristics (d) of a FSuFET structure consisting of 8.8 nm YBCO, 20 nm STO, 1000 nm PZT and 50 nm Au.

4   SUMMARY

The influence of the ferroelectric PZT polarization on normal state and superconducting properties of ultrathin YBCO films was studied. A thin $SrTiO_3$ buffer layer was inserted between the YBCO and PZT layers in order to prevent degradations of the superconductor by the PZT top layer. This buffer layer led to a higher critical temperature of ultrathin YBCO films, while the coercive voltage of the insulating bilayer increased and the remanent polarization slightly decreased.

The changes of the channel resistance R_{DS}, the critical temperature T_C and the critical current density j_C with the applied gate voltage V_G clearly reflect the ferroelectric hysteresis loop of the PZT gate insulator. By reversing the remanent polarization, nonvolatile changes of 7.9 %, 0.65 K and 9.0 % in a 8.8 nm thick channel were measured for $\Delta R_{DS} / R_{DS}$, ΔT_C and $\Delta j_C / j_C$, respectively. The maximum modulations at the two extreme voltages $\pm V_{max}$ are 10 %, 1 K and 15.6 %.

References

Aidam R, Fuchs D and Schneider R 1999 accepted for publication in Physica C
Geerk J, Linker G and Meyer O 1989 Materials Science Reports **4**, 193
Mannhart J, Schlom D G, Bednorz J G and Müller K A 1991 Phys. Rev. Lett. **67**, 2099
Kanno I, Fujii S, Kamada T and Takayama R 1997 Appl. Phys. Lett. **70** 11

Inst. Phys. Conf. Ser. No 167
Paper presented at Applied Superconductivity, Spain, 14-17 September 1999
© 2000 IOP Publishing Ltd

Plasma Optical Emission Studies during HTSC Thin Film Deposition Process Optimisation

V N Tsaneva[1,3], C Christou[1], J H Durrell[1,2], G Gibson[1,2], F Kahlmann[2], E J Tarte[2], Z H Barber[1,2], M G Blamire[1,2] and J E Evetts[1,2]

[1]Dept. of Materials Sci., Cambridge University, Pembroke Street, Cambridge CB2 3QZ, UK
[2]IRC in Superconductivity, Cambridge University, Madingley Rd., Cambridge CB3 0HE, UK
[3]Institute of Electronics, Bulg. Acad. Sci., Tsarigradsko chaussee 72, Sofia 1784, Bulgaria

ABSTRACT: Optical emission spectroscopic measurements were performed during high-pressure dc-diode YBCO thin film sputter deposition process optimization aimed at stabilization of discharge parameters and lowering the deposition temperature. Oxygen dissociation catalytic effect of small He or H_2 additions to Ar/O_2 sputtering gas was not detected at high pressures, but He stabilized the plasma thus yielding higher T_c's. The CuI emission lines were found to be sensitive to small deviations of condensing species flux stoichiometry on H_2 addition, which led to deposition of smooth YBCO films with reduced T_c.

1.   INTRODUCTION

The development of reproducible and stable processes for homogeneous deposition of high quality high Tc (HTS) films is crucial for the fabrication of superconducting devices. High-pressure dc-diode sputtering in oxygen proved to be suitable for the deposition of YBCO with high quality critical parameters (Poppe 1988), and it allows scaling to cover relatively large areas, necessary for microwave applications. However, at pressures around 300 Pa the deposition rate is low and the film surface is often covered with needle-like outgrowths. Higher deposition rates and better surface morphology were obtained when Ar was added to the sputtering gas (Mechin 1997). Nevertheless, due to spatial inhomogeneities and instabilities of the plasma, the control of the process input parameters (pressure p, discharge current I, substrate temperature T_s) does not ensure a stable and reproducible deposition of homogeneous YBCO films. The specific nature of the YBCO targets determines temperature dependent oxygen exchange with the gas environment, which is further complicated by the plasma influence. In its turn this oxygen in/out-diffusion leads to changes of the target electrical and thermal conductivity, of the stoichiometry of the surface altered layer, and hence of the coefficient of secondary ion-electron emission. For a dc discharge running at constant current, these changes are manifested by variation of the cathode voltage, which affects the plasma parameters and the sputtering rate. Moreover, local instabilities can cause long-term plasma variations due to positive feedback between the modified target surface stoichiometry and the secondary electron emission, resulting in inhomogeneous film deposition. These problems become more severe in the presence of an applied magnetic field, i.e. magnetron sputtering, for which Kruger (1993) has achieved plasma stabilization with the help of a computer-controlled process for adjustment of the oxygen partial pressure over the target, based on real-time monitoring O-atom (OI) optical emission at 844.6 nm.

The rate of production of atomic oxygen in oxygen discharges can be enhanced by addition of H_2O, H_2, N_2O (Costa 1979) or He (Tsaneva 1995). Gavaler (1991), Cukauskas (1992) and Fan (1997) have used H_2O and H_2 additions during YBCO sputtering and observed plasma stabilisation and an increase of T_c and j_c of the deposited films, though accompanied by film surface roughening for H_2. Due to the higher activity of oxygen atoms compared with molecules (Westerheim 1991, Sawa 1993), fully oxidized films with improved morphology can be deposited at lower temperatures in more atomized oxygen plasmas.

Small stoichiometry deviations, as well as inhomogeneities, have a large effect on the superconducting properties and morphology of HTS films (Chew 1990), though it is difficult to reveal them by conventional methods, such as XRD, EDAX or AES. Optical emission spectroscopy (OES) of the sputtering plasma is a non-intrusive method which can give *in situ* information about the condensing species flux and, due to the non-linear dependence of different species emission intensity on their concentration, small stoichiometry deviations may yield detectable OES variations. Once established, these sensitive characteristics of the optical emission can be used as feedback for maintaining reproducible deposition of high-quality HTS thin films.

2.   EXPERIMENTAL

YBCO films were deposited in a dc diode high-pressure sputtering system in on-axis geometry. The $SrTiO_3$ and $LaAlO_3$ substrates were placed at a distance of 14 mm facing the target. The heater was noncontact radiative-type (Wagner 1995) which allowed double-side coating. The heater temperature T_h was measured on a dummy metal substrate located within the heater. The substrate temperature T_s, checked by optical pyrometry, differed from T_h within 20°, depending on chamber pressure and heating foil configuration. If not otherwise stated, the sputtering gas mixture was $Ar:O_2=7:3$ at a pressure of 260 Pa, and film thickness was 145 ± 10 nm. After deposition, the chamber was filled with 1 atm pure O_2 and the sample was cooled to room temperature for about 80 min. The films were characterized by optical microscopy, AFM, XRD and Raman spectroscopy, 4-point and inductive T_c measurements.

The optical emission was sampled by a periscope directed at the centre of the sputtering discharge. After reflection from its mirror, the signal was fed through a sapphire window to a quartz waveguide coupled to the entrance slit of Jobin Yvon spectrometer Triax 550 equipped with 1800 gr/mm grating yielding spectral resolution of 0.02 nm, and measured by a CCD.

3.   RESULTS AND DISCUSSION

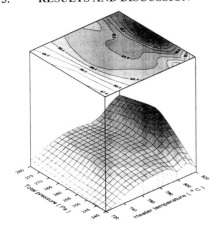

The deposition process parameters were optimized regarding T_c and surface morphology, as a low density of outgrowths is required for multilayer device applications of the YBCO films. The FWHM of the (005) rocking curves ($\Delta\omega$) and the c-lattice parameters were also monitored. Fig.1 shows 3D plot and contour map of the T_c dependence on the total Ar/O_2 pressure and heater temperature.

Entirely c-axis oriented YBCO films with $\Delta\omega < 0.3°$ and $T_c \geq 88.7$ K were obtained down to 750°C at a total pressure of 260 Pa, with 5-fold reduced density of outgrowths. Moderate increase of c-axis lattice parameter to 1.1685 nm suggested non-optimal oxygenation at 750°C. During some

Fig. 1 *3D-plot and contour map of Tc dependence* deposition runs there were pressure drifts, leading
on total Ar/O₂ pressure and heater temperature

to films with $\Delta\omega > 0.3°$ and Tc ≈ 88 K; Raman spectra of these films showed cation disorder, according to Cohen (1995). These pressure drifts, reflecting plasma instabilities, were more often observed with heated substrates, which indicated a connection with the heating of the target. To reveal the mechanism, the plasma optical emission was studied at different heater temperatures.

Both atomic and ionic gas species emission intensity increased with increasing T_h, whilst the intensity of both the ionic and atomic lines of the target elements decreased. The latter result is in contradiction with the results of Nathan (1998), which may be explained by their normalisation to an O_2I line. The line intensity is not proportional to the concentration of the emitting species, but if stringent requirements to the populating and relaxation of the excited states are met, actinometry (Coburn 1980) can been applied, based on emission normalisation to a trace gas (usually Ar) with similar threshold and shape of the electron-impact cross section. Although actinometry has not been validated for our system (Granier 1994), we have used the ratios of different atomic emission lines to the ArI 750 nm-line in order to describe qualitatively the behaviour of ground state concentrations of the respective atoms (Fig. 2).

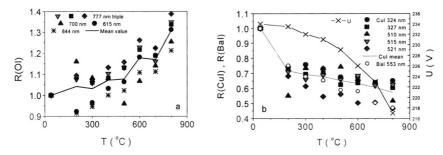

Fig. 2 *Temperature dependence of the ratios of OI (a), CuI and BaI (b) to ArI peak height, normalized to the "room temperature" values; the right Y-axis in (b) refers to the target voltage*

Taking into consideration the small target-substrate distance and the high pressure, it is plausible to assume moderate heating of the gas in the gap between them. The increased thermal velocity of the species causes gas rarefaction (Rossnagel 1988, Donchev 1997), which leads to reduced quenching collision frequency and higher species optical emission intensity. The drop in Cu and Ba actinometric signal can be attributed to reduction in number of Ar ions for sputtering, the reduced target voltage, and the reduced backscattering of the sputtered atoms. This latter effect will limit the drop in deposition rate due to the reduced sputtering rate and lower sticking coefficient, and, in support of this, an increase of the deposition rate has been observed by Nathan (1998).

The observed decrease of target voltage with temperature (Fig. 2b) and its gradual drop during a deposition run in Ar/O_2 (Fig. 3c) may lead to speculation about oxygen outdiffusion from the target (Gavaler 1991, Kruger 1993). T_c depression has been observed for films deposited from a deoxygenated target, which has been "cured" by addition of gases with catalytic dissociative effect in molecular oxygen (Gavaler 1991). We studied the influence of small additions of He and H_2 on the OE spectra in relation to the characteristics of the deposited YBCO films. The pressure and heater temperature were kept constant at 260 Pa and 750°C, respectively, which with Ar/O_2 yielded slightly deoxygenated films according to the relative intensity of (001) and (002) peaks in XRD (Ye 1994); and $\Delta\omega = 0.296°$; $T_c = 88.9$ K. The enhancement of the oxygen actinometric signal observed previously (Tsaneva 1995) for small additions of He at lower pressures was not found for the higher pressures used, but the relatively more stable plasma conditions (without pressure drifts) yielded a slightly higher film quality ($\Delta\omega = 0.243°$; Tc = 89.2 K).

A small addition of hydrogen (5 Pa) to the sputtering Ar/O_2 gas mixture did not lead to gross changes of the oxygen actinometric signal. The YBCO film had a deteriorated c-axis orientation ($\Delta\omega = 0.371°$) and double superconducting transition with onset at 79 K, whilst exhibiting a smooth surface with extremely low outgrowth density. This suggested stoichiometric deviations, which were not detected by the XRD. However, the CuI emission lines, monitored during the deposition, showed different behaviour from all previous measurements. Fig. 3 presents the time evolution of the CuI actinometric signals. One can see that after about 60 min all CuI lines increased, which leads to the conclusion that a presumably thermo-chemical mechanism gave rise to increased copper concentration in the plasma. The reason for the discrepancy with other authors' results (Gavaler 1991, Cukauskas 1992 and Fan 1997) may be the higher target temperature reached in our on-axis system: the first two cited authors have used off-axis configuration, and the third has worked with lower

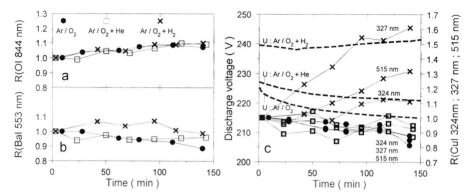

Fig. 3 *Time dependence of the ratios of different OI (a), BaI (b) and CuI: 324, 327 and 515 nm (c) to ArI peak height, normalized to the initial values. Target voltage evolution is also shown in (c) with dashed lines for the three indicated gas mixtures*

heater temperatures (yielding lower T_c's). The effect of smaller H_2 percentages, as well as other gas mixture compositions will be further investigated.

4. CONCLUSIONS

OES was used to gain insight into the optimization of the input discharge parameters and substrate temperature during high-pressure dc-diode YBCO sputter deposition. It has proved to be a sensitive tool for *in situ* monitoring of the optimal stoichiometry of the condensing species flux to the substrate, and for revealing stoichiometry deviations during the deposition run. Small additions of He to the sputtering Ar/O_2 gas mixture helped to overcome plasma instabilities thus yielding higher film quality at lower deposition temperatures, whilst the H_2 additions used led to deterioration of film properties, but improved morphology.

Acknowledgements
Fruitful discussions with Dr. J. Wolfman are gratefully acknowledged.
This work was supported by the UK Engineering and Physical Sciences Research Council and Contract F-819 with Bulgarian Fund "Scientific Investigations".

REFERENCES

Chew N G, Goodyear S W, Edwards J A, Satchell J S, Blenkinsop S E and Humphreys R G 1990 Appl. Phys. Lett. **57**, 2016
Coburn J W and Chen M 1980 J. Appl. Phys. **51**, 3134
Cohen L F, Li Y B, Gibson G, MacManus-Driscoll J L 1995 Proc. MRS. **401**, p.351-356
Costa M D, Zuliani P A and Deckers J M 1979 Can. J. Chem. **57**, 568
Cukauskas E J, Allen L H, Sherill G K and Holm R T 1992 Appl. Phys. Lett. **61**, 1125
Donchev T 1997 Bulg. J. Phys. **24**, 74
Fan S C 1997 PhD Dissertation, Cambridge
Gavaler J R, Talvacchio J, Braggins T T, Forrester M G and Greggi J 1991 J. Appl. Phys. **70**, 4383
Granier A, Chereau D, Henda K, Safari R and Leprince P 1994 J. Appl. Phys. **75**, 104
Kruger U, Kutzner R and Wordenweber 1993 Appl. Phys. Lett. **62**, 1559
Mechin L 1997 Report, Cambridge (unpublished)
Nathan S S, Muralidhar G, Rao G M and Mohan S 1998 Vacuum **49**, 221
Poppe U, Schubert J, Arons R R, Evers W, Freiburg C H and Reichert W 1988 Sol. State Commun. **66**, 661
Rossnagel S M 1988 J. Vac. Sci. Technol. A **6**, 19
Sawa A, Kosaka S, Obara H and Aoki K (1993) IEEE Trans. Appl. Supercon. **3**, 1088
Tsaneva V, Donchev T and Nurgaliev T 1995 Appl. Supercond. **148** IOP Conf. Ser. pp.839-842
Wagner G A, Somekh R E and Evetts J E 1995 Inst. Phys. Conf. Ser. **148**, 855
Westerheim A C, Yu-Jahnes L S and Anderson A C 1991 IEEE Trans Mag. **27**, 1001
Ye J and Nakamura K 1994 Phys. Rev B **50**, 7099

Inst. Phys. Conf. Ser. No 167
Paper presented at Applied Superconductivity, Spain, 14-17 September 1999
© 2000 IOP Publishing Ltd

Dynamic electrical response of YBaCuO thin films as a function of substrate crystallinity for electronic applications

A De Luca, A Dégardin, É Caristan, G Klimek, J Delerue, A Gaugue, A K Gupta[*], J Baixeras and A Kreisler

LGEP, SUPÉLEC, UMR CNRS 8507, Universités Paris 6 et Paris 11, Plateau du Moulon, 91192 Gif-sur-Yvette Cedex, France
[*]National Physical Laboratory, Dr K. S. Krishnan Road, New Delhi 110 012, India

ABSTRACT: The *V-I* characteristics of YBaCuO thin films sputtered on single-crystal substrates have been measured using a pulsed current technique, to avoid ohmic heating effects. The results are discussed in the framework of a statistical model, which provided an empirical approach for the pinning phenomena in YBaCuO films, under application of a weak static magnetic field.

1. INTRODUCTION

Conventional dc four-probe voltage-current (*V-I*) measurements are often complicated by ohmic heating at the resistive contacts between the superconductor and the current leads. In order to circumvent such a problem, we have developed an original pulsed current technique.

In this paper, the detailed description of the experimental set-up is followed by a preliminary report on the *V-I* characteristic investigations in granular YBaCuO thin films deposited on SrTiO$_3$ and MgO single-crystal substrates. We used a statistical model to discuss qualitatively the pinning mechanisms in these films when a weak static magnetic field was applied.

2. EXPERIMENTAL PROCEDURES

YBaCuO thin films were prepared *in situ* by both dc (Chandra et al 1994) and rf (Dégardin et al 1998) on axis magnetron reactive sputtering. In this study, we have used single-crystals of SrTiO$_3$ (small lattice mismatch with YBaCuO) and MgO (despite the 9 % lattice mismatch, because of its interesting dielectric properties). The thickness of films deposited by rf sputtering was about 200 nm whereas the thickness of films deposited by dc sputtering was about 800 nm. Both unpatterned YBaCuO films of 5 mm width and microbridges of 50 to 200 μm width (patterned on YBaCuO samples by a standard photolithography process and ion beam etching), were characterised by electrical transport. To make low resistance ($< 10^{-5}$ Ω cm^2) electrical contacts, four gold pads were e-beam evaporated through metal masks onto the sample, which was then annealed at 475 °C for 15 minutes in pure flowing oxygen and slowly cooled down to room temperature. 50 μm diameter gold wires were ultrasonically bonded on these contacts.

The schematic diagram of the pulsed current system is shown in Fig. 1 (left). The current was delivered by a programmable current source (Keithley Inst., model 2400) as single pulses of duration ranging from 3 ms to several hundred ms and amplitude ranging from 1 μA to 1 A. The voltage appearing between the voltage pads was amplified by a variable gain low noise voltage preamplifier (Stanford Research Systems, model 560) before being measured by a programmable digital oscilloscope (Sefram, model 5920). The temperature *T* was measured using a calibrated Cernox thermometer and stabilised at ± 0.1 K. The whole system was automatically controlled with

LabVIEW™ software by means of IEEE bus. A static magnetic field H could be applied parallel to the *ab*-plane of the film and perpendicular to the transport current.

As the single pulse method is very sensitive to the electrical noise due to the large bandwidth of the oscilloscope, a 12 dB/oct lowpass filter (integrated in the preamplifier box) was interposed between the preamplification function and the voltage measurement by the oscilloscope in order to minimise the high frequency noise. The resulting voltage noise level for these measurements was 7 nV/Hz$^{1/2}$ at 100 Hz. This value is comparable to the nominal noise level of the preamplifier (4 nV/Hz$^{1/2}$ at the same frequency), showing that the noise introduced by the experimental system was very low.

An example of voltage pulse waveform across the sample in response to a current pulse of duration 100 ms and maximal amplitude $I_m = 39$ mA is shown in Fig. 1 (right). The observed initial and final transients of about 0.5 ms in duration can arise from i) imperfect high frequency common mode rejection of the instrumentation amplifier, ii) transient currents induced in connections, or iii) kinetic inductance effects in superconducting material itself. After the initial transient, the voltage increases up to its normal value with a rise time resulting from the filter response. In order to rule out these filtering effects, the measured voltage value was defined as the average over 10 to 200 points taken between the two arrows shown in Fig. 1 (right). The number of points was chosen in function of the current pulse duration which was dependent on the current pulse amplitude and hence on the superconducting quality of the sample. The duration between two pulses allowed the return to equilibrium state; the measurements were averaged over 5 to 20 current pulses at the same fixed I_m, T and H values.

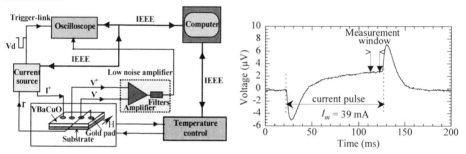

Fig. 1 (*left*) Schematic representation of the pulsed current system. (*right*) Typical shape of the voltage pulse across a sample as a response to a current pulse (see text).

3. STATISTICAL MODEL

In order to introduce the vortex pinning mechanisms, the *V-I* relationship has been studied with a theoretical model, to allow the analysis of the flux flow regime in a superconductor containing structural defects whatever their nature (Baixeras et al 1967). For a particular pinning centre and as soon as the local current density exceeds the critical current density J_c, the vortex which is initially pinned can move freely. In the course of its motion, the vortex can meet other defects and, depending on the depth of the potential well, it can (or cannot) be pinned until the local current density reaches the corresponding critical current density. Therefore if a current density J is applied to the film, the resulting Lorentz force exerted on the vortices is such that some of these vortices are free to move. This implies therefore electrical losses; consequently an electrical field E is developed across the sample under test. The *E-J* relationship has been analysed to give information on vortex pinning. In order to characterize the statistical spreading of pinning forces, a distribution function $f(J_{ci})$ of the critical current density associated to the vortex "i" has been introduced. The expressions for $f(J_{ci})$ and for E are respectively given by:

$$f(J_{ci}) = \begin{cases} 0 & \text{for } J_{ci} < J_{cm} \\ \dfrac{1}{J_\sigma} \exp(-\dfrac{J_{ci} - J_{cm}}{J_\sigma}) & \text{for } J_{ci} \geq J_{cm} \end{cases} \quad \text{and} \quad E = A\left[J - J_{cm} - J_\sigma\left(1 - \exp\left(-\dfrac{J - J_{cm}}{J_\sigma}\right)\right)\right],$$

where $A = \dfrac{B\phi_0}{\eta}$, with B the magnetic induction, ϕ_0 the flux quantum and η the viscosity coefficient assumed to be the same for all vortices. A, J_{cm} and J_σ are adjustable parameters. J_{cm}, which is the critical current density determined with the model, represents the smallest pinning force acting on the assembly of vortices. J_σ corresponds to the width of the distribution function and represents the spreading of the pinning forces.

4. RESULTS AND DISCUSSION

4.1 Film structure and resistive transition in zero magnetic field

Information relative to the samples is collected in Table 1, which contains roughness data from the substrate (r_{sub}) and from the film (r_{film}), both obtained by atomic force microscopy (AFM), as well as resistance ratio R_{300K}/R_{100K}, critical temperature (T_c) and critical current density (J_c) obtained by electrical transport using a 1 μV/cm criterion in zero field. X-ray diffraction analyses showed that all films were predominantly c-axis oriented and AFM revealed a granular morphology of the film surface.

Table 1. Characteristics of YBaCuO samples studied in the framework of this paper (see text).

Sample #	Sputtering type	Substrate type	r_{sub} (nm rms)	r_{film} (nm rms)	R_{300K}/R_{100K}	$T_c(R=0)$ (K)	$J_{c(77K, H=0)}$ (A/cm²)
TM1 (patterned)	rf	unannealed MgO	0.5	7	2.5	78	1.2×10^4
TM2 (patterned)	rf	annealed MgO	0.5	8.5	2.9	80.2	4×10^4
T115-6 (unpatterned)	rf	SrTiO$_3$	0.2	5	2.3	85	8.8×10^4
SKS12 (unpatterned)	dc	SrTiO$_3$	5	8.4	1.5	74.1	–

By comparing YBaCuO films deposited on MgO substrates, we can notice that better T_c and J_c values were obtained for sample #TM2, which could be attributed to the morphology of the MgO substrate surface. Indeed, this substrate underwent annealing at 1000 °C for 5 hours under flowing oxygen, to allow the formation of terrace-like morphology, which is favourable to island growth of YBaCuO. Moreover, the higher value of R_{300K}/R_{100K} for sample #TM2 reflects a better electrical connection between grains. On the contrary, sample #TM2 is rougher than sample #TM1, which can be detrimental for some applications. Nevertheless, a higher roughness might be related with a potentially larger number of pinning centres (in favour of vortex pinning); this could explain better superconducting properties of sample #TM2.

For films deposited on SrTiO$_3$ substrates, T_c of sample #SKS12 is well below T_c of sample #T115-6, which can be attributed to the morphology of the substrate surface on one hand and to a non optimised thickness of the deposit by dc sputtering on the other hand. Indeed a thickness larger than 800 nm was necessary to obtain dc-sputtered films with optimised superconducting properties, whereas the optimal thickness for rf sputtered films has been shown to be about 200 nm.

4.2 Study with weak static magnetic field

Both YBaCuO samples deposited on SrTiO$_3$ substrates were studied under low field applied parallel to ab-plane. The resulting resistive transition curves E-J are displayed in Fig. 2. We can notice the good agreement between the experimental data and the theoretical curves computed from the model. Moreover, we can remark that sample #T115-6 is field dependent to a lesser extent than sample #SKS12, even at temperatures in the vicinity of T_c.

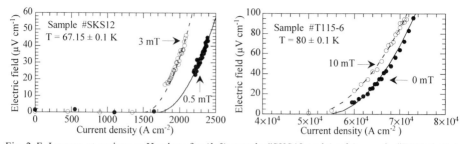

Fig. 2 *E-J* curves at various $\mu_0 H$ values for (*left*) sample #SKS12 and (*right*) sample #T115-6. Note good agreement between experiments (bold or open circles) and theory (continuous or dashed lines).

J_c can be deduced either directly from the experimental data using the 1 µV/cm criterion or from the calculated *E-J* curve using a fitting method. Results are plotted as a function of $\mu_0 H$ in Fig. 3 (left), where we note the satisfactory agreement between experiment and theory. Two regimes are visible in the variation of J_c for sample #SKS12. The very low field region (up to 0.5 mT) corresponds to the transition of weak links formed by grain boundaries. The higher field region corresponds to intrinsic pinning, which induces a nearly constant J_c value. On the contrary, for sample #T115-6, intrinsic pinning seems to be dominant on the full range of *H* values.

The J_σ field dependence is illustrated in Fig. 3 (right). As intrinsic pinning is dominant for *H* parallel to *ab*-plane, the distribution is very sharp. It however broadens at very low fields because the intergrain Josephson behaviour tends to be dominant. By plotting $J_c^{1/2}$ from measurement data as a function of *T* in the vicinity of T_c and in zero field, we find that $J_c(T)$ follows the relationship $J_c(T) = J_c(0)\,[1-T/T_c]^\alpha$, with $\alpha = 2$ as best fit value in line with SNS type for the Josephson barriers.

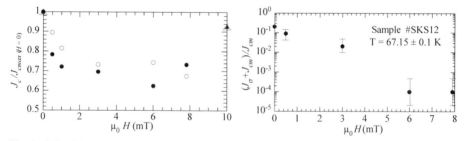

Fig. 3 (*left*) Field dependence of J_c normalised to the zero field case (J_{cmax}) for sample #SKS12 (circles) at 67.15 K and sample #T115-6 (triangles) at 80 K. The open symbols correspond to the measured values and the bold symbols to the calculated values. (*right*) Field dependence of J_σ.

5. CONCLUSION

V-I curves of granular YBaCuO films elaborated on single crystals have been measured using a pulsed current technique. Preliminary results have been discussed with a statistical model that allowed us to make a distinction between weak link effects and intrinsic pinning, for low magnetic fields applied parallel to *ab*-plane. We have shown that the pinning is strongly influenced by the film structure and the substrate nature.

REFERENCES

Baixeras J and Fournet G 1967 J. Phys. Chem. Solids **28**, 1541
Chandra R, Gupta A K and Kumar V 1994 Indian J. Pure & Appl. Phys. **32**, 133
Dégardin A, Bodin C, Dolin C and Kreisler A 1998 Eur. Phys. J. AP **1**, 1

Inst. Phys. Conf. Ser. No 167
Paper presented at Applied Superconductivity, Spain, 14-17 September 1999
© *2000 IOP Publishing Ltd*

Protective coatings for $YBa_2Cu_3O_7$ – thin film devices

W Prusseit

THEVA Dünnschichttechnik GmbH, Hauptstr. 1b, D-85386 Eching-Dietersheim, Germany

ABSTRACT: To guarantee long term performance of $YBa_2Cu_3O_7$ -film based devices and to avoid degradation of films which are in contact with metals, protective encapsulations will become an absolute necessity. This survey evaluates a variety of plastic compounds with respect to their efficiency as cheap cap layers at cryogenic temperatures.

1. INTRODUCTION

The commercial application of $YBa_2Cu_3O_7$ – thin film based devices puts tough requirements on the environmental compatibility and long term stability of such films and the protection against degradation has become a major issue in every product development. Humidity, aggressive ambient, and thermal cycling can slowly degrade the superconducting properties of $YBa_2Cu_3O_7$ - films and hence the long term device performance. However, there is yet another often ignored but very fast degradation mechanism. Whenever $YBa_2Cu_3O_7$ - films are in direct contact with other metals (e.g. in the case of coated conductor tape, electrical contacts etc.) and are exposed to atmospheric moisture the resulting local galvanic cell leads to fast and irreversible corrosion.

Possible preventive measures are either protective layers on top of the $YBa_2Cu_3O_7$ - films or hermetically sealed encapsulations. The latter are expensive, if they can be realized at all. Consequently, appropriate protective layers applied after fabrication and bonding of thin film devices are absolutely necessary. However, the demands on such layers are numerous and very tough. Residual moisture absorption, thermal expansion and compatibility with cryogenic ambient and the delicate material rule out most of the standard encapsulations developed for the protection of semiconductor devices. Looking for alternatives we have studied and practically tested a variety of organic and silicone - based compounds.

2. CORROSION MECHANISMS

2.1 Atmospheric degradation

The chemical reactivity of $YBa_2Cu_3O_7$ under environmental conditions is mainly governed by its Barium component. BaO may react in a two-step process according to the following equations:

$$BaO + H_2O \rightleftharpoons Ba(OH)_2$$
$$Ba(OH)_2 + 2H^+ \rightleftharpoons Ba^{2+} + 2 H_2O$$

In pure water $YBa_2Cu_3O_7$ is very stable. However, the law of mass action implies that even a small concentration of H^+ (acid) will shift the equilibrium and turn BaO into freely soluble Ba^{2+} - ions. This washing out of barium results in a collapse of the $YBa_2Cu_3O_7$ - lattice (exception: sulphuric acid cannot attack $YBa_2Cu_3O_7$ since $BaSO_4$ is insoluble in water). Under environmental conditions it is only the carbonic acid from dissolved atmospheric CO_2 which leads to the long term degradation of $YBa_2Cu_3O_7$. Hence, an effective protection layer has to inhibit the access of water or CO_2 or both.

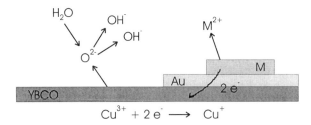

Fig. 1   Local galvanic cell consisting of $YBa_2Cu_3O_7$ and a metal M (e.g. solder)

2.2   Electrochemistry

Since all cations are in their maximum oxidation state (e.g. copper in the +2 and +3 state) $YBa_2Cu_3O_7$ is a very strong oxidizing agent and its electrochemical potential is positive and comparable to that of gold. Consequently, whenever it is brought into contact with less "noble" metals a local galvanic cell builds up and the normal metal is oxidized. In turn, the oxidation state of Cu^{3+} is reduced, $YBa_2Cu_3O_7$ looses oxygen, and becomes an insulator which is eventually destroyed by chemical reactions with water. The underlying chemical process is depicted in Fig. 1.

Some metals build up a passivation layer on their surface and the reaction comes to a standstill immediately. However, most metals will not and $YBa_2Cu_3O_7$ will corrode electrochemically. Unfortunately, this is a very fast process and in a moist ambient a $YBa_2Cu_3O_7$ - film can be destroyed within a few hours. There is nearly no device which does not require electrical contacts and solders usually contain alloying components which keep the corrosion process going. Fig. 2 shows the striking example of three $YBa_2Cu_3O_7$ bridges on the same wafer, which has become wet when moisture condensed after it was removed from liquid nitrogen. Only the upper bridge had solder pads and was destroyed within hours whereas the other bridges with pure gold pads remained intact.

Fig.2   Three $YBa_2Cu_3O_7$ – bridges with contact pads of gold and In-solder (top) which have become wet.

In the case of coated conductors $YBa_2Cu_3O_7$ is even deposited on metal and has electrical contact to the substrate (even if there is a buffer). As a direct consequence of the above mechanism protective encapsulation will become an absolute necessity.

3. ORGANIC CAP LAYERS

There is a variety of established techniques to apply inorganic protective layers like e.g. SiO_2, CeO_2 etc. by vacuum deposition. Although these compounds may have specific advantages, e.g. good adhesion and match to the thermal expansion of the superconductor, they require expensive equipment and processing time and consequently are less economic. Step coverage and tightness are also important issues in this respect and it is questionable if such layers can be reasonably applied on a large scale and complicated geometry at all.

In contrast, plastics in form of resin, glue, or resist are intriguing cheap and easy to handle. However, the requirements for the cryogenic use are very tough. The most important ones are chemical compatibility with the delicate superconductor material, curing, adhesion, thermal expansion mismatch, water absorption and diffusion, dielectric losses, and good insulation. The essential properties of the compounds subject to this survey are listed in Tab 1. They were taken from standard literature Saechtling (1998) or manufacturers' product informations. Some materials are immediately ruled out since they require excessive curing temperatures (polyimide) or release acids (chlorinealcylsilanes, some epoxys) during polymerization.

Compound name	Cure/polymerization	H_2O absorption	Therm Exp. α (ppm/K)	ε	$\tan \delta$ (1MHz)
Chlorinealkylsilanes	$H_2O \rightarrow$ HCl release	—	< 10	—	—
Cyanacrylate	H_2O	1-3 %	80	2.75	—
Epoxy glue	UV $\rightarrow$ acid release	0.7 %	35	3.5-5	0.01
Ethylmethacrylate	UV	1.7 %	104	3.8	—
Novolak resist	UV, 100°C	—	30-50	4.5	0.03
PMMA resist	UV, 100°C	2 %	70	2-3	$4\text{-}40 \cdot 10^{-3}$
Polyester resin	Kat. H_2O_2, 100°C	0.4 %	20-40	5	0.02
Polyimide	350-400°C	3 %	27	3.4	$5 \cdot 10^{-3}$
PTFE (Teflon AF®)	100°C	< 0.01 %	100	1.9	$< 2 \cdot 10^{-4}$
Silicone rubber	100°C	< 0.1 %	200	3.5	0.02
(Poly)-Styrene	H_2O_2	< 0.1 %	60-80	2.5	$4 \cdot 10^{-4}$
(Poly)-Vinylacetate	H_2O_2	3%	160-200	3.2	0.05

Tab. 1 Important properties of plastics compounds

4. EXPERIMENTS AND RESULTS

The resins or precursor monomers were applied by spin coating at 3000 rpm on $YBa_2Cu_3O_7$ test chips half covered with a gold film and indium solder pads. The resulting thickness of the coatings depends on the viscosity (dilution) and could be easily adjusted around 10-30 μm. After polymerization and curing the samples were immersed into de-ionized water at RT for 3-20 hours together with a bare $YBa_2Cu_3O_7$ - reference sample. Another batch of test samples was cycled by immersing into liquid nitrogen (LN-Test) and warming up to RT. After that treatment these samples were immersed into 0.1m HCl for one minute to detect cracks and pinholes (tightness test). The results are summarized in Tab. 2.

Compound name	Adhesion wetting	H$_2$O-Test	LN-Test	Tightness	Applicability
Cyanacrylate	++	+	+	+	Yes
Epoxy glue	–	o	–	–	No
Ethylmethacrylate	+	–	–	–	No
Novolak PF resist	+	–	–	–	No
PMMA resist	++	–	o	–	No
Polyester resin	++	+	–	–	No
PTFE (Teflon AF®)	+	+	+	++	Yes
Silicone rubber	+/o	+	+	+/o	Yes
(Poly)-Styrene	+	–	–	–	No
(Poly)-Vinylacetate	+	–	–	–	No

Tab. 2 Test results (cf. text), (indicators: + good, o moderate, – bad)

Out of about 20 tested compounds of different manufacturers only three survived the LN – cryogenic test in good shape and are appropriate protection layers. Amorphous Teflon AF® which can be dissolved by Flourinert (3M) is a registered trademark of DuPont and protected by U.S. patents. It has already been approved for superconducting microwave devices by Nagai (1991). Elastosil® RVT-1 A-07 silicone rubber is a registered trademark of Wacker – Chemie GmbH and can be diluted with toluene. At ambient conditions it is cured within a few minutes by vulcanization releasing alkaline amines. Although the water diffusion in silicones is relatively high the low residual water absorption seems to be the crucial parameter which renders silicones an effective protection. Cyanacrylate (by Loqtite) is a widely used industrial adhesive and can only be dissolved by a special Debonder (Sichel GmbH). Inductive measurements of the transition temperatures and critical current densities of YBa$_2$Cu$_3$O$_7$ -films prior and after application of the coatings did not reveal any film degradation.

In appropriate dilution Teflon and silicone rubber can be either spin- or spray- coated resulting in good step coverage and tightness. Cyanacrylate is ready for use. However, the cost of Teflon AF® as well as Cyanacrylate is high whereas Elastosil® silicone rubber is very cheap and may become interesting when applied in large volume. Processing of Cyanacrylate is hampered by its short curing time (seconds) which may practically restrict its application to small area chips.

Currently, microwave characterization and long term stability checks of Elastosil® covered YBa$_2$Cu$_3$O$_7$ – films are performed to approve this new type of protective resin for application in RF-filters.

ACKNOWLEDGEMENT

This study has been supported by the BMBF under contract No. 13N7266.

REFERENCES

DuPont U.S. Patent 4,977,297 and 4,982,056
Nagai Y, Suzuki N, Sato M, and Konaka T 1991 Jap. J. Appl. Phys. **30**, 2751
Saechtling H-J 1998 Kunststoff Taschenbuch 27. Ausgabe, Carl Hanser Verlag, München

Inst. Phys. Conf. Ser. No 167
Paper presented at Applied Superconductivity, Spain, 14-17 September 1999
© 2000 IOP Publishing Ltd

Fabrication of low microwave surface resistance $YBa_2Cu_3O_y$ films on MgO substrates by self-template method

M Kusunoki, Y Takano, M Mukaida and S Ohshima

Department of Electrical and Information Engineering, Yamagata University, Yonezawa, Yamagata, 992-8510, Japan

ABSTRACT: In-plane aligned c-axis YBCO films having low R_s were fabricated on MgO substrates by self-template method. To identify the effect of the grain boundaries on R_s, four kinds of films with different densities of $45°$ grains (D_g) were prepared. The R_ss systematically changed corresponding to the D_g. The in-plane aligned YBCO film showed R_s of $0.14 m\Omega$ (20K) and $1.32 m\Omega$ (77K) at 10GHz, respectively. The R_s of this film is comparable to that of in-plane aligned $YBCO/BaSnO_3/MgO$. The results showed that the effect of the lattice mismatch between YBCO and substrate on R_s is negligible.

1. INTRODUCTION

The critical current density (J_c) strongly depends on the angle between grains [1]. This indicates that the surface resistance (R_s) is also influenced by the grain boundaries [2-5]. The $45°$ rotated misaligned grains in the surface often appear in c-axis oriented $YBa_2Cu_3O_y$ (YBCO) films deposited on (100) MgO substrate. Recently, it has been reported that ratios of $45°$ grains to $0°$ grains depend on substrate temperature [6]. This is considered to be due to the thermal expansion of the MgO lattice constant. In order to obtain the strong coupling between the grains, the a- (or b-) axis of YBCO should have the same orientation. However, the growth temperature (T_{in}) necessary to realise perfect in-plane aligned YBCO film does not correspond with the optimum temperature (T_{opt}) required to obtain good values of the c-axis length, critical temperature (T_c) and surface roughness. If we suppose that the orientation of the film is determined at the beginning of the growth, by changing the growth temperature from T_{in} to T_{opt} during the deposition, the in-plane aligned YBCO film can be obtained maintaining the film qualities mentioned above. In this paper, we show the influence of the $45°$ grain boundary on R_s and demonstrate how to prepare the low-R_s film using self-template method. The R_s values of the films will then be compared with the R_s values obtained for YBCO films grown on $BaSnO_3$ buffer layers.

2. SAMPLE PREPARATION

YBCO films were deposited on 20×20 mm^2 (100) single crystalline MgO substrates by ArF excimer laser ablation. Figure 1 is a timing chart of the deposition sequence for the heater temperature. The temperature was monitored by a pyrometer at a heater block located near the substrate. T_t and T_m indicate the temperatures during the template and main layer depositions, respectively. Four values of T_t 670, 688, 713 and 730 °C were used in order to obtain different D_g values. A constant T_m value of 710°C was chosen to get identical characteristics among each sample. The c-axis length, critical temperature T_c and surface roughness are optimised at this temperature of T_m. The thickness of the template layers was 60 nm. Total thickness of the films was 600 nm. XRD φ-scans of the YBCO

(102) peaks are shown in Fig. 2. The (110) peaks of the MgO substrates appeared every $90°n$, where n is an integer. The density of $45°$ grains D_g is defined as:

$$D_g = \frac{I_{45}}{I_0 + I_{45}} \times 100 \qquad (\%) \qquad (1)$$

Here, I_0 and I_{45} are the intensities of the (102) YBCO peaks which are parallel and $45°$ rotated to the a-axis of the MgO, respectively. D_g increased with increasing T_t. The results indicate that the main layers containing 90 % of the film volume grow epitaxially keeping the orientation of the template layers. Using the values of D_g, these samples are named as $S^{2\%}$, $S^{13\%}$, $S^{37\%}$ and $S^{85\%}$, and R_s of the samples are shown as $R_s^{2\%}$, $R_s^{13\%}$, $R_s^{37\%}$ and $R_s^{85\%}$, respectively. Furthermore, the θ-2θ scans of the XRD showed that all films were perfectly c-axis oriented. The T_c values of the films prepared using this method were more than 87 K as measured by the conventional four-point probe method.

3.   RESULTS AND DISCUSSIONS

The R_s values were measured by the sapphire dielectric resonator method at 22 GHz using a cryocooler. A YBCO film deposited on a LaAlO$_3$ substrate as prepared by THEVA [7,8] was used at one end of the resonator as a reference sample. The R_s of the reference sample (R_s^{ref}) has been measured beforehand [9]. The R_s of the object sample (R_s^{obj}) is calculated by

$$R_s^{obj} = 2R_s^{ave} - R_s^{ref} \qquad (2)$$

R_s^{ave} defined as [10]

$$R_s^{ave} = \frac{30\pi\varepsilon_r}{1+w}\left(\frac{2Lf_0}{c}\right)^3 \left(\frac{1+\dfrac{w}{\varepsilon_r}}{Q_u} - \tan\delta\right) \qquad (3)$$

Here, f_0, c, L, ε_r, w and Q_u are the resonant frequency, the light velocity in vacuum, the length of the rod, the relative permittivity, the ratio of electric field energy stored outside to inside the rod and unloaded quality factor, respectively. The loss tangent of the dielectric $\tan\delta$ was neglected in the calculation.

Figure 3 depicts the dependence of the R_s^{obj} on D_g at 70 K. The plotted circles are the obtained R_s values. It is interesting to note that $R_s^{13\%}$ and $R_s^{85\%}$ are almost the same. This is explained as follows. When the $45°$ grains are considered to be the main domains, the density of the $0°$ grains is regarded to be 15 % (=100-85). This indicates that

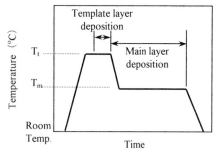

Fig. 1.   Timing chart of the deposition sequence for the heater temperature. The T_t and T_m are the temperatures used in the template and main layer depositions, respectively. Four values of T_t were used in order to obtain different D_g. The main layer was deposited at a constant T_m of 710°C to get identical characteristics for all samples.

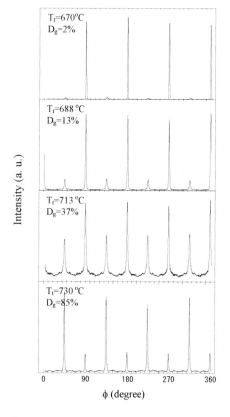

Fig. 2.   XRD ϕ-scans of the YBCO (102) peaks. The (110) peaks of the MgO substrates appeared every $90°n$, where n is an integer. D_g increased with increasing T_t. The main layers having 90 % of the volume in the films grow epitaxially keeping the orientation of the template layers.

the amount of grain boundaries is dominant for R_s. Following this concept, the amount of grain boundaries and R_s take maximum values at 50 % of D_g. The R_s-D_g characteristics can be estimated as a broken line in Fig. 3. The temperature dependence of R_s is shown in Fig. 4. Open circles, triangles, squares and diamonds correspond to $R_s^{2°}$, $R_s^{13°}$, $R_s^{37°}$ and $R_s^{85°}$, respectively. In this figure, R_s-T curves end at around 30 mΩ at higher temperatures. We have to note that these ends do not correspond to the T_c of the films. This is due to the limitation of the measurement of the quality factor, where the bandwidth at – 3dB point could not be measured due to the overlapping of the insertion attenuation to the background level of the network analyser. The films having 0°-45° grain boundaries are characterised by large residual surface resistances. It is therefore evident that the 0°-45° grain boundary is the main cause for the higher surface resistance. Almost perfectly in-plane aligned sample S$^{2°}$ showed R_s of 0.68 mΩ at 20 K and 6.40 mΩ at 77 K. The R_s values scaled with f^2 law to 10 GHz were 0.14 mΩ at 20 K and 1.32

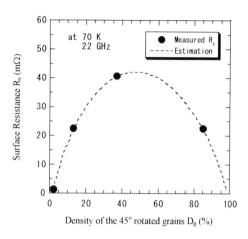

Fig. 3. Dependence of R_s^{obj} on D_g at 70 K. Measured R_s values are shown as closed circles. The estimation curve is shown by broken line. The equivalence of $R_s^{13°}$ and $R_s^{85°}$ indicates that the amount of grain boundaries is dominant for R_s.

mΩ at 77 K, respectively. For comparison, the R_s values (R_s^{BSO}) of the perfectly in-plane aligned (D_g=0) YBCO films deposited on BaSnO$_3$ (BSO) buffer layers [11] are shown as closed circles in Fig.4. Since the lattice constant of BSO lies between that of the a-axis values of YBCO and MgO, it is possible to align the in-plane orientation of YBCO within a wide substrate temperature range. The YBCO growth parameters are the same as those for the main layers of the films made by the self-template method. The R_s^{BSO}-T characteristic agrees with that of S$^{2°}$. From these results it is seen that the effect of lattice mismatch between YBCO film and substrate on R_s is negligible. Moreover, it is possible to prepare low R_s YBCO films on MgO using self-template method.

4. CONCLUSIONS

In summary, we prepared low-R_s YBCO films on MgO using the self-template method and investigated the effect of the 0°-45° grain boundary on R_s. The density of the 45° rotated grain has been successfully controlled by the growth temperature of the template layer. R_s systematically changed corresponding to the amount of the 0°-45° grain boundaries. A conspicuous feature of the films having 0°-45°

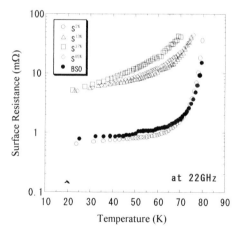

Fig. 4. The temperature dependence of R_s. Open circles, triangles, squares and diamonds correspond to the $R_s^{2°}$, $R_s^{13°}$, $R_s^{37°}$ and $R_s^{85°}$, respectively. The films including 0°-45° grain boundaries are characterized by large residual surface resistances. The closed circles are the R_s values (R_s^{BSO}) of the perfectly in-plane aligned (D_g=0) YBCO film deposited on a BaSnO$_3$ (BSO) buffer layer. The R_s^{BSO}-T characteristic agrees with that of S$^{2°}$. This indicates that the effect of the lattice mismatch between YBCO film and substrate on R_s is negligible.

76

grain boundaries is a large residual surface resistance. On the other hand, the R_s values are drastically decreased in almost 45° grain-free film. R_s of the almost perfectly in-plane aligned YBCO film were 0.68 mΩ at 20 K and 6.40 mΩ at 77 K, on the assumption of tanδ = 0 in the calculations. The R_s values scaled to 10 GHz were 0.14 mΩ at 20 K and 1.32 mΩ at 77 K, respectively. The R_s-T characteristic of the film is in good agreement with that obtained for a YBCO film grown on a BSO buffer layer. This indicates that R_s is mainly dominated by the amount of grain boundaries. Moreover, the lattice mismatch between YBCO and substrate is negligible.

5.   ACKNOWLEDGEMENT

The authors wish to thank K. Develos for her help with the preparation of this manuscript and K. Aizawa for the technical support.

REFERENCES

[1]   Dimos D, Chaudhari P and Mannhart J 1990 Phys. Rev. B**41**, 4038
[2]   Hylton T L, Kapitulnik A, Beasley M R, Carini J P, Drabeck L and Gruner G 1988 Appl. Phys. Lett. **53**, 1343
[3]   Herd J S, Oates D E and Halbritter J 1997 IEEE Trans. Appl. Supercond. **7**, 1299
[4]   Pinto R, Goyal N, Pai S P, Apte P R, Gupta L C and Vijayaraghavan R 1993 J. Appl. Phys. **73**, 5105
[5]   Castel X, Guilloux-Viry M, Padiou J, Perrin A, Sergent M, Paven-Thivet C L 1997 J. Alloys Compd. **251**, 74
[6]   Mukaida M, Takano Y, Chiba K, Kusunoki M and Ohshima S 1999 Jpn. J. Appl. Phys. **38**, 1945
[7]   Utz B, Semerad R, Bauer M, Prusseit W, Berberich P, Kindr H 1997 IEEE Trans. Appl. Supercond. **7**, 1272
[8]   Kinder H, Prusseit W, Semerad R and Utz B 1997 Advances in Superconductivity **IX**, 1011
[9]   Kusunoki M, TakanoY, Mukaida M and Ohshima S 1999 Physica C, **321**, 81
[10]   Kobayashi Y and Katoh M 1985 IEEE Trans. Microwave Theory and Tech. **33**, 586
[11]   Mukaida M, Takano Y, Chiba K, Kusunoki M and Ohshima S, Jpn. J. Appl. Phys. *in printing*

Inst. Phys. Conf. Ser. No 167
Paper presented at Applied Superconductivity, Spain, 14-17 September 1999
© *2000 IOP Publishing Ltd*

Atomic force microscopy with conducting tips: an original and efficient way to probe superconducting film surface

A Dégardin,   O Schneegans,   É Caristan,   F Houzé,   P Chrétien,   A De Luca,
A Wawrzyniak, L Boyer and A Kreisler

LGEP, SUPÉLEC, Universités Paris 6 et Paris 11, CNRS UMR 8507, Plateau de Moulon, 91192
Gif-sur-Yvette Cedex, France

ABSTRACT: Topographical and local electrical contact resistance surface images on 50 to
200 nm thick YBaCuO films sputtered on MgO and $SrTiO_3$ substrates have been obtained
simultaneously by an original technique of AFM using a conducting tip. Electrical images have
revealed a grain structure with terraces of one unit cell height. Moreover, for the thickest films, the
electrically connected areas between grains are clearly visible. This can be correlated to better
electrical transport properties of these films.

1.   INTRODUCTION

In order to study the microstructure and growth mechanisms of YBaCuO thin films, both the
scanning tunnelling microscope (STM) and the atomic force microscope (AFM) have been used
extensively (Moreland et al 1991 and Hawley et al 1991). Recently, we have reported first
observations of YBaCuO films by AFM using a conducting tip (Schneegans et al 1998a). We have
demonstrated the ability of this technique to perform simultaneously image acquisition of both
topography and local electrical contact resistance within a given area of the sample under test.

In this paper, after a brief description of the YBaCuO film preparation and the AFM
experimental set-up with its conducting probe, we report topographical and electrical images
obtained on YBaCuO films of various thicknesses deposited on single-crystal substrates. We then
proceed to a discussion of these results in relation with superconducting film properties.

2.   EXPERIMENTAL PROCEDURES

2.1   Sample preparation

Good *c*-axis oriented YBaCuO films were prepared by on axis rf magnetron sputtering using
an *in situ* process described in detail by Dégardin et al (1998). Films were grown at 300 mtorr total
pressure of a reactive atmosphere ($Ar+O_2+H_2O_{vapour}$), at temperature in the 720–730 °C range and at
120 W rf power. The deposition rate was about 50 nm/h. For this paper, YBaCuO films of thickness
50 to 200 nm were deposited onto MgO single crystals. Before YBaCuO deposition, MgO substrates
were annealed at 1000 °C for 5 h under flowing oxygen, in order to create steps on the MgO surface,
and thus favour the YBaCuO growth (Moeckly et al 1990). To validate our characterization
technique, we have also studied the surface of a YBaCuO film deposited on $SrTiO_3$ single-crystal
substrate, which is known to be among the best substrates for YBaCuO growth.

Some characteristics of YBaCuO samples studied here are given in Table 1. The film
thickness was measured with a Dektak II profilometer. The critical temperature (T_c) and critical
current density (J_c) were measured by electrical transport on 5 mm width unpatterned films, so the
presence of a rather large number of weak links within the film area under test can account for T_c and
J_c depletion. The film roughness was determined from conventional AFM topography.

Table 1. Some characteristics of YBaCuO samples studied in the framework of this paper (see text).

Sample #	Substrate	Film thickness (nm)	$T_{c(R=0)}$ (K)	$J_{c(77 K)}$ (A/cm^2)	Roughness (nm rms)
TM1	MgO	200	86	7.5×10^4	8.5
TM2	MgO	100	83.5	5×10^4	6.3
TM3	MgO	50	60	–	2.5
TS1	SrTiO$_3$	200	85	8×10^4	5.0

2.2   AFM with a conducting tip: experimental set-up

The thin film surface was observed at room temperature with a modified Nanoscope® III AFM using a conducting tip made of doped silicon coated with doped diamond. This laboratory-made system has been described in detail elsewhere (Houzé et al 1996 and Schneegans et al 1998b). The applied tip-sample force was maintained constant at about 50×10^{-9} N. By applying a bias voltage of 1 V between probe and sample (positive electrode), local electrical contact resistance could be potentially measured in the 10^2-10^{11} Ω range. Simultaneous cartographies of both the topographical and electrical features of the surface under study were obtained as grey-scaled pictures.

To obtain electrical images of the YBaCuO films grown on MgO substrates, several scans of the area under test were performed sequentially: indeed, at the beginning of sample examination, the film surface was insulating, due to the presence of a passivation layer. Thus a continuous abrasion by the tip apex for about 10 minutes allowed this layer to be partly removed. We further noticed a transformation of the morphology and a degradation of the electrical performances after successive scans of too long duration ($\approx$ 30 minutes). However, this phenomenon was not observed for the YBaCuO film deposited onto SrTiO$_3$ substrate, probably due to the hardness of this latter material. It should be stressed that the contact resistances measured here are very high due to the small radius of tip apex, and therefore to the tiny tip/sample contact area.

3.   RESULTS AND DISCUSSION

Topographical and electrical images for YBaCuO films elaborated on MgO substrates are shown in Fig. 1.

The first and most striking feature is the granular aspect of the films as revealed by the topographical maps. Most grains seem to present a terrace structure, much better displayed on the corresponding electrical images. For samples #TM1 and #TM2, conducting grains of columnar shape are also visible. Such grain morphology types were observed at other surface locations for all samples, suggesting a general feature of these films, and showing therefore a good reproducibility of the characterization technique.

The second feature concerns both the roughness and the coalescence of terrace shaped grains, which increase with the film thickness. Indeed, the thickest film (sample #TM1) exhibits a higher roughness value (see Table 1), a larger surface grain size and a larger spreading of its conducting terraces, which seem to create a suitable network for the electrical current flow. This can be correlated with the better electrical transport properties measured on it.

By studying these images more closely, we have established that the edges of the steps are more conducting than the terraces themselves. This seems to be well in line with the currently admitted resistance anisotropy of YBaCuO. Moreover, we have determined that the height difference between two terraces is in good agreement with the YBaCuO unit cell constant in the c-axis direction ($\approx$ 1.17 nm).

As a final check of our characterization technique, we have studied the surface of sample #TS1 (Fig. 2 (a)) and compared it with the surface of sample #TM1 (Fig. 2 (b)). The grain structure observed on the electrical image for sample #TS1 exhibits a spiral shape with numerous well-defined terraces. This image also enlightens large terrace areas electrically connected to each other and above all more conducting than areas observed on sample #TM1. This can be correlated with the slightly higher J_c value measured on sample #TS1 (see Table 1).

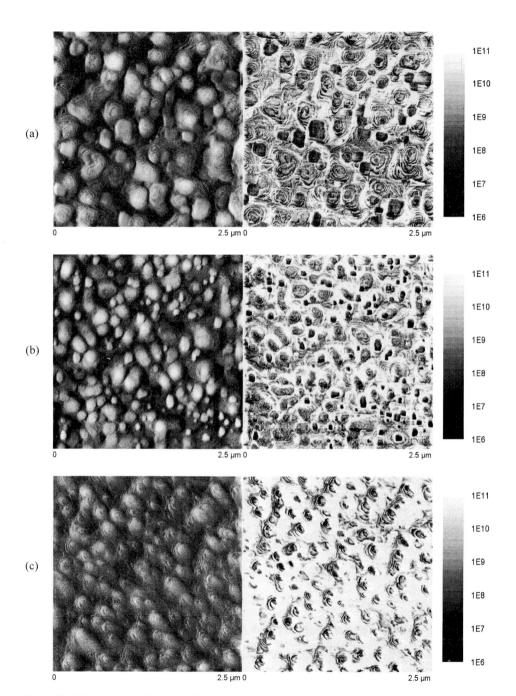

Fig. 1 (*left*) Topographical cartographies in the "illuminate mode" and (*right*) electrical cartographies with their scale in Ω of surfaces of three YBaCuO films deposited on MgO single crystal substrates (see Table 1): (*a*) 200 nm thick; (*b*) 100 nm thick; (*c*) 50 nm thick. Note the evolution of the terrace-shaped grain size and quality of electrical images.

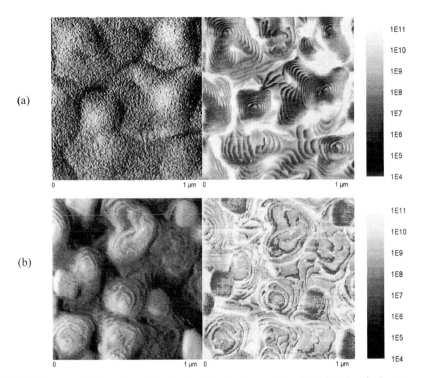

Fig. 2 (*left*) Topographical cartographies in the "illuminate mode" and (*right*) electrical cartographies with their scale in Ω of surfaces of two 200 nm thick YBaCuO films grown on (*a*) SrTiO$_3$ and (*b*) MgO single crystal substrates. Fig. 2(b) is an enlargement of Fig. 1(a). Note the difference in the electrical resistance contrast between the two films.

4.   CONCLUSION

Surfaces of YBaCuO thin films of various thicknesses sputtered on single-crystal substrates have been studied by AFM using a conducting probe. Growth steps have been observed. Moreover, for the thickest films, the electrically connected areas between grains are clearly visible. This can be correlated to better electrical transport properties of these films.

Our electrical images should now be correlated with results obtained by other imaging techniques such as TEM, in order to improve the understanding of YBaCuO growth processes.

REFERENCES

Dégardin A, Bodin C, Dolin C and Kreisler A 1998 Eur. Phys. J. AP **1**, 1

Hawley M, Raistrick I D, Houlton R B and Beery J G 1991 Science **251**, 1587

Houzé F , Meyer R, Schneegans O and Boyer L 1996 Appl. Phys. Lett. **69**, 1975

Moeckly B H, Russek S E, Lathrop D K, Buhrman R A, Li J and Mayer J W 1990 Appl. Phys. Lett. **57**, 1687

Moreland J, Rice P Russek S E, Jeanneret B, Roshko A, Ono R H and Rudman D A 1991 Appl. Phys. Lett. **59**, 3039

Schneegans O, Chrétien P, Caristan É, Houzé F, Dégardin A and Kreisler A 1998a SPIE Proc. **3481** eds D Pavuna and I Bozovic pp. 265-73

Schneegans O, Houzé F, Meyer R and Boyer L 1998b IEEE Trans. Comp. Pack. and Manuf. Technology **21**, 76

Inst. Phys. Conf. Ser. No 167
Paper presented at Applied Superconductivity, Spain, 14-17 September 1999
© 2000 IOP Publishing Ltd

BaZrO₃ as a buffer layer for high T_C superconducting thin film applications

M. Maier, M. Mauer, F. Martin, J.C. Martínez, H. Adrian

Institut für Physik, Johannes Gutenberg-Universität, 55099 Mainz, Germany

ABSTRACT: We investigate the properties of BaZrO₃ as a buffer layer in two different applications: bi-epitaxial Josephson Junctions and deposition of YBa₂Cu₃O₇₋δ on technical substrates like silicon and sapphire. Pulsed Laser Deposition (PLD) was used to grow combinations of epitaxial layers containing BZO, YSZ and CeO₂. The quality of the films was investigated by means of X-ray diffraction, transport and susceptibility measurements.

1. INTRODUCTION

Up to now the best YBa₂Cu₃O₇₋δ (YBCO) single crystals are grown in BaZrO₃ (BZO) crucibles (Erb 1996). Extremely low interdiffusion between both materials is the main reason for the high purity (better than 99.995%). Due to this inertness BZO is a promising material for being used in high quality heterostructures involving YBCO layers. Especially for the development of HTS devices on appropriate substrates BZO can either be used as an interdiffusion barrier or as an intermediate layer to improve or to control the epitaxial growth of YBCO.

In this work we describe the use of BZO in superconducting heterostructures. First we show how BZO can be used for making bi-epitaxial Josephson Junctions. In the second part, we will discuss the use of BZO as a buffer layer for deposition of YBCO on silicon and sapphire.

We used Pulsed Laser Deposition to grow epitaxial oxide layers on different substrates. The process was carried out in oxygen pressure from 0.5 mbar down to 10^{-7} mbar and deposition temperatures up to 850°C, depending on the deposited material. Self manufactured sintered ceramic targets were ablated with energy densities up to $J_{th} = 4.5$ J/cm².

The films were characterised by X-ray diffraction and atomic force microscopy (AFM) as well as scanning electron microscopy (SEM). Inductive and resistive measurements were carried out to determine the electrical properties of the superconductor.

2. BI-EPITAXIAL GRAIN BOUNDARY JUNCTIONS

During the past years a number of techniques have been developed for the construction of Josephson Junctions (JJ) in high temperature superconductors. The two more successful have been grain boundary junctions in bi-crystalline substrates and ramp-type junctions. However, in spite of their better transport properties, these junctions can only be implemented in limited geometries. This factor encouraged us to investigate bi-epitaxial JJ. These junctions present at 4.2K typical critical currents of the order of 10^{-4} Am⁻² and $I_c R_n$ products of only 1mV (Boikov1997). However, with this sort of junctions, it is possible to construct complicated JJ-arrays by simple photo-lithographic processes.

A set of buffer layers is used to generate an in-plane rotation of YBa₂Cu₃O₇₋δ ($a = 3.84$ Å) to create a 45° grain boundary. This can be achieved by continuous lattice match from the SrTiO₃ substrate (STO, a = 3.9 Å) via $BaZrO_3$ (BZO, a = 4.19 Å), Yttrium-stabilised ZrO_2 (YSZ, a = 5.14Å) and CeO_2 (a = 5.41Å) to avoid the 45° in-plane rotation of CeO_2 relative to the STO substrate. Structuring of this three layer stack by Ion Beam Etching (IBE, Ar$^+$, 500V) gives a growth base where the YBCO superconductor grows cube-to-cube on STO and 45° in-plane rotated on the three layer stack.

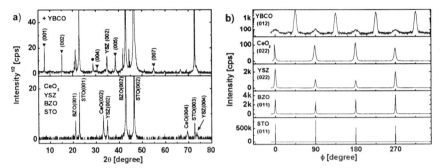

Fig. 1. a) X-ray diffraction pattern of a STO/BZO/YSZ/CeO$_2$ heterostructure (bottom) and the complete heterostructure after YBCO deposition (top). The inverted triangles correspond to YBCO reflections and the ball shows a reflection which could be due to BaCeO$_3$ or CeO$_2$ (111). b) Φ-scan of the YBCO layer deposited on a structured tri-layer (BZO/YSZ/CeO$_2$). As expected both 0° and 45° orientations are present in the YBCO layer. The lower scan shows a Φ-scan of the STO substrate.

X-ray diffraction in Bragg-Brentano geometry shows only film reflections corresponding to (00l)-orientations related to the substrate and the three layer stack (Fig. 1a bottom). A good c-axis orientation of the YSZ layer is achieved only for high deposition temperatures (>800°C) and high energy densities (>3.5 J/cm^2).

After removing the tri-layer in half of the substrate by ion beam etching (IBE), we deposited the YBCO layer. The superconductor shows a clear c-axis orientation (Fig. 1a top) and the (005)-peak shows a full width at half maximum (FWHM) of 0.36°. Due to interface reactions during deposition, the CeO$_2$ (00l)-peaks disappear and an additional peak at 28.8° can be observed. This can be related either to a CeO$_2$-(111) or a BaCeO$_3$-phase. The three-layer stack shows full in-plane orientation checked by Φ-scans with a four circle diffractometer (see Fig. 1b). For the YBCO layer we obtain peaks at Φ = 0° and Φ = 45° on STO and on the three layer stack, respectively. The latter exhibits a in-plane orientation with FWHM of 2.5°. The YBCO layer directly on top of the STO substrate yields less intensive and sharp reflections. This is due to strong surface damaging by IBE as seen in AFM measurements. This epitaxial difference of quality on both sides is as well reflected in the electronic properties of the YBCO layer. The superconducting transition was measured inductively with an AC-susceptometer. We estimated a T_c onset of 85 K. A kink observed at 80K could be related to the YBCO layer on top of the tri-layer.

3. $BaZrO_3$ FOR $YBa_2Cu_3O_{7-\delta}$ DEPOSITION ON TECHNICAL SUBSTRATES

3.1. $BaZrO_3$ as buffer layer for silicon substrates

The integration of HTS devices into semiconductor technologies requires the deposition of high quality HTS thin films on silicon substrates. To avoid interdiffusion usually a YSZ buffer layer is introduced. At deposition temperatures above 700 °C a natural $BaZrO_3$ interdiffusion layer is built which degrades the first mono-layers of the YBCO film and thus the epitaxial and electronic properties (Cima 1988). We show that the introduction of an artificial BZO layer improves the quality of the superconductor. Figure 2a shows a X-ray diffraction measurement of

83

a three layer stack ẎSZ(70nm)/BZO(50nm)/YBCO(70nm). The BZO layer shows reflections corresponding to the (00*l*)- and to (0*ll*)- orientation. Four circle X-ray measurements confirm this 0° and 45° out-of-plane orientation of the BZO c-axis. Additionally a ±9° in-plane rotation of the BZO layer is observed. The YBCO layer on top thus exhibits a 45°±9° in-plane rotation. These results are in good agreement with observations of other groups (Boikov 1994, Schlomm 1996). However the introduction of a very thin (3nm) BZO layer leads to a pure 45° in-plane rotation of the YBCO layer as shown in Fig. 2b. Due to its thickness the BZO layer is not visible in the four circle measurement.

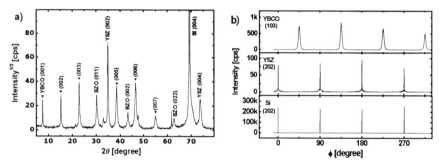

Fig. 2. a) *θ–2θ-scan* of a YBCO thin film deposited on a YSZ/BZO buffer layer on silicon. b) *Φ-scan* of a Si/YSZ/BZO/YBCO heterostructure. As described in the text we used only a 3 nm thick BZO layer.

As a result the c-axis orientation of the HTS could be improved to FWHM<1.3°. As we can see in Fig. 3 resistive measurements give a T_c onset of 89.6 K and a $T_c(R=0)$ = 84.8 K. AFM measurements show a surface roughness of only 1.1 nm (RMS).

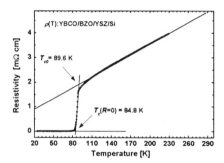

Fig. 3. Superconducting transition of the YBCO thin film measured resistively.

3.2. BaZrO₃ as buffer layer for sapphire substrates

Sapphire (*r-plane* Al₂O₃) is one of the best materials for High Frequency (HF) applications of high temperature superconductors. This is due to its low HF-losses (*tanδ*<10⁻⁷ at 77K) and low cost. To avoid interdiffusion of Al into the YBCO and to improve crystalinity, a buffer layer of CeO₂ is most commonly introduced. However the formation of BaCeO₃ at the interface generates the formation of pinholes at the interface (Boikov 1997a). This inconvenient can be avoided by the use of BZO instead of CeO₂. In the present report, a 50 nm thick BZO buffer layer was introduced in situ as a buffer layer for YBCO. The growth conditions of BZO have been optimised to a deposition temperature of 400°C and a oxygen pressure of 5×10⁻⁶ mbar.

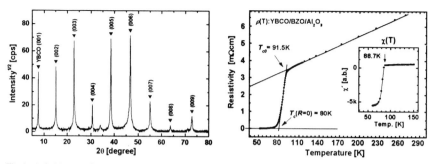

Fig 4. a) *θ–2θ-scan* of a YBCO thin film deposited on sapphire. b) Superconducting transition of the YBCO thin film measured inductively and resistively.

In Fig. 4a we show a $\theta/2\theta$-scan. Only film reflections corresponding to the (00l)-orientation of the YBCO layer can be observed with a well-defined c-axis orientation of FWHM of 0.6° for the (005)-peak. Due to a substrate miscut of more than 1° the substrate peaks are not visible. The peak at $2\theta = 33.9°$ could be related to a reaction product of Al_2O_3 and BZO at our deposition temperature of 840°C. Inductive and resistive measurements show a T_c onset of 91.5 K with a broad transition of about 10 K. These results can be considerably improved by further optimisation of the growth parameters.

4. CONCLUSIONS

In conclusion we showed the possibility of creating a bi-epitaxial grain boundary by the use of BZO, YSZ and CeO_2 as buffer layers between STO and YBCO. A substantial effort has to be invested in order to improve the surface quality of the STO surfaces after ion beam etching.

We show that BZO can be used as interdiffusion barrier for the growth of HTS on technical substrates. By depositing a 3 nm thick BZO layer between YSZ and YBCO it was possible to improve substantially the quality of the superconducting layer.

A potential use of BZO as a buffer layer for sapphire in HF applications, depends still on further investigations of the interface between the substrate and BZO.

ACKNOWLEDGEMENTS

The authors would like to acknowledge the financial support by the TMR-Program (*EG-ERBFMBICT972217*) and the Innovationsstiftung Rheinland-Pfalz (*8036-286261/222*).

REFERENCES

A. Erb et al. 1996, Physica C **258**, 9
Yu. A. Boikov et al. 1997, Phys. Solid State **39**, 1542
M. J. Cima et al. 1988, Appl. Phys. Letters **53**, 710
Yu. A. Boikov et al. 1994, Supercond. Sci. and Techn. **7**, 801
D. G. Schlom et al. 1996, J. of Mat. Res. **11**, 1336
Yu. A. Boikov et al. 1997(a), Phys. Rev. B **56**, 11313

Inst. Phys. Conf. Ser. No 167
Paper presented at Applied Superconductivity, Spain, 14-17 September 1999
© *2000 IOP Publishing Ltd*

Comparison of structural and compositional properties of YBaCuO thin films for microwave applications

A Dégardin[1], S Bourg[1], X Castel[2], C Bodin[1], J Berthon[3], C Dolin[4], É Caristan[1], F Pontiggia[1], A Perrin[2] and A Kreisler[1]

[1]LGEP, SUPÉLEC, UMR CNRS 8507, Universités Paris 6 et Paris 11, Plateau du Moulon, 91192 Gif-sur-Yvette Cedex, France
[2]LCSIM, UMR CNRS 6511, Université Rennes 1, Avenue du Général Leclerc, 35042 Rennes Cedex, France
[3]LPCS, ESA CNRS 8073, Université Paris 11, Bâtiment 414, 91405 Orsay Cedex, France
[4]Unité de Physique UMR 137 CNRS - Thomson-CSF, Domaine de Corbeville, 91400 Orsay, France

ABSTRACT: The influence of MgO substrate thermal annealing prior to YBaCuO deposition has been investigated by studying structural and physico-chemical properties of YBaCuO thin films. An improvement in the crystalline quality of films sputtered onto annealed (vs unannealed) substrates has been established, despite Ca-diffusion from MgO at the film/substrate interface. Moreover, better critical temperature and microwave surface resistance values have been observed for annealed substrates.

1.   INTRODUCTION

Among the factors influencing YBaCuO film properties, the substrate preparation before superconducting material deposition plays a significant role. Norton et al (1990) reported that substrate thermal annealing allowed the formation of steps on its surface, which then favored laser-ablated YBaCuO film growth, by providing preferential sites for island nucleation of YBaCuO grains.

In this paper, after a brief description of the experimental procedures, we present and discuss results about the influence of MgO substrate thermal annealing on the structural and compositional properties of the superconducting films. Other film characteristics like critical temperature (T_c) and microwave effective surface resistance (R_s) are also considered.

2.   EXPERIMENTAL PROCEDURES

YBaCuO films were prepared by on axis rf magnetron sputtering using an *in situ* process described in detail by Dégardin et al (1998). Films were grown under a reactive gas mixture ($Ar+O_2+H_2O$vapour) of 300 mtorr total pressure, at temperature in the 720–730 °C range and at rf power of 120 W. The deposition rate was about 50 nm/h. For this paper, 200 nm thick YBaCuO samples were sputtered on MgO (100) single crystal substrates of two types. Sample #1 corresponds to a film grown on an unprocessed MgO substrate, i.e. as delivered by the manufacturer (Escete BV). Sample #2 refers to a film fabricated under the same conditions on a MgO substrate, which was annealed at 1000 °C for 5 h under pure flowing oxygen before YBaCuO deposition.

Both bare MgO substrates and YBaCuO samples were studied with various characterization techniques. The surface morphology was observed with a Nanoscope® III (Digital Instruments) atomic force microscope. The crystalline structure was studied using X-ray diffraction (XRD). The orientation along the growth direction as well as the spread of the grain out-of-plane tilt angle were determined from standard θ-2θ spectra and rocking curve analyses, using a Philips diffractometer. The in-plane orientation was obtained from φ-scans with a Philips four-circle texture diffractometer. Cu $K\alpha_1$ radiation was used in all cases. The volume composition was investigated qualitatively by secondary ion mass spectrometry (SIMS), with a Caméca IMS4F instrument, using an O^{2+} ion sputtering beam. The surface composition was studied with X-photoelectron spectroscopy (XPS). The XPS analysis was carried out at a base pressure of some 10^{-10} torr with a Vacuum Generator CLAM 4 analyzer, equipped with a 9-channeltron detector and using the non-monochromatic Al $K\alpha$ radiation (1486.6 eV). T_c was determined from ac susceptibility measurements and R_s from measurements performed at 36 GHz using a conical cavity resonator technique (Roelens et al 1998).

3.   RESULTS AND DISCUSSION

3.1   Surface morphology

Fig. 1 shows the AFM topographies of a MgO substrate before and after thermal annealing. Before the thermal process, we can notice a non-uniform surface with numerous circular holes whose average depth is about 3 nm. After annealing, the surface presents sharply curved terraces. However, the thermal treatment produces no variation in the substrate roughness (≈ 0.5 nm rms).

AFM cartographies of YBaCuO films have revealed granular surfaces with large grain size (Schneegans et al 1998). Roughness of sample #2 (8.5 nm rms) is slightly higher than roughness measured on sample #1 (7.3 nm rms).

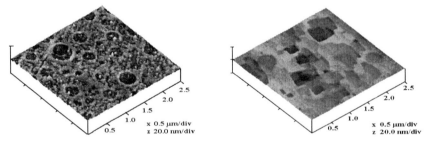

Fig. 1 Surface morphologies of a MgO substrate before (*left*) and after (*right*) thermal annealing. Images were taken in the AFM "topographical mode". Note the changes in the structure.

3.2   Structural properties

θ-2θ XRD analysis has revealed that all YBaCuO samples were highly c-axis oriented. No secondary phases could be detected with this technique. The measured c-axis lattice parameter values ranged from 1.169 to 1.170 nm, with a value closer to that of bulk YBaCuO for sample #2.

The full width at half maximum (FWHM) values measured from rocking curves on the YBaCuO (005) peak were typically 0.4° for sample #1 and 0.3° for sample #2. The FWHM values of the MgO (200) peak were 0.11° for unprocessed substrates and 0.15° for annealed substrates.

X-ray φ-scans of the YBaCuO (103) reflection have provided further information on the in-plane orientation, as shown in Fig. 2. For sample #1, we can notice the presence of two major orientations: in addition to the 0° orientation corresponding to the alignment of the a and b axes of the film with those of the substrate, the φ-scan indicates a high density of predominantly 45° rotated grains. For sample #2 on the contrary, there is no evidence of misoriented grains. This is in good agreement with the results obtained by Moeckly et al (1991). Thus the MgO substrate thermal treatment before YBaCuO deposition seems to be effective in minimizing in-plane misorientation, which could be explained by a reduction of defects in the MgO annealed substrate.

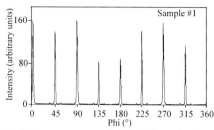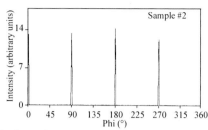

Fig. 2 X-ray φ-scans of the (103) peak for sample #1 (*left*) and sample #2 (*right*). Note the single orientation occurring at 0° for sample #2.

3.3 Compositional properties

SIMS profiles for both YBaCuO samples exhibit a stable Y:Ba:Cu composition, all through the YBaCuO layer. As CaO is a well-known impurity in MgO, we have also studied the profile of the parasitic calcium element. We can notice in Fig. 3(a) a diffusion of Ca from the substrate to the YBaCuO/MgO interface, which has only been observed for sample #2, seemingly due to the high annealing temperature of the substrate. This has been verified by SIMS and XPS analyses performed on bare substrates of both types.

We have also investigated YBaCuO film surface by XPS analyses, because we had previously detected the presence of a passivation layer on the film surface during AFM measurements with a conducting tip (Schneegans et al 1998). Moreover, it is well known that upon exposure to air, a YBaCuO surface reacts to form $Ba(OH)_2$ and $BaCO_3$. Fig. 3 (b) shows the O1s peak spectrum for sample #2. The same behaviour has been observed for sample #1. The peak was deconvoluted into two major components: one at the binding energy of 529.3 eV, which is due to cuprous species, and the other at 531.3 eV, which is generally associated with surface impurities like carbonates, hydroxides or strongly adsorbed oxygen species. The minor line at 533.5 eV is due to C-O impurities. This was confirmed by the observation of the $Ba3d_{5/2}$ peak, found to be split into a line at 777.6 eV and another at 779.9 eV. The low binding energy component is relevant to the superconducting cuprate from the bulk of the film, whereas the high binding energy line can be associated with impurities on the film surface. Further investigations need to be undertaken to complete these results and identify the nature of phases present on the film surface.

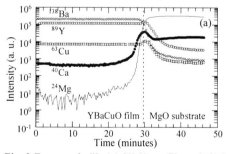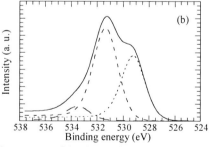

Fig. 3 For sample #2, (*a*) SIMS profile and (*b*) XPS spectrum of the O1s peak.

3.4 ac susceptibility and surface resistance

χ_{ac} inductive measurements were performed on both samples. Results are presented in Fig. 4. The thermal evolution of χ' allowed us to determine T_c, defined as the upper point of the observed transition. We can notice that sample #1 has a broad transition with $T_c \approx 80$ K, whereas sample #2 exhibits a sharp transition with a better value of T_c (83.7 K). Moreover, for the latter film we have calculated a low value for the normalized area under the χ'' peak $A(\chi'') = 0.05$ K. Castel et al (1995) have established a correlation between $A(\chi'')$ and R_S values. Therefore, a low R_S value should be expected for this sample.

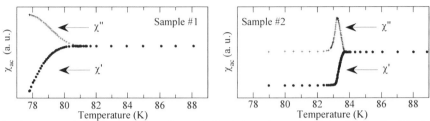

Fig. 4 Superconducting transitions of sample #1 (*left*) and sample #2 (*right*) obtained by inductive measurements. Note the sharpness of the transition for sample #2.

Thermal variations of R_S are shown in Fig. 5 for both samples. We notice that better results are obtained for sample #2, which exhibits an R_S value of 8 to 9 mΩ at 36 GHz below 70 K. All these results can be correlated with the structural properties presented in section 3.1. Indeed, disturbance in the crystallinity of sample #1 such as misorientations within the *a-b* plane should generate more grain boundaries and also degrade the microwave performances of this sample. Thus the substrate thermal annealing seems to result in beneficial effects on the superconducting film properties.

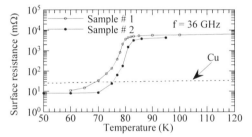

Fig. 5 Effective microwave surface resistance at 36 GHz as a function of temperature for both YBaCuO samples. The dashed line represents R_S evolution for copper.

4.   CONCLUSION

YBaCuO films have been sputtered on non-annealed and annealed MgO substrates. Structural properties as well as ac and microwave performance have been improved for films deposited on the annealed substrates. Further studies are in progress to discuss these results in relation to film electrical transport properties (such as the critical current density) or optical properties (such as the ultra-fast optical-response).

REFERENCES

Castel X, Guilloux-Viry M, Perrin A, Le Paven-Thivet C and Debuigne J 1995 Physica C **255**, 281
Dégardin A, Bodin C, Dolin C and Kreisler A 1998 Eur. Phys. J. AP **1**, 1
Schneegans O, Chrétien P, Caristan É, Houzé F, Dégardin A and Kreisler A 1998a SPIE Proc. **3481** eds D Pavuna and I Bozovic pp. 265-73 (see also this conference, paper #10-98)
Moeckly B H, Russek S E, Lathrop D K, Buhrman R A, Li J and Mayer J W 1990 Appl. Phys. Lett. **57**, 1687
Norton M G, Summerfelt S R and Carter C 1990 Appl. Phys. Lett. **56**, 2246
Roelens Y, Achani M, Bourzgui N, Tabourier P and Carru J C 1998 PIERS Conf. proc. **2**, 644

Inst. Phys. Conf. Ser. No 167
Paper presented at Applied Superconductivity, Spain, 14-17 September 1999
© *2000 IOP Publishing Ltd*

Non-destructive characterisation of HTSC wafer homogeneity by mm-wave surface resistance measurements

R Heidinger and R Schwab

Forschungszentrum Karlsruhe, Institut für Materialforschung I, D-76021 Karlsruhe, Germany

ABSTRACT: A measurement facility for the surface resistance R_S of superconducting films is presented that allows non-destructive R_S mapping of large area wafers at 145 GHz. The quantitative interpretation of the obtained data sets is described for small size defects introduced by 2 MeV α-irradiation. Large scale R_S variations are parameterised based on an cumulative distribution function. The median value and the spread of the distribution are used to group degrees of wafer perfection found in the qualification process of YBCO films on sapphire.

1. INTRODUCTION

Progress in implementing superconducting components into mobile and satellite-based communication systems has driven enhanced requirements for the production of large area wafers for which high-T_c superconductor (HTSC) films are epitaxially grown on dielectric substrates. Non-destructive methods that determine performance relevant materials parameters must provide a characterisation of wafers with typically 3" diameter and more. The geometrical flexibility and the focusing potentials of open resonator set-ups have incited continued experimental efforts in the past (Martens et al 1991, Orbach-Werbig 1994) to bring forward a facility for the determination of the surface resistance R_S. The successful realisation of an open resonator system at 145 GHz that combines spatially resolved and temperature variable R_S measurements (Schwab et al 1998) allows the direct characterisation of wafer homogeneity. In this paper, procedures are introduced to establish a quantitative description for isolated defects and also for large scaled inhomogeneities which are applied to specify wafer perfection representative for a large number of YBCO wafers.

2. MEASUREMENT TECHNIQUE AND EXPERIMENTAL SET-UP

The open resonator was realised in a the (quasi-) hemispherical Fabry-Perot arrangement. It consists of a spherical mirror (radius of curvature R_0) and a plane mirror with a resonator length L slightly below R_0. The linearly polarised TEM_{mnq} modes of the set-up fulfil the resonator equation

$$f_{mnq} = (c/2L)((q+1) + \tfrac{1}{2\pi}(2m+n+1)\arccos(1 - 2L/R_0)) \qquad (1),$$

where c is the speed of light. They are detected by transmission measurements in swept frequency operation with the signals coupled very weakly through holes in the spherical mirror. From the resonance curve, the centre frequency f_0 and the Q-factor Q_0 is obtained by fitting an asymmetric Lorentzian line shape. Apart from the negligible coupling losses and the targeted loss terms related to the surface resistance of the mirror elements, the Q-factor can be affected by diffractive losses due to limited mirror extensions and scattering losses at mirror imperfections, particularly at coupling holes. For the basic TEM_{00q} modes which have a Gaussian field distribution perpendicular to the resonator axis, a proper choice of L allows to focus effectively the beam width w_0 at the plane mirror element while keeping diffraction losses at the spherical mirror negligible. With the HTSC wafer as the front part of the plane mirror element, a R_S mapping is generated by radial and rotational movement of this element (μ_0: vacuum permeability):

$$R_{S,map}(r,\varphi) = 2\left(\tfrac{\pi}{2}\mu_0 f_0(r,\varphi)L(r,\varphi)Q_0^{-1}(r,\varphi) - R_{scat}\right) - R_{S,sp} \qquad (2),$$

where apart from the measured quantities f_0 and Q_0, the exact resonator length is determined from f_0 by solving eq.1 for TEM_{00q} modes. The surface resistance $R_{S,sp}$ of the spherical mirror can be obtained from the conductivity of the normal conducting mirror material (usually Cu) and the scattering loss term R_{scat} from calibration with normal conductors (Schwab et al 1996).

The actual set-up was installed at 145 GHz with the following parameters: $L\approx117.5$ mm, $R=120$ mm, $w_0=3.3$ mm and $R_{scat}=45(\pm5)$ mΩ (cf. Fig.1).Whereas the spherical mirror is fixed on the cold plate of a gas flow cryostat, the plane mirror has flexible thermal coupling. It is kept in position by a thermally insulating rod connected to the translational stages maintained near room temperature. This special arrangement inside a vacuum vessel allows reliable and reproducible positioning of the cryogenically cooled wafer as well as defined warming-up conditions for temperature dependent measurements of $R_S(T)$.

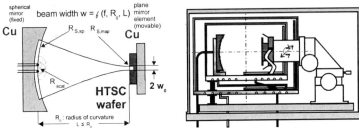

Fig.1 Geometrical parameters of a (quasi-)hemispherical Fabry-Perot resonator and sketch of the mirror arrangement in the vacuum vessel of the established R_S measurement facility

Although $R_S(T)$ measurements, for which in the actual set-up areas of special interest can be selected, bear an outstanding potential for unravelling the underlying loss mechanisms (Dierlamm et al 1999, Schwab 1999) and give additional criteria to judge wafer perfection, this aspect exceeds the scope of the present paper. The R_S mapping at 77 K was performed for wafer characterisation of a large quantity (≈25) wafers with YBCO films grown on both sides of CeO_2 buffered 3" sapphire substrates (thickness: 430 μm) either by the pulsed laser deposition or by sputtering. As the film thickness ($d \approx 300$ nm) is still comparable to the penetration depth λ of the electromagnetic wave, the observed effective $R_{S,eff}$ values are higher than the material specific R_S values. For $d/\lambda \approx 1.2 - 1.4$, the enhancement factor amounts to $1.55 - 1.65$.

3.   QUANTIFICATION OF LARGE SCALE R_S HOMOGENEITY

Typical for the investigated YBCO wafers are R_S variations that extend over scales that exceed the beam width. For these cases, the measured data can be directly interpreted as local $R_{S,eff}(r,\varphi)$ data which indicate two-dimensional structures in the wafer homogeneity. For a compact description, the data are transferred into a normalised distribution function $S(R_{S,eff})/S_0$ by combining the number of data points within a given $R_{S,eff}$ data interval with a weight function for the effective area covered by each data point. This function quantifies the normalised area fraction of $R_{S,eff}$ values observed in the film. For further data reduction, the cumulative sum $\mathcal{S}(R_b)$ is analysed which sums up the normalised distribution function up to an upper bound R_b for $R_{s,eff}$. Thus the D_{50} parameter is defined by $\mathcal{S}(D_{50}) = 0.5$ which represent the median value of the distribution, for which the condition $R_{S,exp} \leq D_{50}$ is fulfilled for half of the film area. Two additional quantities, the onset parameter D_{10} and the terminal parameter D_{90}, are defined in analogy to the D_{50} parameter. With these three parameters, the spread ($\Delta D = D_{90} - D_{10}$) and asymmetry of the distribution ($F = (D_{90}-D_{50})/(D_{50}-D_{10})$) can be quantified (cf Fig.2).

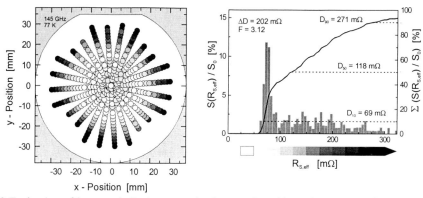

Fig. 2 Evaluation of large scale R_S homogeneity for a wafer with moderate R_S perfection in terms two-dimensional maps, the normalised distribution function and the cumulative sum

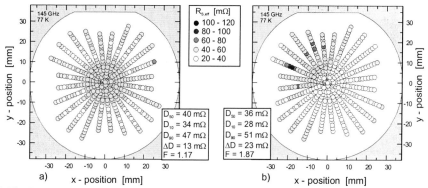

Fig.3 The R_S mapping and the distribution parameters determined for both sides of a wafer with high R_S perfection. Note the local R_S perturbation in sector (-20,10) of side b).

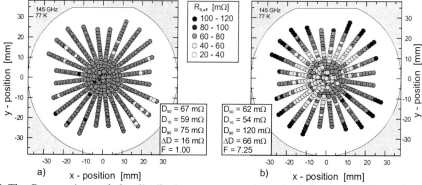

Fig.4 The R_S mapping and the distribution parameters determined for wafer with moderate R_S perfection. Film b) shows enlarged spread values caused by a less perfect wafer edge.

The different degrees of large scale R_S homogeneity that have been found for the investigated YBCO wafers can be attributed to 3 groups. Wafers with high R_S perfection have D_{50} values between 25-50mΩ and ΔD values between 10-20 mΩ (cf. Fig.3). The most frequent group with average R_S perfection shows enhanced D_{50} values (50-75 mΩ). The spread values often stay low (10-20 mΩ), if they are enlarged, the distribution extends to the higher $R_{S,eff}$ tail resulting in high values of the D_{90}

and of the asymmetry parameters (cf. Fig.4). The group with moderate R_S perfection, which has been typically encountered during explorative stages of the wafer production, is characterised by D_{50} values far above 75 mΩ and at least similar ΔD values (cf. Fig.2). For this group, the R_S mapping reflects symmetry patterns corresponding to the wafer movement in the production process indicating effective profiles in the parameters of the deposition process. The large scale R_S homogeneity may be superimposed by contributions from local defects which can become evident in wafers with high and average R_S perfection (cf. Fig.3b).

4. DETECTION AND QUANTIFICATION OF ISOLATED DEFECTS

Even though the measurement facility detects isolated R_S perturbations, they are encountered for a minority of the specimens. To demonstrate the characterisation potential, a model wafer was irradiated with 2 MeV α-particles. A sequence of spots (1mm dia.) were created in two orthogonal rows at intervals of 10 mm. The defects appear in the R_S mapping with a broadened width according to the size of the beam width parameter w_0. In addition, two linear scans were extracted closely along the axis of one row. With these linear scan data, the convolution integral for a linear R_S profile function and the weight function of the mm-wave profile was executed (cf. Fig5). The analysis yields the central positions of the perturbations and a quantitative measure for their strength from which the peak height can be separated for a known defect width and vice versa. The linear analysis is sensitive to the matching of the scanning path, the accuracy can be improved by performing the two dimensional convolution integral.

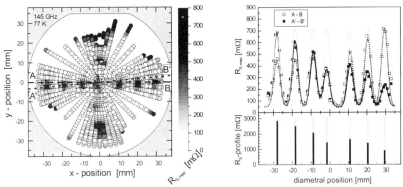

Fig. 5 The R_S mapping and line scans determined for the α-irradiated model wafer and quantifation of the linear R_S profile by calculating the convolution integral with the linear beam profile function.

5. ACKNOWLEDGEMENT

The authors thank J. Geerk, G. Linker and F. Ratzel at the Forschungszentrum Karlsruhe for kindly providing the YBCO wafers produced by the Inverted Magnetron Sputtering method and for producing the model wafer by α-irradiation and M. Lorenz at the University of Leipzig for the YBCO wafers produced by Pulsed Laser Deposition.

REFERENCES

Dierlamm A, Keskin E, Schwab R, Heidinger R and Halbritter J 1999 Proc of this conference
Martens J, Hietala V, Ginley D, Zipperian T and Hohenwarter G 1991 Appl. Phys. Letters. **71**, 1736
Orbach-Werbig S 1994 PhD Thesis Univ Wuppertal (D)
Schwab R and Heidinger R 1996 Dig.21st Int Conf IR + MMWaves, Berlin, 14 –19 July, Paper ATh7
Schwab R, Heidinger R, Spörl R 1998 Dig.23rd Int Conf IR + MMWaves, Colchester, 7-11 Sept, 379
Schwab R 1999 PhD Thesis Univ Karlsruhe (D) + Forschungszentrum Karlsruhe Report FZKA 6331

Inst. Phys. Conf. Ser. No 167
Paper presented at Applied Superconductivity, Spain, 14-17 September 1999
© 2000 IOP Publishing Ltd

Temporal response of high-temperature superconductor thin-film using femtosecond pulses

V Garcés-Chávez, A De Luca*, A Gaugue*, P Georges, A Kreisler* and A Brun

Laboratoire Charles Fabry de l'Institut d'Optique, UMR 8501 CNRS, B.P. 147, 91403 ORSAY Cedex, FRANCE
*Laboratoire de Génie Électrique des Universités Paris 6 & Paris 11, UMR 8507 CNRS, Supélec, Plateau de Moulon, 91192 GIF SUR YVETTE Cedex, FRANCE

ABSTRACT: In order to develop fast infrared detectors, we have used the pump-probe time-resolved spectroscopy technique to study the thermal properties of carriers and phonons in YBaCuO films. With a 800 nm laser source producing 100 fs pulses at 1 kHz repetition rate, we have studied the variation of film reflectivity as a function of time delay between the pump and probe pulses. For a 200 nm thick film, we have observed two optical fast responses of 2.4 ps and 20 ps, followed by a slow response (> 0.5 ns).

1. INTRODUCTION

In the last decade, the availability of high temperature superconductor materials (HTS) had an important impact in fast infrared detector development. Such development, combining superconductor materials and optical technologies, was initiated as soon as the potential for detection of optical transients was exhibited with YBaCuO films (Donaldson et al 1989). Gong et al (1993) studied in detail the femtosecond reflectivity of YBaCuO films in the picosecond range from 12 K to 300 K. Stevens et al (1997) discussed the reflection and transmission ultrafast response in terms of various electronic processes. Lindgren et al (1999) performed femtosecond measurements using an electrooptic sampling technique. We are presenting here our first results in the temporal response of YBaCuO films.

2. SAMPLE FABRICATION

The samples used in this work were 200 nm thick YBaCuO films deposited on single crystal (100) MgO substrate of 10×10 mm^2 area and 500 μm thickness (Dégardin et al 1998). The films were made *in situ* by a radiofrequency magnetron on axis reactive sputtering technique. Structural characterization was made using X–ray diffraction. Although granular, the films exhibited textural orientation with the c-axis perpendicular to substrate plane.

For the electrical characterization we used a standard four-point probe technique with gold contacts e-gun evaporated onto the film through a metal mask. The samples exhibited critical temperature (T_c) in the 80 to 86 K range and critical current density (J_c) in the 5×10^4 to 8×10^5 A/cm^2 range at 77 K. Lower T_c and J_c values are attributed to the granularity of the films.

3. EXPERIMENTAL SET-UP

Fig. 1 shows the pump-probe time resolved spectroscopy set-up. A pump pulse was focused on the film surface producing a variation in its optical properties. A second and less intense pulse (probe pulse) was focused on the same area, without modifying the film properties. We have used a regenerative amplifier source producing 100 fs pulses, with an energy of 0.5 mJ at 800 nm (1.5 eV) and 1 kHz repetition rate. These pulses were amplified using the chirped pulse

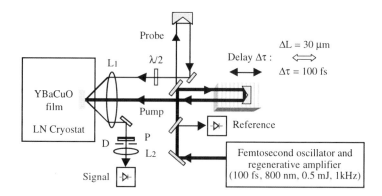

Fig. 1. Pump-probe time-resolved spectroscopy experimental set-up.

amplification (CPA) method (Rullière 1998). In this work, the pump and the probe beams were focused using a lens of 180 mm focal distance leading to 76 μm and 36 μm spot diameter, respectively, with 7.2 μJ/cm^2 and 3 μJ/cm^2 intensity, respectively. Pump and probe polarisations were perpendicular to each other, with an angle of 4° between the beams. For each pump-probe delay position, the amount of light reflected from the probe beam was calculated from the signal delivered by a photodiode detector. This system allowed us to measure reflectivity with a 10^{-3} resolution within the temporal 100 fs to 0.5 ns range. The sample was mounted in a liquid-nitrogen cooled cryostat (DN 1704, Oxford Instruments) and was attached to a special holder in a He exchange gas chamber. A PT100 temperature sensor was inserted into the holder body.

4. RESULTS

4.1. Electrical characterization

Standard four-point probe technique was also used *in situ*. We have measured the voltage values at different temperatures with a high-performance instrument (Source 2400, Keithley Instruments) while a current of 50 μA was injected. Fig. 2 shows typical resistive transitions, before and after the pump-probe experiment. T_c $(R=0) = 81$ K was obtained in both cases for this film, so we could estimate that the laser pulses did not alter significantly the transport properties.

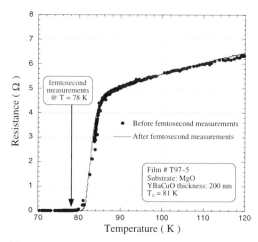

Fig. 2. Resistive transition of a film under test; femtosecond measurements were performed slightly below the superconducting transition.

4.2. Optical fast response

The relative reflectivity variation as a function of the temporal delay between the pump and the probe pulses is shown in Fig. 3. We can see the nonlinear temporal response of a thin YBaCuO film at 78 K (i.e. just below the superconductiong transition). We have measured a maximum variation of 1.5% in the reflectivity. The strong decrease in the reflection probe signal is a consequence of arrival of both the pump and probe pulses at the same time and also at the same position (zero delay in Fig. 3). Three different time constants could be deduced from this plot, with increasing values for increasing delay time. The first one has a value of $\tau_1 = 2.4 \pm 0.5$ ps, the second one of $\tau_2 = 20 \pm 2$ ps and the last one $\tau_3 > 0.5$ ns (with a large uncertainty).

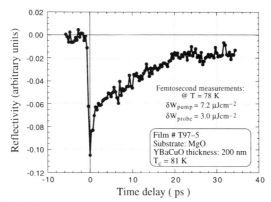

Fig.3. Reflectivity temporal dependence observed with pump-probe system.

5.   DISCUSSION

5.1. Phenomenology and modelling

An optical pulse illuminating a HTS material can destroy Cooper pairs and create high energy quasiparticles (Gilabert 1990). These quasiparticles can release their energy following the general scheme depicted in Fig. 4. The electron population thermalizes very rapidly via electron-electron scattering with characteristic time $\tau_{e\text{-}e}$. These excited quasiparticles recombine themselves to form Cooper pairs with emission of phonons with a characteristic electron-phonon scattering time $\tau_{e\text{-}ph}$. The resulting phonons in excess can release their energy via phonon-electron scattering (characteristic time $\tau_{ph\text{-}e}$) or they can escape to the substrate (characteristic time τ_{es}).

Electron / phonon dynamics in the HTS materials can be explained using the two-temperature model (Lindgren et al 1999). Assuming $\tau_{e\text{-}e}$ is much shorter (≈ 100 fs) than other characteristic times, electron (T_e) and phonon (T_{ph}) temperatures can be considered independent from each other, so the power density is governed by the coupled differential equations:

$$C_e \frac{dT_e}{dt} = \frac{\alpha P_{in}(t)}{V} - \frac{C_e}{\tau_{e\text{-}ph}}(T_e - T_{ph})$$

$$C_{ph} \frac{dT_{ph}}{dt} = \frac{C_{ph}}{\tau_{ph\text{-}e}}(T_e - T_{ph}) - \frac{C_{ph}}{\tau_{es}}(T_{ph} - T_s)$$

where C_e and C_{ph} are the electron and phonon specific heats, α is the radiation absorption coefficient, V is the volume of the illuminated area, T_s is the substrate temperature and $P_{in}(t)$ is the incident optical power. Moreover, thermal balance gives $\tau_{e\text{-}ph} C_{ph} = \tau_{ph\text{-}e} C_e$.

5.2. Measurement results

In the literature we found the following values for YBaCuO: $\tau_{es} = 2.4$ ns (Ghis et al 1994, electrical measurements), $\tau_{e\text{-}ph} \approx 5$ ps and $\tau_{es} > 10$ ns (Stevens et al 1997), $\tau_{e\text{-}ph} \approx 1.1$ ps and $\tau_{ph\text{-}e} = 42$ ps (Lindgren et al 1999). If we compare these values with our experimental results, we

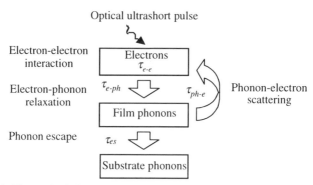

Optical ultrashort pulse

Electron-electron
interaction

Electrons
τ_{e-e}

Electron-phonon
relaxation

τ_{e-ph} τ_{ph-e} Phonon-electron
scattering

Film phonons

Phonon escape τ_{es}

Substrate phonons

Fig. 4. The excited electron relaxation after an optical ultrashort excitation.

can probably associate (see § 4.2) $\tau_1 = 2.4$ ps with τ_{e-ph} and $\tau_3 > 0.5$ ns with τ_{es}. Nobody has directly measured τ_{ph-e} by the pump-probe spectroscopy technique. We can see clearly in Fig. 3 a slope with time constant $\tau_2 = 20$ ps. This value could be associated to τ_{ph-e} which has been obtained indirectly by Lindgren et al (1999). Probably we saw this temporal response because the pump energy was only 2.4 times the probe energy and, in this configuration, the probe pulse could have created some modification in the film properties. However, we find $\tau_{ph-e}/\tau_{e-ph} = 9 \pm 3$ which does not agree with the value of 38 determined by Lindgren et al (1999). Electron-vortex or phonon-vortex interaction could also be invoked to explain time constant τ_2. Moreover, the granular structure of the film (from which result critical temperature and critical current density depletion) implies a non-negligible density of grain boundaries (i.e. weak links) that might play a role in the observed reflectivity.

6.   CONCLUDING SUMMARY

A pump-probe time-resolved femtosecond spectroscopy has been validated by testing YBaCuO film of 200 nm thickness ($T_c \approx 81$ K) deposited on MgO substrate, at 78 K. Our system exhibits a resolution of 10^{-3} in reflectivity and can explore pump-probe delay in the 100 fs to 0.5 ns range. We have obtained an ultrashort response of 2.4 ps which can be attributed to electron-phonon energy relaxation, and a long decay (> 0.5 ns) which can be attributed to phonon escape to the substrate. An intermediate time constant of ≈ 20 ps was also observed, which origin cannot be clearly ascertained.

ACKNOWLEDGEMENTS

This work was supported in part by a MENRT program (ASP 1997). The authors are very grateful to Mr. É. Caristan (LGEP) for film elaboration.

REFERENCES

Dégardin A, Bodin C, Dolin C and Kreisler A 1998 Euro. Phys. J. AP **1**, 1
Donaldson W R, Kadin A M, Ballentine P H and Sobolewski R 1989 Appl. Phys. Lett. **54**(24), 2470
Ghis A, Villegier J-C, Nail M, Gibert Ph and Striby S 1994 SPIE Proc. **2159** eds M Nahum and J-C Villégier pp 55-9
Gilabert A 1990 Ann. Phys. Fr. **15**, 255
Gong T, Zheng L X, Xiong W, Kula W, Kostoulas Y, Sobolewski R and Fauchet P M 1993 Phys. Rev. B **47**, 14495
Lindgren M, Currie M, Williams C, Hsiang T Y, Fauchet P M, Sobolewski R, Moffat S H, Hughes R A, Preston J S and Hegmann F A 1999 Appl. Phys. Lett. **74**, 853
Rullière C (ed) 1998 Femtosecond Laser Pulses (Berlin: Springer) pp 163-8
Stevens C J, Smith D, Chen C, Ryan J F, Podobnik B, Mihailovic D, Wagner G A and Evetts J E 1997 Phys. Rev. Lett. **78**, 2212

Inst. Phys. Conf. Ser. No 167
Paper presented at Applied Superconductivity, Spain, 14-17 September 1999
© 2000 IOP Publishing Ltd

Growth and Superconducting Properties of YBa$_2$Cu$_3$O$_y$/Nd$_{0.7}$Sr$_{0.3}$MnO$_3$ Multilayers

L M Wang†, S S Sung†, B T Su‡, H C Yang‡, and H E Horng§

†Department of Electrical Engineering, Da-Yeh University, Chang-Hwa 515, Taiwan, R. O. C.
‡Department of Physics, National Taiwan University, Taipei 106, Taiwan, R. O. C.
§Department of Physics, National Taiwan Normal University, Taipei 117, Taiwan, R. O. C.

ABSTRACT: Transport properties of superconducting YBa$_2$Cu$_3$O$_y$/Nd$_{0.7}$Sr$_{0.3}$MnO$_3$ (YBCO/NSMO) multilayers were measured. The critical temperatures and the activation energies of the YBCO/NSMO multilayers are decreased as the ferromagnetic NSMO layers become thicker or the superconducting YBCO layers get thinner. The lower activation energy accompanying with a diminishing sign reversal of Hall coefficient in the mixed state implies a decrease of flux pinning in YBCO/NSMO multilayer. The suppressions of T_c and pinning are attributed to the proximity induced ferromagnetic pair breaking in decoupled YBCO layers.

1. INTRODUCTION

Recently, the study of alternately stacked superconducting(S) and ferromagnetic(F) layers is of considerable interest. Techniques for preparation of epitaxial films have been established and thereby make it possible to control the electronic or magnetic properties by varying the layer thickness. Artificial-grown F / low-T$_c$ superconductors multilayers, as for example, Ni/V (Homma et al 1986) and Fe/Nb (Kawaguchi et al 1992) have been prepared to study the interplay of superconductivity and magnetism.

In particular, artificial multilayers consisting of high-T$_c$ superconducting-ferromagnetic materials may attract great interest due to the anistropy in the high-T$_c$ superconductors. The growth of superconducting-ferromagnetic heterostructures based on the superconductor DyBa$_2$Cu$_3$O$_7$ and Sr$_{1-x}$Ca$_x$RuO$_3$ as a barrier material has been presented to study the vortex dynamics in DyBa$_2$Cu$_3$O$_7$/Sr$_{1-x}$Ca$_x$RuO$_3$ multilayers. (Mieville et al 1996) In more recent studies, the fabrication of devices of Nd$_{0.7}$Sr$_{0.3}$MnO$_3$ (NSMO) / YBa$_2$Cu$_3$O$_7$ (YBCO) heterostructures has been reported also. (Dong et al 1997) The NSMO films undergoing a transition from a paramagnetic state to a ferromagnetic state at around 200 K and exhibiting a huge negative magnetoresistance ratio, -$\Delta R/R_H$ ~ 10^6 % at ~ 60 K and a magnetic field 8 T were observed. (Xiong et al 1995) Moreover, NSMO can be indexed as pseudo-cubic with a lattice parameter ~ 3.86 Å, which is close to YBCO a-b lattices parameters. Growth of NSMO/YBCO multilayers is expected to bring interesting information on the study of proximity coupling through a ferromagnetic material as well as on the fabrication of high-T$_c$ superconductor/colossal magnetoresistance heterostructure based devices.

2. EXPERIMENT

High quality YBCO and NSMO thin films have been deposited onto SrTiO$_3$(100) substrates by an off-axis sputtering system. A radio frequency sputtering system with a rotational quartz substrate holder was constructed in this study. Fig. 1(a) shows the configuration of an off-axis rf magnetron sputtering system used in this work. Two sputtering guns were mounted face to face in a stainless-steel vacuum chamber pumped by mechanical and turbo pumps. One target was a stoichiometric superconducting YBCO compound while the other was the stoichiometric NSMO compound. The growth conditions of YBCO films have been described elsewhere. (Wang et al 1997) Typical deposition parameters for NSMO films are the same as that for YBCO films except a larger horizontal distance, ~ 5.3 cm, from the center of target to the substrate.

In the stacking process of multilayers, however, the deposition temperature for YBCO grown onto NSMO was changed to 680℃, while NSMO layers were grown at 720℃. To improve the T_c of YBCO layers grown on NSMO, YBCO/NSMO/YBCO (500 Å / 700 Å / 400 Å) trilayers were first measured using a standard four-probe technique with the silver pastes mounted on the top YBCO layer. Fig. 1(b) shows the resistive transition for YBCO/NSMO/YBCO trilayers deposited at different conditions. The curves show a resistance maximum at around 200 K owing to the contribution of the resistance from NSMO layer. We find an improved T_c of the top YBCO layer and an obvious ferromagnetic-paramagnetic transition at 200 K for the YBCO/NSMO/YBCO trilayer grown at 720℃, 720℃ and 680℃ respectively as showing in Fig. 1(b) and it's inset. The lower deposition temperature, 680℃, for YBCO grown on NSMO than that for YBCO grown on SrTiO$_3$(~ 720℃), may be due to the good lattice match for YBCO(a = 3.88 Å and b = 3.84 Å) and NSMO(~ 3.86 Å).

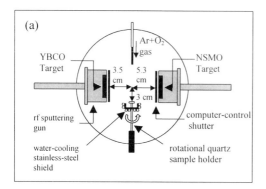
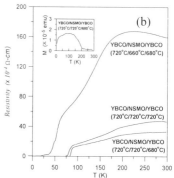

Fig. 1 (a)The Schematic diagram of the rf sputtering system. (b)Resistive transitions for YBCO/NSMO/YBCO (500 Å/700 Å/400 Å) trilayer grown at various temperatures indicated in the parentheses. The inset shows the temperature dependence of magnetization for the YBCO/NSMO/YBCO (500 Å/700 Å/400 Å, 720°C/720°C/680°C) multilayer.

For the growth of YBCO/NSMO multilayers, YBCO and NSMO layers were alternatively deposited until a desired thickness of a multilayer was reached. For transport measurements, films were photothographically patterned to a 2-mm long by 500-μm wide bridge containing two Hall terminals. Six gold dots were then evaporated onto the contact areas to allow simultaneous measurements of the resistive and Hall signals using standard dc techniques. Hall voltages were taken in opposing fields parallel to the c-axis up to 7 T and at a dc current density of J ~ 10^3 A/cm^2.

3. RESULTS AND DISCUSSION

In Figs. 2(a) and (b), we show the resistivities as a function of temperature for YBCO/NSMO (240 Å / 38.5 Å) × 4, (240 Å / 77 Å) × 4, (240 Å / 116 Å) × 4, and (120 Å / 38.5 Å) × 8 multilayers in magnetic fields of 0, 1, 2, 3, 4, 5, 6 and 7 T parallel and perpendicular to the crystal c axis, respectively. The resistivity under magnetic field shows a broadening behavior. The resistive transition gets broader in fields parallel to the c axis compared with the data obtained with fields parallel to the ab plane. The result indicates the anisotropic properties of YBCO/NSMO multilayers which are similar to that observed in YBCO thin films and YBCO/PrBa$_2$Cu$_3$O$_y$ superlattices. (Yang et al 1999) On the other hand, the critical temperature decreases systematically in multilayers with a fixed thickness of YBCO layers and increased thickness of NSMO, or with a fixed thickness of NSMO layers and decreased thickness of YBCO. Notably, for YBCO/NSMO multilayers with thicker NSMO layers, such as the YBCO/NSMO (240 Å / 116 Å) ×4 multilayer, an obviously negative magnetoresistance could be observed in the normal state. The decreased resistivities of

YBCO/NSMO multilayers in magnetic fields are due to the magnetoresistance contributed from the ferromagnetic NSMO. An interesting point in the resistive transition for YBCO/NSMO (240 Å / 116 Å) ×4 multilayers as shown in Figs. 2(a) and (b), is the zero-magnetoresistance point occurring at around 30 K. The presence of zero-magnetoresistance is attributed to the competition between the negative magnetoresistance due to NSMO and the positive magnetoresistance due to the flux motion in YBCO.

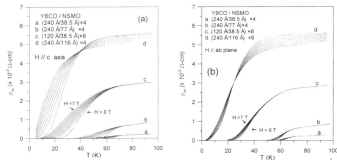

Fig.2 Temperature dependence of the longitudinal resistivity ρ_{xx} for the YBCO/NSNO (240 Å / 38.5 Å) × 4, (240 Å / 77 Å) × 4, (240 Å / 116 Å) × 4, and (120 Å / 38.5 Å) × 8 multilayers in magnetic fields parallel to (a) the c axis and (b) the ab plane. The increment of the field is 1 T in each curve.

Fig. 3 shows the Hall coefficient R_H versus temperature for YBCO/NSMO (240 Å / 38.5 Å) × 4, (240 Å / 77 Å) × 4, and (240 Å / 116 Å) × 4 multilayers in fields of 3 and 7 T. It is known that in the normal state the Hall coefficient of YBCO is positive, increasing as the temperature is lowered and goes to a negative maximum in the mixed state. Moreover, the Hall coefficient of NSMO is negative in the paramagnetic state and goes into a sign change as the temperature is lowered to the ferromagnetic region. (Yang et al Unpublished) The positive Hall coefficient observed in YBCO/NSMO multilayer indicates the transport carriers are governed by YBCO. Otherwise, the Hall resistivity systematically increases in multilayers with a fixed thickness of YBCO layers and increased thickness of NSMO. The observed increase in R_H indicates a decreased carrier density in the multilayers with thicker NSMO layers, which is consistent with higher resistivity data. In addition, the mixed-state sign reversal of Hall coefficient cannot be observed in the YBCO/NSMO multilayer. The diminished sign change of Hall coefficient may be due to the pair breaking effect contributed by NSMO layers, which results in a decrease of pinning force. (Wang et al 1994)

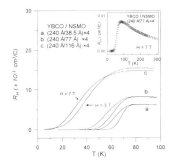

Fig. 3 Temperature dependence of the Hall coefficient for YBCO/NSNO multilayers in magnetic fields of 3 and 7 T. The inset shows the Hall coefficient of YBCO/NSNO (240 Å / 77 Å) × 4 multilayers at temperatures up to 300 K.

The resistive transition under magnetic fields also can be understood as a thermally activated flux motion. Table 1 displays the resistivity ρ, Hall coefficient R_H, $T_c(50\%)$, and the activation energy at field of 1 T for a series of YBCO/NSMO multilayers. The activation energy reveals a systematic decrease in multilayers with thicker NSMO layers or thinner YBCO layer and is 1~2 order smaller than that obtained from YBCO films. The result indicates a poor pinning ability for YBCO/NSMO multilayer, which is due to the decoupling of NSMO layers. The depinning behavior observed in YBCO/NSMO multilayer is consistent with the diminishing sign reversal of Hall coefficient observed in mixed state.

TABLE 1 The resistivity ρ, Hall coefficient R_H, $T_c(50\%)$, and the activation energy at field of 1 T for a series of YBCO/NSMO multilayers

Samples YBCO/NSMO	$\rho(300\ K)$ $(\mu\Omega cm)$	$R_H(300\ K)$ $(\mu cm^3/C)$	$T_c(50\%)$ (K)	$U_0(1\ T)$ (K)
(240 Å / 38.5 Å)	475.7	2867	65.3	3541
(240 Å / 77 Å)	1769	3533	62.6	1007
(240 Å / 116 Å)	6311	7459	25.4	31
(120 Å / 38.5 Å)	3967	5059	41.7	138

4. CONCLUSION

In conclusion, we report the growth and characterization of YBCO/NSMO multilayers. The positive Hall coefficient in the normal state indicates the hole-type carriers dominantly contributed by YBCO in YBCO/NSMO multilayers. The lower activation energy accompanying with a diminishing sign reversal of Hall coefficient in the mixed state implies a decrease of flux pinning in YBCO/NSMO multilayer. The suppressions of T_c and pinning are attributed to the proximity induced ferromagnetic pair breaking in decoupled YBCO layers.

ACKNOWLEDGMENT

The authors thank the National Science Council of the Republic of China for financial support under Grant No. NSC 88-2112-M-212-002 and No. NSC 88-2112-M-002-028.

REFERENCES

H. Homma, C. S. L. Chun, G. G. Zheng, and I. K. Shuller 1986 Phys. Rev. B **33**, 3562
K. Kawaguchi and M. Sohma 1992 Phys. Rev. B **46**, 14722
Ctirad Uher, Joshua L. Cohn, and Ivan K. Schuller 1986 Phys. Rev. B **34**, 4906
Zoran Radovic, A. I. Buzdin, and John R. Clem 1991 Phys. Rev. B **44**, 759
Koichi Kuboya and Kenji Takanaka 1998 Phys. Rev. B **57**, 6022
L. Mieville, E. Keller, J.-M. Triscone, M. Decroux, A. Fischer, and E. J. Williams 1996 Phys. Rev. B **54**, 9525
Z. W. Dong, R. Ramesh, T. Venkatesan, Mark Johnson, Z. Y. Chen, S. P. Pai, V. Talyansky, R. P. Sharma, R. Shreekala, C. J. Lobb, and R. L. Greene 1997 Appl. Phys. Lett. **71**, 1718
G. C. Xiong, Q. Li. H. L. Ju, S. N. Mao, L. Senapati, X. X. Xi, R. L. Greene, and T. Venkatesan 1995 Appl. Phys. Lett. **66**, 1427
L. M. Wang, H. W. Yu, H. C. Yang, and H. E. Horng 1997 Physica C **281**, 325
H. C. Yang, L. M. Wang, and H. E. Horng 1999 Phys. Rev. B **59**, 8956
H. C. Yang, B. T. Su, L. M. Wang, and H. E. Horng Unpublished
Z. D. Wang, Jiming, and C. S. Ting, Phys. 1994 Rev. Lett. **72**, 3875

Inst. Phys. Conf. Ser. No 167
Paper presented at Applied Superconductivity, Spain, 14-17 September 1999
© *2000 IOP Publishing Ltd*

Ferromagnetic and superconducting oxide heterostructures for spin injection devices

L.Fàbrega[1], R.Rubí[1], J.Fontcuberta[1], V.Trtík[2], F.Sánchez[2], C.Ferrater[2] and M.Varela[2]

[1]Institut de Ciència de Materials de Barcelona (C.S.I.C.), Campus de la U.A.B., 08193 Bellaterra (SPAIN)
[2]Dpt. Física Aplicada i Òptica, Universitat de Barcelona, Diagonal 648, 08028 Barcelona (SPAIN)

ABSTRACT: We report on the growth by pulsed laser ablation of $YBa_2Cu_3O_{7-\delta}/SrTiO_3/La_{1/3}Sr_{1/3}MnO_3$ heterostructures in cross-strip geometry, intended to be used for spin polarized quasiparticle injection experiments. The current injected from the ferromagnet produces a suppresion of the critical current of the superconductor; this effect is shown to be independent of the current directions configuration.

1.   INTRODUCTION

The mature stage of epitaxial growth of complex oxide thin films allows nowadays conceiving and obtaining electronic devices made of monolithic oxide heterostructures. Among the most interesting ones are those combining high temperature superconductors (HTS) and ferromagnetic half-metallic manganites (FM), due to the remarkable properties of both types of materials. HTS/I/FM junctions, where I is an insulating oxide, constitute the basis of spin polarized quasiparticle injection (SPQPI) devices: they are based on the reduction of the superconducting gap by the injection of a spin polarized electrical current coming from the ferromagnet, and could be used to build superconducting three terminal devices, as well as ultrafast, high resolution sensors. These devices are still in an early stage of development, due to the difficulty of synthesising homogeneous tunnel junctions in the required scale, and the lack of knowledge of their basic phenomenology. In fact, aside the technological aspects, such heterostructures are very interesting also from a basic point of view, in order to study non-equilibrium superconductivity and interaction between superconductivity and ferromagnetism.

Research on SPQI in oxides has been focussed on $YBa_2Cu_3O_7/LaAlO_3/Nd_{2/3}Sr_{1/3}MnO_3$ (Dong 1998), $YBa_2Cu_3O_7/LaCuO_4/La_{2/3}Sr_{1/3}MnO_3$ and $YBa_2Cu_3O_7/SrTiO_3/La_{2/3}Sr_{1/3}MnO_3$ (Stroud 1998) heterostructures; HTS/FM bilayers have also been studied (Vas'ko 1997). Usually, a decrease of the superconducting critical current has been reported when a current I_{inj} is injected from the ferromagnet, with reported gains $G=\Delta I_c/I_{inj}$ between 1 and 35. Based on the early works with classical superconductors and normal metals (Iguchi 1977), the effect has been attributed to a reduction of the superconducting gap, as a result of the non-equilibrium produced by the quasiparticle injection. This gap reduction is expected to be enhanced due to the polarization of the charge carriers coming from the ferromagnet. Therefore, these reports raise considerable expectations, although further checks need to be performed in order to discern whether the observed phenomena are indeed due to the polarized current or to some extrinsic effects such as heating or inhomogeneities.

We report here on the growth and characterization of HTS/I/CMR heterostructures in cross-strip geometry. V-I measurements on these heterostructures reveal a critical current suppression, which might be attributed to pair breaking induced by the current injected from the ferromagnet to the superconductor. The effect is shown to be independent of the current configuration. The obtained gain is of about 2 at 40K.

2. EXPERIMENTAL

The heterostructures were grown on (100) $SrTiO_3$ and (100) $LaAlO_3$ single crystal substrates by on-axis pulsed laser deposition using a KrF excimer laser (λ=248 nm, τ=34ns). Sintered targets of the stoichiometric compounds $La_{2/3}Sr_{1/3}MnO_3$ (LSMO), $SrTiO_3$ (STO) and $YBa_2Cu_3O_7$ (YBCO) were used. The laser operated at 10Hz and the beam was focused to a fluence of about 3 J/cm^2. The deposition process took place in O_2 atmosphere. The growth conditions (temperature and oxygen pressure) were 750°C, 0.2mbar for LSMO, 800°C, 10^{-3} mbar for STO and 850°C, 0.3mbar for YBCO. The thickness of the three layers was kept to 60nm, 10nm and 60nm, respectively.

The extreme sensibility of the properties of these complex oxides to their microstructure and composition makes it a key issue to obtain reproducible, high quality heterostructures. Another problem concerns the definition and fabrication of the most suitable geometry for the junction: some photolithographic processes may result in a deterioration of the layers, due to the high sensitivity of these materials (specially HTS) to some chemicals. In order to obtain small area, homogeneous junctions, we chose a cross-strip configuration. Heterostructures with such geometry could be obtained by either using shadow masks during film growth or by successive patterning of the different layers, after the deposition. Both techniques have been used. In the first case, narrow (200µm wide) ceramic shadow masks made by laser patterning 0.6 mm thick alumina plates were used. We have studied the microstructure, magnetic and superconducting properties of these strip films; it turned out that, due to the thermal shield effect of the ceramic masks, a decrease of about 50°C was necessary in order to recover the good superconducting and ferromagnetic properties of the thin films deposited without masks

The heterostructures were characterized by X-Ray Diffraction (XRD), Scanning Electron Microscopy (SEM) and Atomic Force Microscopy (AFM). Transport measurements were performed in a closed-cycle cryostat by applying a DC current. The conventional four-probe method was used. Gold contacts were sputtered on the film surface, and wires were attached to them with silver paint.

In the injection experiments, the currents were fed as illustrated in Fig. 1. To extract each V(I) point, several V measurements were averaged after applying the current I. In order to avoid sample heating, the current was switched off after measuring each V(I) point, and some time was left between measurements. The maximum injected power density $P_{inj} \sim J_{inj}^2 R_n A$ –where $R_n A \sim 1.6 \cdot 10^{-3}$ Ωcm^2 is the junction resistance times its area- is of about < 1 W/cm^2, much less than the value 100-1000W/cm^2 required to raise the YBCO film temperature significantly (Bogulavskij 1994); thus, it appears that any heating effect may be ignored in the reported measurements.

3. RESULTS AND DISCUSSION

Fig. 1 is a SEM image of a (001)YBCO/(100)STO/(100)LSMO trilayer in the chosen cross-strip geometry; the channels, 200µm wide, are seen quite flat. We have shown recently (Fàbrega 1999) the high epitaxy and crystal quality of these heterostructures The LSMO and YBCO layers were also reported to display respectively good ferromagnetic ($T_C\approx350K$) and superconducting properties ($T_c\approx85K$).

Fig. 2 displays the V-I_{SC} characteristics of the superconductor recorded at 40K while injecting a current $I_{barrier}$ from the LSMO to the YBCO layer. In this measurement, two currents are fed in the YBCO strip; I_{SC} is the transport current injected through the contacts 1-3, while $I_{barrier}$ is injected between 8 and 3. The voltage is measured between contacts 4 and 2 (Fig. 1). Since $I_{barrier}$ does not flow through all the superconductor strip length, the voltage drop in the dissipative state of the superconductor must be suitably corrected.

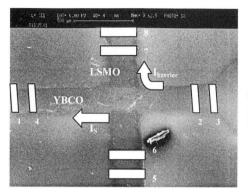

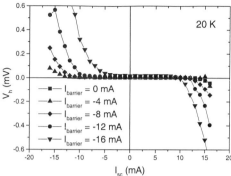

Fig. 1. SEM image of a cross-strip junction and scheme of the contacts and positive sense of the currents.

Fig. 2. $V_h(I_{SC})$ curves as a function of $I_{barrier}$, measured at 20 K.

Assuming that the junction length is much shorter than the superconductor length, the voltage drop in the half of the superconductor strip where $I_{sc}+I_{barrier}$ flows –i.e., between the junction and contact 4- is given by:

$$V_h(I_{SC}, I_{barrier}) \approx V_{42}(I_{SC}, I_{barrier}) - \frac{V_{42}(I_{SC}, I_{barrier} = 0)}{2}$$

This is the voltage value displayed in Fig. 2. The V-I_{SC} curves in this graph reveal a shift of each V-I_{SC} along the I_{SC} axis, when $I_{barrier} \neq 0$. The shift is different for $I_{SC}>0$ and $I_{SC}<0$. Similar features have been previously reported in junctions both with ferromagnetic and normal metals (Wang 1994, Vas'ko 1997, Dong 1998) and are associated to the summation effect of I_{SC} and $I_{barrier}$: their values must be added or subtracted depending on whether they flow parallel or antiparallel.

However, Fig. 2 also reveals another important feature: the shift with $I_{barrier}$ of each V-I branch is much smaller than the value of $I_{barrier}$. This fact can only be explained if the actual injected current $I_{inj,eff}$ responsible of the voltage drop is sensibly smaller than $I_{barrier}$.

An estimate of the current actually injected through the barrier $I_{inj,eff}$ may be obtained from the shift of these V_h-I_{SC} curves with regard to the I=0 axis. Indeed, if we assume a symmetric effect of the injected current, i.e., that any pair-breaking effect is independent of the injection direction, then the V-I curves, where I is the total current flowing in the superconductor, must be symmetric with respect to the origin. Therefore, one can evaluate $I_{inj,eff}$ from:

$$I_{inj,eff}(I_{barrier}) \equiv \frac{I_{c+}(I_{barrier}) - I_{c-}(I_{barrier})}{2}$$

where I_{c+} and I_{c-} are the values of I_{SC} corresponding to the onset of dissipation for a certain $I_{barrier}$ value.

Fig. 3 displays the V-I curves of Fig. 2, but depicting in the current axis the actual value of the current flowing in the superconductor, $I=I_{SC}+ I_{inj,eff}$, where $I_{inj,eff}$ has been evaluated using the above expression. The inset in Fig. 3 displays the obtained values of $I_{inj,eff}$ as a function of $I_{barrier}$. It may be seen that only a fraction of $I_{barrier}$ is effectively contributing to the dissipation in the superconductor. This effect has been reported to occur for other heterostructures with either normal or ferromagnetic metals (Vas'ko 1997, Iguchi 1994). A likely cause is an inhomogeneous current injection. In fact, a homogeneous current injection is expected to occur only when the length of the junction l_j is smaller than the transfer length $l_t \sim (R_n A d_{LSMO}/\rho_{LSMO})^{1/2}$, d_{LSMO} and ρ_{LSMO} are the thickness and resistivity of the LSMO layer (Boguslavskij 1994); in our heterostructures, $l_t \sim 30\mu m$ and $l_j \sim 200\mu m$, therefore non-uniform current injection is likely. Other possible causes of inhomogeneous injection are inhomogeneities in the barrier. This result therefore means that the junction quality needs to be further improved in order to get $I_{inj} \sim I_{barrier}$.

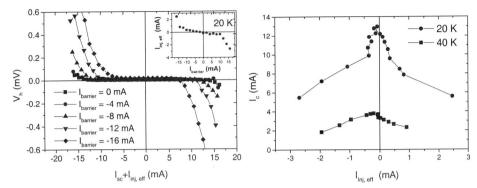

Fig. 3. $V_h(I_{SC}+I_{inj.eff})$ as a function of $I_{barrier}$. The inset shows $I_{inj.eff}$ versus $(I_{barrier})$

Fig. 4. $I_c(I_{inj.eff})$ curves at 20K and 40K

The onset of dissipation of the V-I's on Fig. 3 corresponds to the actual critical current I_c of the superconductor, for a injected current $I_{inj.eff}$. It may be clearly appreciated that there is a critical current suppression induced by the injected current. This is shown in Fig. 4, for two different temperatures. A feature worth remarking on this picture is the fact that the injected current has the same effect for both senses of injection; this is in agreement with the hypothesis used to extract $I_{inj.eff}$.

The observed linear suppression of I_c with the injected current from the ferromagnet, for any current configuration, must be attributed to the reduction of the superconducting gap, induced by the nonequilibrium created in the superconductor by the injection of the charge carriers. The current gain of about 2 is in agreement with the values reported for other HTS/I/CMR junctions (Dong 1998).

4.   CONCLUSIONS

In conclusion, we have grown thin films and heterostructures combining LSMO and YBCO by pulsed laser ablation. The use of ceramic shadow masks during film growth has been proven suitable for the obtaining of small area junctions. Injection measurements reveal a phenomenology similar to some earlier reported data, with a critical current suppression induced by the injected charge carriers. However, indications of non-homogeneous current injection effects have been detected and thus further improvement of the tunnel barrier quality is needed.

ACKNOWLEDGEMENTS

The authors acknowledge financial support from the Spanish CICYT (Projects MAT96-0911 and MAT97-0699), the Catalan DGR (Project GRQ95-8029), and the EU (OXSEN Network)

REFERENCES

Bogulavskij Yu M et al. Physica C **220** 195-202 (1994)
Dong Z W, Pai S P, Ramesh R, Venkatesan T, Johnson M, Chen Z Y, Cavanaugh A, Zhao Y G, Jiang X L, Sharma RP, Ogale S and Greene R L 1998, J. Appl. Phys. **83**, p. 6780
Iguchi I et al. 1997, Phys. Rev. **B 16**, 1954
Stroud R M, Kim J, Eddy C R, Chrisey D B, Horwitz J S, Koller D, Osofsky M S, Soulen R J and Auyeung R C Y 1998, J. Appl. Phys. **83**, p. 7189
Vas'ko V A, Larkin V A, Kraus P A, Nikolaev K R, Grupp D E, Nordman C A, Goldman A M 1997, Phys. Rev. Lett. **78**, 1134
Wang Q and Iguchi I 1994, Physica **C 228**, 393

Inst. Phys. Conf. Ser. No 167
Paper presented at Applied Superconductivity, Spain, 14-17 September 1999
© 2000 IOP Publishing Ltd

Effect of YBa$_2$Cu$_3$O$_{7-x}$ electrodes on the properties of SrTiO$_3$ based varactors

P K Petrov, Z G Ivanov,* and S S Gevorgyan,

Dept. of Microelectronics, Chalmers University of Technology, 412-96 Gothenburg, Sweden,

*Dept. of Physics, Chalmers University of Technology & Gothenburg University, 412-96 Gothenburg and Swedish National Testing and Research Institute, 501 15 Borås, Sweden

ABSTRACT: A set of YBCO/STO based multilayer structures were fabricated and investigated. It was observed that when a YBCO thin film is used as an upper layer, the crystal cell of the STO film becomes orthorhombic and a "butterfly" type hysteresis loop appear in the C-V dependence. The observed performance may be explained by ferroelectric phases in STO films due to stress induced cubic-orthorhombic phase transformation.

1. INTRODUCTION

In recent experiments (Nakamura et al. 1995, Findikoglu et al. 1997) a "butterfly" type hysteresis loop is observed in C-V characteristics of epitaxial YBCO/STO/YBCO capacitors. STO capacitors with SrRuO$_3$ electrodes, however, does not exhibit similar hysteresis loops (Li et al. 1998). A "butterfly" type C-V hysteresis in thin epitaxial film capacitors based on BaTiO$_3$ (BTO) (Hoerman et al. 1998) is attributed to the ferroelectric nature of BTO. On the other hand, it is known that STO is an incipient ferroelectric, i.e. it has only a paraelectric state and does not undergo ferroelectric phase transformation at temperatures at least down to 1K (Kittel 1971). To explain the hysteresis performance of YBCO/STO/YBCO structures we made detailed X-ray analysis of STO films in multilayer structures (Petrov et al 1999). It has been found that STO films become orthorhombic due to the YBCO films grown on top of them. Hence, one may expect that the physical properties of STO films should be changed correspondingly. Particularly, it has been shown theoretically (Fujita and Ishibashi 1997) that a cubic-orthorhombic transition leads to formation of ferroelectric phase in ferroelectric crystals.

In this report we make an attempt to explain the "butterfly" type C-V hysteresis in YBSO/STO/YBCO capacitors observed in our and similar (Nakamura et al. 1995, Findikoglu et al. 1997) experiments.

2. EXPERIMENTAL

Multilayer structures YBCO/STO/LAO, YBCO/CeO$_2$/STO/LAO, and YBCO/MgO/STO/LAO were prepared and investigated. The epitaxial films were on-axis laser ablated from stoichiometrical targets. An oxygen relaxation technique (Petrov et al. 1998) was used during the deposition of STO films with a total thickness of 400 nm for all investigated samples. 30nm CeO_2 or 30nm MgO buffer layers were *in situ* deposited between STO and YBCO layers to improve device properties for varactors applications (Carlsson et al. 1999). The deposition conditions for the all layers are reported in (Petrov et al. 1998).

The crystal structure of the samples was investigated by X-ray diffraction (XRD) performed on a Philips X'Pert PW3098/20 diffractometer with Cu $K\alpha$ radiation. In order to examine the in-plane orientation of the films in the multilayer structures XRD ϕ- scans as well as Grazing-Incidence Diffraction (GID) measurements were performed. The advantage of GID is that when a monochromatic X-ray beam irradiates the sample surface grazingly (with an angle of incidence of about 1^0) a diffraction with considerable high intensity takes place from lattice planes (*hkl*) perpendicular to the sample surface. These measurements were reported in (Petrov et al. 1999).

For examining the electrical properties of the fabricated multilayer structures, a set of planar varactors were patterned using photolithography and Ar ion milling. The C-V measurements were performed on a Precision LRC meter HP 4285A at 77K, 1 MHz.

3.   RESULTS AND DISCUSSION

The XRD patterns for ω-scan along the [001] direction of all investigated multilayers contained only the (001) reflections of the corresponding single layers. ϕ-scans for the (103)LAO, STO, MgO and (109) YBCO confirmed the in plane epitaxial growth of each single layer in the multilayer.

The *a* and *b* lattice parameters for STO and YBCO layers were calculated from the GID patterns of different multilayers (Table 1). A single STO layer grows on LAO substrate with a slightly elongated cubic cell, $a=b\neq c$. When YBCO is grown on top of an STO layer an additional distortion in the *a-b* plane is observed and $a\neq b$.Growth of intermediate CeO_2 or MgO buffer layer however reduces the difference between *a* and *b* parameters of the STO crystal cell. Such a dielectric buffer layer introduced between YBCO and STO films improves multilayers crystallinity, protects the STO surface during processing and reduces the varactor leakage current (Carlsson et al. 1999). The thickness of buffer layer is sufficiently small not to affect the C-V performance of the varactors.

Table 1 X-ray measurement results

Sample	*c* (001)		*b* (010)		*a* (100)	
	Cell parameter Å	Micro strain, $\Delta c/c$	Cell parameter Å	Micro strain, $\Delta b/b$	Cell parameter Å	Micro strain, $\Delta a/a$
STO/LAO	3.915	0.0002	3.851	0.0023	3.851	0.0074
Au/YBCO/STO /LAO	3.908	0.0018	3.875	0.0032	3.843	0.0026
Au/YBCO/CeO₂ /STO/LAO	3.908	0.0018	3.857	0.0029	3.832	0.0108
Au/YBCO/MgO /STO/LAO	3.908	0.0007	3.877	0.0021	3.866	0.0167

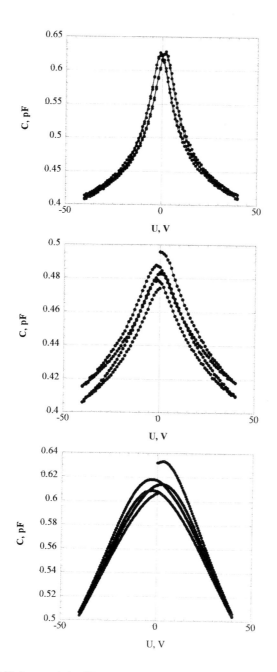

Fig. 1 C-V characteristics (T= 77K, f = 1 MHz) of the following capacitor structures:
a) YBCO/STO/LAO,
b) YBCO/CeO$_2$/STO/LAO,
c) YBCO/MgO/STO/LAO.

C-V dependencies, representative for the investigated multilayer capacitors, are presented in Fig. 1. C-V characteristics of YBCO/STO capacitors have a clearly defined "butterfly" type hysteresis loop that is typical for the ferroelectric materials under applied electric field. Such a behaviour may be explained by the discussed above stress induced cubic-orthorhombic phase transformation of STO films. In the case of capacitors with CeO_2 (Fig. 1b) and MgO (Fig. 1c) buffer layers the "butterfly" shape of the C-V behavior is distorted. A charge accumulation process may explain the distortion. We suggest that the charge accumulation takes place at CeO_2/STO and MgO/STO interfaces, and not in the STO film itself, since such a behavior was not observed in single STO capacitors (Fig. 1a).

In comparison with CeO_2 buffered varactors, the ones with MgO buffer exhibit better crystalline structure (with minimal difference between b and a lattice parameters), higher capacitance and electrical tuning, lower charge accumulation and leakage current. On other hand the distortion of the "butterfly" shape of the hysteresis loop is more pronounced. Such performance may be a result of complex interplay between charge accumulation and ferroelectric domain formation process.

4. CONCLUSIONS

A set of YBCO/STO based multilayers were fabricated and investigated. It was observed that when a YBCO thin film is used as an upper layer, the crystal cell of the STO film becomes orthorhombic and a "butterfly" type hysteresis loop appears in the C-V dependence. The observed performance may be explained by the formation of ferroelectric phase in the STO film due to a stress induced cubic-orthorhombic phase transformation. This process may be accompanied by a charge accumulation at the dielectric/STO interface. Experiments to study these suggestions are in progress.

5. ACKNOWLEGEMENTS

The authors are grateful to Orest G. Vendik and Erik Kollberg for useful discussions and their interest in this work. This work was partly supported by NICOP (Naval Research Lab., USA), CHACH (Chalmers Center for High Speed Technology) and Swedish Materials Consortium on Superconductivity.

REFERENCES

Carlsson E F, Ivanov Z G, Petrov P K et al 1999 Swedish patent Int. Patent Appl. 9901297-3
Findikogly A T, Jia Q X and Reagor D W 1997 IEEE Trans. Appl. Supercond. 7, pp. 2925-8
Fujita K and Ishibashi Y 1997 Jpn. J. Appl. Phys. Part1 9B, pp. 6133-5
Hoernam B H, Ford G M, Kaufmann L D et al. 1998 Appl. Phys. Lett. 73, pp. 2248-50
Kittel C 1971 Introduction to Solid State Physics (John Wiley & Sons Inc.: New York) p. 476
Li H-C, Si W, West A D and Xi X X 1998 Appl. Phys. Lett. 73, pp. 190-2
Nakamura T, Tokuda H, Tanaka S et al. 1995 Jpn. J. Appl. Phys. Part1 4A, pp. 1906-10
Petrov P K, Carlsson E F, Larsson P et al 1998 J. Appl. Phys. 84, pp. 3134-40.
Petrov P K, Ivanov Z G, Gevorgian S S 1999 Paper presented on E-MRS 1999 Spring Meeting, June 1-4, 1999 Strasbourg, France.

Inst. Phys. Conf. Ser. No 167
Paper presented at Applied Superconductivity, Spain, 14-17 September 1999

Raman Spectroscopy as a characterisation tool for FIB damage of HTS thin films.

N. Malde[1], F. Damay[1], L. F. Cohen[1] and M. W. Denhoff[2]

1.Blackett Laboratory, Imperial College, Prince Consort Rd, London SW7 2BZ
2.Institute for Microstructural Sciences, National Research Council, Canada

ABSTRACT: A focused beam of 200 keV Si ions was used to pattern an YBCO film. The fluence were varied from 1×10^{13} ions/cm^2 to 1×10^{16} ions/cm^2 and Raman spectra of the damaged material was studied. The frequency change of the O(4) mode indicated oxygen loss. The intensity of the 340 cm^{-1} phonon mode decreased as a function of damage, relative to the background in a manner which would allow for rapid characterisation of FIB processes. Additionally the background increased across the Raman spectra and a broad peak centred around 570 cm^{-1} appeared.

1.   INTRODUCTION

Focused ion beams have proved a useful tool for milling HTS thin films (Blank et al. 1995), patterning substrates for HTS junction templating (Soulome et al. 1997) and direct bombardment into HTS films for junction definition (Tinchev 1997, Denhoff et al. 1997) . The latter technique although perhaps less controllable than electron beam fabricated junctions (see for example Pedyash et al. 1997) still offers advantages of being a relatively rapid one shot process. Depending on the fluence and energy of incident ions, the damaged material may be superconducting with depressed T_c, non-superconducting but crystalline or non-superconducting and amorphous. In theory, Raman spectroscopy lends itself to characterisation of these different kinds of damage. Exploration of the Raman spectroscopy as a function of FIB fluence for a fixed ion beam energy is reported here.

2.   EXPERIMENTAL

2.1   FIB Damage of HTS thin film

The 160nm thick YBCO film used in this study was grown by pulsed laser deposition on cerium oxide buffered MgO substrates and supplied by the Institute for Microstructural Sciences of the National Research Council of Canada. Patches 4mm x 2mm of various fluences of 200 keV Si ions were written in the film. TRIM simulation of this process was used to model the scattering processes of ions within the film material. The TRIM simulation of a 200nm thick film, shown in figure 1 illustrates that the longitudinal range of the ions incident upon the material surface is about 175 +/- 32nm while the lateral projection in the x-y plane about the point of incidence is 65 +/-

40nm. This means that for a 50nm beam diameter, the damage line is of the order of 200nm. However, the bulk of ion damage actually occurs in a region of material somewhere between the surface and the full range of incident ions. A small layer ($^-$10 nm) of material on the surface remains unaltered by this process while material to a depth of 175nm is damaged. Non-ionising energy loss can be calculated as a function of depth for 200 keV Si ions. It has been found by other researchers that the reduction in T_c depends linearly on this damage. For example, from Tinchev 1997, a dose of 1 x10^{13} cm-^{2} would lower T_c by 28 K at the damage level of 600 eV/ion/nm. Tc can be found as a function of depth into the film as shown in figure 2, for a dose of 2 x10^{13} cm$^-$^{2}. Also from the TRIM calculation we can estimate using the displacement per atom that at the peak of the profile amorphisation will occur at about 1 x 10^{15} cm^{-2}. This implies that the highest fluence will have sputtered material from the surface of the film, and indeed this patch was rendered transparent to visible light.

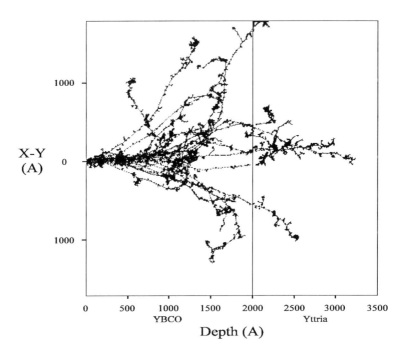

Figure 1: X-Y Longitudinal projection calculated by TRIM simulation. Tracks are shown for twenty Si ions of 200 keV energy incident with an angle of $0°$ upon a 2000Å thick $YBa_2Cu_3O_{7-d}$ film. The ions lose energy during collisions in the crystal lattice, 70% of energy is lost through ionisation, 29% is lost in phonon creation/annihilation leaving 1% of energy in the creation of vacancy sites in the crystal.

2.2   Raman Spectroscopy

Raman spectroscopy was performed using a Renishaw system 2000 Raman microprobe at room temperature. The 514.5 nm line of an Argon - ion laser was focused normal to the sample surface with a lateral spot diameter of 2-3 μm. The incident laser power through a x50 microscope objective was kept at 1.5 mW on the sample surface, to minimise any localised heating effects. The Raman scattered signal was collected onto a CCD detector array in a backscattering geometry for Raman shifts of 100 to 1000 cm^{-1}. The backscattered signal was polarised with a configuration of

(XX). The integrated intensities and the mode frequencies of the phonon lines were extracted by curve fitting using SPSS Peak Fit V.4 software. The A_{1g} O(4) 500 cm^{-1} and 585 cm^{-1} lines were fitted using a Lorentzian function, whereas, a Fano profile (see Fano 1866) was used to fit the O(2, 3) 340 cm^{-1} line.

The Raman signal penetrates approximately 100nm into the film and the spectrum obtained is an average of the film properties over that depth . As the film becomes less conducting the penetration depth will increase and at high fluence the film thickness will be reduced because material will be sputtered off the film surface.

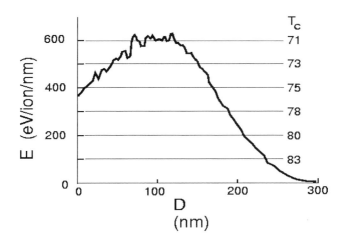

Figure 2. The non ionising energy loss for 200 keV Si ions incident on YBCO. The Tc values are our estimate for a dose of 2 x 10^{13} cm^{-2}, starting with a T_c for the undamaged YBCO of 85K.

3.   RESULTS AND DISCUSSION

Figure 3 shows the spectra for virgin, 2 x 10^{13} cm^{-2}, 1 x 10^{15} cm^{-2} , 1 x 10^{16} cm^{-2} patches. Two patches of each fluence were written in the film and the spectra were found to be reproducible. Also shown for comparison is the spectra from the sapphire substrate. At the two higher fluences, two substrate peaks at 380 cm^{-1} and 420 cm^{-1} become visible. This means that the light is penetrating through the film and the spectra represent an average of properties sampled throughout the film thickness.

Firstly comparing the spectra from the undamaged film to 2 x 10^{13} cm^{-2} damaged area. The only significant change is the O(4) peak position which has shifted from 504.82 cm^{-1} to 503.68 cm^{-1}. Both oxygen disorder and oxygen loss shift the frequency of the O(4) mode down. For a T_c depression of the order of 10K due to oxygen loss, we would expect the O(4) mode to move down by approximately 2 cm^{-1}. This shift also depends on the initial oxygen content of the film substrate material, growth conditions etc.. The observation of a shift of only 1 cm^{-1} is surprisingly low, given the T_c profile shown in figure 2. For significant deoxygenation the 450 cm^{-1} peak should also shift of which there is no evidence from the spectra.

The inset to figure 3 shows intensity of 340 cm^{-1} as a function of dose. At high fluence a broad 340 cm^{-1} mode is still visible suggesting that close to the substrate quasi- crystalline film although oxygen disordered, remains intact. Macfarlane et al., 1988, showed that when oxygen is removed to O_6 and the film is tetragonal, the 500 cm^{-1} mode is absent, and the 340 cm^{-1} mode is enhanced relative to orthorhombic O_7 . The variation of the 340 cm^{-1} intensity suggests that the Raman spectra if correlated with junction properties as a function of fluence, could be used in principle to characterise ion damage.

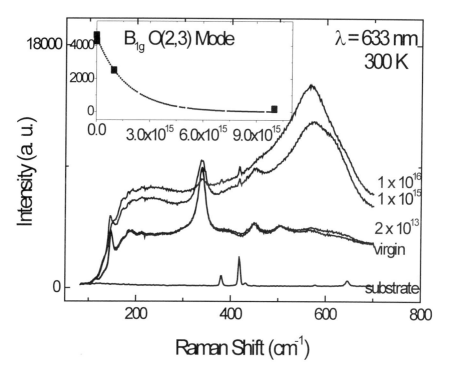

Figure 3 Shows the Raman spectra from the un-implanted virgin film and from the three patches with varying fluence. The spectra from the sapphire substrate is also shown for comparison. Inset shows variation in the integrated intensity of the B_{1g} 340 cm^{-1} O(2,3) mode as a function of fluence.

Two additional features are clear from the spectra shown in figure 3. The flat background which is a signature of HTS Raman spectra and which has previously been interpreted as electronic in origin, increases with increasing fluence. The rise of the background could be related to changes in the electronic scattering cross-section as a function of ion damage (Krantz and Cardona 1997). Both 1 x 10^{15} cm^{-2} and 1 x 10^{16} cm^{-2} show a broad mode centred around 570 cm^{-1}. A mode at this frequency is associated with deoxygenation (Iliev et al. 1993,) and oxygen disorder (Ivanov et al. 1995). We interpret this feature as being due to oxygen disorder, associated with the implanted ions in the remaining quasi-crystalline material.

REFERENCES

Blank D H A et al. 1995 IEEE Transactions on Applied Superconductivity **5**, 2786
Denhoff M W et al. 1997 Inst. Phys. Conf. Ser.**158,** 515
Fano U Phys. Rev. **124**, 1866
Iliev M et al. 1993 Phys. Rev. B **47**, 12341
Ivanov V G et al. 1995 Phys. Rev. B **52**, 13652
Krantz M and Cardona M 1995 Journal of Low Temp. Phys. **99**, 205
Macfarlane R M et al. 1988 Phys. Rev. B **38**, 284
Soulome Y and Okabe Y 1997 IEEE Transactions on Applied Superconductivity **7**, 2311
Tinchev S S 1997 Journal of Applied Physics **81**, 324

Inst. Phys. Conf. Ser. No 167
Paper presented at Applied Superconductivity, Spain, 14-17 September 1999
© 2000 IOP Publishing Ltd

DC Transport and High Frequency Behaviour of c-Axis REBa₂Cu₃O₇₋ₓ Thin Films

W. Schmitt, R. Aidam, J. Geerk, G. Linker and R. Schneider

Forschungszentrum Karlsruhe, Institut für Festkörperphysik, POB 3640, 76021 Karlsruhe, Germany

ABSTRACT: To investigate the influence of the RE element on the electrical transport properties, thin $REBa_2Cu_3O_{7-x}$ (RE = Y, Yb, Gd, Sm) were deposited on $LaAlO_3$ substrates by sputtering of hollow cylinder targets. The growth of the films was characterised by X-ray diffraction measurements. The critical temperature, T_C, the critical current density, j_C and the resistivity, ρ were determined. The surface resistance R_S was determined at 18.9 GHz. It was found that T_C, the c-axis lattice parameter and ρ are increasing with the RE element ion radius while j_C is decreasing. The lowest R_S values were found in $GdBa_2Cu_3O_{7-x}$ films.

1. INTRODUCTION

The relationship between transport properties and the charge carrier concentration is a vital aspect in the investigation of high-temperature superconducting cuprates. A reduced charge carrier concentration causes an increasing normal state resistivity (Wuyts et al 1995a). The influence of the carrier concentration on the critical temperature is manifested in the universal phase diagram (Tallon et al 1995b).

Many approaches have been made to vary the volume charge carrier concentration in $YBa_2Cu_3O_{7-\delta}$, mainly by changing the oxygen concentration (Wuyts et al 1995a), by pressure (Tallon et al 1995b), by an external electrical field (Auer et al 1996a) or by the substitution of elements in the compound. In contrast to other substitutions, the replacement of Y by other rare earth elements (RE) is not connected to a valence change of the substituted atom. However, experiments on bulk material by Guan et al (1996b) showed a significant variation of the critical temperature and the normal-state resistivity. Even a variation of the charge carrier concentration in ceramic $REBa_2Cu_3O_{7-\delta}$ (REBCO) was detected. A possible explanation of this variation might be given by bond valence calculations, which predict a displacement of holes between the CuO chains and the CuO_2 planes with the RE element (Samoylenkoy et al 1996c).

Until now systematic examinations of RE substitution in thin films are still lacking. The following article will report about RE (RE = Yb, Y, Gd, Sm) substitution experiments in thin films with special regard to the deposition parameters of the different systems. The influence of the RE element on the c-axis lattice parameter ,the critical temperature, the normal-state transport behaviour and the critical current density are discussed. Under the aspect of possible high frequency applications special attention is directed on the surface resistance R_S of the different compounds.

2. EXPERIMENTAL

The deposition of thin REBCO films is influenced by several parameters such as the substrate material and the deposition conditions. Because of its high frequency stability, its chemical and mechanical stability and its low lattice mismatch, $LaAlO_3$ was selected as a substrate material.

The films were deposited by reactive sputtering of hollow cylinder targets which enables a material transport in 1:1 stoichiometry (Geerk et al 1988). Therefore ceramic sinter targets with a specific material composition of $REBa_2Cu_3O_{7-\delta}$ were used in the sputtering gun. The deposition parameters in Table 1 enable an epitaxtial growth of the films in the tetragonal $REBa_2Cu_3O_6$ phase. In a second oxygenation step the films are transformed to the superconducting, orthorombic $REBa_2Cu_3O_{7-\delta}$ phase. The deposition process resulted in thin films with excellent dc transport and high frequency properties.

	YbBCO	YBCO	GdBCO	SmBCO
O_2- partial pressure (mbar)	0,2	0,2	0,2	0,2
Ar-O_2- total pressure (mbar)	0,75	0,75	0,5	0,3
Deposition temperature T_s (°C)	780	840	840	780
DC Power (W)	150	150	210	120
Target-substrate-distance (mm)	60	60	60	60
Deposition rate (Å/min)	11	7	19	4

Table 1 Deposition parameters of c-axis REBCO films.

However, YbBCO films could not be grown with suitable quality, since deposition temperatures above 780°C could not be applied. Above 780°C a transparent, isolating film started to grow. This observation might be explained by the lower melting point of YbBCO in comparison to other REBCO systems. To get optimised SmBCO films the sputtering was assisted by a magnetic field during the deposition. However, degradations in T_C occurred in several SmBCO films, possibly due to Sm-Ba substitution (Murakami et al 1996d).

Since the oxygen concentration in REBCO is important for the transport behaviour, the oxygenation of the films was optimised according to the phase diagrams and the diffusion coefficients of each system. All the films deposited and oxygenated under optimised conditions showed a slightly overdoped behaviour according to the nonlinearity in their resistivity.

The c-axis lattice parameter was determined by x-ray diffraction, the thickness and the composition of the films by Rutherford backscattering spectroscopy. In order to determine T_C, the critical current density and the resistivity, the samples were photolitoraphically patterned into stripes and four probe measurements were performed. The surface resistance R_S was examined at 18,9 GHz by using a dielectric resonator technique (Klein et al 1992a). In the following the RE element is represented by its ion radius (IR). The results were reproduced with a large number of samples.

3.   RESULTS AND DISCUSSION

All c-axis oriented $REBa_2Cu_3O_{7-\delta}$ films prepared under optimised conditions had an excellent cristallographic quality with a mosaic spread below 0,3°. The c-axis lattice parameters of optimised samples are represented by the lines in Fig. 1 for all systems. The c-axis lattice parameter rises with increasing RE ion radius which is in good agreement with measurements on bulk material (Kistenmacher 1987, Guillaume et al 1992b).

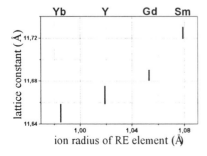

Fig.1 c-axis lattice parameter vs. IR

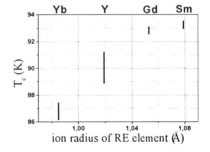

Fig.2 T_C vs. IR

Fig.2 shows resistively measured critical temperatures for the REBa$_2$Cu$_3$O$_{7-\delta}$ materials. The vertical lines represent the variation in T$_C$ of samples prepared under optimised conditions. The critical temperature rises with the RE element ion radius, which agrees with examinations on bulk material (Kistenmacher 1987).

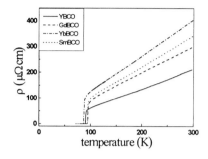

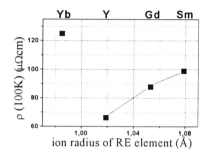

Fig.3 Resistivity of REBCO samples

Fig.4 Resistivity at 100 K vs IR

Measurements of REBa$_2$Cu$_3$O$_{7-\delta}$ films with lowest normal-state resistivity are shown in Fig.3. The resistivity of these samples at 100 K, ρ(100 K), are plotted versus the RE element ion radius in Fig.4. YBCO films showed the lowest ρ(100 K) values of 68$\mu\Omega$cm. Consulting only YBCO, GdBCO and SmBCO samples a clear tendency for ρ(100 K) to rise with the RE element ion radius is detected as it was also found in bulk material (Guan et al 1996b). This increase might be explained by bond valence calculations (BVS). The valence of in-plane copper ,Cu(2), increases with decreasing radius of the RE ion, implying an increase of the carrier concentration in the CuO$_2$ plane (Samoylenkow et al 1996c). Within the model of predominant carrier transport in the CuO$_2$ planes an increasing resistivity might be explained by the decreasing charge carrier concentration. Despite the smallest ion radius of Yb, YbBCO films showed the highest resistivity. This behaviour does not agree with investigations on bulk and might be explained by the deposition problems mentioned in chapter 2.

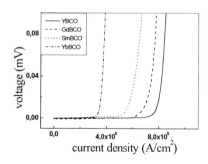

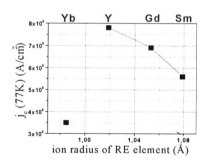

Fig.5 I-V characteristics at 77 K of REBCO

Fig.6 Critical current density at 77 K vs IR

Measurements of the critical current density, j$_C$, at 77 K of REBCO samples are presented in Fig.5. The results are plotted in Fig.6 versus the RE element ion radius. YBCO films showed the highest values of 7.8E10^6 A/cm^2. Apart from YbBCO films, the critical current density decreases with increasing radius of the RE ion. A possible explanation for this behaviour could again be given by the increasing charge carrier concentration with decreasing RE ion radius as calculated by BVS. The pinning force of a point defect is proportional to the charge carrier concentration (Mannhart et al 1991a). Therefore, the critical current density drops with decreasing number of charge carriers. This is in accordance with the results in Fig.6. An explanation within the weak link model can be excluded

116

because of the sharp mosaic spread of the films (Strikovsky 1991b). The behaviour of YbBCO samples might again be explained by the optimised samples.

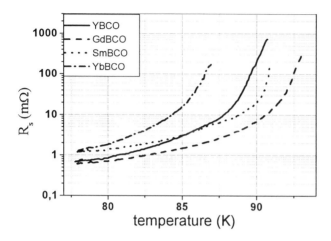

Fig.7
Surface resistance at 18.9 GHz versus temperature for the REBCO systems

The surface resistance R_s of the $REBa_2Cu_3O_{7-\delta}$ films was examined at 18.9 GHz. The samples with lowest R_S at 77 K are presented in Fig.7. The lowest R_S values of 0,65 mΩ of all systems are found for GdBCO samples. A correlation between R_S and the RE element ion radius cannot be not confirmed.

4.   SUMMARY AND CONCLUSION

The influence of the RE element on the properties of REBCO films was examined. The c-axis lattice parameter and the critical temperature rise with the RE ion radius. The normal-state resistivity increases with the radius of the RE ion. This might be explained with the decreasing charge carrier concentration in the CuO_2 planes according to BVS calculations. The decrease of the critical current density with the RE ion radius might consistently be explained by depinning. The low surface resistance and its high T_C make GdBCO films interesting for microwave applications especially if an operation temperature of 77 K and high frequencies were used.

REFERENCES

Auer R and Schneider S 1996a J. Appl. Phys. **81**, 3237
Guan W, Chen J C, Cheng S H and Xu Y 1996b Phys. Rev. B **54**, 6758
Samoylenkoy S V, Gorbenko O Yu and Kaul A R 1996c Physica C **278**, 49
Murakami M, Sakai N, Higuchi T and Yoo S I 1996d Supercond. Sci. Technol. **9**, 1015
Wuyts B , Moshchalkov V V and Bruynseraede 1995a Phys. Rev. B **53**, 9418
Tallon J L, Bernhard C, Shaked H, Hitterman R L and Jorgensen J D 1995b Phys. Rev. B **51**, 12911
Klein N, Dähne U, Poppe U, Tellmann N, Urban K, Orbach S, Hensen S, Müller G and Piel H 1992a J. Superconductivity **5**, 195
Guillaume M, Allenspach P, Mesot J, Roessli B, Staub U, Fischer P and Furrer A 1992b Z. Phys. B **90**, 13
Mannhart J, Schlom D G, Bednorz J G and Müller K A 1991a Phys. Rev. Letter **67**, 2099
Strikovsky M, Linker G, Ganonov S, Mazo L and Meyer O 1991b Phys. Rev. B **45**, 12522
Xi X X, Linker G, Meyer O, Nold E, Obst B, Ratzel F, Smithey R, Strehlau B, Weschenfelder F and Geerk J 1988 Z. Phys. B **74**, 13
Kistenmacher T 1987 Solid State Communications **65**, 981

Inst. Phys. Conf. Ser. No 167
Paper presented at Applied Superconductivity, Spain, 14-17 September 1999
© 2000 IOP Publishing Ltd

Measurements of the microwave properties of DyBa₂Cu₃O₇ films on AO-, LAO-, YAO-substrates in dielectric resonators

K. Irgmaier, S. Drexl, R. Semerad, K. Numssen and H. Kinder

Physics Department E10, Technical University of Munich, D-85747 Garching, Germany

ABSTRACT: We have compared the microwave surface resistance of YbaCuO (YBCO) and DyBaCuO (DyBCO) films on three different substrate materials, namely sapphire (AO), LaAO₃ and YAO₃ (YAO). It has been found that although the critical thickness of DyBCO films on AO is lower, R_s decreases even more, so that a reduction to 70% of YBCO was achieved. The slope of the R_s-curve below 77 K is steeper with Y than with Dy resulting in a crossover of both curves around 60 K again. Finally results of measurements of HTS films on YAO are reported.

1. INTRODUCTION

More than ten years after the discovery of the first High Temperature Superconductors (HTS) compound applications of these new materials start to enter the market. One of the potential applications of HTS is in designing rf-filters. Replacing metallic conductors with HTS materials allows a reduction of size and weight of such filters. This is important for applications in future satellite communication systems (Kässer 1998).

For these thin film applications a reliable and economical YBCO coating technique is essential providing a low microwave surface resistance R_s. The question arises: how can R_s be reduced without increasing the thickness of the film significantly and how can we. To answer this question we replaced the element Y in the YBCO compound by dysprosium to get films with a higher transition temperature based on the lattice strains of the ion size effect (Williams 1996). We have coated DyBCO of various thicknesses on AO, LAO- and YAO-substrates. In order to observe improvement we measure the temperature and power dependent microwave properties of the DyBCO films.

2. EXPERIMENTAL SETUP

The films were produced by thermal reactive evaporation to obtain homogeneous high quality films (Berberich 1994). Especially for the microwave characterisations we coated films on 1'' and 2'' substrates. Measurements of the surface resistance have been done by using dielectric sapphire resonator technique (Kajfez 1986). Our system allows the temperature dependent characterisation based on a dielectric parallel plate resonator for 1'' size and 10.9 GHz in the TE011 mode (Irgmaier 1998). Cooling was provided by a continuous flow cryostat using liquid helium under atmospheric pressure. To determine the power dependent properties we used a TWT amplifier and a dielectric resonator (2'') in pulsed mode (200 µs) at 5.6 GHz and at 77 K. The measurements were evaluated with a network analyser. Important parameters for assessing the film qualification for filter applications, surface resistance R_s at 77 K and the temperature and power dependence of R_s were measured.

3. RESULTS

We have investigated the surface resistance of DyBCO films with different thicknesses on AO substrates. The results show that DyBCO films with a thickness of 250 nm possess, from 75 K on, a lower effective surface resistance than YBCO films with a thickness of 330 nm (Fig. 1). The intrinsic surface resistance decreased to 70% of YBCO (Fig. 2). At lower temperatures, however, the DyBCO films show a higher surface resistance than YBCO films. When using thicker films the expected further decrease of the effective surface resistance of DyBCO films can not be observed. Thicker films show a higher power dependent surface resistance. With the optical microscope, cracks in the film were detected.

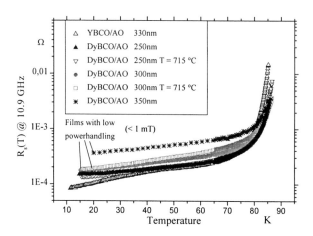

Fig. 1. Surface resistance of DyBCO and YBCO films deposited on sapphire (AO) substrates. Two samples were coated at a temperature of 715 °C (normaly 650 °C).

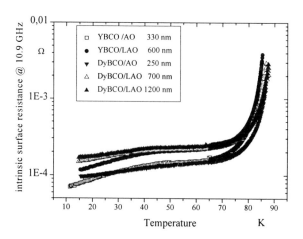

Fig. 2. Intrinsic surface resistance of DyBCO and YBCO films deposited on AO- and LAO substrates.

On LAO substrates, DyBCO films are processed up to the maximum thickness of 1200 nm. Their R_s are, with 330 μΩ at 77 K, smaller than those of YBCO films with a thickness of 600 nm. DyBCO films show a larger remaining R_s at low temperatures than YBCO films as seen in figure 3. The temperature dependent increase of R_s is smaller and a tendency to built a hyperbolic point isnot so pronounced. The measurements show that the intrinsic surface resistance at 77 K can be reduced by about 50 μΩ through substitution with Dy (Fig. 2).

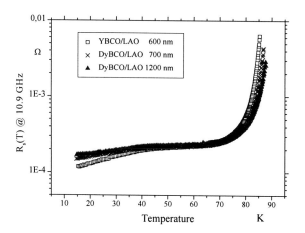

Fig. 3. Surface resistance of DyBCO and YBCO films deposited on LAO substrates.

The power dependent surface resistance at 77 K of DyBCO on LAO (700 nm) is compared to the values of YBCO on AO (330 nm). The measured maximum critical B_{rf}-field of both films is about 15 mT. Figure 4 illustrates the power independent surface resistance below 15 mT.

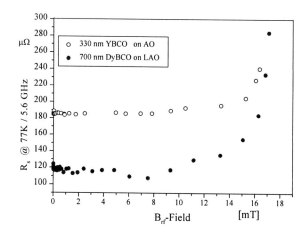

Fig. 4. Power dependent surface resistance of DyBCO and YBCO films deposited on LAO.

120

Temperature dependent measurements of films on YAO substrates which do not contain twin boundaries were performed. The critical thickness of DyBCO and YBCO on YAO is about 400 nm (LAO 600 nm). The qualification of these films for filter applications could be determined by power dependent measurements. The temperature dependent R_s curve is shown in figure 5.

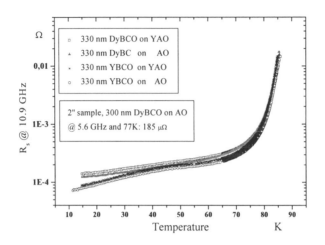

Fig 5. Surface resistance of DyBCO and YBCO films deposited on YAO_3 (YAO).

4. CONCLUSIONS

Measurements of high frequency properties of DyBCO films have shown that one can reduce the intrinsic resistance of HTS-films to about 70 % of YBCO by substituting Y with Dy. On sapphire, the crack formation limits the film thickness to about 250 nm (LAO 600 nm). The power handling on both substrates is the same for YBCO and DyBCO. Coating of HTS layers on twin-free YAO produce no lower surface resistance due to the higher intrinsic surface resistance and the lower critical thickness on YAO.

ACKNOWLEDGEMENT

This work was supported by the BMBF project 13N6827/1 of the German Ministry of Research and Technology.

REFERENCES

Berberich P, Utz B, Prusseit W and Kinder H 1994 Homogeneous high quality YBCO films on 3''and 4'' , Physica C **219,** 497-504
Irgmaier K et al. 1998 Large area RE films, VDI Tech Centre Germany FKZ 13N6827/1
Kässer T et al. 1998 A satellite repeater comprising superconducting filters, IEEE MTT-S 375-378
Kajfez D et al. 1986 Dielectric resonators, Artech House Microwave Library
Williams G and Tallon J 1996 Ion size effect on T_c and interplanar coupling in RBaCuO, Physica C **258,** 41-46

Inst. Phys. Conf. Ser. No 167
Paper presented at Applied Superconductivity, Spain, 14-17 September 1999
© *2000 IOP Publishing Ltd*

$YBa_2Cu_3O_x/CeO_2/NdGaO_3$ epitaxial heterostructures at different substrate orientations.

I K Bdikin[#], I M Kotelyanskii, E K Raksha, A D Mashtakov, P B Mozhaev, P V Komissinskii and G A Ovsyannikov

Institute of Radio Engineering and Electronics RAS, 103907 Moscow, Russia,
[#] Institute of Solid State Physics, Chernogolovka, 142432, Moscow district, Russia.

ABSTRACT: $YBa_2Cu_3O_x/CeO_2$ epitaxial heterostructures were grown on $NdGaO_3$ substrates with surface orientation tilted from the (110) plane. Two types of epitaxial relations were observed for CeO_2 buffer layer, depending on deposition technique and deviation angle of the substrate surface from the (110) $NdGaO_3$ plane. The $YBa_2Cu_3O_x$ thin films on the CeO_2 buffer layer grow either oriented with c axis normal to substrate plane, or following the orientation of the <100> axes of the buffer layer. Observed changes in epitaxy can be explained with different film formation mechanisms at different deposition rates.

1.   INTRODUCTION

Materials with perovskite structure are often used as substrates for $YBa_2Cu_3O_x$ (YBCO) thin film preparation due to their similarity with high-temperature superconductor structure and element composition. Utilization of buffer layers often is necessary to prevent chemical interaction between film and substrate or to provide better lattice match. Cerium oxide CeO_2 is one of the most often applied buffer layer materials, providing excellent lattice match and suppressing chemical interaction at temperatures up to 750 °C (Mashtakov 1997). Fluorite structure of CeO_2 differs substantially from the substrate and film perovskite structure, that can lead to changes in epitaxy of both buffer layer on the substrate and YBCO film on the buffer layer (Kotelyanskii 1996, Mozhaev 1999). Such changes can increase when substrate surface deviates from the standard crystallographic orientations.

In the present paper we study epitaxial relations of the CeO_2 films on $NdGaO_3$ (NGO) substrates with deviation of orientation form the standard crystallographic planes and growth of YBCO films on the obtained CeO_2/NGO heterostructures.

2.   EXPERIMENTAL

(110), (120), (130) and (010) crystallographic orientations of NGO substrates were chosen for buffer layer deposition. This set of orientations is obtained as a result of substrate plane rotation around [001] NGO axis.

CeO_2 films with typical thickness 300-400 Å were deposited by RF-magnetron sputtering and by e-beam evaporation. RF-sputtering of Ce target was performed in $Ar/O_2 = 2/1$ atmosphere at pressure 1.4×10^{-2} mbar and $U_{dc} = 200$ V. Typical deposition rate was 50-70 Å/min. E-beam evapoiration of CeO_2 was performed at pressure $\leq 10^{-4}$ mbar with typical deposition rate 5 Å/min. Substrate temperature during deposition was held at 650-750 °C (Mashtakov 1997).

YBCO films with thickness about 1500 Å were deposited on the buffer layer surface using either DC-sputtering at high oxygen pressure or laser ablation techniques. Typical sputtering parameters were: pure oxygen pressure 4 mbar, discharge voltage 280 V, discharge current density on target 1.5-1.75 A/cm². Deposition rate was about 5 Å/min. Laser ablation was performed at

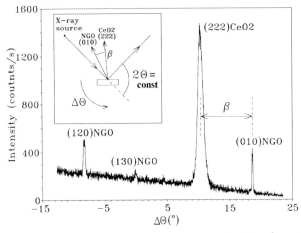

Fig. 1. Rocking curve in a wide angular scanning diapason. On the inset: schematics of the methodics. The angle between reflections is equal to the angle between normals to the crystallographic planes.

0.8 mbar O_2 and laser beam energy density of 1.7 J/cm^2; typical deposition rate was 400 Å/min. Deposition temperature in both cases was 750-780 °C. After deposition the YBCO films were oxygenated for 2 hours in 1 bar of O_2 at 350-450 °C (Mashtakov 1997, Mozhaev 1999).

Crystallographic properties of the obtained heterostructures were studied by X-ray diffractional techniques. For precise evaluation of the angles between crystallographic planes of the layers in the heterostructures the original methodics of rocking curve measurement in a wide angular scanning range was implemented. This methodics is illustrated on Fig. 1 with a sample of the rocking curve and geometry of the measurement on the inset. The 2θ-angle was set constant so, that the characteristic line of the X-ray source provided reflection peak from one of the substrate planes. The sample rotation around the (001) axis resulted in additional peaks on the rocking curve, corresponding to strong reflections from the film and substrate, due to wide-band X-ray radiation spectra. This technique allows direct and high precision measurement of the angles between film and substrate planes (for example β on Fig.1).

3.   RESULTS

Two types of epitaxy were observed in CeO$_2$/NGO heterostructures, each of two variants (Fig. 2 and Table). "Standard" epitaxial relations of CeO$_2$ thin film on a (110) NGO substrate can be written as $(001)_C \| (110)_N$, $[110]_C \| [001]_N$ (Fig. 2, epitaxial type Ia); index 'C' denotes plane or direction in CeO$_2$ buffer layer, 'Y' – in the YBCO thin film and 'N' – in the NGO substrate. For all

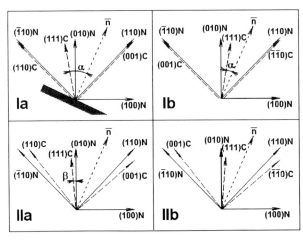

Fig. 2. Scematics of epitaxial growth of CeO$_2$ on the NdGaO$_3$ substrate with normal n. Variants a, b of the epitaxy types result from alignment of the [001] axis of CeO$_2$ along symmetrically equivalent directions [110] and [$\underline{1}$10] of the substrate.

Table. Epitaxial type of CeO_2 growth on $NdGaO_3$ substrate at different deposition techniques and inclination of the [111] CeO_2 axis to substrate normal α and [010] $NdGaO_3$ axis β, correspondingly.

Substrate orientation	Deposition technique	Type of epitaxy	α, degree	β, degree
(110)	RF-sputtering	Ia	55	10
(110)	e-beam evaporation	Ia	55	10
(120)	RF-sputtering	IIa	30	3
(120)	e-beam evaporation	Ia	36	10
(130)	RF-sputtering	Ib	10	-8
(130)	RF-sputtering	IIb (+IIa)	16 (and 22)	-3 (and 3)
(130)	e-beam evaporation	Ia	28	10
(010)	e-beam evaporation	II	0	0

examined substrates the $[110]_C \| [001]_N$ relation remained correct, but CeO_2 orientation in the $(001)_N$ plane depended both on substrate surface orientation and on deposition technique. "Standard" epitaxy was observed for e-beam evaporated CeO_2 films on the (110), (120) and (130)-oriented substrates.

When RF-sputtering technique was used, "standard" epitaxy was observed only for (110) orientation of the substrate. For some of (130) substrates the same epitaxial type was observed (Fig. 2, Ib), but the $[001]_C$ was rotated 90°: $(001)_C \| (1\underline{1}0)_N$. RF-sputtered CeO_2 films mainly showed another type of epitaxy, when the $(111)_C$ plane is close to $(010)_N$ plane (Fig. 2, IIa and IIb). Similar to type I a and b variants of epitaxy, type II show two possible orientations of the $[001]_C$ axis: close to $[110]_N$ direction (IIa, (120) substrates) and close to $[\underline{1}10]_N$ direction (IIb, (130) substrates). Both IIa and IIb variants co-existed on one of (130) NGO substrates (see table). This type of epitaxy was observed only on the (010) substrates when e-beam evaporation technique was utilized.

The YBCO thin film orientation on the CeO_2 buffer layer depended mainly on the deposition technique. Laser ablation resulted in YBCO films oriented with the c axis normal to the substrate plane independent of buffer layer orientation. On Fig. 3, curve b, clear c-oriented film reflections can be seen on a θ/2θ-scan along the substrate normal. No YBCO alignment along the $[001]_C$ direction, inclined 31° from the substrate normal, is observed (Fig. 3, curve d). The X-ray diffractional φ-scans proved that the $<100>_Y$ axes are oriented along and perpendicular to the $[001]_N$ direction of the substrate.

DC sputtering at high oxygen pressure resulted in YBCO films oriented with the c axis along the $<100>_C$ directions of the buffer layer (Fig. 3, curves a and c). Cubic symmetry of CeO_2 results in 3 possible domain orientations of the YBCO film. Experimentally preferential domain formation was observed with axis c close to substrate normal, resulting in monodomain growth for Ia, IIa CeO_2

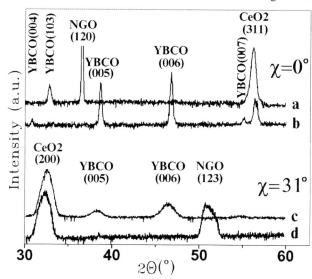

Fig. 3. X-ray diffractional θ/2θ-scans of the YBCO/CeO_2 heterostructures on the (120) $NdGaO_3$ substrates. Curves a, b represent scans along the substrate normal, while c and d curves show scans along the [001] CeO_2 direction, tilted 31 degrees from the substrate normal. Scans a, c were obtained from a sample with DC-sputtered YBCO film; scans b and d from a sample with YBCO film produced using laser ablation technique.

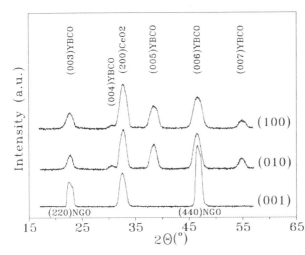

Fig. 4. X-ray diffractional θ/2θ-scans of the YBCO/CeO₂ heterostructure on a (130) NdGaO₃ substrate, deposited using DC-sputtering technique, along the <100> directions of the buffer layer. Only two YBCO domains are present at <100> CeO₂ orientations that are close to the substrate normal.

epitaxial variants and two domains in Ib, IIb cases. On Fig. 4 the two-domain structure of the YBCO film on the (130) NGO substrate is revealed by θ/2θ-scans along the $<100>_c$ directions of the buffer layer. Three YBCO domains were observed in case IIa, when all CeO₂ <100> directions are almost uniformly inclined from the (120) substrate normal.

4. DISCUSSION

The epitaxial relations in the YBCO/CeO₂/NGO heterostructures depend strongly on deposition technique and deviation from crystallographic planes. Rotation of the substrate plane from $(110)_N$ to (010) crystallographic planes of NGO results in change of epitaxy from type I to type II. Utilization of RF-sputtering technique results in change of the type of epitaxy even at deviation angles of 18° ((120) NGO substrate), while when e-beam evaporation is used, epitaxy type changes at deviation angles more than 26°. Difference in deposition rates can be possible reason of this distinction. Structure of the film, grown at low deposition rate, stronger depends on substrate atomic structure, while orientation of the film, grown at high deposition rate, depends on the opening facets surface energy. The $(111)_c$ facets provide minimal surface energy, favoring rotation of the $<111>_c$ axis towards substrate normal Kotelyanskii 1996). Change from type I epitaxy to type II results in such rotation of $[111]_c$ axis, and happens at smaller deviation angle for high deposition rate RF-sputtering, than for low deposition rate e-beam evaporation. Similar reasons can lead to change from variant Ia to Ib epitaxy: the $[111]_c$ axis turns closer to normal.

YBCO thin film deposition on the CeO₂ buffer layer shows similar behavior. High deposition rate during laser ablation results in films with facets of minimal surface energy, i.e. $(001)_Y$-oriented films. At low deposition rate superconductor structure follows structure of the CeO₂ buffer layer. Formation of one or two domains instead of three confirms tendency of the film to grow with the c axis as close to substrate normal, as possible.

Authors would like thank V.A. Lusanov, Yi.A. Boikov for fruitful discussion. The work was supported in part by the Russian Foundation for Basic Research, State Program of Russia "Modern Problems of the Solid State Physics", division "Superconductivity", Swedish Material Consortium and the INTAS program of the EU.

REFERENCES

Kotelyanskii I M et al 1996 Thin Solid Films **280**, 163
Mashtakov A D et al 1997 Inst Phys. Ser No158 (EUCAS'97), p.158
Mozhaev P B et al 1999 Technical Physics **44**, 242

Inst. Phys. Conf. Ser. No 167
Paper presented at Applied Superconductivity, Spain, 14-17 September 1999
© 2000 IOP Publishing Ltd

Phase relations in thin RBCO (R=Lu,Ho,Y,Gd,Nd) epitaxial films prepared by MOCVD

S V Samoilenkov[a], **O Yu Gorbenko**[a], **S V Papucha**[a], **N A Mirin**[a], **I E Graboy**[a], **A R Kaul**[a], **O Stadel**[b], **G Wahl**[b] and **H W Zandbergen**[c]

[a] Chemistry Department, Moscow State University, 119899 Moscow, Russia
[b] IOPW, TU Braunschweig, Bienroder Weg 53, 38108 Braunschweig, Germany
[c] National Centre for HREM, Delft TU, Rotterdamseweg 137, 2628 AL Delft, The Netherlands

ABSTRACT: Phase relations in thin epitaxial RBCO films were found to differ significantly from those of bulk R-Ba-Cu-O ceramics. The appearance of well-oriented inclusions of non-equilibrium-in-bulk secondary phases (R_2CuO_4 (R=Nd and Gd), R_2O_3 (R = Lu,Ho,Y), $Lu_2Cu_2O_5$, Ba_2CuO_3 and $BaCu_3O_4$) in slightly off-stoichiometric films was confirmed by means of XRD, SEM with EDX, and HREM. The formation of coherent or semi-coherent interfaces between embedded inclusions and the matrix is considered to be critical for a stabilization of otherwise non-equilibrium oxide phases in epitaxial RBCO films.

1. INTRODUCTION

Various deposition techniques have been successfully adopted for the growth of $RBa_2Cu_3O_7$ (RBCO, R – rare earth element) thin films with excellent superconducting properties. Metal-Organic Chemical Vapour Deposition (MOCVD) proved to be one of the most promising approaches due to the low requirements to vacuum conditions, its high throughput and resulting quality of the layers achieved (Watson 1997, Becht 1998). However, the composition of the growing film in MOCVD usually deviates from the composition of the vapour phase, due to the complexity of processes involved. Therefore, the MO precursors molar ratio has to be adjusted in an empirical way to achieve stoichiometric films. The evident consequence of such situation is the appearance of secondary phases in films. This issue has been extensively studied for YBCO films grown by MOCVD (Gorbenko et al 1991, Li et al 1992, Hudner et al 1993, Schulte B et al 1993, Waffenschmidt et al 1994, Doudkowsky et al 1995) and various physical vapour deposition techniques (Gong et al 1994, Han et al 1994). The dependence of superconducting properties on a formal cation composition has been considered with no direct reference to the actual phase relations. Meanwhile, there is hardly any evident agreement concerning the composition of secondary phases. Intriguing thermodynamic issues (e.g. the common appearance of Y_2O_3, the non-equilibrium in bulk ceramics secondary phase) gained only a little attention (Weiss 1995). As well, the prominent changes in the phase composition of R-Ba-Cu-O films with the change of R remained beyond the scope of the research made so far. In this report, we present a systematic study of (001)RBCO epitaxial thin films grown by MOCVD. The appearance of non-equilibrium in bulk phases is especially outlined. We show the important role of inclusions' size, structure and orientation in stabilizing particular secondary phases in RBCO films.

2. EXPERIMENTAL SECTION

RBCO (R=Lu,Ho,Y,Gd) films were grown on single crystal (001)$LaAlO_3$ and $SrTiO_3$ substrates at a growth rate of 0.4μm/hr by single source MOCVD (Samoylenkov et al 1996). The films were cooled down in oxygen atmosphere, with an annealing step at 450°C for 1hr. XRD analysis was carried out using SIEMENS D5000 and DRON 3M diffractometers. SEM and EDX element analyses were performed with CAMSCAN electron microscope and an EDAX 9800 system. TEM/HREM of the films cross-sections were performed with Philips CM30ST electron microscope equipped with the field emission gun operating at 300kV and a Link EDX element analysis system. Superconducting properties of RBCO films were determined from ac magnetic measurements.

3.   RESULTS AND DISCUSSION

High epitaxial quality and superconducting characteristics have been achieved for all (001)RBCO films grown under optimized conditions, with exception of R=Nd (Table 1). The optimal deposition conditions followed the line of CuO-Cu_2O-O_2 equilibrium (Lindemer&Specht 1995). The variation of T_c and c lattice parameter of RBCO films for R=Lu-Gd is in a accordance with the dependencies known for the bulk RBCO. The critical current density j_c(77K,100 Oe) decreases somewhat with the increase of $r(R^{3+})$. It is due to this behaviour that LuBCO films, even though LuBCO possesses the lowest T_c value, have the highest j_c at low temperatures. This behaviour could be attributed to the increase of the pinning efficiency with the decrease of $r(R^{3+})$. Interestingly, recent results on high-quality NdBCO thin films (T_c = 93-95 K) prepared by physical methods (Moon& Oh 1997, Li&Tanabe 1998) demonstrate rather low n values of 1.1–1.2 in the $j_c = const \cdot (1$-$T/T_c)^n$ dependence, in a good agreement with our data (see the inset of Fig.1(b)). Probably, the remarkable increase in the single crystal perfection of the epitaxial films in the row LuBCO → Y(Ho)BCO → GdBCO→NdBCO accounts for this (notice, that the typical FWHM of the (005)RBCO rocking curve is about $0.30°$ for the best LuBCO films, $0.20°$ for YBCO and as low as $0.05°$ for NdBCO). Another possible reason is the decrease of the hole doping of CuO_2 planes of $RBa_2Cu_3O_7$ (δ=0) in this row, leading to the lowering of H_{irr} and, consequenlty, to the decrase of j_c.

Deviations of the cation composition from the 1:2:3 ratio caused the appearance of secondary phases (Table 2). The observed number of phases (1, 2, or 3) was in accordance with the Gibbs' rule, provided the (001)-oriented RBCO phase was predominant in the film. We could not detect any products of the chemical interaction between RBCO and these secondary phases by HREM and XRD. Therefore, the phases observed were supposed to be in thermodynamical equilibrium to RBCO.

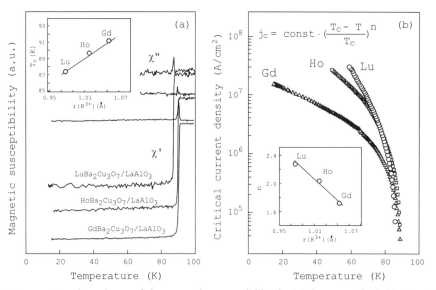

Fig.1 Temperature dependence of the magnetic susceptibility in 0.1 Oe magnetic field (a) and of critical current density measured at 100 Oe of RBCO films grown on $LaAlO_3$ substrates under optimized MOCVD conditions.

However, this phase relations do not correspond to the well-established phase equilibria in bulk R-Ba-Cu-O systems. For instance, no evidence was found for the presence of R_2BaCuO_5 or $BaCuO_2$, which are the equilibrium secondary phases with the bulk RBCO. At the other hand, all the non-equilibrium-in-bulk phases were *well-oriented* as observed by HREM and XRD (Samoylenkov et al 1999) due to their good structural matching with perovskites. Therefore, it must be due to the *low*

Table 1. The optimal deposition conditions found and typical superconducting and structural characteristics of RBCO films grown on (001)LaAlO$_3$ substrates by MOCVD.

R	r(R^{3+}), Å	T$_{dep}$, °C	p(O$_2$)$_{dep.}$, mbar	T$_c$, K	j$_c$(77K,100 Oe), $\times 10^6$ A/cm^2	c, Å
Lu	0.977	800	1.3	88	2.7	11.670(5)
Ho	1.015	820	1.7	90	2.5	11.690(5)
Y	1.018	820	1.7	90	2.0	11.690(5)
Gd	1.053	835	2.5	91	1.0	11.710(5)
Nd	1.109	850	5.0	65	-	11.760(5)

energy of coherent or semi-coherent interfaces formed between the secondary phase inclusions and RBCO matrix (or perovskite substrate), that particular secondary phases become more thermodynamically stable. In contrast, R$_2$BaCuO$_5$ and BaCuO$_2$ have structures rather different from perovskite one, and that prohibits their epitaxial growth. Consequently, their nucleation in the growing film becomes unfavourable and new phase relations appear.

The phase relations change remarkably with the change of R (Table 2). Interesting situation have been observed for Nd-Ba-Cu-O system (Figure 2). Unstable-in-bulk BaCu$_3$O$_4$ phase was found to be in equilibrium with Nd$_{1+x}$Ba$_{2-x}$Cu$_3$O$_{7-\delta}$ solid solutions. Respectively, an excess of copper that does not affect T$_c$ and even j$_c$ of YBCO thin films (Gorbenko et al. 1991) resulted in low-T$_c$ or even non-superconducting Nd$_{1+x}$Ba$_{2-x}$Cu$_3$O$_{7-\delta}$ ($x > 0$) films. To achieve high-T$_c$ films, the deposition of stoichiometric Nd$_1$Ba$_2$Cu$_3$O$_{7-\delta}$ or films with admixture of BaCu$_3$O$_4$ and/or Ba$_2$CuO$_3$ is necessary. We tried to grow such films on LaAlO$_3$ and SrTiO$_3$ substrates. Actually, typical T$_c$ values for those films (50-65K) were the highest of achieved by us, but still too low to correspond to the high-quality Nd$_1$Ba$_2$Cu$_3$O$_7$ material.

Table 2. The secondary phases in non-stoichiometric (001)RBCO films, their structural characteristics, occurrence and orientation.

Phase	Measured lattice parameters, Å	Orientation(s)	Occurrence and size	Bulk material reference
		LuBCO,HoBCO,YBCO,GdBCO,NdBCO		
CuO	d(002) = 2.525 d(202) = 1.583	(001)CuO // (001)RBCO and (101)CuO // (001)RBCO	film surface > 1 μm	JCPDS 5-0661
BaCu$_3$O$_4$	orthorhombic a = 11.01(2), b = 5.50(1), c = 3.923(2)	(001)BCO//(001)RBCO [210]BCO//[100]RBCO and (210)BCO//(001)RBCO [001]BCO//[100]RBCO	film surface 0.1-5 μm	unstable in bulk, as a secondary phase in YBCO single crystals: Bertinotti et al 1989
Ba$_2$CuO$_3$	orthorhombic c = 12.92(3)	(001)BCO//(001)RBCO [100]BCO//[100]RBCO	film surface > 1 μm	JCPDS 39-497
		LuBCO,HoBCO,YBCO		
R$_2$O$_3$	distorted cubic Lu: $a_\perp$* = 10.33(3) Ho,Y: $a_\perp$* = 10.53(3)	(001)R$_2$O$_3$//(001)RBCO [110]R$_2$O$_3$//[100]RBCO	film-substrate interface or film matrix < 20 nm	JCPDS 12-0728 (Lu) JCPDS 44-1268 (Ho) JCPDS 41-1105 (Y)
		GdBCO,NdBCO		
R$_2$CuO$_4$	tetragonal Gd: c = 11.85(1) Nd: c = 11.86 (2)	(001)RCO//(001)RBCO [100]RCO//[100]RBCO	—	JCPDS 24-0422 (Gd) JCPDS 39-1390 (Nd)
		LuBCO		
Lu$_2$Cu$_2$O$_5$	—	(001)LCO//(01$\bar{1}$)LAO (013)LCO//(011)LAO	film-substrate interface < 0.2μm	JCPDS 34-0387

* the lattice parameter perpendicular to the substrate plane

128

Another complication occurs in this case. To avoid the formation of (100)-oriented grains, the temperature for MOCVD growth of NdBCO thin films must be held as high as 850°C (Table 1). At this high temperature, the melting of various eutectics is possible, complicating growth and crystallization mechanisms. Recently, Kumagai et al (1998) reported similar low T_c's for their NdBCO films grown by MOCVD. According to their TEM results, they suggested that the phase separation with the formation of low-T_c $Nd_{1+x}Ba_{2-x}Cu_3O_{7-\delta}$ solid solution took place. Further study is evidently needed to clarify this question.

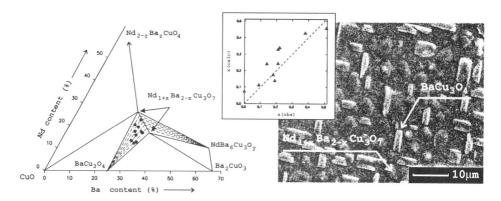

Figure 2. Phase relations in copper-rich part of Nd-Ba-Cu-O system in thin epitaxial films grown on (001) perovskite substrates. Surface morphology of NdBCO film with rectangular precipitates of $BaCu_3O_4$.

ACKNOWLEDGEMENTS

INTAS-RFBR (IR 97-1954), Volkswagen Stiftung (I/73628), RFBR (99-03-32590a) and Stichting Fundamenteel Onderzoek der Materie and the European Commission (contract ERMRXCT 98-0189) are acknowledged for a financial support. S.V.S. acknowledges the support of ISSEP and Robert Havemann Foundation.

REFERENCES

Becht M 1998 Appl Supercond **4**, 465.
Bertinotti A, Hamman J, Luzet D, Vincent E 1989 Physica C **160**, 227.
Doudkowsky M, Santiso J, Figueras A et al 1995 J de Physique IV **5** Coll.C5, 415.
Gong J P, Kawasaki M, Fujito K 1994 Phys Rev **B50**, 3280.
Gorbenko O Yu, Kaul A R, Tretyakov Yu D et al 1991 Physica C **190**, 180.
Han Z, Selinder T I, Helmersson U 1994 J Appl Phys. **75**, 2020.
Hudner J, Thomas O, Mossang E et al 1993 J Appl Phys **74**, 4631.
Kumagai Y, Yoshida Y, Iwata M et al 1998 Physica C **304**, 35.
Li Y Q, Zhao J, Chern C S et al 1992 Physica C **195**, 161.
Li Y and Tanabe K 1998 J Appl Phys **83**, 7744.
Lindemer T B, Specht E D 1995 Physica C **255**, 81.
Moon S H, Oh B 1997 IEEE Trans Appl Supercond **7**, 1185.
Samoylenkov S V, Gorbenko O Yu, Graboy I E et al 1996 J Mater Chem **6**, 623.
Samoylenkov S V, Gorbenko O Yu, Graboy I E et al 1999 Chem Mater **11**, in press.
Schulte B, Maul M, Häussler P, Adrian H 1993 Appl Phys Lett **62**, 633.
Waffenschmidt E, Waffenschmidt K H, Arndt F et al 1994 J Appl Phys **75**, 4092.
Watson I M 1997 Chem Vap Deposition **3**, 9.
Weiss F, Pisch A, Bernard C, Schmatz U 1995 J de Physique IV **5** Coll.C5, 151.

Inst. Phys. Conf. Ser. No 167
Paper presented at Applied Superconductivity, Spain, 14-17 September 1999

The effect of laser beam intensity homogenisation on the smoothness of YBCO films obtained by laser ablation

R A Chakalov[1,2], F Wellhöfer[1,3], S Corner[2], M Allsworth[2], C M Muirhead[2]

[1] Interdisciplinary Laser Deposition Facility, [2] School of Physics and Astronomy, [3] School of Metallurgy and Materials, University of Birmingham, Birmingham B15 2TT, UK

ABSTRACT: In conventional pulsed laser deposition, boulders and grains of off-stoichiometric phases are major contributors to the overall film roughness. These may be caused by a low laser fluence in part of the ablated spot. To overcome this problem we use an optical beam intensity homogeniser. By means of crossed cylindrical lenses the raw beam of the excimer laser is cut into segments and overlaid at a plane of integration. We present evidence that the resulting more uniform and homogeneous spot on the target also significantly improves the overall smoothness of the deposited YBCO films.

1. INTRODUCTION

The excellent electrical properties of thin films obtainable by laser ablation and a high deposition rate have made pulsed laser deposition (PLD) a very popular technique for the fabrication of thin films of high temperature superconductors and other related oxides. But despite significant improvements in the technique over the last decade, epitaxial thin films grown by PLD still rather often suffer from a poor surface morphology. A rough surface, however, is a very serious problem for multilayer structures, which are becoming increasingly important for HTS materials applications.

Some major contributors to an increased surface roughness are boulders and grains of off-stoichiometric phases. There are a number of factors which may cause particulate formation, and by careful process optimisation they can largely be eliminated. However, one of the factors can be a low laser fluence on the target material, which may be insufficient for the initiation of the proper laser ablation process (Chrisey and Hubler 1994, Dam et al 1996). In PLD, the energy density in the focussed laser spot must exceed a certain threshold value (Dam et al 1995). However, due to spatial inhomogeneities of the beam intensity, there can be areas within the spot with quite different laser fluence.

To overcome this problem we used a commercial optical beam homogeniser (MicroLas 1998). By means of crossed cylindrical lenses, the raw beam of the excimer laser is first cut into segments, and then overlaid at a plane of integration. There seems to be a widespread belief that this is highly beneficial for PLD thin films, but there is as yet little supporting evidence in the literature (Wagner et al 1998).

2. THE OPTICAL HOMOGENISER

The raw beam of an excimer laser has a specific intensity profile due to the nature of the energy pumping electrical discharge. Modern lasers are supplied with special electrodes, which are designed to decrease any inhomogeneity. Nevertheless, the beam profile in the horizontal axis can be characterised as "top hat", and in the vertical axis the distribution is nearly Gaussian. A widespread and simple solution to the problem is to cut the peripheral areas by suitable rectangular or circular apertures. This approach has the disadvantage that the size of the homogeneous area is difficult to determine and, furthermore, can vary with the operating conditions. Moreover, since the losses may be high and thus the energy available within the spot low, lasers are often used in regimes close to their limits with frequent change in operating conditions.

A more effective way of improvement can be realised by an external optical system usually called homogeniser. It consists of two arrays of cylindrical lenses. The first one separates the beam cross-section into segments. The second array together with a condensor lens overlay all segments in a plane. Thus the beam is integrated again with a spatial redistribution of the intensity. The more light segments are overlaid the better is the homogeneity. In our case 24 segments were formed by 6×4 element arrays.

3. EXPERIMENTAL DETAILS

A 248 nm excimer laser was used to deposit thin films of YBCO from stoichiometric ceramic targets without and with a homogeniser in the beam path. The focussing elements were aligned in such a way that the area of the spot on the targets was 8 mm^2 in both cases, even though the x and y dimensions differed slightly. A laser fluence of 1.5 J/cm^{-2} was used at a repetition rate of 5 Hz to deposit epitaxial YBCO films on single crystal SrTiO$_3$ (100) substrates. The 'race-tracks' formed by the laser interaction on the targets were investigated by SEM. To allow a direct comparison, the same deposition conditions were used in both cases: 760°C substrate temperature, 0.4 Torr oxygen pressure, 80 mm distance target-substrate, 10°C/min cooling down rate in 700 Torr oxygen. This set of parameters was obtained after optimisation of the process to achieve an acceptable compromise between a high T$_c$ and a relatively smooth surface. No noticeable difference in the critical temperature of the films deposited with and without homogeniser was measured (T$_c$ ~ 88 K). The film surface morphology was examined by AFM in contact mode (ARIS 3300, Burleigh).

4. RESULTS AND DISCUSSION

SEM images of the laser-imprints on the rotating targets are shown on Fig.1. The total number of pulses was 4000, which is also typically used for the deposition of YBCO-thin films preparation. However, since the target was rotated, the estimated number of pulses on any given area was ~250. According to Dam et al (1995) the typical target morphology resulting from ablation with a fluence above the threshold is a flat-topped pillar structure. We observed such a structure over almost the whole area of the 'race-track' obtained from the deposition *with* the homogeniser (Fig.1e-1h). By contrast, the SEM-images of the race-track obtained *without* the homogeniser, show much more pronounced differences between the morphologies observed close to the centre and those in the outer edge of the race-track (Fig.1a-1d). The observed change from hillocks to a pillar-like morphology is consistent with the interpretation that the local laser fluence significantly varies across the race-track and remains partially below the ablation threshold. At low fluence the target surface is subjected to pulsed thermal cycling rather than ablation (Dam et al 1996). During the melting-resolidification processes phase separations take place, and inclusions of second phases are formed within the YBCO matrix. During subsequent laser pulses, parts of these inclusions are also ablated.

A comparison by AFM of the film surfaces deposited under identical conditions either without (Fig.2a), or with the beam homogeniser (Fig.2b), also clearly shows differences in their microstructure. The film deposited without the homogeniser contains a large number of high boulders. These dominate the surface morphology and consequently also lead to a larger film roughness. The form of the boulders does not allow us to assign them to fine droplets from the target or to a-axis oriented particles. But the fact that the boulders are totally absent in the film deposited with the beam homogeniser in line supports our conclusion that they originate from

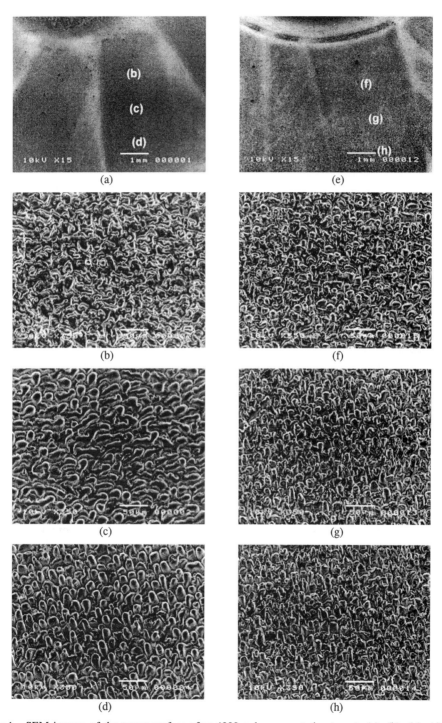

Fig.1. SEM images of the target surface after 4000 pulses on rotating target: (a), (b), (c), (d) – without homogeniser; (e), (f), (g), (h) – with homogeniser.

132

second phases formed in the target as a result of local areas subjected to a fluence below the ablation threshold. On the other hand, the microstructure of the film deposited with the beam homogeniser is free from any boulders, and its roughness determined only by the height and trenches between the growth-islands of YBCO, as is characteristic for the microstructure of high-quality YBCO-films deposited by PLD.

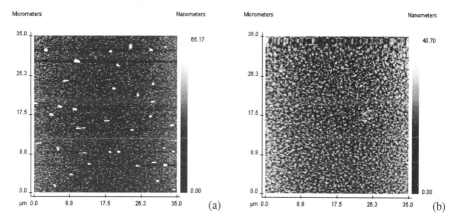

Fig.2. AFM images of thin films deposited (a) without and (b) with homogeniser in the beam line.

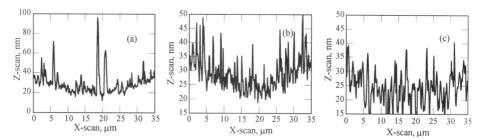

Fig.3. AFM linear scans along x-axis of the images presented on Fig.2: (a), (b) lines on the surface of film deposited without homogeniser and (c) with homogeniser; scan (a) goes over a line containing boulders; scan (b) – between the boulders.

In Fig.3. the linear AFM scans are compared. It can be seen (Fig.3a, 3b) that the higher value of the peak-to-peak roughness ($r_{pp} > 75$ nm) of the film deposited without a homogeniser is caused by the presence of boulders. The film obtained with beam energy homogenisation showed $r_{pp} \sim 15$ nm and average height of the surface $r_a < 5$ nm over 35×35 μm^2 area (Fig.3c). Thus we demonstrate that the elimination of the boulders using the beam homogeniser improves significantly the overall smoothness of the YBCO thin films deposited by PLD.

REFERENCES

Chrisey D B and Hubler G K 1994 Pulsed Laser Deposition of Thin Films (New York: Wiley)
Dam B, Rector J, Chang M F, Kars S, deGroot D G and Griessen R 1995 Appl. Surf. Sci. **86**, 13
Dam B, Rector J, Johansson J, deGroot D G and Griessen R 1996 Mat. Res. Soc. Symp. Proc. **397**, 175
MicroLas Lasersystems GmBH 1998, Göttingen, Germany, web-page http://www.microlas.de
Wagner F X, Scaggs M, Koch A, Endert H, Christen H-M, Knauss L A, Harshavardhan K S and Green S M 1998 Appl. Surf. Sci. **127-129**, 477

Inst. Phys. Conf. Ser. No 167
Paper presented at Applied Superconductivity, Spain, 14-17 September 1999

Ag/ YBCO thin film growth by single resistive evaporation.

A Verdyan, I Lapsker and J Azoulay.

Center for Technological Education Holon, POB 305, Holon, 58102, Israel.

ABSTRACT: It commonly accepted that thin film formation of YBCO on conducting substrate is one of the keys to further development of advanced devices in the microelectronic and high field applications. In this work we report on the preparation of superconducting YBCO thin film deposited on unbuffered silver substrate layer using a simple conventional vacuum system equipped with only one single resistively heated evaporation source. Ag film (500 nm) was first deposited on a polished clean MgO substrate. A pulverized mixture of Y, Cu, and BaF_2 was then inserted in to resistive evaporation boat. The evaporation process lasted 15 minutes thus coating the Ag film with a 1000 nm thickness of amorphous film. Subsequently heat process was carried out under a low oxygen partial pressure. The results of the film evaluation are presented and discussed.

1. INTRODUCTION

Silver is known to be a useful doping element in the improvement of the mechanical and electrical properties of the high T_c oxide superconductors particularly of YBCO (for example see Jin et al. 1992 and Nacamura 1991). By now it is widely accepted that Ag does not enter the YBCO lattice (Joo et al. 1992) and thus does not degrade the critical temperature. However silver doping does affect the microstructure by increasing the density and grain size and by promoting texture crystallization. Furthermore silver is known to improve the ductility of YBCO tapes and the acceleration rate of YBCO synthesis. Recently Budai et al. (1993) have reported on highly textured YBCO thin films grown on Ag (001), (110) and (111) single crystal surfaces.

In this report we present a YBCO thin film preparation on silver thin film deposited on MgO substrate, using a simple conventional vacuum system housing single resistive heated source for evaporation.

2. EXPERIMENTAL.

A simple inexpensive turbomolecular pumping vacuum system with a base pressure better than 10^{-3} Pa and equipped with a single resistively heated source was used for the film preparation. No device of any kind was applied to control or regulate the thickness or stoichiometry of the film during the evaporation process. The distance between the substrate and evaporation source was predetermined, after few calibration attempts, to be 6 cm for which the starting materials were weighed to yield films of the desired thickness and stoichiometry provided evaporation to completion was done. A clean polished MgO substrate glued to a heating element by silver paint was baked out to 450°C for few minutes to outgase the surface. Pure Ag thin film was then deposited onto the MgO surface. During the evaporation the substrate temperature was kept at 250°C which was measured a thermocouple attached to dummy MgO substrate glued next to it. A precursor of yttrium, copper and BaF₂ weighed in the atomic proportion to yield stoichiometry YBCO thin film well ground to a pulverized mixture was than quickly inserted into new tungsten boat. The vacuum system was flushed with oxygen and then pumped down to its base pressure thus leaving the system with a residual pressure mostly due to oxygen, which helped later to defluorinate the film during the heat treatment. Typical duration time of vacuum break between these two stages was about 10 minutes. The MgO substrate coated now with Ag thin film was kept at 200°C during evaporation of the pulverized mixture precursor. This process lasted about 14 minutes, thus coating the Ag film 1000 nm sickness of amorphous BaF₂, Cu and Y uncompounded constituents with a typical rate about 1 nm/s . The solid phase reaction was subsequently carried out at a temperature of 740°C under low oxygen partial pressure of about 1 Pa for about 30 minutes (Azoulay 1996). The films thus obtained were found to have the normal high Tc superconducting phase and were evaluated for their electrical and structure properties.

3. RESULTS AND DISCUSSION

The room temperature normalized values of resistance vs. temperature in zero magnetic field measurement for different samples are compared in Fig. 1.

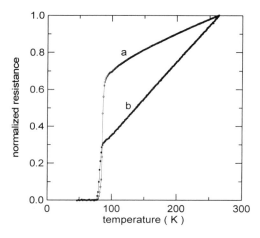

Fig. 1 The room temperature normalized resistance vs. temperature for YBCO
a) deposited on MgO, b) deposited on Ag film evaporated on MgO substrate.

The samples were prepared simultaneously on the same MgO substrate except that sample (a) was deposited directly on the MgO substrate while sample (b) was deposited on Ag film, which was previously evaporated on MgO substrate. The room temperature resistivity and the slope of sample (b) were found to be identical to those of Ag. This quite expected since as long as the temperature is above, T_c where the resistivity of the YBCO film is larger then that of the silver film, the current flows mainly through the silver film. However the onset value and zero resistance of the two samples are the same to within less than 0.5 K. The

X-ray diffraction (XRD) plot carried on the YBCO film (sample (b) in Fig. 1) is shown in Fig. 2 in which all the well documented lines of YBCO show up.

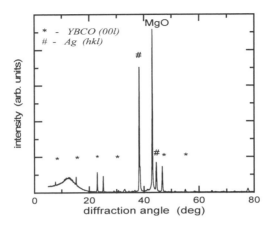

Fig. 2   XRD plot of YBCO film of sample **b** in Fig. 1.

We have studied the topological structure of this sample using scanning electron microscope (SEM). The micrograph is shown in Fig. 3 is taken from a continuous area and features small grains with typical size of 0.5μm. The area in this picture was analyzed by energy dispersive X-ray and was found to be YBCO.

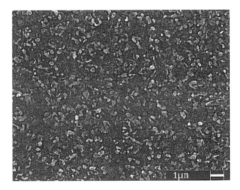

Fig. 3   SEM micrograph of YBCO film of sample **b** in Fig. 1.

It was also found that in spite of the annealing temperature the silver layer does not mix with the YBCO layer. We have also studied the feature of the Ag film (sample (b) in Fig. 1). From SEM measurements it was found that the Ag film was completely continuous. Typical XRD plot of Ag film used for preparing sample **b** is depicted in Fig. 4. This 2θ scan was carried out through a strong filter to reduce the MgO (200) line tail thus showing up the strong Ag (111) line. The rocking curve width of this line was measured to be about 0.5° FWHM thus indicating a mosaic form of the Ag film. This feature is attributed to the relatively low substrate temperature during the deposition and also due to absence of any subsequent high temperature annealing treatment.

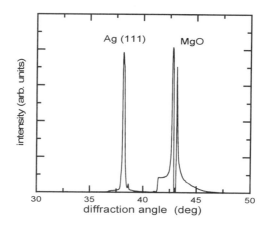

Fig. 4 XRD plot of Ag film of sample **b** in Fig. 1.

4. CONCLUSION

In this study it was shown that YBCO thin film can be grown on silver film directly using a very simple inexpensive vacuum system for resistive evaporation of pulverized mixture Y, BaF_2 and Cu. In spite of the relatively high temperature final heat treatment, the silver layer does not mix with the YBCO film. It is therefore deduced that silver layer acts as a passive substrate during the solid phase reaction between the BaF_2, Cu and Y constituents which was carried out at 740°C. It is believed that the Ag thin film substrate will improve the electrical and physical properties of the YBCO film. Investigation on this issue is being carried out in our laboratory.

ACKNOWLEDGMENTS

The Israel ministry of Science supported this work.

REFERENCES

Azoulay J 1996 J. Am. Ceram. Soc. **79**, 568
Budai J D, Young R T and Chao B S 1993 Appl. Phys. Lett. **62**, 1836
Jin S, Kammlott G W, Tiefel T H and Chen S K 1992 Physica C **198**,333
Joo J, Sigh J P, Poeppel R B, Gangopadhay A K and Mason T O 1992 J. Appl. Phys. **71**, 2351
Nakamura O, Chan I N, Guimpel J and Schuller I K 1991 Appl. Phys. Lett. **59**, 1245

Importance of the interface effects in Au-YBa2Cu3O7-δ sintered composites at low temperature .

C. Lambert-Mauriat, J.M. Debierre and J. Marfaing,

Lab. MATOP-CNRS, Case 151, F- 13397 Marseille Cedex 20.

Abstract. We combine experimental and numerical studies to investigate the variations of the resisitivity in Au-YBCO composites, as a function of the gold volume fraction, Φ, in the temperature range 50K - 300K. Below the superconducting critical temperature, T = 92 K, a systematic shift of the metal-superconductor transition threshold, Φ_s, is observed as T varies and the resistivity curves present an unexpected maximum whose amplitude and position change with T. Our experimental results are well described within a random conductor network model which incorporates porosity and interfaces resistances. A very good agreement between experimental and numerical results is found and a possible interpretation of the temperature dependence of this interfacial resistivity is discussed.

1. Introduction

Electrical transport in high-T_c superconductor ceramics strongly depends on the microscopic properties of the ideal material and on the mesoscopic characteristics, such as the granulometry for example. In the granular composites, the factors which can induce strong anomalies in the resisitivity cannot be analyzed very accurately, mainly because of the nature and importance of the barriers between the supeconducting grains. In this context, we have combined experimental and numerical studies to investigate the behavior of the resistivity of Au-YBCO composites as a function of the gold volume fraction, Φ, in the temperature range 50K - 300K, and this contribution presents the results obtained.

2. Experimental procedure

Our composites were prepared by mixing different volume fractions, Φ and $\Phi' = 1 - \Phi$, of Au and YBCO powders. Then, the two powders were cold pressed under 3.5 x 10^8 Pa at room temperature and slab samples of dimension 9 x 2 x 0.3 mm^3 were sintered. The thermal cycle consisted in a heating up to 910°C (under Argon flow) during 5 min, at a rate of 18°C/min. Then, the sintering temperature was decreased at a rate of 7°C/min, in oxygen atmosphere, with a two-hour step at 450°C. Finally, the samples were cooled down to 200°C, at a rate of 1.5°C/min and air quenched, as detailed elsewhere (Regnier 1996). Different characterizations were carried out on the samples: i) standard four-probe alternative (36Hz) resistance measurements in a classical closed cycle helium PAR cryostat, in the range 20 to 300 K; ii) X-rays diffraction measurements performed onto a D 5000 Siemens equipment with the λ CuK$_\alpha$ radiation Ni filtered; iii) Scanning Electron Microscope study with local energy dispersion spectroscopy analysis (EDS). For each composition, three samples were systematically prepared under identical conditions before measurements. The total error was estimated to about 7% for ρ and below 1% for Φ and Φ'.

3. Results and discussion

The general behavior of the resistivity as a function of temperature is a metallic-like conduction at high temperature, in the whole composition range. For $\Phi < 0.63$, the resistance falls off to zero at a critical temperature T_c^0 which varies with Φ and for $0.5 < \Phi < 0.63$, the transition progressively broadens as Φ increases. While it does not reach zero any longer, the resistivity curve still presents a shoulder when $\Phi > 0.63$. The concentration dependence of the resistivity measured at 273K is plotted as a function of Φ in Fig. 1.

The inflexion point approximately locates the percolation threshold for Au, $\Phi_c \approx 0.2$ and our experimental data are comparable with previous results for Au and YBCO composites (Xiao 1988) and for sintered cylindrical pellets (Streitz 1988). The conduction properties of random composites are usually well interpreted by the percolation theory (Sahimi 1994).

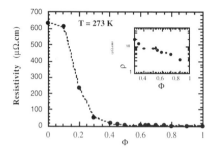

Fig. 1 concentration dependence of the resistivity measured at 273K as a function of Φ

Let us now focus on the superconducting regime. Up to $\Phi = 0.45$, the superconducting transition is not much affected by Au addition as T_c^0 values attest (Fig.2). Our results compare well with previous experiments showing that T_c^0 varies slightly when $\Phi < 0.4$ and significantly more beyond (Imanaka 1990). Due to the sharp decrease of the curve $T_c^0(\Phi)$ for $\Phi \geq 0.6$, the experimental percolation threshold of the superconducting phase tends to $\Phi'_s = 0.35$. The percolation thresholds of Au and YBCO, $\Phi_c = 0.2$ and $\Phi'_s = 0.35$ are thus not equal, contrary to what is expected for grains of equal size.

At low temperature, $60\ K \leq T \leq 200\ K$, the resistivity versus concentration presents a maximum, not predicted by the simple percolation theory and of which position and amplitude varies.

Representing both materials by unit cubes, simulations were performed on a L^3 cubic lattice: each unit cube is either left empty with a probability Π, either filled by gold with a probability $(1 - \Pi)\ \Phi$ or filled by YBCO with a probability $(1 - \Pi)\ (1 - \Phi)$. A direct matricial calculation of the potential in each site gives the total conductance of the system. Effects of porosity and interfacial resistance were first separately examined but the shape of the numerical curves were correctly reproduced only when combining the two effects. In this case we find an excellent agreement at low temperature, as well for the percolation threshold as for the position and magnitude of the resistivity maximum (Fig. 2).

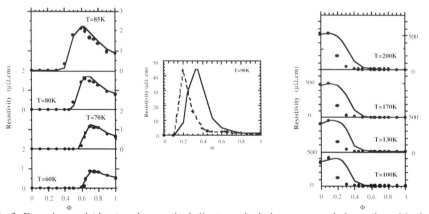

Fig.2: Experimental (dots) and numerical (line) results below, near and above the critical temperature of pure YBCO Tc = 92 K and their temperature and concentration dependence.

Above the critical temperature, the agreement is satisfying, giving a good description of the increase of resistivity (T = 90 K) but the position of the peak is not well adjusted by the model. From the numerics, it is clear that both parameters, porosity and interfacial resistance, R_{int}, must be taken into account to correctly describe the electrical response of the composites. The temperature dependence of these parameters show that two distinct regimes exist, giving insight into the nature of the interfaces.

Below T_c, R_{int} is small and it roughly increases linearly, up to T = 85 K. Thus, its metallic behaviour can be interpreted as the non existence of insulating phases at the Au-YBCO interfaces. The values of the YBCO resistivity, obtained from the calculations, are reported in Fig. 3 and compared with values of the literature.

Above T_c, the interfacial resistance displays a semiconducting character. In Fig. 4, our numerical estimates for R_{int} are compared with the resistivity of deoxygenated YBCO (x = 6.76 and x = 6.78 with T_c^0 = 68 K and 80 K respectively) (Cava 1987). The nice agreeement, concerning the shape of the curves, confirms that YBCO is very likely deoxygenated in our samples, forming ybco-YBCO (deoxygenated) and pure YBCO-YBCO bonds.

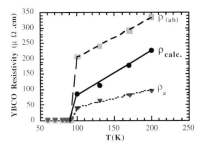

Fig. 3:Numerical values of R_{int} below Tc.

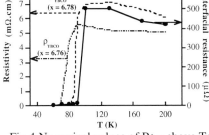

Fig.4:Numerical values of R_{int} above Tc.

From the model it is possible to evaluate the proportions of the two types of interfaces, Au-YBCO and ybco-YBCO which can be partially deoxygenated. Above Tc, this is displayed in Fig.5 with the normalized resistivity at T = 130 K. At the maximum in resistivity, Φ_{max} = 0.15, neither YBCO-YBCO bonds nor Au-Au bonds percolate.

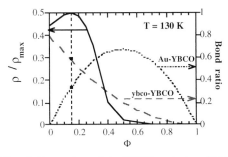

Fig. 5: Proportion of Au-YBCO and ybco-YBCO interfaces and normalized resistivity evaluated from the model at T = 130 K.

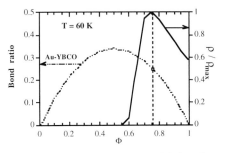

Fig.6: Proportion of Au-YBCO interfaces (dotted line)and normalized resistivity (solid line) evaluated from the model at T = 60 K.

Besides, the number of ybco-YBCO bonds is twice the number of Au-YBCO bonds. Since the resistance of the latters is smaller, R_{int} is dominated by the formers. This is already visible at T = 90 K, where the sharp increase in resistivity is due to ybco-YBCO interfaces.

At T = 60 K (Fig.6) there are no ybco-YBCO bonds and the maximum in resistivity is only due to the Au-YBCO bonds. Thus, R_{int} obtained from the numerics can be attributed to the Au-YBCO interfaces alone. Its metallic behavior can be intrepeted as the non existence of insulating phases at the Au-YBCO interfaces. Their influence is particularly important, since YBCO (or deoxygenated YBCO) is in the superconducting state, which considerably reduces its role. The low value of interfacial resitivity also indicates that no drastic degradation occurs when Au and YBCO grains are in contact.

4. Conclusion

To conclude, let us consider the estimates for ρ_{YBCO} obtained in the numerics. It is satisfying to verify that, as expected, their temperature dependence is linear above T_c (Fig 3). According to the model, porosity and interfaces affect the resistivity of sintered materials. So, it is more accurate to compare our ρ_{YBCO} estimates with measurements both in the (ab)-plane and along the a axis in single crystals (Tozer 1987; Penney 1988; Friedmann 1990). Over the whole temperature range our numerical data are well bounded by the experimental values of ρ_a and ρ_{ab} which, *a posteriori*, validates our model.

References

Cava, R.J., Batlogg, B., Chen, C.H., Rietman, E.A., Zahurak, S.M., and Werder, D.,1987, **Phys Rev B 36** 5719.

Friedmann, T.A., Rabin, M.W., Giapintzakis, J., Rice, J.P., and Ginsberg, D.M., 1990, **Phys. Rev. B 42** 6217.

Imanaka, N., Saito, F., Imai, H., Adachi, G., Yoshiwaka, M., and Okuda, K., 1990, **J. Appl. Phys. 67**, 915.

Penney, T., von Molnak, S., Kaiser, D., Holtzberg, F., and Kleinsasser, F., 1988, **Phys. Rev. B 38** 2918.

Regnier, S., Alfred-Duplan, C., Vacquier, G., and Marfaing, J., 1996, **Appl. Supercond. 4** 41.

Sahimi, M.H., 1994, **Applications of Percolation Theory** (Taylor and Francis, London) Chap. 12.

Streitz, F.H., Cieplak, Xiao G., M.Z., Gavrin, A. Bakhshai, A., and Chien, L., 1988, **Appl. Phys. Lett. 52**, 927.

Tozer, S.W., Kleinsasser, A.W., Penney, T., Kaiser, D., and Holtzberg, F., 1987, **Phys. Rev. Lett. 59** 1768.

Xiao, G., Streitz, F.H., Cieplak, M.Z., Bakhshai, A., Gavrin, A., and Chien, L., 1988, **Phys. Rev. B. 38**, 776.

Inst. Phys. Conf. Ser. No 167
Paper presented at Applied Superconductivity, Spain, 14-17 September 1999
© 2000 IOP Publishing Ltd

Investigation of microstructure of ramp-type YBa$_2$Cu$_3$O$_{7-\delta}$ structures

H Sato*, F J G Roesthuis, A H Sonnenberg, A J H M Rijnders, D H A Blank and H Rogalla

Low Temperature Division, Department of Applied Physics, University of Twente, P.O. box 217, 7500 AE, Enschede, The Netherlands

ABSTRACT: We studied the morphology of ramps in YBa$_2$Cu$_3$O$_{7-\delta}$ films and subsequently the barrier layer. The ramps have been fabricated by Ar ion beam etching using standard photoresist masks. SEM and AFM showed the formation of tracks along the slope of the ramp, originating from the irregular shape of the edge of the photoresist mask. The modified reflowed resist process and pre-annealing process show a smoother ramp surface with possible improvements on the junction characteristics.

1.   INTRODUCTION

All high-T_c superconducting Josephson junctions have attracted extensive attention from the point of view of both fundamental studies and electronic applications of high-T_c superconductors. Trilayer type junctions and especially ramp type junctions are very attractive to realize high-T_c superconducting integrated circuits, since Josephson junctions can be fabricated at a desired location and with device properties determined by the requirements of the application.

One of the difficulties to realize high-T_c Josepshon junctions is the very short coherence length ξ (ξ_{ab}~1.5 nm in *ab* plane and ξ_c <1 nm along *c* axis) of the high-T_c superconductor. The crystalline orientation of high-T_C superconducting films should be intentionally controlled to realize all high-T_C Josephson devices, taking account of anisotropic ξ. A superconductor-normal metal-superconductor (SNS) junction is one of the promising Josephson devices, since the junction shows the Josephson effect due to the proximity effect even if the thickness of the barrier is thicker than ξ. One of the other difficulties to demonstrate high-T_c integrated circuits is a multilayer process. The ground plane is required to realize low inductance, and each Josephson junction should be connected by superconducting interconnections.

Recently, several authors have reported about the trilayer junctions using all *a*-axis oriented YBa$_2$Cu$_3$O$_{7-\delta}$/PrBa$_2$Cu$_3$O$_{7-\delta}$/YBa$_2$Cu$_3$O$_{7-\delta}$ (*a*-YBaCuO/*a*-PrBaCuO/*a*-YBaCuO) (Barner 1991, Hashimoto 1992), *a*-YBaCuO/PrGaO$_3$/*a*-YBaCuO (Tsuchiya 1997), *c*-YBaCuO/*c*-PrBaCuO/*c*-YBaCuO (Maruyama 1999, Akoh 1999), and (103)-YBaCuO/(103)-PrBaCuO/(103)-YBaCuO (Sato 1994, Sato 1996) trilayer films. All junction types showed a resistively shunted junction (RSJ)-like *I-V* characteristic and the clear Fraunhofer-like magnetic field dependence of I_C. The *a*-axis oriented junctions have longest ξ normal to substrate. However, fabrication process of high quality *a*-YBaCuO films is very difficult compared to *c*-YBaCuO film and (103)-YBaCuO film, since cracks are easily formed by the thermal expansion mismatch along *c*-axis of YBaCuO film. For *c*-axis oriented junctions, one of the unsolved questions is the direction of Josephson coupling. The junctions have extremely short ξ_c, but the c-axis oriented junctions with 20-nm-thick PrBaCuO barrier were reported RSJ-like IV characteristics. A coupling direction for (103)-oriented junctions

*On leave from Electrotechnical Laboratory, 1-1-4 Umezono, Tsukuba, Ibaraki 305-8568, JAPAN.

is also an open question, since (103) films showed twining and surface covered by (100) facets and (001) facets. Another problem for a-axis and (103) oriented junctions is superconducting wiring films, since a-axis and (103) oriented wire films will show an anisotopic critical current in substrate plane.

The ramp type junctions was also reported by several authors (Hunt 1999, Verhoeven 1996, Moeckly 1997). Since the ramp type junctions have an epitaxial grown insulating film, it is easier to combine multilayer structures compared to trilayer junctions. In fact, a high-T_c flash type A/D converter (Verhoeven 1996) and high-T_c sampler using ramp type junctions have been reported (Hidaka 1999). Very recently, Satoh *et. al.* (1999) reported a small spread of the critical current I_c with 1-σ=8% in the YBaCuO ramp-type junctions, which had the modified interface of YBaCuO film as a junction barrier. It seems a promising Josephson junction to realize small scale circuits. One of the problems for ramp type junctions is the design method. Since trilayer type junctions are most popular for the low-T_c integrated circuits, many design methods have been developed for trilayer type junctions. This suggests that trilayer junctions are preferred for large scale integrated circuits. But at this moment, the ramp type junctions still have advantages to realize the high-T_c integrated circuits using Josepshon junctions for small circuits.

In this paper, we study the morphology of ramp surface in YBaCuO films and subsequently the barrier layer by Scanning Electron Microscopy (SEM) and Atomic Force Microscopy (AFM). Since the ramp surface is prepared ex-situ, the morphology of ramp surface can be observed during the junction fabrication process. The junctions which were fabricated by our standard process showed RSJ-like I-V characteristics at higher temperature, but showed flux-flow-like I-V characteristics at lower temperature. SEM image showed the formation of tracks along the slope of the ramp, originating from the irregular shape of the edge of the photoresist mask. A modified reflowed resist process and pre-annealing process showed a significant improvement on surface morphology of the ramp surface.

2. CHARACTERISTICS OF RAMP TYPE JUNCTIONS

The YBaCuO ramp type junctions were fabricated by our standard fabrication process (Verhoeven, 1996). Fig. 1(a) shows the typical I-V characteristic at 81.2 K. The barrier thickness is 20 nm and the width of the junction is 10 µm. The ramp angle was estimated to be 24°. Both base and counter electrodes showed T_c of 88 K. The junction shows RSJ-like I-V characteristics with I_c of 50µA and junction resistance R_n of 1 Ω. Fig. 1(b) shows the I-V characteristic of minimum I_c with an external field. As seen in the figure, the junction shows almost completely suppression of I_c at 81.2 K. At lower temperature, however, the junction showed flux-flow like I-V characteristics. I_c was estimated 1.4 mA at 40 K and the response of I_c to external field was less than 10 % of total I_c.

According to our previous results, the junctions with 20-nm-thick PrBaCuO barrier showed lower I_c at 40 K and good response to the external field. This suggests that the junction could have pinholes at the barrier. In order to solve this pinhole problem, we tried to observe the surface morphology of ramp type junctions by scanning electron microscope SEM.

Fig. 2 shows a SEM image of ramp surface. The junction has 100-nm-thick YBaCuO base electrode, 100-nm-thick SrTiO$_3$ insulating film, 20-nm-thick of PrBaCuO barrier layer, and 150-nm-thick YBaCuO counter electrode. The width of junction is 15 µm. As shown in the figure, two important features can be seen. (1) The edge of the ramp region is not straight. This can be explained by an imperfection of the lithography mask and/or the lithography process. (2) Many holes can be seen along the ramp region. This can be explained by (a) a lithography error, (b) a recrystallization of the facet on the pre-annealing process at ramp surface, and (c) a growth mechanism at ramped surface.

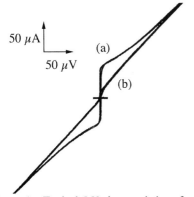

50 μA

50 μV

(a)

(b)

Figure 1. Typical *I-V* characteristics of a 10-µm-width junction with PBCO barrier of 20 nm at 81.2 K (a) without and (b) with external field.

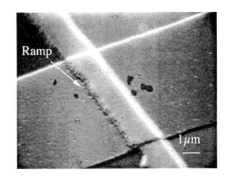

Ramp

1 µm

Figure 2. SEM image of a 15-µm-width ramp covered with 20 nm PrBaCuO barrier layer and 150 nm YBaCuO counter electrode.

3.   RAMP STRUCTURING WITH REFLOWED RESIST

First we checked the resist mask for ramp structuring. Figure 3 shows the SEM images of the resist pattern on the YBaCuO (100 nm)/SrTiO$_3$ (100 nm) bilayer films. All images were observed after 10-nm-thick Au deposition to avoid the charging up effects. As seen in Fig. 3(a), the stripes at the resist edge and the holes at the top of resist edge are can be seen. The stripes would come from interference of incident electrons. The holes could be an imprint of the mask, since a

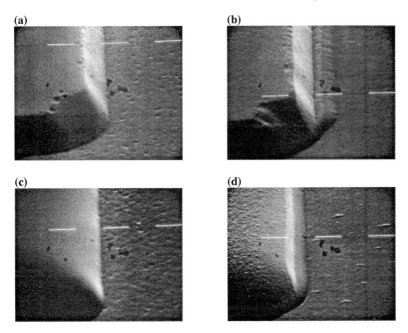

(a)

(b)

(c)

(d)

Figuer 3. SEM images of (a) the edge of normal resist (b) the etched ramp with normal resist, (c) the edge of reflowed resist and (d) the etched ramp with reflowed resist. The dashed line on each image shows a 1-µm-length. Note that the 10-nm-thick Au was deposited on the top of whole sample, in order to avoid charge up phenomena.

contact type mask aligner was used to make resist patterns. Fig. 3(b) shows the SEM image of the structured ramp with resist mask. The Ar ion beam had an incident angle of 35° to the substrate surface. The same sample of Fig. 3(a) was etched. As seen in the Fig. 3(b), we can distinguish each hole at the top edge of the photoresist shown in the Fig 3(a). Figure 3(b) also shows the formation of tracks along the slope of the ramp, originating from the holes at the edge of the photoresist mask. Since the angle of the resist wall was estimated to be 60°, it is concluded that the top edge of the resist is projected to the film sample, and the shape is copied to the ramp surface.

In order to overcome this lithography error, a reflowed resist process was introduced (Hunt 1996). After patterning of the resist, the resist was post baked at 110 °C at hot plate. Post baking makes the resist flow, thus the smooth resist profile was produced since a surface tension of resist. The edge of the reflowed resist on the YBaCuO/SrTiO$_3$ bilayer film is shown in Fig. 3(c). The edge of the resist is not visible and the rounded profile is suggested by the corner of the resist. The etched ramp with reflowed resist is shown in Fig. 3 (d). A smooth ramp surface and a straight boundary of two resist regions can be seen. Two regions are explained by the Ar-ion bombardment. The top part and the bottom part of the resist are corresponding to the Ar-ion bombarded surface and not bombarded surface. It is clear that the boundary of two resist regions is most important part for the ramp structuring.

4.   INFLUENCE OF THE ANNEALING PROCESS

Second, we checked the annealing process before deposition of PrBaCuO barrier layer and YBaCuO counter electrode. Four samples which consist of PLD-deposited YBaCuO (100 nm)/SrTiO$_3$ (100 nm) bilayer films were prepared as film samples. The ramp was formed by conventional Ar-ion etching with reflowed resist. Figure 4 shows AFM images of ramp morphology after different processes. The scan area is 5 μm square on all images. Z divisions are 300 nm/div, except (b) is 400 nm/div. Fig. 4(a) shows a typical AFM image of the ramp after etching and removing photoresist. The Ar ion beam had an incident angle of 30° to the substrate. The ramp angle was 24°. Smooth ramp surface can be seen and the roughness of the ramp region is estimated to be 7 to 8 nm.

In order to reveal the morphology of the ramp surface before depositing the barrier layer, an

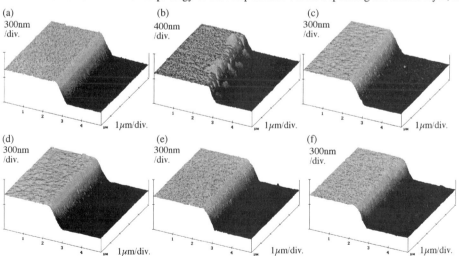

Figure 4.   AFM images of the ramp surface after several processes. (a) After ramp structuring. (b) After annealing at 780 °C/30min. (c) After annealing at 740 °C/30min. (d) After an additional annealing step at 780 °C/30min for the sample shown in (c). (e) After annealing at 740 °C/30min+780 °C/30min. (f) After annealing at 740 °C/30min+780 °C/30min+50-nm-thick PrBaCuO deposition. The scan area is 5 μm square in all images. The z scales are 300 nm/div, except (b) is 400 nm/div.

annealing experiment was carried out. According to the usual fabrication process, the ramp was cleaned by an Ar-ion beam. Then the sample was annealed at 780 °C at 20 Pa of O_2 atmosphere with 30 minutes. The samples are cooled down at 1 atm of O_2 atmosphere. Figure 4(b) shows the AFM images of the ramp after annealing process. As shown in figure, large grains can been seen at the ramp surface. Typical grain height 100 nm. A plausible explanation for the grains is the recrystallization of the amorphized YBaCuO surface layer which originated in the Ar-ion etching process.

In order to avoid the growth of the grains, we tried to reduce annealing temperature, since these grains may enlarge the risk of pinholes in the barrier layer. Figure 4(c) shows the AFM image of the annealed ramp surface at 740 °C. The small particles are visible at the ramp surface, but the overall ramp morphology is quite smooth, which is comparable to Fig. 4 (b). After AFM measurements, the same sample was annealed at 780 °C in 20 Pa of O_2 for 30 minutes. The sample, however, does not show in any visible changes (Fig. 4 (d)). These results suggest the relation between recrystallization and temperature. It is well known that the recrystallization depends on the temperature. At 780 °C, the species at ramp surface can move for long distances. At 740 °C, species can move to reconstruct, but distances should be short. During the recrystallization on 740 °C, the species are thermaly stabilized. Therefore, the ramp surface does not show any visible change after annealing at 780 °C. In order to confirm this phenomenon, a two-step annealing experiment at 740 °C and 780 °C was performed. First, the ramp sample was annealed at 740 °C with 30 minutes. After that, the ramp sample was annealed at 780 °C. During annealing experiment, the gas pressure was fixed to 20 Pa of O_2. The sample was cooled in 1 atm of O_2. Figure 4(e) shows the AFM image of the ramp sample after two step annealing. The ramp surface is still smooth compared to Fig. 4(b).

Other researchers also found grains due to recrystallization at the ramp surface. In order to avoid a rough grains at the ramp surface, Horstman *et. al.* (1998) have used Br-ethanol wet etching to remove the amorphous layer. Another approach used by Inoue *et. al.* (1997) is to supply an active oxygen flux during reheating for the subsequent deposition to prevent rough recrystallization of the surface layer.

Finally, we tried deposition of 50-nm-thick PrBaCuO film on the two-step annealed sample. The PrBaCuO film was in situ fabricated on the annealed ramp sample by PLD method at 780 °C. The O_2 pressure was changed from 20 Pa to 5 Pa just before PrBaCuO film deposition. After deposition, O_2 pressure was increased to 1 Pa, immediately. The process time at 5 Pa was less than 5 minutes. Figure 4 (f) shows AFM images of the ramp surface with 50-nm-thick PrBaCuO. The surface of ramp sample shows a smooth surface with a RMS value of 7.3 nm, which is comparable to the roughness of the etched ramp. We believe the two-step annealing will serve good characteristics of ramp type Josephson junctions.

5. CONCLUSION

We have studied the morphology of ramps in YBaCuO films and subsequently the barrier layer by SEM and AFM. It is comfirmed that the ramp surface structure with conventional resist mask showed the formation of tracks along the slope of the ramp, originating from the irregular shape of the edge of the photoresist mask. Significant improvement has achieved by using reflowed resist, since reflowed resist had a rounded profile. It also confirmed that the annealing temperature before depositon of the barrier layer and counter electrode is very important for suraface morphology of the ramp. Preliminary experiments using the two-step annealing with 50-nm-thick PrBaCuO film showed gentle ramp surface. We believe these processes greatly enhance the characteristics of the ramp-type Josephson junctions.

ACKNOWLEDGEMENT

The authors would like to thank Daniel van Zelst for sample fabication. One of the authors (HS) acknowledges the Science and Technology Agency of Japan for their financial support.

REFERENCES

Akoh H and Sato H, 1999 Extended Abstract of 1999 Internal Workshop on Superconductivity Co-Sponsored by ISTEC and MRS, (Hawai) 39.

Barner J B, Rogers C T, Inam A, Ramesh R, and Bersey S 1991 Appl Phys. Lett. **59**, 742.

Hashimoto T, Sagoi M, Mizutani Y, Yoshida J, and Mizushima K 1992 Appl. Phys. Lett. **60**, 1756.

Hidaka M, Satoh T, Koike M, and Tahara S 1999 Trans of Appl. Supercond, **9** 4081.

Horstmann C, Leinenbach, Engelhardt A , Gerber R , Jia J L, Dittmann R, Memmert U, Hartmann U, Braginski A I 1998 Physica C, **302** 176.

Hunt B D, Forrester M G, Talvacchio J, McCambridge J D and Young R M 1996 Appl. Phys. Lett. **68** 3805.

Hunt B D, Forrester M G, Talvacchio J, Young R M 1999 Trans of Appl. Supercond, **9** 3362.

Inoue S, Nagano T, Hashimoto T, Yoshida J, 1997 Extended Abstract of International Superconductivity Electronics Conference (Berlin) 85.

Maruyama M, Yoshida K, Horibe M, Fujimaki A, Hayakawa H, 1999 Trans of Appl. Supercond, **9** 3456.

Moeckly B H, Char K 1997 Appl Phys Lett. **71** 2527.

Sato H, Akoh H and Takada S 1994 Appl. Phys. Lett. **64** 1286.

Sato H, Nakamura N, Gjøen S R, Akoh H, 1996 Jpn. J. Appl. Phys. **35** L1411.

Tsuchiya R, Kawasaki M, Kubota H, Nishino J, Sato H, Akoh H, Koinuma H 1997 Appl. Phys. Lett. **71** 1570.

Verhoeven M 1996 HIGH-Tc SUPERCONDUCTING RAMP-TYPE JUNCTIONS (Ph.D Thesis).

Inst. Phys. Conf. Ser. No 167
Paper presented at Applied Superconductivity, Spain, 14-17 September 1999
© 2000 IOP Publishing Ltd

a-axis 45° tilt and twist YBa₂Cu₃O artificial grain boundaries: new perspectives of the biepitaxial technique for the realisation of Josephson junctions

F Tafuri

Dip. Ingegneria Informazione, Seconda Università di Napoli, Via Roma 29, 81031 Aversa (CE) and INFM Dip. Scienze Fisiche Università di Napoli Federico II, P.le Tecchio 80, 80125 Napoli, Italy

F Miletto Granozio, F Carillo, F Lombardi and U Scotti di Uccio

INFM-Dip. Scienze Fisiche Università di Napoli Federico II, P.le Tecchio 80, 80125 Napoli, Italy

K Verbist, O Lebedeev and G Van Tendeloo

EMAT, University of Antwerp (RUCA), Groenenborgerlaan 171, B-2020 Antwerp, Belgium

ABSTRACT: YBa₂Cu₃O₇₋δ (YBCO) artificial grain boundary Josephson junctions have been fabricated, employing a recently implemented biepitaxial technique. The grain boundaries can be obtained by controlling the orientation of the MgO seed layer and are characterised by a misalignment of the c-axes (45° c-axis tilt or 45° c-axis twist). High resolution electron microscopy showed the presence of perfect basal plane faced boundaries in the cross sections of tilt boundaries. The phenomenology of the Josephson effect gives evidence of profound differences with asymmetric in-plane 45° (001) tilt bicrystal and biepitaxial grain boundary Josephson junctions. An explanation of these differences could be given in terms of the d-wave nature of the order parameter. The analysis indicates that the biepitaxial technique is suitable for significant improvements and is promising for applications and fundamental studies.

1. INTRODUCTION

YBa₂Cu₃O₇₋δ (YBCO) junctions based on artificial grain boundaries (AGBs), that are characterised by a 45° relative misalignment of the c-axes and are realised by using the biepitaxial technique (Char 1991), have been investigated (Tafuri 1999). The AGB is obtained at the interface between a (103) film deposited over a (110) SrTiO₃ substrate and a c-axis film deposited over a MgO seed layer. Junctions associated with this AGB exhibit promising Josephson properties. Two different AGBs characterised by a 45° tilt or twist of the c-axis across the GB are considered (Fig. 1). We will refer to them in the following as '*tilt*' or '*twist*' AGBs respectively.

The aim of the present work is to investigate some almost unexplored configurations of the wide class of GBs, that are characterised by a misalignment of the c-axes, and to look for a direct correlation between the grain boundary microstructure and superconducting properties. This is of great interest for the understanding of the nature of transport in superconducting GBs. Finally, we will show that the technology we implemented provides a Josephson structure that is able to offer some advantages, in terms of applications, over junctions obtained by other techniques. Some examples are the possibility of complex circuit fabrication and high values of the I$_C$R$_N$ product (I$_C$ being the critical current and R$_N$ the normal state resistance). The achievement of higher values of I$_C$R$_N$ with respect to traditional biepitaxials (Char 1991), or a Fraunhofer-like dependence of I$_C$ on the magnetic field can

148

be understood within the framework of the d-wave order parameter symmetry (Tafuri 1999). Furthermore, the structure considered in this work offers a unique possibility to modify the structure and transport properties of the AGB by controlling the orientation of the interface, which is defined by the fabrication process.

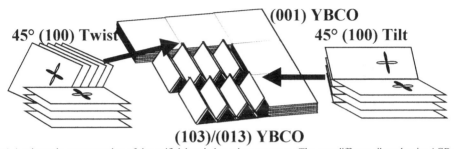

(001) YBCO

45° (100) Twist

45° (100) Tilt

(103)/(013) YBCO

Fig.1 A schematic representation of the artificial grain boundary structure. The two different tilt and twist AGBs are evidenced. The d-wave component of the order parameter expected in our junction configurations is also reported.

2. EXPERIMENTAL

2.1 Fabrication procedure and TEM set-up

The fabrication process involves the deposition of MgO and YBCO thin films and ion-milling procedures (Di Chiara 1996, Tafuri 1999). (110) oriented MgO thin films are deposited by RF magnetron sputtering from a stoichiometric oxide target on (110) SrTiO$_3$ (STO) substrates. A thin MgO seed layer (20 nm) was deposited at a substrate temperature of 600° C. A standard lithographic procedure, employing a Nb mask, ion milling and reactive ion etching, was used to pattern the seed layer. YBa$_2$Cu$_3$O$_{7-x}$ films with a thickness of 120 nm were deposited by inverted cylindrical magnetron sputtering in an Ar/O$_2$ atmosphere (P$_{O2}$=P$_{Ar}$=50 Pa) at a temperature of 780°C. Cross section (CS) as well as plan view (PV) samples for electron microscopy have been prepared by standard mechanical polishing and ion milling. High resolution electron microscopy (HREM) observation were performed in a Jeol 4000 EX microscope with a point resolution of 0.17 nm.

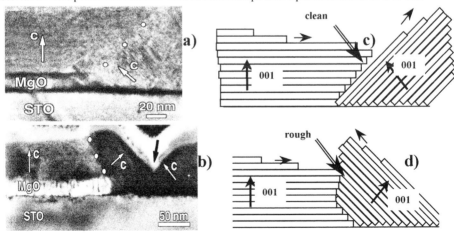

Fig.2 High magnification cross section TEM image recorded along the (001)$_{STO}$ direction of: a) 45° tilt AGB with a basal plane of the (103) film as interface plane; b) 45° tilt AGB with a rough irregular interface; in c) and d) the schematic representations corresponding to the images a) and b) are reported respectively.

2.2 Description of the artificial grain boundaries

The cross section (CS) HREM investigations confirmed the expected nature of the AGBs both for the twist and the tilt cases shown in Fig. 1 (Verbist 1999, Tafuri 1999). In Fig 2 two examples are given. In Fig. 2a, a perfect faceted interface is shown, separating the (001) basal plane (BP) of the (103) oriented film and the (103) plane of the (001) film. In Fig. 2b, the *tilt* AGB exhibits an irregularly stepped interface. This TEM analysis has been performed on a microbridge previously characterised in terms of its transport properties and the sample has been prepared thorough the focused ion beam thinning technique (Verbist 1999). The orientation of the (103) domains at the interface is different in Figs. 2a and 2b, as shown by the arrows indicating the $[001]_{YBCO}$ direction and in the schematic representation in Fig.2c and 2d. The two configurations originate from the different growth direction of the (103) domains with respect to (001) domains. They are different not only because of the roughness which influences the uniformity of the junction but also for the microstructure, which could give rise to different transport properties. The almost perfect interface of Fig. 2a might be suitable for the fabrication of homogeneous Josephson junctions and could be selected by using vicinal (110) substrates (in particular with the normal tilted off [110] axis towards the [010] direction). A (103) growth with a single orientation has been confirmed by X-rays analyses (Scotti 1999) and by preliminary TEM investigations on grain boundary interfaces.

3. TRANSPORT PROPERTIES AND THE JOSEPHSON EFFECT

In Fig. 3, typical current vs voltage (I-V) characteristics of a 45° (100) tilt GB junction are reported for different temperatures. The characteristics present a shape that is typical of the Resistively Shunted Junction (RSJ) model with no excess current. The critical current and the normal state resistance are $I_C = 80$ μm and $R_N = 20$ Ω at T = 4.2 K respectively. These provide a nominal critical current density $J_C \approx 10^4$ A/cm^2, a nominal normal state specific conductance σ_N about 6.3 (μΩ cm^2)$^{-1}$ and $I_C R_N \approx 1.6$ at T = 4.2 K. These values fall in the typical ranges of the tilt case $0.5 \div 10 \times 10^3$ A/cm^2 and $1 \div 10$ (μΩcm^2)$^{-1}$ respectively. *Twist* AGB junctions typically are characterised by higher values of J_C in the range $0.1 \div 4.0 \times 10^5$ A/cm^2 and of σ_N in the range $20 \div 120$ (μΩcm^2)$^{-1}$ (at T = 4.2 K). Deviations from the RSJ model are more remarkable as the critical current density increases and therefore typically for the twist case. In both cases the maximum working temperature T_C of the devices is typically higher than 77 K. The $I_C R_N$ values are high in both cases, of the order of 1-2 mV at T = 4.2 K and $50 \div 100$ μV at T = 77 K respectively. They are larger for the corresponding J_C values than those provided by conventional biepitaxials and are of the same order of magnitude as in

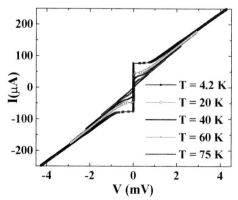

Fig. 3 I-V characteristics of a 45° (100) tilt GB junction at different temperatures.

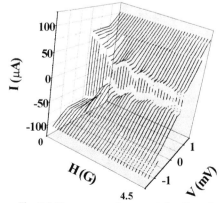

Fig. 4 I-V curves are shown as a function of an externally applied magnetic field at T = 4.2 K.

bicrystal and step edge junctions (Tafuri 1999).
The Josephson nature of the junctions has been verified by applying an external magnetic field H. A typical $I_C(H)$ pattern measured at T = 4.2 K is shown in Fig. 4 where I-V curves are plotted as a function of H. A Fraunhofer-like dependence of I_C on H is evident. The pattern is symmetric around zero magnetic field, and in all samples the absolute maximum of I_C occurs at zero magnetic field (Tafuri 1999). Such a dependence of I_C is quite different from those of asymmetric in-plane 45° tilt bicrystal and biepitaxial junctions, in which the absolute I_C maximum is observed for H $\neq 0$. On the other hand, the two behaviours reflect different GB microstructures and their influence on transport properties. An explanation could be found within the framework of a d-wave symmetry order parameter (Tsuei 1994, Kirtley 1994). As a matter of fact, the order parameter orientations should not produce an additional π phase shift along our junction in contrast with the 45° tilt bicrystal junctions (Hilgenkamp 1996, Tafuri 1999). A schematic representation of the order parameter orientations in our junction is also shown in Fig. 1. On this basis, we can also understand why the $I_C R_N$ values of our junctions are higher than those reported for traditional biepitaxials. We can argue that no unquantised magnetic flux would be expected in our junctions and low frequency $1/f$ noise could be lower than in the asymmetric in plane 45° (001) tilt bicrystal and biepitaxial junctions. This has been confirmed by different measurements, which are in progress. Other details on the barrier nature of 45° (100) tilt and twist AGBs can be also found elsewhere (Tafuri 1998, 1999).

4.   CONCLUSIONS

The Josephson properties   and the microstructure of grain boundary junctions that characterised by a 45° relative misalignment of the c-axes (45° *tilt* and *twist*) have been investigated through transport and HREM measurements respectively.   The observed phenomenology gives evidence of profound differences with asymmetric in-plane 45° (001) tilt bicrystal and biepitaxial grain boundary Josephson junctions. An explanation of these differences could be given in terms of the d-wave nature of the order parameter. A low frequency $1/f$ noise lower than in the asymmetric in plane 45° (001) tilt bicrystal and biepitaxial junctions, that would be expected for this type of GB within the framework of the d-wave, has been confirmed by preliminary measurements The analysis indicates that the biepitaxial technique is suitable for significant improvements and is promising for applications and fundamental studies. We have shown the feasibility of junctions based on atomically clean interfaces that can be reproducibly obtained by realising our type of junction on vicinal substrates

This work has been partially supported by the project PRA-INFM "HTS Devices" .

REFERENCES

Char K, Colclough M S, Garrison S M, Newman N and Zaharchuk 1991 Appl. Phys. Lett. **59**, 733
Di Chiara A, Lombardi F, Miletto Granozio F, Pepe G, Scotti di Uccio, Tafuri F and Valentino M 1996 J. Supercond. **9**, 245
Hilgenkamp H, Mannhart J and Mayer B 1996 Phys. Rev. B **53**, 14586
Kirtley J R, Tsuei C C, Sun J Z , Chi C C, Yu-Jahnes L S, Gupta A, Rupp M and Ketchen M B 1995 Nature **373**, 225
Scotti di Uccio U, Lombardi F and Miletto Granozio F 1999 Physica C
Tafuri F, Miletto Granozio F, Carillo F, Di Chiara A, Verbist K and Van Tendeloo G 1999 Phys. Rev. B **59**, 11523
Tafuri F, Nadgorny B, Shokhor S, Gurvitch M, Lombardi F, Carillo F, Di Chiara A and Sarnelli E 1998 Phys. Rev. B **57**, R14076
Tsuei C C, Kirtley J R, Chi C C, Yu-Jahnes L S, Gupta A, Shaw T, Sun J Z and Ketchen M B 1994 Phys. Rev. Lett. **73**, 1994
Verbist K, Lebedev O, Van Tendeloo G, Tafuri F, Miletto Granozio F and Di Chiara A 1999 Appl. Phys. Lett. **74**, 1024

Inst. Phys. Conf. Ser. No 167
Paper presented at Applied Superconductivity, Spain, 14-17 September 1999
© 2000 IOP Publishing Ltd

Interval deposition: Growth manipulation for use in fabrication of planar $REBa_2Cu_3O_{7-\delta}$ junctions

A J H M Rijnders, G Koster, D H A Blank and H Rogalla

Univ. of Twente, Dept. of Applied Physics, Low Temperature Division, The Netherlands.

ABSTRACT: Planar $REBa_2Cu_3O_{7-\delta}$ junctions are, due to the geometry, very suitable for use in integrated superconducting electronics and devices. A prerequisite are smooth barrier interfaces to avoid mixing of the anisotropic properties of the high-T_c superconductors. We present a growth method, based on a periodic sequence: fast deposition of the amount of material needed to complete one monolayer followed by an interval in which no deposition takes place and the layer can reorganize. Starting with an atomically smooth substrate, we studied the growth, using the interval deposition technique, of SrO, $La_{0.5}Sr_{0.5}CoO_3$ and $DyBa_2Cu_3O_{7-\delta}$, with reflection high-energy electron diffraction and scanning probe microscopy.

1. INTRODUCTION

One of the key-goals in High Temperature Superconducting (HTS) electronics is the development of (tunnel) junctions. Due to the anisotropy of the coherence length in cuprates, i.e., 10-20 Å in the ab-direction and ~1-2 Å in the c-axis direction many groups have directed their research towards the development of junctions in the ab-direction. Different kinds of junctions, including grain boundary, ramp-type, step edge and a-axis sandwich-type have been realized by several groups.

In complex HTS electronic circuits and devices which include a large number of Josephson junctions a planar technology is desirable. Despite research efforts on a-axis junctions and planar microbridges, they have not (yet) been proven to be viable for complex circuits. Planar c-axis junctions with an artificial barrier are expected to yield better results, provided that the junction interfaces can be controlled on an atomic level. The planar geometry of the c-axis junctions combines the *in situ* deposition of both electrodes and barrier with the relative easy integration in electronics.

As already mentioned above, it is necessary to grow smooth layers to avoid the mixing of properties. The roughness of the surfaces and interfaces must be small compared to the barrier thickness. Moreover, the superconducting properties are easily degraded at the interfaces due to the short coherence length in the c-axis direction.

Occasionally, the deposition conditions such as the substrate temperature and ambient gas pressure (oxygen in the case of oxide materials) can be optimised for true two-dimensional (2D) growth, e.g., homo-epitaxy on $SrTiO_3$ (001). True 2D reflection high-energy electron diffraction (RHEED) specular intensity oscillations are observed depositing $SrTiO_3$ with pulsed laser deposition (PLD) at a temperature of 850 °C and an oxygen pressure of 0.04 mbar (Koster 1998). The relatively high temperature in combination with a low oxygen pressure enhances the mobility of the ad-atoms on the surface and, therefore, the probability of nucleation on top of a 2D island is minimized. The as-deposited ad-atoms can migrate to the step edges of 2D islands and nucleation only takes place on fully completed layers.

152

In general, during deposition of different kinds of materials, i.e., metals, semiconductors and insulators, by different deposition techniques, a roughening of the surface is observed. In case of 2D nucleation, determined by the supersaturation, limited interlayer mass transport results in nucleation on top of 2D islands before completion of a unit cell layer. Still, one can speak of a 2D growth mode. However, nucleation and incorporation of ad-atoms at step edges is proceeding on an increasing number of unit cell levels, which is exhibited by damping of the RHEED intensity oscillations.

Several groups have investigated the possibility to apply a form of growth manipulation to promote interlayer mass transport (Rosenfeld 1993). They suggest applying two different temperatures, two different growth rates or periodical ion bombardment to increase the number of nucleation sites and thus decrease the average island size. This will enhance the transport of material on an island to a lower lying level. Usually, for epitaxy of complex oxide materials, the regime of temperatures and pressures is limited by the stability of the desired phases, e.g., $YBa_2Cu_3O_7$ can only be grown in a specific temperature and pressure regime (Hammond 1989). At low temperatures, a-axis oriented films are formed whereas at high temperatures the material decomposes. Periodic ion bombardment is very difficult to realise in view of the stoichiometry of oxide materials. Growth rate manipulation to impose layer-by-layer growth could be a possibility to overcome this problem. In case of PLD, a typical value for the deposition rate within one pulse is of the order of 10 μm/s (Geohegan 1995). Therefore, a high supersaturation is expected when the plume is on and thus the number of very small 2D nuclei can be very high. Subsequently, when the plume is off, larger islands are formed through recrystallisation, exhibited by the typical relaxation of the RHEED intensity of the specular spot during PLD (Karl 1992). Since small islands promote interlayer mass transport, one can utilize the high supersaturation achieved by PLD by maintaining it for a longer time interval and suppress subsequent coarsening.

Accordingly, to circumvent premature nucleation due to the limited mobility of the ad-atoms at a given pressure and temperature, causing a multi-level 2D growth mode, we introduce the possibility of interval deposition. Exactly one unit cell layer is deposited in a very short time interval, i.e., of the order of the characteristic relaxation times, typically 0.5 s (Koster 1999), followed by a much longer interval during which the deposited material can rearrange. During the short deposition intervals, only small islands will be formed due to the high super saturation typical for PLD. The probability of nucleation on the islands increases with their average radius (Rosenfeld 1993) and is, therefore, small in case of fast deposition. The total amount of pulses needed to complete one unit-cell layer has to be as high as possible, to minimize the error introduced by the fact that only an integer number of pulses can be given. Both, a high deposition rate and sufficiently accurate deposition of one unit-cell layer can be obtained by PLD using a high laser-pulse frequency.

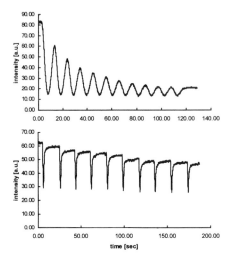

2.   EXPERIMENTAL

For this study, thin films are deposited using PLD in combination with high pressure RHEED (Rijnders 1997). For deposition of SrO a single crystalline target was used, whereas sintered pellets were used for deposition of $REBa_2Cu_3O_{7-\delta}$ and $La_{1-x}Sr_xCoO_3$. The incident angle of the 20 keV electrons was set at 1°, while the intensity of the specular reflection was recorded with a CCD camera.

Fig. 1: Intensity of the specular spot during growth of $La_{0.6}Sr_{0.4}CoO_3$ at 2 Hz (a) and with interval deposition, 20 pulses at 10 Hz for every monolayer (b).

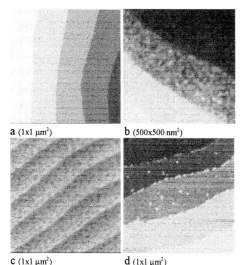

a (1x1 µm²)   b (500x500 nm²)

c (1x1 µm²)   d (1x1 µm²)

Fig. 2: The surface morphology of SrTiO₃ after treatment (a), and subsequent deposition of SrO using standard PLD, 26 pulses at 1 Hz (b), after deposition of SrO using interval deposition with 40 pulses at 10 Hz (c) and 50 pulses at 10 Hz (d).

SrTiO₃ substrates were specially prepared to obtain a single terminated surface with only unit cell steps (Fig. 2a). The substrates were pretreated in H₂O before etching in an NH₄-HF solution followed by an anneal step at 950 °C in 1 bar of oxygen. Using this procedure (Koster 1998) atomically smooth substrates, with TiO₂ as the terminating surface layer, are obtained. The miscut angle of the substrates used in this study is below 0.2°.

3. RESULTS AND DISCUSSION

A possible candidate for barrier material in a superconductor-normal metal-superconductor (SNS) junction is La$_{0.5}$Sr$_{0.5}$CoO₃, due to the small lattice mismatch with REBa₂Cu₃O$_{7-\delta}$. In Fig. 1a the RHEED intensity is given during the growth of La$_{0.5}$Sr$_{0.5}$CoO₃ at 20 Pa of oxygen and a substrate temperature of 650 °C with a continuous pulse frequency of 2 Hz. The surface is transiting from a single level system to a multi level system, as indicated by the damping of RHEED intensity oscillations.

Fig. 1b shows the RHEED intensity during 10 cycles of deposition (at 10 Hz) followed by a time interval of no deposition, using the same oxygen pressure and substrate temperature, according to the new approach. In this case the number of pulses needed per unit cell layer was estimated to be about 20. The decay of the intensity after each unit cell layer is significantly lower compared to the situation in Fig. 1a. The recovery of the intensity after each deposition interval is fast when exactly one unit cell layer is deposited. Note that, besides nucleation on the next level, the decrease in intensity also can be ascribed to the fact that only an integer number of pulses can be given to complete a unit cell layer. A slightly lower or higher coverage causes a longer recovery time. This situation will deteriorate with every subsequent unit cell layer, also indicated by increasing relaxation times.

Deposition of SrO using PLD with a pulse repetition rate of 1 Hz leads to a multi level surface, as can be seen in Fig. 2b. Before completion of a monolayer, nucleation takes place on a 2D island. To overcome this problem interval deposition is applied. Fig. 2c and 2d shows the atomic force micrograph of a SrO monolayer applying 40 and 50 pulses, respectively, at 10 Hz. In these cases the deposition temperature is set to 500 °C. Subsequently, a one-hour *in situ* anneal step at 850 °C is applied. Fig. 2c shows a non-completed monolayer, whereas in Fig. 2d small islands are visible.

The diffraction pattern (Fig. 3a) after deposition (pO₂ 15 Pa, substrate temperature 780 °C and a laser fluence on the target of 1.3 J/cm²), using standard PLD at 1 Hz, of approximately 4 unit cells on a treated SrTiO₃ substrate, shows streaks, indicating a roughened surface. Also clear three-dimensional (3D) spots are visible. These

a   b

Fig 3: RHEED pattern after deposition of 4 unit-cell layers of DyBa₂Cu₃O$_{7-\delta}$ on a treated SrTiO₃ substrate (a), and on an SrO terminated SrTiO₃ substrate (b).

154

spots correspond with a real space distance of 4.27 Å, which can be associated with Cu_2O crystallites. The terminating layer of $REBa_2Cu_3O_{7-\delta}$ is expected to be a BaO layer (Streiffer 1991). The starting layer on a TiO_2 terminated $SrTiO_3$ substrate is also a BaO layer. As a consequence, during deposition of the first unit-cell layer, one CuO_x layer is not incorporated when stoichiometric deposition takes place. When a SrO terminated $SrTiO_3$ substrate is used, only streaks are observed (see Fig. 3b). Now, the interface layer will be a CuO_x layer. The perovskite stacking sequence is preserved. In this case, the formation of Cu_2O crystallites is prevented, as indicated by RHEED.

Nevertheless, streaks in the RHEED pattern indicate a slightly roughened surface. The use of interval deposition is hampered by the number of pulsed needed for completion of one unit-cell layer. In the case of $DyBa_2Cu_3O_{7-\delta}$, using the deposition conditions mentioned above, approximately 12 pulses are needed. To minimize the error in the number of pulses, alternative settings (pO_2 50 Pa, substrate temperature 850 °C and a laser fluence of 8 J/cm^2) are used. Using these settings the number for completion of one unit-cell layer is approximately 40. Clear transmission spots are visible in the RHEED pattern (Fig. 4a) taken after deposition of 50 nm using a laser repetition rate of 1 Hz. The position of these spots corresponds with the lattice parameter of $DyBa_2Cu_3O_{7-\delta}$. Clearly, the surface has become rough. Using the interval deposition technique, only streaks are visible, i.e., no transmission spots, indicating a smoother surface.

4.   CONCLUSIONS

Pulsed Laser *interval* Deposition is a suitable technique to enhance layer-by-layer growth as shown by the heteroepitaxial growth of $La_{0.5}Sr_{0.5}CoO_3$ and SrO on TiO_2-terminated $SrTiO_3$. It enhances the layer-by-layer growth of $DyBa_2Cu_3O_{7-\delta}$ resulting in smooth surfaces as shown by RHEED. Here, on SrO-terminated $SrTiO_3$, the formation of CuO_x precipitates is prevented, as indicated by RHEED. Especially Pulsed Laser Deposition is appropriate for use of interval deposition because of the high deposition rate during every pulse. In combination with high pulse repetition rates a unit-cell layer can be deposited in a very short interval.

REFERENCES

Geohegan D B and Puretzky A A 1995, Appl. Phys. Lett. **67,** 197
Hammond R H and Bormann R 1989, Physica C **162-164,** 703
Karl H and Stritzker B 1992, Phys. Rev. Lett. **69,** 2939
Koster G, Rijnders A J H M, Blank D H A, Rogalla H 1998, Mater. Res. Soc. Symp. Proceedings **526,** 33
Koster G, Kropman B L , Rijnders A J H M , Blank D H A and Rogalla H 1998, Appl. Phys. Lett. **73,** 2920
Rijnders A J H M, Koster G, Blank D H A, and Rogalla H 1997, Appl. Phys. Lett **70** 14
Rosenfeld G, Servaty R, Teichert C, Poelsema B and Comsa G 1993, Phys. Rev. Lett. **71,** 895
Streiffer S K, Lairson B M, Eom C B, Clemens B M, Bravman J C and Geballe T H, 1991, Phys. Rev. B **43,** 13007

 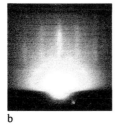

a   b

Fig 4: RHEED pattern after deposition of a 50 nm thick film of $DyBa_2Cu_3O_{7-\delta}$ using standard PLD with a repetition rate of 1 Hz (a), and using interval deposition, 40 pulses at 10 Hz for every unit-cell layer (b).

Inst. Phys. Conf. Ser. No 167
Paper presented at Applied Superconductivity, Spain, 14-17 September 1999
© 2000 IOP Publishing Ltd

Microwave responses of $Bi_2Sr_2CaCu_2O_{8+y}$ intrinsic Josephson junctions

K. Nakajima[1,2]**, H. B. Wang**[1,2]**, Y. Aruga**[1]**, T. Tachiki**[1]**, Y. Mizugaki**[1,2]**, J. Chen**[1,2]**,
T. Yamashita**[2,3] **and P. H. Wu**[4]

[1]Research Institute of Electrical Communication, Tohoku University, Sendai 980-8577, Japan
[2]CREST, Japan Science & Technology Corporation, Japan
[3]New Industry Creation Hatchery Centre, Tohoku University, Sendai, Japan
[4]Department of Electronic Science & Engineering, University of Nanjing, Nanjing 210093, China

Abstract: Mesas with various *a-b* plane sizes have been fabricated by Ar ion milling on $Bi_2Sr_2CaCu_2O_{8+y}$ (BSCCO) single crystals. Current steps at even voltage intervals are induced when intrinsic Josephson junctions (IJJs) are subject to microwave irradiation. Higher order steps are produced with increasing microwave power while neither the power nor temperature affects the step intervals. The intervals are much large, and even greater than those of Shapiro steps assuming all junctions are phase-locked, indicating that electromagnetic waves with very higher frequencies might have been excited in IJJs.

1. INTRODUCTION

Intrinsic Josephson junctions (IJJs) (Kleiner et al 1992) in high-T_C superconductors are important candidates for applications at high frequencies up to terahertz region. So far many interesting experiments have been carried out to investigate the microwave responses and emissions in $Bi_2Sr_2CaCu_2O_{8+y}$ (BSCCO) single crystals or films. Irie et al (1997) and Prusseit et al (1997) observed microwave-induced steps on current-voltage characteristics and explained them in terms of phase-locking fluxon motion in the IJJs. Non-Josephson emission from intrinsic BSCCO junctions reported by Hechtfischer et al (1997) was regarded as Cherenkov radiation. Wang et al (1999) also reported the observation on the microwave responses of stacked BSCCO intrinsic junctions at frequencies around 7 GHz. We observed a series of remarkable microwave-induced current steps on the *I-V* curves in a 10 µm×10 µm sample, and the voltage spacing between neighboring steps was constant at about 4 mV and microwave power-independent. Some preliminary explanations are discussed here, in addition, experimental results on a 5 µm× 5 µm sample are given.

2. SAMPLE FABRICATION

The process for making mesa structure has already reported elsewhere in detail (Wang et al 1999). In brief, conventional photolithography and Ar ion milling were used to define the sizes of a mesa structure in the *a-b* plane and along c-axis. SiO layer was evaporated into the sample to be an insulator. Measuring 10 µm× 10 µm and 5 µm× 5 µm in the *a-b* plane for sample A and B, the mesas are 600 Å high along *c*-axis respectively, implying that there are about 40 intrinsic junctions involved in the two stacks.

3. EXPERIMENTAL RESULTS AND DISCUSSIONS

For sample A we measured the *I-V* curves having about 40 resistive branches, in good agreement with the number estimated from the stack height of 600 Å. Owing to the very small area in the *a-b* plane, heating effect was not serious, resulting in a big voltage jump of about 30 mV for each junction. The normal resistance was 1 kΩ for 40 junctions, say, the normal resistance per each junction $R= 25$ Ω. Together with a critical current of $I_C= 150$ µA, this yielded a characteristic voltage I_CR of about 3.75 mV, or a characteristic frequency of 1.9 THz.

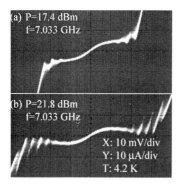
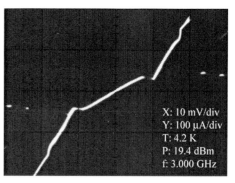

Fig. 1 Typical I-V curves of a 10 μm×10 μm mesa under different microwave power.

Fig. 2 Typical I-V curves of a 5 μm×5 μm mesa under microwave irradiation.

In an attempt to look for finer structures under microwave radiation, we measured the *I-V* curves in low voltage region and showed them in Fig. 1 for different power levels. At a certain power level some interesting structures appeared on the I-V curve (Fig. 1(a)) which, at higher microwave power, could be clearly identified as steps. Further increase of the power resulted in more steps visible over wider voltage range, as shown in Fig. 1 (b). We noticed that the step structures depended on the operating temperature quite remarkably. Temperature rise could easily quench the step structures. Supported by more measurements at different frequencies nearby, we observed that the current steps were stable only within a relatively narrow frequency range. Naturally, one will ask whether there will be step structures again in other frequency ranges if we further increase the frequency. Our preliminary experiments did produce positive evidence.

We notice that our sample sizes (10 μm×10 μm) are much larger than the Josephson penetration depth which is ~ 1 μm. Thus Josephson vortices can be easily generated and driven across the mesa. Collective resonance might have been excited by the microwaves. As the intervals are not changeable, possibly we are encountering geometric resonance of the junction cavity. With the 10 μm sample, to fit our experimental data of 4 mV voltage spacing between neighbouring steps, we need a velocity of $\bar{c} = 1 \times 10^6$ m/s assuming 40 stack junctions in series. It has been noted by Kleiner et al (1994) and Lee et al (1997) that when the common superconducting layer in a stack is much thinner than the London penetration depth, one may find a mode possible which corresponds to a rectangular lattice of vortices moving at a velocity close to that of light in the medium. If we take ε=20 as the dielectric constant of the medium, the velocity of the highest speed mode is about 2.6×10 m/s which is not far from the value we need ($\bar{c} = 1 \times 10^6$ m/s). Thus such a rectangular lattice of moving vortices might be the origin of the step structures observed. To further test this explanation, as shown in Fig. 2, microwave-power-independent current steps at higher voltage were also observed for sample B with smaller size in the *a-b* plane. Obviously, more work should be done on the dependence of the step structures on junction sizes.

REFERENCES

Hechtfischer G, Kleiner R, Ustinov A V and Müller P 1997 Phys. Rev. Letts. **79**, 1365.

Irie A, and Oya G 1997 Physica C **293**, 249.

Kleiner R, Steinmeyer F, Kunkel G, and Müller P 1992 Phys. Rev. Lett. **68**, 2394.

Kleiner R 1994 Phys. Rev. B **50**, 6919.

Lee J U, Guptasarma P, Hornbaker D, El-Kortas A, Hinks D and Gray K E 1997 Appl. Phys. Lett. **71**, 1412.

Prusseit W, Rapp M, Hirata K and Mochiku T 1997 Physica C **293**, 25.

Wang H B, Aruga Y, Tachiki T, Mizugaki Y, Chen J, Nakajima K, Yamashita T and Wu P H 1999 Appl. Phys. Lett. **74**, 3694.

Inst. Phys. Conf. Ser. No 167
Paper presented at Applied Superconductivity, Spain, 14-17 September 1999
© 2000 IOP Publishing Ltd

Experiments on Energy Level Quantization in Underdamped Josephson Junctions

B. Ruggiero, C. Granata, E. Esposito, M. Russo, L. Serio, and P. Silvestrini[a]

Istituto di Cibernetica del CNR, I-80072, Arco Felice (Napoli), Italy
Macroscopic Quantum Coherence Group, Istituto Nazionale Fisica Nucleare,
I-80126, Napoli, Italy

ABSTRACT: We propose a technique to control the effective dissipation by integrating molybdenum resistors with Josephson junctions. The meander shaped resistive lines provide an "in-situ" filtering stage, which strongly reduce the external noise coming into the junction. The extremely low value of dissipation obtained is encouraging in view of new experiments of tunneling between energy levels as Macroscopic Quantum Coherence, basic ingredient of quantum computing. We present experiments on the evidence of energy level quantization at temperatures above the classical-quantum crossover one in underdamped Josephson junctions.

1. INTRODUCTION

Macroscopic quantum effects in Josephson systems have attracted great interest in the scientific community both for the physics involved and in view of applications (Silvestrini 1999). Very recently, quantum devices based on the Josephson effect with a very low dissipation level have been proposed as solid-state qubits under control (Averin 1999). These ideas are based on the phenomenon of a coherent superposition of two distinct quantum states (Leggett 1992, Nakamura et al 1999). Up to now some other macroscopic quantum effects have been observed (Schwartz et al 1985, Martinis et al 1987, Kuzmin and Haviland 1991, Rouse et al 1995, Silvestrini et al 1997, Ruggiero et al 1999a). In this context here we present experiments on Josephson junctions with an improved insulation between the sample and environment showing evidence of energy level quantization (ELQ) at temperature above the classical-quantum crossover temperature $T_0 = h\omega_J/2\pi k_B$, where ω_J is the bias dependent plasma frequency (Martinis et al 1987).

2. JOSEPHSON JUNCTIONS WITH "IN-SITU" FILTERING STAGE

We propose a technique to control the effective dissipation by integrating in a compact way molybdenum resistors wiring with the junction. The resistors are made of 20 μm-wide thin films sputtered during the fabrication process, and located as close as possible to the junction. The meander shaped resistive lines provide an "in-situ filtering stage", which strongly reduces the external noise coming into the $Nb/AlO_X/Nb$ junction. As result, the

158

junction circuit up to high frequency, and the effective dissipation is determined by the value of the molybdenum resistors(R_M=6.5 kΩ). With respect to our previous noise reduction configurations (Silvestrini et al 1997, Ruggiero et al 1998, Ruggiero et al 1999a), the present one is compact and avoids the effect of stray capacitance. This renders the proposed technique useful for low noise applications of Josephson junctions, where highly integrated filters are recommended, as well as in view of new experiments in the quantum limit, where a very low dissipation is required. Details on the device fabrication are reported elsewhere (Ruggiero et al 1999b). A photo of the sample is shown in Fig.1a. To determine the noise condition as well as the dissipation level for our system, we have measured the switching current distribution as a function of the bias current, P(I), in the temperature range 1.2 K ÷ 4.2 K. The measurements were performed by standard time of flight technique (Ruggiero et al 1998) measuring experimental histograms equivalent to the switching current distributions. The results for the two different configurations including (configuration A) or not (configuration B) the meander filtering stages, are reported in Fig.1b. We report the measured distribution width σ= $(<I^2>-<I>^2)^{1/2}$ as function of $T^{2/3}$, and we stress that data were split into two different sets for the two different configurations. The fitting value of the effective resistance at low temperatures was different: R=15 kΩ (configuration A) while R=1 kΩ (configuration B). In fact, the molybdenum meander resistors in the current and voltage lines decouple efficaciously the system from the external circuit and they determine the effective dissipation at low temperature (Ruggiero et al 1999b). The experimental study of the effective dissipation relevant in the thermally-activated supercurrent decay of meander-type filtered Nb/AlO$_x$/Nb Josephson junction, shows a substantial improvement of the dissipation level at low temperature. Work is in progress to further decrease the effective dissipation by suitable choice of the geometrical pattern of the resistors, as well as by the use of different materials, such as AuPd.

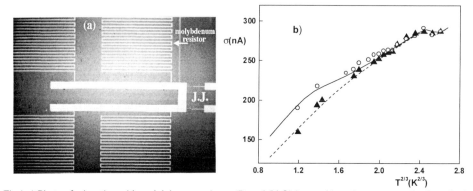

Fig.1 a) Photo of a junction with molybdenum resistors (R_M=6.5 kΩ) inserted into the current and voltage leads. The meander-type filters improve the insolation of the junction from the electromagnetic environment. b) Experimental data for the distribution width σ versus $T^{2/3}$ for configuration A (dots) and configuration B (triangles). Data are compared with the theoretical predictions within the classical theory (Buttiker et al 1983). The relevant junction parameters are independently measured (the junction capacitance C=18 ± 3.6 pF, the critical current I_c=48 ± 0.5 µA at T=4.2 K and I_c=55 ± 0.5 µA at T=1.2 K, and the superconducting gap Δ=1.37 ± 0.01 meV at T=4.2 K). The R vs T dependence is obtained as: $1/R=1/R_0exp(\Delta(T)/k_BT)+1/R_{sat}$, with R_0= 1 Ω

and the saturation resistance R_{sat}=15 kΩ (full curve) and R_{sat}=1 kΩ (dashed line). This dependence confirms that the effective dissipation is dominated by intrinsic mechanisms.

3. EVIDENCE OF ELQ IN JOSEPHSON JUNCTIONS

In our recent experiments on Josephson junctions (Silvestrini et al 1997, Ruggiero et al 1999a), we found that, increasing the sweeping frequency of the external bias, namely in non-stationary conditions for the system (Silvestrini 1991, Silvestrini et al 1996), the switching current distribution, P(I), and the escape rate out of the zero voltage state as function of the current bias, Γ(I), present many oscillations for T>>T_o (Silvestrini et al 1997). Data were consistent with a quantum picture, assuming that the tunneling through the barrier can only occur from quantized energy levels. To induce non-stationary conditions in the junctions our electronics is fast enough to allow high sweeping frequencies of the external bias (resulting in dI/dt up to 100A/s). In building our experimental setup, great care has been devoted to try to have the system dissipation dominated by the intrinsic mechanism, namely, the quasiparticle tunneling, hence with a very low intrinsic damping level at low temperature. We had a 87.3 kΩ SMD resistor located close to the junction, while a great care has been devoted to reduce any stray capacitance, which may determine the real part of the complex impedance at frequencies of the order of the level spacing. The experiment used Nb-AlO$_x$-Nb Josephson junctions with high quality factor V_m>80mV. The junction parameters independently measured at 1.3 K for sample here presented are: the critical current I_c=80 ± 1 μA, and the junction capacitance C=1.2 ± 0.2 pF, corresponding to a critical current density J_c =700 A/cm^2. The system dissipation is determined from the fitting of data in the pure thermal limit (Ruggiero et al 1998), namely from the low frequency measurements, and results in an effective resistance of R=(10 ± 5) kΩ. The plasma frequency ω_j is a very important parameter to fit data and it has been also independently determined by measuring at low temperature (down to 40 mK) the well known transition from the classical to the quantum regimes (Martinis et al 1987). In Fig.2 we report both the experimental histogram P(I) and escape rate

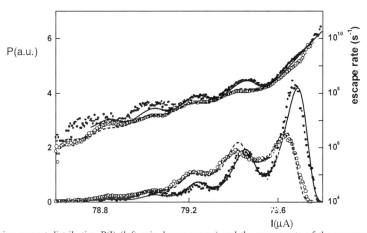

Fig.2. Switching current distribution P(I) (left axis, lower curves) and the escape rate of the supercurrent state as a function of the current I (right axis, upper curves) for a Josephson junction. Dots refer to data taken at T=1.2 K and dI/dt = 102 A/s; open circles refer to data taken at T=1.2 K and dI/dt= 33 A/s. The theoretical predictions from

Silvestrini et al (1996) refer to I_c=80 µA, C=1.2 pF, R= 10 kΩ, and T=1.2 K with two different sweeping frequencies: dI/dt = 102 A/s (solid line) and dI/dt =33 A/s (dashed line).

as a function of the external current I biasing the junction obtained for the sample at T=1.3 K and T= 2 K. The sweeping frequency is high enough to induce the observation of quantum effects, dI/dt>20 A/s. The experimental data already give evidence of energy level quantization. Two different sweeping frequencies are plotted to stress that energy change is in fact the cause of the observed oscillation. Time period of oscillations is quite different in measurements at different sweeping frequencies, but energy spacing is definitely the same.

4. CONCLUSIONS

We have presented a new design of extremely underdamped Josephson junctions which allows to integrate in a compact way molybdenum resistors with junctions. These resistive molybdenum lines do decouple the sample from the external circuit and the electromagnetic environment. This renders the proposed technique useful for devices using the switching properties of Josephson junctions, when integrated filters are recommended, as well as in view of new experiments in the quantum limit, where a very low dissipation is required. We have also presented experimental evidence of ELQ in Josephson junctions at temperature higher than the crossover one. This has been possible by a fast sweep rate of the external bias and having a great care in decoupling the sample from the environment (small dissipation).

REFERENCES
[a] e-mail: silvestrini@fisps.cib.na.cnr.it

Averin D V 1999 Nature **398**, 748.

Buttiker M, Harris E P, and Landauer R 1983 Phys. Rev. B **28**, 1268; and references therein.

Kuzmin L S and Haviland D B 1991 Phys. Rev. Lett. **67**, 2890.

Larkin A I, and Ovchinnikov Yu N 1986 Sov. Phys. JETP **64**, 185 [1986 Zh. Eksp. Teor. Fiz. **91**, 318].

Leggett A J 1992 Quantum Tunneling in Condensed Media, Kagan Yu and Leggett A J eds. (Elsevier Science Pub) pp 2- 36; and references therein.

Martinis J M, Devoret M H and Clarke J 1987 Phys. Rev. B **35**, 4682; and references therein.

Nakamura Y, Pashkin Yu A, and Tsai J S 1999 Nature **398**, 786.

Rouse R, Han S, and Lukens J E 1995 Phys. Rev. Lett. **75**, 1614 .

Ruggiero B , Granata C, Palmieri V G, Esposito A, Russo M, and Silvestrini P 1998 Phys. Rev. B **57**, 134.

Ruggiero B, Castellano M G, Torrioli G, Cosmelli C, Chiariello F, Palmieri V G, Granata C, and Silvestrini P 1999a Phys. Rev. B **59**, 177.

Ruggiero B, Granata C, Esposito E, Russo M, and Silvestrini P 1999b Appl. Phys. Lett. **75**, 121.

Schwartz D B, Sen B, Archie C N, and Lukens J E 1985 Phys. Rev. Lett **57**, 1547.

Silvestrini P 1991 Phys. Lett. A **152**, 306.

Silvestrini P, Ruggiero B, and Esposito A 1996 Low Temp. Phys. **22**, 195 [1996 Fiz. Nik.Temp. **22,** 252].

Silvestrini P, Palmieri V G, Ruggiero B, and Russo M 1997 Phys. Rev. Lett. **79**, 3046.

Inst. Phys. Conf. Ser. No 167
Paper presented at Applied Superconductivity, Spain, 14-17 September 1999
© 2000 IOP Publishing Ltd

Current-Phase Relation in High-T_C YBCO Josephson Junctions

E. Il'ichev, V. Zakosarenko, R.P.J. IJsselsteijn, H.E. Hoenig, V. Schultze, H.-G. Meyer

Institute for Physical High Technology, Dept. of Cryoelectronics, P.O. Box 100239, D-07702 Jena, Germany

ABSTRACT: The current-phase relation (CPR) of YBCO step-edge, and several configurations of bicrystal grain boundary Josephson junctions (JJ) has been investigated experimentally. The CPR was obtained from the measurement of the impedance of the phase-biased junction. The method was tested using well-known Nb/AlO$_x$/Nb tunnel junctions and the expected sinusoidal CPR was obtained. In the case when thermal noise is negligible the only remarkable deviation from a sinusoidal dependence has been found for the CPR of 45 degree grain boundary JJs. Moreover, for asymmetrical 45 degree grain boundary JJs a large π-periodic component of the CPR has been experimentally observed. These results are consistent with d-wave pairing symmetry and the intrinsically shunted tunnel JJs model.

The dependence of the supercurrent I of a superconducting weak link on the phase difference φ of the order parameter is a basic characteristic of the weak link and usually called the current-phase relationship (CPR). In general $I(\varphi)$ is an odd periodic function of φ with a period 2π. Therefore $I(\varphi)$ can be expanded in a Fourier series:

$$I(\varphi)=I_1\sin\varphi+I_2\sin2\varphi+\dots \qquad (1)$$

For weak links based on conventional s-wave superconductors the relation between the coefficients I_n is:

$$I_1 > |I_n|; \; n > 1 \qquad (2)$$

where n is a number of corresponding term in series (1). For instance, for a tunnel junction $I_1 >> |I_n|$ ($n>1$) and we have the well-known sinusoidal CPR. For a junction with normal-metal interlayer the harmonics in series (1) are not negligible (Kulik and Omel'yanchuk 1978), but relationship (2) is valid

In the frame of the d-wave scenario ($d_{x^2-y^2}$) the magnitude and phase of the pair wave function depend on the direction in momentum space (see e.g. Van Harlingen 1995). In particular, there is a difference of π between the phase of Cooper pairs moving along the k_x direction and that of pairs moving along k_y direction. The result is a highly unusual expected CPR for certain configuration of weak links. Let θ_1 (θ_2) denote the angle between the normal to the plane of the junction and the a axis in electrode 1 (2). It can be shown (Walker and Luettmer-Strathmann 1995) that the symmetry of the problem dictates:

$$I_1 = I_{1,1} \cos 2\theta_1 \cos 2\theta_2 + I_{1,2} \sin 2\theta_1 \sin 2\theta_2 \qquad (3)$$

It is clear from Eq. 3 that in series (1) the amplitude of the first harmonic $I_1=0$ for a junction with $\theta_1=0$ $\theta_2=45°$ - so-called asymmetric 45° junction. Therefore for these junctions an anomalous periodicity of the CPR is expected.

More generally many unusual types of the CPR are possible when d-wave symmetry is present (Reidel and Bagwell 1997). There is not only academic interest to investigate the CPR of high-T_c JJs. Recently several configuration of the phase qubit have been proposed (see e.g. Ioffe et al 1999 and Zagoskin 1999). Operation of these devices based on an unusual CPR in order to realize a double degenerated state.

In the present work first we demonstrate the validity of the developed method for the CPR measurements using a well-known Nb/AlO$_x$/Nb tunnel junction. After that we summarize the obtained for step-edge and 24°, 30°, 36°, 45° YBa$_2$Cu$_3$O$_{7-x}$ grain boundary JJs.

To investigate experimentally the CPR, most commonly, the weak link is incorporated in a superconducting loop with a small inductance L (single-junction quantum interferometer). The phase difference φ across the junction in the interferometer can be controlled by the external magnetic flux Φ_e applied to the loop. The quantum interferometer is coupled inductively to a tank circuit with inductance L_T and resistance R_T. The effect of the interferometer on the parameters of the tank circuit can be represented in terms of effective values L_{eff}, R_{eff}. As was shown by Rifkin and Deaver (1976), one can deduce the CPR from a measurement of the effective tank circuit impedance as a function of Φ_e. To restore the CPR in the whole range $-\pi \leq \varphi \leq \pi$ the condition $\beta < 1$ has to be fulfilled.

We measured the phase angle α between the drive current I_{rf} and the tank voltage U at the resonant frequency as a function of Φ_e. The later has been produced by a low-frequency current I_{dc} in the tank coil. As shown by Il'ichev et al (1998b)

$$\tan \alpha \propto dI(\varphi)/d\varphi_e,\qquad(4)$$

where $\varphi_e=2\pi\Phi_e/\Phi_0$.

The low-T_c structures were fabricated using a standard Nb/AlO$_x$/Nb technology. To achieve $\beta < 1$ we have fabricated a interferometer loop with 6 washers, each having an inductance of 55pH, connected in parallel. 10-turns coupling coil is placed on each washer and connected in series. The coupling coils are connected to a large 5x5 mm^2 loop integrated on the chip. This configuration allows to obtain sufficient coupling between the tank coil and the interferometer with a total inductance of $L\approx9$pH.

For high-T_c structures we used [001] oriented bicrystal SrTiO$_3$ substrates with misorientation angles of 24, 30, 36, 45 degree. The YBa$_2$Cu$_3$O$_{7-x}$ films were deposited by laser ablation and interferometers were patterned by Ar ion beam

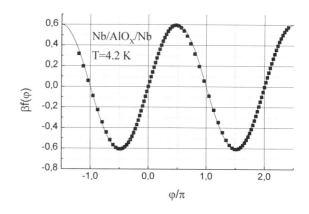

Fig. 1. The current phase relation of a Nb/AlO$_x$/Nb tunnel junction obtained experimentally (squares) and function $\beta f(\varphi) = 0.6 \sin(\varphi)$ (solid line).

etching. The grain boundary junction was incorporated in a 5×5 mm² square washer. The square washer-holes had side-lengths of d = 50 μm leading to a geometrical inductance of L = 1.25 $\mu_0 d$ ≈ 80 pH.

A nearly perfect sinusoidal CPR, obtained from Nb/AlO$_x$/Nb tunnel junction is shown in Fig.1. An example of the CPR obtained for 36° grain boundary JJ is presented in Fig.2. Here deviations from the sinusoidal dependence were also not observed.

Note that the sinusoidal CPR has been obtained for step-edge JJ (Il'ichev et al 1998a), and 24° (Il'ichev et al 1998b), 30°, 36° (Il'ichev et al 1999a) grain boundary JJs. The present measurements have some important consequences for the study of transport mechanisms in high-T_c JJs. The sinusoidal CPR gives evidence that the Cooper pairs do not use ballistic channels, i.e. the super-current flow is mediated by direct tunneling through interfaces of low transparency. This statement is consistent with many other data for HTS JJs and agrees with the phenomenological intrinsically shunted tunnel junction model, which assumes that the quasiparticle current is dominated by resonant tunneling whereas the Cooper pairs can only tunnel directly.

In contrast, for 45° grain boundary JJs deviations from a sinu-soidal CPR have been observed. For symmetrical JJs with $\theta_1=\theta_2=22.5°$ (Il'ichev et al 1998c) as well as for JJs with $\theta_1=19°$

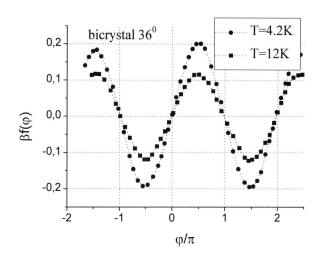

Fig. 2. Current phase relation for YBa$_2$Cu$_3$O$_{7-x}$ bicrystal Josephson junction with misorientation angle of 36 measured at two temperatures.

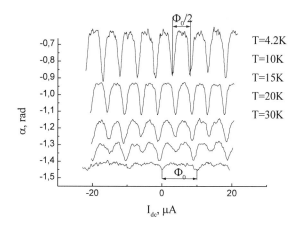

Fig. 3. Phase angle α between the driving current and output voltage for 45 degree grain boundary junction measured at several temperatures as a function of the dc current trough the tank circuit. The curves are shifted vertically for clarity.

and $\theta_2=26°$ experimental results are consistent with a model of an inhomogeneous JJ with randomly distributed alternating current density (Il'ichev et al 1999b). In all these cases a 2π-periodical CPR has been obtained and relationship (2) is valid.

For asymmetrical 45° grain boundary JJs experimental results are shown in Fig. 3. The behavior of this sample at low temperatures is defined by the π–periodic component of $I(\varphi)$ (see relationship (4)). Totally 6 samples with asymmetrical 45° grain boundary JJs have been investigated and only for 2 of them relation $I_2>I_1$ has been obtained. It can be explained by the faceted nature of the grain boundaries (Il'ichev et al 1999c).

In conclusion, we have shown that the CPRs of grain boundary JJs agree with the phenomenological intrinsically shunted tunnel junction model with a pronounced d-wave symmetry component in the pair wave function.

We are grateful to M. Fistul, P. Müller, M. Kupriyanov, H. Hilgenkamp, J. Mannhart, A. Golubov, M. Grajcar, and R. Hlubina for fruitful discussion. Financial support by the DFG (Ho 461/3-1) and INTAS (No 11459) is gratefully acknowledged.

References

Il'ichev E et al 1998a Appl. Phys. Lett. **72**, 731
Il'ichev E et al 1998b Advances in Solid State Phys. **38**, 507
Il'ichev E et al 1998c Phys. Rev. Lett. **81** 894
Il'ichev E et al 1999a IEEE Trans. Appl. Supercond. **9**, 3994
Il'ichev E et al 1999b Phys. Rev. **B59** 11502
Il'ichev E et al 1999c, Phys. Rev. **B60**, 3096
Ioffe L B et al 1999 Nature **398** 679
Kulik I O and Omel'yanchuk A N 1978 Sov. J. Low Temp. Phys. **4**, 142
Reidel R A and Bagwell P F 1997 preprint
Rifkin, R Deaver B S 1976 Phys. Rev. **B13** 3894
Van Harlingen D J 1995 Rev. Mod. Phys. **67**, 515
Walker M B and Luettmer-Strathmann J 1995 Phys. Rev. **B54**, 588
Zagoskin A M 1999 cond-mat 9903170

Inst. Phys. Conf. Ser. No 167
Paper presented at Applied Superconductivity, Spain, 14-17 September 1999
© 2000 IOP Publishing Ltd

A directly coupled Josephson-Fraunhofer-Meissner Gauss-meter

E Sassier, Y Monfort, C Gunther and D Robbes

GREYC UPRES-A 6072 CNRS, ISMRA et Universite de Caen
6 Bd du maréchal Juin, F-14050 Caen Cedex

ABSTRACT: A high sensitivity Josephson-Fraunhofer-Meissner magnetometer (JFM) has been derived from the well-known directly coupled dc SQUID design. The whole sensor is 0.85x0.85-cm² large. It exhibits a transfer coefficient of 550 V/T or 550 A/T as it is respectively current or voltage biased. The measured white noise is 600 fT/√Hz and the 1/f noise corner is around 400 Hz. The white noise level is about 4 times higher than predicted by the standard RSJ theory. We have investigated some effects of thermal fluctuations of the superconducting electrodes in the vicinity of the edges of the junction.

1. INTRODUCTION

The Fraunhofer pattern of HT_c Josephson junctions was recently proposed to be used for magnetometry (Dolabdjian 1996). Though such a magnetometer (Josephson-Fraunhofer-Meissner magnetometer, JFM) is less sensitive than SQUID magnetometers, some of its characteristics make it attractive. Its technology is simple, its dynamic range is potentially large and it can be referenced to zero-field. However, some potential applications such as non-destructive testing still require sensitivity improvements so as to reach values of about 0.1 pT/√Hz. Until now, two ways have been attempted in order to achieve significant sensitivity improvements. The first one consists in operating several junctions connected in series. Krey et al (1999) made a 105 junctions device, obtaining a field-to-voltage transfer rate of 7500 V/T and a sensitivity of 1.2 pT/√Hz. The second solution is to use concentrator structures. V. Martin et al (1996) obtained a transfer rate of 66 V/T (264 A/T) and a sensitivity of 3 pT/√Hz with a grain boundary junction associated with square washers and a focuser in a flip-chip arrangement.

This paper reports the performances of a high field-to-critical-current or field-to-voltage transfer rate JFM magnetometer. It is directly coupled to a large washer, in a design inspired from a usual directly coupled SQUID geometry. We first describe the geometry and the theoretical performances of the device. We detail the measurements of these performances in the second part, and we lastly examine a possible source of the disagreement between the calculated and measured noise levels of the device.

2. DESIGN

A Josephson junction is made of two superconducting electrodes coupled by the tunneling of Cooper pairs through a thin insulating or normal barrier. The critical current of such a structure is sensitive to the magnetic field component parallel to the barrier plane. In the short junction limit with respect to the Josephson length, the field-dependant critical current variation follows the ideal law :

$$I_c = I_0 \, \text{sinc}\left(\pi \frac{B}{B_0} \right) \qquad (1)$$

where B_0 is the first null point of the Fraunhofer pattern. Within the RSJ situation, the white noise is mainly due to thermal effects. Likharev and Semenov (1972) expressed it as follows :

$$S_V(f) = \left(1 + \frac{1}{2}\left(\frac{I_0}{I}\right)^2\right) 4k_B TR_N \left(\frac{R_d}{R_N}\right)^2 \qquad (2)$$

where I is the current crossing the junction, R_d and R_N are respectively the dynamic and the normal resistance of the device.

HT$_c$ junctions are now most commonly patterned in a planar geometry on a grain boundary (GBJ). The first devices of this type were surprisingly found to exhibit a B_0 value within a factor 10 to 100 lower than expected. This phenomenon was soon later found by Rosenthal (1991) and Humphrey (1993) to be due to the current screening the field in the superconducting electrodes, which concentrate the field density in the sensitive zone as a secondary effect. This concentration effect can be reinforced by adding superconducting structures near the junction. For example, V. Martin et al (1996) obtained a nearly-60 concentration factor by patterning junctions between two 630-μm-by-630-μm superconducting squares distant of 5 μm from each other, and by using 5x5-mm flip-chip focuser.

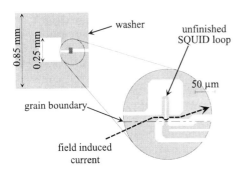

Figure 1 : Design of the device. The current screening the flux changes in the washer loop circulates near the junction. The junction senses the field created by this current, which increases its transfer rate with respect to the ambient field This device is an unfinished SQUID, and the design was not optimized.

Our test design is derived from the well known design of directly coupled dc SQUIDs. The leading idea of the design is to make the screening current of a superconducting loop circulate in a narrow strip as close as possible of the sensitive device. The key elements of the system sensitivity are then, beyond the sensitivity of the device itself, the washer loop inductance and the distance between the device and the coupling strip. The device we used was an unfinished SQUID patterned on a 200 nm thick YBaCuO layer grown on a 24° disoriented bicrystal. The dimensions of the design are shown in figure 1. In this state, the junction had not been divided into the two junctions of the future SQUID, so that the current screening the flux in the washer loop partly circulated close to the junction. The ratio between the two inductances of the unfinished SQUID inner loop arms made the current circulate mostly near the junction.

3.   MEASUREMENTS

The following measurements have been made in a heavily shielded magnetic environment. The device was enclosed in an almost water-proof metallic cylindrical box, which was immersed in liquid nitrogen. Two coils in the Helmholtz arrangement were fixed to the support inside the box, in order to apply magnetic field to the device. We used alternatively two different amplifiers. The simpler was a SSM-2017 directly connected at the output of the device. Its gain was 1000, and its noise voltage e_n was 0.95 nV/√Hz. The second had a cooled transformer as an input stage. The total gain of this amplifier was 25000 at 20 kHz, and its noise voltage was about 60 pV/√Hz.

Using the first amplifier, we found that the junction had a maximum critical current of about 800 μA, a normal resistance R_N of about 0.5 Ω and a dynamic resistance of 1 Ω (figure 2a). We plotted then the field-to-voltage figure of the junction at an optimal bias current of 1000 μA (figure 2b). The maximal transfer rate was found to be about 550 V/T, which corresponds to a critical current modulation of about 550 A/T, with R_d=1 Ω.

We used the second amplifier to measure the noise spectrum of the device. We applied to the device a modulation field of about-B_0 amplitude and 30-kHz frequency, in order to eliminate the 1/f noise component of the amplifier. The output signal was demodulated by a lock-in amplifier (DSP 580, Stanford Research). The figure 2 shows a spectrum obtained with a bias current of about 1000 µA, which set the maximum field-to-voltage transfer rate. We obtained a white noise of about 600 fT/√Hz. The corner of excess noise is at about 500 Hz, and its value reached 10 pT/√Hz at 1 Hz.

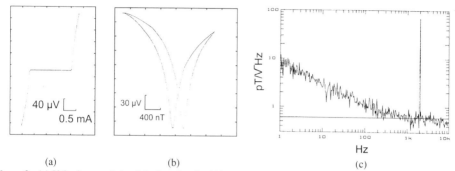

| (a) | (b) | (c) |

Figure 2 : **(a)** V(I) characteristic of the junction. I_c=800 µA, R_N=0.5 Ω and R_d=1.0 Ω. **(b)** V(B) characteristic with a bias current of 1 mA. dV/dB$_{max}$=550 V/T. (c) Spectrum of the device noise, with a field modulation at 20 kHz. The peak is a reference signal at 2kHz.

4.   DISCUSSION

This transfer rate of 550 A/T can be compared to the typical rate of a NKT-company's commercial dc SQUID directly coupled to a 0.8x0.8-cm² pick-up loop, as given by Dolabdjian (1999). It is only a factor 5 higher than our present design, which may be not optimized. The theoretical and measured white noise of such a SQUID are respectively 50 and 60 fT/√Hz, whereas the theoretical white noise of our JFM is 180 fT/√Hz (e.g. only 3.6 times higher). The compared dynamics of the two devices should be roughly the same, if operated in a standard Flux Locked Loop configuration. First measurements of a JFM dynamic range and slew rate were reported in Dolabdjian (1996) (112 dB and 2 mT/s).

However, the theoretical white noise level of 180 fT/√Hz is obviously not in agreement with the measurements. We did several measurements in different bias situations, in order to determine the nature of this noise. We measured the white noise at 20 kHz at the output of the transformer amplifier, without modulation. It became negligible compared to the transformer noise when the junction was biased by a current smaller than the critical value. On the other hand, it remained unchanged when the bias current and the magnetic field were set so as to bias the device at the summit of the field-to-voltage characteristic, where the transfer rate is null. These observations strongly suggest that the noise measured at 20 kHz is due to critical current fluctuations or dynamic resistance fluctuations.

We investigated a possible mechanism linked to thermal fluctuations inspired by Voss (1976) work on semiconductors noise, and by Lepaisant (1985) work on high frequency lines noise. We modelled the device as a simple line coupled to the substrate via a thermal resistance.

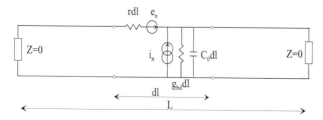

Figure 3 :model of the thermal noise in an YBCO line liying on a substrate. r,C_0,g_{bd} are the linear resistance of the film, the linear capacitance and boundary conductance between the film and the substrate.

Each elementary part of the line is then coupled to the substrate by a linear thermal capacitance C_0, a linear thermal boundary resistance $1/g_{bd}$ and the temperature noise generator related to this resistance (figure 3) The calculation can then be lead as for a high frequency line, the corresponding noise generators being $e_n=2k_BT^2r$ and $i_n=2k_BT^2g_{bd}$. The temperature fluctuation at the middle of the line of length L is the sum of the fluctuations propagating along the line. We did the integration and found a maximum fluctuation of :

$$\overline{T}_n^2 = \frac{\dfrac{\sinh(\alpha L)}{\alpha}\left(\overline{e}_n^2 + |Z_c|^2\overline{i}_n^2\right) + \dfrac{\sin(\beta L)}{\beta}\left(\overline{e}_n^2 - |Z_c|^2\overline{i}_n^2\right)}{8(\cosh(\alpha L)-\cos(\beta L))} \tag{3}$$

where

$$Z_c = \frac{r}{\alpha + j\beta} = \sqrt{\frac{g + jC_0\omega}{r}} \tag{4}$$

Moreover, the critical current density of the junction is related to the temperature by the following formula, in the macroscopic quantum model :

$$J_c = \frac{e\hbar\sqrt{n_1^*(T)n_2^*(T)}}{2m_e\zeta\sinh\left(\dfrac{2a}{\zeta}\right)} \tag{5}$$

where a is the insulator thickness, ζ is the coherence length of the Cooper pairs, and

$$n_1^*(T) = n_2^*(T) = n_0^*\left(1 - \left(\frac{T}{T_c}\right)^4\right) \tag{6}$$

are the super-carrier densities in the two electrodes. With the thermal parameter values given by literature, we obtained a maximum temperature fluctuation density of 121 nK/√Hz. The relative Josephson current fluctuation is then $\Delta Ic/Ic = 7\ 10^{-9}$ which turns into a maximum voltage fluctuation density of $R_d\Delta Ic = 3$ pV/√Hz. Unfortunately, this value is obviously about two orders of magnitude too low to explain the white voltage noise level we measured (320 pV/√Hz).

5.   CONCLUSION

We observed an apparently white noise level up 500 Hz about 4 times higher than expected from theory. This phenomenon was also observed by several other groups (Krey 1999, Martin 1996) with HT_c YbaCuO devices, but it now remains unexplained. Despite this surprisingly high noise level, we achieved a transfer rate of 550 A/T to be compared to the typical SQUID rate (~2500 to 3000 A/T) and a sensitivity of about 600 fT/√Hz, without any design optimization for this use. We now work on especially designed directly coupled junctions.

REFERENCES

Dolabdjian C, Poupard P, Martin V, Gunther C, Hamet JF and Robbes D (1996) Rev. Sci . Inst. **67** (12) 4171-4175

Dolabdjian C, Saez S, Robbes D, Bettner C, Loreit U, Dettman F, Kaiser G and Binneberg A (1999) EUCAS proceedings

Krey S, Brügmann O and Schilling M (1999) App. Phys. Lett. **74** (2) 293-295

Lepaisant M (1985) PhD Thesis, Université de Caen

Likharev K K and Semenov V K (1972) Pis'ma Zh. Eksp. Theor. Fiz. **15**, 625-629 or (1972) JETP Lett. **15** 442-445

Martin V (1996) PhD Thesis, Université de Caen

Rosenthal P A, Beasley M R, Char K, Colclough M S and Zaharchuk G (1991) App. Phys. Lett. **59** (26)

Voss R F and Clarke J (1976) Phys. Rev. B **13** (2) 556-573

Inst. Phys. Conf. Ser. No 167
Paper presented at Applied Superconductivity, Spain, 14-17 September 1999
© 2000 IOP Publishing Ltd

Spatially inhomogeneous temperature effects in Josephson tunnel junctions

D. Abraimov, M. V. Fistul, P. Caputo and A. V. Ustinov

Physikalisches Institut III, Erlangen-Nürnberg Universität, D-91058 Erlangen Germany

G. Yu. Logvenov
OXXEL GmbH, Technologiepark Universität, D-28359 Bremen Germany

ABSTRACT: Low Temperature Scanning Laser Microscopy (LTSLM) has been used to study a two-junction interferometer. Inhomogeneous LTSLM patterns in the form of "rings" around the Josephson junctions have been observed. We have shown that this thermoelectric effect is due to an interplay between a nonlinear current-voltage characteristic of a Josephson junction and inhomogeneous temperature distribution in electrodes. The response of the dc SQUID drastically changes in the presence of magnetic field. We argue that the observed thermomagnetic effect is caused by the temperature gradient across a tunnel Josephson junction.

1. INTRODUCTION

Considerable attention has been devoted to a study of various temperature effects in Josephson coupled systems. Most of these effects appear because the critical current and quasiparticle resistance of a Josephson junction depend on temperature (Likharev 1986). Even a small temperature increase leads to a substantial change of the nonlinear current-voltage characteristic (CVC) of the Josephson junctions (Doderer 1997).

Interesting temperature effects appear in the presence of temperature gradients in Josephson junction (Ryazanov et al 1982, Panaitov et al 1984, Smith et al 1980). In particular, a temperature gradient applied perpendicular to the Josephson junction barrier induces a Josephson current flowing across the junction and therefore, non zero value of the Josephson phase difference. It leads to the thermoelectic analogies of stationary and non stationary Josephson effects that have been observed in low resistive superconductor-normal metal-supercondutor (SNS) Ta-Cu-Ta Josephson junctions (Ryazanov et al 1982, Panaitov et al 1984). These thermoelectric effects are expected to be very tiny in the superconductor-insulator-superconductor (SIS) junctions that have a much higher value of the junction resistance. Until now such effects have not been experimentally observed in SIS junctions. Recently, Guttman et al (1997) proposed to use a two-junction interferometer in order to observe the *thermomagnetic* (thermophase) effects in these junctions.

In this paper we report on the observation of spatially inhomogeneous Low Temperature Scanning Laser Microscopy (LTSLM) response of small *homogeneous* SIS Josephson junctions. We present an analysis of the response that allows us to account for its dependence on the bias current, intensity and depth of laser beam power modulation. Moreover, we observed the dependence of LTSLM response on magnetic field when two junctions are enclosed in a superconducting loop. We argue that the observed thermomagnetic effect caused by the presence of a temperature gradient across a Josephson junction.

2. EXPERIMENTAL SETUP AND MEASUREMENTS

We have studied a two-junction interferometer consisting of two Nb/Al-AlO$_x$/Nb Josephson junctions on Si substrate. The area of each junction is 9 μm^2. The critical current density is about 1kA/cm^2 with the spread of junction parameters of less than 5%. An optical image of two-junction interferometer is shown in Fig. 1(a), where white arrows indicate the junctions. The sample is mounted in vacuum on the copper block that is furnished with a heater and temperature sensor and has direct contact to the liquid He-4 bath. We used a lead superconducting shield to reduce the influence of an external magnetic field. A magnetic field perpendicular to the substrate was applied by using a small coil placed inside the shield. All measurements were carried out at T/T_c=0.92. At this temperature the junctions have the nonhysteretic CVC with the critical current I_c of about 21μA.

The temperature effects in Josephson coupled systems were studied by using the LTSLM (Sivakov et al 1996). LTSLM technique uses a focused laser beam for local heating of the sample. Local heating leads to a decrease of the critical current. The laser beam induced variation of the voltage drop across the sample is recorded versus the beam coordinates. In order to increase the spatial resolution, the intensity of the beam is modulated at a frequency of several kHz. We have systematically LTSLM imaged our two-junction interferometer biased at various values of the bias current I and magnetic field. The latter can be expressed in terms of magnetic flux per interferometer cell Φ. In the absence of magnetic field we observed an *inhomogeneous* LTSLM response in the form of a dark "ring" pattern surrounding each Josephson junction. Typical images are shown in Fig 1(b)-(d) for various values of bias current. The most striking feature of our measurements is that the response has a maximum when the position of hot spot does not coincide with the junction. Moreover, the radius of the ring increases with the bias current.

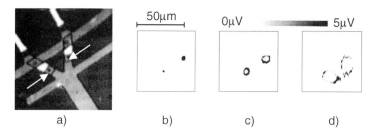

a)   b)   c)   d)

Fig. 1 Optical image (a) and LTSLM images (b-d) of interferometer for different values of bias current I: b) I=16.53 μA, c) I=17.04 μA, d) I=17.74 μA.

In the magnetic field we observed oscillations of the critical current of two-junction interferometer with the field (Fig. 2(a)). Under these conditions we saw an *asymmetric* LTSLM response from the two junctions. This asymmetry changed periodically with magnetic field and reached a maximum at the external flux $\Phi=\Phi_0(2n+1)/4$ (Fig. 2(b)-2(d)), where n is an integer.

3. THEORETICAL ANALYSIS OF LTSLM RESPONSE AND DISCUSSION

We present here a theoretical analyses of LTSLM response of two junction interferometer that allows us to consistently explain the measurements. We use a resistively shunted Josephson junction model (RSJ) to describe the overdamped Josephson junction (Likharev 1986). In this model, the CVC of a Josephson junction (or two junction interferometer) is given by

$$V = R_n \cdot \sqrt{I^2 - I_c^2}, \qquad (1)$$

where R_n and I_c are, accordingly, the normal resistance and critical current of the interferometer. In the presence of the laser beam, the inhomogeneous temperature distribution along the sample occurs. The maximum temperature is reached at the hot spot positioned at the distance r from the junction. The characteristic length of the temperature decay s is determined by the heat transfer in the sample.

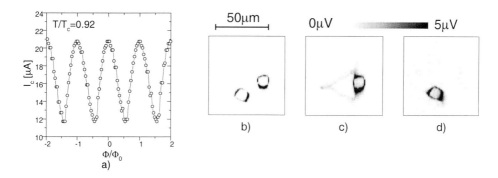

Fig. 2 Magnetic field dependence of the interferometer critical current (a) and LTSLM images for different values of magnetic field: b) $\Phi=0$, c) $\Phi=1/4\Phi_0$, d) $\Phi=3/4\Phi_0$.

The suppression of the junction critical current depends on r. Due to the temporal modulation of the laser beam intensity, the critical current oscillates between its minimum and maximum values $I_{cmin}(r)$ and $I_{cmax}(r)$ (these dependencies are shown in Fig. 3(a)). In the absence of magnetic field, the LTSLM response of two junction interferometer can be approximately written in the form:

$$\delta V(r) = \begin{cases} 0, & if \ I < I_{c\min}(r) \\ R_n \cdot \sqrt{I^2 - I^2_{c\min}(r)}, & if \ I_{c\min}(r) < I < I_{c\max}(r) \\ R_n \cdot \sqrt{I^2 - I^2_{c\min}(r)} - R_n \cdot \sqrt{I^2 - I^2_{c\max}(r)}, & if \ I > I_{c\max}(r) \ . \end{cases} \qquad (2)$$

By using the response function (2) we have calculated the expected LTSLM response for different values of bias current I (Fig. 3(b)). For small values of I the expected maximum of response coincides with the junction. The maximum of response shifts away from the junction when the bias increases; the LTSM response should appear then in the form of a "ring". The appearance of such rings qualitatively well explains the experimental images shown in Fig. 1(b-d). Moreover, we obtain a good fit to experimental data by using the value of $s=3\mu$m. Thus, the interplay between nonlinear CVC of the junction and inhomogeneous temperature distribution leads to the appearance of various inhomogeneous LTSLM patterns.

However, this model alone cannot explain the magnetic field dependent asymmetry in the response of the two-junction interferometer. This thermomagnetic effect can be explained by taking into account the laser beam induced temperature gradient between the junction electrodes. Indeed, the temperature gradient across the junction leads to an additional Josephson phase difference (or magnetic flux) $\delta\Phi$. The sign of induced magnetic flux depends on which junction is illuminated by the laser beam. That leads to the asymmetry of the LTSLM response in the magnetic field. The critical current of symmetrical interferometer can be written in the form (Likharev ,1986):

$$I_c(\Phi \pm \delta\Phi) = \sqrt{I_{c1}^2 + I_{c2}^2 + 2I_{c1}I_{c2}\cos\left(\frac{2\pi(\Phi \pm \delta\Phi)}{\Phi_0}\right)}, \qquad (3)$$

where I_{c1} and I_{c2} are critical currents of junctions.

By assuming a reasonable value of $\delta\Phi \approx 10^{-2}\Phi_0$ and using Eqs. (1)-(3) we have calculated the LTSLM response for the external flux $\Phi = -1/4\Phi_0$ corresponding to the maximum of the observed asymmetry (Fig. 3(c)). We argue that the asymmetry of LTSLM response is a "fingerprint" of the thermomagnetic effect that has been predicted by Guttman et al 1997, caused by presence of temperature gradient across a Josephson tunnel junction.

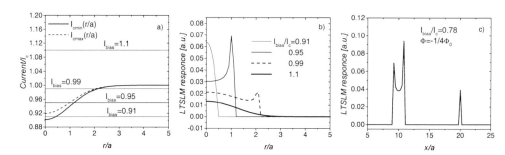

Fig. 3 a) Typical dependencies of the critical currents $I_{cmin}(r)$ and $I_{cmax}(r)$ on the hot-spot position r (a is the size of the junction). b) Calculated LTSLM response of the interferometer for different values of the bias at zero value of magnetic field; c) Calculated LTSLM response of the interferometer in the magnetic field corresponding to $\Phi = -1/4\Phi_0$; x is the hot-spot position along the line connecting the centers of junctions.

REFERENCES

Doderer T., 1997 Int. J. Mod. Phys. B **11**, 1979

Guttman G. D., Nathanson B., Ben-Jacob E., and Bergman D. J.1997 Phys. Rev B **55**, 12691

Likharev K. K.   1986 Dynamics of Josephson junctions and circuits (Gordon and Breach, Philadelphia)

Panaitov G. I., Ryazanov V. V., Ustinov A. V., and Schmidt V. V. 1984 Phys. Lett. A**100**, 301

Ryazanov V.V. and Schmidt V.V. 1982 Solid St. Comm. **42**, 733

Sivakov A. G., Zhuravel' A. P., Turutanov O. G., and Dmitrenko I. M. 1996 Appl. Surf. Sci **106**, 390

Smith A. D., Tinkham M. and Skocpol W. J. 1980 Phys. Rev. B **22**, 4346

Inst. Phys. Conf. Ser. No 167
Paper presented at Applied Superconductivity, Spain, 14-17 September 1999
© *2000 IOP Publishing Ltd*

Efficient multi-Josephson junctions oscillator using a cavity

G Filatrella[1], N F Pedersen[2], and K Wiesenfeld[3]

[1]INFM unit and Science Faculty, University of Sannio, Via Caio Ponzio Telesino 11, I-82100 Benevento Italy
[2]Department of Electric Power Engineering, The Technical University of Denmark, DK-2800 Lyngby, Denmark
[3]School of Physics, Georgia Institute of Technology, Atlanta, GA 30332, USA

ABSTRACT: A theory is presented for a highly efficient microwave oscillator consisting of a large number of underdamped Josephson junctions in a high-Q cavity. The junctions – with a spread in the parameters – interact with each other only through the cavity. Numerical calculations show the following behavior: As the junctions are switched to the voltage state one at a time, at a threshold number of junctions the system enters a coherent state. At this coherent state all the active junctions oscillate in phase at the resonant frequency of the cavity. This sudden phase transition of nonlinear oscillators is somewhat similar to that found in a gas laser. The conversion efficiency from DC power to AC (microwave) power is very high, typically 10-20 GHz. The model explains recent experimental results by Barbara et al. [1999: Phys. Rev. Lett. **82**, 1963] on a multi Josephson junction microwave oscillator as well.

1. INTRODUCTION

To enhance the power delivered by Josephson oscillators it is necessary to achieve coherent phase-lock of the junctions. There have been several types of means suggested to obtain the entrainment of the phases, among them we mention passive loads, two-dimensional circuits, and resonant cavities. Perfect phase-locking would in principle allow to exploit the power of the array that is expected to scale as N^2 with the number N of junctions (Wiesenfeld et al, 1992). Unfortunately it is not easy to achieve such synchronous motion, so that in practice the efficiency was limited well below the theoretical estimated. Recently, Barbara et al. (1999) have been able to produce arrays whose efficiency has raised to the value of 17%, a value that is much higher than the previously reported for instance by Benz and Burroughs (1991), although previous experiments achieved a higher absolute value of the emitted power. The main difference between the Maryland experiment and the previous results is the use of underdamped junctions, but also other features of this experiment are essential: The interaction between the junctions is mediated by a high-Q cavity, and the junctions are activated "row by row". Moreover, the power delivered presents a distinct threshold effect: it increases sharply once a certain number of rows have been activated. On the basis of these characteristics Barbara et al have suggested that one can exploit the formal analogy between the lasers and the arrays of Josephson junctions, proposed as early as 1972 by Tilley (see also Bonifacio, 1982). The purpose of this paper is to propose an explicit coupling scheme between the junctions and the cavity (and so, implicitly, among the junctions themselves) to explain the observed features.

2. THE COUPLING MECHANISM

A schematic model for the interaction between the junctions and the cavity is shown in Fig. 1, where we have tried to retain only the most essential features of the model. In fact the cavity is in this model is reduced to a simple RLC lumped resonator, a simplification with respect to the actual resonator that is better described by a distributed model. The junctions interaction with the cavity is also simplified by assuming that each one interacts solely with the cavity, and that no interaction takes place directly between the junctions.

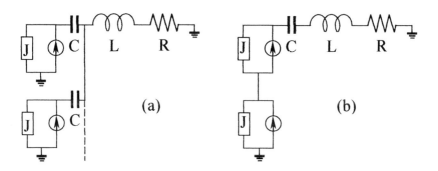

Fig. 1: Schematic electric circuit of the parallel (a) and series (b) array.

This is also a simplification, because the junctions are placed in a two-dimensional network, so that they do interact directly with each other. Still we believe that such an approximation can be used because in two-dimensional arrays the coupling is strongly directional: it is very strong for junctions belonging to the same row, and is rather weak for junctions belonging to different rows (Filatrella and Wiesenfeld, 1994). We can therefore imagine that each row behaves as a single oscillator consisting of perfectly phase-locked junctions, and we concentrate on the problem to lock the rows to each other. This is consistent also with the experimental observations that are always made in terms of rows, considered as the elementary oscillators. The equations for the circuit 1a, where the junctions are coupled in parallel (Monaco et al 1990), read:

$$\beta \ddot{\Phi} + \dot{\Phi} + I_j \sin \Phi_j = I - \ddot{q}_j \qquad (1)$$

$$\ddot{P}i \frac{R}{L\omega_{rc}} \dot{P} + \frac{1}{Lc\omega_{rc}^2} q_j = \frac{1}{\beta_L} \Phi_j \qquad (2)$$

Where j=1,...N, β is the junction capacitance, I_j is the critical current, I is the bias current, q_j is the charge on the jth coupling capacitor, P is the total charge on the load capacitor, c is the coupling capacitance, and L, R, and C are the load inductance, resistance, and capacitance, respectively. To make the equations dimensionless we have used the frequency ω_{RC} and the analogous of the SQUID parameter $\beta_L = h/4pi \; eLI_0$. In the series circuit 1b instead and the equations read (Wiesenfeld et al, 1997):

$$\beta \ddot{\Phi} + \dot{\Phi} + I_j \sin \Phi_j = I - \ddot{q}_j \qquad (3)$$

$$\ddot{P}i \frac{R}{L\omega_{rc}} \dot{P} + \frac{1}{Lc\omega_{rc}^2} q_j = \frac{1}{\beta_L} \sum_{i=1}^{N} \Phi_j \qquad (4)$$

We have introduced in Eq.s (1-4) the critical currents I_j, whose distribution $P(I_j)$ is assumed to be Lorentzian around an average value I: $P(I_j) = (\gamma /\pi)(\gamma^2 + (I-I_j)^2)^{-1}$.

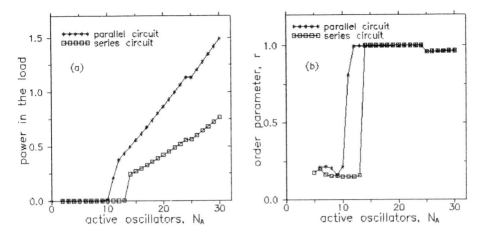

Fig. 1. (a) Load power versus number of active junctions and (b) fraction of the phase-locked junctions (Kuramoto order parameter r) for parallel (stars) and series (squares) array. Parameters of the simulations are: N=30, β=10, γ=0.003, δ=0.006, $β_L$=1, Q=100.

The qualitative picture of what is happening is the following: The resonance curve of the passive RLC load is centered at the frequency Ω, while the natural (i.e., uncoupled) frequencies of the Josephson junctions have a spread ΔΩ around the center of the distribution $Ω_0$= Ω - δ. The key point is that when a new junction is added to the circuit, it oscillates off resonance, and the power in the load increases slowly. At a certain critical level, after a certain number of junctions have been added, the power into the resonator is large enough to pull the junctions frequencies well into the resonance. From now on, since the active oscillators are pumping the cavity at the resonance frequency Ω, the power of the load is large enough to force most (if not all) the junctions activated to phase-lock to the cavity, and so indirectly to the other junctions. The results of the simulations confirm this qualitative analysis; in Fig.2a we can observe a sharp increase of the power when there are 11 or 14 active junctions for the parallel and series coupling scheme, respectively. We want to emphasize that the mechanism just described depends very little on the order on which the oscillators are activated: we have tried to sort the oscillators in several (10) different orders, but the threshold at which the power increases is little changed (at most the curve deviates with one junction).

3. A GENERAL FRAMEWORK

To further check that the qualitative explanation of the origin of the threshold is correct, we have also plotted in Fig.2b the behavior of the fraction of phase-locked junctions as a function of the activated junctions. To do so, we have introduced an "order parameter" analogous to the Kuramoto order parameter often used for coupled nonlinear oscillators (Kuramoto 1975, Strogatz 1994):

$$r = \frac{1}{N_A} \sum_{i=1}^{N_A} e^{i\Phi_i}$$

That is a measure of the number of the junctions locked (normalized to the total number of active junctions N_A). As one can notice, in Fig. 2b the Kuramoto order parameter shows a sharp increase in correspondence of the threshold, jumping instantaneously to 1 (i.e., all active junctions become locked). This "perfect" synchronization is a desired feature for applications, and therefore we believe that the Kuramoto model can be successfully employed for a deeper understanding of oscillators strongly coupled through a resonator (Filatrella et al 1999), as already demonstrated for overdamped,

176

weakly coupled arrays (Wiesenfeld et al, 1998). We believe that the key ingredient is to modify the Kuramoto model by allowing the coupling constant to depend upon the dynamics, i.e. by writing:

$$\dot{\theta} = \omega_j - \frac{1}{N_A} \sum_{i=1}^{N_A} K_i sin(\theta_j - \theta_i + \alpha)$$

In the traditional Kuramoto model the coupling constant K_i is a constant, while we propose it should depend upon the dynamics: it should have a relatively large value if the ith oscillator is phase-locked, and otherwise has a relatively small value. Moreover, the sum is extended only to the active junctions. This general picture can embody in generic dynamical terms the features of Josephson junctions arrays coupled to a resonant cavity.

3. CONCLUSIONS AND AKNOWLEGMENTS

The model presented in this paper describes the experimental results presented by Barbara et al in 1999. It is suggested that the resonant cavity provides the key mechanism for the strong interaction among the junctions, that leads to the very efficient DC to AC power conversion in the experiments.

We gratefully acknowledge P. Barbara and C. Lobb for extensive discussions and for showing us their experimental results prior to publication. We also thank T. Bohr, J. Mygind, and S. Benz for useful discussions, and the DTU Department of Physics for hospitality.

REFERENCES

Barbara P, Cawthorne A B, Shitov S V, and Lobb C J 1999 Phys. Rev. Lett. **82**, 1963
Benz S P and Burroughs C J 1991, Appl. Phys. Lett. **58**, 2162
Bonifacio R, Casagrande F, and Milani M 1982, Lett. Nuovo Cimento **34**, 520
Filatrella G and K Wiesenfeld 1994, J. Appl. Phys. **78**, 1878
Filatrella G, Pedersen N F and Wiesenfeld K 1999, unpublished
Kuramoto Y 1975 Proc. of the Int. Symposium on Mathematical Problems in Theoretical Physics, ed H Araki, Lecture Notes in Physics **39** (Berlin: Springer); Chemical Oscillations, Waves, and turbulence (Berlin: Springer, 1984).
Monaco R, Grønbeck-Jensen R, and Parmentier R D 1990, Phys. Lett. A **151**, 195
Strogatz S H 1994, "*Norber Wiener's Brain Waves*" in Lecture Notes in Biomathematics, **100** (New York: Springer)
Tilley A R 1970, Phys. Lett. A **33**, 205
Wiesenfeld K, Benz S P, Booi P A A 1992, J. Appl. Phys. **76**, 3835
Wiesenfeld K, Colet P and Strogatz S H 1997, Phys. Rev. Lett. **76**, 404
Wiesenfeld K, Colet P and Strogatz S H 1998, Phys. Rev. E **57**, 1563

Inst. Phys. Conf. Ser. No 167
Paper presented at Applied Superconductivity, Spain, 14-17 September 1999
© 2000 IOP Publishing Ltd

Analysis of a hot-spot response of a long Josephson junction in the flux-flow regime

M V Fistul and A V Ustinov

Physikalisches Institut III, Erlangen-Nürnberg Universität, D-91058 Erlangen Germany

ABSTRACT: We theoretically investigate a Low Temperature Scanning Microscopy (LTSM) response of a long Josephson junction in the flux-flow regime. The LTSM response appears here due to the presence of a spatially inhomogeneous interaction between Josephson current wave and linear electromagnetic waves propagating in the junction. We show analytically and verify numerically that the LTSM response displays small oscillations and two pronounced maxima nearby the junction boundaries. The dependencies of LTSM response on the bias current and the size of Josephson junction are found to be in good accord with earlier LTSM experiments.

1. INTRODUCTION

Considerable attention has been devoted to a study of *flux-flow* regime in long Josephson junctions (Kulik and Yanson 1972, Barone and Paterno 1982, Golubov et al 1996, Koshelets et al 1997). In this regime, a long Josephson junction of the size $L \gg \lambda_J$, where λ_J is the Josephson penetration length, is subject to an external magnetic field H applied parallel to the junction plane. It is a well known that under these conditions and in the presence of a bias current flowing through the junction the dense chain of Josephson vortices propagates via the junction and its current-voltage characteristic (CVC) displays a pronounced resonance, the Eck peak (Eck et al 1964, Kulik and Yanson 1972). This resonance appears due to an interaction between the Josephson current density wave and an electromagnetic wave (EW) propagating along the Josephson junction. The voltage position of this resonance V_0 is determined by the spectrum $\omega(k)$ of EWs and is proportional to H in a homogeneous long Josephson junction (Kulik and Yanson 1972). Thus, a long Josephson junction biased on the resonance becomes a tunable source of microwave radiation (Nagatsuma et al 1983, Koshelets et al 1997).

The dynamic states of a long Josephson junction, i. e. the interaction between the propagating Josephson vortices and EWs, can be studied by Low Temperature Scanning Microscopy (LTSM). LTSM technique uses a focused laser or electron beam for local heating of a sample. Local heating leads to an additional dissipation in the small area. A hot-spot induced variation of the voltage drop on the sample is recorded versus the hot spot coordinates. This method allows to visualize the distribution of electromagnetic fields inside a junction (Mayer et al 1991, Quenter et al 1995).

In this paper we present a theory (numerical simulation and analytical approach) of *inhomogeneous* LTSM response of a long *homogeneous* Josephson junction in the flux-flow regime. Our analysis allows to account for the LTSM response dependence on the bias current, magnetic field and the junction size. We also will be able to consistently explain the previously published measurements (Quenter et al 1995).

2. MODEL AND NUMERICAL SIMULATIONS OF LTSM RESPONSE

As a model for our analysis, we consider a one-dimensional long Josephson junction in the presence of magnetic field H and the bias current I. The Josephson phase difference of this system

$\varphi(x,t)$ depends on the coordinate x along the junction and, in the presence of a finite voltage V, on time. This dependence is described by the normalized dynamical equation (Barone and Paterno 1982):

$$\frac{\partial^2 \varphi(x,t)}{\partial x^2} - \frac{\partial^2 \varphi(x,t)}{\partial t^2} - \alpha(x)\frac{\partial \varphi(x,t)}{\partial t} - \sin \varphi(x,t) = -\gamma. \qquad (1)$$

Here, the unit of time is the inverse Josephson plasma frequency and the coordinate x is normalized to λ_J, γ is the normalized density of the bias current and $\alpha(x)$ is the dissipation coefficient that depends on the quasiparticle tunneling resistance of the junction. In the presence of a hot-spot placed at the point x_0 the dissipation coefficient locally increases: $\alpha(x)=\alpha+\varepsilon f(x-x_0)$. The properties of $f(x-x_0)$ function are simple: $f(0)=1$ and this function falls to zero on the distance Δx. The normalized external magnetic field h appears via the boundary conditions:

$$\left.\frac{\partial \varphi(x,t)}{\partial x}\right|_0 = \left.\frac{\partial \varphi(x,t)}{\partial x}\right|_l = h. \qquad (2)$$

The numerical simulations of both the CVC and the LTSM response were performed for a particular set of Josephson junction parameters, namely, the normalized junction length $l=25$, magnetic field $h=4$ and the dissipation coefficient $\alpha=0.1$. For our numerical calculation we used the value of the dissipation increase due to the hot-spot $\varepsilon=0.05$ and the size Δx of hot spot was $0.2\lambda_J$.

By making use of the direct numerical integration of the Eq. (1) we obtain the CVC displaying a sharp resonance. The simulation is shown in Fig. 1(a). The resonance appears due to a strong interaction between the dense chain of Josephson vortices and EWs. Similarly to the previous work by Koshelets et al 1997 we obtain that even in our relatively long Josephson junctions ($l\alpha{\sim}2.5$) the resonance splits into a series of resonant Fiske Steps.

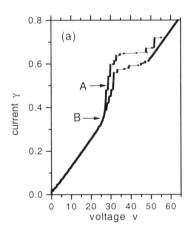

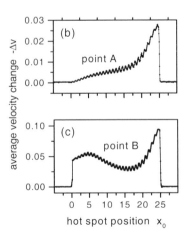

Fig. 1 Numerically simulated current-voltage characteristic (a) and LTSM response (b)-(c) of a long Josephson junction in the presence of an external magnetic field $h=4$. LTSM response are shown for two values of bias current: b) $\gamma=0.5$ (point A); c) $\gamma=0.35$ (point B).

We also numerically simulated the LTSM response of the junction for different values of the bias current (Fig. 1(b)-1(c)). The most striking feature is that the response has a pronounced maximum nearby the boundary where the Josephson vortices enter to the junction, at $x_0 \leq 25$ (Fig. 1b). Moreover, when the bias current decreases the maximum of the response appears also nearby the other boundary (Fig. 1c). We also find small oscillations of the simulated LTSM response along the junction.

3. ANALYTICAL APPROACH AND DISCUSSION

In this Section we derive the LTSM response of a long Josephson junction in the limit of large magnetic field $h>>1$ (flux-flow regime). Note here that the opposite case $h<<1$ has been studied by Malomed and Ustinov 1994. In the flux-flow regime, the solution of the Eq. (1) can be written in the form:

$$\varphi(x,t) = hx + \omega t + \varphi_0(x) + \operatorname{Im}\varphi_1(x)e^{i\omega t}, \qquad (3)$$

where $\varphi_0(x)$ is the time independent perturbation of the Josephson phase induced by the hot-spot, $\varphi_1(x)$ is the amplitude of EW propagating in the junction and $\omega=2eV/\hbar$. These two functions $\varphi_0(x)$ and $\varphi_1(x)$ satisfy the equations:

$$\frac{d^2\varphi_0(x)}{dx^2} = \omega\varepsilon f(x-x_0), \qquad (4a)$$

$$\frac{d^2\varphi_1(x)}{dx^2} + \left[\omega^2 - i\omega\alpha - i\omega\varepsilon f(x-x_0)\right]\varphi_1(x) = e^{i(hx+\varphi_0(x))}. \qquad (4b)$$

By making use of the method elaborated in (Kulik and Yanson 1972, Barone and Paterno 1982) we obtain the *inhomogeneous* LTSM response of the junction that appears due to an interaction between the Josephson current wave and EWs in the presence of hot-spot:

$$\delta V(x_0) = \operatorname{Im}\frac{2}{\alpha}\iint dxdx_1\varphi_0(x_1)G(x,x_1)\sin\left[h(x-x_1)\right] +$$

$$+ \varepsilon\omega\left[\left|\int dxe^{ihx}\operatorname{Re}G(x,x_0)\right|^2 - \left|\int dxe^{ihx}\operatorname{Im}G(x,x_0)\right|^2\right], \qquad (5)$$

where $G(x,x_1)$ is the Green function of the equation:

$$\frac{\partial^2 G(x,x_1)}{\partial x^2} + \left(\omega^2 - i\alpha\omega\right)G(x,x_1) = \delta(x-x_1).$$

The analysis of the Eq. (5) immediately shows that the first term in the right-hand part of Eq. (5) is odd function of the magnetic field and the bias current, and the second term is even function.

We also obtain that the shape of LTSM response crucially depends on the size of the junction L. The crossover occurs in the region of $l \sim l_0 = 1/\alpha$, where the particular length l_0 determines the attenuation of the EWs in the junction. A most interesting result has been found in the limit of large L, when $l >> 1/\alpha$. In this case the interaction between the hot-spot induced EWs and Josephson vortices occurs only on the distance l_0 from the boundaries. It leads to the LTSM response displaying two maxima on the boundaries of the junction:

$$\delta V(x_0) \cong \begin{cases} \dfrac{2\varepsilon}{V_0^2 \alpha^2} e^{\alpha(x_0-l)}, & V - V_0 << \alpha \\[4mm] \dfrac{2\varepsilon}{V_0^2 (V-V_0)^2} e^{-\frac{\alpha l}{2}} \cosh\left[\alpha(x_0 - \dfrac{l}{2})\right], & V - V_0 >> \alpha \end{cases} \qquad (6)$$

The width of maxima is approximately l_0. Moreover, the relative amplitudes of the maxima depend on the bias point. Thus, the amplitude of the maximum associated with the exit of vortices decreases when the bias current approaches to the top of the resonance. Note here, that the Eq. (6) contains only the leading term over parameter $\alpha/\omega << 1$. In the limit ($l >> 1/\alpha$) the small oscillations of $\delta V(x_0)$ have been also found. The amplitude of such oscillations $\alpha/\omega << 1$ and the number of oscillation $n = hl/\pi$. All these analytically described features of LTSM response are in very good agreement with our numerical simulation that have been carried out for not very large value of parameter $l\alpha \sim 2.5$ (Fig. 1). The main result of our theory, i. e. Eq. (6), also allows to well explain the experimentally observed LTSM pattern with two maxima (Quenter et al 1995).

In the opposite limit of small junction ($l << 1/\alpha$) the hot-spot induced interaction between the Josephson vortices and EWs occurs in a whole sample and the inhomogeneous LTSM response linearly depends on the hot-spot coordinate. It also is in a good agreement with previously published numerical simulations and experiments (Quenter et al 1995).

REFERENCES

Barone A and Paterno G 1982 Physics and Applications of the Josephson Effect (Wiley, New York, 1982)

Golubov A A, Malomed B A and Ustinov A V 1996 Phys. Rev. B **54,** 3047

Eck R E, Scalapino D J and Taylor B N 1964 Phys. Rev. Lett **13**, 15

Koshelets V P., Shitov S V, Shchukin A V, Filippenko L V, Mygind J and Ustinov A V 1997 Phys. Rev. B **56**, 5572

Kulik I O and Yanson I K 1972 The Josephson Effect in Superconductive Tunneling Structures (Israel Program for Scientific Translations, Jerusalem, 1972).

Malomed B A and Ustinov A V 1994 Phys. Rev. B **49**, 13024

Mayer B, Doderer T, Huebener R P and Ustinov A V 1991 Phys. Rev B **44**, 12463 (1991)

Nagatsuma T, Enpuku K, Irie F and Yoshida K 1983 J. Appl. Phys. **54**, 3302

Quenter D, Ustinov A V, Lachenmann S G, Doderer T, Huebener R P, Müller F, Niemeyer J, Pöpel R and Weinmann T 1995 Phys. Rev B **51**, 6542

Inst. Phys. Conf. Ser. No 167
Paper presented at Applied Superconductivity, Spain, 14-17 September 1999
© 2000 IOP Publishing Ltd

Enhancement of the supercurrent by coherent tunneling in the double-barrier Nb/Al-AlO$_x$-Al-AlO$_x$-Nb devices

I P Nevirkovets[1,2] and J B Ketterson[1,3]

[1]Northwestern University, Department of Physics and Astronomy, Evanston, Illinois 60208
[2]Institute for Metal Physics, National Academy of Sciences of the Ukraine, UA-252680 Kyiv-142, Ukraine
[3]Northwestern University, Department of Electrical and Computer Engineering, Evanston, Illinois 60208

ABSTRACT: We have carried out experiments on Nb/Al-AlO$_x$-Al-AlO$_x$-Nb double-barrier junctions with a "dirty" middle Al layer. At low temperatures, the devices displayed a critical current larger than that possible for a system considered as a simple, series-connection, of two (Nb/Al-AlO$_x$-Al and Al-AlO$_x$-Nb) junctions, and a novel subgap structure. The behaviour is explained as a manifestation of the Andreev bound states appearing in the double-barrier devices.

1. INTRODUCTION

Recently, the SINIS type junctions have demonstrated their potential as switching elements for superconducting digital circuits (Maezawa and Shoji 1997, Sugiyama et al. 1997), voltage standards (Schulze et al. 1998), and microrefrigerators (Manninen et al. 1997). In particular, the Nb/Al-AlO$_x$-Al-AlO$_x$-Nb devices displayed a characteristic voltage of order 0.1 mV and non-hysteretic behaviour at a temperature of 4.2 K. Up to now, most of the work was devoted to fabrication and application aspects of the devices. However, the lack of experimental and theoretical work on the underlaying physical mechanisms of the electron transport hinders further optimization of the device characteristics.

In this work, we report the results of an experimental investigation of the current-voltage characteristics (IVC's) of the double-barrier Nb/Al-AlO$_x$-Al-AlO$_x$-Nb devices, which suggests that the current through the devices can not be understood as a two-step sequential tunneling process. We have found peculiarities that can be explained by coherent transport of quasiparticles, rather than by the ordinary Josephson effect.

2. DEVICE CHARACTERIZATION

The Nb/Al-AlO$_x$-Al-AlO$_x$-Nb structures were fabricated by a conventional whole-wafer dc magnetron deposition process. Both oxydized Si and R-plane sapphire substrates were used, and 10 μm × 10 μm two-terminal devices were fabricated. The thickness of the external Nb electrodes was 100 nm for all the devices. Usually, in Nb/Al junction technology, little attention is paid to the quality of the Al films. However, we have found that a significant modification of the IVC occurs when using "dirty" Al films in the middle electrode of the double-barrier Nb/Al-AlO$_x$-Al-AlO$_x$-Nb junctions. Here we consider devices in which the Al electrode was deliberately contaminated by introducing 1×10^{-5} torr of O$_2$ during sputtering of the Al. The Al used to form the bottom junction was always "clean" (i.e., deposited without deliberate contamination), whereas for the top junction both "dirty" Al and composite ("dirty" Al covered with "clean" Al) electrodes were used. The tunnel barriers were formed by thermal oxidation of Al in pure oxygen, identically for the bottom and top barriers, to provide a specific tunneling resistance (defined for a single junction from the stack) in the range 10^{-7} - 10^{-6} $\Omega \times cm^2$. The devices were characterized by measuring their IVC's between the bottom and top Nb electrodes.

3. RESULTS AND DISCUSSION

First, we consider a device fabricated on a sapphire substrate with the thickness of the middle ("dirty") Al electrode d_{Al}=7 nm. The IVC of the device is shown in Fig. 1 at T=4.2 K (curve a) and T=1.8 K (curve b). At T=4.2 K, the device has an I_c~20 µA (see inset in Fig. 1), and nonhysteretic behavior. At T=1.8 K, the dc Josephson current is I_c=2.38 mA and the IVC is hysteretic. Measurement of the I_c vs. H dependence has shown a good quality diffraction pattern (Nevirkovets et al. 1999a).

We can estimate the magnitude of I_c assuming the device is a series connection of two identical, independent junctions, in which the superconductivity in the middle Al layer is induced by the proximity with the external Nb. The influence of the proximity effect can be estimated using the tunneling model (McMillan 1968). If we use these assumptions and a straightforward analysis that involves the formula for the critical current (Ambegaokar and Baratoff 1963), then we obtain I_c~31 µA at T=4.2 K (Nevirkovets et al. 1999a), which is in a reasonable agreement with the observed ~20µA.

Estimation of the theoretical I_c at T=1.8 K must account for the possible existence of an intrinsic energy gap in Al. The double-junction device (cf. Fig. 1) does not display any peculiarity corresponding to V_{2d}=2(Δ_{Nb}-Δ_{Al})/e, unlike the devices reported by Blamire et al. (1991) and Nevirkovets (1997). Instead, it reveals a novel subgap structure around the voltage V=Δ_{Nb}/e (see Fig. 1, steps 1 and 2 in the curve b) and the gap-sum step at V_{2s}=2Δ_{Nb}/e (step 3 in the curve b). Our analysis has shown that the structure may be identified as the gap-difference and gap-sum peculiarities at a voltage V_d=(Δ_{Nb}-Δ_{Al})/e and V_s=(Δ_{Nb}+Δ_{Al})/e (Nevirkovets et al. 1999a, Nevirkovets et al. 1999b). Fig. 2 shows the dependence of $I_c(T)$ (solid circles) and $\Delta_{Al}(T)$ (open circles) derived from the temperature evolution of the IVC, and the temperature dependence of the gap-sum voltage (crosses) for the same device (cf. Fig. 1). For ease of comparison with $\Delta_{Al}(T)$, the gap-sum energy vs. T dependence is shifted along the energy axis by 2.51 meV to lower energy. There is a "tail" in the $I_c(T)$ dependence at T>2.5 K; as the temperature decreases, a sharp rise in both the $\Delta_{Al}(T)$ and $I_c(T)$

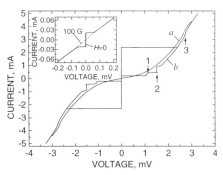

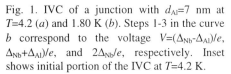

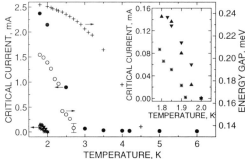

Fig. 1. IVC of a junction with d_{Al}=7 nm at T=4.2 (a) and 1.80 K (b). Steps 1-3 in the curve b correspond to the voltage V=(Δ_{Nb}-Δ_{Al})/e, (Δ_{Nb}+Δ_{Al})/e, and 2Δ_{Nb}/e, respectively. Inset shows initial portion of the IVC at T=4.2 K.

Fig. 2. $I_c(T)$, $\Delta_{Al}(T)$, and 2$\Delta_{Nb}(T)$ dependences (solid circles, open circles, and crosses, respectively) for the same device (cf. Fig. 1). The 2$\Delta_{Nb}(T)$ dependence is shifted along the energy axis to the lower energy by 2.51 meV. Stars and triangles are $I_c(T)$ dependences for SININIS devices (shown in the inset on a magnified scale).

dependences appears in the same temperature interval near $T\sim2.5$ K. This behavior is characteristic of proximitized systems (see, e.g., Gilabert et al. 1971).

The value of $\Delta_{Al}=0.203$ meV derived from the voltage position of steps 1 and 2 in curve b (see Fig. 1) can be used as a measure of the energy gap in the Al electrode at $T=1.8$ K. With $\Delta_{Nb}=1.38$ meV at $T=1.8$ K, we have obtained $I_c=1.70$ mA. This value is considerably smaller than the experimentally measured maximum $I_c=2.38$ mA (Nevirkovets et al. 1999a). Therefore we conclude that our double-barrier devices cannot be considered as a simple, series-connection, of the two junctions.

Such a high value of I_c is unlikely due to the fact that the critical temperature of the "dirty" Al is increased up to $T\sim2.5$ K, at which the sharp raise of the I_c and Δ_{Al} occurs. In order to prove this assumption, we carried out an experiment on SINININIS junctions, where S=Nb, N=Al (the parameters of the S, N, and I components were the same as in ordinary SINIS devices). In the SINININIS junctions, the middle Al film is far away from the two Nb electrodes, so that any induced superconductivity in it should be substantially reduced as compared with the SINIS geometry. As expected, the magnitude of the critical current in SINININIS junctions is considerably lower than in SINIS junctions; moreover, the $I_c(T)$ dependence has a steep raise at $T\sim1.95$-2.0 K, without any noticeable "tail" at higher temperatures (see Fig. 2 and the inset in this figure, where the $I_c(T)$ dependences for SINININIS junctions are shown on an expanded scale). An additional argument is that the temperature dependence of the gap-sum voltage does not reveal any feature at $T\sim2.5$ K (see Fig. 2, crosses); in fact, it equals to $2\Delta_{Nb}/e$, so that Al is not superconducting at this voltage. Therefore, both the dc supercurrent through the SINIS devices and Δ_{Al} are enhanced at $V<2\Delta_{Nb}/e$.

We will assume that Andreev bound states may appear in our double-barrier devices and they provide an additional channel for the supercurrent, which therefore can exceed the maximum I_c value possible in the single-barrier junction (provided the barrier transparency is identical in the two cases) (Nevirkovets and Shafranjuk, 1999c). Next we present clear evidence that coherent transport of quasiparticles can contribute to the electric current through the devices and substantially modify their current-voltage characteristics. In Fig. 3, the IVC of a Nb/Al-AlO$_x$-Al-AlO$_x$-Nb device with less transparent tunnel barriers is shown at $T=4.2$ K (curve a) and $T=1.8$ K (curve b). This device was also fabricated with a "dirty" middle electrode (with the thickness $d_{Al}=14$ nm). The IVC at $T=4.2$ K displays only the gap-sum feature at $V=2\Delta_{Nb}/e$ and is smooth in the subgap region. This behaviour is typical of similar devices reported earlier (see, e.g., Maezawa and Shoji 1997). However, as the temperature is lowered (below $\sim$3K), an unusual feature appears in a narrow energy interval in the vicinity of $V=\Delta_{Nb}/e$, which has a step-like shape. The position of the step moves towards lower

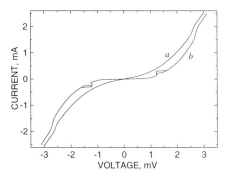

Fig. 3. Typical IVC of a Nb/Al-AlO$_x$-Al-AlO$_x$-Nb device on the Si substrate with a 14 nm "dirty" Al electrode. Curves (a) and (b) are measured at $T=4.2$ K and 1.8 K, respectively.

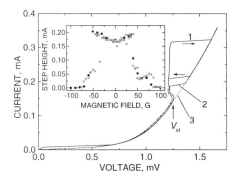

Fig. 4. Initial portion of the IVC of an identical device (cf. Fig. 3) measured at $T=1.82$ K and $H=0$, 60, and 100 G (curves 1-3, respectively). Inset shows step height (measured at a voltage $V_{st}\approx1.25$ mV) vs. magnetic field dependence for two devices.

voltages as the temperature is decreased, while the current increases. Note that the dc Josephson current is much smaller than the step height even at low temperatures (see curve measured at $H=0$ in Fig. 4). The shape of the step is sensitive to a small magnetic field. A magnetic field suppresses the step, so that by $H=100$ G it disappears completely. The inset in Fig. 4 shows the step height as a function of magnetic field for the two identical devices. The step height was measured at $V=1.25$ mV (a voltage at which the step height is maximal at zero field, as indicated by the arrow in the Fig. 4) with respect to the background current at this voltage in a field of 100 G. Note we have a minimum at $H=0$ and an indication of periodic behaviour.

High sensitivity to a magnetic field is strong evidence that the effect is related to phase coherent electron transport through the device, although it is unlikely that it is due to usual Josephson effect, because the dc Josephson current, I_c, is much smaller than the amplitude of the step. We suggest that the observed effect is due to a coherent quasiparticle current between the external Nb electrodes and may be a manifestation of the Andreev bound states.

4.   CONCLUSION

Both the magnitude of the supercurrent at $V=0$ and the novel magnetic-field sensitive subgap features observed on double-barrier Nb/Al-AlO$_x$-Al-AlO$_x$-Nb devices suggest that the devices cannot be considered as a simple, series-connection, of the two junctions. Rather, the AlO$_x$-Al-AlO$_x$ trilayer acts as a single barrier, and it turns out that this fairly thick structure can, under some conditions, maintain almost as large a supercurrent as a single AlO$_x$ barrier in Nb/Al-AlO$_x$-Nb junctions. We suggest that Andreev bound states may provide an additional channel for the coherent transport in the double-barrier junctions.

ACKNOWLEDGMENT

This work was supported by the Northwestern Materials Research Center under the NSF MRSEC program, Grant No. DMR9309061. Fruitful discussions with S. E. Shafranjuk are acknowledged.

REFERENCES

Ambegaokar V and Baratoff A 1963 Phys. Rev. Lett. **10**, 486; Phys. Rev. Lett. **11**, 104
Blamire M G, Kirk E C G, Evetts J E and Klapwijk T M 1991 Phys. Rev. Lett. **66**, 220
Gilabert A, Romagnon J P and Guyon E 1971 Solid St. Communs. **9** 1295
Maezawa M and Shoji A 1997 Appl. Phys. Lett. **70**, 3603
Manninen A J, Leivo M M and Pekola J P 1997 Appl. Phys. Lett. **70**, 1885
McMillan W L 1968 Phys. Rev. **175**, 537
Nevirkovets I P 1997 Phys. Rev. B **56** 832
Nevirkovets I P, Ketterson J B and Lomatch S, 1999a Appl. Phys. Lett. **74**, 1624
Nevirkovets I P, Ketterson J B, Shafranjuk S E and Lomatch S 1999b (unpublished)
Nevirkovets I P and Shafranjuk S E 1999c Phys. Rev. B **59**, 1311
Schulze H, Behr R, Mueller F and Niemeyer J 1998 Appl. Phys. Lett. **78**, 996
Sugiyama H, Yanada A, Ota M, Fujimaki A and Hayakawa H 1997 Jpn. J. Appl. Phys. **36**, L1157

Inst. Phys. Conf. Ser. No 167
Paper presented at Applied Superconductivity, Spain, 14-17 September 1999
© 2000 IOP Publishing Ltd

Transition properties of $YBa_2Cu_3O_7$ step-edge Josephson junctions with various step orientations

Soon-Gul Lee[1], Yunseok Hwang[2], Jin-Tae Kim[3].

[1]Department of Physics, Korea University, Jochiwon, Chungnam 339-800, Republic of Korea.
[2]Department of Physics, Korea University, Sungbuk-ku, Seoul 136-701, Republic of Korea.
[3]Korea Research Institute of Standards and Science, Taedok Science Town, Taejon 305-600, Republic of Korea

ABSTRACT: We have studied transition properties of $YBa_2Cu_3O_7$ step-edge junctions with different step orientations ranging from 0 degree to 165 degrees with respect to one of the major axes of substrate at 15-degree interval. The junctions were prepared on $SrTiO_3$ (100) substrates by pulsed laser deposition and argon ion milling with photoresist mask. We investigated current-voltage characteristics and critical current of the junctions as a function of the angle. The junction critical current showed an angle dependent modulation with maxima near 0 or 90 degree and minima near 45 and 135 degrees. We believe that the critical current variation with the step orientation is associated with the symmetry of high Tc superconductor.

1.   INTRODUCTION

Since the discovery of high temperature superconductors, various types of high Tc Josephson junctions were developed for electronic applications. Among them bicrystal junctions (Dimos et al 1988), step-edge junctions (Simon et al 1990), and ramp-edge junctions (Gao et al 1990) are most widely used. Bicrystal junctions are easy to fabricate, reproducible, and excellent in noise property. However, high cost of the substrate and restricted topological freedom are major obstacles to applications that require many junctions. On the other hand, ramp-edge junctions have large topological freedom, but the fabrication technology of reliable high quality junctions is yet to be developed. Step-edge junctions can be viewed as a compromise of the two.

Step-edge junctions are cost effective, easy to fabricate, and reproducible. In addition, they have good noise properties comparable to bicrystal junctions and relatively large topological freedom. For those reasons, the step-edge junction can be the major junction type for the future high Tc electronics. To maximize the topological freedom of step-edge junctions, junction properties need to be investigated systematically for various step-line orientations with respective to the crystallographic axes of the substrate.

Usually it is recommended that the step-line is oriented parallel to one of the substrate axes to get good step-edge junction properties. Different step-line orientation will induce different microscopic structure and thus affect the critical junction properties. However, how it affects the junction properties is not known clearly, nor has it been studied systematically. In addition, the effects of symmetry of the high T_c superconductor have not been studied in step-edge junctions.

In this work, we fabricated YBCO step-edge junctions with various tilt angles of the step-line with respect to the $SrTiO_3$ substrate axes, and investigated current-voltage characteristics and critical current of the junctions. Discussion of the results was based on the structural analyses and d-wave symmetry.

2.   JUNCTION FABRICATION

12 junctions were made on a 1 cm × 1 cm SrTiO₃ (100) substrate and each junction had a different step-line angle, from 0° to 165° at every 15°, with respect to the crystal axis (Fig. 1). Substrate steps were formed on SrTiO₃ (100) by ion milling with photoresist mask. We deposited about 20-nm-thick Au film on the substrate prior to photolithography. The usage of Au film is known to reduce scattering of ultraviolet under the Cr mask during exposure (Braginski 1996), and the step-line was much better defined according to scanning-electron-microscopic study. To make the ramp-angle of the steps uniform, the substrate was rotated at a few rpm's during milling with the substrate's normal tilted from the beam incidence direction by 20°. The ramp angle of the step was about 60°.

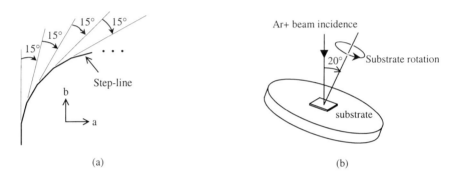

(a)   (b)

Fig. 1. (a) Step geometry on the SrTiO₃ substrate.   (b) Milling conditions for the formation of substrate steps.

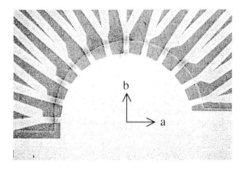

Fig. 2. Scanning electron microscopic image of the step-edge junctions on the SrTiO₃ (100) substrate. Crystal axes of the substrate are oriented along the horizontal and the vertical axes.

YBCO film was deposited by pulsed laser deposition method and patterned by Ar ion milling with a photoresist mask. Step height (h) was 220 nm – 270 nm and the film thickness (t) was 140 nm – 160 nm with t/h ≈ 2/3. During film deposition oxygen pressure was 400 mtorr and the substrate temperature was 810 °C. After deposition the film was annealed for 1 hour at 500 °C in 1 atmospheric pressure of oxygen.

Fig. 2 shows the 12 step-edge junctions with a common electrode. Crystal axes of the substrate are along the horizontal and the vertical directions. Starting from the leftmost junction the step-line angle of each junction in clockwise direction is 0°, 15°, 30°, 45°, 60°, 75°, 90°, 105°, 120°, 135°, 150°, 165°, respectively. Junction width is 5 μm.

3.   EXPERIMENTAL RESULTS AND DISCUSSION

We investigated electrical properties by measuring current-voltage (I-V) curves and critical current of the junctions, and the surface structure by scanning electron microscopy. Junction I-V curves of sample A are shown in Fig. 3 (a). All curves show a sharp transition above the critical currents. According to the thermal fluctuation theory (Ambegaokar and Halperin 1969), $\gamma = \hbar I_c / e k_B T > 10^3$, and thus thermal rounding at the transition is negligible, as observed in Fig. 3 (a). Above I_c, the curves show downward curvature instead of the typical upward curvature of the standard resistively-shunted-junction model (Barone and Paterno 1982). The reason for downward curvature could be flux flow which was self-induced by the junction current.

In Fig. 3 (b) critical currents of the junctions obtained from the I-V curves for 3 different samples are plotted as a function of the step-line angle. Dotted lines connecting data points are drawn as a guide to the eye. As shown in the figure, I_c is maximum near 0° and 75°-90°, and minimum around 30°-45° and 135°. In other words, when the step-line is parallel to the crystal axis of the substrate, critical current is maximum, and when the step is tilted by 45° from the axis, I_c is minimum. Even though some of the data points show slight deviation, the general trend as mentioned above can be easily recognized. Among the samples we have studied, many of them showed similar characteristics as those in Fig. 3 (b).

However, some of our samples showed quite different characteristics. All of those samples had non-RSJ I-V characteristics, i.e., severe rounding at the transition and downward curvature with increasing junction current. In those cases, it was difficult to define critical current and no correlation was found between the critical current and the step-line angle. Surface observed under a scanning electron microscope revealed bumps along the step line.

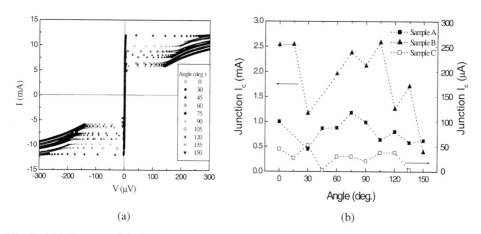

(a)   (b)

Fig. 3. (a) I-V curves of the junctions of sample A. (b) Critical current as a function of the step-line angle for samples A, B, and C. Sample A and B were measured at 77 K and C at 70 K.

Modulation of the junction critical current in angle can be analysed as following. One possibility is microscopic structure along the step line. The atomic arrangement along the step will be jagged and the period of the jag depends on the tilt angle. Growth of $YBa_2Cu_3O_7$ (YBCO) film on the jagged edge will be different from that on the edge without jag, and thus geometry of the grain boundary might be different. Scanning electron microscopic study supported this interpretation. Equally spaced small bumps were observed along the step line and the period was dependent on the tilt angle of the step line. However, it is not clearly understood quantitatively.

Another and a more persuading possibility is d-wave symmetry. Epitaxial growth of $YBa_2Cu_3O_7$ film on the $SrTiO_3$ substrate induces grain boundary junctions at the step-edge and the tilt angle of the grain boundary is equal to the step-line angle. According to the theory of the d-wave symmetry, the critical current of the grain boundary junction is proportional to the cosine of the twice of the tilt angle (Sigrist and Rice 1992). The experimental observation that maxima near $0°$ and $90°$ and minima near $45°$ and $135°$ matches well with the d-wave theory.

According to Mannhart et al (1997) the bicrystal grain boundary is zigzagged and the distribution of the current density is strongly affected by the orientation of the boundary due to d-wave symmetry. Step-edge junctions will be also zigzagged for the same reason as the bicrystal junction and the critical current is the integral of the complicated current distribution. The random nature of zigzagging will lower the reproducibility of the critical current, which is believed to cause the deviation of some data points in Fig. 3 (b). We also believe that, for the same reason, we couldn't observe the modulation as that in Fig. 3 (b) for some of our samples.

4.   CONCLUSION

We have studied current-voltage characteristics and critical current of $YBa_2Cu_3O_7$ step-edge junctions for various tilt angles of the step-line with respect to the crystallographic axes of the $SrTiO_3$ substrate. Junction critical current showed modulation behaviour with maxima near $0°$ or $90°$ and minima around $45°$ or $135°$. The I_c modulation is believed due to difference in the microstructure of the epitaxially grown $YBa_2Cu_3O_7$ film at the step and d-wave symmetry.

ACKNOWLEDGEMENTS

This work was supported by the Ministry of Education, Republic of Korea.

REFERENCES

Ambegaokar V and Halperin B I 1969 Phys. Rev. Lett. **22**, 1364.

Barone A and Paterno G 1982 Physics and Applicaions of the Josephson Effect (New York, John Wiley & Sons) Chap 6.

Braginski A I 1996 SQUID Sensors: Fundamentals, Fabrication and Applications, ed H Weinstock (Dordrecht, the Netherlands: Kluwer Academic Publishers) pp 235-288.

Dimos D, Chaudhari P, Mannhart J and LeGoues F K 1988 Phys. Rev. Lett. **61**, 219.

Gao J, Aarnink W A M, Gerritsma G J and Rogalla H 1990 Physica C **171**, 126.

Mannhart J, Hilgenkamp H and Gerber Ch 1997 Physica C **282-287**, 132.

Sigrist M and Rice T M 1992 J. Phys. Soc. Jpn. **61**, 4283.

Simon R W, Burch J F, Daly K P, Dozier W D, Hu R, Lee A E, Luine J A, Manasevit H M, Platt C E, Schwarzbeck S M, St John D, Wire M S and Zani M J 1990 Science and Technology of Thin Film Superconductors 2, eds R D McConnell and R Noufi (New York: Plenum) pp 549-558.

Inst. Phys. Conf. Ser. No 167
Paper presented at Applied Superconductivity, Spain, 14-17 September 1999
© 2000 IOP Publishing Ltd

Characteristics of step-edge $YBa_2Cu_3O_y$ junctions on various step-angle substrates

S.Y. Yang[1], Chun-Hui Chen[1], J.T. Jeng[1], H.E. Horng[1], and H.C. Yang[2]

[1]Department of Physics, National Taiwan Normal University, Taipei 117, Taiwan
[2]Department of Physics, National Taiwan University, Taipei 106, Taiwan

ABSTRACT: The Josephson coupling was investigated by measuring the characteristics of YBCO Josephson junctions on step-edge substrates with various step angles. Through the etching process developed in this work, the step angle can be varied over a range up to $70°$ to obtain a series of grain-boundary orientations for YBCO Josephson junctions. It was found that the critical current density J_c of the junction decreased with raising the step angle. And also, the I_cR_n's of the junctions with various step angles were observed to fall onto a single scaling line $I_cR_n \propto (J_c)^p$ with $p \sim 0.5$. This implies that the physical mechanism of the Josephson coupling is similar for all various grain-boundary orientations, but differs in the coupling strength.

1. INTRODUCTION

Due to the layered structure of the high-T_c YBCO superconductor, the transport properties are anisotropic and have been demonstrated by many investigators (Tachiki 1989, Song 1991, Horng 1997 and Yang 1999). The anisotropic superconductivity leads to different coupling strengths for the weak links when various interfaces were formed at the intergranular boundaries of high-T_c Josephson junctions (Hermann 1995, Yang 1999). Some researches indicated that the interfaces of the weak links could be adjusted by using step-edge substrates with various step angles (Mitzuka 1993 and Foley 1997). For example, by using $2\tan^{-1}(ma/nc)$ with a = 3.86 Å and c = 11.68 Å (YBCO lattice constants), and m and n are integers, the orientations of the interface resulting from different step angles are <102> for $18.6°$, <101> for $36.2°$, <302> for $52.2°$ and <201> for $66.3°$ etc. Hence, it is very important to be able to control the step angle of the substrate when investigating the anisotropic Josephson coupling. Therefore, in this work, we develop the fabrication technique which can not only reproduce the highly homogeneous step-edge MgO substrates but also achieve various step angles for the step-edge MgO(001) substrates. Furthermore, the properties of the Josephson junctions on the step-edge substrates are studied.

2. EXPERIMENTAL DETAILS

The substrates used for the $YBa_2Cu_3O_y$ (YBCO) Josephson junctions are MgO(001) single crystals. To obtain a step-edge MgO substrate, we developed the following fabrication processes. First, the photoresist AZ1518 was spread onto the substrates followed by annealing the substrates at $80°$ C for 15 minutes. Half of the photoresist was then lifted off through exposure and development. The remained photoresist acted as a mask when the substrate was etched. The substrates were bombarded by Ar ion at a pressure of 0.1 mtorr to etch the uncovered surface of the substrates. The sketch to illustrate the etching process is shown in Fig. 1. The acceleration voltage for the Ar ion beam is 750 V and the beam current is 30 mA. To prevent the photoresist from heat damage during the ion milling, the substrate was attached by using silver paste to a sample holder which was maintained at 17 °C. The sample holder is rotatable and hence the incident angle θ of the Ar ion beam

190

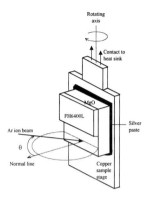

Fig. 1. Configuration of the incident Ar ion beam during the etching process for the MgO substrates. The incident angle θ is made by the propagation of the Ar ion beam with respect to the normal line of the substrate surface.

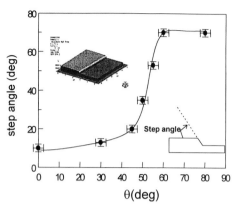

Fig. 2. Step angle versus the incident angle of the Ar ion beam. The inset illustrates the definition of the step angle of a step-edge substrate. A typical AFM image of a 70⁰-step substrate is shown by the left-up inset.

to the substrate can be varied. After ion milling, a step-edge line was formed on the substrates. The step height h of each substrate is around 2000 Å. Then, the step-edge substrates were annealed at 500 °C for 10 hours at atmosphere.

The superconducting YBCO films were deposited onto the step-edge MgO(001) substrates by a homemade off-axis rf magnetron sputtering system with target stoichiometric YBCO. The deposition details were reported elsewhere (Horng 1997). These YBCO films on the step-edge MgO substrates were then patterned to the Josephson junctions through standard photolithographical process (Horng 1998). To obtain a good contact for the electrical measurement, gold was evaporated onto the contact pads of the samples. Aluminum wires were then bound to connect the pads and the electrical board by using a wire bounder. The conventional four-point method was used through the electrical measurement. The criterion voltage for the I_c or J_c is 1 μV.

3. RESULTS AND DISCUSSION

With the rotatable sample holder, the incident angle θ of the Ar ion beam can be varied. The experimental step angles on the MgO substrates for various incident angles θ are shown in Fig. 2. It was found that the step angle can be manipulated by adjusting the incident angle θ. According to the results shown in Fig. 2, the step angle is about 10° when the Ar ion beam is incident normally (θ = 0°) to the substrate. With increasing the incident angle θ up to 50°, the step angle is raised to 35°. When the θ is further increased to 60°, a rapid increase in the step angle up to 70° was resulted. A typical surface image, which was taken by the atomic force microscope (AFM), of the 70⁰-step MgO substrate is shown as the up-left inset in Fig. 2. Then, the step angle remained at 70° as θ ≥ 60°. This empiric relationship between the step angle and the θ shown in Fig. 2 is repeatable in this work. Hence, through the developed etching process, step-edge MgO substrates can be reproduced and the step angle can be varied from 10° to 70°.

To examine the homogeneity of the step-edge line on each substrate, the local surface morphology taken by the AFM, was investigated for three positions which were located along the step-edge line and separated by 1 mm from two neighboring locations. From the AFM images, the maximum variation in the step height along the step line is 75 Å for the 2000 Å-high step. And also, the standard deviation of the step angle is about 2° along each step-edge line. These results reveal that the morphology along each fabricated step-edge line is highly homogeneous.

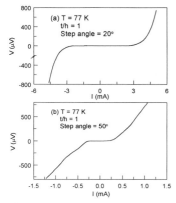

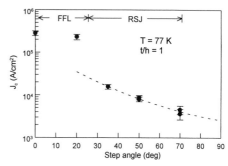

Fig. 3. V-I curves at 77 K of the YBCO films on (a) the 20°-step and (b) the 50°-step MgO substrate. The ratio t/h of the thickness of the YBCO film to the step height is approximately 1.

Fig. 4. Critical current density J_c at 77 K as a function of the step angle. The ratio t/h of the thickness of the YBCO film to the step height is about 1 for each sample. The dashed line is to indicate the decrease in the J_c of the step-edge YBCO Josephson junctions when the step angle is increased.

By using the highly homogeneous step-edge MgO substrates, the YBCO Josephson junctions were manufactured. The ratio t/h of the thickness of the YBCO film to the step height is set at around 1 for the investigated junctions. In Fig. 3, the V-I curves of the 20°- and 50°-step YBCO Josephson junctions at 77 K are plotted. It was found in Fig. 3(a) that the V-I curve of the 20°-step junction exhibited the flux-flow-like (FFL) behavior. This implies that the 20°-step junction acts as a superconducting YBCO micro-bridge instead of a superconducting weak link. However in Fig. 3(b), the V-I curve of 50°-step junction displays the RSJ-like behavior, which gives the evidence that a weak link is formed for this sample. In addition, the two kinks of the V-I curve in Fig. 3(b) imply that the step-edge junction contains two weak links in series. Through a careful inspection, the critical current densities J_c of the 20°- and 50°-step junctions at 77 K are 2.3 x 10^5 A/cm^2 and 8.7 x 10^3 A/cm^2, respectively. The J_c's of other junctions on various step-angle substrates were also measured at 77 K and shown in Fig. 4.

According to the experimental results in Fig. 4, the J_c was observed to almost remain unchanged when the step angle is increased form zero to 20°. As raising the step angle up to 35°, an abrupt decrease in J_c by an order of magnitude was found. Furthermore, the J_c is reduced monotonously with increasing the step angle from 30° to 70°. By comparing the experimental results in Fig. 4 with those in Fig. 3, it was found that the higher J_c's (> 10^5 A/cm^2) correspond to the FFL V-I curves, whereas the V-I curves with lower J_c's (~ 10^4 A/cm^2) are RSJ-like. Hence, it is indicated that the YBCO films were grown epitaxially across the step edges with step angles lower than 35°, but the grain boundaries were formed at the step edges when sputtering the YBCO films onto the substrates with step angles higher than 35°. On the other hand, it is noted that the J_c of the step-edge Josephson junction is reduced when the step angles is raised from 35°

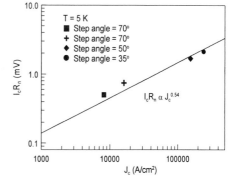

Fig. 5. I_cR_n of step-edge YBOC Josephson junctions at 5 K versus critical current density J_c. Each data point was obtained for an individual junction, as illustrated by the legend. The solid line denotes the scaling line $I_cR_n \propto (J_c)^{0.54}$.

to 70°. Since the Josephson coupling energy of a Josephson junction is proportional to the critical current density, the Josephosn coupling strength of the step-edge YBCO weak link is reduced when the critical current density is lowered by increasing the step angle. The variation in the Josephson coupling strength is believed to be due to the change in the interfaces of the YBCO at the grain boundaries on step-edge substrates with various step angles.

In Fig. 5, the relationship between the I_cR_n and the J_c of the step-edge YBCO Josephson junctions at 5 K is shown. Each point in Fig. 5 was obtained for an individual Josephson junction. It was found that the junction of lower J_c exhibits a lower value of I_cR_n. Besides, the I_cR_n's of the junctions with various step angles were observed to fall onto a single scaling line $I_cR_n \alpha (J_c)^p$ with p ~ 0.5 (denoted by the solid line). This scaling line was also found for other types of grain-boundary Josephson junctions (Hermann 1995, Char 1991 and Vildic 1993) and can be explained by using the intrinsically shunted junction model (Gross 1997) in which the Cooper pairs tunnel through an insulating barrier with localized defects inside. Hence, the observation of the scaling line shown in Fig. 5 implies that the tunneling behaviors of the step-edge YBCO Josephson junctions with various grain-boundary orientations are SIS-like, but differ in the coupling strength when the grain-boundary orientation is changed.

4. CONCLUSION

By developing the etching technique in this work, the step-edge MgO(001) substrates can be reproduced steadily for high-T_c YBCO Josephson junctions. The step angle can be adjusted within the range from 10° to 70°. This provides a significant way to create different orientations of the interfaces for the high-T_c YBCO weak links. It was found that the weak link will be formed when the step angle exceeded a critical value. Furthermore, the Josephson coupling energy is reduced for the junction with a higher step angle. Finally, the observed I_cR_n-J_c curve implies that the tunneling behavior is SIS-like for all various grain-boundary orientations.

5. ACKNOWLEDGEMENT

This work is supported by the National Science Council of R.O.C. under grant Nos. NSC88-2112-M003-008, NSC88-2112-M003-009, NSC88-2113-M002-019, and NSC88-2112-M002-020.

REFERENCES

Char K, Coclough M S, Garrison S M, Newman N and Zaharchuk G 1991 Appl. Phys. Lett. **59** 733
Foley C P, Lam S, Sankrithyan B, Wilson Y, Macfarlane J C and Hao L 1997 IEEE Trans. Appl. Supercond. **7** 3185
Gross R, Alff L, Beck A, Froehlich O M, Koelle D and Marx A 1997 IEEE Trans. Appl. Supercond. **7** 2929
Hermann K, Kunkel G, Siegel M, Schubert J, Zander W, Braginski A I,Jia C L, Kabius B and Urban K 1995 J. Appl. Phys. **78** 1131
Horng H E, Yang S Y, Jeng J T, Wu J M and Yang H C 1997 IEEE Trans. Appl. Supercond. **7** 1177
Horng H E, Yang S Y, Lee W L, Yang H C and Wu J M 1998 Inst. Phys. Conf. Ser. No. **158** 683
Mitzuka T, Yamaguchi K, Yoshikama S, Hayashi K and Enomoto Y 1993 Physica C **218** 229
Song L W, Narumi E, Yang F, Shao H M and Kao Y H 1991 Physica C **174** 303
Tachiki M and Takahachi S 1989 Solid State Commun. **72** 1083
Vildic M, Friedl G, Uhl D, Daalmans G, Kohler H, Meyer H, Bommel F and Saemanns-Ischenko G 1993 IEEE Trans. Appl. Superocnd. **3** 2357
Yang H C, Wang L M and Horng H E 1999 Phys. Rev. B **59** 1
Yang S Y, Chen C H, Horng H E, Lee W L and Yang H C 1999 IEEE Trans. Appl. Supercond. **9** 3121

Inst. Phys. Conf. Ser. No 167
Paper presented at Applied Superconductivity, Spain, 14-17 September 1999
© 2000 IOP Publishing Ltd

On the combination of $YBa_2Cu_3O_{7-d}$ and niobium technology: Material and electrical interface characterization

H J H Smilde[1], D H A Blank, G J Gerritsma and H Rogalla

University of Twente, Department of Applied Physics, Low Temperature Division, P.O. Box 217, 7500 AE Enschede, The Netherlands

ABSTRACT: Using a few different configurations, the interface between $YBa_2Cu_3O_{7-d}$ and niobium has been characterized chemically as well as electrically. In order to avoid a natural insulating oxide layer between both materials to be formed, thin chemical barriers of silver, gold or platinum have been applied.

Supercurrents are observed up to 21 A/cm^2 for the current flowing in the c-axis direction of the $YBa_2Cu_3O_{7-d}$, while for the ramp-type configuration the maximum supercurrent density is 3.1 A/cm^2 at T = 4.2 K. The R_nA values for both types of junctions are of the order of 10^{-6} Ωcm^2.

1. INTRODUCTION

Taking advantage of the attractive aspects of niobium (Nb) as well as $YBa_2Cu_3O_{7-d}$ (YBCO), devices considering the contact between both materials have to be studied in detail. For the application in high frequency and digital electronics, the YBCO material, with its larger superconducting gap, offers in principle the opportunity to make very fast electronic circuitry. On the other hand, the niobium is easier to process and, consequently, more suitable for complicated fast electronics, and offers the possibility for a fast read-out circuit for the HTS electrical circuit.

Also from fundamental point of view these low temperature superconductor (LTS)/high temperature superconductor (HTS) structures are worthwhile for closer consideration. YBCO is an anisotropic HTS, and its charge carriers are assumed to be hole-like. While, on the other hand, niobium is a low temperature superconductor, characterized by its s-type superconductivity and having electron-like charge carriers. Interesting phenomena at the interface concern the charge carrier depletion close to the interface (band bending), the oxygen depletion on the YBCO side and a contact between an s-wave and a d-wave superconductor.

2. PREPARATION

Roughly two types of junctions have been prepared: junctions of which the current flows in the c-axis direction of the YBCO, in our case called planar junctions, and junctions of which the current direction is in the ab-plane of the YBCO, which we call the ramp-type junctions.

For the planar junctions, a bi-layer of YBCO (100nm) and a noble metal (2 to 20 nm) have been deposited in-situ by standard pulsed laser deposition on (100) $SrTiO_3$ substrates.

---

[1] e-mail address: h.j.h.smilde@tn.utwente.nl

The obtained YBCO films have the c-axis oriented perpendicular to the substrate surface, and have $T_{c,0}$'s above 89 K. With argon-ion milling a ramp-edge is fabricated in the bi-layer. And as a final step, a layer of 80 nm of Nb is ex-situ dc-sputtered on to the sample. The T_c of the Nb layer is 9 K. Structuring of the Nb overlap

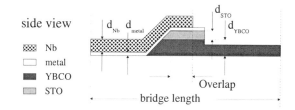

Fig. 1: Ramp-type configuration

is done with a standard lift-off process using photo resist. The definition of the bridge area and the contact paths is performed using argon ion etching. No extra insulating layer is used to separate the YBCO and the Nb at the ramp edge, because, due to the reactivity of Nb with the oxygen of YBCO, a natural barrier will form. The resistance of this barrier is large enough to force the main current to flow in the c-axis direction of the YBCO (~ 10^{-3} Ωcm^2, see section 4.1) and, furthermore, no supercurrent is observed in the YBCO-Nb connections without a chemical barrier.

The fabrication of the second type of junctions starts with the in-situ laser deposition of a bi-layer of 150 nm c-axis-oriented YBCO and 100 nm of SrTiO$_3$. With argon-ion milling a ramp of 20 to 25 degrees is etched in the bi-layer along the crystal axis of YBCO. After removal of the photo resist, the ramp is cleaned with 3 pulses argon ion etching 500 V and 10 pulses at 50 V under an angle of 45 degrees. Then, the sample is in-situ annealed at 780°C in an oxygen ambiance and using pulsed laser deposition a thin chemical barrier of Pt or Au is deposited. Ex-situ, the last step of the earlier mentioned junctions is repeated for the deposition of the Nb and the structuring of the devices. The resulting device is shown in Fig. 1.

3. INTERFACE CHARACTERIZATION

Deposition of Nb on top of YBCO without a chemical barrier between both materials leads to the formation of a thin reaction layer in between them. This layer of approximately 6 nm consists of Nb that has reacted with oxide from the YBCO layer (Usagawa et al. 1998). The reaction layer avoids a supercurrent to be observed from the YBCO to the Nb.

A closed thin layer of a few nanometers of a noble metal between both superconductors prevents this reaction layer to be formed (Ma et al. 1991). In this work Ag, Au and Pt have been used as chemical barriers. From TEM observations of silver deposited in-situ on YBCO some Ag$_2$O particles were observed at the interface. The, for YBCO characteristic, horizontal lattice fringes continued with Pt the closest up to the interface, from all the used barrier materials. Furthermore, as shown in the TEM micrograph, the grain structure in the Pt continued into the poly-crystalline Nb, indicating a good material match also between Pt and Nb (see Fig. 2).

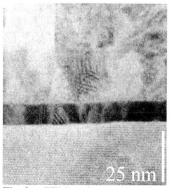

Fig. 2: TEM micrograph of a YBCO/Pt/Nb tri-layer

The observations lead to the conclusion that Pt is the most suitable barrier material from the chemical point of view. It covers the YBCO the most homogeneously and seems the best compatible with YBCO as well as Nb. And it offers the best possibilities to thin the barrier down to the limit of coverage of the YBCO.

4. ELECTRICAL MEASUREMENTS

4.1 Direct Contact YBCO - Nb

The same procedure as for the planar junctions is used for the fabrication of the devices with a direct contact between YBCO and Nb, except that the noble metal has just been omitted. The ex-situ deposition of Nb on YBCO leads to resistance versus temperature characteristics, which increase as the temperature decreases. This indicates that at least one of the interface parts is semi-conducting, which might be due to the reaction layer of NbO_x or oxygen depleted YBCO near the interface. The resistance is of the order of 1 kΩ, resulting in values of $R_nA \sim 10^{-3}$ Ωcm^2 of the normalized contact resistance. No critical current could be observed at T = 4.2 K.

Measurement of the differential resistance dV/dI as a function of the applied voltage shows clear gap features on top of a semi-conducting background (Fig. 3). The values of the clear gap features are determined to be around Δ ~ ± 1.3 mV, which is a little bit smaller than the bulk gap value of Nb and much smaller than the bulk gap value of YBCO.

4.2 Planar Junctions

As a chemical barrier for the planar junctions Ag, Au as well as Pt have been tried. All of them show RSJ-like junction behavior without excess current. In Table 1 the average values of the electrical properties at T = 4.2 K for 15 nm thin barriers are given. In Fig. 4 the I-V characteristic of a planar junction with Pt is depicted, as well as its differential resistance as a function of the bias voltage and current. The constant value of the differential resistance of the junction in the normal state points towards an S'/N/S contact. However a thorough check has not been performed. The planar junctions with a Au barrier show similar behavior.

Fig. 5 presents the average critical current density as a function of the Pt thickness. Decreasing the thickness the critical current density first increases. Then, presumably due to an incomplete coverage of the YBCO by the Pt layer, the value decreases significantly. Assuming the dependence of the coherence length in the normal metal to be exponential, such a fit has been added through

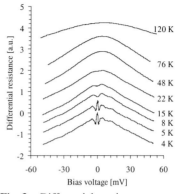

Fig. 3: Differential resistance as a function of the bias voltage of a YBCO/Nb contact at several temperatures (shifted vertical axis)

Table 1: Electrical properties planar junctions at 4.2 K (barrier thickness 15 nm, average values)

barrier	J_c [A/cm^2]	R_nA [Ωcm^2]	IcRn [μV]
Ag	0.5	1·10^{-6}	0.5
Au	4	9·10^{-7}	3
Pt	1	7·10^{-7}	0.8

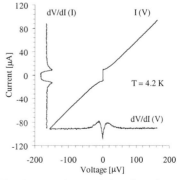

Fig. 4: I-V characteristic of a planar junction and differential resistances dV/dI (in arbitrary units and shifted axis) as a function of the bias current and bias voltage

196

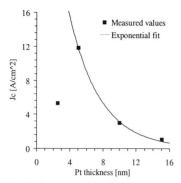

Fig. 5: Average critical current density at T = 4.2 K as a function of the Pt thickness for planar YBCO/Pt/Nb junctions

the points with the thickest Pt layers. Deducing the normal metal coherence length for Pt resulted in a value of $\xi_{n,Pt} \sim 4$ nm at 4.2 K, which is quite small for Pt at this temperature.

The maximum observed critical current was 21 A/cm^2 for the 5 nm Pt barrier junctions, while the average I_cR_n product was 15 μV for these planar junctions.

4.3 Ramp-Type Junctions

In the ramp-type junctions compared to the planar junctions, the total cross-section through which the current is forced to flow is significantly decreased. Another characteristic of the ramp-type configuration is that the junction is a – by ion milling – damaged area, which might induce imperfections of the YBCO material near the ramp. This might result in worsened superconducting properties near the interface with the noble metal and the Nb.

Only ramp-type samples with Pt as a chemical barrier have been prepared. Measurement of the resistance as a function of the temperature showed a slightly increasing resistance with decreasing temperature. Below T ~ 35 K the resistance decreased slightly until the T_c of the Nb. Fig. 5 presents the I-V characteristic of a ramp-type junction. The maximum observed supercurrent density at T = 4.2 K was 3.1 A/cm^2. The I_cR_n product yielded 16 μV, while the normalized contact resistance had a value of $R_nA \sim 6 \cdot 10^{-6}$ Ωcm^2. No excess current was observed. Although the contacts showed a clear junction behavior, the I_c was close to the noise limit, as becomes clear estimating the noise rounding parameter: $\Gamma = 2\pi k_B T/I_0\Phi_0 \sim 0.15$.

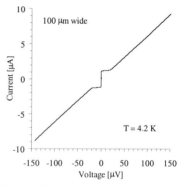

Fig. 6: I-V characteristic of a ramp-type junction of YBCO/Pt/Nb at 4.2 K

5. CONCLUSIONS

In summary, we fabricated junctions between a Nb and a YBCO electrode. A thin chemical barrier of a noble metal has been deposited in between both superconductors, which successfully led to a supercurrent through the junctions. Differential resistance measurements of the planar junctions point towards an S'/N/S contact. However, the research has to be continued to determine more theoretically needed parameters for the description of the junctions. The results on the ramp-type junctions offer the opportunity for interesting work on the d-wave character of the YBCO superconductor.

REFERENCES

Ma Q Y, Schmidt M T, Weinman L S, Yang E S, Sampere S M and Chan S-W 1991 J. Vac. Sci. Techn. A **9 (3)**, 390

Usagawa T, Wen J, Ishimaru Y, Koyama S, Utagawa T and Enomoto Y 1998 Appl. Phys. Lett. **72 (24)**, 3202

Inst. Phys. Conf. Ser. No 167
Paper presented at Applied Superconductivity, Spain, 14-17 September 1999
© 2000 IOP Publishing Ltd

Anomalous resistance peaks in highly transmissive ScS contacts with microwave irradiation

K. Hamasaki, A. Saito and Z. Wang*

Department of Electrical Engineering, Nagaoka University of Technology, Nagaoka 940-2188, Japan
*KARC, Communication Research Laboratory, M.P.T. 588-2, Iwaoka-chou, Nishi-ku, Kobe 651-2401, Japan

ABSTRACT: We report on oscillatory but non-periodic microwave-induced peak structures in the differential resistance dV/dI versus bias voltage V curves which is observed only in highly transmissive superconductor-constrictions-superconductor (ScS) Josephson contacts. The quasi-particle current characteristics of ScS contacts without microwave radiation were well explained by Andreev reflection current. The voltage interval between the peaks changed as the bias voltage was increased, but did not depend on microwave frequency. The observed structures may be associated with the interaction between the microwave and Andreev reflection currents.

1. INTRODUCTION

Microwave properties of superconducting highly transmissive mesoscopic Josephson junctions have been recently the focus of significant theoretical interest (Gunsenheimer 1999). It is the presence of single- or multiple-Andreev reflection phenomena which lead to interest behaviors in such devices. In the past, many authors have studied on the transport and microwave properties of non-tunneling SNS junctions. Almost all of these junctions are two-dimensional (2D) systems, and generally exhibit a broad variety of poorly reproducible current-voltage (I-V) curves. Such 2D-SNS junctions cannot satisfy the main assumption of a ballistic contact model, i.e., all weak link sizes must be small compared to the elastic mean free path and superconducting coherence length. We have developed a field-assisted growth process to fabricate superconductor-constrictions-superconductor (ScS) contacts. By this method, we could continuously vary the I-V characteristics from insulator to tunnel-contact and finally up to highly transmissive contact. The quasiparticle characteristics of these contacts were explained by Andreev reflection current (Hamasaki et al. 1999).

In this paper we report the first experimental study on the microwave properties of highly transmissive ScS contacts. The observed dV/dI peak structures at high bias voltages were oscillatory but non-periodic in bias voltage, inconsistently with the Josephson voltage-frequency relation.

2. DEVICE FABRICATION

The schematic layout of the substrate is shown in Fig. 1(a). An ScS contact is in coplanar waveguide. Figure 1(b) shows the AFM image of a junction area which is patterned using selective niobium anodization process (SNAP). The fabrication procedure of ScS contacts was described in previous papers (Abe et al. 1992, 1994). Briefly, Nb-constrictions were fabricated in a $Nb/Al_2O_3/Nb$ (NbN) insulating sandwich by applying an electric field of <2 V/nm in liquid helium. Figure 1(c) shows the cross-sectional TEM images on the contact before and after the field adjustment process. The constriction diameter is on the order of 10 nm, and close to the electron mean free path of the sputtered Nb films (Hamasaki et al. 1994). The constriction length is equal to the thickness of the insulator.

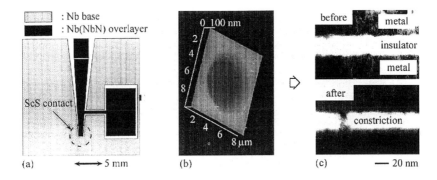

Fig. 1. (a) Schematic layout of the substrate. (b) AFM image of the junction window patterned by SNAP. (c) Cross-sectional TEM micrographs before and after applying field to an insulating SIS sandwich in liquid helium.

3. MICROWAVE PROPERTIES

Figure 2(a) shows a set of dc differential resistance dV/dI vs. V characteristics for a Nb-c-NbN contact at various levels of relative microwave power (full rf power $P_{rf,max}$=10 mW). The critical current I_c of this contact at 4.2 K is 30 μA. Magnetic shielding was achieved using a small Nb can. The dashed lines represents the calculated BTK (Blonder, Tinkham and Klapwijk 1983) plots for Z=0 and P_{rf}=0. The broad dV/dI peak at about 4 mV corresponds the superconducting energy gap. We define here the device quality factor Q as the dynamic resistance ratio R_d(1 mV)/R_N(10 mV), which means deviation from the clean limit of the BTK single Andreev reflection model. Q^{-1} =2 indicates that the Andreev reflection probability A(E)=1. Figure 2(b) shows a plot of Q^{-1} vs. $P_{rf}^{1/2}$. It is found that the Andreev reflection probability rapidly decreases as the microwave power is increased. The physical reason for this effect is due to the breaking of the Andreev reflection process by photon absorption.

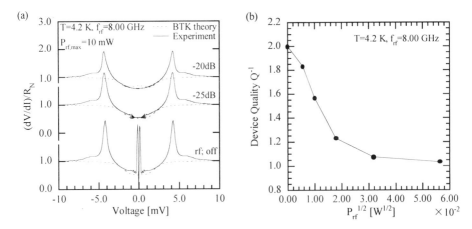

Fig. 2. (a) (dV/dI)/R_N vs. V curves of a Nb-c-NbN at different microwave power level. Dashed lines indicate the predictions of the BTK single-Andreev reflection model without rf fields. (b) Q^{-1} vs. $P_{rf}^{1/2}$ characteristics. Q^{-1} =R_N(10 mV)/R_d(1 mV)

Figure 3(a) shows an enlarged plot of the low voltage region of the dV/dI curve shown in Fig. 2(a). Two types of dV/dI peak structures are observed in this curve. For V<0.4 mV, the voltage interval between the dV/dI peaks is accurately $2*hf_{rf}/2e=32.5$ μV for a microwave frequency of 8.00 GHz, consistently with the Josephson voltage-frequency relation (We adjusted rf power where the magnitude of odd-Shapiro-steps was the largest, based on the Russer model (Russer 1972)). A new feature of the present data is oscillatory but non-periodic peak structures occurring at high bias voltage. Such peaks are indexed by integers n with n=1 corresponding to the last and strongest peak. At voltages of V>0.4 mV, the voltage interval between dV/dI peaks gradually increased with increasing the bias voltage. The peak positions V_n are continuously moved to higher bias voltages with increasing rf power. In Fig. 3(b) only $V_{n=1}$ data is plotted. The dotted lines show the voltages of Shapiro steps calculated from the Josephson voltage-frequency relation, $V_s=m*hf_{rf}/2e$ for 15.7 GHz. In this experiment, the ScS contact was exposed to 15.7 GHz irradiation to obtain a large Shapiro-step interval.

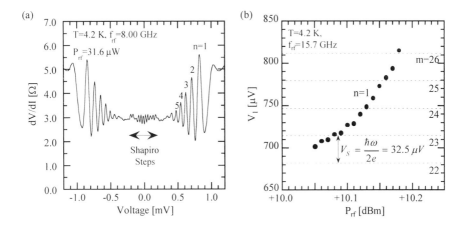

Fig. 3. (a) Enlarged view of the low voltage region of the sample corresponding to Fig. 2(a). The critical current $I_c=30$ μA at 4.2 K. (b) Plots of the V_1 vs. P_{rf} for the n=1 peak. The dashed lines represent the voltages calculated from $V_s=m*hf_{rf}/2e$ for $f_{rf}=15.7$ GHz.

Figure 4(a) shows the power dependence of the peak positions V_n (n=1-5). The dashed lines show the linear dependence of V_n on $P_{rf}^{1/2}$. The voltage interval V_n-V_{n+1} between the dV/dI peaks, however, do not depend on microwave frequency f_{rf} as shown in Fig. 4(b). We also investigated the critical current dependence of the voltage interval with varying the temperature. The peak amplitude is increased with increasing critical current, but the voltage interval V_n-V_{n+1} did not depend on the critical current, and the shape of the superconducting striplines near the junction window. Apparently these qualitative features are different from the behaviors of the usual Shapiro steps or photon-assisted tunneling steps. Also, these structures were not related to the subharmonic energy-gap structures.

Most of the measured clean metallic ScS contacts exhibited similar behaviors shown in Figs. 2-4, but in some samples only the Shapiro steps were observed. The reason for this is still not clear.

As mentioned above, in the absence of applied microwaves the quasiparticle characteristics were well explained by the BTK single-Andreev reflection model. However, the situation changes in the presence of an additional microwave field. In this case electrons and holes moving in the constrictions can absorb photons. As pointed out by Gunsenheimer (1999), such processes cannot be correctly described by the Boltzmann equation which does not contain information about off-diagonal elements of the density matrix. Aberle and Kummel (1996) also discussed the influence of microwave field on the quasiparticle current of mesoscopic SNS systems. However, we have no effective theory which explains the experimental data.

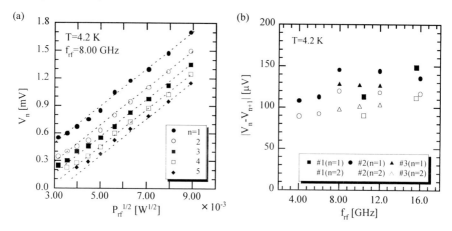

Fig. 4. (a) $P_{rf}^{1/2}$ dependence of dV/dI peak voltages V_n (n=1-5). (b) Frequency dependence of the voltage interval, $V_n - V_{n+1}$ (n=1,2) for three samples.

4. CONCLUSION

Both frequency and power dependences of microwave-induced non-periodic differential resistance peak structures have been studied. Such structures were observed only for highly transmissive ScS contacts. The voltage interval between these peaks did not depend on microwave frequency. The qualitative features of this new structures cannot be explained using the existing theoretical models. We suggested that the structures may be related to the interaction between the microwave and Andreev reflection currents.

ACKNOWLEDGMENTS

Part of this work was done as a Basic Research 21 for breakthroughs in info-communication project supported by the Ministry of Posts and Telecommunications, and Support Center for Advanced Telecommunications Technology Research, Foundation. The authors also wish to thank A. Kawai and T. Ishiguro for AFM and TEM observations.

REFERENCES

Abe H, Hamasaki K and Ikeno Y 1992 Appl. Phys. Lett., **61**, 1131
Abe H, Hamasaki K, Kojima K and Sasaki M 1994 J. Appl. Phys. **33**, 3435
Aberle H and Kummel R 1996 Phys. Rev. Lett. **57**, 3206
Blonder G E, Tinkham M and Klapwijk T M 1982 Phys. Rev., B **25**, 4515
Gunsenheimer U 1999 Physics and Applications of Mesoscopic Josephson Junction
    (The Physics Society of Japan, Edts. Ohta and Ishii) pp.113-133
Hamasaki K and Saito A 1999 ibid, pp.247-266
Hamasaki K and Abe H 1994 IEEE Trans. Appl. Supercon. **3**, 2215
Russer J 1972 J. Appl. Phys., **43**, 2008.

Inst. Phys. Conf. Ser. No 167
Paper presented at Applied Superconductivity, Spain, 14-17 September 1999
© 2000 IOP Publishing Ltd

Effects of He$^+$ irradiation on electromagnetic and transport properties of YBa$_2$Cu$_3$O$_{7-\delta}$ grain boundary Josephson junctions

M. A. Navacerrada, M. L. Lucía and F. Sánchez-Quesada

Departamento de Física Aplicada III, Facultad de Ciencias Físicas, Universidad Complutense, 28040 Madrid, Spain

ABSTRACT: We report a study of electromagnetic and transport properties of YBa$_2$Cu$_3$O$_{7-x}$ grain boundary Josephson junctions irradiated with He$^+$ at 80 keV. Irradiation always induces a decrease of the critical temperature of the electrode. However, more than 65% of the irradiated junctions show an increase of the critical current (I$_C$). An analysis of these results in the framework of the filamentary model suggests that the increase of I$_C$ can be mainly connected with a structural rearrangement in the most disordered resistive regions of the barrier activated by irradiation.

1. INTRODUCTION

Photoexcitation results (Elly 1997), electromigration studies (Moeckly 1993) and annealing experiments (Kawasaky 1992; Sydow 1999) performed in YBa$_2$Cu$_3$O$_{7-x}$ (YBCO) grain boundary Josephson junctions (GBJs) show that the initial effect of a grain boundary is to create an elastic strain gradient which results in the local disordering of oxygen chains and in the segregation of defects to the vicinity of the barrier. These results give evidence of the crucial role that oxygen stoichiometry plays at the origin of the weak-link nature of grain boundaries. On the other hand, a way of inducing alterations in the oxygen sublattice of YBCO thin films is the irradiation with light ions at low energies (Huber 1993). In this framework, irradiation of GBJs may bring new data about the nature of this kind of junctions.

In this paper we present a study of the effects of helium irradiation on the macroscopic parameters of GBJs: I$_C$ flowing across the barrier, normal resistance R$_N$, Swihart velocity c and ratio of the relative dielectric constant to the barrier thickness ε/t. The analysis of these parameters when GBJs are altered by irradiation is a way to obtain information about the microstructure of the barrier (Navacerrada 1999a). We give examples where Josephson coupling can be improved by irradiation although the superconducting properties of the electrodes are always degraded. We show that the effects of irradiation on the junction depend on the dose and on the current distribution in the barrier of the as grown GBJs.

2. EXPERIMENTAL

Josephson junctions were generated using bicrystalline substrates of (100) SrTiO$_3$ with a symmetrical tilt angle of 24°. YBCO films having 500 Å thickness and c-axis orientation were epitaxially grown in a high pressure (3.4 mbar) pure oxygen dc sputtering system (Poppe 1995). In the deposition process the substrate temperature was 900 °C. Films were patterned by wet etching in H$_3$PO$_4$, obtaining junction widths ranging between 2 and 50 μm. Patterned microbrigdes (barriers and electrodes) were irradiated at room temperature with 80 keV He$^+$ ions incident 7° from the surface normal to avoid channeling effects. Ion current was kept small, in the order of 500 nA, to avoid heating of the sample during irradiation. For the selected ion energy the projected range is 3500 Å so that ions have enough energy to stop in the substrate causing irradiation effects on GBJs and on YBCO electrodes. Doses ranged between $D=5\times10^{13}$-10^{15} cm^{-2}. All electrical measurements were made in a double Co-Netic cylinder surrounding the samples for correct magnetic shielding. Current-voltage (I-V) characteristics were registered using an ac Oxford Instruments electronic system. The foot shapes in the resistive transition near T_C were fitted to the thermally activated phase slippage model in order to deduce R_N (Gross 1990).

3.   RESULTS AND DISCUSSION

We have irradiated junctions with widths ranging between 2 and 50 μm at different doses. We have observed Fiske steps in the small GBJs and both Fiske and flux-flow resonances in the long junctions. From the position of resonance steps in the I-V characteristics it is possible to determine c and ε/t (Yi 1996). We present in Table I some representative examples of the evolution of transport and electromagnetic parameters upon irradiation of the junctions.

SAMPLE	w μm	T_C K	R_N Ω	I_C μA	λ_L nm	c ($\times10^{-6}$) m/s	ε/t nm^{-1}
1. B. I.	2	89	17.5	98	140	1.8	33.6
A. I. ($D=5\times10^{13}$ cm^{-2})	2	78	14.5	112	152	2	23
2. B. I.	2	89.6	15.7	127	140	2	27.5
A. I. ($D=5\times10^{13}$ cm^{-2})	2	79	19.3	76	149	1.8	29
3. B. I.	10	89.5	4	761	140	4	6.8
A. I. ($D=10^{14}$ cm^{-2})	10	80	3	845	150	4.1	5.9
4. B. I.	10	89.5	4	761	140	3.8	7.5
A. I. ($D=8\times10^{13}$ cm^{-2})	10	82	4.5	610	150	3.2	9.8
5. B. I.	10	90	5.2	475	140	3.6	8.5
A. I. ($D=3\times10^{14}$ cm^{-2})	10	65	4	302	155	3.7	7.6

B. I. : Before irradiation; A. I. : After irradiation
Table I. Summary of parameters of several representative examples of GBJs before and after irradiation. Data of I_C, λ_L, c y ε/t are quoted at 20 K.

The effect of irradiation on the electrodes is always a decrease of the critical temperature (T_C), along with an increase of the London penetration depth λ_L and the corresponding decrease of the carrier density (Fuchs 1996). Details of the procedure to obtain λ_L after irradiation will be published elsewhere (Navacerrada 1999a). Concerning the properties of the barrier, 65% of our irradiated samples (small and large junctions) showed a decrease of R_N for $D<5\times10^{14}$ cm^{-2}. This decrease of R_N is accompanied by an increase of I_C when the superconducting properties of the YBCO electrodes are not severely degraded as can be observed in examples 1 and 3 of Table II. For $D\geq3\times10^{14}$ cm^{-2} (see example 5) the decrease of the carrier density of the electrodes is too large to observe any increase of I_C although R_N may still decrease. In any case, the decrease of R_N is always accompanied by a decrease of ε/t and an increase of c (Navacerrada 1999b) in agreement with photoexcitation experiments (Elly 1997). For $D>5\times10^{14}$ cm^{-2}, R_N of the barrier always increases in the same line than results obtained by electron irradiation experiments on biepitaxial junctions at doses of 1.5×10^{16} cm^{-2} (Tafuri 1998).

As shown in Fig. 1(a), after irradiation we observe a decrease of the width of the oscillations in the magnetic field dependence of I_C, $I_C(B)$ curves, due to the increase in the value of λ_L of the electrodes. The shape of the $I_C(B)$ curves does not change, indicating that the current distribution is not remarkably altered. Only if irradiation promotes a change in the behavior of the junction from long to small regime we can observe modifications in the modulation of $I_C(B)$. As an example, we show in Fig. 1(b) the modification of $I_C(B)$ at 50 K of a 4 μm wide junction after an irradiation process with D=8×10^{13} cm^{-2}.

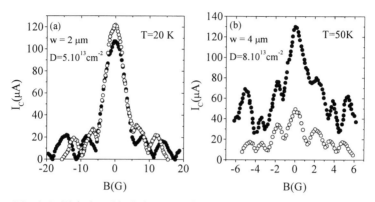

Fig. 1. $I_C(B)$ before (black dots) and after irradiation (white dots) of two junctions. In example (a) I_C increases after irradiation and in the example (b) I_C decreases.

R_N is a good parameter to analyze our barriers since I_C is also dependent on the value of T_C of the electrode which is altered by irradiation. We present in Table II two examples of junctions irradiated with the same doses in which the values of R_N are similar before irradiation ($R_{Nbefore}$). However R_N after irradiation (R_{Nafter}) progress in opposite directions. These results show that, for the same dose, the effect of irradiation on the properties of the barrier is dependent not only on R_N but also on the shape of $I_C(B)$ of the as grown GBJs, i.e. on the particular distribution of the supercurrent in the barrier. The filamentary model (Moeckly 1993) may account for these results. This model proposes that the supercurrent in the barrier is restricted to channels of good crystalline lattice order between dislocations. In this sense, junctions with similar values of R_N may have comparable number of filaments or similar amount of ordered material inside the barrier, but different $I_C(B)$ shapes, pointing to very different current distributions in the barrier (Froehlich 1995). This is the case of the examples chosen for Table II. The final effect of irradiation will depend on the region of the barrier affected by ion impacts. If the ions impact preponderantly the superconducting filaments, the effect can be very similar to the one reported for thin films of YBCO irradiated with light ions at low energies (Huber 1993). We observe an increase of the value of R_N in the same manner than the resistance increases in irradiated thin films and in our electrodes. On the other hand, if ions preferably impact the most disordered non-superconducting regions of the barrier, a diffusion process affecting mainly the oxygen sublattice may be activated. In this sense, He$^+$ irradiation can promote a partial structural rearrangement in these regions and so a decrease of R_N. This behavior can be compared to previously reported results on activation of diffusion processes in grain boundaries by different methods: electron irradiation (Kizuka 1998), electromigration (Moeckly 1993) and annealings in ozone atmosphere (Kawasaki 1992).

We have shown the potential of light ion implantation as a tool to controllably modify GBJs parameters, without significant alteration of critical current spatial distribution in the junction. Helium irradiation may be regarded as a tool to obtain complementary information about the nature of the microstructure of the junctions and opens new possibilities to tailor *a posteriori* barrier performances and hence devices based on Josephson junctions. A remarkable result is the possibility to increase the critical current of GBJs although the T_C of the electrodes is degraded.

w μm	D cm^{-2}	$R_{Nbefore}$ Ω	ΔR Ω
10	3×10^{14}	5.2	-1.2
10	3×10^{14}	4.9	0.6
10	10^{14}	4	-1
10	10^{14}	3.8	0.8

Table II. Examples of junctions of similar values of $R_{Nbefore}$ in which R_{Nafter} progress in opposite directions. Values of $\Delta R = R_{Nafter} - RN_{before}$ are quoted.

This work has been supported by CICYT grant MAT-97-0675. The authors would like to thank C.A. I.de Implantación Iónica from U.C. M.

REFERENCES

Elly J, Medici M G, Gilabert A, Schmidl F, Seidel P, Hoffmann A and Schuller I. K. 1997 Phys. Rev. B **56**, R8507

Froehlich O M, Schulze H, Beck A, Mayer B, Alff L, Gross R and Huebener R P 1995 Appl. Phys. Lett. **66**, 2289

Fuchs A, Prusseit W, Berberich P and Kinder H 1996 Phys. Rev. B **53**, R14745

Gross R, Chaudhari P, Dimos D, Gupta A and Koren G 1990 Phys. Rev. Lett. **64**, 228

Huber R, Schneider M, Wagner U and Ziemann P 1993 J. of All. and Comp. **195**, 255

Kizuka T, Iijima M and Tanaka N 1998 Phil. Mag. A **77**, 413

Kawasaki M, Chaudhari P and Gupta A 1992 Phys. Rev. Lett. **68**, 1065

Moeckly B H, Lathrop D K and Buhrman R A 1993 Phys. Rev. B **47**, 400

Navacerrada M A, Lucía M L and Sánchez-Quesada F 1999a submitted to Phys. Rev. B

Navacerrada M A, Lucía M L, Iborra E and Sánchez-Quedada F 1999b communication at the *International Conference on Physics and Chemistry of Molecular Oxide Superconductors* (Stockholm, July)

Poppe U, Klein N, Dahne U, Solther H, Jia C L, Kabius B, Urban K, Lubig A, Schmidt K, Henger S, Orbach S, Müller S and Piel H 1995 J. Appl. Phys. **71**, 5572

Sydow J P, Berninger M, Buhrman R A and Moeckly B H 1999 IEEE Trans. Appl. Supercond. **9**, 2993

Tafuri F, Nadgorny B, Shokhor S, Gurvitch, Lombardi F, Carillo F, Di Chiara A and Sarnelli E 1998 Phys. Rev. B **57**, R14076

Yi H R, Gustafsson M, Winkler D, Olsson E and Claeson T 1996 J. Appl. Phys **79**, 9213

Inst. Phys. Conf. Ser. No 167
Paper presented at Applied Superconductivity, Spain, 14-17 September 1999
© 2000 IOP Publishing Ltd

Zero-crossing steps in overdamped Josephson junctions

A Takada and A Chiba

Hakodate National College of Technology, Hakodate, 042-8501 Japan

T Noguchi

Nobeyama Radio Observatory, Nobeyama, Nagano 384-1305 Japan

ABSTRACT: The Shapiro step crossing the zero-current axis in the current-voltage characteristic, so called *zero-crossing step*, for the overdamped Josephson junction irradiated with an rf-signal is numerically simulated based on the resistively and capacitively shunted junction model. As an rf-signal, either a pulse train signal or a biharmonic signal is adopted. The heights of these steps are simultaneously examined as a function of the rf-amplitude. We consider that the obtained results can be applied to the conventional unbiased dc-voltage standards without a slow step procedure or a crisis of chaos.

1.   INTRODUCTION

The Shapiro step crossing the zero-current axis in the current-voltage (I-V) characteristic for the irradiated Josephson junction, so called *zero-crossing step*, is made use of a precise voltage measurement based on the Josephson voltage standard. The scheme for use of this step was first proposed by Levinsen *et al* (1977). For the conventional voltage standard, the superconducting tunnel junctions with a relatively large capacitance as well as a large quasiparticle resistance have been required. In this case it is, however, not easy to avoid a slow step procedure to determine the stable point on the *I-V* characteristic. In addition, a careful design of the system is required in order to avoid chaos which makes the junction unstable. It has been known for the junction driven by a sinusoidal signal that the crisis of chaos can be avoided whenever the drive frequency for the junction is well above the Josephson plasma frequency or the specific capacitance of the junction can be negligible, in other words, the junction is overdamped.

In this study, we examine the limit of the junction damping to avoid chaos and the zero-crossing step for the overdamped junction. This step is resulted from the enhancement of ac Josephson effect induced by the biharmonic and pulse train signals as reported by Monaco (1990) and Maggi (1996) who also predicted the presence of the zero-crossing step. However, to our knowledge, the detailed study has not been made yet.

2.   SIMULATION ON *I-V* CHARACTERISTICS AND SURVEY OF CHAOS

2.1   *I-V* characteristic

The resistively and capacitively shunted junction model, sometimes referred to the Stewart-McCumber model, was adopted to calculate the *I-V* characteristics of junctions in terms of the normal resistance R, the specific capacitance C, and the maximum Josephson current I_C. In addition, both the sinusoidal current-phase relation and the constant maximum Josephson current, *i.e.*, the frequency independent I_C, are assumed. Taking into account the reduced rf-frequency Ω introduced by Russer (1970), we obtain the normalized current equation for the irradiated junction as a function of the phase ϕ,

$$\beta\ddot{\phi} + \dot{\phi} + \sin\phi = \alpha_0 + \alpha_1 \sin\Omega\tau, \ (1)$$

where $\beta = 2eR^2CI_C/\hbar$ is the hysteresis parameter (Stewart (1968a) and McCumber (1968b)), α_0 and α_1 are the dc-current and the amplitude of rf-current respectively normalized by I_C, $\Omega = \omega\hbar/2eI_CR$ is the reduced rf-frequency and $\tau = (2e/\hbar)I_CRt$ is the normalized time. In this model, Ω is the prime factor which determines the Shapiro step height of the junction. The time-averaged derivative of phase $\langle\dot{\phi}\rangle$ in eq.(1) is comparable to the junction voltage normalized by I_CR. One should note that the Shapiro step appears at the voltage of $n\Omega$ in normalized unit, where n is integer. We can expand eq.(1) in the case of a sinusoidal drive signal to that of any periodic signal as follows,

$$\beta\ddot{\phi} + \dot{\phi} + \sin\phi = \alpha_0 + \alpha_1 \sum_{n=1}^{\infty} a_n \sin(n\Omega\tau + \theta_n), \ (2)$$

where a_n and θ_n are the relative amplitude and the phase difference for the harmonics of the signal, respectively. Since we are interested in the pulse train consisting of a rectangular pulse having height I_P, width T_P and repetition period T, either a_n or θ_n is considered to be fixed and is a function of the duty cycle(= T_P/T) of the pulse train. For the biharmonic signal the similar equation can be obtained,

$$\beta\ddot{\phi} + \dot{\phi} + \sin\phi = \alpha_0 + \alpha_1\{\sin\Omega\tau + a\sin(2\Omega\tau + \theta)\}, \ (3)$$

where $a = 1/\sqrt{2}$ and $\theta = -\pi/2$ are chosen here to make this signal pulse train-like. Thus the I-V characteristic is still determined by β and Ω even for these non-sinusoidal signals.

2.2 Range of chaotic state

In order to find the lower limit of the desirable junction damping for various rf-signals, we firstly survey the range of chaotic state in β-Ω plane (Kautz (1981) and Cronemeyher et al. (1985)). An electronic analog of the Josephson junction, designed by Magerlein (1986), was used to observe the onset of chaos. In order to complete the model described above, dc and ac bias sources are connected to the junction analog.

Fig.1 shows β-Ω planes in which the onset of chaos in the I-V curves was surveyed for various rf-signals, (a) sinusoidal, (b) biharmonic (pulse train-like) and (c) pulse train with 25% in duty cycle. In this figure,

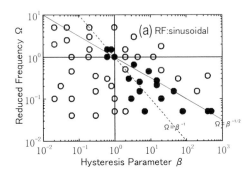

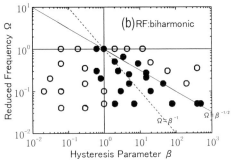

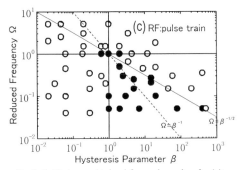

Fig.1 β-Ω planes obtained for various signals: (a) sinusoidal, (b) biharmonic, (c) pulse train ($T/T_P = 0.25$). Closed circles indicate points at which irregular structures appear in the vicinity of the Shapiro steps in I-V curves. Open circle indicate no visible irregular structure.

the closed circles indicate points at which irregular structures appear in the vicinity of the Shapiro steps in the I-V curves and the open circle indicate no visible irregular structure. We characterize the chaotic state in terms of the devil's staircase observed as the irregular structure in the I-V curve at which the power spectrum of the junction voltage has a wide band noise component. In this figure, the normalized dc bias α_0 is varied from -2 to 2, and the normalized rf-amplitude α_1 is from 0 to 2 by the step of 0.1. The range concerning α_0 is seemed to be insufficient for high Ω, especially $\Omega \gtrsim 1$, because the Shapiro step shifts toward a high current region in the I-V characteristic with increase in Ω. However, as discussed in the next section, the appearance of the zero-crossing step is almost restricted to be the first order step and limited in the region of $\Omega \lesssim 1$. It is concluded that the region with respect to a crisis of chaos bounded by $\beta=1$ and $\Omega=\beta^{-1/2}$ is common for the various signals. The latter condition denotes that the junction is driven at the Josephson plasma frequency. It is found that at least $\beta \lesssim 0.6$ seems enough to avoid chaos.

3   ZERO-CROSSING STEP IN OVERDAMPED JUNCTIONS

The heights(widths) of the Shapiro step and the zero-crossing step are numerically calculated as a function of rf-amplitude as shown in Fig.2. In this figure various pulse train(-like) signals, (a) pulse train (T_P/T=0.25), (b)biharmonic signal(pulse train-like), and (c) pulse train (T_P/T=0.1) are considered and all step heights are normalized by I_C. Further, the solid line and the dashed line show the height of nth order step and the step height measured from the zero-current axis to the step edge for positive current direction, respectively. Thus the difference between these two kinds of lines is compatible with the fraction of the nth step crossing over the zero-current axis. It should be noted that Ω=0.44 and β=0.01 are assumed for all plots and the dc component involved in the pulse train itself was eliminated at the beginning of the calculation. In this figure, in order to mutually compare the results obtained for the different duty cycles in addition to the waveforms, the rf-amplitudes α_{1rms} for various signal are common to be in root mean square (rms). For the biharmonic signal $\alpha_{1rms} \approx 0.866\alpha_1$ is obtained from the relation $\alpha_{1rms}=\alpha_1[(1+a^2)/2]^{1/2}$ and $a=(1/2)^{1/2}$. For the pulse train signal $\alpha_{1rms} \approx 0.433\alpha_1$ for T_P/T=0.25 and $\alpha_{1rms}=0.3\alpha_1$ for T_P/T=0.1 are obtained from the relation $\alpha_{1rms}=\alpha_1[(1-\delta)/\delta]^{1/2}$, where $\delta=T_P/T$ and $\alpha_1=I_P/I_C$. In Fig.2(a), it is found that the part of 25% in the first order step which has the maximum amplitude of approximately I_C crosses over the zero-current axis. Quite similar results are obtained for the pulse train-like biharmonic signal as shown in

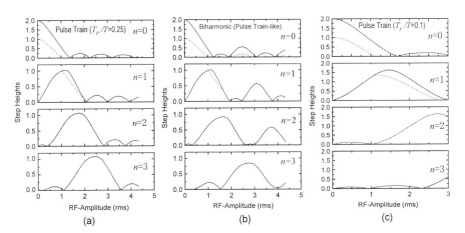

(a)   (b)   (c)

Fig.2. RF-amplitude dependence of step height for various rf-signals: (a) pulse train (T_P/T=0.25), (b) biharmonic signal(pulse train-like), and (c) pulse train (T_P/T=0.1). Solid line shows the height (width) of nth order step. The step heights are respectively normalized by I_C. Dashed line shows the step height measured from the zero-current axis, meaning the difference between the both lines for any rf-amplitude comparable with the part of step crossing the zero-current axis.

208

Fig.2(b). These results suggest that the shape of pulse is not restricted to be rectangular. In the case of the biharmonic signal, θ in eq.(3) is further varied from 0 to $-\pi/4$. Then the zero-crossing step can be scarcely obtained. For the pulse train with 10% in duty cycle the maximum height of the first order step is approximately $1.6I_C$ [Fig.2.(c)]. In this case, the part of 42% of I_C in the first order step crosses over the zero-current axis. Furthermore, the step height for this pulse has a relatively weak dependence of the rf-amplitude when compared to the case of 25% in duty cycle.

For the junction array, the decay of the rf-power is one of serious problems. Therefore, the small duty cycle of the pulse train seems preferable to avoid this problem. With respect to the practical overdamped junction arrays used in programmable voltage standards, the required rf-frequency for $\Omega=0.44$ is, 1.06-6.38GHz for SNS(Nb-PdAu-Nb) junctions having $I_C R$ =5-30μV (Benz (1995)) and 80GHz for SINIS (Nb/Al/AlOx/Al//AlOx/Al/Nb) junctions having $I_C R$ =120μV(Schulze *et al.* (1998)), respectively. One can also estimate the required rise-time t_r of the pulse based on the rough definition of $t_r=T_P/2$. For example, t_r=50ps is needed for $I_C R$ =5μV and 0.6ps for $I_C R$ =120μV when 10% is adopted as the duty cycle of the pulse train. Consequently, we consider that the overdamped junctions can be applied to the unbiased dc-voltage standards without the slow step procedure and a crisis of chaos.

One should note that $\Omega=0.44$ assumed in this study is not always optimum to obtain the large zero-crossing step for the overdamped Josephson junction. For $\Omega \lesssim 1$ the step height is extremely small. On the other hand, for $\Omega \gtrsim 1$ the center of the first step would be located at $\alpha_0 \gtrsim 1$ in current and $\langle \dot\phi \rangle \gtrsim 1$ in voltage. Taking into account that the maximum step height for any cases never exceed $2I_C$, it is presumed that the zero-crossing step is rarely observed for $\Omega \gtrsim 1$ even though still a larger step can be obtained. Therefore, it is roughly estimated that the range in which the relatively large zero-crossing step is limited in the range of $0.1 \lesssim \Omega \lesssim 1$.

4. CONCLUSIONS

The overdamped Josephson junction with hysteresis parameter $\beta \lesssim 0.6$ is not suffered a crisis of chaos even when the junction is driven by the non-sinusoidal signal such as the pulse train as well as the pulse train-like biharmonic. The zero-crossing step in this junction having an effectively large height can be obtained under an appropriate amplitude of the rf-signal. For example, the pulse train signal with 10% in duty cycle induces the zero-crossing step in which the height is comparable with 42% of the maximum Josephson current. The small duty cycle such as 10% of the pulse train seems rather preferable than 25% for the purpose to obtain a large zero-crossing step having a relatively weak dependence on the rf-amplitude. The range of the reduced frequency, $0.1 \lesssim \Omega \lesssim 1$, is considered to be preferable to obtain the effectively large zero-crossing step.

The authors thank to S.Kaechi for his assistance. This work was partly supported from the South Hokkaido Foundation for the Promotion of Science and Technology in Japan.

REFERENCES

Benz S P 1995 Appl.Phys.Lett. **67**, 2714
Cronemeyer D C, Chi CC and Pedersen N F 1985 Phys.Rev.**B31**, 2667
Kautz R L 1981 J.Appl.Phys. **52**, 3528
Levinsen M T, Chiao R Y, Feldman M J and Tucker B A 1977 Appl.Phys.Lett. **31**, 776
Magerlein J H 1978 Rev.Sci.Instrum. **49**, 486
Maggi S 1996 J.Appl.Phys. **79**, 7860
McCumber D E 1968b J.Appl.Phys. **39**, 3113
Monaco R 1990 J.Appl.Phys. **68**, 679
Russer P 1972 J.Appl.Phys. **43**, 2008
Schulze H, Behr R, Muller F, and Niemeyer J 1998 Appl.Phys.Lett. **73**, 996
Stewart W C 1968a Appl.Phys.Lett. **12**, 277

Inst. Phys. Conf. Ser. No 167
Paper presented at Applied Superconductivity, Spain, 14-17 September 1999
© 2000 IOP Publishing Ltd

PREPARATION AND CHARACTERIZATION OF a-AXIS ORIENTED NdBa$_2$Cu$_3$O$_{7-\delta}$/doped PrBa$_2$Cu$_3$O$_{7-\delta}$/NdBa$_2$Cu$_3$O$_{7-\delta}$ TRILAYERS

Mutsumi Sato, Gustavo A. Alvarez, Furen Wang, Tadashi Utagawa, Keiichi Tanabe, and Tadataka Morishita

Superconductivity Research Laboratory, ISTEC, 1-10-13 Shinonome, Koto-ku, Tokyo 135-0062, Japan

ABSTRACT: a-Axis oriented NdBa$_2$Cu$_3$O$_{7-\delta}$ /PrBa$_2$(Cu,Co)$_3$O$_{7-\delta}$/ NdBa$_2$Cu$_3$O$_{7-\delta}$ trilayers were prepared on (100) SrTiO$_3$ substrates using a DC-95MHz hybrid plasma sputtering. In the case of 50 nm thick PrBa$_2$(Cu,Co)$_3$O$_{7-\delta}$ barrier layer, the current-voltage characteristics of the trilayers were consistent with the resistively-shunted-junction model. Well developed Shapiro steps corresponding to 5-20 GHz microwave irradiation were observed. A clear power dependence of the maximum amplitude for the microwave induced Josephson steps was obtained. The Josephson supercurrent under applied magnetic field showed different patterns dependent on the junction areas.

1. INTRODUCTION

For device application in electronics, the reliability and reproducibility of Josephson junctions are necessary for circuit design and fabrication. For this purpose, multilayer Josephson junctions made of superconductor/insulator/superconductor (SIS) structures or superconductor/normal metal/superconductor (SNS) structures have been investigated. In the case of high temperature superconductor, many junctions have been fabricated using a combination of YBa$_2$Cu$_3$O$_{7-\delta}$ (YBCO) and PrBa$_2$Cu$_3$O$_{7-\delta}$ (PBCO) or NdBa$_2$Cu$_3$O$_{7-\delta}$ (NBCO) and PBCO because of their good lattice maching (Hashimoto et al. 1992) (Alvarez et al. 1997). Recently, it was reported that PBCO has localized states leading to hopping conduction

(Yoshida et al. 1997), furthermore PBCO single crystal showed superconducivity (Zou et al. 1996). The doped PBCO, therefore, has preferably been employed as a normal conductor (Inoue et al. 1997) (Horibe et al. 1997). In this paper, the structural properties and current-voltage (I-V) characteristics of the trilayers consisted of a-axis oriented NBCO (a-NBCO) and a-axis oriented Co-doped PBCO (a-PBCCO) are reported.

2. EXPERIMENTAL

The $PrBa_2Cu_{2.8}Co_{0.2}O_{7-\delta}$ (PBCCO) was choiced as a barrier because even though the small variation (only 0.2 %) of its b-axis lattice constant in relation with PBCO (Yang et al. 1992), the barrier resistance was increased four orders of magnitude at 50 K (Sato et al. 1999). The a-NBCO/a-PBCCO/a-NBCO trilayers were deposited on (100) $SrTiO_3$ substrates using a DC-95MHz hybrid plasma sputtering in on-axis configuration. The fabrication conditions of a-NBCO and a-PBCCO films are listed in Table 1. The trilayers were prepared combining these conditions for each layer. The substrate temperature was set in order to avoid the growth of c-axis oriented grains. The growth rate was 0.2 μm/h for a-PBCCO films and 0.5 μm/h for a-NBCO films. After the deposition, the trilayers were transfered to the loading chamber, and cooled down to room temperature in 1 atm oxygen. The crystal orientation, surface morphology and crystallinity were examined by X-ray diffraction (XRD), atomic force microscopy (AFM) and Rutherford backscattering spectrometry (RBS), respectively. Trilayer junctions with areas of 10×10 μm^2, 15×15 μm^2, and 20×20 μm^2 were fabricated using photolithographic and ion milling techniques (Konishi et al. 1995). The temperature dependence of resistance and I-V characterstics of the trilayers were measured by a four-probe method.

Table 1. Preparation conditions of a-NBCO and a-PBCCO films

film	substrate temperature	RF power	DC power	sputtering gas	total pressure	target composition
NBCO	710 ℃	20 W	0.5 A x 150 V	50% Ar + 50% O$_2$	600mTorr	$NdBa_2Cu_3O_{7-\delta}$
PBCCO	670 ℃	20 W	0.2 A x 180 V	50% Ar + 50% O$_2$	600mTorr	$PrBa_2Cu_{2.8}Co_{0.2}O_{7-\delta}$

3. RESULTS AND DISCUSSION

a-Axis oriented NBCO/PBCCO/NBCO trilayers with 50 nm thick PBCCO barrier layer were prepared (Sato et al. 1999). The XRD patterns of the samples showed that the (005)/(200) intensity ratios were less than 0.01. The samples were a-axis oriented films. The mean roughness of 2.32 nm over a 5×5 μm^2 area and χ_{min} of 3.7 % show that samples have smooth surfaces and good crystallinity. The RBS spectrum shows that there are no increases in the scattering yield, i.e. no disorders at the interfaces between the top NBCO layer and PBCCO layer, and between PBCCO layer and bottom NBCO layer, respectively. The trilayers showed superconductive transition at 60 K. The I-V characteristics were resistively shunted junction (RSJ)-like without excess current. Stable Shapiro steps for 5-20 GHz microwave irradiation were observed and these steps were clearly modulated as a function of the microwave power at 4.2 K. Figs. 1 and 2 show the 9 GHz and 15 GHz microwave power dependence of the magnitude of each induced step, respectively. These results are qualitatively consistent with the theory for the magnitude of the n-order of step and expressed as $| J_n(2eV/hf)|$, where J_n is the n-order Bessel function, and V, h, f are the microwave voltage, the Plank constant, the microwave frequency, respectively (Grimes et al. 1968). The dependence of the critical current (I_c) under applied magnetic field (H) was also observed at 4.2 K. The results were different for each junction area. In the 15×15 μm^2 junction, the pattern of I_c-H were Fraunhofer-like, as shown in Fig. 3. On the other hand, in the 20×20 μm^2 junction, the pattern of I_c-H suggested spatial variation of I_c , as shown in Fig. 4 (Barone et al. 1977).

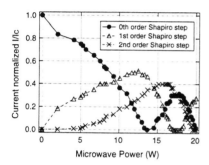

Fig. 1. 9 GHz Microwave power dependence of the maximum amplitude for the Shapiro steps on the 15×15 μm^2 junction area

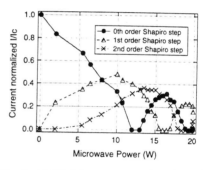

Fig. 2. 15 GHz Microwave power dependence of the maximum amplitude for the Shapiro steps on the 15×15 μm^2 junction area

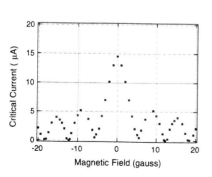

Fig. 3. Magnetic field dependence of the critical current on the 15×15 μm^2 junction area

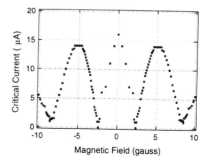

Fig. 4. Magnetic field dependence of the critical current on the 20×20 μm^2 junction area

4. CONCLUSION

a-Axis oriented NBCO/PBCCO/NBCO trilayers were fabricated. The trilayers have smooth surfaces and good crystallinity. The I-V characteristics of the trilayers were RSJ-like. Well developed Shapiro steps corresponding to 5-20 GHz microwave irradiation were observed and showed clear oscillations corresponding to the microwave power. The magnetic field dependence of the critical current showed different patterns dependent on the junction area.

ACKNOWLEDGEMENT

This work was supported by the New Energy and Industrial Technology Development Organization (NEDO) as Collaborative Research and Development of Fundamental Technologies for Superconductivity Applications.

REFERENCES

Alvarez G A, Utagawa T, and Enomoto Y 1997 Phisica C **282-287** pp 1483

Barone A, Paterno G, Ruso M, and Vaglio R 1977 Phys. Stat. Sol. (a) **41** pp 393

Grimes C C and Shapiro S 1968 Phys. Rev. **169** pp 397

Hashimoto T, Sagoi M, Mizutani Y, Yoshida J, and Mizushima K 1992 Appl. Phys. Lett. **60** pp 1756

Horibe M, Kawai K, Ohta T, Fujimaki A, and Hayakawa H 1997 Extended Abstracts of 6th Int. Superconductive Electronics Conf. Vol.2 pp 97

Inoue S, Nagano T, Hashimoto T, and Yoshida J 1997 Extended Abstracts of 6th Int. Superconductive Electronics Conf. Vol.2 pp 85

Konishi M and Enomoto Y 1995 Jpn. J. Appl. Phys. **34** pp 1271

Sato M, Alvarez G A, Wang F, Utagawa T, Tanabe K, and Morishita T 1999 submitted to Physica C

Yang H D, Lin M W, Chiou C K, and Lee W H 1992 Phys. Rev. B **46(2)** pp 1176

Yoshida J, and Nagano T 1997 Phys. Rev. **55** pp 11860

Zou Z, Oka K, Ito T, and Nishihara Y 1996 Proceedings of the 9th International Symposium on Superconductivity pp 365

Inst. Phys. Conf. Ser. No 167
Paper presented at Applied Superconductivity, Spain, 14-17 September 1999

Electrical characteristics of all-NbCN Josephson junctions with TiN films as barriers

H Yamamori, S Kohjiro, H Sasaki and A Shoji

Electrotechnical Laboratory, 1-1-4, Umezono, Tsukuba, 305-8568, Japan

ABSTRACT: Electrical characteristics of NbCN/TiN/NbCN Josephson junctions have been investigated as functions of TiN-film resistivity ρ and thickness d. Prior to fabricating junctions, TiN films were prepared on Si wafers by reactive-magnetron sputtering in mixtures of Ar and N_2 and their resistivities were measured by a four-terminal method at room temperature. Junctions were fabricated using TiN films with different resistivities and thicknesses. Critical current density J_c and normal state resistance R_n for junctions were measured at 4.2 K. It was found J_c for junctions exponentially varied with change in d. On the other hand, variation of ρ for TiN films did not strongly affect J_c for junctions. The I_cR_n product for junctions was varied from 10 μV to 1 mV by changing the TiN film thickness from 65 nm to 30 nm.

1. INTRODUCTION

Studies of overdamped Josephson junctions with SNS or SINIS structures (Kauts *et al.* 1994, Benz *et al.* 1995, Maezawa *et al.* 1997, Frizsch *et al.* 1998) have received much interest because they have potential for use as components of digital to analog (D/A) converters for programmable Josephson voltage standard (Hamilton *et al.* 1995, 1997), superconducting microwave oscillators, and non-latching logic circuits, e.g., rapid-single-flux-quantum (RSFQ) circuits (Likharev *et al.* 1991).

In previous papers, we reported fabrication and electrical characteristics of overdamped Josephson junctions with NbCN/TiN/NbCN, NbCN/MgO/TiN/NbCN, or NbCN/TiN/MgO/NbCN structures (Wang *et al.* 1997, 1998). Among those junctions, we chose NbCN/TiN/NbCN junctions for voltage-standard application because they had relatively low I_cR_n products (< 100μV) and good controllability of critical current density J_c. Furthermore, a high uniformity in critical current density over a 3 inch diam. wafer was achieved (Yamamori *et al.* 1999). In this paper, we report electrical characteristics of NbCN/TiN/NbCN Josephson junctions fabricated using TiN films with various resistivities ρ and thicknesses d.

2. FABRICATION

The preparation of NbCN films and TiN films was carried out in a sputtering system equipped with 6 inch diam. targets of Nb and Ti. The base pressure of the system was below 1 x 10^{-4} Pa. The incident power, the gas flow rate, the total gas pressure, and the sputtering time were automatically controlled using a personal computer. The base and counter NbCN films were deposited in a N_2, Ar, and C_2H_2 gas mixture at a total pressure of 2 Pa. The incident power was 600 W and the deposition rate for NbCN films was 1.3 nm/s. The critical temperature T_c was about 16K. TiN films were deposited in N_2 gas or in N_2 and Ar gas mixture at a total pressure of 1.33 Pa. The incident power was 400 W or 600W. During the deposition of TiN films, the substrate-holder was rotated at 10 rpm to improve uniformity. An average deposition rate for TiN films was about 0.02 nm/s. There was a little dependence of the resistivity on temperature in the range of 4 K to 300 K, that was rather semiconductive, while a single crystal of TiN shows a metallic temperature dependence (Johansson *et al.* 1985). As substrates for film deposition, we used 3 inch diam. Si wafers.

Junctions were fabricated as follows. First, wirings to the base electrodes of junctions were

214

fabricated by depositing a NbCN film on a wafer and by patterning it in a mixture of CF_4 and O_2. Then, a trilayer of NbCN/TiN/NbCN was deposited. Junction areas were defined by reactively-ion etching unnecessary portion of the trilayer film in CF_4. Typical etching rates for NbCN and TiN films were 0.33 nm/s and 0.83 nm/s, respectively. The typical thickness of base and counter NbCN films was 100 nm. Finally, wirings to the counter electrodes of junctions were fabricated by depositing a Nb film and by patterning it in a CF_4 plasma.

3.   EXPERIMENTS AND DISCUSSION

Figure 1(a) shows results of an X-ray diffraction (XRD) measurement for TiN films prepared at different N_2 partial pressures. Open circles in the figure show ratio of intensity of X-rays diffracted from (111) plane and that from (200) plane. Figure 1(b) is a plot of resistivities for TiN films measured at room temperature (RT). The peaks of resistivity seen around 15% and 30% of N_2 pressure may be the result of mixture of (111) and (200) phases. The small variation of resistivity over 40% for films deposited at 400W suggests the content of nitrogen in the films was not increased by the increase of N_2 partial pressure (Wittmer 1985).

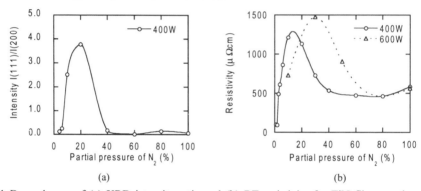

Fig.1 Dependences of (a) XRD intensity ratio and (b) RT resistivity for TiN films on the partial pressure of N_2.

We fabricated junctions using TiN films with RT resistivities of 0.56, 0.86, 1.21, and 1,47 mΩcm. The thickness of TiN films was 50 nm. Figure 2 (a) and 2 (b) show dependences of J_c and product of normal state resistance R_n measured at 4.2K and the junction area A on the resistivity of TiN films. As shown in Fig.2(a), J_c seems to have a tendency to increase with decrease in ρ although there is some spread among the J_c values. On the other hand, $R_n A$ for junctions increased as resistivity of TiN films increased as shown in Fig.2 (b).

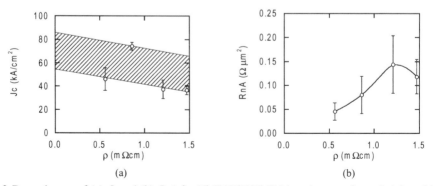

Fig.2 Dependences of (a) J_c and (b) $R_n A$ for NbCN/TiN/NbCN junctions on the resistivity of TiN films ρ. The thickness of TiN films was 50 nm.

Figure 3(a) shows the relation between d and J_c for juctions with two different ρ values of TiN films. As seen in the figure, junctions with the same ρ value showed exponential dependence of J_c on d. From the slope of the lines drawn in the figure, the coherence length for the TiN films with ρ=0.56 mΩcm and 1.21 mΩcm were calculated to be 11 nm and 21 nm, respectively. Figure 3(b) shows the relation between d and R_nA. As seen in the figure, R_nA did not show a propotional dependence on d, which is expected from a simple barrier model. The origin of large R_nA values for junctions with TiN films thinner than 50 nm is not clear at the present stage. But as described below, there is possibility some insulating layers exist at the interface of TiN films and NbCN electrodes..

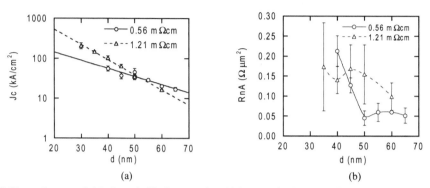

(a)   (b)

Fig.3 Dependences of (a) J_c and (b) R_nA on the thickness of TiN film d for NbCN/TiN/NbCN junctions with different resistivities of TiN films.

Figure 4(a) and 4(b) show I-V characteristics for series of 400 NbCN/TiN/NbCN junctions with TiN thickness of 40 nm and 60 nm, respectively. As shown here, small hysteresis was normally observed on I-V curves for junctions with TiN films thinner than 50 nm. This indicates that thin insulating layers, e.g., Schottkey barrier (Keller et $al.$ 1973), exist at the interface of TiN films and NbCN electrodes.

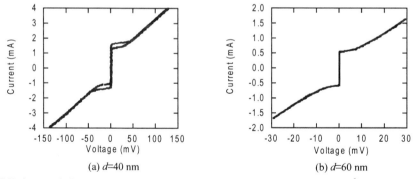

(a) d=40 nm   (b) d=60 nm

Fig.4 I-V characteristics at 4.2 K for series of 400 junctions: (a) J_c is 103 kA/cm^2, I_cR_n 80 μV and junction area 1.8x1.8 μm^2 and (b) J_c is 35 kA/cm^2, I_cR_n 20 μV and junction area 2.0x2.0 μm^2.

Figure 5 is a plot of I_cR_n for the same junctions as in Fig.3 (a) and 3 (b). Reflecting the enhancement of R_nA below d=50 nm for junctions with ρ=0.56 mΩcm seen in Fig. 3 (b), the d dependence of I_cR_n for those junctions changes at d=50 nm, as indicated by solid lines in Fig.5. From Fig.5, it is found that widespread I_cR_n values (10μV-1mV) can be obtained for NbCN/TiN/NbCN junctions by changing the thickness of TiN films (65 nm - 30 nm).

216

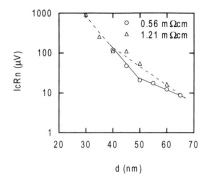

d (nm)

Fig.5 Dependence of $I_c R_n$ products on the thickness of TiN film d for NbCN/TiN/NbCN junctions with different resistivities of TiN films.

4.   CONCLUSION

Electrical characteristics of NbCN/TiN/NbCN junctions with various resistivities and thicknesses of TiN films have been investigated. J_c for junctions showed exponential dependence on TiN film thickness while $R_n A$ for junctions increased with decreasing the TiN film thickness. The dependence of $R_n A$ on the TiN film thickness is different from that expected from a simple barrier model and suggests the existence of insulating layers at the interface of TiN films and NbCN electrodes. By changing the TiN film thickness from 65 nm to 30 nm, $I_c R_n$ products for junctions was varied from 10 μV up to 1 mV.

ACKNOWLEDGMENTS

The authors are grateful to Q Wang and T Kikuchi for technical assistance and discussion. They   thank T Sakuraba, F Hirayama and A Fukushima for management of liquid helium. T Sakamoto is appreciated for continuous support and encouragement.

REFERENCES

Benz S P, Reintsema C D, Ono R H, Eckstein J N, Bozovic I and Virshup G F 1995 *IEEE Trans. on Appl. Supercond.* **5**, 2915.
Frizsch L, Schubert M, Wende G and Meyer H -G 1998 *Appl. Phys. Lett.* **73**, 1583
Hamilton C A, Burroughs C J and Kautz R L 1995 *IEEE Trans Instrum. Meas.* **44**, 223
Hamilton C A, Burroughs C J and Benz S 1997 *IEEE Trans. on Appl. Supercond.* **7**, 3756
Johansson B O, Sundgren J -E and Greene J E 1985 *J.Vac. Sci. Technol. A*, **3**, 303
Kautz R L, Benz S P and Reintsema C D 1994 *Appl. Phys. Lett.* **65**, 1445
Keller W H and Nordman J E 1973 *J. Appl. Phys.*, **44**, 4732
Likharev K K and Semenov V K 1991 *IEEE Trans. on Appl. Supercond.* **1**, 3
Maezawa M and Shoji A 1997 *Appl. Phys. Lett.* **70**, 3603
Wang Q, Kikuchi T, Kohjiro S and Shoji A 1997 *IEEE Trans. on Appl. Supercond.* **7**, 2801
Wang Q, Katsuta J, Kikuchi T, Kohjiro S and Shoji A 1998 *Applied Supercond.* **5**, 339
Wittmer Marc 1985 *J. Vac. Sci. Technol. A*, **3**, 1797
Yamamori H, Sasaki H and Shoji A 1999 *Japanese J. Appl. Phys.* **38**, L734

Inst. Phys. Conf. Ser. No 167
Paper presented at Applied Superconductivity, Spain, 14-17 September 1999
© 2000 IOP Publishing Ltd

Effects of post-ion-milling, pre-annealing, and post-annealing on the characteristics of high-T_c ramp-edge junctions

Gun Yong Sung[1], Chi Hong Choi[1], Kwang-Yong Kang[1], Moon-Chul Lee[2], and Soon-Gul Lee[3]

[1]Electronics and Telecommunications Research Institute, Yusong, Taejon 305-350, Rep. of Korea

[2]Dept. of Physics, Korea University, Seoul, 136-701, Rep. of Korea

[3]Dept. of Physics, Korea University, Jochiwon, Chungnam, 339-800, Rep. of Korea

ABSTRACT: We fabricated high-T_c superconducting $YBa_2Cu_3O_{7-x}$(YBCO)/$YBa_2Cu_{0.79}Co_{0.21}O_{7-x}$(Co-YBCO)/YBCO ramp-edge Josephson junctions on (001) $SrTiO_3$ single crystal substrates and studied the effects of post-ion-milling and pre-annealing of the ramp-edge prior to the top layer and post annealing process. The ion-beam voltage, the ion-beam incident angle, and the photoresist mask angle to yield smooth slopes with an angle of about 30° were optimized. The morphology of the edge was improved by the post-ion-milling, as were the edge-surface-induced epitaxial growth and the small interface resistance between the top YBCO layer and the Co-YBCO barrier. Annealing prior to barrier deposition recovered the ramp-edge surface and increased the Tc of the edge. Annealing assisted an epitaxial growth of the top YBCO layer on the ramp edge. Post-annealing at a temperature above the deposition temperature and cooling at 500 Torr O_2 induced the epitaxial rearrangement of Co-YBCO at a high oxygen vapor pressure. The current-voltage characteristics of the junctions showed RSJ-like behavior.

1. INTRODUCTION

For the realization of high-temperature superconducting digital circuits, a Josephson junction design is necessary that is lithographically definable, compatible with multilayer processes, and has a high design flexibility. These requirement can be fulfilled by a ramp-edge junction, where the bottom electrode is shaped as a smooth ramp-edge and an artificial barrier layer as well as the top electrode grow epitaxially on this ramp-edge(Gao et al 1990, Hunt et al 1991, Char et al 1993). The advantages of this type junction are the free placement of the junction in the chip design, the possibility to choose from a large variety of barrier materials including $PrBa_2Cu_3O_{7-x}$(Gao et al 1990), Ga-$PrBa_2Cu_3O_{7-x}$(Verhoven et al 1995), Co-$PrBa_2Cu_3O_{7-x}$(Yoshida et al 1999), Co-$YBa_2Cu_3O_{7-x}$(Char et al 1994), Ca-$YBa_2Cu_3O_{7-x}$(Antognazza et al 1995), Pr-$YBa_2Cu_3O_{7-x}$(Stolzel et al 1993) $LaSrGaO_4$(Sung et al 1997), interface-engineered $YBa_2Cu_3O_{7-x}$(Moeckly et al 1997). Hunt et al (1999) found that a number of materials and process parameters have a surprisingly large impact on the resistance and I-V characteristic of the ramp-edge junctions. These factors include the bottom electrode material and deposition conditions, the bottom electrode insulator cap material, the edge cleaning technique, and the growth conditions of the normal metal and top electrode. In addition, the morphology and the shape of the ramp-edge are also crucial points in the fabrication of the ramp-edge junctions.

In this paper, we investigated the effect of the post-ion-milling, the pre-annealing of the ramp-edge prior to the top electrode, the post-annealing process, and the RF-plasma cleaning on the junction's I-V characteristic as well as the top and bottom electrode quality.

218

2.   EXPERIMENTAL PROCEDURE

To fabricate ramp-edge junctions, a bilayer consisting of a 200 nm thick $YBa_2Cu_3O_{7-x}$ (YBCO) films for the bottom electrode and a 100 nm thick $SrTiO_3$ insulating layer was deposited *in-situ* by pulsed laser deposition(PLD) (Sung et al 1995). The films were patterned using standard optical lithography processes. The etching mask for the ramp-edges consisted of conventional AZ5214E photoresist, that was baked at 110 °C for 2 min. The ramp-edge was etched by argon ion-beam using an ion-beam current density of 1.5 mA/cm² and voltages from 150 V to 450 V. The sample was mounted on the sample holder with different beam angles (α) and aligned along the PR mask angle (θ) as shown in Fig. 1(c). After the ion-beam milling with $\alpha=30°$, $\theta=90°$ to etch the ramp-edge having an angle of 30°, the ramp-edge was etched again under the different sample geometry. The post-ion-milling geometry was $\alpha=45°$, $\theta=0°$. After the post-ion-milling process, the etching mask was removed by ultrasonic stirring in acetone and rinsing in methanol.   After ion milling the ramp-edge, a 60 min. annealing step at 470 °C and 500 Torr O_2 pressure was carried out in the PLD chamber to recover the surface of the YBCO ramp-edge.   Afterwards, the barrier materials $YBa_2Cu_{0.79}Co_{0.21}O_{7-x}$(Co-YBCO) and another 200 nm thick YBCO film for the top electrode were deposited *in-situ* by PLD mentioned above. Co-YBCO was used as a barrier material, because it is lattice-matched with YBCO and has comparable deposition conditions.

The junctions themselves were defined by a second argon ion-beam etching step. To insure electrical contact to the bottom electrode vias were chemically etched by 1% HF solution. A 200 nm thick gold layer was also deposited by PLD and patterned by a lift-off process to provide electrical contacts. Finally, the junctions were annealed at 800 °C for 30 min. and 470 °C for 120 min. under 500 Torr oxygen atmosphere.

3.   RESULTS AND DISCUSSION

3.1   Effect of Post-ion-milling, Pre-annealing, and Post-annealing

The shape of the ramp-edges and their morphology were investigated by SEM. The morphology of the ramp-edge was improved by the post-ion-milling as shown in Fig. 1 (a) and (b). Because the smooth surface of the ramp-edge is one of the critical parameter to determine the junction resistance and homogeneity, the post-ion-milling is essential to improve   the junction quality. T_c of the patterned bottom electrode increased by an annealing at 470 °C for 60 min. in 500 Torr oxygen atmosphere. We believe that by annealing before depositing YBCO top electrode, the ramp-edge surface could recovered to their equilibrium state, which is helpful for epitaxial growth of barrier layer and top YBCO layer on the ramp-edge.

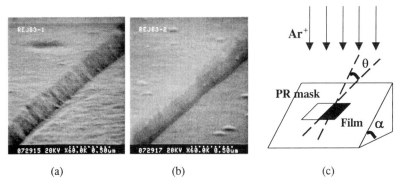

(a)   (b)   (c)

Fig. 1. SEM images for the surface of YBCO bottom layer ramp-edge (a) before and (b) after the post-ion-milling. (c) A schematic of the sample geometry during ion milling.

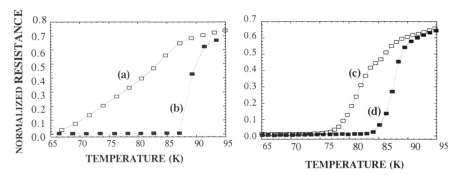

Fig. 2. R-T plot for the junctions (a) without and (b) with post-annealing at 800 °C, 30 min. and 470 °C for 120 min. in 500 Torr oxygen atmosphere. R-T plot for the junctions annealed at 470 °C in 500 Torr oxygen atmosphere for (c) 2 hrs. and (d) 5 hrs.

Dramatic increase of the junctions' T_c by post-annealing at 800 °C, 30 min. and 470 °C for 120 min. in 500 Torr oxygen atmosphere is shown in Fig. 2 (b). In addition to high temperature annealing effect, post-annealing time was also effective to increase the T_c of the junctions as shown in Fig. 2(d). We think that the annealing above deposition temperature and cooling at 500 Torr O_2 atmosphere may induce the epitaxial rearrangement for Co-YBCO with a high oxygen vapor pressure(Sydow et al 1998).

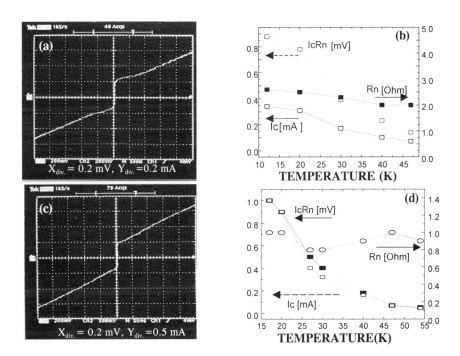

Fig. 3. I-V characteristics of the ramp-edge junctions having a 20 nm thick Co-YBCO barrier measured at 30 K (a) without and (c) with the RF-plasma cleaning. I_c, R_n, I_cR_n variation of the junctions (b) without and (d) with the RF-plasma cleaning as a function of temperature.

220

3.2 RF-Plasma Cleaning Effect

Reduction of the junction resistance was observed by RF-plasma cleaning before the barrier and top electrode layer deposition as shown in Fig. 3. By the RF-plasma cleaning, the junction showed a clear resistively-shunted-junction behavior. The phases of the two superconducting electrodes may be strongly coupled through the weak link by cleaning of the ramp-edge surfaces before barrier layer deposition. I_cR_n and J_c of the junction with a 20 nm thick $YBa_2Cu_{0.79}Co_{0.21}O_{7-x}$ barrier were 1 mV, 2.5×10^5 A/cm^2 at 17 K, 0.32 mV, 1.0×10^5 A/cm^2 at 30 K, and 0.05 mV, 1.25×10^4 A/cm^2 at 54 K.

4. CONCLUSIONS

We have optimized the ion beam voltage, ion beam incident angle, and PR mask angle to yield smooth slope with an angle of about 30°. The morphology of the edge was improved by the post ion milling and RF-plasma cleaning after etching the ramp-edge patterns. T_c increase of the edge surfaces was obtained by an annealing before depositing YBCO top layer. Post-annealed junctions showed an enhanced T_c and a Resistively Shunted Junction(RSJ) current-voltage characteristics. Through the RF-plasma treatment, the current-voltage characteristics of these junctions followed the resistively shunted junction model with excess current. I_cR_n and J_c of junctions with a 20 nm thick Co-YBCO barrier were 1 mV, 2.5×10^5 A/cm^2 at 17 K, 0.32 mV, 1.0×10^5 A/cm^2 at 30 K, and 0.05 mV, 1.25×10^4 A/cm^2 at 54 K.

ACKNOWLEDGMENT

The authors acknowledge the support by the Ministry of Information and Communications and the Ministry of Science and Technology, REP. OF KOREA. The authors would like to thank Dr. Bun Lee, Head of Telecommunications Basic Research Lab., for his supports.

REFERENCES

Antognazza L, Moeckly B H, Geballe T H and Char K 1995 Phys. Rev. B **62**, 4559
Char K, Colclough, Geballe T H and Myers K E 1993 Appl. Phys. Lett. **62**, 196
Char K, Antognazza L and Geballe T H 1994 Appl. Phys Lett. **65**, 904
Gao J, Aarnink W A M, Gerritsma G J and Rogalla H 1990 Physica C **171**, 126
Hunt B D, Foote M C and Bajuk L J 1991 Appl. Phys. Lett. **59**, 982
Hunt B D, Forrester M G, Talvacchio J and Young R M 1999 IEEE Trans. Appl. Supercon. **9**, 3362
Moeckly B H and Char K 1997 Appl. Phys. Lett. **71**, 2526
Stolzel C, Siegel M, Adrian G, Krimmer C, Sollner J, Wilkens, Schulz G and Adrian H 1993 Appl. Phys. Lett. **63**, 2970
Sung G Y and Suh J D 1995 Appl. Phys. Lett. **67**, 1145
Sung G Y and Suh J D 1997 IEEE Appl. Supercon. **5**, 2308
Sydow J P, Buhrman R A and Moeckly B H 1998 Appl. Phys. Lett. **72**, 3512
Verhoeven M A J, Gerrisma G J, Rogalla H and Golubov A A 1995 IEEE Trans. Appl. Supercon. **5**, 2095
Yoshida J, Inoue S, Hashimoto T and Nagano T 1999 IEEE Trans. Appl. Supercon. **9**, 3366

Inst. Phys. Conf. Ser. No 167
Paper presented at Applied Superconductivity, Spain, 14-17 September 1999
© 2000 IOP Publishing Ltd

Dynamics of Josephson vortices in intrinsic Josephson junctions in Bi₂Sr₂CaCu₂O$_y$ single crystal mesas

A Irie[1,2] and G Oya[2,3]

1) Department of Electrical and Electronic Engineering, Utsunomiya University, 7-1-2 Yoto, Utsunomiya 321-8585, Japan
2) CREST, Japan Science and Technology Corporation (JST), Japan
3) Department of Energy and Environmental Science, Utsunomiya University, 7-1-2 Yoto, Utsunomiya 321-8585, Japan

ABSTRACT: Josephson vortex dynamics in intrinsic Josephson junctions in Bi₂Sr₂CaCu₂O$_y$ single crystal mesas has been investigated experimentally and numerically. In a magnetic field, clear vortex-flow branches were observed in addition to multiple quasiparticle branches on the current-voltage characteristics of the mesas at 4.2 K. The vortex velocities were characterized by the highest velocity of electromagnetic waves in the junctions. Simulations also showed that the stable vortex motion with the highest velocity mode could occur even when only some junctions in a stack were concerned in vortex flow.

1.   INTRODUCTION

Recently, much attention has been paid to the investigation of stacks of Josephson junctions both from the point of views of their application in cryoelectronics and of the fundamental physics. They show a variety of new physical phenomena. One of their interesting features is the in-phase locking of Josephson vortices in different junctions in the stack, which was first observed in artificially two-stacked Josephson junctions (Ustinov et al 1993). For applications such as high frequency oscillators, the in-phase locking is useful for improving the impedance matching with free space and the output level. It is well known that highly anisotropic layered high-temperature superconductor Bi₂Sr₂CaCu₂O$_y$ (BSCCO) acts as natural stacks of Josephson junctions, or intrinsic Josephson junctions (IJJs) (Kleiner et al 1992, Oya et al 1992, Yurgens et al 1996 and Irie et al 1997). They are formed atomically by stacking of superconducting CuO₂ layers and insulating BiO and SrO layers in the c-axis direction. For intrinsic Josephson junctions, the magnetic coupling between adjacent junctions is very strong due to the superconducting layers of the thickness t of 3 Å much smaller than the London penetration depth λ_L of ~2000 Å. Such a coupling may lead to a stronger phase-locking phenomenon over many junctions in the stack than that for artificial junctions. Furthermore, the advantage of IJJs is that the number of junctions in the stack can be easily controlled by the thickness of IJJ mesas. Therefore, we have investigated vortex dynamics in IJJs by means of the observation of the Josephson vortex flow characteristics of BSCCO mesas with various sizes and the numerical simulation based on the inductive coupling model for a stack of 9 Josephson junctions.

In this paper, we report the experimental and numerical results on Josephson vortex flow characteristics in intrinsic Josephson junctions.

2. EXPERIMENTAL

Measurements were performed on mesas patterned on the surfaces of BSCCO single crystals. The lateral dimensions ($L \times W$) of the mesas were 160×40 and 320×40 μm^2 and their heights, h, were between 7 and 150 nm. Figure 1 shows a schematic view of the fabricated mesa. An external magnetic field, B, was applied perpendicular to the c-axis and L of the mesa, and the I-V characteristics along the c-axis were measured at 4.2 K by a conventional four-terminal method. Our mesas are considered to be stacks of long Josephson junctions because the Josephson penetration depth λ_J of IJJs in BSCCO is ~0.5-1.0 μm. In simulations, we used the inductive coupling model (Kleiner 1993 and Sakai et al 1993), which well accounts for many dynamical phenomena observed in stacks of conventional long Josephson junctions.

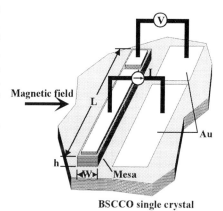

Fig. 1 A schematic view of a fabricated mesa.

3. RESULTS AND DISCUSSION

3.1 *I-V* characteristics

In zero magnetic field, the *I-V* curves of BSCCO mesas show multiple hysteretic resistive branches with a spacing of ~20 mV, which are the quasiparticle branches (QPBs) of the individual intrinsic Josephson junctions (Kleiner et al 1992, Yurgens et al 1996 and Irie et al 1997). The critical currents I_c of IJJs are sensitive to the magnetic field and the field dependence of I_c was similar to that of a conventional long Josephson junction. Moreover, we have successfully observed Josephson vortex-flow branches (JVFBs) in the *I-V* characteristics. An example of the *I-V* curves of a mesa (90 IJJs) with the dimensions of $320 \times 40 \times 0.135$ μm^3 for different values of B is shown in Fig. 2(a). With increasing B, JVFBs appear starting from the $V = 0$ branch and from every QPBs, and shift to higher voltages. In the range of $B < 0.2$ T, all junctions in the mesa are not observed to be in a vortex-flow state. Therefore, the density of vortices in this mesa is considered to be low on account of the applied

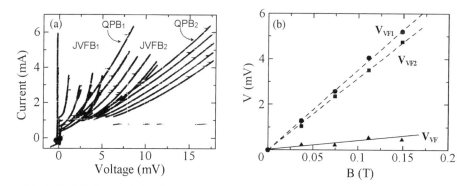

Fig. 2 (a) *I-V* characteristic of a BSCCO mesa with 90 intrinsic Josephson junctions in different magnetic fields (0, 0.03, 0.075, 0.11, 0.15 T) at 4.2 K. (b) Field dependence of maximum voltages of the first and second VFBs and the flow voltage V_{VF} per one intrinsic Josephson junction in the mesa.

field slightly larger than the lower critical field of the mesa, which is estimated from the field dependence of the critical current (Irie et al 1999).

Figure 2(b) shows the voltage V_{VF1} at the top of JVFB$_1$ and the corresponding voltage V_{VF2} of JVFB$_2$ measured from QPB$_1$, as a function of B. It is found that both V_{VF1} and V_{VF2} increase linearly with B. From these, we can determine the vortex-flow voltage V_{VF} per one junction from the difference between V_{VF1} and V_{VF2} because one junction in the stack is already switched to the resistive state at QPB$_1$. The obtained V_{VF} is also plotted in Fig. 2(b). From V_{VF} and dV_{VF}/dB, we found that a maximum number of ten junctions contribute to JVF$_1$ as shown in Fig. 2 (a) and that the flow velocity v_F of vortices was about 3×10^6 m/s. Similar results also were observed for other mesas with 20–100 IJJs. Consequently, we found that vortex flow occurred only in 10-20 % of IJJs in each measured mesa in the presented magnetic field range and the flow velocities were in the range of $(1-4)\times10^6$ m/s in these mesas.

For single long Josephson junction, the maximum velocity of vortex flow is limited by the characteristic velocity of the electromagnetic wave, namely, the Swihart velocity. For a stack of N Josephson junctions, it is given by (Kleiner 1993 and Sakai et al 1993)

$$c_n = c_0 \sqrt{\frac{d}{\varepsilon d'\{1 + 2S \cos[n\pi /(N+1)]\}}}, \qquad n=1,2,\ldots,N, \qquad (1)$$

where c_0 is the light velocity in vacuum, d is the barrier thickness, ε is the dielectric constant, $d'=d+2\lambda_L\coth(t/\lambda_L)$ is the effective magnetic thickness and $S=-\lambda_L/[d'\sinh(t/\lambda_L)]$ is coupling parameter. Eq. (1) indicates that N stacked junctions have N mode velocities of electromagnetic waves. For $n=1$ this gives the highest velocity which is related to a rectangular lattice of moving vortices while for $n=N$ this gives the lowest related to a triangular one. Using typical parameters for BSCCO (Irie et al 1998), the observed vortex flow velocity gives good agreement with the velocity calculated for $n=1$ (5.3×10^6 m/s) and $n=2$ (2.7×10^6 m/s). This indicates that vortices move with the velocity close to the highest one of the electromagnetic wave rather than the lowest which has been observed so far for mesas with large number of junctions, when only some junctions in a stack are concerned in vortex flow.

3.2 Numerical simulations

To acquire a better understanding of the vortex-flow regime in intrinsic Josephson junctions, the system of coupled sine-Gordon equations based on an inductive coupling model has been solved numerically. In normalized units, it is given by (Sakai et al 1993)

$$\begin{pmatrix} \varphi_1'' \\ \vdots \\ \varphi_n'' \\ \vdots \\ \varphi_N'' \end{pmatrix} = \begin{pmatrix} 1 & S & & & \\ & \ddots & & 0 & \\ S & 1 & S & \\ & 0 & & \ddots & \\ & & & S & 1 \end{pmatrix} \begin{pmatrix} \ddot{\varphi}_1 + \alpha\dot{\varphi}_1 + \sin\varphi_1 - \gamma \\ \vdots \\ \ddot{\varphi}_n + \alpha\dot{\varphi}_n + \sin\varphi_n - \gamma \\ \vdots \\ \ddot{\varphi}_N + \alpha\dot{\varphi}_N + \sin\varphi_N - \gamma \end{pmatrix}, \qquad (2)$$

together with the boundary condition for the applied magnetic field

$$\varphi_n'(0,\tau) = \varphi_n'(\ell,\tau) = -b(1+2S) = -\eta \ , \qquad (3)$$

where $\varphi_n(x,\tau)$ is the phase difference in the nth junction, the spatial coordinate x and time τ are normalized to $\lambda_J=[\Phi_0/(2\pi\mu_0 d'J_c)]^{1/2}$ and $\omega_0^{-1}=[\Phi_0\varepsilon/(2\pi d J_c)]^{1/2}$, respectively, J_c is critical current density, α is the dissipation coefficient, γ is the normalized bias current density, $b=B/(\mu_0 J_c\lambda_J)$ is the normalized magnetic field and $\ell=L/\lambda_J$ is the normalized junction length. A typical simulated current-voltage (γ-v) characteristic of a stack of nine Josephson junctions is shown in Fig. 3(a). Here, $v=\Sigma<\partial\varphi_n/\partial\tau>$. Parameters of the stack are $\ell=10$, $S=-0.499997$, $\alpha=0.1$ and $\eta=0.22$. This situation corresponds to the

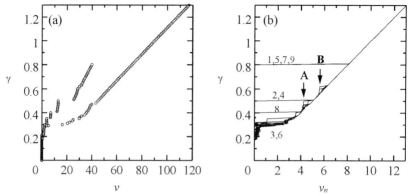

Fig. 3 Numerically simulated *I-V* characteristics of a stack of 9 Josephson junctions. (a) A total junction-voltage and (b) individual junction-voltages of the stack.

low density of vortices in the stack. By applying magnetic field, one can see several discrete resistive branches in the γ-v characteristic due to vortex flow. Figure 3(b) shows the γ-v_n characteristics of the individual junctions in the stack. Notations in this figure indicate the junction index in the stack. The number of junctions switching to the resistive state depends on the bias current. For $\gamma < 0.4$, only two junctions(3,6) are in resistive state. For $0.4 < \gamma < 0.6$, junction 8 and subsequently junctions 2, 4 switch to the resistive state as γ increases. In this region, the current steps (Points A and B) appear remarkably. From the analyses of vortex dynamics in the junctions at these steps, it is found that vortices flow in phase with the $n=1$ mode of the different Fiske number n_x in the layer plane ($n_x=3$ for step A and $n_x=4$ for step B) of the stack. Moreover, from our simulations we found that the $n=1$ mode appeared preferentially in the case of low vortex densities in stacks. These results agrees with our experimental results, and hence indicates that the $n=1$ mode can be also excited without the rectangular lattice of dense vortices in all junctions in the stack.

4.   CONCLUSIONS

Josephson vortex dynamics in stacks of intrinsic Josephson junctions in $Bi_2Sr_2CaCu_2O_y$ single crystal mesas has been investigated experimentally and numerically. Josephson vortex-flow branches were observed on the *I-V* characteristics at 4.2 K and then vortex flow velocity was estimated to be (1-4)$\times 10^6$ m/s. Simulations based on the coupled sine-Gordon equations also showed that the vortex motion with the highest velocity mode was more stable rather than that with the lowest, when not all junctions but some in a stack are concerned in vortex flow.

REFERENCES

Irie A, Hirai Y and Oya G 1998 Appl. Phys. Lett. **27**, 2159
Irie A, Kaneko S and Oya G to be published in Int. J. Mod. Phys. B.
Kleiner R, Steinmeyer F, Kunkel G and Müller P 1992 Phys. Rev. Lett. **68**, 2394
Kleiner R 1994 Phys. Rev. **B50**, 6919
Oya G, Aoyama N, Irie A, Kishida S and Tokutaka H 1992 Jpn. J. Appl. Phys. **31**, L829
Sakai S, Ustinov A V, Kohlstedt H, Petraglia P and Pedersen N F 1994 Phys. Rev. **B50**, 12905
Ustinov A V, Kohlstedt H, Cirillo M, Pedersen N F, Hallmanns G and Heiden C 1993 Phys.Rev. B **48**, 10614
Yurgens A, Winkler D, Zavarisky N V and Claeson T 1996 Phys. Rev. B **53**, R887

Inst. Phys. Conf. Ser. No 167
Paper presented at Applied Superconductivity, Spain, 14-17 September 1999
© *2000 IOP Publishing Ltd*

On Feynman's derivation of Josephson equations

D.-X. Chen, J. J. Moreno, and A. Hernando

Instituto de Magnetismo Aplicado, UCM-RENFE-CSIC,
28230 Las Rozas, Madrid, Spain.

A. Sanchez

Grup d'Electromagnetisme, Departament de Física, Universitat Autònoma de Barcelona,
08193 Bellaterra, Barcelona, Catalonia, Spain.

ABSTRACT: In this work, we have tried to clarify several questionable points in Feynman's treatment on Josephson effects and to give its clear formulation.

1. INTRODUCTION

After Josephson made his microscopic theory on the superconducting tunnelling effect (Josephson 1962) Feynman proposed a macroscopic treatment for deriving Josephson equations (Feynman 1965). Feynman's equations were later justified to some extent on the grounds of BCS theory (Rogovin 1974) and completed to include the effect of the external circuit (Otha 1977). Owing to its simplicity and elegance, his treatment offered a powerful key for the understanding of Josephson effects, and it was cited by many authors till now, especially in textbooks (Mercerau 1969, Barone 1982, Alario 1991, Poole 1995, Kittel 1996). However, a few points in Feynman's derivation were obscure, which caused a big confusion in its later descriptions. In the present paper, we try to clarify these points.

2. FEYNMAN'S DERIVATION OF JOSEPHSON EQUATIONS

Considering two cylindrical superconductors S1 and S2 of the same volume V_s coupled through a thin insulating layer I of area A_J, Feynman first built a quantum-mechanical equation system for superconducting electrons and then used it to Cooper pairs, which were regarded as ideal bosons with charge $-2e$. The resulting two equations are

$$i\hbar \frac{\partial \Psi_1}{\partial t} = -eV_{12}\Psi_1 + K\Psi_2,$$

$$i\hbar \frac{\partial \Psi_2}{\partial t} = eV_{12}\Psi_2 + K\Psi_1, \tag{1}$$

where

$$\Psi_{1,2} = \sqrt{n_{1,2}} \exp(i\gamma_{1,2}) \tag{2}$$

are the wavefunctions in S1 and S2, $n_{1,2}$ and $\gamma_{1,2}$ being the Cooper-pair densities and the phases of wavefunctions in S1 and S2. The voltage applied between S1 and S2 is defined as $V_{12} = V_1 - V_2$, where $V_{1,2}$ are the electric potentials in S1 and S2. Thus, $-eV_{12}$ and eV_{12} in Eqs. (1) are the chemical potentials in S1 and S2 assuming the zero to be halfway between. K is the coupling constant that characterizes the junction.

Substituting Eqs. (2) in Eqs. (1), we get two equations by equating the real parts and two equations by equating the imaginary parts. The dc Josephson equation, i.e., the current density from S1 to S2, is calculated from the first two as

$$J_{12} = 2e\frac{dn_1}{dt} = -2e\frac{dn_2}{dt} = J_c \sin\gamma_{12}^{(1)}. \tag{3}$$

Since n_1 and n_2 should remain constant if the effect of the external battery providing the voltage V_{12} was considered (for a strict treatment on this, see Ohta 1977), we assume

$$n_1 = n_2 = n_0, \tag{4}$$

so that the critical current density J_c in Eq. (3) becomes

$$J_c = 4en_0K / \hbar. \tag{5}$$

The ac Josephson equation is derived from the second two and Eq. (4) as

$$d\gamma_{12}^{(2)} / dt = 2eV_{12} / \hbar. \tag{6}$$

In Eqs. (3) and (6), it is defined that

$$\gamma_{12}^{(1)} = -\gamma_{12}^{(2)} = \gamma_2 - \gamma_1. \tag{7}$$

3. CLARIFICATION OF FEYNMAN'S DERIVATION

3.1 Phase difference

Comparing Eqs. (3) with (6), we see that if J_{12} and V_{12} are clearly defined as made above, the phase differences in the dc and ac equations, $\gamma_{12}^{(1)}$ and $\gamma_{12}^{(2)}$, are opposite in sign. Without substituting the Cooper-pair charge q by $-2e$ in the ac equation, this contradiction in Feynman's treatment was shown implicitly (Feynman 1965), but it was made explicit later by some other authors (Kittel 1996). We should mention that following Feynman, other kinds of combinations of γ_{12} for the dc and ac equations may also be derived (Mercerau 1969, Barone 1982, Vicent 1991, Poole 1995), but the reasons for this variety are all due to unclear definitions of the current and/or voltage and confused definitions of Cooper-pair charge.

Inspecting the above derivations, we find that the ac Eq. (6) can be derived from Eqs. (1) with $K = 0$, i.e., for two mutually isolated superconductors. Therefore, being a direct consequence of the definitions of the chemical potentials in Eqs. (1) and the phases in Eqs. (2), the phase difference $\gamma_{12}^{(2)}$ in Eq. (6) should be questionless. In the following we show that the correct γ_{12} in Eq. (3) should also be $\gamma_{12}^{(2)}$.

With this purpose, we calculate the junction free energy (coupling energy) by following the procedure applied by Josephson (1965). Consider two systems A and B, which are identical except for the fact that A contains a junction like the one treated in Sec. 2 and B is a single superconducting region. Suppose that A and B are connected to generators supplying an identical current $I = A_J J_{12}$ to each. In this case, the change in the free energy of the junction is $dF_J = d(F_A - F_B)$, which should equal the difference between the works done by the generators to A and B, $(V_A - V_B)Idt$. Since the voltage across B, $V_B = 0$, we have $V_A - V_B = V_A = V_{12}$. Adjusting I carefully such that V_{12} remains very small and the process turns out to be reversible, we have from Eqs. (3) and (6) for the surface density of the junction free energy, $f_J = F_J / A_J$, that

$$df_J = V_{12}J_{12}dt = \frac{\hbar J_c}{2e} \sin(\gamma_2 - \gamma_1)d(\gamma_1 - \gamma_2),$$ (8)

whose integration leads to the coupling energy density

$$f_J = f_{J0} \cos\gamma_{12}^{(1,2)} + \text{const.},$$ (9)

where the coupling energy constant is

$$f_{J0} = \frac{\hbar J_c}{2e}.$$ (10)

Compared with the well known formula for the coupling energy density, $f_J = -f_{J0} \cos\gamma_{12} + \text{const.}$ (Josephson 1962), Eq. (9) is incorrect owing to a wrong sign. The correct formula can be obtained only when the phase difference in the dc Eq. (3) is changed to $\gamma_{12}^{(2)}$, which is therefore the right expression of γ_{12}.

Without mentioning it explicitly in Feynman's treatment, K was regarded as a positive quantity just like J_c and e. Thus, in order to change the sign of the phase difference in the dc Eq. (3), the only way is to redefine K as a negative quantity. This modification means that the coupling between S1 and S2 will be stable when the total energy is reduced by the coupling.

3.2 Dimension of J_c

We should mention that a factor of $2e$ was missing in the dc equation derived by Feynman (1965). However, even $2e$ is included as in Eq. (3), this equation is still dimensionally wrong, since in SI units, the dimension of J_{12} is A/m^2 but the dimension of $2edn_1/dt$ is A/m^3. In fact, some authors already noticed this and made relevant modifications to Feynman's derivation (Mercerau 1969, Kittel 1996). Since the time rate of Cooper-pair density is for unit volume and the Cooper-pair current density is for unit area, Eqs. (3) should be changed as

$$J_{12} = 2el_s \frac{dn_1}{dt} = -2el_s \frac{dn_2}{dt} = \frac{-4eK l_s}{\hbar} \sqrt{n_1 n_2} \sin\gamma_{12}^{(2)},$$ (11)

where $l_s = V_s / A_J$ is the common length of S1 and S2, and the phase difference has been changed in accord with the redefinition of K. Correspondingly, Eqs. (5) and (10) become

$$J_c = -4en_0 K l_s / \hbar,$$ (12)
$$f_{J0} = -2n_0 K l_s.$$ (13)

We now discuss the physical meaning of K based on Eqs. (12) and (13).

First, we see that being a coupling constant, K characterizes extrincally the entire S1-I-S2 assembly rather than intrincally its central coupled part. From the microscopic derivation, the intrinsic properties of a junction such as J_c and f_{J0} are functions of the property of superconducting materials of S1 and S2 and the coupling strength between both. Actually, the former is expressed by n_0 in Eqs. (12) and (13) and the latter by $-Kl_S$. Thus, we see that K is not an intrinsic constant and it is inversely proportional to the coupled superconductors' length. As an intrinsic coupling constant, we can define instead

$$K^* = -K\, l_S\, /\, \delta_J ,\qquad(14)$$

where δ_J is the coupling length. Replacing Eq. (14) in Eq. (13), we obtain

$$K^* = \frac{f_{J0}}{2\delta_J n_0} .\qquad(15)$$

This means that considering all the Cooper pairs around the barrier within a given coupling length δ_J, K^* is the coupling energy constant per Cooper pair per unit junction area.

3. CONCLUSIONS

We have examined Feynman's derivation of Josephson equations and clarified several points as critical current density J_c, coupling constant K, coupling energy constant f_{J0}, and the phase difference γ_{12}. The correct expression of $\gamma_{12} = \gamma_{12}^{(2)}$ is consistent with the gauge-invariant phase difference $\theta_{12} = \gamma_{12}^{(2)} - \frac{2e}{\hbar}\int_1^2 \mathbf{A}\cdot d\mathbf{l}$ proposed by Teitel (1983) rather than $\theta_{12} = \gamma_{12}^{(1)} - \frac{2e}{\hbar}\int_1^2 \mathbf{A}\cdot d\mathbf{l}$ originally derived by Josephson himself (1965, 1964). We believe the present work has provided a clear formulation of the elegant Feynman's treatment for the Josephson effects.

REFERENCES

Alario M A and Vicent J L 1991 Superconductivitad, (Madrid: EUDEMA) p. 136
Barone A and Paterno G 1982 Physics and Applications of the Josephson Effect (New York: Wiley) p. 1
Feynman R P 1965 Lectures on Physics Vol. 3 (Reading: Addison-Wesley) Chapt. 21
Josephson B D 1962 Phys. Lett. 1, 251
Josephson B D 1964 Rev. Mod. Phys. 36, 216
Josephson B D 1965 Advan. Phys. 14, 419
Kittel C 1996 Introduction to Solid State Physics, 7th edition (New York: Wiley) p. 366
Mercereau J E 1969 Superconductivity Vol. 1, ed Parks R D (New York: Marcel Dekker) p. 393
Ohta H 1977 Proc. Int. Conf. Superconducting Quantum Devices, eds Hahlbohm H D and Lubbig H p.35
Poole C P Jr., Farach H A and R. J. Creswick 1995 Superconductivity, (San Diego: Academic Press) p. 426
Rogovin D and Scully M O 1974, Ann. Phys. 88, 371
Teitel S and Jayaprakash C 1983 Phys. Rev. Lett. 51, 1999

Inst. Phys. Conf. Ser. No 167
Paper presented at Applied Superconductivity, Spain, 14-17 September 1999
© *2000 IOP Publishing Ltd*

Fabrication and properties of Sr$_2$AlTaO$_6$/YBCO interface-controlled junctions

Gun Yong Sung, Chi Hong Choi, Seok Kil Han, Jeong-Dae Suh, and Kwang-Yong Kang

Electronics and Telecommunications Research Institute, Taejon, 305-350, REP. OF KOREA

ABSTRACT: We fabricated interface-controlled ramp-edge junctions with barriers formed by modifying the ramp-edge surface instead of epitaxially grown barrier layers. Low-dielectric Sr$_2$AlTaO$_6$(SAT) layer was used as an ion-milling mask as well as an insulating layer for the ramp-edge junctions. We optimized the deposition condition of SAT films. The barriers were produced by structural modification at the ramp-edge of YBCO base electrode using plasma-treatment prior to the deposition of YBCO counterelectrode. We investigated the effects of 2-step ion-milling process variables and reflowed photoresist on the ramp-edge angle and morphology. The junction parameters were improved by using *in-situ* RF-plasma cleaning treatment.

1. INTRODUCTION

High temperature superconducting (HTS) Josephson junctions are the important element of superconducting device applications, like superconducting quantum interference device(SQUID) magnetometers, voltage standards, and single flux quantum (SFQ) logic circuits. These applications require the junction that has critical currents (I$_c$) in the range 100~500 μA, normal resistance (R$_n$) values of one to several ohms, and inductances of several pH, with 1-σ spread less than 10%. Bi-crystal junction(Sung et al 1999), step-edge junction, and ramp-edge junction have been widely investigated. Among these junctions, the ramp-edge junction showed several technical advantages compared to others. However, there exist technological problems in the fabrication of these junctions because of using a multilayer structure in ramp-edge junctions. The deposition process is not suitable for the formation of uniform and atomically thin barriers for HTS junctions, because the film growth at high temperature is not homogeneous.

Recently, Moeckly and Char (1997) reported the interfaced-engineered junction (IEJ) which was the ramp-edge type junction fabricated with no barrier deposition. In this junction, the barrier was formed on a ramp-edge surface of YBCO base electrode by structural modification using vacuum annealing and *in-situ* plasma treatment at high substrate temperature prior to the deposition of counter electrode. Hunt et al (1999) reported two ways making the interface-controlled barrier, which were fabricated by using an ion-beam damage at elevated temperature or using a chemical treatment. Satoh et al (1999) suggested the modified interface junctions with LaSrAlTaO insulating layer. This process includes *in-situ* etching process without intentional barrier deposition and vacuum annealing.

Although SrTiO$_3$ is known as a conventional insulating materials for the HTS multilayer process, it has a too large dielectric constant for high frequency applications. In contrast, the dielectric constant reported for SAT films of ε = 23~30 indicates that it is an excellent choice for high-speed digital circuits (Findikoglu et al 1993). In our process, Sr_2AlTaO_6 (SAT) layer was used as an ion-milling mask as well as an insulating layer for the ramp-edge junctions. In 2-step ion milling process using SAT as an etching mask, an ion-milled YBCO ramp-edge was not exposed to solvent through all fabrication procedures. In this study, we have investigated the optimal condition of pulsed laser deposition (PLD) for SAT films on YBCO layer and the effects of the 2-step ion-milling process variables and reflowed photoresist (PR) on the ramp-edge angle and morphology. The junction parameters were improved by using *in-situ* RF-plasma cleaning treatment prior to counterelectrode deposition.

2. EXPERIMENTAL PROCEDURE

Our process for fabricating junctions was based on a 2-step ion milling process, in which the etching mask was not a PR but a patterned SAT layer on the YBCO base electrode. An ion-milled YBCO ramp-edge was not exposed to solvent. YBCO, SAT, and Au for contact pads were deposited by KrF PLD on $LaAlO_3$ (LAO) (100) single crystal. The energy density of the laser beam on the target, the oxygen pressure in the chamber, the substrate temperature, and the substrate to target distance were varied to optimize the thickness, the surface roughness, and the film quality(Sung and Suh 1995). The YBCO layer was deposited in 100 mTorr oxygen at about 820 ℃. The SAT layer was grown at the substrate temperature of 700 ~ 780 ℃ and the oxygen partial pressure of 100 mTorr. The target-substrate distance for YBCO layer deposition was 6.5 cm. The distance for the growth of SAT was 5.5 ~ 6.5 cm. YBCO electrode and SAT insulating layer were deposited sequentially on the substrate.

The SAT layer was patterned using 350V and 30 mA Ar ion-milling with a reflowed AZ5214 PR mask. The patterned PR were reflowed at 130 ~ 160 ℃ for 3min. To create the ramp with various slopes, the incident angle of the Ar ion-beam (α_1) was varied between 30° to 45°. The substrate was rotated during the ion-milling. After the PR was removed, the ramp-edge was etched by second-ion-milling using the SAT mask. The incident angle of the Ar ion beam was $\alpha_2 = 30°$ to 65 ° to the substrate surface. Final angle of the ramp-edge was about 21 ° from substrate surface.

After making the ramp-edge of the base electrode, the sample was transferred to the PLD chamber and heated to between 500 to 700 ℃ in vacuum for 1 hr. We applied plasma by biasing the heater with an RF-source, using Ar as an ionizing gas at pressure of 50 ~ 100 mTorr. This plasma treatment was performed for several minutes with a forward power of 50 ~ 200W. Following the plasma-treatment, the YBCO counterelectrode and SAT insulator were deposited by PLD. The counterelectrode was patterned by ion-milling to define the junction width. Finally, Au was also deposited by PLD and patterned by lift-off for the contact pads.

3. RESULTS AND DISCUSSION

To optimize the deposition condition of the SAT films on 200 nm thick YBCO films on LAO (100) substrate, we varied the substrate temperature in the range of 700 ℃ and 780 ℃ under an optimal oxygen pressure of 100 mTorr. The surfaces of the films deposited at higher energy density were significantly smoother than the ones deposited at lower energy density. The optimal laser energy density incident on the target was 2.0 J/cm^2. To obtain the film with an optimal thickness, the optimal substrate-target distance was determined as 5.5cm. SEM images of the SAT/YBCO bilayer on LAO substrate show a relatively smooth surface and dense cross-section shown in Fig. 1. Ramp-edge was patterned using 2-step ion-milling, to protect the etched YBCO ramp-edge surface from the chemicals. The PR reflowing temperature, the incident angle of ion-beam, and the substrate rotation were varied to obtain a desirable ramp-edge, in which has a smooth surface and clean top and bottom of the ramp and has angles below 30°. Fig. 2 show a wall on top of the ramp. It seems that the wall formed by redeposition of the ion-milled materials in front of the steep edge of the PR during the ion-milling. This

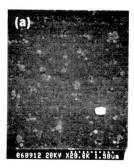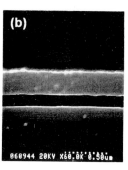

Fig. 1. SEM images of (a) the surface morphology of the SAT films deposited on the YBCO films and (b) the cross-section of the SAT/YBCO bilayer, in which the YBCO layer was chemically etched.

problem can be solved by reflowing the PR as reported from Hunt et al(1996) and Mallison et al (1997). Fig. 3 show the ramp-edge that has a smooth surface, no redeposition wall, and an angle of 20°. This ramp-edge was obtained using the SAT mask. This mask was ion-milled with the reflowed AZ5214 PR mask and rotated during the ion-milling. After the PR was removed, second-ion-milling was carried out by ion-milling using the SAT mask. The incident angle of Ar ion-beam was $\alpha_2 = 30°$ ~ 45 ° to the substrate surface and the substrate was rotated.

The characteristics of the ramp-edge junction depend on the interface state of between the base electrode and counterelectode. A long tail below T_c in R-T curve and low T_c were originated from the contamination and ion-beam damage at the ramp-edge surface during the fabrication process as well as the quality of YBCO electrodes. In order to overcome these problems, we employed an *in-situ* plasma-treatment prior to the deposition of the YBCO counterelectrode. This plasma-treatment was performed for several minutes with forward power of 50 ~ 200 W in 100 mTorr Ar. As shown in Fig. 4 (a), the critical temperature of the junctions and base electrode significantly decreased with increasing the RF-plasma power. In case of the forward power of 200 W, the YBCO counterelectrode degraded severely and the junction showed a tail in R-T characteristic as shown in Fig. 4(b). We believe that the high power plasma-treatment may damage the substrate surface so that the counterelectrode can not be grown epitaxially on the plasma-treated substrate. The RF-plasma cleaning condition was optimized at the forward power of 50 W. The T_c of the junctions was 82 K. At 70 K, the critical current density was 1×10^5 A/cm^2.

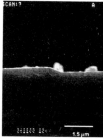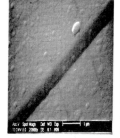

Fig. 2. SEM images of the surface and cross-section of the ramp of the SAT mask, in which was patterned using 350 V Ar ion-milling with no reflowed PR and no rotation.

Fig. 3. SEM images of the surface and cross-section of the ramp of the SAT/YBCO bilayer, in which was patterned using a reflowed PR and rotation during ion-milling.

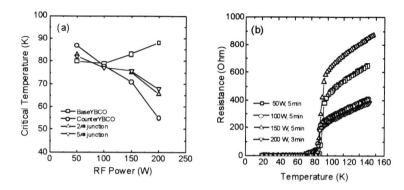

Fig. 4. (a) Critical temperature vs. RF power plots of the base electrode, counterelectrode, 2 μm, and 5 μm junction. (b) Resistance vs. temperature plots of 5 μm junction at various RF-plasma conditions.

4. SUMMARY

The deposition condition of the SAT thin film on the YBCO electrode has been optimized. In the 2-step ion milling process, an ion-milled YBCO ramp-edge was not exposed to solvent through the fabrication process. The ion-beam voltage, ion-beam incident angle, PR mask angle, and PR reflowing temperature were optimized to obtain the ramp-edge with smooth surface and an angle of about 21°. Clean and smooth surface of the ramp-edge was obtained by using the reflowed PR , the liquid nitrogen cooling, and rotation during ion-milling. The junction parameters were improved by an *in-situ* RF-plasma cleaning treatment. The optimized RF power was 50 W. The junction showed a T_c of 82 K and $J_c > 1 \times 10^5$ A/cm^2 at 70K.

ACKNOWLEDGMENT

The authors acknowledge the support by the Ministry of Information and Communications and the Ministry of Science and Technology, REP. OF KOREA. The authors would like to thank Dr. Bun Lee, Head of Telecommunications Basic Research Lab., for his supports.

REFERENCES

Findikoglu A T, Doughty C, Bhattacharya S, Li Q, Xi X X and Venkatesan T 1993 IEEE Trans. Appl. Supercond. **3**, 1425
Hunt B D, Forrester M G, Talvacchio J, McCambridge and Young R M 1996 Appl. Phys. Lett. **68**, 3805
Hunt B D, Forrester M G, Talvacchio J and Young R M 1999 IEEE Trans. Appl. Supercon. **9**, 3362
Mallison W H, Berkowitz S J and Hirahara A S 1997 IEEE Trans. Appl. Supercon. **7**, 2944
Moeckly B H and Char K 1997 Appl. Phys. Lett. **71**, 2526
Satoh T, Hidaka M and Tahara S 1999 IEEE Trans. Appl. Supercon. **9**, 3141
Sung G Y and Suh J D 1995 Appl. Phys. Lett. **67**, 1145
Sung G Y, Suh J D, Kang K Y, Hwang J-S, Yoon S-G, Lee M C, and Lee S G 1999 IEEE Trans. Appl. Supercon. **9**, 3921

Inst. Phys. Conf. Ser. No 167
Paper presented at Applied Superconductivity, Spain, 14-17 September 1999
© 2000 IOP Publishing Ltd

Current transport in Au/YBa2Cu3Ox junctions in *c*-axis and tilted directions of YBa2Cu3Ox thin films

P. Komissinski[1,2] and G.A. Ovsyannikov[2]

[1]Chalmers University of Technology and Gothenburg University, Gothenburg, Sweden
[2]Institute of Radio Engineering and Electronics RAS, Moscow, Russia

ABSTRACT: We report on the fabrication technique and investigation of current transport properties of Au/YBa2Cu3Ox junctions. The junctions were processed in epitaxial YBa2Cu3Ox (YBCO) thin films grown by laser ablation. (001) LaAlO3 and (120) NdGaO3 substrates were used to grow correspondingly (001)-oriented and tilted YBCO films. Current-voltage characteristics, *I-V*-curves, and the dependencies of the differential resistance on the applied voltage, Rd(V), were measured and the anisotropy of the current transport in these junctions was studied.

1.   INTRODUCTION

The resistance between high-temperature superconductor and normal metal, HTS/N, is an important parameter from fundumental and practical point of view. The anisotropy of *c*-axis and *a-b*-plane resistance (R_c and R_{a-b}) has often been observed in the tunneling characteristics of high-temperature superconductors [Divin et. al. 1995]. In the particular case of HTS/N junctions this anisotropy combined with the structural defects in the surface area of the superconductor can produce unexpected results in the work of circuits, containing such contacts. Recently, a number of attempts has been made to minimize the interface resistance in these contacts [Yuzi-Xu 1997, Daly 1997, Terai 1995, Sanders 1995, etc.]. The obtained results are often non-consistent with each other.

This paper is devoted to the experimental investigation of the current transport in Au/YBCO junctions with a current flow along (001)-oriented and tilted high-temperature superconducting YBCO thin films.

2.   YBCO FILMS AND SAMPLE FABRICATION.

The junctions were prepared using the sequence of operations shown in Fig.1. YBCO epitaxial films were grown by laser ablation (Fig. 1a) at temperatures 780-800°C and at 0.8mbar oxygen pressure. We used (001) LaAlO3 substrates to grow *c*-axis oriented YBCO and (120) NdGaO3 substrates to obtain YBCO films with *c*-axis tilted on 18.4° from the normal to the substrate plane (tilted YBCO films). The YBCO film was in situ covered by a thin layer of normal metal, Au, with a thickness of 20nm grown at 100°C by laser ablation

Junction areas and electrodes were defined by using photolithography and Ar ion milling (Fig. 1a). In order to provide an one-directional electrical contact plane of the YBCO film, the region of the junction was insulated on the side by a CeO2 layer with a central window having an area of 10x10μm² (Fig. 1b). The geometry of electrodes allowed for 4-point measurement of the interface resistance. (Fig. 1c). YBCO films were characterized by magnetic succeptibility

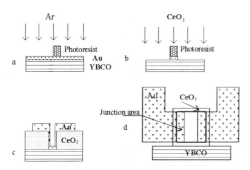

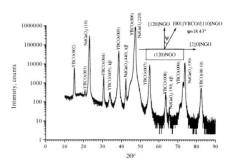

Fig. 1. A fabrication procedure of Au/YBCO junctions:
a) definition of the junction area by photolitography and ion milling;
b) deposition of insulating CeO₂ layer;
c) deposition of the contact pads;
d) top view of the junction;

Fig. 2. X-ray diffraction pattern of Au/YBCO double layer structure on NdGaO₃ (120) substrate. Tilting by the angle $\psi = 18.4°$ to the normal of the substrates provides (00n) reflections of YBCO film.

measurements of the critical temperature (T_c) and the width (ΔT_c) of the superconducting transition. Values of T_c's $> 89K$ and ΔT_c's $< 0.5K$ for c-axis YBCO films and T_c's $> 85K$ and ΔT_c's $< 2K$ for tilted YBCO films were obtained. X-ray θ-2θ scan was performed at the angle $\psi = 18.4°$ and showed that c-axis of YBCO film on (120) NdGaO₃ substrates follows the tilting angle of (120) from (110) NdGaO₃ substrate ($\psi = 18.4°$), see Fig.2

3.   EXPERIMENTAL RESULTS AND DISCUSSIONS.

The dependencies of resistance R versus temperature T, R(T), for Au/YBCO junctions and test planar YBCO microbridges (4μm wide) were measured at temperatures 4.2-300K at 1-5μA bias currents as well as their I-V curves. Parameters of c-axis and tilted Au/YBCO junctions presented in the table 1.

Table 1. Parameters of Au/YBCO junctions.

Sample	ψ,°	T_c, K	R_N(T=T_c, V=0,), Ω	R_d(T=4.2K, V=0), Ω	R_d(T=4.2K, V=0)/R_N	R_NS, 10^{-6}, $\Omega \cdot cm^2$
P32J2	0	89.3	33.2	103.0	3.1	33.2
P32J3	0	89.5	19.5	52.0	2.7	19.5
P32J4	0	89.9	22.9	55.3	2.4	22.9
P34J3	0	89.2	56.1	102.0	2.1	56.1
H2J2	18.4	18.7	1.6	0.7	0.4	1.6
H2J3	18.4	48.2	1.6	1.0	0.6	1.6
H2J4	18.4	40.1	1.8	1.3	0.7	1.8
H5J2	18.4	42.3	0.4	0.2	0.5	0.4
H5J3	18.4	60.3	0.3	0.2	0.7	0.3
H5J4	18.4	61.1	0.5	0.3	0.6	0.5

R(T) dependencies for both types of investigated Au/YBCO junctions are shown in Fig. 3: (a) a solid line for ψ=0, the direction of the current flow coincides with the (001)-direction in YBCO film (a c-axis junction), and (b) a dashed line for ψ=18.4°, the direction of the current flow deviates on ψ=18.4° in relation to the (001)-axis of YBCO film (a tilted junction).

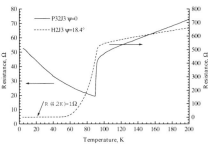
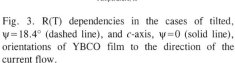

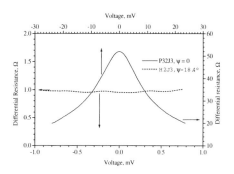

Fig. 3. R(T) dependencies in the cases of tilted, $\psi = 18.4°$ (dashed line), and c-axis, $\psi = 0$ (solid line), orientations of YBCO film to the direction of the current flow.

Fig. 4. A comparison of the Rd(V) dependencies in the case of tilted, $\psi = 18.4°$ (dashed line) and c-axis, $\psi = 0$ (solid line), Au/YBCO junctions.

At $T > T_c$, the $R(T)$ dependencies for both cases are of a metallic type, i.e., the resistance decreases with decreasing temperature, the latter is typical for YBCO film carrying current in the basal plane of YBCO. As a rule, the values of T_c for bridges and investigated tilted junctions were lower than ones of YBCO films measured immediately after deposition of the double-layer Au/YBCO structure. The degradation of superconducting properties of the films was apparently associated with a decrease in the amount of oxygen during ion etching. This effect is strong in the case of tilted YBCO films because of the high rate of the oxygen depletion in a-b basal planes of YBCO film.

It can be seen from the Fig. 3 that the resistance R of tilted junctions decreases monotonously with T at $T < Tc$. For c-axis junctions we have $R \sim 1/T^2$. This different behavior indicates a different type of conductivity of the YBCO/Au interface in tilted and c-axis junctions. In the tilted junctions thecurrent transport is partially in a-b-planes which results in an ohmic type contact with metallic conductivity. In the case of c-axis junctions an oxygen depleted semiconducting surface layer is realized. We have not observed a zero-bias conductance peak probably because of the lower junction resistances compared with [Sanders 1995]. Similar R(T) behaviour and the values of characteristic interface resistance ($R_N S$) were obtained in [Ekin 1988, Daly 1997] in the case of c-axis Au/YBCO interface.

Fig. 4 presents the typical dependencies of differential resistance of Au/YBCO junctions on applied voltage. R_N of the tilted junction is a constant value of 0.3-1.6Ω in the current range below approximately 1mA, which provides a value of $R_N S \sim 10^{-7}$-10^{-6} $\Omega \cdot cm^2$. Taking into account the square of a-b-plane area, involving in the current transport, we have $R_N S(a$-$b) \sim 10^{-7}$ $\Omega \cdot cm^2$. There are four orders of magnitude difference of this value with the theoretically estimated one [Kupriyanov 1995, Kupriyanov 1991, Deutscher 1991]. This difference is possibly caused by the oxygen depletion along a-b planes during the ion milling of Au/YBCO tilted junction. According to our estimations the in-plane superconducting critical current density for proceeded tilted YBCO films is $J_c \sim 10^3 A/cm^2$ at $T = 4.2K$, i.e. a least four orders of magnitude lower than in the case of c-axis YBCO films.

For c-axis structures $R_d(V)$ dependence has a maximum at zero bias current (and voltage) with the value of $\sim 10^{-5}$ $\Omega \cdot cm^2$, see a solid line on Fig. 4. As it was already mentioned, an oxygen depletion effect is a possible reason for the tunnel R(T) dependence and for the high value of $R_N S$. Fig. 5 presents $R_d(V)$ dependencies for c-axis junction, obtained after the annealing of this junction in oxygen atmosphere at 600°C. The $R_d(V)$ dependencies were measured at $T = 4.2K$ immediately after each annealing step. The value of $R_N S$ decreases rapidly with the increase of the annealing time and after 5 hours of annealing is equal to $\sim 10^{-6}$ $\Omega \cdot cm^2$. The subsequent annealing for 15 hours

236

did not affect the R$_N$S value. This means that the oxygen contents in the surface layer of YBCO is close to the maximum value.

4.   CONCLUSIONS.

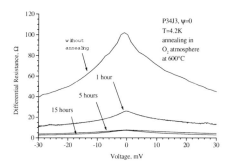

Fig. 5. R$_d$(V) dependencies of c-axis Au/YBCO junction from the annealing time in oxygen atmosphere at 600°C.

We have investigated the current transport properties of c-axis and tilted Au/YBCO junctions. A strong anisotropy of the R(T) dependencies and I-V-curves was observed, caused by the different type of conductivity in a-b-plane and in c-axis direction of YBCO. An oxygen depletion effect and its influence on the contact resistance value weas investigated. The depression of Tc and Jc in tilted YBCO film during the fabrication of Au/YBCO junction makes it questionable the applications of these films in superconducting devices.

Authors are gratefull to Z.G. Ivanov, M.Yu. Kupriyanov and P. Mozhaev for fruitful discussions of the experimental results.

This work has been supported by the ESPRIT project HTS-RSFQ-23429, Swedish Materials Consortium on Superconductivity, Russian Foundation of Fundamental Research and INTAS program.

5.   REFERENCES.

Daly G M, Pond J M, Osofsky G M, Horwitz J S, Soulen R J, Chrisey D B, Auyeung R C Y 1997 IEEE Trans. Applied Superconductivity 7 pp 2153-6.
Deutscher G et.al. 1991 Physica C 185-189 p 216.
Divin Y Y, Poppe U, Shadrin P M, Seo J W, Kabius B and Urban K 1995 Proc. 2nd Eur. Conf. on Applied Superconductivity 2 pp 1359-62.
Ekin J W, Larson T M, Bergen N F, Nelson A J, Swartzlander A B, Kazmerski L L, Panson A J and Blankenship B A 1988 Appl. Phys. Lett. 52 pp 1819-21.
Kupriyanov M Y and Likharev K K 1991 IEEE Trans. Magn. 27 pp 2400-3.
Kupriyanov M Y 1995 Dr. of Sciences degree thesis, NPI MSU edition (Moscow), in Russian.
Sanders S C, Russek S E, Clickner C C, Ekin J W 1995 IEEE Trans. Applied Superconductivity 5 pp 2404-7.
Terai H, Fujimaki A, Takai Y, Hayakawa H 1995 IEEE Trans. Applied Superconductivity.5 pp 2408-11.
Yuzi-Xu, Ekin J W, Russek S E, Fiske R, Clickner C C, Takeuchi I, Trajanovic Z, Venkatesan T, Rogers C T 1997 IEEE Trans. Applied Superconductivity 7 pp 2836-9.

Inst. Phys. Conf. Ser. No 167
Paper presented at *Applied Superconductivity, Spain, 14-17 September 1999*
© 2000 IOP Publishing Ltd

Intrinsic Josephson effects in Tl-2212 thin films

O S Chana[1], D M C Hyland[2], R J Kinsey[3], G Burnell[3], W E Booij[3], M G Blamire[3], C R M Grovenor[4], D Dew-Hughes[2] and P A Warburton[1]

[1] Dept. of Electronic Engineering, King's College London, Strand, London, WC2R 2LS, UK
[2] Dept. of Engineering Science, University of Oxford, Parks Road, Oxford, OX1 3PJ, UK
[3] Dept. of Materials Science, University of Cambridge, Cambridge, CB2 3QZ, UK
[4] Dept. of Materials, University of Oxford, Park Road, Oxford, OX1 3PH, UK

ABSTRACT: We have measured Josephson RSJ-like current voltage characteristics in 2μ-wide bridges patterned in the [cos20° 0 sin20°] direction in 20° mis-aligned thin films of $Tl_2Ba_2CaCu_2O_8$ grown on vicinal $LaAlO_3$ substrates. The temperature dependence of the critical current of the series array of underdamped intrinsic junctions is well described by SIS tunnelling. Consecutive junctions in the array appear to behave independently. Features in the current-voltage characteristics are attributed to a resonant coupling mechanism between infrared-active optical c-axis phonons and oscillating Josephson currents.

1. INTRODUCTION

The development of a technology to fabricate high T_c Josephson junctions has already been important in explaining the underlying physics behind the high T_c materials and also in electronic device applications. At present existing types of junction (Gross *et al.*, 1997) include ramp edge, bicrystal grain boundary and electron beam damaged junctions. In all of these junctions, the direction of current transport is parallel to the copper-oxide planes. In contrast to this type of device, junctions that have current flow normal to the planes can yield intrinsic Josephson effects (Müller *et al.*, 1995). Such effects have been observed in a range of high temperature superconducting single crystals and c-axis oriented thin films.

We present here a novel and simple technique for the manufacture of intrinsic Josephson junctions. Superconducting $Tl_2Ba_2CaCu_2O_8$ (Tl-2212) films are epitaxially grown on vicinally cut $LaAlO_3$ substrates so that the copper-oxide planes make an angle of 20° to the substrate surface. Passing a current through a microbridge patterned in the [cos20° 0 sin20°] direction gives a component of current normal to the copper-oxide planes (Yan *et al.* 1997). Each set of copper-oxide planes acts as a single Josephson junction; hence we have a series array of Josephson junctions.

The temperature dependence of the critical current is well described by the Ambegaokar-Baratoff theory (Ambegaokar and Baratoff, 1963) implying SIS coupling between copper-oxide planes. We have applied a magnetic field of up to 30mT to suppress the critical current (Chana *et al.*, 1999). Insensitivity to this field strongly suggests that the junctions are not grain boundary SNS type junctions. We have also patterned control devices in the orthogonal [0 1 0] direction whose current-voltage (IV) characteristics always exhibit flux-flow behaviour. This is further indication that the observed Josephson effects are not due to grain boundaries.

We have traced out hysteretic IV curves as expected for a series array of underdamped Josephson junctions. The temperature dependence of the critical and return current of at least three junctions has been shown to be identical. Temperature independent features on the IV curves have been linked to a phonon resonance of the Tl-2212 crystal (Schlenga *et al.*, 1998).

2.   EXPERIMENTAL

In our three-stage sputter-pattern-anneal process an amorphous Ba-Ca-Cu-O precursor is r.f. sputtered onto a LaAlO₃ vicinal substrate. The substrate c-axis has a misalignment of 20° to the normal. Wet chemical etching using a standard photolithographic process was used to pattern the microbridge into the precursor. The 5μm long and 2μm wide microbridge was aligned to the [cos20° 0 sin20°] direction as shown in Fig. 1. A control microbridge that was identical in size but orthogonal in orientation to this device was also included in the pattern (see SEM in Fig.1). The etchant used was dilute adipic acid. *Ex-situ* annealing in a sealed crucible with a Tl source at 850°C completes the processing of our patterned phase pure Tl-2212 film. The film thickness is 0.7μm. Four circle XRD analysis, TEM and EDAX have been used to check crystallographic alignment, orientation and stoichiometry (Hyland *et al.*, 1999).

Gold contact pads were evaporated to make four-terminal connections to the device. Cooling down to 4K was achieved by immersing the sample into a liquid helium dewar. Electrical shielding and filtering was used on all measurement electronics. We have used current biasing to record the response of our devices.

From the bridge geometry and the spacing of the copper oxide planes in Tl-2212, we can estimate that there is a series array of 750 intrinsic junctions along a 5μm long microbridge. Junctions can be patterned repeatably and reliably using the method discussed.

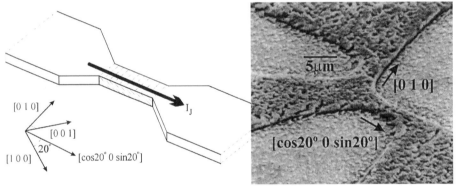

Fig. 1: Schematic of our device geometry (not to scale), showing the direction of the copper-oxide planes in the bridge which is patterned in the [cos20° 0 sin20°] direction. Also shown is an SEM of a typical device – superconductor is dark, substrate surface is light.

3.   RESULTS

3.1   Resistivity

The critical temperature of the films varies from 99 to 108K. The film has resistivities at 273K of 12mΩm and 0.13mΩm in the c- and b- direction respectively. These data show consistency with results from Tl-2212 single crystals (Harshman *et al.*, 1992) and show the high quality of our films even when grown on vicinal substrates. The anisotropy ρ_c/ρ_b (Chana *et al.*, 1999) from T_c to room temperature varies at most by a factor of 1.5 from single crystal data (Duan *et al.*, 1991).

3.2   Current-voltage characteristics

The current-voltage characteristics at 4.2K of a typical intrinsic Josephson microbridge are shown in Fig. 2a. When the bias current exceeds the critical current of the junction with the lowest critical current, I_{c1}, that junction switches into its resistive state. If the current is reduced at this stage, we trace back to the zero voltage state along the hysteretic return path, approaching I_{r1} at V=0 for the first junction. Increasing the current further when the first junction is switched on will cause the junction with the next highest critical current, I_{c2}, to switch into its resistive state. We can continue to trace out up to 6 junctions in this way. No more distinct voltage steps are seen if we increase the bias current beyond the value of I_{c6}, possibly due to a very large number of junctions simultaneously switching. The magnitude of the voltage jump decreases with increasing current due to self-heating. The control [0 1 0] direction microbridges only showed a flux-flow response in the IV curve. The critical current density of the control microbridge was at least an order of magnitude greater then the microbridge in the [cos20° 0 sin20°] direction.

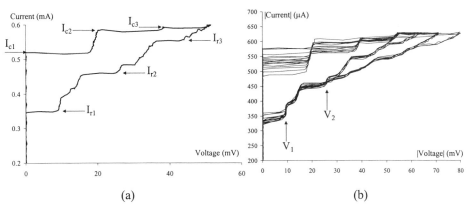

(a)   (b)

Fig. 2: (a) A single IV sweep, and (b) 16 IV sweeps of a microbridge in the [cos20° 0 sin20°] direction at 4.2K.

The normalised temperature dependence of the critical and return currents for the first three junctions (as defined in Fig. 2(a)) are shown in Fig. 3. These data fit very well to the theory of Ambegaokar and Baratoff (1963) for SIS tunnelling. The fit of temperature dependence for consecutive junctions suggests that the subsequent devices are in fact replicas of the first device differing only in their respective I_c. The critical current temperature dependence is fundamentally different from that of SNS type Josephson junctions. SNS-like temperature dependence has been observed in Tl-2212 bicrystal grain boundary junctions by Hu et al. (1996), suggesting that the Josephson effects that we observe in our bridges are not due to the existence of grain boundaries. In addition the control microbridge in the [0 1 0] direction which is adjacent to the [cos20° 0 sin20°] direction bridge (see SEM in Fig. 1) never exhibits RSJ-like characteristics. This is further evidence against the existence of grain boundary junctions.

There is a spread of critical currents in the intrinsic Josephson junctions as shown in Fig. 2(b). This spread has been observed by Mros et al. (1997) in Bi-2212 crystals. They conclude that in a stack of intrinsic Josephson junctions different quasiequilibrium fluxon modes exist and are

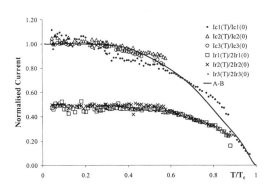

Fig. 3: Normalised temperature dependence of the critical and return currents for the first three junctions. The solid line is a fit to the critical current data using the Ambegaokar-Baratoff theory.

responsible for the observed spread. There is also a spread in the return path to V=0. However there is a step at V_1=9.7mV that appears with consistency on each sweep and also in other microbridges. The step at V_2 also corresponds to a voltage of 9.7mV across the second junction. This voltage has no temperature dependence.

It is possible that this step is due to a Fiske resonance. The expected position of any Fiske step is dependent upon whether the 750 intrinsic junctions oscillate independently (as is suggested by our I-V characteristics) or are phase-locked. In the former case we calculate the Swihart velocity to be $2.8 \times 10^5 ms^{-1}$, leading to a Fiske step at 150μV. In the latter case (Krasnov *et al.*, 1999) the 750-fold splitting of the Swihart velocity leads to Fiske steps which are so closely spaced that they cannot be resolved. It seems unlikely that the features at V_1 and V_2 are due to Fiske resonances.

Schlenga *et al.* (1997) have measured similar voltage steps using a different intrinsic Josephson device structure and attribute this to the Josephson oscillations being at the same frequency as the c-axis longitudinal phonons. Any phonon resonance would not be temperature dependent. The corresponding Josephson frequency for V_1 is 4.7THz. This is in agreement with measurements of longitudinal c-axis phonon resonances in Tl-2212 by grazing incidence infrared spectroscopy (Tsvetkov *et al.*, 1999).

4.   CONCLUSIONS

We have observed intrinsic Josephson effects in microbridges patterned in mis-aligned Tl-2212 thin films grown on vicinal LaAlO₃ substrates. We can trace out RSJ-like junction behaviour in current-voltage curves for up to six of the junctions within the series array. The temperature dependence of the critical currents for the junctions is shown to take the SIS form. The current-voltage characteristics of three consecutive junctions imply that these junctions behave independently. The spread in I_c for a junction is similar to that observed with stacked intrinsic junctions in Bi-2212 and is attributed to different quasiequilibrium fluxon modes. The observed temperature independent voltage steps are due to resonance with longitudinal c-axis phonons.

Acknowledgements

This work is supported by EPSRC grants GR/L50808, GR/M43401 and GR/K93044.

References

Ambegaokar V and Baratoff A 1963 Phys. Rev. **11**, p 486
Chana O S, Hyland D M C, Kinsey R J, Booij W E, Blamire M G, Grovenor C R M,
    Dew-Hughes D, Warburton P A 1999 Physica C in press
Duan H M, Kiehl W, Dong C, Cordes A W, Saeed M J, Viar D L and Hermann A M 1991
    Phys. Rev. B **43**, p 12925
Harshman D R and Mills A P 1992 Phys. Rev. B **45**, p 10684
Hu Q H, Johansson L G, Langer V, Chen Y F, Claeson T, Ivanov Z G, Kislinski Y and
    Stepantsov E A 1996 J. Low Temp. Phys. **105**, p 1261
Hyland D, Chana O S, Grovenor C R M, Dew-Hughes D and Warburton P A 1999 unpublished
Gross R, Alff L, Froehlich O M, Koelle D and Marx A 1997 IEEE Trans. Appl. Supercon. **7**, p 2929
Krasnov V M, Mros N, Yurgens and Winkler D 1999 IEEE Trans. Appl. Supercon. **9**, p4499
Mros N, Krasnov V M, Yurgens A, Winkler D and Claeson T 1997 Phys. Rev. B **57**, p 8135
Müller P 1995 Adv. Sol. St. Phys. **34**, p 1
Schlenga K, Kleiner R, Hechtfischer G, Moessle M, Schmitt S, Müller P, Helm C, Preis C,
    Forsthofer F, Keller J, Johnson H L, Veith M and Steinbeiß 1997 Phys. Rev. B **57**, p 14518
Tsvetkov A A, Dulic D, van der Marel D, Damascelli A, Kaljushnaia G A, Gorina J I,
    Senturina N N, Kolesnikov N N, Ren Z F, Wang J H, Menovsky A A and Palstra T T M 1999
    submitted to Phys. Rev. B.
Yan S I, Fang L, Si M S and Wang J 1997 J. Appl. Phys. **82**, p 480

Inst. Phys. Conf. Ser. No 167
Paper presented at Applied Superconductivity, Spain, 14-17 September 1999
© 2000 IOP Publishing Ltd

Superconducting properties of La-substituted Bi-2201 crystals

Ya G Ponomarev[1], Kim Ki Uk[1], M A Lorenz[2], G Müller[2], H Piel[2], H Schmidt[2], A Krapf[3], T E Os'kina[4], Yu D Tretyakov[4] and V F Kozlovskii[4]

[1] M.V. Lomonosov Moscow State University, Faculty of Physics, 119899 Moscow, Russia
[2] Bergische Universität Wuppertal, Fachbereich Physik, Gaußstr. 20, D-42097 Wuppertal, Germany
[3] Humboldt-Universität zu Berlin, Institut für Physik, Invalidenstr. 110, D-10115 Berlin, Germany
[4] M.V. Lomonosov Moscow State University, Faculty of Chemistry, 119899 Moscow, Russia

ABSTRACT: The superconducting gap Δ_s has been measured in $Bi_2Sr_{2-x}La_xCuO_{6+\delta}$ single crystals in a wide range of temperatures 4.2 K $\leq T \leq T_c$ by point-contact and tunnelling spectroscopy for current in c-direction. The value of Δ_s(4.2 K) was found to scale with the critical temperature T_c in the whole range of doping levels with the ratio $2\Delta/kT_c = 12.5 \pm 2$. The closing of the gap Δ_s at $T = T_c$ has been registered in the underdoped, optimally doped as well as in the overdoped samples.

1. INTRODUCTION

Recently it has been proposed by Deutscher (1999) that for the underdoped copper oxide superconductors there exist two different gap energies Δ_p and Δ_s. The larger gap (pseudogap) Δ_p, measured by angle-resolved photoemission spectroscopy (ARPES) or tunnelling spectroscopy, is supposed to be a half of the energy required to split an incoherent Cooper pair. The smaller (superconducting) gap Δ_s, determined by electron Raman scattering or Andreev spectroscopy, is associated with a superconducting state. According to Deutscher (1999) Δ_s scales with the critical temperature T_c on doping, while Δ_p in underdoped samples continues to grow as $T_c \rightarrow 0$. It remains unclear why tunnelling spectroscopy should register in the underdoped HTSC samples only one excitation energy (Δ_p). Probably one possible explanation comes from the STM tunnelling measurements (NIS contacts) on the underdoped $Bi_2Sr_2Ca_1Cu_2O_{8+\delta}$ (Bi-2212) single crystals (Renner et al 1998). The authors claim that a superconducting gap Δ_s does not depend on temperature and transforms directly into a pseudogap Δ_p of exactly the same magnitude at $T > T_c$. Unfortunately this is in conflict with the results obtained on SIS Bi-2212 contacts (Miyakawa et al 1998 and 1999) which give a convincing evidence that a superconducting gap Δ_s in the underdoped Bi-2212 crystals closes at $T = T_c$. The absence of scaling of Δ_s with T_c on doping in the underdoped Bi-2212 crystals has been reported in several STM studies (Miyakawa et al 1998 and 1999 and Nakano et al 1998). In contrast Matsuda et al (1999) observed that Δ_s passes through a maximum at the optimal doping level in accordance with Deutscher (1999). Finally there exists a wide scatter of the published values of a superconducting gap Δ_s(T = 4.2 K) for the optimally doped Bi-2212 crystals measured by the STM technique in the same geometry ($\mathbf{j} \parallel \mathbf{c}$).

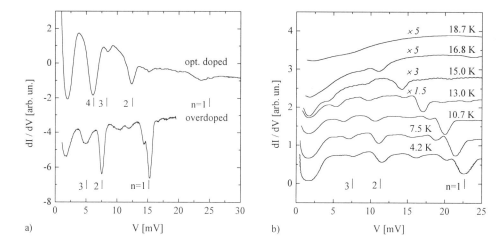

Fig.1 a) Subharmonic gap structure (SGS) in the dI/dV-characteristics of SNS contacts in optimally doped and overdoped Bi-2201(La) single crystals at T = 4.2 K with current in c-direction. The bias voltages $V_n = 2\Delta_s/en$ corresponding to the dips in the SGS are marked by bars. b) Temperature dependence of the SGS in the dI/dV-characteristics of a SNS contact in an underdoped Bi-2201(La) single crystal with T_c = 19.5 K. Some of the characteristics are magnified by indicated factors.

It was interesting for us to verify whether the model proposed by Deutscher (1999) is applicable in the case of single-plane Bi$_2$Sr$_{2-x}$La$_x$CuO$_{6+\delta}$ crystals (Bi-2201(La)) with the value of the coherence length exceeding significantly that for other members of BSCCO-family (Yoshizaki et al 1994). In this system the entire range from overdoped to underdoped samples can be covered by substitution of Sr by La without damaging significantly the crystalline structure (Khasanova et al 1995 and Yang et al 1998). The existence of a pseudogap Δ_p has already been reported for Bi-2201(La) (Harris et al 1997 and Vedeneev 1998) but no detailed studies of the doping dependence of Δ_s have yet been done.

2.   EXPERIMENTAL RESULTS AND DISCUSSION

The Bi$_2$Sr$_{2-x}$La$_x$CuO$_{6+\delta}$ crystals with the actual La concentration $0.1 \leq x \leq 0.5$ and the maximum T_c of about 25 K were grown from copper-oxide-rich melt. The X-ray diffraction pattern of the Bi-2201 phase showed pseudo-tetragonal symmetry with $\mathbf{a}$ = 0.538 - 0.539 nm for all tested powders and crystals. Furthermore the c-axis lattice parameter demonstrated a strong dependence on La doping in accordance with Khasanova et al (1995) and Yang et al (1998).

The superconducting gap Δ_s has been measured in the Bi-2201(La) single crystals in a wide range of temperatures 4.2 K $\leq T \leq T_c$ by point-contact and tunnelling spectroscopy for current in c-direction using a conventional break-junction technique (Aminov et al 1996). After formation of a crack in the crystals at helium temperature it was easy to drive the break junctions into a point-contact regime (contact of a SNS type) with the help of a micrometer screw. In the dI/dV-characteristics of SNS Bi-2201(La) contacts a typical series of sharp dips at bias voltages $V_n = 2\Delta_s/en$ has been registered (Fig. 1a). The dips compose a so called subharmonic gap structure (SGS) caused by multiple Andreev reflections (Devereaux et al 1993). The highly symmetric form of the dips is consistent with a dominant s-wave pattern of the order parameter for current in c-direction (Devereaux et al 1993 and Klemm et al 1999).

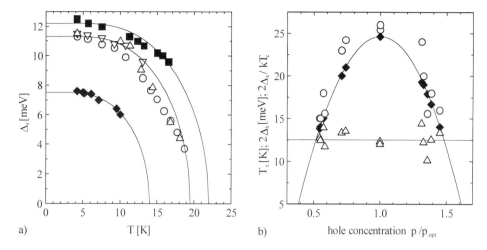

Fig. 2 a) Superconducting gap Δ_s vs temperature T for three underdoped Bi-2201(La) single crystals with different T_c (solid lines - BCS model).   b) Variation of $2\Delta_s(4.2$ K) (open circles), T_c (solid diamonds) and $2\Delta/kT_c$ (open triangles) with the normalized hole concentration p/p_{opt} for the investigated Bi-2201(La) single crystals.

We have found no noticeable difference in the form of SGS for contacts in the underdoped, optimally doped or overdoped Bi-2201(La) crystals which indicates that the symmetry of the order parameter is not changing significantly with doping. At the same time the value of the superconducting gap $\Delta_s(4.2$ K) calculated from the SGS showed scaling with T_c in the whole range of doping as proposed by Deutscher (1999), with the maximum value $\Delta_{s\,max} = (12.7 \pm 0.5)$ meV corresponding to the optimal doping level ($T_{c\,max} = (24.6 \pm 0.5)$ K, $2\Delta_{s\,max}/kT_{c\,max} = 11.8 \pm 0.7$). We have obtained the same value of $\Delta_{s\,max}(4.2$ K) from the gap feature in the current-voltage characteristics (CVCs) of break junctions in the optimally doped Bi-2201(La) crystals in a tunnelling regime (contacts of SIS type). The value of $\Delta_{s\,max}$ determined in the present investigation is in reasonable agreement with $\Delta_{s\,max} = (10 \pm 2)$ meV measured by ARPES at $T = 9$ K in optimally doped Bi-2201(La) crystals (Harris et al 1997).

The most pronounced Andreev ($n = 1$) dip in the SGS ($V_1 = 2\Delta/e$) could often be observed in the dI/dV-characteristics even close to $T = T_c$ (Fig. 1b) which made it possible to study the temperature dependencies of the superconducting gap Δ_s for underdoped (Fig. 2a) as well as for overdoped Bi-2201(La) crystals. The main result of these studies is that the superconducting gap Δ_s definitely closes at $T = T_c$ in the whole range of doping which is again consistent with the model of Deutscher (1999).

To demonstrate the scaling of the superconducting gap $\Delta_s(4.2$ K) with the critical temperature T_c for Bi-2201(La) crystals in the whole range of doping we have plotted $2\Delta_s$ and $2\Delta_s/kT_c$ vs normalized hole concentration (p/p_{opt}) using the empirical expression (Tallon et al 1995) $T_c/T_{c\,max} = 1 - 82.6 \cdot (p - 0.16)^2$ (Fig. 2b). It can be easily seen from Fig. 2b that within experimental errors the ratio $2\Delta_s/kT_c = 12.5 \pm 2$ does not change on doping. Furthermore this ratio exceeds noticeably the reduced-gap ratios for the optimally doped 2212- and 2223-phases ($2\Delta_s/kT_c \approx 7$) (Hudáková et al 1996). The value of $\Delta_{s\,max}$ (4.2 K) for Bi-2201(La) crystals is only two times smaller than that for the 2212-phase, so an unexpectedly low critical temperature $T_{c\,max} \approx 25$ K for the single-plane phase needs some explanation.

We have observed a new specific subharmonic gap structure due to the intrinsic multiple Andreev reflections effect (IMARE) for the current in **c**-direction. We conclude that in contrast to the 2212- and 2223-phases the $Bi_2Sr_{2-x}La_xCuO_{6+\delta}$ crystals could be treated as SNSN... superlattices (Frick et al 1992). The spacers in Bi-2201(La) crystals acquire metallic properties probably due to the presence of La. Multiple Andreev reflections of quasiparticles take place between the adjacent superconducting CuO_2-planes. The SGS for the stacks of n Andreev contacts measured at $T < T_c$ were of extremely high quality and exactly fitted the SGS for a single SNS contact after normalisation of the voltage scale ($V \to V/n$). It was shown by Buzdin et al (1992) that in SNSN... superlattices the proximity effect can significantly increase the ratio $2\Delta/kT_c$ mainly by depressing the critical temperature T_c. This could possibly explain the relatively low T_c in the investigated Bi-2201(La) crystals. The influence of the proximity effect on the $\Delta_s(T)$-dependencies in BSCCO samples containing intergrowths of 2212- and 2223-phases has been reported by Aminov et al (1991) and Ponomarev et al (1993) earlier.

In conclusion, we have found that the superconducting gap $\Delta_s(4.2$ K$)$ in $Bi_2Sr_{2-x}La_xCuO_{6+\delta}$ single crystals scales with the critical temperature T_c in the whole range of doping levels with the ratio $2\Delta/kT_c = 12.5 \pm 2$. The closing of the gap Δ_s at $T = T_c$ has been observed in the underdoped, optimally doped as well as in the overdoped samples. We have obtained an experimental evidence proving that the $Bi_2Sr_{2-x}La_xCuO_{6+\delta}$ crystals could be treated as superlattices of a SNSN... type, for which the ratio $2\Delta/kT_c$ is increased due to the proximity effect.

ACKNOWLEDGMENTS

This work was supported in part by the ISC on High Temperature Superconductivity (Russia) under the contract number 96118 (project DELTA) and by RFBR (Russia) under the contract number 96-02-18170a.

REFERENCES

Aminov B A et al 1991 JETP Lett. **54**, 52
Aminov B A et al 1996 Phys. Rev. B **54**, 6728
Buzdin A I, Damjanović V P and Simonov A Yu 1992 Physica C **194**, 109
Deutscher G 1999 Nature **397**, 410
Devereaux T P and Fulde P 1993 Phys. Rev. B **47**, 14638
Frick M and Schneider T 1992 Z. Phys. B - Cond. Matter **88**, 123
Harris J M et al 1997 Phys. Rev. Lett. **79**, 143
Hudáková N et al 1996 Physica B **218**, 217
Khasanova N R and Antipov E V 1995 Physica C **246**, 241
Klemm R A, Rieck C T and Scharnberg K to be published in Journ. Low Temp. Phys.
Matsuda A et al 1999 Phys. Rev. B **60**, 1377
Miyakawa N et al 1998 Phys. Rev. Lett. **80**,157
Miyakawa N et al 1999 Phys. Rev. Lett. **83**,1018
Nakano T et al 1998 J. Phys. Soc. Jpn. **67**, 2622
Ponomarev Ya G et al 1993 Journ. of Alloys & Compounds **195**, 551
Renner Ch et al 1998 Phys. Rev. Lett. **80**,149
Tallon J L et al 1995 Phys. Rev. B **51**, 12911
Vedeneev S I 1998 JETP Lett. **68**, 230
Yang W L et al 1998 Physica C **308**, 294
Yoshizaki R et al 1994 Physica C **224**, 121

Inst. Phys. Conf. Ser. No 167
Paper presented at Applied Superconductivity, Spain, 14-17 September 1999
© *2000 IOP Publishing Ltd*

Dependence of the gap parameter on the number of CuO2 layers in a unit cell of optimally doped BSCCO, TBCCO, HBCCO and HSCCO

Ya G Ponomarev[1], N Z Timergaleev[1], Kim Ki Uk[1], M A Lorenz[2], G Müller[2], H Piel[2], H Schmidt[2], C Janowitz[3], A Krapf[3], R Manzke[3]

[1] M.V. Lomonosov Moscow State University, Faculty of Physics, 119899 Moscow, Russia
[2] Bergische Universität Wuppertal, Fachbereich Physik, Gaußstr. 20, D-42097 Wuppertal, Germany
[3] Humboldt-Universität zu Berlin, Institut für Physik, Invalidenstr. 110, D-10115 Berlin, Germany

ABSTRACT: We have measured the superconducting gap Δ_s in optimally doped samples of $Bi_2Sr_2Ca_{n-1}Cu_nO_{2n+4}$, $Tl_2Ba_2Ca_{n-1}Cu_nO_{2n+4}$, $HgBa_2Ca_{n-1}Cu_nO_{2n+2}$ and $HgSr_2Ca_2Cu_3O_8$ by Andreev and tunnelling spectroscopy and by ARPES. We have found that the low-temperature value of Δ_s within experimental errors is linearly increasing with the number n of CuO2 layers in the unit cell of the investigated HTSC-families ($n \leq 3$). The variation of the critical temperature $T_{c\,max}$ with n does not obey this simple relation. For a given n the values of Δ_s at $T = 4.2$ K for the Tl- and Hg-families practically coincide but systematically exceed that in the Bi-family.

1.   INTRODUCTION

It is generally assumed that in the layered high-T_c superconductors (HTSC) $Bi_2Sr_2Ca_{n-1}Cu_nO_{2n+4+\delta}$ (BSCCO), $Tl_2Ba_2Ca_{n-1}Cu_nO_{2n+4+\delta}$ (TBCCO) and $HgBa_2Ca_{n-1}Cu_nO_{2n+2+\delta}$ (HBCCO) the superconductivity in a unit cell is located in blocks of n closely spaced CuO2 layers intercalated by Ca (Waldram 1996). These superconducting blocks are separated in **c**-direction by insulating or semiconducting blocks (spacers) with a structure universal for a given HTSC-family. The spacers play an important role in the formation of superconducting properties of the above mentioned compounds serving as charge reservoirs. For a phase with a given number of CuO2 layers n the maximal critical temperature $T_{c\,max}$ (and correspondingly the maximum superconducting gap $\Delta_{s\,max}$ (Deutscher 1999 and Ponomarev et al 1999b)) can be achieved by the variation of the excess oxygen concentration δ in the spacers. The excess oxygen attracts electrons from the CuO2 layers without introducing a significant scattering of charge carriers in the superconducting blocks ("distant" doping). In addition the excess oxygen forms traps in the spacers, thus strongly influencing the transport in **c**-direction due to resonant tunnelling effects (Abrikosov 1999 and Halbritter 1996).

The physical properties of spacers are significantly influenced by doping. Metallization of spacers with the increasing concentration of the excess oxygen changes the **c**-axis transport in HTSC qualitatively causing 2D → 3D transition. At the same time in HTSC crystals with low δ the electron transport is strongly anisotropic and supercurrent in the **c**-direction is essentially of a Josephson nature (Abrikosov 1999 and references therein). The intrinsic Josephson effect (Suzuki et al 1999 and references therein) markedly affects **c**-axis superconducting properties of these crystals and makes it extremely difficult to perform proper **c**-axis tunneling measurements in a single-contact mode.

		T_c [K]	$\Delta_s(4.2\ K)$ [meV]	$2\Delta_s / kT_c$	Reference
BSCCO	$n=1$	25 ± 2	12.7 ± 0.5	11.8 ± 1.1	this work
	$n=2$	86 ± 4	25 ± 1	6.7 ± 0.5	this work
	$n=3$	110 ± 5	36 ± 1.6	7.6 ± 0.5	this work
TBCCO	$n=1$	73	14	4.5 ± 0.4	Tsai et al (1989)
	$n=2$	104 ± 5	34 ± 1	7.6 ± 0.5	this work
	$n=3$	118 ± 5	47.7 ± 1.4	9.4 ± 0.5	this work
HBCCO	$n=1$	97	15	3.6	Chen et al (1994)
	$n=2$	120 ± 5	33 ± 1	6.4 ± 0.4	this work
	$n=3$	124 ± 5	49 ± 1.5	9.2 ± 0.5	this work
HSCCO	$n=3$	107 ± 5	36 ± 1.5	7.8 ± 0.5	this work

Table 1. Superconducting properties of the investigated HTSC compounds.

Samples with the maximum critical temperature $T_{c\ max}$ corresponding to the excess oxygen concentration $\delta = \delta_{opt}$ are defined as optimally doped. For the optimally doped HTSC samples the dependence of the critical temperature $T_{c\ max}$ on the number of CuO_2 layers n is strongly nonlinear (Phillips 1994). Several theoretical models have been proposed to explain the nontrivial dependencies $T_{c\ max}(n)$ in HTSC materials (Phillips 1994, Byczuk et al 1996, Kresin et al 1996, Chen et al 1997 and Leggett 1999). Unfortunately this problem remains unsolved mainly because of the limited and often conflicting experimental data on the variation of superconducting properties of HTSC with the number of CuO_2 planes in the superconducting blocks (Hudáková et al 1996 and Wie 1998).

This paper presents low-temperature measurements of the superconducting gap Δ_s as a function of n ($n \leq 3$) in optimally doped samples of the BSCCO-, TBCCO- and HBCCO-families with identically constructed superconducting blocks. A correlation of the obtained $\Delta_s(n)$-dependencies with the structure of spacers in these families has been found.

2.   EXPERIMENTAL RESULTS AND DISCUSSION

We have measured the superconducting gap Δ_s in optimally doped samples of BSCCO ($n = 1, 2, 3$), TBCCO ($n = 2, 3$), HBCCO ($n = 2, 3$) and $HgSr_2Ca_2Cu_3O_{8+\delta}$ (HSCCO, $n = 3$) by Andreev and tunnelling spectroscopy at $T = 4.2$ K with the current in c-direction using a break junction technique (Aminov et al 1996). The optimally doped samples of BSCCO ($n = 2, 3$) have been studied also by angle-resolved photoemission spectroscopy (ARPES) at $T \approx 30$ K. The details of ARPES measurements have been published elsewhere (Müller et al 1997).

Point-contact and tunnelling spectroscopy studies of the optimally doped single crystals of $Bi_2Sr_{2-x}La_xCu_1O_{6+\delta}$ gave exactly the same value of the gap $\Delta_s(4.2\ K) = (12.7 \pm 0.5)$ meV (Ponomarev et al 1999b) (Table 1), which is in reasonable agreement with the results of the earlier ARPES measurements of $\Delta_{s\ max}(9\ K) = (10 \pm 2)$ meV by Harris et al (1997). Similar measurements have been performed on the optimally doped single crystals of BSCCO ($n = 2$) and polycrystalline samples of BSCCO ($n = 3$). In the dI/dV-characteristics of SNS contacts (point-contact regime) a clearly defined subharmonic gap structure (SGS) caused by multiple Andreev reflections has been observed for all investigated samples (Fig. 1a). In all cases, the dips at bias voltages $V_n = 2\Delta/en$ composing the SGS had a symmetric form which points to the absence of a strong anisotropy of the gap Δ_s in the **ab**-plane (Devereaux et al 1993). The value $\Delta_s(4.2\ K) = (25 \pm 1)$ meV obtained in this work for BSCCO ($n = 2$) from Andreev and tunnelling spectroscopy (Table 1) is well supported by recent studies of intrinsic Josephson effect ($\Delta_s = 25$ meV) (Ponomarev et al 1999a and Suzuki et al 1999) and is in agreement with the ARPES measurements by Müller et al (1997) ($\Delta_{s\ max}(30\ K) = (28 \pm 2)$ meV).

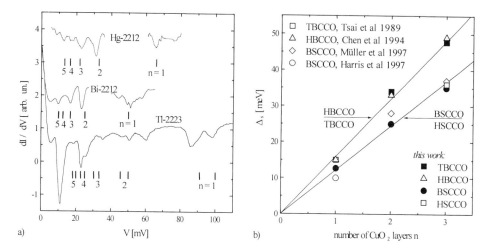

Fig.1 a) Subharmonic gap structure (SGS) in the dI/dV-characteristics of SNS contacts in an optimally doped polycrystalline sample of HBCCO ($n = 2$) and in single crystals of BSCCO ($n = 2$) and TBCCO ($n = 3$) at $T = 4.2$ K. The bias voltages $V_n = 2\Delta_s/en$ corresponding to the dips in SGS are marked by bars. b) The superconducting gap $\Delta_s(4.2$ K) vs the number of CuO$_2$ layers n for the optimally doped samples of BSCCO-, TBCCO- and HBCCO-families. The straight solid lines are guides to the eye.

We have also estimated the values of Δ_s for the TBCCO ($n = 2$, 3) single crystals from Andreev spectroscopy measurements at $T = 4.2$ K. The same was done for the optimally doped polycrystalline samples of HBCCO ($n = 2$, 3). All the resulting values are summarised in Table 1. It should be noted that the experimental data obtained in this work from Andreev and tunnelling spectroscopy corresponded to the current in the c-direction in case of single crystals.

Recently an impressive experimental evidence supported by sound theoretical considerations has been presented which showed that in layered HTSC crystals a s-wave component of the order parameter is dominating for the transport in the c-direction (Klemm et al 1999). Probably the d-wave component is filtered out due to the specific character of this transport. This view is in accordance with recent studies of the intrinsic Josephson effect in BSCCO (Ponomarev et al 1999a). Furthermore it has been pointed out by Devereaux and Fulde (1993) that in the case of an anisotropic s-wave order parameter the SGS in the dI/dV-characteristic of a SNS junction should split into two separate structures corresponding to $V_{n1} = 2\Delta_{min}/en$ and $V_{n2} = 2\Delta_{max}/en$. Actually we have often observed such splitting for the best SNS contacts in all three investigated HTSC-families (see for example the SGS for the TBCCO ($n = 3$) contact in Fig 1a). If the version with the anisotropic s-wave order parameter is correct then the ratio $\Delta_{max}/\Delta_{min}$ in the samples of the investigated HTSC-families does not exceed $(\Delta_{max}/\Delta_{min}) = 1.1$ at $T = 4.2$ K.

Combining our results with the data obtained earlier for HBCCO ($n = 1$) (Chen et al 1994) and TBCCO ($n = 1$) (Tsai et al 1989), we have plotted the gap Δ_s vs n for all three HTSC-families (Fig. 1b). Obviously the gap Δ_s in the investigated HTSC-families is proportional to the number of CuO$_2$ layers n in the range $n \le 3$. It should be noted that the critical temperature $T_{c\ max}$ vs n does not obey this simple relation (Phillips 1994). The most striking result coming from Fig. 1b is the coincidence of the $\Delta_s(n)$-dependencies for TBCCO- and HBCCO-families. This result means that the value of Δ_s in optimally doped samples is totally controlled by the physics in the module consisting of a superconducting multi-plane block sandwiched between two Ba-O planes. Substitution of Ba-O planes by Sr-O planes in HBCCO or TBCCO should cause a transition of $\Delta_s(n)$ to the BSCCO-branch (Fig. 1b). To verify this assumption we have measured Δ_s and T_c for the HSCCO ($n = 3$) compound. As was expected the values of Δ_s and T_c for this material appeared to be practically the same as for BSCCO ($n = 3$) (Table 1). It can be then supposed that a total substitution of Sr by Ba in BSCCO ($n = 3$) could raise Δ_s and T_c in this hypothetical phase close to the typical values for HBCCO- and TBCCO-families.

In conclusion we have found that the low-temperature value of a superconducting gap Δ_s is linearly increasing with the number n of CuO_2 layers in the unit cell of optimally doped samples of the BSCCO-, HBCCO- and TBCCO-families ($n \leq 3$). The variation of the critical temperature $T_{c\,max}$ with n does not obey this simple relation. For a given number of CuO_2 layers n the values of Δ_s at $T = 4.2$ K for TBCCO- and HBCCO-families practically coincide but systematically exceed that in the BSCCO-family.

ACKNOWLEDGMENTS

The authors would like to thank K. Winzer for supplying high-quality TBCCO ($n = 3$) single crystals. The authors are indebted to T.E. Os'kina, Yu.D. Tretyakov and N. Kiryakov for providing the BSCCO ($n = 2$) crystals and the polycrystalline samples of BSCCO ($n = 3$) and HBCCO ($n = 2, 3$). This work was supported in part by the ISC on High Temperature Superconductivity (Russia) under the contract number 96118 (project DELTA) and by RFBR (Russia) under the contract number 96-02-18170a.

REFERENCES

Abrikosov A A 1999 Physica C **317-318**,154
Aminov B A et al 1996 Phys. Rev. B **54**, 6728
Byczuk K and Spalek J 1996 Phys. Rev. B **53**, R518
Chen J et al 1994 Phys. Rev. B **49**, 3683
Chen X, Xu Z, Jiao Z and Zhang Q 1997 Phys. Lett. A **229**, 247
Deutscher G 1999 Nature **397**, 410
Devereaux T P and Fulde P 1993 Phys. Rev. B **47**, 14638
Halbritter J 1996 Journ. Low Temp. Phys. **105**, 1249
Harris J M et al 1997 Phys. Rev. Lett. **79**,143
Hudáková N et al 1996 Physica B **218**, 217
Klemm R A, Rieck C T and Scharnberg K 1999 to be published in Journ. Low Temp. Phys.
Kresin V Z et al 1996 Journ. of Superconductivity **9**, 431
Leggett A J 1999 Phys. Rev. Lett. **83**, 392
Müller A et al 1997 HASYLAB, Annual Report I, 291
Phillips J C 1994 Phys. Rev. Lett. **72**, 3863
Ponomarev Ya G et al 1999a Physica C **315**, 85
Ponomarev Ya G et al 1999b Contribution 10-50, Proc. of EUCAS´99
Suzuki M et al 1999 Phys. Rev. Lett. **82**, 5361
Tsai J S et al 1989 Physica C **162-164**, 1133
Waldram J R 1996 Superconductivity of Metals and Cuprates, IOP Publishing Ltd
Wei J Y T et al 1998 Phys. Rev. B **57**, 3650

Inst. Phys. Conf. Ser. No 167
Paper presented at Applied Superconductivity, Spain, 14-17 September 1999

Three-dimensional Josephson junction networks: General properties and possible applications

R. De Luca

Dipartimento di Fisica and INFM, Università degli Studi di Salerno, I-84081 Baronissi (Salerno), Italy

ABSTRACT: The general properties of three-dimensional Josephson junction networks are briefly summarized. Their application as circuit models of vectorial magnetic field sensors and their use in the study of the magnetic properties of granular superconductors are discussed.

1. INTRODUCTION

The study of the electrodynamic properties of one-dimensional (1D) or two-dimensional (2D) Josephson junction arrays (Tinkham and Lobb 1989, Phillips et al. 1993, Wolf and Majhofer 1993, Chen et al. 1994, Auletta et al. 1994, Paasi et al. 1996) has been of relevance for a twofold reason. A first reason is that a complete knowledge of the behavior of these systems has brought about a deep understanding of the intergranular properties of high-T_c superconductors in the low magnetic field region. A second reason may be found in the large variety of devices which are modelled and studied through Josephson junction networks. The static and dynamic properties of three-dimensional (3D) Josephson junction networks (Yukon and Lin 1995, De Luca et al. 1998), however, have not been investigated in depth yet.

In the present work we present a general approach to the study of the electrodynamic properties of an inductive cubic network of Josephson junctions (JJ's) in the presence of an external forcing agent, such as a bias current, a magnetic field, or both.

2. MODEL AND EQUATIONS

We consider a current-biased inductive cubic network of JJ's in the presence of a constant external magnetic field. A time-dependent gauge-invariant phase difference $\varphi_\xi(\vec{r},t)$, where ξ denotes the direction perpendicular to the junction plane and $\vec{r}$ the position of the junction in space, is associated to each Josephson junction in the network. The JJ's are located on the sides of the cube. To take into account the magnetic energies of the branch currents $i_\xi(\vec{r},t)$, an inductor L is inserted on each branch. Neglecting capacitive effects, the resistively shunted junction (RSJ) model (Barone and Paternò 1982) can be adopted to write the dynamical equations for the twelve phase differences as follows:

$$O_J\left(\varphi_\xi(\vec{r},t)\right) \equiv \frac{\Phi_o}{2\pi R}\frac{d}{dt}\varphi_\xi(\vec{r},t) + I_J\,sin\,\varphi_\xi(\vec{r},t) = i_\xi(\vec{r},t), \qquad (1)$$

where Φ_o is the elementary flux quantum and O_J is a non-linear operator. In Eq. 1 R and I_J are the resistive parameter and the maximum Josephson current of each junction in the network, respectively. In order to solve the above dynamical equations for the phase differences, one needs to express the branch currents in terms of the φ_ξ's.

We first write fluxoid quantization relations by denoting as $\Phi_{\mu\nu}(\vec{r}, t)$ the generic magnetic flux through the cubic face parallel to the $\mu\nu$-plane ($\mu\nu = yz, zx, xy$), so that:

$$\frac{2\pi}{\Phi_o}\Phi_{\mu\nu}(\vec{r}, t) = 2\pi n_{\mu\nu}(\vec{r}) + \varphi_\mu(\vec{r} + a\hat{v}, t) - \varphi_\mu(\vec{r}, t) - \varphi_\nu(\vec{r} + a\hat{\mu}, t) + \varphi_\nu(\vec{r}, t), \quad (2)$$

where $n_{\mu\nu}(\vec{r})$ is an integer, a is the length of the cubic side and $\hat{\eta}$ is the unit vector in the generic η-direction. The fluxes $\Phi_{\mu\nu}(\vec{r}, t)$ can be written in terms of the branch currents and of the externally applied flux $\mu_o \vec{H} \cdot \vec{S}_{\mu\nu}(\vec{r}), \vec{S}_{\mu\nu}(\vec{r})$ being the area vector oriented in the positive ξ-direction ($\xi \neq \mu, \nu$), as follows:

$$\Phi_{\mu\nu}(\vec{r}, t) = L I_{\mu\nu}(\vec{r}, t) + M I_{\mu\nu}(\vec{r} + a\hat{\xi}, t) + \mu_o \vec{H} \cdot \vec{S}_{\mu\nu}(\vec{r}) \quad (3a)$$

$$\Phi_{\mu\nu}(\vec{r} + a\hat{\xi}, t) = M I_{\mu\nu}(\vec{r}, t) + L I_{\mu\nu}(\vec{r} + a\hat{\xi}, t) + \mu_o \vec{H} \cdot \vec{S}_{\mu\nu}(\vec{r}), \quad (3b)$$

where M is a mutual inductance coefficient, for which the ratio M/L can be estimated to be 0.05 (De Luca et al. 1998), and $I_{\mu\nu}(\vec{r}, t) = i_\mu(\vec{r}, t) - i_\mu(\vec{r} + a\hat{v}, t) + i_\nu(\vec{r} + a\hat{\mu}, t) - i_\nu(\vec{r}, t)$

Equations 2-3 can be written in a more compact form by defining the following column vectors:

$$\mathbf{I_S} = \begin{pmatrix} I_{yz}(0, t) \\ I_{yz}(a\hat{x}, t) \\ I_{zx}(0, t) \\ I_{zx}(a\hat{y}, t) \\ I_{xy}(0, t) \end{pmatrix}, \Phi = \begin{pmatrix} \Phi_{yz}(0, t) \\ \Phi_{yz}(a\hat{x}, t) \\ \Phi_{zx}(0, t) \\ \Phi_{zx}(a\hat{y}, t) \\ \Phi_{xy}(0, t) \end{pmatrix}, \Phi_{ex} = \mu_o a^2 \begin{pmatrix} \vec{H} \cdot \hat{x} \\ \vec{H} \cdot \hat{x} \\ \vec{H} \cdot \hat{y} \\ \vec{H} \cdot \hat{y} \\ \vec{H} \cdot \hat{z} \end{pmatrix}, \quad (4)$$

so that Eqs. 3 can be rewritten as follows:

$$\Phi = \mathbf{G} \cdot \mathbf{I_S} + \Phi_{ex}, \quad (5)$$

where $\mathbf{G}$ is a non-singular 5×5 matrix defined according to Eqs. 3a-b. Similarly, the five independent fluxoid quantization conditions can be rewritten as:

$$\Phi = \frac{\Phi_o}{2\pi}(\mathbf{F}\varphi + 2\pi\mathbf{N}), \quad (6)$$

where $\mathbf{F}$ is a 5×12 matrix, φ is a column vector whose components are the twelve phase differences $\varphi_\xi(\vec{r}, t)$, and $\mathbf{N}$ is the column vector whose components correspond to the initially trapped flux numbers $n_{\mu\nu}(\vec{r})$ in the five cubic faces for which the above relation is written. From fluxoid quantization it can be easily argued that the elements of the matrix $\mathbf{F}$ are either zeroes or ± 1. This matrix allows us to write the current vector $\mathbf{I_S}$ in terms of the branch current vector $\mathbf{i}$ as follows: $\mathbf{I_S} = -\mathbf{Fi}$. Moreover, due to charge conservation at the network's nodes, the twelve components of $\mathbf{i}$ can be expressed in terms of five independent branch currents, which we shall gather in a column vector $\mathbf{i_I}$, and of the bias current vector $\mathbf{I_B}$. In this way we may write:

$$\mathbf{i} = \mathbf{T}\mathbf{i_I} - \mathbf{I_B}, \quad (7)$$

where $\mathbf{T}$ is a 12×5 matrix with elements $T_{ij} = 0, \pm 1$. We notice that the particular form of the column vector $\mathbf{I_B}$ depends on the particular bias configuration we choose for the system.

By combining the above vectorial equation, we can write the dynamical equations as follows:

$$O_J(\varphi) + \frac{\Phi_\varrho}{2\pi}\mathbf{B}(\mathbf{F}\varphi + 2\pi\mathbf{N}) = \mathbf{B}\Phi_{ex} + (\mathbf{C} - \mathbf{I})\mathbf{I_B}, \qquad (8)$$

here $\mathbf{I}$ is the 12×12 identity matrix, $\mathbf{B} = \mathbf{T}(\mathbf{FT})^{-1}\mathbf{G}^{-1}$ and $\mathbf{C} = \mathbf{T}(\mathbf{FT})^{-1}\mathbf{F}$.

RESULTS

We integrated the dynamical equations (Eq. 8) by means of Mathematica 3.0. From the knowledge of the time evolution of the phase differences, the branch currents can be derived by Eq. 1. The branch voltages, on the other hand, can be defined as follows:

$$V_\xi(\vec{r},t) = L\frac{di_\xi(\vec{r},t)}{dt} + \frac{\Phi_\varrho}{2\pi}\frac{d\varphi_\xi(\vec{r},t)}{dt}, \qquad (9)$$

here the time derivatives are taken as total, given that, in this context, the position vector plays the role of an index. The voltages can be normalized to RI_J, and , by introducing the normalized time $\tau = Rt/L$ and the normalized branch currents $\tilde{i}_\xi(\vec{r},t) = \frac{i_\xi(\vec{r},t)}{I_J}$, one can set

$$_\xi(\vec{r},t) = \frac{V_\xi(\vec{r},t)}{RI_J} = \frac{d\tilde{i}_\xi(\vec{r},\tau)}{d\tau} + \frac{1}{\beta}\frac{d\varphi_\xi(\vec{r},\tau)}{d\tau}, \text{ where } \beta = 2\pi\frac{LI_J}{\Phi_o}.$$

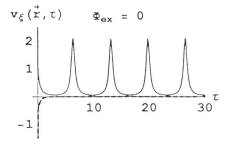

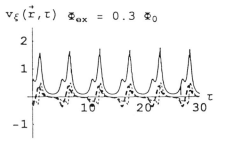

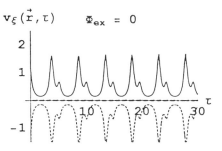

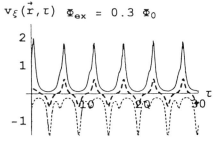

Fig. 1 Time dependence of the normalized voltages $v_x(a\hat{y} + a\hat{z}, \tau)$ (thick dashed line), $_y(a\hat{x} + a\hat{z}, \tau)$ (full line) , and $v_z(a\hat{x} + a\hat{y}, \tau)$ (thin dashed line) for two different bias configurations (a (top) and b (bottom) as in the text), for $\beta = 0.8\pi$ and for $_{ex} = \mu_o Ha^2 = 0$ and $\Phi_{ex} = \mu_o Ha^2 = 0.3\Phi_o$.

From the general form of the dynamical equations for the phase differences one can argue that the system presents stationary solutions which are periodic in τ when the constant bias current exceeds a field-dependent threshold. These features have been extensively studied in the case of d.c. SQUIDs (Barone and Paternò 1982). In the case of 3D inductive cubic networks of Josephson junctions, however, an additional aspect comes into play: The threshold current depends on the field direction. Moreover, for these 3D models there exist 28 different bias configurations or, having fixed a bias direction (i.e., the current is injected in only one face and is drawn only through the opposite face) as in our case, 16 possible bias configurations are present, for which different reponses are expected. In particular, we analyze the cases in which the current bias is injected through the node $(1, 0, 1)$ and is drawn through the nodes $(0, 1, 0)$ and $(1, 1, 0)$, respectively denoted as configuration a and configuration b. For these configurations we might say that the bias direction is the y -axis. For these two cases, in Fig. 1 we report the time dependence of the normalized voltages $v_x(a\hat{y} + a\hat{z}, \tau)$, $v_y(a\hat{x} + a\hat{z}, \tau)$, and $v_z(a\hat{x} + a\hat{y}, \tau)$, all pertaining to branches converging to the $(1, 1, 1)$ node, for fields applied in the z -direction and for $\beta=0.8\pi$. The initially trapped flux in the system is zero and the chosen values of the applied field are such that $\Phi_{ex} = \mu_o Ha^2$ is zero in the first case and is equal to $0.3\Phi_o$ in the second.

4. CONCLUSIONS

We have developed a general approach to study the electrodynamic properties of current-biased inductive cubic networks of small Josephson junctions in the presence of a constant magnetic field. These systems are interesting because they present characteristic responses which may be exploited to model vectorial magnetic field sensors (Di Matteo et al. 1999). These 3D model systems have indeed been used in detailed studied of flux penetration mechanisms in granular superconductors in the presence of an external magnetic field oriented along an arbitrary direction in space (Tuohimaa et al. 1999). Due to these promising perspectives, it might be useful to further investigate the characteristic features of these systems starting from this general analysis.

REFERENCES

Auletta C, Raiconi G, De Luca R and Pace S 1994 Phys. Rev. B **49**, 12 311.
Barone A and Paternò G 1982 Physics and Applications of the Josephson Effect (New York: Wiley).
Chen D X, Sanchez A and Hernando A 1994 Phys. Rev. B **50**, 13735.
De Luca R, Di Matteo T, Tuohimaa A and Paasi J 1998 Phys. Rev. B **57**, 1173.
Di Matteo T, De Luca R, Paasi J and Tuohimaa A 1998 IEEE Trans.on Applied Superc., to be published.
Paasi J, Tuohimaa A and Eriksson J-T 1996 Physica C **259**, 10.
Phillips J R, Van der Zant H S J, White J and Orlando T P 1993 Phys. Rev. B **47**, 5219.
Tinkham M and Lobb C J 1989 Solid State Phys. **42**, 91.
Tuohimaa A, Paasi J, De Luca R and Di Matteo T 1999, to be published.
Wolf T and Majhofer A 1993 Phys. Rev. B **47**, 7481.
Yukon S P and Lin N Chu H 1995 IEEE Trans.on Applied Superc. **5**, 2959.

Inst. Phys. Conf. Ser. No 167
Paper presented at Applied Superconductivity, Spain, 14-17 September 1999
© 2000 IOP Publishing Ltd

Current transport mechanism in YBCO bicrystal junctions on sapphire.

G A Ovsyannikov, I V Borisenko, A D Mashtakov and K Y Constantinian

Institute of Radio Engineering and Electronics RAS, Moscow 103907, Russia

ABSTRACT: YBCO junctions were made on r-cut sapphire bicrystal substrates in which the directions <1120> for both parts of the substrate have the angles ±12° to the plane of the interface of electrodes. The junctions were tested at dc and mm waves. The junctions with 5 μm width have high normal resistance R_N up to 20 Ω, and $I_C R_N$ up to 2 mV, and tolerance of characteristic interface resistance around 30% on a chip. DC, microwave and magnetic characteristics of the junctions have been investigated experimentally. The sinusoidal superconducting current-phase relation $I_S(\varphi)$ and a linear dependence of critical current density vs square root of the barrier transparency have been revealed experimentally. Results are discussed in terms of supercurrent transport via Andreev bound states in symmetric bicrystal junctions with twinned electrodes.

1. INTRODUCTION

The high values of normal-state resistance R_N and critical frequency $f_C=(2e/h)I_C R_N$, as well the nonhysteretic I-V curves of high-T_C superconducting (HTSC) Josephson junctions make them appreciably superior to low-T_C superconducting junctions at liquid-helium temperature (T=4.2K). The high critical temperature and proper superconducting gap give promising opportunities for applications at frequencies higher than those, corresponded to energy gap of ordinary (say, Nb) superconductor. However, the aspects involved in the reproducible fabrication of high quality HTSC Josephson junctions on one hand, and the mechanism, describing current transport, on the other hand are the problems which have not been solved yet. The most reproducible junctions, having a spread of critical current ±12% per chip, are fabricated on $SrTiO_3$ bicrystal substrates (Vale 1997), but because of their high dielectric constant ε>1000 they are unsuitable for high-frequency applications. Sapphire having a relatively low ε≈9-11 and low losses (tan δ≈10^{-8} at 72 GHz), is the traditional material used in microwave electronics. Here we present the results of fabrication and characterization of HTSC Josephson junctions on sapphire bicrystal substrates in a view of determination of current transport mechanism. The high frequency dynamics of those junctions is discussed.

2. EXPERIMENTAL RESULTS

The Josephson junctions were fabricated on the r-cut sapphire bicrystal substrates for which the directions <1120> Al_2O_3 for both parts were misoriented at the angles ±12° to the plane of the interface of electrodes. The $YBa_2Cu_3O_x$ (YBCO) film was deposited by dc sputtering at high oxygen pressure after the CeO_2 epitaxial buffer layer RF magnetron sputtering. The following epitaxial relation: (001)YBCO//(001)CeO_2//(1102)Al_2O_3, [110]YBCO// [001]CeO_2// [1120]Al_2O_3 was fulfilled for the deposited films (Fig.1). Thin film YBCO bridges each 5 μm wide and 10 μm long, crossing the bicrystal boundary, were fabricated by RF plasma and Br_2-ethanol etching (Mashtakov 1999). The angle γ between the normal to the interface and current direction was varied from 0° to 54°. The bicrystal junctions (BJ) with current density I_C/S up to 10^5 A/cm^2 at T=4.2K gave the

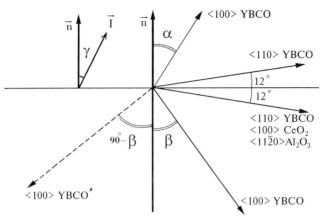

Fig.1. Crystallographic axes orientations of CeO_2 and YBCO films in sapphire bicrystal junction with $\alpha=33°$, $\beta=-33°$. The domain of the film with the direction <100>YBCO' misoriented on the angle $\beta'=90°-\beta$ is the twin to YBCO.

parameters: R_N up to 20 Ω, I_C of order of 100 μA, and $I_C R_N$ up to 2 mV. $I_C(T)$ nearly linear increases with T reduction at $T \ll T_C$ (Andreev 1994). On other hand at $T_C-T \ll T_C$, where thermal fluctuations predominate, the approximated dependence $I_C(T)$ is closer to the quadratic dependence $I_C \propto (1-T/T_C)^2$. Other dc parameters of the junctions discussed elsewhere (Mashtakov 1999). Most unusual feature for investigated junctions was linear dependence of I_C vs square root of barrier transparency $\sqrt{D}$.

Current-phase relation $I_S(\varphi)$ strongly depends on the type of contacts between superconductors. For $T_C-T \ll T_C$ the deviations of $I_S(\varphi)$ from $I_S(\varphi)=I_C\sin\varphi$ are small for any of

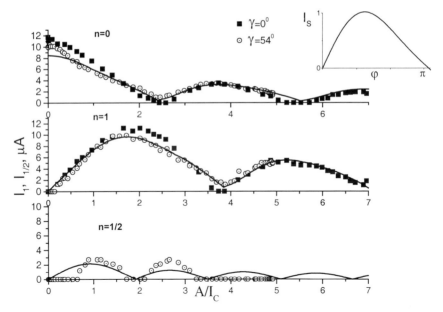

Fig.2. Normalized RF current dependence of the critical current (n=0), first Shapiro step (n=1) and half Shapiro steps (n=1/2) for two BJ $\gamma=0$ (filled squares), and $\gamma=54°$ (opened circles). Solid lines show the calculated curves for fixed $\delta=0.2$ of current-phase relation $I_S(\varphi)=(1-\delta)I_C\sin\varphi+\delta I_C\sin2\varphi$. The current-phase relation for this case is shown in the inset.

superconducting junction, but at $T \ll T_C$ $I_S(\varphi) = I_C \sin\varphi$ remains for SIS junction (Likharev 1979) regardless the transparencies of the barrier $\overline{D} \ll 1$. On the other hand, for junctions of d-wave superconductors (DID) $I_S(\varphi) = I_C \sin\varphi$ takes place only for tunnel junctions in certain range of T, α and $\overline{D}$ (Tanaka 1997, Riedel 1998, Barash 1995).

To estimate the deviation from $I_S(\varphi) = I_C \sin\varphi$ we have measured I-V curves under applied monochromatic mm wave radiation $A \sin(2\pi f_e t)$, $f_e = 40 \div 100$ GHz (Mashtakov 1999). Fig. 2 shows the variation of $I_C(A)$, first $I_1(A)$ and half $I_{1/2}(A)$ Shapiro steps for two BJs with $\gamma = 0$ (symmetrical bias) and $\gamma = 54°$ (nonsymmetrical one). The calculated functions using RSJ model for $f_e > 2eI_C R_N/h$ in the case of $\delta = 0.2$ for $I_S(\varphi) = (1-\delta)I_C \sin\varphi + \delta I_C \sin2\varphi$ are presented on Fig.2. The difference from the case of $I_S(\varphi) = I_C \sin\varphi$ of $I_{C,1}(A)$ is small and fits well to experiment. At the same time, a small deviations $I_S(\varphi)$ from sin-type dependence yield subharmonic (fractional n/m) Shapiro steps. The maximum amplitude of subharmonic steps $I_{m/n}$ are proportional to harmonics of $\sin(n\varphi)$ in $I_S(\varphi)$. The precise measurements of $I_n(A)$ (n=0,1,2), as well $I_{m/n}(A)$ at T=4.2 K ($T/T_C \approx 0.05$) allows us to state the absence of $\sin(2\varphi)$ components in $I_S(\varphi)$ function for all investigated BJs with symmetrical biasing ($\gamma = 0 \div 36°$) with accuracy at least of 5%. For strong asymmetric biasing ($\gamma > 40°$) the contribution of the component $\sin2\varphi$ increases monotonously.

3. DISCUSSION

The $I_S(\varphi)$ can be determined from the energy of bound Andreev levels E_B in the junction since $I_S(\varphi) \infty dE_B/d\varphi$ (Tanaka 1997). For SIS junctions the dependence (Riedel 1998):

$$E_B(\varphi) = \Delta_0 \sqrt{1 - \overline{D} \sin^2 \left(\frac{\varphi}{2}\right)} \tag{1}$$

gives $I_S(\varphi) = I_C \sin\varphi$. For the tunnel junction of two d-wave superconductors with gaps $\Delta_{R(L)} = \Delta_0 \cos(2\theta + 2\alpha(\beta))$ E_B depends on 4 angles: quasiparticle incidence angle $-\theta$, phase $-\varphi$, misorientation angles α (and β). Andreev levels for mirror-symmetric d-wave junctions ($\alpha = -\beta$) at several α are presented at Fig.3a. One can see that in the wide range of $\alpha = 10° \div 45°$ $E_B(\varphi)$ dependence is very close to E_B, given by (2) which corresponds to the case of the fixed $\alpha = \pi/4$:

$$E_B = \pm \Delta \sqrt{\overline{D}} \sin(\varphi/2) \tag{2}$$

Note, the proportionality of $I_C \infty \sqrt{\overline{D}}$, observed in experiment (Ovsyannikov 1999) is directly follows from equation (2).

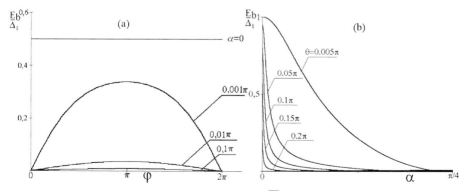

Fig3. (a)- Andreev levels in $D_\alpha ID_{-\alpha}$ junction for several α, $\overline{D} = 10^{-4}$, $\theta = \pi/6$, T=4.2K. $E_B(\varphi)$ confined with equations (1) and (2) for $\alpha = 0$ and $\pi/4$ correspondingly. (b)-Amplitude of Andreev levels in $D_\alpha ID_{-\alpha}$ for several θ, $\overline{D} = 10^{-4}$, T=4.2K.

256

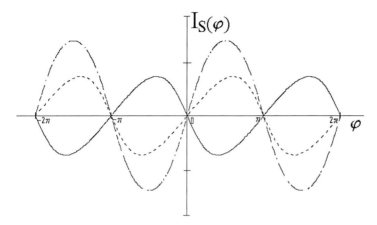

Fig.4. Current-phase relation calculated for symmetrical (45°, 45°)-dotted line and mirror symmetrical (45°,-45°)-solid line bicrystal junctions. Dashed line corresponds to the parallel connection of these two junctions. T==4.2K, $\overline{D}$ =10⁻⁴.

The behavior of the amplitude of Andreev levels with increasing misorientation angle α at several quasiparticle incidence angles (θ) is shown on Fig.3b. For $\alpha=10°\div45°$ the condition max$|E_B|>0.1\Delta_0$ satisfied for small amount of incident quasiparticles in the range $\theta=0\div10°$. Therefore the averaged income of the quasiparticles would be small.

The results of $I_S(\varphi)$ calculation for symmetrical and mirror-symmetrical junction with misorientation angle 45° using the technique (Riedel 1998) are shown on Fig.4. Taking into account the twins in superconducting films, the experimental samples may be considered as a parallel connection of pairs of two similar BJs (in our experiment $D_{33}ID_{-33}$ and $D_{33}ID_{57}$). Even the calculated $I_S(\varphi)$ for $D_{45}ID_{45}$ as well as for $D_{45}ID_{-45}$ are nonsinusoidal for experimental T==4.2K and $\overline{D}$ =10⁻⁴, the resulting current through the parallel connection of these junctions is $I_S(\varphi)\approx I_C\sin\varphi$ (see Fig.4) as we observed in experiment. For larger values of the γ (36°÷54°) the nonsinusoidal parts of the $I_S(\varphi)$ dependence doesn't cancel and subharmonic Shapiro step which appears at the I-V curve under microwawe irradiation due to contributions of two nonsymmetrical type BJs.

4. ACKNOWLEDGEMENTS

The work was partially support by Russian Foundation of Fundamental Research, Russian State Program "Modern Problems of the Solid State Physics", "Superconductivity" division, and INTAS program of EU.

REFERENCES

Andreev A V et al 1994 Physica C **226**, 17
Barash Yu S, Bukrhardt H and Rainer D 1994 Phys. Rev. Lett. **77**, 4070
Barash Yu S, Galaktionov A V and Zaikin A D 1995 Phys. Rev. **B52,** 661
Likharev K K 1972 Rev. Mod. Phys. **51**, 102
Mashtakov A D et al 1999 Technical Physics Letter **25**, 249
Ovsyannikov G A et al 1999 Abstract Book of Int. Conf on Physics and Chemistry of Molecular and Oxide Superconductors (Stockholm) p 154
Riedel R A and Bagwell P F 1998 Phys. Rev. **B57**, 6084
Tanaka Y and Kashiwaya S 1997 Phys. Rev. **B56,** 892
Vale L R et al 1997 IEEE Tr. Appl. Superconductivity **7**, 3193

Inst. Phys. Conf. Ser. No 167
Paper presented at Applied Superconductivity, Spain, 14-17 September 1999
© 2000 IOP Publishing Ltd

Observation of subharmonic gap structures in NbN/AlN/NbN tunnel junctions

Z Wang[1], A Saito[2], and K Hamasaki[2]

1. Kansai Advanced Research Center, Communications Research Laboratory, 588-2 Iwaoka, Iwaoka-cho, Nishi-ku, Kobe, Hyogo 651-2401 Japan
2. Department of Electrical Engineering, Nagaoka University of Technology, 1-1603 Kamitomioka, Nagaoka, Niigata 940-2188 Japan

ABSTRACT: Subharmonic gap structures in NbN/AlN/NbN tunnel junctions were investigated by observation of dV/dI-V characteristics and temperature dependence of dV/dI-V curves. We describe measurements and theoretical analysis on the subharmonic gap structures in the NbN/AlN/NbN junctions with different current density. Subharmonic gap structures were clearly observed in the dV/dI-V curves at voltage $2\Delta/ne$, and the values of n were increasing with the current density in the junctions. We tried to explicate the subharmonic gap structures using the theory due to Octabio, Tinkham, Blonder, and Klapwijk and demonstrated good agreement with the theory for the subharmonic gap structures in NbN/AlN/NbN tunnel junctions.

1.   INTRODUCTION

NbN tunnel junctions are promising for applications of high frequency and high-speed Josephson devices. In order to obtain high performances for the devices, however, the junctions must have large I_cR_N products and small R_NC_J constants. This requires that the junctions have a high current density, because the R_NC_J constants scale with the $1/J_c$. We have recently made an advantage in development of high current density and high quality NbN/AlN/NbN tunnel junctions for submillimeter wave SIS mixers (Wang et al 1994, 1997a, 1997b, Uzawa et al 1998). The junctions showed excellent tunneling characteristics with a current density up to 100 kA/cm^2 and sensitive heterodyne mixing properties in submillimeter wave regions. However, the high current density junctions exhibited significant subgap current, which is interesting to analysis noise properties for SIS mixers (Dieleman et al 1997, 1998). Such subgap current was also observed in many Josephson junctions and was explained for three physical processes: Josephson self-coupling (JSC) (McDonald et al 1976), multiparticle tunnelling (MPT) (Schrieffer et al 1963, Bratus et al 1995), and multiple Andreev reflection (MAR) (Flensberg et al 1989, Kleinsasser et al 1994).

In this paper, we report on the investigations of the subgap current the NbN/AlN/NbN tunnel junctions. Subharmonic gap structures (SGS) were observed in the I-V characteristics at voltage $2\Delta/ne$, where Δ is the superconducting energy gap and n is an integer. We measured dV/dI-V characteristics and temperature dependence of the SGS for the junctions with different current density. We show that multiple Andreev reflection is responsible for the subgap current in our NbN/AlN/NbN tunnel junctions using the theory due to Octabio, Tinkham, Blonder, and Klapwijk (OTBK) (Octabio et al 1983), and demonstrated good agreement with the theory for the SGS in our junctions.

258

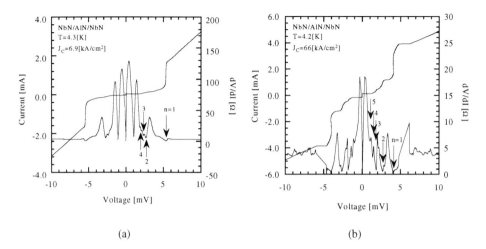

(a)   (b)

Fig. 1. I-V and dV/dI-V characteristics for two NbN/AlN/NbN tunnel junctions with different current density at 4.2 K. The current density is (a) 6.9 kA/cm² and (b) 66 kA/cm².

2.   **EXPERIMENTALS**

NbN/AlN/NbN tunnel junctions with J_c in a wide range (0.1 kA/cm²-100 kA/cm²) were fabricated on single-crystal (100) MgO substrates using the fabrication processes described elsewhere (Wang et al 1994). NbN and AlN films were prepared by rf magnetron sputtering in a load-lock sputtering system at ambient substrate temperatures. Subharmonic gap structures were investigated in the voltage (V) dependence of current (I) and dV/dI for the junctions with different current density. We also carefully measured the temperature dependence of the dV/dI-V curves, and fitted the temperature dependence for the voltage position of SGS at $2\Delta/ne$ with the temperature dependence of gap energy (BCS theory). The subharmonic gap structures were characterized due to MAR by compared with the simulated results using OTBK theory.

3.   **RESULTS AND DISSCUSIONS**

Figure 1 shows I-V and dV/dI-V characteristics measured at 4.2 K for two NbN/AlN/NbN tunnel junctions with (a) J_c=6.9 kA/cm² and (b) J_c=66 kA/cm². The junction size was (a) 4x4 μm² and (b) 1 μm in diameter. The Josephson current I_c was suppressed with a magnetic field. In a very wide range of J_c (100 A/cm² to above 100 kA/cm²) the junctions showed a very good junction quality, and the Josephson currents I_c showed well-behaved Josephson tunneling properties with a BCS-like temperature dependence and ideal magnetic field dependence (Wang et al 1997).

The SGS, as shown in Fig. 1, were observed in the dV/dI-V characteristics at voltage $2\Delta/ne$ for both low-J_c and high-J_c junctions. The number of n grows with increasing current density, i.e., with decreasing the AlN barrier thickness. The SGS at voltage $2\Delta/ne$ up to n=5 was clearly visible in the junction with J_c=66 kA/cm² (Fig. 1(b)). The gap voltage V_g, as shown in Fig. 1(b), decrease with increasing J_c due to the self-heating effect, and the final layer of 66 kA/cm² junction was driven normal above 5 mV.

We think that the subharmonic gap structures in the I-V curve of our junctions is due to multiple Andreev reflection. We try to explicate the subharmonic gap structures using the OTBK theory. In OTBK model, quasiparticle currents I_{qp} through the N-S contact is given by two different nonequilibrium distribution functions $f_{\rightarrow}(E)$ and $f_{\leftarrow}(E)$, where the f(E) is a function of the probability

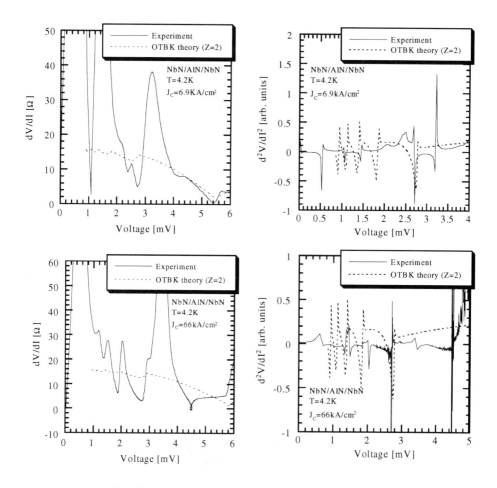

Fig. 2. dV/dI-V and d²V/d²I/-V characteristics of the junctions shown in Fig. 1 with the predications of the OTBK's MAR model. The line is experimental data, and the dash line is calculated by OTBK theory with Z=2.

of Andreev reflection A(E), the probability of ordinary reflection B(E), and the dimensionless barrier strength Z. Figure 2 compares the dV/dI-V and d²V/d²I-V characteristics of the junctions of Fig. 1 with the predications of the OTBK's MAR model. The dimensionless barrier strength Z is 2. The positions of SGS show a good agreement between the experiments and OTBK model, even though the magnitude of the SGS is different.

According to the OTBK theory, SGS which results from MAR are clearly visible in the OTBK (dV/dI)R_N-V curves, as increasing temperature. We also measured the temperature dependence of dV/dI-V characteristics for investigating the MAR process in our tunnel junctions. Figure 3 shows the temperature dependence of the voltage positions of SGS for two junctions of Fig. 1. Also shown are fits to the data using the temperature dependence of superconducting energy gap predicted by BCS theory. SGS with large number of n are clearly observed until the temperature rises up to nearby T_c, and the temperature dependence of voltage positions of SGS agree with the BCS theory. These results suggest that the SGS in our junctions result from MAR process, although further detailed analysis and measurements are needed.

260

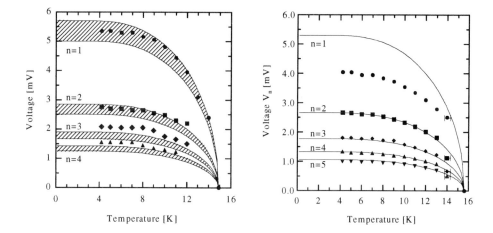

Fig. 3. Temperature dependence of the voltage positions of SGS for the junctions shown in Fig. 1. The lines are fit to the data using the temperature dependence of superconducting energy gap predicted by BCS theory.

4.   CONCLUSIONS

We have made an initial investigation on subharmonic gap structures in NbN/AlN/NbN tunnel junctions with high current density, and demonstrated good agreement with the OTBK's MAR model for the subharmonic gap structures in the junctions. Subharmonic gap structures were clearly observed in the dV/dI-V curves at voltage $2\Delta/ne$, and the values of n were increasing with the current density in the junctions. Similar results were also obtained in Nb/AlOx/Nb tunnel junctions and were concluded due to pinhole defects in the tunnel barriers (Kleinsasser 1994). We are working on to make detailed investigation for SGS in a wide range of current density (0.1 kA/cm^2-100 kA/cm^2), and measuring noise characteristics of NbN/AlN/NbN junctions at voltage below the gap energy for investigating the relation between the SGS and noise properties in the junctions.

REFERENCES

Bratus E N, Shumeiko V S, and Wendin G 1995 Phys. Rev. Lett. **11**, 2110
Dieleman P, Bukkems H G, and Klapwijk T M 1997 Phys. Rev. Lett. **79**, 3486
Dieleman P, and Klapwijk T M 1998 Appl. Phys. Lett. **72**, 1653
Flensberg K and Hansen J B 1989 Phys. Rev. B **40**, 8693
Kleinsasser A W, Miller R E, Mallison W H, and Arnold G B 1994 Phys. Rev. Lett. **72**, 1738
McDonald D G, Johson E G, and Harris R E 1976 Phys. Rev. B **13**, 1028
Octabio M, Tinkham M, Blonder G B, and Klapwijk T.M 1983 Phys. Rev. B **27**, 6739
Schrieffer J R and Wilkens J W 1963, Phys. Rev. Lett. **10**, 17
Uzawa Y, Wang Z, and Kawakami A 1998 Appl. Phys. Lett. **73**, 680
Wang Z, Kawakami A, Uzawa Y, and Komiyama B 1994 Appl. Phys. Lett. **64**, 2034
Wang Z, Kawakami A, and Uzawa Y 1997a Appl. Phys. Lett. **70**, 114
Wang Z, Kawakami A, and Uzawa Y 1997b IEICE Trans. Electron. **E80-C**, 1258

Inst. Phys. Conf. Ser. No 167
Paper presented at Applied Superconductivity, Spain, 14-17 September 1999
© 2000 IOP Publishing Ltd

Effect of trapped Abrikosov vortices on the critical current of YBCO step-edge junctions

K-H Müller, E E Mitchell and C P Foley

CSIRO Telecommunications and Industrial Physics, Sydney 2070, Australia

ABSTRACT: We have developed a theoretical model to describe the effect of Abrikosov vortices on the critical current I_c of a short Josephson junction in planar geometry where the magnetic field H_a is applied perpendicular to the thin-film electrodes. The phase difference along the junction due to the Meissner shielding current is calculated using a model proposed by Humphreys et al (1993) while the phase difference caused by the vortex current is calculated using a superposition of vortex and mirror vortex currents. The model describes well jumps observed in $I_c(H_a)$ of YBCO/MgO step-edge junctions and reveals the likely positions of vortices in the electrodes.

1. INTRODUCTION

Above a certain applied magnetic field Abrikosov vortices form in the electrodes of Josephson junctions. Depending on the positions of these vortices, they modify the critical current I_c of the junction which is undesirable in junction-based devices like SQUID's or RSFQ's. The effect of Abrikosov vortices on I_c has been studied in detail in LTS junctions (i.e. Miller et al 1985) where a junction is formed between two overlapping superconducting strips. In HTS materials a junction can be created by the formation of a grain boundary along two adjacent thin-film strips. When a magnetic field is applied perpendicular to the film strips, a theoretical investigation is intricate due to significant field enhancements at the edges of the strips and in the junction. We have developed a detailed theoretical model for such a junction in planar geometry to describe the experimentally observed jumps in I_c for increasing and decreasing magnetic fields. Fitting the experimental data reveals the likely position of the Abrikosov vortices.

2. EXPERIMENT

Step-edge junctions were fabricated by etching steps of height ~ 400 nm and step angle $\sim 145°$ (see Fig. 1) (Foley et al 1999) into MgO substrates using an Ar-ion beam and depositing YBCO films of thickness $d \simeq 250$ nm onto the substrate by magnetron sputtering. Using optical lithography and ion-beam etching at liquid nitrogen (Foley et al 1999), thin-film strips (electrodes) ~ 1 mm long were patterned over the step. The widths of the junctions were in the range 0.8 - 18 μm. Four terminal current-voltage measurements were made by passing a low frequency (~ 1 Hz) ac current through the junctions and applying magnetic fields in the range 0 - 3 mT perpendicular to the electrodes (Mitchell et al 1999). Values of I_c were obtained by fitting the resistively shunted junction model to the current-voltage curves.

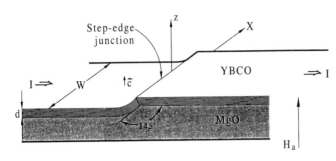

Fig. 1: YBCO thin film strips on MgO which form a step-edge junction. The applied field H_a is perpendicular to the film.

3.   THEORY

We consider a homogeneous Josephson junction in planar geometry of width W and thickness d where a magnetic field H_a is applied perpendicular to the thin-film strip electrodes in z direction as shown in Fig. 1. The critical current of the junction is given by

$$I_c = \frac{I_o}{W} \, | \int_{-W/2}^{W/2} e^{i \, \varphi_{12}(x)} \, dx \, | \quad , \tag{1}$$

where I_o is the maximum critical current and $\varphi_{12}(x)$ the gauge-invariant phase difference in x direction along the junction (Barone 1982). The phase difference $\varphi_{12}(x)$ consists of a Meissner part and an Abrikosov vortex part where

$$\varphi_{12}(x) = \varphi_{12,M}(x) + \varphi_{12,V}(x) \quad . \tag{2}$$

From the 2nd Ginzburg-Landau equation one derives for the Meissner part of $\varphi_{12}(x)$

$$\varphi_{12,M}(x) = \frac{4\pi\mu_o\lambda_{eff}}{\phi_o} \int_0^x J_{M,x}(x, y = 0^+) \, dx \quad . \tag{3}$$

Here $\lambda_{eff} = \lambda^2/d$ where λ is the London penetration depth and $J_{M,x}(x, y = 0^+)$ is the Meissner part of the surface sheet current density along the junction.
According to Humphreys et al (1993) :

$$\varphi_{12,M}(x) = \sum_{n=0}^{\infty} \frac{2 \, \pi^2 \, \mu_o \, K_n \, \sin(q_n x)}{\phi_o \, q_n^2 \, F_n} \quad , \tag{4}$$

where $q_n = (2n + 1)\pi/W$. Expressions for F_n and K_n can be found in Humphreys et al (1993). The function K_n is proportional to the applied field H_a.
The vortex contribution $\varphi_{12,V}$ to φ_{12} is obtained from the 2nd Ginzburg-Landau equation which gives

$$\varphi_{12,V}(x) = \frac{2\pi\mu_o\lambda_{eff}}{\phi_o} \int_0^x [J_{V,x}(x, y = 0^+) + J_{V,x}(x, y = 0^-)] \, dx \quad , \tag{5}$$

where $J_{V,x}(x, y = 0^{\pm})$ is the vortex part of the sheet current density along the junction. The sheet current density of a single vortex in an infinite thin-film is given by (Pearl 1964)

$$\vec{J}_V^S(\vec{r}) = \frac{\phi_o}{2\mu_o} \frac{\vec{e}_z \times \vec{r}}{r(2\lambda_{eff})^2} \left[H_1 \left(\frac{r}{2\lambda_{eff}} \right) - N_1 \left(\frac{r}{2\lambda_{eff}} \right) - \frac{2}{\pi} \right] , \tag{6}$$

where H_1 is a Struve function and N_1 a Neumann function. The sheet current density $\vec{J}_V(x, y, x_o, y_o)$ of an Abrikosov vortex at (x_o, y_o) located in the thin-film strip electrodes is given by a superposition of vortex and mirror vortex currents $\vec{J}_V^S$ where

$$\vec{J}_V(x, y, x_o, y_o) = \sum_{n=-\infty}^{\infty} [\ \vec{J}_V^S(x, y, x_o + 2nW, y_o) \tag{7}$$

$$- \vec{J}_V^S(x, y, x_o + 2nW, -y_o)$$

$$- \vec{J}_V^S(x, y, -x_o + (2n+1)W, y_o)$$

$$+ \vec{J}_V^S(x, y, -x_o + (2n+1)W, -y_o) \ ] \ .$$

4. RESULTS AND DISCUSSION

When an increasing magnetic field is applied to a junction, the Meissner shielding current $J_{M,x}(x, y = 0^{\pm})$ along the junction increases. This causes the phase $\varphi_{12,M}(x)$ of Eq. (3) to vary faster along the junction which, according to Eq. (1), reduces the critical current I_c of the junction. I_c as a function of the applied field H_a then resembles a Fraunhofer diffraction pattern (Barone 1982). In the case of a junction in planar geometry where the field is applied perpendicular to the electrodes, the spacing between the minima in $I_c(H_a)$ is proportional to W^{-2} and also depends on λ and d as shown by Humphreys et al (1993). In planar geometry, the magnetic field at the edges and in the junction is several times larger than the applied field and the magnetic field has a maximum in the center of the junction. If the applied field exceeds a certain strength, Abrikosov vortices are first created near the center of the junction. Three different forces act on a newly formed vortex and move it into its equilibrium position. These forces are the Lorentz force due to the Meissner shielding currents in the film electrodes, the vortex mirror image force due to reflections of the vortex currents at the edges of the electrodes and the pinning force in the film.

Experimentally we have observed the sudden creation and annihilation of vortices when an applied magnetic field is increased or decreased. Figure 2(a) shows the measured I_c versus an increasing applied magnetic field H_a at 77 K where $W = 2.2 \ \mu m$ and $d = 0.25 \ \mu m$. A jump in I_c is observed at $\mu_o H_a = 1.5 \ mT$. Fig. 2(b) shows the corresponding calculated result using Eqs. (1)-(7) assuming that a single Abrikosov vortex is present at position $(x_o, y_o) = (0, 2 \ \lambda)$. The London penetration depth is $\lambda(77K) = 0.2 \ \mu m$. The model prediction is in good agreement with the experimental data. Fig. 3(a) shows the measured I_c versus a decreasing applied field where $\mu_o H_{max} \simeq 3 \ mT$. Compared to Fig. 2(a) the peak in I_c has shifted from 0 mT to 0.5 mT and a large jump in I_c is observed at $\mu_o H_a = 0.1 \ mT$. Fig. 3(b) shows the calculate d result which assumes that two Abrikosov vortices are present. One is located at $(x_o, y_o) = (0, 2.5 \ \lambda)$ and the other at $(x_o, y_o) = (0, -2.5 \ \lambda)$. The jump at $\mu_o H_a = 0.1 \ mT$ in a decreasing applied field occurs when one of the two vortices is annihilated. Here no attempt was made to model the jump at $\mu_o H_a = 2 \ mT$ in Fig. 3(a).

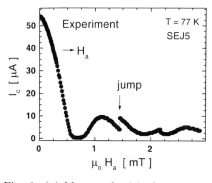 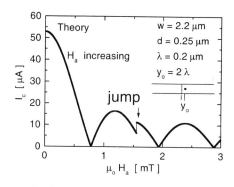

Fig. 2: (a) Measured critical current I_c of a YBCO/MgO step-edge junction versus applied magnetic field H_a at 77 K. The magnetic field is increased from 0 mT to 3 mT. A jump occurs at $\mu_o H_a = 1.5$ mT.
(b) Calculated critical current I_c of junction in planar geometry versus applied magnetic field H_a using Eqs. (1)-(7). At $\mu_o H_a = 1.5$ mT an Abrikosov vortex is assumed to enter one of the electrodes sitting at equilibrium position $(x_o, y_o) = (0, 2\,\lambda)$.

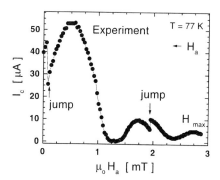 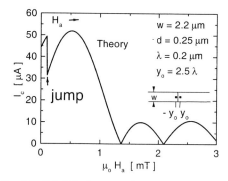

Fig. 3: (a) Measured critical current I_c of a YBCO/MgO step-edge junction versus applied magnetic field H_a at 77 K. The magnetic field is decreased from 3 mT to 0 mT. Jumps occur at $\mu_o H_a = 2$ mT and 0.1 mT.
(b) Calculated critical current I_c of junction in planar geometry versus applied magnetic field H_a using Eqs. (1)-(7). The maximum in I_c has shifted to $\mu_o H_a = 0.5$ mT due to two vortices at positions $(x_o, y_o) = (0, \pm 2.5\,\lambda)$. At $\mu_o H_a = 0.1$ mT, one of the Abrikosov vortices is assumed to leave its electrode.

REFERENCES

Barone A and Paterno G 1982 Physics and Applications of the Josephson Effect (John Wiley & Sons, Inc)
Foley C P, Mitchell E E, Lam S K H, Sankrithyan B, Wilson Y M, Morris S J and Tilbrook D L 1999 IEEE Trans on Applied Superconductivity (in press)
Humphreys R G and Edward J A 1993 Physica C **210**, 42
Miller S L, Biagi K R, Clem J R and Finnmore D K 1985 Phys. Rev. B **31**, 2684
Mitchell E E, Foley C P, Müller K-H and Leslie K E 1999 Physics C (in press)
Pearl J 1964 Appl. Phys Letters **5**, 65

Inst. Phys. Conf. Ser. No 167
Paper presented at Applied Superconductivity, Spain, 14-17 September 1999
© 2000 IOP Publishing Ltd

Fabrication of YBCO step-edge Josephson junctions

J T Jeng†, Y C Liu†, S Y Yang†, H E Horng†, J R Chiou‡, J H Chen‡, and H C Yang‡

† Department of Physics, National Taiwan Normal University, Taipei 116, Taiwan, R.O.C.
‡ Department of Physics, National Taiwan University, Taipei 106, Taiwan, R.O.C.

ABSTRACT: By properly adjusting the exposure time and the incident angle of the ion beam, we fabricated reproducible step-edge substrates with step angles varying from 10° up to 70°. The fabricated step-edge junctions show the resistively-shunted-junction (RSJ) behavior with double grain-boundary junctions when the step angle was about 70°. The value of the thickness-to-height ratio, t/h, was about 1 to 1.4. The feature of the multiple grain boundaries was observed in the voltage-current characteristics. The details of fabrication and characterization of the step-edge junctions are reported.

1. INTRODUCTION

Since the first demonstration of high-T_c Josephson grain-boundary junctions, various technologies have been developed to fabricate controllable grain boundaries for practical applications. Among these technologies, the most appealing devices are the bicrystal grain-boundary junctions (Gross 1991) and the step-edge junctions (Char 1991). Although the bicrystal junctions are accepted as the simplest way to obtain single grain-boundary junctions, the limitations in the circuit layout and the high cost are still existing problems for such junctions. However, the step-edge junction with comparable characteristics is a good choice for its flexibility in the circuit layout and lower cost. Therefore, in attempting to obtain a high quality step-edge junction, various techniques have been explored to make a step edge on the substrate (Hermann 1991, Yi 1996, Francke 1998). The simplest among these methods is the directional etching on the substrate with the aid of photoresist mask.

In this work, the fabrication of a step edge on a SrTiO$_3$ substrate with various incident angles of ion beam was investigated. The edge profile of the photoresist mask was optimized with the proper exposure time. The milling rates on the photoresist and the SrTiO$_3$ substrate were measured at various incident angles in order to correlate the milling rates with the step angle. Typical results of the fabricated step-edge junction are also reported.

2. EXPERIMENTAL DETAILS

The SrTiO$_3$ step edge was patterned with a standard photolithography process using AZ1518 photoresist and an Ar$^+$ ion milling. The photoresist was coated on the substrate with a spinner and then baked at 70 °C for 15 minutes before the exposure. The optimal exposure time was found to be about 30 seconds with the development time of 30 seconds at an ultraviolet intensity of 14 mW/cm^2. To fabricate the step-edge Josephson junctions, high quality YBa$_2$Cu$_3$O$_y$ films (Wang 1996) with thickness of 1500 - 3000 Å were deposited onto the step-edge substrate with a step height of 1500 - 2500 Å. The deposited films were patterned with a standard photolithography process followed by an Ar$^+$ ion milling. The energy of the Ar$^+$ beam was 500 eV while the current density was 1 mA/cm^2. When the films were milled to about 500 Å thick, they were then wet etched with the dilute hydrochloric acid to complete the patterning. In this way, the substrates were not damaged in the ion-milling process. Therefore, the substrates were reusable when the films were removed. When the

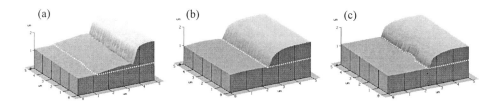

Fig 1. AFM image of the AZ1518 photoresist with different exposure time, (a) insufficient exposure of 20 seconds, (b) proper exposure of 30 seconds, (c) over-exposed for 50 seconds. The scanning area is 5×5 μm^2. The dashed line indicates the interface between the photoresist and the SrTiO$_3$ substrate. The upper part of the dashed plane is the photoresist.

patterns were formed, the contact pads were cleaned with an Ar$^+$ ion milling and then immediately followed by evaporating Au film to reduce the contact resistance in the bonding pads. The electrical properties of the junctions were measured with a four-probe method. When the measurement was completed, the films were removed from the substrate with dilute hydrochloric acid. The recycled step-edge substrates were examined with an atomic force microscope (AFM) to check whether the step edges were damaged or not. It was found that the step-edge profile was still steep and straight after removing the deposited film. The substrates were then reused to fabricate the YBCO junction with another t/h value.

3. RESULTS AND DISCUSSION

First, in order to find the optimum exposure time, the detailed profile of the patterned photoresist edge was examined with AFM. The AFM images of the photoresist edge at different exposure time are shown in Fig. 1. For the sample with an insufficient exposure time, 20 seconds, there is a small tail at the bottom of the photoresist edge as shown in Fig 1(a). On the other hand, when the photoresist was exposed over 50 seconds, the thickness of the photoresist becomes thinner and the edge is wavy as shown in Fig. 1(c). However, with an optimum exposure time of 30 seconds, the photoresist edge was sharp and straight as shown in Fig. 1 (b). With this photoresist edge, a sharp and straight step edge can be formed on the substrate. After the proper exposure and development, the photoresist was baked at 70 °C for 30 minutes to reduce the milling rate of the photoresist. The thickness of the photoresist was about 1.3 μm.

The set-up for the Ar$^+$ ion milling is shown in Fig. 2. The milling angle, θ, is the angle of the incident ion beam makes with respect to the normal of the substrate. The sample holder was water-cooled at about 20°C to prevent heat damage on the photoresist during the milling process. The θ dependence of the step angles of the SrTiO$_3$ step edges is shown in Fig 3(a). We find that with incident angles $70° \geq \theta \geq 60°$, the step angles will be sharper than that with the incident angles $\theta < 60°$. In order to understand this behavior, the milling rates of the SrTiO$_3$ substrate and the photoresist at various incident angles θ were measured as shown in Fig. 3(b).

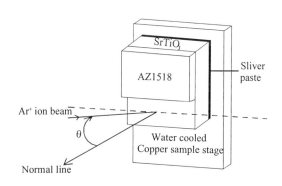

Fig 2. Set-up of the Ar$^+$ ion milling.

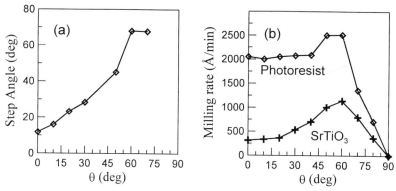

Fig 3. (a) The step angle as a function of the milling angle θ. The fabricated step angle was optimized at 70° ≥ θ ≥ 60°, (b) the milling rate as a function of the milling angle θ for both the photoresist and the substrate.

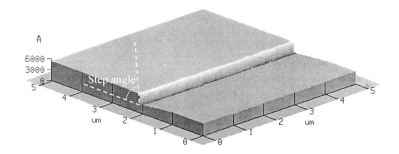

Fig 4. Step-edge SrTiO₃ substrate with sharp step angle.

It was found that the fastest milling rate of the SrTiO₃ was 1200 Å/min at θ = 60°. The incident angle at 70° ≥ θ ≥ 60° creates a sharp step-edge angle as shown in Fig 3(a). When θ ≥ 70°, the milling rate of SrTiO₃ substrates becomes slower. The step-edge substrates used in this work were fabricated with an incident angle θ = 60°. The step-edge profile is shown in Fig. 4. Smooth surface and sharp step angle are demonstrated in the profile.

Generally, the V-I curves of the fabricated step-edge junctions exhibited resistively-shunted-junction (RSJ) behavior for samples with t/h = 0.5 - 1.4. However, the operating temperature of the junctions was found to be lower than 77 K for t/h < 1 while the junction showed a flux-flow-like behavior for t/h > 1.4. Hence, the optimum values of t/h = 1 - 1.4 were chosen. The voltage-current relations at various temperatures for the junction with t/h = 1.2 and junction width

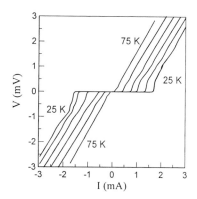

Fig 5. V-I curves of the junction with t/h = 1.2 at various temperatures. All these curves are RSJ-like with $R_n \sim 2\ \Omega$.

~ 2 μm are shown in Fig 5. It was found that the junction was RSJ-like with a temperature independent shunted resistance $R_n \sim 2\,\Omega$. The critical current was found to increase linearly with decreasing temperature. This behavior corresponds to the behavior of superconductor-insulator-superconductor (SIS) junctions (Horng 1999). The I_cR_n product changes from 400 μV at 75 K to 1.4 mV at 4.2 K.

Although the V-I curves are RSJ-like, excess supercurrents and kinks were observed in the V-I characteristics. Excess supercurrents are common features in grain-boundary junctions. They are due to the multiple superconducting filaments in parallel with the Josephson tunneling channels which comprising the junctions (Moechly 1993). In addition, the kinks in V-I curves are due to serial grain boundaries with slightly different critical currents in the step-edge junction. The serial grain boundaries are expected to affect the noise characteristics of the step-edge junctions (Yang 1999).

4. CONCLUSION

The optimized exposure time of the photoresist and the incident direction of the ion beam were manipulated to fabricate step-edge substrates for high-T_c Josephson junctions. Sharp and smooth step edges were reproducibly fabricated with sharp and straight photoresist. Using the step-edge substrates, typical behaviors of the step-edge Josephson junctions are demonstrated.

5. ACKNOWLEDGEMENT

This work is supported by the National Science Council of R.O.C. under grant Nos. NSC88-2112-M003-008, NSC88-2112-M003-009, and NSC88-2113-M002-019.

REFERENCES

Char K, Colclough M S, Garrison S M, Newman N and Zaharchuk G, 1991 Appl. Phys. Lett. **59** 733
Francke C, Offner M, Kramer A, Mex L and Muller J, 1998 Supercond. Sci. Technol. **11**, 1311
Gross R, Chaudhari P, Kawasaki M, Ketchen M B   and Gupta A., 1991 Appl. Phys. Lett, **58**, 543
Herrmann K, Zhang Y, Mück M, Schubert J, Zander W and Braginski A I, 1991 Supercond. Sci. Technol. **4** 583
Horng H E, Yang, S Y, Jeng J T, Wu J M, Chen J H, and Yang H C, 1999 IEEE Trans. Appl. Supercond., **9**, 3986
Koch R H, Clarke J, Goubau, W M, Martinis J M, Pegrum C M, VanHarlingen D J, 1983 J. Low Temp. Phys., **51**, 207
Moeckly B H, Lathrop D K, and Buhrman R A, 1993 Phys. Rev. B, **47**, 400
Wang L M, Yu H W, Yang H C, and Horng H E, 1996 Physica **C 256**, 57
Yang S Y, Chen C H, Horng H E, Lee W L and Yang H C, 1999 IEEE Trans. Appl. Supercond. **9**, 3121
Yi H R, Gustafsson M, Winkler D, Olsson E, and Claeson T, 1996 J. Appl. Phys., **79**, 9213

Inst. Phys. Conf. Ser. No 167
Paper presented at Applied Superconductivity, Spain, 14-17 September 1999
© 2000 IOP Publishing Ltd

On-chip spread of the characteristic parameters in $YBa_2Cu_3O_7$– Josephson junctions with $PrBa_2Cu_3O_7$–barrier on $LaAlO_3$

H Burkhardt, A Rauther, A Kaestner and M Schilling

University of Hamburg, Institute for Applied Physics and Microstructure Advanced Research Centre, Jungiusstrasse 11, 20355 Hamburg, Germany

ABSTRACT: Ramp-edge Josephson junctions from high-temperature superconductors are needed for several applications like digital circuits, magnetometers, and voltage standards because of their circuit design flexibility. In this work ramp-edge junctions with a sandwich structure of $YBa_2Cu_3O_7$ / $PrBa_2Cu_3O_7$ / $YBa_2Cu_3O_7$ are investigated. We investigate the temperature-dependent conductivity and the RSJ-behaviour (*resistively shunted junction* model) in the I(V)-curves. We pay special attention to the on-chip spread of the characteristic parameters critical current I_c, normal state resistance R_n and the excess current I_{ex}.

1. INTRODUCTION

So far, Josephson junctions from high-temperature superconductors (HTS) with artificial barriers lack of a reliable small spread of the characteristic properties of the devices like critical current density $j_c = I_c/A$ and normal state resistivity $\rho_n = R_n A$ with junction area A. To reduce rf-losses the preparation process was transferred from $SrTiO_3$– to $LaAlO_3$–substrates. For a small chip-to-chip variance the process has to be controlled very precisely. The achievement of homogenous films and barriers with respect to thickness, epitaxy and crystal quality is needed on every chip to obtain a homogeneous current flow across the junction which leads to a small spread in the electronic properties. Therefore we use statistical methods for the design of experiments (DOE) for all subprocesses in the multilayer preparation process. This is described in detail by Subke et al. (1997,1999a,b) and Heinsohn et al. (1998), who investigate the influence of the process parameter on the chip-to-chip spread. In addition to a previous paper by Burkhardt et al. (1999), we examine in this work the on-chip spread of the parameters normal state resistance and excess current by investigating the temperature dependent IV-curves and their derivatives dV/dI.

270

2.   FABRICATION, LAYOUT AND MEASUREMENT SETUP

The pulsed laser deposition process used on (100) LaAlO$_3$–substrates is described in detail by Burkhardt et al. (1999). The barrier thickness is chosen as d=30nm of PrBa$_2$Cu$_3$O$_7$ and the width of the junctions as W=4.5µm. For the etching we use a parallel-plate reactor. The angle of the ramp amounts α=20±1°. Our layout consists of two parts of a serial array with 15 Josephson junctions each where every junction can be measured individually. The geometrical extension of one part of the array amounts to 350µm with a distance between both of 4mm. Fig. 1a,b shows a circuit diagram and phase-contrast micrograph of the ramp-edge Josephson junction serial array with 30 Josephson junctions which can be characterised individually. The IV-characteristics are measured by a four-point setup with lock-in technique.

3. RESULTS

3.1.   Critical current, normal state resistance and excess current

By evaluating the peaks in the derivatives d^{2}V/dI^2 of Josephson junctions we determine the I$_c$(T) dependence of each of 25 Josephson junctions on one chip. Between I$_{bias}$=3I$_c$ and I$_{bias}$=10mA we determine the slope of the I(V)-curves as normal state resistance and by extrapolating this straight line to V=0 the excess current I$_{ex}$. The so determined excess-current is in this work only used to get a measure of the RSJ-behaviour of the Josephson junctions. The results of the on-chip average values with error bars are shown in Fig. 1c. In comparison to I$_c$ the R$_n$ and the I$_{ex}$/I$_c$ have negligible spread. The devices have critical temperatures with average T$_c$=86.8±0.2K, our typical value on LaAlO$_3$, indicating homogeneous superconducting layers. The normal state resistance as presented in Fig. 2a is approximately 1Ω and nearly temperature-independent with a slight increase to lower temperatures. This result agrees with Boguslavskij et al. (1992) and Strikovskiy et al. (1996) who prepared *ex situ* respectively *in situ* ramp-edge Josephson junctions with

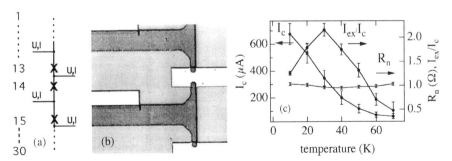

Fig. 1: a) Circuit diagram and b) phase-contrast micrograph of the ramp-edge Josephson junction serial array. The shown part is 85 x 120 µm^2. c) Average values of critical current (left axis), normal state resistance and excess current (right axis) with error bars. Lines are guides-to-the-eye.

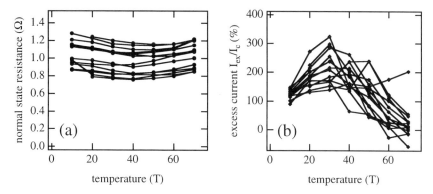

Fig 2: a) Normal state resistance and b) excess current for 14 Josephson junctions on one chip. Lines are guides-to-the-eye.

PrBa$_2$Cu$_3$O$_7$ barrier. In contrast, Verhoeven et al. (1996) found a rising conductivity with temperature G∝T$^{4/3}$ indicating channels of two localised states dominating in the barrier. They use PrBa$_2$Cu$_3$O$_7$ as insulating layer in their ramp-edge Josephson junctions instead of SrTiO$_3$ as in our work (s. Burkhardt et al. 1999). We assume that in our Josephson junctions the mean hopping distance is larger than the barrier thickness d=30nm, since for a thicker barrier d=100nm we also observe variable range hopping in our Josephson junctions as shown by Predel et al. (1997). The determined excess current, as depicted in Fig. 2b, is high and increases even in the middle part of the temperature range to several times of the critical current. This results from the observation that the IV-curves at high temperatures near T$_c$ and at lower temperatures near 4.2K are RSJ-like, but around 50K the IV-curves have flux-flow shape. Fig. 3 gives an impression of the temperature-dependence of the I$_c$R$_n$-product respectively of the spread of the found j$_c$(ρ_n) values at 10K. The solid line in Fig. 3b is a fit with j$_c$∝$\rho_n^{-1.5}$, close to the observation of Boguslavskij et al. (1992) who found j$_c$∝$\rho_n^{-1.3}$.

3.2.   Discussion

As shown earlier by Burkhardt et al. (1999) the spread in the critical current density is not due to a spread in the barrier thickness as expected by Horstmann et al. (1998) but rather originates from the conduction mechanism respectively from the spread in the temperature exponent x in I$_c$(T,d)=I$_c$(T=0,d=0)·exp(-d/ξ_j)·(1-T/T$_c$)^x. We found I$_{c,0,0}$=249.9±9.3mA, x=2.83±0.17, and d=31.7±0.3nm. Our interpretation is that the critical current finds channels in the barrier for Cooper-pair tunneling which can be different in every junction because of inhomogeneities in the heteroepitaxial overgrowth or in the density of localised states in the barrier material. In contrast, the conductivity from the normal conducting quasi-particles is nearly temperature independent as is the spread of the normal state resistance.

272

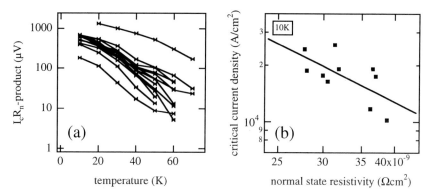

Fig. 3: a) Temperature-dependent on-chip I_cR_n-product for 14 Josephson junctions. b) The statistical spread of the critical current density over normal state resistivity at 10K.

4.   CONCLUSIONS

At a temperature of T=10K the on-chip spread of the parameter I_c amounts to $\sigma \approx 56\%$, of R_n about $\sigma=13\%$, and of $I_{ex}=13\%$. A slight different effective barrier thickness causes a great difference in I_c and R_n because of the exponential dependence $I_c(d)$ and $R_n(d)$. We found no evidence for a systematic on-chip spread originating from inhomogeneity of the barrier film thickness across the chip.

ACKNOWLEDGEMENT

We gratefully acknowledge support by the Deutsche Forschungsgemeinschaft in the Sonderforschungsbereich SFB 508 "Quantum materials - lateral and hybrid structures".

REFERENCES

Boguslavskij Yu M, Gao J, Rijnders A J, Terpstra D, Gerritsma G J and Rogalla H 1992 Physica C **194**, 268

Burkhardt H, Rauther A and Schilling M 1999 Physica C, accepted

Heinsohn J K, Reimer D, Richter A, Subke K O and Schilling M 1998 Physica C **299**, 99

Horstmann C, Leinenbach P, Engelhardt A, Gerber R, Jia J L, Dittmann R, Memmert U, Hartmann U and Braginski A I 1998 Physica C **302**, 176

Predel H, Burkhardt H and Schilling M 1997 Inst. Phys. Conf. Ser. No **158**, 471

Strikovskiy M D and Engelhardt A 1996 Appl. Phys. Lett. **69**, 2918

Subke K O, Krey S, Burkhardt H, Reimer D and Schilling M 1997 ISEC Proceedings, edited by Koch H and Knappe S **2**, 88

Subke K O, Krey S, Burkhardt H, Bartold A, Schilling M 1999a IEEE Trans. on Appl. Supercond. **9**, 3125

Subke K O, 1999b ph d thesis, Hamburg

Verhoeven M A, Gerritsma G J, Rogalla H and Golubov A 1996 Appl. Phys. Lett. **69**, 848

Inst. Phys. Conf. Ser. No 167
Paper presented at Applied Superconductivity, Spain, 14-17 September 1999
© *2000 IOP Publishing Ltd*

Flux flow properties of $Bi_2Sr_2Ca_1Cu_2O_{8+y}$ intrinsic Josephson junctions fabricated by silicon ion implantation

K. Nakajima[1,3], S. Sudo[1], T. Tachiki[1] and T. Yamashita [2,3]

[1]RIEC, Tohoku University, Sendai 980-8577 Japan
[2]NICHe, Tohoku University, Sendai 980-8579 Japan
[3]CREST, Japan Science and Technology Corporation, Japan

ABSTRACT: Flux flow properties have been studied for the $Bi_2Sr_2Ca_1Cu_2O_{8+y}$ intrinsic Josephson junctions fabricated by a silicon ion implantation. The intrinsic Josephson junctions indicate typical current-voltage characteristics identical of conventional mesa type junctions. Flux flow voltages were measured under the magnetic field up to 2T for the junction involving ~90 stacks of elementary Josephson junctions. The maximum flux flow velocity of 1.15×10^6 m/s was obtained at B=1T. The maximum velocity corresponds to a electromagnetic mode with n lower than 3.

1. INTRODUCTION

Intrinsic Josephson effects emerged in high-Tc cuprates like $Bi_2Sr_2Ca_1Cu_2O_{8+y}$ (BSCCO) hold potential for exotic applications. Among them, high-frequency oscillation expected from the collective motion of Josephson vortices attracts our attentions. From the fundamental point of view the intrinsic Josephson effects are argued on the model of stacked a number of Josephson junctions consisting of superconducting CuO_2 layers separated by non-superconducting layers like BiO layers in BSCCO. The stacked junctions model gives rise to a particular electromagnetic dynamics that includes a variety of electromagnetic modes with individual characteristic velocities propagating along the superconducting CuO_2 layers. It is supposed that magnetic field applied parallel to CuO_2 layers stabilize the collective motion of Josephson vortices with high characteristic speeds. This possibility is of interest in connection with the high frequency generation. In this paper, we study on the collective motion of Josephson vortices from the flux flow properties under high magnetic fields up to B=2T.

2. EXPERIMENTAL

Intrinsic Josephson junctions (IJJ) were fabricated by inhibit-ion-implantation of silicon (Si) into BSCCO single crystals grown by the traveling solvent floating method. Prior to the Si ion implantation, the crystals were cleaved in ambient atmosphere then immediately introduced into a vacuum chamber to deposit silver (Ag) and gold (Au) layers for 20nm each by dc sputtering for the surface passivation. 200nm thick Au layers were further deposited and patterned for 10μm×200μm by lift-off. The thick Au layers play a role of stopping layers for Si ions as well as top electrodes. Si ions with the acceleration energy of 80keV were implanted into BSCCO

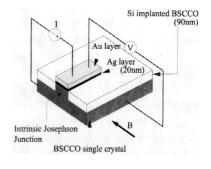

Fig. 1 Schematic view of flux flow measurements.

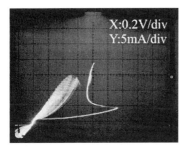

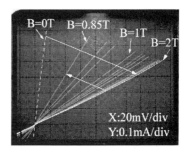

Fig. 2 Typical *I-V* characteristic of a Si-implanted BSCCO intrinsic junction measured at 4.2K

Fig. 3 Flux flow branches for various magnetic filed measured for a BSCCO intrinsic junction at 40.0K.

penetrating the thin passivation layers and inhibited the superconductivity for a certain depth corresponding to the ion penetration range. The dose of Si ions was 6.9×10^{15} ions/cm^2. The current-voltage (*I-V*) properties of the junctions were measured by a three-terminal. A superconducting magnet applied magnetic fields parallel to the CuO$_2$ layers of BSCCO up to 2T. Figure 1 illustrates a schematic view of the measurement under a magnetic field.

3. RESULTS AND DISCUSSION

Figure 2 shows a typical *I-V* curves measured for the IJJ at 4.2K without external magnetic field. The number of branches, which represents the number of junctions involved, counts 60 approximately. Therefore, the superconductivity of BSCCO is inhibited sufficiently by Si ions as deep as 90nm. The depth profile of Si concentrations obtained by a simulation using TRIM codes. Si concentrations within the inhibited layer with the thickness of 90nm ranges from 0.5 to 0.1 atoms per BSCCO unit cell. The critical current I_C for the superconducting branch is about 0.5mA. Magnetic field applied parallel to the CuO$_2$ layers generates the flux flow voltage on the superconducting branch superposing to the apparent voltage due to the contact resistance. Figure 3 shows in the superconducting branches measured at 40.0K under the external magnetic fields with various strengths. As the external field is set in parallel to CuO$_2$ layers, the vortex motion driven by the Lorentz force due to c-axis currents generates the flux flow voltage. Figure 4 shows the magnetic field dependence of the flux flow voltage U_{FF} measured at the bias current of 0.35mA. For the first approximation, the flux flow voltage U_{FF} is expressed as $U_{FF} = Bhv_{FF}$, where B is the magnetic flux density, h the height of IJJ (90nm) and v_{FF} the vortex speed.. Figure 5 shows the magnetic field dependence of the vortex speed v_{FF}. Velocity-constant steps found in Fig.5 imply that different modes of collective vortex motion are excited. The vortex speed v_{FF} increases with increasing B and approaches the maximum value of about 1.15×10^6 m/sec at around B=1T then gradually decreases with further increase of B.

The electromagnetic dynamics in IJJ consisting of N Josephson junctions described by coupled sine-Gordon equations. The equations provide N characteristic velocities of electromagnetic modes, which correspond to N individual electromagnetic modes in the IJJ. The characteristic velocity is expressed as

$$C_n = C_{SW}\left(1 + 2S\cos[n\pi/(N+1)]\right)^{-\frac{1}{2}}, \quad n = 1, 2, \cdots, N, \qquad (1)$$

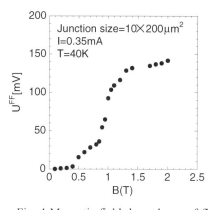

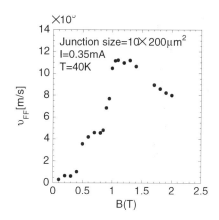

Fig. 4 Magnetic field dependence of flux flow voltages measured at 40.0K.

Fig. 5 Magnetic field dependence of vortex speed measured at 40.0K.

where S is the coupling parameter and C_{SW} the Swihart velocity supposed for a single junction consisting IJJ. The coupling parameter S is expressed by the effective coupling length $s = -\lambda_L/\sinh(t/\lambda_L)$ and the effective magnetic thickness $d' = d + 2\lambda_L \coth(t/\lambda_L)$ as $S = s/d'$, where t is the electrode thickness, λ_L the London penetration depth. The Swihart velocity C_{SW} is given by $C_{SW} = \left(\dfrac{d}{\mu_0 \varepsilon_r \varepsilon_0 d'} \right)^{\frac{1}{2}}$, where Φ_0 is the flux quantum, ε_r the dielectric constant, d the barrier thickness. For BSCCO IJJ, by using $t=3$Å, $d=12$ Å, $\varepsilon_r=20$ and $\lambda_{ab}=2000$ Å instead of λ_L, we have $C_{SW} \cong 1.4\times10^5$ m/s. Therefore, the highest speed corresponding to the mode $n=1$ in the equation (1) is obtained to be $C_1 = 3.9\times10^6$ m/s. Although the flux flow branches in Fig.3 are linear so that no velocity matching steps appear, the maximum velocity of 1.15×10^6 m/s is faster than $C_4 \cong 9.8\times10^5$ m/s. Therefore, the maximum velocity at $B=1$T corresponds to higher speed modes with the order n lower than 3. So far, such a lower order mode has not been achieved except for the case of mesas involving a small number of elementary junctions. The highest speed mode $n=1$ corresponds to the collective motion of Josephson vortices in alignment with a rectangle lattice. It is noticed that $B=1$T is a characteristic field where the Josephson vortices closely packed in layer by layer with the distance of $t+d$ and lie in the line along CuO_2 layers with the distance of $\lambda_J=\gamma(t+d)$, where γ (~1000 for BSCCO) is the anisotropy constant. In the present case, it is supposed that a collective motion of close packed Josephson vortices is stabilized with the rectangle lattice at the characteristic field.

4. CONCLUSIONS

$Bi_2Sr_2Ca_1Cu_2O_{8+y}$ intrinsic Josephson junctions with the size of 10μm×200μm involving ~60 elementary junctions were fabricated by a silicon ion implantation. Current-voltage characteristics measured under magnetic fields up to B=2T have revealed the flux flow branches, as the fields were set parallel to CuO_2 layers. The maximum speed was estimated to be 1.15×10^6 m/s from the flux flow voltage at the bias current of 0.35mA under the magnetic field B=1T. The maximum speed corresponds to the characteristic velocity of an electromagnetic mode with the order n lower than 3.

REFERENCES

Bulaevskii L N, Dominguez D, Maley M P and Bishp A R 1996 Phys. Rev. B 53, 14601

Hechtfischer G, Kleiner R, Schenga K, Walkenhorst W and Müller P 1997 Phys. Rev. B 55, 14638

Irie A, Hirai Y and Oya G 1998 Appl. Phys. Lett. 72, 2159

Kleiner R and Müller P 1994 Phys. Rev. B 49, 1327

Ustinov A V, Kohlstedt H, Cirillo M, Pedersen N F, Hallmans G and Heiden C 1993 Phys. Rev. B 48, 10614

Kleiner R 1994 Phys. Rev. B 50, 6919

Sakai S, Ustinov A V, Kohlstedt H, Petraglia P and Pedersen N F 1994 Phys. Rev. B 50, 12905

Lee J U, Guptasama P, Hornbaker D, El-Kortas A, Hinks D and Gray K E 1997 Appl. Phys. Lett. 71, 1412

Nakajima K, Yamada N, Chen J, Yamashita T, Watauchi S, Tanaka I and Kojima H 1999 IEEE Trans. Appl. Supercond. 9, 4515

Inst. Phys. Conf. Ser. No 167
Paper presented at Applied Superconductivity, Spain, 14-17 September 1999
© 2000 IOP Publishing Ltd

SNS ramp-type Josephson junctions for highly integrated superconducting circuit applications

R Pöpel, D Hagedorn, F-Im Buchholz and J Niemeyer

Physikalisch-Technische Bundesanstalt, Bundesallee 100, D-38116 Braunschweig, Germany

ABSTRACT: At PTB, a fabrication process has been developed to produce SNS Nb/PdAu/Nb Josephson junctions in sub-micron-sized ramp-type configuration. Test circuits of single junctions and junction arrays have been fabricated and investigated. Experiments have been carried out on the normal-state resistance, the temperature dependence of critical currents, and the influence of externally applied magnetic fields. The measurement results are compared with theoretical descriptions.

1.   INTRODUCTION

In the field of low-temperature superconducting integrated circuitry there is a strong interest in using intrinsically-shunted Josephson junctions (JJs) as active circuit elements which offer the possibility of reducing the circuit area down to the physical size of the junction contact area in the sub-micron range.

In SNS technology, increasing activities were reported in the last few years. Large series arrays of Nb/PdAu/Nb JJs were fabricated by Benz (1995) at NIST with characteristics ideally suitable for application in programmable voltage standards and D/A converters. Burroughs et al (1998) have reported on a programmable 1-volt DC voltage standard containing 32 768 SNS junctions, with the smallest cell in the device consisting of 128 junctions. The properties of PdAu barriers of SNS junctions suitable for programmable voltage standards were investigated by Sachse et al (1997) at PTB. At IPHT, single Nb/Ti/Nb junctions have been fabricated by Fritsch et al (1998, 1999). At a thickness of the Ti layer of 20 nm and a junction area of 0.49 μm^2 they obtain $I_C \cdot R_N$ products of 87 μV. Lacquaniti et al from IEN (1999a) report on single Nb/TaO$_x$/Nb junctions with barrier thicknesses of 10 nm. For junction areas from 100 to 1500 μm^2 they obtain $I_C \cdot R_N$ products of about 100 μV. Nb/Al/Nb junctions with an Al barrier thickness of 100 nm and an area of 25 μm^2 fabricated by Lacquaniti et al (1999b) show a non-hysteretical, but non-typical, SNS current voltage characteristic, resulting in an $I_C \cdot R_N$ product of 430 μV. At the University of Cambridge, Moseley et al (1999) fabricated small junction arrays by patterning 50-nm wide spaces by means of a focussed ion beam in a Cu/Nb bilayer. With junction areas of 0.05 μm x 0.5 μm, they measured $I_C \cdot R_N$ products of between 69 and 98 μV.

2.   FABRICATION OF RAMP-TYPE JOSEPHSON JUNCTIONS

At PTB, for the production of sub-micron-sized JJs, a fabrication process has been developed in SNS Nb/PdAu/Nb technology, which is based on a combination of standard non-contact photolithography using an optical wafer stepper and etching techniques (Pöpel et al (1999)). The technology parameters have been set so that JJs can be verified in a specially-designed ramp-type junction (RTJ) configuration and with contact areas of A = 0.25 μm x 1.3 μm, see Fig. 1(a). 3-inch Si wafers have been used as substrates with an Al$_2$O$_3$ layer of 30-nm thickness sputtered on them, which serves as an etch stop during the fabrication process. Within the whole RTJ assembly, the single

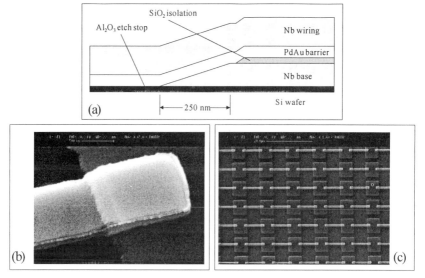

Fig.1: SNS ramp-type Josephson junctions (contact areas: $A = 0.25$ μm x 1.3 μm),
(a): schematic diagram of the cross section (to scale),
(b) and (c): microphotographs (SEM) at different magnifications.

layers and thicknesses are as follows: Nb base electrode (80 nm), SiO₂ isolation (50 nm), PdAu barrier (40 nm), and Nb wiring electrode (100 nm). Test circuits of single JJs and series arrays containing up to 10 000 junctions have been fabricated. Figs. 1(b), (c) show microphotographs (SEM) of RTJs located within a 10 000 JJs series array.

3. MEASUREMENTS AND RESULTS

3.1 Junction Parameters

Fig. 2(a) shows the current-voltage characteristic of a single JJ. From the values of the critical current $I_C = 760$ μA and the normal-state resistance $R_N = 40$ mΩ obtained by measurement, a value of the $I_C \cdot R_N$ product of 30.4 μV is determined. In relation to the size of the bottom Nb electrode of $A = 0.325$ μm², a critical current density of $j_C = 234$ kA/cm² is deduced. However, in view of the fact that the area of the interface of the PdAu layer is larger for the top Nb electrode than for the bottom Nb base electrode, an inhomogeneously distributed current flow must be assumed across the normal-conducting layer, with a higher current density at the base electrode and a lower current density at the wiring electrode. When the measurement data were compared with those

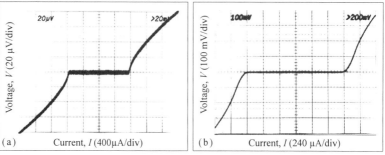

Fig.2: Current-voltage characteristics of ramp-type Josephson junctions,
(a): single junction, (b): series array containing $N = 10\,000$ junctions.

obtained from junctions of conventional window-type design investigated by Sachse et al (1997), with a well-defined size of the junction area and with equal PdAu barrier thicknesses of 40 nm, the presence of an effective junction area of twice the size of the area of the bottom ramp has recently been found (Pöpel et al (1999)). Thus, the effective values of the contact area and the critical current density are determined to be $A_{eff} = 0.65\ \mu m^2$ and $j_{Ceff} = 117\ kA/cm^2$, respectively. Fig. 2(b) shows the I-V curve of a series array containing $N = 10\,000$ RTJs. Here, the junction parameters are as follows: $I_C = 720\ \mu A$, $R_N = 30.8\ m\Omega$, $I_C \cdot R_N = 22.2\ \mu V$, $j_C = 220\ kA/cm^2$, and $j_{Ceff} = 110\ kA/cm^2$. Obtained for several junction arrays, two containing $1\,000$ JJs and three containing $10\,000$ JJs, the mean values of $I_C \cdot R_N$ and j_C are determined to be $I_C \cdot R_N = 21\ \mu V$ and $j_C = 200\ kA/cm^2$. The RTJs exhibit excellent hysteresis-free current-voltage characteristics with a small parameter spread which was investigated at increased voltage magnifications.

3.2 Dependence of Critical Currents on Temperature

The dependence of the critical current I_C on the temperature T, the normal-state resistance R_N, the normal conductor thickness d and the coherence length ξ_n is given by the expression:

$$I_C = \frac{B_1(T)}{eR_N} \frac{d}{\xi_n} \exp\left(-\frac{d}{\xi_n}\right),\tag{1}$$

where e is the electron charge. In different theoretical descriptions $B_1(T)$ takes the following forms, according to:

Zaikin and Ginzburg (1986), applied by Sachse et al (1997): $\quad B_1(T) = 64\pi k_B T / (3 + 2\sqrt{2})$, (2)

Likharev (1979): $\qquad\qquad\qquad\qquad\qquad\qquad\qquad B_1(T) = 4\Delta^2 / (\pi k_B T)$, (3)

and Kupriyanov et al (1999) / Zaikin and Zharkov (1981): $\quad B_1(T) = 64\pi k_B T\, C_0^2(0, T)$. (4)

Here, $C_0^2(0, T) = \Delta^2 / [\pi k_B T + \Delta^* + \sqrt{2\Delta^*(\pi k_B T + \Delta^*)}]^2$, $\Delta^* = \sqrt{(\pi k_B T)^2 + \Delta^2}$, Δ being half the energy gap. As evaluated by Kupriyanov et al (1999), the decay length at the critical temperature T_C, $\xi_{nd} = \xi_n \cdot (T/T_C)^{1/2}$, can be determined from the experimental data by taking two measurements of the critical current I_C at different temperatures into account.

Fig. 3 shows the measured values of the critical current I_C of a single RTJ as a function of the temperature T. For Eq.(1) and Eq.(2), this leads to $\xi_n = 5.35$ nm; for Eq.(1) and Eq.(3) to $\xi_n = 17.3$ nm, and for Eq.(1) and Eq.(4) to $\xi_n = 7.4$ nm. The calculated values of ξ_n fit well with the measured curve shown in Fig. 3 when a constant factor B_2 is introduced by which the calculated critical currents must be divided. For each theoretical description, $B_2 = 2.0$ (Zaikin and Ginzburg (1986)), $B_2 = 61.59$ (Likharev (1979)), and $B_2 = 3.57$ (Kupriyanov et al (1999) / Zaikin and Zharkov (1981)). For the latter, the course of the critical current is shown in Fig. 3.

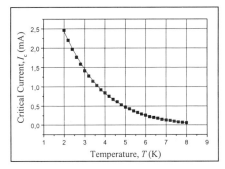

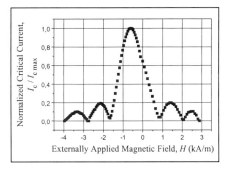

Fig. 3: Dependence of the critical current I_C of a single ramp-type Josephson junction on the temperature T.

Fig. 4: Dependence of the normalized critical current $I_C/I_{C\,max}$ on an externally applied magnetic field H.

280

3.3 Dependence of Critical Currents on Externally Applied Magnetic Fields

Fig. 4 shows the dependence of the critical current I_C of a single RTJ on an externally applied magnetic field H. The ratios of the absolute maximum to the first-order and second-order maxima, as well as distances of the minima, fit the Fraunhofer dependence, which indicates a homogeneous current distribution across the junction. Nevertheless, the absolute maximum is slightly shifted to a negative magnetic field value. This effect may occur due to magnetic fields originating from high critical current densities and a junction configuration which is asymmetric with regard to the direction of the externally applied magnetic field perpendicular to the ramp. Similar shapes were measured by Niemeyer (1979), especially on SNS junctions with thin normal conductor layers of Ag strongly diluted by magnetic impurities of Mn.

4. CONCLUSION AND OUTLOOK

Test circuits containing up to 10 000 SNS Nb/PdAu/Nb ramp-type junctions with contact areas of $A = 0.25\ \mu m \times 1.3\ \mu m$ have been fabricated and investigated. With a thickness of the PdAu normal metal layer of $d = 40$ nm, values of junction parameters of about $j_C = 200$ kA/cm^2 and about $I_C \cdot R_N = 21\ \mu V$ have been achieved for the critical current density and for the product of critical current and normal-state resistance. The junctions exhibit excellent hysteresis-free current-voltage characteristics with small parameter spread. To estimate the coherence length, the temperature dependence of the critical current has been measured and compared with different theoretical descriptions. When external magnetic fields are applied, a homogeneous current distribution across the junction is indicated. The fabrication process fulfils the requirements for an implementation of sub-micron Josephson junctions in highly integrated circuits in the fields of programmable Josephson voltage standards and rapid single flux quantum applications.

ACKNOWLEDGEMENTS

The authors wish to thank F. Müller and H. Schulze for helpful discussions and R. Harke and H. J. Kuntze for technical assistance. The work is partly supported by the European Communities (project No. SMT4-CT98-2239) and by the Deutsche Forschungsgemeinschaft (DFG), Germany (project No. NI 253/3-1).

REFERENCES

Benz S P 1995 Appl. Phys. Lett. **67**, 2714
Burroughs C J, Benz S P, Hamilton C A and Harvey T E 1999 IEEE Trans. Instrum. Meas. **48**, 279
Fritzsch L, Schubert M, Wende G and Meyer H-G 1998 Appl. Phys. Lett. **73**, 1583
Fritzsch L, Elsner H, Schubert M and Meyer H-G 1999 Ext. Abstr. ISEC'99 Claremont 256
Kupriyanov M Yu, Brinkman A, Golubov A A, Siegel M and Rogalla H 1999 to be published: Physica C
Lacquaniti V, Gonzini S, Maggi S, Monticone E, Steni R and Andreone D (1999) to be published: IEEE Trans. Appl. Supercond. (ASC'98)
Lacquaniti V, Gonzini S, Maggi S, Monticone E and Steni R, Int. J. Mod. Phys. (1999)
Likharev K K 1979 Rev. Mod. Phys. **51**,101
Moseley R W, Bennett A J, Booij W E, Tarte E J and Blamire M G 1999 Ex. Abstr. ISEC'99 Claremont 247
Niemeyer J 1979 Thesis, University of Göttingen
Pöpel R, Hagedorn D, Weimann T, Buchholz F-Im and Niemeyer J 1999 to be published: Supercond. Sci. Technol.
Sachse H, Pöpel R, Weimann T, Müller F, Hein G and Niemeyer J 1997 IOP Publ. Ltd: Bristol Inst. Phys. Conf. Ser. **158**, 555
Zaikin A D 1986 Nonequilibrium Superconductivity **174**, ed V L Ginzburg (New York: Nova Science Publ.) pp 57-136
Zaikin A D and Zharkov G F 1981 Fiz. Nizk. Temp. **7**, 375

Inst. Phys. Conf. Ser. No 167
Paper presented at Applied Superconductivity, Spain, 14-17 September 1999
© 2000 IOP Publishing Ltd

Josephson Current along the *c*-Axis of the Mesa-Structured Bi$_2$Sr$_2$CaCu$_2$O$_{2+\delta}$ Single Crystals

K Hirata, S Ooi and T Mochiku

National Research Institute for Metals, 1-2-1 Sengen, Tsukuba 305-0047, JAPAN

ABSTRACT: To study the physical properties of the intrinsic Josephson coupling in Bi$_2$Sr$_2$CaCu$_2$O$_{8+\delta}$ single crystals, we have performed the current-voltage (*I-V*) measurements along the *c*-axis in magnetic fields perpendicular and parallel to the axis. A pronounced effect of magnetic fields to the *I-V* characteristics appears in the multiple-branch behaviour and the Josephson critical current, when the magnitude of the parallel field exceeds the 2D-3D dimensional-crossover field B_{2D}. This is considered as a consequence of missing a long-range order of the correlation of vortex lines along the *c*-axis.

1. INTRODUCTION

It has been recognized as an intrinsic property of high T_c superconductors (HTSCs) that Josephson coupling occurs between the Cu-O layers through the non-superconducting layers, based on the layered crystal structures of HTSCs. Especially, in the strongly anisotropic materials, such as Bi$_2$Sr$_2$CaCu$_2$O$_{8+\delta}$ (Bi-2212). Kleiner et al (1992, 1994) have observed dc Josephson effect and multiple branches, which were considered to arise from the inhomogeneity of the superconducting gap in the sample. Schlenga et al (1996) have found a subgap structure in the multiple branches, caused by the quasiparticle current arising from a subgap structure in the density of states. In magnetic fields perpendicular to the *c*-axis, Cherenkov radiation has been reported by Hechtfischer et al (1997). Microwave response of the intrinsic Josephson junction has been studied by Prusseit et al (1997). These results indicate the intrinsic properties of the Josephson coupling in HTSCs.

The effect of the magnetic field to the intrinsic Josephson junction has been discussed mainly from the points of the Josephson critical current by Luo et al (1995) and Yurgens et al (1999), and the Fraunhofer behavior by Latyshev et al (1997). Most of these measurements have been made to follow the studies on the single Josephson junction. As the Josephson effects in HTSCs have their own properties, the results might have to be considered to reflect the magnetic phase diagram of HTSCs. The magnetic properties of Bi-2212 have been studied in details with the field applied along the *c*-axis, and the magnetic phase diagram has become almost to be established. However, the interlayer coupling between the superconducting layers has not been well understood in the presence of vortices and the current along the *c*-axis. There are few reports to discuss on the interlayer coupling from the results of *I-V* measurements in magnetic fields. The studies on the *I-V* characteristics will be needed also for applications of the intrinsic Josephson junction. So, we have measured *I-V* characteristics in magnetic fields to study the interlayer coupling between the Cu-O layers in presence of vortices and Josephson current, compared with the magnetic phases of Bi-2212.

2. EXPERIMENTAL

Bi-2212 single crystals were grown with a travelling solvent floating zone method by Mochiku et al (1995). Small pieces of single crystals for the mesa-structuring were obtained by cutting and cleaving the ingots into small pieces with dimension of about 4mmx5mm and thickness of 20-50μm. The superconducting transition temperature T_c of the samples is about 87K. A cleaved surface was used for the mesa-structuring. The mesa-structure was made with Ar ion milling on the Au-deposited surface. The detailed fabrication is described by Prusseit et al (1997). Typical size of the mesa structure for the I-V measurements is 10μmx10μm in area and about 50nm in thickness.

Current-biased I-V measurements have been performed with a two-point configuration. The magnetic field was applied parallel and perpendicular to the c-axis with a field cooling process. The top electrode of the mesa-structure was made by depositing gold to the cleaved surface, and was attached to a 25μm Au/Ni wire. The base electrode was taken on the base plane of the sample. The contact resistance was about 0.5Ω at 4.2K, which was not subtracted in the experimental results shown below. The resistivity of the mesa-structured Bi-2212 was 88Ωcm at 290K. The resistivity showed metallic temperature dependence down to 160 K and became semiconductive decreasing the temperature till T_c. In the I-V measurements, it always becomes a great problem that the heating effect may occur in the mesa-structure and the electrodes. In the measurements, the heating effect is negligible, because we will discuss in the small range of current and there is distinct difference in the properties of the multiple branches between parallel and perpendicular magnetic fields with the same current range. The magnetic phase diagram has been obtained from the magnetization measurements with a SQUID (Quantum Design) magnetometer. The sample showed the 2D-3D dimensional-crossover behaviours at 0.05T at lower temperatures of the irreversibility region. The doping level of the sample is considered to be over-doped in taking account of the temperature dependence of the normal resistivity mentioned above, and, of no spin gap features in the I-V characteristics of the normal state.

3.   RESULTS AND DISCUSSIONS

Figure 1 shows the temperature dependence of the Josephson critical current J_c without magnetic field. The Josephson critical current density is estimated as 480 Acm⁻² from the zero-voltage branch (the 0th order branch). Higher order branches show higher critical currents. As the height of the current at zero-voltage depended on the attachment of the thin wire to the surface, degradation of the sample surface might be one of the reasons, pointed out by Irie et al (1998). J_{c0} and J_{ch} have been defined as the values of the current deviating from the linearity of the 0th order branch and of the largest one in the highest order branch, respectively. In the figure, J_{c0} and J_{ch} can be fitted well with the Ambegaokar-Baratoff relation (1963) for the Superconductor/ Insulator/Superconductor (SIS) junctions: $J_C(0,T)=(\pi\Delta(T)/2eR_N)\tanh(\Delta(T)/2k_BT)$, where k_B is the Boltzmann constant. The constant resistivity ρ_n (resistance R_N) of about 200Ωcm in the normal state is assumed for the fitting. The resistivity obtained from the fitting is rather larger than the value from R-T measurements mentioned above. These

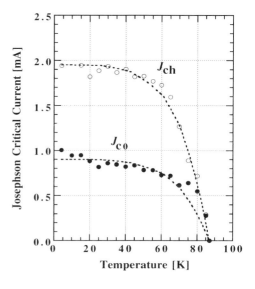

Fig.1 Temperature dependence of the Josephson critical current without magnetic field. Dotted lines are the Ambegaokar-Baratoff relation for the SIS junctions.

facts may indicate that the mechanism differs in conductivity between the superconducting state and the normal state. The value of $\Delta(0)/k_BT_C$ is also obtained as 3.5 from the fitting, which means that Bi-2212 is in the strong coupling regime in superconductivity.

Figures 2 (a) and (b) show the I-V characteristics at 4.2K in magnetic fields applied parallel and perpendicular to the c-axis, respectively. Several cycles of the I-V measurements are plotted at each magnetic field. In the absence of the magnetic field, there are several essential

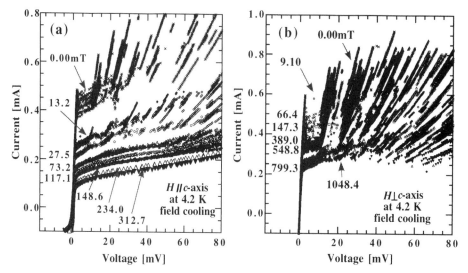

Fig.2 Current-voltage characteristics of mesa-structured Bi-2212 in magnetic fields at 4.2K (a) In parallel fields, multiple branches are clearly seen at lower fields, but significantly diminishes in the lower order of branches with increasing magnetic field. (b) In perpendicular fields, there is no significant change in multiple branching behaviour even at 1048.4mT.

features to be noted. In the higher order branches other than 0th order, each branch does not appear at the same voltage in several cycles of the measurements. This contradicts to the scenario of the earlier explanation for the multiple-branch by Kleiner et al (1992), in which the multiple-branch is caused by the inhomogeneity in the intrinsic Josephson stacks, that is, different values of Josephson critical currents. Instead of that, the recent model taking account of the charging effect proposed by Koyama et al (1996), and the numerical calculations made by Machida et al (1998) based on the model lead to a reasonable explanation for our results. That is, each branch corresponds to the one of the mode of charge density wave.

In the Fig.2 (a), multiple branches and hysteresis are clearly seen at low magnetic fields. With increasing magnetic field, the behaviour of the multiple branching gradually smears out and disappears thoroughly at 73.2mT in the low voltage region, although at high voltage region there still remain the multiple branches. From the magnetization measurements in this field region, the vortices are in an irreversibility region and the value of the magnetization is independent to temperature below 20K. Then, it is considered that the vortices are strongly pinned in a solid phase. There exists also the dimensional-crossover field B_{2D} around 50mT, at which the second peak behaviour was observed. When the field increases crossing the B_{2D}, the vortex state changes from a line-like (three dimensional) structure with the hexagonal vortex lattice (Bragg lattice) to the vortex lattice configuration only in two-dimensional order, observed by the neutron scattering measurements by Cubitt et al (1993). Therefore, it is easily suggested that, before losing the correlation of each vortex line in the c-direction at B_{2D}, the multiple branching can exist for the sake of the strong interlayer coupling. Above B_{2D}, losing the correlation of the lines in the c-direction, the interlayer coupling is reduced and the Josephson critical current decreases. It is not so clear that the multiple branches disappear in the low-voltage region and still remain in the high-voltage region. One of the possible explanations is, that in the high-voltage region, the voltage results from a large phase difference in Josephson current through the Josephson stacks, which is caused by the presence of the Josephson vortices in the 2D vortex configurations, and it allows a Josephson current mode to be formed in the high voltage region.

In contrast to parallel magnetic fields, the multiple branching is observed even in the low-voltage region in Fig.2 (b), and up to 3T in the perpendicular fields, which are not shown here. Lee et al (1995) has not observed so significant multiple-branching behaviours even at 20mT. This might depend on the junction size to enhance the inhomogeneity of the Josephson current flow through the junctions. Furthermore, this appearance of the multiple branches excludes the heating effect in the branching behaviours of the parallel fields mentioned above. This also proves that the multiple branching actually relates to interlayer coupling of the layers in presence of vortices. The magnetic phase diagram for perpendicular fields has not been well established, compared with the parallel case, for the sake of the difficulty in measuring the magnetization with a precise setting of the sample. A little misalignment causes the c-axis component of the magnetic field at high fields. In this perpendicular field and the c-axis current, a characteristic behaviour of the Josephson vortices has been shown by Machida et al (1998). In-phase and out-of-phase configurations of the Josephson vortices are induced with increasing a magnetic field, and the in-phase case enhances the Josephson current. Although we have not observed such an enhancement, these arguments and experiments will be important for realising a Josephson plasma generation excited by the vortex flow method.

4.   SUMMARY

We have measured the I-V characteristics of mesa-structured Bi-2212 single crystals in magnetic fields at 4.2K to study the properties of the intrinsic Josephson junction, reflecting the vortex states. Without a magnetic field, the multiple branching and hysteresis are observed in the I-V curves. The possible origin of the multiple branching is a charging effect in the superconducting layers, which causes the Josephson current mode corresponding to a static charge density wave. The temperature dependence of the Josephson critical current was well explained by the Ambegaokar-Baratoff relation, although it is pointed out that there is a difference in the transport mechanism between the tunneling resistivity through the junction in the superconducting state and in the normal state. Furthermore, a clear feature was found in the multiple branching at the dimensional-crossover field B_{2D} in fields parallel to the c-axis. This is considered as a consequence of the correlation of vortices along the c-direction. No critical field has been found in the I-V characteristics with perpendicular fields.

The authors would like to thank W. Prusseit and M. Rapp for preparing the mesa-structure. We are pleased to acknowledge useful discussions with M. Tachiki on the I-V characteristics.

REFERENCES

Ambegaokar V and Baratoff A 1963 Phys. Rev. Lett. **10**, 486.
Cubitt R, Forgan E M, Yang G, Lee S L, Paul D Mck, Mook H A, Yetraj M, Kes P H, Li T W, Menovsky A A, Tarnawski Z and Mortensen K (1993) Nature (London) **365**, 407
Hechtfischer G, Kleiner R, Ustinov A V and Müller P 1997 Phys. Rev. Lett. **79**, 1365.
Irie A, Hirai Y and Oya G 1998 Appl. Phys. Lett. **72**, 2159.
Kleiner R, Steinmeyer F, Kunkel G and Müller P 1992 Phys. Rev. Lett. **68**, 2394.
Kleiner R and Müller P 1994 Phys. Rev. **B48**, 1327.
Koyama T and Tachiki M 1996 Phys. Rev. **B54**, 16183.
Latyshev Y I, Monceau P and Pavlenko V N 1997 Physica **C293**, 174.
Lee J U, Nordman J E and Hohenwarter G 1995 Appl. Phys. Lett. **67**, 1471.
Luo Sh, Yang G and Gough C E 1995 Phys. Rev. **B51**, 6655.
Machida M, Koyama T and Tachiki M 1998 Physica **C300**, 55.
Mochiku T, Hirata K and Kadowaki K 1997 Physica **C282-287**, 475.
Schlenga K, Hechtfischer G, Kleiner R, Walkenhorst W, Müller P, Jphnson H L, Veith M, Brodkorb W and Steinbeiß E 1996 Phys. Rev. Lett. **76**, 4993.
Prusseit W, Rapp M, Hirata K and Mochiku T 1997 Physica **C293**, 25.
Yurgens A, Winkler D, Claeson T, Yang G, Parker I F G and Gough C E 1999 Phys. Rev. **B59**, 7196.

Inst. Phys. Conf. Ser. No 167
Paper presented at Applied Superconductivity, Spain, 14-17 September 1999
© 2000 IOP Publishing Ltd

Effects induced by microwaves on SINIS junctions

G Carapella, G Costabile, and R Latempa

Unita' INFM and Dipartimento di Fisica, Universita' di Salerno , I-84081 Baronissi, Italy

ABSTRACT: We fabricated and tested double barrier devices consisting of $Nb/AlO_x/Al/AlO_x/Nb$ films, in which the AlO_x barriers are thin ($\approx 2nm$) and the Al film is thinner than the coherence length in Nb (thinner than ≈ 30 nm). The devices exhibit Josephson critical current and their I-V characteristic is highly hysteretic and shows in zero magnetic field evenly spaced resonance's which become very unstable when an external magnetic field is turned on. When the devices are irradiated with microwaves, Shapiro steps are observed. At low rf power, the resonance's are more stable than the zero voltage current. Increasing the rf power, they appear as current branches symmetrical aside a Shapiro step. Increasing further the rf power, they are turned into resistive steps alternated to the Shapiro steps. If the microwave is chosen to have twice the frequency related to the resonance voltage spacing by the Josephson relation, Shapiro steps of half integer order appear in the I-V characteristic.

1. INTRODUCTION

In recent years SINIS junctions have gained considerable attention [1]-[9] (here S, I, and N denote a superconducting, an insulating or a normal metal layer, respectively). The investigations of such devices were focused mainly on non-equilibrium properties, proximity effect, and Andreev reflection.

Much experimental work [2], [5]-[8] was done with $Nb/AlO_x/Al/AlO_x/Nb$ double-barrier devices to explore their potential for application to superconductive electronics. The occurrence of dc supercurrent at $T = 4.2$ K, non-hysteretic behaviour and $I_C R_N \sim 0.5$ mV have been demonstrated. Here we report about new experimental results on $Nb/AlO_x/Al/AlO_x/Nb$ structures. In particular, dc and ac Josephson effect in these structures are clearly demonstrated at $T = 4.2$ K. Moreover, they exhibit current singularities in zero and nonzero external magnetic field, sometimes collateral to rf induced steps, and subharmonic Shapiro steps, indicating a very rich dynamics. However, some effects suggest that the structure cannot be regarded merely as two series biased SIS' junctions.

2. EXPERIMENTS

To fabricate the $Nb/AlO_x/Al/AlO_x/Nb$ structure we used the procedure [10] employed to make stacked $Nb/AlO_x/Nb$ junctions with access to the intermediate electrode. The two devices on which we report here have double overlap geometry [see insets in Fig. 1(a) and Fig. 1(b)], and physical dimensions $L \times W = (600 \times 20)$ μm^2 (device A) and $L \times W = (50 \times 50)$ μm^2 (device B), respectively. For both devices, the outer Nb electrodes were 300 nm thick, while the intermediate Al electrode was ≈ 30 nm.

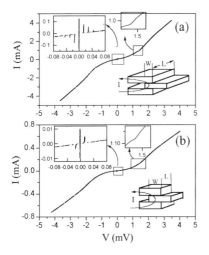

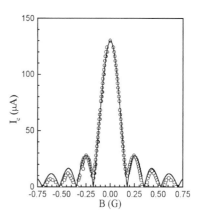

Fig.1 (a) Current voltage characteristic of the SINIS structure. The geometry and biasing condition are shown in the inset. The magnifications show the knee typical of the proximity effect and a series of current singularities around the dc Josephson current. The IVC is recorded at T=4.2 K, and at zero applied magnetic field. (b) Same as in (a), but here the SINIS has a smaller area.

Fig. 2 Critical current of the A device as a function of the magnetic field. The experimental data follow the typical diffraction pattern: $I_C(B) = I_0 |sin(\pi B/B_0)| / (\pi B/B_0)|$ with $I_0 = 126$ µA, and $B_0 = 0.175$ G.

Although our devices provide an electrical access to the intermediate electrode, the configuration of our sample holder did not allow us to perform four contacts measurements of the junctions independently. This is why we reported results obtained series biasing the structure. Figure 1(a) shows the current-voltage characteristic (IVC) of the device A, series biased. Though we used a relatively thick middle Al layer, the IVC shows a typical signature of the proximity effect on the structure, i.e., the knee at voltage $V = 2(\Delta_{Nb} - \Delta_{Al}) / e \approx 1.45$ mV, where Δ_{Nb} is the energy gap of outer Nb electrode and Δ_{Al} is the induced [2], [3], [5], [7], energy gap in the intermediate Al. The enlargement of the voltage region around $V = 0$ shows the dc Josephson current, $I_0 \approx 126$ µA , and other current singularities evenly spaced of $\Delta V_{ZFCS} \approx 15$ µV. Similar IVC is exhibited by the smaller area device B [Fig. 1(b)], with $I_0 \approx 10$ µA and only a zero field current singularity (ZFCS) at $V \approx 13$ µV.

Figure 2 shows the characterisation of the device A in the presence of a magnetic field. The critical current I_0 follows a Fraunhofer-like pattern. This is typical of rectangular, not electrically long, junctions having uniform critical current density J_0. From the pattern we roughly estimate a magnetic thickness [11] $d \approx 200$ nm from $d = \Phi_0/B_0L$, where Φ_0 is the flux quantum, $B_0 \approx 0.175$ G is the first zero in Fig. 2, and L is the junction length. This value of d agrees with the values estimated from pattern in magnetic field of SINIS structures reported in the literature [6]. The Josephson penetration depth can also be estimated as $\lambda_J \approx 1200$ µm from [11] $\lambda_J = \sqrt{\Phi_0 / 2\pi\mu_0 J_0 d}$, where $J_0 \approx 0.8$ A/cm^2 is obtained from the I_0 value in Fig. 1(a).

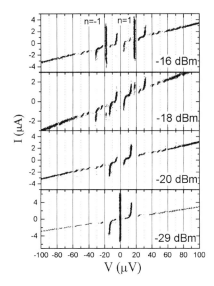 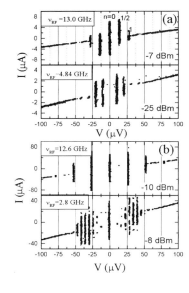

Fig. 3 Evolution of the ZFCS of B device under the effect of a microwave signal at frequency ν_{RF}=8.60 GHz of increasing (from the bottom to the top in the figure) power level P_{RF}. At P_{RF} = -16 dBm also Shapiro steps are induced.

Fig.4 Ordinary and half integer Shapiro steps induced by a microwave signal in the B device [in (a)] and in the A device [in (b)].

In Fig. 3 is shown the behaviour of device B, but similar results are obtained in device A, under the influence of a microwave field at the frequency ν_{RF} = 8.60 GHz. At low microwave level, the ZFCS is only weakly modified. Increasing the amplitude of the rf signal, the dc Josephson current is destroyed, and the ZFCS starts to extend, antisimmetrically, to the opposite side of the resistive branch. A further increase of the rf amplitude induces other branches of the ZFCS's family, and the negative antisymmetric branch grows further. Eventually, one can find (P_{RF} = -16 dBm in Fig. 3) Shapiro steps at voltages $V_n = n\Phi_0\nu_{RF}$, (n = - 1, n =1) coexisting with both the branches of the ZFCS's. We emphasise that this rf induced antysimmetric current branch does not represents a voltage locked state, and that it takes negative current values, a phenomenon that was never observed, neither in junctions nor in junction stacks. Zero crossing, half integer Shapiro steps are also observed in the devices. An example is shown in Fig. 4(a). These steps are enhanced if the rf frequency satisfies the empirical condition $\nu_{RF} = N\Delta_{ZFCS}\Phi_0^{-1}$, where N is an integer. The presence of these half integer steps could be ascribed to a nonuniform current distribution induced at the particular frequency or at the presence of non sinusoidal [1], [9] components in the current-phase relation. However, ordinary Shapiro steps are normally exhibited in the devices, as shown in Fig. 4. Results of Fig. 3 and Fig. 4 are a direct evidence for ac Josephson effect in our SINIS structures.

3. CONCLUSIONS

In summary, the investigated Nb/AlO$_x$/Al/AlO$_x$/Nb structure exhibits, at T = 4.2 K, ordinary dc and ac Josephson effect as well as some new effects. The understanding of these effects requires a further theoretical and experimental insight into the structure that, however, seems promising also from the applicative point of view.

REFERENCES

[1] Kupriyanov M Yu and Lukichev V F 1998 Zh. Eksp. Teor. Fiz. **94**, 139 [Sov. Phys. JEPT 67 1163 (1988)]

[2] Blamire M G, Kirk E C G, Evetts J E and Klapwijk T M 1991 Phys. Rev. Lett. **66**, 220

[3] Helsinga D R and Klapwijk T M 1993 Phys. Rev. B **47**, 5157

[4] Volkov A F 1995 Phys. Rev. Lett. **74**, 4730

[5] Capogna L and Blamire M G 1996 Phys. Rev. B **53**, 5683

[6] Maezawa M and Shoji A 1997 Appl. Phys. Lett. **70**, 3603

[7] Nevirkovets I P 1997 Phys. Rev. B **56**, 832

[8] Nevirkovets I P, Ketterson J B and Lomatch S 1999 Appl. Phys. Lett. **74**, 1624

[9] Nevirkovets I P and Shafranjuk S E 1999 Phys. Rev. B **59**, 1311

[10] Carapella G 1999 Phys. Rev. B **59**, 1407

[11] Barone A and Paterno' G 1982 Physics and Application of the Josephson Effect (Wiley, New York)

Inst. Phys. Conf. Ser. No 167
Paper presented at Applied Superconductivity, Spain, 14-17 September 1999

Highly localised light ion irradiation of $YBa_2Cu_3O_{7-\delta}$ using metal masks

W E Booij, N H Peng*, F Kahlmann, R Webb*, E J Tarte, D F Moore, C Jeynes* and M G Blamire

IRC in Superconductivity, University of Cambridge, Madingley Road, Cambridge CB3 0HE, United Kingdom
* Surrey Centre for Research into Ion Beam Applications, School of Electronic Engineering, Information Technology and Mathematics, University of Surrey, Guildford, GU2 5XH, United Kingdom

ABSTRACT: Irradiation of $YBa_2Cu_3O_{7-\delta}$ with H_2^+ ions with energy in the range of 20-100 keV can be used to fully suppress superconductivity. Simulations suggest that such an irradiation process in combination with a high resolution mask can be used to create highly localised damage regions in thin YBCO films (50-200 nm). This technique has been experimentally verified using an ion beam implantation set-up where low temperature Junction measurements can be made in-situ during implantation. The high resolution metal mask (~40 nm) was made with a focused ion beam, using a newly developed in-situ resistance monitoring technique for endpoint detection. The fabricated devices, which can be characterised during the irradiation process, show Josephson coupling over a limited temperature region. This is a consequence of the relatively long superconducting barriers (~100 nm) that result from the current combination of ion energy and masking.

1. INTRODUCTION

Focused Electron Beam Irradiation (FEBI) of $YBa_2Cu_3O_{7-\delta}$ (YBCO) thin films and related compounds under the right conditions can be used to create high quality Josephson junctions (Pauza et al. (1997), Booij et al. (1997)). Key advantages of the technique are that only a single high quality superconducting layer is required and the film remains unbroken, avoiding many of the complications encountered in multilayer HTS Josephson junction types.

The mechanism by which Josephson coupling arises in FEBI junctions is well understood and is rooted in the sensitivity of the superconducting state in high temperature superconductors to oxygen ordering. Point defects created by various means (e.g. electron and ion irradiation) can result in strong and even complete suppression of the superconducting state. If a sufficiently short barrier (length $L \sim \xi_{nd}$, the dirty limit coherence length) of such material is created in a HTS film, proximity coupling can occur and a Josephson junction is formed (Pauza et al. (1997), Booij et al. (1997), Davidson et al. (1996)). The process by which these defects are created is largely irrelevant (as long the defects are sufficiently uniformly distributed and the barrier is short enough), and a natural extension of the FEBI process is masked ion irradiation (Kahlmann et al. (1998), Katz et al. (1998), Tinchev (1990)). Unlike FEBI, the masked ion damage process (MID) can easily be scaled up to produce large numbers of junctions using existing technology. A number of studies have provided conclusive evidence that Josephson junctions can be created through masked ion irradiation, although the I_cR_n values are still limited by the minimum barrier length that can be obtained (~100 nm) (Kahlmann et al. (1998), Katz et al. (1999)). Consequently such junctions only show Josephson behaviour over a small temperature interval above the T_c of the irradiated barrier (S' barrier).

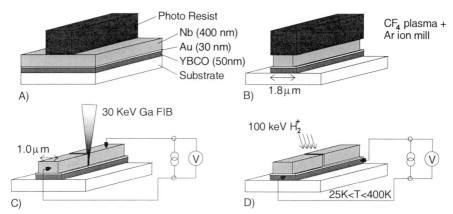

Fig. 1 Schematic representation of the fabrication process to create highly localised barriers

To study whether the barrier length can be reduced we have developed a new in-situ resistance monitoring technique for prototyping high-resolution masks using a focused ion beam. The advantage of this technique is that the process is more flexible and far simpler than electron beam lithography based processing. Furthermore we use metallic masks, which have much greater ion stopping power than the resist masks used until now. By performing low temperature electrical measurements during the implantation process, alignment can be verified. New simulations have been performed which take into account the interaction of ions with the masking layer prior to entering the superconductor.

2.   FABRICATION PROCESS AND MEASUREMENTS

2.1.   Combined track and mask definition

For the experiments we used high quality YBCO films (200 nm) grown by laser ablation on LaAlO$_3$. To open windows for access to the YBCO contact pads, a resist lift off mask was applied prior to deposition of the ion stopping mask, which consists of 30 nm Au followed by 400 nm Nb. The Au layer is necessary to avoid de-oxygenation of the YBCO layer by direct contact with Nb. To ensure good adhesion of the metal masking layer, the Au layer was deposited using DC sputtering at low pressure, whereas the DC sputtering conditions for the Nb layer was optimised for low stress. After lift-off a new photo-resist layer was applied to define tracks in both the mask and the YBCO layer. The pattern was first transferred to the Nb mask by CF$_4$ plasma etching in a RIE. Undercutting of the resist mask occurred due to the isotropic nature of the plasma process used. Using the same resist mask, the exposed Au and YBCO material was removed by Ar ion beam milling at 500 eV on a rotating water-cooled sampleholder. Since Argon ion milling is a highly directional process, the result is that the YBCO tracks are only partially covered by the Nb mask, with as much as 0.4 µm at either edge exposed. However, the exposed part of the tracks will be damaged during the ion implantation process and will simply result in additional shunting. As a final step Au contact pads are deposited on the exposed YBCO pads while making sure that the is no direct electrical contact between Au pads and the Nb mask other than through the YBCO layer. A schematic representation of the processing is shown in Fig. 1.

2.2.   FIB high resolution slit definition

To define the slit in the Nb mask through which the barrier region is formed, we used a focused ion beam system (Philips Electron Optics/FEI FIB 200 workstation) with a Ga source. This system has a milling spotsize of 8 nm for the 4 pA beam current we employed. However, the aspect ratio that can be achieved using direct milling is severely limited (<3) by re-deposition effects. Using

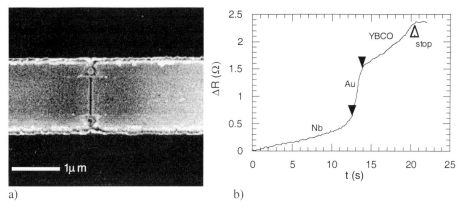

a)   b)

Fig. 2 a) Image obtained in the FIB of a slit approximately 50 nm wide cut into the Nb mask. b) Resistance increase of the track as a function of time during the FIB slit etching process. The points at which the different interfaces are reached can be easily identified. This particular trace shows a calibration run, where the etching went beyond the depth normally used for slits in junction fabrication experiments.

iodine gas injection during the Ga ion mill, which results in the formation of volatile niobium iodide products, can significantly increase the attainable aspect ratio. One disadvantage of using a FIB for this process is that there is no natural endpoint, and detection techniques such as stage current and emitted secondary electron monitoring are simply to inaccurate. We therefor developed a new technique which uses in-situ resistance monitoring of the track whilst etching to monitor the slit formation progress Latif et al.). In the case of our Nb/Au/YBCO tri-layer, the thin Au layer with its high conductivity means that the endpoint for slits as narrow as 40 nm can be found accurately (see Fig. 2)

2.3. H_2^+ irradiation with in-situ low temperature characterisation

For H_2^+ irradiation the sample is transferred to a sample holder with electrical contacts, which is mounted on a closed cycle refrigeration unit. This allows us to measure the electrical characteristics of the device at any temperature between 25K and 400 K whilst irradiating. Through careful shielding and filtering of the signal lines we can accurately measure Josephson characteristics (see inset Fig.3a). By cooling the sample below the T_c we can probe the electrical characteristics of the YBCO region, exposed by the slit, during implantation. Dosing can be controlled accurately by monitoring the resistance of the region at a given temperature and using the resistive transition as an endpoint. An example of a resistance measurement as a function of temperature for a barrier thus formed is shown in Fig. 3a. The difference between the up and down $R(T)$ curve is caused by annealing of defects, during the temperature ramp, which reached a maximum of 100 K.

3. JUNCTION PROPERTIES

After implantation the samples were stored for a couple of days at room temperature and re-measured. As anticipated the characteristics had changed dramatically. The barrier for which the $R(T)$ is shown in Fig. 3b has a low current resistive transition around 25 K directly after implantation whereas as a consequence of defect annealing this has increased to 70 K. The steep $R(T)$ curve is typical for a long barrier whose effective resistive length changes rapidly as a function of temperature. There are two mechanisms to this process: a) An intrinsic effect due to quasi-particle reflection which even occurs in perfectly uniform barriers with a finite T_c (Booij et al. (1999)). b) Scattering of ions causes the barrier to have a graded distribution of T_cs both along the length and thickness of the track (Kahlmann et al. (1998), Pauza et al. (1997)). To get an estimate of the extent of the barrier we can consider the R_nA value just below the T_c of the track; 0.44 $\Omega\mu m^2$. From measurements on long tracks implanted with 100 keV H_2^+ ions, we know that the resistivity of YBCO with a T_c around 60 K is 4 $\mu\Omega m$ (T=89 K) (Booij et al. (1999)). The electrical length of the barrier is therefore approximately

292

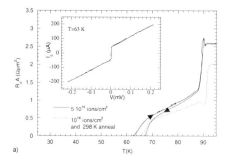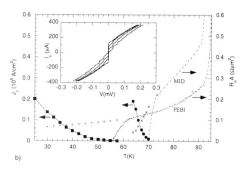

Fig. 3 a) The continuous lines show a up/down $R(T)$, which was measured in-situ, of the same barrier after irradiation with $5 \cdot 10^{15}$ ions/cm^2. The discontinuous line was obtained after further irradiation up to 10^{16} ions/cm^2 and storage at room temperature. The inset shows an IV characteristic measured during implantation. b) J_c and $R_n A$ as a function of temperature for a junction produced by MID after room temperature annealing and a low quality FEBI junction. The inset shows a MID junction under microwave irradiation.

110 nm, which is more than twice as wide as the slit in the mask. As a consequence the J_c of the MID junction rises very rapidly as temperature is reduced and 5-10 K below the resistive transition all RSJ-like features disappear. For comparison the much better J_c and $R_n A$ properties of a low quality FEBI junction, with a known barrier length of 20 nm, are shown.

The reason why the barrier length is twice as wide as the slit width becomes apparent from ion scattering calculations that take proper account of the presence of the slit in the mask. Simple calculations on a continuous masking layer suggest that 400 nm Nb is more than sufficient to stop H_2^+ ions with an energy of 100 keV. However, the presence of the slit means that ions scattered high up in the mask can travel through vacuum and re-enter the mask or YBCO track with sufficient energy and abundance to cause significant widening of the damage profile. To obtain better reproduction of the slit dimensions either the mask has to be increased in thickness (50 %) or the ion energy lowered. The same calculations also show that for 100 keV H_2^+ ions the damage is still quite uniform along the thickness of the film, even for a 200 nm thick YBCO layer. However, widening of the barrier due to lateral straggle does occur, and for future experiment we will resort to thinner films.

We believe that the improved understanding of masking in combination with in-situ measurements will be able to deliver damage regions with a length of 20 nm.

REFERENCES

Booij, W. E., C. A. Elwell, et al. (1999).IEEE Trans. Appl. Superconductivity **9**(2 Pt3), 2886-2889.
Booij, W. E., A. J. Pauza, et al. (1997).Phys. Rev. B **55**(21), 14600-14609.
Davidson, B. A., B. Hinaus, et al. (1996).IEEE Trans. Appl. Superconductivity **7**(2 Pt3), 2518-2521.
Kahlmann, F., A. Engelhardt, et al. (1998).Appl. Phys. Lett. **73**, 2354.
Katz, A. S., A. G. Sun, et al. (1998).Appl. Phys. Lett. **72**(16), 2032-2034.
Katz, A. S., I. S. Woods, et al. (1999).I.E.E.E. Trans. Appl. Supercon. **9**(2), 3005-3007.
Latif, A., W. E. Booij, et al. submitted to Journal of Vacuum Techniques .
Pauza, A. J., W. E. Booij, et al. (1997).J. Appl. Phys. **92**(11), 5612-5632.
Tinchev, S. S. (1990).Superconductor Science & Technology **3**(10), 500-503.

Inst. Phys. Conf. Ser. No 167
Paper presented at Applied Superconductivity, Spain, 14-17 September 1999

Cryogenic dielectric resonators for future microwave communication systems

N Klein, M Winter and H R Yi

Forschungszentrum Jülich, Institut für Schicht- und Ionentechnik, D-52425 Jülich, Germany

ABSTRACT: Recent progress in manufacturing dielectric ceramics and single crystals with high dielectric constant and low microwave losses has turned out to be a challenge for the development of novel devices for satellite communication. From this development device performance is expected to benefit for possible device operation temperatures ranging from cryogenic temperatures around 50 - 150 K (achievable with one-stage close cycle refrigerators) over temperatures from 150 to 200K (in principal achievable with radiation cooling) towards room temperature, if novel dielectric resonator structures with lower loss contribution of the metallic housing would become available.

1. INTRODUCTION

Most ionic crystals exhibit a strong decrease of microwave losses upon decreasing the temperature from room temperature to liquid nitrogen temperature and below. The fundamental intrinsic loss mechanism is the absorption of electromagnetic energy by thermally excited phonons giving rise to power law dependencies of the loss tangent $\tan\delta \equiv \mathrm{Im}\{\varepsilon_r^*\}/\mathrm{Re}\{\varepsilon_r^*\}$, with ε_r^* being the complex permittivity of the dielectric material (Sparks 1982, Gurevich 1991). This loss reduction allows for high quality factors of dielectric resonators at cryogenic temperatures if either modes with strong field confinement inside the dielectric resonator or metallic shielding cavities consisting partially of high-temperature superconducting (HTS) walls are employed. Some of the low-loss polycrystalline microwave ceramics also exhibit a significant reduction of microwave losses upon cooling and a much smaller temperature coefficient of the permittivity in comparison to single crystal materials.

Therefore, novel resonator structures utilising this material potential are worth to be investigated. If the use of cryogenics will become common in microwave communication systems due to the progress in HTS devices, the cryogenic dielectric resonators devices may be of particular relevance for frequencies above 10 GHz where the benefit of planar HTS passive devices is questionable. On the other hand, in application areas like satellite communication where the refrigerator power efficiency is an important issue, cryogenic dielectric resonator devices operating at moderate cryogenic temperatures of 100 – 200 K may be a challenging alternative to HTS systems.

In this article we describe our work on novel dielectric resonator devices with strong potential for applications at cryogenic temperatures. The first example are dielectric dual-mode filters for satellite transponders from C- to Ka-band frequencies using sapphire, single crystal lanthanum aluminate ($LaAlO_3$) or commercial microwave ceramics as dielectric materials (Schornstein 1997, Schornstein 1998, Klein 1999). The second example are sapphire-rutile composite dielectric resonators which exhibits very high Q-values of 10^6 - 10^7 and a turning point at a temperature T_{TP} in the temperature dependence of the resonant frequency f, such that $\partial f/\partial T = 0$ at $T = T_{TP} \approx 50 - 75$ K. These compensated resonators have a strong potential to be used as microwave frequency standards to be operated on a closed cycle refrigerator (Hao 1998).

2.   SINGLE CRYSTAL DIELECTRICS AND MICROWAVE CERAMICS AT CRYOGENIC TEMPERATURES

Among the materials usable for dielectric resonators ($\varepsilon_r \geq 10$) the following ones exhibit a strong potential for operation at cryogenic temperatures:

Al_2O_3 (sapphire and sintered alumina):   Among all dielectrics, high purity single crystals of sapphire exhibit the lowest dielectric losses at cryogenic temperatures. At 10 GHz, the loss tangent drops almost proportional to T^5 from about $7 \cdot 10^{-6}$ at room temperature to about $6 \cdot 10^{-8}$ at 77 K with a slight anisotropy for electric fields aligned along the crystallographic c- direction ($\varepsilon_r = 11.4$) with respect to the a,b direction ($\varepsilon_r = 9.4$) (Braginsky 1987). The frequency dependence of the loss tangent is approximately linear corresponding to a constant values of $Q \cdot f$. Sintered alumina has been optimised with respect to low losses. At 10 GHz $\tan\delta$- values of $2 \cdot 10^{-5}$ at room temperature and $5 \cdot 10^{-6}$ at 77 K were reported (Alford 1998).

Both sapphire and alumina exhibit a strong temperature coefficient of the permittivity of about 100-140 ppm at room temperature and 20 – 30 ppm at 77 K (Dick 1994) resulting in stringent requirements for temperature stabilisation. On the other hand, the high thermal conductivity and the low losses of Al_2O_3 are in favour of operation at high levels of microwave power, such as transmitter filters.

TiO_2: rutile and sintered titania: Rutile is challenging because of its very high permittivity of above 100 and its relatively low loss tangent at cryogenic temperatures. The permittivity is strongly anisotropic (Klein 1995, Tobar 1998) and   exhibits an extremely large   negative temperature coefficient $\partial \varepsilon_r /\partial T = - 1000$ ppm/K at $T = 77$ K. Therefore, TiO_2 is not very useful as dielectric resonator material on its own, but it can be combined with other dielectrics like sapphire in order to provide a turning point in the temperature dependence of the resonance frequency (see section 4). Sintered titania exhibits loss tangent values of $1.5 \cdot 10^{-5}$ at 77K and 10GHz, the temperature coefficient is similar to rutile (Alford 1998). However, titania exhibits an isotropic behaviour of permittivity and loss tangent. Similar to rutile, it is only considered to be useful in case of composite dielectric resonators.

Single crystalline lanthanum aluminate ($LaAlO_3$) and yttrium aluminate ($YAlO_3$): $LaAlO_3$ has become very popular as substrate material for HTS films but it is also a challenging dielectric resonator material due to its relatively high and isotropic permittivity of $\varepsilon_r = 23.4$. The temperature coefficient of the permittivity is similar to sapphire. As discussed in detail in one of our previous publications (Zuccaro 1997), the losses at cryogenic temperatures are governed by a relaxation peak in $\tan\delta (T)$ at 40-70 K. The amplitude of this peak depends on the crystal growth technique and is very small for crystals grown   by the Verneuil technique. Recently, for Crochralzki grown crystals delivered from one particular crystal grower in China the peak was found to be even smaller (Baumfalk 1999). At room temperature, $Q \cdot f = 500.000$ GHz holds true for almost all $LaAlO_3$ crystals, at 77K the lowest reported loss tangent values are around $3 \cdot 10^{-6}$ at 10 GHz for Verneuil material (Zuccaro 1997). $YAlO_3$ has a permittivity of 16 and the loss tangent at 77K is about 10^{-5} at 10 GHz. $YAlO_3$ can also be used as substrate material for HTS films.

Sintered BaMgTaO (BMT): BMT is one of the most popular commercial microwave ceramics. Its permittivity is rather high ($\varepsilon_r = 22-24$) and its temperature coefficient is only a few ppm / K and can be optimised to be zero (Wakino 1989). At room temperature, $Q \cdot f$ can be as high as 350.000 GHz for frequencies above 10 GHz   (see MURATA datasheet), at lower frequencies it drops continuously according to our own measurements (e.g. $Q \cdot f = 150.000$ GHz at $f = 4$ GHz). Fig. 1a shows our experimental results on the temperature dependence of the loss tangent for a BMT ceramic delivered by MURATA for two different frequencies. The employed dielectric resonator technique is described elsewhere (Zuccaro 1997). The observed strong decrease of $\tan\delta$ with temperature (Fig. 1) shows that BMT is a challenging material for cryogenic dielectric resonators with Qs in the range of 10^5 at 4GHz. There are three advantages of BMT with respect to single crystal $LaAlO_3$: First, the temperature coefficient of the permittivity is much smaller from cryogenic temperatures to room temperature. Secondly, BMT is easier to machine than single crystals of $LaAlO_3$. The third advantage is that BMT is already a mass product and therefore much cheaper in comparison to $LaAlO_3$.

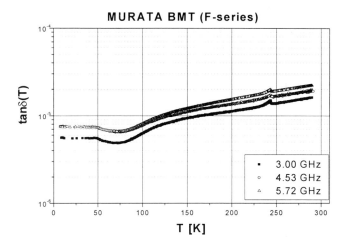

Fig.1: Measured temperature dependence of the loss tangent for a BMT sample provided by MURATA.

3.   CRYOGENIC DIELECTRIC FILTERS FOR SATELLITE COMMUNICATION

There are several reasons to use cryogenic subsystems in satellite payloads: First, a significant weight reduction of input multiplexers (IMUX) at L and C band frequencies can be achieved if cavity or dielectric filters will be replaced by planar HTS filters to be cooled with a space qualified closed cycle refrigerator. In addition, the use of cooled preamplifiers gives rise to about 1 dB of receiver sensitivity.

For the potential use of HTS in output multiplexers (OMUX) at L- and C-band frequencies 2D planar HTS filters operated in an edge current free mode have been developed (Baumfalk 1999). At Ku and Ka-band frequencies, the use of HTS planar filters is questionable at all, because miniaturisation is getting less important in comparison to L-and C-band. In addition, due to the quadratic frequency dependence of the surface resistance of HTS films the performance of planar HTS filters is problematic, in particular at Ka-band frequencies. On the other hand, in particular at Ka band frequencies, the performance of conventional filters is not satisfactory. Therefore, dielectric filters with improved performance are highly desired both for the IMUX and OMUX.

Our approach relies on the use of the fundamental mode of a dielectric hemisphere arranged in a metallic shielding cavity (Schornstein 1997, 1998). Originally, the design was to arrange the hemisphere on an HTS groundplane (Schornstein 1997), however, the feedthrough of coupling and tuning elements through the ground plane was found to be problematic. In the current design, we are using either an HTS endplate on top or no HTS endplate all (Klein 1999). For a quasielliptic four-pole filter, the in-band insertion loss was found to be (-0.04 ± 0.02) dB, which is a factor of 10 lower than for conventional filters (Klein 1999). The unloaded quality factor at 77 K was found to be 111.000 with the upper HTS endplate being employed and 80.000 without HTS endplate. In both cases the hemispheres were machined from single crystal $LaAlO_3$ grown by the Verneuil technique. According to simulations of the electromagnetic fields with the computer code MAFIA (Schmidt 1992) unloaded quality factors of 250.000 would be possible if an upper and lower HTS endplate would be used. However, already at a Q_0 of 100.000 the measured insertion loss is already dominated by losses due to the normal conducting coupling antennae. A further increase of Q_0 would therefore be only challenging for filters with smaller relative bandwidth. On the other hand, the filter with hemispheres machined from MURATA BMT ceramic exhibits a Q_0 of 27.000 at room temperature and 65000 at 77 K. These values are already very attractive for many applications.

The power handling capability of the C-band filter was found to be about 180 watts (with and without HTS endplate, the maximum rf magnetic field on the HTS film was 0.4 mT at that power level). Initially, there was a limitation due to multipacting at about 30 watts. The multipacting threshold could be increased up to about 200 W by a redesign of the coupling antennae performed by our industry partner Bosch Telecom.

In table 1 the projected performance of our filter approach for the three relevant satellite bands is quoted both for room temperature operation and operation at 77K. The numbers quoted in the tables have been calculated using MAFIA simulations and experimental data for the losses of the employed dielectric materials. Note that operation at moderate cryogenic temperatures between 77K and 300K may be of some advantage from the system point of view. This holds true in particular for the OMUX, where the dissipated power in each filter is rather high. However, at T=77K the power efficiency of state-of-the art Stirling coolers is only about 4 – 5 % and increases with increasing operation temperature. According to the numbers quoted in the tables, the use of croygenic dielectric filters is challenging for Ku- and Ka band frequencies. The additional use of HTS components at these frequencies and the use of HTS wall segments of the dielectric filters should be considered in detail taking into account system consideration.

band / downlink-frequencies [GHz]	rel. Bandwidth [%]	Q_0 / P_0 conv.	Q_0 / P_0 HR $T = 300K$	Q_0 / P_0 HR $T = 77K$
C / 3,7 – 4,2	0,9	10.000 / 2,3	30.000 / 0,76	> 100.000 / 0.23
Ku / 10,95 – 12,20	0,3	9.000 / 7,7	20.000 / 3,5	80.000 / 0.87
Ka / 19,7 – 20,2	0,2	6.000 / 17	10.000 / 10	50.000 / 2,0

Table 1a (OMUX): Unloaded quality factor Q_0 and dissipated power P_0 [W] of a quasielliptic 4-pole filter for an input power level of 60 watts for conventional cavity filters (conv.) in comparison to projected values for our hemispherical dielectric filters (HR) at room temperature employing BMT and at 77 K employing LaAlO$_3$ for C-band and sapphire for Ku and Ka-band frequencies.

band / downlink-frequencies [GHz]	rel. Bandwidth [%]	Q_0 / IL conv.	Q_0 / IL HR $T = 300K$	Q_0 / IL HR $T = 77K$
C / 3,7 – 4,2	0,9	10.000 / 0,34	30.000 / 0.11	> 100.000 / 0.032
Ku / 10,95 – 12,20	0,3	9.000 / 1,3	20.000 / 0.5	80.000 / 0.12
Ka / 19,7 – 20,2	0,2	6.000 / 3,6	10.000 / 1.8	50.000 / 0.31

Table 1a (IMUX): Unloaded quality factor Q_0 and insertion loss IL[dB] for a quasielliptic 8-pole filter.

4. CRYOGENIC COMPOSITE WHISPERING-GALLERY MODE RESONATORS FOR LOW-NOISE FREQUENCY STANDARDS

Cryogenic whispering-gallery (WG) mode resonators provide the highest quality factors in the temperature range above liquid helium temperatures. In addition, the high mechanical stability makes sapphire WG resonators extremely challenging as frequency stabilising elements in microwave circuits. It has been demonstrated by some groups that the achievable phase noise values of X-band microwave oscillators based on cryogenic WG sapphire resonators are about 10 - 30 dB smaller than for state-of-the-art quartz oscillators at the same frequency (e.g. Dick 1992).

However, the high temperature coefficient of the resonance frequency of sapphire resonators of about 30 ppm / K results in a very poor short (1 - 1000 s) and long time (> 1000 s) frequency stability presuming realistic temperature stability in the range of a few millikelvins. This drawback can be overcome by using sapphire-rutile composite resonators, where the opposite signs of the temperature coefficients of sapphire and rutile are utilised to generate a turning point in the temperature dependence of the resonance frequency.

We have designed a composite resonator consisting of two ring-shaped thin (350 μm) rutile platelets being arranged at the top- and bottom endplate of a sapphire cylinder (Fig. 2). This composite dielectric puck is hold in place inside a copper cavity via quartz spacers. According to numerical field simulations using MAFIA (Schmidt 1992) the fraction of electromagnetic energy stored in the rutile discs is expected to be 0.001. From experimental data of permittivity and loss tangent of rutile and sapphire this filling factor should correspond to a turning point at 62 K, an unloaded quality factor of

10^7, and a second derivative of resonant frequency with temperature at the turning point of 0.2 ppm / K^2. The latter corresponds to a relative frequency stability of $2 \cdot 10^{-13}$ for temperature deviations of 1 mK from the turning point. This number makes sapphire-rutile composite WG resonators extremely challenging for frequency standard applications.

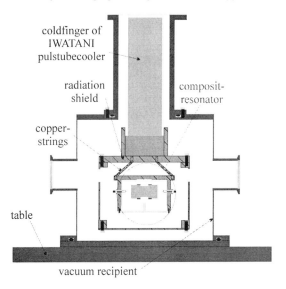

coldfinger of
IWATANI
pulstubecooler

radiation
shield

composit-
resonator

copper-
strings

table

vacuum recipient

Fig.2: Cryogenic setup of a whispering-gallery-mode composite resonator suspended from the cold head of a pulse tube refrigerator.

Experimentally, we have found the turning point at 65 K and $Q_0 = 0.9 \cdot 10^7$. Fig. 4 shows experimental results on the frequency stability of a loop oscillator using our cryogenic composite resonator as a frequency stabilising element. As a reference, we used a HP5352B microwave frequency counter with oven controlled quartz reference. The experimental results on oscillator frequency variations from the nominal value of about 10 GHz and resonator temperature versus time (Fig.3a) and Allan variance (Fig. 3b) calculated from the data in Fig. 3a indicate that at integration times below 500 s the frequency temperature Allan variance mainly follows the temperature Allan variance. This observation indicates, that the current temperature stability of about ± 5 mK still needs to be improved. For longer observation times we observed a periodic frequency variation of about ± 30 Hz with a periodicity of 24 h. This obser-

vation can be explained by thermal expansion of the oscillator loop circuit at room temperature. Within the time interval depicted in Fig.3a, this effect results in slight linear slope of the oscillator frequency and in a slight upturn of the Allan variance at integration times above 500 s.

Further work in this experiment will concentrate on improvement of temperature controlling and the implementation of circuit stabilisation techniques like the Pound method. In addition, measurements of frequency stability employing a hydrogen maser as reference are in progress.

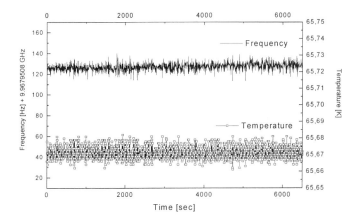

Fig. 3a: Frequency and temperature variations vs. Time.

298

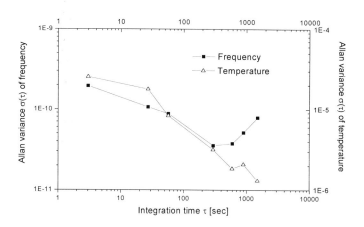

Fig. 3b: Allan variance calculated from the data in Fig. 3a.

ACKNOWLEDGMENTS

The work on the filters has been performed in close collaboration with Bosch Telecom in the framework of a project funded by the German ministry of research an education. The work on the composite dielectric resonators has been performed in close co-operation with J.C. Gallop and L.Hao from the National Physical Laboratory in the UK and has been funded by the European commission in the framework of the BRITE-EURAM project "DiHiMiCo".

REFERENCES

Alford NMcN 1998, Mat. Res. Soc. Symp. Proc. **500**, 183
Aminov B A et al. 1999, IEEE Trans. on Appl. Sup. **9**, 4185
Baumfalk A et al 1999, IEEE Trans. on Appl. Sup. **9**, 2857
Braginsky V B et al. 1987, Phys. Lett. A **120**, 300
Dick G J et al. 1992, Proc. of the 6th European Frequency and Time Form, 35
Dick G J et al. 1994, IEEE Trans IEEE Trans. UFFC **42**, 812
Gurevich V L and Tagantsev A K 1991, Adv. Phys. **40**, 719
Hao L et al. 1999, IEEE Trans. Instr. and Meas. **48**, 99
Klein N et al. 1995, J. Appl. Phys. **78**, 6683
Klein N et al. 1999, IEEE Trans. on Appl. Sup. **9**, 3573
Schmidt D and Weiland T 1992, IEEE Trans. Magn. **28**, 1793
Schornstein S et al. 1997, Inst. Phys. Conf. Ser. **158**, 267
Schornstein S et al. 1998S, IEEE-MTT-S Int. Microwave Symp. Digest, 1319
Sparks M et al. 1982, Phys. Rev. B. **26**, 6987
Tobar M E et al. 1998, J. Appl. Phys. **83**, 1604
Wakino K et al. 1989, Ferroelectrics **91**, 69
Zuccaro C et al. 1997, J.Appl. Phys. **82**, 5695

Inst. Phys. Conf. Ser. No 167
Paper presented at Applied Superconductivity, Spain, 14-17 September 1999
© 2000 IOP Publishing Ltd

YBCO receive coils for low field Magnetic Resonance Imaging (MRI).

A A Esmail*, D Bracanovic*, S J Penn*, D Hight†, S Keevil,* T W Button‡ and N McN Alford*

*South Bank University, 103 Borough Rd. London SE1 OAA, UK
†Surrey Medical Imaging Systems (SMIS), Guildford, Surrey GU2 5YF, UK
‡Birmingham University, IRC in Materials, Birmingham B15 2TT, UK
*Guy's and St Thomas' Hospital London

ABSTRACT: Planar $YBa_2Cu_3O_x$ (YBCO) superconducting surface coils tested in a 0.15T (6.1MHz) MR imager are reported. The coils' performance and sensitivity are compared with identical normal metal coils. A detailed clinical study of the wrist and knee is also evaluated. It was found that superconducting YBCO has a SNR 3 times better compared to a silver coil cooled to 77 K. Clinical images with YBCO are also found to be superior than those taken with copper and silver coils. This paper forms part of an investigation to produce an inexpensive low field prototype MRI imager.

1. INTRODUCTION

MRI is a technique that has now established itself in modern medicine. It provides clinical images with exceptional contrast and resolution. However, in most cases, MRI requires a superconducting magnet which leads to high running costs. These magnets also have high capital costs and space requirements to accommodate the fringe field (Allen 1994).

To reduce the overall cost of a MRI system, superconducting magnets may be replaced with low field permanent magnets. Reducing the magnetic field strength reduces the SNR and hence deteriorates image quality. Nevertheless, NMR imaging has been possible using the earth's magnetic field (Stepsinik 1990). Since lower fields lead to lower SNR the idea is to enhance the SNR by using superconducting pick-up coil. It has already been demonstrated that a pick-up coil made from superconducting YBCO can improve SNR where coil noise is the dominant factor (Hall 1991, Penn 1993, Withers 1993, van Heteren 1994, Wosnik 1996, Penn 1999). The coil noise is the dominant factor in the case of low field systems and surface coils.

2. MATERIALS AND METHODS

2.1. MRI Receiver coils

The preparation of YBCO thick film is discussed elsewhere (Alford 1991). A number of coils with different numbers of turns were made. Silver pads were fired on the ends of the coils to allow tuning capacitors to be attached. The capacitors were attached with indium solder. Copper and silver mimics of similar geometry and design to the YBCO coil were made. The copper mimics were etched on to fibreboard. The silver mimics were produced by printing silver on an alumina substrate and firing appropriately. Both silver and copper coils have been used to compare the sensitivity of superconductivity YBCO coil.

2.2.   Quality Factor (Q) measurements

All coils were tuned to 6.1MHz and matched to 50Ω using a combination of ceramic and PTFE capacitors. Q measurements were performed at room temperature (RT) and 77 K on the tuned circuits with a HP4194A Impedance/Gain Phase analyser. The low temperature Q measurements were made by immersing the coil in liquid nitrogen in a polystyrene container. Two single loops were made to couple with the coil under test. One of the loops excited the coil and the other measured the response. The Q results of three coils are shown in Table 1.

TABLE 1
Q measurements for coils at 6.1MHz and RT and 77 K

Coil type	Q at room temperature	Q at liquid nitrogen 77 K
Copper	108	437
Silver	65	244
YBCO	N/A	3338

2.3.   Imaging System and technique.

Both SNR measurements and clinical imaging were performed using an IMIG 0.15T (6.1MHz) MR imager (Surrey Medical Imaging Systems, England). SNR and clinical wrist and knee (R-wrist and R-knee) imaging was performed with all coils. SNR measurements were obtained by imaging a cylindrical phantom of a solution of 150mM NaCl with 1mM $CuSO_4$ in water. The SNR measurements were made on reconstructed multi spin echo 2-dimensional images. Clinical imaging was performed to examine human R-wrist and R-knee images. Coronal and axial images of the wrist and knee have been acquired using a standard multi-spin echo sequence. Wrist images were performed with a Field Of View (FOV) of 140mm. A fast scout gradient echo (GE) scan has also been acquired using a YBCO coil.

3.   RESULTS/DISCUSSIONS

3.1.   SNR

The SNR results as a function of distance from the coil are shown in Fig. 1. The YBCO receive coil shows improvement in SNR over copper and silver coils. Superconducting YBCO obtained the highest SNR of 176 at 3cm from the receive coil. When compared to cooled silver coil at

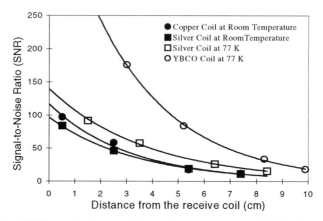

Fig. 1. SNR result of copper, silver and superconducting YBCO coils as function to distance from the coil.

3cm the YBCO coil shows a three fold improvement. The SNR of the YBCO at 8.3cm is also better than that of the cooled silver coil demonstrating that the YBCO coil has a better depth of field range than that of other coils.

3.2.   Wrist/knee images

Figure 2 shows human T_1-weighted R-wrist images. The coronal images are taken using the same slice and position of the R-wrist. The images are taken using a) a copper mimic room temperature coil, b) silver cooled coil at 77 K and c) superconducting YBCO coil at 77 K. A scout scan shown in 2d) has also been performed using the YBCO coil. Scout images of this quality are not normally seen especially with an acquisition time of 17s (compared to 450s for clinical wrist scan).

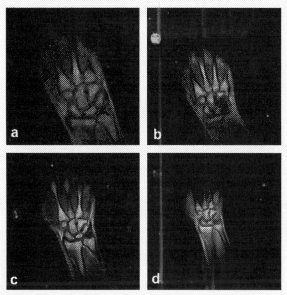

Figure 2. MRI T_1-weighted wrist taken using a) room temperature copper coil, b) silver coil at 77 K, c) superconducting YBCO coil at 77 K and d) scout scan at 77 K. FOV=140mm.

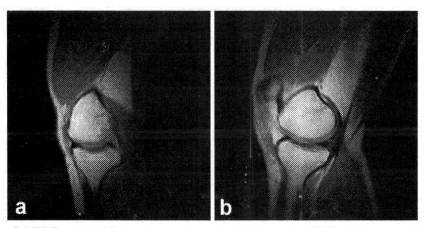

Figure 3. MRI T_1-weighted R-knee images taken using a) silver coil at 77 K and b) superconducting YBCO coil at 77 K. FOV=170mm.

4.   CONCLUSION

The clinical images obtained using all the coils confirms the quality of the design and fabrication. The performance seen from YBCO for scout scanning is one major area that requires further research. The scout scan is taken in 17s and opens work for rapid acquisition. This also suggests the need for optimum MRI sequence application for YBCO scanning. One area of interest may be in paediatric work where fast acquisition scanning is required to reduce patient discomfort and minimise movement artefacts.

ACKNOWLEDGEMENTS

The authors would like to thank the support of EPSRC under the Medlink program, Grant No. GR/L25028.

REFERENCES

Alford N McN, Button T W, Adams M J, Hedges S, Nicholson B and Philips W A, 1991 "Low surface resistance in $YBa_2Cu_3O_x$ melt-processed thick films," Nature, **349**, pp 680-683

Allen D, Elster M D, 1994 "Questions and Answers in Magnetic Resonance Imaging" Mosby Publishing,

Hall A S, Alford N McN, Button T W, Gilderdale D J, Gehring K A and Young I R, 1991 "Use of high temperature superconductor in a receiver coil for Magnetic Resonance Imaging". Magn. Reson. Med., **20**, pp 340-343

Penn S J, Alford N McN, Hall A S, Button T W, Johnstone R, Zammatio S J and Young I R, 1993 "Design of RF receiver coils fabricated from high temperature superconductor for use in whole body imaging equipment," Applied Superconductivity, **1**, pp 1855-1861

Penn S J, Alford N McN, Bracanovic D, Esmail A A, Scott V and Button T W, 1999 "Thick Film YBCO Receive Coils for Very Low Field MRI", In press IEEE Trans. Superconductivity, June

Saint-Jalmes H, Coeur-Joly O, Guilloux-Viry M, PerrinA, Thivet C, Padoiu J, Doussellin G, Pellan Y, 1996 "Low Field Nuclear Magnetic Resonance Imaging using a High Temperature Superconductor YBaCuO Thin Film Receiver Coil Operating at 77K," Journal of Magnetic Resonance Analysis, pp. 53-56)

Stepsinik J, Erzen V and Kos M, 1990 "NMR Imaging in the Earth's Magnetic Field," Magnetic Resonance in Medicine, **15**, pp 386-391

van Heteren J G, James T W and Bourne L C, 1994 "Thin film high temperature superconducting RF coils for low field MRI," **32**, pp 396-400

Withers R S, Liang G C, Cole B F and Johansson M, 1993 "Thin-film HTS probe coil for Magnetic Resonance Imaging" IEEE Trans. App. Supercon., **3**, pp 2450-2453

Wosik J, Nesteruk K, Xie L M, Zhang X P, Gierlowski P, Jiao C and Miller J H. 1996 "Enhanced-resolution Magnetic Resonance Imaging using High-T_c superconducting rf receiver coils," Proceedings of HTS meeting on Physics, Materials and Applications. University of Houston, Texas. March 12-16

Inst. Phys. Conf. Ser. No 167
Paper presented at Applied Superconductivity, Spain, 14-17 September 1999
© *2000 IOP Publishing Ltd*

Demonstration of a power handling of 0.5 kW at 0.4 GHz in a stripline of YBCO on a 3" LaAlO$_3$ wafer

G Koren   N Levy   E Polturak

Physics Department, Technion, Haifa, Israel

Y Koral

Microwave Division, Elisra Electronic Systems Ltd., Bnei-Brak, Israel

ABSTRACT: Thin YBa$_2$Cu$_3$O$_7$ films were prepared by laser ablation deposition on 2"   and   3" wafers of (100) LaAlO$_3$ and patterned into striplines with gold contacts for testing of microwave transmission versus power at 77K. For straight striplines along the diameter of the wafers, with 0.5μm x 5.5mm cross section we found that the power handling at 0.4GHz was 500-1000W for the striplines on the 2" wafers, and about 500W on the 3" wafers. Above these powers the striplines failed catastrophically, mainly near the normal gold contacts.

1.   INTRODUCTION

This study is part of a project to build a multiplexer for power transmission of microwaves simultaneously in 4 channels, and to couple this power into a single antenna with minimum losses. The power requirements are that each channel should withstand up to 40W at 0.2-0.4GHz. To obtain this goal, we planned to use 4 phase shifters made of thin YBa$_2$Cu$_3$O$_7$ (YBCO) films patterned into striplines and assembled in a special geometry which will be described later on. Each one of these shifters must be able to change the phase by up to 180^0 and 4 such shifters are needed. As a first step, large area films of high quality of the high temperature superconductor YBCO were prepared and patterned into straight striplines for testing the feasibility and power handling of the device. The films were prepared and patterned in the Technion, and the power handling experiments were carried out in Elisra.

2.   EXPERIMENTAL

In order to prepare the large area films, we used a modified laser ablation deposition system as shown in Fig. 1. The main changes compared to a standard laser deposition systems are: i. The plume is not perpendicular to the wafer but at 45^0 to it, ii. This allows for a simple scanning of the heater in a horizontal plane which in turn enables, in principle, any size of coated area, iii. We added a commercial bottom heater to relax a bit the heating required from the top heater, which allowed us to use standard heating wires made of Kanthal.   In addition, the top heater was made of ceramics, and its shape made it spread the heat quite evenly on the fragile LaAlO$_3$ (LAO) wafer to avoid breakage.

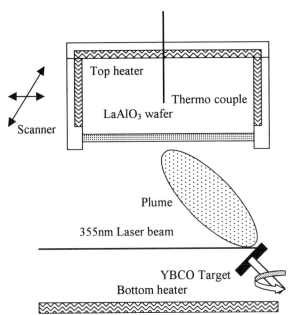

Top heater

Thermo couple

LaAlO₃ wafer

Scanner

Plume

355nm Laser beam

Fig. 1
The laser deposition system

YBCO Target

Bottom heater

The deposition conditions were: 980 °C in the top heater and 1kW in the bottom heater, 0.1 Torr oxygen flow, 1-2J/cm^2 laser fluence on the target, 10Hz laser repetition rate, total of 20,000 laser pulses that produced about 500nm film thickness, and annealing in 20Torr of oxygen pressure. The $T_c(0)$ of the resulting films are in the range of 88-90 K all over the wafer as can be seen in Fig. 2. Fig. 2 was obtained by a direct 4-probe transport measurement of 5 locations along the diameter of the wafers, parallel to the original direction of the plume. Since we used 4x5 spring loaded gold tipped contacts, the films were unharmed and used later on for the power testing. For this, the films were patterned by deep UV photolithography and wet acid etching into straight transmission lines as seen in Fig. 3 (a). Gold contacts were evaporated via shadow masks on the two edges of the stripline and the whole wafer was then reannealed in oxygen at 800^{0}C. The patterned wafers were mounted in a simple testing chamber, and the contacts were connected to the input and output SMA connectors with several gold wires. We used wire bonding to the gold contacts on the stripline and electric welding to the SMA connectors. For the microwave power handling tests, the input SMA was connected to the microwave source and the output SMA to a 50Ω load. The test chamber was evacuated and filled with helium gas for good thermal exchange, and immersed slowly into a dewar filled with liquid nitrogen. The testing temperature was thus 77K, except for the normal gold contacts which were heated by the microwave power to a slightly higher temperature.

3.   RESULTS

In the microwave transmission tests, the power through the stripline at 0.4GHz was increased slowly with dwells of a few minutes after each additional 50W. The transmitted and reflected power were monitored, until a catastrophic failure occurred.

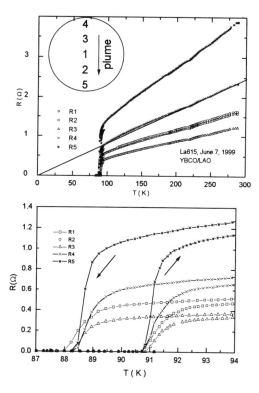

Fig. 2 Resistance versus temperature of 5 areas of the film along the diameter parallel
to the plume.

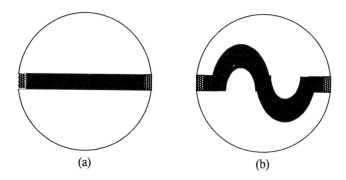

Fig. 3 The stripline for the power testing (a), and a typical stripline of the phase
shifter (b).

We found that the power handling capability of the striplines on the 2" wafers ranged between 500 and 1000W and that of the striplines on the 3" wafers was about 500W. In all, we tested 12 striplines on the 2" wafers and 5 on the 3" wafers. The failure was generally caused by a damaged area in the stripline near the gold contacts, with burning marks and sometimes breakage or cracking of the wafer. We suspect that heating of the normal contacts is the cause of failure and this problem will have to be dealt with in the future. For now, the power handling results are satisfactory for the present project and we are in the stage of building the actual shifter made of YBCO. A schematic diagram of this device is shown in Fig. 4. It contains 4 wafers of 3" diameter two striplines in the form shown in Fig. 3 (b), and two ground planes.

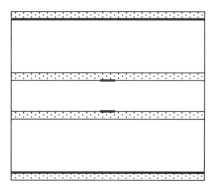

Fig. 4   A schematic cross section of the YBCO based phase shifter on the 3" wafers

So far we built a conventional such shifter with copper striplines and tested it under low power at room temperature. We measured a transmission loss of 0.6dB in the 0.2-0.4GHz range, while the calculated loss was 0.4bB. The calculated loss for the superconducting device is 0.2dB and it remains to be seen what the actual loss at high power will be. Finally, we should note that for 4 such shifters which are necessary for the whole multiplexer, we shall have to use 16 wafers, 8 with patterning and contacts like in Fig. 3 (b), and 8 unpatterned for the ground planes.

Inst. Phys. Conf. Ser. No 167
Paper presented at Applied Superconductivity, Spain, 14-17 September 1999
© 2000 IOP Publishing Ltd

Characterisation of the superconducting and microwave properties of YBCO films simultaneously sputtered on both sides of sapphire wafers

AG Zaitsev[1], R Schneider[1], J Geerk[1], G Linker[1], F Ratzel[1], R Smithey[1], S Kolesov[2] and T Kaiser[3]

[1]Forschungszentrum Karlsruhe, Institut für Festkörperphysik, P.O. Box 3640, D-76021 Karlsruhe, Germany
[2]Fachbereich Elektrotechnik Universität Wuppertal, D-42119 Wuppertal, Germany
[3]Cryoelectra GmbH, D-42287 Wuppertal, Germany

ABSTRACT: The properties of $YBa_2Cu_3O_{7-x}$ (YBCO) films sputtered simultaneously on both sides of CeO_2 coated sapphire wafers of 3 inch diameter met the requirements for their application in passive microwave devices. The films exhibited a high critical current density of 3-5 MA/cm² at 77 K and a low microwave surface resistance, e.g., R_S = 16.5-18.5 µΩ at 1.92 GHz and 77 K, measured on both sides by the disk resonator technique. The films were able to handle a high microwave power corresponding to magnetic field amplitudes (B_{HF}) of 12.5, 15 and 18 mT at 77, 70 and 60 K, respectively. High stability of the films with respect to thermal cycling, patterning, contacting, etc. was observed. The effective R_S of the YBCO films was considerably enhanced by Au contact layers. This effect was not related to a degradation of the YBCO films, but it was explained in terms of the impedance transformation.

1. INTRODUCTION

The implementation of passive microwave devices requires the preparation of high-quality double-sided YBCO films on large-area sapphire substrates (Hein 1997). These YBCO films should be at least 250-300 nm thick, they should exhibit a high critical current density (J_c(77K)>3MA/cm²), a low microwave surface resistance (R_S(77K,2GHz)<25µΩ) and a high power handling capability (expressed in terms of the microwave magnetic field at the film surface B_{hf}(77K)>10mT). In general, gold contacting layers on top of the YBCO films are required. The YBCO film properties should not degrade due to thermal cycling, patterning and preparation of metal contacts. We characterised the double-sided YBCO films prepared by simultaneous sputtering on both sides of sapphire wafers with respect to these requirements. This technique is a promising alternative to the conventional side-by-side film deposition.

2. EXPERIMENTAL RESULTS

The YBCO films were simultaneously sputtered on both sides of 3 inch diameter (3″) *r*-cut sapphire wafers buffered with 30 nm thick (001) CeO_2 (Geerk et al. 1998). The substrate thickness was 0.43 mm. The film thickness was measured by Rutherford backscattering spectrometry. Typically 250 to 500 nm thick YBCO films were prepared. They were crack-free and exhibited a high critical temperature, T_c, ranging from 90K to 91.5 K.

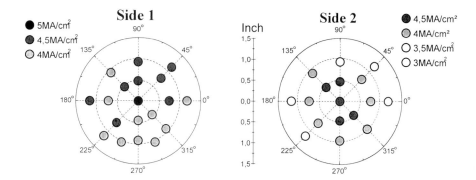

Fig. 1. Distribution of J_c(77K) across the surfaces of 300 nm thick YBCO films on the two sides of a 3″ sapphire wafer.

The critical current density, J_c, was inductively measured using the 3rd harmonic response of the YBCO films. This technique proposed by Claassen et al. (1991) had two advantages relevant to the present work. First, it provided a good resolution between the two sides of the wafer. Second, the measurement was not affected by coating the YBCO film with a gold layer. The spatial resolution across the YBCO film surface was obtained by multiplexing between a number of identical coils. Typical measurement results obtained for a 300 nm thick double-sided YBCO film are shown in fig. 1. The J_c(77K) was 3.5-5 MA/cm^2 on both sides of the wafers.

The R_S of the YBCO films was measured by the disk resonator method, whose fundamentals were described by Kolesov et al. (1997). The disk resonator also appears as an ideal tool for studying the effect of a metallic contact layer on the effective R_S of the YBCO films.

For the disk resonator measurements one side of the double-sided YBCO film was patterned in a circle of 60mm diameter yielding a resonance frequency of ~1.92 GHz for the TM$_{010}$ mode. The other side of the wafer remained unpatterned. The unloaded quality factor, Q_U, of the disk resonator was determined from one-port reflection measurements using a Hewlett Packard 8720C network analyzer. The measurements were performed in the temperature range of 63-80 K. Figure 2 shows the typical temperature dependence of Q_U of two disk resonators made of 300 nm and 500 nm thick 3″ YBCO films, respectively. The Q_U values were independent of the network analyzer output power up to 5 dBm. As shown in fig. 2, the YBCO films exhibited a very good performance, e.g., Q_U = 194000 for the 300 nm thick YBCO film at 77K, and Q_U = 387000, for the 500 nm thick film. These values are remarkably high for large-area disk resonators (Kolesov et al 1997).

Using the Q_U data one can estimate the effective R_S averaged over the entire surface of the YBCO films on both sides of the substrate:

$$R_S = \mu_0 \omega h/(2Q_U) \quad , \qquad\qquad (1)$$

where μ_0 is the permeability of vacuum, ω is the circular resonance frequency, and h is the substrate thickness. Both the dielectric loss in the substrate and the radiation loss are neglected in (1), which was found to be a justifiable assumption for the disk resonators on sapphire (Kolesov et al 1997, Hein et al. 1997).

The dependence of R_S on T is also shown in fig. 2. In particular, we obtained R_S = 16.5 μΩ and R_S = 17.5 μΩ for the 300 nm and 500 nm thick YBCO films at 77 K, respectively. Both values meet the requirements discussed in the introduction.

In order to study experimentally the effect of a metal coating on the microwave losses in the YBCO films we deposited Au layers in a thickness of up to 1.2 μm on the unpatterned YBCO-side of the disk resonator by dc magnetron sputtering at ambient temperature in a pure Ar atmosphere. The dc resistivity, ρ_{Au}, varied linearly from 5.64 μΩcm at 80 K to 5.35 μΩcm at 63 K. These values are about 10 times higher than those of single crystals. Using the measured ρ_{Au} we estimated the skin depth, δ, of our Au films at 1.92 GHz. The resulting δ ranged from 2.73 µm to 2.66 µm, i.e., well above the thickness of Au relevant for the present work.

The deposition of Au did not change the J_c(77K) of the YBCO films, indicating that they did not degrade. However, the Au film reduced dramatically the quality factor of the resonator, as shown in fig.3. For example, Q_U(77K) of 194000, measured for the uncoated resonator made of the 300 nm thick YBCO film, decreased to 49000 and 11500 for Au thicknesses of 0.3 and 1.2 µm, respectively. Such a decrease in Q_U corresponds to a strong enhancement of the effective R_S of the Au coated YBCO side of the resonator, as it is shown in fig. 3. In particular, the increase in the case of the 0.3 µm Au layer is 6-fold and in the case of the 1.2 µm thick Au layer even 20-fold. The experiment was accomplished by etching the Au layer from the YBCO surface. The Q_U of the disk resonator was completely restored. This result indicates the electrodynamical nature of the effect of the Au layer on the effective R_S of the YBCO films. In addition it reveals high stability of the YBCO films on sapphire with respect to various treatments, including repeated thermal cycles, J_c and microwave measurements, patterning, sputtering and etching of Au films. The effect of the Au coating on the quality factor of the disk resonator can be explained by an enhancement of the effective R_S of the YBCO film due to the impedance mismatch at the YBCO-Au interface. The quantitative description of this effect in terms of the impedance transformations is discussed in more detail elsewhere (Zaitsev et al. 1999). It predicts a strong enhancement of the effective R_S of the Au-YBCO bilayers, if the YBCO thickness is comparable to the London penetration depth, λ_L. However, Au layers with thicknesses above 1 µm considerably enhance the effective R_S even for thick YBCO films (700-800 nm). A higher quality of the YBCO films (in terms of shorter λ_l) reduces the effect of a Au layer, while a low ρ_{Au} leads to a further increase of the effective R_s of a Au coated YBCO film.

The microwave power handling capability of the YBCO films was measured in a Nb

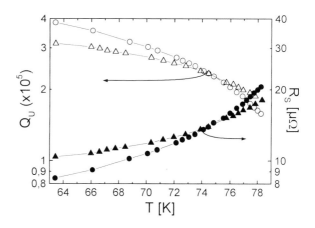

Fig. 2. Unloaded quality factor Q_U (open symbols) and microwave surface resistance R_S (filled symbols) obtained by the disk resonator measurements for 300 nm (triangles) and 500 nm double-sided YBCO films (circles).

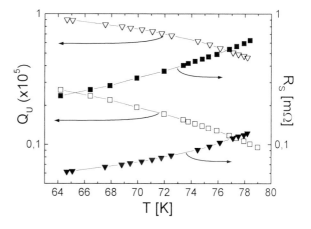

Fig. 3. Q_U (open symbols) and R_S (filled symbols) obtained by the disk resonator made of the 300 nm thick double-sided YBCO film with its unpatterned side coated with 300 nm thick (triangles) and 1200 nm thick (squares) Au film.

310

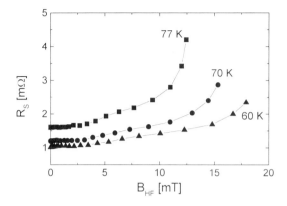

Fig. 4. Microwave surface resistance versus peak magnetic field amplitude at the surface of a 300 nm thick YBCO film. The measurement frequency was 19.15 GHz.

shielded dielectric resonator. This measurement technique was described by Diete et al. (1997). The microwave power applied to the YBCO film was characterised by the amplitude of the magnetic field, B_{HF}, at the film surface. Figure 4 shows typical results obtained for a 300 nm thick YBCO film at the temperatures relevant for the present work. Although R_S exhibited a slight increase with increasing microwave field amplitude, no dramatic enhancement of R_S was observed up to relatively high fields of 12.5, 15 and 18 mT at 77, 70 and 60 K, respectively. These values meet well the range cited in the introduction.

3.   DISCUSSION AND CONCLUSION

The further development of double-sided YBCO film preparation techniques requires the use of non-destructive measurements suitable for the characterisation of the superconducting and microwave properties of the YBCO films on both sides of the substrate in the relevant range of temperature (60-77 K) and frequency (1-4 GHz). The inductive J_c measurements and the disk resonator measurements at microwave frequencies are essential techniques for these purposes. Using these techniques we examined the double-sided YBCO films prepared by the simultaneous sputtering on both sides of 3 inch diameter sapphire wafers. The films exhibited high critical current densities of 3-5 MA/cm² at 77 K and low microwave surface resistances, e.g., average R_S of 16.5-18.5 μΩ at 1.92 GHz and 77 K. The films could handle high microwave power up to B_{HF} values of 12.5, 15 and 18 mT at 77, 70 and 60 K respectively. These properties meet well literature data for high-quality large-area YBCO films that are suitable for applications in passive microwave devices. The films were very stable with respect to thermal cycling, patterning, and contacting. The effective R_S of the YBCO films was considerably enhanced by Au contact layers. This effect was not related to a degradation of the YBCO films and was explained in terms of the impedance transformation.

REFERENCES

Claassen J H, Reeves M E and Soulen R J, Jr. 1991, Rev. Sci. Instr. **62**, 996.
Diete W, Getta M, Hein M, Kaiser T, Müller G, Piel H and Schlick H 1997, IEEE Trans. Appl. Supercond. 7, 1236:
Hein M A 1997 Supercond. Sci. Technol. **10**, 867
Geerk J, Ratzel F, Rietschel H, Linker G, Heidinger R and Schwab R 1998 Presented at Applied Superconductivity Conf., Palm Desert, USA Sept.13-18 (1998). To be published in IEEE Trans on Appl. Supercond. (1999).
Hein M A, Bauer C, Diete W, Hensen S, Kaiser T, Müller G and Piel H 1997 J. Supercond. **10**, 109
Kolesov S, Chaloupka H, Baumfalk A and Kaiser T 1997 J. Supercond. **10**, 179
Zaitsev A G, Schneider R, Geerk J, Linker G, Ratzel F and Smithey R 1999 accepted for publication in Appl. Phys. Lett. Scheduled for **10**, n.26.

Inst. Phys. Conf. Ser. No 167
Paper presented at Applied Superconductivity, Spain, 14-17 September 1999
© 2000 IOP Publishing Ltd

Investigations on the unloaded quality factor of planar resonators with respect to substrate materials and packaging

A Baumfalk, M Reppel, H Chaloupka and S Kolesov

Univ. of Wuppertal, Dept. of Electrical Engineering, Fuhlrottstr. 10, 42119 Wuppertal (Germany)

ABSTRACT: Planar HTS microwave resonators have been developed at Wuppertal University offering an extremely high unloaded quality factor Q_0. For high power applications edge current free TM_{010} mode disk resonators are best suitable, for low power applications a miniaturised double-symmetric hairpin resonator gives excellent results. Due to the high Q_0 of these resonators parasitic influences such as dielectric losses, housing and packaging contributions have become important. These influences have been investigated and reduced, so that quality factors of 360,000 (disk resonator: 50 K, 4 GHz) and 100,000 (double-symmetric hairpin: 50 K, 1.8 GHz) could be reached on $LaAlO_3$ substrates.

1. INTRODUCTION

In actual and future communication systems there is a growing demand in miniaturised low loss microwave resonators and filters. In communication satellites for instance the reduction of size and mass of all components is the main objective. Additionally, a reduced power consumption of improved RF-devices directly translates into a mass reduction in the amplifier section and power supply. In mobile communication systems on the other hand, the available bandwidth must be utilised with maximum efficiency. To reduce the loss of spectral bandwidth due to guard bands for example, highly selective filters must be used. In general, the selectivity of coupled resonator filters is limited to a certain extend due to microwave losses. As an example, Fig. 1 shows the transmission characteristics of an eight-pole quasi-elliptic filter in the ideal (lossless) case, with an unloaded Q of 50,000 and of 10,000 of the utilised resonators. Besides the increased midband insertion loss it can easily be seen that the selectivity is remarkably reduced due to the rounding of the filter skirts. This example illustrates that high Q resonators are absolutely necessary where highly selective filters are demanded.

In general, HTS resonators and filters offer the chance to achieve both, reduction of size and mass as well as very high unloaded quality factors compared to conventional resonators. For extremely high-Q resonators which have been developed at Wuppertal University, parasitic losses such as dielectric losses and contributions of housing and packaging additionally have to be taken into account. These influences have been the topic of investigation in this paper.

2. GENERAL CONSIDERATIONS

The unloaded quality factor Q_0 of microwave resonators denotes the ratio of oscillating power $P_{osc} = \omega_0 W$ (W - stored energy) and dissipated power P_{diss}. The dissipative losses can be divided into separate contributions such as conductive losses, dielectric losses, housing losses and packaging losses. Introducing

$$\frac{1}{Q_0} = \frac{P_{diss}}{P_{osc}} = \frac{1}{Q_c} + \frac{1}{Q_d} + \frac{1}{Q_h} + \frac{1}{Q_p}, \qquad (1)$$

a specific quality factor Q_c, Q_d, Q_h and Q_p can be defined for each loss contribution.

In most high–Q resonator applications the limitation of the unloaded quality factor is mainly influenced by only one of these contributions, depending on the operating environment (resonator design, packaging, temperature etc.). Using temperature dependent analysis of the circuit, it is possible to separate these influences and to identify the most stringent limitation, giving a starting point for further optimisation.

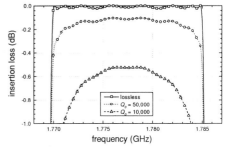

Fig. 1: Effect of losses onto the insertion loss of a highly selective bandpass filter

Conductive Losses: Losses in the conductors can be evaluated by a surface integration of the magnetic field $\vec{H}$ over the conductor surface S_c whereas the stored energy is given by the volume integral over the resonator volume V_r. As described in Chaloupka (1999) one therefore finds the following proportionality:

$$Q_c \propto \frac{l_c}{\lambda_0 R_s} \ , \ l_c = \frac{2 \int\limits_{Vr} \left| \vec{H} \right|^2 dV}{\int\limits_{Sc} \left| \vec{H} \right|^2 dS} , \tag{2}$$

where λ_0 and R_s represent wavelength and surface resistance of the conductor material, and the geometric factor l_c denotes the ratio between volume and conductor surface, weighted by the magnetic field strength of the utilised resonant mode. This factor represents the resonator design, and in general increases with the resonator size.

Conductive losses in HTS structures are highly temperature dependent due to R_s of the thin film materials, leading to a relatively strong influence close to the critical temperature T_c. Temperature dependent R_s measurements excluding all other contributions can help to classify the influence on the resonator structure.

Dielectric Losses: The contribution of dielectric losses Q_d to the dissipated power depends on the loss angle $\tan \delta$ of the utilised dielectric material and the form factor β_d, describing the amount of electrical field energy that is stored inside this material:

$$Q_d \propto \frac{1}{\beta_d \tan \delta} . \tag{3}$$

The temperature dependence of this contribution directly reflects the temperature dependent dielectric loss tangent of the substrate material.

Housing and Packaging Losses: Housing losses can be described as a parasitic coupling of the wanted resonator mode to unwanted modes of the lossy housing structure introducing energy dissipation due to currents in the housing walls. Additional packaging influences are losses due to imperfect electrical contacts to the ground and to the ports of the device. The design goal of all resonators is to reduce these influences to a neglectable value, so that the potential of the utilised resonator can fully be exploited. Whereas the housing contribution can be regarded as temperature independent in most situations, packaging losses can be connected to the field penetration depth in the HTS materials, leading to a temperature dependent behaviour.

3. DISK RESONATORS

For edge current free TM_{010}-mode disk resonators with a substrate thickness h, $l_c=h$ holds, which is the highest value realised for planar resonators and is obtained due to a highly uniform current distribution on the conductor surface. Therefore this resonator type is very well suitable for high Q and high power applications as presented in Kolesov et al (1997). On 1 mm thick LaAlO$_3$ (ε_r=24) substrate the dielectric form factor β_d is close to the ideal value of 1, so that the stored energy is mainly concentrated in the substrate material, showing very low housing contributions.

In Fig. 2, the temperature dependence of Q_0 is displayed for 4 GHz TM$_{010}$ disk resonators on YBCO and DyBCO thin films produced by TU Munich on various different substrate materials (Baumfalk et al 1999). These values were obtained in loss optimised test housings that do not influence the measured quality factor. Whereas the loss dependence on sapphire substrates (circles) directly reflects conductive (R_s) losses, resonators on Czochralski grown LaAlO$_3$ substrates from Lucent Technologies (triangles) show reduced quality at temperatures around 60 K due to high losses in the substrate at these temperatures (see Klein et al (1996)). On Verneuil grown LaAlO$_3$ material (squares) this contribution is reduced but still noticeable. Best results at temperatures of technical interest (> 70 K) have been achieved on Czochralski LaAlO$_3$ from Beijing University (crosses), that do not show high losses at any temperature. Improved thin film materials (DyBCO: filled circles) can give an additional advantage in this R_s limited temperature range because of a higher critical temperature T_c.

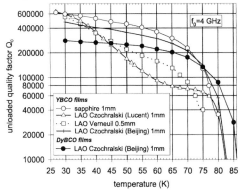

Fig. 2: Q_0 of TM$_{010}$ disk resonators (4 GHz) manufactured from different substrate and thin film materials

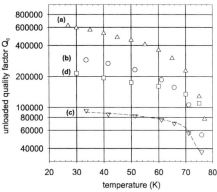

Fig. 3: Influence of the packaging to the unloaded Q factor of TM$_{010}$ mode disk resonators (sapphire substrates)

The influence of packaging loss is illustrated in Fig. 3. In curve (a) the resonator has been measured in a loss-optimised test housing which is relatively large, using a lossless capacitive contact between groundplane and housing. A significant reduction of the measured quality factor can be observed in the miniaturised filter housing (b) that gives a compromise between loss and size. Introducing a lossy contact utilising a one component silver epoxy glue (c) results in a noticeable reduction of Q_0. Assuming a temperature independent packaging loss of 130,000 additionally to (b) reproduces the measured data in the range of measurement error (dashed line). After further improvement of this contact (d), the initial Q_0 values could be reproduced at temperatures of technical interest (70 K and higher), so that an unloaded quality factor of 130,000 could be measured at 70 K.

4. MINIATURISED PLANAR RESONATORS

Miniaturised planar resonators on high-permittivity substrates offer a way to realise highly selective receive filters for communication systems as even at relatively low frequencies (< 2 GHz) a high filter order ($N = 8$-10) can be fabricated on 2" substrates as e.g. shown in Greed et al (1999). As the geometric parameter l_c of these resonators (see (2)) is significantly lower (about 5–10 times) than that of resonators with uniformly distributed current, the achievable unloaded quality factor Q_0 and the power handling capability is restricted. Furthermore, due to the relatively open structure, these resonators are more sensitive to housing losses (Q_h). In order to minimise these housing losses, different planar resonators were investigated with the help of full-wave analysis software (Sonnet *em*) and as result, a double-symmetric hairpin resonator with a significantly higher Q_h was developed

314

(Reppel et al 1997). This theoretical investigation has now experimentally been verified by measurements of the unloaded quality factors of the optimised and of a conventional hairpin resonator whereby the same HTS thin film (YBCO on 0.5 mm thick LaAlO$_3$), the same packaging and the same housing has been used.

First measurements of Q_0 of the developed double-symmetric hairpin resonator resulted in high Q_0-values at low temperatures ($Q_0 > 100,000$ for $T < 38$ K, $f_0 = 1.8$ GHz). However, higher Q_0-values as achieved were expected. The following investigation showed that Q_0 was now not limited by

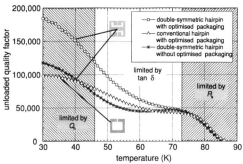

Fig. 4: Measured Q_0 of the developed double-symmetric and of a conventional hairpin resonator ($f_0 = 1.8$ GHz, $P_{in} = -10$ dBm, YBCO on 0.5 mm thick LaAlO$_3$).

housing losses anymore, but limited by the packaging (contact between the ground plane of the resonator structure and the housing, denoted by Q_p). The measurements have been repeated with an improved packaging, and the same Q_0-value of 100,000 could already be achieved at a much higher temperature (50 K). The results of these measurements are shown in Fig. 4. It can be seen that at temperatures close to T_c ($T > 70$ K) the unloaded quality factors are similar as the surface resistance R_s of the HTS thin film is dominating, whereas at lower temperatures ($T < 70$ K) the optimised double-symmetric hairpin resonator displays a significantly higher quality factor. Furthermore, at temperatures around 60 K Q_0 is reduced due to losses in the LaAlO$_3$ substrate.

5.   SUMMARY

Planar microwave resonators have been developed based on thin film HTS materials offering an extremely high unloaded quality factor. The influence of several loss contributions has been investigated in order to improve the unloaded quality factor which is available for the RF filter design. For TM$_{010}$-mode disk resonators on LaAlO$_3$ substrate materials, dielectric losses have been identified as a limitation at temperatures of technical interest. Using a low loss LaAlO$_3$ material Q_0 could be improved by a factor of 2 to 130,000 (70 K, 4 GHz). For practical use of these resonators in miniaturised filter housings additional packaging losses had to be taken into account because of the very high quality factors of the resonators. In a filter set-up a reduction of the unloaded Q to 60,000 had been observed, which could be overcome using an improved contact technology.

The Q-factor of conventional hairpin resonators is at temperatures below 70 K dominated by housing losses. Using an optimised double-symmetric resonator geometry, Q_0 was improved from 70,000 to 100,000 (50 K, 1.8 GHz) due to lower housing losses. Similar to the disk resonator case, this value could only be obtained using a low loss contact technology, indicating the importance of this issue.

REFERENCES

Baumfalk A, Chaloupka H, Kolesov S, Klauda M and Neumann C 1999 IEEE Trans. Appl. Supercond. **9** No. 2, pp 2857-61
Chaloupka H J 1999 to be published in Applications of Superconductivity (NATO ASI series)
Greed R B et al 1999 IEEE Trans. Appl. Supercond. **9** No. 2, pp. 4002-5
Klein N, Schlen A, Tellmann N, Zuccaro C and Urban K W 1996 IEEE Trans. Microwave Theory Tech. **44**, pp 1369-73
Kolesov S, Chaloupka H, Baumfalk A and Kaiser T 1997 J. Supercond. **10** No. 3, pp 179-87
Reppel M, Chaloupka H, Kolesov S 1997 Inst. Phys. Conf. Ser. No 158 pp 323-6
Sonnet *em* electromagnetic analysis software, version 4.0a, Sonnet Software Inc.

Inst. Phys. Conf. Ser. No 167
Paper presented at Applied Superconductivity, Spain, 14-17 September 1999

315

Progress Towards Superconducting Optoelectronic Modulators

J M O'Callaghan[1], J Fontcuberta[2], L Torner[1], R Pous[1], A Cardama[1], S Bosso[3], A Perasso[3].

(1) Universitat Politècnica de Catalunya - Dpt of Signal Theory and Communications - Campus Nord D3 - 08034 Barcelona - Spain
(2) Institut de Ciència de Materials de Barcelona (CSIC). Campus UAB, 08193 Bellaterra - Spain
(3) Pirelli Cavi e Sistemi. Viale Sarca 222. 20126 Milano, Italy.

ABSTRACT: This paper describes the preliminary steps taken to develop Mach-Zehnder integrated modulators with high temperature superconducting (HTS) materials. These include tests to prove: 1) That the performance of standard electrooptical modulators improves when they are cooled down to liquid nitrogen temperatures; 2) That superconducting YBCO with low RF losses can be grown on electrooptic (lithium niobate) substrates; 3) That the reduced RF losses achieved with YBCO electrodes on lithium niobate will have a significant effect on the electrooptical performance of Mach-Zehnder modulators.

1. INTRODUCTION

High-speed, optoelectronic Mach-Zehnder (M-Z) integrated modulators are being widely used in fibre-optic communication systems. They are also candidates to act as interface between cryogenic electronic systems and room temperature electronics. A relevant figure of merit in these devices is the ratio between the bandwidth of the device and the power of the RF modulating source. Since this ratio can be improved by lowering the RF losses in the metal electrodes of the modulator, the use of superconducting electrodes might lead to devices operating at higher modulation speed with weaker electrical signals. Yoshida et al (1999) have already demonstrated working M-Z modulators with Nb and NbN electrodes. This work describes the steps being taken to develop modulators with YBCO electrodes on lithium niobate (LNO) substrates. The estimates of performance enhancements due to the HTS material and due to the low-temperature operation of the device are also described.

2. MACH-ZEHNDER MODULATOR BASICS

A Mach-Zehnder modulator (Alferness, 1982) is made of a LNO crystal, where an optical waveguide pattern has been fabricated by titanium deposition techniques (Hutcheson, 1987). Light is coupled into the titanium waveguides through their ends, at the edges of the LNO crystal, with optical fibres precisely aligned with the waveguides. Two optical waveguides run parallel through most of the wafer and joint near the ends in a single waveguide at the input and output of the device (Fig. 1). Electrical fields of opposite directions can be applied to each of the two parallel waveguides with a coplanar (CPW) electrode. This introduces a phase shift between the optical fields in the two waveguides. Thus, after recombination of the two optical beams at the output waveguide, the envelope of the optical signal is controlled by the voltage in the coplanar electrode and the device works as an amplitude modulator.

In steady-state sinusoidal operation, the relative phase shift $\Delta\phi$ can be expressed as:

$$\Delta\phi(t) = \pi \frac{V_{dc}}{V_\pi} + \pi \frac{Z_m}{Z_m + Z_o} \frac{V_m}{V_\pi} F(f)\cos(2\pi ft) \tag{1}$$

where V_{dc} is a bias voltage, V_m is the applied RF voltage, V_π is the voltage to be applied to shift the phase by 180°, Z_m the characteristic impedance of the CPW electrode, and Z_o the source impedance (usually 50Ω). In the equation above, $F(f)$ is a positive, dimensionless factor that describes the dependence of the phase modulation with the frequency of the RF source (f). At DC $F(0)=1$ and, as frequency increases, $F(f)$ decreases due to the attenuation in the CPW line and due to the mismatch between the velocity of propagation of the electrical fields and that of the optical fields:

$$F(f) = \sqrt{\frac{e^{-2\alpha L} - 2e^{-\alpha L}\cos\theta + 1}{(\alpha L)^2 + \theta^2}} \tag{2}$$

where L is the length of the CPW line, α its attenuation constant and

$$\theta = \frac{2\pi f}{c}(n_m - n_o) \tag{3}$$

being n_m and n_o the effective indexes of the electrical (microwave) and optical fields respectively. By adjusting the DC bias for maximum modulation depth, the envelope of the optical signal that results after combining the fields of the two optical waveguides is proportional to $\Delta\phi$ (as long as $\Delta\phi\ll 1$, which is usually the case) and thus, the variations in the power of the optical signal are a replica of those of the modulating source V_m as long as its frequency f is within the cut-off of $F(f)$.

The value of L in a M-Z modulator represents a trade-off between bandwidth and modulation depth. A large value of L results in a low V_π and a large modulation index but, according to Eq. (2), increasing L reduces the bandwidth, specially if α is large. Therefore, decreasing α improves this trade-off between bandwidth and modulation depth.

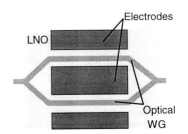

Fig. 1. Diagram of Mach-Zehnder modulator. A CPW electrode introduces a phase shift in the two optical waveguides, which results in an envelope modulation at the output waveguide.

3. OPERATION OF A STANDARD M-Z MODULATOR AT LOW TEMPERATURES

The performance of a commercial, packaged modulator has been measured as a function of temperature from 300K down to 85K. The electrical-to-optical transmission coefficient has been measured using a HP8730A Lightwave Component Analyser, a HP-8168A tuneable laser and other ancillary equipment.

Two main effects were observed as the device was cooled: first, the total optical power at the output of the device decreased and second, the cut-off frequency of the transmission coefficient increased. The first effect is mainly due to optical losses in the coupling between fibres and LNO waveguides; the second is due to the changes of $F(f)$ with temperature (i.e., due to the temperature dependence of α, n_m and n_o). The second effect is summarised in Fig. 2, where the electrical-to-optical transmission coefficient is shown at two temperatures. To

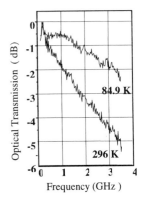

Fig. 2. Electrical-to-optical transmission coefficient in a M-Z modulator with gold electrodes.

account for optical coupling losses, the graphs in Fig. 2 have been normalised with respect to the low-frequency values of the transmission coefficient at room temperature. These results suggest a significant performance enhancement due to the operation at low-temperatures, since the slope of the curve at 85K is about half that of the one at room temperature, and thus the bandwidth of the device roughly doubles at 85K.

4. YBCO THIN FILM DEPOSITION

Superconducting YBCO thin films have been grown on X-cut LNO single crystals using YSZ as buffer layer. This buffer is needed due to the poor structural and chemical compatibility of LNO and YBCO. For Mach-Zehnder modulators, the buffer layer has to allow the epitaxial growth of YBCO, and has to have moderate permittivity at optical wavelengths. YSZ has been chosen as buffer layer since it meets both requirements.

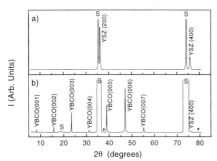

Fig. 3. θ-2θ X-ray diffractograms of a YSZ/LNO film (a), and a YBCO/YSZ/LNO heterostructure (b). Unknown peaks are denoted by a star. Substrate peaks are labeled with a S

The YSZ and YBCO films were deposited by on-axis RF magnetron sputtering, using home-made, stoichiometric targets; the YSZ one contained 16% Y_2O_3. Optimal deposition conditions for YSZ are 50mTorr of total pressure, with 38% O_2 and 62% Ar, substrate temperature 700-750°C and RF power of 50W. For superconducting YBCO, the growth conditions had to be changed to higher temperatures, 780°C, and higher total pressure, 300-400mTorr; the O_2: Ar pressure ratio was kept almost constant. The samples were quenched to 500°C in the growth atmosphere, and then slowly brought to room temperature under 500Torr of O_2.

Fig. 3 shows the θ-2θ X-ray diffraction (XRD) patterns of a YSZ/LNO film and a YBCO/YSZ/LNO heterostructure grown on a substrate at 700-750°C and with 50W of RF power. The YSZ layer grows along the (100) direction; the a-axis spread perpendicular to the film, estimated from the full width at half maximum of the ω-scan for the (400) peak, is of only 0.3°. The optimal thickness of the buffer layer turned out to be around 80 nm. On the top of the (100) YSZ layer, the YBCO film grows along the c-axis direction, i.e. with the superconducting Cu-O planes parallel to the film surface, as intended; ω-scans for the (006) peak indicate a c-axis spread of the order of 1°, for 100nm-thick films. As it can be appreciated in Fig.3, the XRD of the YBCO/YSZ/LNO heterostructure reveals that, after the YBCO growth, the YSZ layer suffers a significant crystalline deterioration, and crystals with (111) orientation are observed. This translates in a noticeable granularity in the YBCO film, which has been observed by AFM.

The superconducting critical temperature Tc of the YBCO films was measured by inductive and transport methods. The best films display a Tc~88K, obtained from the resistance drop, and transition widths of 3-4K.

Overall, the YBCO growth process still has low yields and the results above correspond to the best samples grown. Further work is planned to obtain an improved and repeatable process.

5. PERFORMANCE PREDICTIONS FOR A SUPERCONDUCTING MACH-ZEHNDER MODULATOR

To experimentally asses the attenuation values in the M-Z electrode that could be achieved with a YBCO/YSZ/LNO heterostructure, one-port CPW resonators have been built using this heterostructure and compared with identical resonators made of gold. Figure 4 summarises the results: At 55K and 7 GHz, the attenuation constant α of the YBCO resonators is 2.8 times smaller than that of the gold resonators; furthermore, below 55K losses due to the superconductor are dominated by other types of losses (dielectric and radiation).

From these RF results, frequency-dependent attenuation constants have been inferred for both gold and YBCO electrodes as a function of the electrode length (L). These values of

318

attenuation constants have been used in Eq. (2) with $\theta=0$ to find the 3dB roll-off in the frequency response (i.e., the frequency for which F equals $1/\sqrt{2}$ in Eq. (2)). The election of $\theta=0$ is justifiable since our numerical calculations show that the propagation velocities of the optical and electrical signals can be made equal with the use of a top-plate electrode (Kawano et al, 1991), which has little effect on the overall attenuation constant of the CPW line.

Figure 5 shows the bandwidth vs. L for three types of electrodes: Au at room temperature, Au at 77K, and the YBCO grown for this work (on LNO) at 55K. As found before, a significant improvement is obtained by just cooling a M-Z with Au electrodes. Note however that the two curves for Au decay rapidly as L is increased, indicating a bad trade-off between bandwidth and modulation depth. This does not happen for YBCO which, for $L>1.6$ cm, shows a higher bandwidth and a slower decay with L. Thus, the use of YBCO is advantageous over that of Au except for very short modulators (very broadband but with very small modulation indexes) and would allow high data rates with reduced driving powers of the modulating source.

6. CONCLUSION

Reducing the RF losses in the coplanar electrodes of M-Z modulators can enhance their performance. A sizeable bandwidth improvement has already been observed by just cooling a modulator with Au electrodes to 85K. Significant further improvements could be obtained by substituting Au by YBCO in the modulator electrodes.

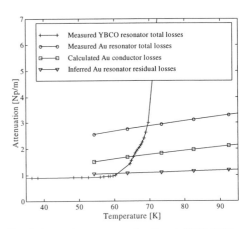

Fig. 4. Attenuation constants of gold and YBCO CPW resonators at 7 GHz. Traces shown are: measured total α for Au electrode (O); calculated α due to Au ($\Box$); α due to dielectric & radiation loss (∇); and total measured α for YBCO (+). Dielectric & radiation losses dominate below 55K.

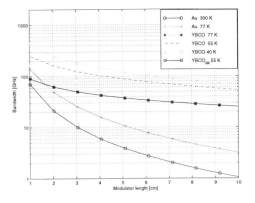

Fig. 5. Bandwidth (GHz) vs. length L (cm) for a M-Z modulator with several different types of electrodes: Au at 300K (O), Au at 77K (+),the YBCO grown on LNO at 55K (■). Other lines correspond to state of the art YBCO on MgO

ACKNOWLEDGEMENTS

This work has been funded by Red Eléctrica de España and Cables Pirelli S.A.

REFERENCES

Alferness R 1982 IEEE Trans MTT, **30** p 1121-36
Hutcheson L 1987 Integrated Optical Circuits and Components (Marcel Dekker) p 169
Kawano K, Nogushi K, Kitoh T, and Miyazawa H, 1991 IEEE Photonics Tech Letters, 3 p 919-21
Yoshida K, Kanda Y, Kohjiro S, 1999 IEEE Trans MTT, **47** p 1201-5

Inst. Phys. Conf. Ser. No 167
Paper presented at Applied Superconductivity, Spain, 14-17 September 1999

Applications of coupled dielectric resonators using HTS coated SrTiO₃ pucks: tuneable resonators and novel thermometry

J C Gallop & L Hao

National Physical Laboratory, Queens Rd., Teddington, TW11 0LW, UK

ABSTRACT: At cryogenic temperatures some single crystal dielectrics possess very low loss and when combined with an HTS shielding enclosure can provide high stability, high Q resonators operable in the temperature range 40 K-70 K which show great promise for frequency standard applications. A number of problems remain to be solved. Here we report implementation of electronic tuning of a sapphire dielectric puck resonator by using a SrTiO₃ tuning element situated in the evanescent field region outside the sapphire puck. In addition the same structure may be used as a very sensitive function of temperature, allowing the possibility of very high resolution thermometry.

1.   INTRODUCTION

A relatively little known but potentially important application of HTS shielded cryogenic dielectric resonators is in realising ultra-stable oscillators with long term stability and very low phase noise performance. We have previously reported on the ability to use combinations of low-loss dielectrics to provide whispering gallery resonances which are temperature compensated, that is the frequency versus temperature variation has a turning point within the useful range of temperature operation, in this case between 40 K and 80 K (Hao et al.). The resonant frequency of a dielectric resonator is determined by its relative permittivity and physical dimensions. To produce a resonator with a specific resonant frequency one must have well characterised and homogeneous single crystal dielectrics and good electromagnetic modelling capabilities. However we estimate that, at best, the machining tolerances (including crystallographic axes alignment), material parameters and modelling accuracy limit the ability to produce a required frequency to an accuracy of no better than ~ 0.01 %. Much greater accuracy is required for local oscillator or frequency standard applications and tuning to an accuracy of perhaps 1 in 10^{10} (1 Hz at 10 GHz) is desirable (1 in 10^6 of the tuning range of 0.01 %).

Various tuning techniques may be attempted. These include temperature tuning, where small controlled temperature changes affect the oscillator frequency via the slowly varying function df/dT. However typically the resonator is operated at a temperature turning point where df/dT=0, to achieve the highest stability, thereby eliminating the possibility of temperature tuning. Mechanically adjustable elements within the resonator housing have also been used but are prone to vibrational instabilities. The most appropriate form of tuning from the point of view of simplicity, speed and noise isolation could be provided by voltage controlled permittivity of a dielectric element in the resonator structure, such as is demonstrated by some paraelectric or ferroelectric high permittivity materials. In the remainder of this paper we explore this latter approach to fine tuning.

2.   RESONATOR TUNING USING SrTiO₃

The single crystal ferroelectric material of choice in this work is the perovskite structured strontium titanate (SrTiO$_3$ or 'STO'). A major reason for this choice is that it is structurally and chemically highly compatible with the cuprate high temperature superconductors and has been used widely as a substrate material for the deposition of HTS SQUIDs and other low frequency thin film superconducting devices. Its relative permittivity is very high even at room temperature and shows considerable crystalline anisotropy. As the temperature is lowered the permittivity rises strongly and bulk crystals of STO enter a paraelectric state before becoming ferroelectric below about 35 K. In both of these states the permittivity can be controlled by an applied electric field. In Figure 1 we report measurements of $\varepsilon_r(T)$ with applied electric field. The results are in reasonable agreement with other measurements (Lacey et al., Gevorgian et al.). The loss tangent $\tan\delta(T)$ of STO at cryogenic temperatures is not as low as for sapphire, being typically in the range 3×10^{-5} to 10^{-3}. It is not clear from the published literature that these losses are intrinsic but their variability suggests this is not the case and that future improvements may be possible.

Microwave losses in the STO puck must not compromise the overall Q of the resonator (since this is essential for high oscillator stability or low phase noise) so the puck is situated some mm from the sapphire puck in its fringing field where the STO stored energy is a small fraction of the total. It is coated with thick film YBCO on both sides to provide electrodes to apply a d.c. electric field E. Microwave power may be coupled in and out through two adjustable loops soldered to the ends of coaxial cables on opposite sides of the copper housing. The resonator is incorporated into a loop oscillator geometry (Gallop et al. 1997) whose frequency is recorded by a microwave counter as the direct voltage applied to the STO tuning element is varied. Figure 2 shows a schematic layout of the loop oscillator. We have demonstrated that modest electric fields of ~50 kVm^{-1} can produce a frequency change of ~0.003%. This should be compared with the design target suggested above of 0.01% so that a tuning voltage of less than 200 V should be adequate to achieve this.

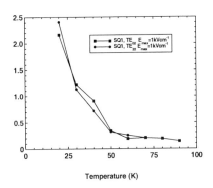

Fig. 1. Temperature dependence of the percentage change in $\varepsilon_r(T)$ of SrTiO$_3$ due to the application of a dc electric field.

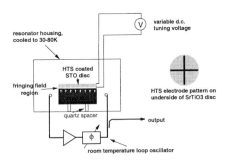

Fig. 2. Schematic of oscillator with HTS/STO fine tuning.

The construction of a tuneable stable resonator requires considerable care in its design. This arises from the fact that minute movements of the STO element with relation to the other dielectric elements which predominantly determine the frequency of the device can cause detectable frequency shifts. Minute movements might arise from thermal inhomogeneities, whether time dependent or spatial gradients in temperature or even creep in the structure which defines the spacing of the STO and main dielectrics. A similar problem has already been observed in copper housings that we presently use in the prototype system outlined here. In future designs we plan to use a low-loss low-permittivity spacer to try to fix the separation of STO and dielectric puck. We are implementing a thin support ring made from single crystal quartz, placed in a region where the fringing fields are relatively low.

3.   COUPLED RESONATORS

The work described in the previous section concerns a configuration where the sapphire puck and the STO puck are rather tightly coupled together. The temperature and voltage change of the composite puck resonant frequency can be reasonably accurately modelled in terms of a voltage V or temperature T dependent perturbation through $\varepsilon_r(V,T)$. A different situation arises when the microwave fields of the sapphire and STO pucks are only extremely weakly coupled. Then the resonant frequency of a particular mode of the sapphire puck f_0 (assumed temperature independent to a first approximation) is quite unaffected by the STO except for a temperature selected resonance condition when the resonant frequency $g_0(T)$ of a mode in the STO comes into close coincidence with that of the selected sapphire mode. In this situation there is a measurable interaction between the two modes so that they may be treated as coupled independent resonators. The following four equations describe a first-order perturbation calculation of the resulting temperature dependent frequencies and linewidths:

$$f(T) = f_0(T) + \mathrm{Re}\left[\frac{A}{(f_0 - g(T)) + iW_{STO}}\right]$$

$$g(T) = g_0(T) + \mathrm{Re}\left[\frac{A}{(g_0(T) - f_0) + iW_{sap}}\right]$$

$$W_{sap}(T) = W_{sap} + \mathrm{Im}\left[\frac{A}{(f_0 - g(T)) + iW_{STO}}\right]$$

$$W_{STO}(T) = W_{STO} + \mathrm{Im}\left[\frac{A}{(g_0(T) - f_0) + iW_{sap}}\right]$$

Here A is the coupling strength between the two modes and is essentially proportional to the overlap of the electromagnetic standing wave patterns of the stored energy of the two field distributions of the modes, integrated throughout the housing. W_{sap} and W_{STO} are the unperturbed linewidths of the sapphire and STO resonances respectively. Which of the two coupled modes is observed in any experiment depends on the nature of the input and output coupling structures, especially their positions.

Figure 3 shows some experimental results for the temperature variation of such a coupled mode (predominantly the sapphire TE_{011} mode) as T is changed over a small range. Note that two STO resonances in turn come in to coincidence with the sapphire resonance, each producing similar frequency and width shifts. The solid curves represent fits to the experimental data for frequency shift and linewidth, using the above equations and treating ε_r as a linear function of T, with coefficient as derived from Lacey et al. (1998). The agreement between experimental and the model is seen to be very good. Note that the coupled resonant frequency varies rapidly over a temperature range of around 50 mK. Simulations of the coupled resonators have been carried out using MAFIA finite difference software and Figure 4 shows the E field distribution for two cases, one far from and the other close to the resonance condition. Note how the E field in the STO element is much enhanced at resonance for the latter.

The coupled resonators provide the possibility of sensitive temperature change measurements to be made. Thus the rate of change of resonant frequency with temperature can be at least as high as 75 MHz/K (see Fig. 3). Since the output frequency of a microwave loop oscillator based on a high Q dielectric resonator can be stable to at least 1 in 10^{11} for an averaging time of 1s (Hao et al.) this thermometer has a potential temperature resolution of ~1.5 nK, comparable with the best high resolution susceptibility or superconducting transition edge thermometers. It has an added advantage that it is non contacting since the temperature change of interest would be occurring in the STO element which is coupled through the fringing electromagnetic field to the main resonator. The latter would require to be separately temperature compensated using, for example a combination of sapphire and rutile elements (Gallop et al.).

322

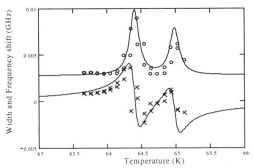

Fig. 3. Experimental results for a coupled mode (predominantly the TE_{011} mode in sapphire) as T is changed over a small range. The solid curves represent fits to the experimental data for frequency shift (crosses) and linewidth (circles).

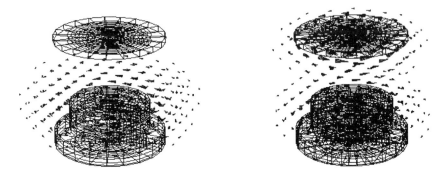

Fig. 4 Finite difference calculations of the E field in a sapphire/STO/quartz resonator: a) far from & b) close to a coupled resonance.

4.   CONCLUSIONS AND FURTHER WORK

We have demonstrated in principle that electronic fine tuning of HTS shielded dielectric resonators may be carried out using low loss tuning elements of single crystal STO to which small voltages may be applied to adjust the relative permittivity and hence the combined dielectric resonator eigenfrequency. Even at cryogenic temperatures the loss tangent in single crystal samples of STO is typically ~10^{-4}. However for the fine tuning application described here the filling factor (that is the fraction of the total stored electromagnetic energy) in the STO element will be only of order 10^{-3} so that overall loaded Q values as high as 10^{7} may be expected for whispering gallery modes. Further work is planned to evaluate the potential of the coupled resonator geometry as an ultra high resolution non-contacting thermometer. Issues of mechanical stability and microwave power dissipation in the STO element will need to be addressed. Further improvements would result from reduction in the loss tangent of single crystal STO.

REFERENCES

M A Hein (1997) *Supercond. Sci. Technol.* **10**, pp. 867-71.

J C Gallop, L Hao, F Abbas and C D Langham (1997) *IEEE Trans. Appl. Supercond.* **7**, pp. 3504-7

S Gevorgian, E Carlsson, P Linner, E Kollberg, O Vendik & E Wikborg (1996) *IEEE Trans MTT* **44**, pp. 1738-41

L Hao, N Klein, W J Radcliffe, J C Gallop and I S Ghosh (1998*) Proc.12th European Frequency and Time Forum*, pp. 112-4

D Lacey, J C Gallop and L E Davis (1998) *Meas. Sci. Technol.* **9**, pp. 536-9

Inst. Phys. Conf. Ser. No 167
Paper presented at Applied Superconductivity, Spain, 14-17 September 1999
© 2000 IOP Publishing Ltd

Bandpass Filter on a Parallel Array of Coupled Coplanar Waveguides.

A Deleniv[a), V Kondratiev[a), I Vendik[a), P K Petrov[b), E Jakku[c), S Lappävuori[c).

[a)Department of Microelectronics and Radio Engineering, St.-Petersburg Electrotechnical University, St.-Petersburg, 197376, Russia.

[b)Department of Microelectronics, Chalmers University of Technology, S-412 96 Göteborg, Sweden.

[c)University of Oulu, FIN-90401, Oulu, Finland.

ABSTRACT. A novel bandpass interdigital filter based on a parallel array of coupled coplanar waveguides (CPW) is presented. A very small size of this filter is superior in comparison with other type filters based on CPW and leads to lower insertion loss. The three-pole Chebyshev bandpass filter with 14% equal-ripple bandwidth at 1.775 GHz and 0.6 dB insertion loss at T=78 K was designed, manufactured and measured. High-Tc superconducting CPW structure is fabricated on 0.5 mm thick $LaAlO_3$ substrate. A good agreement between measured and simulated data was observed. The simulated characteristics of the five-pole Chebyshev filter are presented as well.

1. INTRODUCTION

The use of high-temperature superconducting (HTS) films in microwave applications provides a realization of high performance narrow-band planar filters with low insertion loss and sharp skirts. Coplanar waveguide (CPW) is one of the simplest examples of planar transmission lines and is particularly well suited to single-sided HTS film. The characteristics of CPW do not depend strongly on the substrate thickness, and a wide range of impedances is achievable with a reasonably small cross-section. These facts support an opinion that CPW is very attractive basis for compact filter designing.

Few investigations of filters using CPW have been carried out. Design and performance of the inductively coupled three-pole pass-band CPW filter is discussed in [1]. It should be mentioned that the geometry of a shunt stub providing strong coupling between a feed line and a first (last) resonator was determined experimentally. Dual version of this filter was discussed in [2]. The end coupled half-wavelength sections were used there. The strong coupling realization problem was solved herein with aid of an interdigital capacitor. Meaningfully more simple design procedure is achieved by use of mixed edge and end coupling [3].

In this paper we propose a CPW filter with an interdigital structure which is based on edge coupled quarter-wavelength resonators. This filter is superior in comparison to common filter structures with aligned in series half-wavelength CPW sections or quarter-wavelength sections in the case of using both impedance and admittance inverters.

2. THEORY AND DESIGN

The equivalent diagram of *n* order interdigital filter is presented in Fig.1. It consists of *n+2* coupled quarter-wavelength lines. We use as a basis the theory given in [4] for strip line interdigital filters. It should be pointed that this theory is well suited for CPW based structure on thick substrates providing almost the same effective dielectric constant for all dominant modes. In this case the uniform media with $\varepsilon_{ef}=(\varepsilon_r+1)/2$ can be assumed.

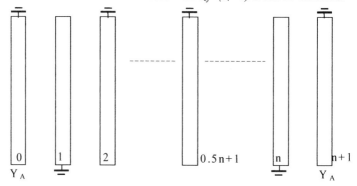

Fig.1 Interdigital filter with short-circuited lines at the ends [4].

Following the guide [4], we use elements of the low-pass prototype to calculate self and mutual capacitances for the cross section of an array of parallel coupled lines. These elements are used to create characteristic matrixes for the structure with a canonical response. The real response will be slightly different due to coupling between nonadjacent lines. We use the spectral domain method for quasi-TEM analysis [5] to calculate both capacitance and inductance matrixes of the structure under simulation. The schematic layout of a three-pole CPW interdigital filter is shown in Fig.2.

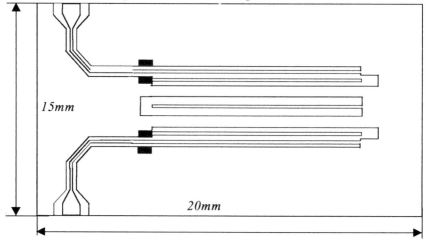

Fig.2 Layout of the three-pole CPW interdigital filter

We assume that multicoupled CPW's present themselves as sequentially placed signal and ground electrodes. It can be seen that the ground electrode was excluded from the filter structure between the feed line and the first (last) resonator. This decision was enforced by a need to achieve a high coupling. In the next step we step by step analyze a characteristic of two CPW neighbors. Their capacitance and inductance matrixes are composed from the elements of corresponding structure matrixes. Such a way allows us to keep information about all cross-couplings in the pair parameters. As the next step we create pair matrixes from the structure with canonical response in the same manner described above.

Further, an iteration process is used to find CPW pair geometry with a response, which is the same as the canonical one. From our experience we can conclude that this procedure provides an elegant treatment of filter synthesis problem being repeated two or three times to each pair of lines sequentially. We truly believe that this readily programmable task can be very useful for CAD oriented design.

We developed two filters with equal-ripple response with a center frequency of 1.775 GHz and 1.8 GHz using mentioned above algorithm (three and five-pole respectively). Two air-bridges were added in each sample to suppress undesirable slot modes (the places are marked as black rectangles in Fig.2). For filter fabrication 15x20mm YBCO film with a thickness of about 200nm, grown on 0.5 mm thick *LaAlO₃* substrate was used. To make a contact pads, a 250nm Au film was electron-beam evaporated in the top of YBCO layer through an *Al* foil shadow mask. The filter patterns were prepared by using photolithography and ion milling.

The three-pole filter was measured at T=78 K using HP 8510C network analyzer and closed cycle cryocooler using standard Willtron test fixture. The simulated and measured characteristics are presented in Fig.3. The filter revealed a maximum insertion loss of 0.6 dB over a 14% pass-band and better than 11 dB return loss. A small pass-band frequency shift and form degradation can be explained as a result of a contribution of open ends and short circuits which have not been taken into account in the simulation procedure. One more possible reason is the dielectric constant, different as compared with one used in simulation.

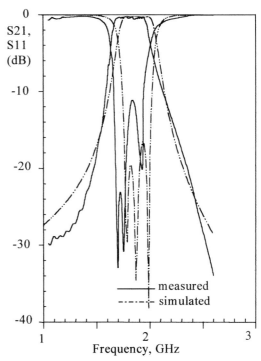

Fig.3 Simulated and measured performance of the three- pole coplanar interdigital filter.

The five-pole version of this filter occupies the same area like a three-pole one. Unfortunately its experimental characteristics are not available at the moment and only simulation results are presented in Fig.4

3. CONCLUSION

An effective CAD-oriented iteration algorithm to synthesize CPW based interdigital filters is proposed. A high-performance HTS filter has been simulated and measured. Good simulation accuracy based on quasi-TEM spectral domain analysis is supported by the experimental results. The filter proposed is suitable for use in miniaturized microwave integrated circuits.

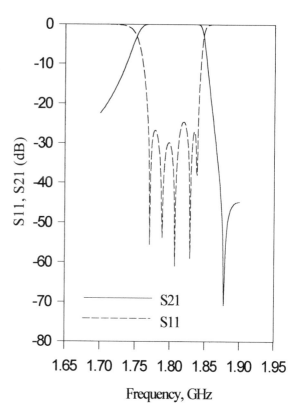

Fig.4 Simulated performance of the five- pole coplanar interdigital filter.

ACKNOWLEDGMENT

One of the authors (Anatoli Deleniv) was supported by the personal INTAS grant YSF 98-116.

REFERENCES

1. Vogt. A. and Jutzi.W. // IEEE Trans., Microwave Theory Tech 1997, **MTT-45**, (4), pp.492-497.
2. Williams D.F. and Schwartz S.E. // IEEE Trans., Microwave Theory Tech 1983, **MTT-31**, (7), pp.558-566.
3. Budimir D., and Robertson I.D. // Microw. Opt. Technol. Lett., 1997, **15**, pp.52-54.
4. Matthaei G.L., Young L., and Jones E.M.T. // "Microwave filters, impedance –matching networks, and coupling structures", Artech House, Norwood, 1980.
   5. Drake E., Medina F., and Horno M. // IEEE Trans., Microwave Theory Tech., 1993, **MTT-41**, (2), pp.492-497.

Inst. Phys. Conf. Ser. No 167
Paper presented at Applied Superconductivity, Spain, 14-17 September 1999
© 2000 IOP Publishing Ltd

Progress in superconducting preselect filters for mobile communications base stations

J S Hong[1], M J Lancaster[1], D Jedamzik[2] and R B Greed[2]

[1]School of Electronic and Electrical Engineering, University of Birmingham, Edgbaston, Birmingham B15 2TT, U.K.
[2]Marconi Research Centre, Great Baddow, Chelmsford, Essex, CM2 8HN, U.K.

ABSTRACT: In this paper we summarise the recent progress of novel HTS preselect bandpass filters for mobile communications applications. The filters with an 8-pole quasi-elliptic function response have been designed to cover a 15MHz sub-band which is centred at different frequencies within a receive band of 1710 to 1785MHz. The HTS filters have been fabricated using double sided $YBa_2Cu_3O_7$ thin films on MgO substrates with a size of 22.5×39mm. The filters have been tested in a vacuum cooler, showing excellent performance. The experimental results including power dependence, temperature dependence and non-linear characteristics are presented

1.   INTRODUCTION

The technology and system challenges of next generation mobile communications have stimulated considerable interest in using high temperature superconducting (HTS) materials for the construction of high performance filters for mobile communication applications (Hong 1998). The implementation of HTS filters consisting of miniature yet very high Q HTS resonators into mobile communications base stations can increase both selectivity and sensitivity, which in return enhances overall system performance to increase capacity and coverage and to reduce infrastructure costs (Greed 1999).

Described here is the recent progress of novel HTS preselect bandpass filters being developed as part of the European Community's Advanced Communications Technologies and Services (ACTS) program on Third Generation Mobile Communications. The project is called as Superconducting Systems for Communications (*SUCOMS*). It involves a number of companies (GEC-Marconi, Thomson CSF and Lybold) and two Universities (Birmingham and Wuppertal).

The preselect bandpass filters have been developed for the DCS1800 standard, which covers a receive band of 1710 to 1785MHz and a transmit band of 1805 to 1880MHz. The filters were designed to cover only a 15 MHz sub-band of the 75MHz receive band. Therefore the typical fractional bandwidth is less than 0.9%. In order to achieve both the low insertion loss and high selectivity, the filters were designed to have an 8-pole quasi-elliptic function response. The HTS filters have been fabricated using double sided $YBa_2Cu_3O_7$ thin films on MgO substrates with a size of 22.5×39mm. The filters have been tested in a vacuum cooler, showing excellent performance. The experimental results including power dependence, temperature dependence and non-linear characteristics are presented.

2.   FILTER DESIGNS AND FABRICATIONS

To start with the filter designs we studied different types of filters against the following specification:

- Centre frequency 1777.5MHz
- Passband width 15MHz
- Passband insertion loss ≤ 0.3dB
- Passband return loss ≤ -20dB
- 30dB-rejection bandwidth 17.5MHz
- Transmit band rejection ≥ 66dB
  (1805-1880MHz)

This leads to identify an 8-pole quasi-elliptic function filter with a pair of transmission zeros at finite frequencies to be designed for the *SUCOMS* project (Hong 1999). We have then developed a novel microstrip filter configuration shown in Fig.1 for realisation of the quasi-elliptic function response. The filter is comprised of eight microstrip meander open-loop resonators. The attractive features of this filter configuration are not only its small and compact size, but also its capability of allowing a cross coupling to be implemented such that the pair of transmission zeros (or attenuation poles) at finite frequencies can be realised. Two filters of this type at different centre frequencies for a low sub-band of 1710 to 1725MHz and a high sub-band of 1770 to 1785MHz were designed with the aid of a full-wave electromagnetic simulator.

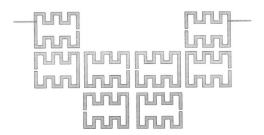

Fig. 1. Developed 8-pole quasi-elliptic function bandpass filter comprised HTS microstrip meander open-loop resonators (Hong 1998).

Several HTS filters have been fabricated using double sided YBa$_2$Cu$_3$O$_7$ thin films on 0.3mm thick MgO substrates with a size of 39×22.5mm and a dielectric constant of 9.65. The fabricated filters were mounted on gold plated titanium carriers. The use of titanium carriers is for the thermal match of the MgO substrates, while the gold plating provides a good electric grounding and low loss. The carriers with the HTS filters were then packaged into filter housings for testing. The housing compartment has a size of 4.25×39.2×30mm. The size of the packaged filters is considerably smaller than similar filters using any other technology.

3.   EXPERIMENTAL RESULTS

The packaged filters were measured in a vacuum cooler. Shown in Fig.2 are measured performance of the two HTS filters at the lowest and the highest channels of the receive band. The measured minimum insertion loss at a temperature of 55K is typically 0.3dB, which includes the losses of two K-connectors and silver-epoxy contacts. The experimental return loss was on average lower than value given by the design specification mainly due to the fabrication tolerance (Hong 1999). Each of the filters showed the similar characteristic of the quasi-elliptic function response with two diminishing transmission zeros near the passband edges. The two transmission zeros result in a sharper filter skirt so as to improve the selectivity of the filter. The measured wide band frequency response of the filter at the lowest channel is plotted in Fig.3, showing very good rejection (almost beyond 80dB) over the entire transmit band.

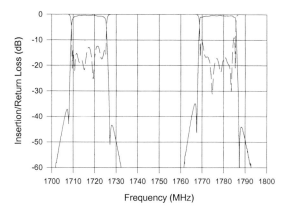

Fig. 2. Measured performance of two fabricated HTS bandpass filters at the lowest Channel (1710 - 1725MHz) and the highest channel (1770 – 1785MHz) of a receive band (1710 – 1785MHz) at a temperature of 55K.

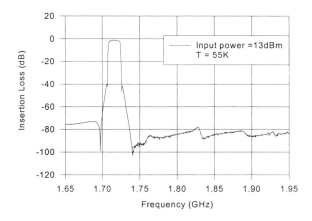

Fig. 3. Measured wide band frequency response of the HTS filter.

The measured temperature dependence of filter is shown in Fig.4. The higher the temperature, the faster the frequency shifts due to the effect of kinetic inductance of the HTS thin films. The centre frequency shifting of the filter was found to be less than 1MHz when the temperature was changed from 45K to 65K. Since the filters will work at a system operating temperature of about 55K, where the temperature stability can be better than 0.5K, the centre frequency shifting would not be an issue for the application. The maximum input power of the filters is specified to be 0dBm. The power dependence of filter was measured. As can be seen from Fig.5, the filter response was almost unchanged even when the input power was increased to 20dBm. The non-linear characteristics of the filter were measured with single tone input. The input tone had a frequency of 1715MHz. The measured outputs of fundamental, second- and third-order harmonics are plotted against the input in a log-log plot of Fig.6. The measured 3rd harmonic was more sensitive to the input power than the 2nd harmonic, and it also followed 3:1 slope against the input, indicating an intrinsic non-linear property of the HTS thin film. The input power was increased up to about 32dBm when the filter showed an insertion loss larger than 3dB. From these measured results a low intermodulation distortion (IMD) can be expected for this type of filters.

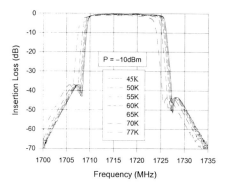

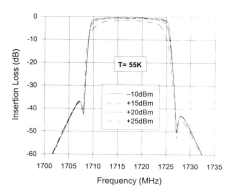

Fig. 4. Measurement of temperature dependence of the HTS bandpass filter.

Fig. 5. Measurement of power dependence of the HTS bandpass filter.

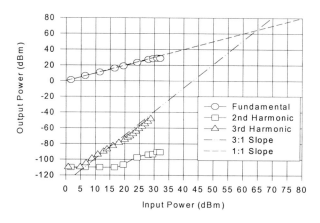

Fig. 6. Measurement of non-linear characteristics of the HTS bandpass filter with single tone input, showing an extrapolated 3rd intercept of +63dBm.

4.   CONCLUSIONS

We have described the progress in the HTS quasi-elliptic function filters that have been developed for mobile communications base stations. It has been demonstrated that the HTS filters can be reliably designed and repeatedly fabricated. The recent experiment results including power dependence, temperature dependence and non-linear characteristics have further verified the excellent performance of this type of filter.

REFERENCES

Hong J S, Lancaster M J, Jedamzik D and Greed R B 1998 *IEEE MTT-S Int. Microwave Symposium,* Baltimore, June 1998
Greed R B, et.al., 1999 *IEEE Trans., Applied Superconductivity, vol.9 no.2, 4002*
Hong J S, Lancaster M J, Jedamzik D and Greed R B 1999 *IEEE Trans. MTT-47, no.9, 1656*

Inst. Phys. Conf. Ser. No 167
Paper presented at Applied Superconductivity, Spain, 14-17 September 1999
© 2000 IOP Publishing Ltd

Frequency dispersion of tunability and losses in ferrite/superconducting structures

A R Kuzhakhmetov[1], A P Jenkins[1], D M C Hyland[1], M Yu Koledintseva[2] and D Dew-Hughes[1]

[1] Department of Engineering Science, University of Oxford, Parks Road, Oxford, OX1 3PJ, UK
[2] Department of Radioengineering, Moscow Power Engineering Institute, 111250 Moscow, Russia

ABSTRACT: The results of experimental research of tunable microwave structures based on $Tl_2Ba_2Ca_1Cu_2O_8$ film resonators combined with ferrite plane elements of various geometry are presented. Tunability by means of external magnetic field and microwave losses in these structures are studied in the frequency range 2-15 GHz. In the structures under investigation the resonators were placed either under the ferrite planes or under ferrite thin films grown on GGG substrates. At 2.4 GHz maximum tunability of about 10% with $Q \approx 450$ was obtained. At 11.4 GHz the maximum Q-factor was about 2500, the best reported recently for tunable structures, with a tunability of 1.3%.

1. INTRODUCTION

There has been an interest for the last several in investigation of tunable high temperature superconducting (HTS) structures by using ferromagnetic (Oates 1999, Tsutsumi 1997) or ferroelectric (Lancaster 1998, Vendik at al 1999) materials. The basic operation of the both groups of devices is changing the permittivity or the permeability by applying an external magnetic or electric field respectively. As a matter of fact all conventional ferroelectrics have quite large losses at microwave frequencies (typically ferroelectric $SrTiO_3$ $\tan\delta_{elec} \approx 10^{-2}$) which prevent large values of quality factor. The best ferromagnetic materials have quite low electrical losses $\tan\delta_{elec} \approx 10^{-3}$--$10^{-4}$ but very big magnetic losses near the ferromagnetic resonance where tunability is a maximum. Our experiments established the possibility to find a compromise between reasonable high tunability and losses in ferrite/superconducting structures in a relatively wide frequency range.

2. EXPERIMENTAL

Superconducting TlBaCaCuO resonators combined with ferrite substrates of various saturation magnetisation were produced to investigate tunability of these structures. $Tl_2Ba_2Ca_1Cu_2O_8$ thin films were prepared using a two stage, ex-situ process, the details of which are published elsewhere (Morley at al 1993). Briefly, precursor films were RF sputter deposited from a 200mm diameter Ba:Ca:Cu composition 2:1:2 target, followed by a post-deposition anneal with Tl-2212 powder as a source of Tl_2O vapour. This process routinely produces phase pure films of around 0.7 μm thickness. The surface resistance of 1cm square films produced using this process was measured using a sapphire dielectric resonator at 24 GHz (Jenkins at al 1996). Shown in the Fig.1 is a plot of surface resistance as a function of temperature (scaled to 10 GHz assuming an f^2 relation). TlBaCaCuO films had the following average parameters: $T_c \cong 103$-105K and $R_s \cong 0.5$ mΩ (at 10 GHz, 77K).

These Tl-based films deposited on LaAlO$_3$ substrates were used to manufacture microstrip resonators. The resonators had a split ring shape with around 5 mm diameter and 0.5 mm line width that corresponded to a fundamental mode at 2.4 GHz, and impedance Z_0=34 Ω. An average tunability factor of such a ring resonator was expected in comparison with transversely and longitudinally wave propagation according to Tsutsumi (1997) where it was found that changing the magnetic field direction from transverse to longitudinal improves magnetic tunability. Direction of applied magnetic field with the electromagnetic wave determines the Faraday rotation which governs tunability. However the split ring geometry was chosen in order to reduce size and radiation losses.

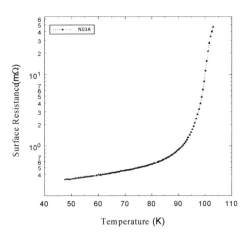

Figure 1. The surface resistance's temperature dependence of typical Tl$_2$Ba$_2$Ca$_1$Cu$_2$O$_8$ thin film

Several superconducting resonators and an identical silver one were produced. Structures were placed in a microwave copper box which allowed good radiation shielding. External magnetic field was applied using either a solenoid (with average diameter 30 mm, 1300 turns and magnetic field strength about 1200 Oe with a current 3 A) or Helmholtz coins. It was established by the experiments that the Q-factor and f_0 of resonators without ferrite was not affected by the magnetic field up to 1200 Oe. The resonators' unloaded Q-factor without ferrite were around 3000 with a copper ground plane and not less than 4500 with a superconducting one at the fundamental mode 2.4 GHz and 77K. In the structures under these investigations the resonators were placed under either the polycrystalline garnet ferrite substrates (Hiltek Microwave Y15A, Y72A) or garnet ferrite thin films (5 μm and 10 μm thick) grown on Gallium Gadolinium Garnet (GGG) substrates. Employed ferrites had saturation magnetisation $4 \cdot \pi \cdot M_s$=300÷1760 G, magnetic linewidth ΔH=1Oe and tgδ_{elec}≈10^{-4} at room temperature. External magnetic field is applied parallel to the ferrite and LaAlO$_3$ planes along the wave propagation direction.

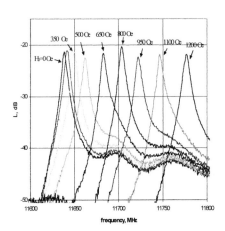

Figure 2. Frequency response at different values of the applied fields.

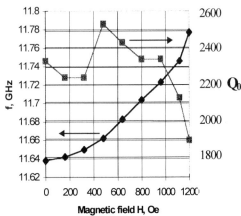

Figure 3. Resonant frequency and Q-factor dependence on applied magnetic field.

3.   RESULTS AND DISCUSSION

Tunability and losses are observed at fundamental and minor modes in the frequency range 2-15 GHz and at temperature down to 50K using an HP8510 Network analyser and Leybold Heraues cryocooler. The typical frequency response of the resonator as a function of applied magnetic field is shown at Fig. 2. The estimated values of resonant frequency and Q-factor are showed at Fig.3. It can be seen that the decrease in Q is not monotonic (Fig.3). We could suppose that this effect was due to nonuniform electromagnetic field distribution in the split ring resonator. However a similar Q-factor dependence on magnetic field was observed by Tsutsumi (1997) and Oates (1999) where linear resonators were used. Thus we probably conclude that it is the effect of dispersion on the coupled microstrip resonator's electromagnetic properties and the ferrite's magnetic properties through Faradey rotation at low magnetic field strength. At 11.4 GHz the maximum Q-factor using ferrite Y75A with $4 \cdot \pi \cdot M_s$=1500 G was around 2500, higher than in recent papers, and the achieved tunability was 1.3%.

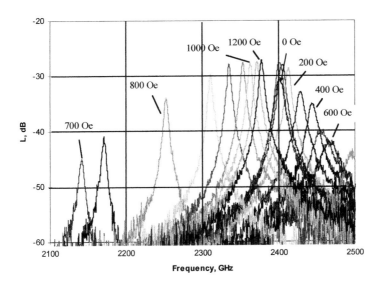

Figure 4. Frequency response of resonator at the fundamental mode near ferromagnetic resonance.

The maximum magnetic field strength of our magnetic system (1200 Oe) allowed the frequency response at the fundamental to be observed (2.4 GHz) before and after ferromagnetic resonance which resonant frequency was given by Clarricoats (1961)

$$f_0 = \mu_0 \cdot \gamma \cdot H_0$$

where H_0 is applied magnetic field, γ - gyromagnetic ratio and $\mu_0 \cdot \gamma \cong 2.8$ MHz/Oe.

Thus it was possible to reach up to 3.3GHz for the ferromagnetic resonance frequency using our equipment. Fig.4 shows the frequency response of the resonator near ferromagnetic with 5 μm thickness YIG grown on GGG. It is quite significant to see the

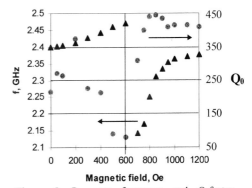

Figure 5. Resonant frequency and Q-factor dependence on applied magnetic field near ferromagnetic resonance.

resonant peak disappear with increasing magnetic field and then reappear at the lower frequency according to the change of complex permeability μ' and μ'' (Clarricoats 1961). This is cleaerer in Fig.5 where the calculated values of resonant frequency and Q-factor are shown. The frequency shift was observed to be rapid when the ferromagnetic resonance frequency was close to the fundamental mode of the resonator (H_0=600-700 Oe) and rather slow where the ferromagnetic resonance frequency was far away from 2.4 GHz. In this experiment the resonant shift of 250MHz and an insertion loss of -20 dB with Q_L=250-450 was obtained. As we can see from Fig. 5 there is a possibility to have magnetic switching between two resonant peaks having the same quality factor. It must be noted that the realatively low values for quality-factor can be explained by an impurity or a polycrystalline structure for the YIG film.

4.   CONCLUSIONS

Magnetically tunable split ring TlBaCaCuO resonators have been demonstrated. Measurements were carried out at different temperatures which may allow an estimate of effective values of real and imaginary parts of the permeability. Measurements in a wide frequency range established the strong dependence of parameters that such superconducting/ferroelectric structures have upon non-linear behaviour of μ' and μ''. The experiments showed that there was a compromise between high tunability and relative good quality factor in magnetically tunable resonant structures by choosing a correct working point on a complex permeability characteristic and at a required frequency.

ACKNOWLEDGEMENT

This work was supported by the Royal Society/NATO grant and in part by the EPSRC under grant no. GR/L08137. The authors would like to thank B.Pedley for her valuable contribution.

REFERENCES

Clarricoats P G Microwave ferrites (London: Chapman & Hall) 1961
Lancaster M J, Powell J, Porch A, Superconductor Science & Technology. 11, no.11; Nov. 1998; pp 1323-34.
Jenkins A P, Dew-Hughes D, Studies in High Tc, ed A.V.Narlikar, 17 (Part I), 1996, pp179-219
Oates D E, Dionne G F, IEEE Trans. on Appl. Superconductivity, 9, 1999, pp 4170-75.
Morley S M, Jenkins A P, Su L Y, Adams M J, Dew-Hughes D and Grovenor C R M, IEEE Trans. on Appl. Superconductivity, 3, 1993, pp 1753-1756.
Tsutsumi M, Fukusako T, Electronics Letters, Apr 10 1997, 33, No.8, pp.687-688
Vendik G, Hollmann E K, KozyrenA B, PrudanA M, Journal of Superconductivity. 12, no.2; 1999; p.325-38.

Inst. Phys. Conf. Ser. No 167
Paper presented at Applied Superconductivity, Spain, 14-17 September 1999
© 2000 IOP Publishing Ltd

Comparative study of YBaCuO thin film microwave surface resistance measurement methods*

A Dégardin[1], A Gensbittel[1], M Boutboul[2], P Crozat[3], M Achani[4], Y Roelens[4], J-C Carru[4], M Fourrier[2] and A Kreisler[1]

[1]LGEP, SUPÉLEC, UMR CNRS 8507, Universités Paris 6 et Paris 11, Plateau du Moulon, 91192 Gif-sur-Yvette Cedex, France
[2]LDIM, EA 253 MENRT, Université Paris 6, Case 92, 4 Place Jussieu, 75252 Paris Cedex 05, France
[3]IEF, UMR CNRS 8622, Université Paris 11, Bâtiment 220, 91405 Orsay Cedex, France
[4]IEMN, UMR CNRS 8520, Université Lille 1, Avenue Poincaré, BP 69, 59652 Villeneuve d'Ascq Cedex, France

ABSTRACT: Microwave surface resistance (R_s) of granular YBaCuO thin films sputtered on MgO single-crystals was investigated. For unpatterned films, R_s was determined from conical cavity resonator measurements. For patterned film structures (coplanar waveguide and microstrip resonators), theoretical models were used for R_s extraction. Moreover, for the microstrip structure, three temperature domains representative of the power non-linear effects were established and discussed by considering intrinsic and extrinsic contributions to the surface resistance.

1. INTRODUCTION

The knowledge of the surface resistance R_s of high critical temperature superconductors plays a major role in the development of passive microwave devices. In order to study devices made of YBaCuO thin films deposited on MgO single crystals, we have undertaken several measurements in the 1 to 40 GHz frequency range involving two aspects. The first one is related to R_s determination from conical cavity resonator measurements on unpatterned films. The second one concerns R_s determination of patterned film structures such as coplanar waveguide resonators and microstrip resonators. Non linear effects with respect to microwave power are also considered.

2. EXPERIMENTAL PROCEDURES

2.1 YBaCuO film elaboration

200 nm thick YBaCuO films were grown by on axis rf magnetron reactive sputtering using an *in situ* process, described in detail by Dégardin et al (1998). Due to its low dielectric constant (ε_r = 9.7), (100) MgO was used as substrate ($10\times10\times0.5$ mm³). Prior to YBaCuO deposition, MgO was annealed at 1000 °C for 5 h under flowing oxygen, in order to create steps on its surface and so favour YBaCuO growth. *c*-axis oriented granular films were obtained, with critical temperature T_c of 86 K and critical current density $J_c = 7.5\times10^4$ A/cm² at 77 K. This rather low J_c value was determined from measurements on unpatterned films; thus the presence of a rather large number of weak links within the film area under test can account for J_c depletion.

* Work supported by a specific funding of Paris 6 University (*PPF Composants supraconducteurs*)

2.2   Cavity resonator technique for surface resistance determination

Before patterning, the temperature dependence of YBaCuO film surface resistance was measured between 50 K and 100 K by end-wall substitution of a conical cavity resonating at 36 GHz in the TE_{011} mode (Roelens et al 1998). The cavity was machined from oxygen-free high-conductivity (OFHC) copper to ensure low ohmic loss. Its length was 10 mm and end-diameters were 9.4 mm and 13 mm, thus allowing to characterize samples with $10{\times}10$ mm^2 or $15{\times}15$ mm^2 area. The cavity was mounted on a cold finger inside a cryogenerator. An indium foil was inserted between the cavity and the cold finger in order to ensure good thermal contact. Microwave transmission measurements were performed using a HP85107 network analyser. The measured surface resistance R_{sm} (also called the effective surface resistance) was determined by the following expression: $R_{sm} = A/Q_0 - BR_{Cu}$, where A and B are geometrical factors, Q_0 the unloaded cavity quality factor measured with a YBaCuO film and R_{Cu} the OFHC copper surface resistance. R_{sm} is film thickness dependent.

2.3   Device technology

Half-wavelength ($\lambda/2$) coplanar waveguide (CPW) structures and microstrip resonators were patterned using a conventional photolithographic process followed by Ar ion-beam milling.

CPW structures were i) transmission lines (of geometrical length = 6.8 mm) of various widths (w = 30 to 100 μm) and ii) 10 GHz resonator (geometrical length = 6.4 mm, w = 70 μm, input/ouput coupling gap length = 30 μm). The CPW to ground plane distance was chosen to obtain Z_0 = 50 Ω as characteristic impedance.

Microstrip resonators were formed of 0.5 mm wide and 11.14 mm long S-shaped lines. The fundamental resonant frequency f_0 was 5.3 GHz. Both resonator ends were open-circuited and coupled to feed lines through gaps of 0.5 mm length. 1 μm thick Au films (with 10 nm thick Cr sticking layers) were deposited onto the backside of the MgO substrate as ground plane.

Au contact metal was deposited by e-beam evaporation through metal masks for the microstrip structures or by a photoresist liftoff for CPW lines.

To compare superconductor with normal metal performances, 1 μm thick Au circuits were manufactured on similar structures.

2.4   Device characterization

For coplanar structures, transmission parameter measurements were performed with a HP8510C network analyser in the 5 to 40 GHz range using a helium flow cryostat (Crozat et al 1994). For contacting samples, two commercial microwave coplanar probes (Picoprobes) were used.

For microstrip resonators, scattering parameters were measured using a HP8720D network analyser operating around 5 GHz. SMA connectors were used to link the metallic package installed in a helium flow cryostat to the network analyser (Boutboul et al 1998).

3.   RESULTS AND DISCUSSION

3.1   Unpatterned films

The determination of unpatterned film R_{sm} values from 36 GHz conical cavity resonator measurements allowed us to study the quality of YBaCuO films and thus to discuss substrate preparation. Some thermal variations of R_{sm} are shown in Fig. 1.

We notice that films elaborated on MgO single-crystals exhibit better results than the film deposited on a polycrystalline YSZ substrate. Moreover they exhibit lower values than for copper below 70 K. Finally, the film deposited on annealed MgO substrate exhibits lower R_{sm} values than those observed for the film deposited on unprocessed MgO. These results demonstrate the influence of the crystalline nature (polycrystal versus single-crystal) and of the preparation (thermal annealing) of the substrate on the microwave quality of YBaCuO films.

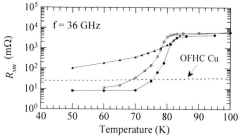

Fig. 1 Effective surface resistance at 36 GHz as a function of temperature for three YBaCuO films deposited on unannealed MgO substrate (open circles), annealed MgO substrate (filled circles) and polycrystalline YSZ substrate (triangles). The dashed line represents R_{sm} for OFHC copper.

3.2 Coplanar waveguide structures

From S_{ij} parameter measurements performed on transmission lines, we have determined the expressions of the resistance per unit length R_u and the propagation constant β respectively given by:

$$R_u = 2 Z_0 \left| S_{21}(f) \right|_{dB} \frac{1}{8.868 \, l} \qquad \text{and} \qquad \beta = \frac{-\,\text{Arg}(S_{21}(f))}{l}$$

where l is the line length. From β, the phase velocity V_{ph} was calculated according to $V_{ph} = 2\pi f / \beta$.

Fig. 2(a) shows R_u as a function of frequency for two line width values. We can notice that R_u, which is representative of the losses in the material, exhibits the usual superconducting f^2 law in the considered frequency range. Besides, we have verified that for metallic lines the evolution followed the expected $f^{1/2}$ law. Moreover, we remark in Fig. 2(a) that R_u is inversely proportional to the line width w. These results validate our technological processes. Finally, the evolution of V_{ph} which is constant through the whole considered frequency range, as shown in Fig. 2 (b), confirms the frequency independence of phase velocity for superconducting lines, which are therefore free from phase dispersion.

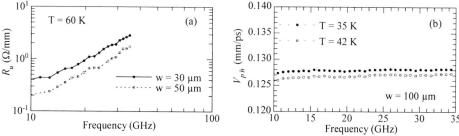

Fig. 2 (a) Evolution of resistance per unit length as a function of frequency, for two widths of CPW lines; note the f^2 dependence. (b) Evolution of phase velocity as a function of frequency for a 100 µm width CPW line; note the non-dispersion.

From the measurements of the temperature-dependent unloaded quality factor Q_0 of the CPW resonator, R_s could be inferred as (Rauch et al 1993): $R_s = \mu_0 2 \pi f_0 \lambda_L L_{tot} / Q_0 L_{kin}$, where μ_0 is the vacuum permeability, λ_L the London magnetic penetration depth, L_{tot} the total inductance per unit length, which is composed of the external inductance L_{ext} and the kinetic inductance L_{kin} of the superconducting line. L_{kin} is dependent on λ_L, which itself depends on temperature. λ_L was obtained by comparison between the measured and the calculated temperature dependence of the resonant frequency. For the calculation, a two-fluid model was used for the temperature dependence as $\lambda_L(T) = \lambda_{L0} / \sqrt{1 - (T/T_c)^\gamma}$, with $\lambda_{L0} = \lambda_L(T = 0K)$. Preliminary results have given $\lambda_{L0} = 500$ nm and $R_s \approx 2$ mΩ at 10 GHz below 35 K. This rather high value of R_s is characteristic of a granular film, which is our case.

3.3 Microstrip resonators

For microstrip resonators, the extraction of R_s values was performed from Q_0 values, as shown in Fig. 3. (a). Dielectric and ground plane losses were taken into account in the calculation of R_s (Boutboul et al 1998). Film granularity can account for the rather high R_s values found at 5 GHz (1 to 2 mΩ).

We have also considered the power dependence of Q_0 and investigated whether the observed non-linearity originated from intrinsic properties of the superconducting material or extrinsic properties such as grain boundaries. As shown in Fig. 3 (b), the non-linearity was characterised by the ratio $r = \Delta Q_0 / Q_{0l}$, with $\Delta Q_0 = Q_{0l} - Q_{0h}$. Q_{0l} represents Q_0 at low input power (-15 dBm) and Q_{0h} is Q_0 at high input power (+ 10 dBm). We have identified three temperature regions, in line with Halbritter's model (1995) who discussed the R_s behavior as a function of temperature. Two components are attributed to R_s: i) an intrinsic component R_{si} which varies according to BCS theory and ii) an extrinsic component R_{sext} related to such defects as grain boundaries. R_{sext} dominates for $T < 0.9 T_C$. In region #1 ($T < T_C/2$), the low (and constant) r value would mean that non linear effects are hidden by dominant extrinsic effects. Region #2 ($T_C/2 < T < 0.9 T_C$), in which r increases rapidly, would correspond to the Josephson vortex penetration in grain boundaries. Finally in region #3 ($0.9 T_C < T < T_C$), R_s is dominated by the intrinsic properties of the material. As the density of Cooper pairs decreases as T is closer to T_C, the non-linear effects would decrease and so explain the decreasing of r in region #3.

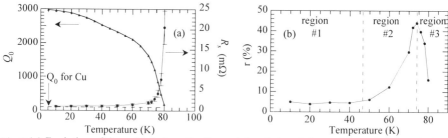

Fig. 3 (a) Evolution of the unloaded quality factor (triangles) and the surface resistance (circles) as a function of temperature; measurements were performed at low input microwave power (-15 dBm). (b) Evolution of the non-linearity as a function of temperature (see text).

4. CONCLUSION

Microwave surface resistance (R_s) was determined on unpatterned YBaCuO thin films as well as on patterned film structures such as coplanar waveguide and microstrip resonators. R_s values were mainly representative of the film granularity. Non linear effects with respect to microwave power were put into light and discussed with relation to the R_s behavior as a function of temperature.

REFERENCES

Boutboul M S, Kokabi H and Pyée M 1998 Physica C **309**, 71
Crozat P, Bouchon D, Hénaux J-C, Adde R and Vernet G 1996 Proc. 183rd Meeting of the Electrochimical Society, eds S I Raider, D P Foly, C Claeys, T Kawi and R K Kirschman, p. 283
Dégardin A, Bodin C, Dolin C and Kreisler A 1998 Eur. Phys. J. AP **1**, 1
Halbritter J 1995 J. Supercond. **8**, 691
Rauch W, Gornik E, Sölkner G, Valenzuela A A, Fox F and Behner H 1993 J. Appl. Phys. **73**, 1866
Roelens Y, Achani M, Bourzgui N, Tabourier P and Carru J C 1998 PIERS Conf. proc. **2**, 644

Inst. Phys. Conf. Ser. No 167
Paper presented at Applied Superconductivity, Spain, 14-17 September 1999
© 2000 IOP Publishing Ltd

Calculation of RF spurious in planar superconducting structures.

J. Parrón, C. Collado, J. Mateu, E. Úbeda, J.M. O'Callaghan, J.M. Rius

Universitat Politècnica de Catalunya - Dept of Signal Theory and Communications - Campus Nord D3 - Jordi Girona, 1-3 08034 Barcelona - Spain

ABSTRACT: A general algorithm has been developed to calculate intermodulation and harmonic spurious generated by planar microstrip superconducting structures. The algorithm combines a numerical tool used in linear electromagnetic calculations (Method of Moments) with Harmonic Balance routines which are normally used for calculating the non-linear response in electronic circuits with lumped components. With these techniques, the non-linear response of planar structures can be calculated without restrictions in their shape. As an example, the fundamental and third-order intermodulation products (IPs) of a half-wave microstrip resonator are calculated.

1.   INTRODUCTION

Planar, high temperature (HTS) superconducting filters are being actively developed to take advantage of their low volume, reduced insertion losses and high selectivity. Even though superconducting filters with coupled-line resonators already offer a significant reduction in volume respect to those based on cavity or waveguide resonators, other types of superconducting miniature resonators are also being used (Hong et al 1996, Hong et al 1997, Zhang et al 1995) to minimise the wafer surface needed and to facilitate packaging and cooling. These miniaturised resonators hold large current densities which enhance non-linear effects such as intermodulation distortion. Thus, depending on the type of resonator chosen, for a given HTS material quality there is a trade-off between filter size and non-linear performance. Predicting this trade-off at the design stage would require CAD packages able to model complicated planar topologies with non-linear electromagnetic algorithms. This paper presents a step towards developing these CAD packages by combining the Method of Moments –widely used in linear electromagnetic simulations– with the Harmonic Balance algorithm used in non-linear simulations of lumped electronic circuits.

2.   LINEAR ANALYSIS

For solving the linear part of our problem, we will work in the spectral domain by applying the boundary condition on the surface of the embedded conductors:

$$\hat{n} \times (\vec{E}_i(\vec{r}) + \vec{E}_s(\vec{r})) = \hat{n} \times Z_s \vec{J}_s(\vec{r}) \qquad (1)$$

that is, tangential components of incident and scattered electric fields are related to tangential components of surface current, $\vec{J}_s(\vec{r})$, through a surface impedance Z_s (zero for perfect electric conductors, complex quantity for superconductors).

As we are interested in obtaining the current induced on the surface of the conductors due to incident fields (those produced by the signal source in absence of metallic or superconducting structures), we express the scattered fields generated by the surface current $\vec{J}_s(\vec{r})$ in terms of vector and scalar potentials. So the equation to solve is (Mosig et al 1989):

$$\hat{n} \times \vec{E}_i(\vec{r}) = \hat{n} \times (j\omega \int_s \vec{G}_A(\vec{r} \mid \vec{r}') \vec{J}_s(\vec{r}') ds' - \frac{1}{j\omega} \nabla (\int_s G_V(\vec{r} \mid \vec{r}') \nabla \cdot \vec{J}_s(\vec{r}') ds') + Z_s(\vec{r}) \vec{J}_s(\vec{r})) \qquad (2)$$

which can be re-written as:

$$\hat{n}\times\vec{E}_i(\vec{r})=L_{EJ}(\vec{J}_s(\vec{r}))$$ (3)

where $L_{EJ}(\vec{J}_s(\vec{r}))$ is a linear operator including the appropriate Green's functions for vector and scalar potentials $\bar{\bar{G}}_A(\vec{r}\,|\,\vec{r}\,')$, $G_V(\vec{r}\,|\,\vec{r}\,')$ (Mosig et al 1989) and the surface impedance Z_s.

Here, we apply the well-known Method of Moments (MoM, Harrington 1968) by decomposing $\vec{J}_s(\vec{r})$ in Rao, Wilton and Glisson (RWG, Rao et al 1982) linear triangle basis functions ($\bar{u}_i(\vec{r})$). Testing functions ($\bar{w}_m(\vec{r})$) are the same RWG triangles (Gallerkin Method). The use of these basis functions allows us in first place, to conform accurately any geometrical surface or boundary, and, in second place, to simplify some computations.

Then, expressing (3) in matrix form:

$$E = Z\,J$$ (4)

where the matrix elements are:

$$J_s(\vec{r}) \approx \sum_i j_i\,\bar{u}_i(\vec{r}) \qquad e_m^I = \langle \bar{w}_m(\vec{r}), \hat{n}\times\vec{E}_i(\vec{r}) \rangle \qquad z_{mi}^E = \langle \bar{w}_m(\vec{r}), L_{EJ}(\bar{u}_i(\vec{r})) \rangle$$

Here $\langle X,Y \rangle$ denotes the Hilbert inner product and j_i are the unknowns to find out.

To solve (4) we can compute, store and invert the impedance matrix Z if the number of unknowns is small or use one of the different available iterative techniques when this number of unknowns becomes larger. As examples of these techniques, we can mention conjugate gradient (CG) or biconjugate gradient (BiCG), which, used together with multilevel matrix decomposition algorithm (MLMDA, Rius et al 1999) or multilevel fast multipole algorithm (MLFMA, Song et al 1997), allow us to obtain good results in large problems with low computational requirements.

3. NON LINEAR ANALYSIS

The Z matrix obtained from the linear analysis of the meshed structure relates the incident field E with the vector of current density J at a given frequency. To take into account the non-linear effects, the surface impedance can be decomposed in two terms, one linear and other nonlinear. The lineal contribution is considered in a linear analysis and is included in the matrix Z (Eq. 4). The non-linear term is calculated in time-domain. It produces a field e_{nl} in each cell of the mesh depending on the current density j crossing through the cell.

$$e_{nl}(\vec{r}) = Rs_{nl}(j(\vec{r}))\cdot j(\vec{r}) + Xs_{nl}(j(\vec{r}))\cdot\frac{dj(\vec{r})}{dt}$$ (5)

For the K+1 frequencies appearing in the spurious signal under analysis, the final solution of the vector of currents J has to meet:

$$E_i^k + E_{nl}^k = Z^k \cdot J^k \qquad k = 0...K$$ (6)

where E_{nl}^k is the *kth* component of the Fourier Transform of the field quantity e_{nl}.

This kind of non-linear problems can be solved using an iterative algorithm based on the harmonic balance technique (Fig.1).

Fig.1. Scheme for the MoM-HB algorithm. Uppercase variables: frequency domain, lowercase: time domain. Direct and inverse Fast Fourier Transform (FFT) are used to convert one to another.

The algorithm begins considering the lineal solution J as initial vector of currents. This current is converted to time domain and using (5) the non-linear electric field e_{nl} is calculated. This field is then converted to frequency domain (E_{nl}^m) and, using (6), the new vector of currents is found and compared with that initially proposed. If the difference is greater than a user-defined error bound, a new estimate of currents is given for the iteration m+1:

$J^{m+1} = p \cdot J^m + (1-p) \cdot J^{m-1}$. Where the parameter p ($0<p<1$) fixes the trade-off between speed and robustness of the convergence towards the solution.

4. EXAMPLE

A two-port, half-wave microstrip resonator with capacitive coupling has been numerically tested. The non linearity has been considered quadratic in order to make comparisons with other authors:

$$Rs_{nl}(j(\vec{r})) = \Delta Rs \cdot |j(\vec{r})|^2$$
$$Xs_{nl}(j(\vec{r})) = \Delta Xs \cdot |j(\vec{r})|^2 \qquad (7)$$

As a first step the lineal response of the resonator is obtained for the two fundamental harmonics f_1 an f_2 (1KHz apart to ensure that they are within the resonance band), and also for the two third order IPs $2f_1$-f_2 and $2f_2$-f_1.

The resonator is then excited with f_1 and f_2 and the HB algorithm is applied to obtain the distribution of current for each frequency. At each iteration, equation (5) is used in time-domain to calculate non-linear electric field vector e_{nl} from the current density vector j. The resulting vector (e_{nl}) depends on the number of frequency components that are being considered (in j and in the rest of vectors and matrices throughout the problem). However, the lowest frequency components in all variables do not depend strongly on the total number of frequency components being considered. To improve computational time, our simulation does not take into account spurious beyond the third-order intermodulation products. Tests performed using simpler, 1-D non-linear analysis software (Collado, 1999) indicate that the loss in accuracy is not significant.

Fig. 2 shows the current density of the third order IP $2f_1$-f_2 both at the surface of the strip, and in a cut along the centre of the resonant microstrip line.

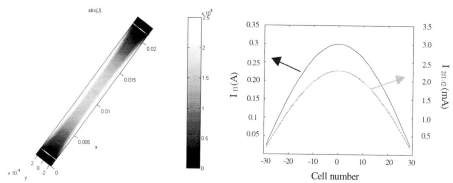

Fig 2: 2D plot of the current density of the 3rd order IP ($2f_1$-f_2) in a microstrip resonator and current distribution of that IP and one fundamental harmonic along the microstrip resonator. The non-linearities in surface impedance are set to: $\Delta Rs=5.1e-13\Omega$ m^2/A^2, $\Delta Xs=2.3e-21$ H m^2/A^2. These values have been obtained from the intrinsic dependence of the density of superelectrons with the current density given in Dahm and Scalapino (1997). The other parameters are: w=0.5 mm, h=0.5 mm, er=9.85, tanδ=1e-5, l=20 mm, Rs=10$\mu\Omega$, λ_L=200 nm

For a two port resonator, Dahm and Scalapino (1997) have derived an equation that relates the maximum current of the third order intermodulation product ($I_{2f_1-f_2}$) with that of the fundamental (I_{f_1}). This equation makes use of the line's non-linear distributed parameters $R(I)$, $L(I)$, C, G, where $R(I) = R + \Delta R \cdot I^2$ and $L(I) = L + \Delta L \cdot I^2$. These parameters depend on the surface impedance and on the distribution of current density in a cross section of the line (Vendik et al 1998):

$$\Delta R = \frac{\int_{-w/2}^{w/2} Rs_{NL}(j) * j^2 dw}{\left(\int_{-w/2}^{w/2} j dw\right)^4} \qquad \Delta L = \frac{\int_{-w/2}^{w/2} Xs_{NL}(j) * j^2 dw}{\left(\int_{-w/2}^{w/2} j dw\right)^4} \qquad (8)$$

Table 1 shows the current levels for the fundamental and third order IP (I_{2f1-f2}) as calculated by our software and compares them with Dahm and Scalapino's equation, wich makes use of ΔL, ΔR in Eq. 8. Differences are smaller than 8%.

	I_{f1} (A)	I_{2f1-f2} (mA)	IL_{f1} (mA)	IL_{2f1-f2} (mA)
Dahm and Scalapino (1997)	0.31	2.44	7.80	6.12e-2
This work	0.30	2.27	7.65	5.81e-2
Error	3.3 %	7.5 %	2 %	5.3 %

Table 1: Current levels for the fundamental and third order IP (I_{2f1-f2}). The second and third columns correspond to currents at the centre of the line, whereas the fourth and fifth are for those at the resistive termination. Simulation parameters: $\Delta Rs=5.1e-13\Omega\ m^2/A^2$; $\Delta Xs=2.3e-21\ H\ m^2/A^2$; $w=0.5\ mm$; $\Delta R=11.4e-3\ \Omega/m\ A^2$; $\Delta L=5.1e-11\ H/m\ A^2$;

5.   CONCLUSIONS

A novel combination of MoM and Harmonic Balance algorithms has proved to be useful for calculating spurious in microstrip HTS resonators. This numerical technique is very general and, beyond these preliminary results, can be used to analyse the non-linear performance of HTS planar circuits with complicated topologies .

ACKNOWLEDGMENTS

This work has been funded by CIRIT throug grant SGR00043. Graduate students J.Parrón and E. Ubeda are supported by the Generalitat de Catalunya, Comissionat per a Universitats i Recerca, under grants 1997 FI 00679 and 1997 00747, respectively. The author would also like to thank Dr. Salvador Talisa for his helpful comments and suggestions.

REFERENCES

Collado C, Mateu J, O'Callaghan J M 1999 European Conference on Applied Superconductivity.
Dahm T and Scalapino D J 1997 J.Appl.Phys **81**, 4, pp 2002-9
Harrington R F 1968 Field Computation by Moment Methods (New York: MacMillan)
Hong J S and Lancaster M J 1996 Electronics Letters **32**, 16, pp 1494-6
Hong J S and Lancaster M J 1997 IEEE Trans. Microwave Theory Tech. **45**, 12, pp 2358-65
Mosig J R et al 1989 Numerical Techniques for Microwave and Millimeter-Wave and Passive
  Structures edited by Itoh T, (New York: John Wiley & Sons) pp 133-213
Rao S M, Wilton D R and Glisson A W 1982 IEEE Trans. Antennas and Propag., **30**, 3, pp. 409-18.
Rius J M, Úbeda E, Parrón J and Mosig J R 1999 Microwave and Optical Technology Letters, **22**, 3
Song J, Lu C C and Chew W C 1997 IEEE Trans. Antennas Propag., **45**, 10, pp 1488-93.
Vendik O G, Vendik I B and Kaparkov D I 1998 IEEE Trans. Microwave Theory Tech. **46**, 5, pp 469
Zhang D, Liang G C, Shih C F, Johansson M E, Lu Z H and Withers R S 1995 App.
  Superconductivity **3**, 7-10, pp 483-96

Inst. Phys. Conf. Ser. No 167
Paper presented at Applied Superconductivity, Spain, 14-17 September 1999
© 2000 IOP Publishing Ltd

Full Wave Analysis of YBCO Coplanar Transmission Lines on Lithium Niobate Substrates

E. Rozan[1], J. O'Callaghan[1], R. Pous[1], J. Byun[2] and F. J. Harackiewicz[2]

1. Universitat Politècnica de Catalunya, Signal Theory and Communications, Campus Nord, 08034 Barcelona - 2. Department of Electrical Engineering. Southern Illinois University at Carbondale, Carbondale, Illinois, 62901-6603

ABSTRACT: The fabrication of a superconductor Mach-Zehnder modulator may suppose a significant increase of its modulation performance. The behavior of this electro-optical device made with YBCO electrodes can be evaluated through numerical modeling based on full-wave methods like Spectral Domain Approach (SDA) and Finite Difference Time Division (FDTD). Analysis are centered on the description of high Tc YBCO coplanar (CPW) transmission lines fabricated on Lithium Niobate/Yttria-Stabilized Zirconia multilayer substrates. Good agreement is observed between the different methods.

1. INTRODUCTION

The information revolution has created a huge demand of high-speed communications. Optical communication systems have been studied and developed because of their huge transmission capability. The electrooptic (EO) devices, as sub-systems, design are a key process element to optimize the transmission velocity. In particular, the Mach-Zehnder EO modulator has been identified as one of the most interesting devices. The presence of both optical and microwave technologies and a multi-layer, highly anisotropic LNO substrate creates a complex component. Different numerical analyses involving frequency dependence have been presented to simulate the microwave parameters of the coplanar traveling-wave (CPW) electrode configuration and to deduce the modulator performance in the works of Kubota et al (1980) or Kawano et al (1991)

The present paper shows a Spectral Domain Approach (SDA), presented by Itoh (1989), and a FDTD approach and compares them with the Point Matching Method (PMM) analysis. In designing electrode configuration, SDA and FDTD algorithms were applied to CPW structures for full-wave analysis. The Spectral Domain Approach has established itself as a predominant numerical method with its high accuracy in spite the method has been limited to relatively simple geometry due to the complexity in pre-processing, specifically, deriving geometry specific Green functions. The FDTD has very little pre-processing, and thus, quickly also led itself into a very popular numerical tool as computing costs continue to decline.

In this paper, the characteristic impedance and the microwave effective index of CPW structures based on anisotropic substrate with thin buffer layer are discussed. In particular, as in EO modulator structures, the dielectric buffer employed is usually SiO_2; its effects on the CPW microwave characteristics are clarified.

2. FDTD FORMULATION

The FDTD algorithm is formulated by discretizing Maxwell's equations both in time and space (Yee,1966). In anisotropic material, the electric flux density is related to the electric field by a permittivity tensor. For this case, due to the strong anisotropy of LiNbO$_3$ layer, a uniaxial permitivity tensor was employed in the simulation. The subcell model eliminates the normal restriction that sets the spatial grid increment to be less than or equal to the smallest physical feature in the solution space. In this case, the thin material sheet technique is used to model the thin buffer layer.

The two fluid model assumes that the electron gas in a superconductor material consists of two gases, the superconductivity electron gas and the normal electron gas.

$$\vec{J} = \vec{J}_n + \vec{J}_s \quad , \quad \vec{J}_n = \sigma_n \vec{E} \tag{1}$$

The superconducting fluid current density is obtained using London equation

$$\frac{\partial \vec{J}_s}{\partial t} = \frac{1}{\mu_0 \lambda_L^2} \vec{E} \tag{2}$$

where λ_L is the London penetration depth. The following equation for E_z is obtained by combining Ampere's law with the two fluid model.

$$\frac{\partial E_z}{\partial t} = \frac{1}{\varepsilon} \left(\frac{\partial H_y}{\partial x} - \frac{\partial H_x}{\partial y} - \sigma E_z - J_{sz} \right) \tag{3}$$

where the superconducting current density Jsz is obtained from (1).

Once the temporal fields are obtained, the line parameters can be calculated from the discrete Fourier transform of these fields.

3.   SPECTRAL DOMAIN APPROACH FORMULATION

In conventional space domain analysis, the structure can be analyzed by formulating the following coupled homogeneous integral equations, which can be solved for the unknown propagation constant β.

$$\int \left[Z_{xx}(x - x', y) J_x(x') + Z_{xz}(x - x', y) J_z(x') \right] dx' = E_x(x)$$
$$\int \left[Z_{zx}(x - x', y) J_x(x') + Z_{zz}(x - x', y) J_z(x') \right] dx' = E_z(x) \tag{4}$$

where E_x, E_z are the field components at the interface where the metal, J_x and J_z the current density

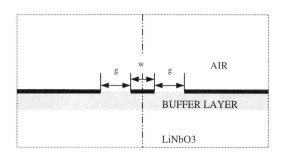

Fig. 1: multi layer CPW configuration

strip is present and Z_{ij} $(i,j=[x,z])$ are the Green's functions calculated in function of β. In order to simplify the Green functions calculation process, it is more convenient to apply a Fourier transform in the x-direction and (4) may written under a matrix system form

$$\begin{bmatrix} \tilde{Z}_{xx} & \tilde{Z}_{xz} \\ \tilde{Z}_{zx} & \tilde{Z}_{zz} \end{bmatrix} \begin{bmatrix} \tilde{J}_x \\ \tilde{J}_z \end{bmatrix} = \begin{bmatrix} \tilde{E}_x \\ \tilde{E}_z \end{bmatrix} \tag{5}$$

The sign ~ denotes for the Fourier transform. Electrical field could be expressed as a superposition of TM and TE modes. For each mode, a close expression of the Green functions is obtained in function of geometric parameters. The resolution of the system is based on the Galerkin procedure.

As the general formulation only deals with isotropic materials, system (5) has to be improved to take in account the LNO anisotropy. The solution exposed by Cai and Bornemann (1992) follows the general formulation and only modifies the closed expressions of Z_{ij}. They also proposed to introduce a complex boundary in the matrix system to include the effect of the superconductor surface impedance. The resulting equations are

$$\begin{bmatrix} \tilde{Z}_{xx} - Z & \tilde{Z}_{xz} \\ \tilde{Z}_{zx} & \tilde{Z}_{zz} - Z \end{bmatrix} \begin{bmatrix} \tilde{J}_x \\ \tilde{J}_z \end{bmatrix} = \begin{bmatrix} \tilde{E}_x \\ \tilde{E}_z \end{bmatrix} \tag{6}$$

Defining λ as the two-fluid model penetration depth for superconductors, Z in (6) becomes $1/\sigma t$ if $t << \lambda$, otherwise $Z_s \coth(t/\lambda)$. In the case of a normal metal, the conductivity σ is real and the skin depth δ is used instead of λ.

4.   POINT MATCHING METHOD

In order to confirm the full-wave methods FDTD formulation, a Point Matching Method (PMM) has been developed. The basic formulation is presented in Marcuse (1982). The algorithm allows calculating the C capacitance per unit of length for the structure in Fig. 1 and for the same structure embedded in vacuum. From the capacitance calculations, the effective propagation index and the characteristic impedance Z_m can be expressed as

$$n_m = \sqrt{\frac{C}{C_o}} \qquad \text{And} \qquad Z_m = \frac{1}{c_o \sqrt{CC_o}} \tag{7}$$

with c_o the speed of the light in the vacuum.

5.   NUMERICAL RESULTS

The cross-sectional view of the analyzed CPW structure is shown in Fig. 1. The characteristic impedance Z_m and the microwave index n_m of the above mentioned model were calculated by employing the FDTD and the SDA and are compared to the PMM. Fig. 2 shows the calculated characteristic impedance Z_m and the microwave effective index n_m, as a function of the buffer layer thickness. The buffer layer thickness can be one of parameter for impedance matching with the external 50 Ω circuit. As shown in the figure, FDTD and SDA agree well with PMM. This confirms that for usual CPW configuration, the line microwave parameters (characteristic impedance and effective index) are dispersiveless and should not be considering in a modulator design process. This may be explained by the small dimension of central electrode width and separation gap regarding to the operating microwave wavelength.

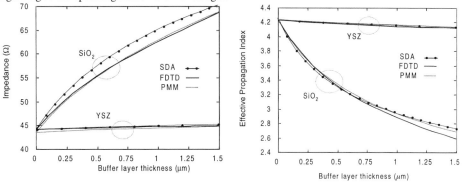

Fig. 2: Impedance and Effictive propagation index for CPW ($w=8$ µm - $g = 16$ µm) for different buffer configuration

346

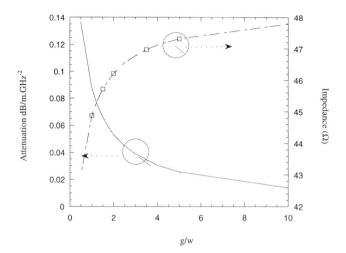

Fig. 3: Attenuation and Impedance for YBCO CPW.

In Fig. 3, representation of attenuation per unit of length and characteristic impedance is shown in function of the ratio w/g for a YBCO CPW line. YSZ thickness has been set at 0.1 μm. RF losses may be reduced when the electrode separation is increased. Also the 50 Ω impedance external line match could be achieve for a large electrode gap. The introduction of YBCO does not affect notably the transmission lines parameters, effective index or characteristic impedance calculated for normal electrode case. Only the attenuation coefficient is sensibly reduced.

6.   **CONCLUSION**

Numerical methods describing CPW electrode design have been presented. The analyses were based on FDTD and SDA with the anisotropic effect in corporate. The characteristic impedance and the microwave effective index were taken into account together with the attenuation per unit of length.

The CPW electrode design will help to obtain an accurate model for electro-optical modulator description. In particular, the improvement of modulator performance could be described from the model analyzed in this work.

ACKNOWLEDGEMENT

This work has been funded by the European Comission through TMR grant ERBFMBICT961112.

REFERENCES

Cai Z and Bornemann J 1992 IEEE Trans. MTT, 40
Itoh T 1989 Numerical Techniques for Microwave and Millimeter-Wave Passive Structures (New York:Wiley & Son) Chap 5
Kawano K, Nogushi K, Kitoh T and Miyazawa H, 1991 IEEE Photonics Tech. Letters, 3
Kubota K, Noda J and Mikami O 1980, IEEE Journal Quantum Electronics, 16
Marcuse D 1989 IEEE Journal of Quantum Electronics, 25
Yee K 1966 IEEE Trans. Antennas Prop., 14

Inst. Phys. Conf. Ser. No 167
Paper presented at Applied Superconductivity, Spain, 14-17 September 1999
© 2000 IOP Publishing Ltd

Simulation of Global Heating Effects on Nonlinear Resonance Curves at Microwave Frequencies.

A. L. Cowie[1], S. Thiess[1], N. A. Lindop[2], J.C. Gallop[3] and L.F. Cohen[1]

1.Blackett Laboratory, Imperial College Prince Consort Rd, London SW7 2BZ
2.Dept. of Chemistry, Kings Buildings, University of Edinburgh, EH9 3JZ
3.National Physical Laboratory, Queens Rd, Teddington, Middlesex.

ABSTRACT: The resonance curve shapes of a thin film parallel plate HTS resonator were observed as a function of microwave power. Skewed resonance curves and a pronounced shift of resonance frequency were observed at input powers above 1mW. An iterative model was set up in order to explore whether the skewed shape and frequency shift observed originated from global heating.

1. INTRODUCTION

At low microwave power the surface resistance is independent of power and the resonance curve is a symmetric Lorentzian curve. At high microwave power one manifestation of nonlinear behaviour is a skewed resonance curve, see Chin et al., 1992 and Oates et al., 1993. The skewness can be associated with vortex nucleation or heating. In the present study an iterative model was set up to predict how global heating would affect the resonance curve shape as a function of microwave power and delay time at each frequency step.

2. EXPERIMENTAL METHOD

The experiments were performed with a parallel plate resonator (two 5mm x 10mm x 200nm thick $YBa_2Cu_3O_{7-\delta}$ films on 1mm thick $LaAlO_3$ substrate with a 0.2mm thick sapphire spacer), excited in a TE01 mode. The resonant structure was clamped together in a copper microwave housing and placed in a continuous flow cryostat. The microwave CW source was amplitude modulated at 10 kHz in order for phase sensitive detection of the output signal. The frequency of the microwave source is slowly stepped across the resonance. The time spent at each frequency is used as an experimental variable although not dealt with further in this paper. The coupling was varied at 21K and minimum power so that the transmission coefficient, $T(=S_{21}^2)$, defined as the ratio between input and output power was always $\sim 1 \times 10^{-3}$ at resonance. The input power to the resonator housing was varied between 10μW and 0.1W.

3. 1D GLOBAL HEATING MODEL

Considering the heat balance equation for the total system:

$$Q_{k+1} = Q_k + Q_{diss} + Q_{transfer} \qquad\qquad [1]$$

where, Q_{k+1} represents the heat energy in the thin film after a step in frequency across the resonance frequency and Q_k is the heat energy present in the film before the step in frequency is taken. Q_{diss} is the extra heat created in the films from the microwave loss and $Q_{transfer}$ is the heat energy transferred away to the heat bath.

For simplicity, it is assumed that all the dissipated power in the resonator results in excess heat in the film. The microwave signal to the resonator is modulated at $f_{mod} = 10kHz$, and the microwave frequency is held at a particular value for a certain delay time t_{delay}. The variation in all the parameters were calculated over a cycle of 100 1μs steps. This cycle corresponded to a single period of the amplitude modulation of the lock-in where power was on for 50μs and off for 50μs. The cycle is then repeated $n = t_{delay}/100μs$ times before plotting a point on the resonance curve. The heat and corresponding temperature rise created by dissipated power when the signal is on for a time period $dt = 1μs$ at resonance frequency are written:

$$Q_{diss} = 2S_{21}(1-S_{21})P_{in}.dt \qquad and \qquad T_{rise} = Q_{diss}/(c_1.m_1) \qquad [2]$$

where c_1 and m_1 are the specific heat capacity and mass of the film. To describe the quantity of heat energy transferred away from the film to the heat bath in each iterative frequency step some simplifying assumptions are made. First it is assumed that the transfer of heat from the film to the bath is a 1D path through the film and substrate. The parallel plate resonator is treated as two infinite planes about a sapphire spacer. Heat transferred laterally through the spacer or across the film surface to the edges is neglected. It is assumed that the film is heated homogeneously to temperature T_k and that there is no thermal barrier at the film/substrate interface. Consequently, the transfer of heat through a layered 3-D structure has been reduced to heat transfer in 1-D across a slab of dielectric substrate with one surface held at the film temperature T_k, and the other in thermal equilibrium with the cryogenic heat bath at $T_{heat\ bath}$. The heat transferred away and the temperature fall bath in time period dt are written as
:

$$Q_{transfer} = \kappa.2A. (T_k - T_{heat\ bath})dt/d \qquad and \qquad T_{fall} = Q_{transfer}/(c_2.m_2) \qquad [3]$$

where κ is the thermal conductivity of the dielectric substrate, A, d, c_2 and m_2 are the area, thickness, specific heat capacity and mass of the substrate, respectively [1]. The factor of 2 is because the area of heat transfer twice the area of a single film.

The heated Lorentzian curve is produced using an iterative calculation across a frequency span which covers the resonance frequency of the resonator. For each step in the frequency of the microwave signal, the output power, temperature, line-width, penetration depth and resonance frequency is calculated.

We have used the dirty d-wave expression for the temperature dependence of the London penetration depth and we take into account the thin film thickness in our calculation of the new resonance frequency. However, these calculations do not take into account the temperature dependence of the thermal conductivity and the specific heat capacitance of the film or the substrate. Realistic values were found for all the thermal parameters. The sum of the values for electron and phonon specific heat capacity of $YBa_2Cu_3O_{7-\delta}$ is estimated to be $c_1 \sim 162.8\ JK^{-1}kg^{-1}$ in agreement with Manhart et al. 1996. From Michael *et al.* 1992, the 21K specific heat capacity of $LaAlO_3$ is estimated as $c_2 \sim 75\ JK^{-1}kg^{-1}$ and the thermal conductivity of $\kappa \sim 20WK^{-1}m^{-1}$, in agreement with Pukhov 1997. It was assumed that $\lambda(0) \sim 150nm$ and the experimentally determined $R_s(T)$ function was used.

4. RESULTS

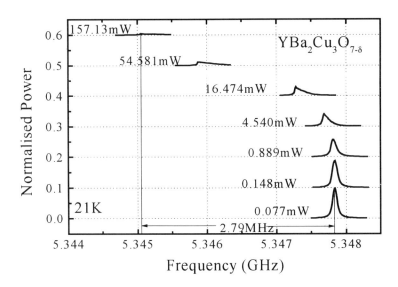

Figure1: The measured resonance curves at 21K for fixed delay time of 1.8sec, as a function of the microwave power varied from 77 μW to 157 mW for film set Y1 (YBa$_2$Cu$_3$O$_{7-\delta}$) at 21K. Curves are normalised by scaling with the input power. If the transmission coefficient remains unaffected by input power the height of each curve is 0.1 on this scale. Curves are offset from each other for ease of viewing.

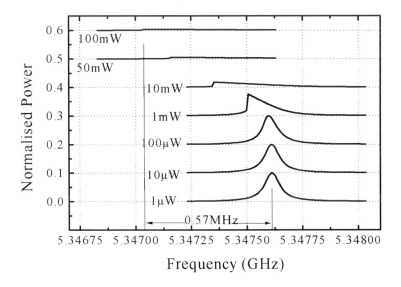

Figure 2: The results of the model calculation for a 1.8 sec fixed delay time, κ = .0.22 WK^{-1}m

350

Fig. 1 shows the resonance curves at 21K for fixed delay time of 1.8sec, as a function of the microwave power (varied from 77 μW to 157 mW). At low power the resonance curves are a simple Lorentzian curve shape. Upon crossing a threshold power, at about 1mW, when the frequency is swept across the resonance a pronounced tail off in the signal occurs and the full cavity resonance is not mapped out. As the full resonance curve is not mapped out once the threshold power is crossed, the true centre frequency is not known. Nevertheless using the peak of the resonance curve as a guide only, the peak frequency shifts down by 2.79 MHz across this power range. From the temperature dependence of the resonance frequency (not shown), a 3 MHz change at 21K could only be produced by a global temperature rise of 28 K.

The results of the model calculation for a 1.8 sec fixed delay time and variable input power are shown in Figure 2. Key features of the experimental data appear to be reproduced by the model. The simulated skewed curves resemble those measured in experiment, the output power is reduced at high power and the frequency shifts downwards. The curve shape is in reasonable agreement with experiment. The accuracy of the comparison of absolute power between experiment and simulation depends on our ability to correctly describe the heat transfer. In order to reproduce the skewness at approximately the same input power as in the experiment, the value of κ had to be changed from the realistic value of 22 $WK^{-1}m^-$ to 0.2 $WK^{-1}m^-$. Otherwise the simulated input power at which the output power is reduced to zero is of the order of 5W. The unrealistic values of κ or input power indicates that an important thermal resistance has been overlooked such as thermal barrier between film and substrate or heating of the sapphire substrate.

4. CONCLUSIONS

A model calculation was set up to explore whether global thermal effects could generate skewed resonance curves and large frequency shifts observed experimentally. The model ignores frequency shifts due to any other mechanism. The simulation produces consistent curve shapes but fails to explain the huge frequency shifts adequately even when the heat transfer parameters have been corrected for the unknown thermal barrier. In experiment, if the delay time is reduced to 0.2s, large frequency shifts are observed without the resonance curves becoming skew for input powers greater than 150 mW. These are associated with an intrinsic nonlinear mechanism which is not global heating related, as reported elsewhere by Thiess et al. 1999. At low temperatures, even for short delay times, all films fail catastrophically in our system as a result of global heating. This work has enabled a better understanding of the behaviour of a parallel plate resonator in a continuous flow cryostat at high microwave power.

REFERENCES

Chin C C et al. 1992 Phys. Rev. B **45**, 4788
Mannhart J 1996 Supercon. Sci. and Tech. **9**, 49
Michael P C 1992 Journal of Appl. Phys. **72**, 107
Oates J H et al. 1993 IEEE Trans on Appl. Supercon. **3**, 17
Pukhov A A 1997 Supecon. Sci. and Tech. **10**, 82
Thiess S, Cowie A, Gallop J C and Cohen L F, preprint

Inst. Phys. Conf. Ser. No 167
Paper presented at Applied Superconductivity, Spain, 14-17 September 1999
© 2000 IOP Publishing Ltd

351

Forward coupled high temperature superconducting filters for DCS1800 mobile phone base stations

A. Andreone, A. Cassinese, M. Iavarone, P. Orgiani, F. Palomba, G. Pica, M. Salluzzo, and R. Vaglio

INFM and Dipartimento Scienze Fisiche, Universita' di Napoli *Federico II*, Napoli, ITALY

C. Oliviero and G. Panariello

Dipartimento Ingegneria Elettronica, Universita' di Napoli *Federico II*, Napoli, ITALY

R. Monaco

Istituto Cibernetica, C.N.R., Arco Felice, ITALY

A. Guidarelli Mattioli and S. Cieri

Omnitel Pronto Italia, Ivrea, ITALY

A. Matrone and E. Petrillo

Ansaldo C.R.I.S., Napoli, ITALY

ABSTRACT: High temperature superconducting (HTS) bandpass filters for operation in European DCS1800 cellular base stations have been fabricated. A d.c. sputtering system has been developed to grow *in situ* $YBa_2Cu_3O_{7-\delta}$ (YBCO) double side films on 2" diameter $LaAlO_3$ substrates in a high oxygen pressure. YBCO planar filters spanning the frequency range 1710 to 1785 MHz have been successfully obtained.

1. INTRODUCTION

The wireless industry has seen an explosive growth in the last years, especially in the field of mobile phone communications. Expansion of cell capacity as well as optimisation of the quality coverage are critical issues for mobile phone operators. As wireless penetration increases, more and more sites are needed to manage both the customers growth and their expectations on good quality cellular coverage on the territory. It is therefore important for operators to optimise power link budget between *Base Transceiver Station* (BTS) and *Mobile Station* (MS). An improvement in the link budget can be achieved increasing the sensitivity of the MS receiver. The standard solution is the insertion of a *Band Pass Filter* (BPF) and of a *Low Noise Amplifier* (LNA) between the antenna and the main receiver. The preselector filter defines the usable spectrum for the operator based on allocated spectrum rejecting or attenuating "out-of-band" signals. Recently, the use of HTS filters integrated with cryogenic LNAs has been proved to be an attractive alternative to conventional technology (Liang et al, 1995; Hong et al, 1999). HTS-based solutions have been under development for the last five years and recent advances in design, materials, and cooling technologies have finally made such subsystems feasible (Greed et al, 1999). The low loss of HTS microwave components can increase base stations sensitivity as well as decrease noise and therefore expand coverage. This translates also in sharp skirt filters with improved selectivity, reducing adjacent channel interference. These advantages can come with a significant size reduction, even including the cooler, which allows superconducting subsystems to be installed on top of the antenna mast.

We present here the fabrication and test of planar L-band HTS filters as first step towards the development of a hybrid superconducting/semiconducting *Mast Head Amplifier* (MHA) receiver front-end for use in cellular systems. The receiver is based on a YBCO band-pass microstrip filter and on a very low noise cryogenic amplifier. A special low vibration cryocooler is under development to house the filter and the amplifier. The system follows the European DCS1800 specifications to be used as a first filtering mast-mounted stage in a mobile phone base station. A first trial in an Omnitel Italia test facility is scheduled for year 2000.

2.   FILM GROWTH AND PROPERTIES

The system designed in our Laboratory to produce large area, double-side, *in situ* superconducting films is schematically illustrated in Fig. 1a. In the main vacuum chamber, the upper plate allocates a 4" zero length sputtering source, which can be used in diode and/or unbalanced magnetron configuration, while on the lower plate a linear feedthrough supports a cylindrical 3" heater. The feedthrough can bring the heater up to close contact with the cathode surface. The sputtering gas is fed close to the target through a circular multi-hole pipe. The films are grown in high oxygen pressure ($P \approx 2 \cdot 10^2$ Pa) using a stoichiometric target. To avoid film contamination in the double-side process, a thin (0.25 mm) $LaAlO_3$ separator is placed in contact with the heater surface. An appropriate frame is used to position both the separator and the substrate concentric to the heater. A load-lock system and a wobble-stick with twisters allow to transfer and rotate the 2" substrates (Fig. 1b). The intro-chamber is equipped with a 2" zero length magnetron source and a second 3" heater to deposit gold contacts and to coat the ground plane.

The structural properties of the films were characterized by X-ray diffraction and ϕ-scan measurements. The samples are truly epitaxial with typical FWHM values for the rocking curve on the (005) reflection of $0.2°$.

The superconducting properties were inductively measured. Critical temperatures mapping revealed a high degree of uniformity on both sides of the sample with Tc values ranging between 87 and 88 K. J_C values at 77 K were typically between 1.5 and 2.5 MA/cm^2.

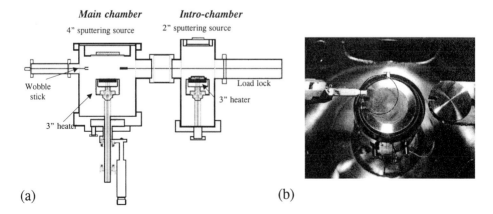

Fig. 1: (a) schematic view of the sputtering system and (b) detail of the manipulation inside the main chamber

3.   HTS MICROWAVE FILTERS

The performance of a microwave filter depends on the intrinsic (microwave losses, power handling, etc.) and extrinsic (housing, external coupling, etc.) characteristics of the device. In general, a band-pass filter can achieve a higher selectivity by increasing the degree of poles, namely the number of resonators. However, because the quality factor of resonators is finite, the insertion losses of the filter increase as the number of resonators increases. On the other hand, for a given number of poles, it is possible to select a certain type of filter characteristics that not only meets the selectivity requirement but also results in minimum insertion losses for a given quality factor of resonators. In particular, the capability of placing attenuation poles near the cut-off frequencies of pass band improves the selectivity using less resonators.

For a filter designer, therefore, conventional filter technology imposes trade-offs between selectivity (number of poles) and efficiency (insertion loss) and limits the range of practical designs. The use of superconducting materials removes these constraints. HTS in fact, due to their lower surface resistance at relatively high cryogenic temperatures, allows the fabrication of low loss and highly selective band-pass planar filters with compact dimensions (< 2" diameter).

However, the performance of superconducting filters degrades drastically at high power, representing one of the main limiting factors in telecommunications applications. Besides non linear behaviour in the signal transmission, HTS filters can generate also intermodulation distortion, leading to the production of spurious output signals. The limitation of the power handling is mainly due to the crowding of the r.f. current at edges of each microstrip resonator. At a fixed input power, the peak current depends on the width of the microstrip, i.e. wider strip corresponds to lower peak current. Therefore, using the appropriate design, it is possible to develop filters capable to handle several watts of input power with a linear response.

We have designed a series of forward-coupled planar superconducting filters with a Tchebischev response which meet high selectivity criteria, low pass band insertion losses, and high power handling in a compact size. A full-wave analysis of the electromagnetic response of the devices has been carried out using an advanced commercial simulation package. These filters utilise coupling due to the even- and odd-mode velocity difference of coupled microstrip linear resonators (Matthaei et al, 1980). The design parameters were obtained by a full-wave analysis method (Davino et al, 1999). Their in-band response have the whole 75 MHz band or a 25 MHz wide sub-band of the DCS1800 frequency allocation spanning the range 1710 to 1785 MHz. The number of poles ranges from 5 to 7 to fit on a 50 mm diameter and 5 mm thick LaAlO$_3$ single crystal substrate.

The layout of a five poles interdigital filter and the simulated response are shown in Figs. 2a and 2b respectively. The filter consists of parallel coupled $\lambda/2$ resonators with low internal impedance (10 Ω), characterised by wide microstrips (w = 2.92 mm), in order to lower the peak current at the edges. The signal is transmitted through 50 Ω microstrip lines (w = 0.17 mm) parallel coupled to each resonator. This configuration minimises current crowding at the feed lines, therefore improving the power response of the overall planar structure. The filter has central frequency at 1.76 GHz and 1.5 % fractional bandwidth.

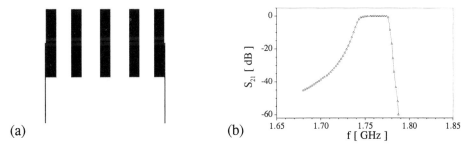

(a)   (b)

Fig. 2 : the layout of a five poles interdigital L-band filter (a) and its simulated response (b).

354

4.   RESULTS AND DISCUSSION

In this section the response of high power handling five poles band-pass filters in Nb and YBCO is shown. Niobium has one of the lowest surface resistance amongst superconductors at 4.2 K and therefore is ideal for a test of the performance of the device. The 600 nm thick Nb films (T_c = 9.2 K) are deposited using conventional rf sputtering on the two sides of a 0.5 mm thick, 2" diameter single crystal $LaAlO_3$ substrate. The filters are fabricated by a standard photolithography process and wet etching. Gold pads are sputtered on the input feed lines followed, in the case of YBCO filter, by an annealing step at 500 °C in 400 Torr of pure oxygen in order to lower the contact resistance. A vectorial network analyzer is used to measure the scattering parameters.

In Fig. 3 the measured transmission response (S_{21}) of a niobium and a YBCO based filter prototypes is shown at T/T_c = 0.5. No effect of the temperature on the overall response of the filter is observed up to 0.9 T_c. The comparison between the two materials is quite satisfactory, apart for a small shift of the center frequency due the difference between the values of the dielectric constant of the two $LaAlO_3$ substrates. The filters efficiency and selectivity fulfil the design requirements, with passband insertion loss lower than 0.3 dB, in-band ripple less than 1 dB, and out-of-band rejection better than 50 dB @25 MHz in the upper frequency region. Using maximum input power (10 dBm), the filter response was studied at various operating temperatures. No change in the scattering parameters increasing the input power was observed.

The onset of nonlinearities was monitored using also a two-tone intermodulation configuration. No degradation of the filter electrical characteristics and no presence of third-order spurious products were observed up to a reduced temperature T/T_c = 0.95. This test guarantees that the filter structure is robust and designed to handle much higher input power. Further measurements using a high power (input power ≤ 50 dBm) L-band amplifier are under progress to evaluate the maximum power such interdigital filters can handle in the linear regime.

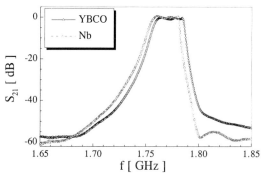

Fig. 3: measured response of a Nb (circles) and a YBCO five poles band-pass filter (upper triangles) at T/T_c = 0.5

REFERENCES

Davino D, Oliviero C, Panariello G and Verolino L 1999 Il Nuovo Cimento, submitted

Greed R B, Voyce D C, Jedamzik D, Hong J S, Lancaster M J, Reppel M, Chaloupka H J, Mage J C, Marcilhac B, Mistry R, Hafner H-U, Auger G and Rebernak W 1999 IEEE Trans. Applied Supercond. **9**, 4002

Hong J S, Lancaster M J, Greed R B, Voyce D C, Jedamzik D, Holland J A, Chaloupka H J and Mage J C 1999 IEEE Trans. Applied Supercond. **9**, 3893

Liang G-C, Zhang D, Shih C-F, Johansson M E, Withers R S, Oates D E, Anderson A C, Polakos P, Mankiewich P, de Obaldia E and Miller R E 1995 IEEE Trans. Applied Supercond. **5**, 2652

Matthaei G, Young L and Jones E M T 1980 Microwave Filters, impedance matching networks and coupling structures (Dedham: Artech House)

Inst. Phys. Conf. Ser. No 167
Paper presented at Applied Superconductivity, Spain, 14-17 September 1999
© 2000 IOP Publishing Ltd

Performance of passive microwave devices made from thin films of YBa$_2$Cu$_3$O$_{7-\delta}$ superconductor

E. Moraitakis, M. Pissas, D. Niarchos and G. Stratakos†

Institute of Materials Science, NCSR Demokritos, 153 10 Ag. Paraskevi, Athens, Greece
†Institute of Communications and Computer Systems of the National Technical University of Athens, Zografou 15773 Athens, Greece

ABSTRACT: High quality thin films of YBa$_2$Cu$_3$O$_{7-\delta}$ superconductor were deposited on LaAlO$_3$(100) substrates with sizes up to 1 in x 1 in, with a simple sputtering technique. The results show that these films are of good quality with perfect c-axis orientation, epitaxial growth, critical temperature of $\geq$ 90 K, critical current density of $\geq$ 1 x 10^6 A/cm^2 at 77 K and thickness uniformity better than 5 % over an 1inch x 1inch substrate. A two-pole bandpass filter with centre frequency of 13 GHz and 3 dB fractional bandwith of about 15%, was designed to test the microwave performance of these films.. A YBCO filter exhibits at 77 K an insertion loss of more than 5 dB lower than an equivalent Au filter at the same temperature.

1. INTRODUCTION

Promising developments in high-T_c epitaxial films and cryogenic technology have made it possible to achieve attractive microwave devices such as resonators, filters, channelizers, antennas and delay lines (Gallop 1997, Hein 1997). These devices will be used in future satellite and wireless communication systems, where better performance and higher compactness in device properties are required. In order to achieve these goals, high quality superconducting films with low surface resistance (R_s) are required over at least 1inch x 1inch substrates.

In this paper we demonstrate a simple homemade sputtering apparatus for the deposition of high quality YBCO thin films over large areas. We report on the properties of the films deposited with this technique and the performance of microwave devices made from them.

2. PREPARATION AND CHARACTERISATION

The magnetron source was designed to accept a large YBCO disk-shaped target of 10 cm diameter, in order to achieve uniform deposition over large area and to operate in an unbalanced magnetic field configuration (Savvides 1993), in order to eliminate resputtering effects (Selinder 1991). In such a magnetic field configuration the plasma is directed away from the substrate (Fig.1). Less bombardment of the growing film by energetic species is expected and as a consequence the deposition of stoichiometric YBCO films is more straightforward at moderate pressures. Furthermore we found that a simple vacuum system pumped only by a roughing pump (to approximately 5 mTorr) is sufficient for the deposition of high quality YBCO films without the need for high vacuum systems (Koren 1993). Details of the deposition technique can be found elsewhere (Moraitakis 1998). Briefly, prior to each deposition, an Ar-O$_2$ mixture was introduced at a fixed partial pressure ratio P(O$_2$)/P(Ar)=1/10 and a total pressure around 400 mTorr was stabilized. DC power P= 55-60 W was fed to the sputtering source while the deposition temperature was stabilized to T$_S$= 730^{0}C with a temperature controller. In

356

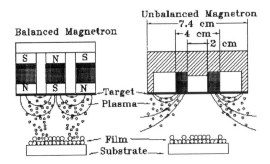

Figure 1. Schematic cross-section drawing of the magnetic field configurations in a (conventional) balanced magnetron and our unbalanced magnetron. In the latter case the inner magnet is a permanent magnet of ring-shape while the outer one is a soft magnet.

this work we present results for single crystal LaAlO₃(100) substrates with sizes up to 1inch x 1inch, put on the heater block with silver paste. The substrate to target distance was fixed to 3.5- 4 cm. Under these conditions the deposition rate is about 15 A/min. After each deposition the temperature was lowered to 450°C at a rate of about 10°C/min in 1 Atm of pure O_2, where the film was kept for 30 min to ensure maximum oxygenation.

Figure 2 shows a typical X-ray diffraction pattern using the θ-2θ geometry (Bragg-Brentano diffractometer) of a 300 nm thick film on LaAlO₃(100) substrate. Only the (00ℓ) peaks of the YBCO structure are observed, which means that the films are oriented with the c-axis normal to the substrate. The c-axis of the films has a normal value of about 11.68 Å. Furthermore the epitaxial growth of the films on the LaAlO₃(100) substrates was examined by means of pole-figures experiments using a 4-cycle diffractometer (Kroman 1992, Holiastou 1997).

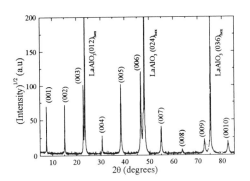

Figure 2. X-ray diffraction pattern of a typical 300 nm thick film on LaAlO₃ (100) substrate.

The superconducting properties of the films were examined by a home-made ac-susceptometer. Figure 3(a) shows a typical ac-susceptibility curve for an YBCO film on 1cm x 1cm LaAlO₃(100) substrate, with T_c= 91.8 K and ΔT_c= 1 K. Figure 3(b) shows the same for an YBCO film on 1inch x 1inch LaAlO₃(100) substrate, with T_c= 91.8 K and ΔT_c= 2.5 K. The ac-susceptibility probes the whole volume of the sample and the above results indicate that YBCO films with homogeneous superconducting properties can be deposited up to at least 1inch x 1inch with this sputtering technique. The critical current of YBCO films deposited was measured with standard four-probe geometry on 50 μm wide and 1 mm long bridges. The patterning of the films was done with typical photolithographic and wet etching techniques. The critical current of the films with thicknesses ~300 nm was 1.3 – 2 x 10⁶ A/cm² at a temperature of 77 K and zero magnetic field, using an electric field criterion of 1 μV/cm. Furthermore the thickness uniformity in a 1 inch x 1 inch patterned YBCO film measured using a Dektak IIA profilometer was better than 5%. The results encourage us to believe that with proper modification of the heater block we may extent the deposition of YBCO films with good properties on substrates with larger sizes and moreover good-quality double-side covered YBCO films may be prepared. This is important for the fabrication of microwave devices with eliminated losses in the ground plane (Foltyn 1991, Jenkins 1997, Yeliang 1996).

2.   MICROWAVE DEVICES

In order to test the microwave properties of the as prepared YBCO films on LaAlO$_3$ substrates, a two-pole bandpass filter was designed which operates at 13 GHz with 15% 3dB fractional bandwidth. The geometrical design parameters of the YBCO filter were obtained and simulated by the HP-Eesof Series IV commercially available computer-aided design (CAD) program. In the simulation, high T$_c$ superconductors has been specified as perfect conductors with extremely high conductivity. The LaAlO$_3$ substrate was considered as a dielectric with a dielectric constant of 23 and low dielectric loss (tanδ=1 x 10^{-4}). The input and output impedance of the filter is set to 50 Ω. Figure 4 shows the layout of the two-pole filter. YBCO filters were constructed with standard photolithographic and wet etching process that were described above. For comparison the same filter was constructed on a gold (Au) film 2.5 μm thick deposited on LaAlO$_3$(100) substrate by sputtering. In both cases (YBCO and Au filters) the metallization of the second side of the LaAlO$_3$ substrates was done by covering them with Au layers while the electrical contacts to the filters were done by applying silver pads on the films. The filters were put successively in a copper cavity which was specially designed and electrical connections with the 50 Ω input and output feed lines were done.

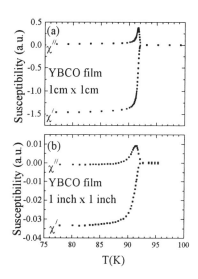

Figure 3. Ac-susceptibility measurements of YBCO films on LaAlO$_3$ with dimensions 1cm x 1cm (a) and 1inch x 1inch (b). Both films were about 300 nm thick. The measurements were done using a frequency f= 997 Hz and ac-field H_{ac}= 0.5 Oe.

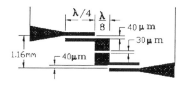

Figure 4.   Design of the two-pole band-pass filter with a resonant frequency of 13 GHz on a LaAlO$_3$ substrate 0.5 mm thick.

Figure 5(a) shows the frequency response of transmission (S$_{21}$) and reflection (S$_{11}$) coefficients for the YBCO filter measured using a HP8510C vector network analyzer at 77 K. The corrected value for the losses in the cables of the insertion loss IL = 20log S_{21} at 13 GHz, is IL= -0.985 dB for the YBCO filter at 77 K, while the theoretical one is IL= -0.856 dB, for this 2-pole filter design. Figure 5(b) shows the frequency response of the coefficients for the Au filter. The IL of the YBCO filter is 5.2 dB lower than the same filter made from Au at 13 GHz. This result proves the successful performance at microwave frequencies of the YBCO films made with this sputtering technique.

358

2. CONCLUSIONS

We have reported on the preparation of YBCO films with a simple sputtering technique, which combines an unbalanced magnetron configuration and a large YBCO target. YBCO films of good quality with perfect c-axis orientation, epitaxial growth, critical temperature of $\geq$ 90 K, critical current density of $\geq$ 1 x 10^6 A/cm^2 at 77 K and thickness uniformity better than 5 % were deposited up to 1inch x 1inch substrate. A two pole YBCO filter, which operates at 13 GHz, exhibits an insertion loss of more than 5 dB lower than an equivalent Au filter on LaAlO$_3$ substrate, at 77 K. The results are promising for the deposition of good quality double-sided YBCO films over large areas. This is important for more potential applications in microwave superconducting devices.

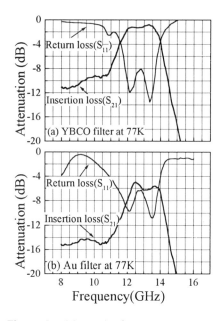

Figure 5. Measured frequency response of the transmission (S_{21}) and reflection (S_{11}) coefficient at 77 K of the: (a) YBCO filter and (b) gold filter.

REFERENCES

Gallop J C 1997 Supercond. Sci. Technol. **10** A120.
Hein M A 1997 Supercond. Sci. Technol. **10** 867.
Savvides N and Katsaros A 1993 Appl. Phys. Lett. **62** 528.
Selinder T I, Larsson G, Helmersson U, Rudner S, 1991 J. Appl. Phys. **69** 390.
Koren G 1993 Physica C **209** 369.
Moraitakis E et al, 1998 Supercond. Sci. Technol. **11** 686.
Kroman R, et al 1992 J. Appl. Phys. **71** 3419.
Holiastou M et al, 1997 Supercond. Sci. Technol. **10** 712.
Foltyn S R et al, 1991 Appl. Phys. Lett. **59** 1374.
Jenkins A P et al, 1997 IEEE Trans. Appl. Supercond. 7 2793.
Yeliang Z et al, 1996 J. Supercond. **6** 625.

Inst. Phys. Conf. Ser. No 167
Paper presented at Applied Superconductivity, Spain, 14-17 September 1999

Surface wave high-temperature superconducting resonators

G A Melkov[1], Y V Egorov[1], A N Ivanyuta[1], V Y Malyshev[1] and A M Klushin[2]

[1]Kiev Taras Shevchenko University, 64,Vladimirskaya St., Kiev, 252017, Ukraine
[2]Institut für Schicht - und Ionentechnik, Forschungszentrum Jülich GmbH, Leo-Brandt-Str. 1, 52425, Jülich, Germany

ABSTRACT: We investigate the new type of quarter-wavelength surface wave resonators which may be effectively used in microwave cryoelctronics. The design usually consist of a base high-conducting plate (metal or superconductor) of definite width and length placed in the middle of waveguide and one or two supplementary dielectric layers. We use standard method of partial regions for calculating of resonance frequencies on resonator's length, width and temperature. Experimental investigations in the 3-cm waveband have shown a good agreement with theory.

1.   INTRODUCTION

For the investigation of the microwave properties of different materials and the design of microelectronic devices parallel-plate resonators and microstrip constructions are often used. In the special case of superconducting electronics the manufacturing of two superconducting layers on the same substrate is rather difficult technical problem especially in the high part of mm-waveband. Moreover, it is important to design input circuit with lowest losses based on microstrips or coaxial cables.

We have solved some of these problems using only a single metal or high-temperature superconducting (HTS) plate to design a new type of microwave resonators. The fundamental mode is a surface wave formed by currents of high magnitude over the conducting surfaces of metal films on dielectric substrates. The concrete geometrical construction is determined by using.

2.   THEORETICAL ANALYSIS OF SURFACE WAVE RESONATOR

We are analyzing the example of surface wave resonators (SWR) that consists of conducting film on dielectric substrate with outer dielectric layer excited fundamental mode or rectangular waveguide (Fig.1) using a partial wave synthesis developed by Harrington (1968). The significant simplification is due to the fact that fundamental SWR mode, which we are interested, is an even one only and the symmetrical conditions give a possibility to analyze a half of system introducing magnetic wall in the symmetry plane z=0. A half of SWR ($z \geq 0$) was divided into a four part region:

I.	$0 \leq x \leq D$	$0 \leq y \leq l$	$0 \leq z \leq w/2$
II.	$D \leq x \leq a$	$0 \leq y \leq l$	$0 \leq z \leq w/2$
III.	$0 \leq x \leq a$	$l \leq y \leq b$	$0 \leq z \leq w/2$
IV.	$0 \leq x \leq a$	$0 \leq y \leq b$	$z \geq w/2$

360

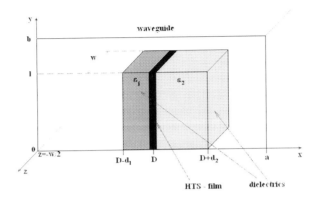

Fig.1 Surface wave resonator in waveguide

The electric and magnetic fields in each region are, without restriction of generality, described by the so-called electric and magnetic longitudinal (LE and LM) section waves in the x direction, which result from a Hertz vector consisting only of x component:

$$\vec{A}(x,y,z,t) = \vec{e}_x \cdot A(x,y,z) \cdot e^{-i \cdot \omega \cdot t} \qquad \text{(LE waves)},$$
$$\vec{F}(x,y,z,t) = \vec{e}_x \cdot F(x,y,z) \cdot e^{-i \cdot \omega \cdot t} \qquad \text{(LM waves)}.$$

The Helmholtz equations for these waves are:

$$\Delta \vec{A} + k^2 \varepsilon \cdot \vec{A} = 0,$$
$$\Delta \vec{F} + k^2 \varepsilon \cdot \vec{F} = 0.$$

Here ε - dielectric permeability, $k = \omega / c$, c - light speed. Each field component results from a superposition of LE waves and LM waves. Lewin (1975) has shown that:

$$\vec{E} = -\nabla \times \vec{F} + \frac{i}{k \cdot \varepsilon} \nabla \times \left[\nabla \times \vec{A} \right],$$
$$\vec{H} = \nabla \times \vec{A} + \frac{i}{k} \nabla \times \left[\nabla \times \vec{F} \right].$$

We used the following boundary conditions for the tangential electric and magnetic fields:

$$E_y(0,y,z) = E_z(0,y,z) = E_y(a,y,z) = E_z(a,y,z) = E_x(x,b,z) = E_z(x,b,z) =$$
$$= E_x(x,0,z) = E_z(x,0,z) = 0;$$

$$\frac{E_y(D \pm 0, y, z)}{H_z(D \pm, y, z)} = \pm Z_s \quad \text{for } y \leq l \text{ and } z \leq w - 2,$$

where Z_S is the surface impedance of HTS film. Coffey and Clem (1991) have shown that

$$Z_s = ik\widetilde{\lambda} \cdot \coth(h/\widetilde{\lambda}).$$ Here h – thickness of HTS film, $\widetilde{\lambda} = \left[\dfrac{\lambda^2(T)}{1+2j\lambda^2(T)/\delta_n^2(T)} \right]^{1/2}$ -

complex penetration depth, T – temperature, $\lambda(T)$ - London penetration depth. $\delta_n^2(T) = 2\rho_n/k$; ρ_n – normal resistivity. Vendik et al (1998) have shown that a simple equation for London penetration depth $\lambda(T) = \lambda(0)/\sqrt{1-(T/T_c)^4}$ can be used, were T_c – critical temperature of the HTS film. On the boundaries of regions the tangential components of electric and magnetic field were compared.

3.   EXPERIMENTAL RESULTS AND DISCUSSIONS

We have investigated the microwave SWR made from cooper films grown on poly-crystalline Al_2O_3 substrates and $YB_2C_3O_7$ (YBCO) films grown on single crystalline yttria stabilized zirconia oxide (YSZ) substrates. The YBCO films were about 0.35 μm thick. The Al_2O_3 substrate with $\varepsilon_1 = 9.8$ and the YSZ substrate with permeability $\varepsilon_1 = 26$ were exploited as the dielectrics in the SWR. The second dielectric was vacuum with $\varepsilon_2 = 1$.

We report the dependencies of the fundamental mode frequency F of SWR on the resonators length, resonators width and the temperature. F was measured at a fixed temperature by the standing wave coefficient technique, described by Ginston (1957). To increase the measurement precision, F was measured using a match-terminated waveguide. A matching load was located behind the resonator at $z > w/2$ (see Fig.1).

The typical results of F (l), F (w) and F (T) obtained by investigating the SWR reso-nator situated in a 3-cm rectangular waveguide are shown in Fig. 2, Fig. 3 and Fig. 4 respectively.

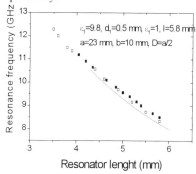

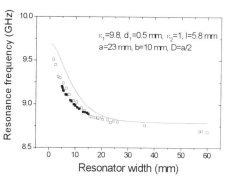

Fig.2 Dependence of fundamental resonant frequency of the cooper SWR on resonators length. Solid line– theory (solid film), open squares- solid film, solid squares – film consist of two parts:

$l \times w_1$ and $l \times (w - w_1 - \delta)$, where

$w_1 = 2.5\, mm$, gaps width $\delta = 0.15\, mm$.

Fig.3 Dependence of fundamental resonant frequency of the cooper SWR on resonators width. Solid line– theory (solid film), open squares- solid film, and solid squares – film consist of two parts:

$l \times w_1$ and $l \times (w - w_1 - \delta)$, where

$w_1 = 2.5\, mm$, gaps width $\delta = 0.15\, mm$.

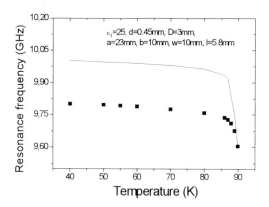

Fig.4. Measured (squares) and simulated dependence of resonant frequency of the YBCO surface wave resonator on YSZ substrate on temperature.

The parameters of HTS film used for obtaining the theoretical curves were: $T_c = 90.5$ K, $\lambda(0) = 150$ nm, $\rho_n(T_c) = 0.33 \cdot 10^{-7}\ \Omega \cdot m$.

The resonant wavelength of fundamental mode $\lambda = c/F$ can be approximately represented as $\lambda = 4(l + \Delta)$. Δ is determined by the fields outside the resonator. In the case of no dielectrics, we have as $w \rightarrow 0$, $\Delta \rightarrow 0$. Whereas in the case of resonator with dielectric plates, the resonant wavelength becomes $\lambda = 4(l_\varepsilon + \Delta_\varepsilon)$ with conditions of: $l_\varepsilon = l/\sqrt{\varepsilon_{eff}}$; $\varepsilon_{eff} = \varepsilon(d_1, d_2, \varepsilon_1, \varepsilon_2)$-effective dielectric permeability, $1 < \varepsilon_{eff} < \max\{\varepsilon_1, \varepsilon_2\}$; and $\Delta_\varepsilon = \Delta(l_\varepsilon, w)$, $\Delta_\varepsilon < \Delta$.

The results demonstrate a good agreement between the measured and simulated data. Fig. 2 and Fig. 3 document that the slit does not change SWR fundamental resonance frequency more than 1%. At the same time it effectively suppresses the appearance of the higher order oscillations in the resonator.

We can say, that there are two main peculiarities of the microwave surface wave resonators. Firstly, they are simple in structure. Secondly, a high current density and homogeneity in the HTS film can be obtained intrinsically. Therefore these resonators can be potentially used in devices with Josephson junction array, for measuring the microwave properties of the HTS films, and in the tunable microwave filters, especially at mm waveband range.

REFERENCES

Coffey M W and Clem J R 1991 Phys. Rev. Lett. **67**, 386
Ginston E L 1957 Microwave measurements (New York, Toronto, and London: McGraw-Hill Book Company, Inc.)
Harrington R F 1968 Field Computation by Moment Method (New York: Macmillan)
Lewin L 1975 Advanced Theory of Waveguides (London: Butterworth and Co Ltd.)
Vendik O G, Vendik I B and Kaparkov D I 1998 IEEE Trans. on MTT **5**, 469

Inst. Phys. Conf. Ser. No 167
Paper presented at Applied Superconductivity, Spain, 14-17 September 1999
© 2000 IOP Publishing Ltd

TBCCO based HTS Channel Combiners for Digital Cellular Communications

A P Jenkins, D Dew-Hughes, E G Edwards, D Hyland and D J Edwards

The Department of Engineering Science, University of Oxford, Parks Road Oxford OX1 3PJ

ABSTRACT: This paper reports results from the testing of disk resonator structures fabricated from TBCCO based 2212 thin films and illustrates their potential as channel combiners for cellular base station applications. The R_s of these films has been measured at 5.5GHz using a sapphire dielectric resonator and shown to be less than 0.5 mΩ at 80K. The power handling characteristics of devices designed for operation in the UMTS frequency bands have been investigated, using power levels as high as 4W. In addition, a dual channel combiner will be demonstrated that operates in a cryocooler at 70K with spectrally correct UMTS 3rd Generation signals and high power levels.

1. INTRODUCTION

Linearity and spectral re-growth are issues that are becoming of increasing importance in cellular base station transmitter subsystems. With the imminent roll out of Third Generation mobile systems (3rd Gen) as discussed by Street(1999), filter structures that can enhance spectral efficiency and therefore make better use of the resource available to service providers (on both the receive and transmit side) will be of great interest to those seeking to exact the greatest return from their investment in 3rd Gen spectrum licences. If HTS technology is to have an impact on this application area then devices operated under realistic signal conditions and power levels need to be demonstrated, such as those produced by Liang(1995). In an effort to address part of this issue, disk like resonant structures have been designed and fabricated from HTS material that are designed to act as single channel (5MHz) combiners for use in UMTS base station transmitters. Due to the non-constant envelope of the modulation scheme used for the proposed new system, extreme requirements for linearity will be placed on power amplifier systems. By combining very narrow and low insertion loss devices the performance burden can be shared between the analogue combiners and the linearised power amplifier.

2. DESIGN AND SIMULATION

2.1 Disk Structure

Disk structures are well known and have been used for HTS power applications by Jenkins(1996) and Aminov (1999). However, in order to extract the best performance from the HTS structure, careful consideration needs to be paid to the coupling/feed structure. For a combiner application, the resonant structure needs to be tightly coupled to yield a small insertion loss while maintaining a high RF current density at the feed points. In order to achieve these requirements, a half-wavelength low impedance structure is used by Jenkins (1998) that couples to the TM_{011} mode of the disk resonator over a quarter wavelength. The low impedance of the line allows a larger RF current to flow and the quarter wavelength allows a high degree of coupling. A hexagonal structure was used as it was found that this made the following simulations much more computationally efficient with little observable impact on the current distribution when compared to a circular structure.

2.2   Simulation Results

The design goal was for a 3dB band width of 5MHz with an insertion loss of 0.5dB (5MHz is the channel occupancy for a 4Mcp/s UMTS signal and the insertion loss simulation is based on an R_s figure of $1m\Omega$). The device described was simulated using an HP EEsof electromagnetic simulator. The frequency response for the mode of interest is shown in Fig. 1 and in Fig. 2 the current distribution is given. No large current peaks are observed in the current distribution, indicating that for a given superconducting J_c, this design will have a larger power handling capability than a comparable microstrip structure, as suggested by Jenkins.

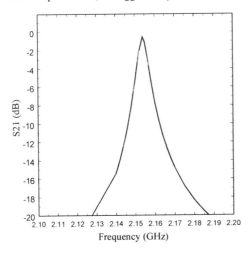

Fig. 1 Simulated Frequency Response of Combiner Structure

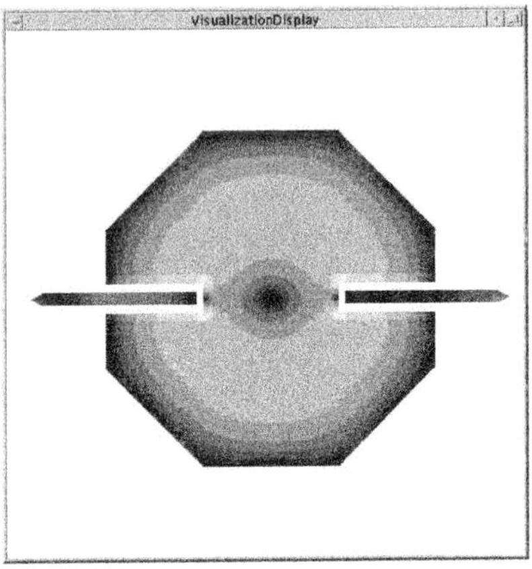

Fig. 2 Simulated Current Distribution of the TM_{011} Mode

3.   MEASUREMENTS

A number of structures were fabricated from Tl-Ba-Ca-Cu-O 2212 HTS material produced in house. The details of the process are given by Morley (1993). Thin films were produced on 50mm diameter $LaAlO_3$ substrates and characterised in terms of surface resistance at 5.5GHz using a sapphire dielectric resonator. R_s values of less than $0.5m\Omega$ were measured (at 80K) for such films. Subsequently, such films were patterned into devices using wet etching.

3.1   Frequency and Insertion Loss Performance

The devices were assembled into suitable enclosures and measured using an HP8510 vector network analyser. The frequency response at a base temperature of 56K was recorded, as well as the insertion loss at three temperatures (calibration was carried out at each temperature). Fig. 3 shows the measured wide-band response of a typical device, with Fig. 4 showing a screen dump of the close in response of a device.

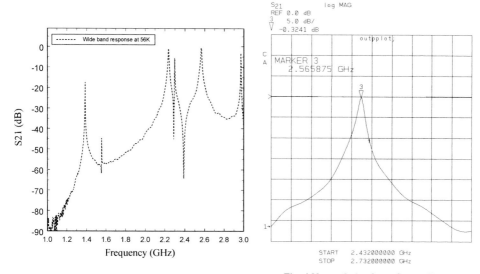

Fig. 3 Measured Frequency Response　　**Fig. 4 Network Analyser Screen Dump.**

Given in the table below are calibrated insertion loss measurements at three temperatures. Owing to the very low figures that need to be measured, calibration (response only) was carried out at each temperature. The overall uncertainty associated with this measurement is at least +/- 0.1dB, given such factors as flexible cable being used in the experimental arrangement and only a response calibration being possible at low temperatures.

Temperature (K)	Frequency (GHz)	Insertion Loss (dB)	3dB Bandwidth (MHz)
56	2.566	0.22	8.8
80	2.566	0.82	8.8
95	2.564	0.94	9.0

As can be seen, the 3dB bandwidth is larger than the design bandwidth and the frequency is higher. The larger measured bandwidth and reduced insertion loss can be attributed to greater coupling due to the increased frequency. The device can still be operated at 95K with only a small penalty in performance.

3.1　Power Handling Characteristics

A device was then integrated with a commercially available cryocooler supplied by the Hymatic Company and operated at 70K. Power was applied to the device and measured using a spectrum analyser. The assembled cooler is shown in Fig. 6 with three devices assembled on the cold head. Fig. 5 shows the measured response of the HTS device under maximum available power from the power amplifier used. The HTS device showed no compression at a power level of over 3.5W.

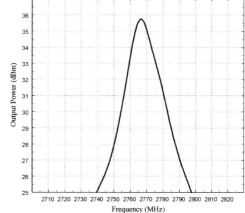

Fig. 5 Power and Frequency Response at 70K.

366

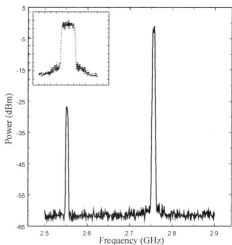

Fig. 6 Cryocooler and Microwave Package.

Fig. 7 Output Spectrum of Dual Channel Combiner. Inset shows close up of the UMTS Signal.

Fig 7 shows the output spectrum from the UMTS dual channel combiner when being feed with two 4MHz band width spectrally correct UMTS signals. One channel is run at a much higher power level to simulate asymmetric power profiles due to power control. The inset shows a close up of one of the spectrally correct UMTS signals.

4.   CONCLUSIONS

Large area TBCCO 2212 thin films on LaAlO$_3$ substrates have been produced. A number of HTS disk structures have been fabricated and tested from these films that have been shown to handle high power levels. These have been integrated as a cryo-package and cooled in a commercial cryocooler. Successful operation of a Dual Channel UMTS combiner has been demonstrated.

5.   ACKNOWLEDGEMENTS

The authors would like to thank Dr Street for his software generation of the UMTS test signals, Dr Grovenor of the Materials Department for supply of HTS thin films and the Hymatic Engineering Company for the loan of a closed cycle cryocooler. This work was supported by the EPSRC under grant no. GR/L08137

REFERENCES

Aminov B A et al 1999 IEEE.Trans.Appl.Supercond **9**, 2 pp4185-4188
Jenkins A P et al 1996 IEE Colloq. Superconducting Microwave Circuits **96/094** pp3/1-3/5
Jenkins A P et al 1998 J. Superconductivity **11**, 1 pp5-8
Liang G C 1995 IEEE MTT-S Digest, pp191-194
Morley S M et al 1993 IEEE.Trans.Appl.Supercond **3** pp1753-1756
Street A M 1999 Presented at NCAPP'99

Inst. Phys. Conf. Ser. No 167
Paper presented at Applied Superconductivity, Spain, 14-17 September 1999
© 2000 IOP Publishing Ltd

Thermal Breakdown of Superconducting State in HTS Devices Based on S/N Transition

E Loskot, M F Sitnikova, V Kondratiev

Department of Microelectronics and Radio Engineering, St.-Petersburg Electrotechnical University, St.-Petersburg 197376, Russia

ABSTRACT. A special problem of application of devices based on the S/N transition consists in the prevention of thermal spontaneous switching. A modified technique for analysis and evaluation of parameters determining thermal breakdown of S-state in an HTS switching element is proposed. The theoretical investigation based on the one-dimensional heat flow model is presented. We obtained the closed form equation for the minimum current of the normal phase expanding and equation for the velocity of the S/N boundary movement.

1. INTRODUCTION

Controlling HTS devices (microwave switches, power limiters and phase shifters) are based on a phase transition from the superconducting state (S - state) to the state of the normal electrical resistivity (N - state) in a thin superconducting film strip. To minimize the control power and current, the switching elements are implemented as narrow bridges. Owing to heterogeneity of the HTS material, the local inclusions of the normal phase into the HTS film being in the S-state can appear under a passing transport current (Fig. 1). The temperature distribution in the film is shown in Fig. 2 and corresponds to a thermal domain. The N-phase inclusions can result in instability of the film state followed by spontaneous switching the whole film to the N – state. The thermal breakdown of the S-state in such structures is caused by a movement of a thermal domain boundaries, if the heat generation in the domain exceeds the heat dissipation. It should be noted that the thermal destruction can take place when the current value is less than the critical one. Thus, it is important to know specific parameters characterizing this phenomenon, which is undesirable in a practical application of the devices using S/N transition.

2. THERMAL PROCESS MODELING

In order to evaluate the effect of the transport current on the state of the S-N bridge, the theoretical investigation of a thermal destruction of the S-state in the HTS microwave switching element has been performed. We have applied the procedure based on algorithm developed in [1] for a simple one-dimensional heat flow model. The switching element made as a transmission line section is based on the HTS film on a dielectric substrate of thickness h. The film of thickness d and width B is controlled by dc current I. The substrate is kept at the constant temperature of the coolant T_0. We have studied the equilibrium state and expanding the local inclusion both for the case of uniform material of the bridge and of nonuniform structures.

The heat power flow per unit volume generated by the current in thermal domain is determined by Joule heating:

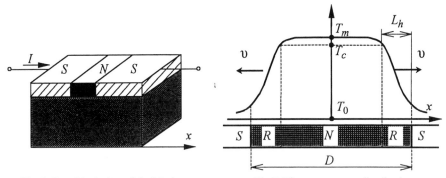

Fig. 1. Local inclusion of the N-phase

Fig.2. The temperature distribution corresponding to a thermal domain

$$Q(T,I) = \frac{I^2 \cdot R_S(T)}{B^2 \cdot d} \ , \tag{1}$$

The surface resistance $R_S(T)$ of the bridge is described by a step function. The height of the step increases with increasing temperature above a critical one (T_c).

Let us suppose that the heat from the thermal domain is effectively removed only to the substrate and $B<<h$. The heat power flow per unit volume escaping from the bridge into the substrate is determined as:

$$W(T) = \frac{\kappa}{d \cdot d_{\text{eff}}} \cdot (T - T_0), \tag{2}$$

where d_{eff} is the effective thickness of a substrate layer where the temperature change occurs, κ is the thermal conductivity of the substrate. In this discussion the thermal conductivity is isotropic and temperature independent.

From the thermal balance between Joule heating and heat transfer away from the bridge the stability and size of the normal inclusion can be defined. If the heat balance equation has several solutions (Fig. 3), the co-existence of the phases with different temperatures is possible. Solutions, which don't meet the following condition

$$\left. \partial W/\partial T \right|_{T(I)} > \left. \partial Q/\partial T \right|_{T(I)} ,$$

are unstable with respect to a small disturbance of temperature. In Fig.3 one can see that, when the flowing current exceeds the current I_m, the temperature distribution of the superconducting and normal phases will take place. I_m is called the minimum current of the hotspot existence.

Under some conditions the interphase S/N boundaries move independently with a constant velocity υ. The movement of the temperature wave can be described by the expression for the energy conservation low of a thermal domain:

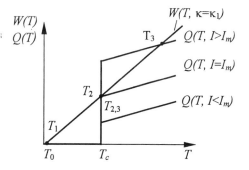

Fig. 3. Graphical representation of the heat balance equation

$$\frac{1}{2}\cdot\left(K\frac{\partial T}{\partial x}\right)^2 + \upsilon\cdot\int_x^\infty \nu\cdot K\cdot\left(\frac{\partial T}{\partial x}\right)^2 dx - S(T,I) = 0 \tag{3}$$

with the boundary conditions:

$$T(\infty) = T_0, \qquad T(-\infty) = T_3, \qquad \left.\frac{dT}{dx}\right|_{\pm\infty} = 0, \qquad D \gg L_h.$$

Here ν and K are the heat capacity and the thermal conductivity of the film correspondingly, $S(T,I)$ is the "potential energy" of the thermal domain

$$S(T,I) = \int_{T_0}^{T}[(W(T) - Q(T,I))\cdot K]dT. \tag{4}$$

The movement of the S/N boundary comes from the compensation of the potential energy difference between S and N phases by the work of friction force of thermal domain.
This statement follows by the equation for the velocity of the interphase S/N boundary

$$\upsilon(I) = -(I - I_p)\cdot\left[\frac{\partial S[T_3(I),I]}{\partial I}\right]_{I_p} \bigg/ \sqrt{2}\cdot\int_{T_0}^{T_3} \nu\cdot S(T,I)^{1/2}dT. \tag{5}$$

The current I_p for $\upsilon=0$ (S/N boundaries are immovable) is determined as the minimum current of the normal phase expanding.
For a simulation of the thermal process, the closed form equation for the minimum current of the normal phase expanding and equation for the velocity of the S/N boundary movement were obtained

$$\upsilon = (I_p - I)\cdot I_p\cdot M_1\cdot\left[T_c^2 - \frac{(M_2\cdot T_0)^2}{(M_2 - M_3\cdot I_p^2)^2}\right], \tag{6}$$

$$(I_p^2)^2 + a\cdot(I_p^2) + b = 0. \tag{7}$$

where a, b, M_1, M_2 and M_3 are variables dependent on parameters of the thermal model of the bridge (geometrical dimensions of the model, film and substrate thermal conductivity, critical and bath temperature, heat capacity of the film).
Fig. 4 illustrates the calculated dependence of the S/N boundaries velocity on dc current for

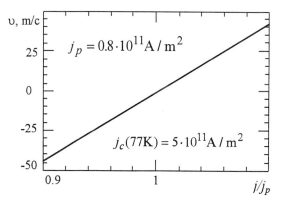

Fig. 4. Calculated velocity of the S/N boundary movement

typical HTS YBa$_2$Cu$_3$O$_7$ film of 25µm width and 0.2µm thickness on sapphire substrate of 0.5mm thickness.

In the case of nonuniform structure, due to complexity of the problem, the numerical simulation is necessary. In given consideration the structure discontinuity can be described as a domain with an increasing (decreasing) the thermal power generation in the domain or the thermal power elimination to the substrate. If the discontinuity can be considered as a point thermal source, the thermal conductivity equation can be represented as follows:

$$\frac{\partial^2 T}{\partial x^2} K + Q(T) - W(T) + F(T) \cdot \delta(x) = 0 . \tag{8}$$

The function F(T) characterizes the additional power which is dissipated in the local structure discontinuity:

$$F(T) = \int_{-l}^{l} dx [Q(T,x) - Q(T) - W(T,x) + W(T)] , \tag{9}$$

where Q(T,x) and W(T,x) are the characteristics of the discontinuity, x is the coordinate along which the current flows, l is the dimension of the local discontinuity.

Different solutions to (8) show that several kinds of localized thermal domains may exist in the nonuniform structure instead of a single domain in the uniform structure. The results of the numerical simulation demonstrate that a local discontinuity does not change remarkably the parameters of the thermal breakdown of the superconducting state in the HTS film. A visual demonstration of all peculiarities associated with the localization of the thermal domain on the discontinuity can be represented by *I-V* curve (Fig. 5). In figure j_r is the current density of the superconductivity restoration, $j_c{}^*$ is the critical current density at the area of the discontinuity.

The experimental verification of the model is in progress. The results of modeling the thermal processes allow to select a regime of control of the switching element preventing spontaneous switching.

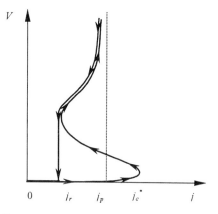

Fig. 5. *I-V* characteristic of a superconducting bridge with a thermal domain

REFERENCES

[1] Gourevitch A.V., Mints R.G.// Uspeskhi Fiz. Nauk in Russia, 1984, Vol. **142**, No. 1, pp. 61-98.

[2] Skocpol W. J. et al. // J. Appl. Phys., 1974, Vol. **45**, No. 9, pp. 4054-4066.

[3] Buznikov N.A., Pukhov A.A.// Pis'ma Zh. Tekh. Fiz (Tech. Phys. Lett.), 1996, Vol. **22**, No.12, pp. 45-49.

Inst. Phys. Conf. Ser. No 167
Paper presented at Applied Superconductivity, Spain, 14-17 September 1999
© 2000 IOP Publishing Ltd

HTS Switchable Diplexer

V Kondratiev, A Svishchev, I Vendik

St.-Petersburg Electrotechnical University, 197376, St.-Petersburg, Russia

E Jakku, S Leppävuori

University of Oulu, FIN-90401, Oulu, Finland

ABSTRACT. The paper demonstrates the feasibility of a design of fully integrated HTS diplexer using microwave switches in each channel. The switchable diplexer consists of two 3-pole filters having 1% bandwidth and two switches based on S-N transition. Two versions of the switchable diplexer are under consideration: with the S-N switch connected in series and in shunt.

1. INTRODUCTION

Microwave multiplexers are of interest for many applications. Different groups have published results of a design and investigation of diplexers and multiplexers [1-3]. We discuss in this paper a possibility of a design of a switchable diplexer using HTS components only. As a microwave switch, the HTS switch based on the S-N transition under dc control current can be used. Small insertion loss and high isolation in two different states of the S-N element of the switch have been demonstrated recently [4]. Small switching time of the S-N switch was measured as a function of the control current [5]. It was shown that the switching time for the transition to the on and off states of the S-N switch can be small enough (~ 10 ns). The main advantage of the diplexer with S-N switches is a homogeneous technological process of manufacturing all components of the microwave device including filters and switches.

2. DESIGN OF THE DIPLEXER

The diplexer consists of two band-pass filters and two S-N switches. Using method of a design of HTS filters [6] which takes into account the HTS film parameters, we have designed two 3-pole pass-band filters based on half-wave-length resonators with end coupling. The central frequency of the filters is 8.94 GHz and 9.06 GHz and the pass-band is 1%. Two versions of the switchable diplexers have been simulated and designed. Both types use the same filters and different S-N switches. The layout of the diplexer a) with the S-N switch connected in series is shown in Fig. 1. The S-N switch consists of the switching element and two impedance transformers (Fig. 2). A narrow strip of 10 um length and 350 um width is used as the HTS switching element. The transformers are used in order to optimize the characteristics of the switch [4]. The optimization allows obtaining the same values of the transmission coefficient in the on state and the reflection coefficient in the off state:

372

$$\left|S_{21}^{(on)}\right| = \left|S_{11}^{(off)}\right| = 1 - \frac{1}{\sqrt{K}}, \tag{1}$$

where K is the commutation quality factor determined in [4,7]

$$K = \frac{R_{sur}^{N}}{R_{sur}^{S}}, \tag{2}$$

R_{sur}^{N} and R_{sur}^{S} are the surface resistances of the HTS film in the N-state and in the S-state correspondingly.

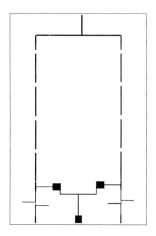

Fig.1 Layout of the diplexer
with the switch in series.

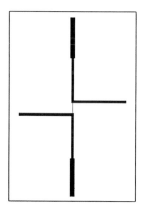

Fig.2 Switching element.

The characteristics of the S-N switch are shown in Fig. 3.

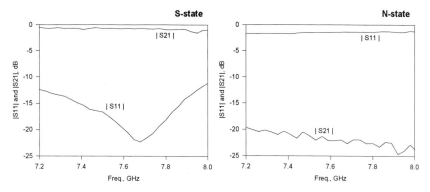

Fig.3 The results of measurement
of the S-N switch

The second type b) of the diplexer with the S-N switches connected in parallel is shown in Fig. 4. Comparison of these two versions of the HTS diplexer allows concluding the following:

Diplexer a) with the S-N switch connected in series has a lower size as compared with the diplexer b); it can be used as a power limiter; the design procedure is more simple because the filters and the switches can be designed separately and then connected by a transmission line section.

Diplexer b) with the S-N switch connected in shunt occupies a larger area as compared with the diplexer a); but it allows obtaining a higher value of the reflection coefficient in the off state and can handle a higher microwave power level.

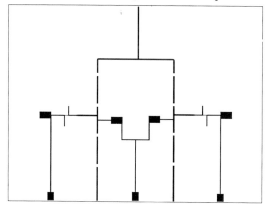

Fig.4 Layout of the diplexer
with the switch in shunt.

3. RESULTS

The simulation performance of the diplexer type a) is depicted in Fig. 5. As one can see, there is no channel influence onto each other, thus it is possible to operate with channels separately. Such diplexer was manufactured using double-sided YBCO film on LaAlO$_3$ substrate of 0.5 mm thickness. Unfortunately due to nonhomogeneous HTS film, the narrow switching elements burned under low level of the control current.

374

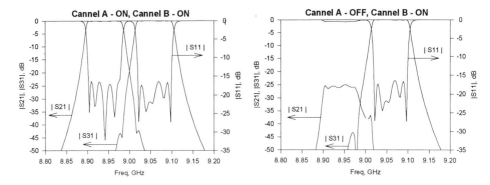

Fig.5 Performance of the diplexer

CONCLUSION

Using the S-N transition in HTS films gives a possibility to develop switchable diplexers and multiplexers, which are fully integrated. The performance of the switchable device depends on the S-N switch characteristics determined by the commutation quality factor of the HTS film. The presented characteristics support the feasibility of development of a fast switchable diplexer with a low insertion loss and a low reflection coefficient in both states.

ACKNOWLEDGMENT

The work was supported by the Science Council on Physics of Condensed Matter (Project № 98063) in Russia. One of the authors (V.K.) was supported by the grant of Finnish Academy of Sciences.

REFERENCES

[1] R.R. Mansour, Van Dokas, G. Thomson, Wai-Cheung Tang, Ch. M. Kudsia, IEEE Trans. MTT, 1994, Vol. **42**, No. 12, pp. 2472-2479

[2] S. Talisa, M.A. Janocko, D.L. Meier, J. Talvacchio, C. Moscowitz, D.C. Buck, R.S. Nye, S.J Pieseski, G.R. Wagner, IEEE Trans. MTT, 1996, Vol. **44**, No. 7, pp. 1229-1239

[3] F.A. Miranda, G. Subramanyam, F.W. Van Keuls, R.R. Romanofsky, IEEE Trans. Appl. Superconductivity, 1999, Vol. **9**, No. 2, pp. 3581-3584

[4] A. Svishchev, I. Vendik, A. Deleniv, P.Petrov, A. Zaitsev,R. Wördenweber,El. Letteres, 1998, Vol. **34**, No. 13. Pp.1329-1330

[5] I. Vendik, D. Kaparkov, M. Gaidukov, S. Razumov, V. Osadchiy, V. Sherman, O. Buslov, B. Chan, S. Bolioli, B. Dirassen, Proc. 25th EuMC, Bologna, 1995, pp. 931-936

[6] I.B. Vendik, V.V. Kondratiev, D.V. Kholodniak,S.A. Gal'chenko,A.N. Deleniv, M.N. Goubina, A.A. Svishchev, S. Leppävuori, J. Hagberg, E. Jakku, IEEE Trans. Appl. Superconductivity, 1999, Vol. **9**, No. 2, June, pp. 3577-3580

[7] I.B. Vendik, O.G. Vendik, E.L. Kollberg, V.O. Sherman, IEEE Trans. MTT, 1999, Vol. **47**, No. 8, pp. 1553-1562

Inst. Phys. Conf. Ser. No 167
Paper presented at Applied Superconductivity, Spain, 14-17 September 1999
© 2000 IOP Publishing Ltd

Optical Modulator with Superconducting Resonant Electrodes for Subcarrier Optical Transmission

K. Yoshida[1], H. Morita[1], H. Kanaya[1], Y. Kanda[2], T. Uchiyama[3], H. Shimakage[3], and Z. Wang[3]

[1] Graduate School of Information Science and Electrical Engineering, Kyushu University, Fukuoka 812-8581, Japan.
[2] Faculty of Engineering, Fukuoka Institute of Technology, Fukuoka 811-0295, Japan.
[3] KARC Communications Research Laboratory, Kobe 651-2401, Japan.

ABSTRACT: We have studied a low voltage optical modulator employing high Tc superconducting resonant-type electrodes for use in the sub-carrier optical transmission system in which microwave signals are carried by intensity modulated lightwave. It is shown that microwave modulation of lightwave can be made efficient by using superconducting electrodes with high Q-factor. A preliminary experiment using LiNbO₃ (LN) optical modulator with YBa₂Cu₃Oₓ (YBCO) electrode demonstrates dc and microwave modulation of the optical wave.

1. INTRODUCTION

In the conventional optical transmission system, which transmits digital data by optical pulses, optical modulators with extremely broad bandwidth are necessary to broaden its bandwidth of the base-band transmission system. Recently, sub-carrier transmission system in which microwave or millimeter wave signals are carried by intensity modulated lightwave has been intensively studied (Sueta and Izutsu 1990), (Dagli 1999). In this system, optical modulators are not required to have such broadband modulation characteristics. Resonant type optical modulators (Izutsu et al. 1988), (Yoshida et al. 1994) are suited as the devices because they can drive with high modulation efficiency at the expense of the narrowing the bandwidth.

We have been studied so far the application of the superconducting electrodes to travelling-wave-type LN optical modulators with broad bandwidth (Yoshida et al. 1997, 1999), and it has been demonstrated that the optical modulator with superconductors which have low loss and low dispersion is superior to that using normal-conductors. In this paper we have studied the resonant type LN optical modulator with high Tc superconducting electrodes. Theoretical performance of resonant type optical modulators is presented in Sec. 2. The design of resonant type optical modulators with impedance matching circuits is presented. YBCO thin films on MgO substrate were employed as the electrodes. In Sec. 3 we present a preliminary experiment on the optical modulator which was realized by flip-chip bonding of LN and MgO substrates.

2. THEORETICAL PERFORMANCES OF THE MODULATOR WITH SUPERCONDUCTING ELECTRODES

Figure 1 shows the schematic of a resonant type optical modulator, which has Mach-Zehnder type Titanium diffused optical waveguides on LN substrates. Input light is split into the two waveguides. The splitted lightwaves are phase modulated by the signal voltage at each waveguide, and they are added interferentially at the output end, giving rise to intensity modulated optical wave. Since resonant type optical modulators employ the standing-wave voltage singal, high efficient modulation can be obtained even at microwave or millimeter wave frequencies if we introduce a high Q resonance circuit composed of high Tc superconducting electrodes.

The electrode consists of a feeder line, a matching circuit, resonance circuits, and a dc bias line as shown in Fig. 2. In order to obtain high efficiency, the circuit is designed so as to satisfy the impedance matching condition. A stub line is added for impedance matching. The impedance of the feeder line is designed as 50Ω, but that of the resonance line is not 50Ω because the size of the coplanar waveguide (CPW) is determined by the shape of the Mach-Zehnder type optical waveguides. Therefore, by adding a stub with a proper length L_2, the input impedance can be adjusted to be 50Ω. When the width and the gap of the resonance line are 10μm and 15μm, respectively, the length of the impedance matching stub is adjusted to be 20μm in the case of Q=10000 at 1GHz. In this case the resonance frequency was 12.45GHz when impedance matching is satisfied for the resonance line of 2.5mm length. In the calculation, the loss of the resonance circuit was represented by the Q-factor which includes conductor, dielectric and radiation losses.

Figure 2 shows a pattern of an electrode, which was designed by CPW including the stub for impedance matching. The stub and resonance lines were terminated by the shorted end. In addition, the dc bias line was introduced to adjust the operating point.

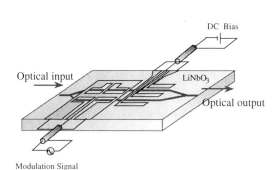

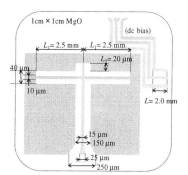

Fig. 1 Schematic of a resonant type optical modulator

Fig. 2 A pattern of an electrode of a resonant type optical modulator by CPW transmission lines

The modulation depth of resonant type optical modulators is given by (Yoshida et al. 1994)

$$F(\omega) = T(\omega) \cdot \frac{\cosh\{\gamma(\omega)L_1\} - \cosh(\beta_0 L_1)}{(\gamma(\omega)L_1)^2 - (\beta_0 L_1)^2} \cdot \frac{\gamma(\omega)L_1}{\sinh\{\gamma(\omega)L_1\}} \qquad (1)$$

where $\gamma(\omega)$ is the propagation constant, and $T(\omega)$ is the ratio of the voltage of incident wave to that at the feeder point. When a stub is used as the matching circuit, $T(\omega)$ is given as

$$T(\omega) = \frac{2}{1 + y_1 + y_2} \qquad (2)$$

where y_1 and y_2 are the admittances of the resonant electrode and the stub, respectively, which are normalized by the characteristic admittance.

The frequency dependence of modulation depth was numerically obtained by changing Q-factors as free parameters. In the calculation, the best-suited stub lengths were used for impedance matching for a given Q-factor. Figure 3 shows the results of the calculation. It has become clear that modulation depth increases as Q-factor increases. It is also seen that the peak frequencies change with Q-factors, because the frequencies satisfying impedance matching condition are different for each Q.

Figure 4 shows that the relationship between Q-factor and the driving voltage which is defined by the amplitude of the applied ac voltage necessary for π radian phase shift of lightwave. This result indicates that the driving voltage can be decreased by employing high Q superconductor electrodes.

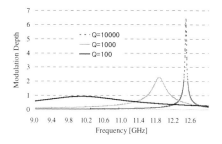

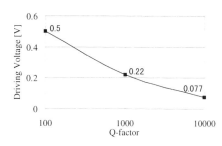

Fig. 3 Frequency dependence of modulation depth using stub circuit

Fig. 4 Relationship between Q-factor and driving voltage

3.   EXPERIMENTAL

In our previous studies (Yoshida et al. 1997, 1999), modulation experiments employing low Tc superconductors (Nb and NbN) for resonant type optical modulators were carried out. In the case of low Tc superconductors, the fabrication of a thin film electrode directly on an LN substrate was possible. In the present experiment, high Tc superconductor was fabricated on a MgO substrate. Subsequently, as shown in Fig. 5, the MgO substrate was glued by flip-chip bonding to the LN substrate in which optical waveguides are fabricated. The photo of the setup is shown in Fig. 6. Optical fibers were connected to both ends of the LN substrate. DC bias voltage and microwave signal were supplied to electrodes on the MgO substrate by micro probes, respectively. This module was installed into the cryostat. The modulator was then cooled down by a refrigerator. A block diagram of the measurement system is shown in Fig. 7.

From the dc modulation experiment, the threshold voltage was obtained. Using this value, the overlap integral (Γ) between optical and microwave fields (Izutsu 1988) are estimated as 0.21, which compares favorably with the result of numerical simulation: $\Gamma = 0.20$. This means that the electric field was effectively applied to optical waveguides even by the flip-chip bonding of LN and MgO substrates. Secondly, the return loss (S_{11}) of the feeder line was obtained at 26K as shown in Fig. 8. It is seen that the resonance frequency was 13.2GHz, and that impedance matching was also satisfied at this frequency.

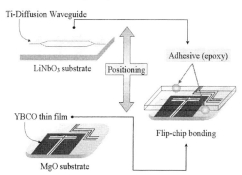

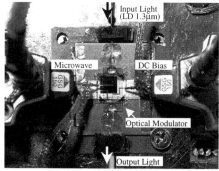

Fig. 5 Schematic of the flip-chip bonding of LN and MgO substrates

Fig. 6 Photo of the setup of the resonant type optical modulator

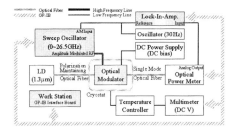

Fig.7 Experimental System

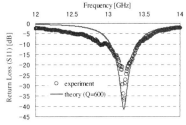

Fig. 8 Resonance characteristics at 26K

In Fig. 9 microwave modulation depth was obtained by the envelope detection method (Izutsu et al. 1988), whose measurement system is shown in Fig. 7. Modulation characteristic was successfully observed around 13GHz. This result shows that optical wave was modulated effectively by the microwave signal, which was supplied through high Tc superconducting electrode. Detailed studies are in progress.

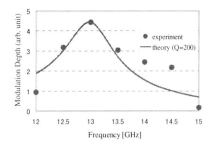

Fig.9 Modulation characteristics at 26K

4.   CONCLUSION

The design and experiment of a resonant type optical modulator with high Tc superconducting electrodes was presented. The impedance matching of the electrode was realized by using a stub line. It is shown by numerical calculation that the modulation efficiency increases when high Q superconductors are used as the electrodes. The experimental optical modulator was realized by a flip-chip bonding of LN and MgO substrates. Based on the design and calculation, modulation experiments were carried out at low temperatures. It was demonstrated that the optical modulator with high Tc superconducting electrode by the flip-chip bonding method operated effectively.

REFERENCES

Izutsu M, Murakami H and Sueta T 1988 IEICE Trans. Electron **J71-C**, 5, pp653-658
Dagli N 1999 IEEE Transactions on Microwave Theory and Thechnics **47**, 7, pp1151-71
Sueta T and Izutsu M 1990 IEEE Trans. Microwave. **38**, 5, pp477-82
Yoshida K, Kanda Y, and Kohjiro S 1999 IEEE Transactions on Microwave Theory and Thechnics **47**, 7, pp1201-05
Yoshida K, Nomura A and Kanda Y 1994 IEICE Trans. Electorn. **E77-C**, pp.1181-84
Yoshida K, Minami A, and Kanda Y 1997 IEEE Trans. Appl. Sureprcond. **7**, 2, pp3508-11

Inst. Phys. Conf. Ser. No 167
Paper presented at Applied Superconductivity, Spain, 14-17 September 1999
© 2000 IOP Publishing Ltd

YBa$_2$Cu$_3$O$_{7-\delta}$ thin films grown by RF sputtering on buffered LiNbO$_3$ substrates

L.Fàbrega[1], R.Rubí[1], F.Sandiumenge[1], J.Fontcuberta[1], C.Collado[2] and J.O'Callaghan[2]

[1]Institut de Ciència de Materials de Barcelona (C.S.I.C.), Campus de la U.A.B., 08193 Bellaterra (SPAIN)
[2]Dpt. Teoria del Senyal i Comunicacions, Campus Nord UPC, Jordi Girona 1-3, 08034 Barcelona (SPAIN)

ABSTRACT: We report on the microstructural and superconducting properties (critical current, RF losses) of epitaxial YBa$_2$Cu$_3$O$_{7-\delta}$ thin films grown by RF magnetron sputtering on LiNbO$_3$ with a buffer layer of yttria-stabilized zirconia. We show that the use of this buffer layer allows to obtain superconducting thin films with high enough critical temperatures and critical currents, and RF performances which make them interesting for fabrication of electrodes for RF feeding of a variety of devices.

1. INTRODUCTION

LiNbO$_3$ (LNO) is an emblematic ferroelectric, electrooptic material commonly employed for a variety of optoacoustic and electrooptical devices, such as components in telecommunication networks. The growth of HTS thin films on single crystals of this material is interesting because, for instance, it would allow the substitution of the normal metal electrodes used at present to feed the RF signal required to excite the LNO crystal; the use of HTS, with, lower surface resistance, would imply a decrease of the excitation power and an increase of the device speed and bandwidth.

However, LNO has a bad structural matching with the layered, perovskite-based HTS, which makes it difficult the growth of good superconducting thin films directly on LNO crystals. Additional problems have been claimed to arise from cationic interdiffusion at the interface -observed by Hashiguchi (1992) for YBa$_2$Cu$_3$O$_{7-\delta}$- and the instability of the LNO surface at the temperature and atmosphere conditions required for the growth of HTS (Wu 1995). In spite of these problems, superconducting thin films of YBa$_2$Cu$_3$O$_{7-\delta}$ (YBCO) have been grown directly on LNO substrates with different cuts, by laser ablation and sputtering techniques; however, the microstructure and superconducting properties of these films is usually rather poor, which makes them unsuitable for most applications. Remarkable exceptions seem to be the films obtained by Lee (1989) and Tiwari (1996), both of them grown on Y-cut substrates.

The use of a suitable buffer layer might be expected to result in an increase of the microstructural quality of the films, and therefore in better superconducting properties. Several works have been already carried out, mainly on Y- and Z-cut LNO substrates using YSZ and MgO buffer layers (Imada 1991 and Hashiguchi 1992), because of their simplicity. Among these, MgO latter must be disregarded because of the existence of an extended interdiffusion.

Here we report on the growth of YBCO thin films on X-cut LNO single crystals using YSZ as buffer layer. This material, which has been shown to allow the epitaxial growth of YBCO, is most suitable due to its moderate dielectric constant and refractive index. The obtained layers are

380

characterized structurally by X-ray diffraction and electron and atomic force microscopy. Superconducting properties were analyzed by transport, inductive and RF losses measurements.

2. EXPERIMENTAL AND RESULTS

The films were deposited by on-axis RF magnetron sputtering, using homemade, stoichiometric targets; the YSZ one contained 16% Y_2O_3. Optimal deposition conditions for YSZ are 50mTorr of total pressure, with 38% O_2 and 62% Ar, substrate temperature 700-750°C and RF power of 50W.

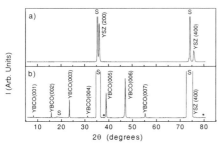

Fig. 1 XRD patterns (θ-2θ scan) of (a) a YSZ film and (b) a YBCO/YSZ bilayer. Labels S and * denote substrate and unknown peaks, respectively.

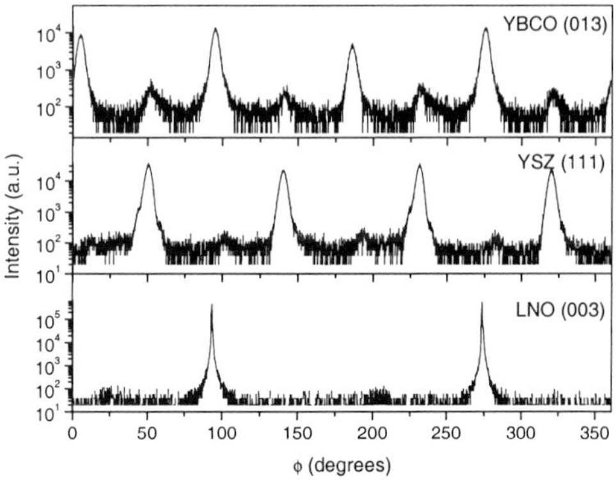

Fig. 2 φ-scans for a YBCO/YSZ/LNO heterostructure.

For superconducting YBa$_2$Cu$_3$O$_{7-\delta}$, the growth conditions were slightly changed to higher temperatures, 780°C, and higher total pressure, 300-400mTorr; the O_2: Ar pressure ratio was kept almost constant. The samples were quenched to 500°C in the growth atmosphere, and then slowly brought to room temperature under 500Torr of O_2.

Fig. 1a shows the θ-2θ diffraction patterns of a 800Å-thick YSZ layer, obtained at optimal deposition conditions. The YSZ grows along the (100) direction, as revealed by the dominant presence of the (h00) reflections: only a minor residual peak, corresponding to the (111) reflection, is observed at 2θ≈30Å. The rocking curve around the (400) peak has a full width at half maximum (FWHM) of 0.5°, indicating that films are indeed highly oriented. The lattice parameter inferred from the (h00) peaks positions is 5.04Å, slightly smaller than that obtained for bulk YSZ, a≈5.14Å. Transversal electron diffraction patterns provide a similar lattice parameter of 5.06Å, both in the plane and along the growth direction. This is suggestive of a dominant cubic YSZ phase, as expected for the used Y_2O_3 content, growing (100) oriented. The off-angle XRD analyses (φ-scan, Fig. 2) confirm the epitaxial growth of the YBCO and (h00) YSZ layers. From the φ-positions of the (111)YSZ and (003)LNO reflections, it may be seen that there is a 7° misorientation between the a-axis of the YSZ cubes and the (00l) direction of the LNO substrates, likely to better accommodate the lattice mismatch between both materials.

The θ-2θ scans of the YBCO/YSZ bilayers (Fig. 1b) indicate that (i) the superconductor grows along the c-axis direction, with a c-axis spread of 1.0° (FWHM of the rocking curve around the (006) peak); and (ii) the YSZ peaks weaken and broaden, as compared to those of a YSZ thin film. This latter effect reveals a deterioration of the YSZ crystallinity, due to the change of the deposition conditions required to grow the YBCO. This point has been checked by in-situ annealing an optimal YSZ film, at the deposition conditions of YBCO; the XRD pattern of the YSZ film after this treatment is similar to that of the bilayer.

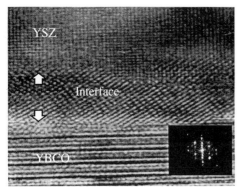

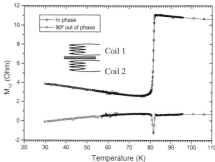

Fig. 3 Cross-sectional HRTEM image YBCO/YSZ interfaces in a bilayer.

Fig. 4 Superconducting transition of a YBCO/YSZ/LNO heterostructure, measured with the mutual inductance technique. The inset shows the film and the two coils.

Fig. 2, displaying the φ-scans of a YBCO/YSZ/LNO heterostructure, reveals the epitaxial growth of the YBCO cube on cube over the YSZ layer.

Transverse High Resolution Transmission Electron Microscopy (HRTEM) images of the bilayers reveal a rough structural interface between the LNO and the YSZ, which might be due to deterioration of the LNO surface prior to the YSZ deposition (Wu 1995), as a consequence of the adverse deposition conditions required for the growth of the latter. Due to this rough interface, the first nm of YSZ grow highly distorted. Far enough from the interface, however, the YSZ lattice rearranges and the film displays a structure with few defects. Because of this improvement of growth quality, the surface of these films is quite smooth, as revealed by Atomic Force Microscopy. Typical r.m.s. roughness values of about 35Å are obtained for optimal YSZ films, 800Å thick.

HRTEM images reveal also (Fig. 3) a thin (~40Å) YSZ/YBCO interface, with a structure different from that of both compounds. This interface might arise from cationic interdiffusion or as a consequence of the deterioration of the YSZ film at the deposition conditions of YBCO as previously discussed. Above this interface, and parallel to it, the CuO_2 planes are clearly appreciated, evidencing the c-axis growth of the HTS. However, the growth seems to deteriorate as it proceeds further, and many structural defects become apparent at higher thickness. However, the surface topography of the YBCO films, as observed by AFM, reproduces that of the YSZ layer.

Finally, the superconducting properties of the grown YBCO films were analyzed. First, a basic characterization was done measuring the superconducting transition by the mutual inductance technique. In this technique the thin film is placed between two small coils -1 and 2 in Fig.4- and the mutual inductance is defined as $M_{12} \equiv V_2/I_1$. V_2 is the voltage induced in coil 2 by a current I_1 circulating in coil 1. Since the mutual inductance depends on the field screening produced by the superconductor, as we change the temperature of the film we are able to measure its critical temperature.

A critical temperature of 85K, with a transition width of about 2K, was obtained (Fig. 4). In order to determine the transport critical current (J_C), a 50μm-wide bridge was photolitographically patterned onto the YBCO layer. The conventional four-probe method was used to measure V-I curves. Gold contacts were sputtered on the film surface, and wires were attached to them with silver paint. The critical current as a function of the temperature is shown in Fig. 5; values of $1.2 \cdot 10^5 A/cm^2$ were reached at 77K.

Two identical waveguide resonators were patterned from Au/YSZ/LNO and YBCO/YSZ/LNO heterostructures. Their attenuation constants are shown in Fig. 6. It may be seen that below 60K the resonator with YBCO has much lower RF losses than that with gold. Comparison

382

of other Au resonators with and without YSZ also evidenced that the YSZ layer does not introduce any additional losses.

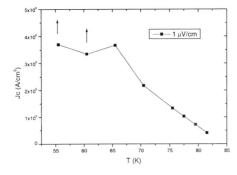
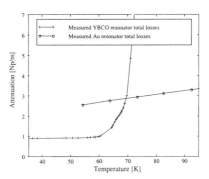

Fig. 5 J_C curves measured on a patterned YBCO/YSZ/LNO heterostructure. The two lower temperature points may be underestimated because of heating problems.

Fig. 6 Attenuation constant for YBCO/YSZ/LNO and Au/YSZ/LNO resonators.

3. CONCLUSIONS

We have demonstrated the epitaxial growth of YBCO thin films on X-cut LNO substrates using a thin YSZ buffer layer. The films show good superconducting properties. Measurements reveal that the YSZ layer does not introduce RF losses and that below 60 K the losses in the heterostructure are lower than those associated to Au. These results are extremely encouraging for the substitution of metallic electrodes by YBCO on LNO-based devices.

ACKNOWLEDGEMENTS

Financial support from Pirelli Cavi and Red Eléctrica de España is acknowledged.

REFERENCES

Collado C et al. (1998), IEEE Trans. on Appl. Supercond.,
Guptasarma et al. (1992), Physica **C 203**, 129
Hashiguchi et al. (1992), Jpn. J. of Appl. Phys. **31**, 780
Hohler et al. (1989), Appl. Phys. Lett. **54**, 1066
K.Imada K et al. (1991), Supercond. Sci. and Technol. **4**, 473
Lee S G et al. (1989), Appl. Phys. Lett. **55**, 1261
Tiwari et al. (1996), J. of Electronic Mat. **25**, 131
Wu N J et al. (1995), J. Mater. Res. **10**, 3009

Inst. Phys. Conf. Ser. No 167
Paper presented at Applied Superconductivity, Spain, 14-17 September 1999
© 2000 IOP Publishing Ltd

383

Tl-2212 films for microwave devices

H Schneidewind, M Zeisberger, H Bruchlos, M Manzel

Institut für Physikalische Hochtechnologie, D-07702 Jena, Germany

T Kaiser

Bergische Universität Wuppertal, D-42097 Wuppertal, Germany

ABSTRACT: High quality $Tl_2Ba_2Ca_1Cu_2O_8$-films (Tl-2212) were deposited on 2 inch diameter $LaAlO_3$ substrates and CeO_2 buffered sapphire by off-axis magnetron sputtering followed by a special annealing process. The best film on $LaAlO_3$ exhibits a surface resistance of R_s = 135 μΩ at 10 GHz and 77 K up to microwave field levels of about 2 mT. R_s-values of 250 μΩ were measured for films on CeO_2 buffered sapphire. Two-pole bandpass filters with a centre frequency of 4 GHz and a bandwidth of 17.5 MHz using λ-dual-mode ring resonators have been fabricated from double-sided Tl-2212 films on $LaAlO_3$.

1. INTRODUCTION

In satellite communication systems the application of high temperature superconducting (HTS) materials with critical temperature as high as possible can improve the efficiency of the applied cryocooler significantly. Up to now $YBa_2Cu_3O_7$-films (YBCO) are the choice for planar microwave devices because of the well established YBCO technology, but the critical temperature of these films is too low to operate on cryocoolers meeting the severe requirements of satellite payload (volume, weight, power consumption).

The Tl-2212 films with critical temperatures noticeably above 100 K offer an alternative to the YBCO films consequently allowing an essential improvement of the performance of HTS communication subsystems. The higher critical temperature allows an operation at 77 K sufficiently away from the superconducting transition leading to a minimization of the influence of operating temperature fluctuations. A further advantage is to operate the Tl-based devices at a higher temperature compared to the YBCO material leading to higher cryocooler efficiency.

In this paper we present the preparation of high-quality Tl-2212 thin films on both sides of 2" $LaAlO_3$ substrates as well as the processing and characterization of microstrip bandpass filters on $LaAlO_3$. However $LaAlO_3$ as HTS substrate material exhibits some disadvantageous microwave properties. Consequently a further improvement in microwave performance of planar HTS filters is expected by using sapphire substrates with excellent microwave properties. We show first results of the preparation of Tl-2212 on 2" CeO_2 buffered sapphire substrates.

2. FILM DEPOSITION AND PROPERTIES

Epitaxial Tl-2212 films deposited on $LaAlO_3$ as well as on sapphire were prepared by means of the standard two-step method. However the growth of high quality Tl-2212 films on R-plane sapphire always requires the use of a buffer layer to avoid chemical reactions and to ensure a low lattice mismatch. We use 50 to 100 nm thick CeO_2 buffer layers deposited by planar rf magnetron sputtering in an Ar/O_2 gas mixture at 1 Pa, rf power 100 W, and about 800 °C substrate temperature. The resulting films show excellent (100) orientation without any (111) components and a nearly perfect in-plane epitaxy. The half-width of the rocking curve of the (200) peak amounts to < 0.6 °.

In the first fabrication step Tl-free Ba-Ca-Cu-O-precursor films are deposited from a planar dc or rf driven 4" $Ba_2Ca_1Cu_2$-alloy target by means of reactive high-rate magnetron sputtering in an off-axis geometry. Precursor films of 400 nm thickness are prepared in 40 % Ar + 60 % O_2 at 2 Pa absolute pressure in dc mode or at a mixture of 85 % Ar + 15 % O_2 at 5 Pa absolute pressure in rf mode, respectively. The substrate temperature is kept at about 250 °C to avoid water contamination and to enhance adhesion of the films. In order to passivate the highly reactive precursor films against moisture und CO_2 in the air the deposition process is followed by a rf excited post oxidation process in pure O_2. The as deposited films are amorphous, electrically insulating, and appear mirror smooth.

In the second preparation step the thallium oxide is incorporated into the precursor films during crystallization to the Tl-2212 phase. For that purpose the precursors are annealed in Tl-oxide rich atmosphere and subsequently cooled down under flowing oxygen. This annealing process is carried out in special tube furnaces where the precursor film together with a sintered Tl-Ba-Ca-Cu-O pellet acting as the Tl_2O-source are enclosed in a Au/Pt-container.

The conditions for the Tl-oxide incorporation are completly different for processing Tl-2212 on $LaAlO_3$ or on CeO_2 buffered sapphire, respectively. The reasons are possible reactions between the CeO_2 buffer and the growing Tl-2212 film (O'Connor 1996) in comparison to the chemical stable $LaAlO_3$. In order to prevent degrading chemical reactions of the CeO_2 buffer layer we annealed the films on sapphire at lower temperatures (860 °C in maximum) compared to the standard process on $LaAlO_3$ (Huber 1995). The annealing time was decreased to 30 min or less. Furthermore we changed the composition of the Tl-Ba-Ca-Cu-O pellets to enable a sufficient Tl_2O formation at lower temperatures. In addition the chemical interaction between the growing Tl-2212 film and buffer layer or substrate material respectively, turned out to be strongly dependend on buffer layer quality. Therefore we focussed our interest on the developement of nearly perfect epitaxial buffer layers. Not only the c-axis orientation but especially an outstanding in-plane orientation is the key for receiving stable buffer layers. After intensive optimization our CeO_2 buffer layers resist not only the severe conditions of annealing at 860 °C for 30 min but they also allow to produce high quality Tl-2212 films.

The epitaxial Tl-2212 films are characterized by XRD, microanalysis, SEM, ac-susceptometry, and microwave surface resistance measurements. The superconducting properties of films prepared on both substrate materials are summarized in the following table.

	$LaAlO_3$	CeO_2 buffered sapphire
Main phase	2212	2212
Grain size	10 μm	1 μm
Critical temperature	105 K	103 K
Transition width (10 % to 90 %)	1 K	1 K
Surface resistance at 10 GHz and 77 K	135 μΩ	250 μΩ

Fig. 1 shows the $\chi'(T)$-curves for Tl-2212 films on CeO_2 buffered sapphire with various buffer film thicknesses (indicated at curves). The measurements were carried out at two different magnetic field strengths (6 and 600 μT). A decrease in superconducting transition temperature with increasing CeO_2 layer thickness is clearly obvious. This degradation is caused by the formation of microcracks during annealing and subsequent cooling due to thermal expansion coefficient mismatch. SEM micrographs show the quantity and length of the microcracks increasing with enhanced buffer layer thickness. In the case of buffer layers thicker than about 250 nm we observe closed rectangular patterns of cracks whereas below 250 nm down to 100 nm we see single cracks at a rather lower quantity. Therefore we chose buffer layer thickness to values below 100 nm to prevent degrading microcracks in the Tl-2212 films.

For application of HTS thin films in passive microwave devices not only a low microwave surface resistance is important but also the power dependence of the microwave losses to ensure a high power stability of the devices. Therefore the power dependence of the surface resistance was measured by means of a sapphire resonator at 19 GHz for different temperatures (seen in Fig. 2). At low fields a constant level of microwave surface resistance R_s is visible. Depending on temperature the microwave surface resistance increases considerable above field levels of about 2 mT. This result indicates the improved power handling capability of Tl-2212 films compared to the Tl-2223 material.

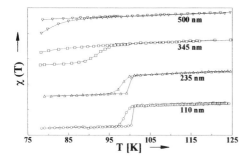

Fig. 1   Influence of buffer layer thickness on $\chi'(T)$ characteristics (curves are shifted in vertical direction for clearness)

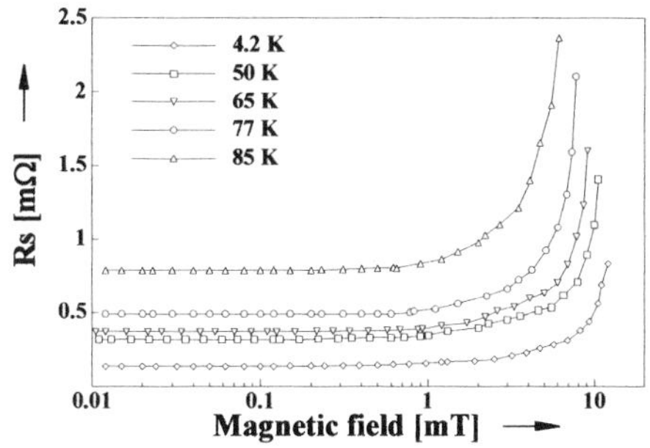

Fig. 2   Surface resistance of a Tl-2212 film on 2" LaAlO$_3$ at 19 GHz for different temperatures

3.   FILTER DESIGN AND PREPARATION

Fig. 3 shows the patterned filter structures using λ-dual-mode ring resonators to realize an elliptic filter function. The 2-pole bandpass filters with a centre frequency of 4 GHz and a bandwidth of 17.5 MHz have been fabricated from double-sided Tl-2212 films on 2" LaAlO$_3$ substrates. The operation principles of the dual-mode ring resonator have been explained by Guglielmi (1990).

Patterning of the Tl-2212 filters was carried out using a standard photolithographic technique and wet etching process. Further details of the filter preparation are given by Zeisberger (1998).

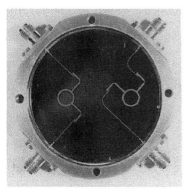

Fig. 3 Copper housing with installed symmetric (left) and asymmetric (right) 2-pole filter based on λ-dual-mode ring resonators.

4.   FILTER RESULTS

The filters mounted in a copper housing with SMA connectors and tuning elements located above the transmission lines were cooled down to 77 K in an open dewar arrangement which allows access to the tuning elements during operation. The measurement was done using a HP network analyzer and a S-parameter set. The setup was calibrated to substract the losses of the input and output cables. But the losses caused by the SMA-connectors and the silver paint to contact the filters are still included in the measurement results.

The measured frequency response of a symmetric and an asymmetric 2-pole filter is shown in Fig. 4. The centre frequency amounts to 4.006 GHz and 4.002 GHz, respectively. The insertion loss for both filter configurations is better than 0.31 dB. The filters were tuned to show a return loss S_{11} below -23 dB over the hole bandwidth of 17.5 MHz. The transmission characteristics of the symmetric filter turned out clear transmission zeros at each side of the passband whereas no zeros were observed in the case of the asymmetric filter.

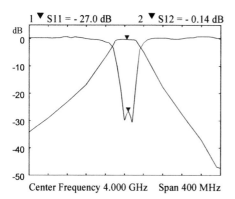 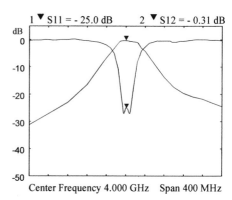

Fig. 4   Measured insertion loss S_{12} and return loss S_{11} of a symmetric (left) and asymmetric (right) 2-pole filter with superconducting ground plane after tuning at 77 K

The shown results were obtained after several redesigns of the filter layout. In the case of $LaAlO_3$ substrates this successive procedure of experimental trial and redesign after measurement is essential because the simulation and numerical design of filter performance is complicated by the rather high permittivity of $\varepsilon_r \approx 24$. Therefore it is an important step forward to prepare Tl-2212 filters on substrates more suitable for microwave application, for instance sapphire.

5. CONCLUSIONS

The presented results demonstrate the successfull implementation of 2-pole Tl-2212 filters based on λ-dual-mode ring resonators on double-side coated $LaAlO_3$ substrates. The fabrication of 4- or 8-pole filters on $LaAlO_3$ is rather difficult because of the unfavourable microwave properties of the $LaAlO_3$ material. Furthermore filter performance will be complicated by the substrate twinning and the low thermal conductivity. Therefore we optimized the technology to prepare 2" Tl-2212 films on CeO_2 buffered sapphire. The results are encouraging to continue improvement of the quality of Tl-2212 films on sapphire, to establish a technology for double-sided deposition on 2" sapphire substrates, and in future on 3" sapphire substrates. Filters will be prepared from Tl-2212 films on CeO_2 buffered sapphire.

ACKNOWLEDGEMENTS. This work is supported by the German BMBF under Grant No. 13 N 7386. The authors thank B. Conrad, G. Bruchlos, and K. Kirsch for technical support.

REFERENCES

Guglielmi M and Gatti G 1990 Proc. 20th European Microwave Conference pp 901-906
Huber S, Manzel M, Bruchlos H, Hensen S, Müller G 1995 Physica C **244**, 337
O'Connor, Dew Hughes D, Bramley A P, Grovenor C R M et al. 1996 Appl. Phys. Lett. **66**, 115
Zeisberger M, Manzel M, Bruchlos H, Diegel M, Thrum F et al. 1998 presented at the ASC´98

Inst. Phys. Conf. Ser. No 167
Paper presented at Applied Superconductivity, Spain, 14-17 September 1999

Spectral-domain Modelling of Superconducting Microstrip Structures

Z. D. Genchev[1], A. P. Jenkins[2], D. Dew-Hughes[2]

[1] Institute of Electronics, Bulg. Ac. Sci.,1784 Sofia, Bulgaria(zgenchev@iegate.ie.bas.bg)
[2] Department of Engineering Science, University of Oxford, OX1 3PJ UK

ABSTRACT: We present an analysis of thin parallel plate microstrip superconducting structures in which the effects of the superconducting material, of finite complex conductivity and finite thickness, are taken into account through the concept of sheet resistance. A new analytical solution for the propagation properties of parallel plate transmission line with three substrates is given.

1. INTRODUCTION

The phase velocities and attenuation constants of very thin superconducting strips were analysed by Pond (1989) by means of full-wave analysis consisting of a combination of the sheet resistance and the spectral-domain approaches. Kuo et al (1991) extended this method using pulse functions, while Cai et al (1994) determined the resonant frequencies of patch resonators using the same approach. Calculation of the Green dyadics, by taking into consideration resistive boundary conditions and the unequal distribution of the surface current density on the two sides of the conductor, was given by Amari et al (1996). In all of these treatments the electromagnetic field outside the substrate decreases monotonically to zero at infinity. It is of practical and theoretical importance to consider the field patterns and propagation properties for a multiple layer transmission line composed of several thin superconducting films separated by dielectrics, i.e., to consider vertical combination of several double-sided deposited substrates. This is the aim of this paper.

2. TRANSMISSION LINE ON ONE DOUBLE SIDED SUBSTRATE

The resistive boundary condition (Pond 1989, Senior 1975) can be introduced for a TM electromagnetic wave ($\frac{\partial}{\partial y} \equiv 0$) propagating in the positive z direction of the form $\exp\{i(\omega t - az)\}$ with Re(a)>0 and Im(a)<0 as follows:

$$E_z(x = \frac{d}{2}) = R\{H_y(x = \frac{d}{2} + 0) - H_y(x = \frac{d}{2} - 0)\} \tag{1}$$

$$E_z(x = -\frac{d}{2}) = R\{H_y(x = -\frac{d}{2} + 0) - H_y(x = -\frac{d}{2} - 0)\} \tag{2}$$

In Eqs. (1,2) we suppose that the tangential electric field (E_z) is continuous at the interfaces x $= \pm \frac{d}{2}$ separating the vacuum regions ($\frac{d}{2} \le x \le \infty$), ($-\infty \le x \le -\frac{d}{2}$) from the dielectric substrate $|x| \le \frac{d}{2}$ (which has a relative scalar complex permittivity ε). The jump of the tangential magnetic field models the existence of thin superconducting layer of thickness t and complex conductivity $\sigma = \sigma_n - i(\omega\mu_0\lambda^2)^{-1}$. The resistance of this sheet is given by:

$$R = \frac{1}{\sigma t} \tag{3}$$

The thickness t is presumed to be small compared to any characteristic field penetration length i.e., $t \ll \lambda; t \ll \delta = [\frac{2}{(\omega\mu_0 \sigma_n)}]^{1/2}$ (3a)

δ is the classical skin depth. Let k_b and k denote the following wave numbers

$$k_b = [a^2 - k_o^2]^{1/2}, \qquad k = [a^2 - k_o^2 \varepsilon]^{1/2} \tag{4}$$

$\operatorname{Re}(k_b) > 0; \operatorname{Re}(k) > 0; k_o = \dfrac{\omega}{c_o} = \omega\sqrt{\varepsilon_0\mu_0}$

Then, the magnetic field has the following structure:

$$H_y(x > d/2) = C\exp\{-k_b(x - d/2)\} \tag{5a}$$

$$H_y(|x| < d/2) = Ach(kx) + Bsh(kx) \tag{5b}$$

$$H_y(x < -d/2) = D\exp\{k_b(x + d/2)\} \tag{5c}$$

In Eqs. (5) A,B,C,D are arbitrary constants, for which from Eqs. (1,2) we find four linear relations. The dispersion relation which follows from these relations is a product of two different transcendental functions:

$$F_1(s,c)F_2(s,c) = 0, \tag{6}$$

where

$$s \equiv sh(\frac{kd}{2}), c \equiv ch(\frac{kd}{2}) \tag{7}$$

and

$$F_1(s,c) = \frac{ks}{k_o^2\varepsilon} + \frac{iR}{\omega\mu_0}(c + \frac{ks}{\varepsilon k_b}) \tag{8}$$

$$F_2(s,c) = F_1(c,s) = \frac{kc}{k_o^2\varepsilon} + \frac{iR}{\omega\mu_0}(s + \frac{kc}{\varepsilon k_b}) \tag{9}$$

Let us note that if Hy (x)=Hy(-x) the dispersion relation is given by Eq. (8) F_1=0 and that if H_y is an odd function of x the dispersion relation is $F_2 = 0$. On the other side, multiplication $F_1 F_2$=0 leads exactly to Eq. (21) given by Pond et al (1989), namely:

$$\tanh(kd) = -\frac{2fg}{g^2 + f^2}; g = \frac{k_o^2\varepsilon}{k}; f = -\frac{i\omega\mu_0}{R} + \frac{k_o^2}{k_b} \tag{9b}$$

3. PROPAGATION PROPERTIES OF MULTIPLE- LAYER TRANSMISSION LINE

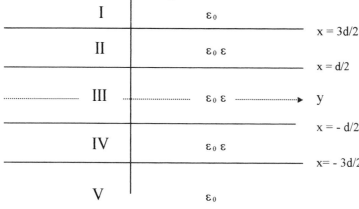

Fig.1. Parallel plate transmission line with three substrates.

Fig. 1 shows a parallel plate transmission line consisting of three identical substrates (regions II,III,IV). These three substrates have thicknesses d and relative dielectric permittivities ε. The substrate (region III) is without metallization, whereas the substrates in the IInd and IVth regions are double-sided coated with superconducting (or conventional) material. The sheet resistance of interfaces $x = \pm \frac{3d}{2}, x = \pm \frac{d}{2}$ is given by formula (3) with restrictions (3a) satisfied. The regions I (x>3d/2) and V (x<-3d/2) are vacuum halfspaces. Instead of Eq.(5) now the general solution (TM-mode) of Maxwell equations contains eight arbitrary constants ($H_1, C, C_1, A, B, D, D_1, H_2$) for wich we choose to have dimensions of magnetic field. This analytical solution is given by the following expressions:

$$H_y^{(1)}(x > \frac{3d}{2}) = H_1 e^{-k_b(x-\frac{3d}{2})} \tag{10a}$$

$$H_y^{(2)}(\frac{d}{2} < x < \frac{3d}{2}) = Cch(k(x-\frac{d}{2})) + C_1 sh(k(x-\frac{d}{2})) \tag{10b}$$

$$H_y^{(3)}(|x| < d/2) = Ach(kx) + Bsh(kx) \tag{10c}$$

$$H_y^{(4)}(-3d/2 < x < -d/2) = Dch(k(x+d/2)) + D_1 sh(k(x+d/2)) \tag{10d}$$

$$H_y^{(5)}(x < -3d/2) = H_2 \exp[k_b(x+\frac{3d}{2})] \tag{10e}$$

We introduce the following three notations

$$\frac{\omega \mu_0}{k_0} = \sqrt{\frac{\mu_0}{\varepsilon_0}} = Z_o = 377\Omega \tag{11a}$$

$$R' = \frac{R}{R - iZ_0 \dfrac{k_b}{k_0}} \tag{11b}$$

$$\rho_{(3)} = \frac{R' ch(kd) + \dfrac{k}{k_b \varepsilon} sh(kd)}{R' sh(kd) + \dfrac{k}{k_b \varepsilon} ch(kd)} \tag{11c}$$

By using all boundary conditions at $x = \pm d/2, \pm 3d/2$ we derive the dispersion equation which we will write by a common formula (valid for both cases j=3-this section and j=1- one substrate case-section 2):

$$F_{1j}(s,c)F_{2j}(s,c) = 0 \tag{12}$$

where:

$$F_{1j}(s,c) = \frac{ks}{k_0^2 \varepsilon} + \frac{iR}{\omega\mu_0}(c + \frac{s}{\rho_{(j)}}) \tag{13a}$$

$$F_{2j}(s,c) = F_{1j}(c,s) = \frac{kc}{k_0^2 \varepsilon} + \frac{iR}{\omega\mu_0}(s + \frac{c}{\rho_{(j)}}) \tag{13b}$$

Obviously in the one-substrate case:

$$\rho_{(1)} = \frac{k_b \varepsilon}{k} \tag{14}$$

whereas in the three substrates case we must use $\rho_{(3)}$, i.e. Eq. (11c). Let us note that in the limit $|k|d \to 0$ we have for the fundamental mode the approximate solution:

$$a^2 = k_0^2 \varepsilon (1 - \frac{iC_{(j)}R}{d\omega\mu_0}) \tag{15}$$

where $C_{(1)} = 2$, $C_{(3)} = 0,438$. In many applications the quality factor Q of a waveguide or cavity resonator is given by Collin (1994) as $Q = \dfrac{\operatorname{Re}(a)}{2|\operatorname{Im}(a)|}$. This result demonstrates an enhancement of Q in the case of three substrate construction over that of the single substrate case.

REFERENCES

Amari S and Bornemann J 1996 IEEE Trans. Microwave Theory Tech. 44, 967.

Cai Z and Bornemann J 1994 IEEE Trans. Antennas Prop. 42, 1443.

Collin RE 1992 Foundations for Microwave Engineering (McGraw-Hill).

Kuo CW and Itoh T 1991 IEEE Microwave and Guided Wave Lett. 1, 172.

Pond JM, Krowne CM and Carter WL 1989 IEEE Trans. Microwave Theory Tech. 37, 181.

Senior TBA 1975 Radio Science 10, 645.

Inst. Phys. Conf. Ser. No 167
Paper presented at Applied Superconductivity, Spain, 14-17 September 1999
© 2000 IOP Publishing Ltd

Highly reproducible double-sided 3-inch diameter YBCO and YBCO / SrTiO₃ / YBCO* thin films for microwave applications

M Lorenz, H Hochmuth, D Natusch, K Kreher

Universität Leipzig, Fakultät für Physik und Geowissenschaften, D-04103 Leipzig, Germany

T Kaiser

Cryoelectra GmbH, D-42287 Wuppertal, Germany

R Schwab, R Heidinger

Forschungszentrum Karlsruhe, Institut für Materialforschung I, D-76021 Karlsruhe, Germany

C Schäfer, G Kästner, D Hesse

Max-Planck-Institut für Mikrostrukturphysik, D-06120 Halle/Saale, Germany

ABSTRACT: A large-area pulsed laser deposition process for high-quality $YBa_2Cu_3O_{7-\delta}$ (YBCO) thin films on both sides of R-plane sapphire substrates with CeO_2 buffer layer is used routinely to optimize planar microwave filters for satellite and mobile communication systems. With the experience of more than 700 double-sided 3-inch diam. YBCO:Ag films a high degree of reproducibility of j_c values above 3.5 MA/cm^2 and of state of the art R_s values is reached. YBCO / SrTiO₃ / YBCO* / CeO_2 film systems on R-plane sapphire wafers show as microwave resonators encouraging electrical tunability.

1. INTRODUCTION

High-T_c superconducting (HTS) $Y_1Ba_2Cu_3O_{7-\delta}$ (YBCO) thin films on low dielectric loss substrates are suitable candidates for applications as passive microwave devices in future communication systems. There is a huge number of activities in many countries to develop microwave devices using HTS thin films, because in this field real market applications of HTS subsystems seem to be possible in the nearest future. Very recently, the physical and technological aspects of HTS thin films at microwave frequencies were summarized in the excellent book by Hein (1999).

A highly reproducible PLD process for large-area 3-inch diameter and double-sided YBCO films on sapphire substrates for microwave applications is developed and continuously improved by Lorenz et al (1996, 1997a, b, 1999). The electrical and microwave performance of this PLD-YBCO films is comparable to high-quality films deposited by other techniques as e. g. thermal coevaporation developed by Kinder et al (1997). Reproducible deposition processes are base technologies for the successful application of HTS microwave devices in commercial products and several processes are now ready to fulfill the demands of the device developing community.

This paper describes the state of the art of a highly reproducible large-area PLD technique for HTS-YBCO and dielectric SrTiO₃ thin films on buffered sapphire wafers by comparing maps of the critical current density j_c and of the microwave surface resistance R_s.

392

2. EXPERIMENTAL

YBCO:Ag thin films, dielectric SrTiO₃ layers, and CeO₂ buffer layers were deposited by PLD at very similar conditions, using a KrF excimer laser operating at 248 nm wavelength, on both sides of 3-inch diameter R-plane sapphire wafers of 430 μm thickness as described in more detail by Lorenz et al (1996, 1997a, b). The double-sided films are deposited subsequently at substrate temperatures around 760°C. During deposition of several hundreds of 3-inch diameter sapphire wafers a high degree of film quality and reproducibility is reached .

Most of the PLD-YBCO:Ag films are used for development of passive microwave filters for the advanced satellite and communication technique at Robert BOSCH GmbH Stuttgart. Mapping of j_c at 77 K, employing the inductive and side-selective method by Hochmuth and Lorenz (1996), is used as a simple and fast routine control of the stability of PLD process and of the resulting YBCO film quality. R_s-mapping using an open resonator technique by Schwab et al (1998) at 145 GHz provides sensitive information on lateral R_s homogeneity over the 3-inch diam. film surface. This information is very important for control of lateral homogeneity of the deposition process. The dependence of R_s on the microwave surface magnetic field B_s at 8.5 GHz and 77 K was measured with a 16 mm diam. sapphire resonator in the center of the 3-inch diam. films by Kaiser et al (1997), giving direct information on the power handling capability of the films. TEM cross sections were prepared at MPI for Microstructure Physics in Halle/Saale as recently described by Kästner et al (1999).

3. REPRODUCIBILITY OF PLD-YBCO FILMS

Fig. 1 demonstrates the obtained lateral homogeneity and reproducibility of the critical current density of the double-sided 3-inch diam. YBCO films on CeO₂ buffered R-plane sapphire. In more detail, Fig. 1 shows the j_c-scans of 8 consecutively deposited double-sided YBCO films with sample No. S 100 to S 108. High critical current densities of 4 to 5.5 MA/cm² at 77 K with a YBCO thickness of about 250 nm are obtained routinely by the large-area PLD technique. The variation of j_c from sample to sample is of nearly the same quantity as for one and the same 3-inch diam. YBCO film. Fig. 1 is a prove for the very good reproducibility of the large-area PLD, which is obtained after more than 5 years optimization work on the process.

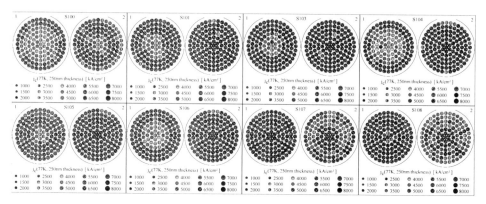

Fig. 1. Reproducibility of the large-area PLD process: j_c maps of 8 consecutively deposited double-sided 3-inch diam. YBCO:Ag thin films on CeO₂ buffered R-plane sapphire.

More important than electrical film properties at low frequencies of some kHz is the film performance at microwave frequencies of some GHz. Therefore, Fig. 2 shows R_s maps of typical double-sided 3-inch diam. YBCO films at 145 GHz and 77 K as measured at Research Center Karlsruhe. Because of the application of the PLD-YBCO films as microwave devices the sensitive microwave properties are definitive for optimization of the PLD process. Fig. 2 demonstrates very good lateral homogeneity of R_s over single 3-inch diam. films which is within the statistical variation of R_s values due to the measuring system. However, from side one to side two of the same sample and from sample to sample there are higher differences of the levels of R_s which are due to the sequential deposition of both YBCO:Ag films on side 1 and 2. However, the planar stripline structures of the microwave filters are patterned always on side 2 whereas side 1 which is the ground plane affects less sensitively the total filter performance. Indeed, C-band filters with very good bandpass performance which is suitable for satellite communication subsystems have been structured from PLD-YBCO:Ag thin films at Robert BOSCH GmbH Stuttgart.

The microwave power handling capabilities of selected PLD - YBCO:Ag films show similar $R_s(B_s)$ characteristics and a good reproducibility. The R_s values at 8.5 GHz remain constant below 400µΩ up to a microwave surface magnetic field of about 10 mT as shown by Lorenz et al (1999).

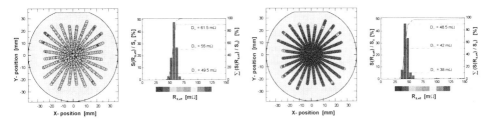

Fig. 2. Lateral homogeneity of microwave surface resistance: R_s-mapping with histograms, taken at 145 GHz and 77 K, of a typical double-sided PLD-YBCO:Ag film sample No. S 070. For side 1 (left) and side 2 (right) were found average R_s values of 55 and 42 mΩ, respectively.

4.   YBCO / SrTiO₃ FILMS FOR TUNEABLE MICROWAVE DEVICES

In order to develop electrically tuneable microwave filters and phase shifters there is a need for combinations of dielectric $SrTiO_3$ or ferroelectric $Sr_{1-x}Ba_xTiO_3$ films at one hand and of HTS-YBCO films on the other hand. For example, several microstrip-based phase shifter designs were proposed by van Keuls (1998). In order to deposit YBCO / $SrTiO_3$ bilayers on R-plane sapphire substrates, there are necessary additional CeO_2 buffer and YBCO* seed layers in between the sapphire substrate and the $SrTiO_3$ film as proposed e.g. by Boikov et al (1995). Only by introducing a more than 30 nm thick c-axis oriented YBCO* film on the CeO_2 buffer layer we found a nearly perfect (100)-texture of the $SrTiO_3$ film as proved by XRD phi-scans.

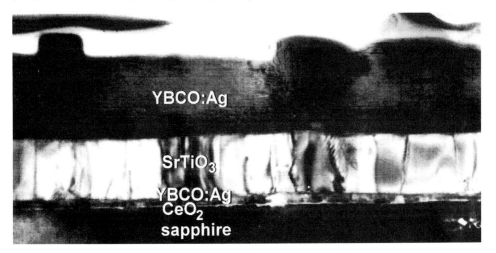

Fig. 3. TEM cross-section of sample No. G 877 with YBCO / SrTiO₃ / YBCO* / CeO₂ films on R-plane sapphire showing two a-axis oriented grains within the c-axis oriented YBCO:Ag film on top. The right a-axis oriented grain nucleates at a step of the dielectric SrTiO₃ layer.

Fig. 3. shows a TEM cross section of one of the first deposited 4-layer film systems on sapphire. All films were deposited in-situ in one deposition run which appears as a main advantage of the flexible PLD technique. Obvious in Fig. 3 are the a-axis oriented grains in the top YBCO film on $SrTiO_3$ which are found with much higher probability compared to YBCO on CeO_2 buffer layers. Because of this structural imperfectness, the j_c at 77 K of the YBCO films on $SrTiO_3$ is only about 2 MA/cm^2 up to now, which is about a factor of two smaller than the j_c of PLD-YBCO films on CeO_2 buffered sapphire as shown in Fig. 1. Correlation of microstructure and microwave properties of HTS-YBCO films were discussed for example by Kästner et al (1999) and Lorenz et al (1997b and 1999). A-axis oriented grains are known to degrade both j_c and R_s. Test resonators made from the described YBCO / $SrTiO_3$ / YBCO* / CeO_2 film system on sapphire wafers showed encouraging tunability of bandpass frequency.

CONCLUSIONS

A PLD technique is presented which allows the fully reproducible double-sided coating of 3-inch diameter sapphire wafers by thin YBCO/CeO_2 films with laterally homogeneous j_c and R_s values. Hundreds of PLD-YBCO:Ag films were already used as microwave bandpass filters for future communication systems by Robert BOSCH GmbH Stuttgart, Germany. The flexible PLD-technique seems to be advantageous compared to other deposition techniques, particularly if more complicated combinations of HTS and dielectric films as e. g. YBCO / $SrTiO_3$ films on sapphire substrates with CeO_2 buffer and YBCO* seed layer for use as electrically tuneable microwave filters have to be deposited in-situ.

ACKNOWLEDGEMENTS

We thank M. Klauda, C. Neumann, T. Kässer, F. Schnell and R. Schmidt (Robert BOSCH GmbH) for the long-term friendly cooperation. We are indebted to the German BMBF and to Robert BOSCH GmbH for continuous support within „Leitprojekt: Supraleiter und neuartige Keramiken für die Kommunikationstechnik der Zukunft", and to the Saxonian Ministry of Science and Art.

REFERENCES

Boikov Yu A, Ivanov Z G, Kiselev A N, Olsson E, Claeson T 1995 J. Appl. Phys. **78**, 4591
Hein M 1999 High-Temperature-Superconductor Thin Films at Microwave Frequencies
    (Berlin, Heidelberg, New York: Springer Tracts in Modern Physics 155)
Hochmuth H and Lorenz M 1996 Physica C **265**, 335
Kaiser T, Bauer C, Diete W, Hein M, Kallscheuer J, Müller G, Piel H 1997
    Inst. Phys. Conf. Series **158**, 45
Kästner G, Schäfer C, Senz S, Kaiser T, Hein M, Lorenz M, Hochmuth H, Hesse D 1999
    Supercond. Sci. Technol. **12**, 366
Kinder H, Berberich P, Prusseit W, Rieder-Zecha S, Semerad R, Utz B 1997
    Physica C **282 – 287**, 107
Lorenz M, Hochmuth H, Natusch D, Börner H, Lippold G, Kreher K and Schmitz W 1996
    Appl. Phys. Lett. **68**, 3332
Lorenz M, Hochmuth H, Natusch D, Börner H, Thärigen T, Patrikarakos D G, Frey J, Kreher K, Senz
    S, Kästner G, Hesse D, Steins M, Schmitz W 1997a IEEE Transact. Appl. Supercond. **7**, 1240
Lorenz M, Hochmuth H, Frey J, Börner H, Lenzner J, Lippold G, Kaiser T, Hein M and
    Müller G 1997b Inst. Phys. Conf. Series **158**, 283
Lorenz M, Hochmuth H, Natusch D, Lippold G, Svetchnikov V L, Kaiser T, Hein M,
    Schwab R, Heidinger R 1999 IEEE Transact. Appl. Supercond. **9**, 1936
Schwab R, Heidinger R, Geerk J, Ratzel F, Lorenz M, Hochmuth H 1998 Proc. 23rd Int.
    Conf. on Infrared and Millimeter Waves, University of Essex, Colchester, U.K.,
    September 7 - 11, 1998
Van Keuls F W, Romanovsky R R, Miranda F A 1998 Integrated Ferroelectrics **22**, 373

Inst. Phys. Conf. Ser. No 167
Paper presented at Applied Superconductivity, Spain, 14-17 September 1999

395

Voltage tunable $YBa_2Cu_3O_7$ -$BaTiO_3$ - microwave ring resonator processing and characterization

A Lacambra, M Rossman, J Blanco, G Dimarco, D Dixon, F Leon, Yu Vlasov[*] and G L Larkins Jr

Future Aerospace Science and Technology Center for Space Cryoelectronics, Florida International University, Miami, FL, USA
[*] On leave from the Ioffe Physico-Technical Institute, St. Petersburg, Russia

ABSTRACT: We have integrated tunable coplanar capacitors with a $YBa_2Cu_3O_7$ ring resonator on $LaAlO_3$ substrate allowing the control of the odd resonant frequency with the application of an external bias. $BaTiO_3$ and $SrTiO_3$, well-known ferroelectric materials, have been selected as the dielectric material. The electrodes are made by creating an interdigitated gap on the $YBa_2Cu_3O_7$ ring over which the dielectric thin film has been deposited. Both the dielectric and the $YBa_2Cu_3O_7$ thin film were grown by the pulsed laser deposition technique.

1. INTRODUCTION

Fabrication of continuously electronically tunable microwave elements with low loss, which are compatible with high T_C superconductors is of widespread interest. Many investigators have or are currently working on this problem (Chrisey et al 1993, Galt et al 1993, Beall et al 1993, Chakalov et al 1998, Karmanenko et al 1998). All of these efforts to solve the problem involve ferroelectric materials used as dielectric elements. Only few researchers have actually studied and reported results from such devices in actual circuits (Chrisey et al 1993, Beall et al 1993, Lancaster 1998).

This paper reports on work done on a ring resonator design which incorporates $BaTiO_3$ and $SrTiO_3$ filled interdigitated coupling capacitors in the ring. $BaTiO_3$ was selected as our starting material because its compatibility issues on our superconductor of choice, $YBa_2Cu_3O_7$, had already been solved in prior work (Larkins 1995). We expanded our study to $SrTiO_3$ because this ferroelectric material does not easily react with $YBa_2Cu_3O_7$ and its dielectric constant has a greater increase at low temperature than $BaTiO_3$. The last factor is the most important to achieve better tunability.

In the paper we present the design, fabrication procedure, testing and results obtained from these structures.

2. DESIGN

The basic device is a microstrip ring resonator with $\lambda/8$ coupling lines. Two identical interdigitated capacitors were inserted at the 90° points of the ring (Fig. 1). All lines were nominally 50 Ω (0.169 mm width) and small triangular shaped contact pads were provided at the feedpoints of the resonator. Chokes were added to allow the introduction of dc bias to the interdigitated capacitors. Based on the ring diameter (14 mm), the literature value of 25 for

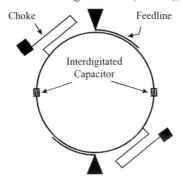

Fig. 1, ring resonator artwork (not to scale)

the dielectric constant of the substrate (LaAlO₃), and the simple effective dielectric constant equations given in texts for microstrip lines, the designed fundamental resonant frequency of this circuit was calculated as 1.7 GHz. The dimensions of the capacitors were a 0.06 mm gap with a finger length of 0.6 mm. The spacing between the feedlines and the ring is 0.1 mm.

3.   FABRICATION

The resonators described in this paper were fabricated on 20 mm × 20mm × 0.5 mm (100) oriented LaAlO₃ attached with colloidal silver paste to a heater capable of routinely attaining 850 °C. An initial YBa₂Cu₃O₇ thin film of approximately 700 nm thickness was deposited at 780-820 °C in 500 mTorr of O₂ using PLD-system equipped with a Q-switched frequency tripled Nd-YAG laser operating at 355 nm. Pulse energies were 150 mJ/pulse with an on-target beam size of 1 mm². Typically films grown in this fashion exhibited a $T_{C(R=0)}$ of 86-88 K with an onset of ~89 K (Vlasov et al, 1999).

After the initial YBa₂Cu₃O₇ deposition the substrate was mechanically removed from the heater, the YBa₂Cu₃O₇ side was covered with black wax and the remaining silver paste was removed in a 1:1 HNO₃:H₂O etch. The substrate was then rinsed in deionized water (DI water), blown dry using nitrogen and the black wax removed in a series of solvents (trichloroethylene, acetone, methanol) and once again blown dry using nitrogen.

Then the sample was reattached to the heater using colloidal silver paste (YBa₂Cu₃O₇ film covered face toward the heater) and a second film of YBa₂Cu₃O₇ was put on the other side of the substrate using the previously described deposition procedure.

The substrate was then removed from the heater and the side with silver paste was coated with black wax. Photoresist was then spun onto the YBa₂Cu₃O₇-film side, exposed and developed. The film was patterned using a 3% acetic acid in DI water solution, rinsed, blown dry, and the black wax removed as before.

At this stage the patterned resonator was installed in the microwave test system and characterized prior to the growth of the BaTiO₃ or SrTiO₃ thin film.

The ferroelectric thin film was grown at high temperature (700 °C for BaTiO₃, 750 °C for SrTiO₃) in 100 mTorr of O₂ using the same laser parameters as used in the YBa₂Cu₃O₇ growths. This was followed by a photolithographic step where the BaTiO₃ or SrTiO₃ was removed from the entire surface of the sample except for two small patches over the interdigitated capacitors in the ring resonator. The photoresist was then stripped in acetone and methanol and the entire circuit was blown dry in nitrogen.

At this stage the device was complete and the final testing of it initiated.

4.   TEST SETUP

The resonators were tested in a specially designed test fixture affixed to a closed cycle refrigerator of Janis Research Co. model CCS-1020. The refrigerator is capable of operation as low as 10 K. Temperature measurement was done using a Lakeshore model 330 temperature controller. Microwave measurements from 300 kHz to 6 GHz were performed using a Hewlett-Packard 8753C network analyzer and an 85047A s-parameter test set where s_{21}, the forward power transfer s-parameter, was measured. Data was collected with a computer through GPIB interface.

Cable and connector losses were compensated for at room temperature. No calibration at low temperature was done, since most of the losses were observed to come from the co-axial cables outside of the refrigerator.

All microwave measurements were performed at an incident power of +20 dBm (100 mW). Cable losses reduced this value to about 12 dBm at the sample at 2 GHz at room temperature.

4.   RESULTS

At room temperature (290 K) there were no prominent resonances observed. Fig. 2 shows the entire spectral region from 1 GHz through 6 GHz at a temperature of 10 K for the same resonator both with and without the $BaTiO_3$ material over the interdigitated capacitors. The fundamental resonances (peaks a and b) are at 1.65 GHz and 1.83 GHz. These are well

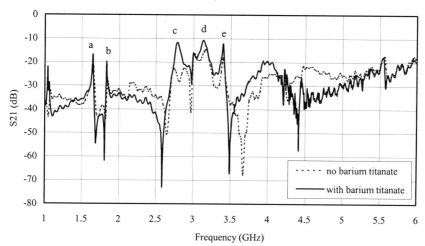

Fig. 2, T = 10 K. Note fundamental resonances at (a) 1.65 GHz and (b) 1.83 GHz. Solid line characteristic for the resonator with $BaTiO_3$ exhibits values of Q for indicated peaks: (a) – 180; (b) – 231; (c) – 48; (d) – 42; (e) – 177.

within process tolerance of the design frequency of 1.7 GHz. The higher frequency resonances in the spectra are harmonics of the fundamental resonances and resonances caused by the coupling of the feedlines to the resonator and reflections at the coupling capacitors.

398

Also note that the addition of the BaTiO$_3$ has significantly improved the coupling coefficient as can be seen in Fig. 3. This is indicated by both the higher Q and the lower loss of the peaks in the spectrum of the resonator after the addition of the BaTiO$_3$.

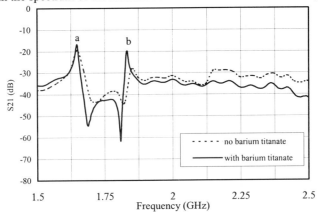

Fig. 3, T = 10 K. Note the enhanced Q and peak heights after the BaTiO$_3$ was added.

5.   CONCLUSIONS

High quality ring resonators with BaTiO$_3$ and SrTiO$_3$ filled coupling capacitors have been fabricated. The effects of the increased coupling due to the BaTiO$_3$ fill have been observed. The device is ready for testing for voltage tunability and quality factor modulation.

More work remains to be done to optimize the ferroelectric response at low temperatures. Mixture of SrTiO$_3$ and BaSrTiO$_3$ filling materials need to be tested and the design still requires optimization in terms of the reduction of the gap in the interdigitated capacitors.

6.   AKNOWLEDGMENTS

This work has been supported by the United States Air Force's Office of Scientific Research under grant number F49620-95-1-0519 and by the Office of Naval Research under grant number N00014-99-190315.

REFERENCES

Beall J A, Ono R H, Galt D, Price J C 1993 IEEE MTT-S Intl. Microwave Symp. Digest **3**, 1421
Chakalov R A, Ivanov Z G, Boikov Y A, Larsson P, Carlsson E, Gevorgian S, Claeson T 1998 Physica C: Superconductivity **308**, 279
Chrisey D B, Horwitzs J S, Pond J M, Carrol K R, Lubitz P, Grabowski K S, Leuchtner R E, Carosella C A 1993 IEEE Trans. Appl. Supercond. **3**, 1528
Galt D, Price J C, Beall J A, Ono R H 1993 Appl. Phys. Lett. **63**, 3078
Karmanenko S F, Dedyk A I, Barchenko V T, Chakalov R A, Lunev A V, Semenov A A, Ter-Martirosyan L T 1998 Supercond. Sci. Technol. **11**, 284
Lancaster M J, Powell J, Porch A 1998 Supercond. Sci. Technol. **11**, 1323
Larkins G L Jr, Avello M Y, Fork D K 1995 IEEE Trans. Appl. Supercond. **5**, 3049
Vlasov Yu, Lacambra A, Soto R, Larkins G L Jr, Stampe P and Kennedy R 1999 IEEE Trans. Appl. Supercond. **9**, 1642

Inst. Phys. Conf. Ser. No 167
Paper presented at Applied Superconductivity, Spain, 14-17 September 1999

399

Modeling and fabrication of a microelectromechanical switch for H-Tc superconducting applications

I O Hilerio[1], J R Reid[2], J S Derov[2], T M Babij[1], and G Larkins[1]

[1]Florida International University/ECE Department, Miami, FL, USA
[2]Air Force Research Laboratory/SNHA, Hanscom AFB, MA, USA

ABSTRACT: Microelectromechanical systems (MEMS) technology has demonstrated significant advantages when used as capacitive switches. Our experience in MEMS technology, coupled with that in superconducting thin films, will allow us to obtain better performance parameters. This paper describes the modeling and fabrication of microelectromechanical (MEM) membrane capacitive switches suitable for RF high temperature superconducting applications. The fabrication and theory of operation of these switches is discussed. Models of the electromechanical and microwave properties are presented. Simulation analysis show the switches are capable of attaining low insertion loss (less than 0.2 dB at 20 GHz), and high isolation (greater than 20 dB at 20 GHz).

1. INTRODUCTION

Superconducting applications are in need of a small, cheap, and reliable switch. MEM switches offer all three characteristics. Our previous work, as reported by Hilerio (1999), focused on MEM switches fabricated on gallium arsenide substrates. This work facilitates the application of these devices to high temperature superconducting applications. Regardless of the variety of topologies and materials available, all MEM switches share key characteristics which make this technology so captivating. Miniaturization is one of the most appealing characteristics of the switches. Dimensions are typically in several hundreds of microns. In addition, standard photolithography based processing is used in fabricating the switches. This allows batch fabrication, resulting in a per unit price reduction. Further, as with integrated circuits, the switches can be monolithically fabricated eliminating the cost of hybrid assembly. In addition, the devices rely on electrostatic actuation providing the switches with high efficiency and high operating frequency. These characteristics ease the integration of these devices into areas such as superconducting applications that can greatly benefit from such devices.

2. MEM CAPACITIVE SWITCHES

Having successfully modeled and developed MEM switches on semi-insulating substrates, similar results are expected for the switches when constructed with superconducting thin films. The devices described herein are capacitively coupled MEM switches where the active element is a thin metallic membrane. As illustrated in Fig. 1, the switch is fabricated on top of a gallium arsenide substrate. A gold coplanar waveguide (CPW) serves as the transmission path for the RF signal to be switched. The contact area between the top membrane and the CPW center conductor is covered with a thin dielectric film. The dielectric layer helps avoid any direct mechanical contact between the metal plates. This results in only electrical switching of the RF signal, eliminating any DC current flow and reducing possible stiction problems between the plates.

The switch is triggered by applying a DC potential difference between the membrane and the waveguide. When the potential is high enough the air bridge collapses and comes into mechanical contact

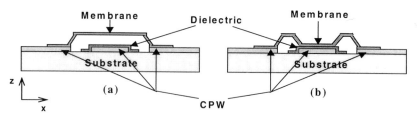

Fig. 1. Basic layout of a shunt-type capacitive membrane switch in (a) unactuated and (b) actuated positions.

with the dielectric above the center conductor. This creates a high capacitance between the plates and shunts the RF signal to ground. When the applied potential is dropped below the minimum pull-down voltage required, the membrane returns to its natural state. In the up state there is only a small parasitic capacitance and the signal is thus transmitted without any significant loss. Pull-down voltages between 20 and 40 volts were recorded. These voltages are comparable by those reported by Goldsmith *et al* (1996) and Yao and Chang (1995).

The RF switches were fabricated using a surface micromachining process with a total of four masking steps. The starting substrate is a semi-insulating gallium arsenide wafer. The process flow is implemented as follows: (1) a gallium arsenide (GaAs) substrate is chemically cleaned; (2) a thin layer of gold is deposited by thermal evaporation and patterned by a lift off process, resulting in the CPW structure; (3) a passivation layer of silicon nitride (Si_3N_4) is deposited and patterned using a lift off step; (4) a sacrificial layer of a 3 μm thick polymer (PMGI SF 15) is then spin-deposited and patterned, serving as the support structure for the membrane; (5) a second layer of gold is deposited and patterned on top of the structural layer; (6) finally, the polymer below the membrane is removed by wet etching, releasing the membrane. Fig. 2 shows an example of a typical membrane capacitive switch.

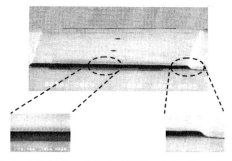

Fig 2. SEM micrograph of a typical fabricated GaAs MEM switch. Note the air bridge formed below the membrane.

A critical parameter for these switches is the ratio of the capacitance in the actuated and unactuated states, defined respectively as C_{ON} and C_{OFF}. These values define the microwave performance, with C_{ON} determining the electrical isolation and C_{OFF} driving the insertion loss. Approximations of these values can be made by assuming the switch is a parallel plate capacitor. When the switch is unactuated the OFF capacitance formed is determined by the area of the membrane and the separation between the plates (h_g). When the membrane snaps down the ON capacitance is given by the membrane area, the dielectric thickness (h_d), and the dielectric constant (ε_{die}). The ON/OFF capacitance ratio, which also represents the ON/OFF impedance ratio, is then given by,

$$\frac{C_{ON}}{C_{OFF}} \approx \frac{Z_{ON}}{Z_{OFF}} \approx \frac{\varepsilon_{die} h_g}{\varepsilon_0 h_d}$$

where ε_0 is the permittivity of air.

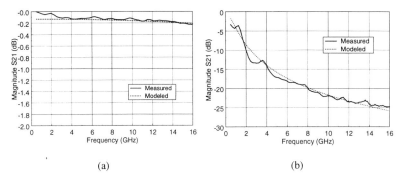

(a)   (b)

Fig 3. (a) Measured shunt switch insertion loss (up position). The dashed line corresponds to a modeled C_{OFF} of 0.06 pF. (b) Measured switch isolation (down position). The dashed line corresponds to a modeled C_{ON} of 7.6 pF.

Since h_g can be a few microns, h_d can be as low as 0.1 μm, and ε_{die} is about 7.5 times larger than ε_0, the ON/OFF ratio is roughly 50-150. This value is high enough to allow good switching isolation. The simulations in Fig. 3 demonstrate the switch performance, corroborating that the capacitance ratio is sufficient. For this particular device, the ON capacitance is estimated at 7.6 pF, while the OFF capacitance was calculated at 0.06 pF, resulting in a ON/OFF ratio of 126. In the transmission state, the switch possesses an insertion loss measured to be less than 0.25 dB up to 16 GHz. Meanwhile, it shows that the measured isolation increases from a low frequency measurement-limited minimum of 4dB at 500 MHz to 25 dB at 16 GHz. Mechanical characterization demonstrated switching times of 10-12 μs.

3.   MEM SWITCHES FOR HT-C SUPERCONDUCTIVE APPLICATIONS

Having demonstrated the advantages and performance of conventional MEM switches, we are now ready to proceed with the fabrication of the switch for superconducting applications. The devices will be similarly modeled and fabricated. We are currently developing shunt and series type switches. Insertion loss and isolation parameters have been simulated as seen in Fig. 4. Membrane widths of 50, 100, and 150 μm are shown. As expected, wider membranes result in better isolation and increased parasitic capacitance. In general, any of these dimensions will result in acceptable switching performance.

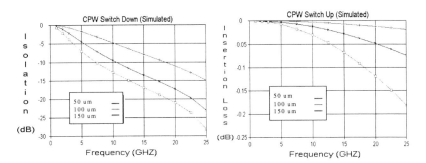

Fig 4. Simulated measurements of MEM switch for HT-c superconductive applications.

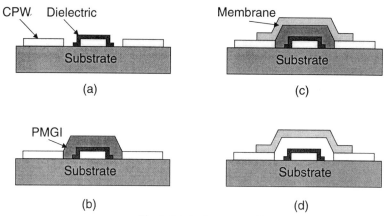

Fig. 5. Fabrication process flow.

The fabrication process for these devices is similar to that of the conventional switch. Again, a four-mask process is used. The fabrication sequence, shown in Fig. 5, starts with a lanthanum aluminate substrate covered with a 0.35 μm film of YBCO on one side. The CPW structure is formed by coating the YBCO layer with a thin photo resist and exposing it with the CPW mask using a ultra violet light mask aligner. The resulting structure is used to etch the YBCO utilizing a 35:1 de-ionized water:phosphoric acid solution. Next, a thin dielectric layer of silicon nitride, approximately 1500 Å, is reactively sputtered and patterned (using lift off) over the contact area below the membrane. A sacrificial layer of PMGI SF 15 polymer is then spin-deposited, exposed, and developed to serve as the support structure for the membrane. Subsequently, a 5 μm thick coat of photo resist is spun on top of the sacrificial layer and patterned to define the mask for the membrane structure. The membrane is then formed by the thermal evaporation of a 0.05 μm thick layer of chromium followed with a 0.6 μm thick layer of gold. Removing the photo resist leaves behind the patterned membrane. Finally, the sacrificial layer is removed by immersing the device in a PMGI stripper solution for about 1 hour, followed by individual rinses in de-ionized water and methanol. The methanol is then evaporated by placing the device on a hot plate set at 100°C. At this point we are currently iterating the membrane deposition step.

4. CONCLUSION

Microelectromechanical capacitive membrane switches suitable for RF superconducting applications have been described. Simulation analysis shows insertion loss less than 0.2 dB and high isolation of 25 dB at 25 GHz. These switches are fabricated using well-known fabrication techniques borrowed from the integrated circuit industry. Of the many potential applications, the switches can be used in superconductive applications such as antennas, phase shifters, and phased arrays. Fabrication of the devices is currently underway and expected performance should be attained. This work was funded by the Air Force Office of Scientific Research grant number F49620-95-1-0519.

REFERENCES

Goldsmith C, Randall J, Eshelman S, Lin T H, Denniston S, Chen S, and Norvell B, 1996 IEEE Microwave Theory Tech. Symp. Dig. **6**, 1141-4
Hilerio I 1999 Master's Thesis, Florida International University
Yao J J and Chang M F 1995 Proc. 8th Int. Conf. On Solid States Sensors and Actuators pp 384-7

Inst. Phys. Conf. Ser. No 167
Paper presented at Applied Superconductivity, Spain, 14-17 September 1999
© 2000 IOP Publishing Ltd

Planar electrooptical phase Modulator: An intermediate step towards superconductor Mach-Zehnder Integrated Modulators

E Rozan, C Collado, J Prat and J M O'Callaghan

Universitat Politècnica de Catalunya - Dept of Signal Theory and Communications - Campus Nord D3 - Jordi Girona, 1-3 08034 Barcelona - Spain

ABSTRACT: The fabrication of electro-optical modulators with superconducting electrodes requires the introduction of a specific buffer layer, like Yttria-Stabilized-Zirconia (YSZ), to obtain good YBCO quality. Conventional modulators use SiO_2 as buffer layer to optimize the electro-optical performance. Thus, the characterization of a gold electro-optical resonant phase modulator including YSZ can be used to demonstrate that the presence of this material is not disturbing the electro-optical modulation properties. A resonant slot-line planar electrode layout has been used to experimentally assess these properties. The measurement protocol proposed is operational from room to cryogenic temperatures. The improvement of the modulated phase measurement of a gold slot-line resonant modulator, observed when temperature decreased, is presented.

1.   INTRODUCTION

Even though the use of high temperature superconductors (HTS) deposited onto Lithium Niobate (LNO) substrates might lead to a substantial improvement in the performance of integrated, traveling-wave Mach-Zehnder (MZ) HTS modulators, no published reports are easily found on successful attempts to fully build and test such devices. This is partly due to the many requirements that have to be simultaneously met while the HTS deposition processes are still under development, such as the need for good quality HTS material over large surfaces, a custom-made optical-waveguide pattern in the LNO, etc. Therefore, test structures and procedures have to be devised to assess the performance of the various components of the M-Z modulator without having to assemble a whole device.

A previous work (Rozan et al, 1998) indicates that coplanar YBCO structures can be grown onto LNO substrates showing RF losses significantly smaller than identical structures made of gold. However, to achieve good YBCO quality for these structures, the deposition of a Yttria Stabilized Zirconia (YSZ) buffer layer was required due to the poor structural and chemical compatibility of LNO and YBCO. While the work by Rozan et al (1998) showed that the YSZ buffer had little effect on the RF performance of the electrodes, an experimental assessment of the effect of this buffer on the electrooptical performance of the modulator was still missing. This assessment is presented here through the measurement of the optical phase modulation produced with a resonant planar structure on YSZ/LNO.

The use of a resonant phase modulator instead of a MZ one has many of the simplifying advantages mentioned above: requires much less substrate area and is much less sensitive to the optical insertion losses that might appear at the connection between the optical fibers and the LNO substrate when the device is cooled.

2.   PLANAR ELECTRODE DESIGN CONSIDERATIONS

Typical electrooptical devices are made on a LNO wafer where optical waveguides have been fabricated by titanium deposition and a set of planar electrodes placed above the substrate. Their role is to apply the electrical field that will control the optical signal phase. Different electrode configurations may be employed, but popular designs are based on CPW (Fig. 1a) or slot-line (Fig. 1b) electrodes. CPW configuration is ideal for amplitude modulation because the optical phase is changed simultaneously in both optical waveguides, which reduces the voltage required for complete optical cancellation by a factor of two. However, CPW electrodes require a more complex optical waveguide pattern, including two Y-branches. Also, precise alignment technique is necessary to match the electrode gaps with the optical waveguides. For a direct phase modulation, only one optical waveguide is involved, which does not require the fabrication of an optical waveguide pattern adapted to match the size of the structure to be tested.

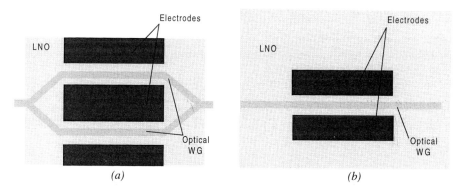

Fig. 1: Coplanar (a) and slot-line (b) electrode configuration

To keep the resonator within a 2.5 cm YSZ/LNO substrate, the resonant frequency has been set to 2 GHz to obtain a half-wave transmission line of 2 cm long. The effective resonator length has been evaluated taking into account the effective propagation index of the line. Electrode separation has been set to 25 μm. Larger electrode separation results in lower RF losses and a higher resonator quality factor, whereas a small separation facilitates the penetration of the electrical field into the optical waveguide, achieving a high electrooptical interaction. Not shown in Fig.   1b is a CPW access line and a taper that couples the center of the resonator and makes the transition to a CPW line compatible with a 3.5mm coaxial connector. To improve electrical coupling, a SMD capacitor has been placed across the coupling gap.

Also, overlap parameter should be improved with respect to the conventional design, where buffer layer material is SiO_2, as YSZ is characterized by a high dielectric constant (similar to the LNO) which allows a better electrical field penetration.

3.   RESONANT PHASE MODULATOR MEASUREMENT

The slot-line resonator described in the previous section has been fabricated with gold electrodes, and its modulation performance has been tested from room temperature down to almost 100 K. Optical fiber pigtail connections have been realized using a precise fiber manipulation and alignment system. To assure a good light transmission, careful LNO edge polishment was done prior to connecting the fibers to the optical waveguide in the LNO. The full measurement set-up is presented in Fig. 2. From the various possible methods available to measure phase modulation, the protocol used for this experiment is based on heterodyne demodulation technique, being the least sensitive to optical insertion losses. This makes the measurement robust to possible degradations in the performance of the optical LNO-fiber connections, which may occur when the device is cooled.

The multielectrode DFB laser, emitting at 1550 nm, is connected to the electro-optical modulator to take advantage of its good frequency stability, coherence, and its appropriate optical power. The other light source is tuned at 1550.03 nm to achieve an intermediate heterodyne frequency of 3.75 GHz, measurable on the spectrum analyzer.

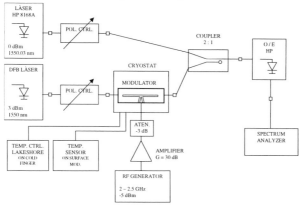

Fig. 2 Full measurement set-up

The phase modulation β is extracted from the ratio between the power of the modulating signal and the carrier signal (P_m/P_c). Using narrowband PM approximation (Carlson, 1986), the modulated signal amplitude is directly βA_c where A_c is the carrier signal amplitude. Thus, β can be related to P_m/P_c through

$$\left(\frac{\beta}{2}\right)^2 = \frac{P_m}{P_c} \tag{1}$$

It can observed in (1) that modulation index does not depend on absolute signal levels, which makes this technique very appropriate for cryogenic measurement.

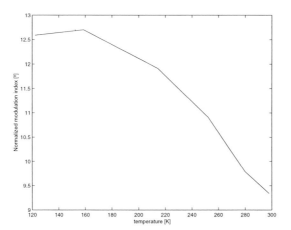

Fig. 3: Normalized modulation index $\beta\left/\sqrt{1-\rho^2}\right.$

Measurements of carrier and modulating signal amplitude have been done as a function of the temperature. An optimized value of the modulation index is extracted by tuning the RF frequency of the modulating signal to its optimum value within the resonance band. Electrical reflection coefficient temperature dependence is also measured. The quality factor achieved by this modulator ranges from 40 at room temperature to 80 at 100 K. From these data, an estimation of the modulation index, assuming perfect impedance matching can be done. This normalized modulation index is shown in Fig. 3 and reflects the combined temperature dependence of the Pockels coefficient and the losses in the electrode. The measurement has been done down to 110 K. a 30% improvement is achieved between room temperature and the lowest temperature.

4. RESONANT PHASE MODULATOR PERFORMANCE

The phase change per unit of length depends on the electrode separation G (Alferness, 1982)

$$\overline{\Delta\Phi} = \frac{\pi r n_o^3 \Gamma}{\lambda G} \tag{2}$$

where

$$\Gamma = G \iint E \left| O' \right|^2 dxdy_x \tag{3}$$

is the overlap parameter. In (3), O' represents the normalized optical field distribution and E the applied electrical field. Other parameters are the optical wavelength λ, the Pockels coefficient, r and the optical waveguide refractive index n_o. The total phase will be obtained by summing (2) along the interaction zone l.

In a resonator, the electrical energy stored is defined by

$$U = \frac{1}{2} \iiint \varepsilon \left| E \right|^2 dV \tag{4}$$

In a slot-line half-wave planar resonator, field distribution can be evaluated by analytical and simulation methods which permits the calculation of the overlap parameter and the electric energy and to obtain a relation between both parameters. Knowing the resonant quality factor can be expressed in function of the energy stored through

$$Q = \frac{\omega_o U}{(1-\rho^2)W_{av}} \tag{5}$$

where $\omega_o = 2\pi f_o$ (being f_o the resonating frequency), ρ the electrical reflection coefficient and W_{av}, the available power, a relation of proportionality will exist between the resonant quality factor and the phase shift

$$\Delta\Phi \propto \frac{\pi r n_o^3}{\lambda G} \cdot \sqrt{(1-\rho^2)Q} \cdot \sqrt{W_{av}} \tag{6}$$

In YBCO electrode resonator quality factor may be expected to reach value close to 400, which supposes improvement of phase by a factor of two.

4. CONCLUSION

These measurements have put in evidence a correct operation of the optical components (fibers, connections, LNO waveguides) in low temperature conditions, showing the viability of cryogenic photonic components. The same protocol measurement would be employed for a superconducting modulator. Measurable phase shift obtained proves buffer layer YSZ is compatible with electro-optical modulation properties. YBCO resonant modulators become then operational device, first step toward superconductor MZ modulators.

REFERENCES

Alferness R 1982 IEEE Trans. MTT, 30 pp 1121-1136
Carlson B 1986 Communications Systems (McGraw-Hill) pp 236
Rozan E, Collado C, Garcia A, O'Callaghan J M, Pous R, Fabrega L, Rius J, Rubí J, Fontcuberta J, Harackiewicz F, 1999 IEEE Trans. App. Superconductivity

Inst. Phys. Conf. Ser. No 167
Paper presented at Applied Superconductivity, Spain, 14-17 September 1999
© 2000 IOP Publishing Ltd

Determination of non-linear parameters in HTSC transmission line using a multiport harmonic balance algorithm.

J. Mateu[1], C. Collado[1], L.M. Rodriguez[2], J.M. O'Callaghan[1]

[1] Universitat Politècnica de Catalunya - Dept of Signal Theory and Communications - Campus Nord UPC, D3 - Jordi Girona, 1-3 08034 Barcelona – Spain.
[2] Centro de Investigación Científica y de Educación Superior de Ensenada (CICESE) Km 107 Carretera Tijuana-Ensenada Ensenada, B.C. Mexico 22860

ABSTRACT: The use of the Multiport Harmonic Balance (MHB) algorithm in the determination of the non-linear parameters of superconducting transmission lines is investigated. MHB is capable of calculating current distributions of fundamental, harmonic and intermodulation spurious in transmission lines with current-dependent inductance and resistance per unit length $(L(I),R(I))$. Since MHB accepts any closed-form expression for $L(I),R(I)$, non-linear parameter extraction can be done by iteratively adjusting $L(I),R(I)$ and comparing the MHB results with measured data. Numerical tests using expressions of $L(I),R(I)$ adequate for HTS lines are performed and indicate that $L(I),R(I)$ can be unambiguously determined.

1.   INTRODUCTION

Passive microwave filters for cellular and space communication systems are one of the most promising applications of High Temperature Superconducting (HTS) materials. Their low RF losses allow for planar HTS filters to be built with selectivities and insertion losses comparable to those obtained with normal metal filters based on cavity resonators. However, the non-linear properties of HTS materials give rise to spurious harmonics and intermodulation products that may limit the dynamic range of systems in which these filters are used. These spurious are generated by local non-linear effects (non-linearities in conductance or surface impedance of the HTS), which depend strongly on the microstructure of the HTS film and hence, on its fabrication process. Thus, a high variability exists on the non-linear performance of HTS devices given by the various existing HTS fabrication techniques, and by the many parameters that control these techniques. Due to this high variability, acquiring an accurate knowledge of the local non-linear effects is difficult. This requires both experimental measurements, and non-ambiguous procedures to relate the experimental observables with the parameters of the local non-linear model.

Several authors have used similar microstrip or stripline resonators to experimentally characterise HTS materials (among others: Oates et al, 1993; Vendik et al, 1997; Willemsen et al, 1997). Despite the similarity among the resonators being tested, the models proposed for the local non-linearities change from an author to another, but they are all adjustable and fit the measured data. While most of the models proposed may be flexible enough to be able to unambiguously describe any HTS sample (that is, keep a one-to-one relation between the parameters in the model and all the experimental observables) by just changing the numerical values of the parameters in the model, the number and variability of the models is so large that is not clear that all of them are applicable to any HTS sample; it is also not clear that all the models proposed keep a one-to-one relationship between the model and the observables, i.e. are unambiguous (this does not necessary apply to the works cited in this paper).

This paper presents a tool that might clarify this situation by contributing to develop (or sort out) models that are both unambiguous and applicable to a wide-range of HTS samples. It is based on the Multiport Harmonic Balance (MHB) algorithm (Collado et al, 1999). This numerical algorithm calculates the voltage and current distribution of the fundamental and spurious tones along non-linear transmission lines. The local non-linearities in the transmission line are set by defining a dependence of the inductance and resistance per unit length on the current flowing through them ($L(I),R(I)$). MHB poses no restrictions on the functions defining $L(I)$ and $R(I)$, and is very fast regardless of the definitions of $L(I),R(I)$. This speed and versatility make MHB ideal for parameter adjustment, that is, $L(I)$ and $R(I)$ can be defined as a function of several parameters and these parameters can be found by matching the calculations made by MHB with the results of experimental measurements.

2. MHB NON-LINEAR PARAMETER FITTING

Figure 1 shows the iteration scheme followed. Starting from a tentative choice of $L(I),R(I)$, the MHB code is executed and its results are compared with the experimental measurements. Usually these are spectrum analyser measurements of the power delivered at the output port of a two-port resonator by the fundamental signal(s), their harmonics, and their intermodulation products. The comparison shown in Fig. 1 is based on an error function that takes into account the relative amplitude errors in several of the tones. As shown below, the selection of the tones that are measured and included in this error function has a significant influence on the fitting process.

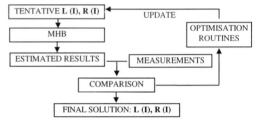

Fig. 1: Iteration scheme for the experimental determination of non-linear parameters ($L(I)$, $R(I)$) in superconductive lines.

3. FITTING WITH SQUARE-LAW NON-LINEARITIES

Many even functions can be postulated for $L(I),R(I)$. In our case, to illustrate the use of MHB in non-linear parameter determination, a square-law dependence has been assumed (i.e., $L(I)= L_0 + \Delta L \cdot I^2$ and $R(I)= R_0 + \Delta R \cdot I^2$). This is applicable whenever non-linearities are weak and $L(I),R(I)$ can be expanded in a Taylor's series. Furthermore, several authors have postulated a square-law dependence (Vendik et al 1997, Hammond et al 1998), and among them, Dahm and Scalapino (1997) show that this dependence occurs if the non-linearities are only due to the intrinsic behaviour of the YBCO.

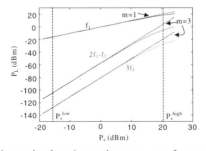

Fig. 2. Power of the fundamental and spurious at the output port of a two-port line resonator as a function of available power from de source, as calculated by MHB with ΔL=8.2 pH A^{-2} m^{-1} and ΔR=0.097 Ω A^{-2} m^{-1}

Figure 2 shows the power of the fundamental, third harmonic and third-order intermodulation product of a line resonator with quadratic $L(I)$ and $R(I)$ as a function of the power available from the source, calculated with MHB.   At low powers, the slope of the curve corresponding to the fundamental is one, and that of the spurious is three.   Also, there is no measurable dependence of the resonant frequency and quality factor on the source power.   At sufficiently high powers, the slope of all curves decreases and both the resonance frequency and quality factor vary when the power of the source is changed.

Our experience indicates that measurements as a function of source power only give information of the joint dependence of $L(I)$ and $R(I)$ with I. The difficulty resides in separating $R(I)$ and $L(I)$ at a given value of I (i.e., at a given source power), since their effect on some of the spurious is very similar.

As discussed below, the election of the spurious to measure and fit is critical for a correct determination of the values of $R(I)$ and $L(I)$. This election of observables will be done at two different power levels, as marked on Fig. 2.

Figure 3 shows a defective fitting process obtained by fitting to the data in Fig. 2. The contours on that figure are loci of (ΔL, ΔR) pairs for which the fitting error is constant. The shapes of the contours show that there is not a single minimum, and thus, no unambiguous determination of ΔL, ΔR is possible. This figure also shows a couple of optimisation trajectories that, starting from different initial values of ΔL, ΔR, end up at a different solution, both of them having equal fitting error. The error function used in this fitting process is the normalised difference between the simulated and measured power of the third order intermodulation product ($2f_2$-f_1) (this is done at low source power, P_S^{low}, but a similar situation occurs at P_S^{high}). This ambiguity is to be expected, since we are trying to adjust two parameters with a single data point.

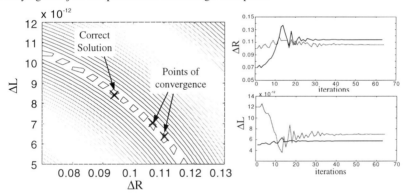

Fig. 3. *(Right)*: Contours of constant fitting error when only the third order intermodulation product is considered in the error function. The point of convergence depends on the initial estimate of ΔL and ΔR. *(Left)*: Evolution of ΔL and ΔR.

To solve this, the contribution of the relative fitting errors of the fundamental (f_1) was added to the error function (by taking the square root of the sum of the relative errors squared).   Similar results were obtained at P_S^{low}, but at P_S^{high} the determination was unambiguous as shown by the fitting error contour plots and the optimisation trajectory in Fig. 4. Thus, the determination of ΔL, ΔR using the fundamental and a third-order intermodulation product is unambiguous only for certain power levels.

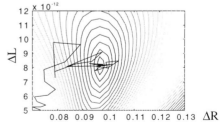

Fig. 4: Fitting error contour plots when both the fundamental and third order intermodulation spurious are considered.   Determination is unambiguous as shown by the single, well-defined minimum in the plots. This plots correspond to P_s^{high}; the equivalent plots for P_s^{low} are similar to those in Fig. 3 and thus, give rise to ambiguity in ΔL, ΔR.

410

Unambiguous determination of $\Delta L, \Delta R$ at both $P_s^{\ low}$ and $P_s^{\ high}$ was achieved when the relative error in the third harmonic $(3f_1)$ was added to the error function of the previous case (as before, by taking the square root of the sum of all relative errors squared). In this case the contour plots of the fitting error look similar to those in Fig. 4, with a single, well-defined minimum. Comparison of all three cases is done in Fig. 5, showing the evolution of $\Delta L, \Delta R$ as they are adjusted.

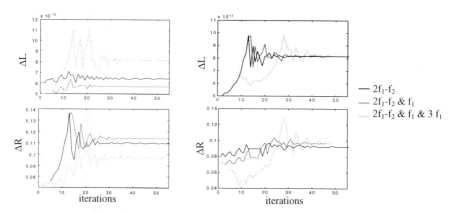

Fig. 5: Evolution of the values of ΔL and ΔR being fitted with the three different error functions being considered. Graphs on the left are for P_s^{high} and those on the right for P_s^{low}.

4.   CONCLUSION

The Multiport Harmonic Balance algorithm has proved to be useful for non-linear parameter extraction. Successful numerical extraction tests have been performed on non-linear lines whose parameters are close to what can be expected in a HTS microstrip resonator. This is expected to contribute towards the advancement of non-linear HTS models that can be applied to many types of HTS materials without ambiguity.

ACKNOWLEDGMENTS

This work has been funded by CIRIT through grant SGR00043 and by CONACYT through scholarship number 118305 for one of the authors (LMR). The authors would also like to thank Dr. Salvador Talisa for his helpful comments and suggestions.

REFERENCES

Collado C, Mateu J and O'Callaghan J M 1999 European Conference on Applied Superconductivity.
Dahm T and Scalapino D J 1997 J. Appl. Phis. **81**, 4, pp 2002-9
Hammond et al 1998 J. Appl. Phis. **84**, 10, pp 5662-7
Oates J H et al 1993 IEEE Trans. Applied Superconductivity **3**, 1, pp 17-22
Vendik O G and Vendij I B 1997 IEEE Trans. Microwave Theory Tech. **45**, 2, pp 173-8
Willemsen B A, Dahm T and Scalapino D J 1997 Appl. Phys. Lett. **71**, 26, pp 3898-900

Inst. Phys. Conf. Ser. No 167
Paper presented at Applied Superconductivity, Spain, 14-17 September 1999
© 2000 IOP Publishing Ltd

Computer simulation of the non-linear response of superconducting devices using the Multiport Harmonic Balance algorithm.

C. Collado, J. Mateu, J.M. O'Callaghan

Universitat Politècnica de Catalunya - Dept of Signal Theory and Communications - Campus Nord D3 - Jordi Girona, 1-3 08034 Barcelona – Spain.

ABSTRACT: The Harmonic Balance algorithm has been adapted to be used in non-linear transmission lines. This is used to calculate the generation of harmonic and intermodulation spurious in circuits based on superconducting transmission lines, which present distributed non-linear effects. The results of the algorithm developed -Multiport Harmonic Balance (MHB)- are in close agreement with theoretical predictions made for a microstrip YBCO half-wave resonator. The intermodulation performance of a coupled-line filter is also calculated as an example and the results are in good qualitative agreement with published measurements.

1. INTRODUCTION

Passive microwave filters for cellular and space communication systems are one of the most promising applications of High Temperature Superconducting (HTS) materials. Their low RF losses allow for planar HTS filters to be built with selectivities and insertion losses comparable to those obtained with normal metal filters based on cavity resonators. However, the intrinsic non-linear properties of HTS materials give rise to spurious harmonics and intermodulation products. The level of these spurious depends on the power of the input signal and this may limit the dynamic range of systems in which these filters are used.

There is no widely available CAD software able to efficiently treat these non-linear effects, and filter developers have to rely on empirical trial and error. To improve this, this work describes a computation approach that efficiently predicts the harmonic and intermodulation spurious of structures made of single and multiple transmission lines, and could be the base for non-linear computer-aided design of HTS filters. No restricting assumptions are made either on the order or the non-linearities, or on the dominance of reactive non-linearities over resistive ones (or vice-versa). Our approach is based on the application of the Harmonic Balance (HB) algorithm (Maas, 1988) to distributed structures and. –when applied to simple, specific cases– it is found to agree closely with the theoretical predictions of non-linearities in HTS lines made by Dahm and Scalapino (1997).

2. METHOD OF ANALYSIS

The analysis is based on the assumption that the non-linear effects in a HTS transmission line can be modelled as non-linear perturbation of the inductance and resistance per unit length:

$$L(I) = L_0 + \Delta L(I)$$
$$R(I) = R_0 + \Delta R\ (I) \tag{1}$$

where I is the current through the line, R_0, L_0 are the linear components of L, R and $\Delta L(I)$, $\Delta R(I)$ the non-linear ones. Note that, unlike other works in HTS non-linearities, our algorithm does not assume that $L(I)$, $R(I)$ can be approximated by a Taylor's series expansion, thus allowing for strong non-linear effects in the transmission line.

In our analysis, a transmission line is modelled as a cascade of identical unit cells (Fig. 1). Thus, any circuit containing a HTS line can be modelled as a linear network with N+1 ports loaded with N non-linear one-port elements and an excitation port (Fig. 1).

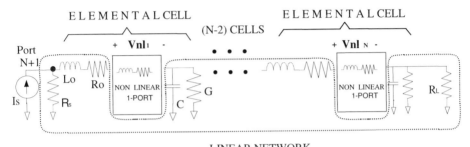

LINEAR NETWORK
(N+1 PORTS)

Fig 1: Discretized transmission line.

As shown in Maas (1988), the whole circuit can be analysed using MHB, in which the response of the linear network is calculated in frequency domain using $v_L = Z\, i_L$, where v_L, i_L are vectors containing all voltages and currents at the N+1 ports of the linear network. This system of (N+1)x(N+1) equations can be reduced to an NxN one involving only the ports where non-linear elements are connected (that is, excluding the port of the current source). To do this, the original system is modified to $v_L = Z\, i_L + v_S$, where v_S accounts for the voltage contribution of the current source. This matrix equation can be expanded to:

$$
\begin{bmatrix} v_{L1} \\ v_{L2} \\ \vdots \\ v_{LN-1} \\ v_{LN} \end{bmatrix} =
\begin{bmatrix} Z_{11} & Z_{12} & \cdots & Z_{1N-1} & Z_{1N} \\ Z_{21} & Z_{22} & \cdots & Z_{2N-1} & Z_{2N} \\ \vdots & \vdots & \ddots & \vdots & \vdots \\ Z_{N-11} & Z_{N-12} & \vdots & Z_{N-1N-1} & Z_{N-1N} \\ Z_{N1} & Z_{N2} & \cdots & Z_{NN-1} & Z_{NN} \end{bmatrix} \cdot
\begin{bmatrix} i_{L1} \\ i_{L2} \\ \vdots \\ i_{LN-1} \\ i_{LN} \end{bmatrix} +
\begin{bmatrix} v_{S1} \\ v_{S2} \\ \vdots \\ v_{SN-1} \\ v_{SN} \end{bmatrix}
\tag{2}
$$

where the subvectors v_{Li}, i_{Li}, v_{Si}, and the submatrices Z_{ij} are evaluated for the K+1 frequencies appearing in the spurious signals:

$$v_{Li} = \begin{bmatrix} v_{Li}^0 & v_{Li}^1 & \cdots & v_{Li}^k \end{bmatrix}^T;\quad i_{Li} = \begin{bmatrix} i_{Li}^0 & i_{Li}^1 & \cdots & i_{Li}^k \end{bmatrix}^T;\quad v_{Si} = \begin{bmatrix} v_{Si}^0 & v_{Si}^1 & \cdots & v_{Si}^k \end{bmatrix}^T;$$

$$Z_{ij} = diag\left(Z_{ij}^l\right) \quad l = 0,1,2,...k \quad i,j = 1,2,...N$$

and the elements of the source vector v_S are calculated using $v_{Si}^l = Z_{i\,N+1}^l \cdot I_{L\,N+1}^l$,where $I_{L N+1}^l$ is the lth frequency component of the current source I_s (Fig. 1). Z results in a sparse N(K+1)x N(K+1) matrix which is inverted to obtain the vector of currents flowing into the lineal network

$$i_L = Y \cdot \begin{bmatrix} v_L - v_S \end{bmatrix} \quad ; \quad Y = Z^{-1}
\tag{3}$$

The MHB algorithm begins assuming zero initial conditions for v_L. Eq. (3) is then used to obtain the vector i_L. These currents cause voltages across the N non-linear one-ports (v_{NL}), which are computed in time domain using $L(I)$ and $R(I)$ given in Eq. (1). By comparing the resulting voltages at the interface between the linear (N+1)-port and the N non-linear one ports, the state variables (voltages) of iteration $m+1$ (v_L^{m+1}) are updated using those of the previous iteration (v_L^m , v_{NL}^m). This is done following the procedure proposed by Hicks et al. (1980): $v_L^{m+1} = v_L^m \cdot p + v_{NL}^m \cdot (1-p)$, where the parameter p $(0 < p < 1)$ fixes the trade-off between speed and robustness of the convergence towards the solution. Convergence is achieved when the components of v_{NL} match those of v_L. If this is not achieved at a given iteration, the currents are re-calculated using Eq. (3) with the updated version of v_L and a new iteration is started.

The software developed with the proposed method has been tested against a commercial time-domain circuit analysis software (PSPICE). To limit the computing time of the time-domain analysis, the number of cells had to be restricted to two. For a two-tone excitation (1.00 and 1.05GHz) the agreement was better than 0.5% for the fundamental signals and 0.1% for the third-order intermodulation products. Computation times on a Pentium PC at 120 MHz were such that, if extrapolated to a calculation of a half-wave resonant line using 400 cells, the time analysis would spend up to 5000 hours, in front of only 120 s using our Multiport Harmonic Balance method.

3.   EXAMPLE 1: YBCO MICROSTRIP RESONATOR

To check for the accuracy of its results and its applicability real situations, our software has been used to analyse a half-wave YBCO microstrip resonator capacitively coupled to a matched generator and a matched load (Fig. 2). The non-linear properties of this structure have been derived by Dahm and Scalapino (1997) from basic theoretical considerations, providing closed-form equations to calculate the currents of harmonic and intermodulation spurious under a series of simplifying assumptions (a square-law dependence of $L(I)$, $R(I)$ in Eq. (1), and negligible losses due to the dielectric substrate). When the resonator is excited with two tones ω_1 and ω_2, both within the resonance band, a close agreement (within 0.2%) between theory and simulation is found for both the peak current of the third order intermodulation $I_{2\omega1-\omega2}$ product and for that of the fundamental mode (I_0)

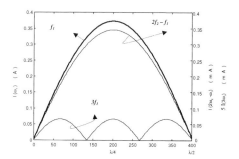

Fig.2. Current distribution of the fundamental (ω_1), third-order spurious ($2\omega_1$-ω_2) and third harmonic, as calculated by our Multiport Harmonic Balance algorithm with $L=L_0+\Delta L_2\ I^2$; $R=R_0+\Delta R_2\ I^2$; $\Delta L_2=8.18e^{-12}$ H/A^2m and $\Delta R_2=2.42e^{-3}$ Ω/A^2m. Peak currents are in close agreement with those predicted theoretically by Dahm and Scalapino (1997).

414

4. EXAMPLE 2: COUPLED-LINE FILTER

The non-linear response of a third-order coupled-line filter was also analysed by properly coupling three half-wave line resonators and using a two-tone source. Fig. 3 shows the peak current in each resonant line at one of the fundamental frequencies and one of the third order intermodulation products. These are in good qualitative agreement with the measurement made by Yoshitake T et al. (1995), which show that the intermodulation distortion is highest at the band pass edges. Quantitative agreement will be possible whenever the non-linear terms $\Delta L(I)$, $\Delta R(I)$ can be determined from the experiments. Efforts are currently under way to define and set-up such experiments

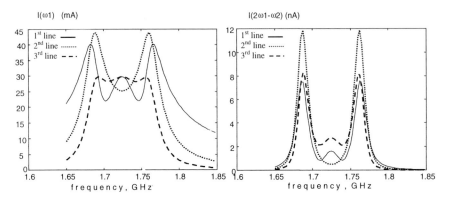

Fig.3. Peak current in each resonating line at one of the fundamental frecuencies (ω_1) and one of the third order intermodulation products ($2\omega_1-\omega_2$).

5. CONCLUSIONS

The Multiport Harmonic Balance code developed for the analysis of non-linearities in HTS lines has proved to be a fast and robust algorithm, capable of modelling any HTS circuit with transmission lines. It has also proved its usefulness in calculating the intemodulation distortion of coupled-line filters without restricting assumptions made neither on the order of the nolinearities, nor on their resistive or reactive nature.

ACKNOWLEDGMENTS

This work has been funded by CIRIT throug grant SGR00043. The authors would also like to thank Dr. Salvador Talisa for his helpful comments and suggestions.

REFERENCES

Dahm T and Scalapino D J 1997 J.Appl.Phys **81**, 4, pp 2002-9
Hicks R G and Khan P J 1980 Electronic Letters **16**, 10, pp 375-6
Maas S A 1988 Nonlinear Microwave Circuits (Artech House)
Yoshitake T and Tahara S 1995 Appl. Phys. Lett. **67**, 26, pp 3963-5

Inst. Phys. Conf. Ser. No 167
Paper presented at Applied Superconductivity, Spain, 14-17 September 1999
© *2000 IOP Publishing Ltd*

Weakly Coupled Grain Model for High T_c Superconducting Thin Films Taking Account of Anisotropic Complex Conductivities

K.Yoshida[1], H.Takeyoshi[1], H.Morita[1], H.Kanaya[1], T.Uchiyama[2], H.Shimakage[2] and Z.Wang[2]

[1]Department of Electronic Device Engineering, Graduate School of Information Science and Electrical Engineering, Kyushu University, 6-10-1 Hakozaki, Higashi-ku, Fukuoka 812-8581, JAPAN

[2]KARC Communications Research Laboratory, 588-2 Iwaoka, Iwaoka-cho, Nishi-ku, Kobe 651-2401, JAPAN

Abstract: An analytical solution of the London equation for the weakly coupled grain model of high-T_c superconducting thin films has been obtained in the case of finite thickness by taking full account of anisotropic conductivities. Using the solution, we provide general expressions for the transmission-line parameters of high-T_c superconducting transmission lines. Dependences of the resistance on the grain size, coupling strength and film thickness have been numerically evaluated and discussed.

1. Introduction

Intensive studies have been made of applications of high-T_c superconducting transmission lines, which have low loss and low dispersion characteristics, to microwave devices for mobile and satellite communications. In these applications the precise evaluation of the transmission-line parameters of superconducting thin films such as the resistance and the kinetic inductance, which result from the complex conductivity, is essential. It has been pointed out by Hylton et al (1988) (1989a)(1989b) that these basic parameters are greatly influenced by the weaklinks(Josephson junctions) inevitably existing in high-T_c thin films, i.e., the so-called "weakly coupled grain (WCG) model". As a result of this model, it is supposed that the weaklink gives rise to the excess surface resistance (Attanasio et al 1991)(Miller et al 1988)(Nguyen et al 1993)(Oates et al 1993) and tends to enhance the kinetic inductance of polycrystalline films compared with that of the single crystal (Porch et al 1993)(Yoshida et al (1996)(1998)). Detailed studies of this model, however, have not been made, especially in terms of anisotropic properties of high T_c superconducting films.

In our previous papers (Yoshida et al (1996)(1998)) we have carried out experimental studies on the WCG model neglecting the anisotropy of high T_c superconducting films. In this paper we make, for the first time, a general theoretical formulation of the effects of anisotropy on the WCG model. In Sec.2 an analytical solution of the London equation for the weakly coupled grain model of high T_c superconducting thin films has been obtained in the case of finite film thickness by taking full account of anisotropic conductivities. Using the solution, we obtained the expressions for the transmission-line parameters of high T_c superconducting transmission lines in Sec.3. Dependences of the resistance on the grain size, coupling strength and film thickness have been numerically evaluated and discussed.

2. Expression for the Magnetic Field Distribution in the Anisotropic Superconducting Thin Film with Grain Boundaries

In Fig.1 we show the schematic figure of the weakly coupled grain (WCG) model proposed by Hylton et al (1988) for a c-axis oriented high T_c superconducting film, where the c-axis of the superconducting film is perpendicular to the substrate. The average grain size is assumed to be **a**, and the film thickness is **d**.

In Fig.2 we show the cross section of the superconducting film representing the WCG model, where the external current K [A/m] per unit length in the y direction is assumed to be flowing in the z direction.

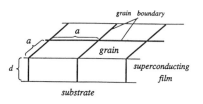

Fig. 1 Schematic of weakly coupled grain (WCG) model.

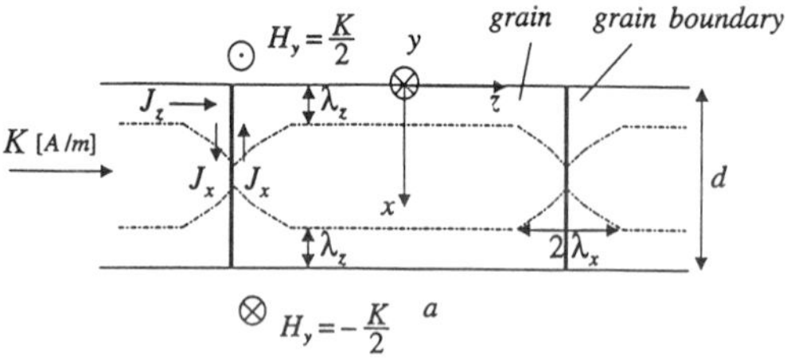

Fig.2 The cross section of the superconducting thin film

The London equation for the anisotropic superconductor is given as (Hylton et al 1989)(Van Duzer et al 1981)

$$\lambda_x^2 \frac{\partial^2 H_y}{\partial z^2} + \lambda_z^2 \frac{\partial^2 H_y}{\partial x^2} = H_y \qquad (1)$$

with the boundary condition at the grain boundary :

$$\frac{\partial^2 H_y}{\partial x^2} \mp \frac{\lambda_x}{\lambda_J^{(0)2}} \frac{\partial H_y}{\partial z} = 0 \quad at \quad z = \pm \frac{a}{2} \qquad (2)$$

with

$$\lambda_J^{(0)} = \sqrt{\frac{\Phi_0}{2\pi \mu_0 2\lambda_x J_c}}$$

where λ_x and λ_z are the magnetic penetration depth in the x and z direction, respectively. J_c is the critical current density for the Josephson junction and $\lambda_J^{(0)}$ corresponds to the penetration depth for an isolated single Josephson junction. According to the symmetry and periodicity of the present problem, other boundary conditions for a single grain located in the region $-a/2 \leq z \leq a/2$ can be expressed as

$$H_y(x=0) = -\frac{K}{2} \qquad (3)$$

$$H_y(x=d) = \frac{K}{2} \qquad (4)$$

The solution of Eq.(1) satisfying boundary conditions Eqs.(2)-(4) is obtained as :

$$\frac{H_y(x,z)}{(K/2)} = \frac{\sinh \frac{1}{\lambda_z}(x - \frac{d}{2})}{\sinh (\frac{d}{2\lambda_z})} \quad -$$

$$\sum_{n=1}^{\infty} \frac{8n\pi}{d^2} \frac{1}{[1+\lambda_z^2(\frac{2n\pi}{d})^2]} \frac{\sin(\frac{2n\pi}{d}x)\cosh(\frac{\sqrt{1+\lambda_z^2(\frac{2n\pi}{d})^2}}{\lambda_x}z)}{[(\frac{2n\pi}{d})^2\cosh(\frac{a}{2\lambda_x}\sqrt{1+\lambda_z^2(\frac{2n\pi}{d})^2}) + \frac{\sqrt{1+\lambda_z^2(\frac{2n\pi}{d})^2}}{\lambda_z^3}\sinh(\frac{a}{2\lambda_x}\sqrt{1+\lambda_z^2(\frac{2n\pi}{d})^2})]} \quad (5)$$

It is shown that in the limit of $d = +\infty$, Eq.(5) coincides with the solution given by Hylton et al (1989a) valid only for the case of a semi-infinite superconductor. Using this solution, the current distribution can be obtained from the Maxwell equation $\mathbf{J} = \nabla \times \mathbf{H}$.

3. The Equivalent Circuit for the High T_c Superconducting Transmission Line with Grain Boundaries

For the case of an applied alternating current as $I e^{j\omega t}$, we obtain the expression for the total current density $\mathbf{J}$ as

$$\mathbf{J} = \mathbf{J}_N + \mathbf{J}_S = ([\sigma_1] - j[\sigma_2])\mathbf{E} \quad (6)$$

with

$$[\sigma_1] = \begin{pmatrix} \sigma_{1x} & 0 & 0 \\ 0 & \sigma_{1y} & 0 \\ 0 & 0 & \sigma_{1z} \end{pmatrix} = \begin{pmatrix} \frac{1}{\rho_c} & 0 & 0 \\ 0 & \frac{1}{\rho_{ab}} & 0 \\ 0 & 0 & \frac{1}{\rho_{ab}} \end{pmatrix}$$

$$[\sigma_2] = \begin{pmatrix} \sigma_{2x} & 0 & 0 \\ 0 & \sigma_{2y} & 0 \\ 0 & 0 & \sigma_{2z} \end{pmatrix} = \begin{pmatrix} \frac{1}{\omega\mu_0\lambda_c^2} & 0 & 0 \\ 0 & \frac{1}{\omega\mu_0\lambda_{ab}^2} & 0 \\ 0 & 0 & \frac{1}{\omega\mu_0\lambda_{ab}^2} \end{pmatrix}$$

where $\mathbf{J}_N$ is the normal-conducting current density, $\mathbf{J}_S$ is the superconducting current density, $\mathbf{E}$ is the electric field, μ_0 is the vacuum permeability, $\lambda_x = \lambda_c$ is the magnetic penetration depth along the c-axis and $\lambda_y = \lambda_z = \lambda_{ab}$ is the magnetic penetration depth in the a-b plane, $[\sigma_1]$ is the normal conductivity tensor, $\sigma_{1x} = 1/\rho_c$, $\sigma_{1y} = \sigma_{1z} = 1/\rho_{ab}$, ρ_c is the resistivity in the c direction and ρ_{ab} is the resistivity in the a-b plane, $[\sigma_2]$ is the superconductivity tensor, ω is the angular frequency.

In Fig.3 we show the equivalent circuit for the transmission line with a unit length made of anisotropic high T_c superconducting films including grain boundaries. In this figure L_m represents the conventional magnetic inductance per unit length, which is almost independent of the conductor material, and L_{kG} and L_{kJ} denote the kinetic inductance of the superconductor per unit length for the grain and the grain boundary, respectively. R_G and R_J represent the resistance per unit length resulted from the grain and the grain boundary, respectively.

The expressions for the kinetic inductance and the resistance of the grain and the junction can be obtained by calculating the kinetic energy of the superconducting electrons and the balance of the power consumption, respectively :

$$\frac{1}{2}L_K|I|^2 = \int \text{Re}\left[\frac{1}{2}\mu_0[\lambda^2]\mathbf{J}_s \cdot \mathbf{J}_s^*\right] dv \quad (7)$$

$$\tfrac{1}{2}R\,|I|^2 = \int \mathrm{Re}\,[\tfrac{1}{2}(J \cdot E^*)]dv \tag{8}$$

where I is the total current as shown in Fig.3 and the volume integral extends over unit length in the z direction. The obtained results are :

$$R_G = \frac{1}{D}\left(\frac{\omega L_{kx}}{Q_x\left(1+Q_x^{-2}\right)} + \frac{\omega L_{kz}}{Q_z\left(1+Q_z^{-2}\right)}\right) \tag{9}$$

$$R_J = \frac{1}{D}\left(\frac{\omega L_{kJ}}{Q_J\left(1+Q_J^{-2}\right)}\right) \tag{10}$$

$$L_{kG} = \frac{1}{D}\left(\frac{L_{kx}}{1+Q_x^{-2}} + \frac{L_{kz}}{1+Q_z^{-2}}\right) \tag{11}$$

$$L_{kJ} = \frac{1}{D}\frac{L_{kJ}^{(0)}}{\left(1+Q_J^{-2}\right)} \tag{12}$$

with

$$L_{kx} = \int_0^d dx\,\frac{1}{a}\int_{-\frac{a}{2}}^{\frac{a}{2}} dz\;\mu_0\lambda_x^2 f_x^2(x,z)$$

$$L_{kz} = \int_0^d dx\,\frac{1}{a}\int_{-\frac{a}{2}}^{\frac{a}{2}} dz\;\mu_0\lambda_x^2 f_z^2(x,z)$$

$$Q_x = \frac{\sigma_{2x}}{\sigma_{1x}}$$

$$Q_z = \frac{\sigma_{2z}}{\sigma_{1z}}$$

$$Q_J = \frac{R_N}{\omega L_J^{(0)}} = \frac{I_c R_N}{f\,\Phi_0}$$

$$L_J^{(0)} = \frac{\Phi_0}{2\pi I_c}$$

$$L_{kJ}^{(0)} = \int_0^d dx\,\mu_0\lambda_J^2\left|J_z(x,\tfrac{a}{2})\right|^2$$

$$\lambda_J = \sqrt{\frac{\Phi_0}{2\pi\mu_0 a\,J_c}}$$

where Q_x and Q_z represent the quality factor for the current component J_x and J_z in the grain, Q_J represents the quality factor of the junction, R_N is the junction resistance, $L_J^{(0)}$ is the kinetic inductance for a single junction, I_c is the junction critical current, L_{kx} and L_{kz} are the kinetic inductance per unit area (sheet inductance) associated with the current component J_x and J_z, respectively, and λ_J corresponds to the Josephson penetration depth in the small grain limit (Hylton et al 1989a). The quantities f_x and f_z represent the current density for the case of a unit applied current, i.e, K=1 : They are defined by

$$J_x(x,y,z) = f_x(x,z)\,K(y) \tag{13}$$

$$J_z(x,y,z) = f_z(x,z) K(y) \qquad (14)$$

Equations (9), (10) indicate that the resistances are proportional to the respective kinetic inductances.

The quantity D in Eqs.(9)-(12) is defined by

$$D = \frac{|I|^2}{\int_{-\infty}^{\infty} K^2(y)dy} \qquad (15)$$

which corresponds to the geometrical factor representing the characteristic length for a particular transmission line geometry. The characteristic length D defined by Eq.(15) depends on the geometry of the transmission line. If we assume that the film thickness is sufficiently thin and that the characteristic impedance is $50\,\Omega$ on MgO substrate with the permitivity $\epsilon_r = 9.4$, D is determined by the geometrical configuration of the transmission line almost independent of the internal structure of the conductor, and we can obtain numerically following the procedure given in Sheen et al (1991) :

$$D = 0.4\ W \qquad \text{for coplanar waveguide}$$
$$D = 0.5\ W \qquad \text{for micro-stripline}$$

where W is the width of the signal electrode.

In order to evaluate the values for the resistance for various parameters, we first introduce the following normalized parameters ; $\alpha = a / (2 \lambda_x)$, $\beta = \lambda_z^2 / \lambda_J^{(0)2}$, $\gamma = d / (2 \lambda_z)$, where α is the normalized grain size, $\beta = \lambda_z^2 / \lambda_J^{(0)2} = (4 \pi \mu_0 / \Phi_0) \lambda_x \lambda_z^2 J_c$ represents the strength of the coupling and γ is the normalized thickness.

In Figs.4,5 and 6 we show the dependence of the resistances per unit length on α, β and γ, respectively. In the calculation we used the typical experimental values for YBa$_2$Cu$_3$O$_x$ superconductors (Friedmann et al 1990)(Hylton et al 1989) : $\lambda_{ab} = 0.15$ [μm], $\lambda_c = 0.5$ [μm], $\rho_c = 2.5 \times 10^{-5}$ [Ωm], $\rho_{ab} = 4.0 \times 10^{-7}$ [Ωm], which leads to $Q_x = 1.3 \times 10^4$ and $Q_z = 2.3 \times 10^3$ at $\omega = 2\pi \times 10^9$ [Hz]. Junction parameters are : $Q_J = 4.8 \times 10^2$ for it $I_c R_N = 1$ [mV] and f $=10^9$ [Hz]. It must be mentioned that Eqs.(9) and (10) lead to the resistances proportional to ω^2 in the case of Q_x, Q_z, Q_J $\gg 1$. The resistances shown in Figs.4,5 and 6 are calculated at $\omega = 2\pi \times 10^9$ [Hz] in the case of D=100 [μm].

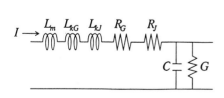

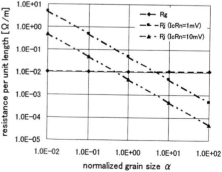

Fig.3 The equivalent circuit for the transmission line with a unit length

Fig.4 The dependence of the resistance per unit length on α for $\beta = \gamma = 1$.

420

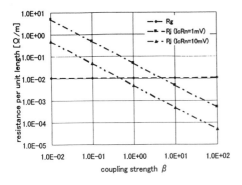

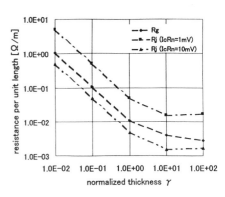

Fig.5 The dependence of the resistance per unit length on β for $\alpha = \gamma = 1$.

Fig.6 The dependence of the resistance per unit length on γ for $\alpha = \beta = 1$.

4. Conclusions

In the present paper we obtained general expressions for the the resistance and the kinetic inductance of the superconducting thin films with grain boundaries by taking full account of anisotropic conductivities, which include the normalized grain size α, coupling strength β and normalized thickness γ as free parameters. Detailed discussions with specific samples are to be our next work.

References

Attanasio C, Maritato L and Vaglio R 1991 Phys. Rev.B, **43**, no.7, pp.6128-6131
Friedmann T A, Rabin M W, Giapintzakis J, Rice J P and Ginsberg D M 1990 Phys. RevB, **42**, no10 , pp.6217-6221
Hylton T L, Kapitulnik A and Beasley M R 1988 Appl.Phys.Lett., **53**, no.14, pp.1343-1345
Hylton T L and Beasley M R 1989a Phys. Rev. B., **39**, no.13, pp.9042-9048
Hylton T L, Beasley M R, Kapitulnik A, Carini J P, Drabeck L and Gruner G 1989b IEEE Trans.Magn., **25**, no.2, pp.810-813
Miller D, Richards P.L, Etad S, Inam A, Venkatesan T, Dutta B, Wu X D, Eom C B, Geballe T H, Newman N and Cole B G 1988 Appl. Phys. Lett., **59**, no.18, pp.2326-2328
Nguyen P P, Oates D E, Dresselhaus G, and Dresselhaus M.S 1993 Phys. Rev. B, **48**, no.9, pp.6400-6412
Oates D E, Nguyen P P, Dresselhaus G, Dresselhaus M S, and Chin C C 1993 IEEE Trans. Appl. Supercond., **3**, no.1, pp.1114-1117
Porch A, Lancaster M J, Humphreys R G and Chew N G 1993 IEEE Trans. Appl. Supercond., **3**,no.1, pp.1719-1722
Sheen D M, Ali S M, Oates D E, Withers R S and Kong J A 1991 IEEE Trans. Appl. Supercond., **1**, No.2, pp.108-115
Van Duzer T and Turner C W 1981 Principles of Superconductive Devices and Circuits, Elsevier North Holland Inc., New York
Yoshida K, Onoue T, Kiss T, Shimakage H and Wang Z 1996 IEICE Trans. Electron., **E79-C**, no.9, pp.1254-1259
Yoshida K, Adou T, Nishioka S, Kanda Y, Shimakege H and Wang Z 1998 IEICE Trans.Electron., **E81-C**, no10 , pp.1565-1572

Inst. Phys. Conf. Ser. No 167
Paper presented at Applied Superconductivity, Spain, 14-17 September 1999
© *2000 IOP Publishing Ltd*

SQUID-NDE of semiconductor samples with high spatial resolution

J Beyer, Th Schurig, A Lüdge*, H Riemann*

Physikalisch-Technische Bundesanstalt Berlin, Abbestr. 2-12, D-10587 Berlin, Germany
*Institute of Crystal Growth, Rudower Chaussee 6, D-12489 Berlin

ABSTRACT: We developed a SQUID-based, noninvasive method for the investigation of semiconductor wafers. It is based on the detection of excited photocurrents via their magnetic field by means of a highly sensitive SQUID system. Our method allows the visualization of small growth-related fluctuations of the doping level of the semiconductor with a spatial resolution of a few tens of a micrometer determined by the excitation spot size. Numerical simulations of the magnetic signals have been performed the results of which show a reasonable agreement with the experiments.

1. INTRODUCTION

A significant issue for the characterization of semiconductor wafers for device fabrication is the determination of the doping homogeneity as yield and device performance can be severely affected by doping variations. Inhomogeneities of the doping level can be caused by spatial variations of the distribution of dopands or impurities. We have proposed and basically demonstrated a completely noninvasive SQUID-based method for the visualization of gradients in the doping level in semiconductor wafers (Schurig 1997, Beyer 1999). Our approach is based on the magnetic detection of photogenerated currents excited in the semiconductor utilizing sensitive SQUID sensors and it extends the range of application of SQUID magnetometry for nondestructive evaluation to the investigation of semiconductor samples. In this paper we describe the principles and the experimental realization of the SQUID photoscanning technique. We report on investigations of silicon wafers with small growth-related doping fluctuations and present results of numerical simulations of the magnetic signals.

2. SQUID PHOTOSCANNING TECHNIQUE

The basic principle of the SQUID photoscanning technique is illustrated in Fig.1. In a semiconductor sample with a spatial variation of the effective doping level N (for instance caused by artificial doping profiles, doping fluctuations or defects such as grain boundaries) there is an internal electric field **E** associated with the gradient of N. The sample is illuminated locally by light with a photon energy hv larger than the bandgap energy E_{GAP} of the semiconductor which results in the generation of nonequilibrium charge carriers in the sample. In a region of a doping gradient the interaction of the nonequilibrium charge carriers with the internal electric field leads to the occurance of a net photocurrent. Highly sensitive SQUID magnetometers are capable of measuring the magnetic field originating from the net photocurrent and thereby detecting the existance of doping gradients. By moving the sample relative to both the illumination and the magnetometer (the position of both is fixed with respect to each other) a scanning of the sample is performed. In contrast to most of the conventional SQUID NDE systems (Wikswo 1995) and SQUID microscopes (Hibbs 1992, Lee 1997)

422

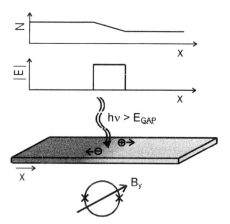

Fig.1: Scheme of SQUID Photoscanning detection. Photogenerated nonequilibrium charge carriers interact with the internal electric field **E** associated with a spatial variation of the doping level N in the semiconductor sample. SQUID magnetometers detect the magnetic field arising from the net photocurrent.

in our approach the spatial resolution is determined neither by the size of the SQUID sensor nor the sample-to-sensor distance. It is limited by the size of the excitation spot which can easily be focused to a few μm and by the minimum incremental motion of the translation stage moving the sample which is ≤ 1 μm in the system we used. The scheme of our SQUID Photoscanning measurement system is depicted in Fig.2. An integrated low-critical temperature (T_C) SQUID magnetometer of our type W9M (Drung 1999) with a chip size of 3.6 x 3.6 mm^2 and effective area of 1 mm^2 is operated in a liquid helium cryostat having a cold-to-warm distance of about 7 mm. The magnetometer is placed at the inner bottom of the cryostat and vertically arranged to sense the horizontal magnetic field.

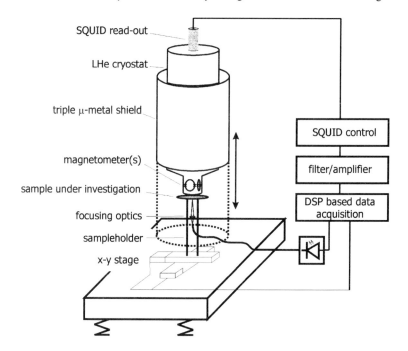

Fig.2: Scheme of SQUID Photoscanning measurement system.

The semiconductor samples typically 0.3 – 1 mm in thickness are placed at the outer bottom of the cryostat at room temperature. The polished sample bottom side is illuminated by laser light of 675 nm wavelength generated by a commercial diode laser and fed via an optical fiber to the focusing optics which is mounted about 9 cm underneath the sample. The optics realizes a laser spot of about 20 μm in diameter and about 7 mW in power on the sample and is aligned on axis with the magnetometer center. The triple μ-metal cylinder used for magnetic shielding can be moved vertically in order to open the setup for sample loading and positioning. The x-y translation stage on which the sample is mounted is placed underneath the shielding and allows one to scan the sample by a lateral translation of the sample with respect to both the optics and the magnetometer. The inner dimension of the μ-metal shielding limits the size of samples which can be fully scanned with the system to about 80 mm in diameter. The whole system is placed on a vibration isolated workstation. The SQUID magnetometer is operated in a flux-locked loop; the system white magnetic field noise at frequencies above about 100 Hz up to the system bandwidth of about 500 kHz amounts to 8 fT/√Hz being dominated by noise contributions of the cryostat and the μ-metal shielding. The magnetometer signal is fed to the data acquisition system controlling the amplitude modulation of the diode laser (typically in the range of several hundreds of Hz to several kHz) and the operation of the x-y scanning stage. A proper phase sensitive detection of the magnetometer signal component at the laser modulation frequency by the DSP based 12 bit data acquisition necessitates a conditioning of the SQUID output. This is realized by appropriate signal filtering, amplifying and phase adjustment. Heavy magnetic shielding and vibration isolation are not stringent prerequisites for the operation of the SQUID Photoscanning system. Electromagnetic and mechanical interferences have to be reduced to an extend ensuring a locked-loop operation of the magnetometer and avoiding an increase in the system magnetic field noise in the relevant laser modulation frequency range. A single μ-metal shield and/or an eddy current shield may be sufficient especially in conjunction with the implementation of reference magnetometers for the suppression of external interferences. The spectral analysis of the magnetometer signal in a narrow effective bandwidth of typically 1 – 12 Hz centered at the laser modulation frequency makes a cryo cooling of the SQUID sensor potentially applicable as cooler interferences typically occur at discrete frequencies.

3.   MEASUREMENTS

The objects of analysis from which the results presented below have been obtained were nitrogen doped silicon wafers. Silicon single crystals typically show doping variations of the order of a few percent of the mean doping level caused by segregation processes during crystallization. In wafers cut from such crystals lateral doping inhomogeneities – so called striations - occur and in the case of wafers vertically cut from the single crystal core the doping distribution contains valuable information about the shape of the crystallization phase boundary. Fig.3 depicts the results of scanning measurements of Si:N vertical cut wafers. The horizontal magnetic field component B_y

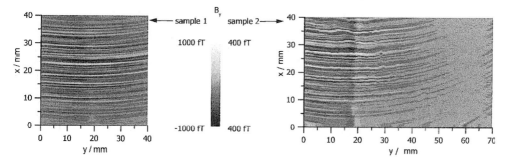

Fig.3: Grey scale map of horizontal magnetic field amplitude B_y obtained from partial scanning of two Si:N vertical cut wafers. The growth direction of the single crystals was in x-direction. The scan step width was 100 μm in x-direction and 1 mm in y-direction. Sample 1: Si:N 3 Ωcm, 45 x 75 mm^2, measurements time: 27 min; Sample 2: Si:N 4.3 Ωcm, 45 x 100 mm^2 measurement time: 40 min

measured with the SQUID magnetometer is shown as a grey scale map with respect to the position of the laser spot on the sample. The growth direction of the crystals of which the wafers were cut was in parallel to the x-direction. The left plot in Fig.3 (sample 1) shows a 40 x 40 mm^2 area scan of the mid region of a 45 x 75 mm^2 Si:N vertical cut wafer with a mean resistivity of 3 Ωcm. The right plot (sample 2) has been obtained from a 40 x 70 mm^2 area scan of the mid and edge region of a 45 x 100 mm^2 Si:N wafer with a mean resistivity of 4.3 Ωcm. The quasiperiodic patterns of the horizontal magnetic field component qualitatively resemble striation patterns which can be found from such wafers with conventional methods such as spreading resistance technique and conventional photoscanning analysis (Lüdge 1997). Maximum field amplitudes of about1 pT for sample 1 and 400 fT for sample 2 and typical period lengths in x-direction of abut 1 mm have been measured. The magnetic field patterns visualize the lateral variation of the doping level in the samples. A magnetic signal is obtained when the laser spot is positioned in a region of a lateral doping gradient whereas no net photocurrent and consequently zero magnetic signal occurs when the photoexcitation takes place in a locally homogeneously doped region. The change in sign of the magnetic signal indicates an inversion of the direction of the net photocurrent distribution which is associated with an inversion of the orientation of the local doping gradient. In both patterns pictured in Fig.3 aquiline regions of the same magnetic signal can be found which display the shape of the crystallization phase boundary. In the plot of sample 2 the contrast smears in the edge region of the sample. This is partially due to the fact that in the edge region the orientation of the local doping gradients and therewith the direction of the net photocurrent distributions is no longer in parallel with the growth direction as it is the case in the mid region of the wafers. Consequently, the monitored y-component of the magnetic field caused by the net photocurrent diminishes. The implementation of a second horizontally sensitive magnetometer arranged to measure B_x is planed in order to measure both horizontal magnetic field components.

4. SIMULATIONS OF THE MAGNETIC SIGNALS

Beyond the noninvasive qualitative visualization of the lateral doping variation it is desirable to quantitatively interpret the detectable magnetic signal. To ascertain sample properties determined by the doping distribution such as the resistivity from the magnetic signal it is necessary to solve a two-step *inverse problem*: firstly to infer the net photocurrent distribution from the magnetic signal and secondly to conclude the doping distribution from the photocurrent distribution. As a first approach we tackled a two-step *forward calculation* of the magnetic signals by performing numerical simulations in the following way. Starting from an assumed doping distribution and a given optical excitation the semiconductor drift-diffusion equations have been numerically solved in two spatial dimensions using the finite elements program package ToScA (Gajewski 1992). ToScA output the net photocurrent density distribution $\mathbf{j}^{net} = (j_x^{net}, j_y^{net})$ which then has been used to calculate the magnetic field $\mathbf{B} = (B_x, B_y, B_z)$ at the position of the magnetometer with a multidipol model and Bio-Savart's law. The input parameters for the simulations are the assumed doping distribution, the location, wavelength and power of the laser illumination, the quantum efficiency η, the lifetime τ_{ne} of the photogenerated nonequilibrium charge carriers and the number and distribution of nodes used for the discretisation of the considered area. An example simulating the situation in an inhomogeneously doped n-Si sample is explained in the following. The considered area of 2 mm in lateral dimensions x and y and 1 mm in thickness is schematically shown in Fig.4a. A linear variation of the net doping level along the x-direction of 10 % of the mean net doping level with a transition width of 400 μm (x = 0.8 mm to x = 1.2 mm) has been assumed. The mean net doping level of 1×10^{15} cm^{-3} corresponds to a mean resistivity of about 4 Ωcm. In the center of the doping variation region a Gaussian shaped distribution (FWHM = 20 μm) of nonequilibrium charge carriers was placed to simulate the electron-hole pairs photogenerated by the laser illumination of 675 nm wavelength , 7 mW of power and a spot diameter of 20 μm (Fig.4b). The additional simulation parameters were $\eta = 1$ assuming that every incident photon produces one electron-hole pair and a nonequilibrium charge carrier lifetime of $\tau_{ne} = 500$ μs which is a reasonable value for crystalline silicon. In Figs.4c and 4d the calculated net photocurrent components j_x^{net} and j_y^{net} are shown as grey scale maps with the same scaling for the

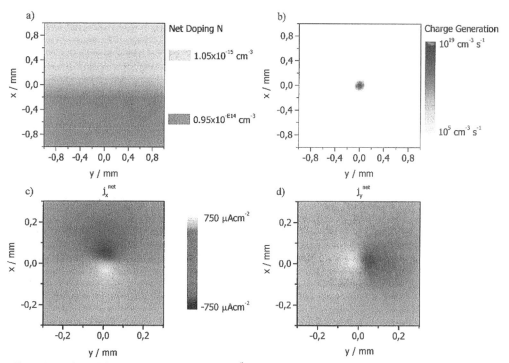

Fig.4: a) top view of simulation area of $2 \times 2 \times 1$ mm^3 with linear doping gradient of 10 % of the mean net doping level N = 1e15 cm^{-3} in x direction; b) Gaussian distribution of the generated charge simulating the photogeneration of nonequilibrium charge carries due to illumination with laser light of 675 nm wavelength and 7 mW power and a spot diameter of 20 μm in the center of the doping gradient region; c) and d) calculated net photocurrent distribution components j_x^{net} and j_y^{net} in the central region of the simulation area

central region of the area under consideration. Note different length scale of Figs.4a, 4b and Figs.4c, 4d. Obviously, the simulated photocurrent distribution is not simply dipolar like. The y-component j_y^{net} is highly symmetric along the axis defined by x = 1 mm but there is a slight asymmetry in j_x^{net} along the direction of the doping gradient. The photocurrent distribution can be described as of warped quadrupolar shape displaying the symmetry breaking caused by the doping gradient. From the calculated photocurrent distribution a distribution of current dipoles was generated and the

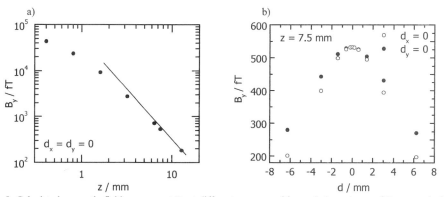

Fig.5: Calculated magnetic field component B_y at different sensor positions: a) dependence of B_y on vertical distance z, line represents distance dependence of B_y for a current dipole in parallel with x direction b) dependence of B_y on lateral sensor displacement $d = (d_x^2 + d_y^2)^{1/2}$ for fixed vertical distance z = 7.5 mm

contribution of each of those to $B = (B_x, B_y, B_z)$ at the sensor position was calculated applying Bio-Savart's law. In Fig.5 the calculated horizontal magnetic field component B_y is plotted for various sensor positions. The dependence of B_y on the vertical distance z from the location of the laser spot and on lateral displacements d_x and d_y for a fixed z = 7.5 mm are shown. At z values of approximately the sample-to-sensor distance (z = 7.5 mm) the calculation of B_y yields a values of 530 fT. This is of the order of the experimental maximum signals of 250 fT to 1000 fT. For z ≥ 3 mm B_y falls off approximately with $1/z^2$ (dipolar behavior) whereas for z ≤ 2 mm it falls off less strongly. The corresponding x component of the magnetic field has been calculated to be $B_x \approx 10^{-4} B_y$. Hence, when measuring the magnetic field at z values of about the experimental sample-to-sensor distance the net photocurrent distribution appears like a dipolar current distribution with an orientation in the direction of the doping gradient. This far field behavior is reasonable as z ≥ 5 mm corresponds to a distance of about ten times the dimension of the calculated current distribution. A test calculation for a sample without a variation of the doping level resulted in B_y = 0.02 fT at a vertical distance z = 7.5 mm. This is an estimation of the uncertainty of the numerical calculation as exact zero magnetic signal is expected when the optical excitation is placed in locally homogeneously doped region. Fig.5b shows that a slight deviation from a precise axial alignment of the magnetometer and the laser spot is not crucial; a lateral displacement d_x or d_y of ± 1 mm only results in a reduction of the magnetic signal of about4 %. Further refined simulations for different sample parameters, for instance transition shape and width and mean doping level, as well as for various experimental parameters as laser power and wavelength are under way. The predictions will be checked experimentally.

5. SUMMARY AND OUTLOOK

In this paper we reported on the concept and the experimental realization of the SQUID Photoscanning technique for noninvasive detection of doping inhomogeneities in semiconductor wafers. Our method combines a high spatial resolution realized by a local illumination of the sample with the high magnetic field sensitivity of low-T_C SQUID magnetometers. With the setup described we are able to detect small growth-related doping level fluctuations and visualize the doping homogeneity of semiconductor wafers of up to 3" in diameter. Forward calculations of the magnetic signals based on the numerical analysis of the drift-diffusion equation in semiconductors have been performed the results of which are qualitatively and quantitatively in fair agreement with the experimental findings and confirm our understanding of the signal origin. Refined simulations in conjunction with calibration measurements of test samples with defined doping profiles will be performed in order to infer quantitative information about the sample properties from the magnetic signal. The instrumentation will be further developed with regard to a second horizontal sensitive SQUID channel, the application of high-T_C SQUID magnetometers and a cryogen-free cooling. The application of our technique for the analysis of semiconductor structures and devices, for instance photovoltaic devices, may be promising and is set on.

REFERENCES

Beyer J, Matz H, Drung D and Schurig T 1999 Appl.Phys.Lett. **74** (19), 2863

Drung D, Knappe S, Aßmann C, Peters M, Wenzel K and Schurig T 1999 Extended Abstracts of 7th Int.Supercond.Electr.Conf. ISEC'99, Berkeley, CA, USA, June 21-25, p 543

Gajewski H, Heinemann B, Langmach H, Telschow G, Zacharias K 1992 "ToScA - Two-Dimensional Semiconductor Analysis " Weierstrass Institute for Applied Analysis and Stochastics Berlin

Hibbs A D, Sager R E Cox D W Aukerman T H, Sage T A and Landis R S 1992 Rev.Sci.Instrum. **63** (7), 3652

Lee T, Chemla Y R, Dantsker E and Clarke J 1997 IEEE Trans.Appl.Supercond. **7**, 3147

Lüdge A, Riemann H 1997 Inst.Phys.Conf.Ser. **160**, 145

Schurig T, Matz H, Drung D, Lüdge A and Riemann H 1997 Inst.Phys.Conf.Ser. **160**, 149

Wikswo J P 1995 IEEE Trans.Appl.Supercond. **5**, 74

Inst. Phys. Conf. Ser. No 167
Paper presented at Applied Superconductivity, Spain, 14-17 September 1999
© 2000 IOP Publishing Ltd

Superconducting sensors for weak magnetic signals in combination with BiCMOS electronics at 77 K for different applications

P Seidel[1], L Dörrer[1], F Schmidl[1], S Wunderlich[1], S Linzen[1], R Neubert[1], N Ukhansky[1,2], S Goudochnikov[2]

[1]Institut für Festkörperphysik, Friedrich-Schiller-Universität Jena, Helmholtzweg 5, D-07743 Jena, Germany
[2]Instite of Terrestrial Magnetism, Ionosphere and Radio Wave Propagation, Russia

ABSTRACT: Different sensors for weak magnetic signals were realised using thin high temperature superconducting films on different substrates including buffered silicon. Superconducting quantum interference devices (SQUIDs), magnetometers and planar gradiometers based on them as well as a new type of a Hall-effect sensor with a superconducting antenna were tested with respect to signal resolution, band width and spatial resolution. To realise adapted systems for biomagnetic research or non-destructive evaluation a common room temperature electronic have some disadvantages. Thus we tested discrete elements as well as special adapted integrated BiCMOS circuits placed near the sensors at 77K. We demonstrate the status of the development of such an electronic.

1.   INTRODUCTION

Weak magnetic signals in the range of some pT down to fT can be detected by superconducting sensors. Superconducting quantum interference devices (SQUIDs) are often coupled to antenna structures to enhance the flux detecting area. Such magnetometer or gradiometer structures were prepared in thin film technology using high-T_C superconducting $YBa_2Cu_3O_{7-x}$ realising the Josephson junctions by bicrystal or step-edge technology. Such sensors can be used for different applications like biomagnetism or non-destructive evaluation (NDE) even in a magnetically unshielded environment. To enhance the signal to noise ratio these sensors can be coupled to a low noise semiconducting amplifier working near the sensor in liquid nitrogen. Even more components of the SQUID-electronics like the standard flux locked loop scheme can be arranged on a costumer chip working at low temperatures. Besides the lower thermal noise the shorter connecting lines for low voltage signals have advantages because for instance environmental noise is not coupled in. Besides all progress in SQUID sensors their operation in unshielded envrionment is still a problem. In some cases where the signals are in a range between the sensitivity of common Hall sensors and very high resolution SQUIDs we propose the use of a new sensor type. The sensitivity of a Hall structure is enhanced by an incoupling antenna prepared out of a high-T_C thin film. This hybrid sensor has a wide range of linearity and requires a much simpler electronics. Thus it offers the possibility of one- or two-dimensional arrays of sensors for NDE in unshielded environment.

2.   HIGH-T_C SQUID SENSORS

We used for our investigations two kinds of Josephson junctions, step-edge and bicrystal junctions. Step-edge junctions were prepared on ion beam etched $SrTiO_3$ substrates with a ratio between step height and YBCO film thickness (d<120 nm) of about 1.3. The bicrystal Josephson junctions were made on $SrTiO_3$ substrates with 24°, 30° or 36.8° grain boundaries. In both cases we used laser deposited films with a critical current density $j_C \geq 2 \cdot 10^6$ A/cm² (at 77 K) and a critical temperature $T_C \approx 90$ K. All structures were patterned with Ar ion beam etching on a cooled sample holder to avoid degradation of the junction properties during the patterning process. All devices were passivated with a double layer of amorphous insulating YBCO and CeO₂ sputtered on to the whole substrate surface.

The values of flux resolution $\sqrt{S_\Phi}$ of dc-SQUIDs with a SQUID inductance of about 30 pH based on step-edge junctions ranged between 7 $\mu\Phi_0/\sqrt{Hz}$ and 2.9 $\mu\Phi_0/\sqrt{Hz}$ (white noise region, 77 K) depending on the preparation conditions during the laser deposition (Fig.1). The Josephson junctions on bicrystal substrates show comparable superconducting properties. The reproducibility of the bicrystal junctions is better while the white flux noise of dc-SQUIDs with same layout parameters is about two times higher at 77 K. The values of the $I_C R_N$ product of bicrystal junctions depend on grain boundary angle as well as on the film thickness of the deposited YBCO film. Bicrystal dc-SQUIDs (d = 150 nm) on 24° grain boundaries reach the highest $I_C R_N$ products of about 400 μV (77K) which is four times higher compared to values of step-edge dc-SQUIDs.

Different layouts of dc-SQUIDs serving as a current sensor in galvanically coupled planar gradiometers and magnetometers were prepared on 10·10 mm² substrate area (Fig.2). Depending on the applications in the field of biomagnetism and non-destructive evaluation the superconducting sensors need a matched field resolution and sensitivity as well as the ability to operate in unshielded environment.

The influence of the layout of the galvanometer dc-SQUID (Fig.2a) and antennas (Fig.2b) on the gradient sensitivity and the balance of the gradiometer was studied. The variation of the geometrical parameters of the dc-

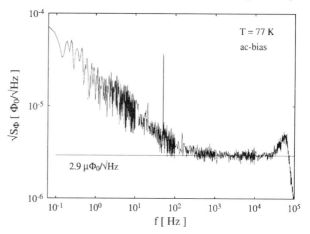

Fig. 1 Flux noise of step-edge dc-SQUD at 77K

SQUID like loop length, loop width and width of the incoupling line allows us to change the mutual inductance and in such a way the sensitivity of the dc-SQUIDs in a wide range. We used values for the mutual inductance between 30 pH and 200 pH and for the width of the superconducting antennas between 2 mm and 8 mm to adjust a field gradient resolution of the sensors.

In such a way we obtained for our gradiometer structures a gradient resolution in magnetically shielded environment of about $\sqrt{S_G} \approx 310$ fT/(cm$\sqrt{Hz}$) in the white noise region and $\sqrt{S_G} \approx 700$ fT/(cm$\sqrt{Hz}$) (at 1 Hz) at 77 K. In unshielded environment we achieved comparable results in the white noise region but increased values of $\sqrt{S_G} \approx 3$ pT/(cm$\sqrt{Hz}$) at 1 Hz. This is even low enough for their application in a heart monitoring system for non-invasive measurements.

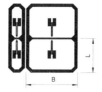

Fig. 2 Layout of a) dc-SQUID   b) gradiometer   c) magnetometer

3. HYBRID HALL SENSOR WITH HIGH-T_C ANTENNA

We have developed and investigated thin-film Hall magnetometers with a high-T_C super-conducting pick-up antenna (Kaiser 1998). The results of first investigations on $SrTiO_3$ substrates have shown a promising performance of these devices. The sensor consists of a superconducting thin-film pick-up antenna with a dimension of $4 \cdot 4$ mm^2 and a structure width of 0.5 mm. Close to the Hall detector the superconducting loop forms an incoupling line with a length of 100 µm and a width of 10µm, see Figs. 3 and 4.

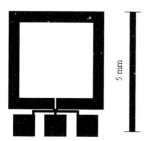

Fig. 3 Layout of the superconducting film structure of the magnetometer

Fig. 4 Inner sensor structure with YBCO line for coupling and Hall detector Bi line

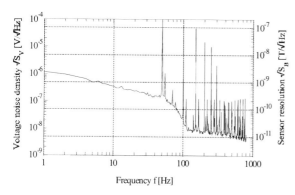

Fig. 5 Voltage noise and resolution of the sensor at 77K

At room temperature we have measured a maximum magnetometer sensitivity of 839 mV/T. At T_C we have found a steep increase of the sensitivity by a factor of about 100 indicating a sufficiently good operation of the superconducting pick-up system. From the layout data a magnetic field amplification factor of 120 has to be expected. At 77 K a maximum sensitivity of 474 V/T was observed increased by a factor of 565 compared to the room temperature value. With this value and the result of the voltage noise measurement (Fig. 5) we calculated a resolution of the magnetometer of about 8 pT/√Hz for frequencies above 0.6 kHz. The noise in the frequency range below 100 Hz is dominated by the laboratory background. In a current project we want to use the hybrid sensor for some NDE applications. For these applications it is not necessary to have the sensitivity of a SQUID magnetometer but it should be better than commercial MR sensors. With optimised layout for the superconducting antenna and Hall sensor this range in sensor resolution will be reached. Furthermore, the fabrication of the hybrid sensor or a sensor array is much more reproducible than for SQUIDs because no Josephson junctions have to be realised. For larger arrays we prefer silicon wafers as substrates using our buffer technology (Schmidl 1998). The linear output characteristics allows a more simple amplification electronics. We plan to cool this electronics together with the sensor by use of a small cryocooler.

4. SEMICONDUCTOR ELECTRONICS AT 77 K

There are existing a lot of different schemes and variants for sensor electronics. Most of these are working at room temperature. There are some problems concerning the long wires between room temperature electronics and sensor at lower temperature (e.g. 77 K). Mainly the relative large capacitance and inductance, the delay time and the possibility of incoupled disturbances lead to difficulties. Because of the moderate sensor temperature we are able to use specially designed semiconductor circuits for a sensor electronics at the same temperature. Additionally to shorter wires with corresponding advantages a smaller Nyquist noise of the electronic scheme can be expected.

4.1   Discrete LNA

We investigated the application of a commercial low noise amplifier (LNA) to improve the performance of a SQUID-electronics. This amplifier was designed to work at low temperatures. As an example figure 6 shows the current and voltage noise of the LNA1815. This type of preamplifier is used in a room temperature ac-bias SQUID electronics with 200 kHz modulation as a directly coupled, cooled amplifier. This system (Ukhansky 1997) was used to measure the noise spectrum of a step-edge SQUID shown in Fig.1. In the case of the directly coupled preamplifier the minimum detectable flux noise is given by $S_{\Phi,min} = S_{V, amp} / (dV/d\Phi)_{SQUID}$. For the given LNA1815 and an expected flux noise of 10 $\mu\Phi_0/\sqrt{Hz}$ the transfer function of the SQUID has to be larger than 17 $\mu V/\Phi_0$.

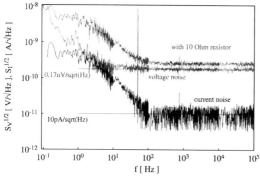

4.2   Adapted costumer chips

In the next step we developed together with the MAZeT GmbH Erfurt, the IPHT Jena and the PTB Berlin adapted costumer chips (ASICs) in standard BiCMOS technology to create complete dc-SQUID electronics at 77 K (except the power supply and the control elements). We investigated two principle schemes, a direct coupled (very useful and simple for SQUIDs with high transfer function) and a more complicated scheme with flux modulation. The former one is presented in the reference (Kunert 1999).

Fig. 6   Current and voltage noise of the LNA 1815 measured with shorted input and with 10 Ω resistor at the input

The principle sketch of the flux modulated electronics based on the developed ASIC DILA04 is shown in Fig. 7. The real system is build of 1 oszillator, 16 flip-flops, 22 gatters, 14 OPA, 8 analog switches and some resistors on chip and some resistors and capacitors out of the chip. In figure 8 the system response for a sinusoidal signal (upper curve) in dc+ , dc- and ac-bias mode is shown for open loop electronics. The complexity of the system leads to some problems, so that the closed loop not worked till now. The amplification on chip is not large enough for the closed loop mode or (if one tries to enhance the amplification) the crosstalk is too large. The crosstalk is the reason for the additional ripple seen in Fig.8 especially in the ac-bias signal.

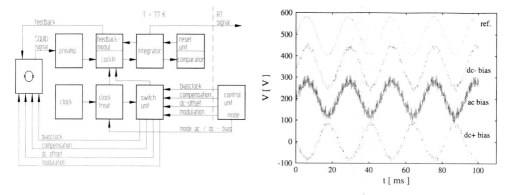

Fig. 7   Simplified circuit of the (flux modulated)
        adapted costumer chip

Fig. 8   Input (reference) and output signal of
        the flux modulated SQUID electronic
        (open loop) in different bias modes

5. APPLICATION SYSTEMS FOR UNSHIELDED ENVIRONMENT

Besides a biomagnetic system for bedside cardiology in unshielded clinical environment (Seidel 1999) we used our sensors for non-destructive evaluation. For investigations in NDE we use a testing system (Fig. 9) consisting of a x-y-positioning system, the dc SQUID or dc SQUID gradiometer in a fibre composite liquid nitrogen dewar from Conductus (24h stand by time, 5 mm distance between inner and outer wall at sensor position) and an adapted dc SQUID electronics. The positioning system allows the adjustment of the sample position in the x-y- plane managed by two motors in 150 cm distance from the dewar with the dc SQUID. The motion of these DC-motors is translated by a system of ceramic rods on the sample moving under the dc SQUID sensor. This experimental setup was chosen to avoid electromagnetic disturbances from the DC-motors. It allows a scanning area of 600 mm in x-direction and 400 mm in y-direction. The maximum scan speed is 30 mm/s enabling the application of the system for the examination of magnetic impurities or eddy current techniques in industrial dimensions. Furthermore the rotation of samples shown on intake valves for motors of car engines (Wunderlich 1998) is possible.

In the dewar different sensor positions can be realized. For the applications introduced here a sensor position with the substrate normal parallel to the x-y-scanning plane is used. Without numerical treatment the measured spatial resolution of the system limited by the distance of the sensor above the measurement object.

As an example Fig. 10 shows the magnetic field gradients of different electronic elements (e.g. SMD resistors and capacitors) measured by this system (distance 15 mm). The elements are not biased and without previous magnetisation. The strong magnetic signal caused by ferromagnetic parts can be seen and should be taken into account if SQUID sensors are placed near such elements.

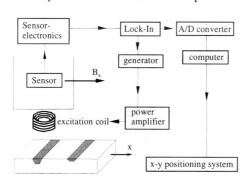

Fig. 9   Experimental setup for investigations in NDE

432

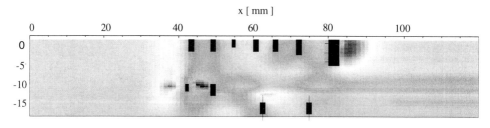

x [mm]

y [mm]

Fig. 10 Magnetic field gradient of different electronic elements (indicated by black boxes). Maximum values are +970 and –900 nT/cm.

5. SUMMARY

Different types of superconducting sensors for the resolution range of 1 to 10 pT/√Hz were fabricated. The adapted electronics can be improved by semiconducting components working at low temperatures. This gives some advantages for application systems in unshielded environment e.g. for non-destructive evaluation or biomagnetism.

ACKNOWLEDGEMENT

We like to thank G. Kaiser, F. Nitsche, J. Kunert, V. Zakosarenko, D. Hieronymus for contributing to this work.

Partially supported by German BMBF (contact Nos. 13N6864, 13N6808A and 13N7235)

REFERENCES

Kaiser G, Linzen S, Schneidewind H, Hübner U and Seidel P 1998 Cryogenics **38** , 625
Kunert J, Zakosarenko V, Schultze V, Gross R, Nitsche F, Meyer H-G these proceedings, No. 6-82
Schmidl F, Linzen S, Wunderlich S, and Seidel P 1998 Appl. Phys. Lett. **72**, 602
Seidel P, Schmidl F, Wunderlich S, Dörrer L, Schneidewind H, Weidl R, Lösche S, Leder U, Solbig O and Nowak H 1999 IEEE Trans. Appl. Supercond. **9**, 4077
Ukhansky N N, Gudochnikov S A, Weidl R, Dörrer L, Seidel P 1997 Ext. Abstr. 6th Int. Superconductive Electronics Conf. , eds. Koch H and Knappe S, pp 80-82
Wunderlich S, Schmidl F, Specht, H, Dörrer L, Schneidewind H., Hübner U and Seidel P 1998 Supercond. Sci. Technol. **11**, 315

Inst. Phys. Conf. Ser. No 167
Paper presented at Applied Superconductivity, Spain, 14-17 September 1999
© 2000 IOP Publishing Ltd

Practical low noise dc SQUIDs for a Cryogenic Current Comparator

E Bartolomé, J Flokstra and H Rogalla

Low Temperature Division, Department of Applied Physics, University of Twente, P.O. Box 217, 7500 AE Enschede, The Netherlands

ABSTRACT: We are developing sensitive washer type Nb-Al/AlO$_x$/Al-Nb dc SQUIDs directly coupled to the inductive loop of a Cryogenic Current Comparator (CCC) for the measurement of very small currents. We investigated the effect of McCumber parameters β_c close to one, and variations in the number of input coil turns n on the SQUID's characteristic curves and noise properties. The SQUIDs reach a sensitivity $\varepsilon \sim 100\,\hbar$, and can directly be coupled to a 100nH CCC loop through only n~15-30 turns. Thus, a CCC current resolution around 1pA/√Hz is expected.

1. INTRODUCTION

The Cryogenic Current Comparator is the most accurate device to compare two currents. It was shown (Sesé et al 1999) that in a so-called type I CCC, the ideally expected current resolution $<i_P^2>^{1/2}$ can only be achieved under a special configuration, in which the superconducting overlapped tube with self-inductance L_{OV} is directly connected to the input coil L_i of the readout SQUID. Assuming that the SQUID is the only noise source of the system, and that the inductances of the two sides of the flux transformer match, $L_{OV}=L_i$, the current resolution per turn CCC can be expressed as: $<i_P^2>^{1/2}=(8\varepsilon/k_{SQ}^2 L_{OV})^{1/2}$, where ε is the intrinsic energy resolution of the SQUID. The upper limit of L_{OV} is determined by the size of the cryostat neck. Thus, in order to enhance the current resolution, ε must be reduced. However, a very small SQUID inductance L_{SQ}, favorable to improve the sensitivity, deteriorates the coupling factor k_{SQ}. Moreover, since perfect coupling requires: $L_{OV}=L_i=n^2 L_{SQ}$, a large number of input coil turns n is necessary to fulfill the matching requirement. This causes the appearance of resonances in the SQUID, which degrade the noise performance. We describe the design and test of SQUIDs aimed to achieve low ε values and to be directly coupled to the CCC system.

2. DESIGN AND CHARACTERISATION

A CCC system read-out by a commercial dc SQUID with $\varepsilon \sim 10^4\,\hbar$ would achieve a current resolution of only ~10pA/√Hz. In practice, this value is further degraded due to the use of a sensing coil, to overcome the mismatch between the high input coil inductance ($L_i \sim 1$-2μH) and the low L_{OV} (<100nH). On the other hand, available nearly quantum limited SQUIDs with a sensitivity of a few $\hbar$ at 4.2K would attain $<i_P^2>^{1/2} \sim 0.1$pA/√Hz. However, L_{SQ} of those SQUIDs is so small (~1-2pH) that a large number of turns (n~300) would be necessary in order to couple the SQUID directly to the overlapped tube. Our goal is to achieve a compromising design. We optimized washer type dc SQUIDs with tightly coupled input coil and Nb-Al/AlO$_x$/Al-Nb tunnel junctions to reach a sensitivity around ~80$\hbar$ at 4.2K (implying a current resolution of ~1pA/√Hz), and to allow direct coupling to L_{OV} with a reasonable number of turns (15-30). The screening parameter was set $\beta_L=2L_{SQ}I_0/\Phi_0=1$, the noise parameter $\Gamma=2\pi k_B T/I_0\Phi_0<0.1$, and the McCumber parameter $\beta_c=2\pi I_0 R^2 C/\Phi_0\le1$, where I_0 is the critical current, C the capacitance, R the total resistance shunting each junction, and Φ_0 the flux quantum. For a critical current density of $J_0=40A/cm^2$, this results in an energy resolution in the white noise region of $\varepsilon/\hbar=1.903d[\mu m]D[\mu m]^{1/2}\beta_c^{-1/2}$, expressed in terms of the washer hole size D, the junction size d and β_c at the work temperature 4.2K. In addition, the operational condition, $\Gamma=6.7\times10^4D[\mu m]<0.1$ should be fulfilled. Usually, β_c is taken ~0.3; however from the expression above it is clear that the sensitivity can be improved by increasing the McCumber parameter up to unity. Besides, Fig. 1 shows that for a given aimed ε the fabrication parameters relax when $\beta_c\rightarrow1$.

This approach requires knowing accurately the capacitance per unit area $C'=C/A$ of the junctions. For this purpose we fabricated 6 series of junctions with areas ranging from 2x4 to 10x10 μm^2, and shunt resistors R_{sh} varying such that the McCumber parameter passes from the hysteretic to the non-hysteretic regime. β_c is deduced from: $\beta_c=[2-(\pi-2)(I_m/I_0)]/(I_m/I_0)^2$ (Zappe 1972), where I_m (the return minimum current) and I_0 are measured from the hysteretic IVC's. The capacitance is then calculated for each series. The linear fit of C as a function of the area gives a specific capacitance of 0.03pF/μm^2 at 4.2K and 0.034pF/μm^2 at 1.6K (Fig. 2). Accuracy was gained at lower temperature thanks to thermal noise reduction.

Two series of SQUIDs were fabricated using our standard Nb/Al technology (Adelerhof et

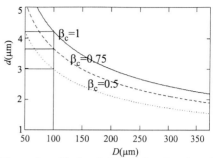

Figure 1 – *Washer hole and junction size of SQUIDs with designed $\varepsilon\sim80\hbar$, for increasing McCumber parameter.*

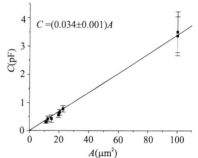

Figure 2 – *Junction capacitance as function of the junction area at 1.6K.*

al 1993). The first includes SQUIDs with β_c tending to one (0.5, 0.75 and 1), junction areas 5x5, 4x4, 2x4μm^2 and a single input turn on top. Large β_c parameters imply high value shunt resistors ranging from 6 to 21Ω. These were made in PdAu, for which a sheet resistance of $R_\square$ (66nm)=9Ωm/m was found. The second series comprises SQUIDs with $\beta_c=0.5$, 5x5μm^2 junctions and increasing number of turns n, in order to achieve the desired coupling. We avoided the use of known techniques generally applied to prevent resonances because they introduce extra noise (Ono et al 1996). Only a damping resistor R_d designed to minimize the noise contribution was placed shunting the washer. Table 1 shows the results of the characterization measurements on SQUIDs with junction area 5x5μm^2. The critical current values were as expected. In some cases (i.e SQ#5,12), only one of the shunt resistors made contact, resulting in a measured R around twice the expected value. The SQUIDs with the smallest junction sizes were poorly reproducible.

SQUID	dxd (μm^2)	D (μm)	n turns	C (pF)	$L_{SQ}+L_{slit}$ (pH)	I_0 (μA)	R_{sh} (Ω)	β_c
#1	5x5	66	0	0.75	104+47	20 *18.8*	5 *3.9*	0.5 *0.3*
#2	5x5	66	1	0.75	104+47	20 *18.4*	5 *4.4*	0.5 *0.4*
#5	5x5	66	10	0.75	104+47	20 *18.5*	5 *8.9*	0.5 *1.7*
#6	5x5	66	20	0.75	104+47	20 *17.5*	5 *2.9*	0.5 *0.2*
#7	5x5	66	50	0.75	104+97	20 *17.9*	5 *4.1*	0.5 *0.3*
#8*	5x5	66	20	0.75	104+47	20 *17.8*	5 *5.7*	0.5 *0.7*
#10	5x5	66	1	0.75	104+47	20 *18.5*	6 *7.6*	0.75 *1.2*
#12	5x5	66	1	0.75	104+47	20 *17.3*	7 *11.4*	1 *2.6*

Table 1 – *Main parameters of SQUIDs with junction area 5x5μm^2. Regular: designed values, italic: experimental values. * SQUID #8 was made on purpose without damping resistor.*

3. RESULTS AND DISCUSSION

3.1 McCumber parameter close to one: $\beta_c \to 1$

SQUIDs with small β_c value showed quite regular characteristics (see also 3.2). In contrast, $\beta_c \approx 1$ SQUIDs presented peculiar I-V and V-Φ curves. These anomalous characteristics could be also calculated back with the help of the junction simulation program JSIM (Fig. 3). We recorded very large values for the transfer function $\partial V/\partial\Phi$, as also reported by Polushkin et al (1999).

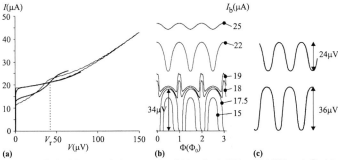

Figure 3 – (a) Measured (thick line) and JSIM simulated (thin line) IVCs of SQ#10 with β_c~1.2. (b) Simulated V-Φ curves at different bias points. (c) Measured V-Φ curves at $2I_0=18.8\mu A$, and at the resonance (I_b~$25\mu A$). The characteristic shapes of the curves at I_b~$18\mu A$ and $19\mu A$ were also observed.

The usual expression for the SQUID's transfer function $\partial V/\partial\Phi$ is linearly dependent on $\beta_c^{1/2}$:

$$\frac{\partial V}{\partial\Phi} = \frac{R_{sh}}{L_{SQ}}\left(\frac{2\beta_L}{1+\beta_L}\right) = \frac{1}{\beta_L}\left(\frac{2I_0}{\pi C\Phi_0}\right)^{1/2}\left(\frac{2\beta_L}{1+\beta_L}\right)\beta_c^{1/2}$$

This holds only for $\beta_c<<1$. When β_c increases, the measured values are much larger than the ones predicted by this formula and follow an exponential dependence (Fig. 4). The JSIM simulations suggest an even stronger exponential behavior.

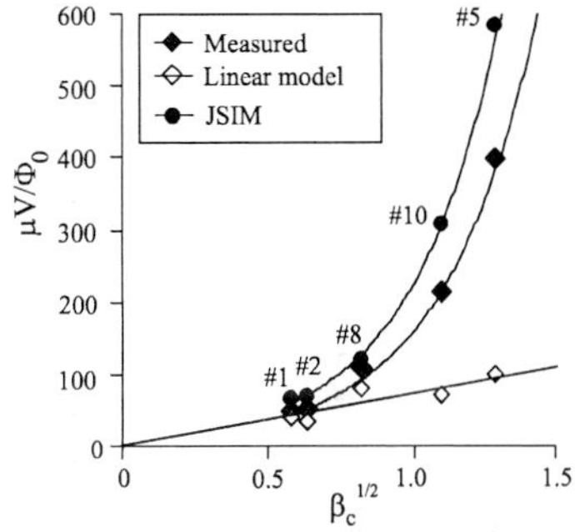

Figure 4 -- Transfer function dependence on $\beta_c^{1/2}$. (SQUIDs with A=5x5 μm^2).

3.2 Resonances

The following consequences were observed as the number of turns increased. Even in the absence of an input coil, resonances appear in the parallel tank circuit formed by the inductance of the SQUID and the series capacitance of the junctions. The resonant voltage V_r measured at the point where the 0 and $\Phi_0/2$ flux IVC's cross each other coincides with the expected value $V_r=\Phi_0/2\pi(L_{SQ}C/2)^{1/2}$, when the slit inductance is taken into account in L_{SQ} and a ~10% junction area reduction due to rounding is assumed. The damping resistor R_d shunting the washer did not seem to suppress resonances sufficiently. The measured IVC's resembled rather the non-damped JSIM curves. A single turn input coil on top of the washer does not affect appreciably the SQUID characteristics. As n becomes larger, additional resonance bumps appear in the IVC's, due to EM $\lambda/2$ standing wave resonances in the long transmission strip line formed by the input spiral coil and the washer (Fig. 5a). The frequency of the mth resonance is: $f_m=m(c/2l)(1/\varepsilon_r)^{1/2}$, where m is an integer, c is the speed of light, l is the input coil length and ε_r is the dielectric constant of the insulator between coil and washer, $\varepsilon_r(SiO_2)$~4. I.e. for SQUID #8 with a 20-turns input coil measuring l=18.6mm, one has: $f_m=m.4GHz$. Indeed resonance bumps coincide with m=5, 8, 11, 13, 14, 15 multiples. An RC filter shunting the input coil was unable to damp the microwave resonances.

436

3.3 Noise measurements

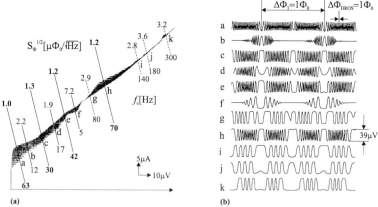

Figure 5– *(a) IVC of SQUID #7 (n=50). White noise (up) and corner frequencies (down). (b) V-Φ_{sig} curves of the two stage system at different bias currents (see text).*

Noise measurements of SQUIDs #1, 6, 7 and 8 were performed in a recently developed two stage configuration using a Double Relaxation Oscillation SQUID (DROS) with reference junction as second stage (Podt et al 1999). The bias current of the first stage $I_{b,1}$ was varied such that the noise could be measured at different operational points in the IVC's (Fig. 5a). The second stage was biased at $I_{b,2}=75\mu A$; at this point the modulation depth was $52\mu V$ and the noise of the DROS was $\sim6\mu\Phi_0/\sqrt{Hz}$. We comment the results for SQUID #7 with the largest number of turns (n=50). At the bias points labeled g, i, j, k the system flux gain $G_\Phi=\partial\Phi_{DROS}/\partial\Phi_{sig}$ is insufficient and the noise is dominated by the DROS. At the resonance points b, d, f the SQUID couples efficiently noise into the DROS, as proven by the two stage V-Φ_{sig} degradation (Fig. 5b). In consequence $\partial V/\partial\Phi_{DROS}$ decreases, $S_{\Phi,pre}^{1/2}\approx(1.8nV/\Phi_0)/(G_\Phi.\partial V/\partial\Phi_{DROS})$ becomes large and the system noise is limited by the electronics. At the bias points a, c, e, h, although relatively large $1/f$ noise is introduced, the white noise levels $1.0\text{-}1.3\mu\Phi_0/\sqrt{Hz}$ are very close to the theoretical value of $0.9\mu\Phi_0/\sqrt{Hz}$ for this SQUID.

4. CONCLUSIONS

A CCC system using one of the measured $\sim1.0\mu\Phi_0/\sqrt{Hz}$ SQUIDs (corresponding to an energy resolution $\varepsilon\sim100\hbar$) directly coupled to the CCC loop through a relatively small number of turns (n$\sim$15-30) would attain a current resolution of $\sim1pA/\sqrt{Hz}$. This represents about one order of magnitude improvement compared to existing systems. For the CCC application, the SQUID must be used in the very low frequency region. Therefore, next efforts will be focused on reducing the $1/f$ noise and/or applying modulation methods to shift the SQUID working point to the optimized white noise region.

REFERENCES

Adelerhof D J, Bijlsma M E, Fransen B P M, Weiman T, Flokstra J and Rogalla H 1993 Physica C **209**, 477

Ono I, Koch J A, Steinbach A, Huber M E and Cromar M W 1996 IEEE Trans. Appl. Supercond. **7**(2), 2538

Podt M, van Duuren M J, Hamster A W, Flokstra J and Rogalla H 1999 Appl. Phys. Lett. **75**(15), 2316

Polushkin V, Glowacka D, Hart R and Lumley J 1999 Rev. Sci. Instr. **70**(3), 2

Sesé J, Camon A, Rillo C and Rietveld G 1999 IEEE Trans. Instr. Meas. (to be published)

Zappe H H 1973 J. Appl. Phys. **44**, 1371

Inst. Phys. Conf. Ser. No 167
Paper presented at Applied Superconductivity, Spain, 14-17 September 1999
© 2000 IOP Publishing Ltd

High-T_c dc SQUID magnetometers in YBa$_2$Cu$_3$O$_{7-\delta}$ thin films with resistively shunted inductances

F Kahlmann[1], W E Booij[1], M G Blamire[1], P F McBrien[1], N H Peng[2], C Jeynes[2], E J Romans[3], C M Pegrum[3] and E J Tarte[1]

[1]IRC in Superconductivity, University of Cambridge, Madingley Road, Cambridge CB3 0HE, United Kingdom
[2]Surrey Centre for Research into Ion Beam Applications, School of Electronic Engineering, Information Technology and Mathematics, University of Surrey, Guildford GU2 5XH, United Kingdom
[3]University of Strathclyde, Department of Physics and Applied Physics, 107 Rottenrow East, Glasgow G4 0NG, United Kingdom

ABSTRACT: Previously, we have used Focused Electron Beam Irradiation (FEBI) to fabricate the resistors in resistively shunted high-T_c dc SQUIDS. In this study we demonstrate that Masked Ion Damage (MID) can successfully replace FEBI. This technique reduces the complexity of the resistor fabrication process considerably. Resistively shunted dc SQUID magnetometers based on bicrystal junctions with a 24° misorientation angle were fabricated using MID. Their inductances varied from 50pH to 200pH, and the measured maximum voltage modulation depths ΔV were in good agreement with theoretical predictions by Enpuku et al.

1. INTRODUCTION

Usually, an autonomous Superconducting Quantum Interference Device (SQUID) is very sensitive to changes in magnetic flux but is not initially a very sensitive magnetic field sensor due to its limited size. This is because the total SQUID inductance L, which increases with increasing SQUID size, degrades the maximum voltage modulation depth ΔV and therefore increases the magnetic flux noise $S_\Phi^{1/2}$ (Enpuku 1993). The design of a highly sensitive dc SQUID magnetometer is therefore usually a matter of finding the right balance between minimising its magnetic flux noise $S_\Phi^{1/2}$, which requires a small inductance L, and maximising its coupling efficiency to a magnetic pickup circuitry, which requires a large inductance L.

Enpuku and Doi (1994) have proposed a solution to this problem. They showed that in the case of low-T_c SQUIDs connecting a shunt resistor R_s in parallel with the SQUID inductance L compensates for the effect of a large SQUID inductance. It has also been proposed that this should not only be true in the case of low-T_c SQUIDs, but also for high-T_c SQUIDs.

The following approximate expression, obtained by fitting an analytical expression to numerical simulations, was given for the maximum voltage modulation depth ΔV:

$$\Delta V = \frac{4}{\pi} \frac{I_c R_N}{1+\beta_{eff}} \left[1 - 3.57 \frac{\sqrt{k_B TL}}{\Phi_0} \right] \qquad (1)$$

$$\beta_{eff} = \frac{\beta_L}{\sqrt{1+(\gamma\beta_L)^2}}$$

Here, I_c is the critical current per junction, R_N the normal resistance per junction, L the SQUID inductance, T the temperature, k_B the Boltzmann constant and Φ_0 the flux quantum. The parameter $\beta = 2LI_c/\Phi_0$ is the modulation parameter and $\gamma = R_N/R_s$ is the damping parameter.

Previously, we have reported the first high-T_c dc SQUIDs with resistively shunted inductances (Kang 1998, Tarte 1999), where the resistors were fabricated by Focused Electron Beam Irradiation (FEBI). It was shown that the resistive shunt leads to an increase of the maximum voltage modulation depth ΔV compared to an unshunted SQUID. Furthermore, the observed normalised maximum voltage modulation depth $\Delta V/I_c R_N$ could be very well described by the expression given in (1).

In this study we demonstrate that MID, instead of FEBI, can be successfully implemented to fabricate the resistors. The intention of replacing high-energy electron irradiation by this technique is to reduce the complexity of the resistor fabrication process, and thereby widen access to the fabrication of resistively shunted high-T_c SQUIDs in general.

2. FABRICATION PROCESS AND MAGNETOMETER DESIGN

The YBCO films used during the course of this work were all grown by off-axis pulsed laser deposition on 1×1cm^2 (100)-SrTiO$_3$ bicrystal substrates with a 24° misorientation angle. In the "as grown" state, they show a transition temperature $T_{c,0}$ of 90K and a critical current density j_c at 77K exceeding 2×10^6A/cm^2. The 200nm thick films were patterned by standard optical lithography and Ar$^+$ ion milling at 500eV on a water-cooled rotating sample stage.

The design used for our SQUID magnetometers is the one introduced by Lee et al. (1995). The square pickup loop has an outer dimension of 2mm and a width of 750μm. The SQUID loop itself is located in the middle of the pickup loop (see Fig. 1). The linewidth of the loop is 4μm. The strips are separated by a distance of 5μm, and the length is varied to get a total SQUID inductance L (kinetic plus geometric inductance) of 50, 100, 150 and 200pH. The junctions are realised by 2μm wide and 6μm long tracks crossing the bicrystal line.

The resistors were fabricated on 4μm wide resistor tracks by 100keV H$_2^+$ implantation through 1.5μm wide slits in a photoresist mask. The implantation dose was 7×10^{15} ions/cm^2, giving resistance values of about 13Ω. For further details of the resistor fabrication process by MID see Kahlmann 1999.

a)

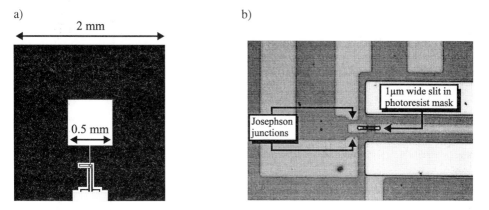

2 mm

0.5 mm

b)

1µm wide slit in photoresist mask

Josephson junctions

Fig. 1 Schematic drawing of the SQUID magnetometer design (a) and a photograph of the SQUID, also showing a 1µm wide slit in the photoresist mask across the resistor track (b).

3. RESULTS AND DISCUSSION

After the implantation, the I-V characteristics and the V-Φ curves of the resistively shunted dc SQUID magnetometers (fabricated all on the same substrate) were measured at 77K. The parameters derived from these measurements are summarised in Table 1. Examples of the measured V-Φ curves for the 200pH dc SQUID magnetometer for different bias currents are given in Fig. 2. It should be noted that, according to (1), the theoretically expected value of ΔV for this magnetometer in the unshunted case is only 2µV.

The effective flux capture area A_{eff} (i.e. the ratio of the flux quantum Φ_0 to the applied magnetic field $B_{\Phi 0}$ necessary to generate one flux quantum captured by the magnetometer, $A_{eff} = \Phi_0/B_{\Phi 0}$) of each SQUID magnetometer was measured using a Helmholtz coil. The results are shown in Fig. 4. We also calculated the effective area according to $A_{eff} = 0.637a[\text{mm}]L_{sl}[\text{nH}]\text{mm}^2$ (Ketchen 1985), where a is the outer dimension of the pickup loop (2mm in our case) and L_{sl} is the slit inductance of the SQUID loop. The measured and calculated effective areas are in very good agreement, as shown in Fig. 3.

Finally, we compare the normalised voltage modulation depth $\Delta V/I_c R_N$ of the SQUID magnetometers to the theoretical values based on (1), which were multiplied by a factor of 0.6 (see Fig. 4). This factor was introduced by Enpuku et al. (1995) to compensate for

L [pH]	I_c [µA]	R_N [Ω]	ΔV [µV]	$\gamma = R_N/R_s$	β_L
50	105	2.6	22	0.2	5
100	105	2.7	13	0.2	10
150	140	2.1	10	0.15	20
200	135	2.1	5	0.15	26

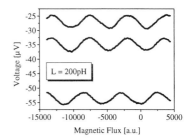

Tab. 1 Parameters of the resistively shunted dc SQUID magnetometers at 77K.

Fig. 2 Measured voltage vs. flux curves of the 200pH resistively shunted SQUID magnetometer.

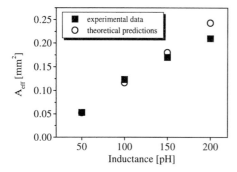

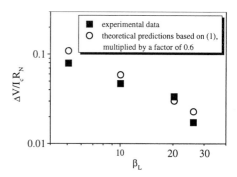

Fig. 3 Measured and calculated effective areas A_{eff} of the SQUID magnetometers.

Fig. 4 Comparison of the normalised voltage modulation vs. β_L for the SQUID magnetometers and the theoretical values based on (1), multiplied by a factor of 0.6.

discrepancy between theory and experiment, the origin of which is not yet fully understood. Using this "fudge" factor of 0.6, the theoretical values describe our experimental data well.

4.   CONCLUSIONS

In conclusion, we have demonstrated that Masked Ion Damage (MID) can be successfully implemented to fabricate the resistors in resistively shunted high-T_c dc SQUID magnetometers. This reduces the complexity of the resistor fabrication process, which was formerly based on Focused Electron Beam Irradiation (FEBI), and enhances the manufacturability of these devices.

REFERENCES

Enpuku K, Shimomura Y and Kisu T 1993 J. Appl. Phys. **73**, 7929

Enpuku K and Doi H 1994 J. Appl. Phys. **73**, 1856

Enpuku K, Tokita G, Maruo T and Minotani T 1995, J. Appl. Phys. **78**, 3498

Kahlmann F, Booij W E, Blamire M G, McBrien P F, Peng N H, Jeynes C and Tarte E J 1999, "Integrated resistors in YBa$_2$Cu$_3$O$_{7-\delta}$ thin films suitable for high-T_c SQUID magnetometers with resistively shunted inductances", this conference

Kang D J, Booij W E, Blamire M G and Tarte E J 1998 Appl. Phys. Lett. **73**, 3929

Ketchen M B, Gallagher W J, Kleinsasser A H, Murphy S and Clem J R 1985, in *SQUID'85 Superconducting Quantum Interference Devices and their Applications*, Hahlbohm H and Luebbig H Eds. Berlin, Springer-Verlag, p. 865-871

Lee L P, Longo J, Vinetskiy V and Cantor R 1995, Appl. Phys. Lett. **66**, 1539

Tarte E J, Kang D J, Booij W E, Coleman P D, Moya A, Baudenbacher F, Moon S H and Blamire M G 1999 IEEE Trans. Appl. Supercond. **9**, 4432

This work was supported by the UK Engineering and Physical Sciences Research Council.

HTS SQUID and cryogenically operated magnetoresistive sensors - an open competition?

C Dolabdjian, S Saez and D Robbes

GREYC-CNRS, UPRES-A 6072, ISMRA et Université de Caen
6, Bd. du maréchal Juin 14050 Caen Cedex, France

C Bettner, U Loreit and F Dettmann

Institut für Mikrostrukturtechnologie und Optoelektronik
Im Amtmann 6, D-35578, Wetzlar, Germany

G Kaiser and A Binneberg

Institut für Luft und Kaeltetechnik
Bertolt -Brecht-Allee 20, D-01309, Dresden, Germany

ABSTRACT: A comparison between an HTS dc SQUID ($\approx 1 \text{ cm}^2$ pick-up loop) and a magnetoresistive (MR) sensor with an additionally magnetic focuser is given. The latter device consists of an amorphous soft ferromagnetic strip 3 mm wide and 18 mm long, with a barber pole structured MR detector placed in its middle and operated in a simple half Wheatstone bridge configuration. The field amplification structure areas of both sensors have nearly the same size. The directly coupled SQUID has a direct white noise level of 86 fT/√Hz above 1 kHz. The MR performances were investigated both at 300 K and 77 K. Direct measurements performed at an effective temperature of 90 K (self heating) led to single MR half bridge resistor of 2*46 Ω at a bias voltage of 2*3.45 V, a maximal voltage sensitivity of 3876 V/T and a noise level of 192 fT/√Hz at frequencies above 100 kHz. All these measurements were made in a shielded environment. A reference Flux Gate Sensor magnetometer was simultaneously operated.

1. INTRODUCTION

Although high T_C Superconducting Quantum devices have proved most of their promises to detect very weak magnetic field variations, the question of their large development is still an issue. As a matter of fact, operating them in non-shielded conditions, with large disturbances at most of frequencies of the electromagnetic spectrum may lead engineers to choose some other sensitive magnetometers. This paper shows that cooled Magneto-resistive sensors combined with magnetic focuser (Dettmann 1999), the size of which is not larger than that of a standard directly coupled SQUID, is a possible candidate for application above 100 fT/√Hz and at high frequencies. This holds even at T = 300 K.

The paper is organized as follows: section 2 give the sensors descriptions, expected and measured sensitivities; in section 3 experimental set up and noise spectra are gathered. Section 4 concludes this paper.

2. SENSOR OVERVIEW

2.1 HTS dc SQUID

A classical HTS directly coupled dc SQUID (Fig. 1-a), made by NKT company, was used as a magnetometer chip. The focuser-loop is electrically coupled to the dc SQUID inductance (L_S). The SQUID had the following characteristics at 77K: L_S = 70 pH, critical current 2 I_C = 15 µA, normal resistance $R_N/2$ = 2 Ω and a maximum peak to peak SQUID voltage of 9 µV. The measured SQUID sensitivity is of 5.3 nT/ϕ_0, in close agreement with the theory of the directly coupled SQUID of Mastuda (1991), and depends on the size of the focuser:

$$S_{SQUID} = \frac{1.25\,\mu_0\phi_0}{d_{Ext}\,L_S}\ [T/\phi_0], \qquad (1)$$

where d_{Ext} is the concentrator external size (m), L_S the dc SQUID inductance (H), μ_0 the permeability of free space and ϕ_0 the flux quantum. Enpuku (1993) expresses the flux sensitivity at the optimal bias point of the dc SQUID by:

$$S_{\phi,SQUID}^{1/2}(f) = \frac{4\sqrt{k_B\,T R_N}}{\frac{7 R_N I_C}{\pi(1+\beta)}\left[1 - 3.57\sqrt{\frac{k_B\,T R_N}{\phi_0}}\right]}\ [\phi_0/\sqrt{Hz}] \quad (2)$$

where R_N and I_C are respectively the resistance (Ω) and the critical current (A) of one Josephson Junction, T is the operating temperature (K), k_B is the Boltzman constant and $\beta = 2L_S I_C/\phi_0$. The voltage to field transfer factor of the directly coupled SQUID and the theoretical field sensitivity are:

$$T_{SQUID} = \frac{\partial V}{\partial \phi}\frac{\phi_0}{S_{SQUID}} = \frac{2I_C R_N}{(1+\beta)}\frac{1}{S_{SQUID}}\ [V/T]\ (3),\ S_{B,SQUID}^{1/2}(f) = S_{\phi,SQUID}^{1/2}(f)S_{SQUID}\ [T/\sqrt{Hz}]\ (4).$$

A theoretical resolution of 50 fT/$\sqrt{Hz}$ can be expected at 77 K with a voltage to field transfer factor of 5300 V/T.

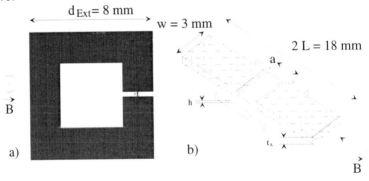

Figure 1: HTS dc SQUID a) and magnetoresistive b) sensors design ($\vec{B}$ is the applied external magnetic field)

2.2 Magnetoresistive Sensors

The design of the magnetoresistive (MR) is shown in Fig. 1-b. The active structure consists of a 24 nm thick (t) permalloy layer with a width of (w_{MR}) 80 µm and a total length of 2.4 mm prepared on top of a silicon substrate. It has a so-called barber pole structure in order to have maximum sensitivity in zero magnetic field. The magnetic flux concentrator of the device consists of two (t_A) 25 µm soft ferromagnetic foil antennas with a width of 3 mm (w), a length (L) of 9 mm. They form a gap with a width (a) of 100 µm in which the detector is placed. Finally, the substrate has step edges with a depth (h) of about 15 µm parallel to the detector strip. Then, external magnetic flux is concentrated by means of the soft ferromagnetic foil antennas and fed into the detector structure. The barber pole structure of the detector is designed to enable a device operation as a Wheatstone half bridge. The sample design and the magnetic parameters lead to the sensitivity of the device that can be expressed by the following equation presented by Dettmann (1999):

$$S_{MR} = \frac{1}{2} \frac{\frac{\Delta R}{R}}{H_x + H_k + \frac{t}{a-s} M_s} \left(\frac{\frac{L}{a}}{2\left(1 + \frac{L-a}{2t_A \mu_{eff}}\right)} + 1 \right) (V/V)/(A/m) \quad (5)$$

where $\Delta R/R$ is the magnitude of the maximum magnetoresistive change, H_x the magnetic field component in strip direction (A/m), H_k the anisotropic magnetic field (210 A/m), M_s the saturation magnetization (10^6 A/m), μ_{eff} the effective permeability and $s = a-w_{MR}$. μ_{eff} is given by the relative permeability of the antenna material ($\mu \approx 10^5$) and the antenna dimensions:

$$\mu_{eff} = \frac{\mu-1}{1+(\mu-1)\frac{2t_A}{L-a}} + 1 \quad (6).$$

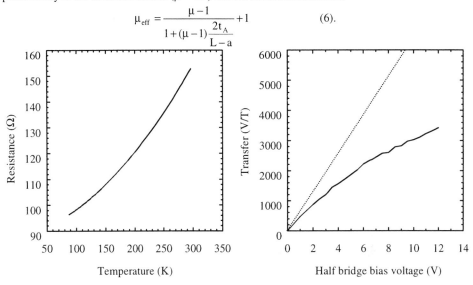

Figure 2: Measured half bridge resistance variation versus the temperature

Figure 3: MR transfer versus bias voltage around 90 K (dashed lined: theoretical transfer (calculated with Eq. 7) - dotted lined: Example of measured transfer)

Using equation (5), one can find that the sensitivity of the device is nearly proportional to the ratio of L/a. The voltage to field transfer coefficient of the magnetoresistive sensor, scaling with the voltage biasing $2*U_{bias}$ of the Wheatstone half bridge, is expressed by:

$$T_{MR} = \frac{S_{MR} U_{Bias}}{\mu_0} = 3.4 \; 10^4 \frac{\Delta R}{R} U_{Bias} [V/T] \quad (7).$$

The white noise of the device is determined by the Nyquist fluctuations. The corresponding resolution is given by the following relations:

$$S_{B.MR}^{1/2}(f) = \sqrt{4k_B \frac{R}{2} T} \frac{1}{T_{MR}} = \mu_0 \sqrt{\frac{2k_B T}{P}} \frac{1}{S_{MR}} [T/\sqrt{Hz}] \quad (8)$$

where k_B is the Boltzman constant, R the mean resistance of the half bridge (Ω), T the effective sample temperature (K) and P the power dissipation in the MR strip (W).

Using equation (8) one can derive that the resolution and the signal to noise ratio can be improved by increasing the power dissipation and by decreasing the device temperature. In order to investigate these facts experimentally measurements of the device sensitivity and noise were performed. The performance of the magnetoresistance sensors was tested, in the best case, at both temperatures 300 K and around 77 K. The figure 2 presents the mean half bridge resistance variation versus the temperature. The resistance value is 150 Ω at 300 K and decrease to 94 Ω at 77 K. Then, a maximal resolution of 140 $fT/\sqrt{Hz}$ and 50 $fT/\sqrt{Hz}$ can be expected, at respectively 300 K and 77 K, with theoretical transfer of around 7800 V/T, at a bias voltage of 2*10 V and a $\Delta R/R \approx 2.3$ %.

3.   NOISE MEASUREMENT

For a safe comparison, we have made the noise measurement of the dc SQUID and the magnetoresistance sensors mounted in the same dewar, surrounded by the same shielding environment (fours µmetal and two soft-iron cylinders). Further more, a low noise fluxgate is used to provide a reference peak signal (23 Hz frequency). The magnetic field was applied through precalibrated Helmoltz coils inside the cylinders. In the both case, we used an ultra low noise room temperature amplifier, noise performance of which are: $e_n = 320$ $pV/\sqrt{Hz}$, $i_n = 5$ $pA/\sqrt{Hz}$, 1/f noise corner around 0.5 Hz for low impedance device, gain 26000. The results are presented in figure 4.

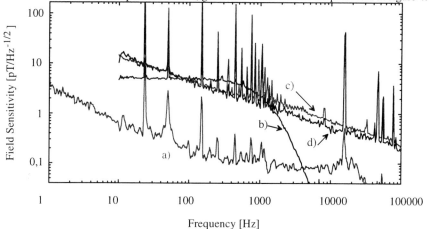

Figure 4: Measured magnetic noise a) SQUID at 77 K: $T_{SQUID} = 5300$ V/T, b) Fluxgate, c) MR at 300 K (bias voltage 2*3.65 V), d) MR at 90 K (self heating): bias voltage 2*3.45 V, $T_{MR} = 3876$ V/T

4.   CONCLUSION

These preliminary measurements show that the performance of MR, combined with magnetic focuser, can compete with SQUID systems for high frequency applications. The MR bandwidth is more than 1 MHz (Bettner 1999) and they are very simple to implement. 1/f noise limits currently their applications in low frequency, however some improvements are thinkable.

REFERENCES

Mastuda M, Murayama Y,Kiryu S, Kasai N, Kashiwaya S, Koyanagi M, Endo T, Kuriki S 1991 IEEE Trans. On Magnetics **27** pp 3043
Empuku K, Shimonura Y, Kisu T 1993 J. Appl.Phys. **73**(11) pp 7929
Dettmann F 1999 Proc. Magnetoresistive Sensoren V, eds Wetzlar pp 09
N.Smith N, Jeffers F,Freeman J 1991 J.Appl. Phys. **69** pp 5082
Bettner C, 1999 (Personal communication)

Inst. Phys. Conf. Ser. No 167
Paper presented at Applied Superconductivity, Spain, 14-17 September 1999
© 2000 IOP Publishing Ltd

HTS SQUID current comparator for ion beam measurements

L Hao, J C Gallop, J C Macfarlane* and C Carr*

Centre for Basic, Thermal and Length Metrology,
National Physical Laboratory, Teddington TW11 0LW, UK
*Dept. of Physics and Applied Physics, Strathclyde University, Glasgow G4 0NG, UK

ABSTRACT: Direct current comparators are unique superconductivity-based devices which allow extremely accurate direct current ratios to be established. Up to the present they have been used only with low temperature superconductors and SQUIDs. We have begun investigations into HTS current comparator designs, with a readout SQUID also made from HTS. The longer term aim of this work is to produce a system capable of high accuracy non-invasive measurement of an ion beam current (in the range 1μA to 1mA). Here we report preliminary measurements on the current ratio accuracy achievable with a simple HTS current comparator and present exploratory experimental results on the measurement of ion beams.

1. INTRODUCTION

The application of superconducting technology to the precise measurement of currents has been extensively developed in the liquid helium temperature regime [e.g. Harvey 1972; Sullivan and Dziuba 1974]. There, it is possible to exploit the properties of "user-friendly" materials such as indium, lead, niobium and alloys to manufacture optimally-wound toroidal transformers incorporating superconducting jointing and nested shields. The HTS materials available to date do not offer similar desirable features and the design of a superconducting current comparator is severely constrained by the inflexibility of these ceramics, and by the difficulty in realising superconducting connections between the various components. The original concept of the cryogenic current comparator was nevertheless based on the use of a straight cylindrical tube of superconductor [Harvey 1972] and it is certainly feasible to replicate this simple geometry with the HTS materials. In practice, there remain serious problems both in establishing sufficiently good coupling to the detector SQUID, and in minimising the errors due to end-effects and imperfect shielding. In this paper we outline our progress towards the realisation of a HTS current comparator, estimate its inherent accuracy limitations, and demonstrate its potential application for the measurement of charged-particle (e.g. Argon-ion) beams.

2. THEORETICAL MODEL

In the absence of a superconducting shield, currents which flow in both the axial and radial directions strongly affect the flux density at the SQUID. This is illustrated in Fig. 1 for the case of axially-flowing currents. The presence of a sufficiently long superconducting shield, which in this simple realisation is in the form of a cylindrical tube, ensures identical coupling between the current and the SQUID, located outside the tube, irrespective of the path taken by the current as it passes through the tube.

The current flow in an ion beam of finite width can be simulated by the simplified case of two wires carrying the same or opposite currents. One of the wires is displaced from the centre line to model a beam of finite width. In this way we can test the predicted insensitivity of the SQUID signal to the position of the wires, and by implication, to the width of the beam.

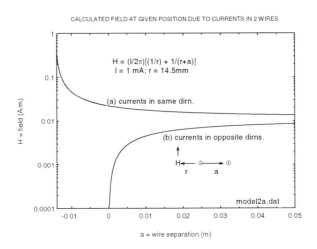

Fig. 1 Calculated field at the SQUID position due to a 1 mA current in two parallel wires separated by distance **a**, for currents in (a) the same and (b) opposite directions.

3. EXPERIMENT

3.1. Superconducting current comparator

The superconducting tube (75 mm long x 23 mm outer diameter) consists of a hollow cylinder of yttria-stabilised zirconia (YSZ) coated on both inner and outer surfaces by a thick film of YBCO, which is continuous around the ends of the tube. The flux detector is an HTS gradiometer SQUID, which is located outside, but as close as possible to, the mid-point of the tube. A flux concentrator consisting of a slotted mu-metal sleeve is tightly fitted around the YBCO tube. Coupling to longitudinal currents flowing in the tube wall is achieved by optimally locating the SQUID close to the slot, such that fringing fields couple additively to the two gradiometer loops. A second YBCO-coated tube 270 mm x 73 mm diameter enclosing the entire assembly acts as a magnetic shield against external fields.

The sensitivity of the detector to the total current flow and the degree of cancellation for equal and opposite currents were verified in a series of 2-wire tests as shown in Fig. 2. The unbalanced flux measured for ramped dc currents (Fig. 2(a)) and for 35 Hz alternating currents (Fig. 2(b)) was less than 1% of the reading recorded for a single wire carrying the same current.

Measurements of the unbalanced flux were also recorded as a function of the angular position of the displaced current, and again a maximum deviation <1% was found. By comparison with the unshielded situation (Fig. 1) it is clear that the superconducting tube effectively reduces the flux discrepancy due to two oppositely-flowing but radially-displaced currents by a factor ~100. Although the degree of balance is undoubtedly degraded in this experiment by end effects due to the relatively small length-to-diameter ratio of the present tube, it appears this simple geometry is adequate for further exploration of the ion-beam monitoring application described below.

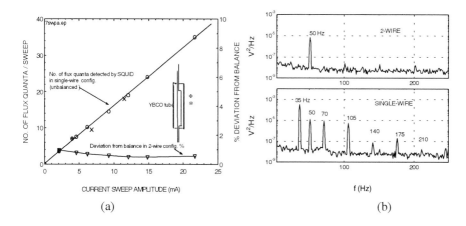

Fig. 2 Current comparator SQUID response in single wire and 2 wire configurations during application of (a) a dc current sweep and (b) a 35 Hz alternating current. Both the dc flux in (a) and the 35 Hz flux together with its harmonics in (b) are suppressed in amplitude in the 2-wire configuration by a factor ~100. The 50 Hz peak is due to mains interference.

3.2. Ion beam measurements

In order to test the feasibility of monitoring an ion beam with the HTS superconducting current comparator, a magnetron plasma discharge ion source was attached to an evacuated glass tube passing through the YBCO cylinder. Note that in principle the beam tube may remain at ambient temperature while it passes co-axially through the superconducting cylinder which is located within an annular nitrogen vessel. A wire was also threaded through the YBCO cylinder for calibration purposes. Argon ion beam currents in the range 0 - 10 μA were generated and the corresponding SQUID signal recorded. An example of the experimental data is shown in Fig.3. The range of adjustment of the ion beam current was limited by instabilities in the plasma source but from the data obtained it is clear that changes of the order 1μA in the beam current can be readily detected. No serious interference between the plasma source and the SQUID was observed; the gradiometric configuration of the detector undoubtedly contributed to its stable operation in a relatively noisy environment.

448

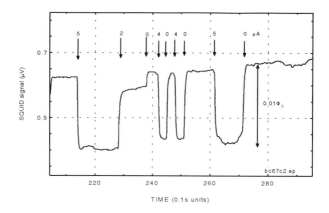

Fig. 3 Changes in SQUID signal corresponding to ion beam currents of 0, 2, 4 and 5 μA as indicated.

4. CONCLUSION

A high-temperature-superconductor cryogenic current comparator has been realised in a simple cylindrical geometry. The critical component, a thick-film coated YSZ tube, was combined with a gradiometric HTS SQUID and a high permeability flux concentrator which enhanced both the sensitivity of the device and its immunity to external magnetic disturbances. The linearity of the SQUID signal has been verified for currents between 1 μA and 25 mA. The degree of balance achieved for equal and opposite currents is better than 1%, a figure which can be improved by increasing the length:diameter ratio of the superconducting cylinder. The application of the current comparator for measurement of an argon ion beam has been successfully demonstrated in the range of beam currents 1 - 20 μA. Ongoing work will focus on reducing the influence of stray fields and end effects, and installing a more controllable ion source which will enable the technique to be extended to ion beams ranging from tens of nano-amps to several milliamps.

ACKNOWLEDGEMENT

The YBCO-coated YSZ cylinder was supplied by Dr Tim Button, University of Birmingham. The gradiometric SQUID was made by Dr Ed Romans, University of Strathclyde. Early trials of the comparator relied on HTS SQUIDs supplied by Dr Y. Shen , NKT (Denmark) and Dr Sven Wunderlich, University of Jena, Germany.

REFERENCES

Harvey IK (1972) Rev. Sci. Instrum. 43, pp. 1626-9.
Sullivan DB and Dziuba RF (1974) Rev. Sci. Instrum. **45**, pp. 517-519.

Inst. Phys. Conf. Ser. No 167
Paper presented at Applied Superconductivity, Spain, 14-17 September 1999
© 2000 IOP Publishing Ltd

An HTS SQUID magnetometer using coplanar resonator with vector reference for operation in unshielded environment

G Panaitov, Y Zhang, S G Wang*, N Wolters, L H Zhang**, R Otto, J Schubert, W Zander, H Soltner***, M Bick, H-J Krause and H Bousack

Institut für Schicht- und Ionentechnik, Forschungszentrum Jülich (FZJ), D 52425 Jülich, Germany
* Department of Physics, Peking University, Beijing 100871, China
**Institute of Physics, Chinese Academy of Sciences, Beijing 100080, China
*** Institut für Festkörperforschung, Forschungszentrum Jülich (FZJ), D 52425 Jülich, Germany

ABSTRACT: we developed an electronic first-order axial gradiometer using an HTS SQUID vector reference to cancel external disturbances. To improve the performance, low noise SQUIDs with coplanar resonators were used for the baseline SQUIDs of the gradiometer. The technique of adaptive disturbance cancellation and possible applications of the gradiometer in magnetocardiography and in non-destructive evaluation are discussed.

1. INTRODUCTION

High temperature superconductors (HTS) SQUIDs have achieved a remarkable magnetic field resolution in magnetic shielding. Many applications, however, require SQUID systems providing high-resolution measurements in unshielded environment. To reduce the interference of the environmental noise, electronic gradiometers are constructed by subtracting the outputs of independent SQUIDs. Great efforts have been undertaken to improve the balance, i.e., common mode rejection (CMR), of the gradiometer system. Tavrin *et al.* (1993) and Borgmann *et al.* (1997) applied mechanical adjustment techniques with a complex cardanic suspension or superconducting plates to tune the balances. In order to avoid such complex mechanical systems, He *et al.* (1999) developed a first-order SQUID gradiometer with a vector reference. All these HTS SQUID gradiometers were constructed on the basis of rf SQUIDs coupled to conventional rf tank circuits.

The development of the superconducting coplanar resonators has greatly improved the performance of our HTS rf SQUIDs. A magnetic field resolution better than 30 fT/√Hz has been achieved in white noise region for the rf SQUID magnetometer with a washer area of 8×8 mm² (Zhang *et al.* 1998). The present paper will describe the performance of the first-order vector reference gradiometer outside magnetic shielding based on these SQUID sensors. Applications of the gradiometer in magnetocardiography (MCG) and non-destructive evaluation (NDE) are presented.

2. SQUID GRADIOMETER SYSTEM

The fabrication of HTS rf SQUIDs, the principle of coplanar resonators, the coupling of rf SQUIDs to the resonator, as well as the readout electronics for rf SQUIDs were described by Zhang *et al.* (1998). We shall focus on the SQUID gradiometer system technique in unshielded environment.

We constructed an electronic first-order gradiometer by using an HTS SQUID vector reference. As the noise performance of gradiometer depends mainly on the quality of the z-oriented SQUIDs, highly sensitive SQUIDs with coplanar resonators were used as sensor (Z_0) and reference (Z_1) magnetometers in this direction. The parameters and performance of the magnetometers in a magnetic shielding are listed in Table I.

450

The distance between the z-SQUIDs (base line) is 7.5 cm. Two SQUIDs with washers of 3.5 mm in diameter coupled to conventional tank circuit were used as the x- and y- references. The performance of the x- and y- reference SQUIDs is similar as those used by He et al. (1999). The orthogonally mounted X- and Y-SQUIDs, together with the Z_1-magnetometer, represent a SQUID vector reference designed to cancel environmental disturbances.

The output signal of the gradiometer with vector reference is a linear combination of the four magnetometer outputs: Z_0 - α Z_1 - βX - γY. The balance parameters α, β, and γ were

Table I. Parameters and performance of SQUID magnetometer with integrated coplanar resonator in a magnetic shielding.

Parameter	Value
SQUID hole area (μm^2)	150×150
SQUID inductance (pH)	225
Washer area (mm^2)	8×8
Pumping frequency (MHz)	830
Field/flux coeff. $\partial B/\partial \Phi$ (nT/ Φ_0)	2.5
White field noise ($fT/\sqrt{Hz}$)	40
Slew rate (mT/s)	1.3
Dynamic range (μT)	> ±1
Total harmonic distortion (THD)	< 10^{-5}

adjusted by three potentiometers to minimise the gradiometer output during the balancing procedure. The balancing of the system was performed in homogeneous magnetic fields created by three pairs of orthogonal Helmholtz coils with diagonals of 2.5 m. For optimal balance, we achieved a CMR of more than 10^4 in z-direction and about 3×10^3 in the x- and y-directions. Details for the balance tuning were described by He et al. (1999).

In Fig. 1. traces (a)-(c) are the magnetic field background noise in one of our laboratories, measured by the magnetometers Z_0, X, and Y, respectively. The disturbances in the unshielded laboratory are high in the low frequency region, about 40 pT/√Hz at 10 Hz in z-direction (trace (a)). This low frequency noise is not only due to the fluctuation of the geomagnetic field but also to the influence of buildings and artificial sources. Usually, this sort of noise has a relatively stable level. The dominating disturbance is the power line with a frequency of 50 Hz, which has a peak-

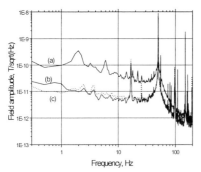

Fig.1. Spectra of the background field measured without shielding in z- (a), x- (b) and y-directions (c) at the FZJ laboratory.

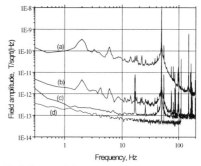

Fig.2. Spectra of the z-magnetometer without (a) and with shielding (d; the 'best balanced' (b) and 'best noise' adapted gradiometer (c).

to-peak value about 30 nT, and its harmonics. However, the amplitude and the waveform of the output signal are changing with time. The noise levels in x- and y-directions (traces (b) and (c)) are about one order of magnitude lower than that in the z-direction. This implies that for disturbance cancellation the z-reference is more important than those of other directions in our laboratory.

Fig. 2 shows the noise cancellation with different balance arrangements. Trace (c) shows the magnetic noise after the suppression of the background (a) with a vector-referenced gradiometer. In this noise measurement, the balance parameters were adjusted to obtain the lowest noise level. The disturbance rejection rate is about 500 at low frequencies, which results in a noise level of 120 fT/√Hz at 10 Hz and white noise level of 70 fT/√Hz. This may be limited by the noises of the two Z-SQUIDs. Trace (d) shows the noise of the Z_0-sensor measured in a magnetic shielding. According to the noise superposition principle, the expected limit of the system noise for a first-order gradiometer without shielding should be $\sqrt{2}$ times the Z_0 magnetometer contribution in shielding, provided that both Z SQUIDs have the same noise. Therefore, this limit can be regarded as the ultimate criterion for an optimally adjusted gradiometer in an unshielded environment. By comparison with the Z_0 spectrum (d) measured in shielding, the gradiometer system noise is found to be close to the theoretically expected limit.

It is important to notice that to obtain the 'best noise' trace (c), the balance parameters α, β and γ differ from those of the 'best balance', found for a homogeneous field in a Helmholtz coil. Fig. 2. clearly demonstrates the difference between the spectra measured with these two sets of balance parameters. One can see that the spectrum, trace (b), measured with the gradiometer optimally balanced in Helmholtz coils is substantially higher than the best noise curve, trace (c). The 'best balanced' gradiometer rejects only the homogeneous component of the background disturbances, (a). The uncompensated signal in trace (b) corresponds to the gradient component of external disturbances, which can be further suppressed, with an appropriate set of balance parameters, to reach the lowest noise level, (c). This adaptive compensation represents an advantage of the first-order electronic gradiometer over the conventional hardware gradiometers with gradient pick-up coils (Vrba 1996).

The function of the X- and Y-SQUIDs is only to improve the orthogonality of the system. As the reference parameters β and γ are normally less than 0.05, the contributions of the X- and Y-sensors to the system noise is negligible. Moreover, the background noise in x- and y-directions in our laboratory was about one order of magnitude lower than in z-direction (see Fig.1). In this case, the CMR for the horizontal directions better than 100 is enough to reduce the disturbances down to a few percent of the system noise.

3. MCG MEASUREMENTS

The typical MCG spectrum of a healthy human is dominated by the beat frequency of about 1 Hz and its harmonics below 40 Hz. The spectral density is about 5 to 10 pT in this frequency range, drops rapidly as the frequency increases, and becomes negligible above 100 Hz (Zhang *et al.* 1999). In our laboratory, the 50 Hz signal from the power line dominates the outputs from the magnetometer and gradiometer. Although the gradiometer suppresses the 50 Hz signal from about 30 nT_{p-p}, measured by magnetometer, to 500 pT_{p-p}, it is still larger than the QRS complex of the heart signal of usually about 50-100 pT. A low pass filter with a cut-off frequency of 30 Hz was employed to reject the disturbance. We tested filters of Bessel, Butterworth or Chebyshev types. They display different characteristics as to magnitude and impulse response over frequency. The Chebyshev filter has a sharp characteristic at the cut-off frequency, but it causes oscillations in the impulse response. We adopted the Bessel filter with a slower varying magnitude-frequency characteristic to avoid such a distortion of the MCG signal. Fig. 3 shows a real-time MCG trace for a healthy person, recorded in the system frequency bandwidth of 0.01 Hz – 30 Hz.

In order to obtain a signal with complete information of amplitude and phase from the components of the heart signal, the bandwidth of 30 Hz of the system is not sufficient. This may invoke the second or higher order gradiometer. We showed (Zhang *et al.* 1999) that the further reduction of the power line disturbance with a second order gradiometer makes it possible to carry out the real time measurement in a frequency band of 130 Hz.

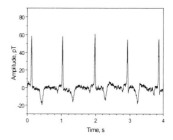

Fig. 3. Real time MCG recorded outside of magnetic shielding in a laboratory at FZJ.

452

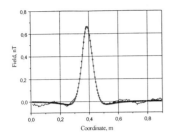

Fig. 4. Field profile of a ferromagnetic particle (line) and dipole fit (circles).

4. NDE MEASUREMENTS

Another possible application of gradiometry in NDE was demonstrated by measuring the signal of small ferromagnetic particles. The samples with small particles of different sizes and magnetic moments were placed on a scanning table and moved underneath the gradiometer sensor. The z-component of the magnetic field was recorded in a frequency bandwidth from dc up to 30 Hz. As the size of the particle is much smaller than the distance between the particle and the SQUID sensor, the model of the localised magnetic dipole can be applied for the particle data analysis. By fitting the field profile $H(x)$ measured for the particle, the amplitude of the magnetic moment P and the distance from the sensor to the particle z can be determined. Fig. 4. shows an example of the measured particle field profile (solid line). From the best dipole fit to the trace (circles), the magnetic moment of the particle was found to be about $2.6 \times 10^{-6} A \cdot m^2$, and the distance to particle about 87 mm, which is in a good agreement with the physical locations of the particle (~ 82 mm). The noise level of the trace determines the magnetic moment resolution to be $\sim 1.3 \times 10^{-7} A \cdot m^2$, at the distance of about 82 mm. By means of the proposed method, one can detect magnetic defects inside nonmagnetic materials.

5. SUMMARY

We have constructed a first-order gradiometer with a vector reference, using low noise rf SQUIDs with coplanar resonators. The system noise was reduced down to a white noise level of 70 fT/√Hz and 120 fT/√Hz at 10Hz. The x- and y-SQUID sensors, which were used to improve the orthogonality of the system, have an only minor contribution to the system noise.

To provide the lowest noise and the best disturbance cancellation, the balance parameters should be adapted to the real environment, rather than to the balance in the homogeneous field. The cancellation parameters should be chosen to minimise the noise in the frequency band actually needed for the measurement. The first order gradiometer is not able to provide a sufficient suppression of the 50 Hz disturbance, because the power line signal contains strong, time-depending spatial gradients.

We have demonstrated an MCG recording in unshielded environment. The data provided the information about low frequency components of the heart signal, since the frequency bandwidth was limited to 30 Hz. It was also demonstrated that the detection of magnetic defects inside a nonmagnetic material is a promising application of first-order gradiometer for non-destructive evaluation (NDE) outside magnetic shielding.

REFERENCES

Borgmann J, David P, Ockenfuss G, Otto R, Schubert J, Zander W and Braginski A I 1997 Rev. Sci. Instrum. **68**, 2730.

He D F, Krause H-J, Zhang Y, Bick M, Soltner H, Wolters N, Wolf W and Bousack H 1999 IEEE Trans. Appl. Supercond., **9**, 3684.

Tavrin Y, Zhang Y, Mück M, Braginski A I and Heiden C 1993 Appl. Phys. Lett. **62**, 1824.

Vrba J 1996 SQUID Sensors: Fundamentals, Fabrication and Applications, ed H Weinstock (Dordrecht: Kluwier Academic Publishers) pp 117-197

Zhang Y, Wolters N, Zeng X H, Schubert J, Zander W, Soltner H, Yi H R, Banzet M, Rüders F and Braginski A I 1998 Appl. Supercond. **6**, 385

Zhang Y, Panaitov G, Wang S G, Wolters N, Otto R, Schubert J, Zander W, Soltner H, H-J Krause and H Bousack 1999, submitted to Appl. Phys. Lett.

Inst. Phys. Conf. Ser. No 167
Paper presented at Applied Superconductivity, Spain, 14-17 September 1999
© 2000 IOP Publishing Ltd

A second-order HTS electronic gradiometer for nondestructive evaluation with room-temperature differencing

C Carr, EJ Romans, A Eulenburg, AJ Millar, GB Donaldson and CM Pegrum

Department of Physics and Applied Physics, University of Strathclyde, Glasgow G4 0NG, UK.

ABSTRACT: We describe a HTS second-order electronic gradiometer formed from two first-order HTS single layer gradiometers with room-temperature differencing. We report on the spatial response of the system, discuss the noise performance in an unshielded environment and present magnetic mapping results.

1.   INTRODUCTION

For SQUID applications in unshielded environments it is desirable to employ gradiometry to reduce the effect of unwanted magnetic fields. Recently 2nd-order gradiometers have been demonstrated in HTS made from either a single HTS layer (Lee 1998) or from a large single-layer flux transformer 'flip-chipped' to a directly coupled magnetometer (Kittel 1998). In the former the typically small substrate size limits the loop areas and the baseline, whereas in the latter very careful alignment is required. Here we have chosen to form a 2nd-order electronic gradiometer by differencing the room-temperature outputs of two 1st-order single layer gradiometers (SLGs) aligned in a planar arrangement.

We have previously reported on the performance of small HTS single layer gradiometers (SLGs) fabricated on $10 \times 10 \, \text{mm}^2$ bicrystal substrates (Carr 1998) and larger devices fabricated on $30 \times 10 \, \text{mm}^2$ bicrystals (Pegrum 1999). The smaller devices have a baseline b (defined as the distance between the centres of the pickup loops) of approximately 4.5 mm, which limits the gradient field sensitivity to $\sim 350 \, \text{fT/(cm} \sqrt{\text{Hz}})$ for the best devices. For the larger devices, $b \sim 13 \, \text{mm}$, and our best measured gradient field sensitivity is $50 \, \text{fT/(cm} \sqrt{\text{Hz}})$. We have made two separate second-order gradiometer systems: the first from two (reasonably well matched) small SLGs (Fig. 1a) and the second from two (reasonably well matched) larger SLGs (Fig. 1b). For the larger system the SQUID separation, s, was slightly larger than ideal as the devices had been previously encapsulated separately. Initial device testing was performed at 77 K with the cryostat inside one layer of mumetal shielding. The electrical and noise characteristics of the individual devices are given in Table 1.

TABLE 1: ELECTRICAL AND NOISE CHARACTERISTICS OF THE INDIVIDUAL SLGS IN THE TWO SYSTEMS

	junction	I_c (μA)	R_n (Ω)	ΔV (μV)	$\sqrt{S_\Phi}$ (1Hz) ($\mu\Phi_0/\sqrt{\text{Hz}}$)	$\sqrt{S_\Phi}$ (1kHz) ($\mu\Phi_0/\sqrt{\text{Hz}}$)
$10 \times 10 \text{mm}^2$ s=12mm	step edge	5	10	20	43	9
	step edge	20	6	16	17	10
$30 \times 10 \text{mm}^2$ s=40mm	24° bicrystal	8	6	5	40	23
	24° bicrystal	10	7	4	75	36

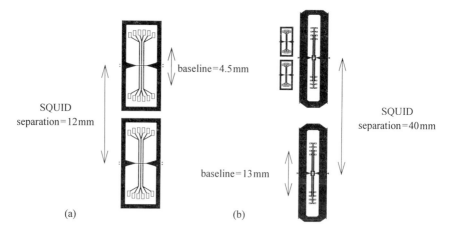

Fig. 1. Schematic representation of the two HTS electronic gradiometer systems, incorporating (a) two devices fabricated on $10\times10\,\text{mm}^2$ substrates, and (b) two devices fabricated on $30\times10\,\text{mm}^2$ substrates. In (b), the small gradiometer system has been shown at the same scale for comparison.

2.   ELECTRICAL CHARACTERISATION

The spatial gradient response of a gradiometer can be characterised using an electromagnetic source with a known response. In this work we have employed an "infinite" current-carrying wire, with a $1/r$ field dependence. The SLG will not perfectly reject uniform fields and has a small magnetometric response due to the presence of the SQUID at its centre: the intrinsic balance is ~1/100 for the small SLG and ~1/1000 for the large SLG. For our 2nd-order system with identical devices we would therefore expect an additional small 1st-order response (the difference in the 2 magnetometric signals).

The gradiometer systems were operated with each SLG in a separate flux-locked loop using Conductus PCI-100 electronics. The differencing was performed using a simple buffered op-amp circuit with one input weighted by an adjustable factor, λ, to allow for slight differences in the modulation/feedback coil couplings between the devices. Initially λ was set to the ratio of the measured transfer functions (V/Φ_0) for the 2 individual SLGs. This method of room-temperature differencing was chosen due to its simplicity and we are currently investigating software-based differencing. Fig. 2a shows the individual responses of the two small SLGs and also the electronically differenced response as the source is moved further away from the system in the plane of the substrates. Within experimental error, the individual SLG responses follow a $1/r^2$ dependence (r measured to centre of the SLG) and the differenced response follows a $1/r^3$ decay (r measured to centre of the system) for the range of standoffs investigated. These dependencies agree with the modelled responses, with the finite balance of each SLG having negligible effect in this standoff regime.

We made similar gradient response measurements for the larger system. But here to remove the need for a much larger current path, we instead used a different source configuration with two 1.25 m long wires carrying current in opposite directions. For the standoffs used and with a wire separation of 10 mm (in the same plane as the gradiometer), the field has a $1/r^2$ dependence (Kittel 1998). The individual, and differenced responses are shown in Fig. 2b. As expected we obtain a $1/r^3$ response for the individual SLGs and a $1/r^4$ response for the 2nd-order system.

In the above analysis we have ignored the fact that in practice the 2 individual SLGs will have slightly different magnetometric effective areas (expected from fabrication tolerances) and therefore the above system will have a small residual response to uniform fields. We measured the effective areas of each small SLG using a pair of 1.2 m diameter Helmholtz coils to which an ac signal was applied, with the resulting (flux-locked loop) SQUID output detected by a lock-in amplifier. The measured values for the two small SLGs were $820\,\mu\text{m}^2$ and $770\,\mu\text{m}^2$, with the electronically differenced signal $50\,\mu\text{m}^2$ as expected. This residual effective area is too small to have any measureable effect on the data in Fig. 2, but

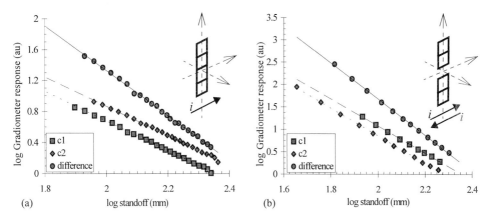

Fig. 2. (a) Gradient response of the small gradiometer system and its individual SLGs to a current-carrying wire. (b) Gradient response of the large gradiometer system and its individual SLGs to opposite-current-carrying wires in the plane of the gradiometer. The curves have been offset for clarity. For the individual curves the standoff is measured to the centre of the SLG; for difference the standoff is measured to the centre of the system.

in any case it can be effectively reduced to zero by slightly adjusting the value of λ.

The flux noise spectra of the small gradiometer system together with the noise spectra of the individual SLGs are shown in Fig. 3. The devices were operated in ac-bias mode with modulation at 256 kHz and bias reversal at 64 kHz. The value of λ was chosen to minimise the 50 Hz mains peak in the differenced signal, in this case resulting in a reduction compared to the SLG peak by a factor of ~18, with no increase in the white noise level above that expected from the two SLGs. Most of the reduction is attributable to the decrease in the residual magnetometric effective area, but clearly a proportion of the mains interference is non-uniform: we see the 50 Hz peak in our lab with highly balanced 2nd-order LTS gradiometers with gradiometric SQUIDs.

3. MAGNETIC MAPPING

To assess the performance of the $30\times10\,mm^2$ system in an unshielded environment, we constructed a turntable on which was placed pieces of paper containing laser-jet ink dots of various diameters. The stand-off between the sample and the lower SQUID was approximately 45 mm. The stepper motor was shielded by one layer of mumetal and was located directly below the gradiometers. Fig. 4 shows the response of the individual gradiometers and the differenced output to two 20 mm diameter

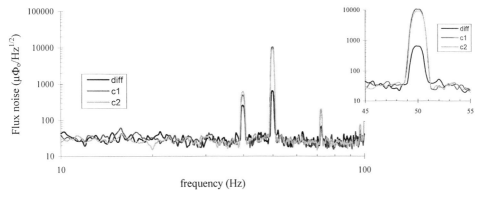

Fig. 3. Unshielded flux noise spectra of the small gradiometer system and the individual SLGs (not recorded simultaneously). The inset shows the region around the 50 Hz mains noise peak in more detail.

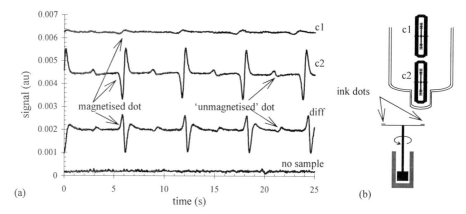

Fig. 4. (a) Response of the large gradiometer system to a magnetised dot and a nominally unmagnetised dot (both of diameter 20mm) rotated at ~17rpm below the gradiometer. For clarity, vertical offsets have been introduced into the curves which were not recorded simultaneously. A schematic of the experimental set-up is shown in (b).

dots, one of which was magnetised (achieved by moving a magnetised screwdriver over it for <1 sec) with the other nominally unmagnetised (although weak magnetisation from the Earth's field is visible). For comparison, the differenced output when only the motor is running with no sample is also shown. The devices were fairly stable and remained flux-locked for considerable periods in the open laboratory, despite the present fairly small $I_c R_n \sim 50$-$70\mu V$ of our larger devices.

4. CONCLUSIONS

In summary, we have shown that simple room-temperature differencing of the individual outputs of two single layer gradiometers results in a system that has the expected response of a 2nd-order electronic gradiometer. The devices have been demonstrated to be suitable for magnetic mapping applications in unshielded environments.

REFERENCES

Carr C, Eulenburg A, Romans EJ, Pegrum CM and Donaldson GB 1998 Supercond. Sci. Technol., **11**, 1317
Kittel A, Kouznetsov KA, McDermott R, Oh B and Clarke J 1998 Appl. Phys. Lett., **73**, 2197
Lee S-G, Hwang Y, Nam B-C, Kim J-T and Kim I-S 1998 Appl. Phys. Lett., **73**, 2345
Pegrum CM, Eulenburg A, Romans EJ, Carr C, Millar AJ and Donaldson GB 1999 to appear in Supercond. Sci. Technol.

Inst. Phys. Conf. Ser. No 167
Paper presented at Applied Superconductivity, Spain, 14-17 September 1999
© 2000 IOP Publishing Ltd

HTS RF SQUID planar gradiometer with long baseline for the inspection of aircraft wheels

M Maus, M Vaupel, H-J Krause, Y Zhang, H Bousack, R Kutzner, and R Wördenweber

Institut für Schicht- und Ionentechnik, Forschungszentrum Jülich, D-52425 Jülich, Germany

ABSTRACT: Recently, an automated aircraft wheel inspection system using a planar gradiometric HTS SQUID sensor was presented by Hohmann et al (1999). As the cracks to be detected are typically located 10 mm beneath the wheel surface, an enlargement of the baseline of the planar gradiometer to about 10 mm is needed for optimum performance. We report on the fabrication of planar gradiometers with enhanced baseline. The parasitic slit inductance was reduced by covering the slit with a superconducting ground plane. For a gradiometer with 5 mm baseline, a reduction of the white flux noise from 144 $\mu\phi_0/\sqrt{Hz}$ to 73 $\mu\phi_0/\sqrt{Hz}$ was observed when covering the slit. Narrowing the gradiometer slit using electron beam lithography proved less effective.

1. INTRODUCTION

The wheels of an aircraft are subject to enormous mechanical stress and heat during take-off and landing. The material fatigue causes hidden cracks inside the wheel, so they have to be checked thoroughly. It was shown by Hohmann et al (1999) that our HTS-SQUID wheel testing system can detect these cracks reliably. To find smaller cracks in a larger depth, a gradiometer with a higher baseline is necessary. However, due to increasing inductance, a longer baseline leads to increasing white noise, which eventually prevents operation of the SQUID in flux-locked loop mode. Thus it is necessary to limit the inductance when enhancing the baseline.

2. CALCULATION OF THE OPTIMUM BASELINE

In this chapter, the optimum baseline for aircraft wheel inspection is estimated. The magnetic signals to be detected are supposed to be generated by a magnetic dipole M. Two noise contributions are taken into account: the SQUID noise and the environmental noise. For simplicity, both are assumed to be white and independent of each other, as proposed by Garachtchenko et al (1999):

$$B_N(b) = \sqrt{\left[B_{hom}/CMRR + B_{grad} \cdot b\right]^2 + \left[b \cdot G_N \cdot f\right]^2} \tag{1}$$

The environmental noise consists of homogeneous magnetic disturbances B_{hom} and gradient disturbances B_{grad}. G_N denotes the gradient sensitivity, b the baseline, CMRR the common mode rejection ratio and f the frequency bandwidth. The SNR (signal-to-noise-ratio) is the quotient of the signal ΔB, measured by the gradiometer and the noise B_N (fig. 1a). The baseline, for which the SNR becomes maximum, can be taken as the optimum baseline.

Fig 1b shows the optimum baseline plotted over the dipole distance for three different values of the CMRR. The optimum baseline goes up with increasing dipole distance and with decreasing CMRR. For conditions as in the inspection of aircraft wheels (d ≤ 15 mm), a rule of thumb *optimum baseline ≈ dipole distance* is confirmed. For a typical dipole moment M = 10^{-7} Am^2, the optimum baseline is 6 mm for a gradiometer with CMRR = 1000 and 14 mm for CMRR = 100. Thus, the goal for improvement of our planar gradiometer is achieving a baseline of approximately 10 mm.

458

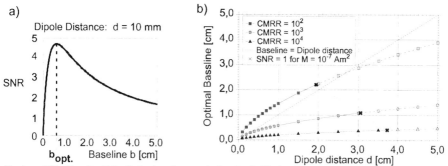

Fig. 1:   a) The Signal-to-Noise Ratio plotted across the baseline b. The baseline where SNR becomes maximum can be regarded to be the optimum baseline b_{opt}. b) The calculated optimum baseline b_{opt} as a function of the dipole distance, for different values of the CMRR, for a dipole moment $M = 10^{-7}$ Am^2. The part of the curves with SNR <1 are skeched dashed because they are experimentally not relevant.

3. THE PLANAR RF-SQUID GRADIOMETER

When using extremely sensitive magnetic field sensors such as SQUIDs for non-destructive evaluation, the challenge is to resolve small fields in the presence of large ambient fields. These environmental fields require a large dynamic range (maximum field amplitude) and a high slew rate (maximum change of field in time). Fig. 2 shows the principle and the layout of the double hole gradiometer, shown by Zhang et al (1997), which has been used by Hohmann et al (1999) for SQUID testing of aircraft wheels. Because only the difference of the shielding currents I_S crosses the junction, the demands to the slew rate and the dynamic range are reduced significantly, as compared to commonly used electronic SQUID gradiometers. The system can be operated in unshielded environment much more reliable, even in motion and with varying orientation.

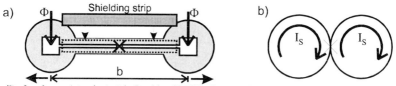

Fig. 2:   Layout (a) and principle (b) of the rf-SQUID planar gradiometer, shown by Zhang et al (1997).

A straightforward enhancement of the baseline of the gradiometer in this layout, however, enhances the area of the slit and thus the inductance of the SQUID L_S increases. One technique limiting the SQUID inductance is the introduction of a superconducting strip on top of the slit of the gradiometer. This strip shields the slit against the magnetic field and reduces the pick-up area of the SQUID and, correspondingly, the SQUID inductance.

4. CALCULATION OF THE SQUID INDUCTANCE

The effect of the shielding strip on the inductance of the planar gradiometer was calculated using the numerical simulation program of Chang (1981) for the calculation of cross section inductances. This approximation is valid because the length of the slit exceeds the other dimensions of the structure by far. The total inductance of the planar gradiometer SQUID is derived from:

$$L_S = \frac{L_{Loop} + L_{Slit}}{2} \tag{2}$$

using the formula of Jaycox and Ketchen (1981) to estimate the loop inductance $L_{Loop} = 126$ pH.

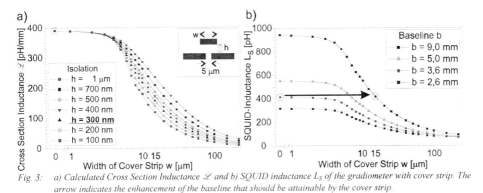

Fig. 3: a) Calculated Cross Section Inductance $\mathscr{L}$ and b) SQUID inductance L_S of the gradiometer with cover strip. The arrow indicates the enhancement of the baseline that should be attainable by the cover strip.

Fig. 3a shows the calculated cross section inductance $\mathscr{L}$ for the planar gradiometer being plotted against the width of the shielding strip w, for different values of the thickness of the isolation layer h. Without shielding, $\mathscr{L}$ is calculated to be 390 pH/mm. This corresponds to the value of 400 pH/mm calculated by Zhang et al (1997). For h = 300 nm, which corresponds to the thickness of the isolation layer being realized in the experiment, $\mathscr{L}$ is reduced by the shielding strip (w=15 µm) to 157 pH/mm.

Using (2), the SQUID inductance L_S was calculated. Fig 3b shows L_S plotted against the width of the shielding strip. A higher baseline causes a rising of the SQUID inductance L_S, whereas a wider shielding strip lowers L_S. For aircraft wheel inspection, currently a gradiometer with 3.6 mm baseline is used. From Fig. 3b, it is expected that it should be possible to enhance the baseline to 9 mm by introducing the covering strip without increasing the SQUID inductance, as indicated by the arrow.

5. FABRICATION

The SQUIDs were fabricated using standard thin film technology (fig. 4), with a step-edge junction and sputtered YBCO thin films. For fabrication of multilayer devices, it is important that the surface of the bottom layer is very smooth, as shown by Ockenfuß et al (1995). The structure is manufactured by photolithography and ion beam etching.

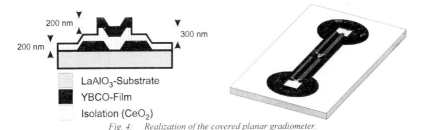

Fig. 4: Realization of the covered planar gradiometer.

6. CHARACTERIZATION

Fig. 5 shows the flux noise $S_\Phi^{1/2}$ of a gradiometer with and one without shielding strip. The level of the white noise decreases from 144 $\mu\Phi_0/\sqrt{Hz}$ to 73 $\mu\Phi_0/\sqrt{Hz}$, nearly by a factor of 2. Thus, the white noise is reduced by the shielding. From the calculation, it was expected to get a white noise lower than the noise of a 3.6 mm baseline gradiometer. A reason may be that the values for the short baseline gradiometers are reproducible best values of a very high number of samples (n>100).

The gradient-to-flux-coefficient rises from 15.1 nT/(Φ_0 cm) to 33.4 nT/(Φ_0 cm). This result confirms that the reduction of the white noise is actually caused by the shielding strip.

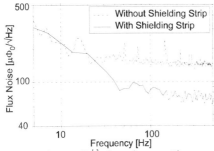

		b= 2.6mm	b= 3.6mm	b= 5.0mm	b= 5.0mm with strip
$S_\Phi^{1/2}$	$\left[\dfrac{\mu\phi_0}{\sqrt{Hz}}\right]$	35	65	**144**	**73**
$\dfrac{\partial^2 B}{\partial\Phi\,\partial x}$	$\left[\dfrac{nT}{\phi_0\ cm}\right]$	56	33	15.1	33.4
CMRR		>1000		664	10
# samples		>100		7	1

Fig. 5: Flux noise $S_\Phi^{1/2}$ of a gradiometer with cover strip and of an uncovered sample. The uncovered one has got a white noise of 144 $\mu\Phi_0/\sqrt{Hz}$, the covered one shows a reduction to 73 $\mu\Phi_0/\sqrt{Hz}$.

Tab. 1: Result of the electrical characterization. For comparison results for short baseline gradiometers are given.

The CMRR of an uncovered planar gradiometer is typically >1000 because of the good symmetry of the layout. However, we determined the CMRR of the gradiometer with cover strip to be only 10. This value of the CMRR is not sufficient for our application. On the one hand, the reduced CMRR can be attributed to the position of the shielding strip, which is displaced by about 0.9 μm. This leads to different large pickup areas of the gradiometer. On the other hand, the quality of the upper YBCO thin film is not homogeneous over the length of the strip. As shown by SEM-photos, the film at one side is very smooth, whereas the surface at the other side shows a rougher morphology, which may be attributed to some extent of a-axis-growth of the film. This finding was confirmed by x-ray measurements. Because of the different shielding effect on both sides of the gradiometer, the inhomogeneity contributes to the observed unbalance.

7. NARROWING THE GRADIOMETER SLIT

An alternative technique to reduce the gradiometer inductance is narrowing the gradiometer slit. A calculation using Chang (1981) yielded the same inductance for a 5 mm baseline gradiometer with 0.5 μm slit width as for our 3.6 mm baseline gradiometer with a 5 μm slit. Using electron beam lithography, a gradiometer with a 550 nm wide slit was prepared. The gradient-to-flux coefficient was measured to be 24.2 nT/(Φ_0 cm). Thus, the method works, however, the use of a shielding strip is more effective.

8. CONCLUSION AND OUTLOOK

A long baseline gradiometer with a magnetic shielding strip was successfully manufactured. The white noise is reduced by the cover strip by a factor of 2. The rising gradient-to-flux-coefficient $\partial^2 B/(\partial\Phi\ \partial x)$ shows that this is in fact based on a reduced pick-up area for the flux. The displacement of the strip as well as an inhomogeneity of the upper YBCO layer contribute to the observed unbalance of the gradiometer. The technological issue of improving the CMRR has to be solved before the gradiometers can be used in aircraft wheel inspection. Then, further enhancement of the baseline to 10 mm is straightforward.

ACKNOWLEDGMENT

Helpful discussions with P. David and A. I. Braginski are gratefully acknowledged.

REFERENCES

Chang W H 1981 IEEE Trans on Mag 17, 764.
Garachtchenko et al 1999 IEEE Trans Appl Supercond 9, 3676.
Hohmann R et al 1999 IEEE Trans Appl Supercond 9, 3801.
Jaycox J M, Ketchen M B 1981 IEEE Trans. on Mag. 17, 400.
Ockenfuß G et al 1995 Physica C 243, 24.
Zhang Y et al 1997 IEEE Trans Appl Supercond 7, 2866.

Inst. Phys. Conf. Ser. No 167
Paper presented at Applied Superconductivity, Spain, 14-17 September 1999
© *2000 IOP Publishing Ltd*

Direct measurement of vortex motion in Nb variable-thickness-bridges

S. Hirano[1], S. Kuriki[1], M. Matsuda[2], T. Morooka[3], S. Nakayama[3]

[1] Research Institute for Electronic Sciences, Hokkaido University, Sapporo,060-0812, Japan
[2] Muroran Institute of Technology, Muroran, 050-8585, Japan
[3] Seiko Instruments Inc., Matsudo, Chiba, 270-2222, Japan

ABSTRACT: We have measured flux noise characteristics of variable-thickness-bridges (VTBs) made on Nb thin films at various temperatures by the direct flux detection method. The short VTB fabricated using focused-ion-beam (FIB) on an epitaxial Nb film showed switching noise with minimum step height of 0.1 Φ_0. For a polycrystalline VTB, long time period observation showed a sequence of steps of different heights at fixed current bias. The distribution of the step heights was roughly approximated by the power law, which may be a signature of the self-organized criticality.

1. INTRODUCTION

It has been found that vortex systems in type-II bulk superconductors exhibit avalanches when slowly driven toward the threshold of instability, with the size distribution showing a power-law behavior which is considered as a signature of the self-organized criticality (SOC) (Bak et al 1987, Bak et al 1988, Bassler and Paczuski 1998). Such avalanches have been observed experimentally by slowly increasing the magnetic field applied to a thin superconducting tube (Field et al 1995). In our previous study, we investigated the flux noise characteristics of Nb variable-thickness-bridges (VTBs) by the direct flux detection method (Hirano et al 1999). It was suggested that the vortex dynamics in the VTBs were different depending on the quality of films. In this report, we studied the flux noise characteristics of a short VTB fabricated using focused-ion-beam (FIB) on an epitaxial Nb film and of a polycrystalline-film VTB in the long time period observation. The epitaxial VTB showed steady switching noise above the critical current. The polycrystalline VTB showed the flux noise with a step height distribution roughly approximated by a power law which may be the evidence of self-organized criticality.

2. EXPERIMENTAL

Epitaxial Nb films (thickness d = 100 nm) were deposited by electron beam evaporation on a polished sapphire substrate at elevated temperature. A microstrip was made by photolithography and Ar plasma etching. A VTB (length l = 100 nm, width w = 47 μm) was made using Micrion 2100 Focused Ion Beam system. Scanning electron microscopy (SEM) showed that the VTB had rounded V-shape cross-section, the width of which determined the bridge length. The critical temperature Tc of the VTB was 8.80 K. The polycrystalline Nb VTB used in this study was the same sample (l = 5.5 μm, w = 450 μm) as in the previous study, where the polycrystalline film was deposited at room temperature. The critical temperature Tc of the VTB was 5.89 K.

462

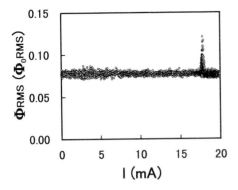

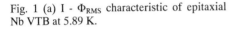

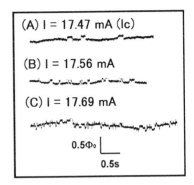

Fig. 1 (a) I - Φ_{RMS} characteristic of epitaxial Nb VTB at 5.89 K.

Fig. 1 (b) Real-time traces of the magnetic flux noise of epitaxial Nb VTB at bias currents of 17.47 mA (= Ic, A), 17.56 mA (B), and 17.69 mA (C).

The thin film VTB sample was mounted on a copper sample holder and suspended in a vacuum chamber immersed in liquid He. The temperature of the sample was controlled with a heater mounted on the sample holder. The thin film sample was set at a distance of 200 - 300 μm from the planar gradiometer of a Nb dc SQUID which detected the flux in the monopole approximation scheme. For measurements at 4.2 K, the sample and the dc SQUID/gradiometer were directly immersed in liquid He.

3.   RESULTS

Fig. 1 (a) shows the I - Φ_{RMS} characteristic of the epitaxial VTB at 5.89 K. Here, Φ_{RMS} was measured at a fixed bias current for a certain period of time, and subsequently the current was increased slowly. The critical current Ic (17.47 mA) was determined from the onset of Φ_{RMS}, where Φ_{RMS} varied in time at a fixed current at about Ic. At higher currents the Φ_{RMS} decreased to the intrinsic level, thus forming a peak within a narrow range of current. Fig. 1 (b) shows real-time traces at 5.89 K at the bias currents almost at the peak position in the I - Φ_{RMS}. At the Ic (A), random telegraph noise (RTN), i.e., switching among two discrete levels, was observed. With increasing current to 17.56 mA; (B), the RTN appeared more frequently. At 17.69 mA (C), switchings among more than two levels appeared. In these time traces, the minimum step height was equal to a flux change of 0.1 Φ_0. The maximum number of

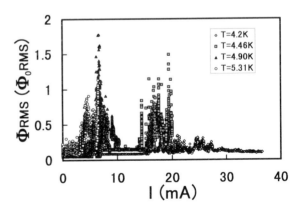

Fig. 2. I - Φ_{RMS} characteristic of polycrystalline Nb VTB at temperatures of 4.2 K, 4.46 K, 4.90 K, and 5.31K.

levels observed was 5.

For the polycrystalline VTB, the static I - V (not shown here) exhibited asymmetric voltage rises for repeated current biases of positive and negative currents. Fig. 2 shows the I - Φ_{RMS} characteristics observed at different temperatures from 4.2 to 5.31 K. At 4.90 K , the largest flux fluctuation was observed. At 5.31 K, fluctuations of Φ_{RMS} was also observed at zero bias current. In general, the number of peaks in the I - Φ_{RMS} was reduced and broader peaks appeared in narrower range of current as the temperature was increased, as seen from the figure.

Fig. 3 shows real-time traces of the flux noise taken over long time period at 4.46 K for different current values at Ic (14.03 mA), at the bottom between peaks (14.26 mA), and at a peak

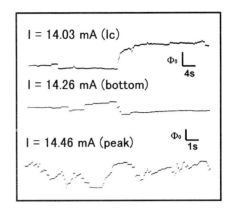

Fig. 3. Real-time traces of the magnetic flux noise of polycrystalline Nb VTB at 4.46 K.

of Φ_{RMS} (14.46 mA). Sudden appearance and disappearance of large fluctuations frequently occurred at widely distributed time intervals of 0.1 - 10 s, without changing the bias current at a fixed temperature. This observation is a clear contrast to the epitaxial VTB which showed steady switching noise. The minimum step height was 0.1 Φ_0.

4.   DISCUSSION

The single peak in the I - Φ_{RMS} characteristic and steady switching noise observed in the short epitaxial VTB, which is similar in the epitaxial VTB with larger length (l = 2.3 μm) in the previous study. It was suggested that the flux fluctuations were generated by the successive random entry of single vortices into the bridge (Hirano et al 1998, Hirano et al 1999). The fact that the obseved maximum number (five) of discrete levels in the switching noise suggests that at most 4 vortices entered simultaneously in the VTB.

Multiple peak structure in the I - Φ_{RMS} and asymmetric I - V in the polycrystalline VTB suggest existence of a large number of trapped vortices, anchored by distributed pinning potentials. Sudden appearance of the flux fluctuation at a fixed current may reflect that the vortex system, which experienced a repetition of pinning and depinning processes, is at the threshold of instability and very susceptible to a small perturbation, where vortices already in steady motion, which gave pulses outside the measurement bandwidth, trigger the subsequent correlated motion of many vortices as a bundle (Pla and Nori 1991). This may indicate the occurrence of vortex avalanche.

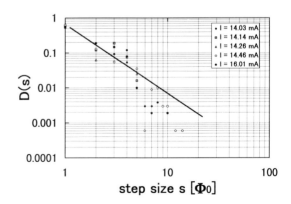

Fig. 4. The probability distribution D(s) of step heights s of the magnetic flux noise versus step height at various bias currents at 4.46 K. A straight line in the figure shows $1/s^2$ dependence.

Fig. 4 shows the distributions of step heights s of the flux noise at various bias currents obtained at 4.46 K, derived from the real-time traces such as in Fig. 3. Here, the D(s) represents the probability distribution of the occurrence of flux switchings of step height s during the observation period. The distributions were roughly approximated by the power law of $1/s^2$, as shown in the figure. Deviations from the power-law behavior at large s show cutoff step heights for the flux fluctuations, which may be due to the limited size, in particular, length of the VTB. The largest step height observed was 14 Φ_0. Beside this finite size effect, the obtained power-law suggests the evidence of the self-organized criticality exhibited by the vortex system in the VTB, driven by the transport current.

6. CONCLUSION

The flux noise characteristic of a short variable-thickness-bridge (VTB) fabricated with focused-ion-beam (FIB) technique on an epitaxial Nb film showed switching noise with a step height of 0.1 Φ_0 above the critical current. This steady switching noise may be generated by the successive entry of single vortices into the bridge. In contrast, the flux noise of a polycrystalline VTB at a fixed temperature showed switching noise with distributed step heights, where probability of the step height is approximated by the power law. The power-law distribution can be thought of a signature of self-organized criticality, exemplified by a dynamical system of vortices driven by the current.

ACKNOWLEDGMENT

The authors wish to thank H. Takahashi and T. Shibayama, Center for Advanced Research of Energy Technology, Hokkaido University, for their help in use of FIB system and Y. Nodasaka, School of Dentistry, Hokkaido University, for help in scanning electron microscopy (SEM). This work is supported by a Grant-in-Aid for Scientific Research on Priority Area "Vortex Electronics".

REFERENCES

Bak P, Tang C, and Weisenfeld K 1987 Phys. Rev. Lett. **59**, 381
Bak P, Tang C, and Weisenfeld K 1988 Phys. Rev. **A 38**, 364
Bassler K E and Paczuski M 1988 Phys. Rev. Lett. **81**, 3761
Field S, Witt J, Nori F, and Ling X 1995 Phys. Rev. Lett. **74**, 1206
Hirano S, Hirata Y, Matsuda M, Morooka T, Nakayama S, and Kuriki S 1998 Proc. 11th Int. Symp. on Superconductivity eds N. Koshizuka and S. Tajima (Springer-Verlag) pp 1205 - 1208
Hirano S, Hirata Y, Matsuda M, Morooka T, Nakayama S, and Kuriki S 1999 J. Appl. Phys. **85**, 7819
Pla O and Nori F 1991 Phys. Rev. Lett. **67**, 919

Inst. Phys. Conf. Ser. No 167
Paper presented at Applied Superconductivity, Spain, 14-17 September 1999
© *2000 IOP Publishing Ltd*

Thermal noise in digital dc SQUIDs

G H Chen, H Du and Q S Yang

Institute of Physics and Center for Condensed Matter Physics,
Chinese Academy of Sciences, Beijing 100080, China

ABSTRACT: The intrinsic noise expression of a digital dc SQUID can be obtained by means of numerical calculation. It shows that the thermal noise has temperature dependence of $T^{6/5}$ and is inversely proportional to the pulse frequency ω_b of the bias current.

1. INTRODUCTION

A digital dc SQUID containing two hysteretic Josephson junctions was firstly proposed by Drung (1986). After that, some new configurations of the SQUIDs have been invented (Fujimaki *et al.* 1988). These inventions provide a great convenience in circuit construction and operation. Nowadays digital SQUIDs as second stage in SQUID circuits can transform signals at low temperature from analog form into pulse one and make the SQUID operation with "on chip flux locked loop" possible (Fujimaki 1991; Yuh and Rylov 1995; Radparvar and Rylov 1995, 1997).

But can the devices be operated as a sensor with a resolution comparable to the ordinary SQUIDs? To answer this question the intrinsic noise in digital dc SQUIDs governed by thermal fluctuations is estimated in this paper.

2. CURRENT FLUCTUATIONS

2.1 Equations

Most digital dc SQUIDs consist of two hysteretic Josephson junctions and work under the condition of pulse current bias with pulse amplitude less than the critical current of the devices and the condition of an appropriate dc-flux bias.

In order to get a higher flux response and a larger open loop dynamical range, we set a flux bias at $\Phi_0/4$, taking the screening parameter $\beta \equiv 2I_c L/\Phi_0 = 1$, where Φ_0 is the quantum flux, I_c the critical current of the junctions and L the effective inductance of the SQUID loop. During the period of a bias pulse, a round way transition back and forth between zero and finite voltage states in the SQUID may take place due to thermal fluctuations. The transition probability depends not only on the noise level, but also on the wave form and amplitude of the bias current. For a certain wave form, say triangle wave with pulse frequency $f_b = \omega_b/2\pi$, we can adjust the pulse amplitude to make the transition probability 1/2. That means the average number of the SQUID output pulses per second is $f_b/2$.

For simplicity, we assume below that the SQUID is symmetrical and the noise is of Johnson type. Then equations for the two junctions in a dc SQUID can be written as

$$I_i = \frac{\hbar c}{2e}\ddot{\varphi}_i + \frac{\hbar}{2er}\dot{\varphi}_i + I_c \sin\varphi_i + \tilde{I}_i(t)$$

$$\langle\tilde{I}_i(t)\rangle = 0 \tag{1}$$

$$\langle\tilde{I}_i(t)\tilde{I}_j(t+t')\rangle = \frac{2k_B T}{r}\delta_{ij}(t') \qquad (i,j=1,2)$$

where $\tilde{I}_i(t)$ and $\tilde{I}_j(t+t')$ are the noise current in the ith junction at time t and that in the jth junction at time (t+t') respectively. All other notations in above equations have their ordinary meanings.

As it is well known that the thermal fluctuations in dc SQUID can be treated equivalently with total current noise $\tilde{I}_t$ and circulating current noise $\tilde{I}_s$ instead of $\tilde{I}_{1,2}$ via a transformation

$$\begin{aligned}\tilde{I}_t &= \tilde{I}_1 + \tilde{I}_2 \\ \tilde{I}_s &= (\tilde{I}_1 - \tilde{I}_2)/2\end{aligned} \tag{2}$$

2.2 The noise due to circulating current fluctuation

The flux noise spectral density of the SQUID due to circulating noise current can be written as

$$\langle\tilde{\Phi}_s^2\rangle = L^2\int \exp(-i\omega t')\langle\tilde{I}_s(t)\tilde{I}_s(t+t')\rangle dt' \tag{3}$$

From equation (1) we get $\langle\tilde{I}_s(t)\tilde{I}_s(t+t')\rangle = k_B T\delta(t')/r$. It gives a contribution to the energy resolution in connection with a white noise spectrum

$$\varepsilon_s = k_B TL/2r \tag{4}$$

Under the low damping condition, ε_s is negligibly small comparing even with quantum noise level for typical SQUID parameters at 4.2K.

2.3 The noise due to total current fluctuation

In dealing with the noise due to total current fluctuation the SQUID can be equivalently considered as a Josephson junction with normal resistance of r/2, junction capacitance of 2c and critical current I_c^s periodically modulated by the normalized effective flux $\phi = \Phi/\Phi_0$ The modulation can be described alternatively by the normalized external flux $\phi_{ex} = \Phi_{ex}/\Phi_0$ together with the parameter β, or by numerical simulation program developed by Fang (JSIM 2.1). The problem has been generally solved by De Bruyn Ouboter and De Waele (1970), Landman (1976) and others. The solution shows that the modulation depth in I_c^s is $\Delta I_c^s = I_c$ for $\beta = 1$.

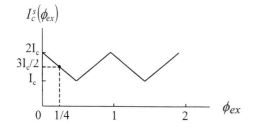

Fig. 1 Linear approximation of critical current modulation with $\beta = 1$; Flux biased at $\phi_{ex} = \phi_0/4$.

For simplicity we take the linear approximation with $\beta = 1$, which gives

$$I_c^s = 2I_c\left(1 - |\ \phi_{ex} - n\ |\right) \tag{5}$$

where n is the integer nearest to ϕ_{ex}.

In operation we flux bias the SQUID at $\Phi_0/4$, as shown in Fig.1, corresponding to a critical current of $I_c^s = 3I_c/2$, so as to get the highest unlocked dynamic range for the device.

In the thermal fluctuation regime, the transition from the zero voltage state to the finite voltage state in a Josephson tunnel junction can be equally treated as a problem for a particle escaping from a metastable state in a potential well. According to Büttiker, Harris and Landauer (1983) and Hänggi, Talkner and Borkovec (1990) the escape rate at low damping limit is given by

$$\tau^{-1} = a_t \frac{\omega}{2\pi}\exp\left(-\Delta U/k_B T\right) \qquad (\tau - lifetime) \tag{6}$$

The prefactor a_t is of order unity. The barrier height ΔU of the potential and the oscillation frequency ω of the particle near the bottom of the well in Eq. (6) are given by (Fulton and Dunkleberger 1974)

$$\begin{aligned} \omega &= \omega_p\left(1 - \alpha^2\right)^{1/4} \\ \Delta U &= \Delta U_0(1-\alpha)^{3/2} \end{aligned} \tag{7}$$

respectively, where ω_p is the plasma frequency (in the range of 10^{11} -10^{12} Hz) and $\Delta U_0 = 2\sqrt{2}I_c^s\phi_0/3\pi$ is the barrier height with the bias current $\alpha = I/I_c^s$ being zero. In the following calculation we simply take $\alpha_t = 1$ in Eq. (6), as given by the transition state theory (see Haggi et al. 1990).

We assume that the bias current is a pulse sequence of triangle wave with period $\tau_0 = 2\pi/\omega_b$ and amplitude α_0 less than one. A transition from zero voltage state to finite voltage state may take place during a pulse period. The probability for the SQUID remaining in its zero voltage state after a period time can be written as

$$W(\alpha_0) = \exp[-\int_0^{\tau_0}\tau^{-1}(\alpha)dt] \tag{8}$$

where $W(\alpha_0)$ is a function of the pulse amplitude α_0, the ratio between plasma frequency and the bias pulse frequency $\Omega = \omega_p/\omega_b$ and the ratio between the energy of Josephson coupling and that of thermal excitation $B = \Delta U_0/k_B T$.

The probability density of the transition is simply given by

$$P(\alpha_0) = -dW(\alpha_0)/d\alpha_0 \tag{9}$$

where $P(\alpha_0)$ is also a function of α_0, Ω and B.

In order to get the highest sensitivity, the device should be locked at the maximum point of $P(\alpha_0)$, $P^{max}(\alpha_0)$, correlated with $W(\alpha_0)$ near 1/2, which has a binomial distribution with root mean square value

$$\sigma_\omega = \sqrt{\frac{W(1-W)}{\omega_b/2\pi}} \tag{10}$$

with $W=1/2$, we have $\sigma_\omega = \sqrt{\pi/2\omega_b}$, corresponding to α_0 uncertainty $\sigma_{\alpha_0} = \sigma_\omega/P^{max}(\alpha_0)$, and further, to a flux noise $\sigma_\Phi = \sigma_{\alpha_0}|d\Phi_{ex}/d\alpha_0|$. At the flux bias point, we have $|d\Phi_{ex}/d\alpha_0| \approx 3\phi_0/4$.

468

Therefore

$$\sigma_\phi = \frac{3}{4}\sqrt{\frac{\pi}{2\omega_b}}\frac{\phi_0}{P^{\max}(\alpha'_0)} \tag{11}$$

The denominator $P^{\max}(\alpha'_0)$ as a function of B and Ω can be numerically calculated from Eq. (6), (7), (8) and (9). The calculation turns out that $P^{\max}(\alpha'_0)$ can be well fitted by

$$P^{\max}(\alpha'_0) \approx 2.2B^\delta \log_{10}\log_{10}\Omega \tag{12}$$

in a very wide range of Ω and B, where δ =0.612. For simplicity, we take δ =3/5 and neglect Ω dependence in Eq.(12) by taking $\Omega=10^3$ for all Ω. As a rough estimate, we get

$$P^{\max}(\alpha'_0) \approx B^{\frac{3}{5}} \tag{13}$$

with $\varepsilon_t = \sigma_\phi / 2L$,finally we get the energy resolution contributed by the total current fluctuation

$$\varepsilon_t = \frac{3^2 \pi\phi_0^2}{2^4 \omega_b L}\left(\frac{4\pi L k_B T}{3\phi_0^2}\right)^{6/5} \tag{14}$$

The term ε_t dominates the thermal noise in digital dc SQUIDs at 4.2K. The energy resolution of the devices could be reduced to a level below 100 times of Plank's constant above the bias frequency of 100MHz and for the typical inductance, say 100pH, of dc SQUIDs.

3.   CONCLUSION

The major contribution to the thermal noise in a digital dc SQUID comes from the total current fluctuation. The energy resolution of the device has ω_b^{-1} behavior with bias pulse frequency and $T^{6/5}$ temperature dependence. Our calculation turns out that digital dc SQUID can be used as a sensor with an intrinsic resolution within a hundred times of Plank's constant at a bias frequency higher than 100MHz.

REFERENCES

Drung D 1986 Cryogenics **26** 623

Fujimaki N, Tamura H, Imamura T and Hasuo S 1988 IEEE Trans. Electron Devices **35** 2412

Fujimaki N 1991 Fujitsu Sci. Tech. J. **27** 59

Yuh P F and Rylov S V 1995 IEEE Trans. Appl. Supercond. **5** 2129

Radparvar M and Rylov S 1995 IEEE Trans. Appl. Supercond. **5** 2142

Radparvar M and Rylov S 1997 IEEE Trans. Appl. Supercond. **7** 3682

Fang E S Simulation software: JSIM2.1 University of California, Berkeley, CA 94720, U.S.A.

De Bruyn Ouboter R and De Waele A T A M 1970 In *Progress in Low Temperature Physics*(Gorter C J Ed.) , North-Holland, Amsterdam, Chap. 6, pp. 243-290

Landman B S 1976 Appl. Supercon. Conference **76** 871

Büttiker M, Harris E P and Landauer R 1983 Phys. Rev. B **28** 1268

Hanggi P, Talkner P and Borkovec M 1990 Rev. Mod. Phys. **62** 251

Fulton T A and Dunkleberger L N 1974 Phys. Rev. B **9** 4760

Inst. Phys. Conf. Ser. No 167
Paper presented at Applied Superconductivity, Spain, 14-17 September 1999
© 2000 IOP Publishing Ltd

Simulations and experiments on HTS resistive SQUIDs

L Hao, J C Gallop, J C Macfarlane,* D A Peden* and C M Pegrum*

Centre for Basic and Thermal Metrology, National Physical Laboratory, Teddington, TW11 0LW, UK
*Dept. of Physics and Applied Physics, Strathclyde University, Glasgow G4 0NG, UK

ABSTRACT: The design and operation of HTS resistive SQUIDs for potential applications in Josephson noise thermometry above helium temperatures are reported. The effects of loop inductance on the operation of two types of R-SQUID, having a single or a double Josephson junction respectively, are modelled by mathematical simulation. Experimental measurements of Josephson radiation generated by the latter device are reported for the first time.

1. INTRODUCTION

Josephson noise thermometry has provided a method of primary thermometry in the temperature region below 1K for a number of years (Kamper and Zimmerman 1971). Following the discovery of the High Temperature Superconductor (HTS) materials, an HTS Josephson noise thermometer, which is expected to extend the temperature range of application of such primary thermometers to as high as 50K, has been proposed (Gallop and Petley 1995). An HTS Josephson junction shunted by a metallic resistor forms the resistive SQUID (R-SQUID) (Hao et al. 1998). With a fixed bias current applied to the R-SQUID, the Josephson radiation is broadened due to thermal noise in the shunt resistor R. A measurement of the linewidth $\Delta f = 4\pi k T R / \Phi_0^2$ in principle provides the absolute temperature. We have described elsewhere the design and fabrication of the YBCO thin film Josephson junction and the noble metal shunt resistor for the case of a single-junction device (Peden et al. 1999). Because the device is inductively coupled to a SQUID pre-amplifier located some 3-5mm from the device, the value of L must be ~1 nH for efficient coupling. Calculations of the circulating supercurrent in the single-junction SQUID indicate that if $\beta_L = 2\pi L I_c / \Phi_0 >> 1$ the output is reduced in amplitude and has a high harmonic content, both of which effects impede the detection of the Josephson oscillation and measurement of its linewidth. Consequently we are carrying out a comparative study of 2-junction HTS R-SQUIDs, similar in design to those which were originally demonstrated using niobium technology at liquid helium temperatures (Krivoy et al. 1995). Our mathematical simulations of these 2-junction R-SQUIDs indicate that their operation is much less affected by the magnitude of the loop inductance. In this paper we summarise results of numerical simulations of single-junction and 2-junction R-SQUIDs, and report first direct experimental observations of the heterodyne Josephson oscillation for the 2-junction HTS R-SQUID.

2. NUMERICAL SIMULATIONS OF RESISTIVE SQUIDS

2.1 Single-junction R-SQUID

The R-SQUID (Fig. 1) with critical current I_1, resistance R_j and capacitance C is shunted by a resistor R in a loop with inductance L. It is required that $R/R_j << 1$ (Hao et al 1998). The circuit can be expressed in terms of the Josephson phase θ by means of Equation (1):

$$-\frac{LC\Phi_0}{2\pi}\frac{d^3\theta}{dt^3} - \frac{\Phi_0}{2\pi}\left[\frac{L}{R_j} + RC\right]\frac{d^2\theta}{dt^2} - \left|\frac{\Phi_0}{2\pi}(1 + R/R_j) + LI_c\cos\theta\right|\frac{d\theta}{dt} + R(I_T - I_c\sin\theta) = 0$$

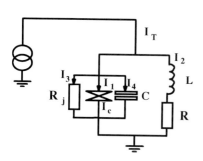

Fig. 1 Schematic diagram of single-junction R-SQUID

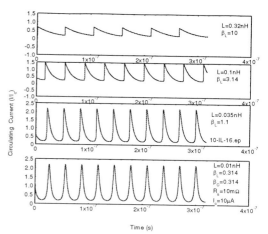

Time (s)

Fig. 2 The circulating current vs. time for single-junction R-SQUID. Inductance values range from 0.01nH to 0.32nH .

2.2 Results of the single-junction R-SQUID simulation

The observable quantity in the magnetically-coupled single junction R-SQUID is the time-dependent circulating current, rather than the junction voltage. Equation (1) was integrated numerically using the adaptive interval version of the basic Runge-Kutta method. In practice, HTS junctions have capacitance $C < 10$ pF, and we have found that so small a value has a negligible effect on the simulations. The simulations indicate that for low inductance ($\beta_L < 1$) the circulating current is a regular, smooth function of time (Fig. 2 lower graph). As the inductance is increased, the signal strength decreases until at $\beta_L \gg 1$ (L=0.32 nH) the Josephson oscillation appears as a very weak and distorted signal (see Fig. 2 top graph). These results were independently verified (Pegrum and McGregor 1999) using the J-SIM software (Fang and van Duzer 1997). It is consequently difficult both to achieve adequate coupling of the Josephson oscillation to the read-out SQUID, and to measure its linewidth accurately.

2.3 Two- junction R-SQUID

A two junction R-SQUID, is shown schematically in Fig.3. The bias currents are independently adjusted such that each junction oscillates at a relatively high Josephson frequency (10's of GHz), whereas the heterodyne (difference) frequency at a much lower value (kHz-MHz) is determined by the dc voltage across R.

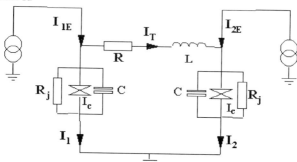

Fig.3 Circuit diagram of 2-junction R-SQUID showing shunt resistor R, loop inductance L, intrinsic junction resistance R_j and capacitance C.

We assume that both junctions have identical critical currents I_c and shunt resistance R_j. (Slight differences in a real device can be compensated by adjustment of the bias currents). The shunt capacitance C, as discussed above, can be neglected . The currents are related by the two equations:

$$I_1 = I_{1E} - I_T \; ; \; \text{and} \; I_2 = I_{2E} + I_T$$

The voltages are given by

$$V_1 = \frac{\Phi_0}{2\pi} \cdot \frac{d\theta_1}{dt} = R_j(I_1 - I_c \sin\theta_1) \; ; \qquad V_2 = \frac{\Phi_0}{2\pi} \cdot \frac{d\theta_2}{dt} = R_j(I_2 - I_c \sin\theta_2)$$

$$\text{It follows that} \quad V_1 - V_2 = \frac{\Phi_0}{2\pi} \cdot \frac{d(\theta_1 - \theta_2)}{dt} = RI_T + L\frac{dI_T}{dt}$$

2.4 Results of the two junction R-SQUID simulation

The first derivatives of each of the 5 unknown quantities are expressed in terms of the 5 (undifferentiated) variables I_1, I_2, I_T, θ_1 and θ_2 thus

$$\frac{dI_T}{dt} = -\left(\frac{R}{2L}\right)\left(I_{1E} - I_{2E} - I_1 + I_2\right) + \left(\frac{R_j}{L}\right)\left(I_1 - I_2 - I_c\left(\sin\theta_1 - \sin\theta_2\right)\right)$$

$$\frac{dI_1}{dt} = -\frac{dI_T}{dt} = -\frac{dI_2}{dt}$$

The results of the numerical simulation of these differential equations describing the circuit (fig. 3) are shown for a range of β_L values in fig. 4. Note that **even for very large β_L values the amplitude of the inductor current is not significantly reduced and its essentially sinusoidal character is maintained.**

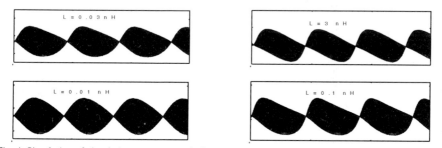

Fig. 4 Simulation of circulating supercurrent in 2-junction R-SQUID for inductance values 0.01, 0.03, 0.1 and 3 nH. The Josephson oscillation is modulated at a frequency corresponding to the voltage across R.

3. EXPERIMENT

A two junction R-SQUID of the type shown in Fig.3 has been fabricated using thin-film YBCO bi-crystal junctions. The resistor R is formed by a Au film 400 nm thick, which bridges a gap in the underlying YBCO film. Deposition and fabrication techniques were optimised such that resistor values of a few 10s of micro-ohms were achieved with negligible Au-YBCO interface resistance (Peden et al.1999). Currents I_{1e} and I_{2e} are applied from external stable current supplies such that the voltage difference (~20 nV) between the junctions determines the frequency of the heterodyne signal (~10 MHz) which is amplified and recorded. The Josephson linewidth can then be measured.

4. RESULTS

The device was installed in a helium-flow cryostat and carefully shielded against magnetic and radio-frequency interference. Its temperature was controlled within 10 mK between 19K-90K. A cryogenic co-axial cable coupled the R-SQUID to a room temperature amplifier. When both junctions were biassed slightly above their respective critical currents, Josephson oscillations were readily observed at heterodyne frequencies between 5 and 35 MHz, in the temperature range 19K - 74K. The oscillations were stable over periods of several hours, and could be reproduced following temperature cycling of the device. An example of the signal recorded on a spectrum analyser for the device temperature near 26K is shown in Fig.5.

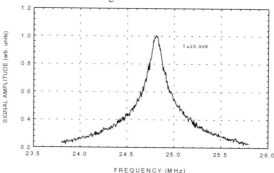

Fig. 5 Spectral recording of Josephson heterodyne oscillation generated by a 2-junction R-SQUID.

5. CONCLUSION

Mathematical simulations of the single-junction and 2-junction R-SQUIDs have been done for a range of inductance values. In the former type of device, large inductance values (L~ 1 nH) which are necessary for efficient coupling to a read-out SQUID cause severe attenuation and distortion of the Josephson signal. In the 2-junction R-SQUID, however, the heterodyne Josephson signal is not similarly affected. First experimental observations of heterodyne Josephson oscillations in a 2-junction HTS R-SQUID are reported in the temperature range 19K-74K. Ongoing work is directed at the measurement of the linewidth as a function of temperature and resistance, and accurate comparison with the theoretical dependence.

ACKNOWLEDGEMENT

Early samples of thin YBCO films and single-junction R-SQUIDs were kindly provided by Dr T. Schurig (PTB Berlin) and Dr Y.Q. Shen (NKT Denmark). 2-junction R-SQUIDs were prepared by Dr E. Romans (University of Strathclyde) .

REFERENCES

Fang, E S and van Duzer T (1997) J-SIM User Manual, Univ. of California, Berkeley, USA
Gallop J C. and Petley B W (1995) *IEEE Trans. Instrum. Meas.* **44**, pp.234-237.
Hao L, Gallop J C, Reed R P, Macfarlane J C and Romans E (1998) Applied
     Superconductivity, **5**, pp.297-301
Kamper R A and Zimmerman J E (1971) J. Appl. Phys. **42**, pp.132-136
Krivoy G S, Abmann C, Peters M and Koch H (1995) J. Low Temp. Phys. **99**, pp. 107-120
Peden D A, Macfarlane J C, Hao L, Reed R P and Gallop J C (1999) EUCAS'99 (submitted).
Pegrum C M and McGregor K 1999 (University of Strathclyde, unpublished Report).

Inst. Phys. Conf. Ser. No 167
Paper presented at Applied Superconductivity, Spain, 14-17 September 1999

473

Integrated resistors in $YBa_2Cu_3O_{7-\delta}$ thin films suitable for superconducting quantum interference devices with resistively shunted inductance

F Kahlmann[1], W E Booij[1], M G Blamire[1], P F McBrien[1], N H Peng[2], C Jeynes[2] and E J Tarte[1]

[1]IRC in Superconductivity, University of Cambridge, Madingley Road, Cambridge CB3 0HE, United Kingdom
[2]Surrey Centre for Research into Ion Beam Applications, School of Electronic Engineering, Information Technology and Mathematics, University of Surrey, Guildford GU2 5XH, United Kingdom

ABSTRACT: Previously, we have demonstrated that the performance of high-T_c dc SQUIDs can be significantly improved by resistively shunting their inductance. During the course of this former work the integrated resistors were fabricated by focused electron beam irradiation. In this study we demonstrate that the combination of ion implantation and conventional optical lithography can be successfully applied to get the same resistance values. This technique of masked ion damaging reduces the complexity of the resistor fabrication process considerably.

1. INTRODUCTION

The design of a dc SQUID is usually a compromise between minimising its magnetic flux noise $S_\Phi^{1/2}$ on the one side and maximising its coupling efficiency to a magnetic pickup circuitry on the other. While a large SQUID inductance L can significantly enhance the coupling efficiency, it also degrades the maximum voltage modulation depth ΔV and therefore increases the magnetic flux noise $S_\Phi^{1/2}$ (Enpuku 1993).

Enpuku and Doi (1994) have proposed a solution to this problem. They showed that connecting a shunt resistor R_s in parallel with the SQUID inductance L compensates for the effect of having a large SQUID inductance. It has also been proposed that this should not only be true in the case of low-T_c SQUIDs but also for high-T_c SQUIDs

Resistors are standard components in low-T_c superconducting circuits, but the same techniques cannot be used for the high-T_c superconducting oxides because of their surface chemistry. However, $YBa_2Cu_3O_{7-\delta}$ (YBCO) is very sensitive to disorder, and we have shown that using high-energy electron irradiation can form resistive regions. Furthermore, this technique was successfully applied to increase the maximum voltage modulation depth ΔV compared to an unshunted SQUID without changing its magnetic flux noise $S_\Phi^{1/2}$ (see Fig. 1, Kang 1998).

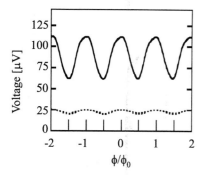

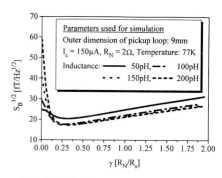

Fig. 1. Voltage modulation curves of an un-shunted bicrystal SQUID (broken line) and a resistively shunted one (solid line), both taken at a temperature of 77K.

Fig. 2. Calculated magnetic field sensitivity $S_B^{1/2}$ in the white noise region. Simulations based on the theoretical predictions by Enpuku et al. (1995).

In this study we demonstrate that the combination of ion implantation and conventional optical lithography can be successfully applied to get the same results. The intention of replacing high-energy electron irradiation by this technique is simply to reduce the complexity of the resistor fabrication process.

To determine the resistor values needed for optimum performance (i.e. the best magnetic field sensitivity $S_B^{1/2}$ in the white noise region), we have performed simulations based on the theoretical predictions by Enpuku et al. (1995). The parameters used in the simulation, shown above in Fig. 2, represent typical values for our SQUID magnetometers. The magnetometers are fabricated from 200nm thick YBCO films on 1cm x 1cm SrTiO$_3$ bicrystal substrates with a 24° misorientation angle. The square pickup loop has an outer side length of 9mm, and the junction tracks are 2μm wide (for further details, see Kahlmann 1999). The critical current I_c per junction is usually of the order of 150μA and the normal resistance R_N is about 2Ω. We find that for optimum performance $\gamma=R_N/R_s$ has to be approximately 0.25, so R_s should be about 8Ω.

2. FABRICATION PROCESS

The YBCO films used during the course of this work were all grown by off-axis pulsed laser deposition (Santiso 1998) on (100)-SrTiO$_3$ substrates. In the "as grown" state, they show a transition temperature $T_{c,0}$ of 90K and a critical current density j_c at 77K exceeding 2×10^6A/cm^2. The 200nm thick films were patterned into 4μm wide tracks using standard optical lithography and Ar$^+$ ion milling at 500eV on a water-cooled rotating sample stage.

On top of the patterned tracks, photoresist (AZ5214) with a thickness of 1.1μm was deposited and slits were defined by means of standard optical lithography. Finally, H$_2^+$ ions were implanted at an energy of 100keV. The implantation was carried out at room temperature and the plane of the substrate is tilted by 14° with respect to the incoming ion beam to avoid channelling effects. Given the low dissociation energy of H$_2^+$, the performed irradiation is expected to be indistinguishable from 50keV proton (H$^+$) irradiation. According to Monte Carlo simulations performed with the TRIM code (Biersack 1984), this should lead

a)

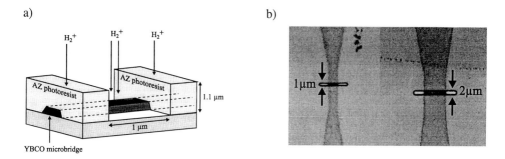

b)

Fig. 3. Schematic drawing of a 1μm wide slit in a resist mask across a YBCO microbridge (a) and photograph of two masked tracks (b).

to a homogeneous defect profile throughout the whole thickness of the YBCO films. The maximum range of the implanted ions within the AZ photoresist is about 550nm.

Fig. 3 shows a schematic drawing of a slit opened in a resist mask across a YBCO microbridge (a) as well as a photograph of two masked tracks (b).

3.   RESULTS AND DISCUSSION

Given the geometry of the resistor tracks (200nm film thickness, 4μm wide), the minimum resolution of the masking technique (1μm) and the desired resistance value of about 8Ω, a resistivity ρ of the irradiated YBCO at 77K of about 640μΩcm is required. Therefore, the $\rho(T)$ curves of 100μm long and 50μm wide tracks (unmasked) were measured after irradiation with different H_2^+ doses ranging from $6x10^{15}$ to $1.9x10^{16}$ ions/cm^2. The results are shown in Fig. 4. The transition temperature $T_{c,onset}$ of the irradiated material decreases approximately linearly with increasing dose, which is in agreement with other studies (Booij 1999, Kahlmann 1998). Complete suppression of T_c is achieved for an implantation dose of $1.8x10^{16}$ ions/cm^2. At this point, the $\rho(T)$ behaviour of the irradiated YBCO changes metallic to insulating. The resistivity exceeds 1600μΩcm, which is close to the maximum resistivity for metallic conductivity of 1500μΩcm, calculated according to the

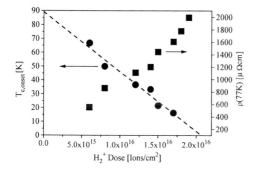

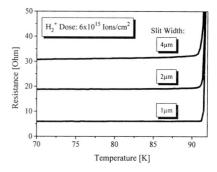

Fig. 4. $T_{c,onset}$ and ρ(77K) versus implantation dose. $T_{c,onset}$ versus dose fits very well to a linear dependence (broken line).

Fig. 5. Resistance versus temperature for three 4μm wide tracks, irradiated through slits with different widths.

Ioffe-Regel criterion (Tolpygo 1996). For an implantation dose of 6×10^{15} ions/cm^2 the resistivity at 77K of 550$\mu\Omega$cm is close to the target value.

Using this implantation dose, masked tracks with varying slit width were irradiated and their $\rho(T)$ curves were measured. Below the transition temperature of the unirradiated material (≈ 91K), the resistance of the tracks shows almost no variation with temperature around 77K (see Fig. 6). The resistance values at 77K are 6Ω, 19Ω and 31Ω for a slit width of 1μm, 2μm and 4μm, respectively. It is obvious that the resistance values do not scale linearly with the slit width. A possible explanation for this behaviour could be the lateral shape of the implantation profile but this needs some further investigation. The result in general makes clear that the resistivity value at 77K cannot be much further reduced because otherwise the transition temperature of the irradiated material would still exceed the operating temperature of 77K (see Fig. 6). Hence, a 1μm wide resolution of the masking technique is vital to get the desired resistance value.

4.   CONCLUSIONS

In summary, we have demonstrated that masked ion implantation can be successfully used to fabricate resistors in YBCO thin films. The combination of standard optical lithography with the implantation of H_2^+ ions at 100keV results in resistance values of a few Ohms at a temperature of 77K. Furthermore, the resistivity of the irradiated YBCO around 77K shows almost no variation with temperature. This property makes it more straightforward to achieve the desired resistance value.

The range of resistance values we obtained is exactly in the range required to optimise the performance of resistively shunted SQUID magnetometers. Further investigations will have to show if integrating these resistors into SQUID magnetometers leads to an improved magnetic field sensitivity of these devices.

REFERENCES

Biersack J P and Eckstein W G 1984 Appl. Phys. **A34**, 73

Booij W E, Elwell C A, Tarte E J, McBrien P F, Kahlmann F, Moore D F, Blamire M G, Peng N H and Jeynes C 1999, IEEE Trans. Appl. Supercond. **9**, 2886

Enpuku K, Shimomura Y and Kisu T 1993 J. Appl. Phys. **73**, 7929

Enpuku K and Doi H 1994 J. Appl. Phys. **73**, 1856

Enpuku K, Tokita G, Maruo T and Minotani T 1995, J. Appl. Phys. **78**, 3498

Kahlmann F 1998 PhD thesis, University of Cologne

Kahlmann F, Booij W E, Blamire M G, McBrien P F, Peng N H, Jeynes C, Romans E J, Pegrum C M and Tarte E J 1999, "High-T_c SQUID magnetometers in YBa$_2$Cu$_3$O$_{7-\delta}$ thin films with resistively shunted inductances", this conference

Kang D J, Booij W E, Blamire M G and Tarte E J 1998 Appl. Phys. Lett. **73**, 3929

Santiso J, Moya A and Baudenbacher F 1998, Supercond. Sci. Technol. **11**, 462

Tolpygo S K, Lin J-Y, Gurvitch M, Hou S Y and Phillips J M 1996 Phys. Rev. B **53**, 12462

This work was supported by the UK Engineering and Physical Sciences Research Council.

Inst. Phys. Conf. Ser. No 167
Paper presented at Applied Superconductivity, Spain, 14-17 September 1999
© 2000 IOP Publishing Ltd

Effect of the thermally activated phase slippage on characterizations of step-edge YBCO SQUIDs

H.E. Horng[1], S.Y. Yang[1], Chun-Hui Chen[1], J.T. Jeng[1], and H.C. Yang[2]

[1]Department of Physics, National Taiwan Normal University, Taipei 117, Taiwan.
[2]Department of Physics, National Taiwan University, Taipei 106, Taiwan.

ABSTRACT: In this report, we investigate the effect of the thermally activated phase slippage (TAPS) on the characterizations of a step-edge YBCO SQUID. It was observed that the V-I curves at temperatures from 10 K to 77 K show nonzero resistance with the bias current close to zero. This fact is attributed to the TAPS in the SQUID and reveals that TAPS is present over a considerable temperature range. Owing to the nonzero resistance of the SQUID, the voltage of the SQUID was found to be modulated by an external magnetic field when the bias current is nearly zero. And also, the bias current corresponding to the maximum V_{pp} in the V-Φ curve was observed to be smaller than that predicted from RSJ model. This deviation suggests that the TAPS reduces definitely the effective critical current of Josephson junctions. All the details will be discussed.

1. INTRODUCTION

Owing to the great potential in applications, the characterizations of superconducting Josephson junctions and SQUIDs have received increasing studies (Lukens 1970, Newbower 1972, Miklich 1995, Haller 1997 and Seidel 1997). Some researches indicated that the magnitude of the superconducting order parameter is reduced definitely at interfaces of the junctions because of the short coherence length of the high-T_c superconductors (Lukens 1970, Newbower 1972, Deutscher 1987). As a result, the disruption of the Josephson coupling caused by the thermal fluctuations becomes significant for high-T_c Josephson junctions. Ambergaokar and Halperin have given a detailed kinetic theory to the effect of the thermal fluctuations (Ambegaokar 1969). They concluded that a nonzero averaged voltage is attributed to the thermal fluctuation for a dc Josephson current. Gross et al. observed the nonzero voltages in the V-I curves for the $YBa_2Cu_3O_{7-y}$ grain-boundary single Josephson junction (Gross 1990). In addition, the contribution from the thermal fluctuations to the Shapiro steps in a YBCO single Josephson junction was investigated by Kautz et al. (Kautz 1992). According to these experimental results, the thermal activated voltage is present over a considerable temperature range.

Although the thermal fluctuations have been analyzed theoretically and found experimentally for the high-T_c single Josephson junction, the study of its influence on high-T_c SQUIDs is still rare. Therefore in this work, the effects of the thermal fluctuations on the characterizations of a step-edge $YBa_2Cu_3O_{7-y}$ SQUID were investigated.

2. EXPERIMENTAL DETAILS

The sample used in this work is a step-edge $YBa_2Cu_3O_{7-y}$ SQUID on a MgO(001) substrate. The step-edge substrate was fabricated by ion milling with the aid of the photoresist of AZ1518. The step height is 2000 Å and the step angle is 70°. Next, a 2000 Å-thick YBCO film was sputtered on the step-edge MgO substrate. Through a standard photolithography process, the film was patterned to be a SQUID with a hole area of 800 μm^2 and a junction width of 4 μm. Then gold was evaporated

478

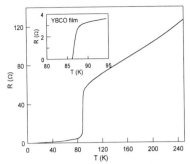

Fig. 1. The R-T curve of the step-edge YBCO SQUID under zero magnetic field. The inset shows the R-T curve for the YBCO film with a bias current being 1 μA.

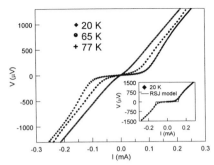

Fig. 2. The V-I curves of the step-edge YBCO SQUID at various temperatures. The V-I curve at 20 K fitted to the RSJ model is shown in the inset.

onto the contact points of the sample for good electric contact. A conventional four-probe method was used for the electric measurement. The temperature was detected by a diode thermometer and controlled by a temperature controller (LakeShore Model 330). To measure the magnetic field modulated voltage of the SQUID, the sample was subjected into a solenoid which provided a magnetic field perpendicular to the plane of the SQUID.

3. RESULTS AND DISCUSSION

In Fig. 1, the R-T curve is shown for the step-edge YBCO SQUID with the bias current 1 μA under zero magnetic field. This curve exhibits a very sharp drop at $T_{c,offset}$ = 88 K. This drop is due to the transition of the YBCO film forming the grain-boundary Josephson junctions. With lowering the temperature T, a broad foot structure was observed for the resistance transition. Furthermore, the resistance becomes zero at T = 10 K. The broadened resistive transition may be attributed to several causes, such as the poor quality of the YBCO film or the weak coupling between the two superconductors of the SQUID. To clarify this, the R-T curve of the YBCO film was measured and shown in the inset of Fig. 1. The resistance of the YBCO film vanishes definitely at $T_{c,zero}$ = 86 K. This result ensures a good quality for the YBCO film and indicates that the broadened resistive transition of the SQUID is not due to the weak coupling between the two superconductors of the SQUID to the broadened resistive transition, the voltage-current (V-I) curves of the fabricated SQUID were investigated at temperatures from 20 K to 77 K, as shown in Fig. 2.

The V-I curve at 20 K exhibited a RSJ-like behavior and was fitted to the RSJ model shown in the inset of Fig. 2. The critical current I_c and the normal resistance R_n used for the fitting curve are 0.12 mA and 5.9 Ω, respectively. A good match was found between the experimental data and the theoretical curve as the bias current $I > I_c$. However, as $I < I_c$, the experimental data deviates from the theoretical curve. And also, a nonzero resistance R_p was observed even at I close to zero. With raising temperature up to 77 K, the I_c was reduced while the R_n almost remained unchanged. Besides, the R_p became higher when the temperature was increased, as represented by the data points in Fig. 3. The measured R-T curve (denoted by the solid line) within the temperature range from 10 K to 80 K was also plotted in Fig. 3. It is worthy to note that the $R_p(T)$ data points coincides with the R-T curve. This implies that the broadened resistive transition in the R-T curve and the R_p's in the V-I curve are attributed to the same origin. To realize the origin, the dynamics of the phase difference φ of the order parameters of two superconductors forming the Josephson junction was investigated.

Under zero magnetic field, the SQUID acted as a single Josephson junction. The equation to describe the dynamics of the phase difference φ for a current-driven Josephson junction with critical

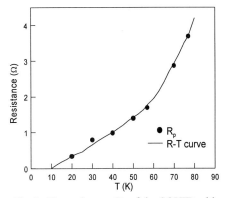

Fig. 3. The resistance R_p of the SQUID with bias current close to zero as a function of the temperature. The solid line denotes the R-T curve of the SQUID.

Fig. 4. Magnetic field modulated voltage oscillations for the YBCO SQUID under bias currents of (a) 10 μA and (b) nearly zero.

current I_c, normal resistance R_n and capacitance C is resulted from $d\varphi/dt = 2eV/\hbar$ and $CdV/dt = I - I_c\sin\varphi - V/R_n + L(t)$, here V denotes the voltage between the two superconductors, I represents the bias current and $L(t)$ stands for a fluctuating noise current induced by thermal noise. By combining these two equations, the equation of motion for the phase difference φ was obtained as $\hbar R_nCd^2\varphi/dt^2 + 2eR_nI_c\sin\varphi - 2eIR_n + \hbar\, d\varphi/dt - 2eR_nL(t) = 0$. This indicates that the movement of the phase difference φ is equivalent to the Brownian motion of a mass $M = (\hbar/2e)^2 C$ in a so-called tilted washboard potential $U = -E_J\cos\varphi - (\hbar I/2e)\varphi$, where $E_J (=\hbar I_c/2e)$ is the Josephson coupling energy. The first term of the potential U exhibits a periodic potential well with a period 2π for the phase difference φ. When the thermal energy k_BT is much lower than $2E_J$ under $I \approx 0$, the phase difference φ oscillates back and forth in a certain well with a frequency $(2eI_c/\hbar C)^{1/2}$. Thus, the time-averaged $<d\varphi/dt>$ is zero, where $<>$ denotes the time average. This leads to a zero time-averaged voltage $<V>$ for the junction bias with a dc Josephson current $(I < I_c)$ via the relation $<V> = (\hbar/2e)<d\varphi/dt>$. However, when the thermal energy is comparable with $2E_J$ and a small I is applied, the washboard potential becomes tilted and the phase difference can be activated thermally to escape from one potential minimum over the barrier to the next minimum. Thus, the phase difference slips randomly by 2π and the variation of the phase difference is no longer periodic. Hence, a nonzero voltage V can be observed for a Josephson current when the thermal energy is comparable with the Josephson coupling energy. Therefore, the thermal fluctuations, which cause the thermally activated phase slippage (TAPS), contribute to the nonzero resistance R_p in the V-I curves and also to the broadened resistive transition in the R-T curve of the SQUID.

Ambegaokar and Halperin (Ambegaokar 1969) indicated that the resistance R_p caused by TAPS under bias current I close to zero or much smaller than I_c is $R_p|_{I\to 0} = R_n[I_0(\gamma)]^{-2}$, where I_0 denotes the modified Bessel function of zero order and increases with $\gamma (=\hbar I_c/2ek_BT)$. Since

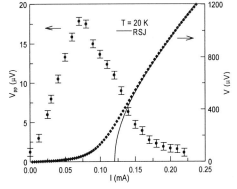

Fig. 5. The peak-to-peak voltage data with errors versus the bias current. The V-I curve (denoted by the rhombus) of the SQUID with the RSJ-fitted curve (denoted by the solid line) is also shown.

the I_c of the SQUID is modulated periodically by the external magnetic flux Φ, the R_p is varied with the applying Φ. Thus, the voltage V of the SQUID is predicted to oscillate periodically with Φ under a bias current close to zero via the relation $V(\Phi) = IR_p(\Phi)$. To check this, the V-Φ curve of the SQUID was measured at 20 K under a bias current 10 μA, which is much smaller than the I_c (= 0.12 mA) obtained from the RSJ model. The observed result was shown in Fig. 4(a). A clear oscillation was found for the voltage of the SQUID when the Φ was varied. This is consistent with our above prediction. The data in Fig. 4(b) were detected when the current source was disconnected from the SQUID. The only electronics, which may apply a current to the sample, connected to the SQUID is the picovoltmeter. By using a galvanometer with a resolution of picoamphere, the current provided by the picovoltmeter can not be sensed. This implies that the bias current is very close to zero. In Fig. 4(b), the V-Φ curve was found for the SQUID. This result gives a strong evidence that the TAPS can generate a magnetic modulation for the voltage of a SQUID bias with Josephson currents ($I < I_c$).

The V_{pp}'s in the V-Φ curves of the SQUID under various bias currents at 20 K were measured and represented by the points with errors in Fig. 5. It was found that the V_{pp} increases and reaches to a maximum value of 18 μV as raising I from nearly zero to 0.07 mA. The V_{pp} is then reduced when the bias current is further increased. According to the RSJ model, the bias current $I_{Vpp=max}$ corresponding to the maximum V_{pp} should be in the neighborhood of the critical current I_c of the SQUID. For comparison, the V-I curve (denoted by the rhombus) of the SQUID at 20 K is shown in Fig. 5 and fitted to the RSJ model (denoted by the solid line) with I_c being 0.12 mA. The experimental $I_{Vpp=max}$ (= 0.07 mA) of the SQUID is about one half of that (= 0.12 mA) obtained from the RSJ model. This suggests that TAPS in the junctions can reduce the effective critical current of the SQUID and thereby gives rise to the reduction in $I_{Vpp=max}$.

4. CONCLUSIONS

The thermally activated phase slippage (TAPS) of the Josephson junctions was observed for the step-edge YBCO SQUID over a considerable temperature range. According to the experimental results, the TAPS becomes dominant to the resistance transition of the grain-boundary Josephson junction. And also, the TAPS gives rise to the nonzero resistance R_p of the SQUID with bias current close to zero. Owing to the R_p, the voltage of the SQUID can be modulated by the external magnetic field under Josephson currents.

5. ACKNOWLEDGEMENT

This work was supported by the National Science Council of R.O.C. under grant Nos. NSC88-2112-M003-008, NSC88-2112-M003-009, NSC88-2112-M002-019 and NSC88-2112-M002-020.

REFERENCES

Ambegaokar V and Halperin B I 1969 Phys. Rev. Lett. **22** 1364
Deutscher D and Moller K A 1987 Phys. Rev. Lett. **59** 1745
Gross R, Chaudhari P, Dimos D, Gupta A and Koren G 1990 Phys. Rev. Lett. **64** 228
Haller A, Tavrin Y and Krause H J 1997 IEEE Trans. Appl. Supercond. **7** 2874
Kautz R L, Ono R H and Reintsema C D 1992 Appl. Phys. Lett. **61** 342
Lukens J E, Warburton R J and Webb W W 1970 Phys. Rev. Lett. **25** 1180
Miklich A H, Koelle D, Ludwig F, Nemeth D T, Dantsker E and Clarke J 1995 Appl. Phys. Lett. **66** 230
Newbower R S, Beasley M R and Tinkham M 1972 Phys. Rev. B **5** 864
Seidel P, Schmidl F, Weidl R, Brabetz S, Klemm F, Wunderlich S, Durrer L and Nowak H 1997 IEEE Trans. Appl. Supercond. **7** 3040

Inst. Phys. Conf. Ser. No 167
Paper presented at Applied Superconductivity, Spain, 14-17 September 1999
© 2000 IOP Publishing Ltd

YBa$_2$Cu$_3$O$_{7-\delta}$ dc SQUID array magnetometers and multichannel flip-chip current sensors

J Ramos, V Zakosarenko, R IJsselsteijn, R Stolz, V Schultze, A Chwala, H E Hoenig, and H-G Meyer

Institute for Physical High Technology, Dept. of Cryoelectronics, P. O. Box 100239, D-07702 Jena, Germany

ABSTRACT: We have prepared arrays of single layer high-T$_c$ dc SQUIDs on 10mm×10mm bicrystal substrates. SQUID arrays with field sensitivities of 85 nT/Φ$_0$ and 105 nT/Φ$_0$ have been prepared on separate chips. A field resolution of ≤2 pT Hz$^{-1/2}$ has been obtained for at least 80% of the devices for frequencies down to 1 Hz. By using multiturn YBa$_2$Cu$_3$O$_{7-\delta}$ (YBCO) input coils and mounting them in a flip-chip configuration we have built a three-channel current sensor. The current resolution of the devices is about 80 pA Hz$^{-1/2}$ down to 10 Hz.

1. INTRODUCTION

Multichannel systems with very sensitive dc SQUIDs magnetometers are of special interest in biomagnetism. On the other hand, dc SQUIDs with moderate resolution values, e.g. of the order of some pT Hz$^{-1/2}$ can already be used for applications in non-destructive evaluation (NDE). Additionally, the use of SQUID arrays can improve the spatial resolution in the investigated area. SQUID arrays can also be operated as current sensors, provided a multiturn input coil is inductively coupled to every SQUID in the array. Further, if every input coil of the SQUID array is connected to a superconducting or a normal-conducting pickup antenna, it would be possible to fabricate sensitive SQUID magnetometers or gradiometers with long baselines. Certainly the application of high-T$_c$ dc SQUID arrays can only be realized if a high yield of operating SQUIDs and multiturn input coils is warranted.

In this article we report on the preparation and characterisation of YBa$_2$Cu$_3$O$_{7-\delta}$ (YBCO) dc SQUID array magnetometers and a three-channel flip-chip current sensor. Compared to our previous work (Ramos et al 1998, 1999a) we have now integrated a feedback loop for every SQUID so that a separate operation of every channel is possible. We also show a possible application of the current sensors in geophysical measurements.

2. EXPERIMENT

Square washer SQUIDs with dimensions 400μm×400μm (referred in the following as SQW4) and 500μm×500μm (referred as SQW5) have been prepared for the experiments. They are prepared on SrTiO$_3$ (STO) bicrystals with a symmetric misorientation angle of 30° (Ramos et al 1998). The thickness of the YBCO layer is 200 nm. The junction width is 2 μm and the coupling hole is a 3μm×200μm slit for the SQW4 and 3μm×250μm for the SQW5 devices. The calculated SQUID inductance is equal to 80 pH for the former and 90 pH for the latter device. Each SQUID is equipped with one integrated feedback coil. We place 11 SQW4 and 9 SQW5 devices on

10mm×10mm substrates. The separation between the SQUIDs is 725 μm and 925 μm, respectively. The cross talk between the SQUIDs has been determined by driving a current through the feedback loop of one SQUID and measuring with the neighbouring SQUID.

For the experiments with the flip-chip current sensors we prepared five superconducting input coils with 19 turns each on one 10mm×10mm single crystalline STO substrate polished on both sides. The line width of the strips was 5 μm and the separation between them 3 μm. The calculated inductance L_{IN} was approximately 14 nH. The chip with the input coils was mounted in a flip-chip configuration with the SQUID arrays. The chips are fixed together with plastic clamps. The separation between the input coils and SQUID sensors is defined by a 10 μm thick mylar foil. The relative positioning of the chips was performed under an optical microscope within an accuracy of better than 20 μm.

The field sensitivity of the dc SQUIDs was measured by using a Helmholtz coil system. The noise characteristics were taken at 77 K, using a flux-modulated electronics with a 3 kHz bandwidth. The SQUIDs were immersed in a plastic dewar filled with liquid nitrogen. The dewar was placed in a magnetic shielding, which consists of three concentric μ-metal cylinders at room temperature.

3. RESULTS AND DISCUSSION

3.1 Dc SQUID array

The superconducting properties of several SQUID arrays have been investigated. The typical critical current density of the junctions, j_c, equals 0.5 - 1.5x10^4 A cm^{-2} at 77 K. The normal state resistance, R_N, of 2 μm wide junctions is typically 4 - 8 Ohms and the critical current - normal state resistance product of the junctions, $I_c R_N$, at 77 K is 180 - 330 μV. Peak-peak modulation voltage values larger than 15 μV are routinely obtained for all SQUIDs on a chip. The most relevant parameters of all SQUIDs in the arrays SQW4 and SQW5 are shown in Table 1. The spread in I_c is typically a factor of 3 over all SQUIDs on one chip. The mutual inductance has been measured by driving a current through the feedback loop and taking the voltage-flux characteristic (Ramos et al 1999b). The relative weak coupling between the SQUID and the feedback coil gives origin to cross talk between one feedback coil and a neighbouring SQUID of about 5 %. By operating every second SQUID in the array the cross talk was only about 0.8 %. These cross talk levels can be understood by roughly estimating the magnetic field of the feedback coil, considering it as the field of a point dipole.

SQUID	SQW4	SQW5
Washer area (μmxμm)	400x400	500x500
Calculated SQUID inductance (pH)	80	90
Mutual inductance of feedback loop (pH)	44	54
SQUID critical current (μA)		40 – 120
SQUID normal resistance (Ω)		2 – 4
Peak-peak SQUID voltage (μV)	> 20	> 15
Field sensitivity (nT/Φ_0)	105	85
Typ. Field resolution @ 1 kHz (pT Hz$^{-1/2}$)		< 1.8
Typ. Field resolution @ 1 Hz (pT Hz$^{-1/2}$)		< 2.1

Table 1. The main characteristics of two SQUID arrays.

The equivalent flux noise of two particular SQUIDs in different arrays is shown in Fig. 1. The different white noise levels are related to their different field sensitivities (see Tab. 1). The field resolution equals 1.5 pT Hz$^{-1/2}$ for both SQUIDs. The white noise level of the SQUIDs extends down to about 2 Hz. Field resolution values better than 3 pT Hz$^{-1/2}$ down to 1 Hz can be obtained on almost all SQUIDs in an array. An example of the field resolution for one 9- and one 11-SQUID array is shown in Fig. 2. The field resolution in the white noise region for all SQUIDs is less than 2 pT Hz$^{-1/2}$. The mean resolution value for all SQUIDs at 1 Hz is about 2 pT Hz$^{-1/2}$. However, the low frequency noise of some devices deviates considerably from the mean value as for SQUIDs number 4 and 8 in the SQW5 array (see Fig. 2). The excess noise at low frequencies for these two SQUIDs could arise, for instance, from microstructural defects in the SQUID washer.

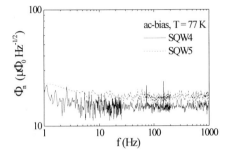

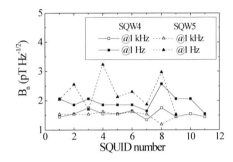

Fig. 1. The spectrum of the equivalent flux noise of SQUIDs in two different arrays. The white noise extends down to about 2 Hz.

Fig. 2. The field resolution of dc SQUIDs in two magnetometer arrays. Although the field sensitivity of both arrays is different, the resolution @1 kHz is very similar.

3.2 Multichannel flip-chip current sensor

The high yield in the fabrication of multiturn input coils (Ramos et al 1988) opens the possibility for preparing multichannel current sensors. In principle, our present concept offers the possibility of accessing up to 11 channels on one flip-chip device. However, due to the 80 % yield in the fabrication of the components only 6 - 7 channels could be realized. The cross talk requirements put a further constriction to the number of channels. We have presently build a three-channel current sensor aiming to obtain a cross talk level of about 0.1 %. Fig. 3 shows the output voltage of the three channels as a function of the current driven through the corresponding input coils. The modulation period of the SQUID voltage is about 3.5 μA/Φ_0. Thus, the mutual inductance between the SQUID and the input coil equals 570 pH. The peak-peak modulation voltage for all channels is > 20 μV. Fig. 4 shows the spectrum of the current resolution of one SQUID current sensor. It equals 80 pA Hz$^{-1/2}$ in the white noise region. On the other hand, the noise at 1 Hz is about 150 pA Hz$^{-1/2}$.

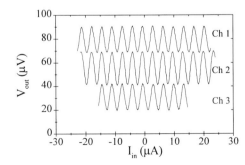

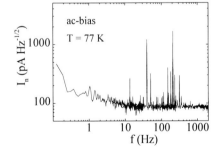

Fig. 3. The SQUID output voltage of a three-channel current sensor as a function of the current through the input coil.

Fig. 4. The current noise spectrum of a SQUID current sensor.

A possible application of these current sensors can be made in geophysical measurements. The so-called Radiomagnetic Sounding (RMS) method, developed by Drung et al (1997) aims at measuring the conductivity of the subsoil down to some 10 m depth. The method makes use of transmitter signals in the frequency range of 10 kHz - 1 MHz. For this purpose, a normal-conducting pickup coil is connected to the input coil of the current sensor. The lowest cut-off frequency of the measurement system, f_c, is given by the relation: $f_c = (1/2\pi)(R/L)$, where R is the resistance of the input-pickup coil circuit and L its total inductance. The response of such pickup circuit to an external magnetic field will be restricted by its resistance at low frequencies. At higher frequencies the current induced in the pickup circuit is proportional to the external magnetic field, i. e. the pickup circuit behaves like a superconducting loop. The highest working frequency is limited by the slew rate of the system, which is typically 1 - 1.5 MHz for small signals in our case. This is the frequency range of interest for the RMS measurements. It is certainly desirable to have the cut-off frequency as low as possible in order to increase the measurement frequency range. This can be achieved by decreasing the series resistance in the pickup circuit and by increasing the pickup loop inductance (e. g. increasing the loop dimensions). The series resistance arising from the gold pad-YBCO contact resistance and the aluminum bonding wires, R_S, has been measured for several input coils and found to be 10 - 60 mΩ. A considerable part of this value belongs to the bonding wires. The resistance of the bonding wires has been found to be 7 mΩ/mm at 77 K.

The cut-off frequency for a system with a pickup antenna can be now estimated. We consider a pickup antenna as a pair of circular loops with a radius r = 2 cm made of copper wire. If this wire is thick enough (radius a $\geq$ 0.5 mm), we could neglect its resistance at 77 K compared to the resistance of the bonding wires. The inductance of the loop can be calculated from the relation (Ramo et al 1965): $L_{PU} \cong \mu_0 r [\ln (r / a)]$. The total inductance of the pickup circuit including L_{IN} and some stray inductance will be L ~ 250 nH. Thus, the cut-off frequency equals 6 - 34 kHz for the given interval of R_S. These estimations, based on experimental data, show the possibility of building a high-T_c SQUID measurement system for exploiting the RMS method.

4.   CONCLUSIONS

We have fabricated single layer SQUID arrays designed for multichannel operation. The crosstalk between channels was 0.8 % when operating only 6 out of 11 channels prepared on a chip. A field resolution value of about 3 pT Hz$^{-1/2}$ down to 1 Hz has been measured for all magnetometers. A current resolution of 80 pA Hz$^{-1/2}$ in the white noise region has been obtained by operating the dc SQUIDs as current sensors. We have shown, that sensitive gradiometers operating at high frequencies can be built by connecting the input coil of the current sensors to a normal-conducting pickup antenna. A possible application in geophysics is given for the RMS method.

We acknowledge the help from K. Kandera, W. Morgenroth and B. Steinbach in the fabrication process. This work was supported by the German BMBF under contract 13N7111.

REFERENCES

Drung D, Radic T, Matz H, Koch H, Knappe S, Menkel S, and H Burkhardt 1997   *IEEE Trans. on Appl. Supercond.* 7 3283
Ramo S, Whinnery J R, and Van Duzer T 1965 *Fields and Waves in Communication Electronics* (New York, John Wiley & Sons, Inc) p 311
Ramos J, IJsselsteijn R, Stolz R, Zakosarenko V, Schultze V, Chwala A, Meyer H-G, and Hoenig H E 1998 *Supercond. Sci. Technol.* **11** 887
Ramos J, Chwala A, IJsselsteijn R, Stolz R, Zakosarenko V, Schultze V, Hoenig H E, and Meyer H-G 1999a ASC-98 (Palm Desert, CA, 1998) *IEEE Trans. on Appl. Supercond.* at press
Ramos J, Zakosarenko V, IJsselsteijn R, Stolz R, Schultze V, Chwala A, Hoenig H E, and Meyer H-G 1999b *Supercond. Sci. Technol.* **12** 597

Inst. Phys. Conf. Ser. No 167
Paper presented at Applied Superconductivity, Spain, 14-17 September 1999
© 2000 IOP Publishing Ltd

Smart DROS sensor with digital readout

M. Podt, D. Keizer, J. Flokstra and H. Rogalla

Low Temperature Division, Department of Applied Physics, University of Twente, P.O. Box 217, 7500 AE Enschede, The Netherlands

ABSTRACT: A 100 MHz prototype of a Double Relaxation Oscillation SQUID (DROS) with the complete flux locked loop circuitry on one single chip, the Smart DROS, has been realized. We measured the power spectral density and the time sequence of the output pulses. The concept of a Smart DROS with digital readout has the potential for a high measurement bandwidth together with a slew rate up to 10^8 Φ_0/s. In this paper, the properties of the Smart DROS will be discussed.

1. INTRODUCTION

The flux locked loop (FLL) electronics used to linearize the output of a SQUID is conventionally implemented with semiconductor electronics at room temperature. Maximum slew rates that can be achieved with conventional dc SQUIDs based on flux modulated FLL readout electronics are of the order of 10^6 Φ_0/s (Koch 1996). Partly motivated by the need to increase the slew rate for the readout of cryogenic particle detectors, e.g. based on superconducting tunnel junctions or superconducting transition-edge microcalorimeters, digital SQUIDs are being developed. In such digital SQUIDs, the sensing SQUID and the complete feedback circuitry can be integrated on one single chip (Radparvar 1994, Fath 1997). As will be discussed in this paper, our Smart DROS allows slew rates up to 10^8 Φ_0/s, about two orders of magnitude larger than the maximum slew rate that can be achieved with conventional dc SQUID systems.

Another main advantage of the digital SQUID concept with on-chip feedback circuitry is that in case of multi-channel systems a cryogenic multiplexer can be used for the readout of several SQUIDs, which greatly reduces the number of wires and simplifies room temperature electronics. Therefore, together with its high slew rate, the digital SQUID is a promising candidate for the readout of particle detector arrays as for example used in x-ray spectroscopy.

2. OPERATION PRINCIPLE OF THE SMART DROS

At the Fujitsu laboratories, Fujimaki (1988) developed a single-chip digital SQUID based on a comparator and an up-down counter. The comparator is a hysteretic dc SQUID biased with a bipolar alternating clock current, such that the output of the comparator is a pulse train of voltage pulses. In our concept of the low T_c digital SQUID, we use a Double Relaxation Oscillation SQUID (DROS) as the comparator. The main advantage of using a DROS (Adelerhof 1995) is that it does not require an external clock, since its relaxation oscillations generate an on-chip clock signal. This enables smooth operation at high clock frequencies.

The output pulses of the DROS serve as the input of a digital counter, which has the role of the integrator of a conventional FLL. The output signal of this up-down counter is fed back to the signal SQUID of the DROS, supplying the feedback flux Φ_{fb}, as is shown in Fig. 1. During each relaxation oscillation cycle, the critical currents of both hysteretic SQUIDs of the DROS, the signal SQUID and the reference SQUID, are compared with each other. Only the SQUID with the smallest critical current I_c will participate in the relaxation oscillation and a voltage pulse will appear across it,

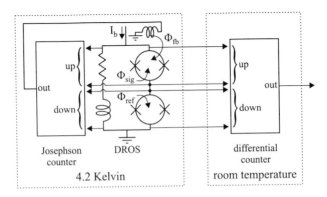

Fig. 1. Schematic overview of the Smart DROS with digital readout.

whereas the other remains superconducting. Each relaxation oscillation of either the signal SQUID or reference SQUID causes the feedback flux to be increased respectively decreased by an amount of δ_{fb}, the quantization unit of the feedback flux. The feedback loop automatically locks the systems into the dynamic equilibrium where both SQUIDs of the DROS have a switching probability of 50%.

To prevent the system from being limited by quantization errors, the quantization unit of the feedback flux should be chosen such that it does not exceed the broadband flux noise of the DROS:

$$\delta\Phi_{fb} < \sqrt{S_\Phi \cdot f_{RO}} \, . \tag{1}$$

However, the quantization unit should not be made too small, since the maximum slew rate of the Smart DROS depends on it through:

$$\frac{\partial\Phi_{sig}}{\partial t} = f_{RO} \cdot \delta\Phi_{fb} \, . \tag{2}$$

Consequently, both the quantization unit and the relaxation oscillation frequency f_{RO} should be made as large as possible. From numerical simulations it turned out that the interference between the plasma oscillations and the relaxation oscillations starts to limit the sensitivity when f_{RO} exceeds ~1% of the plasma frequency (Adelerhof 1995). For our present devices, the plasma frequency is about 100 GHz, which implies an optimum relaxation oscillation frequency of $f_{RO} \approx 1$ GHz (Adelerhof 1994).

The up-down counter of the Smart DROS is implemented in superconducting electronics. It is a semi-digital Josephson counter (Van Duuren 1999), which consists of a superconducting storage loop interrupted by two write gates that are connected to the DROS, as is schematically shown in Fig. 1. These write gates convert the voltage pulses of the DROS to flux pulses and they are dimensioned such that each voltage pulse generates one flux quantum Φ_0. The flux quanta that are added to or extracted from the storage loop causes a circulating current I_{st}, which supplies the feedback flux to the DROS via a feedback coil with a mutual inductance M_{fb}. The quantization unit of the feedback flux is given by:

$$\delta\Phi_{fb} = M_{fb} \cdot \delta I_{st} = \frac{M_{fb}}{L_{st}} \cdot \Phi_0 \, . \tag{3}$$

In the digital readout scheme of the Smart DROS, the readout is performed by a differential counter at room temperature, which counts the number of pulses that are generated across both the reference and the signal SQUID of the DROS, as is shown in Fig. 1. The difference between the number of pulses across both SQUIDs corresponds proportionally to the amount of applied flux.

3.   EXPERIMENTAL RESULTS

Previously, fully operational prototypes of the Smart DROS based on Nb/Al technology have been designed, fabricated and characterized. Experimental transfer functions were measured, which proved the proper operation of these Smart DROSs with readout SQUID (Van Duuren 1999).

As a first step towards a digital readout scheme, we measured the output pulses of a separate DROS with the same properties as the Smart DROS, but without the FLL circuitry. Fig. 2 shows the power spectral density S of the voltage across the reference SQUID. If $I_{c,sig}(\Phi_{sig}) > I_{c,ref}$, only the reference SQUID participates in the relaxation oscillations at a frequency $f_{RO} \approx 100$ MHz, as is shown in Fig. 2a, and the signal SQUID stays superconducting. The broadening of the peaks over a few MHz is due to spectral impurity of the relaxation oscillation frequency. However, since in the digital readout scheme the number of pulses across both SQUIDs of the DROS are counted and the signal corresponds to the difference between the number of pulses across both SQUIDs, the spread in the relaxation oscillation frequency does not affect the proper operation of the Smart DROS.

Fig. 2b shows the power spectral density at $I_{c,sig}(\Phi_{sig}) = I_{c,ref}$. At that point, both SQUIDs have a switching probability of 50%, such that the voltage pulses across the reference SQUID appear at half the relaxation frequency, about 50 MHz. The additional broadening of the peaks compared to Fig. 2a, around −70 dBm/MHz, is caused by the fact that due to noise both SQUIDs of the DROS do not oscillate exactly in turn. This can also be concluded from Fig. 3, which shows a voltage pulse train of 1.2 µs measured across the reference SQUID of a DROS at $I_{c,sig}(\Phi_{sig}) = I_{c,ref}$. If one would use a counter with an input trigger level of about 6 mV, the occurrence of voltage pulse could clearly be distinguished. However, the input trigger level of commercial available counters is typically 20 mV, which requires appropriate amplification of the output pulses.

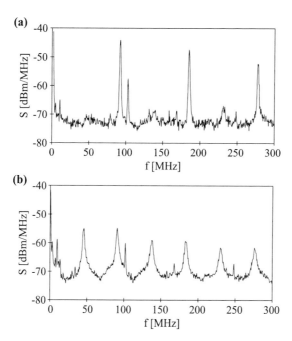

Fig. 2. Spectra of the voltage pulses measured across the reference SQUID of a 100 MHz DROS at (a)
$I_{c,sig}(\Phi_{sig}) > I_{c,ref}$ *and (b)* $I_{c,sig}(\Phi_{sig}) = I_{c,ref}$. *The vertical axis has not been corrected for the gain of the microwave amplifier that was used. The small peak slightly above 100 MHz is caused by environmental noise.*

488

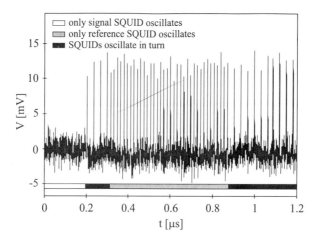

Fig. 3. Voltage pulse train measured across the reference SQUID of a 100 MHz DROS at $I_{c,sig}(\Phi_{sig}) = I_{c,ref}$. The vertical axis has not been corrected for the gain of the microwave amplifier.

4. DISCUSSION AND DISCUSSION

A prototype of the Smart DROS, i.e. a digital SQUID based on a Double Relaxation Oscillation SQUID, has been developed and first experiments towards a digital readout scheme have been performed. The major advantage of using a DROS instead of a conventional dc SQUID is that it does not require an external clock. The Smart DROS has the potential for a very high measurement bandwidth and slew rate. For the prototype, the quantization unit of the feedback flux is $\delta\Phi_{fb} = 0.18$ mΦ_0. The relaxation oscillation frequency is $f_{RO} \approx 100$ MHz, which results in a maximum slew rate of $\partial\Phi_{sig}/\partial t \approx 1.8 \cdot 10^4$ Φ_0/s [from Eq. (2)]. Theoretically, this slew rate can be enhanced by increasing the relaxation oscillation frequency up to a few GHz and by increasing the quantization unit. Since the intrinsic flux noise of the Smart DROS is about $\sqrt{S_\Phi} \approx 5$ $\mu\Phi_0/\sqrt{\text{Hz}}$ (Van Duuren 1999), according to Eq. (1) the quantization unit of the feedback can be increased by a factor of ~100. Consequently, maximum slew rates up to 10^8 Φ_0/s might be possible.

In the prototype, the signal flux is directly applied to the DROS, which limits the dynamic range. By redesigning the Smart DROS and applying the signal flux to the storage loop, the circulating current is kept zero and the dynamic range will be virtually infinite. Thus, further development of the Smart DROS offers the perspective of a very fast, sensitive SQUID sensor, as for example needed for the readout of cryogenic particle detector arrays.

REFERENCES

Adelerhof D J, Nijstad H, Flokstra J and Rogalla H 1994 J. Appl. Phys. **76**, 3875
Adelerhof D J, Van Duuren M J, Flokstra J and Rogalla H 1995 IEEE Trans. Appl. Supercond. **5**, 2160
Fath U, Hundhausen R, Fregin T, Gerigk P, Eschner W, Schindler A and Uhlmann F H 1997 IEEE Trans. Appl. Supercond. **7**, 2747
Fujimaki N, Tamura H, Imamura T and Hasuo S 1988 IEEE Trans. Electron Devices **35**, 2412
Koch R H, Rozen J R, Wöltgens P, Picunko T, Goss W J, Gambrel D, Lathrop D, Wiegert R and Overway D 1996 Rev. Sci. Instrum. **67**, 2968
Radparvar M 1994 IEEE Trans. Appl. Supercond. **4**, 87
Van Duuren M J, Brons G C S, Flokstra J, Rogalla H 1999 IEEE Trans. Appl. Supercond. **9**, 2919

Inst. Phys. Conf. Ser. No 167
Paper presented at Applied Superconductivity, Spain, 14-17 September 1999
© 2000 IOP Publishing Ltd

A low-noise S-band dc SQUID amplifier

G V Prokopenko, S V Shitov, D V Balashov, V P Koshelets and J Mygind*

Institute of Radio Engineering and Electronics RAS, Mokhovaya 11, 103907 Moscow, Russia
*Department of Physics, Technical University of Denmark, B 309, DK-2800 Lyngby, Denmark

ABSTRACT: A completely integrated RF amplifier based on a dc SQUID (SQA) has been designed, fabricated and tested in the frequency range 3.3 - 4.1 GHz. A new launcher system (SQA-unit) has been developed in order to improve RF coupling between coaxial connectors and coplanar lines on the chip. A negative feedback loop has been implemented to increase the dynamic range of the SQA. The following parameters have been measured for the single-stage device at 3.65 GHz in the feedback mode: gain of (11.0 ± 1.0) dB, 3 dB bandwidth of about 210 MHz, and noise temperature (9.0 ± 1.0) K that corresponds to the flux noise $S_\Phi^{1/2} \approx 0.7\mu\Phi_0/\sqrt{Hz}$ and energy sensitivity $\varepsilon_i \approx 160\ \eta$ ($2.0 \cdot 10^{-32}$ J/Hz).

1. INTRODUCTION

A completely RF amplifier based on a dc SQUID (SQA), proposed by Koshelets et al (1996), looks a good choice for an IF amplifier integrated with a SIS mixer and a flux-flow oscillator (FFO) in a fully superconducting sub-mm receiver which can be used for radio astronomy at sub-mm. A SQA has a number of advantages over traditional coolable semiconductor low noise amplifiers due to its ultra-low power consumption and natural compatibility with both a SIS mixer and a FFO. Following the concept of a completely integrated superconducting receiver (Koshelets et al 1997) the study of an integrated SQA is a logical step towards a densely packed imaging array of integrated receivers demonstrated by Shitov et al (1999a). Recently Muck et al (1998) have shown that RF amplifiers based on a niobium dc SQUID can achieve gain of about 18 dB and a system noise temperature in the ranged from 0.5 ± 0.3 K (at 80 MHz) to 3.0 ± 0.7 K (at 500 MHz). However for a real radio astronomy application the intermediate frequency bandwidth of at least 4 GHz is required. Recently the advanced design of a SQA has been developed by Prokopenko et al (1997, 1999) at about 4 GHz. This paper presents recent results on the study of a single-stage 4 GHz SQA with a negative feedback loop, which demonstrates its feasibility as an intermediate frequency amplifier for the PLL sub-mm integrated receiver (Shitov et al (1999b)).

2. DESIGN OF SQA

A microwave design of the 4 GHz SQA with a novel input resonant circuit has been developed and described by Prokopenko et al (1997). Fig. 1 presents a single-stage SQA that consists of the double washer SQUID, which has two square holes of the same size. The input coil consists of two identical connected in series four-turn turn sections, which are positioned inside the corresponding holes in the washer. The capacitors C_1, C_2 are chosen to tune a resonance of the input coil (L_{COIL}) at the signal frequency $f_s \approx 3.7$ GHz. The low-pass filter, based on the two coplanar lines with a cut-off frequency of about 50 GHz, is used to transmit the *dc* bias and the signal at f_s, but prevents the Josephson current $f_J \gg f_s$, from leaking out of the SQUID. The strip between two shunted (R_{sh}) micron-size Nb-AlO$_x$-Nb SIS tunnel junctions is used as an integrated control line (±) for magnetic bias of the SQUID. The resistors R (about 500 Ω each) designed as a high

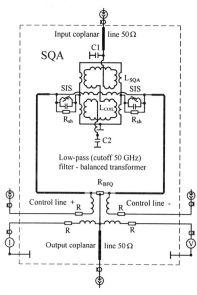

Fig. 1 Equivalent diagram of a single-stage S-band SQA integrated in one chip.

Table 1. Main parameters of the SQA #2_6(2)

PARAMETERS:	MEASURED VALUE:
1. SQUID inductance	$L_{SQA} \approx 70$ pH
2. Coupling coefficient	$k^2 \approx 0.6$
3. Area of SIS junction	≈ 1.0 μm^2
4. Capacitance of SIS junction	$C_{SIS} \approx 0.1$ pF
5. Critical current	$I_C \approx 21$ μA,
6. McCamber parameter	$\beta_C \approx 0.1$
7. Inductance parameter β_L	$\beta_L \approx 1$
8. Shunt resistance per junction	$R_{sh} \approx 8$ Ohm
9. Time constant of the SQUID	$\tau = L_{SQA}/R_{sh} \approx 10^{-1}$ sec
10. Inductance of input coil	$L_{COIL} \approx 3$ nH
11. Input circuit capacitance	$C_1 = C_2 \approx 1$ pF
12. Size of two-stage SQA chip	6×6 mm
13. Central operating frequency	$f_c \approx 3.7$ GHz
14. Dynamical resistance at bias point	$R_D \approx 23$ Ohm,
15. Voltage at the bias point	$V_B \approx 18$ μV,
16. Power gain	$G_1(max) \approx 11.0 \pm 1.0$ dB
17. Noise Temperature	$T_N(min) \approx 9.0 \pm 1.0$ K
18. Intrinsic flux noise	$S_\Phi^{1/2}(min) \approx 0.7 \mu\Phi_0 Hz^{-1/2}$
19. Frequency bandwidth of the SQA	$\Delta f|_{3dB} \approx 210$ MHz

impedance coplanar line and used to prevent leak of RF signals. The resistor $R_{BFQ} = 0.1-1.0$ Ω is used to prevent a flux trapping in the loop of the output circuit. This configuration of the output circuit is designed especially to cancel possible signal leakage (common mode) from the input of the SQA. We call this configuration Balanced Output SQUID Amplifier. The main parameters of the SQA are listed in the Table 1.

We have designed and tested an SQA-unit (see Fig. 2) - the new launcher system with improved RF coupling between input/output coaxial SMA-connectors and coplanar lines of the SQA chip. Although the chip includes two SQA's it is possible to measure each SQA separately. An intermediate (side) SMA-connector of the

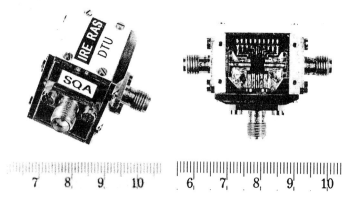

Fig. 2 View on the SQA-unit.

SQA-unit is intended for the single-stage measurement (see Fig. 2).

3.   MEASUREMENT SET-UP

A block diagram of the measurement set-up is shown in Fig. 3. The sample was placed in the SQA-unit inside a liquid ^{4}He cryostat shielded by two external μ-metal cans. The combination of a solid state noise source (Noise Com, NC 3208-A, $T_{NS} \approx 2.0 \cdot 10^5$K at 4.0 GHz) and precise step attenuator was used to supply a calibrated signal to the input of the SQA. A stainless steel cable followed by a 20-dB attenuator was placed at 4.2 K to reduce influence of the 300 K noise at the input of the SQA. A coolable HEMT amplifier (G = 30 dB, T_N = 20 K) and a room temperature FET amplifier (G = 34 dB, T_N = 120 K) are used in front of the spectrum analyzer HP-8563A.

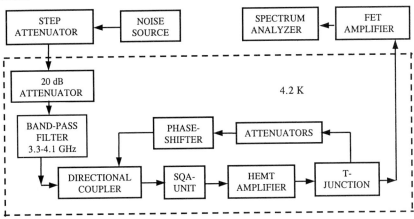

Fig. 3. Block diagram of the measurement set-up for the SQA.

In previous experiments Prokopenko et al (1999) have used a room temperature narrow band-pass filter (Δf = 40 MHz) at the input of SQA to avoid saturation that appeared as a noticeable shift of the working point. To increase the dynamic range of a SQA, the negative feedback loop is implemented, and preliminary results are reported in this paper. The loop is formed by a directional coupler (-16 dB) placed at the SQA input, an attenuator, a phase-shifter, and T-junction. To adjust the feedback signal, coolable attenuators in the range of 23 - 26 dB and a tuneable phase-shifter are used. The implementation of the feedback loop resulted in locking the operation point of the SQA for input signals up to about 100 K.

4.   EXPERIMENTAL RESULTS AND DISCUSSIONS

A noise temperature and a gain of the SQA have been evaluated by a standard Y-factor technique using hot/cold response of the system by reading data from the spectrum analyzer. The "cold" signal was estimated as 5.75±0.25 K for the cold input attenuator of 20 dB and the room temperature step attenuator set at 110 dB. Several different settings of the step attenuator (15 dB, 11 dB and 5 dB) have been used that corresponds to the "hot" signal of 23 K, 55 K, and 122 K at the input of the SQA.

The data set presenting the experimental noise temperature of the system SQA+HEMT, and HEMT alone is shown in Fig. 4. The upper plot is measured for open feedback, the lower plot is for closed one. One can see that the closed feedback mode allows to get the almost linear response up to input signals of about 100 K, while in the open feedback mode the SQA saturates at the input signal as low as ≈55 K. The noise temperature and the gain for these two cases are shown in the Fig. 5. In the case of an open feedback (upper plot): the gain is 8-9 dB, T_N ≈ 7-10 K; for a closed feedback (lower plot): the gain is ≈ 9-11 dB, T_N ≈ 9-12 K. The increase of T_N for the closed feedback can be explained by extra noise from the following HEMT amplifier (T_N = 20 - 25 K) introduced to the input of the SQA via the feedback path, while the effective gain can, in principle, be larger due to improvement of match between the SQA and the signal source. One can

492

see that the feedback mode gives also some improvement of a frequency band of the amplifier (see Fig. 5).

4. CONCLUSION

The experimental S-band amplifier based on a dc SQUID with negative feedback is designed and tested. The main parameters of SQA referred to the SMA connectors of the amplifier in the feed

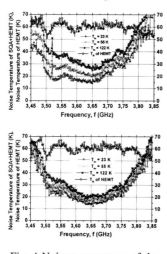

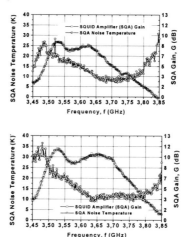

back mode: gain 11.0±1.0 dB, 3 dB bandwidth 210 MHz, noise temperature 9.0±1.0 K at the center frequency of 3.65 GHz. The saturation level at the input is increased up to 100 K that makes the feedback SQA applicable as an IF amplifier for a SIS mixer. The SQA of present design is planned for use with a superconducting integrated receiver as a preamplifier of a coolable IF chain.

Fig. 4 Noise temperature of the system SQA + HEMT and for HEMT only: open feedback (top), closed one (bottom).

Fig. 5 Noise temperature and gain for $T_{in} \approx 122$ K: open feedback (top) and closed one (bottom) The error bar is 1.0 K and 1.0 dB respectively.

5. ACNOWLEDGEMENTS

The work was supported in parts by the Russian SSP "Superconductivity", INTAS project 97-1712, the Danish Research Academy, and the Danish Natural Science Foundation.

REFERENCES

Koshelets V P, Shitov S V, Filippenko L V, Baryshev A M, Shchukin A V, Prokopenko G V, Litskevich P, de Graauw T, Luinge W, v.d.Stadt H, Golstein H, Schaeffer H, Klapwijk T, Gao J-R, Lehikoinen P and Mygind J 1996 Proc.30th ESLAB Symp.ESTEC ESA SP-388 pp 193-202
Koshelets V P, Shitov S V, Filippenko L V, Baryshev A M, de, Luinge W, Golstein H, van de Stadt H, Gao J-R and de Graauw T 1997 IEEE Trans. Appl. Superc. 7, 2, 3598
Muck M, Andre M-O, Clarke J, Gail J and Heiden C 1998 Appl. Phys. Let. 72, 22, 2885
Prokopenko G V, Shitov S V, Koshelets V P, Balashov D B and Mygind J 1997 IEEE Trans. Appl. Superc. 7, 2, 3496
Prokopenko G V, Balashov D V, Shitov S V, Koshelets V P and Mygind J 1999 IEEE Trans. Appl. Superc. 9, 2, 2902
Shitov S V, Koshelets V P, Ermakov A B, Filippenko L V, Baryshev A M, Luinge W, and Gao J-R 1999a IEEE Trans. Appl. Superc. 9, 2, 3773
Shitov S V, Koshelets V P, Filippenko L V, Dmitriev P N, Vaks V L , Baryshev A M, Luinge W,, Whyborn N D and Gao J-R 1999b this conference.

Inst. Phys. Conf. Ser. No 167
Paper presented at Applied Superconductivity, Spain, 14-17 September 1999
© 2000 IOP Publishing Ltd

Noise study of high-T_c YBa$_2$Cu$_3$O$_y$ magnetometers under microwave irradiation

H C Yang†, J T Jeng‡, J H Chen†, S Y Yang‡, and H E Horng‡

† Department of Physics, National Taiwan University, Taipei, Taiwan, R.O.C.
‡ Department of Physics, National Taiwan Normal University, Taipei, Taiwan, R.O.C.

ABSTRACT: The effects of microwave irradiation and static magnetic field on the noise characteristics of dc SQUIDs and magnetometers were investigated. It was found that low static magnetic fields cause large increase in $1/f$ noise. The microwave irradiation reduces the depth of modulation, V_{pp}, distorts the transfer function of V-Φ curves, and increases the $1/f$ and white noises. The V_{pp} shows a local minimum while the corresponding flux noise reveals a local minimum for samples under microwave irradiation.

1.   INTRODUCTION

In many applications, for example, biomagnetism (Romani *et al* 1996) and megnetotelurics (Clarke *et al* 1983), one requires low level of noise to extend down to frequencies of 1 Hz or even lower. Unfortunately, the low frequency $1/f$ noise, which is observed in low T_c-SQUIDs but is not a serious issue, is generally a serious problem in high-T_c SQUIDs and a great effort has been expended in attempting to understand its origin and reduce its magnitude.

A barrier that remains for many applications is that most SQUIDs have to operate in electromagnetically shielded enclosures. For some practical applications, it is necessary to operate SQUIDs in unshielded environments, therefore it is important to understand the effects of radio frequency radiation and static magnetic fields on the SQUID characteristics. In this work, we reported the correlation of the V-Φ curves and noise characteristics of high-T_c SQUIDs and magnetometers in radio frequency radiation and static magnetic fields.

2.   EXPERIMENTS

The samples investigated were the bi-crystal YBa$_2$Cu$_3$O$_y$ (YBCO) SQUIDs and directly-coupled SQUID magnetometers with an effective area of $\sim 1\times10^{-3}$ mm^2 and ~ 0.28 mm^2 respectively. The external magnetic field was shielded with one layer of mu-metal cylinder and one layer of high-T_c superconducting oxide. Under such a shielding, the earth field could be reduced by a factor of 1×10^{-6}. The microwave applied to the SQUIDs or SQUID magnetometers was via a dipole antenna. The noise spectrum was measured with conventional dc and ac bias reversal techniques in rf shielded and magnetic shielded cryostat.

3.   RESULTS AND DISCUSSION

Fig. 1 shows typically V-Φ curves of a 24° bicrystal SQUID at 77.4 K. The inductance of the SQUID was about ~ 50 pH. The I_c value is ~ 20 µA. The normal state resistance of this SQUID was 5.6 Ω. The modulation depth of the V-Φ curve was ~ 28 µV at dc bias current of 52.5 µA at 77.4 K.

From the numerical simulation, an expression for the power spectrum of the flux noise S_Φ has been obtained as (Enpuku *et al* 1993, 1994 and Foglietti *et al* 1989)

$$S_\Phi = 1/(2f) [(LI_c)^2(S_{Ic}/I_c^2)], \qquad (1)$$

for the *1/f* noise. Here L is the loop inductance and I_c is the junction critical current. In Eq.(1) the main contribution of *1/f* noise is from the fluctuation of the critical current. The S_{Ic}/I_c^2 represents the power spectrum of the fluctuation of the critical current. We note Eq.(1) gives the *1/f* noise for the case when the SQUID is operated in the flux-locked loop with dc bias and ac flux modulation.

In Fig. 2 we show the flux spectrum of the SQUID as a function of $LI_c/(2f)^{1/2}$ in the low frequency range. The noise spectrum was measured in the dc bias and ac flux-lock loop. A behavior of *1/f* noise was observed in the low frequency range. Using Eq.(1), we obtained the fluctuation critical current, $S_{Ic}^{1/2}/I_c \sim 6.3 \times 10^{-4}$ $Hz^{-1/2}$ at 10 Hz.

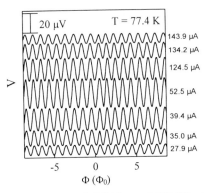

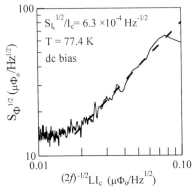

Fig. 1 V-Φ curves of 24° bicrystal SQUID at varied bias currents.

Fig. 2 Noise spectrum of 24° bicrystal SQUID as a function of $LI_c/(2f)^{1/2}$

Work on low-T_c SQUIDs (Koch *et al* 1983) showed that there are generally two separate sources of *1/f* noise. One arises from the motion of vortices in the body of the SQUIDs while the other arises from the critical current fluctuations. The second source of *1/f* noise is fluctuations in the critical current of junctions. Resistance can also contribute *1/f* noise. However, at the low voltages where SQUIDs are usually operated at critical-current fluctuations dominate. Therefore, the contribution of the resistance fluctuation can be neglected. Besides, the critical current fluctuation can be reduced by the bias reversal technique (Koch *et al* 1983 and Koelle *et al* 1999)

Fig. 3 shows the noise spectrum of a bare SQUID in a applied magnetic field of ~1.0 μT for a bare SQUID with an effective area of 1×10^{-5} cm^2. The flux noise was measured with the bias reversal technique, so the *1/f* noise due to the critical fluctuation is eliminated. The SQUID did not show *1/f* noise in the low frequency regime above 1 Hz and the noise level was 5 $\mu\Phi_o/Hz^{1/2}$ in zero magnetic field. When a magnetic field of 1.0 μT was applied, the flux noise at 1 Hz is increased. We note that the vortex usually enters the SQUID when the applied field B $\geq$ 10 μT. Our data show that a magnetic field of 1.0 μT can induce *1/f* noise. It can be that the pinning of the film is particularly weak at the edge and the weak pinning is responsible for the *1/f* noise.

In Fig. 4 we show the $1/V_{pp}$, the $S_\Phi^{1/2}$ and the corresponding magnetic field noise, $S_B^{1/2}$, at 1 kHz of a directly-coupled SQUID magnetometer at varied bias currents without and with the microwave irradiation, where V_{pp} is the depth of modulation of the V-Φ curves. The noise spectrum

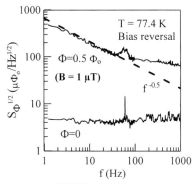

Fig. 3 Flux noise of a bare SQUID with and without an applied field of 1 μT

was measured with the dc bias and flux modulation technique. The frequency of the radiation was f = 7 GHz while the irradiated power was 4 dBm. The critical current and resistance of the magnetometer was ~18 μA and 5.6 Ω respectively at 77.4 K. Several features were observed. First, the $1/V_{pp}$ oscillates at varied bias currents under microwave irradiation (Fig.4a) while the $1/V_{pp}$ reveals a local minimum without the irradiation. Since the I-V curves of SQUID magnetometer are modulated both by microwave irradiation and magnetic fields. Hence, the V_{pp} will oscillate at different bias currents. We did not observe special correlation between the V_{pp} of the SQUID magnetometer without and with the microwave irradiation. Secondly, the $S_B^{1/2}$ (or $S_\Phi^{1/2}$) shown in Fig. 4b increases when the magnetometer was irradiated with the microwave radiation. At a bias current of 75 μA, the magnetic field noise at 1 kHz was ~0.5 pT/Hz$^{1/2}$ and the noise level was increased to ~1.4 pT/Hz$^{1/2}$ when the magnetometer was irradiated with the microwave. Finally, the V_{pp} shows a local maximum while the corresponding flux noise reveals local minimum.

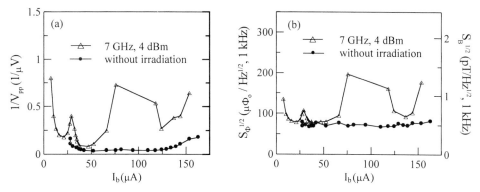

Fig. 4 (a) $(1/V_{pp})$, (b) $S_\Phi^{1/2}$ and $S_B^{1/2}$ at 1 kHz as function of bias currents

In most cases, the physical dimension of SQUIDs is much smaller than the wavelength of the microwave irradiation. In this case the radiation applied to the SQUID can be considered as to be a radio frequency (rf) flux bias and rf current bias. The major effect of the direct coupling is that the magnetic field component of the electromagnetic wave is directly sensed by the SQUID loop as a rf flux bias. The exact geometry and the structure attached to SQUID will determine the frequency dependent transfer function and therefore the noise characteristics.

Studying the effects of rf radiation on the dc SQUID, Koch *et al* (1994) considered rf signals to dc SQUIDs as both a flux or a bias current via the input and output circuitry of SQUID. They showed that rf interference distorts the V-Φ characteristic by both reducing its amplitude and creating asymmetry in the V-Φ curve about the $\Phi_o/2$ point. The first effect can increase the white noise of SQUID while the second effect can lead to a large increase in the level of low frequency noise. The amplitude of the V-Φ curve is reduced when the sample is irradiated with the microwave. In the present study it was observed that the white noise was increased significantly under microwave irradiation. A noise spectrum of $S_\Phi^{1/2}$ (or the corresponding $S_B^{1/2}$) at 1 kHz of the SQUID magnetometer measured with a standard dc bias and flux-lock loop was shown in Fig. 5. The rf field coupled to the SQUID was ~10 nT. Without the microwave irradiation the white noise level at 1 kHz was 80 $\mu\Phi/Hz^{1/2}$ and the V_{pp} was 28 μV. The noise level at 1 kHz was increased to 180 $\mu\Phi/Hz^{1/2}$ and the V_{pp} was reduced to 1.5 μV when the sample was irradiated with the microwave. The increased white noise level was mainly due to the decreased depth of modulation in the V-Φ curve under microwave irradiation. The $1/f$ noise is not increased significantly due to the asymmetry of the V-Φ is not serious and the critical current fluctuation dominates the noise level in the dc bias measurement technique.

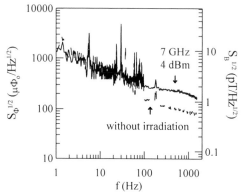

Fig. 5 Noise spectrum of a SQUID magnetometer with a bias current of about 50 μA.

4.   CONCLUSION

We characterized the noise spectrum of SQUIDs and magnetometers. It was found that low static magnetic fields cause large increase in $1/f$ noise. The rf radiation distorts the V-Φ characteristic by both reducing its amplitude and creating asymmetry in the V-Φ curve. The reduced amplitude increases the white noise of SQUIDs while the distorted V-Φ curve leads to an increase in the level of low frequency noise.

REFERENCES

Clarke J, Gamble T D, Goubau W M, Cough R H and Miraky R F 1983 *Geophys. Pros.* **31**, 149
Enpuku K, Shimomura Y and Kisu T 1993 *J. Appl. Phys.* **73**, 7929
Enpuku K, Kotika G and Maruo T, 1994 *J. Appl. Phys.* **76**, 8180
Foglietti V, Gallagher W J, Ketchen M B, Kleinssasser A W, Koch R K and Sandstrom L 1989 *Appl. Phys. Lett.* **55**, 1451
Koch R H, Clarke J, Goubau W M, Martinis J M, Pegrum C M and Harlingen D J Van 1983 *J. Low Temp. Phys.* **51**, 207
Koelle D, Kleiner R, Ludwig F, Dantsker E, and Clarke J 1999 *Rev. Mod. Phys.* **71**, 631
Koch R H, Foglietti V, Rozen J R, Stawiasz K G, Ketchen M B, Lathrop D K, Sun J Z and Gallagher W J 1994 *Appl. Phys. Lett.* **65**, 100
Romani G L, Gratta C Del and Pizzella V 1996 *SQUIDs Sensor: Fundamentals, Fabrication and Applications* (NATO ASI Series E: Applied Sciences – Vol.329) ed H Weinstock (Dordrecht: Kluwer Academic Publishers) pp 445-490

Inst. Phys. Conf. Ser. No 167
Paper presented at Applied Superconductivity, Spain, 14-17 September 1999
© 2000 IOP Publishing Ltd

HTS DC SQUID Flux Locked Loop based on a cooled integrated electronics

J Kunert, V Zakosarenko, V Schultze, R Groß*, F Nitsche*, and H G Meyer

Institute for Physical High Technology, Magnetics/Cryoelectronics Division, P.O.B. 100 239, D-07702 Jena, Germany
* MAZeT GmbH, In den Weiden 7, D-99099 Erfurt, Germany

ABSTRACT: Using a monolithic BiCMOS technology we developed and fabricated a directly coupled flux locked loop electronics at 77 K for high T_C dc-SQUIDs. It has an equivalent input noise of 240 pV/√Hz and a small power consumption. Together with the SQUID the electronics shows a slew rate of up to 2×10^6 Φ_0/s. The SQUID-system operates stable also without any magnetic shielding.

1. INTRODUCTION

The application of the superconducting quantum interferometer (SQUID) for measurement of a small magnetic field without magnetic shielding is limited by high levels of magnetic disturbances. In most cases a SQUID is operated with electronics which supplies a feed back signal to the SQUID compensating the variations of the external magnetic fields and locking the magnetic flux in the SQUID (FLL-electronics). In the presence of large disturbances the dynamic range and slew rate of the whole system, SQUID with FLL-electronics, must be sufficiently high. These parameters depend on the performance of both SQUID and electronics. Recently, the transfer function of high T_c SQUIDs was remarkably enhanced by using 30-degree bicrystal substrates [Minotani, Ramos]. A large bandwidth and low noise values have been obtained when some parts of SQUID electronics were placed in the vicinity of the SQUID in liquid nitrogen [Ukhansky]. Our approach is to integrate all parts of the SQUID electronics on one Si-chip designed to operate in liquid nitrogen.

2. CIRCUIT CONFIGURATION, REALISATION AND CHARACTERISTICS

To create semiconductor devices working at 77 K we have to adapt the design rules, because the parameters of basic devices are changed by cooling them to 77 K. This was described in our previous works [Kunert, Zakosarenko]. We use the hardware description language SPICE with adapted parameters for circuit design and simulation [Simeon, Lu]. We chose a 1.2 μm BiCMOS technology to have many possibilities for the circuit design.

Our FLL-electronics has a simple layout with a directly coupled low noise amplifier, integrator and a feed back loop. A CMOS switch is used to reset the integrator. Thereby the electronics operates as an amplifier with a gain of approximately 4000. The "RESET" modus was also used to test the SQUID characteristics. All these elements were integrated on the chip. The Si-chip was mounted on a printed circuit board (PCB) with some external resistors

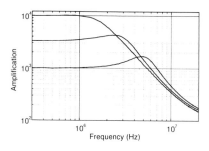
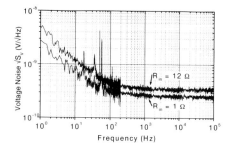

Fig. 1. Frequency response of the FLL-electronics amplifier at 77 K for various coefficients of negative feed back.

Fig. 2. Equivalent input noise figure of the FLL-electronics at 77 K depending of source resistance

and capacitors. Changing the connections on the PCB we can choose between a few circuit variants. The PCB was covered with a cap to prevent a direct contact of the Si-chip with boiling nitrogen and to ensure mechanical stability of the whole electronics. The former is important for reduction of the low frequency noise of the electronics. The encapsulated FLL-electronics has dimensions of 12 mm × 16 mm × 5 mm.

The low-noise amplifier is a differential amplifier on the base of two specially designed low noise n-p-n transistors. It has a voltage amplification of 45 dB, an input impedance of about 300 Ω, an input bias current of 0.1 mA, and an equivalent input white noise of 240 pV/$\sqrt{Hz}$ with 1/f-noise contribution at frequencies lower than 40 Hz. An operational amplifier (OA) following the low-noise amplifier was used in the FLL-electronics as an integrator. This OA has n-p-n transistors at the input and CMOS components in the following stages. This results in a small current consumption of the whole circuit and rail-to-rail operation of the output stage. As a drawback of CMOS devices we have a remarkable contribution of 1/f noise at frequencies up to 2 kHz. The OA has an amplification of more than 86 dB and a bandwidth of 6 MHz. The low-noise amplifier and the OA can be used as a multi-purpose low-noise operational amplifier. Fig. 1 shows the measured transfer function of this amplifier. By these measurements the gain was limited in a standard way introducing a negative feed back through external resistors. It corresponds to the gain factor measured at low frequencies. It is clear from the figure that the -3 dB bandwidth is about 1.8 MHz for a gain factor of 10^4.

The equivalent input noise spectrum of this amplifier is shown in Fig. 2. Due to 1/f noise of the OA the 1/f-corner of the amplifier is shifted to 60 Hz as compared to 40 Hz for the differential amplifier alone. The upper and lower curves in Fig. 2 correspond to the input loaded with a 1 Ω or a 12 Ω resistor, respectively. From these data we can obtain the main parameters of the equivalent input circuit consisting of a voltage noise source $\sqrt{S_V}$, a current noise source $\sqrt{S_I}$, and a noise resistor R_N [Wade, Leach]. The experimental data can be fitted with $\sqrt{S_V} = 230$ pV/$\sqrt{Hz}$, $\sqrt{S_I} = 30$ pA/$\sqrt{Hz}$, $R_N = 10$ Ω and a correlation $\gamma = 0.48$ between $\sqrt{S_V}$ and $\sqrt{S_I}$. The resistance R_N corresponds to the base spreading resistance [Leach] of the input-stage transistor but not to the directly measured input impedance of the differential amplifier.

The whole electronics has a current consumption of 13 mA at a supply voltage ± 2.2 V. This corresponds to a power dissipation of 57 mW in liquid nitrogen.

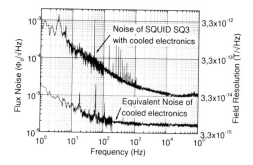

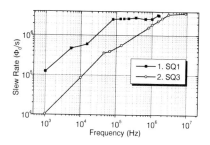

Fig. 3. The noise spectrum of SQ3 with cooled FLL-electronics and the noise of FLL-electronics at 77 K normalised to the flux in the SQUID.

Fig. 4. Slew rate of the FLL-electronics with two SQUIDs. Different integrator capacitance was used for different SQUIDs.

3.   SQUID-SYSTEM MEASUREMENTS

We tested our FLL-electronics with three SQUIDs: with a planar SQUID gradiometer and two different SQUID magnetometers. The gradiometer (SQ1) was a dc SQUID with a grain-boundary Josephson junctions directly coupled to a planar gradiometric pick up loop. The structure was patterned in the $YBa_2Cu_3O_{7-x}$ film deposited on bicrystal $SrTiO_3$ substrate. The SQUID has a flux-to-voltage transfer function $V_\Phi = 60$ $\mu V/\Phi_0$ and an intrinsic noise of 10 $\mu\Phi_0/\sqrt{Hz}$, as measured with our standard room-temperature electronics with flux modulation. The SQUID was placed at a distance of 5 cm from the electronics in common magnetic shielding. There was no shielding between the SQUID and the electronics. The test shows no influence of the electronics on the SQUID noise figure. Taking into account the value of V_Φ one can estimate the contribution of the electronics noise to be not more than 8 % of total noise, which is within the experimental error. The small-signal bandwidth of the whole system was approximately 10 MHz and the slew rate was 2.6×10^6 Φ_0/s. The system SQ1 with FLL-electronics has a dynamic range of 60 Φ_0.

The magnetometer SQ2 used for testing of the system was also fabricated as a single layer structure on a bicrystal substrate. The pick up loop of SQ2 consist of a single flat square turn. The field-to-flux conversion factor of the pick up loop was $B_\Phi = 17$ nT/Φ_0. SQ2 has $V_\Phi = 90$ $\mu V/\Phi_0$ and a white noise level of 10 $\mu\Phi_0/\sqrt{Hz}$. As for gradiometer SQ1 there was no detectable influence of the electronics on the SQUID.

The magnetometer SQ3 was a flip-chip magnetometer with high field sensitivity. A square pick up loop and multi-turn input coil were prepared on a single crystal substrate with a multilayer technology. The input coil was inductively coupled to a single layer washer SQUID deposited on a bicrystal substrate. The magnetometer SQ3 has following parameters: $B_\Phi = 3$ nT/Φ_0, $V_\Phi = 150$ $\mu V/\Phi_0$, and a white noise of 10 $\mu\Phi_0/\sqrt{Hz}$. The system SQ3 with FLL-electronics has a dynamic range of 5 Φ_0. The noise spectrum of this system is shown in Fig. 3. The noise of the electronics normalised to the flux in the SQUID using the measured value V_Φ is also shown for comparison. It is clear, that the noise figure of the system is determined by the intrinsic noise of the SQUID magnetometer. The calculated contribution of the FLL-electronics to the noise is only 1.3 % of the total noise. Relatively large 1/f noise is well known for HTS SQUIDs and is originated by the fluctuation of the critical current of the Josephson junctions. It can be reduced if a bias reversal scheme is used. In our FLL-electronics we can not use bias reversal at the present time.

One of the advantages of our FLL-electronics is the potentially high slew rate of the system. To determine the slew rate we put the signal from a triangle-wave generator with a fixed frequency and variable amplitude to the feed back coil of the SQUID. The slew rate is defined as the maximal rate of the flux change at which the FLL-electronics holds the flux in SQUID without jumps. The experimentally determined dependence of the slew rate on the frequency of the signal is shown in Fig. 4. For lower frequencies the slew rate is limited by the dynamic range of the FLL because the output signal of the FLL-electronics can not exceed ± 2.2 V. For high frequencies the slew rate exceeds 2×10^6 Φ_0/s for both SQUIDs. Large slew rate and dynamic range allow the stable operation in laboratory without magnetic shielding even for the system with the sensitive SQUID magnetometer.

4. CONCLUSIONS

The liquid nitrogen cooled integrated electronics for the high T_C dc-SQUID flux-locked-loop was tested with several SQUIDs. For system characterization a very sensitive SQUID magnetometer was used. In our simple dc-bias scheme the SQUID noise dominates the flux noise of the whole system. For better low frequency noise behavior an ac-bias scheme would be necessary. The slew rate at high frequencies is better than 2×10^6 Φ_0/s and the operation is stable without magnetic shielding.

This FLL electronics is convenient for SQUID application at high frequencies, for instance for Radiofrequency Magnetic Sounding (RMS) measurements.

ACKNOWLEDGEMENT

The work is supported by the German BMBF under contract No. 13N6863. The authors are grateful to R. Stolz (IPHT Jena), N.N. Ukhansky (MSU Moscow) L. Dörrer (FSU Jena) and M. Scheiner (PTB Berlin) for technical assistance and for useful discussions.

REFERENCES

Kunert J, Zakosarenko Z, Schultze V, Meyer H G, Nitsche F Wenske H 1998 J. de Physique IV **8**, Pr3 205-8

Leach W M 1994 Proc. IEEE **82**, 1515-38

Lu W L, Hwang Z W, Lu L S 1996 J. de Physique IV **6**, C3 199-206.

Minotani T, Kawakami S, Kiss T, Kuroki Y and Enpuku K 1997 Jap J of Appl Phy **36(8B)**, L1092-5

Ramos J, Ijsselsteijn R P J, Stolz R, Zakosarenko V, Schultze, Chwala A, Hoenig H E, Meyer H G, Beyer J, and Drung D 1999 IEEE Trans. Appl. Supercond. 9 No. 2, 3392-5

Simeon E, Claeys C, Martino J A 1996 J. de Physique IV **6**, C3 29-42.

Wade T E, Van der Ziel A, Chenette E R, and Riog G A 1976 IEEE Trans. **ED-23**, 998-1011

Ukhansky N N, Gudoshnikov S A, Weidl R, Dörrer L, and Seidel P 1997 ISEC'97 Ext. Abst. 3, 80-2

Zakosarenko V, Kunert J, Schultze V, Meyer H G, Wenske H, and Nitsche F 1999 IEEE Trans. Appl. Supercond. 9 No. 2, 3283-5

Inst. Phys. Conf. Ser. No 167
Paper presented at Applied Superconductivity, Spain, 14-17 September 1999
© 2000 IOP Publishing Ltd

Alternative structures in washer-type high-T_c dc SQUIDs

A B M Jansman, P G Jeurink, A I Gómez, M Izquierdo, J Flokstra and H Rogalla

Low Temperature Division, Department of Applied Physics, University of Twente, P.O. Box 217, 7500 AE Enschede, The Netherlands

ABSTRACT: We present half-slotted SQUIDs with the outermost strip provided with holes and various slot configurations. They are supposed to have high effective areas while their narrow strip lines should reduce the low-frequency noise. A parameterization in current paths is difficult. This complicates the use of a finite-element model. The effective areas of YBaCuO SQUIDs on a SrTiO₃ bicrystal substrate compare to those of completely slotted SQUIDs. Neither the half-slotted nor the new structures show excess low frequency noise when cooled in fields up to 50 µT.

1. INTRODUCTION

Applications of high-T_c dc SQUIDs outside a shielded environment require good performance in the earth magnetic field. Unfortunately, when a SQUID is cooled through its transition temperature in a background field then vortices will be trapped in the superconducting strip lines. Hopping of these vortices leads to excess $1/f$ flux noise in the SQUID (Miklich et al. 1994).

For strip line of width w penetration does not occur for background fields smaller than $B_T=\pi\phi_0/4w^2$ (Clem 1997). For 4 µm wide strip lines, B_T is as large as 100 µT. Slotted SQUIDs consisting of concentric narrow strip lines combine suppressed $1/f$ noise with preserved effective area (Dantsker et al. 1997). For the half-slotted case the effective area even shows a maximum. The dependence of the effective area on the number of slots was well described with a finite element method (Jansman et al. 1998).

In this paper we present a set of alternative washer structures consisting of 4 µm wide strip lines. Qualitative and quantitative expectations are presented for their effective areas. The measured values of the effective areas are compared to the expectations. Noise spectra measured with and without a background field are presented and discussed.

2. DESIGN AND EXPERIMENTS

Below, we will construct washers out of narrow strip lines, designed to have the effective area of the half-slotted SQUID. Jansman et al. (1998) suggest a rule of thumb for the coupling of flux into the SQUID hole. When a current flows through a wide structure it couples less flux than it would if the structure were split up into narrow strip lines. In the half-slotted SQUID, central currents flow through narrow strip lines and couple effectively to the SQUID hole. The negative flux contribution from the opposite currents at the outside is small because they flow through a wide unslotted structure.

Some alternatives for the original slotted SQUIDs are shown in Fig. 1. The remaining strip lines may be oriented radially or tangentially, *i.e.* either their long or their short axis may be directed to the central hole of the washer, respectively. Tangential currents couple much more flux into the central SQUID hole than radial currents do.

502

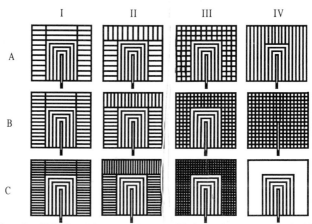

Fig. 1 Alternative four-slots SQUIDs with various designs of holes in the outermost strip. The outer dimensions are 186×204 μm². The junctions are 4 μm wide. They are situated at the grain boundary just below the washer. All washer strip lines are 4 μm wide.

Let us try to estimate the effective areas of these washers. In an optimistic approach, we suppose that for all new structures the current distribution in the central four slots may be assumed to be the same as for a four-slots SQUID. Then variations in effective area have to be due to the current distribution far from the center. The outermost part of the washers of the type II SQUIDs consists of radial holes only. Only the surrounding outermost tangential current couples negative flux efficiently, but presumably less than in the eight-slots case. The resulting effective areas for the SQUIDs with radial holes (type II and to a less extent type I) are expected to be higher than for the SQUIDs with eight slots. The holes in the type III SQUIDs are only small perturbations on the situation of a normal four-slots SQUID, especially when the holes become small. Likewise, the performance of SQUID IV-B should resemble that of a solid SQUID, although even small modifications of the innermost structure may have large consequences. The current distribution and thus the effective area of SQUID IV-A will be the same as for a completely slotted SQUID. SQUID IV-C, is a small directly coupled SQUID with a small effective area. Since in practice the current in the central four slots may well be affected by the fine structure outside, the actual effective areas of the new washers may be somewhat smaller. The inductance seen by the junctions, governing transfer function and flux noise, is mostly determined by the innermost structure of the washer, and can be considered the same for all structures, except IV-B (Jansman et al. 1999).

For a quantitative understanding, the SQUIDs are implemented in the finite-element method of Jansman et al. (1998). The method requires a parameterization of the structure in 200 nm wide

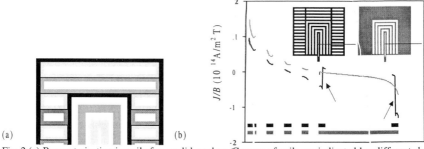

Fig. 2 (a) Parameterization in coils for a solid washer. Groups of coils are indicated by different shades of gray. The central coils are concentrically defined around the slots. Around the holes closed rectangular coils are defined and at the exterior of the washer there is a group of concentric coils.(b) Current distributions J in SQUID I-B and a four-slots SQUID, subject to a magnetic field B. The current distributions apply to the cross section displayed as thin lines in the inset. The arrows indicate discontinuities in the calculated current distributions.

coils of constant current. For the new structures the parameterization is not as obvious as for the standard slotted SQUIDs. The parameterization used below is shown in Fig. 2a. Around the central four slots and at the outer edge of the washer concentric coils are defined. The currents in the outside of the washer are parameterized in terms of groups of coils around the holes. The sum of concentric currents is zero. In Fig. 2b, the calculated current distributions for SQUID I-B and for a four-slots SQUID are shown. Clearly, the distribution in the innermost strips changes with variations far from the SQUID hole. The changed current distribution indicates that the qualitative approach was indeed too optimistic. The distribution in the center of SQUID I-B is representative for all SQUIDs of types I and II. Therefore, the effective area of SQUID I-B can be expected to be considerably lower than for the four-slots SQUID. The calculated effective area values, presented below, can be considered as a bottom value. In reality, some (negative) concentric current will flow around the holes. Since the sum of concentric currents is zero, this results in extra positive current close to the center, increasing the effective area. The limitations of the parameterization become clear from Fig. 2b. The parameterization leads to discontinuous current patterns in strips separating the holes, which is physically improbable. Both the qualitative approach and the calculations can be of qualitative use in the analysis of actual measurements on the alternative structures.

The SQUIDs of Fig. 1 have been fabricated on a $SrTiO_3$ 30° bicrystal substrate. The film, fabricated by pulsed laser deposition, is 100 nm thick. Its critical temperature is 89 K. The film was patterned by standard photolithography. Gold contacts were provided by rf sputtering. As a reference, one standard solid SQUID, three four-slots SQUIDs and one eight-slots SQUIDs were added. They are distributed along the grain boundary line to account for possible inhomogeneity over the film. The critical currents of the 4 μm wide junctions are between 5 and 10 μA at 77 K.

Both the experimental and some calculated effective areas are displayed in Fig. 3. The realistic London penetration depth value λ_L=290 nm at 77 K leads to best agreement for the standard slotted SQUIDs, as is seen in Fig. 3b. In Fig. 3a, the theoretical effective areas of the new structures are all quite low. Furthermore, all predictions of type I and II SQUIDs overlap, since their current distributions are alike. The qualitative expectations were too optimistic with respect to the effective area, because the current distribution in the center is not the same as for the four-slots SQUID. On the other hand, the calculated values turn out to be too pessimistic. As was discussed above, some extra positive current flows around the central slots. This extra contribution is compensated by some net current along the holes. The used parameterization cannot account for such a net current. For SQUID IV-C the model works fine because the parameterization is easy. The best new design (type III-C) has an effective area which is comparable to the completely slotted case.

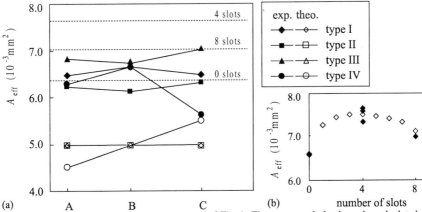

Fig. 3 (a) Effective areas of the alternative structures of Fig. 1. The open symbols show the calculated values with λ_L=290 nm, the closed symbols show the experimental values at 77 K. The calculations on types III-B and C and IV-B were too complex to carry out. As a reference, the theoretical values for the solid SQUID, four-slots SQUID and eight-slots SQUID are indicated by dotted lines. (b) Effective areas of the reference SQUIDs. For these SQUIDs the experimental (♦) and the theoretical (◊) values correspond very well. The values for the alternative structures in (a) are higher than calculated.

504

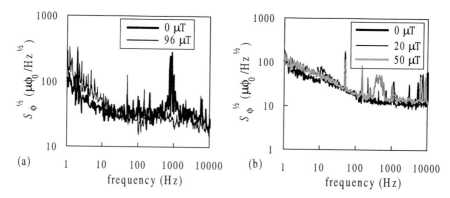

Fig. 4 Noise spectra at 77 K of (a) SQUID III-C in Fig. 1 and (b) a four-slots SQUID that was fabricated on the same sample as a reference. During the noise measurements no current bias reversal was applied. For the four-slots SQUID no increase of low frequency noise is seen for cooling in fields up to 50 μT. SQUID III-C shows a small increase when the background field is 96 μT. For both SQUIDs interference signals at 50 Hz can be seen. The origin of the broad peaks around 1 kHz for SQUID III-C and around 400 Hz for the four-slots SQUID is unknown.

For some of the SQUIDs, the noise was measured in background fields. Unfortunately, the sample degraded considerably before the noise measurements were carried out. The critical currents of the junctions decreased down to a few microamperes. As a consequence the modulation depth of the SQUIDs decreased and the white noise floor rose. Often, when the SQUIDs were cooled down in a field, the flux locked loop circuit could not lock. For some of the SQUIDs, therefore no reliable noise measurements could be performed. In Fig. 4 noise spectra of two SQUIDs on this sample are shown for various fields. Even without a background field, the noise figures contain a $1/f$ component that cannot be reduced by the bias current reversal scheme. The low-frequency noise for field-cooled SQUIDs again does not show the dependence on the strip line width that is expected. For the four-slots SQUID no excess low frequency noise is found for background fields up to 50 μT. Although some effect may be hidden by the intrinsic noise spectra of these SQUIDs, which have already a strong $1/f$ contribution, a $1/f$ noise contribution due to hopping vortices should be well visible for a four-slots SQUID in such high fields. The $1/f$ noise of these SQUIDs in a background field does not relate to the width of the strip lines in the washer. Recently, various SQUIDs with wide strip lines have shown noise spectra which were invariant to the background field during cooling (Krey et al. 1998, Borgmann et al. 1999).

In conclusion, alternative SQUIDs with narrow strip lines were designed. The effective areas are close to those of the completely slotted SQUID. Qualitative arguments and a finite-element method lead to two extreme cases for the current distribution and the effective area in the alternative structures. The actual situation seems to be an intermediate case. Field cooled SQUIDs had the same $1/f$ noise as shielded SQUIDs, independent of the strip line width.

REFERENCES

Miklich A H, Koelle D, Shaw T J, Ludwig F, Nemeth DT, Dantsker E, Clarke J, McN. Alford N, Button T W and Colclough M S 1994 Appl. Phys. Lett. **64**, 3494
Clem J, 1997 Effects of trapped vortices in SQUIDs, presented at the University of Twente
Dantsker E, Tanaka S and Clarke J 1997 Appl. Phys. Lett. **70**, 2037
Jansman A B M, Izquierdo M, Flokstra J and Rogalla H 1998 Appl. Phys. Lett. **72**, 3515
Jansman A B M, Izquierdo M, Flokstra J and Rogalla H 1999 IEEE Trans. Appl. Supercond. **9**, 3290
Krey S, David B, Eckart R, and Dössel O 1998 Appl. Phys. Lett. **72**, 3205
Borgmann J, David P, Otto R, Schubert J and Braginski A I 1999 Appl. Phys. Lett. **74**, 1021

Inst. Phys. Conf. Ser. No 167
Paper presented at Applied Superconductivity, Spain, 14-17 September 1999
© *2000 IOP Publishing Ltd*

Superconductive and traditional electromagnetic probes in eddy-current nondestructive testing of conductive materials.

M. Valentino, A. Ruosi, G. Pepe, G. Peluso

Istituto Nazionale per la Fisica della Materia (INFM), Unita' di Napoli
Università di Napoli "Federico II", Piazzale Tecchio 80, 80125 Napoli, Italy

Abstract: We present herein a comparative study of non-destructive testing performed by an induction coil, with a sensitivity of $24\mu V$ at 70Hz, a flux-gate and a HTc-SQUID-based magnetometer with a field noise density of $20pT/\sqrt{Hz}$ and $0.3pT/\sqrt{Hz}$ respectively. Single-layered and multi-layered aluminium-alloy structures with artificial defects commonly encountered in the aircraft industry, i.e. holes, slots and cracks, have been investigated. Experimental data are discussed in terms of signal-to-noise ratio in the frequency range 10-1000 Hz. The effect of stand-off variation and probe tilting on the output signals is also investigated.

1. INTRODUCTION

With respect to the conventional electromagnetic sensors used in Eddy-Current NonDestructive Evaluation (EC-NDE), SQUID (Superconductive QUantum Interference Device) magnetometers, are attractive for their extremely high magnetic field sensitivity over a large frequency range of the eddy-current distributions and high spatial resolution (Weinstock 1985, Wikswo 1995). Currently, many different kinds of electromagnetic probes are used for EC-NDE measurements, for example induction coils and flux-gate magnetometers. Each one has advantages and drawbacks and hence it is difficult to state which of them is the most effective for a particular application. For aircraft industry applications such as quality tests during the fabrication process or integrity assessments of structures already in use, it is essential to have a high accuracy in the localization of both surface and deeply embedded defects (depth >few mm) in metallic structures. Since the skin penetration depth of a planar electromagnetic wave incident on a conductive plate is given by $\delta=(\sigma\mu_0\pi f)^{-1/2}$ (where μ_0 is the magnetic permeability of vacuum, σ the conductivity of the material and f the frequency of the incident wave), probes having a high sensitivity at low frequencies are required.

In EC-NDE measurements, noise and signal depend on the particular detected defect and it is difficult to measure them separately. Moreover, due the intrinsically different working principle of the three probes, each measures a different physical characteristic. The flux-gate, SQUID and induction coil detect the component of magnetic field parallel to the core, the magnetic flux threading the loop and the inductance variation, respectively. The sensor-independent parameter chosen to perform such a comparison is a type of Signal-to-Noise ratio (S/N) defined as the ratio between the maximum signal variation along the measured line-scan and the standard deviation of the N data points b_i collected along the same line scan, i.e. $\sigma_n = [\sum_i (b_i - b_m)^2/N]^{1/2}$, where b_m is the average value of N points. Since in our measurements line-scans cross a consistent area of the sample located far away the defect, σ_n can be considered a good approximation of the noise in each measurement.

506

An experimental characterization based on a quantitative comparison between the above mentioned sensors is required in order to optimize their performance in relation to the problem under investigation. In this context, a comparative study of nondestructive testing performed by an induction coil, a flux-gate and a SQUID-based magnetometer is discussed here. Five different aluminum-alloy structures with artificial defects such as holes, slots and cracks, modeled after those commonly encountered in the aircraft industry, have been examined (Valentino 1999).

2. EXPERIMENTAL SET-UP

A schematic of the eddy-current NDE system based on HTc-SQUIDs developed at the INFM laboratory at the University of Naples is shown in Fig.1 (a). More details are reported by Ruosi (1998).

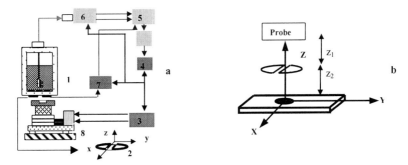

Fig.1. (Left) A scheme of the eddy-current SQUID-based NDE prototype. (1) HTc SQUIDs into a fiber-glass cryostat; (2) Double–D shaped induction coil; (3) X-Y controller; (4) PC; (5) Lock-in amplifier; (6) SQUID controller; (7) Current source; (8) X-Y scanning system. (Right) The probe is located at a distance z_1 above the center of the differential double-D shaped coil which is at a distance z_2 above the sample.

Eddy currents are induced in the sample by a differential double D-shaped coil shown in Fig. 1 (b). Two exciting coils having a different diameter have been used.

The system uses low noise iMAG-3 HTc SQUIDs (Conductus). Since measurements are performed in an electromagnetic unshielded environment, to reject noise disturbances from distant field sources the outputs of two SQUIDs are electronically subtracted. The optimised baseline used in this work is 29.5 mm. Due to the package of the probe, the minimum distance between the bottom SQUID and the inducing coil for measuring the B_z component is $z_1=12$ mm. The field-noise density of our system is less than 0.3 pT/√Hz. The effective field sensitivity for a signal with 6 Hz bandwidth is about 3 pT/√Hz. The operating frequency of the SQUID-based magnetometer ranges from 10 Hz to 20 kHz.

A commercial low-noise three-axial flux-gate magnetometer (Bartington) has been used in this work to compare the performance of different probes. The electronics of the flux-gate, operating at room temperature, is similar to that described in Fig. 1 (a) except for the voltage-optical converter that is not required. The magnetic field noise density of the flux-gate in the gradiometric configuration is 20 pT/√Hz and the measurement range is 10 Hz-1kHz . The optimize baseline of the gradiometer is 30 mm.

The induction coil test was performed using the absolute probe type *PKA 22-6* (Rohmann) with a diameter of 38 mm and a thickness of 40 mm, the operating frequency of this device ranges from 50 Hz to 1 kHz and the maximum voltage sensitivity is 24 μV at 70 Hz. This particular kind of coil, optimized for operating at low frequency, has been chosen for a detection of deep flaws. It can be positioned on the surface of the specimen at a minimum distance $z_2 = 0.3$ mm. The characteristics of the aluminum-alloy specimens with various artificial defects examined in this work are given in Tab. 1. The experimental results shown below were obtained moving the sample with a speed of 1 mm/s beneath the stationary probes. For each measurements the system acquire 1024 samples during a line-scan of 80 mm.

Test	Samples	σ (MS/m)	Overall volume (mm^3)	Defect	t (mm)	w (mm)	L (mm)	D (mm)	d (mm)
1	Al7071Ti651 1 layer	32±20%	100x200x4	Circular hole	1			8	3
2	Al7071Ti651 1 layer	32±20%	150x1000x15	Multiple slots		0.6	40		11 12
3	Anticorodal 2 layers	56±20%	200x200x23	Crack	3	<0.1	400		20
4	Anticorodal 3 layers	56±20%	200x200x43	Crack	3	<0.1	400		40
5	Al7071Ti651 5 layers	32±20%	100x200x16	Circular hole	1			6	13

Table 1. Physical and geometrical characteristics of the samples used in the experiments. The parameters σ, t, w, L, D, d represent the electrical conductivity of the sample, and the thickness, width, length, diameter and depth of the particular defect, respectively.

3. RESULTS AND DISCUSSIONS

To compare the probe sensitivity to deep flaws, output signals as a function of the line-scan coordinate were recorded for the selected samples. Each of the structures used in this study and described in Tab.1 was chosen to test the sensitivity of the sensors in different circumstances. Sample 1 enables the detection of a millimeter-sized hole in single-layered structure at a depth of only a few mm. Sample 2 gives information on the detection of multiple slots at depths greater than 10 mm in a single-layer structure. Measurements on samples 3 and 4 give information on the detection of a long crack in a three-layer structure at even larger depths (20 mm and 40 mm), which is severely testing the sensitivity of the probes. Finally, sample 5 is used for the detection of a millimeter-sized defect in a 5-layer structure at depths larger than 10 mm. The presence of many layers and of insulating gaps between them results both in a significant reduction of the conductivity of the whole structure and in complex wave fields reflected at the interfaces. The attenuation of the output signal can be significant in this case.

As a preliminary step the effect of the distance variation between the probe and the sample, and of the probe tilting on the output signals are investigated. The effect of larger probe lift-off distances (representing, e.g., an insulating layer at the surface of the structure) is less negative in the case of the SQUID and the flux-gate compared to the induction coil. Studies of the effect of a probe tilt angle reveal that the SQUID can be considered, to a very good extent, as a point sensor, which is simplifying both the measurements and the numerical simulations. Measurements of the frequency dependence of the S/N response demonstrate the superior performance of the SQUID and the flux-gate in detecting subsurface and deep defects at depths greater than 10 mm in a single-layer structure. Moreover, the flatter response up to 1 kHz for the SQUID is desirable for the detection of both surface and deeply embedded flaws.

Tab. 2 summarizes the maximized S/N ratios for the measurements performed on the test samples. The diameter and the frequency of the exciting coil and the depth of the defect are also shown. The overall superiority of the SQUID sensor can be appreciated from the table. Indeed, while the coil and the fluxgate fail the test *4* and *5* respectively, the SQUID is always equally or more sensitive to the other sensors in each individual case, it is the only sensor which succeeds in identifying the defects in all structures examined here. In the presence of voids and insulating layers in multi-layered samples, the conductivity is significantly reduced and wave-field reflections at the interface become quite complex. As demonstrated in test *5*, the SQUID is the only sensor capable of detecting these anomalies with sufficient sensitivity. Emphasize that this study has compared the S/N response of each probe in their optimized operating conditions where the SQUID and flux-gate lift-off ($z_2 > 10$ mm) is large if compared to the lift-off of the induction coil ($z_2 = 0.3$ mm).

Probe	Coil diameter (mm)	Frequency (Hz)	Defect depth (mm)	S/N #1	#2	#3	#4	#5
Coil		1000	3	2.58				
	38	70	11		0.65			
		70	12, 13		0.25		-	
		60	20			0.92		0.16
Fluxgate	25	580	3	6.97				
	58	133	20			2.07		
	25	77	11		2.46			
	25	77	12		0.64			-
	58	13	40				0.56	
SQUID	25	580	3	5.21				
	25	133	20, 13			2.54		3.57
	25	77	11		4.20			
	25	77	12		1.95			
	58	13	40				2.78	

Table 2. Maximized S/N ratios for the different probes in the examined samples

4. CONCLUSIONS

An experimental characterisation based on a quantitative comparison between different eddy-current probes in terms of S/N ratio in the frequency range 10-1000 Hz has been successfully carried out. In particular, the performance of an induction coil, a flux-gate and a HTc-SQUID-based magnetometer have been investigated on the basis of comparative experiments on different Al-alloy test samples with artificial defects such as holes, slots and cracks. Experimental results prove the superiority of the eddy-current NDE system based on HTc-SQUIDs in low frequency measurements for the detection of deep defects in particular in multi-layered conductive structures.

ACKNOWLEDGMENTS

This work has been carried out in the frame of the project: "Eddy Current Nondestructive Evaluation using Superconducting Devices" with the support of the Istituto Nazionale per la Fisica della Materia (INFM).

REFERENCES

H. Weinstock 1985 SQUID'85 eds H. D. Halbohm and H. Lubbig (Berlin: de Gruyter) pp. 853-858.
J Wikswo Jr. 1995 IEEE Trans. Appl. Superconductivity 5
M. Valentino 1999 Electromagnetic Nondestructive Evaluation 1999 eds D. Lesselier and A. Razek I Press pp. 159-170
A. Ruosi 1999 IEEE Trans. Appl. Superconductivity 9, pp. 3499-3502

Inst. Phys. Conf. Ser. No 167
Paper presented at Applied Superconductivity, Spain, 14-17 September 1999
© 2000 IOP Publishing Ltd

HTS dc-SQUID with gradiometric multilayer flux transformer

M I Faley, H Soltner, U Poppe, and K Urban

Institut für Festkörperforschung, Forschungszentrum Jülich GmbH, D-52425 Jülich, Germany

D N Paulson, T N Starr, and R L Fagaly

Tristan Technologies inc., 6350 Nancy Ridge Drive, Suite 102 San Diego, CA 92121 USA

ABSTRACT: Planar HTS dc-SQUID gradiometers with a gradiometric configuration pick-up coil and a multi-turn multi-layer flux transformer in a flip-chip configuration with a dc-SQUID using quasiplanar Josephson junctions were produced and tested. The gradiometers have a baseline of about 1 cm, a sensitivity of about 1.6 nT/cm·Φ_0, and a gradient noise of about 40 fT/cm√Hz. Potential applications for these gradiometers in nondestructive evaluation and biomagnetism are discussed.

1. INTRODUCTION

Because of their high sensitivity, SQUID-based measuring systems are used for many applications, *e.g.*, in biomagnetism and nondestructive evaluation (NDE). In contrast to conventional systems with induction coils or fluxgates as sensors, high temperature superconductor (HTS) SQUID systems offer a better sensitivity especially at low (<10 Hz) frequencies. To realize this sensitivity in unshielded environments a gradiometric planar SQUID-structures which subtracts high parasitic background noise by the sensor due to gradiometric configuration of the pick-up coil can be used.

HTS SQUID gradiometers can be made from single $YBa_2Cu_3O_7$ (YBCO) layer with directly coupled pick-up loops or from multilayer structures with multi-turn input coils. On 10 mm x 10 mm substrates HTS single layer gradiometers with directly coupled pick-up loops having a sensitivity of about 500 fT/cm√Hz were demonstrated (see, *e.g.*, Seidel et al. (1999)). A flip-chip arrangement of such a gradiometer with a single layer gradiometric flux antenna on a Si wafer of about 5 cm in diameter permits one to achieve a field gradient sensitivity of 73 fT/cm√Hz in the white noise region and 596 fT/cm√Hz at 1 Hz (Tian et al. (1999)). With an inductive coupling of a magnetometer to a single layer flip-chip flux antenna a second order gradiometer with large baseline was realized (see Kittel et al. (1998) and a reference for a first order gradiometer on a similar basis therein).

A better inductive coupling between the large gradiometric pick-up loops and the SQUID loop can be realised with a multi-turn input coil in close contact with the SQUID washer. In this case even gradiometers with a smaller baseline can provide a better resolution in comparison with a single layer gradiometers. Then increase of baseline could additionally improve the sensitivity of the multilayer gradiometers. A major problem is the preparation of the high quality HTS multilayer structures on large substrates. We have developed this technology on $LaAlO_3$ wafers up to 3 cm in diameter with the help of nonaqueous Br-ethanol etching - the same technique, that we had developed earlier for the preparation of the ramp-type HTS Josephson junctions and dc-SQUIDs (see, *e.g.*, Faley et al. (1997) and references therein). Here, we describe the preparation, properties, and some of the potential applications of HTS flip-chip gradiometers having multilayer gradiometric flux antennas on such large wafers.

510

2. FABRICATION OF THE GRADIOMETERS

The high-oxygen-pressure dc-sputtering technique was used for the deposition of the YBCO and $PrBa_2Cu_3O_7$ (PBCO) films. The c-oriented films of YBCO show a typical critical current density above 10^6 A/cm^2 at 77 K and zero-resistance transition temperature of about 91 K. It was found to be very important that high quality ceramic films are used for the SQUIDs and flux antennas to guarantee good operation parameters of the whole system. Good reproducibility, low noise and stability of the gradiometers can be achieved only with well oxidized c-axis oriented ceramic films with a very small amount of structural defects. Up to 24 dc-SQUIDs with SQUID washers 1 mm in diameter were prepared on single 10 x 10 mm^2 $LaAlO_3$ (100) or $SrTiO_3$ (100) substrates and separated by a diamond disk saw. The main details of the SQUID preparation procedure have been described in previous publications (see, *e.g.*, Faley et al. (1999) and references therein).

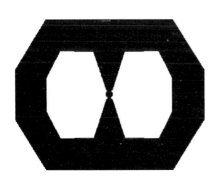

Fig.1. The layout of the planar gradiometric flux antenna with a multi-layer multi-turn flux transformer of about 1 mm in diameter (● in the middle).

Fig.2. Encapsulated planar gradiometers ($\partial B_{x,y}/\partial z$ and $\partial B_z/\partial x,y$) fixed on standard dc-SQUID packages designed for operation together with iMAG© electronics.

The planar gradiometric flux antennas (see Fig.1) with multi-layer multi-turn flux transformers 1 mm in diameter were produced on $LaAlO_3$ (100) wafers 3 cm in diameter. The effective baseline of the gradiometer is about 1 cm. The wide lines of the pick-up loops lead to a stable operation of the gradiometer in changing homogeneous fields. These wide lines can be used for a coupling to a much larger single layer gradiometric flux antenna (with a baseline > 4 cm) in a double flip-chip configuration. In the present study, only the flip-chip configuration of the dc-SQUID and the planar gradiometric flux antenna with a multi-layer multi-turn flux transformer shown in Fig. 1 is considered.

To protect the gradiometers from degradation in ambient atmosphere, we enclosed them together with a feed-back coil and a resistive heater in glass-fibre epoxy capsulations. The capsulations shown in Fig.2 are fixed on standard dc-SQUID packages designed for operation at 77 K together with iMAG© electronics. The gradiometers are being produced in small series and are commercially available.

3. EXPERIMENTAL SETUP AND RESULTS

The experimental setup used for calibration of the planar gradiometer is shown in Fig.3. The gradiometer is placed in the plane z = h above the two parallel wires fed with the same current I. The vertical component of the magnetic field $B_z(x)$ from two wires is given by the following equation:

$$B_z(x) = -\frac{\mu_0 I}{2\pi}\left(\frac{x+d}{(x+d)^2 + h^2} + \frac{x-d}{(x-d)^2 + h^2}\right),$$

where $\mu_0 = 4\pi \ 10^{-7}$ T·m/A. At some optimum height h from the wire's plane z = 0 the field gradient $\partial B_z(x)/\partial x$ is nearly constant (see Fig.3) along the length of the gradiometer. By comparison with the calculated values we determined that the field gradient sensitivity $\partial B_z/\partial x \cdot \Phi_0$ is about 1.6 nT/cm·Φ_0 for the present gradiometer. The gradiometer layout like the one presented in Fig.1 yields a gradient sensitivity of about 1.6 nT/cm·Φ_0. The measured field gradient sensitivity of a gradiometer near fully using the area of the wafer of the diameter 3 cm is better than 1 nT/cm·Φ_0.

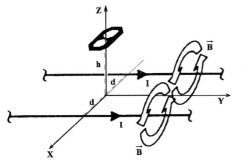

Fig.3. The experimental setup used for calibration of the planar gradiometer.

Fig.4. The calculated field gradient $\partial B_z(x)/\partial x$ at the position of the planar gradiometer with I = 1 mA; h = 3.7 cm; and d = 1.6 cm.

The gradiometer operates without any problems even when being moved in an unshielded environment. But in that case, the corresponding noise spectra mainly represent environmental magnetic field gradient noise. To measure the intrinsic noise of the gradiometer some magnetic shielding is necessary. In Fig.5 the noise of the gradiometer measured in a weak (shielding factor is about of 100 at 50 Hz) magnetic shielding is presented. To our knowledge the observed field gradient noise of about 40 fT/cm√Hz is, at the moment, the lowest for the HTS planar gradiometers at 77K.

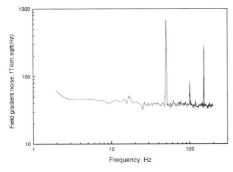
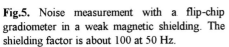
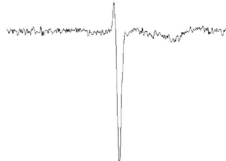

Fig.5. Noise measurement with a flip-chip gradiometer in a weak magnetic shielding. The shielding factor is about 100 at 50 Hz.

Fig.6. A magnetocardiogram obtained with a flip-chip gradiometer by averaging 60 heart beats in a frequency band from 0.03 to 150 Hz. .

512

An example of an MCG-measurement with the flip-chip gradiometer is shown in Fig.6. The magnetocardiogram was obtained in a frequency band from 0.03 Hz to 150 Hz by averaging 60 heart beats with ECG signals recorded simultaneously for triggering. The gradiometer was placed at about 10 cm above the heart of a volunteer with the x-axis normal to the chest and with the z-axis of the gradiometer directed along the body. An MCG of similar quality is expected for the operation in unshielded laboratory conditions with the electronic suppression of the environmental noise with the use of a reference gradiometer and by averaging of about 120 heart beats. An estimated magnetic moment of the heart during the R-peak is about 1 μAm^2.

For the application in NDE, the performances of the magnetometers and gradiometers in unshielded environment are important. The ac-bias reversal scheme effectively reduces the low frequency noise of the gradiometer also in unshielded environment. The laboratory environmental noise is significantly larger than the intrinsic noise of the gradiometer, but can practically be eliminated with the help of a reference gradiometer. The gradiometers can be moved and/or rotated in Earth's field during operation. A homogeneous high quality film on the circumference of the gradiometer can carry a high screening supercurrent. This allows to apply a large change of the homogeneous external field without any flux penetration into the pick-up loops of the gradiometer.

The detection of small ferrous inclusions in non-magnetic materials with the help of a HTS-SQUID gradiometer becomes a standard method in NDE (see, *e.g.*, Tavrin et al. (1999)). The ferrous inclusions can have a specific magnetic moment "p" of about 0.07 $A \cdot m^2/g$ (SmCo$_5$). If we assume a bandwidth of about 10 Hz and a signal-to-noise ratio of about 10 for such measurements, then the acceptable gradient signal from the inclusions is about 0.3 nT/m. The magnetic field gradient $\partial B_z/\partial x$ at the distance "h" from the inclusion of mass M is about $0.64\mu_0 pM/\pi h^4$. For the available gradiometer sensitivity the mass of the detectable ferrous inclusion is about 3 ng on the distance 2 cm from the centre of the gradiometer and about 2 μg on the distance of about 10 cm. The magnetic inclusions may have about 10 times (paper clip) smaller remnant specific magnetic moment than SmCo$_5$ and, correspondingly, about 10 times larger detectable masses.

4. SUMMARY

We have fabricated flip-chip dc-SQUID planar gradiometers including dc-SQUIDs with a quasiplanar HTS Josephson junctions and a multi-turn multi-layer flux transformers with gradiometric configuration of the pick-up coils. These gradiometers can be moved and/or rotated in Earth's field during operation. The gradiometers have a baseline of about 1 cm, a sensitivity of about 1.6 nT/Φ_0, the white noise of about 40 fT/cm√Hz and about 100 fT/cm√Hz at 1 Hz. This magnetic field gradient resolution of the gradiometers seems to be sufficient for the discussed applications for biomagnetism or NDE.

REFERENCES

Faley M I, Poppe U, Urban K, Krause H - J, Soltner H, Hohmann R, Lomparski D, Kutzner R, Wördenweber R, Bousack H, Braginski A I, Slobodchikov V Yu, Gapelyuk A V, Khanin V V and Maslennikov Yu V 1997 IEEE Transactions on Appl. Supercond. 7, No.2, pp 3702-3705
Faley M I, Poppe U, Urban K, Zimmermann E, Glaas W, Halling H, Bick M, Paulson D N, Starr T and Fagaly R L 1999 IEEE Transactions on Appl. Supercond. 9, No.2, pp 3386-3391
Kittel A, Kouznetsov K A, McDermott R, Oh B and Clarke J, 1998 Appl.Phys.Lett. 73, No.15, pp 2197-2199
Tavrin Y, Siegel M, Hinken J, 1999 IEEE Transactions on Appl. Supercond. 9 No.2 pp 3809-3812
Tian Y J, Linzen S, Schmidl F, Dörrer L, Weidl R, and Seidel P 1999 Appl.Phys.Lett. 74, No.9, pp 1302-1304
Seidel P, Schmidl F, Wunderlich S, Dörrer L, Schneidewind H, Weidl R, Lösche S, Leder U, Solbig O and Nowak H, 1999 IEEE Transactions on Appl. Supercond. 9, No.2, pp 4077-4080

Inst. Phys. Conf. Ser. No 167
Paper presented at Applied Superconductivity, Spain, 14-17 September 1999
© 2000 IOP Publishing Ltd

Detection of Internal Defects Using HTS-SQUID for Non-destructive Evaluation

Y Hatsukade[1,2], N Kasai[2], F Kojima[3], H Takashima[2] and A Ishiyama[1]

Waseda University, 3-4-1 Ohkubo, Shinjuku, Tokyo 169-8555, Japan
Electrotechnical Laboratory, 1-1-4 Umezono, Tsukuba, Ibaraki 305-8568, Japan
Kobe University, 1-1, Rokkodai, Nada, Kobe 657-8501, Japan

ABSTRACT: The experimental method to detect shape of internal defects in metals using HTS-SQUID gradiometer for non-destructive evaluation by changing frequency of injection current was investigated. A stair-stepped hole in copper plank sample was scanned by the SQUID changing the frequency from 10Hz to 120Hz. The amplitudes of the signal peaks due to the hole discontinuously changed at 20Hz and 60Hz. The skin depths of these frequencies corresponded to the depths where the shape of the hole varies. It indicated this method has a possibility to detect profile of internal defect.

INTRODUCTION

In recent years, detection of deep-lying flaws in metals by non-destructive evaluation using SQUID (SQUID-NDE) is prospective because the SQUIDs have extremely high sensitivity for magnetic fields at frequencies from dc to MHz. The effectiveness of SQUID-NDE for such flaws was reported on several studies (Valentino 1999, Hohmann 1999, Kreutzbruck 1999, Ruosi 1999). Because skin depth of metal changes according to frequency of flowing current, it should be possible to detect configuration of the deep-lying flaws by changing the frequency. There is a report that the depth of subsurface flaws in multi-layer structures was determined using dual frequency eddy current technique (Carr 1997). Here, we investigated a method to reconstruct a configuration of an internal crack which aperture varied according to the depth in metals by changing the frequency of injection current using a SQUID-NDE system.

For practical use of the SQUID-NDE, robustness to environmental noise, easy handling and low cost are required. High temperature superconductor SQUID (HTS-SQUID) gradiometer satisfies the needs well. So we used a HTS-SQUID gradiometer in this work.

EXPERIMANTAL

.1   SQUID-NDE system

We have developed a NDE system using a HTS-SQUID (Suzuki 1997, Kasai 1999). The system is composed of a HTS-SQUID gradiometer, a dewar with 3.5 mm thick bottom, a scanning table, a lock-in amplifier and a personal computer to acquire data and control the scanning table. The sketch of the system is shown in Fig. 1. A sample fixed on the scanning table is moved in the x direction by compressed air and in the y and z directions by mechanical gears. The microscopic photograph of the HTS-SQUID gradiometer fabricated is shown in Fig. 2. The gradiometer is a planar type one which pick-up coils are directly coupled to a SQUID arranged at the centre of the pick-up coils. The dimension of the pick-up coil and the baseline are 3.5 x 1.75 mm^2 and 1.75 mm, respectively.

514

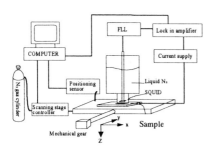

Fig. 1 The schematic figure of the NDE system used in this work.

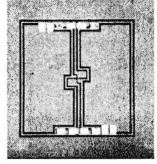

Fig. 2 The microscopic photograph of the HTS-dc SQUID gradiometer. The SQUID loop is located at the centre.

The dimension of the SQUID loop is 25 x 60 μm^2. The flux noise of the gradiometer measured in F operation in a magnetically shielded room was about 0.5×10^{-5} $\Phi_0 / \sqrt{Hz}$ at the white noise regi (10Hz – 300Hz), where $\Phi_0 = 2.07 \times 10^{-15}$ Wb is the flux quantum.

2.2 Samples

A copper plank sample with an artificial defect of a stair-stepped hole, which shape var depending on the depth, was prepared. The dimension of the sample is 100 x 200 x 20 mm^3. T configuration of the hole is shown in Fig. 3. The hole has 3 steps and the diameters of the holes at e, stage are 5, 10 and 20 mm. The depth ranges of the stages are 0–8 mm, 8–14 mm and 14–20 mm fr its surface, respectively. The centre axis of the hole is along the z-axis as shown in Fig. 4. Two ex same dimensional copper samples were prepared for comparison. One of the samples has a circu through hole of 10mm diameter and another has that of 20mm diameter.

2.3 Method

The injected current method was used in this work. The return-current pad without flaw which had the same surface area of the copper sample, was prepared. The pad was put under t sample inserting a thin insulating film between the sample and itself. The sample and the pad w connected using the silver adhesive sheet at one end of these. The ac current of 15 mA at freque range from 10Hz to 120Hz was flowed in the y direction. The sample was scanned in the x direction the HTS-SQUID gradiometer along a line over the centre of the hole shown in Fig. 4. The gradiom, detected the first order gradient component in the x direction of the z component of the magnetic density, dB$_z$/dx. The data were acquired with the PC using lock-in detection technique. The sampl interval was 1.5 mm in the x direction. The lift-off distance was 5 mm in every measurement. measurements were carried out in the magnetically shielded room.

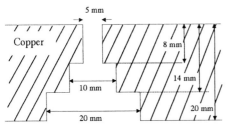
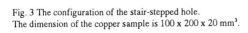

Fig. 3 The configuration of the stair-stepped hole. The dimension of the copper sample is 100 x 200 x 20 mm^3.

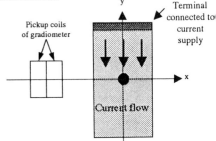

Fig. 4 The scanning direction. A sample set on the scanning table is moved in the x direction.

3. RESULTS AND DISCUSSION

The results of the stair-stepped hole are shown in Fig. 5. The centre peak is due to the hole and the peaks on both sides are due to the side edges of the sample. The relationship between the frequency of the injected current and the centre peak amplitude is shown in Fig. 6. Abrupt changes appeared at the frequencies between 20Hz and 30Hz, 60Hz and 70Hz. The data at 50Hz and especially at 100Hz included uncertainties caused by magnetic noise of the power-line and its harmonic frequencies. The depth of the current penetration into metal was limited by the skin depth effect. The skin depth is described by the following equation : $\delta = 1/\sqrt{\pi\sigma\mu f}$, where σ, μ and f are the electric conductivity, the magnetic permeability and frequency, respectively. The σ and μ of copper are 5.8×10^{8} /m and $4\pi \times 10^{-7}$ H/m, respectively. The same results were rewritten in Fig. 7 as a function of the skin depth. The skin depths at frequencies where abrupt changes appeared correspond to the depth where the diameter of the hole changed. It suggests that the abrupt changes should be due to the change of the shape of the hole. For confirmation, the through holes of 10mm diameter and 20mm diameter were also scanned using the same method. The relationships between the frequencies and the centre peak amplitudes of these through holes are shown in Fig. 8. No discontinuous change appeared in the case of the through holes, and these dependencies of the peak amplitudes on the frequency clearly differed from that of the stair-stepped hole.

We verified the relationship between the frequency and the peak amplitude of the stair-stepped hole by a simple computational simulation approximating the disturbance of the current due to the hole

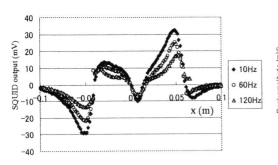

Fig. 5 The experimental results of the stair-stepped hole. The frequencies of the injected current are 10, 60 and 120Hz in this figure.

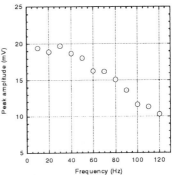

Fig. 6 Frequency vs. centre peak amplitude. Abrupt changes appeared between 20Hz and 30Hz, and 60Hz and 70Hz.

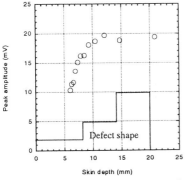

Fig. 7 Skin depth vs. centre peak amplitude rewritten from the same data of the Fig. 6. The shape of the stair-stepped hole is shown at the bottom of the figure.

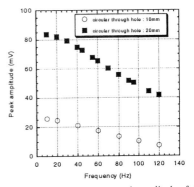

Fig. 8 Frequency vs. centre peak amplitude of the through holes of 10mm diameter and 20mm diameter. The peak amplitudes simply decreased as the frequency increased.

516

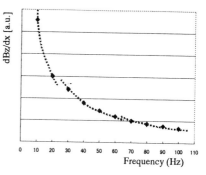

Fig.9 The simulated result on frequency vs. peak amplitude.
Discontinuous change clearly appears at frequency between 20Hz and 30Hz, and slightly between 60Hz and 70Hz.

to a current dipole. We supposed the amplitude of a current dipole should be in proportion to the volume of the hole and the current density of the injected current. The amplitude of a current dipole at z, $p(z)$ was presented as follows: $dp \propto d(z)^2 J(z, f)dz$, where d is the diameter of the hole and J is the current density of the injection current at z. The results of the through holes supported that the peak amplitude was in proportion to a square of the diameter. Because it was supposed the injection penetrated from both the upper surface and the back surface, the current density at z and at frequency f should be described as follows: $J \propto e^{-z\sqrt{\pi\sigma\mu f}} + e^{-(t-z)\sqrt{\pi\sigma\mu f}}$, where t is the thickness of the sample. The peak amplitude at frequencies adopted in the experiments was calculated by the following equation: $\dfrac{dB_z}{dx} = \int_0^t \dfrac{\mu_0}{4\pi} \dfrac{d^2 J}{z^3} dz$. The result is shown in fig.9. Discontinuous change clearly appeared at frequency between 20Hz and 30Hz, and slightly appeared at frequency between 60Hz and 70Hz.

4. SUMMARY

The stair-stepped hole on a copper plank sample was scanned using a NDE system with a HTS-SQUID gradiometer by changing frequency of the injected current from 10Hz to 120Hz. In the relationship between the centre peak amplitudes due to the hole and the frequency, abrupt changes appeared at the frequencies between 20Hz and 30Hz, and between 60Hz and 70Hz. The skin depths of these frequencies corresponded to the depth where the diameter of the hole changed. The circular through hole with 10mm diameter and 20mm diameter were also measured for comparison. The results of the through holes differed from that of the stair-stepped hole in the dependency of the peak amplitude on the frequency. This result shows that the method used in this work has a capability to detect the profile of the internal defects in metals.

REFERENCES

Carr C, McKirdy D M, Romans E J, Donaldson G B and Cochran A, 1997 IEEE Trans. on Appl. Supercond. 7, 3275

Hohmann R, Maus M, Lomparski D, Grueneklee M, Zhang Y, Krause J H, Bousack H and Braginski A I, 1999 IEEE Trans. on Appl. Supercond. 9, 3801

Kasai N, Suzuki D, Takashima H, Koyanagi M and Hatsukade Y, 1999 IEEE Trans. on Appl. Supercond. 9, 4393

Kreutzbruck M V, Baby U, Theiss A, Mueck M and Heiden C, 1999 IEEE Trans. on Appl. Supercond. 9, 3805

Ruosi A, Pepe G, Peluso G, Valentino M and Monebhurrun V, 1999 IEEE Trans. on Appl. Supercond. 9, 3499

Suzuki D, Kasai N, Oomine R, Koyanagi M and Kurosawa I, 1997 HTSED Workshop '97 Sponsored by FED pp 96-98

Valentino M, Ruosi A, Pepe G P and Peluso G, 1999 to be published in AEM 15, eds D Lesselier and A Razek (Amsterdam: IOS Press)

Inst. Phys. Conf. Ser. No 167
Paper presented at Applied Superconductivity, Spain, 14-17 September 1999
© 2000 IOP Publishing Ltd

Development of an HTS cryogenic current comparator

M D Early, K Jones and M A van Dam

The Measurement Standards Laboratory of New Zealand, Industrial Research Limited
P O Box 31-310, Lower Hutt, New Zealand

ABSTRACT: We have developed a design of a toroidal cryogenic current comparator (CCC) that should allow high accuracy current ratios to be established. A high degree of shielding is obtained using split toroidal shells that are alternated in their orientation. Using LTS materials and a DC SQUID, we have investigated a prototype of this geometry and found performance only slightly inferior to that obtained with continuous shielding. Based on these results we expect that a ratio accuracy of 10^{-8} is feasible for an improved construction.

1. INTRODUCTION

Low temperature cryogenic current comparators (CCC's) can be used to establish dc current ratio's to an accuracy of better than 10^{-10}. The usual overlapped-tube CCC consists of a set of toroidal ratio windings enclosed by a thin lead shield that overlaps itself two or three times (often referred to as being like a snake swallowing its tail). The success of this construction is due to the ease with which lead sheet can be shaped and soldered.

The practical difficulties of Helium systems have restricted the use of LTS CCC's, but an HTS CCC could be expected to be extremely advantageous in a wide range of applications (for example in a resistance bridge). However the complex three-dimensional geometry and the difficult mechanical properties of HTS materials have so far thwarted the construction of a similarly accurate HTS version of the device. Elmquist (1999) has achieved performance of better than 10^6 in a CCC based on HTS coated parallel tubes. Arri *et al* (1998) have considered a design where the windings are located near the bottom of an open toroidal gutter.

Based on numerical models of the current distributions in toroidal CCC's we propose to extend the ideas of Arri *et al* (1998) to a more traditional toroidal shape where the windings are enclosed by a series of split shells which are oriented in an alternating fashion. This configuration should achieve improved accuracy and sensitivity. In this paper we describe the results of an experiment to verify the efficacy of using such shells in a geometry that could be successfully constructed using a bulk HTS material. For the purposes of this work the prototype is constructed from LTS materials and a LTS DC SQUID is used.

2. HTS CCC DESIGN

Since the insulation on the ratio windings would be damaged by exposure to the high temperatures used to process HTS materials, the shielding must be formed from finished components. This is possible in a toroidal geometry if the shield is split along a diagonal across the cross-section of the bore. In this way multiple layers of shielding can be formed. Generally it is considered that such breaks in the shielding of CCC's would be highly deleterious. However by alternating the orientation of the diagonal split, errors due to the split in one layer are attenuated by the continuous shielding of the next layer at the same corner. This is confirmed in our numerical modelling work (Section 4). A cross section of a CCC with this feature is shown in Fig. 1. While a split can prevent the establishment of an 'equilibrium' current distribution (that is the distribution when the shielding is continuous), the alternating location of the split allows this current to be redistributed on successive layers.

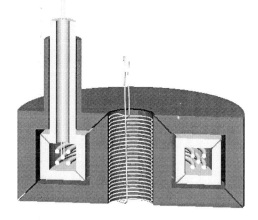

Fig. 1: CCC cross-section. For clarity only one set of external connections to the ratio windings is shown. The dimensions are given in the text.

A second difficulty in emulating the overlapped tube construction is that successive layers cannot be connected in a way that gives rise to a continuous overlap. However, unconnected concentric layers of shielding still provide a large degree of attenuation as long as the position of the gap in each layer is diametrically opposed. This gap (which is in a radial direction) is required so that the internal screening currents are forced to return on the outer surface. An analysis due to Seppa (1990) suggests that the attenuation provided by the spacing between successive toroidal layers of radius r and width w is $\sim e^{-2r/w}$. Hence the main effect of having unconnected layers is that the length of the attenuation path is half as long and we would expect a factor of about 7 times less attenuation in this case.

The exit tube where the windings enter and leave the toroid will also need to be superconducting and we propose to use closely fitting tubes pressed into the shells. The pickup coil shown in Fig. 1. is not appropriate for an HTS device and we expect that an HTS SQUID (with or without a flux transformer) will be able to be directly mounted in this region with good coupling to the toroid.

3. MEASUREMENT

3.1 Construction of the LTS Prototype

The cross section of the CCC shape used for the prototype is as shown in Fig. 1. The inner layer is 10 mm by 10 mm with a wall thickness of 2 mm. The outer layer is 20 mm by 20 mm with a wall thickness of 4 mm. The axial radius of both layers is 15 mm. The shells are machined from a 50:50 tin/lead alloy

The seams around the toroidal shells are machined but have not been further treated to improve the finish. A 0.5 mm cut is made through each layer at diametrically opposed locations to form the gap. Five 10-turn windings are set into the four corners and near the centre of a Teflon former. This maximises the ratio error of the windings. The pickup coil consisted of 26 turns of 0.25 mm diameter niobium wire spread over 20 mm (axial radius 4.7 mm) connected to a commercial DC SQUID. Polyimide tape was used to ensure the insulation between all conducting components is maintained.

3.2   Measurement Method

We made measurements of the ratio error in three configurations: firstly with the shells touching, secondly with the shells soldered together along the seams, and finally with several layers of polyimide tape between the shells to provide about 0.5 mm spacing. The entire CCC was disassembled, modified and reassembled for each configuration.

4.   RESULTS

4.1   Measured Values

The ratio error of each corner winding relative to the central winding for the three measurement configurations is shown Fig. 2. All other ratio errors (e.g. one corner relative to another) can be calculated from these four values. From the measured positions of the windings the ratio error for the central winding relative to the geometric centre of the former can be calculated. In the case of the configuration with the large gap, the 0.4 mm offset of the central winding leads to a -110×10^{-6} error, showing how sensitive the ratio error is to the position of the wires.

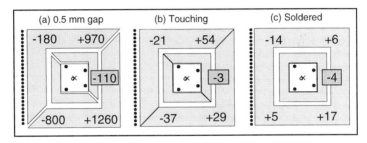

Fig. 2: Measured ratio errors ($\times10^{-6}$) of each corner winding (solid circles) relative to the central winding (open circle). The calculated ratio error of the central winding to the centre of the former (cross) is shown in the box.

There is a substantial reduction in the ratio error by about a factor of 20 as the shells are brought into contact. The ratio errors for the soldered configuration are a further factor of four smaller and indicate the level of error associated with presence of the gap in the shields. For this shape the calculated attenuation factor of the spacing between the layers (Seppa, 1990) is ~ 10^{-4}. We would expect that the difference between the soldered and the touching ratio errors indicate the sort of difference the use of split HTS shields might cause compared with an overlapped tube CCC. We would also expect that using a technique to bring the shells into more intimate contact (such as lapping the seams) would reduce this difference even more.

4.2   Calculation of the Ratio Error

We have previously developed a matrix method for calculating the current distribution on the external surface of an overlapped tube (Early and van Dam, 1999). We have generalised this method to deal with the split shields where the current flowing on the external surface is now unknown. Instead we use the requirement that because of the gap, the net current through the cross section of each shell must be zero. The current distributions we obtain for the two layers of split shells are shown in Fig. 3 for a current of 1 A in the central winding. The external current distribution for the second (outer) layer has the same form as that found for an overlapped tube CCC (Early and van Dam, 1999).

520

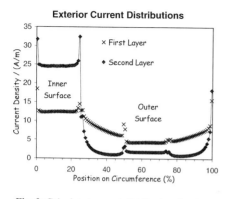
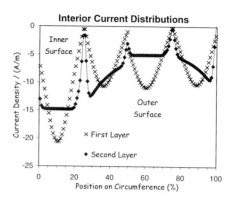

Fig. 3: Calculated current distributions for the prototype CCC for a gap of 0.5 mm and a current of 1 A in the central winding. The circumference for the exterior of the shells is 40 mm for the inner layer and 80mm for the outer.

By calculating the output for pairs of windings with opposing currents we obtain the ratio error. The results for the CCC with a large gap are shown in Table 1 for two values of the shell spacing. We also show the effect of a small offset in the position of the former relative to the shells. Compared to measured values, the 1mm calculated values are of the right order but are not of the same pattern or sign. The values obtained for a 0.5mm shift in the position of the whole former relative to the shells show a change of 268×10^{-6}. We believe that the poor agreement is associated with sensitivity to precise relative locations of the windings.

Table 1: Comparison of measured and calculated ratio errors for the CCC with the large gap.

Spacing (mm)	0.5	0.5	1.0	1.0
Offset of Former (mm)	0	0	0	0.5
Position of Winding	**Measured Error** $(\times10^{-6})$	**Calculated Error** $(\times10^{-6})$		
Inner & Lower	-180	+221	+862	+945
Inner & Upper	-800	+15	-272	-540
Outer & Lower	+1260	-13	-660	-601
Outer & Upper	+970	-91	-1028	-938

5.   CONCLUSION

With the use of alternating symmetry to cancel out the imperfections of successive layers, we have shown that the performance of unconnected shielding with split seams can approach that of a continuous superconducting shield. Careful attention paid to attaining closely mating surfaces may further reduce the error. With an extra layer of shielding we would anticipate further significant reduction in the ratio error and that errors of 10^{-8} would be achievable.

REFERENCES

Arri E, Boella G, Pavese F, Negro A, Vanolo M, Dagnino and Lamberti P 1998 Conference on Precision Electromagnetic Measurements Digest, pp 221-222
Early M D and van Dam M A 1999 IEEE Trans. Instrum. Meas. **48**, pp 379-382
Elmquist R E 1999 IEEE Trans. Instrum. Meas. **48**, pp 383-386
Seppa H 1990 IEEE Trans. Instrum. Meas. **39**, pp 689-697

Inst. Phys. Conf. Ser. No 167
Paper presented at Applied Superconductivity, Spain, 14-17 September 1999
© 2000 IOP Publishing Ltd

Development of HTS Resistive SQUIDs for Noise Thermometry

D A Peden[1] J C Macfarlane[1] L Hao[2] and J C Gallop[2] [3]

[1]Department of Physics and Applied Physics, Strathclyde University, Glasgow
G4 0NG, UK.
[2]Centre for Basic and Thermal Metrology, National Physical Laboratory, Teddington,
TW11 0LW, UK.

ABSTRACT: The use of a High-T_c resistive SQUID (R-SQUID) for Noise Thermometry is currently under investigation. This technique has the potential to provide realisation of the absolute temperature scale in the range 10 K – 50 K. An R-SQUID consists of a noble metal resistor connected as a shunt across a YBCO grain boundary junction, where the shunt resistance R_s is significantly smaller than the junction normal state resistance R_n. Resistors based on a YBCO/Au bilayer intersected by a meandering gap have been fabricated. A second ex-situ Au layer forms the resistive path, and $R_S \simeq 26$ $\mu\Omega$ has been measured at 77 K. Resistors have been used in both single junction and double junction devices. One dimensional current flow simulations and experimental measurements of the resistors in the temperature range of 30 K to 77 K are presented.

1   INTRODUCTION

Josephson noise thermometry using a low-T_c R-SQUID is established for < 1 K (Kamper and Zimmerman 1971). We are currently developing high-T_c R-SQUIDs with the aim of extending this absolute thermometry technique to temperatures of $10\,K-50\,K$. These devices incorporate either one or two bicrystal Josephson junctions in parallel with a noble metal shunt resistor R_s. The R-SQUID is biased so the resistor thermal noise broadens the Josephson radiation of the junctions. The Josephson linewidth is directly related to the device temperature by

$$\delta f(T) = \frac{4\pi\, k_B\, T\, R_s}{\Phi_0^2} \qquad (1)$$

consequently the R-SQUID is an *absolute* thermometer (Peden *et al.* 1999). The 2 junction R- SQUID design (Fig. 1) incorporates two Josephson junctions in a small loop of inductance $\simeq 7.1$ pH.

It is desirable that $R_s \ll R_n$, so that the thermal noise in the shunt resistor dominates over the junction resistance R_n which may be non-linear and temperature dependent, implying that $R_s \leq 0.1$ mΩ (Gallop and Petley 1995). The design of R-SQUID devices was facilitated by investigating a number of resistor test structures in an effort to reduce R_s. Theoretical models have been applied to the YBCO/Au bilayers and meandering resistor gaps employed to promote current flow in the *a-b* plane. Josephson radiation measurements and simulations with two junction R-SQUIDs are presented elsewhere (Hao *et al.* 1999).

---

[3]To whom correspondence should be addressed.

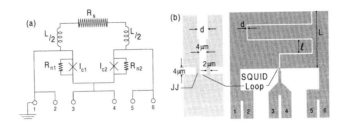

Fig. 1: (a) Schematic diagram of a double junction (2J) R-SQUID; (b) A sketch of the device design.

2 DEVICE FABRICATION

The HTS Resistive SQUID devices and test structures have been fabricated from PLD c- axis YBa$_2$Cu$_3$O$_{7-\delta}$ thin films on 10 mm × 10 mm 24° SrTiO$_3$ bicrystal substrates. Device patterning was by standard photolithography and Ar ion beam milling. To prevent contamination of the YBCO surface and to reduce the contact resistance, R_c, a layer of *ex-situ* Au was deposited *before* photolithographic patterning of a 2J R-SQUID. A second *ex-situ* Au layer was patterned by lift-off photolithography to produce resistor and contact pads. The ion mill was then used to remove the first *ex-situ* Au layer from the junctions and the connections between the contact pads. The 2J R-SQUID design shown in Fig. 1 incorporates 2 μm × 4 μm grain boundary junctions and the resistor is formed by a $d = 6$ μm meander gap across a large 4.0 mm × 7.0 mm YBCO pad.

3 CURRENT FLOW SIMULATIONS

A transmission line model (TLM) was applied to model the current flow in a resistor structure of width L, finger width ℓ, gap width d, and Au thickness t and length b. The model uses the expressions,

$$R_c = \frac{1}{L} \left(\frac{\rho_c \rho_m}{t} \right)^{1/2} \coth \left[b \left(\frac{\rho_m}{\rho_c\, t} \right)^{1/2} \right] \tag{2}$$

$$R_m = \frac{d\, \rho_m}{L\, t} \tag{3}$$

for the contact resistance R_c and the resistance of the metal across the gap R_m. The c-axis current is assumed to flow into the Au over a transfer length,

$$b_{eff} = \left(\frac{\rho_c\, t}{\rho_m} \right)^{1/2} \tag{4}$$

such that the effective gap width $d_{eff} = d + 2b_{eff}$. Table 1 shows data for single junction R-SQUIDs where planar and meander gap models have been used. The latter assumes that the width L is equivalent to the perimeter of the meander gap. A lower limit of (ℓ/L) is shown in the case of the Strathclyde/PTB device, where the model provides a better fit when a planar gap structure is assumed (Peden *et al.* 1999). Consequently, the 2J R-SQUID devices reported here used meander gap resistors with wide fingers of YBCO, $(\ell/L) \sim 0.25$.

Table 1: TLM simulation results of shunt resistance (R_{sim}) on single junction R-SQUID devices. The simulations assume a sample temperature of 50 K, $\rho_c = 9.2 \times 10^{-12}$ $\Omega.m^2$, and $\rho_m = 2.0 \times 10^{-9}$ $\Omega.m$ (Peden *et al.* 1999) and $\rho_c = 5 \times 10^{-12}$ $\Omega.m^2$ for the 2J R-SQUID.

Device	NKT		Strathclyde/PTB		2J R-SQUID
Device Gap	Planar	Meander	Meander		Meander
Simulation	Planar	Meander	Meander	Planar	Meander
d_{eff} (μm)	112.9	72.5	58.5	58.5	106
b_{eff} (μm)	21.4	21.4	26.3	26.3	50
t (nm)	100	100	150	150	1000
(ℓ/L)	N/A	0.1^a	0.0005	0.02	0.25
R_{sim} (mΩ)	1.558	0.247	0.049	1.975	26.6 $\mu\Omega$
R_{exp} (mΩ)	1.5	0.24	1.2–2.0		18 − 26 $\mu\Omega$

^{a}estimated parameter from device design.

4 DOUBLE JUNCTION R-SQUID EXPERIMENTAL RESULTS

4.1 Resistor Structure

To permit a measurement of the resistance across the noble metal shunt, the Josephson junctions of the first such device were left open circuit. Figure 2 shows the variation in R_s with temperature, measured from 18 K to 77 K. The resistance was measured as $\simeq 18$ $\mu\Omega$ at 18 K increasing to $\simeq 26$ $\mu\Omega$ at 77 K. This low value of R_s is an order of magnitude smaller than any results previously reported (Peden *et al.* 1999).

4.2 Double Junction R-SQUID

With bias currents applied to the R-SQUID, the dependence of the Josephson frequency on bias current allowed the resistance of the 2J R-SQUID to be measured

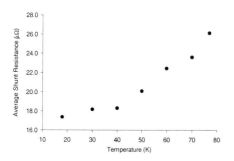

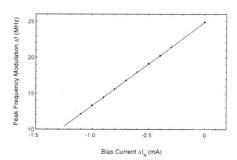

Fig. 2: Variation in R_s with T, with $R_s \simeq 18$ $\mu\Omega$ at 18 K increasing to $R_s \simeq 26$ $\mu\Omega$ at 77 K.

Fig. 3: Variation in Δf with ΔI_R indicating $R_s \simeq 24$ $\mu\Omega$ at 30 K.

from the gradient of $\Delta f/\Delta I_R$ where I_R is the current through R_s,

$$R_s = \Phi_0 \frac{\Delta f(T)}{\Delta I_R} \simeq 24 \ \mu\Omega \tag{5}$$

as shown in Fig. 3. This device was fabricated in an identical method to the test resistor structure described in section 4.1, and shows that YBCO/Au/Au resistors can be reproducibly made in the range $\sim 18 - 30 \ \mu\Omega$.

5 CONCLUSION

Development of high-T_c R-SQUIDs requires the reproducible fabrication of a small shunt resistance R_s so that the thermal noise is dominated by R_s rather than R_n. Meander gap devices have been fabricated in the c-axis films to promote current flow in the a-b plane. The current flow through resistors has been modeled in order to predict R_s before device fabrication. The modeling indicates a lower limit to (ℓ/L) below which meander gaps offer no advantage over planar gaps. A number of resistor test structures have been fabricated in an effort to reduce R_s, and preliminary results show good reproducibility of $\sim 18 - 30 \ \mu\Omega$ YBCO/Au/Au resistors. The temperature dependence of R_s is $< 2\%$ per K in the range $18 - 70$ K. This ex-$situ$ resistor process approaches the lowest contact resistivity values achieved with in-$situ$ processing (Ekin $et.$ al 1993). The resistance at 30 K is independent of bias current up to $I_R > 1$ mA . The value of R_s under operating conditions can be accurately measured from the bias current dependence of the Josephson oscillation frequency.

ACKNOWLEDGEMENTS

We are grateful to E.J. Romans, University of Strathclyde for PLD YBCO films. Also to F. Ludwig, PTB Berlin, and Y. Shen, NKT, Denmark for exchange of samples and useful discussions.

REFERENCES

Ekin J Russek S E Clickner C C and Jeanneret 1993 Appl. Phys. Lett. **62** (4) 369-371.
Gallop J C and Petley B W 1995 IEEE Trans. Instrum. & Meas. **44** 234-237.
Hao L, Gallop J C, Macfarlane J C, Peden D A, and Pegrum C M 1999 EUCAS '99 Proc., Barcelona (submitted).
Kamper R A and Zimmerman J E 1971 J. Appl. Phys. **42** 132-136.
Peden D A, Macfarlane J C, Hao L, Reed R P, and Gallop J C 1999 IEEE Trans. App. Supercond. **9** (2) 4408-4411.

Inst. Phys. Conf. Ser. No 167
Paper presented at Applied Superconductivity, Spain, 14-17 September 1999
© 2000 IOP Publishing Ltd

Application of SQUID based position detectors for testing the Weak Equivalence Principle at the drop tower Bremen

W Vodel, S Nietzsche, J v Zameck Glyscinski, H Koch, and R Neubert

Friedrich Schiller University, Institute of Solid State Physics, Max-Wien-Platz 1, 07743 Jena, Germany

ABSTRACT: Due to their exceptional sensitivity and universality SQUIDs can be applied even in position detectors providing an unusually high position resolution on the order of 10^{-14} m/$\sqrt{\text{Hz}}$. Using this sensor technique gravitational experiments can be performed such as the test of the Weak Equivalence Principle on earth using drop tower facilities or in orbit like in the current Satellite Test of the Equivalence Principle (STEP) of NASA/ESA.

This contribution describes an experimental set-up which is presently tested at the Drop Tower Bremen, Germany, providing a free-fall height of 109 m which corresponds to a measurement time of 4.7 sec. In principle, this apparatus consists of two free falling test bodies, a 2-channel LTS DC SQUID based position detector, magnetic bearings, caging mechanisms, and a liquid helium cooling. All systems are qualified for operation in the drop tower and capable of withstanding the high-deceleration phase of 50 g at the end of each drop. Finally some recent results are discussed.

1. INTRODUCTION

Gravity seems to enjoy a remarkable universality property because all bodies have been found experimentally to fall approximately at the same rate. Einstein developed this principle of universality of free-fall to the level of a grand hypothesis that he termed the "Equivalence Principle" (EP). Einstein used the EP as the basic postulate of General Relativity. As such, it has always deserved to be tested with the highest possible precision.

It was long ago, in the 16th Century, that Galileo performed his famous free-fall experiments at the Leaning Tower of Pisa. Newton, using pendulums, determined the validity of the EP to one part in 10^3, whereas, Eötvös in 1896 achieved a sensitivity of 5×10^{-8} using a torsion balance. Since then others, using progressively more refined experiments, have achieved a sensitivity of about 10^{-11}.

Today physicists are still going at it. But things have become rather sophisticated since then: In our current experiments we use the Drop Tower at Bremen University with an evacuated steel tube having a drop length of 109 m and a flight time of 4.7 s for our free-fall experiments. This drop tower as an earth bound microgravity facility offers the unique possibility of a pseudo-Galilean test of the EP in which a true relative measurement can be performed. The drop capsule is equipped with position detectors of extremely high resolution on the basis of DC SQUIDs for the precise measurement of movement of the two free-falling test bodies.

2. EXPERIMENTAL EQUIPMENT

2.1 DC SQUID System

In order to realize the precision measurement of tiny displacements of free falling test bodies a high sensitive DC SQUID measuring system is required. Such high performance DC SQUID based measurement systems are developed and manufactured at the Institute of Solid State Physics of the Friedrich Schiller University, Jena. All systems use the sensor *UJ 111* (Vodel and Mäkiniemi 1992) designed in a gradiometric configuration based on Nb-NbO$_x$-Pb/In/Au window-type Josephson tunnel junctions with dimensions of 3 μm × 3 μm. To achieve a low total inductance the SQUID loop consists of eight sub-loops connected in parallel. To couple a signal into the gradiometer-type SQUID an input coil system is integrated onto the chip consisting of two coils of 18 turns each, connected in a gradiometric configuration. The input inductance of the SQUID is about 0.8 μH suitable for an optimal matching to all common signal sources.

In contrast to other sensors the SQUID UJ 111 was designed for universal applications in precision measurement technique and works at an extremely low noise level also in a magnetically unshielded environment. The long term stability of the SQUID parameters during a time period of several years is remarkable even though the SQUIDs are not encapsulated hermetically. According to our experience there is no influence on the SQUID parameters caused by more than 100 cooling down cycles. It should be pointed out that the SQUID sensors are also very insensitive to mechanical shocks tested by free-fall experiments from a height of up to 120 m causing a deceleration of up to 50 g at the end of the flight.

For the optimal choice of bias and flux modulation point, a white flux spectral density of 2×10^{-6} Φ_o /√Hz for the SQUID system was found. Using optimal electric and magnetic shielding of the sensor no 1/f knee could be measured down to 10 mHz (Vodel and Mäkiniemi 1992).

2.2 SQUID position detector

The principle components of a SQUID position detector consist of a superconducting pick-up coil and a superconducting diaphragm covering the test body. When this pick-up coil is connected across the input coil of the SQUID and a definite current is trapped in this closed superconducting loop any small movement of the test mass results in an output signal of the SQUID. The superconductive coated test body acts as a tuning slug changing the inductance of the two pick-up coils.

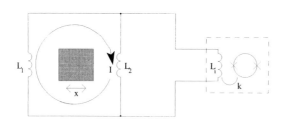

We started our investigations with a position detector using one pick-up coil only (Vodel et al 1997). An improved version of a position detector with two pick-up coils is seen in Fig. 1. The most important advantage of this detector, compared with a detector having a single pick-up coil only, consists in the fact that the current I is not limited by the critical current of the input coil of the SQUID (Nietzsche 1996). According to our calculations (see Vodel et al 1997), a higher current I will provide a much better resolution δx of the detector.

Fig. 1. SQUID position detector with two pick-up coils.

The designing of the position detector, enabling highest resolution, involved testing various configurations of pick-up coils and test bodies and measuring the dependence of inductance on displacement for each one. A special measurement equipment has been developed integrating a micrometer screw and a piezoelectric element providing still finer displacements in the μm-range and capable also of measuring the inductance. The displacement was measured using a commercial detector with a resolution of 100 nm. The best detector configuration we found was a cylindrical pick-up coil and a concentrically aligned cylindrical test body providing a high sensitivity over a wide working range of 1–2 mm (Vodel et al 1997). The detector described offers the following advantages over a single pick-up coil detector used in earlier experiments: no limitation of the persistant current I caused by the SQUID, reduced sensitivity to magnetic disturbances, and higher resolution.

After performing many free-fall experiments with a simplified system having only one single test body, our new system with two test bodies was completed and has been successfully tested in first free-fall tests. Fig. 2 represents a simplified drawing of the new two-body system. The two concentric test bodies (a) and (b) are made of lead and aluminium (coated with niobium), respectively. They are stabilized in the radial

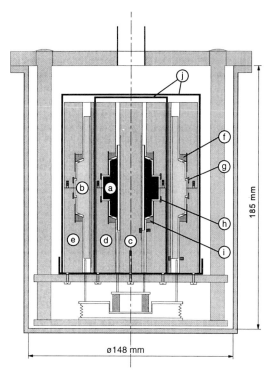

direction by meander-shaped coils and can be controlled in the vertical direction by levitation coils. These coils are supported by special carriers (c), (d), (e). For instance, (f) denotes the upper levitation coil for the outer test body and (i) the lower levitation coil of the inner test body. The pick-up coils required for the position measurement are also attached close to the test bodies. For each body two pick-up coils are needed as described above. (h) shows the lower pick-up coil of the inner test body and (g) the upper pick-up coil of the outer test body. Furthermore all parts are shielded by superconducting screenings (j). The whole system is housed in a vacuum chamber immersed in liquid helium at 4.2 K. The cryostat (20 l LHe), the SQUID electronics, the data processing unit, the control unit and the power supply are mounted in a special structure, the so-called free-flyer, which resists the vibrations and shocks at the beginning and at the end of the drop. The free-flyer itself falls freely inside of the outer fall capsule, which contains a computer-controlled caging mechanism. Despite the weightlessness and the rapid deceleration of about 500 m/s² (50g) after numerous drop experiments neither the cryogenic equipment nor the SQUID electronics have been damaged.

Fig. 2. Simplified assembly drawing of the cryogenic part of the measurement system for testing the Equivalence Principle including two test bodies, levitation coils, and pick-up coils for position measurement (see text for further explanations).

3. RECENT EXPERIMENTAL RESULTS

Our last experiments were done to improve the test body position control unit. This unit fixes the test bodies at the beginning of the drop. Before release, the hanging capsule is under tension from its own weight resulting in a contraction upon release. This would result in an upwards acceleration of the test bodies. After vibrations of the system fade, the control unit has to center the test bodies in their working position quickly. For this purpose the control unit measures the position and the velocity of the test bodies and sends short rectangular shaped pulses of a calculated length through the upper and the lower levitation coils, respectively. These coils situated at both ends of each test body can repel the superconducting test bodies. Theoretically, two pulses from opposite directions are sufficient to attain centering even given an initial velocity. However, after the two pulses there are small deviations from the aspired position and velocity due to errors in determination of the initial velocity and position or due to limitations of the control unit. Therefore, it will be necessary to repeat the process several times. For future precision experiments the final velocity of the test bodies relative to the measurement system should be lower than 0.25 μm/s to avoid exceeding the working range.

Fig. 3 shows a preliminary result of a free-fall test. The test body was preaccelerated for 10 ms. After that the control unit measured the velocity and stopped the test body with a pulse of a calculated length. This process was repeated 4 times with 10 ms left between the pulses. The spikes in the curve are caused by the crosstalking of the levitation coils and the small bandwidth of the A/D converter. After two pulses the velocity became so small, that the next pulses were very short. At this time the current amplitude of the pulses should be reduced and the sensitivity of the control unit has to be increased.

528

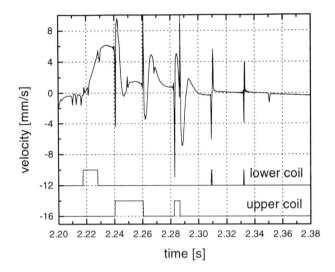

Fig. 3. Velocity of one of the test bodies during the control process recorded during a drop. The lower traces show the compensating pulses of the levitation coils.

CONCLUSIONS

The application of high performance DC SQUIDs allows repeated measurement of the relative postition of a superconducting covered test body with an exceptional noise limited resolution of 4×10^{-14} m/$\sqrt{\text{Hz}}$. The best detector configuration was found to be a cylindrical pick-up coil and a concentrically aligned cylindrical test body providing a high sensitivity over a wide working range of 1–2 mm.

On the basis of these SQUID position detectors it will be possible to prove the validity of Einstein's Equivalence Principle with an accuracy of at least 10^{-13} at the Drop Tower Bremen. This would be by one order of magnitude better than the most accurate experiments performed by Roll, Krotkow and Dicke (1964) who used highly precise modern torsion balance equipment. In the near future, the ongoing STEP project (Satellite Test of the Equivalence Principle) of NASA/ESA will send such a device into space allowing an even higher accuracy than can be achieved on earth (Barlier et al 1991, Everitt et al 1997).

ACKNOWLEDGMENTS

We express our special gratitude to H. Dittus (Center of Applied Space Technology and Microgravity, Bremen University) for his support during the preparation and realization of the free-fall experiments at the Bremen Drop Tower. This work was supported in part by DLR, Germany, contract no. 50WM94383.

REFERENCES

Barlier F, Blaser J P, Cavallo G, Damor T, Decher R, Everitt C W F, Fuligni F, Nobili A, Nordtvedt K, Pace O, Reinhard R, Worden P 1991 STEP Assessment Study Report, **SCI(91)4**
Everitt C W F, Worden P, Ryans-Culclager L, Farnsworth R 1997 STEP NASA Research Announcement: NRA-96-HEDS-03
Nietzsche S 1996 Diploma Thesis Friedrich Schiller University Jena Germany
Roll P G, Krotkov R V, Dicke R H 1964 Ann. Phys. **26** 442-517
Vodel W, Koch H, Nietzsche S, von Zameck-Glyscinski J, Dittus H, Lochmann S and Mehls C 1997 IEEE Transactions on Applied Superconductivity **7** no.2 3343-6
Vodel W, Mäkiniemi K 1992 Measurement Science and Technology **3** no. 2 1155-60

Inst. Phys. Conf. Ser. No 167
Paper presented at Applied Superconductivity, Spain, 14-17 September 1999
© 2000 IOP Publishing Ltd

Magnetic characteristics and noise properties of asymmetric dc SQUID configurations

G Testa, S Pagano, M Russo and E Sarnelli

Istituto di Cibernetica del CNR, Via Toiano 6, 80072 Arco Felice, Napoli, Italy

ABSTRACT: Asymmetric dc-SQUID configurations have been investigated. In comparison to symmetric SQUIDs, asymmetric configurations show an increase of the flux to voltage transfer function V_Φ and a decrease of the magnetic flux noise. The effect of a damping resistance leads to a smaller flux noise for large SQUID inductance L and to an increase of V_Φ with L, up to a maximum value depending on the absolute temperature and the circuit parameters. These results indicate the possibility of using asymmetric dc-SQUID to improve device performance both for small and large inductances.

1.   INTRODUCTION

Several SQUID configurations have been proposed in order to improve the device sensitivity and to reduce the magnetic flux noise. Tesche and Clarke (1977) first studied the characteristics of SQUIDs with asymmetry in the critical current and in the shunt resistance of the two junctions, or in the self inductance of the two arms. Their analysis was mainly concerned with static noise-free properties, like current-voltage and voltage-flux behaviors, as a function of the device parameters. Voltage noise sources were then included in the model to optimize noise properties of symmetric SQUIDs. In particular, an optimal value of the parameter $\beta=2LI_C/\Phi_0=1$ was found, where L is the loop inductance of the SQUID, I_C the critical current of each Josephson junction and Φ_0 the magnetic flux quantum.
Enpuku(1984) proposed a new configuration with a damping resistance shunting the SQUID loop inductance. He showed that both the flux to voltage transfer function and the noise characteristics are considerably improved in the case of large β. As a result SQUIDs with large inductances could better match external flux transformers, necessary for magnetometric applications, therefore increasing the magnetic field sensitivity.
In the present paper we numerically investigate, for the first time, the effect of asymmetry on SQUID performance in the presence of noise. An asymmetry of the critical current of the two junctions normally reduces the flux to voltage transfer function $V_\Phi = dV/d\Phi$ and increases the flux noise. However other asymmetric configurations can improve both voltage and noise performance of the SQUIDs. Such an improvement can be enhanced by the use of a damping resistance. In this case we show that the transfer function $dV/d\Phi$ not only is not degraded for large β as in symmetric configurations, but increases up to a maximum value depending on the absolute temperature and the circuit parameters. In section 2 we show numerical simulations for symmetric and asymmetric dc SQUIDs, with and without damping resistance, and we briefly discuss the results.

530

2. NUMERICAL SIMULATIONS

The equivalent circuit for the dc SQUID has already been shown elsewhere (Teshe 1977 Enpuku 1985). The normalized equations for the SQUID voltage and currents that we have considered are slightly different from the usual ones since we have taken into account asymmetries in the critical current ($I_{c1}=I_c(1-\alpha)$ and $I_{c2}=I_c(1+\alpha)$) and in the normal resistance ($R_1=R/(1-\rho)$ and $R_2=R/(1+\rho)$) of the two Josephson junctions:

$$(1+2\gamma-\rho^2)v_1/\gamma=\{i_b+i_{n1}+i_{n2}-(1-\alpha)\sin\theta_1-(1+\alpha)\sin\theta_2-[i_b-2J+2i_{n1}+2i_{n3}-2(1-\alpha)\sin\theta_1]\,(1+\rho)/2\gamma\} \quad (1)$$
$$(1+2\gamma-\rho^2)v_2/\gamma=\{i_b+i_{n1}+i_{n2}-(1-\alpha)\sin\theta_1-(1+\alpha)\sin\theta_2-[2(1+\alpha)\sin\theta_2-i_b-2J-2i_{n2}+2i_{n3}]\,(1-\rho)/2\gamma\} \quad (2)$$
$$J=(\theta_1-\theta_2-2\pi\phi)/\pi\beta \quad (3)$$
$$v_1=d\theta_1/dt \quad (4)$$
$$v_2=d\theta_1/dt \quad (5)$$

where θ_1 and θ_2 are the junction phase differences, $v_1=V_1/RI_C$ and $v_2=V_2/RI_C$ the normalized voltages, $\gamma=R/R_d$ with R_d the damping resistance, $i_b=I_b/I_C$ and $\phi = \Phi/\Phi_0$ the normalized bias current and applied flux respectively. The time t is normalized by $\Phi_0/2\pi RI_C$.
We have assumed that the normalized noise currents i_{n1}, i_{n2} and i_{n3} are due to Johnson noise generated by the junction and damping resistances with spectral densities $S_1=4KT/R_1$ $S_2=4KT/R_2$ and $S_3=4KT/R_d$ respectively, where K is the Boltzman constant and T the absolute temperature. By normalizing S_i with $I_C\Phi_0/2\pi R$ we obtain $S_{in1}=4\Gamma(1-\rho)$, $S_{in2}=4\Gamma(1+\rho)$, $S_{in3}=4\Gamma\gamma$, where $\Gamma = 2\pi KT/I_0\Phi_0$. The procedure to obtain the low-frequency voltage spectral density is reported elsewhere (Enpuku 1985, Voss 1981).
In the following calculations we choose a small noise parameter $\Gamma=0.01$ in order to reduce the thermal noise and enhance the effect of asymmetries on the device behavior. This value corresponds to a typical low Tc SQUID configuration. Fig.1a and Fig.1b show the results of numerical simulations performed for three different SQUID configurations without any damping resistance.

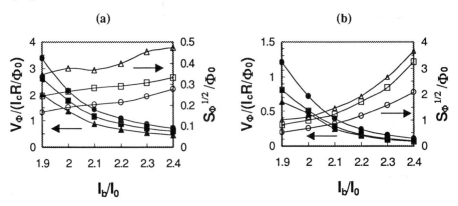

(a)   (b)

Fig.1. Flux to voltage transfer function V_Φ (filled symbols, on the left Y axis) and magnetic flux noise spectral density $S_\Phi^{1/2}$ (open symbols, on the right Y axis) for three SQUID configurations with $\beta=1$ (Fig.1a) and $\beta=5$ (Fig.1b): $\alpha=\rho=0$ (squares); $\alpha=0.6$, $\rho=0$ (triangles) $\alpha=\rho=0.6$ (circles). The damping parameter is $\gamma=0$ in all cases.

In the first configuration the two junctions have the same shunt resistance but different critical currents. The value of the current asymmetry α has been fixed to 0.6 which means $I_{C2}=4I_{C1}$. In the second configuration the two junctions differ for both critical currents and normal resistances. We report here only the case $\alpha=\rho=0.6$ for which the two junctions have the same characteristic voltage I_CR and $I_{C2}/I_{C1}=R_1/R_2=4$. The third configuration is fully symmetric and is characterized by $\alpha=\rho=0$. For all three configurations we have considered a symmetry in the inductances of the two SQUID arms.

Flux to voltage transfer function V_Φ (left axis) and magnetic flux noise $S_\Phi^{1/2}$ (right axis) are reported for the three cases with $\beta=1$ (Fig.1a) and $\beta=5$ (Fig.1b). An asymmetry in the critical current only (triangles) leads to a decrease of V_Φ of about 20% with respect to the symmetric case (squares) and an increase of the magnetic flux noise. This can be a problem, for example, for high critical temperature SQUIDs, in which the uniformity of the critical current along the grain boundary is still one of the main problems to solve. However we have shown that it is possible to exploit asymmetries in the Josephson junctions to improve the performance of the devices. For example, dc SQUIDs with an asymmetry both in the critical current and in the shunt resistance have larger V_Φ values and lower flux noises with respect to conventional symmetric SQUIDs. This improvement occurs for both $\beta=1$ and $\beta=5$. To study the effect of a damping resistance on asymmetric configurations we made numerical simulations with a damping parameter $\gamma=1$ and SQUID parameters $\alpha=\rho=0$ and $\alpha=\rho=0.6$. Results are reported in Fig.2.

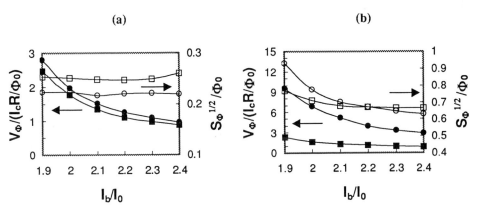

Fig.2.Flux to voltage transfer function V_Φ (filled symbols, on the left Y-axis) and magnetic flux noise spectral density $S_\Phi^{1/2}$ (open symbols, on the right Y-axis) for two SQUID configurations with $\beta=1$(Fig.1a) and $\beta=5$(Fig.1b): $\alpha=\rho=0$ (squares); $\alpha=\rho=0.6$ (circles). The damping parameter is $\gamma=1$ in all cases.

In dc SQUIDs with $\alpha=\rho=0$ the transfer coefficient is almost constant with β, at least for small thermal noise. There is just a small decrease due to the current noise which cannot be shunted by the damping resistance. The SQUID configuration with $\alpha=\rho=0.6$ shows a different behavior. In this case the V_Φ value for $\beta=5$ is more than 3 times larger than the one for $\beta=1$. The effect of a damping resistance on the noise characteristics (on the right Y-axes) is similar to what reported in the literature for symmetrical SQUIDs (Enpuku 1985). The magnetic flux noise is smaller than the case without damping resistance for large β and is almost constant for the whole bias current range. The configuration $\alpha=\rho=0.6$ is confirmed to be the best one also in terms of noise, but for $\beta=5$ the $S_\Phi^{1/2}$ value results larger than the case $\alpha=\rho=0$ for $i_b < 2.2$.

In Fig.3 we compare the behavior of the symmetric ($\alpha=\rho=0$) and asymmetric ($\alpha=\rho=0.6$) configurations as a function of β for $i_b=2.0$. Simulations are referred to a circuit with a damping parameter $\gamma=1$. We can see that V_Φ in the asymmetric configuration is an increasing function of β. The magnetic flux noise of asymmetric SQUIDs is lower than the one typical of the symmetric case up to a value β_{max}; above this value the asymmetric configuration is not convenient for practical applications. Such a maximum β is a function of several factors as, for example, the noise parameter Γ, the asymmetry parameters α and ρ, the damping parameter γ.

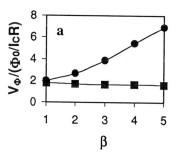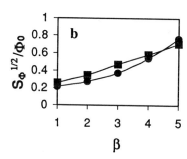

Fig.3. Flux to voltage transfer function V_Φ (Fig.3a) and magnetic flux noise spectral density $S_\Phi^{1/2}$ (Fig.3b) for two SQUID configurations as a function of β: $\alpha=\rho=0$ (squares) and $\alpha=\rho=0.6$ (circles). The damping parameter is $\gamma=1$ in both cases.

The effect of the asymmetric configuration $\alpha=\rho=0.6$, with a damping resistance, on the flux-locked-loop (FLL) operation mode is now discussed. The advantage of a flux noise almost constant over the whole range of bias current has already been discussed elsewhere (Enpuku 1985). Moreover, a larger flux to voltage transfer coefficient increases the open-loop gain $G(\omega)$ of the FLL system, improving the small-signal frequency response of the system. The slew rate, proportional to the linear part of the voltage-flux characteristics, is expected to decrease for large β with respect to the symmetric case.

In conclusion we have presented numerical simulations on the voltage characteristics and noise properties of both symmetric and asymmetric SQUID configurations. We have shown that asymmetric dc SQUIDs with the two junctions differing for both critical currents and shunt resistances have the best performance. We have also shown that for an asymmetric SQUID the use of a damping resistance further improves both the flux to voltage transfer function and the flux noise for large β, useful for a better matching with an input or pick-up coil. Further analysis and comparison with experimental data will be performed in the near future.

This work has been partially supported by the projects PRA-INFM "HTS Devices" and SUD-INFM "Analisi non distruttive con correnti parassite tramite dispositivi superconduttori".

REFERENCES

Enpuku K, Sueoka K, Yoshida K, and Irie F 1984 J. Appl. Phys. **57**, 1691
Enpuku K, Muta T, Yoshida K, and Irie F 1985 J. Appl. Phys. **58**, 1916
Tesche C D and Clarke J 1977 J.Low Temp. Phys. **29**, 301
Voss R F 1981 J. Low Temp. Phys. **42**, 151

Inst. Phys. Conf. Ser. No 167
Paper presented at Applied Superconductivity, Spain, 14-17 September 1999
© 2000 IOP Publishing Ltd

Design and fabrication of the high-T_c radiofrequency amplifier with the microstrip input coil based on bicrystal dc SQUIDs

M A Tarasov[1], O V Snigirev[2], A S Kalabukhov[2], S I Krasnosvobodtsev[3], E A Stepantsov[4]

[1] Institute of Radio Engineering and Electronics, RAS, 103907 Moscow, Russia
[2] Department of Physics Moscow State University, 119899 Moscow, Russia
[3] Lebedev Institute of Physics, Russian Academy of Sciences, 117924 Moscow, Russia
[4] Institute of Crystallography, Russian Academy of Sciences, 117333 Moscow, Russia

ABSTRACT: We have designed and tested a topology of a HTS dc SQUID amplifier with a microstrip-type input coil on bicrystal substrate. A process to fabricate the integrated multilayer YBCO dc SQUID amplifier was developed. The manufactured SQUIDs showed no degradation after input coil fabrication and revealed high performance properties.

1.   INTRODUCTION

In recent years there is a growing interest in radio-frequency amplifiers based on dc SQUIDs (SQA). Potential applications of such devices include the nuclear magnetic resonance, intermediate frequency (IF) amplifiers, preamplifiers for bolometers and others (C.Hilbert et al, 1985). One of the problems arising in the design of the SQA based on low-T_C (LTS) and high-T_C (HTS) superconductors is the effective matching of the microwave signal to the SQUID loop. Additionally, the SQA design involves complicated multilayers technology that is difficult to realize in HTS.

An elegant method of coupling a low-inductance, low size SQUID loop with the relatively high 50 Ω convention impedance was suggested and realized recently in LTS SQA where the input coil was used as a microstrip resonator (M.Mueck et al, 1998). This let to reduce the SQUID loop inductance and a capacitance between the input circuit and the loop. The typical layout of the SQA with the microstrip line includes the open-ended single-layer input coil separated from the HTS film by isolator, what strongly simplify the fabrication process. Nevertheless the degradation of the HTS film and Josephson junctions (JJ) during the insulator deposition and subsequent coil fabrication is still a serious problem. It is important also to avoid shorts between the coil and HTS film. A good quality of the HTS film and the isolating interlayer is the main issue of the technology.

In previous work (M. Tarasov et al,1999) we have presented an equivalent circuit and numerical simulations of the signal coupling mechanism in HTS dc SQUID amplifier with the microstrip input coil. It was shown that even in the simplest version of the developed model gives the good fit to known experimental data and allows to explain experimentally measured parameters. In this report we present the developed fabrication technology process of the integrated HTS radiofrequency amplifier based on bicrystal dc SQUIDs.

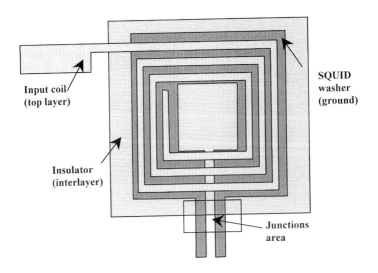

Fig. 1. A typical layout of high-Tc SQA with the microstrip input coil

2. DEVICE FABRICATION.

The SQA layout is shown in Fig 1. The SQUID square washer has dimensions D of 60 μm and 90 μm and the slot in the washer has dimensions l of 120 μm length and d 10 μm width. Using approximate expressions for inductances (Jaycox J.M. et al, 1981) one can estimate that the total SQUID loop inductance L_s is about 170 and 220 pH for each loop dimension.

The YBCO dc SQUIDs were made on 5×5 mm (100) YSZ bicrystal substrates with 37° misorientation angle. The substrates quality was examined by both X-ray and atomic-force microscopy (AFM) methods. The AFM image of the YSZ bicrystal substrate after mechanical polishing is shown in Fig. 2. One can see that the surface roughness is less than 10 nm. Using this technique we can control both substrate surface roughness and bicrystal boundary quality. Together with the X-ray diagnostics this lets us to select substrates with optimal parameters.

The YBCO film of 200 nm thick was deposited by the pulsed laser ablation using Nd:YAG lasers from the two non-stoichiometric targets. This technique let us to produce high quality films with very small amount of droplets. Films have critical temperatures T_C varying in range of 89 - 91 K and $\Delta T_C < 1$ K defined by the magnetic shielding measurements. The quality of the film surface was examined with a scanning electron microscope (SEM) and X-ray techniques. The X-ray analysis showed that film had predominantly c-axis orientation with less than 0.5% a,b-axis component. A very narrow range of the deposition parameters such as laser energy, temperature and oxygen pressure provides a high quality YBCO film suitable for subsequent growing of the insulator layer. A standard photolithography method with wet-etching in 0.1% HNO_3 was used to pattern the SQUIDs. The JJ width was 3÷4 μm. A pair of SQUIDs was placed on each bicrystal substrate.

An amorphous Y-ZrO_2 film of 100÷200 nm thick was used as insulator. The film was deposited with the same technique as YBCO films at room temperature and $6×10^{-2}$ mbar oxygen pressure. The AFM image of the step-edge of YSZ film on YBCO made by lift-off technique is shown on Fig. 3. It was found that the quality of the insulator layer (surface roughness, adhesion, hardness) is strongly depend on the YBCO film properties. The presence of small amount of the a-oriented outgrowths in YBCO leads to the weak adhesion and low hardness of the Y-ZrO_2 film.

The Au film of about 100 nm thick was deposited by dc magnetron sputtering at room temperature and $1.5×10^{-2}$ mbar argon pressure. No buffer layers were used to improve adhesion

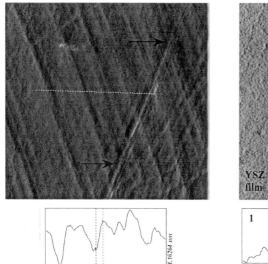

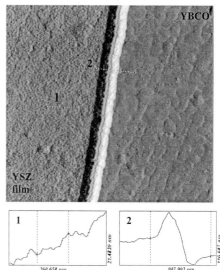

Fig. 2. *AFM* image of the *YSZ* bicrystal substrate after mechanical polishing (upper picture) and cross -section of the local area (at bottom). Scan size *2,5×2,5* µm. A bicrystal boundary marked out by the black arrows.

Fig. 3. *AFM* image of the *Y-ZrO₂* amorphous film on *YBa₂Cu₃O₇₋δ* layer and cross-sections of the local film area (1) and step-edge made by lift-off technique (2).

of the Au film. The 4 ÷ 9-turn input coils of 10 µm width were patterned further by the lift-off technique.

3. MEASUREMENTS.

The SQUIDs were tested at each technology stage. After the SQUIDs structure patterning the critical current I_C and normal resistance R_n were about 15 - 30 µA and 1 - 4 Ohms, respectively, which gives a I_cR_n product of about 40 - 70 µV typical for the SQUIDs on bicrystal substrates with 37° misorientation angle. The summarized experimental data are presented in Table I. An objectionable result is that the voltage modulation depth ΔV_{p-p} vs applied magnetic field for all samples was found small (at most 3÷5 µV, see table), whereas the theoretical predictions for the measured SQUIDs parameters provide voltage modulation depth about 15÷20 µV. Indeed, $\Delta V_{p-p} = \Delta I_C \times R_d$, here $\Delta I_c \approx I_c/(1+\beta)$, $R_d \approx 1.4 \times R_n$ – dynamic resistance, $\beta = I_c \times L_s/\Phi_0$, Φ_0 - flux quantum (K.Enpuku et al, 1994). Hence for the typical values I_c and R_n equal to 20 µA and 2 Ohm, respectively, and L_S = 150 pH we can obtain β = 1.5, ΔI_C = 8 µA and $\Delta V_{p-p} \approx$ 22 µV. Typical values of the leakage resistance between Au coil and YBCO film was found of about 10^6 Ohms. Measurements both after the Y-ZrO₂ film growing and Au deposition and coil patterning in founded ranges of the technological parameters did not indicate any degradation of the SQUIDs.

4. DISCUSSION

We have developed a technology process for the fabrication of the integrated high-Tc dc SQUID amplifiers on bicrystal substrates. This technique involves the insulator interlayer and the single-layer planar Au input coil. This technology allows us to fabricate high-T_c SQA acceptable for further investigations of their microwave properties.

TABLE I. Parameters of the YBCO dc SQUIDs on 37° YSZ bicrystal substrates.

Sample:	Junction width w (μm)	Washer Square D (μm)	Slot length l (μm)	Critical current I_C (μA)	Normal resistance R_n, Ohms	I_cR_n, (μV)	Voltage mod.depth, $\Delta V_{p\text{-}p}$ (μV)
1	3,5	60	120	15	2,3	35	4
2	3,5	60	120	15	1,5	23	3
3	3	60	120	10	4	40	5
4	3,5	90	120	20	2,5	50	5
5	3,5	60	120	30	2	60	5

Nevertheless, the measurements of the YBCO bicrystal dc SQUIDs parameters revealed discrepancy between the experimental and predicted values of the voltage modulations depth. One possible explanation of the discrepancy is that the real inductance of the SQUID loop is much higher than was estimated above. To verify this, we have numerically calculated the inductance of the loops using a two-dimensional computer simulation program (M.M. Khapaev, 1998). This calculations provides values closed to estimated. We can conclude that the observed discrepancy of the modulation depth is caused either by the part of the inductance not included in our calculations or by the asymmetry of the JJ critical currents. To reveal the origin of such SQUIDs behavior we are going to fabricate test chips with various SQUIDs topologies, different from the described above and to use another bicrystal substrates materials such as SrTiO$_3$.

ACKNOWLEDGMENT

The authors are thankful to I.A. Yaminsky and S.N. Polyakov for AFM and X-ray tests and to A.S. Artemov for polishing of the bicrystal substrates. This work was supported in part by the INTAS foundation under grants No. 97-731 and No. 96-268 and the Russian Scientific Council on Superconductivity under grant No. 96-081.

REFERENCES

Enpuku K, Tokita G, and Maruo T 1994. J.Appl.Phys. **76** (12) pp.654-59
Hilbert C and Clarke J, 1985 *Journal of Low Temp. Phys.* **61** Nos. 3/4, pp 237-251
Jaycox J M , Ketchen M B 1981 *IEEE Trans. on Magn.*, **17**(1) pp 400-3.
Khapaev M M 1998 Supercond. Sci. Technol. **10** pp. 389-94,
Mueck M, Andre M-O, Clarke J, Gail J, Heiden C 1998 *Appl. Phys. Lett.* **72** pp 2885 - 7.
Tarasov M et al 1999 Proc. 7th Int. Superc. Electr. Conf. (Berkeley USA) pp 540-2.

Inst. Phys. Conf. Ser. No 167
Paper presented at Applied Superconductivity, Spain, 14-17 September 1999
© 2000 IOP Publishing Ltd

Noise Characteristics of a Two Stage dc SQUID-based Amplifier

D E Kirichenko[1], A B Pavolotskij[2], I G Prokhorova[1], O V Snigirev[1], R Mezzena[3], S Vitale[3], Yu V Maslennikov[4], V Yu Slobodtchikov[4]

[1]Department of Physics, Moscow State University, 119899 Moscow, Russia; [2]Nuclear Physics Institute, Moscow State University, 119899 Moscow, Russia; [3]Department of Physics, University of Trento, 38050 Trento, Italy; [4]Cryoton Co. Ltd., 142092 Troitsk, Moscow Region, Russia.

ABSTRACT: The noise of a two stage dc SQUID-based amplifier has been measured in the temperature range 1.2 - 4.2 K. The white noise level was found close to 0.3 $\mu\Phi_0/\mathrm{Hz}^{1/2}$ at 4.2 K @ 1 kHz for the amplifier's basic version and close to 0.6 $\mu\Phi_0/\mathrm{Hz}^{1/2}$ at 4.2 K @ 1 kHz for the version with an additional damping system. The spectral densities of both amplifier versions have scaled linearly with the temperature and have shown an almost zero residual value at 0 K.

1.   INTRODUCTION

The internal energy sensitivity, ε_v, of a dc superconducting quantum interference device (dc SQUID) is one of the most important figures of a low frequency amplifier based on the SQUID (Koch et al 1981, Danilov et al 1983). It has been shown in theory (Danilov et al 1983) and confirmed by experiment (Van Harlingen et al 1981, Awshalom et al 1988) that the quantum limit can be approached with a low value for the SQUID inductance L. But two serious problems arise in this case for SQUID practical applications in several fundamental physics experiments, including gravitational waves detection. One of them is the problem of effective, resonances-free coupling with the usually relatively high inductance signal source. The second one is the problem of the matching between the low SQUID output voltage noise and the noise of a room temperature electronics.

Formerly we have reported on our investigations in this field (Maslennikov et al 1995, Kirichenko et al 1997). The last version of our two stage dc SQUID-based amplifier with double transformer coupling scheme has been designed and implemented recently. The effective input inductance of the amplifier and the input mutual inductance were close to 2.9 μH and 2.6 nH correspondingly, the loop inductance of the first stage SQUID was of about 11 pH, and the effective coupling k^2 was close to 0.24. The details of the amplifier design and its signal characteristics at temperature 4.2 K has been reported by Kirichenko et al (1999). In the present paper we report the results of noise measurements obtained for this version in the temperature range from 1.2 to 4.2 K.

2.   AMPLIFIER DESIGN

The amplifier consists of a Nb-shielded capsule housing two SQUIDs, a cryogenic probe, a twin channel electronics box and a control unit. The operation of the amplifier can be controlled manually or by a computer. The circuit diagram of the SQUIDs and analog part of room temperature electronics is shown on Fig. 1. The basic parameters of the SQUIDs are listed in the Table.

The first stage SQUID (SQ1) was coupled to the amplifier's input using the double transformer coupling scheme. The input coupling transformer T1, second transformer T2, and the SQUID SQ1 were integrated in a single chip. The coupling transformer T1 was divided into two sections (see Fig. 2). Each part of the transformer has the secondary coil formed as square washer with the hole size of 580 μm and 52 turns primary coil. The primary coil has 5 μm line width and 5 μm spacing. The washers of the transformer's secondary coils are connected to form the first order planar gradiometer. Such a configuration significantly reduces the sensitivity to interference signals.

Parameters of the amplifier	Stage 1	Stage 2
Loop inductance LS (pH)	11	120
Critical current per junction I_0 (µA)	100	15
Shunting resistance per junction RS (Ohm)	1.9	4.0
$\beta_L = 2 \times LS \times I_0/\Phi_0$	1.1	3.6
Effective input inductance L_{IN} (µH)	2.6	0.25
Effective mutual inductance, M_{IN} (nH)	2.6	2.3

The loop of the first SQUID (secondary coil of the T2) was formed in a self-shielded configuration. It consists of two square washers connected in parallel in order to reduce the SQUID's loop inductance. The holes of the washers are 10 µm size. The two 5 turn input coils connected in series were placed over the SQUID washers to form the primary of the T2 transformer.

Another version of the 1st stage SQUID was also designed and realized. The basic configuration is the same described previously, apart from the addition of distributed microwave damping system. Such a system was proposed and tested by Jin et al (1997a) and Jin et al (1997b) in order to suppress the SQUID's structure resonances. We have realized it in different design as two sets of the long resistive stripes made of a thin titanium film over the washers of T1 and SQ1.

The T1 damping set consisting of 9 resistive stripes with 20 µm line width and the 20 µm separation has been deposited and patterned over each washer simultaneously with the Josephson junctions shunt resistors. The stripes were coupled to resonators formed by the input coil windings and T1 washers via vortexes trap slots in the washers. Also two squares of the same resistive film have been placed in the holes of the T1 transformer washers.

The SQ1 damping set contents 5 resistive stripes with 5 µm line width and the 5 µm separation over each washer. These stripes have been deposited and patterned simultaneously with the stripes of the T1, but in direct ohmic contact with the SQUID input coil windings.

The second stage SQUID (SQ2) was placed on a separate chip. Its loop was formed as a square washer with a hole size of 70 µm. The 30-turn SQUIDs input coil was wound with a niobium wire of 42 µm diameter, placed on separate fiberglass substrate and attached to the SQUID's chip.

3.   MEASUREMENT RESULTS

We have measured the noise of the amplifier versus temperature for three different 1st stage SQUIDs. One of them (marked as #1.1.8) had the basic circuit configuration, and two of them (marked as #1.1.9 and #3.1.9) have been fabricated with the distributed microwave damping system. The resonant steps in the current-voltage and flux-voltage curves of the SQUID #1.1.8 were sharper then for the damped SQUIDs #1.1.9 and #3.1.9 (Kirichenko et al 1999).

For all SQUIDs measurements were carried out as follows. A test signal of magnitude Φ_0 was applied to the first SQUID via its feedback coil. The bias current of the first channel was adjusted to obtain the maximum amplitude of the SQUID output. Then the test signal was reduced to

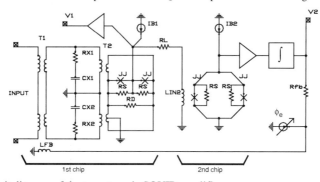

Fig. 1. The circuit diagram of the two stage dc SQUID amplifier.

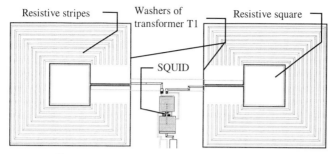

Fig. 2. The sketch of the 1st stage SQUID with the distributed damping system. The primary coils of the coupling transformer T1 and the resistive stripes in the SQUID are removed for clarity.

$0.01\Phi_O$ and the operating point of the first SQUID was set for the maximum slope of the V-Φ_e characteristic by adjusting the dc magnetic flux. We can remark here that typical values for a voltage swing ΔV of the V-Φ_e characteristic was in the range 150 — 200 μV (Kirichenko et al 1999).

The output signal of the first SQUID was used as a test signal for the adjustment of the second SQUID. Also the bias current of the second SQUID was adjusted in order to obtain the maximum amplitude of its output signal. In our experiments one-tenth of Φ_O at the input of the first SQUID has corresponded to 6 — 9 Φ_O at the second SQUID. Thus, the gain factor of the SQ1-R_L-L_{IN2} circuit was about 60 — 90. This factor was optimized by the adjustment of R_L providing the minimum of the whole amplifier noise and to ensure stable operation of the magnetometer in the locked feedback mode.

Further for noise measurement the amplifier was switched to operation in the locked loop mode. To do this the output of the integrator of the second electronic channel was connected to the feedback coil of the first SQUID. In this case the second SQUID operated as a low noise amplifier on the steep linear part of it signal characteristics. The feedback transfer factor was adjusted within the range 40 — 70 V/Φ_O.

The flux noise versus the temperature has been measured reducing the vapor pressure of the liquid helium bath. A vacuum regulator valve was employed to obtain good pressure and temperature stabilization. Before every noise measurements the working point of both SQUIDs have been slightly readjusted in order to compensate the shift of the SQUID's parameters with temperature and minimize the output noise. Figure 4 shows the temperature dependence of the amplifier flux noise spectral density. One can see that this dependence is approximately linear. The linear fit made with the least root mean square method shows that the residual noise extrapolated to T = 0 K is close to

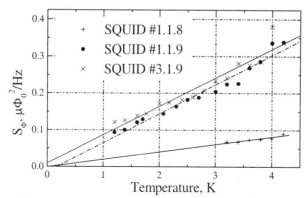

Fig. 3. White flux noise spectral density of the amplifier with various 1st stage SQUIDs versus temperature.

zero for all three SQUIDs.

The noise measurements for the undamped SQUID #1.1.8 have been carried out from T = 4.2 K to 3.2 K. We have found that decreasing the temperature the resonant steps in the signal characteristics of the SQUID became sharper and larger. The working point of the 1st SQUID became too close to one of the resonances and the operation of the SQUID becomes disturbed by this resonance. Thus the measurements of this SQUID's noise with temperature were stopped near 3.2 K.

4. DISCUSSION

The noise graphs (Fig. 3) clearly show the difference between the basic version of the 1st stage SQUID and the damped version. It can be seen that despite of the similarity of all the SQUID's parameters of both versions the noise level of the damped version is significantly larger than the level of the basic one. At the same time the spectral density of this excessive noise linearly depends on the temperature.

We suppose that the only possible sources of the temperature dependent noise are the resistive elements of the distributed microwave damping system. These elements generate the Johnson noise currents, which are magnetically coupled to the 1st stage SQUID. We have estimated the noise introduced by the stripes in the transformer T1 and SQUID SQ1 to be of the order of 7×10^{-8} $\Phi_0/Hz^{1/2}$. The noise introduced by the resistive squares in the holes of the transformer T1 has been estimated as 3×10^{-7} $\Phi_0/Hz^{1/2}$. This value agrees with the measured noise in order of magnitude.

Thus, the sensitivity of amplifier of present design was limited by Johnson noise in the resistive squares inserted in washer holes. Nevertheless these squares suppress also the microwave resonances (Jin et al 1997b) in the transformer structure. Their size can be readjusted in the future design and implementation.

Besides the noise properties previously discussed, it is reasonable to mention also a very high voltage swing ΔV of the voltage-to-flux curve and the corresponding transfer factor $\partial V/\partial \Phi_e$ the ordinary values for which were in range 150 - 200 μV and 300 - 500 μV/Φ_0 respectively. This circumstance together with the sensitivity well below 1 μΦ_0 and the stable operation of the damped SQUIDs in whole temperature range of measurements makes the developed amplifier very promising as a base unit for different SQUID applications at low temperatures.

ACKNOWLEDGEMENT

This work was supported by the NATO Linkage Grant No. 960856 and the project "Intersquid" of the Russian Ministry of Science.

REFERENCES

Awshalom D D, Rosen J R, Ketchen M B, et al. 1988 Appl. Phys. Lett. **53** pp 2108–2110

Danilov V V, Likharev K K and Zorin A B 1983 IEEE Trans. Magn. **19**, 2, pp 572-574

Jin I, Amar A and Wellstood F C 1997a Appl. Phys. Lett. **70** pp 2186-2188

Jin I, Amar A, Stewenson T R, Wellstood F C, et al. 1997b IEEE Trans. Appl. Supecond. **7**, 2 pp 2742-2745

Kirichenko D E, Pavolotskij A B, Prokhorova I G, et al. 1997 in Inst. Phys. Conf. Ser. **158**, pp 727-730 (IOP Publishing Ltd.)

Kirichenko D E, Pavolotskij A B, Prokhorova I G, et al. 1999 IEEE Trans. on Appl. Superc. **9** (2) pp. 2906-08

Koch R H, Van Harlingen D J, and Clarke J 1981 Appl. Phys. Lett. **38** pp 380-382

Muhlfelder B, Johnson W and Cromar M W 1983 IEEE Trans. Magn. **19** pp 303-307

Maslennikov Yu V, Beliaev A V, Slobodtchikov V Yu, et al. 1995 in Inst. Phys. Conf. Ser. **148**, pp 1569-1572 (IOP Publishing Ltd.)

Van Harlingen D J, Koch R H and Clarke J 1981 Physica **B108** pp 1083-1085

Inst. Phys. Conf. Ser. No 167
Paper presented at Applied Superconductivity, Spain, 14-17 September 1999
© 2000 IOP Publishing Ltd

A scanning high-Tc dc SQUID microscopy of the YBa$_2$Cu$_3$O$_{7-\delta}$ thin-film samples at 77 K

S A Gudoshnikov[1], M I Koshelev[2], A S Kalabukhov[2], L V Matveets[1], M Mück[3], J Dechert[3], C Heiden[3] and O V Snigirev [2]

[1] The Institute of Terrestrial Magnetism, Ionosphere and Radio Wave Propagation, Russian Academy of Sciences, (IZMIRAN), Troitsk, Moscow region 142092, Russia

[2] The Department of Physics, Moscow State University, Moscow 119899 GSP, Russia

[3] Institut für Angewandte Physik Justus Liebig Universität Gießen, 35392 Gießen, Germany

ABSTRACT: A scanning high-Tc dc SQUID microscope with a spatial resolution of about 20 μm and a magnetic field sensitivity close to 100 pT/Hz$^{1/2}$ was used for mapping a magnetic flux distribution B$_z$(x,y) in the YBCO thin film samples at 77 K. The images were taken in a perpendicular geometry in a low magnetising fields ranging from 0.1 to 70 A/m for the cases of zero field-cooled (ZFC) and field-cooled (FC) samples. Images of individual vortices trapped in the inspected film have been obtained.

1. INTRODUCTION

In recent years the local behaviour of the magnetic flux in high-T$_c$ superconductors (HTS) has been extensively studied by the different methods. This investigations have both fundamental and practical aspects. The thermally activated motion of the vortices trapped in thin-film washers of the HTS dc SQUIDs produces in it an excess low-frequency noise. The motion of vortices due to the external magnetic field change is a cause of the hysteresis of the SQUID output signal in unshielded environment. A recently developed scanning SQUID microscope (SSM) can provide a detailed map of the magnetic field in a close proximity to the sample surface and seems to be a convenient instrument for investigations of the magnetic flux behaviour in HTS thin film samples.

At T = 4.2 K magnetic images of the single flux vortices trapped in the HTS thin film have been obtained by Kirtley et al (1995) with the low-Tc superconductor (LTS) dc SQUID. Recently Koelle et al (1999) obtained such images at T = 77 K using a low-temperature scanning electron microscope. In our paper we present the images of the magnetic vortices in the HTS thin-film samples obtained at 77 K with the SSM based on HTS dc SQUID. The behaviour of the HTS zero-field cooled (ZFC) and field cooled (FC) thin film samples in week magnetic field is also reported.

2. EXPERIMENTAL SETUP AND SAMPLES CHARACTERIZATION

The design of our SSM based on high-T$_c$ dc SQUID has been described previously by Gudoshnikov et al (1997a). An improved SSM version was used for measurements (Gudoshnikov et al 1999b), which let us to achieve a high spatial resolution close to 20 μm. Measured effective area of the SQUID, A$_{SQ}$, was equal to 660 μm^2, thus magnetic field 3.0 μT corresponds to one flux quantum Φ_0 ($\Phi_0 = h/2e$). A magnetic field sensitivity of the sensor was close to 100 pT/Hz$^{1/2}$ at frequencies higher than 100 Hz. The SQUID was operated in a standard flux-locked-loop mode using the feedback wire coil of 3 nm outer diameter. All measurements were performed at temperature 77.8 K. Cryostat was placed inside of the double-layer mu-metal shield, which provides a residual magnetic field less than 0.2 μT. This value was estimated when SSM was rotated inside shield. The SSM probe was oriented in such way that the direction of the radial component of the residual magnetic field vector was parallel to a sample surface. The applied magnetic field, H_a, created by field coil was perpendicular to a film plane so that H_a was parallel to the c axis of the measured film.

The investigated thin films of $Y_1Ba_2Cu_3O_{7-x}$ were deposited by eximer pulsed laser ablation on 10×10 mm *(100)* SrTiO_3 substrates. The critical temperature, T_c, of the films were determined both by magnetic and resistivity measurements and varied between 87 K and 92 K. Measured critical current density, j_c, was close to $2.5×10^6$ A/cm^2. The squares with side of 1 mm, 0.6 mm and 0.4 mm were patterned by conventional photolithography. Both surface roughness and edges of the obtained films and structures were tested using an atomic force microscopy (AFM) and a scanning electron microscopy (SEM). Typical AFM and SEM images of the samples are shown in Fig. 1 and Fig. 2. One can see that the thickness of film, d, was close to 200 nm, the edges have sharp slope (about 45 degrees) and the deviations from the straight line with average scale of the order of 0.2 μm, comparable with the London penetration depth λ at 77 K.

3. RESULTS

Measurements were carried out both for the case of applying an external field to a zero-field-cooled (ZFC) and field-cooled (FC) superconductors. First measurements were carried out over the central part of the 1 mm × 1 mm square specimen. Obtained image is shown in Fig. 3. A few flux peaks of magnetic field with equal amplitude ΔB_z of 0.8 μT was observed with a signal-to-noise ratio more than 1000. The magnitude of magnetic flux for each of measured peaks was calculated in accordance with the estimation of Kirtley (1995) and was found to be corresponded to one flux quantum. The position of the peaks and their magnitude did not change during the several measuring cycles in a particular cooldown. A density of vortexes, n, can be estimated as close to $2×10^7$ 1/m^2 and magnetic field induction in the sample, $B = n\Phi_0$, is close to $4×10^{-8}$ T. This situation seems reasonable if we suppose the precision of the sample surface alignment in parallel to the shield residual field of the order of 5 degrees.

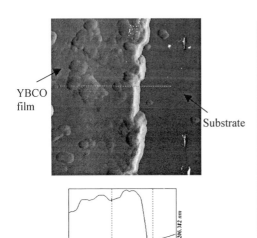

YBCO film

Substrate

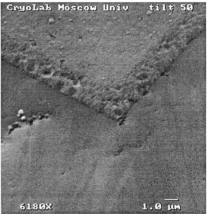

Fig.1. AFM image of the surface and the edge of the YBCO thin film structure fabricated on SrTiO_3 substrate (top).Typical cross-section of the surface and edge taken along the dot line on the picture (bottom).

Fig.2. SEM image of the surface and the edges of the YBCO thin film sample patterned by wet photolithography.

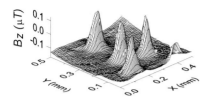 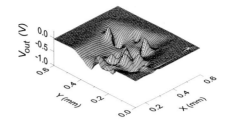

Fig. 3. Magnetic image of the central fragment of the 1mm × 1 mm ZFC YBCO thin film sample at 77 K.

Fig. 4. Magnetic image of the 400 μm × 400 μm ZFC YBCO thin film taken at 77 K

The external magnetic fields values up to 70 μT were applied to a film both perpendicular and parallel and then removed. The obtained images demonstrated that the vortices location were stable. If we use a simple estimation of a vortex magnetic moment, $m = 2\lambda\Phi_0/\mu_0$, vortex pinning energy is much higher than 4×10^{-20} Joule. The upper magnetising field limit for imaging of separate vortexes in developed SSM can be estimated as 5 μT when the average space separation between vortexes will be about 20 μm.

The magnetic image of 400 μm × 400 μm size ZFC specimen is shown in Fig. 4. Here one can see clearly the edges of the sample and vortex-antivortex pairs which generation in the HTS thin films is intensively discussed presently (Melnikov et al 1998). While the SQUID transfer coefficient varies drastically at the measured film edges (Gudoshnikov et al 1997c), the output voltage of the SQUID was given in the ordinate axis.

Fig. 5 shows the magnetic image of the central part (0.5 mm× 0.5 mm) of 1 mm × 1 mm-size thin film specimen, which was cooled in 60 μT perpendicular magnetic field. The image was done in the presence of same external magnetic field thus the variations of the near-surface field were measured with the respect to this level. It is visible that all film area is filled with the chaotically distributed peaks. The similar images were obtained with the samples cooled in the lower fields. In the field 60 μT the average vortex-to-vortex separation is about 0.2 μm. Nevertheless, the typical distance between the peaks is of the order of 50 - 100 μm what is much larger than the SSM resolution. We can suppose only that the density of vortexes in high-quality HTS thin films is not uniform. Such image can be produced by the clusters of the disentangled vortexes, specific for a

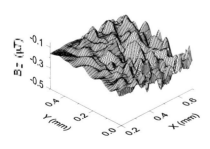 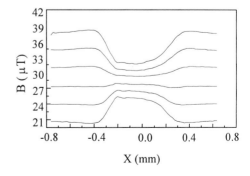

Fig.5 Magnetic image of the FC (60 μT) YBCO thin film sample at 77 K.

Fig.6 . Magnetic field profiles across the film vs applied field varying in small range.

vortex liquid in low magnetic field near the sample critical temperature (Grabtree and Nelson 1997).

Fig. 6 illustrates the example of the behaviour of the FC YBCO thin film strip of the width $w = 0.6$ mm in the magnetic field ranged near 30 μT. The strip was cooled in the field close to 33 μT and the scan across it was taken. The changes of the mutual inductance between SQUID and feedback coil at the strip edges were taken in account. One can find small variations of the measured field over the sample in this scan (third from the bottom) with the scale corresponding to one given in Fig. 5. Before each new scan we opened the SQUID feedback loop, added external field equivalent one flux quanta and then closed feedback loop. One can see that in spite of very high quality of the deposited film there is no visible classic Meissner effect. The magnetic field as low as 30 μT was not expulsed from the sample in cooling procedure. Nevertheless, if the field was increased the shielding effect appears (scans 4 – 6 from the bottom). The changing of the shape of well-similar structure in the scans indicates the additional flux penetration inside the sample. The scans 1 – 2 demonstrate the flux trapping in the sample in the case of decreasing of the applied field.

4.   DISCUSSION

Beginning from the technique, we can say that the high-Tc dc SQUID-based SSM has been developed which magnetic field sensitivity and space resolution allow to obtain images of the individual vortexes trapped in high-Tc thin films at T=77 K in the magnetic field up to 5 μT. This device can give additional insights into the mechanisms which are responsible for the flux penetration and its behaviour in the HTS thin films.

The more striking results is that despite the extensive efforts to cool sample slowly in almost zero field, we always observed some flux trapped in our thin films often of the both signs. If we take in the account the enhancement (Benkraoda and Clem 1996) of the magnetic field near the sample edges which is of the order of $(d/w)^{1/2}$ for rectangular approximation of the sample edges form, the transverse penetration field for our samples will be of about 2 - 3 μT. This value is much higher than remnant magnetic field in used shield. it seems more realistic that some parts of the thin film sample edges have so high curvature for instance due to S-shaped edge (not detectable by the AFM needle) that penetration is observed in all available fields, how low it may be (Trofimov and Kuznetsov 1994).

ACKNOWLEDGMENT

The work is supported in part by the German BMBF under Grant No. 13 N 6898, the INTAS Grant No. 97-0894 and the Russian Program on Superconductivity, Grants Nos. 60199, 96093.

REFERENCES

Benkraouda M, Clem J R 1996 Phys. Rev. B **53** (9), pp 5716-26
Grabtree G W, Nelson D R 1997 Physics Today April pp 38-45
Gudoshnikov S A et al. 1997a Proc. 6th Int. Supercond. Electr. Conf. - (Berlin, Germany) pp 392-4
Gudoshnikov S A et al. 1999b IEEE Trans. Appl. Supercond. **9** (2) pp 4385 - 8
Gudoshnikov S A et al. 1997c Proc. 3rd Eu. Conf. on Appl. Supercond. (The Netherlands) pp 755-8
Kirtley J R K, Ketchen M B et al. 1995 Appl. Phys. Lett. **66** (9) pp 1138-40
Koelle D, Gross R, Keil S, et al 1999 Proc. 7th Int. Superc. Electr. Conf. (Berkeley USA) pp 557-62
Melnikov A S, Nozdrin Yu N et al Phys. Rev. B 1998 **58** (10) pp 1-4
Trofimov V N, Kuznetsov A V 1994 Physica C **235-240** pp 2853-54

Inst. Phys. Conf. Ser. No 167
Paper presented at Applied Superconductivity, Spain, 14-17 September 1999
© 2000 IOP Publishing Ltd

A 40-channel double relaxation oscillation SQUID system for biomagnetic applications

Y H Lee, H C Kwon, J M Kim, Y K Park and J C Park

Korea Research Institute of Standards and Science, P. O. Box 102, Yusong, Taejon, 305-600, Republic of Korea

ABSTRACT: We developed a 40-channel SQUID system based on double relaxation oscillation SQUIDs (DROSs). The DROSs were fabricated from Nb/AlO$_x$/Nb junctions. The large flux-to-voltage transfer of DROSs, typically about 1 mV/Φ_0, enabled direct readout by room-temperature dc preamplifiers, resulting in simple flux-locked loop electronics. The pickup coil is an integrated first-order planar gradiometer with a baseline of 40 mm. The 40-channel system has an average noise level of 1.2 fT/cm/√Hz at 100 Hz, corresponding to a field noise of 5 fT/√Hz, and was applied to measure auditory-evoked neuromagnetic fields.

1. INTRODUCTION

For the superconducting quantum interference devices (SQUIDs) to be useful in a biomagnetic multichannel system, simplicity and compactness of SQUID sensors, including pickup coils and readout electronics, are required. To simplify readout electronics, a SQUID should have a large flux-to-voltage transfer coefficient to enable direct readout by a room-temperature dc preamplifier (Drung 1994). Since DROSs provide large flux-to-voltage transfer and large modulation voltage, it is possible to use simple flux-locked loop electronics in SQUID operation (Adelerhof et al 1994).

To simplify the pickup coil, a planar gradiometer where the thin-film pickup coil is integrated on the same substrate as the SQUID is desirable. In this paper, we constructed a compact 40-channel SQUID system by combining DROSs and integrated planar pickup coils. The operation characteristics and the neuromagnetic application of the system are described.

2. CONSTRUCTION OF 40-CHANNEL SYSTEM

2.1 DROS Planar Gradiometer

A DROS functions as a comparator of two critical currents; the critical current of the signal SQUID and the critical current of a reference SQUID. If the two critical currents are equal, a very large flux-to-voltage transfer coefficient can be obtained (Adelerhof et al 1994). In order to make the DROS more useful in a multichannel system, the reference SQUID was replaced by a reference junction (Lee et al 1998). By using a reference junction, the wires for the reference flux are not necessary and the possibility of flux trapping by the reference SQUID is eliminated. The signal SQUID is a hysteretic DC SQUID type with two 100 μm x 100 μm square holes connected in parallel. The inductance of signal SQUID was designed to be 113 pH. The size of each Josephson junction in the signal SQUID is 4 μm x 4 μm, while the reference junction has a size of 5 μm x 5 μm (Lee et al 1999).

The input coil consists of two serially connected coils, 15-turns each, integrated on each SQUID loop and its inductance is calculated to be 87 nH. Both the SQUID and the input coil are gradiometric. The pickup coil is a first-order gradiometer with two 12 mm x 12 mm coils connected in series. The linewidth of the pickup coil is 0.5 mm and the baseline is 40 mm. The field-to-flux transfer of the flux transformer circuit is calculated to be 0.95 nT/Φ_0.

The sensors were fabricated on 3-inch Si wafers by a simple 4-level process based on Nb/AlO$_x$/Nb junction technology. The junctions were defined by reactive ion etching and the insulation between metal layers was made by SiO$_2$ film deposited by plasma-enhanced chemical vapour deposition. The resistor is a reliable thin film of Pd. The overall size of the sensor is 12 mm x 52 mm and 4 sensors were fabricated on each wafer. The fabricated sensors were glued onto printed circuit board (PCB) holders of 16 mm x 75 mm, where wiring copper leads were printed. At the end of each PCB holder, a 6-pin socket was soldered so that the holder can be quickly replaced from the epoxy block.

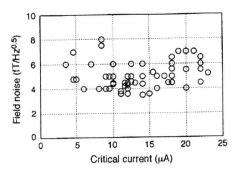

Fig. 1. Magnetic field noise of DROS planar gradiometers versus the critical current of a

Characterization of the sensors was done inside a moderately shielded room with two layers of Mu-metal and one layer of Al, which has shielding factors of about 60 dB at 1 Hz and 90 dB at 100 Hz. Fig. 1 shows the magnetic field noises of 55 sensors versus critical currents of the reference junction. The fabricated DROSs had reference critical currents of 5~25 μA and the maximum modulation voltages of around 110 μV. The flux-voltage curves of DROSs showed almost a step function of the applied field. The maximum flux-to-voltage transfer is around 2 mV/Φ_0, which is about 10 times larger than the transfer coefficient of dc SQUIDs. The average value of field noise is about 5 fT/$\sqrt{\text{Hz}}$ within the reference critical current range of 5~20 μA, but the spread in noise is minimum for the critical current of 9~17 μA.

2.2   Readout Electronics

Because of the large transfer, the DROS output could be connected directly to a room-temperature dc preamplifier without an intermediate matching circuit. Fig. 2 shows the schematic circuit diagram of the flux-locked loop (FLL) electronics. The FLL circuit uses a dc bias current. The preamplifier is an instrumentation amplifier type made of LT1028 (Linear Technology) operational amplifiers and has an input voltage noise of 1.7 nV/$\sqrt{\text{Hz}}$ at 100 Hz and 5 nV/$\sqrt{\text{Hz}}$ at 1 Hz. With a typical transfer coefficient of 2 mV/Φ_0, the preamplifier contributes an equivalent flux noise of 0.85 μΦ_0/$\sqrt{\text{Hz}}$ at 100 Hz. In the frequency range 1~100 Hz, the preamplifier noise contribution is about 10 % of the total SQUID system noise (Lee et al 1999).

Because of the large modulation voltage of DROS, the DROS was quite stable against the offset voltage drift of the amplifier chain. The operation margin for the offset drift is about ±15 μV around the center of modulation voltage, this operation margin for offset voltage is about 3 times larger than the dc SQUID

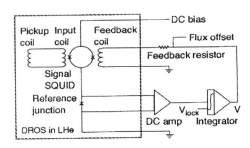

Fig. 2. Schematic circuit drawing of the flux-locked

with additional positive feedback (Ryhanen et al 1991).

The 40-channel system consists of 6 head boxes for FLL circuits and 3 sub-racks for SQUID controls. Each head box has 8 FLL circuits and were fixed on the ceiling of the shielded room. The control electronics are within an RF-shielded Al cabinet located outside the shielded room. The FLL outputs were passed through high pass, low pass and 60-Hz notch filters, and gain-adjustable amplifiers.

2.3   40-Channel Insert

The 40-channel insert consists of 16 rectangular epoxy rods, and each rod has two to four gradiometers in the x and y direction. The distance between two parallel gradiometers is 25 mm. The gradiometers were arranged to measure field components tangential to the body surface; dB_x/dz and dB_y/dz, where z-axis is normal to the body surface. A tangential gradiometer sensitive to off-diagonal derivative has a magnetic field peak just above the current dipole when the detection coil is arranged perpendicular to

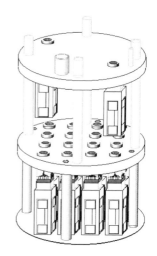

Fig. 3. Structure of 40-channel

the dipole direction. Thus, a less extensive sensor array is needed to get the essential field distribution (Tsukada et al 1995).

In addition to the 40 channels, 4 reference channels were added to pickup background noises and to apply adaptive filtering (Shimogawara et al. 1995). Fig. 3 shows the assembled structure of the SQUID insert. To minimize thermal load, special care was taken in the choice of wire material and radiation shields. Low-thermal-conductivity manganin wires of 127 μm diameter were used for bias current and flux feedback. For the voltage lines, phosphorous bronze wires of 127 μm were used. Since the resistance of the phosphorous bronze wire is about 9 Ω/m, and the input current noise of LT1028 is 2 pA/√Hz at 100 Hz, the voltage noise due to the wire resistance is about 18 pV/√Hz. This noise level is negligible compared with the input voltage noise of LT1028. The connection between the SQUID assembly and the radiation shields were made of low-thermal-conductivity fiber-reinforced plastic tubes. The radiation shields were made of multiple layers of Cu/styrofoam, and the cold evaporating gas was used to cool the signal lines.

The liquid helium dewar has flat tail with an inner tail diameter of 135 mm (CTF Systems Inc.). The dewar has a liquid capacity of 33 L and the boil-off rate without any insert is 2.9 L/day. Because of the low thermal load, the boil-off rate of the complete 40-channel system could be as low as 3 L/day.

3.   40-CHANNEL OPERATION

The magnetic field noise of the 40 channel sensors was in the range of 4~7 fT/√Hz at 100 Hz, with an average value of 5 fT/√Hz at 100 Hz and 10 fT/√Hz at 1 Hz. When several channels were operated simultaneously in the FLL mode, the noise of any channel did not increase noticeably by the operation of other adjacent channels, implying that the cross-talk between channels, due to the relaxation oscillations, is sufficiently low (Lee et al 1998).

To demonstrate the usefulness of the

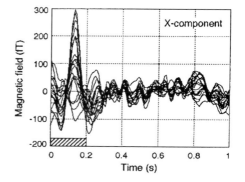

Fig. 4. Auditory-evoked magnetic field signals.

developed system as a biomagnetic multichannel system, the 40-channel system was applied to measure neuro- magnetic fields. Fig. 4 shows the x-component of 40 channel traces of auditory evoked responses. Here the direction from the temporal lobe to the ear is defined as the x-axis. Auditory stimuli of a 1 kHz tone burst, 200 ms duration and 70 dB normal hearing level were applied to the left ear of a normal subject, and the measurement were done over the right temporal lobe. The traces were obtained after averaging 128 times, using 0.3 Hz high pass filter, 100 Hz low pass filter and 60 Hz notch filter. A clear N100m peak, about 100 ms after the stimulus onset, was obtained.

4. CONCLUSIONS

We constructed a compact 40-channel SQUID system based on DROSs and integrated planar gradiometers, and operated to measure neuromagnetic fields. The large flux-to-voltage transfer of the DROSs simplified FLL circuits and the large modulation voltage allowed stable FLL operation against the offset drift of amplifier chain. Because of the low thermal load, the 40-channel insert had very low boil-off rate. The average noise level of the 40-channel planar gradiometer system is 1.2 fT/cm√Hz at 100 Hz, operated inside a magnetically shielded room. This corresponds to a magnetic field noise of 5 fT/√Hz, which is low enough for neuromagnetic measurements. By using the developed system, auditory-evoked fields were successfully measured.

ACKNOWLEDGEMENTS

This work was supported by the Ministry of Science and Technology, Republic of Korea.

REFERENCES

Adelerhof D J, Nijstad H, Flokstra J and Rogalla H 1994 J. Appl. Phys. **76,** 3875
Drung D 1996 SQUID sensors : Fundamentals, Fabrication and Applications, ed H Weinstock (Dordrecht : Kluwer Academic Pub.) pp 63-116
Lee Y H, Kwon H C, Kim J M, Park Y K and Park J C 1998 Appl. Supercond. **5,** 413
Lee Y H, Kwon H C, Kim J M, Park Y K and Park J C 1999 Supercond. Sci. Technol. (in press)
Ryhanen T, Cantor R, Drung D and Koch H 1991 Appl. Phys. Lett. **59,** 228
Shimogawara M.and Kado H. 1995 Biomagnetism: Fundamental Research and Clinical Applications, ed. C. Baumgartner, L. Deecke, G. Stroink and S.J. Williamson (Elsevier) pp. 536-541.
Tsukada K, Haruta Y, Adachi A, Ogata H, Komuro T, Ito Takada Y, Kandori A, Noda Y, Terada Y and Mitsui T 1995 Rev. Sci. Instrum. **66,** 5085

Inst. Phys. Conf. Ser. No 167
Paper presented at Applied Superconductivity, Spain, 14-17 September 1999

YBa₂Cu₃O₇ step-edge dc SQUID magnetometers on sapphire substrates

H-R Lim[1,2], I-S Kim[1], D H Kim[2], and Y K Park[1]

[1]Korea Research Institute of Standards and Science, Yusong PO Box 102, Taejon 305-600, Republic of Korea
[2]Department of Physics, Yeungnam university, Kyungsan, Republic of Korea

ABSTRACT: $YBa_2Cu_3O_7$ step-edge Josephson junctions and dc SQUID magnetometers on sapphire substrates have been fabricated. CeO_2 buffer layer and $YBa_2Cu_3O_7$ films were deposited *in situ* on low angle (~30°) steps formed on sapphire substrates. Large $I_C R_N$ product of 250 µV with junction resistance $R_N \sim 5\ \Omega$ and critical current $I_C \sim 50$ µA at 77 K could be obtained reproducibly with 5 µm microbridges. Direct coupled SQUID magnetometers utilising the step-edge junctions exhibit voltage modulation depth of 16 µV and magnetic field noise of 170 fT/.Hz at 100 Hz for the 70 pH-SQUID.

1. INTRODUCTION

An important technological goal in the development of high temperature superconductors (HTS) for electronic applications such as SQUID (Superconducting QUantum Interference Device) sensors is the fabrication of high quality thin films on large size substrate at low cost. Sapphire (single crystal Al_2O_3) is of considerable interest as a substrate in view of its modest dielectric constant and its commercial availability in large diameter substrate at low cost, as compared with the other oxide single crystals commonly used for HTS thin films depositions. The major problems in using the sapphire as substrates for $YBa_2Cu_3O_7$ (YBCO) thin films are interdiffusion, lattice mismatch, and large difference of the thermal expansion coefficients. In order to overcome these problems, oxide buffer layers have been deposited between sapphire and YBCO. A number of materials, e.g., CeO_2 (Denhoff et al, Zaitsev et al, and Cole et al), MgO (Cole et al), $SrTiO_3$ (Char et al, and Fork et al), and YSZ (Fork et al) have been investigated for the buffer layers. CeO_2 has been shown to be excellent *r*-plane cut sapphire buffer layers. During the last few years encouraging results have been reported on the preparation of high quality YBCO thin films on the CeO_2 buffered sapphire substrates. Josephson junctions are the most important active components in SQUIDs fabrications. Several approaches to fabricating Josephson junctions of YBCO thin film have been investigated in the area of bicrystal junctions, ramp edge junctions, biepitaxial junctions, and step-edge junctions (Koelle et al). Step-edge junctions have many attractive advantages as compared with other type junctions, i.e., easier process, high integration, low cost, and low 1/f noise (Yi et al). Numerous studies have been studied on the fabrication of step-edge junctions on $SrTiO_3$, $LaAlO_3$, MgO, Si (Linzen et al), and sapphire substrate (Adam et al). In spite of lots of potential advantages in sapphire substrates for HTS device applications, only few limited researches have been reported on the fabrication of SQUIDs using the sapphire substrates. In this paper the preparations and characterizations of YBCO step-edge junctions and dc SQUID magnetometers on *r*-plane sapphire substrates are reported.

550

2.   EXPERIMENTAL

Step-edge junctions were fabricated from pulsed laser deposited YBCO films on *r*-plane sapphire substrate. The steps on the sapphire substrate were fabricated by Ar-ion milling through photoresist masks. The step height and the step angle were adjusted by the milling time and the angle of incident Ar beam onto the substrate. After removing the photoresist, sapphire substrates were annealed at 1050 °C for 3 h. CeO_2-buffered YBCO films were grown by pulsed laser deposition technique on the stepped sapphire substrate. The deposition conditions of laser energy and substrate temperature were optimised for best YBCO thin film deposition, under oxygen pressure of 400 mTorr. 15-nm-thick CeO_2 buffer layer was *in situ* deposited prior to the deposition of YBCO thin film. The step height was typically 135 nm while thickness of YBCO films was 120 nm. The step-edge junctions were obtained by patterning 5-μm-wide microbridges positioned across the steps. Au film was deposited by rf-sputtering for electrical pads via lift-off patterning. The dc SQUID magnetometers were fabricated on 10 mm x 10 mm sapphire substrates.

3.   RESULTS AND DISCUSSIONS

Patterned YBCO thin films on CeO_2//sapphire exhibit excellent superconducting properties (Kim et al 1999), i.e., T_C □ 89 K, ΔT_C □ 0.5 K, and J_C . 3×10^6 A/cm². The optimised pulsed laser deposition conditions were substrate temperature of 810 °C, oxygen pressure of 50 mTorr, and laser fluence of 1.5 J/cm² for 15-nm-thick CeO_2 films, and substrate temperature of 790 °C, oxygen pressure of 400 mTorr, and laser fluence of 1.2 J/cm² for 120-nm-thick YBCO films, respectively. Step formations on sapphire substrate were performed by Ar-ion milling through photoresist masks. The step height and the step angle were adjusted by the milling time, the rotation angle and the tilt angle of substrates. The control of step angles in the wide range could be easily obtained by varying the rotation angle (maximum shadow effect at 90° rotation angle). Fig. 1 shows a typical result of step angle dependence on the substrate rotation angle. As shown in the figure, the step angles determined from cross sectional view of SEM image were in the range 70° - 20° when the rotation angles were changed from 0° to 90° with fixed tilt angle of 60°. After removing the photoresist, sapphire substrates were annealed at 1050 °C for 3 h under oxygen gas flow. With the heat treatment the ion beam damaged substrates recovered to sharp step edges and clean surface state. The sharp step-edges as observed from SEM cross sectional images were rounded when covered with CeO_2 buffer layer. 15-nm-thick CeO_2 films and 120-nm-thick YBCO films were *in situ* deposited on the stepped sapphire substrates with different deposition conditions described above. The CeO_2 buffered YBCO thin films were patterned using standard photolithography and Ar ion milling.

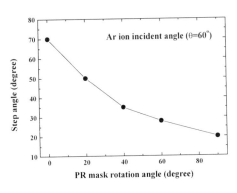

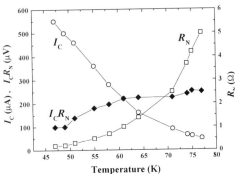

Fig. 1. Step angle dependence on the mask rotation angle.

Fig. 2. Temperature dependence of I_C, R_N and $I_C R_N$ product of the step-edge junctions.

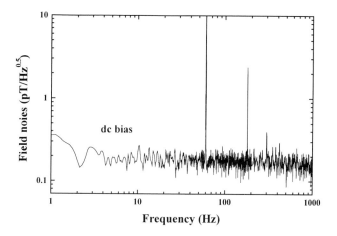

Fig. 3. Magnetic field noise of the SQUID magnetometer with static bias measured in a magnetically shielded room.

According to the *I-V* measurements at 77 K, no junctions were formed on the high angle (.50°) steps. RSJ-like junction-behaviours could be observed only in the CeO_2-buffered-YBCO microbridges on the low angle (. 40°) steps. Temperature dependence of the junction parameters is shown in Fig. 2. A typical 5-μm-wide junction has R_N of 5 Ω and I_C of 50 μA at 77 K. Very high values of R_N and $I_C R_N$ products obtained from the step-edge junctions of CeO_2 buffered YBCO films on sapphire substrates are promising results to realise high performance SQUID and digital circuits (Minotani 1998 and Hunt 1999).

SQUID magnetometers have been fabricated on 1 cm x 1 cm sapphire substrate with the step-edge junctions. The SQUID inductance was designed to have 70 pH as the sum of the slit inductance, parasitic inductance, and kinetic inductances. This SQUID inductance value is 30% lower than the standard 1 cm^2 single layer dc SQUID design (Beyer et al 1998) utilising 30° grain boundary junctions

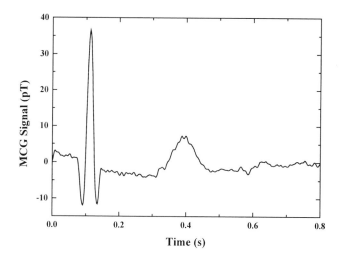

Fig. 4. Record of the time averaged magnetocardiograms using the SQUID magnetometer with static bias measured in a magnetically shielded room.

with 10 Ω resistance. The contact pads were covered by a gold layer which was deposited by rf sputtering through a photoresist lift-off stencil, ultra sonic bonded for external leads.

For the measurement of SQUID properties, the magnetometers were mounted onto PCB boards and the packages were immersed in liquid nitrogen in a fiberglass epoxy dewar. The measurements were carried out in a magnetic shielded room. The SQUID is operated in a flux-locked loop using 100 kHz flux modulation and either static or ac current bias. The SQUID has very high $I_C R_N$ product over 250 μV with 5 Ω resistance per junction. The maximum voltage modulation $\Delta V =$ 16 μV was observed for the device, which is almost half value of the predicted voltage modulation from Enpuku's relation (1993).

The magnetic field noise of the magnetometers measured using dc current bias exhibit almost frequency independent behaviours up to 1 kHz, as shown in Fig. 3. The magnetic field noise at 100 Hz was 170 fT/.Hz. The magnetic field noise spectrum exhibits no significant changes with ac bias at 1.5 kHz but decreases as the frequency increased with a $1/f$ dependence above 200 Hz. The origin of the excess noise at low frequencies is not clear, but is believed problem with ac-bias electronics due to the induced environmental noise. For the demonstration of the performance, Fig. 4 shows time averaged magnetocardiograms measured in magnetically shielded room.

4.   SUMMARY

$YBa_2Cu_3O_7$ step-edge Josephson junctions and dc SQUID magnetometers on sapphire substrates have been fabricated. CeO_2 buffer layer and $YBa_2Cu_3O_7$ films were deposited *in situ* on the low angle (~30°) steps formed on the sapphire substrates. Typical 5-μm-wide junction has R_N of 5 Ω and I_C of 50 μA with large $I_C R_N$ product of 250 μV at 77 K. Direct coupled SQUID magnetometers fabricated on the 1 cm^2 sapphire substrate utilising the step-edge junctions exhibit voltage modulation depth of 16 μV and magnetic field noise of 170 fT/.Hz at 100 Hz for the 70 pH-SQUID.

REFERENCES

Adam R, Benacka S, Chromik S, Darula M, Strbik V, Gazi S, Kostic I, and Pincik E V 1995 IEEE Trans. Appl. Supercon **5**, 2774

Beyer J, Drung D, Ludwig F, Minotani T, and Enfuku K V 1998 Appl. Phys. Lett. **72**, 203

Char K, Ponce F A, Trmontana J C, Newman N, Phillips J M, and Geballe V 1991 Appl. Phys. Lett **58**, 2432

Cole B F, Liang G-C, Newman N, Char K, Zaharchuk G, Martens J S V 1992 Appl. Phys. Lett. **61**, 1727

Denhoff M W, and McCaffrey V 1991 J Appl. Phys **70**, 3986

Enpuku K, Shimomura Y, and Kisu T V 1993 J. Appl. Phys **73**, 7929

Fork D K, Ponce F A, Sloggett G J, Mueller K-H, Lam S, Savvides N, Katsaros A, and Matthews D N V 1995 IEEE Trans Appl. Supercon **5**, 2804

Hunt B D, Forrester M G, Talvacchio J, and Young R. M V 1999 IEEE Trans Appl. Supercon. **9**, 3362

Kim I-S, Lim H-R, Kim D H, and Park Y K V 1999 IEEE Trans. Mag. (to be published)

Koelle D, Kleiner R, Ludwig F, Dansker E, and Clarke J V 1999 Rev. of Mod. Phys. **71**, 631

Kotelyanskii M, Mashtakov A D, and Ovsyannikov V 1997 Inst. Phys. Conf. Ser. No 158, 28

Linzen S, Schmidl F, Doerrer L, and Seidel P V 1995 Appl. Phys. Lett. **67**, 2235

Minotani T, Kawakami S, , Kuroki Y, and Enpuku K V 1998 Jpn. J Appl. Phys **37**, L718

Yi H R, Gustafsson M, Winkler D, Olsson E, and Claeson T V 1996 J. Appl. Phys. **79**, 9213

Zeitsev A G, Ockenfuss G, and Woerdenweber V 1997 Inst. Phys. Conf. Ser. No 158, 25

Inst. Phys. Conf. Ser. No 167
Paper presented at Applied Superconductivity, Spain, 14-17 September 1999

Study of the magnetic recording media using a scanning dc SQUID microscope

S A Gudoshnikov[1], O V Snigirev[2], and A M Tishin[2]

[1] Institute of Terrestrial Magnetism, Ionosphere and Radio Wave Propagation, Russian Academy of Sciences, (IZMIRAN), Troitsk, Moscow region 142092, Russia
[2] Department of Physics, Moscow State University, Moscow 119899 GSP, Russia

ABSTRACT: A scanning high-Tc dc SQUID microscope operated at liquid nitrogen temperature was used for imaging a magnetic flux distribution over a surface of different fragments of standard 3.5-inch floppy disks, achieving a spatial resolution close to 20 μm. Magnitudes of the magnetic field normal component produced by magnetic recording media with in-plane oriented magnetic moments of magnetic particles were measured. The magnetic images of fragments of unformatted, formatted and recorded media of the floppy disks have been obtained.

1. INTRODUCTION

Nano-scale particles and thin magnetic films can obviously contain only a small amount of magnetic ions. For example, a single layer of a previously investigated rare-earth Langmuir-Blodgett film can accommodate up to about 10^{14} magnetic ions (Tishin 1997). Therefore, very sensitive techniques are needed to investigate the magnetic properties of such objects. In the last few years it has been demonstrated that a dc SQUID-based magnetic scanning microscope became a useful tool for investigation of ultra-thin structures (Snigirev 1997a). It is shown that the SQUID-based technique can be considered as the promising method for writing and reading of protected information media (Bohr 1999). Due to extremely large sensitivity (to few $pT/Hz^{1/2}$), a SQUID scanner is able to detect an information contained on ultra-thin magnetic films or other types of nano-sized objects, which are protected by optically non-transparent cover. It is assumed that the information consisting of micro-size magnetic bits can be detected by scanning surfaces of a few mm-scale objects.

Presently, magnetic recording on video and audio tapes, on computer disks and diskettes, credit and ATM cards is the largest and fast growing field of use of magnetic materials. High-density magnetic recording systems require a high technology in development of recording and reading heads and magnetic recording media. A two-dimensional magnetic image produced by the surface of a magnetic material can be used in order to estimate the spatial variation of magnetization, which is of interest in magnetic recording media.

This work was performed to test the feasibility of using a scanning SQUID microscope for reading information from magnetic recording media and to estimate the value of the magnetic signals. We plan to study the applicability of novel SQUID-based method to test the distribution of magnetic field over the different magnetic recording media. As a first step we have imaged the surface of commercially available 3.5" floppy disk.

554

2. EXPERIMENTAL SETUP AND OBJECT CHARACTERIZATIONS

The design of our SSM based on high-T_c dc SQUID has been described previously (Gudoshnikov et al. 1997a). For measurements reported here, we used an improved version, which permitted to achieve a spatial resolution of about 20 μm. YBCO thin-film dc SQUID with an inner hole size of 20 μm ×20 μm and outer size of 40 μm×40 μm was used as a sensor. The SQUID was located within 300 μm from the edge of a $SrTiO_3$ bicrystal substrate polished to a corner. A magnetic field sensitivity of the sensor was close to 100 pT/Hz$^{1/2}$ at frequencies higher than 100 Hz. A minimal SQUID-sample separation was of about 20 μm. In these experiments SQUID was operated in a standard flux-locked loop mode and a normal component of magnetic field over a surface of the sample was measured. All measurements were performed at liquid nitrogen temperature in magnetic shielded environment.

As magnetic recording media objects we used the fragments of commercially available standard 3.5-inch magnetic floppy disks. The magnetic recording media of a floppy disk presents, as a rule, a thin layer of Fe-Co-based microparticles with sizes of about 1 μm. These magnetic particles have in-plane oriented magnetic moments and the coercivity of such magnetic materials has a value of about 30-70 kA/m.

Magnetic moments are initially in-plane randomly oriented. During operation each floppy disk is exposed to formatting process in which the magnetic tracks are created. Each track is a circumference consisted of the magnetic moments oriented along it. Here is North-South (N-S) (or S-N) magnetic poles of magnetic moments orientation state along the track. This is equivalent to a "zero" (0) state of data. In practice, each 3.5-inch floppy disk has 80 tracks of width 0.115 mm with period of 0.2 mm. During recording process some magnetic moments are remagnitized in opposite direction. The situations North-North (N-N) or South-South (S-S) magnetic poles orientation are equivalent to "unit" (1). Thus data recorded on a track represent lineups of magnetic moments of different length and opposite orientation, which lie alone the track.

3. RESULTS

During measurements we tested the following fragments of 3.5-inch floppy disks: unformatted before measurements; formatted; carrying some recorded data; carrying a special recorded data.

The fragment of unformatted sample was taken from a new commercially available floppy disk. We broke the diskette, cut a few parts of magnetic media and tested them. After cool down to liquid nitrogen temperature the magnetic image of a portion of fragment was taken. The first measurements were carried out at a SQUID-sample separation close to 150 μm. Fig. 1, a shows the 3D magnetic image across a fragment of unformatted floppy disk. Dashed lines indicate the edges of the sample. Sharp maximum and minimum of the B_z component of large magnitude ± 10 μT were found near the sample edges whereas there was no noticeable signal over the sample surface at this separation.

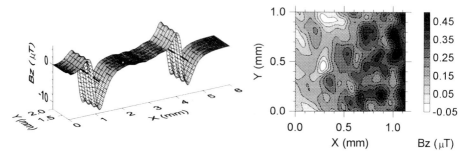

Fig 1. Magnetic images of a fragment of unformatted floppy disk; a) across the fragment, b) grey scale magnetic map of a central portion of the fragment.

To observe the more detail structure of magnetic field over the central part of the sample we reduced the SQUID-sample distance to 50 μm and took SSM image of the 1 mm × 1 mm area in the central part of the sample. The grey scale map of B_z magnetic field component is shown in Fig. 1, b. Observed randomly located small maxima and minima of magnetic signal have had a typical value of ± 0.25 μT.

The next fragment was cut from a new formatted floppy disk of the same pack. For imaging at separation of 150 μm the anomalies of the same type (like above mentioned) have been observed near the sample edges. A 3D magnetic image of the central part of this fragment taken at SQUID-sample distance 50 μm is shown in Fig. 2, a. Magnetic image has a periodic structure looked as a "washer desk". A period of this structure along X-axis is of about 200 micrometers. The magnetic signal amplitudes (a value of the magnetic field from maximum to minimum) were close to 3 μT. The wave-shaped distribution of a magnetic field is smooth and has no any peculiarities. This magnetic structure corresponds to magnetic tracks, which were created due to formatting procedure.

The next sample measured was a fragment of a floppy disk with some recording data, corresponded text information. Fig. 2, b presents a gray scale magnetic image of the fragment portion taken at a SQUID-sample distance of about 20 μm. The gray scale varies from about - 3 μT (white) to + 2.5 μT (black). The diagonal strips with the same period of about 200 μm are distinctly seen.

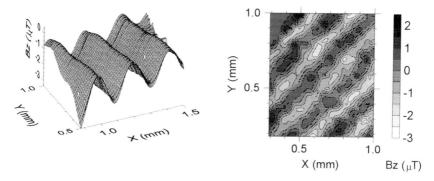

Fig 2. Magnetic images of the fragments of formatted floppy disk; a) 3D image of disk without any information; b) gray scale magnetic map of disk with some recorded information.

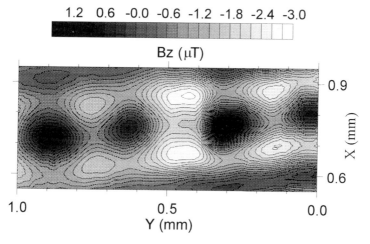

Fig 3. Magnetic image of the individual floppy disk track with special information.

More dark strips correspond to magnetic tracks with recorded data clusters looking as oval black spots. The amplitudes of the individual spots are varied in the range 0.5-1 µT. The minimal size of the spot is about 20 µm, corresponding to the space resolution of the SSM. Thus the SSM space resolution wasn't enough to detect the magnetic flux corresponded to an individual bit of information ("0" or "1").

To see the individual bits we created a special computer program, which allow to record onto a floppy disk some information as an alternating of large consequences of the "units" and the "zeros". About one hundred of "0" follows for only a few "1". For such data, as mentioned above, the consequences of "0" correspond to long single domains. The start and the end of this domains correspond to "1". The maximum of B_z magnetic field should correspond to only one "1". For two "1" signal was slightly less, three "1" we can't find, because two identical oriented magnetic domains (equal "0" consequences) separate very small each time opposite oriented domains (correspond "1").

An example of the grey scale image of the single track with such type of recorded information is demonstrated in Fig. 3. The line-up of well-defined maxima and minima of the magnetic field along one track can be clear seen. The positions of each maximum and minimum on the track (Fig. 3) correspond to the ends of one direction orientated "0" bits. Note, that the values of surface magnetic field have been ranged from −3 µT to +1.6 µT (-0.03 - + 0.016 G).

4. CONCLUSION

Using standard floppy disks as samples we consider the possibility of the scanning SQUID microscopy to investigate magnetic recording media and to read recorded data. Our experiments have demonstrated clearly that the scanning SQUID technique has the potential to image a character of distribution and determine a magnitude of the B_z magnetic field component over the surface of magnetic recording media.

The advantages of using this technique are the sensitivity of SQUID and high dynamic range. These parameters have enough to measured magnetic signals from nano-scale particles and layers from the one side and to fix different levels of magnetic signals with good signal-to-noise ratio from the other side. As an example different value of "1" signal can be detect in our images.

Our SSM space resolution achieved is 20 µm, making it impossible to resolve the single magnetic domain of recording media (one bit of information). But the spatial resolution, at our opinion, can be improved to 0.1 µm by a utilization of soft ferromagnetic tips to pick a surface signal permitting it possible to read the separate bits of information. Future development of this technique will allow creating a new method for characterization of surface magnetic properties of magnetic storage systems.

5. ACKNOWLEDGMENT

The INTAS Grant N 97-0894 and the Russian Program on Superconductivity, Grants Nos. 96093, support the work.

REFERENCES

Bohr J, Gudoshnikov S A et al. 1999 Patent of Russian Federation N 2124755
Gudoshnikov S A et al. 1997a Int. Supercond. Electr. Conf. - 6 (Berlin, Germany) p 392
Snigirev O V, Andreev K E et al. 1997a Phys. Rev. B 55 p 14429
Snigirev OV, Tishin A M, et al. 1997b Phys. Rev. B 55 p 14429
Tishin A M, Bohr J et al. 1997 Phys. Rev. B 55 p 11064

Inst. Phys. Conf. Ser. No 167
Paper presented at Applied Superconductivity, Spain, 14-17 September 1999
© *2000 IOP Publishing Ltd*

Magnetic Detection of Mechanical Degradation of Low Alloy Steel by SQUID-NDE system

N Kasai[1], S Nakayama[2], Y Hatsukade[3], and M Uesaka[4]

[1]Electrotechnical Laboratory, Tsukuba, Ibaraki 305-8568, Japan
[2]Seiko Instsruments Inc., Takatsuka-Shinden, Matsudo, Chiba 3270-2222, Japan
[3]Waseda University, Ohkubo, Shinjuku, Tokyo 169-8555, Japan
[4]Nuclear Engineering Research Lab., Univ. of Tokyo, Tokai, Naka, Ibaraki 319-1106, Japan

ABSTRACT: We have measured the magnetic property changes in low alloy steel A533B specimens caused by tensile deformation and by fatigue using a SQUID-NDE system to investigate the potential of advanced monitoring system for pressure vessels in nuclear power plants. The SQUID with a concentric second-order gradient pick-up coil was used. The large magnetic change by tensile deformation appeared near the Luders band and the magnetic change appeared even in the case of 50 fatigue cycles.

1. INTRODUCTION

Nondestructive evaluation (NDE) is very important to ensure the structural safety and reliability of nuclear power plants. The possibility of NDE by using SQUID (Superconducting Quantum Interference Device) is expected because SQUID has superior magnetic sensitivity. The ability of SQUID-NDE for cracks and corrosion in metals was shown in several studies (Donaldson 1993, Hohmann 1997, Haller 1997). On the other hand, Weinstock and Nisenoff (1985) first detected the magnetic change of stainless steel induced by tensile deformation using SQUID. The magnetic changes on a stainless steel duplex structure by fatigue (Otaka, 1993) and on a steel cylinder by hardening (Schmidl, 1997) were also observed. We measured the magnetic changes on structural carbon steel induced by application of tensile load (Kasai, 1997). And the strong correlation between appearance of Luders bands on the specimen surface and the magnetic field changes on the specimen was found (Kasai, 1998). These results revealed that the degradation of ferromagnetic materials by stress caused the change on magnetic properties in addition to the change on mechanical properties. That is, there is a possibility to detect magnetically the degradation before appearance of cracks using SQUID.

The A533B low alloy steel is a ferromagnetic and the typical material of light water reactor pressure vessels. We have measured the spatial distribution of magnetic property changes in the A533B specimens caused by tensile deformation and by fatigue using a SQUID-NDE system with a concentric second order gradient pick-up coil to investigate the potential of an advanced monitoring system using SQUID for nuclear power plants.

2. SQUID-NDE SYSTEM AND A533B

The NDE system using a Nb based SQUID with a concentric second order gradient pick-up coil was used in this study (Kasai, 1997). Fig. 1 shows the schematic diagram of the system and the coil configuration. The outer coil is 2 mm in diameter and one turn, and the inner coil is 1 mm in diameter and 4 turns. The SQUID mainly detects spatial curvature of z component of magnetic flux density, d^2B_z/dr^2. Then the SQUID is suitable for detection of the area where magnetic property changes occur on ferromagnetic materials. The system was set in a simple

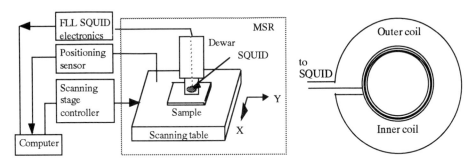

Fig.1 Schematic diagram of SQUID-NDE system and consentric seconed order gradient pick-up coil

magnetically shielded room. No magnetic field was applied to the specimens before and during the measurements by SQUID. The lift-off distance was 5 mm in this work.

A533B is a typical material of light water reactor pressure vessels. It is a low alloy steel. The chemical compositions are 97.1%Fe, 0.17%C, 0.24%Si, 1.37%Mn, 0.6%Ni, 0.46%Mo and small amounts of P, S, and Cu. The ingot of the material was quenched by water-cooling after heating at 860~890°C for 2hrs and 25 min and then annealed by air-cooling after heating 650~665°C for 2 hrs and 19 min.

3.   TENSILE TEST

Four A533B specimens, T1, T2, T3 and T4 were prepared for the tensile test. The configuration and the dimension of the specimen are shown in Fig.2. The plate specimens were subjected to A. C. demagnetization after the shaping. The specimens were scanned by the SQUID-NDE system before applying tensile loads to confirm the demagnetization.

After the measurement for confirmation, the tensile load of 462MPa, 0.378% residual strain, 0.43% residual strain and 580MPa were applied to the T1, T2, T3 and T4 respectively, where the upper yield stress was 575MPa. The stress-strain curve obtained by using an identical extra specimen prepared from the same material is shown in Fig.3. The speckle pattern was measured during the loadings using a laser speckle interferometer by Toyooka (Uesaka, 1997).

The specimens were scanned by the SQUID-NDE system after application of load. The result of T3 is shown as a contour map in Fig.4(a). The Fig.4(b) shows the area where Luders bands were observed on the speckle pattern measurement. The area where large magnetic change appeared corresponds to the area where Luders bands were observed. The correspondence agreed with the result on a carbon steel (Kasai, 1998). The Fig. 5 shows the results along the midline of the specimen. The curve for T1 is a straight line with a constant slope at whole range. The curves for T2 and T3 have the same slope as T1 at the range where the Luders bands do not appear. The curve for T4 becomes again a straight line with the same slope. It shows that we can know the work-hardening area by tensile load from the SQUID measurement.

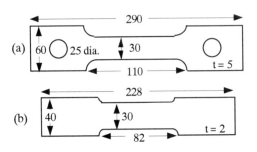

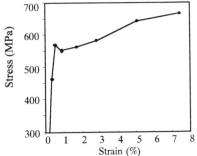

Fig.2 Shape and dimension of specimen
(a)for tensile test and (b) for fatigue test (unit:mm)

Fig.3 Stress-strain curve of A533B specimen

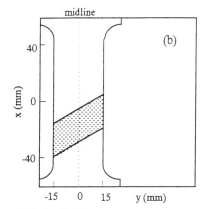

Fig.4 Example of results for tensile test
(a) Contour map of SQUID output for T2, (b) the shaded area is the one where the Luders bands appeared. It was obtained by the laser speckle interferometer.

4. FATIGUE TEST

Another four A533B specimens, F1, F2, F3 and F4 were prepared for the fatigue test. The configuration and the dimension of the specimen are shown together in Fig.2(b). These specimens were not demagnetized. The F1 was not applied any load. The loads with a sinusoidal waveform of 500MPa maximum and 50 cycles, 480MPa maximum and 10^5 cycles, and 500MPa maximum and 10^4 cycles were applied to the F2, F3, F4 respectively. The stress did not exceed the upper yield stress.

Firstly, B-H curves were measured using a Hall probe to obtain an index of fatigue. The residual magnetic flux density B_r on each specimen is shown as a function of the fatigue cycle in Fig.6. The results show that the fatigue on the F3 is low level and that on the F2 is moderate level and F4 is in advanced stage of fatigue.

After the B-H curve measurements, the specimens were scanned by the SQUID-NDE system. The result of F3 on the parallel part of specimen except near plate side edges is shown in Fig.7. The image of the SQUID output for F1 was almost plain. A large positive peak and a negative peak appeared like a sinusoidal waveform in the image for F2. Many upward and downward peaks unsystematically appeared in the image for F3. The image for F4 was rather flat as compared with that of F2 and F3, though there were many irregular small peaks and a large positive peak was at the right end in the image. These images are different with that for T2 and T3. The Fig.7(b) shows the results along the midline of the specimen. The results showed that even slight fatigue induced the magnetic change in several places on the specimen. Large change of curvature of B_z was induced over the whole when fatigue advanced. Finally, the change of the curvature decreased. It shows that we can know the fatigued level by continuously monitoring the transition of the curvature.

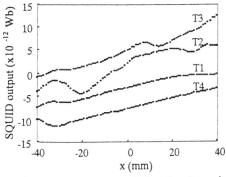

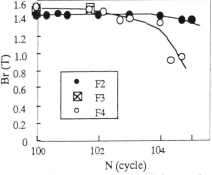

Fig.5 Results along the midline of each specimen   Fig.6 Br versus number of fatigue cycles

560

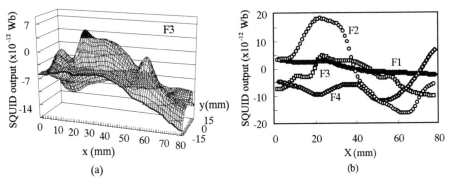

Fig. 7 Results for fatigue test. (a) The magnetic image for the specimen
of F3 and (b) The results along the midline of each specimen.

5. CORRELATION BETWEEN MAGNETIC AND MECHANICAL PROPERTY

The crystal deformation of ferromagnetic material by external stress induces the
magnetization change. It is well known as the inverse effect of magnetostriction. Around the
crystal imperfection such as dislocation, the internal stress field exists and the magnetic
moments change the direction from the axis of easy magnetization. Then the imperfection is an
obstacle for the movement of a magnetic domain. In this work, the tensile load or external field
lined up the magnetic moments. After removal of the load or field, the moment direction in a
local area depends on the state of the imperfection in the area. The SQUID detected
magnetically and macroscopically the spatial change of the state of the imperfection.

6 SUMMARY

We have measured the magnetic property changes induced by tensile load and cyclic
fatigue load on a low alloy steel of A533B by using the SQUID-NDE system with a concentric
second-order gradient pick-up coil. In the tensile test, the large magnetic changes appeared at the
area where Luders band appeared. In the fatigue test, the magnetic changes appeared even in the
case of low fatigue cycle. It is concluded that the magnetic property is very sensitive to the
mechanical degradation and there is the possibility of SQUID-NDE for nuclear power plants. It is
also necessary to investigate the sensitivity of SQUID to the thermal embrittlement and the
radiation embrittlement for the development of the SQUID-NDE system.

REFERENCES

Donaldson G B, Cochran A and Bowman R M 1993 The New Superconducting Electronics, eds H
Weinstock and R W Ralston (Dordrecht: Kluwer Academic) Chap. 6
Haller A, Tavrin Y and Krause H-J 1997 IEEE Trans. on Appl. Supercond. **7**, 2874
Hohmann R, Krause H-J, Soltner H, Zhang Y, Copetti C A, Bousack H, Braginski A I and Faley
M I 1997 IEEE Trans. on Appl. Supercond. **7**, 2860
Kasai N, Ishikawa N, Yamakawa H, Chinone K, Nakayama S and Odawara A 1997 IEEE Trans.
on Appl. Supercond. **7**, 2315
Kasai N, Chinone K, Nakayama S, Odawara A, Yamakawa H and Ishikawa N 1998 Jpn. J. Appl.
Phys. **37**, 5965
Otaka M, Enomoto K, Hayashi M, Sakata S, and Shimizu S 1993 Trans. Jpn Soc. Mech. Eng. **59**,
C-2041 [in Japanese]
Schmidl F, Wunderlich S, Doerrer L, Specht H, Linzen S, Schneidewind H and Seidel P 1997 IEEE
Trans. on Appl. Supercond. **7**, pp.2756-9
Weinstock H and Nisenoff M 1985 SQUID'85-Superconducting Interference Devices and Their
Applications, eds H D Hahlbohm and H Lubbi (Berlin: Walter de Gruyter) 853

Inst. Phys. Conf. Ser. No 167
Paper presented at Applied Superconductivity, Spain, 14-17 September 1999
© 2000 IOP Publishing Ltd

Design aspects of the FHARMON fetal heart monitor

A P Rijpma, H J M ter Brake, J Borgmann, H J G Krooshoop and H Rogalla

Low Temperature Division, Department of Applied Physics, University of Twente, PO Box 217, 7500 AE Enschede, The Netherlands

ABSTRACT: A promising application of SQUID magnetometers is the magnetic measurement of fetal heart activity. A fetal heart monitor based on high-T_c SQUIDs is under development at the University of Twente. In order to operate outside a shielded environment, active compensation is applied to suppress the earth's field. Next, to improve the signal-to-noise ratio, a 2nd order gradiometer is formed. For the gradiometer, a baseline of 6 cm is used, based on fetal magnetocardiograms that were recorded with a low-T_c system, of which the baseline could be varied.

1.   INTRODUCTION

Several groups are presently studying the recording of the magnetic signal of the fetal heart: the fetal magnetocardiogram (fMCG) (Quinn 1994, van Leeuwen 1996). From these studies, it is clear that the recording of fMCG-signals offers medically relevant information. However, these measurements are performed with low-T_c systems in a shielded environment. The costs and complexity of these systems are a serious drawback when considering fMCG-recordings as a routine procedure. Therefore, it would be advantageous to lower the threshold for recording fMCG-signals. To this end, the FHARMON project aims at a turnkey system for use in a clinical environment (i.e. without the use of a shielded room) (Rijpma 1997).

The FHARMON system is planned to make use of high-T_c SQUIDs, of which the white noise level should be below 50 fT/$\sqrt{Hz}$ from 1 Hz on. This is necessary, as the amplitude of the fetal heart signal is only 1-10 pT, whereas the bandwidth of interest is about 1-100 Hz. As the system is to be used outside a shielded room, it has to deal with the environmental magnetic fields. The largest field is the earth's field, which is about 50 μT strong. A bigger problem, however, is caused by the environmental disturbances overlapping the signal frequency bandwidth. Environmental noise levels exhibit a spread by two orders of magnitude to a typical maximum of 10 nT/$\sqrt{Hz}$ at 1 Hz decreasing to roughly 1 nT/$\sqrt{Hz}$ around 10 Hz (Vrba 1995). In the local hospital, we measured a noise level in the order of 100 pT/$\sqrt{Hz}$ in the frequency band of 1-100 Hz. An exception is the (50 Hz) power line frequency, for which the amplitude can easily reach 100 nT. The field gradients are estimated at 30 pT/m$\sqrt{Hz}$ and 10 pT/ m$^2\sqrt{Hz}$.

2.   SENSOR HEAD DESIGN

Fig. 1 shows the sensor head and, more detailed, the SQUID holder. High-T_c SQUIDs are used to form the gradiometer, as they may be cooled by a small turnkey cryocooler. Possibly, a thermal buffer will be used to allow time separation of the cooling process and recording of the fMCG. The Al_2O_3 (sapphire) holder containing the three gradiometer SQUIDs, is the heart of the system. It will be cooled through a short flexible copper link connected to the thermal buffer or, if no buffer is used, to the cold tip of a cooler. The whole cryogenic part will be placed in vacuum. The cryogenic design is discussed in more detail elsewhere (Rijpma 1999).

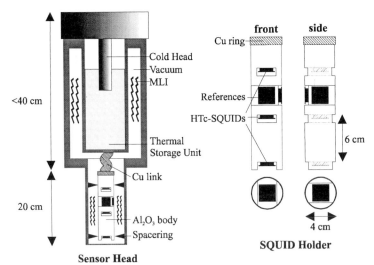

Sensor Head

Fig. 1. Preliminary design of the FHARMON sensor head

Next to the outer wall of the vacuum space, an rf-shield will be placed to shield the sensors from high-frequency magnetic noise fields that negatively influences SQUID operation (Koch 1994). Because the system contains only one measurement channel, it should be possible to move the sensor head over the maternal abdomen in order to search for a good measurement position. Both for reducing the required dynamic range resulting from this operation, as well as to be able to cool the SQUIDs in low background fields, a compensating field will be applied. As we want to suppress the DC background field before cooling, a room temperature reference sensor will be used to control the active compensation coil, which supplies this field.

3.   ACTIVE COMPENSATION COIL

A liquid nitrogen cooled 2nd order axial gradiometer was constructed to carry out tests on noise suppression techniques (Rijpma 1997). This system also included an active compensation coil. The coil was designed to fit on the LN_2 cryostat and produce the same field at all sensor positions. The coil was designed in such a way, that, at the sensor positions, variations in coil factor due to construction tolerances should not exceed 1%. In Fig. 2 the gray area shows the region, in which the coil factor is within 1% of the theoretical factor. The design places the SQUIDs at the peaks of the graph where the gradient of the coil factor is zero. As planned, this makes the system relatively insensitive to errors in sensor position.

The coil factor at several positions inside the cryostat was measured by moving a fluxgate sensor along the axis. The measurements, represented by the circles, are scaled and translated over the axial position to fit the calculated values. Due to the scaling no conclusion can be drawn about the absolute accuracy of the coil. However, the resulting figure shows that between the three SQUID sensor positions the differences in coil factor are well within the 1% limit.

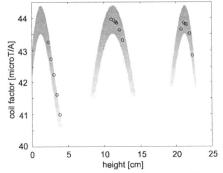

Fig. 2. The coil factor as a function of the distance to the cryostat bottom. The gray area shows were the coil factor differs less than 1% from the theoretical value. The circles represent the measurements.

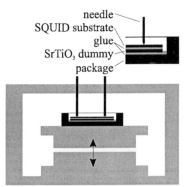

needle
SQUID substrate
glue
SrTiO₃ dummy
package

Fig. 3. Z-table system for sensor
alignment

Additional measurements were performed with the insert containing three SQUID sensors. The insert was placed in the cryostat, which was in turn placed at the center of a Helmholtz coil set. Successive measurement of the sensor response to Helmholtz set and compensation coil, confirm that the coil factor for the three SQUID sensor positions indeed differ less than 1%.

4.   SQUID ALIGNMENT

Tilt of the SQUID sensors leads to an imbalance contribution. This contribution can be compensated for by reference sensors. However, this compensation is limited by two factors: the reference noise level and the field gradient between (tilted) sensor and reference. Therefore, the aim is to align the sensors to within about 0.1 degree; corresponding to an imbalance $C_B=10^{-3}$. Unfortunately, the procedure that we currently use to package the SQUIDs (ter Brake 1997) already results in tilts of about 1 degree, when comparing the alignment of the pick-up-loop with the package bottom. By use of the set-up shown in Fig. 3, which basically consist of three needles poised above a z-translation table, we plan to improve the alignment of the SQUID sensor relative to the package. Initial experiments using makeshift packages and 0.14 mm glass squares show an alignment better than 0.05 degree. The same results are expected when performing the procedure with an actual package and SQUID. It remains to realize the Al₂O₃-holder with adequate accuracy.

5.   GRADIOMETER LAYOUT

In order to determine an appropriate gradiometer order and baseline for the FHARMON system, measurements were performed in shielded environment to estimate the spatial dependence of the fMCG-signal. The system used for these measurements consists of two low-T_c BTi rf-SQUIDs with wire-wound pick-up coils (d=38 mm, N=6). The bottom coil has a fixed position relative to the cryostat. The top coil is connected to the insert for the 2nd SQUID and can be moved from the outside. The available range corresponds to a baseline of 4 to 16 cm. Recorded were the two magnetometer channels and a gradiometer signal formed by electronically taking the difference between the channels. The bandwidth was limited to 100 Hz by the SQUID electronics output filters.

With this system, measurements were carried out on a fetus with a gestational age of about 35 weeks. At two positions above the abdomen, the baseline was varied and recordings were made. Next, an average fetal heartbeat was determined for each recording. In all recordings over 50 beats were available. These beats were located by threshold detection aided by manual corrections. The amplitude of the beat was then calculated from the averaged beat. Assuming an r⁻² spatial dependence as from a current dipole, a reasonable fit to the measurements is found:

$$A_{fetal}(r) = A_0 \left(\frac{r_0}{r} \right)^2 \qquad (1)$$

Here, A_0=2.8 pT is the amplitude of the heart signal at the location of the fixed sensor and r is the sensor location measured along the axis, with r_0=0.08 m the location of the fixed sensor. In Fig. 4 both the measured amplitudes and the fit is shown.

We combined this rough description of the fetal signal with the expected environmental noise and the intrinsic SQUID noise to find the appropriate gradiometer order. We found that a 1st order gradiometer will not sufficiently suppress the environmental noise in unshielded environment. A 3rd order gradiometer was found to be an improvement over the 2nd order gradiometer only for noise levels well above the expected. However, the SQUID holder would have to be significantly increased in length to establish this improvement. Thus, to keep the system relatively compact, a 2nd order gradiometer will be used in the FHARMON system.

564

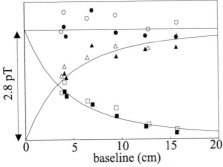

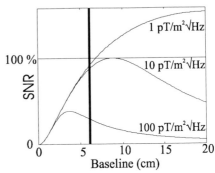

Fig. 4. Amplitude of fetal signal in fixed sensor (●), variable sensor (■) and gradiometer (▲). The lines corrrespond to Eq. (1); The open markers to the second recording location

Fig. 5. The signal-to-noise-ratio as a function of baseline for three environmental noise levels. The middle line corresponds to the expected level of 10 pT/m²√Hz.

Next, the 2nd order gradiometer is evaluated in more detail. Assumed is that, next to the intrinsic noise of 50 fT/√Hz per sensor, only the second order gradient of the environmental noise is relevant. In Fig. 5 the SNR is shown as a function of the system baseline. The SNR is shown relatively, where the maximum obtainable SNR is appointed a value of 100%. A 10 times better and a 10 times worse environmental noise situation are also included in the figure. From these, we see that for low environmental noise levels, a larger baseline can give a significant improvement. However, in case the environmental noise turns out higher than expected we would like to lower the baseline. Therefore, we chose the baseline between the optimum values for the worst case and expected noise level, which leads to a baseline of 6 cm.

6.   CONCLUSION

Several aspects of the FHARMON sensor head were examined. Measurements were carried out with a LN_2 cooled system to show the feasibility of constructing a compact compensation coil, which is designed to be part of the movable sensor head. Using glass squares, a method for controlled packaging of the SQUID sensors was successfully tested. Finally, based on low-T_c fMCG recordings a baseline of 6 cm was chosen for the sensor head.

ACKNOWLEDGEMENTS

The project is supported by the Dutch Technology Foundation (STW), the Institute for Biomedical Technology (BMTI), Philips Medical Systems and Signaal USFA.

REFERENCES

ter Brake H J M, Dunajski Z, van der Mheen W A G and Flokstra J 1989 , Journal of Physics E: Scientific Instr. 22 pp 560-564
ter Brake H J M, Karunanithi R et al. 1997 Meas. Sci. Technol. 8 pp 927-931
Koch R H et al. 1994 Appl. Phys. Lett. 65 pp 100
van Leeuwen P, Lange S, Schüßler M and Lajoie-Junge L 1996 Biomag96, Santa Fe
Quinn A, Weir A, Shahani U, Bain R, Maas P and Donaldson G 1994 British Journ. of Obst. And Gyn. 101 pp 866-870
Rijpma A P, Seppenwoolde Y, ter Brake H J M, Peters M J and Rogalla H 1997 Proc. EUCAS '97, Applied Supercond. 2 eds Rogalla H and Blank D H A (Bristol: IOP Publishing) pp 771-774.
Rijpma A P, ter Brake H J M, Peters M J, Bangma M R and Rogalla H 1999 Cryog. Eng. Conf.
Vrba J 1995 SQUID sensors: fundamentals, fabrication and applications eds Weinstock H (Dordrecht: Kluwer) pp 117

Inst. Phys. Conf. Ser. No 167
Paper presented at Applied Superconductivity, Spain, 14-17 September 1999
© 2000 IOP Publishing Ltd

Field response measurements on direct coupled HTS dc SQUID magnetometers

P.R.E. Petersen[a,b,c], Y.Q. Shen[a], T. Holst[a], and J. Bindslev Hansen[b]

[a]NKT Research Center, Priorparken 878, DK-2605 Brøndby, Denmark.

[b]Department of Physics, Technical University of Denmark, DK-2800 Lyngby, Denmark.

[c]Department of Clinical Neurophysiology, Rigshospitalet, Blegdamsvej 9, DK-2100 Copenhagen, Denmark.

ABSTRACT: Experimentally, we have investigated the magnetic field response of direct coupled $YBa_2Cu_3O_{7-x}$ dc SQUID magnetometers. By progressively etching the pick-up loop of the magnetometers, we changed the inductance and effective area of the magnetometer pick-up loop and thereby changing the effective area of the magnetometers. Results are presented from a series of field response measurements on two magnetometers with different values of the SQUID frame inductance. Furthermore, the measured values are compared to calculated values.

1. INTRODUCTION

For many SQUID magnetometer designs, the aim is to maximise the effective area, A_{eff}, given the substrate dimensions. Normally, this is done by using a larger pick-up loop to focus the magnetic flux into the smaller SQUID body. In order to avoid the more complicated processing of HTS multilayers, many designs are nowadays based on a direct coupling scheme, where the pick-up loop is galvanically coupled to the SQUID loop. The advantage is that this coupling scheme can be realised in a single layer of $YBa_2Cu_3O_{7-x}$. The aim of the present work was to determine how B_0 of a direct coupled dc SQUID magnetometer changed when the inductance and effective area of the pick-up loop changed. $B_0 = \Phi_0/A_{eff}$, is the applied field required to generate one flux quantum, Φ_0 in the SQUID. The SQUID loop itself was formed as a frame since the inductance of a frame structure is less sensitive to the small but unavoidable process variations in the width of the strip forming the SQUID loop than a more slit-like structure traditionally used (Petersen). Two magnetometers with different values of the SQUID frame inductance were made for these measurements.

2. MEASUREMENTS

The magnetometers were patterned in an approximately 200 nm thick $YBa_2Cu_3O_{7-x}$ thin film deposited by pulsed laser deposition on a 10 mm × 10 mm MgO substrate. The resist masks for patterning were made with electron beam lithography, and the patterning was made by Ar^+ ion milling. The SQUIDs were based on step-edge Josephson junction technology (Feng),(Lam). To vary the pick-up loop inductance and effective area of the magnetometers, the pick-up loop was trimmed for each new measurement. The trimmings were done by partly covering the pick-up loop with a simple paper mask and then remove uncovered material by Ar^+ ion milling. The outer dimensions of the pick-up loop were fixed at 8.8 × 8.8 mm² for all the different pick-up loop layouts. $YBa_2Cu_3O_{7-x}$ was only removed from the inner part of the pick-up loop, meaning that we decreased the strip width, w, and thereby changing the pick-up loop layout from being washer-like to become increasingly more frame-like. The advantage of this trimming procedure was that the SQUID characteristics were the same through all the measurements.

For each pick-up loop layout, we measured the current needed to produce one flux quantum, Φ_0, in the SQUID, both by applying a modulation current, I_m, directly to the SQUID frame and by applying a homogeneous external magnetic field, B_{ext}. To generate B_{ext} we used 5-turn coil with a diameter of 79 mm. A schematic drawing of the two cases is shown in Fig. 1.

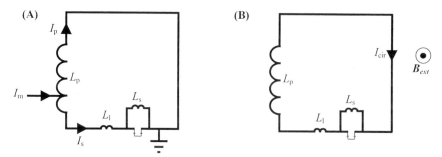

Fig. 1A: Schematic drawing of the direct modulated case. I_m is the modulation current applied directly to the magnetometer. I_s is the part of the modulation current flowing through the SQUID frame and I_p is the part flowing through the pick-up loop of the magnetometer. L_s is the inductance of the square frame of the SQUID and L'_s represents the inductance of the fraction of the SQUID frame that the modulation current flows through i.e. the upper part of the SQUID frame. L_l is the inductance of the current leads that connects the pick-up loop and the square frame of the SQUID, and L_p is the inductance of the pick-up loop.

Fig. 1B: Schematic drawing of the externally modulated case. I_{cir} is the circulating current induced by an externally applied magnetic field B_{ext}. B_{ext} is generated by an external modulation coil.

2.1 Direct modulation

Due to the inductive load of the SQUID frame by the pick-up loop, the direct modulation current I_m will split in two parts. One part, I_s, flows through the SQUID frame, and the remaining part, I_p, flows through the pick-up loop. Since I_m is injected somewhere along the pick-up loop, I_s also passes through a fraction of the pick-up loop. The parameter κ is introduced to represent this fraction. The direct modulation current is given as:

$$I_m = I_s + I_p = I_s + I_s \frac{\kappa L_p + L_l + L'_s}{(1-\kappa)L_p}, \tag{1}$$

where all currents and inductance's are defined as explained in Fig. 2. We measured the value of the directly applied modulation current I_m needed to generate one flux quantum in the SQUID, i.e. the value of I_m where $I_s = I_{\Phi_0}$. I_{Φ_0} denotes the current that generates one flux quantum in the SQUID when flowing through the upper part of the SQUID frame (for details see (Petersen)). Hence:

$$I_m = \frac{I_{\Phi_0}}{1-\kappa}\left(1 + \frac{L_l + L'_s}{L_p}\right). \tag{2}$$

The geometric inductance of a square frame can be determined as in (Drung):

$$L_{frame} = \frac{2\mu_0 h}{\pi}\left(\ln\left(\frac{h}{w}+4\right)+\tfrac{1}{2}\right), \tag{3}$$

where μ_0 is the vacuum permeability, h is the width of frame hole and w is the width of the strips forming the frame. Eq. (3) is valid for any ratio of h/w. Using this expression to determine the pick-up loop inductance L_p, we plot $1/L_p$ vs. the measured I_m as shown in Fig. 2 for magnetometer B197.

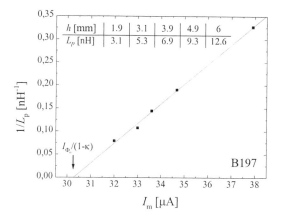

h [mm]	1.9	3.1	3.9	4.9	6
L_p [nH]	3.1	5.3	6.9	9.3	12.6

Fig. 2: Plot of the calculated value of $1/L_p$ vs the measured I_m for magnetometer B197 and the linear fit to these data points. A straight line can connect these points, which indicate that the inductance of the pick-up loop is well estimated by Eq. (3).

The plot in Fig. 2 shows a linear relation between $1/L_p$ and I_m in agreement with Eq. 2. This indicates that the inductance of the pick-up loop is well estimated by Eq. 3. Therefore we determine $I_{\Phi_0} / (1-\kappa) = 30.3$ µA and $L_l + L'_s = 0.77$ nH from the linear fit to the data series. For the second magnetometer, B191, we determined $I_{\Phi_0} / (1-\kappa) = 21.2$ µA and $L_l + L'_s = 0.74$ nH.

2.2 Modulation through an external coil

When an external magnetic field, B_{ext}, is applied to the magnetometer, the flux enclosed by the pick-up loop and upper part of the SQUID frame will be $\Phi_{ext} = B_{ext} A_{eff}$, where A_{eff} is the effective area of the pick-up loop, (see Fig 1(B)). The circulating current, I_{cir}, generated in the pick-up loop and the upper part of the SQUID frame by Φ_{ext} is then:

$$I_{cir} = \frac{\Phi_{ext}}{L_p + L_l + L'_s} = \frac{B_{ext} A_{eff}}{L_p + L_l + L'_s}. \tag{4}$$

B_0 is defined as the value of the applied magnetic field, B_{ext} that generates one flux quantum in the SQUID. When $B_{ext} = B_0$ the circulating current, I_{cir} equals the current needed to generate one flux quantum in the SQUID, I_{Φ_0}. Hence:

$$B_0 = \frac{\Phi_0}{A_{eff}} = \frac{I_{\Phi_0} (L_p + L_l + L'_s)}{A_{eff}} \tag{5}$$

The effective area may be estimated as in (Drung):

$$A_{eff} = h(l - \tfrac{1}{3} w) \tag{6}$$

Using Eq. (5), and Eq. (6), we can express B_0 as function of the pick-up loop hole width. Fig. 3 shows the calculated and measured B_0 vs. the width of the pick-up loop hole, h.

As seen from Fig. 3 we achieved optimum B_0 for a pick-up loop hole width of $h = 3.5$ mm for a square pick-up loop with $l = 8.8$ mm. This is in reasonable agreement with the result found by (Cantor).

B_0 calculated using the formulas of (Drung) to determine the inductance and the effective area of the square pick-up loop is in agreement with the measured values of B_0.

Although B_0 changes as the inductance and effective area of the pick-up loop are changed, the overall level of B_0 is set by the mutual inductance between the pick-up loop and the SQUID frame. We determined the mutual inductance between the pick-up loop and the SQUID frame to be 109 pH

568

and 76 pH for B191 and B197, respectively. The optimum values of B_0 found for B191 and B197 are 4.7 nT/Φ_0 and 6.7 nT/Φ_0, respectively.

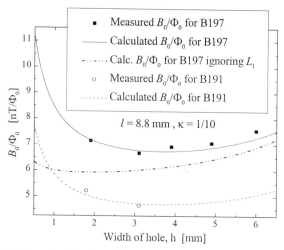

Fig. 3: Measured and calculated B_0 vs. the pick-up loop hole width for the magnetometers B191 and B197. The width of the pick-up loop was $l = (h + 2w) = 8.8$ mm for all measurements. The calculated B_0 curve for B197, in the case where L_l is neglected, is also shown. As seen this curve does not fit the measured data.

There is a considerable inductance contribution from the leads connecting the pick-up loop and the SQUID frame. For B197 we determined $L_l + L'_s$ to be 770 pH (cf. Fig 2). In Fig. 3 we plotted the two curves for B_0 of B197, one with the contributions from both L_l and L'_s, and one only with the contribution from L'_s. As seen, when neglecting L_l the value of the hole width that yields optimum B_0 changes from 3.5 mm to ~2 mm.

3. CONCLUSION

We have made a series of measurements of B_0 for two direct coupled magnetometers. B_0 was measured for different pick-up loop inductance's and effective areas. We found that the optimum B_0 of the square pick up loop of width $l = 8.8$ mm was achieved with a hole width of $h = 3.5$ mm. The inductance of the current leads connecting the pick-up loop and the SQUID frame has significant influence on B_0. This contribution must be included when calculating the strip width yielding optimum B_0.

4. ACKNOWLEDGEMENTS

We thank Dietmar Drung, PTB Berlin, for sharing his unpublished expressions for the frame inductance and the effective area and for useful comments and discussions. We thank Rolf Jensen for technical assistance. Peter R.E. Petersen thanks Prof. Christian Krarup for guidance and the Danish Research Councils for funding through the Danish Centre for Biomedical Engineering.

REFERENCES

Cantor R et al. IEEE Trans Appl. Supercond. **5**, 2927 (1995).
Drung D, unpublished.
Feng Y J, Shen Y Q, Mygind J, Pedersen N F, and Wu P H. Physica C **282-287** p. 2459 (1997).
Lam S and Foley C P. Proc. Intl. Supercond. Electronics Conf. (ISEC'95), Nagoya, Japan (1995).
Petersen P R E, Shen Y Q, Sager M P, Holst T, Larsen B H, and Bindslev Hansen J. (ISEC'99) To appear in Supercond. Sci. Technol. **12** (1999).

Inst. Phys. Conf. Ser. No 167
Paper presented at Applied Superconductivity, Spain, 14-17 September 1999
© *2000 IOP Publishing Ltd*

Single layer high-T_c SQUID gradiometer with large baseline

A Eulenburg, C Carr, A J Millar, E J Romans, G B Donaldson and C M Pegrum

Department of Physics and Applied Physics, University of Strathclyde, Glasgow G4 0NG, UK.

ABSTRACT: We describe a dc-superconducting quantum interference device (SQUID) first order gradiometer fabricated from a single layer of $YBa_2Cu_3O_7$ (YBCO) on a $30 \times 10mm^2$ bicrystal substrate. The device has a baseline of 13mm and an intrinsic balance of $\sim 10^{-3}$. The gradient sensitivity at 77K and 1kHz is $50\,fT/(cm\sqrt{Hz})$ in magnetic shielding and $260\,fT/(cm\sqrt{Hz})$ when operated unshielded in our laboratory. The sensor was used for unshielded magnetocardiography.

1. INTRODUCTION

First order gradiometers based on high-T_c SQUIDs enable the measurement of small, localised magnetic fields in unshielded environments. There are two main techniques for forming such a device. In electronic gradiometry, the output signals of two separate SQUID magnetometers are subtracted. The advantage of this approach is that it allows large baselines and a high degree of balance (Borgmann 1997). However, the electronic subtraction imposes stringent requirements on the system linearity, slew rate and the synchronisation of the multichannel electronics (Ludwig 1998, Espy 1998). The second approach is to form a planar gradiometer by coupling two symmetric pickup loops to a SQUID. Such a device can be made simply from a single layer of high-T_c thin film with the pickup loops directly coupled to the SQUID. However, the size of such single layer gradiometers (SLGs) demonstrated to date - e.g. (Knappe 1992, Wunderlich 1998) - is restricted by the common use of $10 \times 10mm^2$ substrates, limiting the baseline to typically 4mm and the area of the two pickup loops to small values, resulting in best gradient sensitivities of about $400\,fT/(cm\sqrt{Hz})$. As a more sensitive alternative to the autonomous SLG, a large flip-chip flux transformer can be coupled to a magnetometer (Dantsker 1997) or a SLG (Tian 1999) but this requires very accurate manual alignment. Here we describe a SLG with a longer baseline and larger pickup loops which are realised by fabricating a single device on a $30 \times 10mm^2$ substrate.

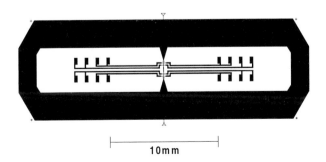

10mm

Fig. 1: Layout of the gradiometer with four dc-SQUIDs in the centre.

2. DESIGN AND FABRICATION

The gradiometer is shown in Fig. 1. It consists of two symmetric pickup loops to which four identical dc-SQUIDs are directly coupled. The estimated inductance of one pickup loop is $L_{loop} \sim 15$nH. The layout results in a baseline $l = 13$mm. The linewidth of the SQUIDs was restricted to 4μm to prevent magnetic flux penetration when the device is operated unshielded (Dantsker 1997b) and to minimise the SQUIDs' effective area for a given inductance. The width of the junctions was 3μm and the SQUID slit width was 4μm. The slit length of 108μm was chosen to set the estimated SQUID inductance to the optimised value for field sensitivity of a directly coupled device of $L_{SQ} \sim 100$pH (Koelle 1999).

The SLG was fabricated on a 30×10mm^2 24° SrTiO$_3$ bicrystal substrate from a YBCO thin film deposited by pulsed laser deposition at a substrate temperature of 820°C, an oxygen pressure of 0.15mbar and a target-substrate distance of 68mm. The film homogeneity along the 30mm long axis of the substrate was optimised by focusing the laser beam at the target to a narrow spot with a horizontal and vertical dimension of $\sim$10mm and $\sim$0.5mm respectively. The resulting plume expands little in the horizontal direction but significantly in the vertical direction, parallel to which we align the longer side of the substrate during deposition. For YBCO films with a thickness of 200nm and $T_c = 90$K at the centre, the thickness was 180nm and $T_c = 89$K at the ends of the substrate. The device was patterned using standard photolithography and argon ion milling.

3. RESULTS AND DISCUSSION

All device testing was performed in liquid nitrogen in a flux locked loop (FLL) with 64kHz ac- bias using Conductus PCI-100 electronics. The individual effective areas A_{eff} of several SQUIDs were measured in the open laboratory using a calibrated pair of 1.2m diameter

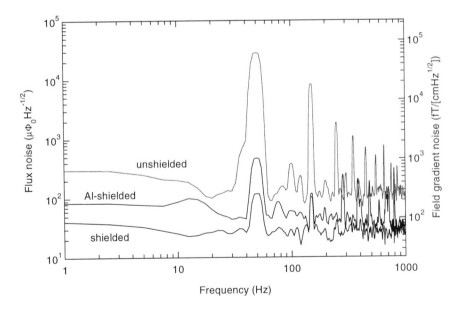

Fig. 2: Closed loop noise spectra of a gradiometer measured with 64kHz ac-bias at 77K measured inside two layers of mumetal shielding (lower curve), inside an aluminium shielded room (middle curve) and unshielded (upper curve).

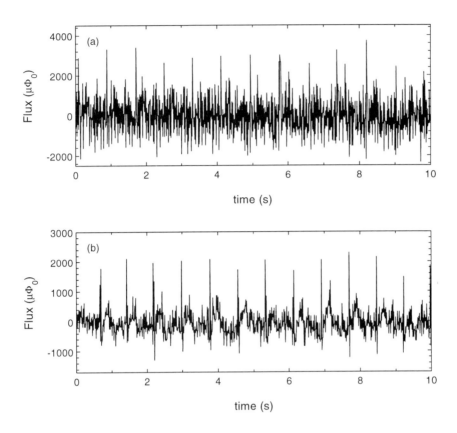

Fig. 3: Real time measurement of the magnetic field originating from the heartbeat of a healthy male volunteer measured (a) unshielded and (b) inside an aluminium screened room.

Helmholtz coils. A 320Hz current corresponding to an RMS magnetic field strength of 250nT was passed through the coils and the resulting SQUID response was lock-in detected. The effective area of individual SQUIDs on the SLG is about $650 \pm 50\mu m^2$. This is of the order we might expect for the SQUID alone as predicted by Ketchen et al. (Ketchen 1985) (though strictly for a square washer) of $A_{eff} \sim \sqrt{A_w A_h} = 775\mu m^2$, where A_h and A_w are the areas of SQUID slit and washer. The effective area of a gradiometer on which one of the pickup loops had been cut (operating as a magnetometer) was measured as $A_{mag} = 0.73mm^2$. Defining the gradiometer balance as $b = A_{eff}/A_{mag}$, we obtain $b \sim 10^{-3}$ which is approximately three times better than the values we had measured previously on smaller SLGs (Carr 1998). We have also employed a parasitic effective area compensation scheme with two neighbouring SQUIDs coupled in an opposite sense to the same gradiometer loop. Details on the achieved balance of about 3×10^{-5} will be published elsewehere (Eulenburg 1999).

The flux noise of one particular SQUID on the SLG measured inside two layers of mumetal shielding is shown in Fig. 2. The critical current and normal state resistance per junction were $8\mu A$ and 5.7Ω respectively and the peak-to-peak voltage modulation was $5\mu V$. The flux noise of $23\mu\Phi_0/\sqrt{Hz}$ at 1kHz corresponds to a gradient sensitivity $S_g^{1/2} = S_\Phi^{1/2}/(lA_{mag})$ of $50fT/(cm\sqrt{Hz})$ at 1kHz. To our knowledge, this is almost one order of magnitude lower than the best values reported for high-T_c SLGs. With the removal of the mumetal shielding, the device's flux noise level in the range 0–1kHz increased approximately fivefold and the magnetic

572

field sensitivity was $260\,\mathrm{fT}/(\mathrm{cm}\sqrt{\mathrm{Hz}})$ at 1kHz. For comparison, using the cut gradiometer, we measured an environmetal field noise level in our laboratory of $\sim 43\mathrm{pT}/\sqrt{\mathrm{Hz}}$ at 1kHz. We also recorded the noise inside an aluminium shielded room that provides high frequency screening but no attenuation at dc. The measured field gradient noise values of $76\,\mathrm{fT}/(\mathrm{cm}\sqrt{\mathrm{Hz}})$ at 1kHz and $293\,\mathrm{fT}/(\mathrm{cm}\sqrt{\mathrm{Hz}})$ at 1Hz demonstrate that low noise operation can be achieved in the earth's magnetic field.

The sensor was used for the detection of the magnetic field of the human heart. The healthy male volunteer was resting on a bed and the orientation of the gradiometer was such that its long axis was normal to the bed and its short axis parallel to the spine of the subject. Here, only preliminary measurements are presented as the data was not acquired with a resolution adequate for medical measurements but instead with a spectrum analyser at a sampling frequency of 80Hz and a signal resolution of only $\sim \pm 1\%$. The spectrum analyser performs low pass filtering above 40Hz to prevent aliasing. As shown in Figure 3(a), the heartbeat can be identified clearly for the unshielded measurement with an approximate signal to noise ratio of 2. The magnetocardiogram of the same volunteer measured inside the screened room is shown in Figure 3(b). Here, the magnetic screening leads to an improvement in the signal to noise ratio to about 6 and the T-wave component of the heartbeat can now be identified well in real time.

4. CONCLUSIONS

In summary, we have demonstrated a high performance first order gradiometer for operation at 77K with low field gradient noise. The device has been used successfully for unshielded magnetocardiography. Finally, we point out that for unshielded measurements, a considerable further performance improvement may be expected when two such devices are operated together thus forming a second order gradiometer with long baseline.

REFERENCES

Borgmann J, David P, Ockenfuss G, Otto R, Schubert J, Zander W and Braginski A I 1997 Rev. Sci. Instrum. **68**, 2730
Carr C, Eulenburg A, Romans E J, Pegrum C M and Donaldson G B 1998 Supercond. Sci. Technol. **11**, 1317
Dantsker E, Froehlich O, Tanaka S, Kouznetsov K, Clarke J, Lu Z, Matijasevic V and Char K 1997 Appl. Phys. Lett. **71**, 1712
Dantsker E, Tanaka S and Clarke J 1997b Appl. Phys. Lett. **70**, 2037
Eulenburg A, Romans E J, Carr C, Millar A J, Donaldson G B and Pegrum C M 1999 submitted to Appl. Phys. Lett.
Espy M A, Kraus R H, Flynn E R and Matlashov A 1998 Rev. Sci. Instrum. **69**, 123
Ketchen M B, Gallagher W J, Kleinsasser A W, Murphy S and Clem J R 1985 in *SQUID'85, Superconducting Quantum Interference Devices and their Applications*, eds H D Hahlbohm and H Lübbig (Walter de Gruyter, Berlin), p. 865
Knappe S, Drung D, Schurig T, Koch H, Klinger M and Hinken J 1992 Cryogenics **32**, 881
Koelle D, Kleiner R, Ludwig F, Dantsker E and Clarke J 1999 Rev. Mod. Phys. **71**, 631
Ludwig F, Beyer J, Drung D, Bechstein S and Schurig T 1998 to be published in IEEE Trans. Appl. Supercond.
Tian Y J, Linzen S, Schmidl F, Dörrer L, Weidl R and Seidel P 1999 Appl. Phys. Lett. **74**, 1302
Wunderlich S, Schmidl F, Specht H, Dörrer L, Schneidewind H, Hübner U and Seidel P 1998 Supercond. Sci. Technol. **11**, 315

Inst. Phys. Conf. Ser. No 167
Paper presented at Applied Superconductivity, Spain, 14-17 September 1999
© 2000 IOP Publishing Ltd

Operation of a YBa₂Cu₃O₇ dc SQUID magnetometer with integrated multiloop pick-up coil in unshielded environment and in static magnetic fields

K–O Subke, C Hinnrichs, H–J Barthelmeß, M Halverscheid, C Pels and M Schilling

Institut für Angewandte Physik und Zentrum für Mikrostrukturforschung, Universität Hamburg, Jungiusstraße 11, D-20355 Hamburg, Germany

ABSTRACT: We investigate the behavior of a dc SQUID magnetometer with integrated multiloop pick-up coil (IMPUC) in unshielded environment and in static magnetic flux densities of up to $220\,\mu$T. The magnetometer is based on the high-temperature superconductor YBa₂Cu₃O₇ with bicrystal Josephson junctions. With shielding we obtain a low frequency flux density noise of $\sqrt{S_B}(3\,\mathrm{Hz}) = 50\,\mathrm{fT}/\sqrt{\mathrm{Hz}}$ at 77 K in zero magnetic field. The noise level at 3 Hz exhibits a moderate linear increase with the static magnetic field of $4.7\,\mathrm{fT}/(\sqrt{\mathrm{Hz}}\mu\mathrm{T})$, confirming the high epitaxial quality of our optimized YBa₂Cu₃O₇ films.

1 INTRODUCTION

Integrated dc SQUID magnetometers based on the high-temperature superconductor YBa₂Cu₃O₇ achieve sufficiently low flux density noise levels for biomagnetic applications. Drung et al reach $53\,\mathrm{fT}/\sqrt{\mathrm{Hz}}$ at 1 Hz with an integrated flux-transformer coupled magnetometer operated at 77 K (1996). However, these results can only be obtained inside heavily shielded rooms. Without shielding the magnetometers are subjected to external noise sources interfering with the SQUID operation and the earth's magnetic field of $50\,\mu$T. The latter leads to vortices pinned in the superconducting films. Their motion causes excess low-frequency noise contributions as described by Ferrari et al (1994) and Miklich et al (1994). Dantsker et al (1997) demonstrated that patterning SQUIDs with slits or holes helps to maintain the noise levels obtained in zero field up to a threshold field determined by the maximum superconducting linewidth. This concept was used by Krey et al (1999) to investigate the influence of patterning on a directly coupled magnetometer. Investigations of integrated flux-transformer coupled magnetometers and multiloop SQUIDs by Krey et al (1998) show that a high epitaxial quality of the superconducting films leads to an only moderate linear increase in excess noise of the magnetometers operated in static magnetic flux densities of up to $110\,\mu$T. In this work we investigate the behavior of a dc SQUID integrated multiloop pick-up coil

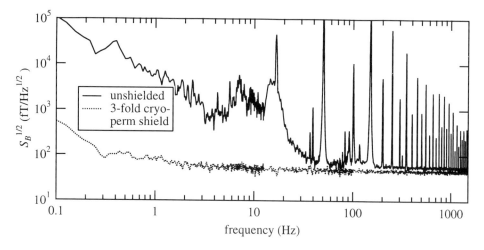

Figure 1: Flux density noise of the IMPUC magnetometer with a threefold cryoperm shield and without shielding in a rural environment at 77 K.

(IMPUC) magnetometer with bicrystal Josephson junctions operated without shielding and in static magnetic fields.

2   PREPARATION

The IMPUC magnetometer, introduced by Scharnweber and Schilling (1996), is prepared on a 10 mm × 10 mm × 1 mm, 24° bicrystal SrTiO₃ (100) substrate. We employ a pulsed KrF Excimer laser for the deposition of the superconducting $YBa_2Cu_3O_7$ and insulating $SrTiO_3$ films. Heinsohn et al (1998) describe the optimization of the $YBa_2Cu_3O_7$ films. First a 100 nm $YBa_2Cu_3O_7$ film with a protective $SrTiO_3$ cap layer for good overgrowth conditions of 25 nm is deposited in situ. It is patterned by conventional photolithography and Argon-ion etching in a parallel-plate reactor, thus defining the washer SQUID with the Josephson junctions and the four pick-up loops. A 50 nm $SrTiO_3$ film is deposited to insulate the $YBa_2Cu_3O_7$ ramp edges from crossovers, and windows for via contacts are etched. The four-turn input coil is structured from a 200 nm $YBa_2Cu_3O_7$ film, with the washer serving as the back contact to the pick-up loops.

With the magnetometer size of 8 mm × 8 mm and a SQUID inductance of 80 pH an effective area of 1.4 mm² is achieved. The bicrystal Josephson junctions of the magnetometer exhibited a maturing behavior over the first six month after the preparation, presumably due to oxygen loss at the grain boundary. As a consequence, the modulation swing was almost doubled to 17 μV and the flux density noise was reduced accordingly.

3   NOISE PROPERTIES

3.1   Unshielded operation

Figure 1 depicts the noise spectra of the IMPUC magnetometer measured in a rural environment at 77 K. For all the measurements the magnetometer was operated with a bias reversal scheme, thus eliminating noise contributions from critical current fluctuations

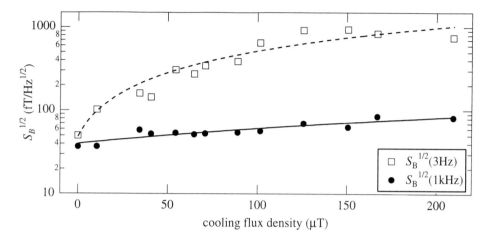

Figure 2: Flux density noise levels at 3 Hz and 1 kHz in dependence of the static magnetic flux density in which the IMPUC magnetometer was cooled and operated at 77 K. The lines are linear fits to the data.

in the Josephson junctions (Dössel 1991). The spectrum obtained inside a threefold cry-operm shield is almost frequency independent down to 3 Hz, below which the shielding admits external noise. For this reason all the spectra are evaluated at 3 Hz for the low-frequency noise levels. In the unshielded measurement a $1/f$ excess noise contribution appears due to external noise sources and vortex motion in the films. Additionally, a broad structure appears around 10 Hz due to wind moving the magnetometer in the earth's magnetic field. The peak at $16\frac{2}{3}$ Hz origins in the railway operation frequency, while the dc/ac converter feeding the SQUID electronics operated close to the magnetometer causes the sharp 50 Hz harmonics. Despite these unfavorable conditions the magnetometer remained in the flux-locked loop operation during all the measurements for more than 30 minutes.

3.2 Operation in static magnetic fields

The investigations of the behavior in static magnetic fields are carried out in a threefold cryoperm shield in liquid nitrogen at 77 K. As described by Krey et al (1999), a pair of Helmholtz coils provide the magnetic field in which the magnetometer is cooled and operated. These measurements were obtained after the end of the maturing process mentioned in Section 2. At zero field, a low-frequency noise level of $\sqrt{S_B}(3\,\mathrm{Hz}) = 50\,\mathrm{fT}/\sqrt{\mathrm{Hz}}$ is reached, a further improvement even compared with the shielded measurement in a rural environment two month after the magnetometer fabrication, as depicted in Fig. 1.

Figure 2 descibes the dependence of the low-frequency noise at 3 Hz and the white noise at 1 kHz on the applied flux density B_0 between zero and $210\,\mu\mathrm{T}$. For the low-frequency noise we obtain a linear increase of $\frac{\partial\sqrt{S_B}}{\partial B_0}(3\,\mathrm{Hz}) = 4.7\,\mathrm{fT}/(\sqrt{\mathrm{Hz}}\mu\mathrm{T})$ due to the $1/f$ excess noise from the motion of vortices. In the white noise regime the noise levels are only weakly affected with a slope of $\frac{\partial\sqrt{S_B}}{\partial B_0}(3\,\mathrm{Hz}) = 0.2\,\mathrm{fT}/(\sqrt{\mathrm{Hz}}\mu\mathrm{T})$. The increased white noise levels can partly be attributed to the observed linear reduction of the effective magnetometer area. This is obviously caused by higher slit inductances between the pick-up loop spokes, resulting in a reduced inductance adaption between the pick-up loops and the SQUID. This parallels the findings of Scharnweber and Schilling (1997) about the temperature dependence of the effective area in this magnetometer type. The deviations from the linear behavior in the low-frequency noise are attributed to measurement errors caused by external noise sources of unidentified origin from the laboratory environment, penetrating the shielding at frequencies below 20 Hz. The moderate linear increase of the low-frequency noise is comparable to the findings of Krey et al (1998). From the linear behavior we conclude that the excess noise contribution is entirely due to the motion of vortices. Their density is approximately linearly dependent on B_0, while the quality of the films determines the extend to which they can contribute to the magnetometer noise.

4 CONCLUSION

We have fabricated a $YBa_2Cu_3O_7$ dc SQUID magnetometer with integrated multiloop pick-up coil based on bicrystal Josephson junctions. In shielded environment a flux density noise of $\sqrt{S_B}(3\,\text{Hz}) = 50\,\text{fT}/\sqrt{\text{Hz}}$ at 77 K is achieved. The magnetometer remains stable in unshielded operation. Investigations of the excess noise caused by static magnetic fields in which the magnetometer is cooled and operated yield a linear slope of $4.7\,\text{fT}/(\sqrt{\text{Hz}}\mu\text{T})$ at 3 Hz due to the motion of vortices trapped in the superconducting films. We attribute this moderate increase to the high epitaxial quality of our optimized $YBa_2Cu_3O_7$ film deposition and multilayer process.

5 ACKNOWLEDGMENT

We acknowledge financial support by the Bundesministerium für Bildung, Wissenschaft, Forschung und Technologie, Federal Republic of Germany, under contract number 13N7323/8.

REFERENCES

Dantsker E, Tanaka S and Clarke J 1997 Appl. Phys. Lett. **70**, 2037

Dössel O, David B, Fuchs M, Kullmann W H and Lüdeke K-M 1991 IEEE Trans. Magn. **27**, 2797

Drung D, Ludwig F, Müller W, Steinhoff U, Trahms L, Koch H, Shen Y Q, Jensen M B, Vase P, Holst T, Freltoft and Curio G 1996 Appl. Phys. Lett. **68**, 1421

Ferrari M J, Johnson M, Wellstood F C, Kingston J J, Shaw T J and Clarke J 1994 J. Low Temp. Phys. **94**, 15

Heinsohn J-K, Reimer D, Richter A, Subke K-O and Schilling M 1998 Physica C **299**, 99

Krey S, David B, Eckart R and Dössel O 1998 Appl. Phys. Lett. **72**, 3205

Krey S, Barthelmeß H and Schilling M 1999 accepted for publication in J. Appl. Phys.

Miklich A H, Koelle D, Shaw T J, Ludwig F, Nemeth D T, Dantsker E, Clarke J, Alford N McN, Button T W and Colclough M S 1994 Appl. Phys. Lett **64**, 3494

Scharnweber R and Schilling M 1996 Appl. Phys. Lett. **69**, 1303

Scharnweber R and Schilling M 1997 IEEE Trans. Supercond. **7**, 3485

Inst. Phys. Conf. Ser. No 167
Paper presented at Applied Superconductivity, Spain, 14-17 September 1999
© 2000 IOP Publishing Ltd

577

Technique for compensation of background magnetic field for high-Tc dc SQUID systems operating in unshielded environment

S A Gudoshnikov[1], L V Matveets[1], S V Silvestrov[1], P E Rudenchik[1], O V Snigirev[2], L Dörrer[3], F Schmidl[3], R Weidl[3] and P Seidel[3]

[1] Institute of Terrestrial Magnetism, Ionosphere and Radio Wave Propagation, Russian Academy of Sciences, (IZMIRAN), Troitsk, Moscow region 142092, Russia

[2] Department of Physics, Moscow State University, Moscow 119899 GSP, Russia

[3] Institut für Festkörperphysik, Friedrich-Schiller-Universität Jena, Germany

ABSTRACT: For application of SQUID-based systems in nondestructive testing, a method of compensation of background magnetic field is suggested. A high-Tc dc SQUID with direct readout electronics based on a liquid-nitrogen-cooled preamplifier is used as a reference magnetometer for compensation of environment disturbances acting on the high-Tc dc SQUID measuring sensor. The reference SQUID has a small effective area making it possible to move the system in the Earth magnetic field without unlocking of the SQUID electronic. A compensation factor of about 30 dB was achieved.

1. INTRODUCTION

In recent years there has been rapid growth in development of instrumentation based on high-Tc Superconducting Quantum Interference Devices (SQUIDs), which are used in non-destructive evaluation and biomagnetism (Wiksvo (1995)). These practical applications require SQUID systems operating in an unshielded environment. Operating without shielding, the SQUID is subjected to the magnetic environment including dc magnetic field of the Earth, power line signals and RF interference.

A few special methods have been developed to maintain the SQUID sensitivity in the presence of strong magnetic fields and electromagnetic interferences. The gradiometric configuration of the sensor is most generally employed. In low-Tc systems wire-wound superconducting pick-up coils are used. Present high-Tc technologies make it possible to fabricate only planar thin film gradiometric structure. So when axial configuration is needed the gradiometers to be made electronically by combining separate SQUID magnetometer outputs.

Koch et al. (1993) developed a compensation technique for an unshielded environment using a reference SQUID to compensate the disturbances in two magnetometers forming an axial electronic gradiometer. Two compensation schemes suitable for first- and second-order gradiometer rejection were developed and tested with high-Tc rf SQUID magnetometers by Borgmann et al. (1997). Ter Brake et al. (1995) suggested the schema of individual flux compensation.

We used a high-Tc dc SQUID with direct readout electronics based on a liquid-nitrogen-cooled preamplifier as a reference magnetometer to compensate environmental magnetic disturbances acting on the measuring high-Tc dc SQUID sensor. This technique is adapted for use in high-resolution nondestructive testing systems operating in an unshielded environment.

2. OPERATING PRINCIPLES

The proposed compensation technique is similar to the second compensation scheme developed by Borgmann at al. (1997). A simplified scheme of the circuit is shown in Fig. 1. The reference channel (1) includes the high-Tc dc SQUID, the SQUID electronics and the cylindrical solenoid coil (S). The reference SQUID (1) is placed at the top end of the solenoid, which acts as a feedback coil of the reference channel. The measuring channel (2) consists of the measuring high-Tc dc SQUID and SQUID electronics. The measuring SQUID (2) is positioned at the bottom end of the solenoid in a point symmetrical to the site of the reference SQUID. Both channels can utilize an ac bias electronics.

The SQUIDs are cooled down in the Earth's magnetic field. During operation the output of the flux-locked loop reference channel (1) is fed to the solenoid (S), which produces a feed-back magnetic field opposite to the external magnetic field variations. Because of similarity of the field near the ends of the solenoid the common mode disturbances at the location of measuring SQUID (2) should be compensated.

In order to define the SQUIDs positioning accuracy and magnetic field homogeneity through the SQUID area it is necessary to calculate the magnetic field distribution near solenoid ends. As fixed parameters R = 5 mm (solenoid radius) and d = 50 µm (diameter of wire) were chosen. It is assumed that maximum magnetic field near the solenoid end should be about 50 µT. This magnetic field distribution was obtained by direct calculation using Biot-Savart law. The calculation accuracy was about 0.01%.

The magnitude of the magnetic field normal component in points lying along the solenoid axis near its ends in dependence on the solenoid length is shown in Fig. 2. a, when solenoid current I is equal to 4 mA. The middle curve (1) represents this dependence for a point located in a plane of solenoid end. The upper (2) and lower (3) curves correspond to the points displaced from the solenoid end by 100 µm inside and outside of solenoid, accordingly.

The calculations have shown that a length of solenoid practically does not effect the homogeneity of a z-component of magnetic field over the relevant areas (i.e. the difference ΔB_z between these curves for each length L of solenoid actually remains constant). Hence, the choice of solenoid's length is rather arbitrary. For our measurements a length of the solenoid of 60 mm and a radius of 5 mm were chosen.

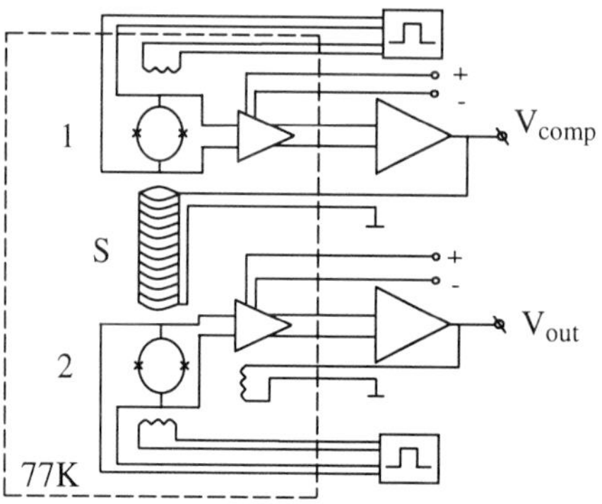

In Fig. 2, b the calculated distribution of the z-component of magnetic field near the end of the 60-mm solenoid is shown. The point with coordinates {0,0} corresponds to the end of the central axis of the solenoid and has 50 µT magnetic field at the same parameters R, d, L, I. The magnetic field is rather homogeneous along the solenoid diameter in the shown area, however it varies fast along the solenoid central axis (h-coordinate). In order to compensate magnetic variation with error less than 1% it is necessary to keep the SQUIDs positions near the solenoid ends with accuracy ± 20 µm.

Fig.1 Compensation scheme for a high-Tc SQUID magnetometer with an additional reference SQUID.

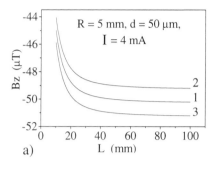
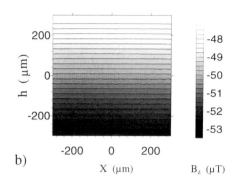

a)

b)

X (µm)   B$_z$ (µT)

Fig. 2. The normal component of magnetic field near the end of solenoid at fixed parameters R, d, and I. a) Dependence of magnetic field on the solenoid length, L, for points located along the solenoid axes: curve 1 – for point located in the plane of solenoid end, curve 2 – for point located in 100 µm outside of the solenoid, curve 3 – for point located in 100 µm inside of solenoid. b) Distribution of the magnetic field z-component near the end of solenoid, where sensor can be placed.

3. TEST MEASUREMENTS

To test the compensation technique a two-channel stick was prepared. The 60-mm solenoid was fixed in a special plastic holder. The measuring SQUID was mounted to the bottom end of the solenoid. The reference SQUID was mounted on a special pivot, which can move near the top end of the solenoid using room temperature micro-screw. Two liquid nitrogen amplifiers were mounted 20 mm away from the reference SQUID. The cryogenic part was 23 mm in diameter and 180 mm in length.

A high-Tc dc SQUID that had a square washer with 30-µm inner hole and 250-µm outer dimension was used as a reference SQUID. A field-to-flux transfer coefficient was estimated as 300 nT per flux quantum ($\Phi_0 = 2\times10^{-15}$ Wb). At 77 K, the SQUID's critical current was about 15 µA, and the voltage modulation was 15 µV.

To get a high slew rate and high frequency bandwidth the SQUID electronics (Gudoshnikov et al. 1999) was optimized. We increased the bias reversal frequency from 20 kHz up to 2 MHz. The new *emitter-connected logic* generator was developed producing both bias current and bias flux square wave signals. It is not necessary to compensate 2 MHz frequency square-wave voltage, for the dynamic range of the LNA was large enough and it was suppressed well by room-temperature amplifiers (based on AD797) and integrator (τ = 0.05 ms). As a result, our reference channel had a slew rate of 5×10^5 Φ_0/s, a 350-kHz bandwidth (for a small signal), a dynamic range of 150 dB. This electronics was operating in a flux-locked loop mode without unlocking in the range from zero magnetic field up to the Earth's magnetic field (close to 50 µT).

Fig. 3, a shows the noise spectra of the reference channel taken under magnetic shielding (lower curve) and in an unshielded environment (upper curve). In magnetic shield the white-noise level is less than 10 $\mu\Phi_0$/Hz$^{1/2}$ (3 pT/Hz$^{1/2}$) at frequencies higher than 800 Hz. We propose that a high level of the low-frequency noise (100 $\mu\Phi_0$/Hz$^{1/2}$ at 10 Hz) is determined by flux motion in the SQUID's washers. The upper curve presents a magnetic situation in the laboratory environment. Low frequency magnetic field variations (close to

580

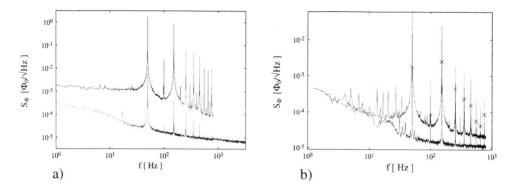

a)

b)

Fig. 3. a) Noise spectra of the reference channel for magnetic shield (lower curve) and unshielded environment (upper curve). b) Noise spectra of the measuring channel for unshielded environment (upper curve) and unshielded environment with compensation (lower curve).

320 pT/Hz$^{1/2}$) were measured up to 50 Hz. The main spectral signals took place at 50 Hz and its harmonics, which fell down from 320 nT (50 Hz) to 3 nT (800 Hz).

A small-square high-Tc dc SQUID with 20 μm inner hole and 40 μm outer size intended for scanning SQUID microscope was used as a signal-measuring sensor during the first test measurements. Fig. 3, b shows a spectral noise density of this small-area SQUID with and without compensation in an unshielded environment. The upper spectrum of Fig. 3, b presents noise of a small measuring SQUID located at 60 mm far from the reference SQUID in the laboratory environment. Different scales of upper curves of Figures 3. a, b are due to different SQUID effective areas. The lower spectrum of Fig. 3, b shows the flux noise with operating compensation system. Crosses at 50 Hz and its harmonics correspond to the same amplitudes in the lower curve. Maximum compensation close to 30 dB was observed at 50 Hz. High harmonics suppression was varied from 20 dB (100 Hz) to 10 dB (800 Hz). Magnetic shield noise measurements gave the same spectral density as in compensation mode, but with less power signals.

The obtained data showed some virtues of this compensation system using liquid nitrogen amplifier. Due to an increase of bias reversal frequency up to 2 MHz system had a high slew rate and a high dynamic range and operated stable in stationary and moving states. We hope to achieve a compensation factor higher than 40 dB in the nearest future.

ACKNOWLEDGMENT

We would like to thank Alexey Kalabukhov for SQUID preparation. The work is supported in part by the German BMBF under Grant No. 13 N 7235, the INTAS Grant N 97-0894 and the Russian Program on Superconductivity, Grant N 96093.

REFERENCES

Borgmann J et al. 1997 Rev. Sci. Instrum. **68** pp 3082-84
Gudoshnikov S A et al 1999 IEEE Trans. on Appl. Supercond. **9** N2 pp 4397-9
Koch R H et al. 1993 Appl.Phys. Lett. **63** pp 3-6
Ter Brake H J M et al. 1995 Inst. Phys. Conf. pp 1503-6
Wikswo J P, Jr. 1995 IEEE Trans. on Appl. Supercond. **5** N2 pp 74-120

Inst. Phys. Conf. Ser. No 167
Paper presented at Applied Superconductivity, Spain, 14-17 September 1999
© 2000 IOP Publishing Ltd

SQUID and Hall-probe microscopy of superconducting films

A Ya Tzalenchuk[1,2], Z G Ivanov[1,3], S V Dubonos[4], and T Claeson[1]

[1]Department of Physics, Chalmers University of Technology and University of Göteborg, S-412 96 Göteborg, Sweden
[2]Institute of Crystallography Russian Academy of Sciences, 117333 Moscow, Russia
[3]Swedish National Testing and Research Institute, S-501 15 Borås, Sweden
[4]Institute for Microelectronics Technology, Russian Academy of Sciences, 142432 Chernogolovka, Russia

ABSTRACT: We have developed a variable temperature scanning micro-magnetometer, incorporating an HTS SQUID and a ballistic Hall-probe sensor. The microscope features non-magnetic temperature regulation and topographic imaging capabilities. The performance of the microscope was demonstrated in measurements of a flux distribution in patterned HTS films and around deliberately introduced defects in a temperature range 5-100 K.

1. INTRODUCTION

A variety of methods has been used to visualise the distribution of the magnetic field in superconducting samples, including Bitter decoration, magnetic force microscopy, electron holography, magneto-optics, scanning SQUID and scanning Hall-probe microscopy (for an excellent review see deLozanne (1999)). The last two devices (Scanning Micro-Magnetometers) offer the best sensitivity, while the spatial resolution is of the order of microns, being limited by the active sensor area and the sensor-to-sample separation. A number of successful SMM designs has been suggested in recent years (Chang et al. (1992), Mathai et al. (1992); Black et al. (1993), Gudoshnikov et al. (1994), Kirtley et al. (1995); Oral et al. (1996), Morooka et al. (1999). We have developed a simple, reliable, and versatile variable temperature SMM, which can work with both a SQUID sensor and a Hall probe.

2. EXPERIMENTAL

2.1 Description of the SMM

A non-magnetic continuous flow cryostat with a variable temperature insert (VTI) is designed and built for the microscope. Both the sensor and the sample are placed in the vacuum chamber of the VTI. The sensor is suspended on a cantilever, which is attached to a cold finger cooled directly by LHe or LN_2. The sample-holder is attached to a rod fixed on a stepper motor driven XYZ translator with an attainable resolution of 0.5 micrometers. This holder is connected by soft copper braids to a variable temperature heat exchanger. To avoid magnetic interference, the sample temperature is regulated by the balance of two fluxes of He gas: cold, directly from the bath, and warm, passing by the room temperature. The balance is maintained by automatic valves placed outside the cryostat. The sensor is operated in the contact mode for best resolution. The cantilever is equipped with a simple capacitive sensor, which allows for rough monitoring of the sample topography. The contact pressure and thus the heat transfer is minimised. The SQUIDs are based on 0.5 μm step-edge Josephson junctions in a $\varnothing$ 8 μm YBa$_2$Cu$_3$O$_{7-\delta}$ (YBCO) loop. The design of the cryostat, cantilever, scanner, and the SQUID sensors used in the microscope is described in more details by Tzalenchuk et al. (1999). The Hall probes are made in a GaAs/GaAlAs heterostructure with a high-mobility two-dimensional electron gas (2DEG). The Hall cross has a 2 μm^2 wide conductive channel. It is characterised by the series resistance of about 10 Ω after illumination with visible light at 4 K through a window in the cryostat. To avoid overheating of the 2DEG the bias currents below 10 μA have been used.

2.2   Sample preparation

We have chosen two examples of practical importance to demonstrate magnetic imaging with the SMM: visualisation of defects in HTS samples and properties of patterned HTS film. Correspondingly, two samples were prepared. The first sample is a 200 nm thick YBCO film deposited by laser ablation on a LaAlO$_3$ substrate. Lithographically defined cross-shaped marks were etched in each corner of the substrate prior to film deposition. During the thermal cycling in the deposition chamber the substrate was intentionally constraint in expansion/contraction. As a result the substrate underwent plastic deformation, which was localised in several narrow zones (shear bands) (Berezhkova and Sustek (1997)). Accordingly, the distance between those of the marks across the shear bands has changed by several micrometers (Fig.1a). The second sample is a YBCO film laser deposited on a SrTiO$_3$ substrate and patterned by conventional lithography and Ar ion milling. Both positive and negative patterns were created on the same chip. The smallest pattern feature is 6 μm in diameter.

3.   EXPERIMENTAL RESULTS

3.1   Properties of a YBCO film in the vicinity of substrate defects

The properties of the superconducting film grown on a substrate around the localised shear bands were studied using SMM equipped with the SQUID sensor. The film was zero-field cooled down to 5 K and a magnetic field of 2 G was applied perpendicular to the film surface. The magnetic images were measured at different temperatures in the range 5-100 K. The exact position of the shear bands was located using the topographic mode of the SSM. Two of the images taken at 18 and 77 K and superimposed on the optical image of the film surface are presented in Fig. 1b. No magnetic contrast was observed below 10 K. The magnetic images at higher temperatures reveal a big density of magnetic flux around the shear bands where superconductivity is suppressed. While at 18 K we observe only separate defects, they form a continuous line about 100 μm wide above 70 K.

3.2   Magnetic field distribution in a patterned YBCO film

Field distribution in a patterned HTS film was measured using the Hall probe. Images measured at zero external magnetic field and at 13 G are presented in Fig. 2a and 2b. The magnetic field penetrates all the way into the film. A finite slope of the magnetic field profile complicates the resolution of the smaller structures on the raw image given the limited greyscale depth of the printout. However, structures as small as 10x10 μm were distinguished by differentiating the raw image. The images obtained with and without external magnetic field are almost exactly the negative of each other in the film region.

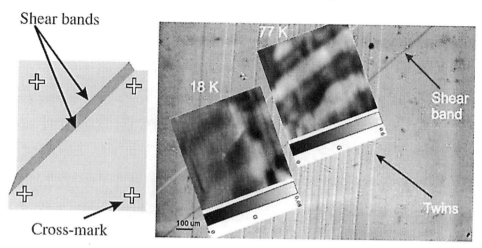

Fig. 1. a) A sketch of the localised plastic deformation in the substrate; b) A magnetic field distribution in the YBCO film around the shear bands superimposed on the optical image of the same region. Note different B scale on the magnetic images: 0.08 G for the 18 K image and 0.6 G for the 77 K image.

A single 600 μm long scan line across the edge of the superconductor was measured above 55 K in the magnetic field (Fig. 2c). It is clearly seen how the slope of the magnetic profile gradually decreases as the temperature approaches T_c. Some features in the curve, however, are still visible at 85 K.

4. DISCUSSION

4.1 Defects in HTS films

The magnetic contrast observed in the YBCO film in the vicinity the localised shear bands in the substrate suggests a strong disorder in the film. A reduced superconducting transition temperature might be a consequence of oxygen off-stoichiometry induced by disorder in the plastically deformed region. Indeed a microbridge patterned across the shear bands has shown the transport properties in agreement with the observed magnetic pattern. The critical current at low temperatures was strongly suppressed compared to the defect-free regions of the film and the I_c vs. B dependence was very non-uniform. The critical current turned to zero above 70 K far below the macroscopic T_c of the film.

4.2 Patterned HTS films

The observed distribution of the magnetic field in the patterned YBCO film is in general explained by the Bean model. The magnetic profile measured in the film bulk represents the density of vortices integrated over the active area of the sensor. Its steepness is determined by the demagnetisation factor and the critical current density proportional to the pinning force at a given temperature. It follows from the images that the 500 μm wide YBCO strips are completely in the mixed state at elevated temperatures. The remanent magnetic profile is determined by the same factors as the one measured in the field, but its slope is opposite representing a zero field on the strip edge and an increasing density of pinned vortices from the edge to the centre.

The contrast from features with a smallest diameter of 10 µm is traceable on the images. The limitation on spatial resolution comes from the large distance between the Hall cross and the chip corner in our non-optimised design of the Hall sensor.

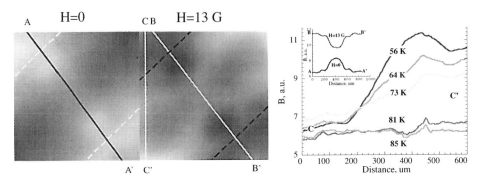

Fig. 2. Magnetic images of a patterned YBCO film at 55 K: a) in zero magnetic field; b) in a magnetic field of 13 G perpendicular to the surface. Dashed lines indicate the strip edges. Rounded spots in the image come from the holes Ø 50 µm etched in the film; c) Cross-sections measured at different temperatures along the line denoted C-C' in Fig. 2b. Inset: cross-sections along A-A' and B-B' corresponding to H=0 and H=13 G.

5. CONCLUSIONS

Scanning SQUID / Hall-probe microscopy has been shown to be very helpful in routine investigations of the quality of both as-grown and patterned HTS films. The simple and versatile microscope developed in this work is capable of imaging the local magnetic properties over a large area in the temperature range of 5-100 K. The demonstrated spatial resolution is about 10 µm for both sensors. This is comparable with the SQUID loop diameter, but can be pushed down to about 1 µm for the Hall probe.

REFERENCES

Berezhkova G V and Sustek V 1997 Crystal. Rep. **42**, 327-40
Black R C et al. 1993 Appl. Phys. Lett. **62**, 2128-30
Chang A M et al. 1992 Appl. Phys. Lett. **61**, 1974-6
deLozanne A 1999 Supercond. Sci. & Tech. **12**, R43-R56
Gudoshnikov S A et al. 1994 Cryogenics **34**, 883-6
Kirtley J R et al. 1995 Appl. Phys. Lett. **66**, 1138-40
Mathai A et al. 1992 Appl. Phys. Lett. **61**, 598-600
Morooka T et al. 1999 IEEE Trans. on Appl. Supercond. **9**, 3491-4
Oral A et al. 1996 J. of Vac. Sci. & Tech. B **14**, 1202-5
Tzalenchuk A Ya et al. 1999. IEEE Trans. on Appl. Supercond. **9**, 4115-8

Inst. Phys. Conf. Ser. No 167
Paper presented at Applied Superconductivity, Spain, 14-17 September 1999
© 2000 IOP Publishing Ltd

Sub-Femtoamp dc SQUID based Cryogenic Current Comparator

C Rillo, J Sesé, A Camón

ICMA, CSIC-Universidad de Zaragoza, 50009 Zaragoza, Spain.

E Bartolomé, J Flokstra

University of Twente, P.O. Box 217, 7500 AE Enschede, The Netherlands.

G Rietveld

NMi Van Swinden Laboratorium, P.O. Box 654, 2600 AR Delft, The Netherlands.

ABSTRACT: The dc SQUID based Cryogenic Current Comparator (CCC) is the most sensitive device for precision measurements (<0.01 ppm) of the ratio of very small dc currents (<1μA). When ideal magnetic flux coupling between the different CCC elements is obtained, current resolutions in the sub-femtoamp range can be attained. There are several uncertainty sources limiting the precision with which the current ratio can be determined. A study of those sources as well as the conditions for ideal coupling between the flux transformer and the toroidal CCC tube are given in this paper. Finally we derive strategies for the construction of optimal CCCs.

1. INTRODUCTION

The need for measuring sub-femtoamp currents when testing electronic devices has triggered industry to develop state of the art instruments, as the Keithley 6430 (www.keithley.com), a room temperature electronic apparatus that is able to detect a direct current as small as 100 aA (about 700 e/s). Using instruments like that, the current generated by Single Electron Tunnelling (SET) devices (which is of the order of 10 pA) could be measured with a resolution of parts in 10^5 and an absolute precision of 1%.

Very sensitive room temperature electronic ampere meters will be adequate to characterize SET devices but they do not meet high precision electrical metrology needs. In fact, the proposed, but no yet realized, electrical quantum current standard, based on SET devices, does not need a current meter. Instead it needs a cryogenic dc to dc converter (current comparator), ideally with a resolution of parts in 10^8 in the current ratio. Up to now there is not any commercial instrument for such purpose.

A few laboratories in the world have started projects aiming to build very sensitive CCCs (i.e.: SETamp project, SMT-CT96-2049 3156 proposal) as one essential step towards the development of the quantum current standard. The heart of the very sensitive CCC is a dc SQUID detector. SQUIDs are poor current resolution devices (about 1 pA for best dc SQUIDs operating at 4.2 K), therefore the CCC primary winding ratio should be of the order of 10^7, for this application, which is unpractical.

As a consequence the desired CCC requires : 1st) better current resolution SQUIDs, i.e.: (nearly) quantum limited SQUIDs, to decrease the winding turn ratio to practical values in the range 10^4-10^5; 2nd) ideal magnetic flux coupling between the different elements of the CCC i.e: coupling coefficients near to 1.

The authors are working on both approaches. The first approach is treated by Bartolomé (1999). SQUIDs of 80 h at 4.2 K (2h at 0.1 K expected) are used. The present status of the second approach is reviewed in this paper.

2.   CCC SENSITIVITY

The Type I CCC (Sullivan 1974) is essentially composed of several primary windings inside a superconducting tube that is overlapped like a snake swallowing its own tail. When the CCC is operated, two opposite currents I_1 and I_2 are passed through the N_1 and N_2 turns of the primary windings and a Meissner current $I_E = I_1 N_1 - I_2 N_2$, equal to the magnetomotive force imbalance, appears in the inner surface of the superconducting overlapped tube and returns through the external surface. As it is shown in Fig. 1, in which the primary windings are not shown for simplicity, the current I_E induces a current I_S in the superconducting flux transformer. This transformer consists of a sensing coil having N_S turns with nominal self-inductance L_S (effective self-inductance L_{Seff} due to the presence of the overlapped tube) and the SQUID input coil with self-inductance L. The SQUID acts as null detector for I_E. We shall define the normalised sensitivity of the CCC, S_{CCC}, as the square root of the ratio between the energy in the SQUID input coil $(L I_S^2/2)$ and the available energy in the CCC $(A_{OV} I_E^2/2)$ which, due to flux conservation in the flux transformer, is given by

$$S_{CCC} = \frac{I_S}{I_E} \sqrt{L/A_{OV}} = \frac{M_{OV,S}}{L_{Seff} + L} \sqrt{L/A_{OV}} \qquad (1)$$

where $M_{OV,S}$ is the mutual inductance between the superconducting overlapped tube and the sensing coil. A_{OV} is the effective inductance of the overlapped tube when surrounded by the superconducting closed shield (shown in Fig. 1). This shield protects the CCC and sensing coil against external noise.

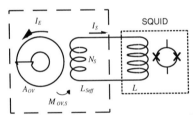

Fig. 1: Schematic representation of an overlapped tube Type I CCC. Indicated are the overlapped tube with effective self-inductance A_{OV}, the flux transformer sensing coil with N_S turns and effective self-inductance L_{Seff}, and the SQUID with input inductance L. The discontinuous lines indicate superconducting closed shields. Primary windings are not shown for the sake of clarity.

It has been deduced theoretically and corroborated experimentally by Sesé (1999a) that ideal magnetic flux coupling can be obtained either, by direct connection of the CCC superconducting overlapped tube with L (Fig. 2),

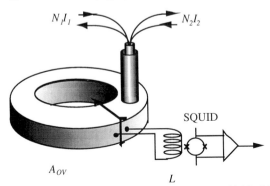

Fig. 2: Schematic representation of a CCC in which the overlapped lead shield of the windings is used as sensing coil.

or by the use of ferromagnetic cores. In both cases expression (1) takes the form

$$S_{ccc} = \frac{x}{x^2+1} \tag{2}$$

where $x = N_S\sqrt{(A_{OV}/L)}$ is a dimensionless parameter. Then, for ideal coupling the normalised sensitivity is a universal function of x and has an optimum value, $(S_{CCC})_{OPT} = 0.5$, for $x = 1$ (i.e.: $(N_S)_{OPT} = \sqrt{(L/A_{OV})}$). This is illustrated by the continuous line in Fig. 3.

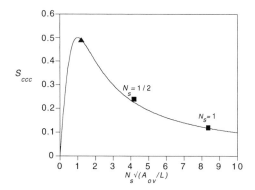

Fig. 3: Normalised sensitivity of the CCC as a function of the dimensionless parameter $N_S\sqrt{(A_{OV}/L)}$. The solid line is the theoretical curve for ideal coupling between overlapped tube and sensing coil (equation (2)). The squares correspond to experimental points with the use of ferromagnetic cores for configurations with $N_S=1$ and $N_S=1/2$, implemented as shown in Fig. 4. The triangle corresponds to the experimental point of the direct connection of the overlapped tube to the SQUID.

When ferromagnetic cores are used $A_{OV}>>L$ and the optimum condition requires $N_S<1$. To do that the implementation of fractional turn loop configurations in the sensing coil has been ideated by Sesé (1999c). This is illustrated in Fig. 4.

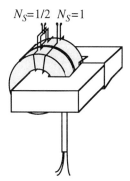

Fig. 4: Schematic representation for a fractional turn loop configuration, $N_S=1/2$, in the secondary coil of a type I CCC. The overlapping tube has a square shape with two ferromagnetic torii inside. Each torus has a one turn winding pick-up coil. They can be connected in series or in parallel to get $N_S=1$ or $N_S=1/2$ respectively.

588

3. CCC CURRENT RESOLUTION

Sesé (1999c) demonstrated that even though very low noise ferromagnetic cores can be used (i.e.: VITROVAC 6025F) the current resolution is dominated by the $1/f$ noise properties of those materials. Therefore, unless new materials with lower $1/f$ noise are discovered, the design of very current sensitive CCCs will be restricted to the direct coupling configuration of Fig. 2. In that case, under optimum conditions, the mean square value of the equivalent input current noise, at the input of a primary winding of N_I turns, is given by the following expression (Sesé 1999b).

$$\langle I_1^2 \rangle = \frac{4L\langle i_{SQ}^2 \rangle}{N_1^2 k_{SQ}^2 A_{OV}} \tag{3}$$

Using a quantum limited dc SQUID operating on the white noise region and with an input coil inductance of about 100 nH (CCC mean diameter about 10 cm), an equivalent SQUID input coil current noise of about 0.1 pA could be attained. Therefore with N_I in the range 10^4-10^5, current resolutions well below the femtoamp range (~10 aA) could be obtained.

4. CONCLUSIONS

The construction of big CCCs with large turn numbers (about 10^5) and A_{OV} of the order of 100 nH has already been tried (Hartland 1993). Drawbacks of these trials where non-ideal coupling due to the use of wire based sensing coils instead of direct coupling, and drifts or excess noise attributed to thermal instabilities or vibrations.

Our design does not use a sensing coil therefore the noise from vibrations does not occur. We have also verified that thermal instabilities of the bath are not relevant at least for standard size CCCs (A_{OV} about 10 nH). We then propose the construction of CCCs of about 100 nH, directly coupled to (nearly) quantum limited SQUIDs. The operation temperature may be in the mK-range using the practical SQUIDs, which have an energy sensitivity of about 100 h at 4.2 K. To avoid drifts and $1/f$ noise the possibility of incorporating a modulation technique of the flux transformer current, similar to those proposed by Park (1995) will be analized.

ACKNOWLEDGEMENTS

The financial support given by the Spanish Ministry of Education and Science (CICYT, MAT98-0668), by the Government of Aragón (scholarship to J.S., BIT04/93) and by the Dutch STW programme is greatly acknowledged.

REFERENCES

Bartolomé E et al. 1999, this Conference
Hartland A 1993 Proc. International Conference on Electromagnetic Measurements, 18-1
Park G S et al. 1995 IEEE Trans. Appl. Supercon. 5, 3214
Sesé J et al. 1999a IEEE Trans. Appl. Supercon. 9, 3487
Sesé J et al. 1999b IEEE Trans. Instrum. Meas., 48, 370
Sesé J et al. 1999c IEEE Trans. Instrum. Meas., 48, in press.
Sullivan D B et al. 1974 Rev. Sci. Instrum. 45, 517

Inst. Phys. Conf. Ser. No 167
Paper presented at Applied Superconductivity, Spain, 14-17 September 1999
© 2000 IOP Publishing Ltd

a-axis tilt biepitaxial YBa$_2$Cu$_3$O$_{7-x}$ dc-SQUIDs with low magnetic flux noise

E Sarnelli,[*†] **G Testa,**[*] **F Carillo,**[†] **and F Tafuri**[†+]

[*]Istituto di Cibernetica del CNR, Via Toiano 6, Arco Felice I-80072 Italy
[†]INFM Dip. Scienze Fisiche Università di Napoli Federico II, P.le Tecchio 80, Napoli I-80125 Italy
[+]Dip. Ingegneria dell'Informazione, II Università di Napoli, Via Roma 29, Aversa I-81031 Italy

ABSTRACT: High temperature dc-SQUIDs based on the biepitaxial technique have been fabricated and measured. [100] tilt boundaries instead of the traditional [001] tilt junctions have been realized. The two types of grain boundaries behave differently from each other, probably as a consequence of the absence of π-junctions in the [100] tilt structures. Magnetic flux-to-voltage transfer functions and magnetic noise characteristics are suitable for using such devices in applications. Finally, the energy resolution of these SQUIDs is the lowest reported in the literature for biepitaxial grain boundary junctions.

The possibility to fabricate devices by high-temperature superconductors represents an important topic for practical use of superconductivity in the next future. Basic requests for such devices are the feasibility, i.e. a relatively high yield and a reproducibility of electrical characteristics, the capability to make complex circuitry and, finally, a low level of intrinsic electrical and/or magnetic noise. Since the introduction of SrTiO$_3$ bicrystal (BC) substrates (Dimos 1988) for the fabrication of Josephson junctions, the strong potentiality of high T$_c$ devices, in particular regarding superconducting quantum interference devices (SQUIDs) and magnetometers, appeared evident. In fact, by varying the misorientation angle of the substrate, different critical currents I$_c$ and characteristic voltages I$_c$R$_n$ can be achieved, allowing the apriori determination of these parameters (here R$_n$ is the "normal state" resistance of the junction). With the bicrystal technique low noise magnetometers (Koelle 1995) have been fabricated. In some cases (Cantor 1995), magnetometers present sensitivities very close to what achievable with low-T$_c$ devices. Moreover, prototype flip-chip first and second order planar gradiometers (Kittel 1998), able to measure cardiac pulses in external environments, have been fabricated. These devices are mainly based on [001] tilt grain boundary junctions (GBJs), even though [100] tilt and twist can be fabricated. Unfortunately, BCs do not represent the ideal solution for devices requiring complicated designs and junctions with different characteristics, unless recurring to very sophisticated multi-bicrystal substrates, which should increase the already high costs of these samples. Alternative solutions have been proposed in the past years, like [001] 45° biepitaxial junctions (Char 1991) and step-edge junctions (Simon 1990). The former are suitable for designing complex (planar) circuitry, but are characterized by low performance considering the high level of electric and magnetic noise associated with them; the latter, in principle, represent a good compromise between the necessity of complex designs and high electrical performance, even though the yield and the reproducibility of step-edge junctions is still poor. In this work we have fabricated and investigated electrical and noise characteristics of [100] 45° tilt biepitaxial YBa$_2$Cu$_3$O$_{7-x}$ (YBCO) dc-SQUIDs. High electrical SQUID performance at T = 77 K and good magnetic noise characteristics make these devices eligible for deep investigation for their use in applications as, for instance, superconducting magnetometry and complex device configurations.

a-axis tilt [100] biepitaxial junctions have been fabricated by the deposition of YBCO films with inverted cylindrical magnetron on a seed MgO layer previously deposited on [110] SrTiO₃ (STO) substrates. MgO films grow along the [110] direction following the substrate orientation, while the YBCO films grow predominantly along the [103] or [013] directions on the STO substrate, and along the [001] orientation on the MgO seed layer respectively (Di Chiara 1997). Typically 30 nm thick MgO films are deposited by single target with an rf planar magnetron spattering. During the process the substrate temperature is kept at 600 °C. During the deposition process an Ar+O₂ atmosphere (P(O₂) = P(Ar) = 50 Pa) is present and the substrate temperature is usually at 780 °C. Details on the fabrication process can be found elsewhere (Di Chiara 1997). Finally, a 100 nm thick gold film is evaporated in the region of contact pads in order to reduce contact resistance of electric connections made by 25 aluminum wedge-bonded wires. The geometries of both the MgO seed layers and the YBCO devices are obtained by standard photolithographic techniques and ion milling processes. During the etching process the sample is cooled down to -80 °C to prevent overeating of the photoresist. Using the procedure described above, we have fabricated [100] 45° tilt biepitaxial YBCO dc-SQUIDs, with a design addressed to the realization of discrete vortex-flow transistors (DVFTs). A DVFT is constituted by an array of hole-like dc-SQUIDs with a loop inductance L_S of the order of few tens of a pH. This is because the discreteness parameter $\Lambda_j = (\Phi_0/2\pi L_s I_{cj})^{1/2} = (1/\pi\beta)^{1/2}$, separating the discrete operation of the device from the continuos one, must be $\Lambda_j < 1$. Here Φ_0 is the magnetic flux quantum and I_{cj} is the critical current of the jth junction of the array. The optimized value of the discreteness parameter for a symmetric geometry (Berman 1994, Gross 1995, Koelle 1995) is $\Lambda_j = 0.6$, which implies $\beta = 0.9$ and $I_c L_s = 900 \times 10^{-18}$ Weber. Considering $I_{cj} \approx 50$ μA, results $L_s \approx 20$ pH. Moreover, a high value of the transresistance $r_m = \partial V/\partial I_g$, where V is the output device voltage and I_g is the gate control current is desirable for a good operation of a DVFT. In an optimized device operation r_m is approaching the limit value R_n/N, where R_n is the junction normal resistance and N is the number of junctions forming the device. Biepitaxial junctions seems to be good candidates for their use in DVFTs, since high values of normal resistance are usually obtained at 77 K with junction dimensions and film thickness in the range of 2-4 μm and 100-200 nm respectively. Junctions with R_n in the range 3-10 Ohm with $I_c R_n$ characteristic products of the order of 10-50 μV at 77 K and 1-2 mV at T = 4.2 K are routinely obtained (Tafuri 1999, Testa 1999).

In Fig. 2 the current-Voltage (I-V) characteristics of two [100] 45° tilt YBCO biepitaxial dc-SQUIDs with low and high inductances L_s = 13 pH (a) and L_s = 74 pH (b) at T = 77 K

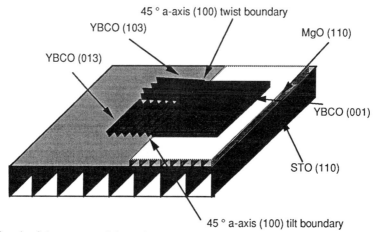

Fig. 1 Sketch of the two possible grain boundary junctions which can be fabricated by using MgO seed layers on (110) SrTiO₃ substrates.

are reported respectively. Thickness and junction dimensions of the two SQUIDs are the same, and the critical current densites J_c are comparable. In fact, for SQUID (a) $J_c = 5.2 \times 10^2$ A/cm^2, while for SQUID (b) $J_c = 3.6 \times 10^2$ A/cm^2. The spread in the junction parameters is the main problem for the fabrication of reliable HTS devices, although such a difficulty is often overcame by adequate annealing procedures for monitoring the junction characteristics. The normal state resistances of the samples scale proportionally to the critical current densities, leading to similar values of the characteristic voltages, actually 16 μV for SQUID (a) and 13.3 μV for SQUID (b) at 77 K respectively. However, even though these values seem quietly low, they are relative to samples with maximum critical temperature of 82 K, and have to be compared with the ones measured at 4.2 K which are 1.3 mV and 0.7 mV respectively. We are confident that an increase of the working critical temperature would increase the I_cR_n values and also, as it will be presented in the next, improve the noise performance of our SQUIDs. The 77 K magnetic field dependencies of the voltage for different bias currents are reported in Fig. 3. The values of the screening parameters $\beta = 2L_sI_c/\Phi_0$ are 0.03 and 0.18 for SQUID (a) and (b) respectively. Considering VFT applications, the β values should be increased. This can be achieved by increasing the critical current of the device as a result of an increasing of the critical temperature. The maximum voltage modulation amplitudes of the two SQUIDs are 10.4 μV and 1.4 μV, corresponding to transfer functions $V_\Phi = \partial V/\partial \Phi$ 36.9 μV/Φ_0 and 5.0 μV/Φ_0 respectively. Comparing these values to the theoretical predictions computed by Enpuku et al. (1994) we observe a discrepancy. In fact, the expected values are $V_\Phi = 49.2$ μV/Φ_0 and $V_\Phi = 23.1$ μV/Φ_0 for SQUID (a) and SQUID (b) respectively.

Magnetic flux noise spectral densities $\sqrt{S_\Phi}$ of the SQUID (a) is reported in Fig. 4. In this measure the SQUID is biased with a standard flux locked loop readout electronics with dc bias current. On the right Y axis the SQUID energy resolution $\varepsilon = S_\Phi/2L_s$ is also reported. At low temperatures a white noise value $\sqrt{S_\Phi} = 3$ $\mu\Phi_0/\sqrt{Hz}$ corresponding to an energy resolution $\varepsilon = 1.6 \times 10^{-30}$ J/Hz has been measured. This value is the lowest reported in the literature relative to YBCO biepitaxial SQUIDs (Miklich 1991). At 77 K, due to the low critical temperature ($T/T_C = 0.95$), the magnetic noise rises to 40 $\mu\Phi_0/\sqrt{Hz}$, corresponding to $\varepsilon \approx 1.0 \times 10^{-28}$ J/Hz. The low-frequency noise performance can be characterized by the estimation of the critical current fluctuation ratio S_{Ic}/I_c^2 (Enpuku 1995). In the SQUID characterized, $S_{Ic}/I_c^2 = 1.5 \times 10^{-4}$ Hz^{-1} at 77 K. Such a high value (typically $S_{Ic}/I_c^2 \approx 10^{-8}$ Hz^{-1}) can be once again due to the reduced critical temperature of the sample. In fact, considering that at low temperature $S_{Ic}/I_c^2 = 6.7 \times 10^{-9}$ Hz^{-1}, the value measured at 77 K is probably coming from thermally activated flux motion since the low-energy of pinning centers.

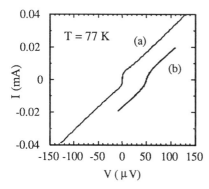

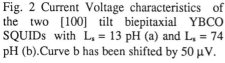

Fig. 2 Current Voltage characteristics of the two [100] tilt biepitaxial YBCO SQUIDs with $L_s = 13$ pH (a) and $L_s = 74$ pH (b).Curve b has been shifted by 50 μV.

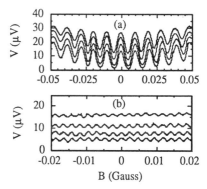

Fig. 3 Magnetic field dependence of the SQUIDs of Fig. 2.

592

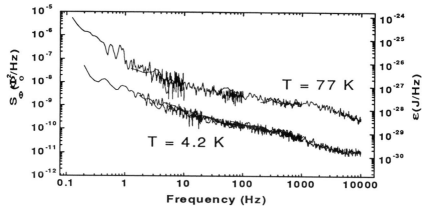

Fig. 4 Noise characteristics of SQUID (a) measured at 77 K and 4.2 K with a 500 kHz flux-locked loop electronic readout.

CONCLUSIONS

YBCO dc-SQUIDs based on the biepitaxial technique with 45° [100] tilt orientation have been fabricated and tested. Their electrical transport characteristics are suitable for the fabrication of complex circuits as VFTs. Noise characteristics are comparable to the behaviors observed in traditional YBCO biepitaxial SQUIDs. In particular, at low temperatures, the [100] tilt SQUIDs present the lowest energy resolution reported in the literature for biepitaxial junctions. A better performance is expected by increasing the critical temperature of the device. This can be achieved by a post annealing treatment and/or introducing vicinal substrates which allow a correctly oriented growth since the first layers. Biepitaxial dc-SQUIDs fabricated on vicinal substrates will be the argument of next papers.

This work has been partially supported by the projects PRA-INFM "HTS Devices" and SUD-INFM "Analisi non distruttive con correnti parassite tramite dispositivi superconduttori".

REFERENCES

Berman D, van der Zant H S J, Orlando T P, and Delin K 1994 IEEE Trans. Appl. Supercond. 4, 161
Cantor R, Lee L P, Teepe M, Vinetskiy V, and Longo J, 1995 IEEE Trans. Appl. Sup. 5, 2927
Char K, Colclough C, Garrison S M, Newman N, and Zaharchuk G 1991 Appl. Phys. Lett. 59, 773
Di Chiara S, Lombardi F, Miletto Granozio F, Scotti di Uccio U, Tafuri F, and Valentino M 1997 IEEE Trans. on Appl. Superc. 7, 3327
Dimos D, Chaudhari P, Mannhart J , and LeGoues F K 1988 Phys. Rev. Lett. 61, 219
Enpuku Doi H, and Maruo T 1994 Jpn. J. Appl. Phys. 33, L722
Enpuku K, Tokita G, Maruo T, and Minotani T 1995 J. Appl. Phys.78, 3498
Gross R, Gerdemann R, Alff L, Bauch T, Beck A, Froelich O M, Koelle D, and Marx A 1995 Appl. Supercond. 2, 443
Kittel A, Kouznestov K A, McDermott R, Oh B, and Clarke J 1998 Appl. Phys. Lett. 73, 2197
Koelle D, Kleiner R, Ludwig F, Miklich A H, Dantsker E, and Clarke J 1995 Appl. Phys. Lett. 66, 640
Miklich A H, Kingstone J J, Wellstood F C, Clarke J, Colclough M S, Char K, and Zaharchuk G 1991 Appl. Phys. Lett. 59, 988
Simon R W, Burch J F, Daly K P, Dozier W D, Hu R, Lee A E, Luine J A, Manasevit H M, Platt C E, Schwarzbek S M, John D St, Wire M S, and Zani M J 1990 Science and Technology of Thin Film Superconductors 2, McConnel R D and Noufi R Eds, Plenum N Y, p 549
Tafuri F, Miletto Granozio F, Carillo F, Di Chiara A, Verbist K, and Van Tendeloo G 1999 Phys. Rev. B 59,11523
Testa G, Sarnelli E, Carillo F, and Tafuri F 1999 Appl. Phys. Lett. accepted for publication

Inst. Phys. Conf. Ser. No 167
Paper presented at Applied Superconductivity, Spain, 14-17 September 1999
© 2000 IOP Publishing Ltd

A Squid Switch for a Macroscopic Quantum Coherence Experiment

M.G. Castellano[1,2], **F. Chiarello**[1,2], **C. Cosmelli**[1,3], **G. Torrioli**[1,2]

[1] Istituto Nazionale di Fisica Nucleare, 00185 Roma, Italy.
[2] Istituto di Elettronica dello Stato Solido, C.N.R., via Cineto Romano 42, 00156 Roma,Italy.
[3] Dipartimento di Fisica, Università di Roma La Sapienza, 00185 Roma, Italy.

ABSTRACT: An experiment of Macroscopic Quantum Coherence can be realized by using a set of SQUIDs cooled at very low temperatures. One of the main problems in realizing such an experiment is the construction of a system performing a Non Invasive Measurement on the SQUID state. Such a device can be realized using a hysteretic SQUID working as a two-state superconducting switch. In this paper we will describe briefly the scheme of the experiment, the theoretical behavior of the SQUID switch, and finally the experimental results on a real switch, obtained at a temperature of 4.2 K.

1. INTRODUCTION

In 1932 Einstein Podolsky and Rosen in their famous paper posed the question whether Quantum Mechanics Theory were a complete theory or not. The problem of incompatibility between Quantum Mechanics predictions and Classical Mechanics was then the subject of a long debate till 1964, when Bell proposed a way to test the predictions of Classical mechanics on a microscopic system. More than 20 years after, Aspect (1982) succeeded in realizing such an experiment using a pair of entangled photons to test the Bell inequalities.

In 1980 Leggett proposed to realize a similar experiment on a macroscopic system to perform a test of Macroscopic Quantum Coherence. The proposed system was an rf SQUID cooled at very low temperature. In 1995 the Rome group begun to work on the realization of the Leggett test, proposing essentially the same experimental set up but using a slightly different test procedure (Cosmelli 1998). In this paper we will describe very briefly the system to be used for a MQC experiment with SQUIDs and the essential features (both from the theoretical and experimental point of view) of the SQUID switch to be used in the experiment.

2. THE MQC EXPERIMENT

The MQC experiment can be realized by using a set of three different SQUIDs: a rf-SQUID (the source of the state to be measured), a SQUID Switch (the analogous of the polarizer in an EPR experiment) and a standard linear amplifier reading the output of the Switch. The source of the state, the rf-SQUID, can be described by a single dynamical variable, the total magnetic flux ϕ linked to the ring, which is subjected to a potential $U = (\phi - \phi_x)^2 / 2L - (I_0 \phi_0 / 2\pi) \cos(2\pi\phi/\phi_0)$, where ϕ_x is the applied flux, L is the SQUID inductance, I_0 is the junction critical current and $\phi_0 = h/2e = 2.07 \times 10^{-15}$ Wb is the flux quantum (Barone 1982). The system equation is the same of a particle of mass C (the junction capacitance) with friction coefficient 1/R, subjected to the same potential. If the parameter $\beta_L = 2\pi L I_0/\phi_0$ is greater than one, this potential is a succession of wells, corresponding to metastable flux states of the SQUID, superposed to a quadratic term. If the SQUID is biased by half flux quantum, the potential has the form of a double well with two degenerate energy levels in the left and right wells (Fig. 1).

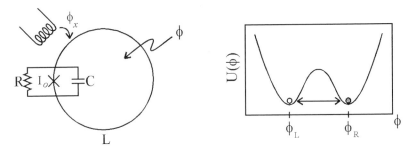

Fig. 1: The rf-SQUID and its symmetric double well potential.

Once the barrier between the two wells is low enough, and if the temperature is so low to avoid thermal activation over the barrier, the flux can have Rabi oscillations between the two wells. This two flux states, representing two circulating current identical in modulus but opposite in versus, are macroscopic in the sense that they differ from a macroscopic number of particles (all the Cooper pairs present in the SQUID loop). The experiment will consist of preparing the SQUID flux state in one of the two wells, leaving then the system to its free evolution, and "measuring" the SQUID state at different times to evaluate the occupation probabilities of the SQUID states.

3.   THE SQUID SWITCH.

The main problem in reading the SQUIDs state is the practical realization of a Non Invasive Measurement (NIM). In fact we cannot use a standard amplifier to read the state of the SQUID. A standard amplifier will have intrinsically a back action towards the rf SQUID. This back action acts as a "force" on the system changing its evolution and making impossible to follow the free evolution of the SQUID dynamics. The theoretical solution to this problem came by C. Tesche (1986) who proposed to use a hysteretic dc SQUID to realize NIM measurements of rf-SQUID flux.

3.1   SQUID Switch: operating principles.

The SQUID Switch is a standard dc-SQUID without the two small resistances usually placed across the Junctions to eliminate the hysteresis in the I/V characteristics. Once the SQUID is hysteretic its behavior is similar to a Josephson junction having a variable critical current. In this case the SQUID critical current is modulated by the external flux.

The Switch operation is the following (Fig. 2): in normal conditions the SQUID is biased by an external constant flux ϕ_{bias} reducing the critical current Ic to a typical value of the order of half the maximum critical current. The SQUID however is not current biased, so it remains in the superconducting state for every value of the input flux. Then the bias current is switched on for a short time Δt at a value slightly below the SQUID critical current. In this condition the SQUID can be linked to one of the two fluxes related to the two values of the current circulating in the rf SQUID (clockwise and counterclockwise). Since these two currents are opposite, their effect will be to lower or to raise the value of the critical current below or above the value of the actual current bias. So the SQUID will switch to the voltage state in the former case, while it will remain in the superconducting state in the second. In this second case the SQUID will perform a Non Invasive Measurement of the rf SQUID state, while in the first case the back action related to the resistive state of the SQUID will modify the SQUID dynamics. All the measurements giving these results will be neglected, keeping only the NIM measurements to calculate the desired quantities. So this technique allows performing NIM measurements on the rf SQUID state at the expense of loosing 50% of the measurements.

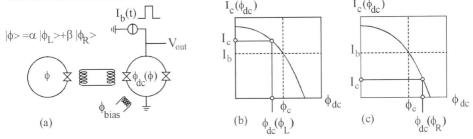

Fig. 2: dc-SQUID switch operating principles. (a) The rf SQUID is in a quantum superposition of left and right flux (current) states. (b) The left state causes an increasing of the dc SQUID critical current that remains above the bias current, and there is no switching (NIM). (c) The right state decreases Ic below the bias current, and there is a transition to the voltage state.

4. THEORETICAL BEHAVIOR AND EXPERIMENTAL RESULTS

The switching probability of the SQUID switch can be evaluated from the expression of the escape rate of the SQUID from the superconducting to the voltage state. In the case of a low dissipation system we can use the usual expression of the escape rate out of a potential well given by Kramers applied to a cubic potential. The flux sensitivity, defined as the interval $\delta\phi \equiv \phi_{dc}^{ext}(P = 90\%) - \phi_{dc}^{ext}(P = 10\%)$ of the applied flux ϕ_{dc}^{ext} between the values corresponding to 10% and 90% for the switching probabilities P, is approximately given by:

$$\delta\phi \cong \frac{\phi_0}{\pi} \frac{\alpha^{1/3}}{\left(1-\alpha^2\right)^{1/2}} \left(\frac{3}{2}\frac{K_B T}{E_j}\right)^{2/3} \left\{\ln\left[\frac{\omega_j \tau}{2^{3/4}\pi}\alpha^{1/4}\left(1-\alpha^2\right)^{1/8}\right]\right\}^{-1/3} \tag{1}$$

were I_0 and C are respectively the critical current and the capacity of a single junction, $\alpha \equiv I_b / 2I_0$ is the reduced bias current, τ is the duration of the bias pulse, $E_j \equiv I_0\phi_0 / \pi$ and $\omega_j \equiv \sqrt{4\pi I_0 / C\phi_0}$. Using this definition the Switch efficiency is set equal to 90%. To test the SQUID behavior we have cooled at a temperature of 4.2 K a typical chip that should be used for the final MQC set up.

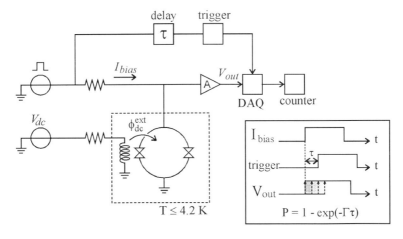

Fig. 3: Experimental set-up for the study of the dc SQUID switch.

With the set up shown in Fig. 3 we made two set of measurements, the first by keeping constant the time τ during which the Switch is on, and varying the amplitude of the external flux, the second by keeping constant the external flux and varying the time τ. The results are shown in Fig. 4 where the experimental points are shown together with the theoretical predictions. The experiment has be done with the following values for the system parameters: $I_b = 29.6\,\mu A$, $I_0 = 25.9\,\mu A$, $C = 0.4\,pF$, $L \approx 3\,pH$.

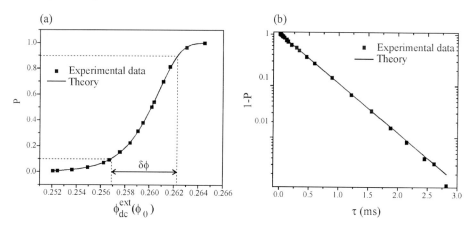

Fig. 4: Experimental results: (a) for τ fixed and ϕ_{dc}^{ext} variable; (b) for τ variable and ϕ_{dc}^{ext} fixed.

We can see that the agreement is very good; this allows to extrapolate the value of $\delta\phi$ below 300mK (the crossover temperature between classical and quantum regime). This sensitivity leads to a value of $\delta\phi$ (as defined in eq.1) of the order of $1\,m\phi_0$, a value that should permit to detect the flux coming from the rf-SQUID.

REFERENCES

Aspect. A., Dalibard J., Roger G. 1982,Phys. Rev. Lett. **49**, 1804.
Barone A. and G. Paterno' 1982, *Physics and Applications of the Josephson Effect*, (J. Wiley & Sons Pub., New York).
Bell J.S. 1964, Physics **1**, 195.
Cosmelli C. 1998, Proc. Int. Conference on Macroscopic Quantum Coherence, E. Sassardi, Y. Srivastava, J. Swain, A. Widom (Boston, World Scientific), 252.
Einstein A., Podolski B. and Rosen N.1935, Phys. Rev. **47**, 777.
Leggett A. 1980, Suppl. Prog. Theor. Phys. **69**, 80.
Tesche C. 1990, Phys. Rev. Lett. **64**, 2358.

Inst. Phys. Conf. Ser. No 167
Paper presented at Applied Superconductivity, Spain, 14-17 September 1999
© 2000 IOP Publishing Ltd

Read-out electronics for a 9-channel high-T$_c$ dc SQUID system

S Bechstein, D Drung, M Scheiner, F Ludwig and Th Schurig

Physikalisch-Technische Bundesanstalt, Abbestr. 2-12, D-10587 Berlin, Germany

ABSTRACT: A read-out electronics for operating a 9-channel high-T$_c$ dc SQUID system has been developed. This electronics consists of two parts – the flux locked loop (FLL) unit mounted on the top of the liquid nitrogen dewar and the control unit (CU). The FLL unit consists of three triple modules each containing three direct-coupled FLL electronics boards with bias reversal. The amplifier voltage noise is 0.4 nV/√Hz with a 1/f corner at about 0.3 Hz. Each module is equipped with a voltage-controlled oscillator for providing the bias reversal clock and a circuit for driving the thick-film heaters implemented in each SQUID package for releasing trapped flux. The working points of the SQUIDs are adjusted using analogue voltages between −10 V and +10 V produced in the CU. In order to minimise the number of wires between FLL and CU, a serial two-line bidirectional (I^{2}C) bus is used to transfer the control signals. An in-system programmable micro-controller is used to control the main features of the CU. The new read-out electronics was successfully tested on a system equipped with 9 simultaneously operated low-noise high-T$_c$ direct-coupled SQUID magnetometers.

1. INTRODUCTION

For operating a high-T$_c$ SQUID system in magnetically disturbed environment highly balanced gradiometers are needed to suppress environmental noise. An elegant approach for multichannel systems is to use only one reference system for all sensing channels and to perform the balancing via software (for a review see Vrba 1996). In order to study the software balancing with high-T$_c$ SQUIDs we are currently building a prototype with a single sensor channel and a reference system involving 8 high-T$_c$ SQUIDs to perform software gradiometry of up to second order.

In this paper we report the development of the read-out electronics for this 9-channel high-T$_c$ dc SQUID system.

2. SYSTEM DESCRIPTION

For all applications using a magnetically shielded room, the SQUID sensors have to be controlled from outside the place of measurement. Therefore, we realised the electronics in two parts – the flux locked loop (FLL) electronics mounted directly on the top of the liquid nitrogen dewar containing the SQUIDs and the control unit (CU) which is placed outside the shielded room. Special cables the length of which can be chosen arbitrarily connect the CU with the FLL electronics.

Minimising the space in favour of a high sensor density, three FLL boards were combined to a FLL triple module (Fig. 1). Each FLL triple module contains:

- 3 x basic direct-coupled FLL circuits (preamplifier voltage noise: 0.4 nV/√Hz, 1/f corner at about 0.3 Hz),
- 1 x voltage-controlled oscillator for providing the bias reversal clock up to 500 kHz,
- 1 x circuit for driving the thick-film heaters implemented in each SQUID package,

598

Fig. 1: Photograph of FLL triple with aluminium cover removed.

1 x circuit controlling the state of the oscillator and the heaters (a serial to parallel input/output
port PCF8574 by Philips)

Each of the three FLL boards of a triple module consists of a basic FLL circuit and one part of the
circuits described above. The FLL electronics are upgraded versions of that described by Drung
(1997). The three boards are placed in an aluminium cylinder with an outer diameter of 47 mm.

The CU supplies and controls the FLL triples in the following manner: The working points
(bias current, bias voltage, bias flux, offset voltage) of the 9 SQUIDs and the VCO are adjusted by
analogue voltages in the ±10V range. All other functions are controlled by a serial two-line
bidirectional bus to minimise the number of wires between FLL and CU. The data transfer between
CU and FLL electronics and the main features of the CU are controlled by an in-system
programmable micro-controller (AT90S2313 by Atmel) with a programmable flash of 2K bytes. The
controller is switched into the sleep mode during SQUID measurements in order to avoid
interference from the controller clock, produced by a 4 MHz resonator. To support mobile
experiments the CU is supplied with overload detectors for all channels. An internal flux generator
(frequency can be switched between two values adapted to the particular application) and low-pass
filter are implemented for monitoring the V-Φ characteristics. The CU is compatible with all of our

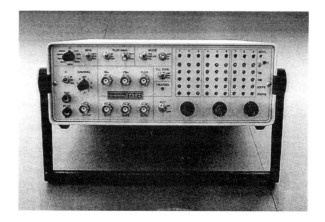

Fig. 2: Photograph of 9-channel control unit.

FLL electronics versions including our 15 MHz bandwidth low-T_c SQUID read-out electronics. A photograph of the completed CU is depicted in Fig. 2. The dimensions of the housing are only 25 cm x 9 cm x 20 cm.

3. MEASUREMENTS

Our electronics system was developed to read-out a sensor array of 9 high-T_c dc SQUIDs to perform software gradiometry up to second order for operation in unshielded or moderately shielded environment. Fig. 3 shows the magnetic field noise spectra of 9 direct-coupled high-T_c SQUID magnetometers mounted on the dewar inset and simultaneously operated in the Berlin magnetically shielded room (BMSR). Six of the devices were produced with the process described by Ludwig et al. (1999) using a novel pickup loop layout that is highly suited for low-noise operation in environmental magnetic fields Ludwig and Drung (1999). The other three magnetometers were provided by the University of Hamburg. Five magnetometers (4 x PTB, 1 x Hamburg) have a noise level of <50 fT/√Hz down to frequencies below 10 Hz. Comparing the sensor noise with that measured for the same SQUIDs with our previous, analogue two-channel electronics, we did not observe differences. In addition, it was demonstrated on a 3-channel SQUID module developed for geophysical applications that the 9-channel CU is also suited for high bandwidth systems.

Preliminary measurements of the crosstalk between SQUID channels belonging to the same FLL triple were performed. The three corresponding SQUIDs devices were arranged to form a vector magnetometer so that they were oriented orthogonally to each other. A signal current of known value was fed into the feedback coil of one channel being in the reset mode and the response of the SQUID magnetometers of the two other channels being in the FLL mode was measured. Crosstalk values of a few tenth of a percent were typically measured for signal frequencies below 5 kHz in spite of the close proximity of the magnetometer sensors (centre-to-centre distance of ≥ 13 mm).

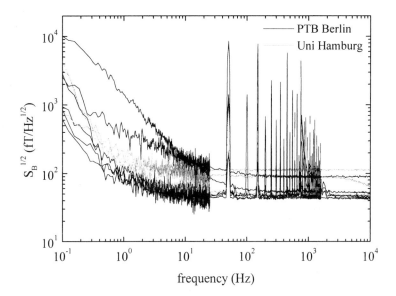

Fig. 1: Magnetic field noise spectra of 9 direct-coupled high-T_c SQUID magnetometers (six fabricated at PTB Berlin and three at the University of Hamburg). All sensors were simultaneously operated with 9-channel electronics and 100 kHz bias reversal. The lines in the spectra are due to power line interference (50 Hz and harmonics).

4. SUMMARY

A new read-out electronics for a 9-channel high-T_c dc SQUID system was developed and successfully tested by noise measurements on a prototype system. In spite of using an in-system programmable micro-controller with a clock frequency of 4 MHz we did not observe differences in the SQUID noise level using our analogue two-channel and our new digitally controlled 9-channel electronics. Furthermore, during the measurements provided so far, the control unit was used as stand-alone device but it could be operated computer-controlled via a RS232 interface.

ACKNOWLEDGEMENT

This work was supported by the German BMBF under contract number 13N7326.

REFERENCES

Drung D 1997 Rev. Sci. Instrum. **68**, 4066

Ludwig F, Beyer J, Drung D, Bechstein S, and Schurig Th 1999 IEEE Trans Appl. Supercond. **9**, 3793

Ludwig F and Drung D 1999 Appl. Phys. Lett. (to appear in November 1 issue)

Vrba J 1996 SQUID Sensors: Fundamentals, Fabrication and Applications (Dordrecht: Kluwer) pp 117-78

Inst. Phys. Conf. Ser. No 167
Paper presented at Applied Superconductivity, Spain, 14-17 September 1999

The role of the geometry in superconducting tunnel junctions detectors

R. Cristiano, M. P. Lissitski, C. Nappi

Istituto di Cibernetica del C.N.R., Arco Felice, Napoli, Italy

ABSTRACT: Recently, experimental results on superconducting tunnel junctions as high energy resolution radiation detectors open perspectives for applications and make urgent to solve problems related to the efficiency and simplification of the working conditions. The junction geometry enters various issues in solving such problems, particularly when arrays of junctions are used. Aspects related to the junction detector geometry are discussed, reporting on an overview of our experiments on Nb-based junctions with various geometry. In particular, results referring to the annular geometry, which has an interest beyond the particular application, are presented.

1. INTRODUCTION

Among cryogenic detectors, Superconducting Tunnel Junctions (STJs) are characterised by high energy resolution and high counting rate capability. The time is mature now for their use both in basic physics experiments as well as in analytical instrumentation like EDX-spectrometers or mass-spectrometers for macromolecules (1997a). In these applications aspects like imaging, geometrical efficiency (the capability to cover large areas to collect the radiation from the source) and simplification of the working conditions enter in a relevant way the problem of providing a suitable detector configuration. Imaging and geometrical efficiency are obtained by increasing the number of junctions, by using for instance junctions arrays. Besides the non trivial task of fabricating many high quality STJs, one of the possible obstacle towards the goal of realising efficient working devices is the achievement of a stable working bias voltage simultaneously in each junction of the array. It is well known that STJ detectors operate in the presence of an external magnetic field, H, which is necessary to suppress the Josephson critical current, I_c, and the Fiske steps in the same time such that a stable bias voltage is obtained. A complete suppression is difficult and is obtained only at relatively high magnetic fields (hundreds of Gauss). Often, the residual amplitudes of both the critical current and the Fiske steps are slightly larger than the quasiparticle current which is typically several (4-6) orders of magnitude smaller than the Josephson critical current. Large magnetic field in turns create regions with depressed energy gap where excited quasiparicle can be trapped and lost, leading to a reduction of the detection signal. For this reason it is common to compromise by applying weaker magnetic fields and choosing the bias point between Fiske steps. In a very recent and interesting work Friedrich et al. (1999a) measured the influence of the presence of Fiske steps on the performance of a STJ detector. At bias voltages corresponding to Fiske modes they found a dramatic drop of the energy resolution, in spite of the fact that the signal magnitude was unchanged. This drop was not due solely to a decrease in the STJ dynamic resistance, as demonstrated by their biasing the STJ on top of a Fiske resonance. For this bias point the dynamic resistance is infinite nevertheless they still observed the worst energy resolution. They also found that the maximum counting rate is strongly affected by the presence of Fiske steps. It is clear that this problem becomes more complex when a large number of STJs are used. Since position and amplitude of Fiske modes depend on the junction geometry, many efforts have been

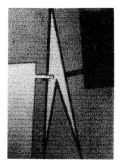

Fig 1. Photograph of various Nb-AlOx-Nb STJ with quadrangular geometry: Rhombus (left), Arrow (center), Rhomboid (right).

devoted to searching for an optimum geometry, where a fast suppression of the I_c vs H pattern side lobes and no steps in the bias voltage region can be obtained. This problem was already investigated several years ago in the context of digital applications. Peterson (1991) proposed various junction shapes and calculated explicitly their I_c vs H pattern. In this paper we report on an overview of our efforts in the search of an optimal geometry and our recent experiments made on Nb-based junctions with various geometry. In particular, results referring to the annular geometry, which have an interest beyond the particular application for radiation detection, are presented and discussed.

2.   QUADRANGULAR JUNCTIONS.

The "diamond" (square junction, magnetic field parallel to the diagonal) is certainly the most popular geometry used in the context of STJ detectors. For this geometry the I_c vs H pattern exhibits an H^{-2} fall-off of the maxima with the field, which is considered convenient in terms of a fast side lobe suppression. The analytical expression for the I_c vs H pattern is given by the following expression:

$$\frac{I_c(H)}{I_c(0)}=2\left[1-\cos\ \pi\frac{H}{H_0}\right]\bigg/\left(\pi\frac{H}{H_0}\right)^2 \tag{1}$$

where $H_0=\Phi_0/\mu_0 Dd$, Φ_0 is the quantum flux, d is the effective magnetic penetration depth and D is the length of the diagonal. Single junctions and arrays of "diamond" junctions have been fabricated and tested by several groups. For this geometry is possible to predict the position and the amplitude of the Fiske steps (Neremberg 1976), with a good agreement with experiments. At small values of the applied magnetic field, several Fiske steps, corresponding to simple and mixed modes of the junction electromagnetic cavity, are present and their amplitudes are far from be negligible. They disturb the biasing of the STJ and lead to the problems in the detection performances that were mentioned in the introduction (Friedrich et al. 1999a). In order to overcome such drawbacks we have investigated different quadrangular geometries. We found (Nappi et al 1996) that a whole class of quadrangular geometries has the same I_c vs H pattern of Eq. (1). A chip containing Nb junctions with various quadrangular geometries like "rhombus", "rhomboid" and "arrow" has been fabricated (see Fig.1). In Fig. 2a the theoretical and the experimental I_c vs H pattern of a rhombus junction is shown. The agreement between theory and experiment is excellent. The advantage of the rhombus with respect to the diamond geometry with the same area is that a faster suppression of the critical current is obtained if the magnetic field is applied parallel to the smaller diagonal. This allows the use of weaker magnetic fields and the reduction of the risk of trapping magnetic vortices. For the rhombus geometry, however, is not possible to derive an analytical expression for the position and amplitudes of Fiske steps.

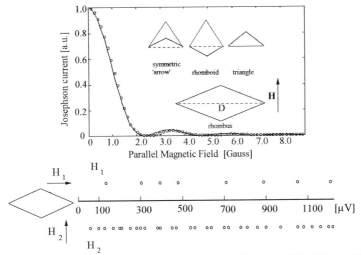

Fig. 2 a) Critical current vs magnetic field pattern for a rhombus junction. b) Positions of Fiske resonances for the above rhombus junction. Field along the diagonal as indicated.

In Fig. 2b the experimental voltage positions of the Fiske steps corresponding to the two different orientations of the external magnetic field are shown. It can be seen that the application of the magnetic field in the direction parallel to the smaller diagonal has the disadvantage that the number of steps is large, with a high step density in voltage region (100-500 μV) where typically the bias of the detector is chosen. Thus, it can be concluded that the rhombus geometry does not have significant advantages with respect to the diamond geometry. We have also measured junctions with the "arrow" and the "rhomboid" geometry (see insert in Fig.1). These geometries were designed in the attempt to have junctions in which the building up of standing waves was hampered as much as possible. Although, the respective Ic vs H patterns agree with the theoretical prediction of Eq.(1), the situation with respect to the presence of Fiske steps does not simplify, so that these geometries do not exhibit an advantageous behavior.

3.   ANNULAR GEOMETRY

We proposed a new configuration of STJ detector based on the use of annular junctions, namely on circular junctions with a small hole in the counter electrode (Nappi and Cristiano 1997b). It has been shown that in presence of n-trapped fluxons threading the non connected electrode, this configuration has the convenient properties that the critical current is suppressed and the resonances with significant amplitude reduce to only one. The flux trapping can be obtained by cooling the junction from $T > T_c$ to $T < T_c$ in a perpendicular magnetic field. The advantage of the annular geometry is that, once the trapping has occurred, the magnetic field can be switched off and there is no need to further sustain any parallel magnetic field during the detector working time. Moreover, the quantum nature of the trapped field guarantees its stability ($\Phi = n\,\Phi_0$ and $\Phi_0 = h/2e$ is the flux quantum).

Annular Josephson tunnel junctions have received a continuous attention, in the course of the last two decades, because of their particularly attractive phenomenology. For instance, investigations on the soliton propagation in the absence of disturbing boundary reflections can be carried out exclusively in the annular geometry. The term *annular* has been reserved actually to a geometry in which the width of the ring electrode is small in comparison with the ring radius and the circumference of the ring is long in comparison with the Josephson penetration length λ_j. In this case, it is possible to reduce the investigation of the junction electrodynamics to a one-dimensional problem. For STJ-detector applications, annular junctions with large width of the ring have to be considered in order to

TABLE I. Positions of the first and second Fiske steps X_{n1}, X_{n2} (m =1,2) with n=1 or 2 trapped fluxons as function of δ.

δ	X_{11}	X_{12}	X_{21}	X_{22}
10^{-6}	1.841	5.331	3.054	6.706
1/7	1.766	5.024	3.048	6.633
1/5	1.705	4.960	3.034	6.494
1/3	1.540	5.273	2.932	6.270
0.5	1.354	6.564	2.681	7.062
0.7	1.182	10.591	2.362	10.798
0.9	1.053	31.446	2.106	31.500
0.99	1.005	–	2.010	–
0.99999	1.0000	–	2.0000	–

guarantee geometrical efficiency and such a reduction is impossible: the problem remains essentially two-dimensional. On the other hand, when the circumference of the junction is smaller than λ_j, the trapped fluxons do not localise and, if no external magnetic field is present, a simple configuration of the field with a cylindrical symmetry sets in. The magnetic field lines have only radial component and they are quite uniformly spread all around the junction circumference. This highly symmetric configuration, as well as the presence of trapped fluxons in itself, has remarkable influence over the behavior, static and dynamic, of the junction. For example, the critical current is completely suppressed and the Fiske step picture undergoes important changes. The theory of two-dimensional annular junctions, under the above simplifying assumptions, has been developed by Nappi et al (1998). It is possible to show that the expression for the Ic vs H pattern takes the following form:

$$\frac{I_c}{I_0} = \left| \frac{2}{(1-\delta^2)} \int_{\delta}^{1} x J_n \left(x \frac{H}{H_0} \right) dx \right| ; \quad H_0 = \frac{\Phi_0}{2\pi R_e \mu_0 d}, \quad I_0 = j_c \pi (R_e^2 - R_i^2), \quad (n = 0,1,2,..) \quad (2)$$

where R_e and R_i are the external and the internal radius of the ring, respectively, $\delta=R_i/R_e$, n is the number of trapped fluxons, j_c is the maximum Josephson critical current density and J_n is the nth-order Bessel function of integer order. The theory predicts also the voltage position of Fiske steps, which depends on two integer numbers (k=0,1,2..., m=1,2,3...) which correspond respectively to the azimuthal and radial modes:

$$V_{km} = (c\Phi_0/2\pi R_e) X_{km} \quad (3)$$

c is the Swihart velocity and X_{km} are, for each k, the m-th solutions of the following equation:

$$J_k'(X_{km}\delta) - \frac{J_k'(X_{km})}{N_k'(X_{km})} N_k'(X_{km}\delta) = 0 \quad (4)$$

Table II. Values of the relative step amplitudes as a function of (n,m); $\delta=1/3$

n	m=1	m=2	m= 3
1	1	3.08×10^{-4}	9.71×10^{-7}
2	1	8.88×10^{-3}	7.00×10^{-5}
3	1	3.67×10^{-2}	9.00×10^{-4}

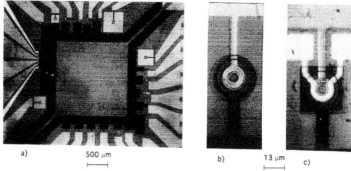

Fig.3. a) Photograph of the chip; b) detail of the annular junction of the type I; c) detail of the annular junction of the type II with the Au control line

where, J'_k and N'_k are the derivatives of the kth-order Bessel functions of the first and second kind respectively. In the presence of n trapped fluxons and no external field the series ($k=n$, m) is selected. In Table I the positions of the first and second Fiske step (m =1,2) are given when one (n=1) or two (n=2) flux quanta have been trapped and for increasing inner hole radius. The position of the second Fiske step (and necessarily higher order steps) moves towards very high voltage as the inner radius increases. Even in the case in which many resonances are present at finite voltages, especially when R_i → 0 (which is the case of interest for STJ detectors), the relative amplitudes of the steps following the first (m =1) are very small. Table II gives the values of the relative step amplitudes as a function of (n,m). Subsequent Fiske steps (m=3,4...) are even smaller in amplitude for this case. At fixed n, Fiske resonances start at X_{nm} , m=1,2,3... and their amplitude is a quickly decreasing function of the resonance order m. From the practical point of view as n increases only one peak is expected at larger and larger voltage X_{n1}.

We have fabricated various *island* annular junctions with different hole diameters and the same external diameter to extensively investigate this configuration for its intrinsic interest, but also to demonstrate the potentiality of this geometry for particle detection (Cristiano et al. 1999b). Two types of annular junctions were fabricated: without (Fig.3b) and with (Fig.3c) a Au thin film control line to generate the perpendicular magnetic field to trap fluxons in the hole. Here we report on the results of measurements of an annular junction of type I which had the internal and external diameters of 7.5 μm and of 16 μm, respectively ($\delta = R_i/R_e$ = 0.47). In the absence of trapped fluxons, we have observed Fiske resonances maximising their amplitude by the application of a suitable parallel magnetic field. The first Fiske resonance was at V_1=550±10 μV, the second and the third resonance were registered at V_2=1080±20 μV and V_3=1440±20 μV, respectively. They correspond to the (1,1), (2,1) and (3,1) resonance modes. By means of a thermal cycle from $T>T_c$ down to T=4.2 K in a residual perpendicular magnetic field we were able to trap a single magnetic fluxon in the annular junction. The I_c vs H pattern at T=4.2 K after the trapping procedure is shown in Fig.4. The experimental I-V characteristic, recorded *at zero external magnetic field* after single flux trapping, clearly exhibited only (1,1) resonance mode . The presence of the (2,1) and the (3,1) resonances, corresponding to the case of n=2 and n=3 trapped fluxons, was also observed, but their amplitudes were absolutely negligible in comparison with the first one. Their presence is to be attributed to a deviation from the ideally symmetric configuration assumed by the theory.

4. CONCLUSIONS

The choice of a suitable geometry can lead to an overall simplification of the detector operations. Diamond junctions at the moment are the simplest and more convenient solution, although the presence of Fiske resonances still pose problems. Annular geometry is certainly of interest. Further

606

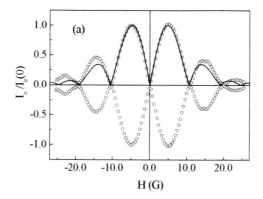

Fig. 4. Ic vs H experimental dependence (open dots) and theoretical curve (solid line)

work is necessary in this direction to better understand various aspects. In fact it is not clear now if trapped vortices in the hole are stable under radiation and if they do have a negative effect on the detector performances. Moreover, the technique to trap the vortices in the hole has to be improved. Nevertheless, in spite of a relatively more complex structure, the annular geometry is very attractive because it offers the possibility to reduce the number of Fiske resonances, to control their voltage positions and to work in a ultra stable magnetic field which is self-sustained.

ACKNOWLEDGEMENTS

This work has been partially supported by INFN under the STJ-Detector Project and by the EC-TMR Programme "Cryogenic Detectors", Contract No. ERBFMRXCT 980167. M. L. is supported under an INFN fellowship. The authors wish to thank G. Ammendola E. Esposito, L. Frunzio, L. Parlato for their help in STJ measurements, D.V. Balashov for his help in the fabrication technology and A. Barone for useful discussions.

REFERENCES

(1997a) Proc. 7th Int. Workshop of Low Temperature Detectors, LTD-7, Ed. S. Cooper (Max Planck Institute of Physics, Munich)
Cristiano R. et al. (1999b) Appl. Phys. Letts **74**, 3389
Friedrich S. et al. (1999a) Proc. 8th Int. Workshop of Low Temperature Detectors, LTD-8 in press
Nappi C. et al (1996) J. Appl. Phys. **80**, 3401
Nappi C. and Cristiano R. (1997b) Appl. Phys. Letts **70**, 1320
Nappi C., Cristiano R. and Lisstskii M.P. (1998) Phys. Rev. **B58**, 11685
Neremberg M.A., Forsyth P.A., Blackburn J.A., (1976) J. Appl. Phys. **47**, 4148
Peterson R.L. (1991) Cryogenics, **31**, 132

Inst. Phys. Conf. Ser. No 167
Paper presented at Applied Superconductivity, Spain, 14-17 September 1999
© 2000 IOP Publishing Ltd

Superconducting YBCO nanobridges for submillimeter-wave detectors

A Gaugue[1], **C Ulysse**[1], **D Robbes**[2], **C Gunther**[2], **J Gierak**[3], **A Sentz**[4], **G Beaudin**[5] and **A Kreisler**[1]

[1] LGEP, UMR 8507 CNRS (Universités Paris 6 & Paris 11), Supélec, Plateau de Moulon, 91192 GIF SUR YVETTE Cedex, FRANCE

[2] GREYC, ESA 6072 CNRS, ISMRa et Université de Caen, 6 Boulevard du Maréchal Juin, 14050 CAEN Cedex, FRANCE

[3] L2M, UPR 20 CNRS, 196 Avenue Henri Ravera, B.P. 107, 92225 BAGNEUX, FRANCE

[4] LDIM, EA 253 MENRT, Tour 12 E2 (case 92), Université Paris 6, 4 place Jussieu, 75252 PARIS Cedex 05, FRANCE

[5] DEMIRM, UMR 8540 CNRS, Observatoire de Paris, 61 Avenue de l'Observatoire, 75014 PARIS, FRANCE

ABSTRACT: Among the detectors that are actually operational in the submillimeter wave range, the superconducting bolometers offer interesting characteristics. Indeed, their inherent slow response can be overcome by developing an integrated planar antenna to couple the incident radiation to the detector on the one hand, and by developing hot electron bolometer technologies based on nanostructures on the other. We describe the design of such a detector and its implementation with thin YBaCuO film. As a preliminary result, a 1×0.3 μm^2 bolometer has been realized by focused ion beam milling, coupled to a log-periodic antenna.

1. INTRODUCTION

Studies of electromagnetic radiation in the submillimeter-wave/terahertz range is important in a number of scientific research fields such as radioastronomy (to observe star formation or galactic structure), atmospheric physics (to observe radiation emitted from a number of molecules in the Earth's atmosphere such as HF and OH at 1.3 THz and 2.5 / 3.5 THz, respectively) or climatology. Moreover, military interest should be found in the analysis of submillimeter wavelength radiation to measure radar signatures.

Two kinds of detectors are classically used for this spectral range. Firstly, Schottky diodes have poor sensitivity and require high local oscillator (LO) power (a few mW) for heterodyne detection. Nevertheless, they are the only devices actually used in receivers at frequencies beyond one THz, and they can operate at room temperature (Röser et al 1994). Secondly, superconductor-insulator-superconductor (SIS) detectors have good performance (Bin et al 1996), close to the quantum limit for frequencies lower than the superconducting energy gap (about 750 GHz for Nb), but they are intrinsically frequency limited (at about twice the gap frequency).

In recent years, a new class of detectors, based on hot electron effects in a superconducting bolometer, has been developed. It represents a very attractive candidate for heterodyne mixing because there is no intrinsic frequency limitation due to its purely thermal sensing principle. Besides, only low LO power is required, in the same range as for SIS mixers (about 100 nw). Both low T_c and high T_c superconductor materials (LTSC and HTSC, respectively) can be used for realizing hot electron bolometers (HEB). Two classes of LTSC HEBs have appeared these recent years, that differ by the relaxation path that allow photon-excited electrons to return to a state of equilibrium. In the phonon-cooled HEB, hot electrons transfer their energy to the phonons, whereas in the diffusion cooled HEB hot electrons transfer their energy by diffusion to a normal metal (electrical contacts or arms of a planar antenna). Their performances are similar and have already shown to be competitive with those of SIS detectors at frequencies around one THz (Yagoubov et al 1999, Wyss et al 1999). Up to 2.5 THz, their noise temperature is 2-3 times lower than that of Schottky diodes.

HTSC HEBs cannot be separated into two classes; they are mainly of the phonon cooled type because electron diffusion mechanisms are negligible in HTSC films (Lee et al 1998, Harnack et al 1999). The analysis of HTSC HEBs is quite different from their LTSC counterparts: phonon dynamics play a major role due to the relatively high operating temperature (Cherednichenko et al 1999) and introduce some excess noise in these devices. Thus, they are not expected to reach the sensitivity of LTSC HEBs. Nevertheless HTSC HEBs exhibit some advantages. HTSC films have a very short electron-phonon relaxation time, which is about 1-2 ps at 80-90 K in YBCO so the output bandwidth in these devices should be an order of magnitude higher (the fundamental limitation is around 130 GHz) than that of LTSC phonon cooled HEBs. Moreover, they require cooling temperatures between 70 K and 80 K, which can be reached with self contained light weight cryocoolers, well suited to space missions. However this kind of detectors has not yet reached the state of technological maturity.

In order to design competitive bolometers for the submillimeter wavelength range - in terms of sensitivity, bandwidth and noise - three aspects have to be investigated i) electromagnetic coupling between the incident radiation and the active detector region, ii) thermal balance inside the active region and iii) electrical coupling between the active region and the readout electronics circuitry. This last point will not be considered in the following.

2.   DETECTOR FABRICATION AND DESIGN

2.1   Incident wave to detector coupling

One of the major problems arising in the submillimeter domain is to couple efficiently the incident radiation to the active detector region. The conventional bolometric structures (monolithic bolometers) where the absorbing and detecting functions are realized by the same surface is inconceivable here, because YBCO film exhibits an extremely low absorptivity in this wavelength range. Two solutions are currently adopted to couple the incident radiation to the thermally sensitive port of the device: absorbing layer coating or planar antenna electrical coupling. The latter solution is preferred to improve the response time versus sensitivity compromise. This coupling method consists in transferring power to a superconducting transition edge thermal sensor by means of a receiving antenna that delivers high frequency currents to the superconductor. Joule heating so arises from the current drawn by the normal electrons.

Since the early eighties, planar antennas have been commonly in use to couple thin-film sensitive elements to radiation fields, from far infrared to millimeter wavelengths. Among the possible geometries we have chosen the self-complementary log-periodic structure, because this kind of antenna can exhibit frequency-independent properties over a very broad band. Our antenna was designed to operate in the 30 μm to 1 mm wavelength range; for our substrate choice (MgO, $\varepsilon_r = 10$), its impedance is about 80 Ω (Gaugue et al 1997).

2.2   The active region: the nanobolometer

In traditional superconducting bolometers, the temperature rise due to incident radiation concerns both electrons and phonons of the film. In hot electron bolometers, the electron temperature can be increased above the phonon temperature of the film. This temperature rise (and consequently the sensitivity) depends on parameters that characterize the thermal link between the superconductor and the thermostat. At high signal frequencies, the device thermostat should be considered from a microscopic point of view, i.e. in terms of phonons. The detector performances thus strongly depends on the heat removal mechanism by phonons. If these are dominantly substrate phonons, the bolometer operates in a traditional mode; if these are dominantly film phonons, the bolometer operates in a hot electron regime.

The major obstacle to obtain a HEB regime is to render negligible any traditional bolometric effect; for this, the phonon temperature of the film should not increase above the phonon temperature of the substrate. Two main conditions should therefore be fulfilled: i) make use of a very thin superconducting film (a few nm) and ii) minimize the film to substrate thermal boundary resistance (i.e. avoid acoustic mismatch between the film and the substrate). Moreover, the substrate must act as a thermostat. This can be achieved by making use of a small bolometer fabricated on a high thermal conductivity substrate. High conductivity (κ) materials on which high quality YBCO films can be grown are magnesium oxide (MgO, $\kappa = 3.4$ WK^{-1}cm^{-1}) and sapphire (Al$_2$O$_3$, $\kappa = 6.4$WK^{-1}cm^{-1}). Moreover, MgO provides a low boundary specific resistance of about 5×10^{-4} KW^{-1}cm^2, whereas Al$_2$O$_3$ provides 10^{-3} KW^{-1}cm^2 and requires a buffer layer to avoid contamination. Modelling of the performance of HTSC HEB was worked out by Karasik et al (1997), who have evaluated the effects of device size and heat sink temperature. The results show that submicron-size devices made from a very thin film (e.g. 10 nm) could reach optimum performance (DSB noise temperature of a few thousand K for 1 μW LO power).

We have developed a processing method to obtain YBCO microstructures of nanometer scale. Three processing schemes are actually operational for patterning YBCO nanobridges: processes based on conventional electron beam lithography combined with wet chemical etching (Kamm et al 1998) or with reactive ion etching (De Nivelle et al 1993) and a process based on regular photolithography followed by focused ion beam (FIB) milling (Ben Assayag et al 1995). We have used this latter process. YBCO film with a thickness of 100nm was produced on a MgO substrate by *in situ* RF magnetron sputtering. The microbridge and the leads to perform two-point probe measurements were patterned by photolithography and ion milling. Then, a log-periodic antenna was formed from a 100 nm thick gold layer produced by RF sputtering and patterned again by photolithography and ion milling. The resulting microbridge structure have an active area of 7 μm $\times$ 10 μm with a transition width of 5 K and a 100 Ω resistance at the midpoint of the superconducting transition, as shown in Fig. 3. Preliminary experiments, performed with the 119 μm wavelength radiation of a discharge H$_2$O laser, have evidenced the main features of a bolometric response. We have also recorded the antenna radiation pattern (Gaugue et al 1998). Finally, the length and width of the bridge were reduced by focused ion beam milling (Fig. 1); as a preliminary result, a bridge of 300 nm width and 1 μm length has been obtained (Fig. 2).

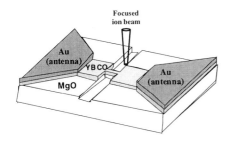

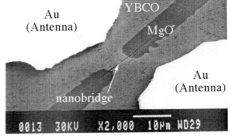

Fig. 1. The milling principle by focused ion beam to reduce width and length of the bridge.

Fig. 2. Geometrical view of an YBCO nanobridge of 0.3 μm width and 1 μm length.

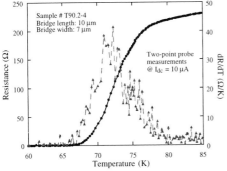

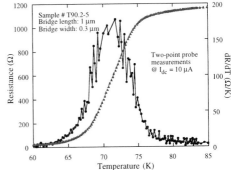

Fig. 3. Resistance and dR/dT as a function of temperature for an antenna-coupled YBCO microbolometer.

Fig. 4. Resistance and dR/dT as a function of temperature for a microbolometer narrowed down to 0.3 μm x 1μm by FIB milling.

The advantage of a such process is to control the nanobridge geometrical dimensions. with a good accuracy. Moreover we have the possibility of monitoring *in situ* and accurately the nanobridge resistance during the FIB milling (Ben assayag et al 1995). This should be required to obtain a good impedance matching between the antenna and the nanobridge. On the other hand, this process - in its present state - does not allow to obtain a nanobridge directly connected to the antenna arms; as shown in Figs. 3 and 4, the superconducting nanobridge is still in series with two microbriges (on the left and right sides). These parasitic microbridges slow down the heat removal process and alter the impedance matching. However, it should be noticed that the FIB technological step does not alter the superconducting transition width.

3. CONCLUSION

The hot electron bolometer is a most promising detector above 1 THz, where SIS junctions are frequency limited and Schottky diodes are rather noisy. We have suggested a new type of process for realizing HTSC HEBs, where the size of the active region is adjusted by focused ion beam milling. YBaCuO microstructures of nanometer scale have been obtained. Additional experiments with different device geometries and further optical measurements are required to investigate in detail the phenomena that occur inside the active region of the HTSC HEB.

ACKNOWLEDGMENTS

The authors are greatly indebted to E. Caristan (LGEP) for YBCO film elaboration, to Pr M Fourrier and Pr G Alquié (Université Paris 6) for laser and cryogenic facilities being made available to us for FIR measurements. This project was partly funded by the Observatory of Paris through a CNES R&T agreement.

REFERENCES

Ben Assayag G, Gierak J, Hamet JF, Prouteau C, Flament S, Dolabdjian C, Gire F, Lesquey E, Günther C, Dubuc C, Bloyet D and Robbes D 1995 J. Vac. Sci. Technol. B **13**(6), 2772
Bin M, Gaidis M, Zmuidzinas J, Phillips T and LeDuc H 1996 Appl. Phys. Lett. **68**(12), 1714
Cherednichenko S, Rönnung F, Gol'tsman G, Gershenzon E and Winkler D 1999 Proc. 10th Int. Symp on Space Terahertz Technology pp 181-9
De Nivelle M J M E, Gerritsma G and Rogalla H 1993 Phys. Rev. Lett. **70**(8), 1525
Gaugue A and Kreisler A, Robbes D and Gunther C, Sentz A, Hamet J-F 1997 Proc. 6th Int. Superconducting Electronics Conf. (ISEC'97), eds H Koch & S Knappe **3** pp 132-4
Gaugue A, Caristan E, Robbes D, Gunther C, Sentz A and Kreisler A 1998 Proc. 3rd European Workshop on Low Temperature Electronics (WOLTE 3), J. Phys. IV France **8**Pr3, 263
Harnack O, Karasik B, McGrath W, Kleinsasser A and Barner J 1999 Proc. 10th Int. Symp on Space Terahertz Technology pp 169-79
Kamm F-M, Pletti A and Ziemann P 1998 Supercond. Sci. Technol. **11**, 1397
Karasik B, McGrath W, Gaidis M, 1997 J. Appl. Phys. **81**(3), 1581
Lee C-T, Li C, Deaver B, Lee M, Weikle R, Rao R and Eom C 1998 Appl. Phys. Lett. **73**, 1727
Röser H, Hubers H, Crowe T and Peatman W 1994 Infrared Phys. Technol. **35**(2-3), 451
Skalare A, McGrath W, Bumble B, LeDuc H, Burke P, Vereijen A, Schoelkopf and Prober D 1996 Appl. Phys. Lett. **68**(11), 1558
Wyss R, Karasik B, McGrath W, Bumble B and LeDuc H 1999 Proc. 10th Int. Symp on Space Terahertz Technology pp 215-28
Yagoubov P, Kroug M, Merkel H, Kollberg E, Schubert J and Hubers H 1999 Proc. 7th Int. Superconducting Electronics Conf. (ISEC'99) pp 450-2

Inst. Phys. Conf. Ser. No 167
Paper presented at Applied Superconductivity, Spain, 14-17 September 1999
© 2000 IOP Publishing Ltd

Methods of submillimeter wave Josephson spectroscopy

M Tarasov E Stepantsov Z Ivanov

Chalmers University of Technology, Gothenburg, Sweden S41296

A Shul'man O Polyansky A Vystavkin

Institute of Radio Engineering and Electronics RAS, Mokhovaya 11, Moscow, Russia 103907

M Darula O Harnack

Research Center Juelich, Juelich, Germany 52425

E Kosarev

P.Kapitza Institute for Physical Problems, Moscow, Russia 117973

D Golubev

P.Lebedev Physical Institute of RAS, Moscow, Russia 117924

ABSTRACT: A HTS Josephson spectrometer has been designed, fabricated and experimentally studied. The spectrometer circuit consists of a YBCO bicrystal Josephson junction integrated with a double-slot or log-periodic antenna and connected in parallel with a gold low-inductance shunt. The YBCO films were deposited by laser ablation on sapphire and MgO bicrystal substrates with misorientation angle of 24 degrees. The selective detector response and RF response at intermediate frequency IF=1.4 GHz were measured in the signal frequency range 60-1250 GHz and the linewidth of Josephson oscillations (LJO) was determined by using three different methods. The RF response corresponds to the frequency down-conversion in a self-pumped Josephson mixer mode. For processing of RF response in this case we suggest a novel method of extracting the signal spectrum.

1.   INTRODUCTION

The Josephson effect is attractive for application in millimeter and submilimeter-wave spectroscopy. The selective Josephson detector response brings possibility to realize Hilbert transform spectrometer suggested by Divin, Polyansky and Shul'man (1980). Such spectrometers were developed using Josephson point contacts, shunted SIS tunnel junctions, HTS Josephson junctions. Another suitable for spectroscopy effect is a self-pumped Josephson mixing (JM). In such converter the input signal is mixed with internal Josephson oscillations. The minimum value of a double side-band (DSB) noise temperature in JM with self pumping (see Likharev and Migulin (1980)) equals to the physical temperature T for $f<0.2f_c$, and increases as $8(f/f_c)^2$ for $f>f_c$.

2. JOSEPHSON JUNCTIONS AND INTEGRATED RECEIVING STRUCTURE

The integrated receiving structure consists of a YBaCuO Josephson junction formed on a bicrystal MgO or sapphire substrate and an Au complementary log-periodic or double-slot antenna. The YBaCuO film 80-100 nm thick was deposited by laser ablation. The 2 μm wide junction has a normal state resistance of 10 Ω and a critical current of 300 μA, measured at 4.2 K. The remarkable feature of IV curves is the low excess current, a clear Fraunhoffer pattern in I_c dependence on magnetic field and correct oscillations of the critical current and Shapiro steps with applied LO power. For low-inductive shunting of Josephson junctions we use both hybrid resistive shunts made by bonding the loop of Au wire 30 μm thick and 5 mm in diameter or integrated thin-film loop deposited on the substrate together with the contact pads. The shunting resistive loop about 5 mm in diameter brings the resistance below 0.1 Ω at 4.2 K and inductance that does not shunt sufficiently the IF signal at 1.4 GHz. The substrate with Josephson junction was attached to the MgO extended hyperhemisphere lens placed in the LHe cryostat with optical window. Backward Wave Oscillators (BWO) in the frequency range 350-650 GHz and 880-1250 GHz were used as a LO source for mixer measurements and receiving structure microwave evaluation. The IF signal from the junction was connected to a matching circuit and amplified by cooled down to 4.2 K amplifier with a cooled circulator at the input.

3. EXPERIMENTAL RESULTS.

We have measured the selective detector response and IF noise under signal radiation at frequencies up to 1250 GHz (Fig. 1). Measurements of the IF noise and the selective detector response can be used for evaluation of the Josephson radiation linewidth. The calculation of the spectrum from detector response of Josephson junction is known as Hilbert spectroscopy and the proposed estimation from IF noise dependence is its modification that allows to simplify measurement technique and improve sensitivity and frequency resolution. The configuration of the measurement setup in this case is the same as for JM with self-pumping. With increasing the intermediate frequency, or decreasing the JO linewidth the voltage position of IF maxima is changed from the position of dynamic resistance maxima to the voltage positions corresponding to the sidelobes of signal frequency mixed with JO frequency at $V_s \pm V_{JO}$. In general we have three dependencies suitable for spectrum recovery:

1. Selective detector response measured at low modulation frequency.
2. IF detector response measured at few GHz that coincides with (1) when IF<LJO
3. IF self-pumped mixer response measured when IF>LJO

The first two cases can be processed using conventional Hilbert Transform method, or its modification, and the latter requires different method.

4. RESPONSE PROCESSING FOR SPECTRUM RECOVERY

The spectrum at 1 THz extracted by Hilbert transform method from detector response is presented in Fig. 2. The RF response dependencies have several features common to detector response, but they do not coincide with the latter. We suggest the following procedure for deducing the incident spectrum. The procedure includes extraction of autonomous noise $Na(v)$ from pumped $Np(v)$ noise $Ne(v)=Np(v)-Na(v)$, preparation of two dependencies shifted in bias voltage by $+v_{if}$ and $-v_{if}$, where $v_{if}=f_{if}\Phi_0$. These two shifted curves are used to deduce sum $Ns(v)=Ne(v+v_{if})+Ne(v-v_{if})$ and difference $Nd(v)=Ne(v+v_{if})-Ne(v-v_{if})$ dependencies. The two last brings the required incident spectrum $S(v)=Ns(v)-|Nd(v)|$. The example of such calculation is presented in Fig. 3.

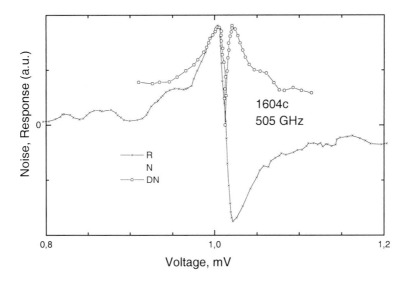

Figure 1. Detector response R and noise DN, latter after deduction of autonomous noise and extraction the square root.

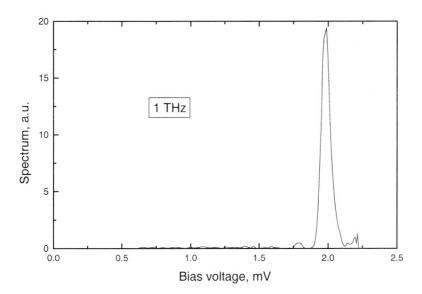

Figure 2. The spectrum extracted from direct detector response using Hilbert transform method.

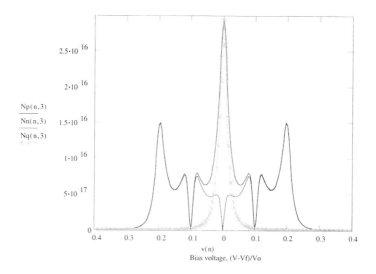

Figure 3. Recovered spectrum Nq, sum Np and difference Nn of shifted dependencies.

5. DISCUSSION

The sensitivity of both mentioned methods depends on the amplifier sensitivity. At low frequencies we can take the amplifier noise about 5 nV/Hz$^{1/2}$. For RF response the amplifier noise temperature can be below 10 K. Taking into account the measured input noise temperature for self-pumped mixer about 1000 K, and detector sensitivity up to 10^6 V/W we can estimate the detector type spectrometer noise S_{det} and self-pumped mixer spectrometer noise S_{spm} as follows:

$S_{det}=5\cdot10^{-9}/10^6=5\cdot10^{-15}$ W/Hz$^{1/2}$

$S_{spm}=k\cdot T_n=1.4\cdot10^{-23}\cdot1000=1.4\cdot10^{-20}$ W/Hz$^{1/2}$.

The frequency resolution corresponds to the Josephson oscillation linewidth, that can be improved by means of low-inductance shunting. Such shunting does not affect significantly the sensitivity of RF response, but reduces proportionally the detector sensitivity. Another advantage of RF method is that spectral resolution does not depend on the step size and it can have much wider dynamic range.

CONCLUSION

We have fabricated a HTS Josephson spectrometer and tested it in mm and submm-wave range. In processing the low frequency or detector response we utilized a Hilbert transform method. Measurement of RF response is equivalent to the self-pumped mixer mode. For extracting spectrum from the RF response a novel method of Josephson spectroscopy from the IF noise dependence has been suggested. This method allows simplifying the measurement technique, improving the sensitivity, dynamic range and increasing spectral resolution by low-inductance shunting without reduction of sensitivity.

REFERENCES

Divin Yu Ya, Polyansky O Yu and Shul'man A Ya 1980 Sov. Tech. Phys. Lett. **6**, 454
Likharev K K and Migulin V V 1980 Radio Engineering and Electron Physics **25**, 1

Inst. Phys. Conf. Ser. No 167
Paper presented at Applied Superconductivity, Spain, 14-17 September 1999

Single photon imaging spectrometers using superconducting tunnel junctions

L Frunzio, CM Wilson, K Segall, L Li, S Friedrich, MC Gaidis and DE Prober

Yale University, Applied Physics, P.O. Box 208284, New Haven, CT 06520-8284, USA

ABSTRACT: We are developing single photon 1-D and 2-D imaging spectrometers using superconducting tunnel junctions for astrophysical applications. They can operate in the energy range from X-ray to NIR. Our devices utilize a lateral trapping geometry. They have superconducting tunnel junctions on the sides of the absorber far apart from it. Energy information is obtained by the total collected charge. Position information is obtained from the fraction of the total charge collected by each junction. These device perform intrinsic imaging with many more pixels than read out channels.

1. INTRODUCTION

Superconductive tunnel junctions (STJ) have been extensively studied in the development of single photon spectrometers in a large variety of configurations (Booth 1996). These detectors have been operated at energies ranging from the X-ray down to the Near Infrared (NIR) (Frank 1998, Lumb 1995, Verhoeve 1997). In the last years, they have been integrated in different devices which also perform imaging (Kraus 1989, Jochum 1993, Friedrich 1996, Rando 1998a). The general approach pursued to date is large format arrays of single pixel detectors up to about hundred (Rando 1998b).

We have developed STJ-based detectors with intrinsic 1-D imaging, meaning that the detectors have many more pixels than read out channels. We have done this employing lateral trapping and band gap engineering. In Fig. 1 is a schematic of the devices.

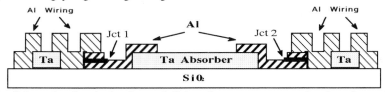

Figure 1. Schematic of a single photon 1-D imaging spectrometer using lateral trapping band gap engeneering and backtunneling. Not shown is an insulating SiO layer between the junction and the wiring. In X-ray devices Ta plugs are not present.

2. OPERATING PRINCIPLE

Many physical processes are involved in the operation of these STJ-based detectors. First, an incident photon is absorbed in the central Ta film breaking Cooper pairs and creating quasiparticles. The quasiparticles relax down to the absorber gap energy and diffuse until they reach the Al. In the Al, they can scatter inelastically, losing energy until they reach the Al gap energy. Once the quasiparticles scatter below the gap of Ta by emission of phonons, they are "trapped" in the Al electrode. This trapping process also produces charge multiplication due to the fraction of emitted phonons with energy larger than twice the Al gap energy. Then, the quasiparticles tunnel and are read out as an excess subgap current. The current pulses are then integrated to obtain the charge collected from each junction.

The total collected charge is proportional to the ratio of the impinging photon energy and the effective energy required to excite a quasiparticle. The latter is proportional to the Ta absorber gap energy. The fraction of charge collected in each junction tells us the location of the absorption event. If the photon is absorbed in the center, then the charge divides equally. If the photon is absorbed at one edge of the absorber, then most of the charge is collected by the closest junction.

Another important process in our optical/UV detectors is backtunneling. Excited quasiparticles enter the Al base electrode from the Ta absorber and scatter down in energy. The higher energy gap of the Ta then confines the quasiparticles near the tunnel barrier. A Ta plug is added interrupting the Al wiring, confining the quasiparticles near the barrier in the counterelectrode. Quasiparticles can then circulate, tunneling and backtunneling. Because both tunneling and backtunneling transfer a charge in the forward direction, we measure an integrated charge many times greater than the number of quasiparticles. This effect gives the junctions charge gain, which is very important for the smaller signals in optical/UV devices.

The energy resolution is the sum in quadrature of many components taking into account creation statistics, the trapping process, the backtunneling, the incomplete cooling of the quasiparticles and the electronic noise. A full explanation of the noise mechanisms in these devices has been presented elsewhere (Segall 1999a)

In the limit of negligible loss in the absorber and complete quasiparticle trapping in the STJ base electrode, the number of pixel is proportional to the resolving power of the spectrometer (Wilson 1999) allowing an intrinsic 1-D imaging capability with only two read out channels.

3. EXPERIMENTAL CONDITIONS

All devices are fabricated at Yale in a high vacuum deposition system with in-situ ion beam cleaning. We start with a wet oxidized Si substrate. The Ta absorber and plugs are then dc magnetron sputtered at 750 °C. Next a Nb ground contact is sputtered. The Al trilayer is then evaporated in one vacuum cycle. An SiO insulating layer is evaporated and finally Al wiring is evaporated. An in-situ ion beam cleaning is performed before each metal deposition to ensure good metallic contact. All layers are patterned photolithographically using either wet etching or lift-off (Gaidis 1994).

Measurements are made in a two stage ^{3}He dewar with a base temperature is 220 mK (Friedrich 1997a). To measure the photo-response of our junctions, we used a room temperature JFET current amplifier. We use an Amptek A250 amplifier with a 2SK146 input transistor. Extra circuitry is added that allows the A250/2SK146 to be dc coupled to the junctions. The amplifier thus provides an active voltage bias for the junction (Friedrich 1997b).

For X-ray measurements, we irradiate the sample with a ^{55}Fe source that emits two Mn lines at 5.89 keV (K_α) and 6.49 keV (K_β). In the optical/UV range, we illuminate the detectors using a small Hg lamp calibration source. A bandpass filter is used to select one photon energy at a time. We bring light into the dewar using an optical fiber, which is UV grade fused silica. The fiber is Al coated to enhance UV transmission up to energies of about 6 eV (200 nm). The filtered light passes through a fiber splitter that divides the light equally between two fibers. One of these fibers is fed into the dewar. The other fiber is fed into a photomultiplier tube which simultaneously measures the intensity.

4. RESULTS

We have made high quality Ta films with a residual resistance ratio of 17. The film are about 600 nm thick and are calculated to absorb about 28% of the incident energy at 6 keV. We have measured a quasiparticle loss time in the absorber of 450 μs at 220 mK. This loss time can be compared to a time of ~10 μs needed for a quasiparticle to diffuse across the absorber and be trapped in the Al junction electrode. We have made high quality Al films with a residual resistance ratio of 12 and a quasiparticle loss time of 57 μs at 220 mK. This loss time can be compared to the ~2 μs tunnel time in a typical junction.

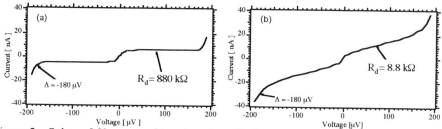

Figure 2. Subgap I-V curves of two junctions. Both measurements are made at T=220 mK. The junction parameters are: (a) Area=400 μm², R_N=2.3 Ω; (b) Area=100 μm², R_N=13.8 Ω.

We have also made high quality junctions. Two characteristics are important for low noise. First, junctions should have a low subgap current to minimize the shot noise. They should also have a large subgap resistance to minimize the contribution of the amplifier voltage noise. Fig. 2a shows the subgap I-V curve of a 400 μm² junction with a normal state resistance R_N=2.3 Ω. At 220 mK we measure a subgap current of 5 nA and a subgap resistance of 880 kΩ. Unfortunately, we have not been able to completely reproduce this quality in the imaging devices. Fig. 2b shows the subgap curve of a 100 μm² junction from a different fabrication run. This junction has R_N=13.8 Ω and the subgap resistance is only 9 kΩ. We do not understand if this difference is a fabrication issue or an experimental setup problem.

We have detected single X-ray photons using many devices. A typical absorber has an active area of 160 x 100 μm². We use a junction area of either 560 μm² or 1860 μm². The latter have different wiring sizes and different trapping layer geometries (Friedrich 1997c). In some samples there are Ta plugs. More detailed results will be presented elsewhere. In any case, we have achieved an excellent energy resolution δE_{FWHM}=26 eV at 5.89 keV in a limited absorber length of about 34 μm. The best energy resolution obtained on a full active absorber length is δE_{FWHM}=60 eV at 5.89 keV. This energy resolution implies a spatial resolution of about 1.6 μm, thus providing 100 pixels with just two read out channels (Segall 1999b).

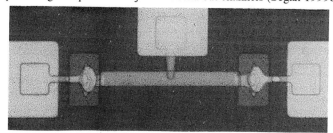

Figure 3. Photograph of the 1-D STJ-based imaging spectrometer tested in optical/UV range.

We have tested other detectors in the optical and ultraviolet region (Wilson 1999). Fig. 3 shows a typical device. The subgap curves of both its junctions look like the one in Fig. 2b. This device has a 100 x 10 μm² Ta absorber. Each Al trap overlaps the absorber by 5 μm. The presence of Ta plugs confines the quasiparticles near the barrier. We measure that the average number of times a quasiparticle tunnels is 23. The energy resolution, measured over the full absorber, is δE_{FWHM}=1.0 eV (δλ=240 nm) at E=2.27 eV (λ=546 nm) green line and δE_{FWHM}=1.6 eV (δλ=83 nm) at E=4.89 eV (λ=253 nm) UV line. For the UV if only a selected range of the absorber is chosen, we obtain δE_{FWHM}=1.1 eV (δλ=57 nm). We have measured the noise spectra of both junctions with no illumination and they are consistent with the measured resolution. However, the noise spectra contain excess noise that we cannot explain.

An energy resolving power of about 3 in the UV implies that the detector can resolve at least 4 spatial pixels at that energy. This particular detector has an active absorber area 90 μm long by 10 μm wide. So, the detector has 4 pixels with dimensions 22 x 10 μm². This is achieved with only two readout channels.

618

5. 2-D IMAGING

We are also developing single photon 2-D imaging spectrometers. For X-ray, we are studying the degrading of energy and spatial resolution of a 2-D absorber with traps at each side (Li 1999). We have begun fabricating these devices, one of those is shown in Fig. 4. Our Ta absorbers have very long quasiparticle loss time making feasible the realization of devices with 1 mm^2 absorber. Devices with large absorbers could resolve about 1000 of 20 x 20 μm^2 pixels with only four read out channels. For optical/UV, the limitations in the count rate resulting from this type of solution could be bypassed using array of strips, as shown in Fig. 4. In fact, this seems to be the best compromise between the needs of a large number of pixels with a simplified read out and the requirement on the count rate. Further improvements are feasible introducing an RF-SET based read out electronics (Schoelkopf 1998)

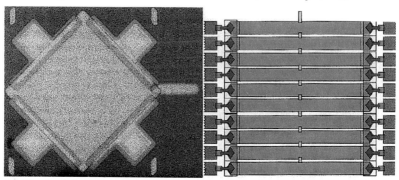

Figure 4. On the left: photograph of a 2-D X-ray imaging spectrometer fabricated at Yale before the wiring layer is deposited. On the right: diagram of a 2-D optical/UV imaging spectrometer using ten absorber strips.

6. CONCLUSIONS

We are developing and testing STJ-based single photon imaging spectrometers. Our devices use a lateral trapping and bangap engineering. We have detected single X-ray, optical and UV photons with these first detectors.

REFERENCES

Booth N E and Goldie D J 1996 Supercond. Science and Technology **9**, 493
Frank M et al. 1998 Rev. Sci. Instrum. **69**, 25
Friedrich S et al. 1996 Nucl. Instrum. Meth. in Phys. Res. **A370**, 44
Friedrich S 1997a Ph.D Thesis, Yale University
Friedrich S et al. 1997b IEEE Trans. Appl. Supercond. **7**, 3383
Friedrich S et al. 1997c Appl. Phys. Lett. **71**, 3901
Gaidis M C 1994 Ph.D Thesis, Yale University
Jochum J et al. 1993 Ann. Physik **2**, 611
Kraus H et al. 1989 Phys. Lett. **B231**, 195
Li L et al. 1999 Proc. LTD8 (in press)
Lumb D H et al. 1995 Proc. SPIE **2518**, 258
Rando N et al. 1998a Proc. SPIE **3435**
Rando N et al. 1998b Proc. SPIE **3445-25**
Schoelkopf R J et al. 1998 Science **280**, 1238
Segall K et al. 1999a Appl. Phys. Lett. (submitted)
Segall K et al. 1999b IEEE Trans. Appl. Supercond. (in press)
Verhoeve P et al. 1997 IEEE Trans. Appl. Supercond. **7**, 3359
Wilson C et al. 1999 Proc. LTD8 (in press)

Inst. Phys. Conf. Ser. No 167
Paper presented at Applied Superconductivity, Spain, 14-17 September 1999
© 2000 IOP Publishing Ltd

Realisation of a flux-flow dc-transformer using high temperature superconductors

S Berger, K Bouzehouane, DG Crété and JP Contour

Unité Mixte de Physique CNRS/Thomson-CSF, Domaine de Corbeville, F91404 Orsay CEDEx

ABSTRACT: The DC transformer, proposed in 1965 by Giaever, involves flux-flow vortex motion in two superimposed thin superconducting films, separated by a thin insulating layer. This paper reports on the first realisation of this device using high temperature superconducting films. It involves pulsed laser deposition of $YBa_2Cu_3O_7$, $PrBa_2Cu_3O_7$ and $SrTiO_3$, with a ramp based multilayer technology. Transport measurements exhibit coupling and decoupling between vortices in the adjacent films. Widening the temperature operating range of this device, mandatory for applications such as RF detection, seems possible with low pinning materials.

1. INTRODUCTION

The DC transformer is a superconductor/insulator/superconductor planar heterostructure where the insulator is thick enough to prevent tunnelling between the superconducting films. Giaever first explained in 1965 the operation of the device, involving vortex flow and magnetic interaction of the vortices. Berger *et al.* made in 1999, a brief review of the experimental work made since then. The principle of operation is briefly described : when an external magnetic field is applied normal to the surface of the structure, regular arrays of vortices are present in both films, with the same flux line density. If the insulator is thin enough, magnetic interaction can couple the arrays of vortices via their magnetic moment, and govern their dynamics. Decoupling occurs at high driving current in the primary: this non linear effect is sensitive to RF signals similarly to a Josephson junction, as predicted by Gilabert *et al* in 1994. It could be used as a Josephson spectrometer, with a voltage to frequency ratio much larger than for a single Josephson junction. High temperature superconductors should make the structure more viable for applications. But due to the drastic deposition conditions associated to the difficulty in controlling the morphology, no device elaboration has already been achieved using these materials.

We report on the first elaboration of a DC-transformer from high temperature superconducting (HTSC) films, using a $YBa_2Cu_3O_{7-\delta}$ / insulator / $YBa_2Cu_3O_{7-\delta}$ (SIS) double heterostructure. We also observed 3 distinct regimes of operation due to the magnetic coupling.

2. EXPERIMENTAL PROCEDURE

The SIS heterostructure is grown by pulsed laser deposition, using the procedure which was previously developed for ramp based multilayer technology (see for example Gao et al, 1992) in order to minimise the surface where the two $YBa_2Cu_3O_{7-\delta}$ (YBCO) layers are superimposed. This procedure includes a first step of ion milling, followed by a second deposition with epitaxial relationship on the etched surface as reported by Bouzehouane in 1996.

The dimensions of the device (240 μm×10 μm excluding connection lines) allow fabrication by regular photolithography. Argon ion etching is carried out using a Kaufmann source, with a beam of incidence of 50° with respect to the normal to the film, in order to define edges of the film with a slope of about 20°. This prevents the formation of grain boundaries in the top YBCO layer. After in-situ cleaning of the surface by an ion beam at low energy (50eV), we deposit i) an insulating bilayer formed by $PrBa_2Cu_3O_{7-\delta}$ (10nm) and $SrTiO_3$ (30nm); ii) the 100nm thick YBCO top electrode. The insulating bilayer helps in reducing the roughness with respect to $SrTiO_3$ grown directly on the YBCO, as observed by Contour *et al.* in 1999. Two other iterations of photolithography are needed to define the complete device. Gold contacts (400x400μm^2) are then sputtered on the two YBCO layers. Film roughness, outgrowth density and ramp profile are controlled by atomic force microscopy and scanning electronic microscopy throughout the fabrication. Fig. 1 shows the final device geometry.

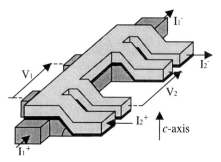

R(T) and V(I) have been measured using three configurations i) current fed in the primary layer only, voltage measured at the primary (subscripts 11); ii) current fed in the secondary layer only, voltage measured at the secondary (subscripts 22); iii) current fed in the primary layer only, voltage measured at the secondary (subscripts 21). No external magnetic field is applied, but we assume that the bias current induces sufficiently high vortex concentration, penetrating from the edges of the bridge.

Fig. 1: Geometry of the device and transport measurement configuration. The insulating layer is represented in dark. The angle α of the ramp is measured by Atomic Force Microscopy (α =20°).

3.   RESULTS AND DISCUSSION

The critical temperatures of the primary and the secondary layers measured on the device are 85.5K and 87.7K respectively. Current leakage through the insulating layer is evidenced by the low interface resistance of 3.4×10^{-4} Ω.cm^2 below the critical temperatures.

Fig. 2 presents the V-I characteristics for T=84.5K. The critical currents Ic of the primary and secondary layers are 2.3 mA and 10.5 mA respectively. These values are read on the V_{11} and the V_{22} plots, since these measurements are equivalent to single layer measurements as long as $I \leq Ic$ because of the resistive nature of the insulating layer.

The plot of V_{21} displays the coupling effect. The voltage appearing in the secondary layer can not be explained by a Lorentz force mechanism since the total bias current remains much smaller than the critical current of the secondary. Moreover, similar critical currents displayed by the V_{11} and V_{21} plots, and a perfect proportionality between the two curves up to the decoupling current Id prove that the two vortex arrays obey the same dynamic laws and justify the conclusion that the magnetic coupling is efficient.

For $I > Id$, the two curves are no longer proportional. While V_{11} still behaves like a typical flux-flow characteristic, the V_{21} curve reaches a maximum value and then begins to drop. Measurements have been interrupted before heating effects arise to prevent thermal runaway. One can note that this decoupling effect occurs very progressively when increasing the bias current: even above the decoupling point, the primary vortices continue to drag the secondary ones, even if the two sets of flux lines drift with different velocities. Clem (1974) analyses this mechanism as the time averaged measurement of increasing alternate motion of the secondary vortices in the direct and the opposite directions compared to the primary array, when the velocity difference between the two arrays is enlarged. These two opposite displacements of the secondary vortices, around the pinning sites, occur once per relative drift (of the arrays) of one vortex lattice parameter.

An important feature is the much lower value of the voltage in the secondary than in the primary, even for current smaller than Id. This tends to demonstrate that fewer vortices are moving in the secondary film than in the primary. The hypothesis that the vortices would move slower in the secondary from the critical current must be dismissed: in this case, the speed (and then the voltage) in the two layers would not increase proportionally with the bias current, just like above

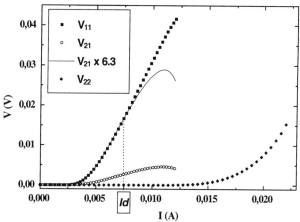

Fig. 2: $V(I)$ characteristics at $T=84.5$K for V_{11}, V_{21} and V_{22} configurations. The shape is sign independent, reproducible in the time, without any hysteresis. The critical current density in the secondary film is much higher than in the primary one. The solid line shows the perfect proportionality between V_{11} and V_{21} for $I < Id \approx 6.9$ mA. Id is defined as the current involving a 5% increase of the proportionality factor between V_{11} and V_{21}.

Id. The reasons why much less flux lines are involved in the secondary layer may be the following : i) there is still an efficient pinning force in this film at the working temperature which is much lower than the Tc of the secondary; ii) a lower vortex concentration in the secondary. They both imply the observed lower voltage.

Preliminary measurements under magnetic field normal to the films did not seem to influence this effect, indicating that the difference in vortex concentration is not the main reason. Therefore, we believe that vortex pinning is the origin of the voltage difference between Ic and Id.

Subsequently, we investigated on the pinning in HTS using AC susceptibility. The maximum value of the dissipative component of the susceptibility, χ''_{max}, is a measure of the pinning as it takes energy to pull a vortex out of its potential well. The smaller is χ''_{max}, the lower is the pinning force. As shown on Fig. 3, there is a sensible correlation between χ''_{max} and the outgrowth density. This indicates that the outgrowth have a significant contribution to pinning.

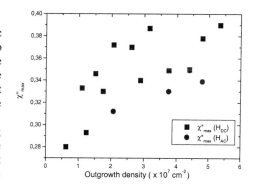

Fig. 3: Correlation of χ''_{max} with the outgrowth density for films with identical thickness (200 nm). Square symbols are used to report data taken for different values of the AC magnetic field amplitude, and circles for different values of the DC magnetic field.

The thickness of the secondary layer might also be a relevant parameter with respect to pinning : a thinner secondary film may facilitate moving the vortices in the secondary. Indeed, the coupling force, acting essentially on the lower part of the secondary vortices, will more easily overcome the force due to the collective pinning or punctual pinning sites present in the film all along the flux line. The relative strength of these opposite effects also depends on the magnetic field and the temperature (through the magnetic coupling length between the two superconductors, and the lengths characteristic of the pinning strength related to thermal energy or vortex line energy, such as the Larkin length). For this study, we choose to reduce the AC susceptibility data in term of critical exponent n, which is defined by the relation between the current density and the electric field : $E = E_0.(J/J_c)^n$. Large values of n indicate strong pinning, while $n \approx 1$ corresponds to low pinning. Fig. 4 shows the dependence of n on temperature for 3 classes of films : on the right part of the chart are YBCO films with thicknesses from 5 nm to 300 nm, from the line connecting circle symbol (for 5 nm) to the right onwards, essentially in increasing thicknesses. For these samples, the outgrowth density was $\approx 4.10^7 cm^{-2}$. Two samples with lower outgrowth density exhibit lower n values, as expected, and correspond to the data plotted with crosses. The sample giving the best result (*i.e.* the lowest n values, dashes) is a NdBa$_2$Cu$_3$O$_7$ sample with typical thickness and outgrowth density. This result may be interpreted in term of a smaller lattice mismatch between NdBa$_2$Cu$_3$O$_7$ and SrTiO$_3$ than with YBCO : a hypothesis to be confirmed is that this translates into fewer defects in the material.

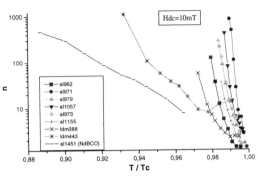

Fig. 4: Critical exponent n of various samples versus reduced temperature. The samples are YBCO when not specified, and described in the text. At low temperature, the vortices are frozen (large values of n); n decreases while the vortex lattice is melting and tends to unity when approaching the critical temperature.

4. CONCLUSION

We have fabricated the first HTSC flux-flow DC-transformer, displaying strong magnetic coupling between vortices across a 40 nm insulating layer. The measurement of the V-I characteristics gives experimental evidence of this effect : they are proportional up to the decoupling current and then we observed a progressive drop of the voltage across the secondary as the current in the primary is increased. The device is viable for applications since a stable decoupling effect occurs at reasonable bias current and temperature. We expect that using NbBa$_2$Cu$_3$O$_7$ superconducting thin films can greatly reduce intrinsic pinning of vortices and consequently, widen the temperature operating range of the device.

REFERENCES

Berger S, K Bouzehouane, D Crété and JP Contour 1999 Eur. Phys. J. AP **6**, 111

Bouzehouane K 1996 PhD thesis Univ. Paris XI France

Clem JR 1974 Phys. Rev. B **9**, 898

Contour JP, Ayache J, Chenu C, Drouet M, Durand O, Magis M and Maurice JL (submitted to Eur. Phys. J. AP)

Gao J, Boguslavskij Y, Klopman B D, Vijbrans R, Gerritsma GJ and Rogalla H 1992 J Appl. Phys. **72**, 575

Giaever I 1965 Phys. Rev. B **15**, 825

Gilabert A, Schuller IK, Moshchalkov VV and Bruynseraede Y 1994 Appl. Phys. Lett. **64**, 2885

Inst. Phys. Conf. Ser. No 167
Paper presented at Applied Superconductivity, Spain, 14-17 September 1999
© 2000 IOP Publishing Ltd

623

Far-infrared Hilbert-transform spectrometer based on Stirling cooler.

O Y Volkov and V V Pavlovskii

The Institute of Radioengineering & Electronics of RAS, Moscow 103907, Russian Federation

Y Y Divin and U Poppe

IFF-IMF, Forschungszentrum Jülich GmbH, Jülich, D-52425, Germany

ABSTRACT: The design and the characteristics of the first Hilbert-transform spectrometer operating with Stirling cooler are reported. The principle of operation of Hilbert-transform spectrometer is based on the ac Josephson effect. High quality $YBa_2Cu_3O_{7-x}$ thin-film bicrystal Josephson junctions have been used in the spectrometer. A temperature of the spectrometer as low as 34 K has been reached with the help of a Stirling cryocooler. The operation of the spectrometer has been demonstrated in the frequency range from 0.16 THz to 3.1 THz with a resolution of 4 GHz in a large part of the range.

1. INTRODUCTION

Hilbert-transform spectroscopy (HTS) of electromagnetic radiation is based on the square-law detection of this radiation by a Josephson junction, where voltage-controlled Josephson oscillations are responsible for a specific selectivity of the detection mechanism. It was shown by Divin et al (1980), that the spectrum $S(f)$ of electromagnetic radiation detected by a Josephson junction can be recovered by Hilbert transformation of the junction response $H(f)$, which is measured as a function of the Josephson frequency $f_j = (2e/h)V$, where V is a voltage across the junction.

Measurements of millimeter- and submillimeter-wave radiation by HTS have been carried out using both low-T_c and high-T_c Josephson junctions (Divin et al 1980, Stumper et al 1984, Larkin et al 1996, Divin et al 1996). The laboratory prototypes of Hilbert-transform spectrometers cooled by cryogenic liquids have been developed (Divin et al 1983, Hinken et al 1988, Divin et al 1993, Tarasov et al 1995, Divin et al 1998).

A necessity to use cryogenic liquids for cooling is considered as a main obstacle on the way of superconducting electronics into the market, and a replacement of them by cryocoolers is required (Rowell 1999). Here, we present the design and characteristics of a Hilbert-transform spectrometer based on a Stirling cryocooler.

2. EXPERIMENTAL DETAILS

A general view of the Hilbert-transform spectrometer based on a Stirling cooler is shown in Fig. 1. The Stirling cryocooler (Model SL200, AEG INFRAROT-MODULE GmbH) consists from a compressor, a pressure-transferring line and a separate component with a coldfinger. The compressor was attached to the mounting plate through elastic supports. The separate component with a coldfinger was fixed directly on the mounting plate. A special vacuum chamber with a polyethylene window was

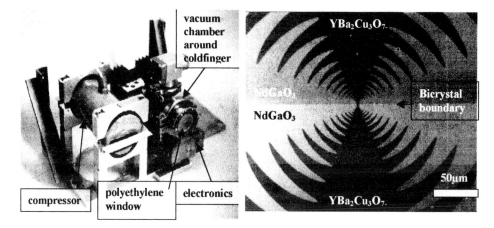

Fig. 1. Photograph of a laboratory prototype of the Hilbert-transform spectrometer, based on a Stirling cryocooler.

Fig. 2. Micrograph (transmitted polarised light) of a YBa₂Cu₃O₇-x bicrystal Josephson junction with an integrated log-periodic antenna.

mounted around the coldfinger. The hot part of the separate component of the cryocooler was in contact with a ripped heat sink. The compressor was magnetically shielded by several layers of mu-metal tape. The same type of shielding has been used around the vacuum chamber. To stabilise the temperature of a cold finger, a very efficient air cooling of the compressor was found to be required. Coldfinger temperatures in the range from 30 to 90 K have been achieved in this cryogenic environment by controlling the ac power of the compressor. Measurements at any temperature in the range 30-90 K were available during several hours with a reasonable drift of 1-2 K.

High-quality $YBa_2Cu_3O_{7-x}$ grain-boundary junctions fabricated on untwinned $2 \times 14°$ (110) $NdGaO_3$ bicrystal substrates (Divin et al 1997) have been used in this prototype of a Hilbert-transform spectrometer. The micrograph of one of our bicrystal junctions is shown in Fig. 2. The widths of the junctions were around 2-3μm. The I_cR_n-products of these junctions were up to 330 μV at 78 K, and the values of resistances R_n ranged from 0.25 to 8 Ohm. A broadband $YBa_2Cu_3O_{7-x}$ log-periodic antenna has been integrated with each junction on the substrate. The substrate with the Josephson junction was mounted in the vacuum chamber on the coldfinger of the Stirling cooler. The electrical connections to Josephson junctions have been made through vacuum connectors and an electronics box was fixed on the vacuum chamber.

An optically pumped far-infrared laser and a backward-wave oscillator with a multiplier were used as sources of monochromatic radiation for the characterisation of the spectrometer. Monochromatic radiation from the sources was focused by a parabolic mirror to the junction antenna through a polyethylene window and a hyperhemispherical Si-lens. With this combination we were able to deliver radiation in the frequency range from 60 GHz to 4.25 THz to the Josephson junction.

3. CHARACTERIZATION OF THE SPECTROMETER

The characterisation of the spectrometer has been made at a temperature of (34 ± 1) K by measuring the voltage dependences of the current response $\Delta I(V)$ of the spectrometer to monochromatic signals with the frequencies f_i from 60 GHz up to 4.25 THz. A $YBa_2Cu_3O_{7-x}$ grain-boundary Josephson junction with $R_n = 1.1$ Ohm and $I_cR_n = 1.5$ mV has been used for these measurements. A combined set of the individual responses $\Delta I(V)$ of the Josephson junction to these monochromatic signals is shown in Fig. 3. Each response curve $\Delta I(V)$ was normalised on its value at low voltages, and all these curves $\Delta I(V)$ in Fig. 3 have got an initial value equal to one in the low-

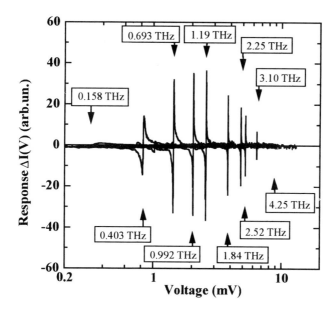

Fig. 3. A combined set of the individual responses $\Delta I(V)$ of the Hilbert-transform spectrometer to monochromatic far-infrared radiation with different frequencies. The $YBa_2Cu_3O_{7-x}$ bicrystal Josephson junction at 34 K on a Stirling cryocooler was used in this spectrometer.

voltage range. Each of the individual responses to radiation with the frequencies f_i in the range from 158 GHz to 3.10 THz shows an odd-symmetric resonance around the voltages $V_i \approx hf_i / 2e$, which is due to the interaction of the internal voltage-controlled Josephson oscillations with the external signal.

As it can be seen from Fig. 3, the frequency range of the spectrometer exceeds one frequency decade in the far-infrared range. The response of the junction with $R_n = 1.1$ Ohm to 70.5μm-radiation ($f_i = 4.25$ THz) has showed only a suppression of the critical current and a gradual decrease at the higher voltages.

In the RSJ model of Josephson junctions, the amplitude of the normalised response near the resonance $V_i \approx hf_i / 2e$ should increase from zero up to some saturation level with an increase of V_i from zero to above $I_c R_n$. The saturation in the amplitude of a resonance response reflects the frequency independent behaviour of the amplitude of Josephson oscillations in the RSJ model. As it can be seen from Fig. 3, in our case the amplitude of the resonance response $\Delta I(V)$ is growing from 1 to 36 with the increase of the frequency f_i from 0.158 THz to around 1 THz and than approaching zero with a further increase of the signal frequency f_i to 4.25 THz. The first trend is in accordance with the RSJ behaviour, and the second one might be attribute to the gradual decrease of the amplitude of Josephson oscillations with the increase of Josephson frequencies.

The width of the resonant response to the highest frequency of 3.1 THz was around 8-9 μV, which corresponds to the Josephson linewidth of 4-4.5 GHz. At lower frequencies, the linewidth first goes down to 3 GHz at frequencies between 1 and 2 THz, and than goes up with the further decrease of the frequency. The Josephson linewidth determines the spectral resolution δf of the technique, and, according to our measurements, the resolving power $\delta f / f$ of around 10^{-3} might be achieved in the terahertz range with Hilbert-transform spectroscopy. An example of application of the developed Hilbert-transform spectrometer is demonstrated in Fig. 4. Radiation to the spectrometer came from a

626

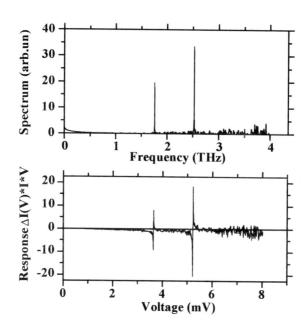

Fig. 4. Response $\Delta I(V) \cdot I(V) \cdot V$ of Hilbert spectrometer to radiation of optically pumped CH_3OH laser (below) and spectrum of laser radiation recovered by Hilbert-transformation of the response (above).

far-infrared CH_3OH laser, pumped by 9P36 line of CO_2 laser. The length of the FIR laser cavity was slightly changed from the optimum one for 2.523 THz line. As the result of this detuning, two odd-symmetric resonances appeared at the response $\Delta I(V) \cdot I(V) \cdot V$ of the spectrometer (below). An application of Hilbert-transformation to this response, according to the principles of HTS, gives the spectrum of incident radiation (above). Two lines, the main at 2.523 THz and the competing at 1.758 THz, are clearly visible in the spectrum. The intensities of laser lines are inside the dynamic range of the spectrometer and no artificial line at the difference frequency has appeared in the spectrum.

REFERENCES

Divin Y Y, Polyanskii O Y, Schulman A Y 1980 Sov. Tech. Phys. Lett. **6**, 454
Divin Y Y, Polyanskii O Y, Schulman A Y 1983 IEEE Trans.Magn. **19**, 613
Divin Y Y, Larkin S Y, Anischenko S E, Khabaev P V, Korsunsky S V 1993 Int. J. Infrared & Millimeter Waves, **14**, 1367
Divin Y Y, Schulz H, Poppe U, Klein N, Urban K, Pavlovskii V V 1996 Appl. Phys. Lett. **68**, 1561
Divin Y Y, Kotelyanskii I M, Shadrin P M, Volkov O Y, Shirotov V V, Gubankov V N, Schulz H, Poppe U 1997 Proc. EUCAS 97, eds H Rogalla and D H A Blank (IOP Publishing) pp 467-70
Divin Y Y, Volkov Y Y, Shirotov V V, Pavlovskii V V, Poppe U, Schmueser P, Tonutti M, Hanke K, Geitz M 1998 Proc. SPIE, **3465**, 309
Hinken J H et al 1988 Proc. 18th Europ. Microwave Conf. pp 177-82
Larkin S Y, Anischenko S E, Kamyshin V V, Khabayev P V 1996 Proceedings SPIE, **2842**, 607
Rowell J M 1999 IEEE Trans. Appl. Supercond. **9**, 2837
Stumper U, Hinken J H, Richter W, Schiel D, Grimm L 1984 Electronics Lett. **20**, 540
Tarasov M A, Shul'man A Y, Prokopenko G V, Koshelets V P, Polyanski O Y, Lapitskaya I L, Vystavkin A N 1995 IEEE Trans. Appl. Supercond. **5**, 2686

Inst. Phys. Conf. Ser. No 167
Paper presented at Applied Superconductivity, Spain, 14-17 September 1999
© 2000 IOP Publishing Ltd

Development and Characterization of Phonon-Cooled NbN Hot-Electron Bolometer Mixer at 810 GHz

C Rösch, F Mattiocco and K-F Schuster

Institut de Radio Astronomie Millimétrique (IRAM), 300 rue de la Piscine, Domaine Universitaire de Grenoble, 38406 St. Martin d'Hères, France

ABSTRACT: We report about the investigation of NbN hot-electron bolometers integrated in two different types of a double dipole structure. Uncorrected double sideband receiver noise temperatures of 1200 K and 1900 K were obtained at 810 GHz. The gain bandwidth at the intermediate frequency (IF) was determined by an impedance measurement with a HP8510 network analyser. In order to allow for rapid measurements the whole setup was integrated in a cryogenic dipstick. Depending on the particular operating point bandwidths between 0.65 GHz and 1.36 GHz were obtained.

1. INTRODUCTION

While superconductor-insulator-superconductor (SIS) mixers have almost replaced Schottky diode mixers in astronomical studies up to 1 THz superconducting hot-electron bolometers (HEB) seem to become the technology of choice for heterodyne detection above 1 THz. Unlike the diffusion cooled HEB mixer where the small length of the superconducting film determines the size of the mixer bandwidth , the dimensions of the microbridge are less critical parameters for phonon-cooled NbN devices. For a given film thickness, the area of the mixer can be adjusted in order to give a wide range in impedance and optimum local oscillator (LO) power level. However, for phonon cooled HEBs the optimal LO power simply scales with the volume of the microbridge. Therefore it is important to use e-beam lithography to reduce the size and to reach low LO power levels which can be generated by solid state sources.

The sheet resist R_s=880 Ω for 5 nm NbN films is relatively well adapted to a variety of antenna structures. After successful fabrication of NbN HEB integrated in a single bowtie dipole antenna (Lehnert 1998, Rösch 1998) a double dipole structure with two different integration schemes of the HEB was investigated (Rösch 1999).

2. DOUBLE DIPOLE ANTENNA

Single bowtie dipoles used for first device tests are appropriate for simple mixer tests in the laboratory but due to large sidelopes they are not suited for real astronomical observations. A double dipole structure was investigated by Skalare (1997). We used a slightly modified structure which was designed for a working frequency of 810 GHz assuming a fused quartz substrate. As shown in Fig. 1 the two antennas are connected by coplanar striplines and the mixer element is placed in the centre of the dipoles. Intermediate frequency (IF) output and DC bias are connected via coplanar rf-filters on either side of the antennas.

628

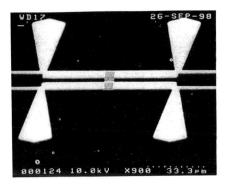

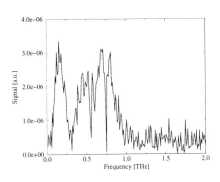

Fig. 1: *Double dipole with a single HEB mixer element in the centre*

Fig. 2: *FTS measurements of a double dipole with central HEB*

3. MEASUREMENTS AND RESULTS

To investigate the response of this antenna as a function of frequency FTS measurements were performed using the HEB mixer in direct detection mode. Therefore a Martin-Puplett interferometer with a chopped liquid nitrogen cold load was used. Since atmospheric absorption was not suppressed the measured response in Fig. 2 is only a preliminary result. However, the structure seems to be reasonably well matched to 810 GHz, but the results have to be confirmed by a setup in a dry nitrogen atmosphere.

The double sideband (DSB) receiver noise temperature was determined by the usual Y-factor method. A typical result for mixing is given in Fig. 3 where the IF conversion curves for hot and cold load are plotted together with the pumped I-V-curve of the device. The best uncorrected DSB receiver noise temperature of 1200 K was found for an LO frequency of 813 GHz and an IF of 1.25 GHz. Since the critical current of the device was relatively high the beamsplitter coupling for the LO power had to be increased causing higher losses. To compare the results with earlier measurements noise temperatures were corrected for beamsplitter losses. The results summarised in Fig. 4 are comparable to the results obtained with a simple dipole. The desired improvement of the beam shape still has to be shown.

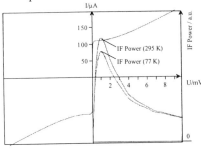

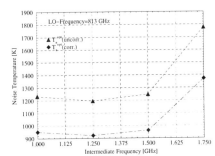

Fig. 3: *Heterodyne measurement*

Fig. 4: *Noise temperatures*

4. NEW ANTENNA STRUCTURE FOR VERY HIGH FREQUENCIES

The antenna pattern of the double dipole can be improved by omitting the coplanar connections to the mixer element. These lines do not only disturb the antenna pattern but also introduce ohmic losses which become particularly important for very high frequencies. Therefore the wide range of achievable device impedances for NbN HEBs suggest another solution for the double dipole coupling (see Fig. 5). In this structure each antenna is furnished with its own mixer and the signals of both antennas are combined in the IF path. In order to conserve the symmetry of the antenna such a structure requires identical mixer elements, what is quite a challenge for device fabrication.

The conversion curve for such a device is shown in Fig. 6. The noise temperature of 1900 K is clearly higher than for the other devices. Probably this is due to the fact that the two mixer elements are not exactly identical. The particular structure of the conversion curves is not understood at the moment.

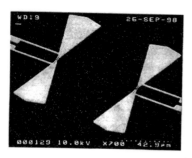

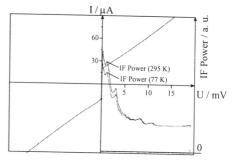

Fig. 5: Double dipole with individual HEB   *Fig. 6: First heterodyne measurements*

5.   IMPEDANCE MEASUREMENTS

In order to estimate the IF bandwidth inherent in the HEB mixer we used a method proposed by Karasik (1997). It was verified experimentally (Ekström 1995) that the impedance of the HEB changes from a high differential resistance at low frequencies to a low ohmic resistance at high frequencies. Since the crossover is related to the intrinsic relaxation of the electron temperature this relaxation time τ_θ can be determined from the impedance $Z(\omega)$.

For these measurements one of the devices with a central HEB mixer element was bonded in a gap of a microstrip transmission line as shown in Fig. 7. Instead of mounting the line in a dewar we installed it in a dipstick which could be immersed into liquid He. The dipstick was connected to a HP8510 network analyser and the DC bias was fed directly by its internal bias supply.

Fig. 7: Microstrip mount of a HEB device for impedance measurements in a dipstick.

IF impedance measurements were performed within a frequency range of 0.2 GHz to 6 GHz at a power level of -60 dBm. Averaging the signal we did not find any particular need for an additional cryogenic amplifier.

The calibration was done by biasing the HEB first to the superconducting state ($Z \approx 0$) and then to the normal state ($Z = R_n$). After de-imbedding the IF impedance from the microstrip fixture the result was fitted simultaneously to the real and to the imaginary part of the theoretical model

$$Z(\omega) = R \frac{1+C}{1-C} \frac{1 + i\omega \frac{\tau_\theta}{1+C}}{1 + i\omega \frac{\tau_\theta}{1-C}}$$

where R is the device resistance. The self-heating parameter C and τ_θ were used as free fitting parameter. The fits for different operating points are shown in Fig. 8 and the results are summarised in Tab. 1. The IF bandwidths are in reasonably good agreement with the results of a similar device obtained by superposing two LOs.

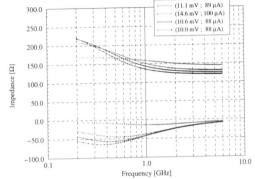

U / mV	I / μA	f / GHz
14.6	102	1.36 +/- 1.24
11.3	87	0.75 +/- 0.27
10.6	85	0.73 +/- 0.19
10.2	85	0.65 +/- 0.17

Tab. 1: Different operating points

Fig. 8: Nonlinear fits of the complex impedance for different operating points.

While the fits are good for the real parts of the impedance they are only acceptable for the imaginary parts. Calculating the uncertainty of the fitting parameters shows that they are quite high and increase with increasing bias voltage. Using higher bias voltage causes the differential resistance at the operating point to decrease. As a result the frequency dependence of the impedance becomes less pronounced and the uncertainty of the fit increases.

The IF bandwidth seems to increase with increasing bias voltage but due to the large error bars this can not be concluded.

6. CONCLUSION

NbN HEB mixer elements were integrated in two different types of double dipoles antenna structures. Noise temperatures as low as 1200 K were obtained for a double dipole with a single central HEB. For the structure with an individual HEB for each dipole noise temperatures are only slightly higher (1900 K). Therefore this type of antenna is promising for higher frequencies but the fabrication process still has to be improved.

Determination of the IF bandwidth by measuring the impedance was possible but special care has to be taken with regard to the uncertainties.

REFERENCES

Ekström H, Karasik B S, Kollberg E, Gol'tsman G N and Gershenzon E M, 1995
    Proc. of the 6th Int. Symp. on Space Terahertz Technology, Pasadena, USA, pp 169-283
Karasik B S, Gaidis M C, McGrath W R, Bumble B and LeDuc H G, 1997
    Proc. of the 8th Int. Symp. on Space Terahertz Technology, Boston, USA
Lehnert T, Rothermel H and Gundlach K H, 1998
    J. Appl. Phys. 83 (7), p 3892
Rösch C, Lehnert T, Schwoerer C, Schicke M, Gundlach K H and Schuster K F, 1998
    Proc. of the 9th Int. Symp. on Space Terahertz Technology, Pasadena, USA
Rösch C, Mattiocco F, Gundlach K H and Schuster K F, 1999
    Proc. of the 10th Int. Symp. on Space Terahertz Technology, Charlottesville, USA, pp 208-214
Skalare A, McGrath W R, Bumble B, LeDuc H G, 1997
    Proc. of the 8th Int. Symp. on Space Terahertz Technology, Boston, USA

Inst. Phys. Conf. Ser. No 167
Paper presented at Applied Superconductivity, Spain, 14-17 September 1999
© 2000 IOP Publishing Ltd

Heterodyne type response in SIS direct detector

A. Karpov*, J. Blondel*, P. Dmitriev**, V. Koshelets**.

*Institut de Radioastronomie Millimétrique, St. Martin d'Hères, France
(present address : Caltech, Pasadena, USA)
**Institute of Radio Electronics and Engineering, Moscow, Russia

ABSTRACT. We present a first observation of single photon mixing in an SIS detector, providing a heterodyne type response in a direct detection experiment. A broad band (40%) SIS direct detector is studied under the effect of the black body radiation. A 1.5 GHz intermediate frequency low noise amplifier is connected to the detector. No local oscillator power is applied in this experiment. At the same time a typical periodical heterodyne response appeared in the intermediate frequency power versus bias voltage dependence. Tien - Gordon like steps are observed in the current voltage characteristic (CVC) of the SIS junction. The Y factor of about 2 is measured in a standard experiment with liquid nitrogen cooled and ambient temperature loads. The heterodyne response may be explained as single photon mixing within the detector band The mode of operation presented here may serve to build a sensitive detector for the millimeter and sub millimeter bands using a SIS junction.

1. INTRODUCTION

The direct detection of radiation at millimeter and submillimeter wavelengths via quantum assisted tunneling in a quasiparticle tunnel junction is a well known effect, as well as the mixing of the signals with a local oscillator in a quasiparticle tunnel junction [1]. These effects have allowed the building of practical ultra-low noise instruments for the detection of weak radiation of distant objects in radio astronomy [2].
Below we describe a new possible mode of operation of an SIS device based on the mixing of the single photons. This regime is intermediate between the detection and heterodyne reception and may be useful for the development of a broad band SIS continuum receiver.
First we describe the SIS detector used in our experiment and the test technique. Then we present the heterodyne type response observed in the direct detection experiment.

2. SIS DETECTOR DESIGN

The waveguide detector design is presented in Fig. 2. The SIS detector comprises a microstrip Nb circuit with SIS junction printed on a 0.2 mm thick quartz substrate and a detector block. The detector block uses a single backshort in a full height 2.3×1.15 mm waveguide. The printed circuit is coupled to the waveguide with a matching probe in the 70 - 130 GHz band [3]. The non-contacting backshort position once adjusted remains fixed in the waveguide.
The printed circuit of the detector is optimized for the junction normal resistance of about 50 Ohm and $R_N \omega C \approx 4$. The Nb-Aloxide-Nb SIS junction has an area of 1.5 μm^2. The coplanar

inductive tuning circuit is connected to the junctions as part of the top electrode layer in the junction fabrication process (Fig. 1). The coplanar design gives a wider operation band and a better tolerance of the manufacturing imperfections of Nb printed circuits [4]. Coplanar line parameters have been calculated with the GPLINES program and the circuit was developed using the EESOF Libra program. The bandwidth of the SIS junction match in the detector is expected to be about 50 GHz.

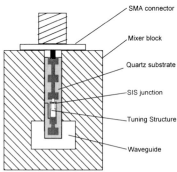

Fig. 1. The SIS detector design. It is a single backshort block with a full height 2.3×1.15 mm waveguide. The printed circuit is coupled to the waveguide with a matching probe in the 70- 130 GHz band. The non-contacting backshort position once adjusted remains fixed.

3. TEST SET

The test set comprises a liquid helium cryostat, the SIS detector, a cooled HEMT IF amplifier, and an ambient temperature amplifier. The signal radiation was coming to the detector horn antenna through the cryostat window. The receiver input window with anti-reflection grooves is made of polyethylene and an infrared filter of expanded polystyrene foam was fixed to the 77 K shield. The detector block was at a temperature of about 4.4 K. The cooled intermediate frequency amplifier used in the experiment had about 5 K noise temperature in the frequency range of 1.2-1.8 GHz [5]. An isolator was fitted between the SIS detector and the first IF amplifier.

In the direct detection experiments a lock-in amplifier was used for the measurement of the junction current variation on shifting the receiver beam between a liquid nitrogen temperature load (77 K) and an ambient temperature (295 K) load.

4. SIS DIRECT DETECTOR OPERATION

A. Detector characterization

The CVC characteristic of the SIS detector measured at 4.4 K temperature is presented in the Fig. 2. The leakage current at 2 mV is 2.5 µA. The response to black body irradiation is illustrated in Fig. 3, which gives the dependence of detected current on bias voltage. One can note the quantum steps above and below the gap voltage. A small contribution of the two-photon process to the tunneling is visible as a second quantum step below the gap voltage. The shape of the quantum steps is similar to the quantum steps produced by monochromatic radiation. This behavior is due to the filter properties of the SIS detector circuit, limiting the detector band [6]. The width of the quantum steps corresponds to the central frequency of the detector band of 90 GHz. The maximum variation of detected current with 77 K and 295 K loads is 0.26 µA. This SIS detector responsivity corresponds to a detector bandwidth of no less than 35-40 GHz [6]. The detector noise in our experiment is limited by the bias supply noise of 4-5 nV/√Hz. A detector sensitivity of ΔT=30 mK√s has been measured with 10 Hz modulation frequency. ΔT may be reduced down to the limit of 1-0.5 mK√s with improvement of the DC amplifier. The last limit is fixed by the shot noise of the detector dark current.

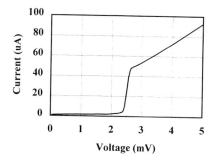

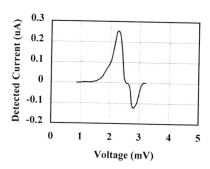

Fig. 2. Current-voltage characteristic of the SIS detector.

Fig. 3. The detected current in the SIS detector as a function of bias. The detector beam is switched at 10 Hz frequency between the 77 K and 295 K loads. Two quantum steps are visible below and one step above the gap voltage.

B. Photon mixing in the SIS detector

In this experiment no local oscillator power is applied to the SIS junction. At the same time a typical heterodyne response is observed through the IF amplifier chain. The IF power versus bias of the SIS detector is presented in Fig. 4. The curves from upper to lower are measured respectively with the 295 K load in front of the receiver, with the 77 K load and with the detector backshort tuned to disconnect the SIS junction and antenna (no radiation at the junction). The output power maximum at 2.33 mV has a shape typical for a heterodyne response in a SIS mixer. The ratio of the IF power levels with 295 K and 77 K loads is close to 2.

The observed effect may be explained as the mixing of single photons within the band defined by the SIS junction match to the detector antenna. This supposition may be confirmed using the data of the Fig. 4. The conversion gain of the SIS mixer at low levels of the local oscillator power (Plo) is proportional to the Plo [1]. In our experiment the Plo is the power of the incident broad band radiation of a black body and Plo is proportional to the black body temperature (Tbb). The signal is also proportional to the incident power and to the Tbb. So, the output IF power resulting from the mixing of the incident non coherent photons should be proportional to the square of the black body temperature $\Delta Pif(Tbb){\sim}Tbb^2$.

The contribution to the IF power from the described mixing process $\Delta Pif(Tbb)$ may be determined as the difference of the output power with the black body in front of the detector and the output power with a detuned detector, disconnected from antenna. The last regime gives a reference, as the power at the receiver IF output port comprises only the noise power of the IF chain and the amplified SIS junction shot noise.

At the bias voltage of 2.33 mV the ratio of the IF contributions from the 77 K and 295 K black bodies is $\Delta Pif(295K)/\Delta Pif(77K)$=12.2. This gives the equivalent temperature of the incident radiation from the 77 K load of $Tins$=295K/√12.2=84 K. The increase of the temperature of incident radiation by 7 K is related to the radiation of the cryostat window and that of the infrared filter. This 7-8 K optics noise was measured in this cryostat in an independent heterodyne

634

receiver experiment. This good agreement of the measured data confirms our hypothesis on the single photons mixing.

The described regime may be interesting to create a broad band low noise detector using a SIS junction. This mode of operation allows one to detect the radiation in a broad band and with a strong variation of the output power at the IF frequency.

In our experiment with the SIS direct detector the stability of available bias sources was putting the limit of sensitivity 10-100 times above the limit related to the device shot noise (see section IV a). A good gain stability of the IF amplifier (10^{-5}), gives an other possibility. One can employ the detection through the regime of the mixing of single photons using the stability of IF chain and not that of the bias circuit.

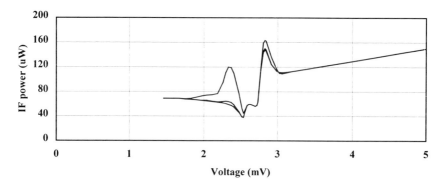

Fig. 4. The IF power versus bias of the SIS detector. The curves from upper to lower are measured respectively with the 295 K load in front of the receiver, with the 77 K load and with the detector backshort tuned to disconnect the SIS junction and the antenna (no radiation at the junction). The output power maximum at 2.33 mV has a shape typical for a heterodyne response in a SIS mixer. The maximum ratio between the IF power levels with 295 K and 77 K loads is close to 1.9.

5. CONCLUSION

We present the mixing of the single photons in a SIS detector. In our experiment no local oscillator power is applied to the SIS junction. Never the less a typical heterodyne response is observed through the IF amplifier chain. The observed IF power level dependence on the incident radiation temperature is well explained with the single photon mixing model.

The described regime may be interesting to create a broad band low noise detector using a SIS junction. The radiation of the receiver optics may be characterized using the described regime.

We developed also a SIS direct detector for the 80 – 120 GHz band with the 30 mK√s sensitivity. With improvement of the DC amplifier, the direct detector noise may be reduced to ~1 mK√s.

REFERENCES

1   J. R. Tucker, IEEE J. of Quantum Electronics, **QE-15**, 11, 1234, 1979.
2   J. E. Carlstrom, J. Zmuidzinaz, "Review of radio science 1993-95", Oxford Un. Press, 1996.
3   A. Karpov at al, Proc of 17 Int. Conf. IR and MM Waves, SPIE-P/1929, 212, 1993.
4   A. Karpov at al, IEEE Tr. on App. Superconductivity, **AS-7**, 2, 1073, 1997.
5   J. D. Gallego, M. W. Pospieszalski, El. div. Int. Rep. **286**, NRAO, 1990.
6   A. Karpov, J. Blondel, and K. H. Gundlach, "App. Superconductivity 1995", **2**, 1741, 1995

Inst. Phys. Conf. Ser. No 167
Paper presented at Applied Superconductivity, Spain, 14-17 September 1999
© *2000 IOP Publishing Ltd*

Metal-high T_c superconductor point contact response to millimeter wave radiation

A Laurinavičius, K Repšas, R A Vaškevičius and A Deksnys

Semiconductor Physics Institute, A.Goštauto 11, 2600 Vilnius, Lithuania

ABSTRACT: Experimental results of the response of a metal - high T_c superconductor (BSCCO-2212) point contact to millimeter wave radiation both at room and liquid nitrogen temperatures are presented. Measurements were carried out with and without bias current. An unexpectedly large response signal was observed when the bias current was applied. This effect is explained by the heating of the interface area below the tip of the metallic probe by the alternating current of the millimeter wave. The response signal was compared with that obtained using a Schottky diode.

1. INTRODUCTION

Many papers have been published describing studies of the response of high-T_c super-conducting films to electromagnetic radiation. These investigations cover a very wide spectrum range and include the optical (Danerud et al 1994, Hone-Zern Chen et al 1996, Berkowitz et al 1996, Shim et al 1998, Shinho Cho et al 1998, Kaila et al 1998), millimeter wave (Ngo Phong et al 1993, Laurinavičius et al 1997) and microwave (Grabow et al 1994) bands. The mechanisms of the responses have been broadly divided into two classes: bolometric and non-bolometric. The bolometers described in the most of the papers were fabricated using patterned YBCO films.

In the present paper we have investigated the metal-high T_c superconductor (BSCCO-2212) point contact response to millimeter wave radiation.

2. EXPERIMENTAL DETAILS

The experiment was carried out at frequencies of 25-30 GHz and in the incident power range of 0,1-32 mW. Superconducting BSCCO-2212 layers with thickness of about 30 μm were used. Samples were prepared on cleaved MgO(100) faces using melt processing technology, i.e. by melting and recrystalyzing coatings of superconducting powder. The fabricated samples showed epitaxial structure. The thickness of the substrate was 0,5 mm.

The schematic diagram of the experimental set-up is shown in Fig. 1. The sample used for the investigation was placed on the wide wall of the rectangular waveguide. Bias voltage V was applied to the point contact metal probe-superconducting layer through the load resistance R. The other silver-superconducting sample contact was connected to the metallic waveguide wall. The diameter of the metallic probe tip was about 20 μm.

636

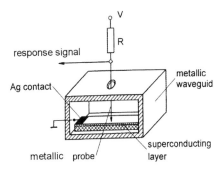

Fig. 1. Schematic diagram of the experimental set-up.

3.   RESULTS AND DISCUSSION

A response signal appeared when the millimeter wave radiation was switch on. The response measurements were carried out with and without bias current. The measurement results are shown in Fig. 2. As it is seen from the figure, the response had a weak dependence on temperature even in the region of the transition from the normal to the superconductive state. This response was also observed at room temperature. Moreover, the point contact and superconductive layer resistance dependencies on temperature were absolutely different (Fig. 3). Compared to the superconductive layer resistance, the point contact resistance decreased slightly with temperature.

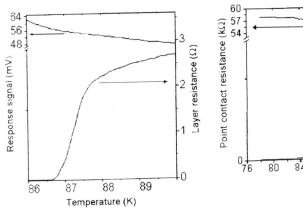

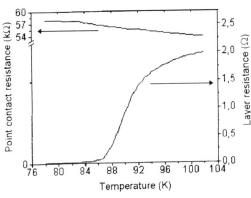

Fig. 2. Point contact response signal and superconducting layer resistance versus temperature.

Fig. 3. Point contact and superconducting layer resistance versus temperature.

From these facts, we conclude that this response is not related to the superconducting phenomenon. We also assume that this response can be associated with the degraded metal-high-T_c superconductor interface. Such assumption was confirmed by Tulina (1997) who investigated the normal metal point contact dc properties of the high-T_c superconductor (BSCCO-2212). In our case, the large point contact resistance and its increase when the

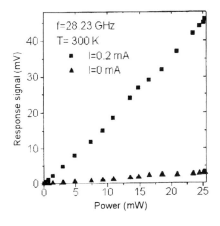

Fig. 4. Response signal versus millimeter wave power with and without bias current.

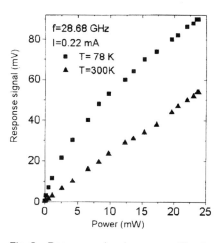

Fig. 5. Response signal versus millimeter wave power at different temperatures.

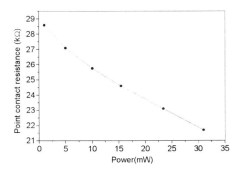

Fig. 6. Point contact resistance versus millimeter wave power.

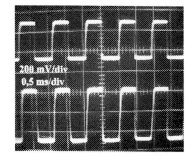

Fig. 7. Point contact response signal (bottom) and Schottky diode detected signal (top) oscillograms at room temperature. Bias current I=30 μA. Incident power P=35 mW.

temperature is decreased (Fig. 3) shows that the nature of the interface layer is semi-conductive. This conclusion is in agreement with that made earlier in a study of the electric histeresis loop of the semiconducting Bi-O_2 layer on the cleaved surface of a BSCCO-2212 single crystal (Xiaolin Wang et al 1993).

In Fig. 4 and Fig. 5, the voltage-power characteristics at different temperatures and bias currents are presented. A small response signal without bias current (Fig. 4) showed that the voltage - current characteristic of the point contact is very close to symmetric. On the other hand, a bias current significantly increased the response signal. This effect can be explained by the heating of the interface area below the tip of the probe by the alternating current of the millimeter wave. The temperature change of the interface layer causes a change in its resistance and a voltage drop on the point contact. The correctness of such reasoning is confirmed by Fig. 6 where the point contact resistance versus the incident millimeter wave power is presented. According to this point of view, the response signal should increase when the temperature of the point contact decreases. This is clearly demonstrated in Fig. 5, where a significant increase of the response signal is observed at temperature 78 K. Moreover, we

638

observed a decrease of the response signal when the bias current was increased. This effect is related to the heating of the interface layer by the dc bias current. The largest response signal was observed at a relatively small bias current (Fig. 7). For comparison, the detected signal of a Schottky diode is presented in the same oscillogram.

4.   CONCLUSION

The observed response signal of the metal high T_c superconductor point contact to millimeter wave radiation is related to the degraded superconductor surface layer which has a semiconductive nature. The origin of the response is associated with a change of the resistance of the point contact by the millimeter wave alternate current, which heats the interface area below the tip of the metallic probe. The obtained results can be useful in the design of cryo-electronic components, which integrate both superconducting and nonsuperconducting elements.

REFERENCES

Berkowitz S J, Hirahara A S, Char K, Grossman E N 1996 Appl. Phys. Lett. **69** 2125
Danerud M, Winkler D, Lindgren M, Zorin M, Trifonov V, Karasik B S, Gol'tsman G N, and Gershenzon E M 1994 J. Appl. Phys. **76** 1902
Grabow B E and Boone B G 1994 J. Appl. Phys. **76** 5589
Hone-Zern Chen, Chen Y C, Wu Z S, Jia S C, Lin C H, and Hsiung Chou 1996 Jpn. J. Appl. Phys. **35** 5308
Kaila M M, Cochrane J W, and Russell G J 1998 J. Supercond. **11** 463
Laurinavičius A, Repšas K, Vaškevičius A R, Lisauskas V, Čepelis D 1997 Inst. Phys. Conf. Ser. No **158** 393
Ngo Phong L, and Shih I 1993 J. Appl. Phys. **74** 7414
Shim S Y, Kim D H, Lim H R, Hwang J S, Park J H, Kim C H, Choi S S, Hahn T S 1998 Appl. Supersond. **6** 37
Shinho Cho, Deok Choi, Harold R. Fetterman 1998 J. Appl. Phys. **84** 5657
Tulina N A 1997 Physica C **991** 309
Xiaolin Wang, Zhuo Wang, Hong Wang and Minhua Jiang 1993 Physica C **208** 259

Inst. Phys. Conf. Ser. No 167
Paper presented at Applied Superconductivity, Spain, 14-17 September 1999

639

Performance of Inhomogeneous Distributed Junction Arrays

M Takeda and T Noguchi

The Graduate University for Advanced Studies, Nobeyama, Minamisaku, Nagano 384-1305, Japan
Nobeyama Radio Observatory, Nobeyama, Minamisaku, Nagano 384-1305, Japan

ABSTRACT: Mixing properties of a new type of distributed SIS junction arrays, which consists of different dimensions of junctions and lengths between every two junctions, has been theoretically investigated using quantum theory of mixing. A set of dimensions of junctions and lengths between every two junctions have been determined so as to minimize a reflection coefficient of an S-parameter, S11, at the input port in the array using microwave CAD. We report on the performance of an SIS mixer with distributed junction array designed to cover the frequency range from 90 GHz to 180 GHz.

1. INTRODUCTION

Tuneless or fixed tuned SIS mixers are highly desirable for complex systems such as multibeam receivers and interferometer arrays at millimeter and submillimeter wavelengths. The bandwidth of such tuned SIS mixers is mainly governed by the $\omega R_n C_j$ product, which is approximately equal to the Q-factor of the resonance circuit, and is strongly dependent on the critical current density of a junction, J_c. In order to achieve a broader bandwidth higher value of J_c is generally required. Unfortunately, high J_c may bring various disadvantages such as large sub-gap leakage current and reduction of yields of junctions in the fabrication (Kleinsasser *et al.* 1993). Recently, Shi *et al.* (1997) proposed an interesting SIS mixer composed of distributed junction arrays, which has a number of identical junctions equally separated by superconducting micro-striplines that act as tuning inductances. It has been shown that the critical current density required to achieve a reasonable bandwidth can be lowered in the mixer with the distributed junction arrays in contrast to the conventional single-junction SIS mixers. This type of distributed junction arrays, however, shows large periodic increase of noise at certain frequencies. These increase of noise will become an obstacle for the broad band operation.

In this paper, we propose a new type of distributed junction arrays, or inhomogeneous distributed junction arrays, which has different dimensions of junctions and lengths between every two junctions, to reduce amplitude of the increase of noise discovered in the conventional distributed junction arrays, or homogeneous distributed junction arrays.

2. THEORY AND OPTIMIZATION OF DISTRIBUTED JUNCTION ARRAYS

The inhomogeneous distributed junction array proposed here consists of a number of junctions with different dimensions distributed on a transmission line as illustrated in Fig.1. In the inhomogeneous distributed junction array the dimensions of junctions and the lengths between every two junctions must be determined for a given current density prior to the calculation of mixing properties. A simplified equivalent circuit of the inhomogeneous array, which consists of N-junctions represented by a combination of linear resistance R_n and geometrical capacitance C_j connected in parallel, is shown in Fig.2. The specific capacitance of an SIS junction is assumed to be $90.0 \, \text{fF}/\mu\text{m}^2$. The reflection coefficient of S-parameter, S_{11}, seen from the input port of the inhomogeneous distributed junction array is not only minimized over the given frequency band but also as independent on frequency as possible, using a commercial microwave software (HP-MDS).

Once a set of optimized dimensions of junctions and spacing between every two junctions, we established an equivalent large and small signal model for the inhomogeneous arrays following the method derived by Shi *et al.* (1997) for homogeneous arrays and then calculated mixing properties of the inhomogeneous arrays using Tucker's quantum theory of mixing (Tucker *et al.* 1985). In the analysis, we have adopted the quasi-five port approximation in which five sidebands are allowed, but the LO voltage is assumed to be sinusoidal (Kerr *et al.* 1993). Only the difference in the analysis for the inhomogeneous arrays from the homogeneous ones is that the conversion admittance and correlation matrices must be derived to every junction, since they depend on I-V curve of individual SIS junction.

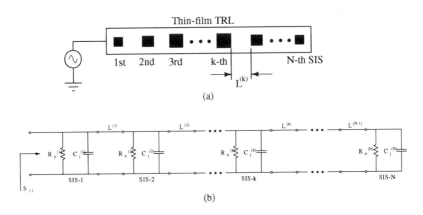

Fig.1 (a) Schematic representation and (b) simplified equivalent circuit of an inhomogeneous arrays

3.   SIMULATION RESULTS

Using the quantum theory of mixing together with the equivalent circuit model established in the previous section we have calculated the mixing properties of the inhomogeneous distributed junction array with an optimized set of dimensions of junctions and lengths between every two junctions in the frequency range from 90 GHz to 180 GHz, which corresponds to a full band of the WR-8 waveguide. IF frequency was assumed to be 1.5 GHz. The number of junctions in the array was set to five so that the array has a moderate input impedance which makes it easy to match the source well. The RF termination normalized to the equivalent normal-state resistance of a junction array, R_{RF}/R_n, has been assumed to be unity over the frequency band. The conversion gain is strongly dependent on the magnitude of the IF termination, which is approximately equal to the dynamic resistance of the pumped junction. We have selected IF termination of 10Ω, since the large IF termination, for example 50Ω which is typical IF termination, seriously degrades the conversion gain in the band.

In the calculations, all the junctions were assumed to have the same critical current density of 2.0 kA/cm^2 for both the homogeneous and inhomogeneous arrays. The dimensions of junctions and the lengths between every two junctions in the inhomogeneous array used in the calculation are listed in Table 1. In the homogeneous array, the area of $2.0\,\mu m^2$ of a junction was adopted, which is a typical area of SIS junctions used in the conventional SIS mixers in the same frequency band. The lengths between every two junctions was determined so as to minimize the receiver noise temperature in the frequency band in the homogeneous array and we adopted the length of $90.0\,\mu m$ in the calculation for the homogeneous array. The width of stripline was assumed to be $8.0\,\mu m$ for both arrays, which corresponds to the characteristic impedance of 8.6Ω.

Table 1.   Simulation parameters

$A^{(1)}$	$A^{(2)}$	$A^{(3)}$	$A^{(4)}$	$A^{(5)}$	$L^{(1)}$	$L^{(2)}$	$L^{(3)}$	$L^{(4)}$
3.0	7.0	9.0	9.0	7.0	50.0	40.0	40.0	50.0

Units of junction area and length of stripline are μm^2 and μm, respectively.

Figure 2 shows a reflection coefficient of S-parameter, S_{11}, and input coupling efficiency seen from the input port for the homogeneous and inhomogeneous arrays. In the homogeneous array, it is found that S_{11} degrades at certain frequencies. On the contrary, such degradations of S_{11} are considerably improved in the well-optimized inhomogeneous array. The improvement of S_{11} at input port results in the improvement of input coupling efficiency. From this result excellent mixing properties can be expected in the inhomogeneous array.

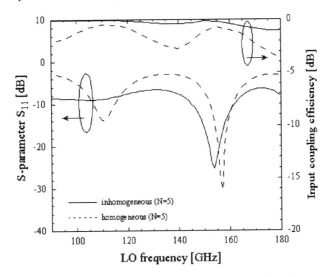

Fig.2   The reflection coefficient of S-parameter, S_{11}, and input coupling efficiency for inhomogeneous (solid lines) and homogeneous (dashed lines) with 5-junctions in frequency range from 90 GHz to 180 GHz.

Figure 3 shows the calculated mixer noise temperature and conversion gain for the homogeneous and inhomogeneous arrays. Note here that the dc bias and LO voltage across the last junction were optimized with respect to the receiver noise temperature at each frequency. As shown in Fig.3, broad-band operation has been accomplished in both arrays in spite of fairly low critical current density of 2.0 kA/cm². In the homogeneous array with five junctions, however, very large increase of noise are observed at several frequencies and the conversion gain is considerably decreased near those frequencies. If the number of junction is increased, for example, the number is set to ten as shown in Fig.3, amplitude of the increase of noise can be significantly suppressed in the homogeneous array. In the inhomogeneous array with five junctions, it is found that amplitude of the increase of noise at those frequencies are considerably suppressed in contrast to the homogeneous ones. The fluctuation of the conversion gain over the frequency band is also very much reduced in the inhomogeneous ones. Degradation of the conversion gain at lower frequency in the inhomogeneous array is caused by degradation of output coupling. If it is possible to make smaller IF termination, or if IF termination normalized to normal-state resistance, $R_{IF}/R_n \leq 1$, the conversion gain at lower frequency will be significantly improved.

From the viewpoint of noise as a function of frequency, it is obvious that the inhomogeneous array with five junctions is nearly competitive with the homogeneous array with ten junctions. The reduction of number of junctions in the array will bring the following advantages. At first, it is easy to achieve good impedance match to the source, since effective normal-state resistance increases. Secondly, the dimensions of junctions can be larger in the array with a small number of junctions, if the impedance match at the input port can be allowed to be sacrificed to a little extent. This brings a great benefit for the fabrication of junctions and makes the fabrication of the inhomogeneous arrays easier than those of the homogeneous arrays. In spite of the restricted optimization, however, it is found that amplitude of the increase of noise has been extremely suppressed in the inhomogeneous array. It is noted here that the SSB (Single Side Band) mixer noise temperature and conversion gain of the inhomogeneous array are less than 30 K and larger than −5.0 dB over the frequency range from 90 GHz to 180 GHz, respectively, which is competitive with those in conventional single-junction SIS mixers.

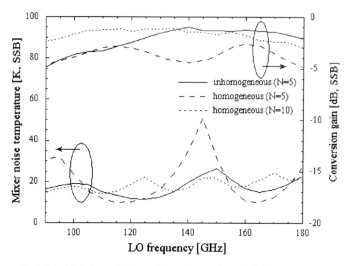

Fig 3 Calculated mixer noise temperature and conversion gain for inhomogeneous and homogeneous arrays in frequency range from 90 GHz to 180 GHz. N is a number of junctions in the array.

4.   CONCLUSION

We have investigated mixing properties of distributed junction arrays with different dimensions of junctions and lengths between every two junctions. If the dimensions of junctions and lengths between every two junctions in an array are well optimized in term of the reflection coefficient of S-parameter, S_{11}, seen from the input port of an array, it is demonstrated that the inhomogeneous distributed junction arrays can achieve low-noise SIS mixers in the frequency range from 90 GHz to 180 GHz. It is also found that critical current density for the inhomogeneous arrays can be very low compared with conventional single-junction SIS mixers as far as same frequency band is concerned.

REFERENCES

A W Kleinsasser, F M Rammo and M Bhushan 1993 Appl. Phys. Lett. **62**, 1017
S C Shi, T Noghuchi and J Inatani 1997 Proc. 8th Int. Symp. on Space Terahertz Tech. pp 81
J R Tucker and M J Feldman 1985 Rev. Mod. Phys. **57**, 1055
A R Kerr, S K Pan and S Withington 1993 IEEE Trans. Microwave Theory Tech. **41**, 590

Inst. Phys. Conf. Ser. No 167
Paper presented at Applied Superconductivity, Spain, 14-17 September 1999
© 2000 IOP Publishing Ltd

HTSC-Bolometer for IR-Spectrometer
- Processing and Operation -

K-S Roever[1], T Heidenblut[1], B Schwierzi[1], T Eick[2], W Michalke[2], E Steinbeiss[2]

[1]Institut fuer Halbleitertechnologie und Werkstoffe der Elektrotechnik,
 Universitaet Hannover, Appelstr. 11a, D-30167 Hannover
[2]Institut fuer Physikalische Hochtechnologie e.V., Winzerlaer Str. 10, D-07745 Jena

ABSTRACT: $GdBa_2Cu_3O_{\{7-x\}}$-superconducting bolometers with T_c around 90 K and very high detectivity D* were fabricated and characterized. A Silicon-On-Nitride (SON) substrate was used to create $c\text{-}Si/Si_xN_y$-membranes with very low thermal conductance G. As a result a remarkable high D* of $2\cdot10^{10}$ $cmHz^{1/2}/W$ at a wavelength of 10 μm was achieved.

Furthermore a new operational mode for bolometers was developed, the so called resistance-bias. Simulations and measurements show electrical time constants τ_{el} below 1 ms. This makes even bolometers with very high D*, which are usually slow, usable for measurements in the frequency range up to 200 Hz.

1.   INTRODUCTION

HTSC bolometers are thermal detectors in which a high-T_c superconducting film is used as an electrical resistance thermometer. The very sharp superconducting transition enables changes in temperatures to be measured with high sensitivity. Possible applications of HTSC bolometers exist in e.g. IR-spectrometers, due to their better spectral broadband high sensitivities than any sensor operating at or above liquid nitrogen temperatures (77 K).

The present work aimed at improvement of the detectivity $D* = Area^{1/2}/NEP$ by reduction of the thermal conductance G through the use of Si_xN_y-membranes. Furthermore the reduction of time constant was investigated by the use of other biasing schemes than the regular current bias.

2.   PROCESSING AND NOISE PROPERTIES

The fabrication process of our bolometers with a silicon-nitride membrane is shown in Fig. 1 and was described before by Sánchez et al. (1998). We tried to improve the noise equivalent power (NEP) of the detector device by optimizing the thermal conductance G which strongly depends on bolometer dimensions.

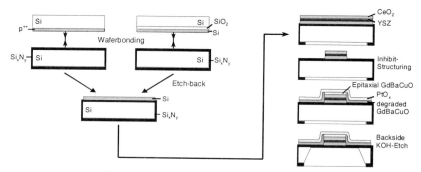

Fig. 1: Processing sequence for bolometerfabrication.

For closed membrane bolometers the thermal conductance G can be estimated as a function of thickness d and sidelength L of the membrane by solving the differential equation of heat diffusion. The resulting G for our bolometers is shown in Fig. 2.

When electrothermal feedback is taken properly into account, the noise terms can be written as:

$$NEP_{Phonon} = \sqrt{4k_B T^2 G} \quad ; \quad NEP_{Johnson} = \sqrt{\frac{4k_B T}{P} \frac{G}{\alpha}} |1 + i\omega\tau| \quad ; \quad NEP_{1/f} = \sqrt{\frac{1}{f} \frac{\gamma_H}{n_C Vol} \frac{G}{\alpha}} |1 + i\omega\tau|$$

in which α is the temperature coefficient, γ_H the Hooge parameter, n_C the number of charge carriers and Vol the volume of the superconducting film. These formulas are independent of the biasing mode. With typical values for these parameters and substituting G of Fig. 1 yields the NEP shown in Fig. 3. The minimum NEP is reached at a membrane size of 1.6x1.6 mm². The resulting detectivity D* for this bolometer structure is determined in Fig. 4.

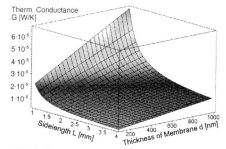

Fig. 2: Thermal conductance G as a function of the membrane layout

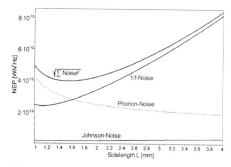

Fig. 3: Noise properties (d = 500 nm; ω = 1000 Hz)

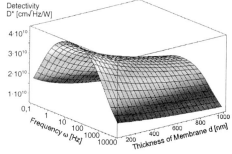

Fig. 4: Detectivity D* (L = 1.6 mm)

As a result a remarkable high D* of $2 \cdot 10^{10}$ cmHz$^{1/2}$/W at a wavelength of 10 µm was achieved with optimized dimensions by the prozessing sequence shown in Fig. 1. But the improvement of D* and therefore NEP led to an increase of time constant which has already been reported elsewere (de Nivelle et al. 1997).

3. OPERATION

3.1 Operational Modes for Bolometers

For IR-spectrometer applications very sensitive but also fast detectors are necessary. Our bolometers with high detectivity have time constants in the order of hundreds of milliseconds. Therefore, we needed to increase speed. We first tried by using voltage-bias which serves a lower electrical time constant τ_{el} (Lee 1996) (de Neville et al. 1997), as can be seen in the theoretical Response.

$$\underbrace{\frac{dV}{dP_{rad}} = \frac{V_B}{\left(G + j\omega C\right)/\alpha - P_{Bel}}}_{current-bias\ response} \quad \underbrace{\frac{dI}{dP_{rad}} = -\frac{I_B}{\left(G + j\omega C\right)/\alpha + P_{Bel}}}_{voltage - bias\ response}$$

Herein C denotes the thermal capacitance and P_{rad} and P_{Bel} are radiation and electrical bias power. Due to temperature inhomogenities this behavior is restricted to an increase in speed to approximately a factor of 5.

During the work we tried to improve the advantage of speed due to negative feedback. We applied a new operational mode for bolometers, the so called resistance-bias. With this mode the radiation power can be compensated almost completely by reducing the electrical power dissipation. Therefore, temperature and bolometer resistance will almost be unchanged by radiation power. In an ideal case thermal characteristics like conductance G and capacitance C have no influence on response. This can also be seen in the mathematical formulation where no τ_{el} can be calculated.

$$\underbrace{\frac{dI}{dP_{rad}} = -\frac{I_B}{2P_{Bel}}}_{current\ read-out} \quad \underbrace{\frac{dV}{dP_{rad}} = -\frac{V_B}{2P_{Bel}}}_{voltage\ read-out}$$

3.2 Simulation and Measurement of Different Modes

We examined possible realizations in electrical circuits by simulations with PSPICE™ Circuit Analysis Version 5.0 from MicroSim Corporation. These Simulations were made with an electrical emulation of thermal bolometer characteristics with the substitute model shown in Fig. 5.

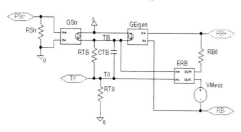

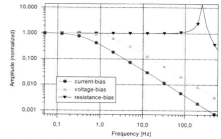

Fig. 5: Substitute model of bolometer for SPICE simulation

Fig. 6: Results of simulations of current-, voltage- and resistance-bias

As can be seen in Fig. 6 the simulations confirm the advantage of voltage- and especially resistance-bias over current-bias. With the new mode the speed of bolometers can be increased to a factor of greater than 100.

When we really built up the circuits it became obvious that voltage and resistance-bias modes had a small response that inevitably lead to high noise. Especially resistance-bias was affected. The reason can be deduced from the mathematical formulations. With increasing electrical power dissipation the response decreases. Therefore, we decided to create a kind of mixed-mode circuit. We used resistance bias for stabilizing bias-points and a mixture of current- and voltage-bias to achieve high response magnitudes. The circuit that obeys these demands is shown in Fig. 7.

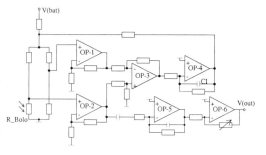

Fig. 7: Circuit for mixed-mode operation

R_Bolo symbolizes the bolometer resistor. The operational amplifiers OP1 and OP2 serve as impedance transformer, OP3 evaluates the difference and OP4 serves as an integral device with a relatively high time constant, which controls the maximum frequency where resistance-bias is dominating. OP5 and OP6 are correction devices to serve an almost constant magnitude from 1 Hz to 100 Hz.

Measurements show the advantage of using resistance-bias only for stabilization. With higher time constants realized by the integral capacitance CI connected to OP4 the signal to noise ratio can be improved in the low to mid frequency regime, as can be seen in Fig. 8.

So it can be deduced from the results with input signal of 30 nW, that radiation down to 30 pW can be measured with this bolometer configuration up to frequencies around 200Hz. That makes the bolometers we produced suitable for measurements in FTIR.

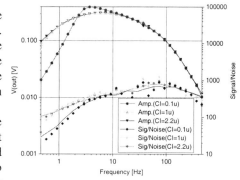

Fig. 8: Results of measurement of circuit in Fig. 7

ACKNOWLEDGEMENT

This work is supported by German Federal Ministry of Education, Science, Research and Technology under Grants 13N6956/2 and 13N6955.

REFERENCES

de Nivelle M J M E, Bruijn M P, de Vries R, Wijnbergen J J, de Korte P A J, Sánchez S, Elwenspoek M, Heidenblut T, Schwierzi B, Michalke W and Steinbeiß E 1997 J. Appl. Phys. **82**, 4719

Lee A T, Richards P, Nam S W, Cabrera B and Irwin K D 1996 Appl. Phys. Lett. **69**, 1801

Sánchez S, Elwenspoek M, Gui C, de Nivelle M J M E, de Vries R, de Korte P A J, Bruijn M P, Wijnbergen J J, Michalke W, Steinbeiß W, Heidenblut T and Schwierzi B 1998 J. Microelectromechanical Syst. **7**, 62

Inst. Phys. Conf. Ser. No 167
Paper presented at Applied Superconductivity, Spain, 14-17 September 1999
© 2000 IOP Publishing Ltd

A superconducting integrated receiver with phase-lock loop

S V Shitov[1], V P Koshelets[1], L V Filippenko[1], P N Dmitriev[1], V L Vaks[2], A M Baryshev[1, 3], W Luinge[3], N D Whyborn[3], J-R Gao[3, 4]

[1]Institute of Radio Engineering and Electronics (IREE), Russian Academy of Sciences, Mokhovaya street 11, GSP-3, 103907 Moscow, Russia
[2]Institute of Physics of Microstructure (IPM), Russian Academy of Sciences, GSP-105, 603600 Nizhny Novgorod, Russia
[3]Space Research Organization of the Netherlands (SRON), 9700 AV Groningen, the Netherlands
[4]Department of Applied Physics and Materials Science Center, University of Groningen, the Netherlands

ABSTRACT: An integrated circuit of the receiver contains a quasi-optical SIS mixer, a FFO as a local oscillator and a twin-SIS harmonic mixer for a PLL loop. The complete integrated circuit is fabricated in a single technological run from a trilayer Nb/AlO$_x$/Nb with J$_c$ = 8 kA/cm^2. A LO power of about 0.15 µW and 0.35 µW is *coupled* from FFO to the receiver and harmonic mixers respectively at about 350 GHz that is enough to pump both mixers optimally. A *rf* scheme and a layout of the chip are discussed along with numerical analysis of the first experimental data.

1. INTRODUCTION

Integration of a *free-running* flux-flow oscillator (FFO) and a SIS mixer into a single-chip quasi-optical Superconducting Integrated Receiver (SIR) is a proven technique. A receiver noise temperature below 100 K (DSB) and a good beam pattern of the *double-dipole antenna* with sidelobes below −16 dB have been reported by Shitov et al (1999) at 500 GHz. To estimate an ultimate performance of the receiver, a detailed test of the reference chip containing only the lens-antenna SIS mixer was performed by Baryshev et al (1998) that showed a superior receiver noise temperature of about 40 K at 470 GHz. A densely packed imaging array receiver for radio astronomy can be a natural niche for SIR that has been demonstrated by Shitov et al (1998). However, to be a practicable narrow-band *spectrometer*, SIR has to have a narrow linewidth of its internal local oscillator (LO). A technique to stabilize superconducting integrated oscillators was developed by Koshelets et al (1996) using room temperature electronics. Recently an externally phase-locked FFO was realized by Koshelets et al (1999a), and a linewidth as narrow as 1 Hz was demonstrated in Fiske mode for the frequency range of 230 – 440 GHz (Koshelets et al 1999b). Taking into account that Fiske steps can be separated closely and almost overlapping, complete band coverage can be achieved. In summary, it looks possible to use the phase-locked FFO as a LO for radio astronomy now already. Here we report on recent study on combining the sensitive SIS mixer and the phase-locked FFO into a quasi-optical superconducting PLL receiver at 350 GHz which is one of the most complex *integrated receiver* circuits made from superconductors (see also Kerr et al 1998).

2. SCHEMATIC OF THE CHIP AND DESIGN CONCEPT

To combine a quasi-optical SIS mixer and a phase-locked FFO in a single circuit, the new chip has been developed; its simplified scheme is presented in Fig. 1.

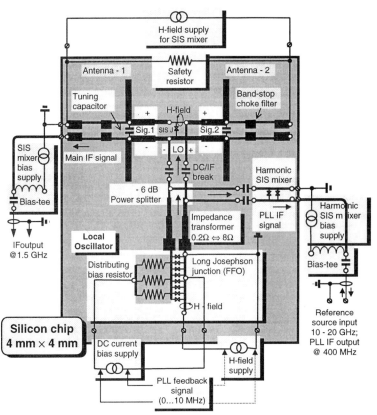

Fig. 1 Simplified equivalent diagram of external supplies and connections necessary to operate the chip receiver. The PLL electronics are not shown.

A high-frequency portion of a PLL circuit that contains a harmonic twin-SIS mixer (Koshelets et al 1996, Shitov et al 1998) and a power distributing circuit has been added to the chip.

Since one FFO feeds two mixers, an exact balance of LO power is very important. The output power from the FFO is split unequally. In order to avoid significant loss of signal towards the LO, coupling level of 25 % seems to be a reasonable value for the receiver SIS mixer (about - 6 dB). Taking into account that the receiver SIS junction and the FFO are coupled effectively to the antenna and the harmonic mixer respectively, the portion of the FFO power available at the receiver SIS junction is further reduced to about 12-15 %. This is twice as low in comparison with the previous SIR design (Koshelets, Shitov et al, 1996). To make the LO power sufficient for operation of the PLL receiver, either a more powerful FFO has to be used or the receiver SIS junction has to be smaller. Nevertheless, this two-device feed has an advantage of low VSWR that can be achieved at an output port of FFO since the harmonic mixer consumes the main portion of the LO power playing a role of the wide-band termination.

To increase the output power of FFO, one can try a wider junction. Since characteristic impedance of such a Josephson transmission line (JTL) reduces with increase of its width, the coupling circuit has to provide higher transformation ratio that may narrow the coupling bandwidth. The increase of width of FFO can, in principle, cause loss of *rf* power to a transverse mode of the JTL. This can occur, according to our estimation, above 420 GHz for a 9-micron wide junction.

A harmonic twin-SIS mixer for a superconducting PLL circuit has been demonstrated experimentally by Koshelets (1996). It was found that an efficient regime of the high-harmonic SIS mixer (up to 50th harmonic and even higher) can be realized in a Josephson mode. Since the harmonic mixer is pumped by a strong reference signal of 10 GHz, a proper isolation is necessary for both the FFO and the receiver SIS mixer. This reference signal can also leak through a common mode, which can be excited due to an essential inductance of wire bonds connecting the chip at 10 GHz.

3. CHIP LAYOUT AND EXPERIMENTAL DATA

To minimize an unavoidable reduction in the fabrication yield due to a growing complexity of the circuit, a special attention was paid to exclude the micron-size elements that were present in earlier SIRs. The narrowest strip width is about 4 micrometers and the smallest gap is about 3 micrometers while the area of SIS junctions is 1.5 μm^2 (R_N = 18 Ω) and 2 x 2 μm^2 (R_N = 6.5 Ω) for the receiver mixer and the harmonic mixer respectively.

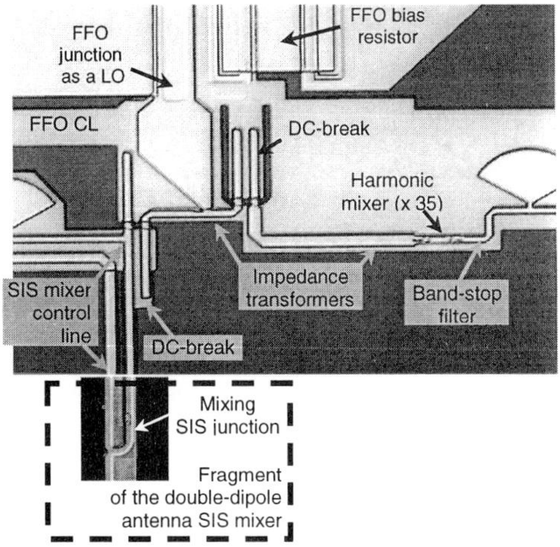

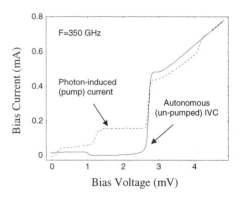

Fig. 2 Composite photo of a central part of the chip. All main elements present. The receiver mixer connection not in scale.

The most complex portion of the circuit is presented in Fig. 2. It contains FFO output terminal, a LO power splitter, a twin-SIS harmonic mixer, *dc*-breaks and a part of the SIS mixer. The output *rf* power is coupled from the FFO via a tapered high-ratio transformer. At the T-junction the LO power splits towards the two mixers being proportional to their input admittance.

A special combination of dc-breaks is used for rejection of the 10 GHz reference signal. It differs from one described by Koshelets, Shitov et al (1996). The harmonic mixer has the *dc*-break of inside type which provides an effective short at 10 GHz, while the receiver mixer uses an inside-outside *dc*-break to prevent the common mode (open at 10 GHz). The suppression factor better than –40 dB was estimated for such dc-breaks at 10 GHz. The control line of the receiving SIS mixer is folded to reduce magnetic field interference to the FFO. All *dc*-connections to both mixers contain *rf*-filters to prevent leak of the signal at 350 GHz. The filters are optimized for coupling of IF port at 1 - 4 GHz and of PLL IF port at 400 – 1000 MHz.

To be sure that at least one Fiske step is close to 350 GHz and the voltage separation of neighboring steps is small enough for continuous coverage using a 4 GHz IF amplifier by picking different side-bands, the length of FFO is varied in the range of 600 ± 50 micrometers. To check a possibility of improvement in LO power, three different widths of FFO were chosen: 4, 6 and 9 micron. The power available from experimental FFO is clearly related to the width of the junction. It was estimated as 0.45 µW in summary for the 9-micron FFO that is about twice more than we got from 4 micron FFO at 500 GHz (see Fig. 3).

To make careful analysis of a circuit under the test, the ratio of the pump levels for the two mixers were used. This ratio depends directly on parameters of passive elements of the *rf* circuit, but not on the FFO power (see Fig. 4). The Fourier transform spectrum presented in Fig. 4 confirms an accurate tuning of the receiver mixer. The power ratio at 350 GHz is about 3.5-

Fig. 3 Experimental IV-characteristics (IVC) of the twin-SIS *harmonic mixer*: autonomous (solid) and pumped (dash). The harmonic mixer plays the role of a matched termination for the LO (FFO). This mixer consumes about 75-80% of the available LO power.

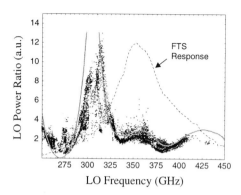

Fig. 4 Experimental data (dots) and calculated best fit (solid) on the power split between receiver mixer (RM) and harmonic mixer (HM), P_{HM}/P_{RM}. The experimental values of the pump power for each mixer P_{HM} and P_{RM} were estimated, being about proportional to a photon-induced current in the mixers. Experimental data on FTS response characterize the instantaneous *rf* bandwidth of the receiver (dash).

4 that approximately corresponds to the design value of –6 dB (See Fig. 1). The asymmetry of the split figure is, according to our numerical simulation, caused by tolerance in the fabrication process and it has a minor effect on both sensitivity and bandwidth of the receiver.

4. CONCLUSION

The concept of a superconducting integrated receiver at 350 GHz with phase-lock loop is developed on the basis of recent progress on quasi-optical SIS mixers and phase-locked FFO; it has passed extensive numerical simulation and is implemented in experimental devices. To our knowledge this is one of the most complicated *integrated receiver* circuits for sub-mm range made from superconductors. A few tests of experimental samples demonstrated reasonable yield: the circuit parameters are close to the design values including accurate tuning of the receiving SIS junction. The LO power obtained from FFO at about 350 GHz is sufficient to pump both the receiver and the harmonic mixers that is a key issue to obtain phase-locked regime of the FFO. The detailed experimental data on performance of the PLL circuit are expected soon. The spectral line measurement is planned.

5. ACKNOWLEDGMENTS

Authors thank Th. de Graauw for his initiation and promotion of the study and V. D. Nguyen for precise mounting of chips. The work was supported in parts by the Russian Program for Basic Research, the Russian SSP "Superconductivity" and ESA TRP contract No. 11653/95/NL/PB/SC.

REFERENCES

Baryshev A M, Luinge W, Koshelets V P, Shitov S V, Klapwijk T M, 1998 unpublished

Kerr A R, Pan S-K, Leduc H G, An Integrated Sideband Separating SIS Mixer for 200-280 GHz, 1998 *Proc. 9th Int. Symp. Space Terahertz Tech.* **9**

Koshelets V P, Shitov S V, Shchukin A V, Filippenko L V, and Mygind J., 1996 *J. Appl. Phys. Lett.* 69 (5) 699

Koshelets V P, Shitov S V, Shchukin A V, Filippenko L V, Dmitriev P N, Vaks V L, Mygind J, Baryshev A M, Luinge W, Golstein H, 1999a *IEEE Trans. on Appl. Supercond.* 9, 4133

Koshelets V P, Baryshev A M, Mygind J, Vaks V L, Shitov S V, Filippenko L V, Dmitriev P N, Luinge W, Whyborn N, 1999b this conference

Shitov S V, Ermakov A B, Filippenko L V, Koshelets V P, Luinge W, Baryshev A M, Gao J-R, Lehikoinen P, 1998 *Proc. 9th Int. Symp. Space Terahertz Tech.* **9** 263

Shitov S V, Koshelets V P, Ermakov A B, Filippenko L V, Baryshev A M, Luinge W, Gao J-R, 1999 *IEEE Trans. on Appl. Supercond.* 9 3773

Inst. Phys. Conf. Ser. No 167
Paper presented at Applied Superconductivity, Spain, 14-17 September 1999

Numerical simulation based on a five-port model of the parallel SIS junction array mixer

M-H Chung and M Salez

DEMIRM, Observatoire de Paris, 61, avenue de l'Observatoire, 75014 Paris, FRANCE

ABSTRACT: We have calculated the harmonic performance of a parallel SIS junction array mixer using a five-port model and Tucker's quantum theory of mixing to investigate and predict its properties. To do this, we perform a large signal analysis in the frequency domain by using the voltage update method to determine the local oscillator voltage waveform across the SIS junction array. Preliminary results of simulations of mixer gain are presented. A comparison of these with simulations of the same array-mixer in the three-port approximation is also reported.

1.   INTRODUCTION

Superconducting-Insulator-Superconducting (SIS) junction mixers have been so far the most sensitive heterodyne detectors in the millimeter and submillimeter wavelength range (Tucker et al 1985). Most mixers use end-loaded single-junction(Blundell et al 1995) or twin-parallel junction(Zmuidzinas et al 1994) designs, where the inductive circuit tuning out the geometric capacitance is a superconductive microstrip line. Recently, a new kind of SIS junction mixer was proposed by Shi et al (1997) to increase the coupling bandwidth, in which a larger number (N>2) of parallel-connected junctions act almost like a bandpass filter. The very encouraging experimental results reported by the same group (Shi et al 1998) partly motivated our interest in this simulation work, in order to investigate the properties of this type of mixers for broadband applications. The theory of quantum mixing developed by Tucker et al(1985) is a powerful tool for analyzing the properties of SIS mixers. However, the validity of the often-used three-port approximation strongly depends on the capacitance of the SIS junctions. Generally it is assumed that the three-port approximation is justified by the large capacitance, short-circuiting the tunneling currents at harmonic frequencies. In this work, we analyze the parallel SIS junction array mixer with a five-port model and compare the results with the three-port approximation. For sake of simplicity, we only present here simulations with N=3.

2.   FORMULATION OF NUMERICAL ANALYSIS

2.1   Large Signal Analysis

To determine the LO voltages across the SIS junctions of the mixer array by using a harmonic balance method, it is necessary to calculate first the quasi-particle currents in the spectral domain flowing through the junctions with the given junction voltages. Withington et al (1989) developed an efficient method to determine the Fourier components of the quasi-particle current in a junction by comparing the Fourier series of this current with Werthamer's relationship. The Fourier series of the voltage and current can be expressed as :

$$V(t) = V_0 + \mathrm{Re}(\sum_{n=1}^{\infty} V_n e^{jn\omega_{LO}t}) \quad \text{and} \quad I(t) = I_0 + \mathrm{Re}(\sum_{n=1}^{\infty} I_n e^{jn\omega_{LO}t}) \tag{1}$$

where V_0 and I_0 are respectively the dc voltage and dc current and ω_{LO} the angular frequency of the local oscillator. The complex I-V characteristic of an SIS junction is $I(V) = I_{kk}(V) + jI_{dc}(V)$ where $I_{dc}(V)$ is the measured dc current and $I_{kk}(V)$ its Kramers-Kronig transform, which is defined as

$$I_{kk}(V) = \frac{1}{\pi} P \int_{-\infty}^{\infty} \frac{I_{dc}(V') - V' / R_N}{V' - V} dV' \tag{2}$$

where P denotes the Cauchy principal value and R_N is the normal state resistance. The quasi-particle current in the spectral domain can then be calculated using:

$$I_n = -j \sum_{k=-\infty}^{\infty} \left[C_k C_{k-n}^* I(V_0 + \frac{kh\omega_{LO}}{e}) - C_k^* C_{k+n} I^*(V_0 + \frac{kh\omega_{LO}}{e}) \right] \tag{3}$$

The complex phase factor, C_k for five-port model is defined as $C_k = \sum_{m=-\infty}^{\infty} A_{(k-2m)1} A_{m2}$ where $A_{mn} = J_m(\alpha_n)e^{-jm\phi_n}$, $\alpha_n = e|V_n| / \hbar\omega_{LO}$ and ϕ_n are respectively the normalized amplitude and the argument of the complex voltage phasor V_n. Hence, if we assume the junction voltage V_n at the fundamental frequency and first harmonic, the quasi-particle current of the junction can be fully derived in the spectral domain (Tong et al 1990).

We now apply the method of Withington et al (1989) to a parallel SIS junction array. In the following, we will limit ourselves to a moderate number of junctions (N=3) although the method is applicable to any number. The circuit model of the large signal analysis is shown in Fig. 1. V_S^n and $Z_S(\omega)$ are LO source voltage and impedance respectively. Each portion of superconducting microstrip line between two adjacent junctions i and $i+1$ is represented by an inductive impedance $Z_{i,i+1}(\omega)$ which can easily be calculated (Chang 1979). The SIS junction labeled i is modeled by its non-linear quasiparticle current conductance $Y_{qp,i}$ and by the susceptance $j\omega C_i$, where C_i is its geometric capacitance (as often done in similar problems, the Josephson tunneling current is here omitted: although this assumption is reasonable for single-junction mixers where a magnetic field can easily suppress the dc Cooper pair current, this might not be as plain in a parallel array mixer due to possible quantized flux in the transmission lines).

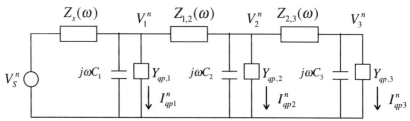

Fig. 1 Circuit model of the 3 parallel SIS junction array for large signal analysis.

V_i^n and I_{qpi}^n are the amplitudes of the Fourier components of the voltage and quasi-particle current as defined above, in the SIS junction i at frequency $n\omega_{LO}(n=1,2)$. When starting the voltage update method, one needs to know the initial voltages across the junctions. By replacing the nonlinear quasi-particle current admittance by the normal state conductance, the voltages of the different junctions can be easily calculated using linear circuit theory. With these roughly estimated voltage values used as the initial voltages for the voltage update method, we can calculate the quasi-particle currents in the spectral domain using the method of Withington et al (1989). The next step is to apply the linear circuit theory to obtain the new junction voltage solutions, different from the previous values. One must then repeat this iteration and each time replace the previous set of junction voltages by the newly computed one, until the desired convergence criterion is reached (Hicks et al 1982).

2.2   Small Signal Analysis

Once the LO voltages across the SIS junctions of the array are determined, one can easily calculate the conversion efficiency of the SIS parallel array mixer. In the five-port model assumption, a 5×5 admittance matrix, of which the elements are given in (35) of Withington et al (1989), represents the linear relationships between the small signal sideband voltages and currents across any lumped SIS junction mixer. Let us call Y_{Ci} such admittance matrix for the ith junction of the array. Due to the non-zero phase of the LO voltages, the arguments of the admittance matrices are modified (Zmuidzinas et al 1994). Using these five-port admittance matrices, the single equivalent five-port matrix representing the whole array can easily be obtained by linear circuit theory. Fig. 2 shows the circuit model in the particular case of the 3-junction array mixer considered here. i_S^n and $Y_S(\omega)$ are small signal source current and admittance respectively.

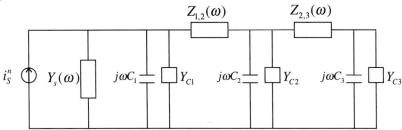

Fig. 2 Model of the parallel SIS junction array mixer for small signal analysis; N=3.

3.   PRELIMINARY SIMULATION RESULTS

For all simulations, we used the measured dc current-voltage characteristic of a typical Nb/AlOx/Nb SIS junction fabricated in our laboratory. The gap voltage is 2.75 mV and the normal resistance is around $12\,\Omega$. The measured dc current of the junction, $I_{dc}(V)$, and its Kramers-Kronig transform, $I_{kk}(V)$, are shown in Fig. 3. The following input parameters are used : $\omega C R_N = 2$ and 6, $\omega L / R_N = 0.5$ and 0.17 at $\omega / 2\pi = 300$ GHz. The values chosen for the normalized inductance correspond to a resonance at this frequency. Intermediate Frequency (IF) is 4 GHz and the IF impedance is $50\,\Omega$. The RF source admittance, at the fundamental and first harmonic frequencies, is set to be equal to the normal conductance of the SIS junction.

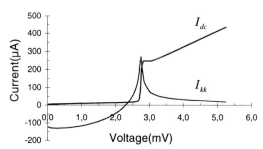

Fig. 3 Measured dc I-V characteristic of a physical Nb/AlOx/Nb junction and its Kramers-Kronig transform.

Fig. 4 shows the calculated maximum single sideband conversion gain using both the 3 port model and 5 port model by varying LO voltage amplitude. The bias voltage V_0 is kept constant, at half of the first photon step below the gap voltage for the calculation.

As expected, all curves seem to show the designed optimal coupling near 300 GHz as a conversion gain maximum. We note a strong qualitative difference in the shape of this maximum for the five-model calculations. Furthermore, while the two curves for $\omega CR_N = 2$ and 6 in the three-port model clearly converge for frequencies higher than the resonance frequency, the five-port simulation results do not follow this tendency. We cannot confirm at this time whether this disparity arises from the harmonic effect.

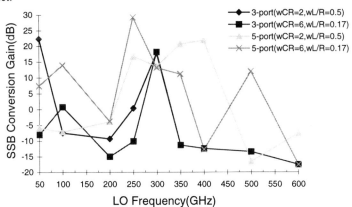

Fig. 4 Calculated maximum SSB conversion gain of the 3 parallel SIS junction array mixer.

4. CONCLUSION

We have performed a mixer analysis of the parallel-connected SIS junction array mixer proposed by Shi et al(1997), based on a five-port model using the voltage update method (Withington et al 1989). An example of simulation results is provided for three SIS junctions, yet this method can be extended to any number of junctions. A preliminary comparison with the three-port model analysis clearly shows discrepancies in the maximum single sideband conversion gain. More analysis work currently under progress, using various numbers of junctions and various geometrical and technological parameters, will be reported on later.

REFERENCES

Blundell R, Tong C-Y E, Barrett J W, Kawamura J, Leombruno R L, Paine S, Papa D C, Zhang X, Stern J A, LeDuc H G and Bumble B 1995 Proc. Sixth International Symposium on Space Terahertz Tech. 123
Chang W M 1979 J.Appl. Phys. vol. **50**, 8129
Hicks R G and Khan P J 1982 IEEE Trans. Microwave Theory Tech. vol. **30**, 251
Shi S C, Noguchi T and Inatani J 1997 Proc. Eighth International Symposium on Space Terahertz Tech. 81
Shi S C, Noguchi T, Inatani J, Irimjiri Y and Saito T 1998 Proc. Ninth International Symposium on Space Terahertz Tech. 223
Tong C-Y E and Blundell R 1990 IEEE Trans. Microwave Theory Tech. vol. **38**, 1391
Tucker J R and Feldman M J 1985 Rev. Mod. Phys. vol. **57**, 1055
Withington S and Kollberg E L 1989 IEEE Trans. Microwave Theory Tech. vol. **37**, 231
Zmuidzinas J, LeDuc H G, Stern J and Cypher S R 1994 IEEE Trans. Microwave Theory Tech. vol. **42**, 698

Inst. Phys. Conf. Ser. No 167
Paper presented at Applied Superconductivity, Spain, 14-17 September 1999
© 2000 IOP Publishing Ltd

Surface resistance of NbN and NbC$_x$N$_{1-x}$ films in the frequency range of 0.5-1.5THz

S Kohjiro and A Shoji

Electrotechnical Laboratory, 1-1-4 Umezono, Tsukuba, Ibaraki 305-8568, Japan

ABSTRACT: Surface resistance R_S of NbN and NbC$_x$N$_{1-x}$ films has been evaluated at 4.2K in the frequency range of 0.5-1.5THz using the Fiske resonant modes in NbC$_x$N$_{1-x}$ (NbN)/MgO/NbC$_x$N$_{1-x}$(NbN) Josephson junctions. At 1THz, typical R_S values of NbN and NbC$_x$N$_{1-x}$ films are 100 and 50mΩ, respectively. These values are lower than theoretical R_S of high-quality Cu and Nb. It indicates the superiority of NbN and NbC$_x$N$_{1-x}$ as an electrode material for SIS mixers and Josephson local oscillators operating above 0.7THz.

1. INTRODUCTION

Superconducting heterodyne receivers consisting of low-noise SIS (Superconductor/Insulator/Superconductor) mixers and tunable Josephson local oscillators are a great candidate for radio astronomy and global monitoring of atmosphere pollution. Imaging array receivers of 9 pixels at 0.5THz are now under the development based on Nb/AlO$_x$/Nb junctions and Nb/SiO$_x$/Nb strip transmission line for rf signal (Shitov *et al* 1999, Koshelets *et al* 1999). The upper frequency limit of such receivers is determined by the gap frequency f_g of Nb, say 0.7THz. To push the operating frequency above 0.7THz, one requires a low-loss normal metal such as Al or, preferably, a superconductor with f_g>0.7THz. Recently, NbTiN was proposed as a candidate of the latter (Kooi *et al* 1998). Properties of 14K<T_C<16K(f_g≈1.2THz) and ρ<70$\mu\Omega$cm were reported for NbTiN films grown on several kinds of substrate materials without intentional heating (Stern *et al* 1998), where T_C is the critical temperature and ρ is the normal state resistivity at a temperature just above T_C. It was also shown SIS mixers consisting of NbTiN/SiO$_x$/NbTiN tuning circuits revealed the double-side-band noise temperature of 520K at 0.83THz (Kawamura *et al* 1999) and 830K at 0.93THz (Jackson *et al* 1999). However, NbTiN-based SIS receivers with f_g≈1.2THz are not suitable for low-noise detection of emissions from some important molecules such as HF (1.23THz). In addition, it is difficult to find the optimum composition of NbTiN (Bumble 1998). By contrast, f_g>1.4THz and ρ<60$\mu\Omega$cm were obtained for NbN and NbC$_x$N$_{1-x}$ films grown epitaxially on (100)MgO substrates (Shoji *et al* 1992a, b, Kohjiro *et al* 1993). Noise temperature of an SIS mixer consisting of NbN/AlN/NbN junctions and a NbN/SiO/Al tuning circuit was as low as 457K at 0.78THz (Uzawa *et al* 1998) . Furthermore, optimization of film-growth is easier since the film quality depends weakly on the concentration of the carbon. It indicates epitaxial NbN or NbC$_x$N$_{1-x}$ has an enormous potential for electrode material of low-noise receivers operating up to 1.5THz. For their further study, surface resistance R_S above 0.7THz should be measured.

In this paper, we have evaluated R_S of NbN and NbC$_x$N$_{1-x}$ films in the frequency range of 0.5-1.5THz using the Fiske resonant modes (Kulik 1967) in NbC$_x$N$_{1-x}$(NbN)/MgO/NbC$_x$N$_{1-x}$ (NbN) junctions.

2. FABRICATION OF JUNCTIONS

NbC$_x$N$_{1-x}$(NbN)/MgO/NbC$_x$N$_{1-x}$(NbN) junctions were fabricated as follows: junction sandwiches were deposited in an rf-magnetron sputtering system on (100)MgO substrates with a dimension of 20 x 20 x 0.5mm^3. In the sputtering system, we used a 6-inch-diam Nb target with a purity of 99.9% and rf power density of 3.3W/cm^2 in the mixture of 91-94 vol.% Ar,

Table 1. Parameters of measured junctions. DC properties of NbN and NbC$_x$N$_{1-x}$ electrodes are those of films deposited on (100)MgO substrates heated up to 250-300°C.

Sample	J_c [kA/cm^2]	V_g [mV]	Base	T_c [K]	ρ [$\mu\Omega$cm]	λ [nm]	Counter	T_c [K]	ρ [$\mu\Omega$cm]	λ [nm]
A	0.42	6.0	NbN	17.2	20	110	NbN	17.2	20	110
B	1.3	6.0	NbN	17.2	20	110	NbCN	16.7	60	180
C	3.2	5.9	NbCN	16.7	60	180	NbN	17.2	53	170
D	1.9	6.3	NbCN	16.7	60	180	NbCN	16.7	60	180
E	2.1	6.0	NbCN	16.7	60	180	NbCN	16.7	70	200
F	2.5	6.5	NbCN	16.7	60	180	NbCN	16.7	60	180
G	2.7	6.2	NbCN	16.7	60	180	NbCN	16.7	70	200

6-9 vol.% N$_2$ and 0-1 vol.% C$_2$H$_2$. The substrate was heated up to 250-300°C during the deposition of both electrodes, while it was cooled down to less than 100°C during the barrier formation. Table 1 shows dc properties of prepared NbN and NbC$_x$N$_{1-x}$ films as well as junction parameters of gap voltage V_g and critical current density J_C. In Table 1, magnetic penetration depth λ was estimated from T_C and ρ (Orlando *et al* 1979, Shoji *et al* 1992a). Typical thickness of base electrode, tunnel barrier, and counter electrode was 250nm, 1nm and 250nm, respectively. After the sandwich formation, individual junctions were patterned using a contact aligner and a reactive ion etching (RIE) technique. Junction area was 35 x 20μm^2. After patterning base electrodes, 3nm-thick MgO and 450nm-thick SiO$_2$ films were deposited on a whole substrate for an etching stopper and an insulation layer, respectively. Contact holes on junctions and those on bonding pads were formed by RIE in a mixture of 68 vol.% CF$_4$ and 32 vol.% H$_2$, followed by a removal of the MgO etching stopper by Ar ion-beam-sputtering. Finally, the sample was completed by depositing a 700nm-thick Nb wiring layer, patterning, and etching in a CF$_4$ plasma.

3. MEASUREMENT

To evaluate the frequency dependence of R_S, the following parameters of the fabricated junctions are required (Broom and Wolf 1977, Kohjiro *et al* 1993): the current height of n-th Fiske step as a function of its voltage $I_S(V_n)$, the voltage of the first Fiske step V_1, critical current of the junction I_C and the dynamic resistance due to the quasiparticle tunneling R_d. Here, the angular frequency ω is related to V_n as $\omega=2\pi V_n/\Phi_0$ due to the ac Josephson effect, where Φ_0 is the flux quantum. To obtain these parameters, I-V characteristics of junctions at 4.2K were measured by conventional point probe method. The magnetic field $B_e(V_n)$ ($0<B_e(V_n)<10$mT) was applied perpendicular to the length of junctions ($L=35\mu$m) with a cooled solenoid coil. The magnetic field is adjusted so that the height of each Fiske step $I_S(V_n)$ becomes maximum, i.e., $B_e(V_n)\sim V_n/(dc_j)$, where $d=\lambda_1+\lambda_2+t_i$ is the effective penetration depth in the junction, λ_1 and λ_2 the magnetic penetration depth of the base and counter electrodes, respectively; t_i the barrier thickness of the junction; and c_j the phase velocity of electromagnetic wave in the junction. I_C was estimated from V_g and the normal resistance R_n of junctions as $I_C\approx0.6V_g/R_n$ which is valid for NbC$_x$N$_{1-x}$/MgO/NbC$_x$N$_{1-x}$ junctions (Shoji *et al* 1992b). V_1 was calculated from d and the specific capacitance of the junction $C_s=8\mu$F/cm^2 (Kohjiro *et al* 1993) as $V_1=\Phi_0/(2L\sqrt{\mu_0 dC_s})$, where μ_0 is the permeability of free space.

4. RESULTS AND DISCUSSION

Figure 1(a) and 1(b) show I-V characteristics of a NbN/MgO/NbC$_x$N$_{1-x}$ ($J_C=1.3$kA/cm^2 and $V_g=6.0$mV) and a NbC$_x$N$_{1-x}$/MgO/NbC$_x$N$_{1-x}$ ($J_C=2.5$kA/cm^2 and $V_g=6.5$mV) junctions, respectively, under several values of the magnetic field of $1.2<B_e(V_n)<3.7$mT. Fiske steps are clearly observed due to several orders of resonance in the junction. When the bias voltage of junctions increases from zero, the step height increases rapidly, reaches the maximum value, and decreases gradually to zero at $\approx V_g/2$. A difference of observed steps between Fig. 1(a) and (b) is as follows: in Fig. 1(a), $I_S(V_n)$ decreases rapidly with increase of V_n for $V_n>2$mV and $I_S(V_n)\approx0$ for $V_n>2.8$mV, while $I_S(V_n)$ decreases gradually for $V_n>2$mV and can be clearly seen at $V_n=3.3$mV in Fig. 1(b). It suggests the following: first, the electrode loss of NbC$_x$N$_{1-x}$ is less than that of NbN above 1THz. Next, the NbC$_x$N$_{1-x}$/MgO/NbC$_x$N$_{1-x}$ junction acts as a low-loss cavity for electromagnetic wave up to 1.6THz.

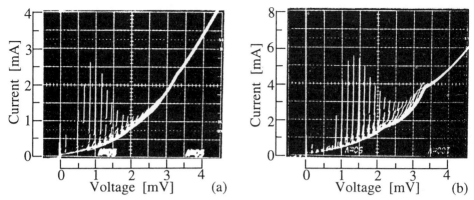

Fig. 1. Fiske steps in I-V characteristics of (a) a NbN/MgO/NbC$_x$N$_{1-x}$ and (b) a NbC$_x$N$_{1-x}$/MgO/NbC$_x$N$_{1-x}$ junctions under the magnetic field of 1.2-3.7mT.

Figure 2(a) shows experimental and theoretical frequency dependence of the magnetic penetration depth $\lambda(\omega)$. Here $\lambda(0)$ is the penetration depth in the low frequency limit. Experimental data of samples A(NbN) and F(NbC$_x$N$_{1-x}$) was calculated as $\lambda(\omega)/\lambda(0)=(nV_1/V_n)^2$, which is valid when base and counter electrodes have the same value of $\lambda(\omega)$ and when C_s is independent of frequency. Theoretical $\lambda(\omega)$ for f_g=1.5THz was derived from the frequency-dependent phase velocity of a superconducting stripline at a reduced temperature of T/T_c=0.11 and 0.46 (Kautz 1978). As shown in Fig. 2(a), $\lambda(\omega)/\lambda(0)$ of NbC$_x$N$_{1-x}$ is slightly larger than theoretical value above 1THz, while, in the wide frequency region, $\lambda(\omega)/\lambda(0)$ of NbN deviates considerably from theoretical one. At 1THz, the deviation from the theoretical value is as large as 10% of $\lambda(0)$. This frequency-dependent penetration depth may be attributed to the imperfection of films, resulting in R_S values larger than theoretical ones (Kohjiro et al 1993).

Figure 2(b) shows the frequency dependence of the surface resistance R_S of NbN and NbC$_x$N$_{1-x}$ electrodes calculated from the measurement of samples A-G. As shown in Fig. 2(b), R_S of NbC$_x$N$_{1-x}$ is about half of R_S of NbN. At 1THz, 40mΩ<R_S<60mΩ for NbC$_x$N$_{1-x}$ and 80mΩ<R_S<120mΩ for NbN. It may correspond to the difference of frequency-dependent penetration depth $\lambda(\omega)$ between NbN and NbC$_x$N$_{1-x}$ films shown in Fig. 2(a). The origin is not clear why R_S of NbC$_x$N$_{1-x}$ film is lower than R_S of NbN, though NbC$_x$N$_{1-x}$ has lower T_C and higher ρ than NbN. It may be attributed to a result of improved structural ordering due to the incorporation of carbon into the film (Cukauskas 1983).

For the comparison, theoretical (de Lange et al 1995) and experimental R_S of Nb and theoretical R_S of high-quality Cu (de Lange et al 1995) are also shown. Experimental data of Nb was obtained from Fiske steps in a Nb/AlO$_x$/Nb junction with area of 40 x 3μm^2, where electrical

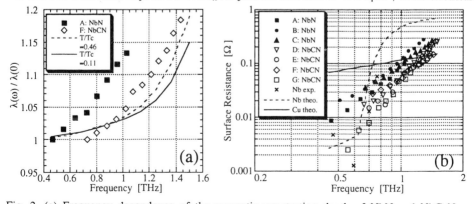

Fig. 2. (a) Frequency dependence of the magnetic penetration depth of NbN and NbC$_x$N$_{1-x}$ electrodes. Solid and dashed lines are theoretical ones for f_g=1.5THz. (b) Frequency dependence of the surface resistance of NbN (closed marks), NbC$_x$N$_{1-x}$ (open marks), Nb (cross: experiment, dashed line: theory) and theoretical Cu (solid line).

properties of both Nb electrodes were T_C=9.0K and ρ=5.3$\mu\Omega$cm. Experimental R_S of Nb film agrees reasonably with its theoretical R_S. The agreement of experimental R_S with theoretical one for Nb films was also reported (Javadi et al 1992). R_S of Nb films is less than 40mΩ below 0.7THz, while it increases rapidly with the frequency above 0.7THz. It is due to the gap frequency of Nb films. As shown in Fig. 2(b), both NbN and NbC$_x$N$_{1-x}$ have much lower R_S than Nb above 0.8THz. The theoretical R_S of Cu was calculated by assuming the mean free path of the electron of 19μm and resistivity of 3.3nΩcm, both of which are corresponding to the ideal bulk. R_S of Cu is proportional to $\omega^{2/3}$ and less than R_S of Nb above 0.8THz. R_S of NbN and NbC$_x$N$_{1-x}$ is proportional to ω^k, where theoretical value is k=2 but 3<k<4 is experimentally observed above 0.7THz. The discrepancy may be due to the imperfection of films and may correspond to the frequency-dependent penetration depth shown in Fig. 2(a). In spite of non-ideal surface resistance, R_S of NbN and NbC$_x$N$_{1-x}$ is lower than R_S of high-quality Cu below 1.0 and 1.3THz, respectively.

5. CONCLUSION

We have measured R_S of NbN and NbC$_x$N$_{1-x}$ films using Fiske resonant modes in NbC$_x$N$_{1-x}$(NbN)/MgO/NbC$_x$N$_{1-x}$(NbN) Josephson junctions. The junctions fabricated on (100)MgO substrates showed 5.9mV<V_g<6.5mV and steep Fiske steps up to 2.8-3.3mV. Above 0.7THz, R_S depends on frequency as ω^k with 3<k<4. At 1THz, we have demonstrated R_S of NbN and NbC$_x$N$_{1-x}$ films are 80mΩ<R_S<120mΩ and 40mΩ<R_S<60mΩ, respectively. These values are lower than theoretical R_S of high-quality Cu and Nb. It indicates NbN and NbC$_x$N$_{1-x}$ are suitable for an electrode material of superconducting heterodyne receivers operating above 0.7THz.

ACKNOWLEDGMENTS

The authors are grateful to M Watanabe and S Segawa for technical assistance of sample preparation. They would like to acknowledge H Yamamori for the measurement system of I-V characteristics and M Maezawa for preparation of Nb/AlO$_x$/Nb junctions. They thank T Sakuraba, F Hiarayama and A Fukushima for management of liquid helium. T Sakamoto is appreciated for continuous support and encouragement.

REFERENCES

Broom R F and Wolf P, 1977 *Phys. Rev. B* **16**, 3100
Bumble B, 1998 private communication
Cukauskas E J, 1983 *J. Appl. Phys.* **54**, 1013
de Lange G, Kuipers J J, Klapwijk T M, Panhuyzen R A, van de Stadt H and de Graauw M W M, 1995 *J. Appl. Phys.* **77**, 1795
Jackson B D, Iosad N N, Leone B, Gao J R, Klapwijk T M, Laauwen W M, de Lange G and van de Stadt H, 1999 *Proc. 10th Int. Sym. Space&THz Tech.* pp 144-56
Javadi H H, McGrath W R, Bumble B and LeDuc H G, 1992 *Appl. Phys. Lett.* **61**, 2712
Kautz R L, 1978 *J. Appl. Phys.* **49**, 308
Kawamura J, Miller D, Chen J, Kooi J W, Zmuidzinas J, Bumble B, LeDuc H G and Stern J A, 1999 *Proc. 10th Int. Sym. Space&THz Tech.* pp 398-404
Kohjiro S, Kiryu S and Shoji A, 1993 *IEEE Trans. on Appl. Supercond.* **3**, 1765
Kooi J W, Stern J A, Chattopadhyay G, LeDuc H G, Bumble B and Zmuidzinas J, 1998 *Int. J. IR and MM Waves* **19** 373; 1998 *Proc. 9th Int. Sym. Space&THz Tech.* pp 283-94
Koshelets V P, Shitov S V, Shchukin A V, Filippenko L V, Dmitriev P N, Vaks V L, Mygind J, Baryshev A B, Luinge W, and Golstein H, 1999 *IEEE Trans. on Appl. Supercond.* **9**, 4133
Kulik I O, 1967 *Sov. Phys.-Tech. Phys.* **12**, 111
Orlando T P, McNiff E J Jr, Forner S and Beasley M R, 1979 *Phys. Rev. B* **19**, 4545
Shitov S V, Koshelets V P, Ermakov A B, Filippenko L V, Baryshev A B, Luinge W and Gao J R, 1999 *IEEE Trans. on Appl. Supercond.* **9**, 3773
Shoji A, Kiryu S and Kohjiro S, 1992a *Appl. Phys. Lett.* **60**, 1624
Shoji A, Kiryu S, Kashiwaya S, Kohjiro S, Kosaka S and Koyanagi M 1992b *Superconducting Devices and Their Applications - Springer Proceedings in Physics* **64** (Berlin: Springer-Verlag) pp 208-13
Stern J A, Bumble B, LeDuc H G, Kooi J W and Zmuidzinas J, 1998 *Proc. 9th Int. Sym. Space&THz Tech.* pp 305-13
Uzawa Y, Wang Z and Kawakami A, 1998 *Appl. Phys. Lett.* **73**, 680

Inst. Phys. Conf. Ser. No 167
Paper presented at Applied Superconductivity, Spain, 14-17 September 1999
© 2000 IOP Publishing Ltd

Development of Compact High-resolution X-ray Detector System using STJ and a SQUID Amplifier

T Ikeda[a], H Sato[a], H Kato[a], K Kawai[a], H Miyasaka[a], T Oku[a], W Ootani[a], C Otani[a], H M Shimizu[a], Y Takizawa[a], H Watanabe[a], H Nakagawa[b], H Akoh[b], M Aoyagi[b] and T Taino[c]

[a] RIKEN (The Institute of Physical and Chemical Research), 2-1 Hirosawa, Wako, Saitama 351-0198, Japan
[b] Electrotechnical Laboratory (ETL), 1-1-4 Umezono, Tsukuba, Ibaraki 305-8568, Japan
[c] Department of Applied Quantum Physics and Nuclear Engineering, Kyushu University, Hakozaki, Fukuoka 812-8581, Japan

ABSTRACT: A compact system for high-resolution X-ray detection, incorporating STJ and a SQUID amplifier, was developed. STJs and the SQUID amplifier with a specially designed magnetic shield were installed in an OFC container of $\phi 90 \times 70$ mm^3 capacity. The magnetic shield for the SQUID amplifier worked well when the external field for STJ operation was applied. Cooling down to 4.2K for SQUID and to 0.3K for STJ was successfully achieved. The time response for the system was achieved up to 2 MHz, the SQUID amplifier itself dominating the response.

1. INTRODUCTION

The usual semiconductor-based X-ray detectors are well-established and achieve an energy resolution of about 120 eV at 6 keV. It is nearly theoretical limit. However, high energy-resolution detectors for infrared, visible, ultra-violet and X-ray regions are now required. Our goal in the development of an X-ray detector is to obtain an energy resolution of < 30 eV at 6 keV, a count rate of more than 10,000 cps per STJ pixel and a response of more than 10 MHz. Compactness and a general-purpose style are also necessary in the case of mounting on a beam line for particle physics or on a telescope for astronomy. The detector system employs a superconducting tunnel junction (STJ) and a superconducting quantum interference device (SQUID) amplifier for X-ray detection and for its readout, respectively. The merits of the SQUID amplifier are as follows: (1) Current-sensitive amplifier suitable for current output from STJ. (2) High-speed: STJ generates a rapid pulse by X-ray. (3) Low-temperature applications: In the case of STJ experiment, a low-temperature stage will soon be available in a cryostat. (4) Small: The amplifier is supplied as a small chip of 5 mm square. This means that an integrated system including STJ can be easily established. In this study, we show the practical use of the superconducting readout devices for high-accuracy radiation detection. The STJ used here was fabricated at RIKEN and operated at 0.35K. It has a pentalayer of Nb(200nm)/Al(50)/AlO$_x$/Al(50)/Nb(150). Its dynamic resistance and capacitance for the size of 100 μm square are 1 kΩ and about 400 pF, respectively. The SQUID amplifier fabricated by HYPRES Inc. is a two-stage type and consists of an input coil whose inductance of 0.3 μH and a series array of 200 SQUIDs. An on-chip heater is available for detrapping the magnetic flux.

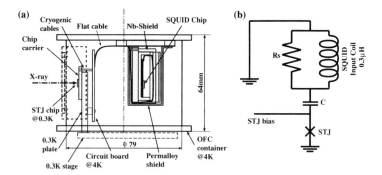

Fig. 1: (a) Configuration of the X-ray detector system; (b) Electrical circuit between an STJ and the SQUID amplifier

The X-ray signal from an STJ is amplified by the SQUID amplifier at 4K, and then amplified again by a 100-gain room temperature module. The combination of the SQUID device and the 100-gain amplifier has responses of 2 MHz and a rise time of 0.4 μsec, respectively. In this paper, the setup and configuration of each part are described. The noise level and frequency dependence are reported, and the present performance is finally discussed.

2. SETUP

The configuration of the X-ray detector is shown in Fig. 1(a). An STJ chip, which includes some junctions, was glued and wired on to a chip carrier 25 mm square at 0.3K. The connection between the STJ and SQUID was adjusted on a circuit board just behind the 0.3K plate. The circuit board was thermally disconnected from the 0.3K stage but thermally connected with the 4K container. Thus, the temperature of the board is expected to be $\sim$ 4K. The board was wired to the STJ chip carrier with cryogenic cables so that heat flow could be avoided. The SQUID amplifier chip was glued on to a small copper plate to obtain powerful cooling. In order to cut the direct thermal noise from the copper plate, a thin fiberglass layer (500 μm) was installed between the SQUID chip and the plate. The wire pattern was designed onto a fiberglass layer and is connected to the circuit board with a 9-pin flat cable. This flat cable has two layers. One layer includes signal and bias lines (9 channels). The other is a thin ground plane. The SQUID chip carrier is attached in a superconducting niobium box at 4K for magnetic shielding. The niobium shield has a size of $36 \times 20 \times 12$ mm^3 and has a multi-layer structure. Any external magnetic flux or radiation can not directly reach the chip. Then, the niobium shield is covered with a cryogenic permalloy shield, which has a single layer. Finally, 0.3K stage for STJ, the circuit board and cryogenic permalloy shield are mounted together inside an OFC container with a volume of $\phi 90 \times 70$ mm^3. X-ray goes through a thin beryllium window of the cryostat and aluminized tape of the OFC container. All parts are assembled in a ^{3}He cryostat which has a shape of $\phi 300 \times 550$ mm^3.

In order to optimize the electric circuit, a simulation was carried out. The circuit used here is shown in Fig. 1(b). The capacitance C of 100 nF was determined so that the signal would be free from deformation. The capacitance makes the current component ($>$ several Hz) pass and reach the SQUID amplifier. The shunt resistance, R_s should be less than 10 Ω to suppress oscillation at the peak point of a signal. However, the noise current, $< i_{noise} > \propto \sqrt{kTR_s}/R_s$, increased with small R_s. Consequently R_s was set to 0.5 $\sim$ 5 Ω. With R_s of 0.5 Ω, a rise time of 1 μsec was obtained at the input coil of the SQUID amplifier. This is consistent with the experimental results. The detail results of the simulations have been

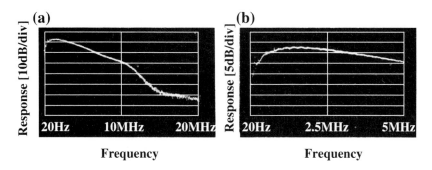

Fig. 2: (a) Response vs. frequency in wide range (b) Response vs. frequency in narrow range

reported elsewhere (Oku 1999).

3.   EXPERIMENTS AND RESULTS

After the SQUID amplifier had been carefully tuned, the noise level and time response were obtained. The characteristic of SQUID was reported by Ikeda et al (1998). Fig. 2 shows the frequency dependence of the total system with R_s of 10 Ω. The sensitivity in the region below 0.5 MHz is very low due to capacitance, C. The highest response was obtained at $1-2$ MHz. At 5 MHz and 10 MHz, losses of -7 dB and -20 dB were observed. The contribution of the co-axial cables was also checked. The cryostat uses co-axial cables which were designed to minimize the external heat flow from outside. The bandwidth of the cables was up to 20 MHz with a small loss of less than -4 dB at around 16 MHz. In the X-ray measurements, the rise time of 0.3 μsec was observed as shown in Fig. 3. It was delayed comparing with the simulated results of 0.1 μsec. Under the same conditions, the noise level as a function of frequency was measured. In the white noise region, the level was 3.2 pA/$\sqrt{\text{Hz}}$ at 1 kHz. In the $1/f$ region, 19 pA/$\sqrt{\text{Hz}}$ at 1 Hz was found. These values were evaluated with a SQUID amplifier gain of 1.8×10^6 V/A, including a room temperature gain of 100.

As a calibration of the linearity to X-ray energy, simulated pulses generated by a pulse generator were used. These signals were applied to the SQUID amplifier in the same way as the STJ signals. Fig. 4(a) depicts the linearity. The vertical axis corresponds to the output signal integrated over time. For this measurement, R_s was set to a small value of 0.5 Ω so that the rise time matched the SQUID bandwidth. The linearity co-efficient was calibrated by a

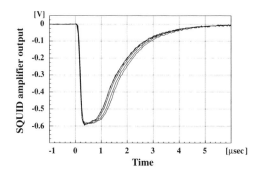

Fig. 3: Typical X-ray signals of the SQUID amplifier output with the room temperature module

662

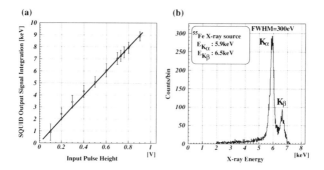

Fig. 4: (a) Linearity in the energy and (b) Energy spectrum for X-ray

real X-ray signal corresponding to 5.9 keV. This plot shows the linearity in a wide range from 1 to 10 keV, with the STJ signal proportional to the X-ray energy. The energy resolution was typically about 940 eV in FWHM for the whole energy region. In better cases such as those shown in Fig. 4(b), the deviation from linearity becomes serious. The spectrum in Fig. 4(b) shows a result from X-ray measurements. The X-ray photons from a ^{55}Fe source make two peaks of 5.9 keV (K_α) and 6.5 keV (K_β). The peak corresponding to K_α has a resolution of 300 eV in FWHM corresponds to the K_α events. The intrinsic resolution of the amplifier was found to be about 100 eV by using a pulse generator. During the measurement, the magnetic field of 10 ~ 50 mT was applied continuously. And its magnitude was frequently changed to obtain the optimum STJ condition. However, any change in V-Φ curve of the SQUID due to the field was not detected.

4. SUMMARY

A compact system for high-resolution X-ray detection using STJ and a SQUID amplifier was developed and works. The magnetic shield suppressed the field around the SQUID, when the field for STJ operation of 10 ~ 50 mT was applied. The response up to about 2 MHz was obtained. The performance of the amplifier was tested in terms of energy resolution, linearity to the X-ray energy and a bandwidth. The energy resolution was measured to be 300 eV in FWHM for 6 keV X-ray. In order to obtain a better energy resolution, the peripheral circuit has been modified to suppress the noise through the SQUID bias cables or accidental oscillation due to unexpected capacitance or inductance of the cables. Also other methods of STJ biasing (Mears 1997) have been checked.

Acknowledgment

This work was performed through Special Coordination Funds for promoting Science and Technology Agency of the Japanese Government. This work was supported in part by Grant-in-Aid under the program numbers 08554002, 09490038, 10740136, 11740163, 11874035, 11780136 and 11740276. T.I., T.O., H.S., W.O. and Y.T. are grateful to the Special Postdoctoral Researchers Program for support of this research.

REFERENCES

Ikeda T et al. 1999 IEEE Transactions on Applied Superconductivity **9**, 2, pp 3858
Mears C A et al. 1997 IEEE Transactions on Applied Superconductivity **7**, 2, pp 3415
Oku T et al. 1999 submitted to Nucl. Inst. and Meth. A as Proceedings of LTD-8

Inst. Phys. Conf. Ser. No 167
Paper presented at Applied Superconductivity, Spain, 14-17 September 1999
© 2000 IOP Publishing Ltd

Broad-band terahertz NbN hot-electron bolometric mixer

J Schubert[1], A Semenov[2], H-W Hübers[1], G Gol'tsman[2], G Schwaab[3], B Voronov[2] and E Gershenzon[2]

[1] DLR Institute of Space Sensor Technology and Planetary Exploration, 12489 Berlin, Germany
[2] Physical Department, State Pedagogical University, 119435 Moscow, Russia
[3] Physical Chemistry Department II, Ruhr University Bochum, 44801 Bochum, Germany

ABSTRACT: We present data on the noise temperature of a NbN lattice-cooled hot-electron bolometric mixer in the spectral range 0.7 THz to 5.2 THz. The bolometer, a 3.5 nm thick film having in-plane dimensions of $1.7 \times 0.2 \ \mu m^2$, was integrated in a planar complementary logarithmic spiral antenna. The antennas were combined with either a 6 mm or a 12 mm diameter extended hemispherical lens. Sensitivity measurements performed at seven discrete frequencies resulted in double side-band receiver noise temperatures of 1500 K (0.7 THz), 2200 K (1.4 THz), 2600 K (1.6 THz), 2900 K (2.5 THz), 4000 K (3.1 THz) 5600 K (4.3 THz) and 8800 K (5.2 THz). The radiation pattern of the antenna at 2.5 THz was found to be almost diffraction limited irrespective of the lens diameter used.

1. INTRODUCTION

High resolution heterodyne spectroscopy in the frequency range 1 THz to 6 THz yields important information on astronomical objects as well as on the chemical composition of the earth atmosphere. Examples are the CII fine structure line at 1.6 THz and the OI fine structure line at 4.75 THz which are major coolant lines of the interstellar medium. The OH rotational transitions at 2.5 THz and 3.5 THz allow the determination of the OH volume mixing ratio which gives important information on the chemistry of the ozone destruction in the stratosphere. A number of on-going astrophysical and atmospheric research programs aimed at these goals require a receiver with the noise temperature close to the quantum limit. Recent studies have shown that superconducting hot-electron bolometric (HEB) mixers are able to satisfy such requirement. Generally, the superconducting film of a HEB is so small that only electrons are heated by the absorbed radiation. Therefore the sensitivity of the device is relatively large; it approaches that of slow conventional bolometers. Depending on the length of the device, speed of the detector response may be either of the order of the reciprocal electron-phonon interaction time in the film (lattice-cooled HEB) or may be dominated by the diffusion of non-equilibrium electrons from the film into metal contacts (diffusion-cooled HEB). Both approaches, however, result in a response time of few tens of picoseconds. In the heterodyne regime the above features provide low noise temperatures as well as a large (several GHz) intermediate frequency bandwidth. A further attractive feature of HEB mixers is the small power requirement from the local oscillator. So far, at frequencies below 2.5 THz, both lattice-cooled and diffusion-cooled HEB mixers have demonstrated receiver double side-band noise temperature of about twenty quantum limits. In this paper, we report noise temperature measurements in the frequency range 0.7 THz to 5.2 THz, which were performed with the same device, a lattice-cooled NbN HEB mixer. We also present 2.5 THz data on the radiation pattern of a planar complementary logarithmic spiral antenna which is combined with extended hemispherical lenses of different diameter.

2.   MIXER DESIGN

Devices were fabricated from 3.5 nm thick NbN films which typically had a room temperature resistivity of 870 μΩ·cm and a superconducting transition temperature of about 10 K. Films were deposited in a nitrogen atmosphere by dc reactive magnetron sputtering of Nb on 350 μm thick optically polished Si substrates of 5 kΩ·cm resistivity. Details of the process are described elsewere (Cherednichenko et al 1997). A planar two-arm complementary logarithmic-spiral antenna was used to couple both the signal and the local oscillator (LO) radiation with the mixer. The log-spiral antenna belongs to a family of frequency independent antennas, i. e. impedance and beam pattern are to a large extent frequency independent. The central part of the antenna was patterned from 250 nm thick gold film using electron beam lithography while the outer part was defined by conventional UV photolithography. The layout of the antenna is shown in Fig. 1. The circle inside which the antenna arms formed inner terminals and sized to represent a spiral, had a diameter of 2.2 μm. The diameter of the circle that circumscribed the spiral structure was 130 μm. Between these circles, the antenna arms inscribe 2.15 full turns. The spiral structure terminated a coplanar line, which was lithographed on the same substrate and had an impedance of 50 Ω. The substrate carrying the HEB, the planar

Fig. 1   Layout of the logarithmic spiral antenna. The gap in the centre of the antenna has the height 1.7 μm and the width 0.2 μm, which defines the length of the HEB.

antenna and the co-planar line, was glued with its rear surface onto the flat side of an extended hemispheric lens. The lens was cut off from an optically polished sphere that was made from high resistivity (>10 kΩ cm) silicon. Spheres of 6 and 12 mm diameter were used for this purpose. The extension of the lens together with the substrate yields a total extension length of 1.2 mm and 2.4 mm for the smaller and the larger lens, respectively. These values are very close to the extension length corresponding to the synthesised elliptic lens for which the beam pattern of the hybrid antenna is expected (Büttgenbach 1993) to be diffraction limited.

3.   NOISE TEMPERATURE MEASUREMENTS

The noise temperature was measured at discrete frequencies from 0.7 THz to 5.2 THz. An optically pumped ring far infrared (FIR) laser (Hübers 1996) and a transversely excited FIR laser (Hübers 1998) were used as a local oscillator in the frequency ranges 0.7 THz to 2.5 THz and 2.5 THz and 5.2 THz, respectively. Results from measurements of the noise temperature at 2.52 THz were identical irrespective of which laser system was used.

The double side-band (DSB) receiver noise temperature was determined by the Y-factor method making use of Ecosorb as the hot and cold load at temperatures of 293 K and 77 K, respectively. To derive the receiver noise temperature from the measured Y-factor at all frequencies above 1 THz, the dissipation-fluctuation theorem in general form was used.

Figure 2 shows the DSB receiver noise temperature as a function of LO frequency in the range 0.7 THz to 5.2 THz. Also shown is the DSB receiver noise temperature corrected for losses (presented below in tabular form) in the optical elements.

	Loss (dB)						
Frequency	0.7 THz	1.4 THz	1.6	2.5	3.1	4.3	5.2
Beamsplitter	0.1	0.2	0.3	0.6	0.7	1.2	1.2
TPX window	0.4	0.4	0.5	0.6	0.6	0.8	0.9
Quartz filter	1.1	1.8	1.9	1.2	1.3	1.5	1.9
Si lens (refl.)	1.5	1.5	1.5	1.5	1.5	1.5	1.5
Si lens (absorp.)	0.1	0.1	0.1	0.1	0.1	0.1	0.1
Total	3.2	4.0	4.3	4.0	4.2	5.1	5.6

The optical losses in the TPX window and quartz filter were determined from transmission curves supplied by the manufacturer. Losses in the beamsplitter and the Si lens were calculated separately. The corrected receiver noise temperature follows closely the 10 $h\nu/k_B$ line. This is 20 times the quantum limit $h\nu/2k_B$ for a DSB receiver operated in the continuum detection mode. The LO power absorbed in the HEB was evaluated from the pumped and unpumped current-voltage characteristics

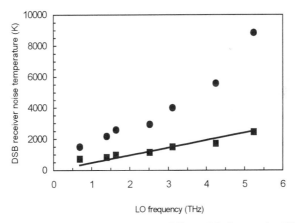

Fig. 2 Double side-band receiver noise temperature for different LO frequencies (filled circles) and the DSB receiver noise temperature corrected for losses in the coupling optics (filled squares). Solid straight line corresponds to the 10 $h\nu/k_B$ limit.

by means of the isothermal method. An absorbed LO power of about 100 nW was determined for the optimal operation regime at all frequencies. The optimum bias current and voltage were about 22 µA and 1.8 mV; no significant dependence of the optimal operation regime on the frequency was observed. The result suggests that the planar logarithmic spiral antenna of the mixer is almost frequency independent from 0.7 THz to 5.2 THz. Knowing the geometry of the antenna it is possible to estimate the wavelength range where the characteristics of the antenna should be frequency independent. The upper boundary corresponds to the wavelength $\lambda \leq 2D$, where D is the diameter of the circle circumscribing the spiral. The shortest wavelength at which the antenna still works properly is about

666

10 times the inner radius at which the spiral deviates from the ideal shape. In our case this yields an upper wavelength of 884 μm (0.34 THz) and a lower wavelength of 44 μm (6.8 THz); this interval encompasses our experimental wavelength range. According to Kraus (1988), such an antenna should have a frequency independent RF impedance of about 75 Ω when suspended in free space. We expect an even smaller value for our antenna since it is supported by the dielectric half-space represented by a substrate which has a thickness much larger than the wavelength.

4.   BEAM PATTERN OF THE ANTENNA

In order to have a frequency independent beam pattern the effective aperture of the antenna should scale with the wavelength. As an example, we present in Fig. 3 the E-plane beam pattern of our hybrid antenna measured at 2.5 THz when using 6 and 12 mm diameter lenses. Solid lines show the diffraction limited pattern that was simulated for the physical diameter d of the lens according to the

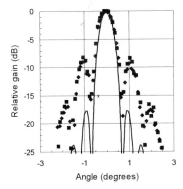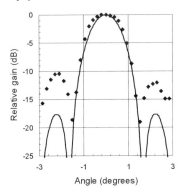

Fig. 3   E-plane beam pattern of the hybrid antenna measured at 2.5 THz for the 12 mm lens (left panel) and the 6 mm lens (right panel). Solid lines show diffraction limited pattern corresponding to the physical diameter of lenses.

expression $(2J_1(v)/v)^2$, where $v=(\pi \tan(\theta)\ d)/\lambda$, θ is the angle and J_1 is the Bessel function of the first kind. The measured full width at half maximum was 1.65° and 0.75° while the calculated profile yields 1.19° and 0.59° for the 6 mm and the 12 mm lens, respectively. Using experimental values of the width of the beam we estimated the Gaussian coupling efficiency to be about 70 % in both cases. Experimental data are in good agreement with the diffraction limited pattern in terms of the main lobe beam-width and the position of the side lobes. It is worth noting, however, that the first side-lobes occur approximately at the −10 dB level which is 7 dB higher than simulations suggest. We attribute the discreapancy to possible deviation of the actual extension length from the optimal value corresponding to the lowest level of side-lobes.

This work was supported by the the German Ministry of Science and Education trough the WTZ grant RUS-149-97. The authors thank U. Bartels for his assistance during the measurements.

REFERENCES

Büttgenbach T H 1993 IEEE Trans. on Microwave Theory Tech. **41**, 1750.
Cherednichenko S, Yagoubov P, Il'in K, Gol'tsman G and Gershenzon E 1997 Proc. of the 8th Int. Symp. on Space Terahertz Technology, p 245.
Hübers H-W, Schwaab G W, Röser H P 1996 Proc. 30th ESLAB Symp. "Submillimetre and Far-Infrared Space Instrumentation", ESA SP-388, p 159.
Hübers H-W, Töben L, Röser H P 1998 Rev. Sci. Instr. **69**, 290.
Kraus J D 1988 Antennas (McGraw-Hill, Inc.).

Inst. Phys. Conf. Ser. No 167
Paper presented at Applied Superconductivity, Spain, 14-17 September 1999
© 2000 IOP Publishing Ltd

Fabrication of Diffusion-Cooled Nb Hot-Electron Bolometer Mixer for Terahertz Applications

M. Frommberger, M. Schicke, P. Sabon, K.H. Gundlach and K.F. Schuster

Institut de RadioAstronomie Millimetrique IRAM, Domaine Universitaire,
38406 St. Martin d'Heres, France

ABSTRACT: We report on heterodyne measurements with a Nb diffusion-cooled hot-electron bolometer (HEB) in a quasi-optical mixer at 800 GHz. The device is a 400 nm long and 12 nm thick Nb microbridge patterned by e-beam lithography in a self-aligned process. Its normal state resistance is 50 Ω. We have measured the double side band (DSB) receiver noise temperature using the Y-factor method with a 300 K and 77 K blackbody load. At 813 GHz, a bath temperature of 4.2 K, and an intermediate frequency of 1 GHz, the DSB receiver noise temperature is 6300 K. We expect significant improvement of the performance by further reduction of the device's length.

I. INTRODUCTION

So far IRAM has succesfully produced NbN HEBs (Rösch1998) with good noise temperatures. The cooling mechanism in the NbN HEBs is based on electron-phonon interaction. To obtain a good performance with NbN HEBs one has to produce thin films with a thickness of about 5 nm or less, to ensure a short phonon escape time τ_{phe} into the substrate, which is necessary for a high IF. The preparation of those thin superconducting films is difficult to control. After all, the performance of the phonon-cooled NbN HEB is ultimately limited by τ_{phe} leading to an maximum IF of about 4 GHz (Karasik1999).

In Nb the electron-phonon interaction is too slow for practical applications. Therefore, Nb HEBs are cooled by fast out-diffusion of the hot electrons into high conductivity cooling pads. For the diffusion-cooled HEB we are not restricted in film thickness. Since in this case $f_{IF} \sim l^{-2}$ a high IF can be achieved by preparing short and narrow superconducting microbridges (Karasik1998), reproducibly patterned with e-beam lithography, which is the big advantage of the diffusion-cooled Nb HEB.

II. NIOBIUM FILM PROPERTIES

Basic research on the properties of Nb thin films gives us the possibility to make a proper choice of the antenna feed impedance and device geometry.

The Nb films have been deposited on fused quartz substrates by dc-magnetron sputtering at a background pressure better than 5×10^{-8} mbar. The transition temperature T_c and the sheet resistance R_{sq} have been obtained by a four point contact measurement in a dip stick. The film thickness has been varied from 2 nm to 300 nm and verified by anodization (Kroger1981), spectroscopy and ellipsometry (Sabon1999).

Figure II.1 shows a logarithmic dependence of T_c on the film thickness (Wolf1976), expected from the BCS expression for the energy gap (Cooper1961). An exponential fit of R_{sq} shows the dependence on the film thickness like it has been predicted by Graybeal et al. (1984).

For the standard 120 nm thick Nb layers sputtered at 1.2 Pa we obtained a T_c of 9.1 K and a resistance ratio R_{300K}/R_{10K} of 5.5, both results are close to literature values (Broom1980, Hinken1989).

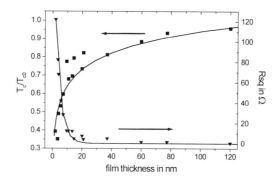

Fig. II.1 Transition temperature T_c and sheet resistance Rsq versus film thickness.

III. DEVICE FABRICATION

The essential part of the fabrication of Nb HEBs is the definition of the length of the Nb microbridge, which is determined by the contact pads (see Fig III.1). Furthermore the resistance of the bridge should be equal to the antenna feed impedance (e.g. 20-150 Ω). This can be achieved by making the film thin (~12 nm) to get a high sheet resistance (±10-20 Ω/□) and by choosing a proper length-width ratio. To define the microbridge, electron-beam lithography (EBL) is required.

The process starts with the deposition of a 12 nm thick Nb film by dc-magnetron sputtering over the hole area of a 200 μm thick fused quartz wafer. Antenna and RF-choke structure are defined by optical lithography and lift-off of a Nb/Al/Au trilayer (Nb: 100 nm, Al: 8 nm and Au: 50 nm). Then the 80 nm thick Au cooling pads are defined in a first EBL step by a standard lift-off process with PMMA 950 K. In the last step the microbridge is defined via reactive ion etching (**RIE**, CF_4, O_2 20:1- plasma). The Al etch-mask is defined by EBL and lift-off.

Figure III.1 shows the remaining Al microbridge connecting the cooling pads. Between the upper and lower contact pad one can recognize the 400 nm long and 150 nm wide microbridge. The Al etch mask will later be removed by a basic solution.

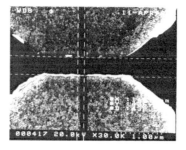

Fig. III.1 REM picture of the 400×150 nm HEB defined by EBL.

IV. EXPERIMENTAL RESULTS

All devices have been characterized by measuring their I(V)-curves in a dip stick at 4.2 K. We obtained a critical current I_c of about 300 μA and normal resistances R_n of about 50 Ω. One of these devices was mounted into our quasi-optical receiver which was initially designed for the 350 GHz frequency band (Rothermel1994). It was glued onto the hyperhemispherical crystal-quartz lens and contacted with a well conducting glue.

The LO was a commercial solid state source consisting of a Gunn oscillator followed by a varistor doubler and tripler, providing about 50 µW in a range from 795 GHz to 813 GHz (Radiometer Physics GmbH, Germany). The oscillator output, radiated from a dual moded horn, was focused onto the dipole antenna via a HD-polyethylene lens and the hyper-hemisperical crystal-quartz lens. The LO and the signal frequencies were combined via a beamsplitter arrangement.

Figure IV.1 shows the I(V)-curve and the conversion curves for the 300 K and 77 K loads. The junction was biased at the saddle point, minimum IF output power in the transition regime. The measurements have been carried out for different IFs and beam splitters. The best receiver noise temperature of 6300K was achieved at 813 GHz and 1 GHz IF. The IF output power was measured with a HP power meter.

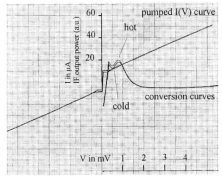

Fig. IV.1 Pumped I(V)- and conversion curves for the hot and cold load.

The heterodyne mixer response was first proven by means of two superposed oscillator signals. We varied the IF between 800 MHz and 2.2 GHz by tuning the "non-signal" LO consisting of a 110 GHz-120 GHz Gunn followed by a varistor-type frequency multiplier providing harmonics up to the 7th with the proper waveguide filter and a rectangular horn mounted to the output. The change of the frequency and the power was observed with a Tektronix spectrum analyzer.

Figure IV.2a shows the obtained results. Unfortunately our cold IF amplifier is limited to frequencies between 800 and 1700 MHz. Figure IV.2b shows the dependence of the receiver noise temperature on the IF-filter position. The data has been measured at 798 GHz, 4.2 K bath temperature and a fixed bias point. We have been using the mylar beam splitter RN19 and a tunable bandpass filter for 1-2 GHz. The plot shows, that the noise temperature increases significantly with the intermediate frequency.

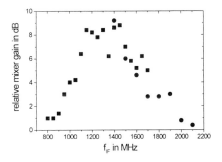 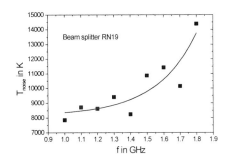

Fig. IV.2 a) Mixing at 798 GHz. Variation of the intermediate frequency by tuning the "non-signal" LO. b) Receiver noise temperature versus intermediate frequency at 798 GHz and 4.2 K physical temperature.

From theory we can estimate the upper frequency limit of our HEB to be 700 MHz (Burke1999, Gershenzon1990). Therefore our measurements show probably the spectral characteristic of our IF amplifier superposed with the 700 MHz cut-off frequency of our HEB.

V. CONCLUSIONS

We succeeded in fabricating a diffusion-cooled superconducting hot-electron bolometer. Since the process is free of difficult alignments, a high degree of reproducibility can be attained. The obtained results are very encouraging for further investigations. This means, improving the e-beam lithography step to obtain shorter devices and related with this, a better impedance matching to the antenna.

Also, with another antenna design, an antireflection coated lense and windows optimized for the operating frequency, considerably better noise temperatures can be achieved at 800 GHz.

VI. ACKNOWLEDGEMENTS

The authors like to thank H. Rothermel, F. Schäfer and K.H. Gundlach for helpful discussions. The authors are grateful to P. Pasturel for substantial technical support.

REFERENCES

R.F. Broom, S.I. Raider, A. Osenbrug, R.E. Drake and W. Walter, *Niobium oxide-barrier tunnel junction*, IEEE Transactions on electronic devices, Vol. 27, No. 10, 1998, October 1980

P.J. Burke, R.J. Schoelkopf, D.E. Prober, A. Skalare, W.R. McGrath, B. Bumble and H.G. LeDuc, *Length scaling of bandwidth and noise in hot-electron superconducting mixers*, Appl. Phys. Lett **68** (23), 3 June 1996

L.N. Cooper, *Superconductivity in the neighborhood of metallic contacts*, Phys. Rev. Lett. **6** (12), June 15, 1961

E.M. Gershenzon, G.N. Gol'tsman, A.M. Lyul'kin, A.D. Semenov and A.V. Sergeev, *Electron-phonon interaction in ultrathin Nb films*, Sov. Phys. JETP **70** (3), March 1990

J.M. Graybeal, M.R. Beasley, *Localization and interaction effects in ultrathin amorphous superconducting films*, Phys. Rev. B, **29** (7), 1 April 1984

J.H. Hinken, *Superconductor Electronics*, Springer Verlag, 1989

B.S. Karasik, A. Skalare, R.A. Wyss, W.R. McGrath, B. Bumble, H.G. LeDuc, J.B. Barner and A.W. Kleinsasser, *Low-noise and Wideband Hot-Electron Superconductive mixers for THz Frequencies*, 6th IEEE Int. Conf. on THz Electronics, Leeds, UK, 1998

B.S. Karasik, W.R. McGrath and R.A. Wyss, *Optimal choice of material for HEB superconducting mixers*, IEEE Transactions on applied Superconductivity, Vol. 9, No. 2, June 1999

H. Kroger, L.N. Smith and D.W. Jillie, *Selective niobium anodization process for fabricating Josephson tunnel junctions*, Appl. Phys. Lett. **39** (3), 1 August 1981

C. Rösch, T. Lehnert, C. Schwoerer. M. Schicke, K.H. Gundlach and K.F. Schuster, *Low-noise NbN phonon-cooled hot-electron bolometer mixers ar 810 GHz*, Proc. of the 9th Int. Symp. on Space Terahertz Technology, Pasadena, CA, USA, 1998

P. Sabon, *memoire CNAM*, to appear oct.1999

S.A. Wolf, J.J. Kennedy and M. Nisenhoff, *Properties of superconducting rf sputtered ultrathin films of Nb*, J. Vac. Sci. Technol., Vol. 13, No. 1, Jan./Feb. 1976

Inst. Phys. Conf. Ser. No 167
Paper presented at Applied Superconductivity, Spain, 14-17 September 1999
© 2000 IOP Publishing Ltd

Anodic oxidation for NbN film thickness measurements and fabrication of NbN thin film resistors

M Schicke[†], K Mizuno[‡], K H Gundlach[†], and K F Schuster[†]

[†] Institut de Radioastronomie Millimétrique, 38406 St. Martin d'Hères Cedex, France
[‡] Central Research Laboratories, Matsushita Electric Industrial Co., Ltd., 3-4 Hikaridai, Seika-Cho, Soraku-Gun, Kyoto, 619-02 Japan

ABSTRACT: Anodic oxidation is a known technique to determine the thickness of certain metallic layers. We show that this method is also applicable for thin NbN layers (2-13 nm). NbN Hot-Electron Bolometers need the preparation of 3-10 nm thick NbN films, which now can be measured before continuing the processing. Secondly, we report on the fabrication of resistive NbN thin films. Our films have a resistivity of $210\,\mu\Omega\cdot$cm, which is almost constant within a temperature range of 4 K to 300 K. Therefore, these films can be used as integrated resistors in Nb or NbN based mixer technology.

1. ANODIC OXIDATION FOR NBN FILM THICKNESS MEASUREMENTS

It is well known that the thickness of some metallic layers (e.g. Nb, Al, Ta, Ti) can be evaluated by anodic oxidation [Young 61]. Under appropriate conditions they anodize at a constant rate (Nb: 0.85 nm/V; Al: 0.88 nm/V [Imamura 91]) up to thicknesses above 100 nm. The thickness is determined from the voltage span in the anodization profile. Previous anodization experiments with NbN films were generally done for thicknesses > 20 nm. These films do not anodize at a constant rate and the anodization profile is not reproducible. During the development of Hot-Electron Bolometer (HEB) mixers with thin (3 to 10 nm) NbN microbridges we have re-examined the anodization of thin NbN films and found a reproducible constant anodization rate of approximately 1.3 nm/V for 2 to 13 nm thick films. This can be of interest for various applications including the investigation of superimposed layers or the passivation of the edges of NbN tunnel junctions. As for NbN HEB mixers the film thickness is an important parameter for the speed, the impedance, and the transition temperature. Its simple evaluation by anodic oxidation before the actual definition of the microbridge can be very useful. We also discuss a model describing the anodization-behavior of NbN films.

1.1 MEASUREMENT SETUP

The sample is connected to the anode of a current source and partially submerged into the electrolyte (see Fig. 1). The cathode consists of a gold wire. The electrolyte is a mixture of 156 g ammoniumpentaborate, 1120 ml ethylene glycol and 760 ml H_2O and is originally used for anodization of Al and Nb [Kroger 89]. Using a constant current density, typically about 0.5 mA/cm², the NbN surface on the substrate is anodized and the voltage drop across the growing oxide layer increases. An RC differentiating circuit is set up to measure dV/dt during the anodization and plot it versus V. To determine the thickness of a metallic film, the entire film thickness is anodized and the endpoint detection is given by the derivative, which becomes infinity when the electrical contact through the film is interrupted (see Fig. 2).

672

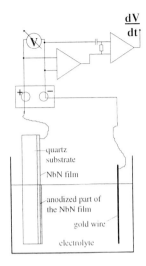

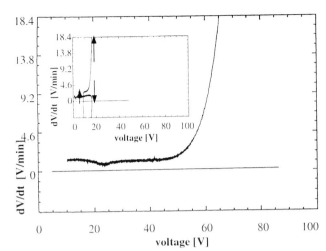

Fig. 1: Schematic measurement setup.

Fig. 2: Anodization curve of a 5 nm thick NbN film. The end point voltage is 6.5 V. The inset shows the curve of a 15 nm thick NbN film. The arrows indicate the direction of plotting.

Film thicknesses up to 13 nm could be measured reproducibly (see Fig. 3). For thicker NbN films the anodization becomes unstable (see inset of Fig. 2), which is tentatively ascribed to the columnar structure of the NbN films. The oxide, which is formed at the surface of the columns is a good insulator, but there might be unstable channels through which the electrolyte could still exchange ions with the film. These channels, once closed by only a few oxide molecules, might be reopened by electrostatic forces between the ions in the electrolyte and the film, which occurs at a threshold voltage of 15 - 20 V. Hence the voltage drops back (negative dV/dt) and then increases again, often at another speed dV/dt.

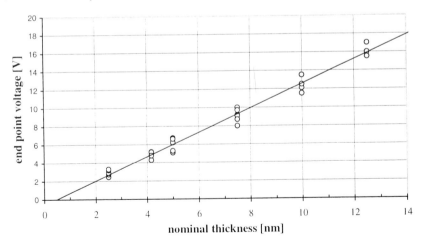

Fig. 3: Relationship between the end point voltage and the anodized film thickness. The offset of 0.5 nm is due to enhanced anodization at the surface of the electrolyte. There the thin film is disconnected from anodization current before the rest of the sample in the liquid is anodized. The nominal thickness was given by the sputter rate.

Whereas Al and Nb film thicknesses can be measured with this method up to a few hundred nanometers, NbN cannot be oxidized profoundly, which is seen when anodizing NbN films thicker than approximately 30 nm. Here, at the end of anodization the oxide layer becomes completely insulating before the whole film is anodized. Thus, finally the voltage does not change anymore and its derivative fluctuates around zero.

1.2 CONCLUSIONS

The measured data show that anodization is a practical method for thickness determination of our thin NbN films. Anodization becomes unstable for thicknesses larger than 13 nm, which therefore is the maximum measurable thickness for NbN with this method. For thickness verification on bolometer substrates, small quartz samples were put around the bolometer substrate during deposition, in order to anodize them later on.

2. FABRICATION OF NBN THIN FILM RESISTORS

Resistive NbN thin films are of interest not only for high frequency applications used in radio astronomy but also for integration in other microelectronic circuits based for instance on Nb or NbN superconducting elements. The films presented here were fabricated by rf-sputtering of Nb at 600 W in a gas mixture of 2.5 sccm N_2 and 98 sccm Ar at 1.35 Pa. Films of different thicknesses have a resistivity of 220 μΩ·cm, which is constant within a temperature range of 4 K to 300 K (see Fig. 5). At a thickness of 90 nm our films have a sheet resistances around 25 Ω/□ (see Fig. 6). Commonly used PdAu or TiAu films have much lower resistivities and therefore need to be long which creates inductivity problems for high frequency applications and space consumption problems for microelectronic circuits.

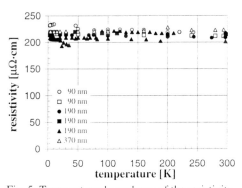
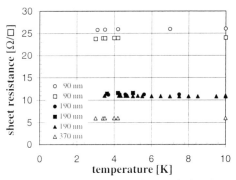

Fig. 5: Temperature dependence of the resistivity.   Fig. 6: Temperature dependence of the sheet resistance around 4 K.

2.1. ADJUSTMENT OF THE RESISTANCE VALUE

The value of a resistor needed for a certain application can be adjusted by etching down the film thickness of the resistor with Reactive Ion Etching (RIE). To determine the etch rate of our NbN films we etched several samples of known film thickness for different times and deduced from the change in the resistance the thickness reduction of the pure NbN layer as a function of the etch time (see Fig. 7). The offset at the time scale corresponds to the time needed to start the etch plasma and to etch the native oxide layer which grows at the film surface as a function of time during which the sample was stored under ambient conditions (see Figs. 7 and 8). Correcting all measurements for the time offset, an etch rate of 0.4 to 0.5 nm/s was obtained (see Fig. 9).

674

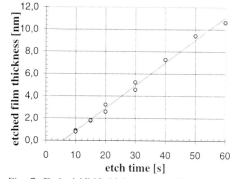

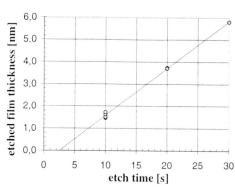

Fig. 7: Etched NbN thickness as a function of etch time for films with an native oxide layer due to 10 days of storage under ambient conditions.

Fig. 8: Etched NbN thickness as a function of etch time for films with an native oxide layer due to 4 hours of storage under ambient conditions.

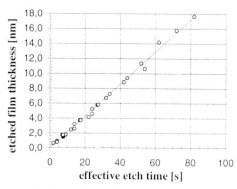

Fig. 9: Etched NbN thickness as a function of the effective NbN etch time for samples of different NbN layer thicknesses and different thicknesses of the native oxide layer.

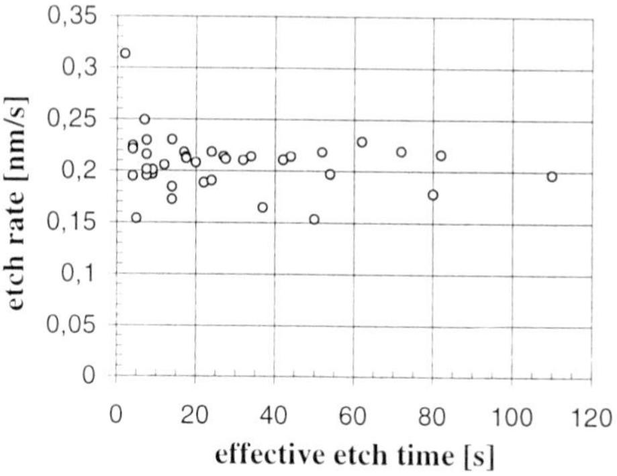

Fig. 10: NbN etch rate as a function of the effective NbN etch time evaluated for the same samples shown in Fig. 9.

2.2. CONCLUSIONS

We have fabricated resistive NbN thin films which are applicable for integration in low temperature superconducting electronics. In radio astronomy this is of great interest since it allows to integrate parts of the signal treating electronics on the same chip as the highly sensitive frequency mixer elements. This reduces the number of cables and connectors and therefore losses and noise. Further examinations have to be made on the formation of the native oxide layer in order to tune the resistance values more accurately.

ACKNOLEDGMENTS

We appreciated very much many interesting and helpful discussions with P. Sabon.

REFERENCES

[Young 61]:   L.Young, *Academic Press*, London and New York (1961)
[Imamura 91]:   T. Imamura, S. Hasuo, *IEEE Trans. Magn.* **27**, 3172 (1991)
[Kroger 89]:   H. Kroger, L. N. Smith, and D. W. Jillie, *Appl. Phys. Lett.* **39**, 280 (1989)

Inst. Phys. Conf. Ser. No 167
Paper presented at Applied Superconductivity, Spain, 14-17 September 1999

675

A Superconducting Tunnel Junction with Superconducting Microstrip Coil for X-ray Detector

T. Taino[1], H. Nakagawa, M. Aoyagi, H. Akoh, K. Maehata[1], K. Ishibashi[1]
H. Sato[2], T. Ikeda[2], C. Otani[2], W. Ootani[2], T. Oku[2], H. Kato[2], K. Kawai[2],
H. M. Shimizu[2], Y. Takizawa[2], H. Miyasaka[2], H. Watanabe[2]

Electrotechnical Laboratory (ETL), 1-1-4 Umezono, Tsukuba, Ibaraki 305-8568, Japan
[1]Kyushu University, 6-10-1 Hakozaki, Higashi, Fukuoka, Fukuoka 812-8581, Japan
[2]The Institute of Physical Chemical Research (RIKEN), 2-1 Hirosawa, Wako, Saitama 351-0198, Japan

ABSTRACT: We have proposed and demonstrated a superconducting tunnel junction (STJ) with microstrip coil for X-ray detectors. The microstrip coil consists of multiple superconducting microstrip lines, and applies magnetic field into the STJ to suppress dc Josephson current. The STJ devices were fabricated by Nb/Al/AlOx/Nb integration technology using 2 μm design rule. From the measurements of the magnetic field dependence of the Josephson current, the magnetic field of 13 mT can be applied to 200×200 μm^2 STJ by supplying current of 49 mA into the microstrip coil.

1. INTRODUCTION

Superconducting tunnel junctions (STJs) are one of the promising candidates for future high performance X-ray detectors, because it has advantages of a high energy resolution, a high counting rate and wide energy range. The energy resolution is mainly determined by the number of quasi-particles exited by irradiation of X-ray. Since superconductor has a smaller energy gap of a few meV than semiconductor one, many quasi-particles can be produced in the STJ. The theoretical energy resolution of 4 eV is expected by irradiation of 5.9 keV X-ray for Nb-based STJ. Energy resolution has been achieved to be 22 eV in the STJ of Ta/Al/AlO$_x$/Al/Ta structure by Verhoeve et al., (1998). The high counting rate of more than 1000 CPS is reported in Nb-based STJ by Frank et al., (1998). In these experiments, a magnetic field of about 10 mT is used to suppress dc Josephson current of STJ.

Development of Josephson IC technology based on the Nb-based STJ held the potential to provide high-speed signal processing circuits by Rylov et al. (1997), Nakagawa et al (1991) and Nagasawa et al., (1997). By combining these Josephson signals processing circuits and STJ detectors, a high performance detector system can be expected, particularly in the case of array detectors. However, the Josephson device utilises the Josephson current, which should be suppressed during use with the STJ detectors. A superconducting quantum interference device (SQUID) had already been introduced in the STJ detector system as an ultra-low noise readout amplifier for the X-ray signal by Frank et al., (1998). In this system, an electromagnet of relatively large size is utilised near the STJ chip to apply a magnetic field. The SQUID readout of the Josephson circuit was in a magnetic shield isolated from the STJ, which resulted in a large system size. Though similar study had been tested by Koshelets et al. (1996), the magnetic field for suppression of Josephson current in this research is small for X-ray detection.

In addition, recently Taino et al. (1998) reported that a small directional fluctuation of the magnetic field in the STJ affected the Fiske-step current which should be also suppressed for the X-ray detector operation. This indicates that the direction of the magnetic field has to be fixed precisely for a high performance X-ray detection.

In this paper, we have proposed and demonstrated a new STJ with a microstrip coil, which can suppress the Josephson current without an external electromagnet and can precisely control the direction of the magnetic field. This new STJ has a potential to realise a small size detector system with a combination of STJs and Josephson circuits.

2. SUPERCONDUCTING MICROSTRIP COIL

When a magnetic field is generated by a flowing current I in a superconducting microstrip line of width W located at a distance h above a superconducting ground plane, magnetic field can be estimated by $\mu_0 I/W$, where μ_0 is the permeability of the free space. The line width is assumed to be larger than the distance or thickness of the microstrip line in order to neglect the fringing effect. The magnetic field at the STJ which is inserted between the microstrip line and the superconducting ground plane can also be estimated by the above relation. In the case of 200×200 μm^2 STJ, a current of 1.6 A is needed for a single microstrip line. It is difficult to handle this large current level in the STJ-IC chip. In order to decrease the necessary current level, the microstrip line is divided in multiple microstrip lines and each line is serially connected in a planar spiral microstrip coil. A schematic illustration of this STJ with the microstrip coil is shown in Fig.1. Utilising a line and space design rule of 2 μm, the microstrip line can be divided into 47 microstrip lines for the 200×200 μm^2

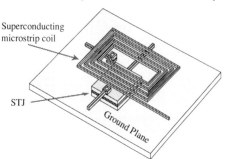

Fig.1 Schematic configuration of STJ with superconducting microstrip coil.

STJ. The current level of the microstrip coil is calculated to be 34 mA to generated the magnetic field of 10 mT. A critical current of a Nb thin film line with 2 μm width and 500-nm thickness was reported to be 90 mA by Imamura et al., (1990), and sufficiently guarantees the above necessary coil current.

3. FABRICATION OF STJ

Figure 2 shows a process flow for the fabrication of a Nb/Al/AlOx/Nb STJ with a Nb microstrip coil. First, a 20-nm MgO film as a stopping layer of an etching process was evaporated on 3 inch Si-wafer by an electron beam gun sputtering. A 400-nm Nb film was deposited on the MgO layer by a dc magnetron sputtering and formed to the ground plane pattern by a photolithography technique and a reactive ion etching (RIE) in CF_4 plasma. A 200-nm SiO_2 film was deposited as a ground plane insulation by a rf sputtering, followed by the deposition of a MgO (20-nm) stopping layer (Fig.2 (a)).

A 150-nm Nb underlayer was deposited by the dc magnetron sputtering and formed to the base-electrode pattern by the RIE in CF_4 plasma. This underlayer is used to reduce stress in the Nb(50-nm)/Al(10-nm)-AlO$_x$/Nb(200-nm) trilayer film, which was subsequently fabricated on a whole wafer by the dc magnetron sputtering without any breaking vacuum (Fig.2 (b)). The

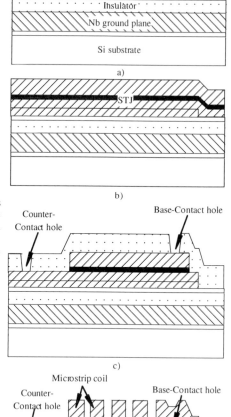

Fig.2 Process flow for the fabrication of STJ.

AlO$_x$ barrier was formed by a thermal oxidation of Al film after deposition. The Josephson current density J$_c$ can be controlled by adjusting the oxidation time and the O$_2$ pressure. In the present time, we choose the O$_2$ pressure of 2660 Pa and the oxidation time of 30 min to obtain a J$_c$ of 200 A/cm^2.

The base- and counter-electrodes of the STJ were formed by RIE in CF$_4$ plasma. After the formation of the STJ, the insulation layer of a 350-nm SiO$_2$ was deposited over the entire STJ by the rf sputtering. Contact holes for the base- and counter-electrodes were formed by the RIE process (Fig.2 (c)).

Finally, a Nb film (800-nm) was deposited and patterned for making the connection lines of the STJ and the superconducting microstrip coil (Fig.2 (d)).

The fabricated STJs had various shapes such as square, circle and diamond ones. A photograph of a 200×200 μm^2 STJ with a microstrip coil fabricated by the above process is shown in Fig.3. The whole 47-turns microstrip coil is 635 μm x 430 μm, which is located on the ground plane. A portion of the coil is overlapped on the STJ area except where the contact hole of 4×4 μm^2 is exposed for wiring.

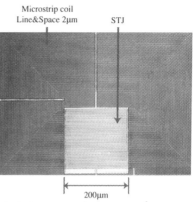

Fig.3 Photograph of 200×200 μm^2 STJ with superconducting microstrip coil.

4. EXPERIMENTAL RESULTS

In order to calibrate the magnetic field generated by the Nb microstrip coil, the magnetic field dependence of the Josephson current I$_c$ was measured by an external electromagnet with 300-turns NbTi wire which was located around the STJ-chip. Fig.4 shows that the magnetic field dependence of I$_c$ for a 50×50 μm^2 STJ at 4.2 K. It is found from this figure that the STJ shows an ideal Fraunhofer pattern, indicating that the fabricated STJ has a good quality. In fact, the fabricated STJ had a Vm value of 80 mV and a gap voltage of 2.8 mV. From the relationship between the critical magnetic field and the dimension of STJ, we can precisely estimate the magnetic field generated from the current in electromagnet, which is given by 1.01×10^5 T/mA.

Figure 5 (a) and (b) show the magnetic

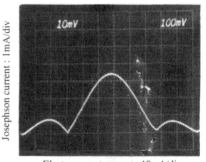

Electromagnet current : 10mA/div
Magnetic field : 1.01×10^{-4}[T/div]

Fig.4 Fraunhofer characteristic of 50×50 μm^2 STJ by using external electromagnet.

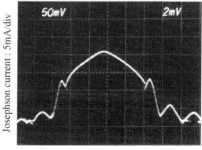

Electromagnet current : 5mA/div
Magnetic field : 5.06×10^{-5}[T/div]
a)

Microstrip coil current : 0.2mA/div
Magnetic field : 5.27×10^{-5}[T/div]
b)

Fig.5 Fraunhofer characteristics of 200×200 μm^2 STJ by using a) superconducting microstrip coil and b) external electromagnet.

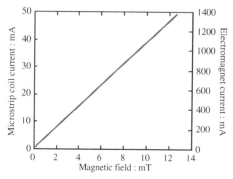

Fig.6 Magnetic field performance of superconducting microstrip coil and electromagnet.

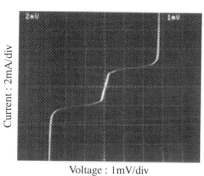

Voltage : 1mV/div

Fig.7 Current-voltage characteristic of 200×200 μm^2 STJ with superconducting microstrip coil feeding 49mA.

Fraunhofer characteristic due to the large junction effect. In comparison between Fig.5 (a) and (b), it is found that we can obtain the same Fraunhofer characteristics of the STJ by applying the magnetic field from the Nb microstrip coil and the electromagnet. By using the calibration of the magnetic field in the electromagnet, the magnetic field generated from the current in the Nb microstrip coil can be estimated to be 0.26 mT/mA. In this STJ, the maximum current of the Nb microstrip coil was 49 mA, which corresponds to the maximum magnetic field of 13 mT.

The magnetic field performance of the Nb microstrip coil and electromagnet is shown in Fig.6. The Nb microstrip coil can apply the magnetic field to the STJ by supplying the 1/25 smaller current than the electromagnet.

Figure 7 shows a current-voltage characteristic of 200×200 μm^2 STJ with a microstrip coil feeding a current of 49 mA into the coil at 4.2 K. It is confirmed that both the Josephson current and Fiske-step current are suppressed successfully.

5. SUMMARY

We have demonstrated a STJ with a microstrip coil comprised of multiple superconducting microstrip line for use as X-ray detector. The Josephson current of the STJ detector was sufficiently suppressed by feeding an appropriate current into the microstrip coil. The use of superconducting microstrip coil in the STJ detector system is expected to coalesce the STJ detector and the Josephson signal processing circuits at the chip level.

REFERENCES

Frank M, Hiller L J, Grand J B, Mears C A, Labov S E, Lindeman M A., 1998 Rev. Sci. Instrum., **69**, pp25-31
Imamura T, Ohara S, Hasuo S., 1990 IEICE Tech Report, **SCE90-14**, pp31-36
Koshelets P V, Shitov V S, Filippenko L V, Baryshev M A, Golstein H, Graruw T, Luinge W, Schaeffer H, Stadt H., 1995 Appl. Phys. Lett, **68**, pp1273-1275
Nagasawa S, Numata H, Hashimoto Y, Tahara S., 1997 ISEC'97 Extended Abstracts, **2**, pp290-292
Nakagawa H, Kurosawa I, Aoyagi M, Kosaka S, Hamazaki Y, Okada Y, Takada S., 1991 IEEE Trans. Appl. Supercond., **1**, pp37-47
Rylov S V, Bunz L A, Gaidarenko D V, Fisher M A, Robertazzi R P, Mukhanov O A., 1997 IEEE Trans. Appl. Supercond., **7**, pp2649-52
Taino T, Ishibashi K, Maehata K, Yoshida S, Povinelli M M, Nakagawa H, Akoh H, Joosse K, Takada S, Kishimoto M, Katagiri M., 1998 Jpn. J. Appl. Phys., **37**, pp25-30
Verhoeve P, Rando N, Peacock A, Dordrecht A, Taylor B G, Goldie D J., 1998 Appl. Phys. Lett., **72**, pp3359-61 (in Japanese)

Inst. Phys. Conf. Ser. No 167
Paper presented at Applied Superconductivity, Spain, 14-17 September 1999

Influence of the junction edge losses on the spectral response of Superconducting Tunnel Junction Detectors

E Esposito[a][c], R Cristiano[a], L Frunzio[a], C Nappi[a], S Pagano[a], M Lisintski[a], G Ammendola[b], L Parlato[b][c], G Pepe[b], G Peluso[b], H Kraus[d], P Valko[d] and A Barone[b]

[a] Istituto di Cibernetica del CNR, I-80072 Arco Felice, Napoli, Italy
[b] INFM-Dipartimento di Scienze Fisiche dell'Università Federico II, I-80128 Napoli, Italy
[c] Osservatorio Astronomico di Capodimonte, I-80131 Napoli, Italy
[d] Particle and Nuclear Physics, University of Oxford, Oxford, United Kingdom

ABSTRARCT: A model to describe the spectral response of a superconducting tunnel junction detector in the presence of the quasiparticle diffusion, edge losses and backtunnelling is presented. Analytical solutions of the model for rectangular and circular junctions are discussed. The predictions of the theory are compared with an experimental spectrum obtained for a Nb-based junction.

1. INTRODUCTION

The efficiency of the charge collecting mechanism of Superconducting Tunnel Junction (STJ) detectors depends strongly on the competition between tunneling and losses of quasiparticles in the electrodes of the junction. In this article we discuss a model describing the influence, on the spectroscopic performance of a STJ detector, of the quasiparticle diffusion process in the presence of bulk and edge losses. Our model takes into account explicitly the possibility of multiple tunnelling of quasiparticles (i.e. the back-tunnelling effect). We assume that quasiparticles can escape tunnel in two ways: trapped at the external perimeter of the junction or lost in the bulk electrodes (Luiten et al 1997). The model is solved in two cases: rectangular junctions and circular junctions. The influence of the geometry on the spectral response is commented. A comparison of the theoretical prediction with an experimental spectrum of a Nb/Al$_2$O$_3$/Nb junction under X-ray irradiation is reported.

2. THE MODEL

The two electrodes of the junction are modelled as rectangular flat boxes of dimensions $a \times b$ or as disks of radius R. Since the thickness electrodes is very small in comparison with the size dimension of the junction area, we assume that the diffusion takes place predominantly in the junction plane. We also assume uniform loss capabilities in the volume of the electrodes and along the perimeter of the junction, and refer to them separately through two different parameters. We treat the situation in which a point-like event at r_o is absorbed in one of the two electrodes producing a number of excess quasiparticle N_0 while the other is initially unaffected. The model equations can be written as:

$$\frac{\partial n_1}{\partial t} = D\nabla^2 n_1 - \gamma_{12}n_1 - \gamma_{loss}n_1 + \gamma_{21}n_2$$

$$\frac{\partial n_2}{\partial t} = D\nabla^2 n_2 - \gamma_{21}n_2 - \gamma_{loss}n_2 + \gamma_{12}n_1$$

where n_1 and n_2 are the non-equilibrium excess two-dimensional quasiparticle densities in electrode 1 and 2 respectively, D is the diffusion coefficient and γ_{loss} is the bulk quasiparticle loss rate, $\gamma_{12}n_1$ is the flux of quasiparticles which tunnels from 1 to 2, while $\gamma_{21}n_2$ is the reverse flux (referred as the back-tunnel current). We restrict the treatment to the linear regime conditions (Ivlev et al 1991). The initial conditions for equations (1) and (2) are represented by a delta function for n_1 and zero for n_2. The boundary condition is the following:

$$\nabla n_i \cdot \hat{u}_n \big|_C = -\kappa n_i \qquad i = 1, 2$$

where C is the contour of the junction area (rectangular or circle), $\hat{u}_n$ is the unitary normal vector exiting the boundary, κ is a coefficient for the edge-quality, which can be related to the edge reflectivity R and to the mean free path of the quasiparticles: $\kappa = (1-R)/2l$ (Luiten et al 1997). Quasiparticles reaching the electrode edges have a finite probability $1-R$ to be lost there, getting out of the tunnelling area. We solved analytically equations (1) and (2) for n_1 and n_2 for both circular and rectangular geometry and used the results to evaluate the total number of collected charge carriers through the integral:

$$N_t = \int_0^\infty dt \int_S dS (\gamma_{12}n_1 + \gamma_{21}n_2)$$

S being the junction area. Analytical expressions of the above integral can be found in Cristiano et al (1999) for both the geometries and for any value of the ratio γ_{21}/γ_{12}. The main result of assuming loss at the edges in the presence of diffusion is a position dependent response $N_t(r_o)$ of the detector. In the presence of loss at the edges the number of quasiparticles tunnelling depends critically on the proximity of the absorption position to the edges. The closer absorption point is to the edge, the lower the number of quasiparticles tunnelling is. In addition to κ, three parameters influence the response of the detector in this model. They can be identified in the diffusion length of one of the two electrodes eg $\Lambda_1 = (D/(\gamma_{12} + \gamma_{loss}))^{1/2}$, the tunnelling probability $\gamma_{12}/(\gamma_{12} + \gamma_{loss})$ and the ratio γ_{21}/γ_{12}. The effect of back-tunnelling is represented by the average number of times each quasiparticle is transferred across the barrier $\langle n \rangle = \gamma/\gamma_{loss}$ (Goldie et al 1994). Once fully determined the STJ response $N_t^{(1)}$ and $N_t^{(2)}$ for each electrode, the spectral response under homogeneous irradiation and Gaussian broadening is:

$$s(N) = \sum_{i=1}^2 \frac{N_{tot}^{(i)}}{\sigma\sqrt{2\pi}} \frac{1}{S} \int_S dS \exp\left(-\frac{\left(N - N_t^{(i)}(\vec{r}_o)\right)^2}{2\sigma^2}\right)$$

where $N_{tot}^{(1)}$ and $N_{tot}^{(2)}$ are the numbers of total events occurring in the electrode 1 and 2 respectively and σ is the standard deviation.

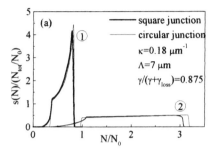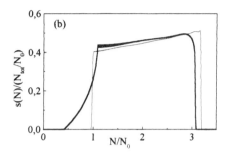

Fig 1 (a) Calculated spectra for a square junction 50x50 μm^2 (thicker line) and circular junction (thinner line) with equal area; label 1 refers to case without backtunnelling ($\gamma_{21}=0$) and label 2 to case of equal rate of direct and back-tunnelling; (b) same curves of (a) labelled 2 on larger scale.

Fig 1 shows for a particular set of a parameter values, corresponding to a considerable presence of edge loss, the effect of multiple tunnelling of quasiparticles. Multiple tunnelling manifest itself shifting the spectrum toward higher charge values and lowering the level of energy resolution. The difference on the spectral response due to different geometries is as expected negligible (Fig 1(a)), even though differences can be appreciated on a magnified scale (Fig 1(b)). Since for the same area the circular junction has less perimeter, a slightly better charge response is obtained in comparison with the square junction. The geometry has negligible influence on the determination of the fitting parameters. In fact, to achieve a good agreement between the two fits it is sufficient to decrease the tunnelling probability $\gamma_{12}/(\gamma_{12}+\gamma_{loss})$ parameter by not more than 1%.

3.   EXPERIMENT AND DISCUSSION

We have used the above model to evaluate the level of edge-quality in our junctions. $Nb/Al_2O_3/Nb$ junctions have been realised with the electrodes of equal thickness (200 nm) (Cristiano et al 1998). The spectrum presented in Fig 2, refers to a junction with a rhombus geometry with area of 50x50 μm^2. The junction was irradiated with a ^{55}Fe X-ray source at a base temperature of about 50 mK. The electronic noise of our set up was estimated to be $\sigma_{cal} = 37\times10^3$ e$^-$ (Esposito et al 1999) The experimental spectrum of Fig 2, shows the characteristic tail toward lower energy that in principle is due to the presence of losses at the edges. The solid curves in Fig 2, represent the fit of the theoretical model. The solid line is calculated using the solutions for circular geometry, while the dashed line is for the square one. Both simulation were obtained using the same value of the parameters: $\kappa=$ 0.2 μm^{-1}, $\Lambda=53$ μm and $\gamma_{12}/\gamma_{loss}=0.230$. The fits reproduce reasonably well the main features of the experimental spectrum, that is the presence of the two ^{55}Fe emission lines in both the right ratios of counts and the energy amplitudes and the low-charge part of the spectrum. The energy resolution was $\sigma=5\times10^3$ e$^-$ including the Fano factor and the tunnelling noise. As for the case illustrated in Fig 1, the shapes of the theoretical spectra exhibit the same principle features with only minor modifications in detail. Although the geometry does have an effect on the pulse height spectra, the most important parameters governing the spectroscopic response are the edge loss parameter, the diffusion length, and the tunnel probability. The parameters of the model have a different effects on the shape of the spectrum best fit: Λ is inversely proportional to the count level in the peak spectrum, whereas the peak position on the charge axis is strongly influenced both by κ and γ_{loss}. Assuming the value of the mean free path $l \sim 10$ nm for our polycrystalline Nb films, and $\kappa = 10$ we can estimate that the parameter 1-$R \sim 267 \times 10^{-3}$, indicating a strong loss at the edges compared with values of about 10^{-4} (Luiten et al 1997).

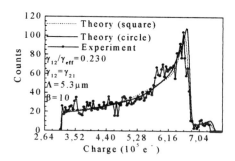

Fig 2 Experimental X-ray spectrum (dotted line) of a Nb/Al$_2$O$_3$/Nb STJ compared with theoretical curves in the case of circular geometry (solid line) and square one (dashed line).

4   CONCLUSION

We have developed a theoretical model to describe the influence of the loss of quasiparticles at the edges of STJ detectors under X-ray irradiation. Our model takes into account the multiple tunnelling of quasiparticles and can be used in the general case of asymmetrical rates of tunnelling $\gamma_{12} \neq \gamma_{21}$, for any value of the ratio γ_{12}/γ_{21}. We have shown that the geometry can slightly influence the spectral response of an STJ. Finally the comparison with an experimental situation has also been shown, demonstrating the usefulness of our model for extracting diagnostic information on our Nb-based STJ detectors.

ACKNOWLEDGEMENT

This work has been partially supported by INFN under the STJ-Detector Project and by the EC-TMR-Network "Cryogenic Detectors", Contract No ERBFMRXCT 980167.

REFERENCES

Cristiano R, Esposito E, Frunzio L, Pagano S, Barone A, Parlato L, Peluso G, Pepe G, Esposito A, Aoyagi M, Akoh H, Nagakawa H, Takada S 1998, JPhys IV France, **8** Pr3-275
Cristiano R Esposito E, Frunzio L, Nappi C, Ammendola G, Parlato L, Pepe G, Kraus H and Valko P 1999, Jour of Appl Phys To be publ october 1999
Esposito E, Cristiano R, Frunzio L, Lisintski M, Nappi C, Pagano S, Parlato L, Peluso G, Pepe G Barone A 1999, Int J of Modern Phys **13**, 1247
Luiten OJ, van den Berg M L, Gòmez Rivas J, Bruijn M P, Kiewiet F B, and de Korte P A J 1997, Proc, *Proceedings of the seventh international Workshop on Low Temperature Detectors (LTD-7), Munich 1997*, edited by S Cooper, pp A12, 25
Goldie DJ, Brink PL, Patel C, Booth N E, Salmon GL 1994, Appl PhysLett **64**, 3169
Ivlev B, Pepe G and Scotti di Uccio U 1991, Nucl Instr And Meth **A300,** 127

Inst. Phys. Conf. Ser. No 167
Paper presented at Applied Superconductivity, Spain, 14-17 September 1999
© 2000 IOP Publishing Ltd

683

Development of superconducting tunnel junctions with Al trapping layers for X-ray detectors

H Sato,[a] T Ikeda,[a] H Kato,[a] K Kawai,[a] H Miyasaka,[a] T Oku,[a] W Ootani,[a] C Otani,[a] H M Shimizu,[a] Y Takizawa,[a] H Watanabe,[a] H Nakagawa,[b] H Akoh,[b] M Aoyagi[b] and T Taino[c]

[a] RIKEN (The Institute of Physical and Chemical Research), 2-1 Hirosawa, Wako-shi, Saitama 351-0198, Japan
[b] Electrotechnical Laboratory (ETL), 1-1-4 Umezono, Tsukuba-shi, Ibaraki 305-8568, Japan
[c] Dept. of Applied Quantum Physics and Nuclear Engineering, Kyushu University, 6-10-1 Hakozaki, Higashi-ku, Fukuoka 812-8581, Japan

ABSTRACT: Nb-based superconducting tunnel junctions (STJs) with Al trapping layers have been developed for X-ray detectors. Some modifications were applied to our fabrication process to reduce the leakage current. The energy resolution was measured by using an ^{55}Fe X-ray source, and for 5.9 keV X-rays, 41 eV (FWHM) with a 20×20 μm^2 junction and 58 eV with a 100×100 μm^2 junction were achieved. According to the number of quasiparticles contributing to the tunneling process, the effective ε value, ε_{eff}, was 0.9 meV, and the charge amplification due to multiple tunneling was about 3.

1.   INTRODUCTION

Superconducting tunnel junctions (STJs) have the potential for good radiation detectors which can achieve high energy resolution and a high photon counting rate. The energy resolution mainly depends on the statistical fluctuation in the number of quasiparticles produced when a photon is absorbed in a STJ and on the electrical noise of the readout system. One of the solutions for relatively reducing the influence of this electrical noise is to increase the charge output by gathering quasiparticles near a tunnel barrier as much as possible. STJs having Al layers on both sides of the tunnel barrier ($Nb/Al/AlO_x/Al/Nb$) have been produced for this purpose and their validity has been demonstrated by other workers (for example, Frank et al. (1998) and Verhoeve et al. (1998)). Since the band gap energy of Al (0.17 meV) is smaller than that of Nb (1.5 meV), quasiparticles that have diffused into the Al layers are trapped in them. This property makes it possible to increase the tunneling rate and tunnel many times before their recombination into Cooper pairs (multiple tunneling), and the increased charge output can thus be measured.

In the early stage of our development, Nb-based STJs ($Nb/Al-AlO_x/Nb$) were produced by using the fabrication techniques studied at ETL (Joosse et al. 1996). During experiments on X-ray detection, we found that not so many ($\sim$ 30 %) of the quasiparticles produced in the Nb layers were detected in our Nb junctions. Therefore, we started the fabrication of STJs having Al trapping layers to increase the charge output and measurement of X-ray spectra. In this article, the fabrication technique and preliminary results of measuring the response for X-rays are reported.

2.   FABRICATION

STJs were fabricated at a facility in RIKEN. The fabrication process was based on that for Nb-based junctions that has already been reported (Sato et al. 1999). Some modifications were applied to this process in order to suit junctions having thick Al layers.

An Al_2O_3 layer (50 nm) as a buffer for phonons was deposited on a sputter-cleaned substrate. A multi-layer of Nb(200 nm)/Al(50 nm)/AlO_x/Al(50 nm)/Nb(150 nm) was deposited by DC magnetron sputtering without breaking the vacuum. Each junction was then structured by reactive ion etching. Before etching the Al layers, a thin SiO_2 layer was deposited to protect the subsequent Al layers from any damage caused by the developer during the photoresist treatment. After isolating the junctions from each other, the edges of the Al layers were oxidized in O_2 plasma to prevent any micro-shorting between them which may occur during the etching process. The entire wafer was next covered with an SiO_2 layer (400 nm). After making the contact holes, an Nb wiring layer (600 nm) was deposited and structured. The STJs were in a square shape having sizes of 20×20 μm^2, 100×100 μm^2, 200×200 μm^2, and 500×500 μm^2.

The typical properties of the STJs obtained with 100×100 μm^2 junctions were as follows: critical current density $J_c = 134$ A/cm^2, normal resistance $R_N = 88$ mΩ, subgap current $I_{sg} = 1.4$ nA at 0.2 mV and 0.35 K. I_{sg} obtained at 0.2 mV and 0.35 K was one of the parameters used to measure the barrier quality of the junctions. I_{sg} normalized by the value measured at 4.2 K is shown in Fig. 1, which was obtained for junctions made without oxidation of the edge of the Al layers (a) or with oxidation (b). By comparing (a) and (b), it can be seen that oxidation of the edge of the Al layers prevented micro-shorting and reduced the leakage current. High-quality junctions could thus be produced with this oxidation process.

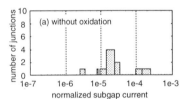 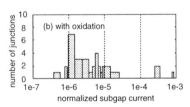

Fig. 1: Distribution of the subgap current obtained with 100×100 μm^2 junctions at 0.2 mV and 0.35 K. The horizontal axis shows the value normalized by I_{sg} measured at 4.2 K.

3.   EXPERIMENTS

3.1   Setup

The effect on charge amplification by the Al trapping layers was tested by measuring an X-ray signal with a ^{55}Fe X-ray source. A 5×5 mm^2 chip consisting of a number of STJs was mounted on a gilded copper plate. This plate was attached to the copper cold stage of a cryostat and cooled to 0.35 K by using liquid ^{3}He. A magnetic field of about 10 mT was supplied by two coils placed near the cryostat to suppress the dc Josephson current. In the case of high-quality junctions, the subgap current was very sensitive to the magnetic field trapped in the junctions. Consequently, special attention was paid to not trap the magnetic field before measuring the X-ray spectra. The signal from the STJ was passed to a multi-channel analyzer via a charge-sensitive or a current-sensitive preamplifier and a shaping amplifier, all being operated at room temperature.

3.2   Results

By integrating the output signal from the current-sensitive preamplifier, the charge contributing to the tunneling process, Q_t, was determined. Typically, $Q_t = 1.1$ pC was obtained for an $E = 5.9$ keV X-ray absorption. This corresponds to the number of tunneling quasiparticles of $N_t = 6.6 \times 10^6$. From N_t, the effective ε value, $\varepsilon_{\text{eff}} = E/N_t = 0.9$ meV was calculated. By comparing ε_{eff} with $\varepsilon = 1.7\Delta_{\text{Nb}}$, the charge amplification due to the multiple-tunneling effect was 3 in our STJs. Here, $\Delta_{\text{Nb}} = 1.5$ meV is the gap energy of Nb. Our Nb junctions which didn't have Al trapping layers typically gathered only $Q_t \sim 0.1$ pC for the absorption of a 5.9 keV X-ray. This result indicates that the Al layers worked properly for enhancing the charge by multiple tunneling.

Pulse height spectra for 5.9 keV X-rays measured at 0.35 K are shown in Fig. 2. These spectra were obtained with junctions fabricated on the same substrate (Si), i.e., they were made in the same batch, so the only difference between them was their size. The energy resolution (FWHM) was 41 eV for 20×20 μm^2, 58 eV for 100×100 μm^2, 65 eV for 200×200 μm^2 and 129 eV for 500×500 μm^2. Since the shape, the pulse height and the energy resolution of each peak strongly depended on the bias voltage and the external magnetic field, the spectra shown in Fig. 2 were obtained after adjusting the bias voltage and the magnetic field to sharpen one peak among two Kα peaks. In each spectrum, two sets of two peaks due to absorption in the bottom layer or in the top layer were observed. It was not easy to determine whether the set was peaks from the bottom layer or from the top layer. In our case, identification was made by using the difference in the counting rate of the X-rays and by an analysis of the rise time of the output signal (Fig. 3).

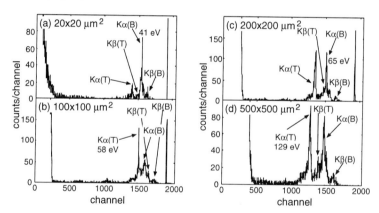

Fig. 2: Pulse height spectra for 5.9 keV X-rays from a ^{55}Fe source. Kα(T) means the signal of the Mn Kα line from the top layer, Kα(B) means that from the bottom layer, and so on. In (a), (b) and (c), the peak from the electric pulser is also shown in the 1900 channel. FWHM for each peak is 9 eV, 20 eV and 25 eV, respectively.

4.   DISCUSSION

The multiple-tunneling time can also be determined from the ratio of the tunneling rate and the recombination rate of quasiparticles. With our 100×100 μm^2 junctions, the inverse of the tunneling rate, $\Gamma_t^{-1} = 400$ ns, was calculated by using equation (1) from Mears et al. (1993), and the inverse of the recombination rate $\Gamma_r^{-1} = 1$ μs was measured from the output signal of the current-sensitive preamplifier. We could derive a multiple-tunneling time of 2.5 from these values, which is consistent with the result obtained from the charge output shown in previous section. Γ_t^{-1} is proportional to R_N. Therefore, in order to increase the multiple-tunneling time

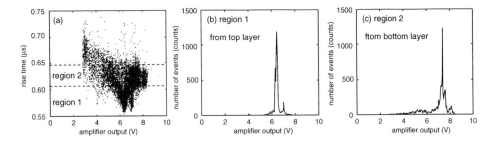

Fig. 3: Example of the rise-time analysis. (a) is a scatter plot of 5%-40% rise time as a function of output of the amplifier obtained with a $200 \times 200 \ \mu m^2$ junction. (b) and (c) are spectra obtained by using pulses in region 1 and 2, respectively.

to obtain better energy resolution for $\sim$keV X-rays, R_N should be smaller.

In Figs. 2 (c) and (d), an unexpected peak can be seen in the 1550 channel. According to the rise-time analysis, it was found that this peak accompanied a peak from the bottom layer. The origin of this peak has not been identified and it should be removed to obtain a clear spectrum for the application of STJs.

5.   SUMMARY

STJs having Al trapping layers were fabricated to a high level of quality. X-ray measurements confirmed the effect of the trapping of quasiparticles and multiple tunneling. Due to this charge amplification effect, values for the energy resolution for a 5.9 keV X-ray of 41 eV with a $20 \times 20 \ \mu m^2$ junction and of 58 eV with a $100 \times 100 \ \mu m^2$ junction were obtained. It is necessary to optimize the thickness of the tunnel barrier and Al trapping layers for even better energy resolution.

ACKNOWLEDGMENTS

H.S., T.I., T.O., W.O. and Y.T. are grateful to Special Postdoctoral Researchers Program for support of this research. This study was performed thorough Special Coordination Funds for promoting Science and Technology Agency of the Japanese Government. This work was supported in part by Grant-in-Aid for Encouragement of the Ministry of Education, science, Sports and Culture under the program numbers 10740136, 08554002 and 09490038.

REFERENCES

Frank M, Hiller L J, Ie Grand J B, Mears C A, Labov S E, Lindeman M A, Netel H, Chow D and Barfknecht A T 1998 Rev. Sci. Instrum. **69**, 25

Mears C A, Labov S E and Barfknecht A T 1993 Appl. Phys. Lett. **63**, 2961

Joosse K, Nakagawa H, Akoh H, Takada S, Maehata K and Ishibashi K 1996 Jpn. J. Appl. Phys. **35**, 2633

Sato H, Ikeda T, Kato H, Kawai K, Miyasaka H, Oku T, Ootani W, Otani C, Shimizu H M, Watanabe H, Nakagawa H, Akoh H, Aoyagi M and Taino T 1999 Proc. Applied Superconductivity Conference 1998, pp 4475

Verhoeve P, Rando N, Peacock A, van Dordrecht A, Poelaert A, Goldie D J and Venn R 1998 J. Appl. Phys. **83**, 6118

Inst. Phys. Conf. Ser. No 167
Paper presented at Applied Superconductivity, Spain, 14-17 September 1999
© 2000 IOP Publishing Ltd

Hot Electron Bolometric mixers based on NbN films deposited on MgO substrates

P Yagoubov, M Kroug, H Merkel and E Kollberg
Department of Microwave Technology, Chalmers University of Technology, S-412 96 Gothenburg, Sweden

J Schubert and H W Hübers
DLR Institute of Space Sensor Technology, D-12489 Berlin, Germany

S Svechnikov, B Voronov and G Gol'tsman
Department of Physics, Moscow State Pedagogical University, Moscow 119435, Russia

Z Wang
Communications Research Laboratory, Kobe, 651-2401, Japan

ABSTRACT: We report on heterodyne measurements with phonon-cooled type Hot Electron Bolometric (HEB) quasioptical mixers. The devices are based on thin NbN films deposited on MgO substrates. MgO as a substrate material has been shown to provide a good acoustic coupling to the film and hence a fast response time of the bolometer. Gain bandwidth measurements have been performed at 0.65 THz. An IF bandwidth of 3.7 GHz was measured for a mixer made from a 5 nm film. The noise temperature performance is investigated in the 0.6-2.5 THz frequency range. The best DSB receiver noise temperatures are: 480 K at 0.6 THz and 1,400 K at 2.5 THz.

1. INTRODUCTION

The aim of the research is to design and fabricate sensitive wideband heterodyne receivers for operation in the terahertz frequency range. These receivers are required for atmospheric observations and radio astronomical applications. Superconducting Hot Electron Bolometric (HEB) mixers represent an attractive candidate for providing high sensitivity heterodyne front-ends at frequencies beyond 1 THz where the sensitivity of SIS mixers degrades drastically. The key advantages of HEB mixers are: operating frequencies from millimeter waves up to far infrared, very low intrinsic noise, IF bandwidth of several GHz and low local oscillator power requirement. There are two types of HEBs under development: phonon-cooled version (Gershenzon 1990, Yagoubov 1998) and diffusion-cooled one (Prober 1993, Wyss 1999). The difference between the two is the cooling mechanism of the hot electrons. We are developing the phonon-cooled HEBs based on thin NbN films. Our earlier devices used Si and sapphire as the substrate material. The intermediate frequency bandwidth of mixers based on 3 nm thick films is 3-4 GHz. It is limited by both electron-phonon interaction time and phonon escape time (Yagoubov 1996). This paper addresses the question of an increase of the bandwidth by shortening the phonon escape time by using MgO as a substrate and we report corresponding receiver sensitivity measurements.

2. MIXER FABRICATION

The NbN films on MgO substrates are deposited by magnetron sputtering. This is done in laboratories in Moscow and Kobe (Japan), the details of the processes are documented in (Yagoubov 1996) and (Wang 1996), respectively. Based on the experience with NbN films on Si and sapphire, the deposition in the Moscow lab is done with substrate heating up to 700 C. In the Kobe lab the substrates are maintained at room temperature, a process originally developed for fabrication of SIS NbN mixers (Wang 1994). It is not clear yet whether the substrate temperature is a critical parameter for the quality of thin NbN films on MgO.

The results in terms of the superconducting properties of the films are comparable. The T_c for 5 nm films is about 9 K for the film, dropping down to 8-8.5 K after patterning. These numbers are lower than those for NbN films on sapphire substrates, about 11 K (Yagoubov 1996). However, only a few films with this thickness have been made so far and a further optimization of the film deposition process is needed.

The film is patterned by e-beam lithography to form a submicron size strip (typical dimensions are: 0.1-0.2 μm long and 2-4 μm wide) across the center gap of an Au logarithmic spiral antenna. The details of the fabrication process are described by Kroug (1997).

3. MEASUREMENT SETUPS

The HEB mixer chip is clamped to a 12 mm diameter elliptical Si lens. The mixer block is mounted in a LHe-cooled vacuum cryostat equipped with a 380 μm Zitex G115 IR radiation filter. For receiver sensitivity measurements we use two different experimental setups. In the frequency range 0.7-2.5 THz the local oscillator is a FIR laser with an output power of about 1 mW. Hot/cold (295/77 K) black body loads are used as a signal source in noise temperature measurements. A simple beamsplitter made from 6 μm Mylar foil couples the signal beam and the local oscillator onto the mixer. The device IF output is connected through a bias-T to a 1.3-1.7 GHz cooled HEMT amplifier with a noise temperature of ≈5 K. The output of the HEMT amplifier is filtered with a 1.5±0.02 GHz band pass filter, further amplified and detected in a power meter. At 0.6 THz the LO source is a BWO.

The IF bandwidth is measured at 0.65 THz with two BWOs serving as LO and signal sources. The radiation from the LO is focused by a Teflon lens and combined with the signal by a 12-μm-thick Mylar beamsplitter. The signal power can be varied by means of a quasioptical attenuator. The IF signal is amplified by a two stage 0.1-12 GHz room temperature Miteq amplifier and registered by a spectrum analyzer.

4. INTERMEDIATE FREQUENCY BANDWIDTH

The ultimate bandwidth of a HEB with phonon cooling mechanism is set by the material electron-phonon interaction time $\tau_{e\text{-}ph}$. NbN is a superconductor with a very short $\tau_{e\text{-}ph}$. This time constant depends on the temperature of the electron gas Θ, with $\tau_{e\text{-}ph} \sim \Theta^{-1.6}$ (Gousev 1994). Since $\Theta \leq T_c$, a film with a high T_c is desirable. At $\Theta = 10$ K $\tau_{e\text{-}ph}$ is equal to 12 ps, corresponding to 13 GHz maximum IF bandwidth. This bandwidth can be realized when non-equilibrium phonons escape the film fast enough without being reabsorbed by the electrons. However, the phonons may dwell in the film which results in decreased bolometer speed. In the ballistic model the phonon escape time τ_{esc} is proportional to the film thickness, coefficient of the acoustic mismatch between film and substrate and inversely proportional to the velocity of sound in the film. Therefore, in order to obtain a wide mixer IF bandwidth one must use ultrathin films and provide a good matching between the film and substrate.

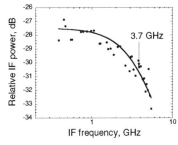

Fig. 1. IF bandwidth measurements with the mixer based on a NbN film deposited on a MgO substrate.

TABLE. DEVICE PARAMETERS AND RESULTS OF NOISE TEMPERATURE MEASUREMENTS.

Device number	M2_1	M2_2	M2_8	J2_1	J2_4
R_{20K}, Ohm	85	165	120	100	60
Bol. length, μm	0.4	0.9	0.25	0.2	0.1
width, μm	4	4	2	2	2
I_c, μA (at 4.2 K)	470	470	230	240	310
DSB NT, K*					
0.6 THz	530	480	-	-	530
0.7 THz	-	-	750	650	-
2.5 THz	-	-	1400	1500	-

*Measurements are performed at 2.2 K ambient temperature.

A systematic investigation of the bandwidth of phonon-cooled NbN HEB mixers on sapphire substrates has been done by Cherednichenko (1997). The measurements were performed at 140 GHz in a waveguide setup. The gain bandwidth varies from 0.9 GHz for a 10 nm thick NbN film to 4 GHz for a 2.5-3 nm film. The results can be explained by the two-temperature model of Perrin and Vanneste (1983). This model describes the situation where not only electrons, but also phonons are being heated up. Fitting experimental data into the model gave τ_{esc} proportional to the film thickness and equal to 13 ps per 1 nm. Hence, even for 3 nm films $\tau_{esc} > \tau_{e-ph}$ which results in a bandwidth being far smaller than its theoretical limit.

A larger bandwidth might in principle be possible by using even thinner films. However, with current film technology a smaller film thickness is accompanied by a decrease of the critical temperature T_c (Cherednichenko 1997). This, in turn, makes τ_{e-ph} longer (since $\Theta \leq T_c$) resulting in a lower speed of the bolometer.

A way to increase the mixer bandwidth, which we propose in this work, is to shorten the phonon escape time by utilizing a substrate with better acoustic coupling to the film, e.g. MgO, which has the same crystalline structure as NbN. This may allow epitaxial growth of the film (Wang 1996) and could provide a better acoustic matching.

We have measured the bandwidth of mixers with two different film thicknesses, 15 nm and 5 nm. The measurements are performed with optimum LO power and dc bias with respect to a maximum IF signal. For the device based on the 15 nm film the bandwidth is 1.3 GHz, for the 5 nm film it is 3.7 GHz, Fig. 1. Both numbers are considerably higher than values we have measured previously using NbN films with the same thickness and T_c on other substrates such as Si or sapphire. The estimate in the framework of the two-temperature model shows that for both devices the phonon escape time is approximately 6 ps per 1 nm film thickness. This can only be attributed to the better acoustic coupling between the film and the substrate.

Better quality (higher Tc) ultrathin (down to 3 nm) films on MgO would allow for at least 5-6 GHz gain bandwidth with this type of HEB.

5.   RECEIVER NOISE TEMPERATURE

The noise temperature measurements were performed with five devices, three from batch M2 and two from batch J2. The first batch (M2) was made from a NbN film deposited on a MgO substrate at 700 C. The second batch (J2) was deposited on MgO at room temperature. The device parameters and the results of the receiver noise temperature measurements are presented in the table. All results are DSB receiver noise temperature, no corrections have been made for optical and mismatch losses as well as the noise contribution from the IF chain.

Mixers from both batches show excellent and similar noise performance. The results are comparable with the best results obtained with devices based on NbN films on Si substrates at lower frequencies and better at 2.5 THz (Yagoubov 1999).

Since the film T_c of both batches is quite low, about 8 K, the noise temperature measurements were performed at 2.2 K ambient temperature. This was achieved by pumping on the He chamber of the cryostat. At 4.2 K ambient temperature the receiver sensitivity is about 15-20% worse.

6. CONCLUSION

We have fabricated and tested NbN phonon-cooled HEBs in the frequency range 0.6-2.5 THz. The devices showed excellent results in terms of sensitivity. The noise temperature at 2.5 THz represents state-of-the-art performance of heterodyne receivers. An improvement in IF bandwidth has been made with utilization of MgO as the substrate material. Optimization of the ultrathin film deposition process will lead to further increase of the bandwidth.

REFERENCES

Cherednichenko S, Yagoubov P, Il'in K, Gol'tsman G and Gershenzon E 1997 Proc. 8th Int. Symp. on Space Terahertz Technology, Cambrige, MA, 245

Gershenzon E M, Gol'tsman G N, Gogidze I G, Elant'ev A I, Karasik B S and Semenov A D 1990 Sov. Phys. Superconductivity, **3**, 1582

Gousev Yu P, Gol'tsman G N, Semenov A D, Gershenzon E M, Nebosis R S, Heusinger M A and Renk K F 1994 J. Appl. Phys., **75**, No 7, 3695

Kroug M, Yagoubov P, Gol'tsman G and Kollberg E 1997 Proc. the 3rd. European Conference on Applied Superconductivity, Veldhoven, Netherlands, (Inst. Phys. Conf. Ser. No 158, p.405)

Perrin N and Vanneste C 1983 Phys. Rev. B, **28**, 5150

Prober D 1993 Appl. Phys. Lett. **62**, 2119

Wang Z, Kawakami A, Uzawa Y and Komiyama B 1994 Appl. Phys. Lett., **64**, No 15, 2034

Wang Z, Kawakami A, Uzawa Y and Komiyama B 1996 J. Appl. Phys., **79**, No 10, 7837

Wyss R, Karasik B, McGrath W, Bumble B and LeDuc H 1999 Proc. 10th Int. Symp. on Space Terahertz Technology, Charlottesville, VA, 214

Yagoubov P, Gol'tsman G, Voronov B, Seidman L, Siomash V, Cherednichenko S and E. Gershenzon 1996 Proc. 7th Int. Symp. on Space Terahertz Technology, Charlottesville, VA, 290

Yagoubov P, Kroug M, Merkel H, Kollberg E, Gol'tsman G, Svechnikov S and Gershenzon E 1998 Appl. Phys. Lett., **73**, 2814

Yagoubov P, Kroug M, Merkel H, Kollberg E, Schubert J, Huebers H.-W, Schwaab G, Gol'tsman G and Gershenzon E 1999 Proc. 10th Int. Symp. on Space Terahertz Technology, Charlottesville, VA, 214

Inst. Phys. Conf. Ser. No 167
Paper presented at Applied Superconductivity, Spain, 14-17 September 1999
© *2000 IOP Publishing Ltd*

Particle detection with geometrically-metastable type I superconducting detectors

M.J. Gomes and T.A. Girard

Centro de Física Nuclear, Universidade de Lisboa, 1649-003 Lisboa, Portugal

C. Oliveira

Physics Department, Instituto Nuclear e Tecnologia,, 2685 Sacavem, Portugal

V. Jeudy

Groupe de Physique des Solides (UMR CNRS 75-88), Universités Paris 7/6, 75251 Paris, France

ABSTRACT: Recent results in the electron irradiation of geometrically-metastable tin lamina by Jeudy et al. suggest the use of rhenium in detector development as a result of its intrinsically higher stopping power. We report a similar response study with lamina of 25 and 50 micron thick rhenium lamina at 330 mK under electron and X-ray irradiation, using a fast-pulse induction system which records the irreversible flux transfer associated with the particle interactions. Experiments yield fully-resolved energy spectra, with a linear response and high efficiency. The results additionally provide information on the thermal evolution of the geometrical energy barrier, and associated flux instabilities.

1. INTRODUCTION

Low energy, high resolution detectors are of increasing interest in particle and astrophysics. Superconducting detectors operated well-below their critical temperatures can in principle obtain improved resolutions over semiconductors by virtue of the order meV binding of the Cooper pairs: Two types of devices have been pursued over the last several years: tunnel junction (STJ) and superheated grains (SSG). The STJ has been extensively investigated (Booth 1996): energy resolutions of 0.7-1% for 5.9 keV X-rays (Verhoeve 1998) with a detection efficiency ~50% and timing resolutions of 20 ms have been obtained; similar results have been achieved in the UV-near IR (Peacock 1996). STJ's are generally of small volume (100-300 nm thickness, 40 x 40 μm^2) and must be operated in the few mK range; low production yield, extremely nonlinear response, and vibrational sensitivity remain practical limitations to their application. SSG devices have received relatively less attention (Turrell 1997): X-ray spectra, with about 20% $\Delta E/E$ have been reported (Heres 1993), but the response is again extremely nonlinear.

Recently Jeudy et al. (1996) reported the thermally-induced destruction of the geometric barrier in Sn lamina by the decay electrons of ^{35}S and ^{109}Cd. In both cases, the response was found to be linear in energy to within 1% over the energies involved. In the ^{109}Cd experiments, first spectra of the monoenergetic internal conversion decay electrons were obtained with ~15% resolution at 60-80 keV, but with an efficiency estimate of < 10%.

Although the geometric barrier in Type I structures is a well-recognized phenomenon (Jeudy 1991), there is no theoretical description of its thermal destruction. In principle, irradiation of the material produces a local heating that induces, as in the magnetic nucleation case, the penetration of flux bundles into the strip volume. The inhomogeneity of the global flux structure in the strip implies that the thermal nucleation efficiency should depend on the position of the heated zone -- the particle must release heat near the strip edges, where the energy barrier is present, to induce flux penetration (Jeudy 1996). To address the question of efficiency and whether the full diamagnetic volume of the strip is sensitive, rather than only the near-edge region, we have investigated the geometric barrier in a rhenium strip. Here, the electron source is internal: rhenium is 62% naturally-occuring [187]Re, a beta decay with decay rate (Firestone 1996) of order 20 Bq/mm[3]. The maximum decay energy is ~2.6 keV, implying an electron range of ≤ 1 μm (Sternheimer 1984) so that the natural decay constitutes a pointlike, homogeneously-distributed irradiation source.

2.   EXPERIMENTAL DISPOSITION

Measurements were performed on 99.99+% pure, annealed, pinhole-free rhenium polycrystalline ribbons of 25-50 μm thickness (L_y), 670-900±20 μm width and 1.5±0.10 cm length. The strips were mechanically fastened to the cold plate of a single shot [3]He refrigerator, as shown in Fig. 1: a 1.5 μm thick mylar sheet was intercalated between the strip and a U-shaped copper pickup coil supported by an epoxy board; the strip was in direct contact with the refrigerator cold plate, which was centered on the axis of an external magnet coil positioned coaxially with the refrigerator. The relative stability of the field was better than 2×10^{-4}, with a homogeneity of better than 1% over the strip location.

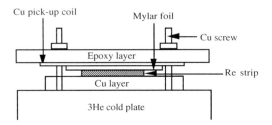

Fig. 1: Sample mounting within the working surface of the refrigerator.

Flux penetration was detected by the copper loop surrounding the strip, which was transformer-bridged to a room temperature, LeCroy HQV810-based, preamplifier-amplifier chain. Flux variations in the strip are detected only if they occur in a time shorter than the few nanoseconds risetime of the preamplifier (Jeudy 1991). The signal was discriminated with a LeCroy MVL407 fast voltage comparator; pulse-height analyses was achieved by repeated cycles with mV variations of the discrimination threshold under otherwise identical conditions.

The device was zero-field cooled to 330 mK. The magnetic field was then ramped upward to 250 G, well-above H_c (210 G), and returned to zero. The signal in this protocol is generally a mixture of magnetically- and thermally-induced events: to isolate those associated with the irradiation, a pause of 20 s was inserted in the field upramp, indicated by the dotted line in Fig. 2. Each measurement cycle required 50 s to execute.

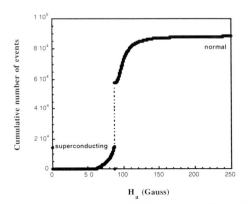

Fig. 2: Typical signal acquisition with magnetic field increase, at T = 330 mK, with pause.

The transitions during the pause were recorded separately. They are typically exponential as seen in Fig. 3(a). A typical pulse height distribution, obtained from successive differences in threshold variation measurements, is shown in Fig. 3(b).

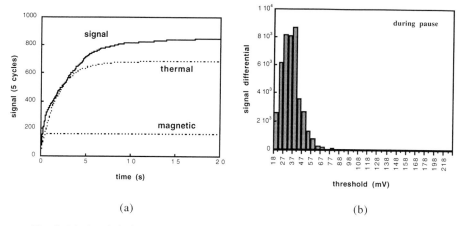

Fig. 3: (a) signal timing curve during a pause; (b) typical pulse-height distributions at T = 330 mK during a pause.

3. DISCUSSION

The results of Fig. 3(a) can be described as the sum of two exponential functions of the form $N(t) = N_M + N_T$, with

$$N_j(t) = n_j \left(1 - e^{-t/\tau_j} \right) , \qquad (1)$$

where N_M corresponds to the magnetic contribution, and N_T, to the thermal. Fitting of the data yields the typical decomposition shown in Fig. 3(a), with $\tau_M = 0.174 \pm 0.01$ s, $\tau_T = 3.18 \pm 0.01$ s. The τ_M corresponds approximately to the decay of the applied magnetic field damping in the coils (t = 130 ms), and a penetration of flux tubes into the superconducting strip induced by the energy deposition of the incident radiation.

The saturation of the timing curve in Fig. 3 results from regeneration of the geometric barrier by new magnetic flux in the strip central region; this implies a time-dependent measurement efficiency (ε), which can be expressed as

$$\varepsilon (t) = \tau_T (\Delta t)^{-1} \left[1 - e^{-\Delta t/\tau_T} \right] , \qquad (2)$$

where Δt is the pause time, yielding $\varepsilon \sim 16\%$ with the present experimental values. The thermal contribution is ~6 Hz, in agreement with estimates of the decay rate anticipated from the volume distribution of decays (Firestone 1996), implying a full volume sensitivity to the irradiation. Measurements made at various pause fields further yield a saturation amount which varies in a Gaussian fashion over the range of first/last penetration fields with a centroid at the midpoint, indicating control by the barrier presence.

Since the time variation of the magnetic flux in the loop (dϕ/dt) is integrated by the electronics, a signal pulse-height is proportional to the flux contained in the thermally-nucleated normal tube,: pulse-height $\sim \int(\partial\phi/\partial t)dt = \int \mathbf{B}\cdot d\mathbf{S} = H_cS$, with S the surface area of the nucleated flux tube, The diameters of the bundles can be estimated from the amplitude distribution of Fig. 3(b), assuming that the surface area of the strip is approximately equal to the sum of the surface areas of each bundle. The bundle diameters are distributed within the range 9-100 μm. The smallest detected flux bundle, limited by the 17.5 mV noise level in these measurements, contains approximately 600 ϕ_0 ($\phi_0 = 20.7$ G·μm²), the largest, a factor of about 200 more.

Energy resolution in this device is thus predicated on the ability to resolve flux bundle size. The measurement step-size was 5 mV, corresponding to 120$\phi_0 \sim 135$ eV, insufficient to resolve the K_α, K_β lines of the ^{55}Fe decay used in calibration. The minimum step-size permissable with present equipment is 0.1 mV, implying an ultimate step-size of $2.4\phi_0$ ~ 2.7 eV.

The observed linearity of the spectra in energy (Jeudy 1996) is also observed in these experiments. This implies that, at a given T, the nucleated volume L_yS is :

$$\Delta E / [L_yS(T,\Delta E)] = \text{constant (T)} , \qquad (3)$$

where ΔE is the deposited energy. Eqn. (3) is consistent with the heating per unit volume, $\int C_x dT$, being only material-dependent.

4.   CONCLUSIONS

The full volume of the strip is sensitive to energy deposition, rather than only the edge regions. This implies that the flux is internally generated by the irradiation, rather than induced from the intermediate state edge structure by heating. The efficiency of such devices, operated in a "pause mode", is however seen to be only 16% of the total decay during the measurement period; a "no pause" measurement mode demonstrates an efficiency of ~100%, but would be limited to energies > 3 keV.

Energy resolution in this device is predicated on the ability to resolve flux bundle size: with present equipment, the minimum step-size is 0.1 mV, implying an ultimate detectable flux variation with the current system ~2 ϕ_0, or ~2.7 eV.

The current timing resolution of these devices is ~10 μs owing to the transformer presence in the electronics; resolutions of order 10 ns have been previously demonstrated (Furlan 1991).

Eqn. (3) implies that increase of L_y should yield smaller pulse amplitudes for the same ΔE, so that the strip thickness itself may act as an variable amplifier. With some twenty type I superconductors, there is a large range of detector design flexibility.

We acknowledge fruitful discussions with Prof. G. Waysand (Université de Paris) and Prof. Grzegorz Jung (Beer-Sheva University) in the completion of this work. The activity was supported in part by Programa CERN grants FAE/1154/97 and FAE/1211/98 of the Ministry of Science & Technology of Portugal.

REFERENCES

Booth N, Cabrera B and Fiorini E 1996, Ann. Rev. Nucl. Part. Science, 46.
Firestone RB 1996 Table of Isotopes (New York, Wiley).
Furlan, M et. al. 1991 Low Temperature Detectors for Neutrinos and Dark Matter-III (Gif-sur-Yvette, Ed. Frontiers).
Heres O, Dubos H, Girard TA, Limagne D, Perrier P, Torre JP and Waysand G. 1993 Jour. Low Temp. Phys. **93**, 449.
Jeudy V, Limagne D and Waysand G 1991 Europhys. Lett. **16**, 491 and references therein.
Jeudy V, Collar JI, Girard TA et. al. 1996 Nucl. Instr. & Meth. **A373**, 65.
Peacock A, Verhoeve P, Rando N, Perryman MAC, Taylor BG and Jacobsen P 1996 ESTEC preprint 28.7.
Sternheimer RM, Berger MJ and Seltzer SM 1984 At. Dat. Nucl. Dat. Tabl. **30**, 261.
Turrell BG 1997 Proc. 7th Intern. Workshop on Low Temperature Detectors (Munich, Max Planck), p. F1.
Verhoeve P, Rando N, Peacock A and Van Dordrecht A 1998 Appl. Phys. Lett., 2072.

Inst. Phys. Conf. Ser. No 167
Paper presented at Applied Superconductivity, Spain, 14-17 September 1999
© 2000 IOP Publishing Ltd

(YBCO/STO)$_n$ multi-layers : a way to avoid substrate micro-machining of microbolometers ?

N Cheenne[*+]**, D Robbes**[*]**, J P Maneval**[°]**, S Flament**[*]**, B Mercey**[+]**, J F Hamet**[+]** and L Mechin**[*]

[*] GREYC-Instrumentation – UPRES-A 6072 – ISMRA – 6, Bd M^{al} Juin – 14050 CAEN – France
[+] CRISMAT – UMR 6508 - ISMRA – 6, Bd M^{al} Juin – 14050 CAEN – France
[°] LPMC – ENS Paris – 24, rue Lhomond – 75232 PARIS Cedex 05 – France

ABSTRACT: We have refined thermal conductance measurements for two distinct types YBCO/STO buffer/MgO structures : YBCO single layer and YBCO decoupled by a superposition of ultra-thin YBCO/STO bilayers. Excitations of the films were performed using a modulated laser beam at 780 nm that heats a reference sample and the multilayered one. Dispersion in the measured values is thought to be associated with local defects, as suggested using an OBIV imaging technique based on confocal microscopy. In addition, we propose a novel method, using the transient response to critical current steps, for measuring the phonon escape times at 77 K.

1.   INTRODUCTION

The sensitivity of microbolometers made with thin films is highly dependant on the heat escape from the film to the substrate. Méchin *et al.* (1997) have used micromachining techniques to decrease this heat flow.   However, these techniques lead to fragile detectors because of the suspended structure they occur. A solution to this problem based on work of Sergeev *et al.* (1994) has been investigated.   Using the high value of thermal boundary resistance (R_{bd}) between YBCO/STO and increasing the number of interfaces between the top YBCO film and the substrate an increase of the thermal resistance (R_{eq}) between the latter is expected. In the first part, we resume the description of films growth that can be found elsewhere (Robbes *et al.* (1999)) with more details and give the result of our first measurements based on electric response of strip under light illumination. In section 3, a way to reveals local material defects using confocal microscopy is presented.   Section 4 presents results obtained by a new method, using transient response to critical current step, giving access to the phonon escape times at 77 K.

2.   MATERIAL GROWTH AND CHARACTERISATION

2.1   Material description

The films were grown by pulsed laser deposition except for the gold terminal film used as electrical contact areas.   The substrate used was MgO (001) with an SrTiO$_3$ buffer layer. Fig. 1 shows schematically the structures of the reference and of the multilayered films. The upper YBCO films had Tc around 88K after patterning of device on the film. Expected X Ray Diffractogram was found for the superlattice, but its study revealed a poor oxygenation of the YBCO in the superlattice : the calculated mean c parameter was found equal to 11.82 Å. TEM study confirmed this defect and

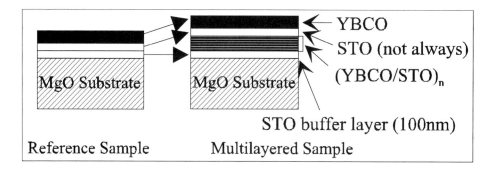

Fig. 1 : Schematic structure of the specimens

let also appear short-cuts between two consecutive YBCO layers across STO layer. Further detail may be found in Robbes *et al.* (1999).

The microbridges on the specimen were patterned by classical photolithography. First dimension of the bridges was 600 μm x 40 μm but because of the heterogeneity of the strip (see § 3) smaller microbridges were etched in the strips (length less than 10 μm and less than 5 μm wide depending on specimen).

2.2 Thermal boundary conductance measurement

The basic principles of our method have been previously described by Langlois *et al.* (1994).

Table 1 gathers the differences between films that vary by the number of bilayers YBCO/STO, by the thickness of films and by the presence of an STO layer between the superlattice and the top YBCO film. The values of the thermal conductance between the top YBCO film and its substrate (G_{eq}) for these specimen are also given. The increase of G_{eq} with the number of bilayers in the superlattice is well observed, except for Ley 4579, which possesses a low Tc (6 K under the other films), most likely a result of the poor quality of the film.

An effect of the thickness of bilayers YBCO/STO appears clearly, comparing the two first specimen to the others. The effect of the thermal insulation associated to the superlattice is greater in the case of Ley 3704 (a factor more than 10 is found for only 10 bilayers approximately 11.3 nm thick) than in the case of Ley 4584 (a factor around 2.5 is found for 20 bilayers 8.0 nm thick) if Ley 3657 is chosen as the reference. A manifestation of phonon wavelength could be seen in this result. Its value could draw near that of the bilayer thickness and then the phonon could become "blind" to the boundary.

Name of the film	Number of bilayers	Thickness of one bilayer (nm)	Thickness of YBCO top film (nm)	STO layer between YBCO and superlattice	Temperature measurement (K)	Thermal conductance between YBCO film and substrate (W/K/cm²)
Ley 3657	0		100	No	86	1000
Ley 3704	10	11.3	200	No	86	60-100
Ley 4583	0		100	Yes	86	500
Ley 4579	10	8.0	100	Yes	80	480
Ley 4584	1	8.0	100	Yes	86	380-440
Ley 4581	5	8.0	100	Yes	86	280-340
Ley 4582	20	8.0	100	Yes	86	160-180

Table 1 : Main characteristic of the specimens

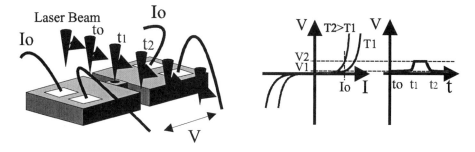

Fig. 2 : Principe of OBIV imaging

3. OBIV IMAGING REVEALS HETEROGENITY OF THE SPECIMENS

The value of G_{eq} found for Ley 4583 is lower than the expected value by a factor of 2. This result could be explained by the relative non-homogeneity of the films. This fact has already been shown by our previous work (Robbes *et al.* (1999)) by displacing the laser spot (our excitation source) along the strip. The electric response was not constant revealing more sensitive area on the strip.

Divin and Shadrin (1994) have used laser scanning microscopy for visualising electrical inhomogeneities of high Tc superconducting films but only at room temperature. Flament has developed a similar system enabling measurements down to 80 K. During the experiments, the current biased samples are placed in a freezing stage equipped with a glass window. The scanning focused laser beam of a confocal microscope heats locally the cooled samples and induces an increase of the sample voltage (see Fig. 2). By storing the value of the voltage change at each scanned point, we obtain a grey scaled image consisting of 256 x 256 points with 8 bit signal resolution showing the homogeneity of the active area. Figures 3 to 5 have been obtained with this non destructive technique we call OBIV imaging for Optical Beam Induced Voltage imaging.

Fig. 4 is a high magnified OBIV image of a 40 x 600 µm² strip on LEY 4582. A more sensitive part is clearly seen on the left part close to the contact area. We must mention at this point that this hot spot has been observed at the same position in the strip whatever was the bridge orientation or position in the scanned area. The measured signal is then actually due to a local extreme sensitivity and not to any parasitic electrical resonance. In Fig. 5, the OBIC image of an other strip of Ley 4582 is superposed (in the middle) on the optical one. In that case, both scanning speed and amplifying system bandwidth have been properly chosen so to avoid any lateral shift of the OBIV image from the optical one. Although LEY 4582 is one of the most uniform specimen, both active and passive parts are still visible. This non homogeneous response is more evident in Fig. 6 showing the response of LEY 4584 after some degradation of the film. Only the lower edge of the strip is active.

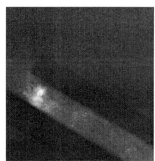

| Fig. 3 : Ley 4582 | Fig. 4 : Ley 4582 | Fig. 5 : Ley 4583 |

This technique could be very interesting to estimate the effective sensitive area of a strip and correct the G_{eq} given in section 2. For example, the G_{eq} value calculated for Ley 4583 is expected to be at least 2 times greater. This small value could be explained by a bad estimation of the absorbing surface of the strip. To visualise local active zone could also be used to etch sensitive structures (like junction, SQUID…) precisely in this special areas. Hence, OBIV imaging is then of great interest.

4. A NOVEL DETERMINATION OF YBCO FILM THERMAL COUPLING

An important information about heat transfer characteristics between a film and its substrate may emerge from a seemingly purely electronic effect. In narrow superconducting bridges, a larger than critical current ($I>I_c$) induces dissipation by PSC (Phase-Slip Centers) (Skocpol et al. (1974)) rather than by vortex flow. The point is that PSC's nucleate in a finite time t_d governed by an equation derived from Ginzburg-Landau by Pals and Wolter (1979) :

$$t_d \, (I > I_c) = \tau_\Delta \int_0^1 \frac{2x^4 dx}{(4/27)(I/I_c)^2 - x^4 + x^6} \qquad \text{(Eq. 1)}$$

in which the gap relaxation time τ_Δ is the only adjustable parameter. (Another version of this equation is necessary when T is not in the vicinity of T_c). Gap relaxation is usually related to electron-phonon interactions ; however, in many practical situations, it is limited by the slower process of phonon escape from the film. The corresponding time τ_γ, instead of τ_Δ, becomes the relevant prefactor of the integral, as it has been shown for YBCO films (Jelila et al. (1998)). One passes from τ_γ to the thermal conductance g_{bd} by Eq. 2 :

$$g_{bd} = \frac{C'd}{\tau_\gamma} \qquad \text{(Eq. 2)}$$

(d : film thickness ; C' (specific heat per unit volume) = 1.27 J/K.cm^3 around 80 K).

To establish experimentally the proposed procedure, we have submitted a number of epitaxial YBa$_2$Cu$_3$O$_{7-x}$ films to current steps and measured t_d on the nanosecond scale as a function of the ratio I/I_c in order to eventually obtain τ_γ from a fit to Eq. 1. Our τ_γ's turn out to be temperature-independent in a wide range (1 to 83 K), and proportional to film thickness from 30 to 150 nm. We found a rather uniform value of the ratio $\tau_\gamma/d = 75$ ps/nm (Harrabi and Ladan (1999)) for epitaxial YBCO deposited on crystalline MgO, whatever the temperature of the measurement. This is consistent with the model of mismatch-limited phonon radiation.

From this value, we obtain an interface conductance g_{bd} around 1670 W/K.cm^2 in semi-quantitative agreement with the above quoted values. We stress that the present method provides the coefficient of heat transfer *at any temperature, even in the superconducting state.*

REFERENCES

Divin Y Y and Shadrin P M 1994 Physica C **232**, 257-262

Harrabi Kh and Ladan F-R 1999, EUCAS Satellite Conf. SMART'99

Jelila F S, Maneval J P, Ladan F-R, Chibane F, de-Ficquelmont A, Méchin L, Villégier J C, Aprili M and Lesueur J 1998 Phys. Rev. Lett. **81**, 1933

Langlois P, Robbes D, Lam Chok Sing M, Gunther C, Bloyet D, Hamet J F, Desfeux R and Murray H 1994 J. Appl. Phys. **76**, 6

Méchin L, Villégier J C and Bloyet D 1997 J. Appl Phys. **81**, 10

Pals J A and Wolter J 1979 Phys. Lett. **70A**, 150

Robbes D, Cheenne N, Hamet J F and Mercey B 1999 Physica C Special issue Digital Applications, Josephson Junctions and Sensors

Sergeev A V, Semenov A D, Kouminov P, Trifonov V, Goghidze I G, Karasik B S, Golt'tsman G N and Gershenzon E M 1994 Phys. Rev. B **49**, 13

Skocpol W J, Beasley M R and Tinkham M 1974 J. Low Temp. Phys. **16**, 145

Inst. Phys. Conf. Ser. No 167
Paper presented at Applied Superconductivity, Spain, 14-17 September 1999
© *2000 IOP Publishing Ltd*

A Self-Consistent Hot-Spot Mixer Model for Phonon-Cooled NbN Hot Electron Bolometers

Harald Merkel, Pourya Khosropanah, Pavel Yagoubov, Erik Kollberg

Department of Microelectronics, Chalmers University of Technology, S-412 96 Göteborg, Sweden

ABSTRACT—A device model for superconducting hot electron bolometers (HEB) is set up using a hot spot mixing concept based on a coupled system of nonlinear one dimensional heat transport equations for electrons and phonons. Closed form solutions for the electron and phonon temperature profiles are available for a polynomial expansion for the electron thermal conductivity and for strictly localized electron-phonon interaction. The hot spot length is used in a small signal model predicting IV curves, conversion gain and noise performance for NbN HEB.

1. INTRODUCTION

Since feasibility of Hot-Electron Bolometric (HEB) mixers at THz frequencies has been shown, the objective of bolometer research focuses now on optimization of the IF bandwidth, the required LO power and noise by improving mixer geometry, topology and new materials. Therefore self-consistent physics-based device models are needed to reliably predict and extrapolate device parameters. Here a device model valid for phonon- and diffusion-cooled HEB is presented. Nonlinear one-dimensional (1D) heat transport equations for electrons and phonons serve as a model basis including spatially non-homogenous heating. These relations are purely 1D unless the bridge's width does not exceed twice the lateral penetration depth (about 1 μm for NbN). Electron-phonon-interaction is the dominant cooling channel for NbN HEB. In order to describe the transition to normal state correctly, linearization is not possible. Nevertheless the nonlinear heat balances reduce to a single nonlinear relation where closed form analytic solutions are available under the following conditions: The electron thermal conductivity in the superconductor λ depends exponentially on the electron temperature. Using $\lambda \propto T_e^{2.6}$ instead does not introduce a significant error. Assuming the phonons moving purely randomly, almost no power is transported by phonons in film direction and the term in the heat balance vanishes. Solving these heat balances (see section 2), HEB mixing is explained by a hot spot changing in length due to RF, LO and dc bias heating changes (c.f. Wilms-Floet (1999) and Merkel (1999)). The resulting time varying resistance change gives rise to a parametric oscillation (Moon (1992)) treated in section 3.

2. NONLINEAR LARGE SIGNAL MODEL

For unpumped IV curves the device is in Meissner state for small voltages. Exceeding a critical current, relaxation oscillations are observed and power is dissipated in the HEB. Above a critical heating power electrons will reach the critical temperature, a normal conducting hot spot is created and stable operation is possible. Increasing the device voltage widens the hot spot until the device is in normal state. The electron and phonon temperature profiles along the HEB bridge are determined by a system of coupled power balances. Electrons are heated by RF and LO radiation and dc bias. For frequencies above the quasiparticle bandgap RF heating is uniform whereas bias heating is confined to the normal conducting regions. The electron-phonon coupling depends on the electron and phonon temperature to an exponent of 3.6 (Gershenzon 1991). Its efficiency is $\sigma \propto T_e^{-1.6}$ (Gousev 1994) is linearized in an average electron temperature (here: T_c).

$$\frac{d}{dx}\lambda(T)\frac{d}{dx}T(x) - \sigma(T_c)\left[T^{3.6}(x) - T_{phonon}^{3.6}(x)\right] + p(x) = 0 \qquad (1)$$

$$\frac{d}{dx}\lambda_{phonon}\frac{d}{dx}T_{phonon}(x) - \sigma_{phonon}(T_c)\left[T_{phonon}^{4}(x) - T_{substrate}^{4}\right] = -\sigma(T_c)\left[T^{3.6}(x) - T_{phonon}^{3.6}(x)\right] \qquad (2)$$

Specifying the heating power density $p(x)$ a single equation is left after some manipulation:

$$\frac{d}{dx}\lambda_0 T^{2.6}(x)\frac{d}{dx}T(x) - \sigma(T_c)\left[T^{3.6}(x) - T_{substrate,eff}^{3.6}\right] + \left[\frac{1}{2L}P_{LO} + P_{DC}\cdot\frac{(u(x+H)-u(x-H))}{2H}\right] = 0 \qquad (3)$$

Here $\lambda_0 T^{2.6}$ is the electron thermal conductivity fitted to experimental data (Poole 1995). The electron cooling efficiency is given by $\sigma \propto \dfrac{c_{el}}{\tau_{e\to ph}}D$, where D is the film cross section and $c_{el} = 1.6\times10^{-4}\cdot T_e\dfrac{J}{K^2 cm^3}$ the thermal capacity of the electrons. The electron-phonon interaction time for NbN is set to $\tau_{e-ph} = 440K^{1.6}ps\cdot T^{-1.6}$ (c.f. Gousev (1994)). The bridge length is 2L where the hot spot is 2H. The heating power due to RF (dc) is denoted P_{LO}, (P_{DC}) using the unit step $u(x)$. Ideal cooling in the antenna pads and specifying interface conditions at the hot spot ends yields:

$$T(-L) = T(+L) = T_{bath} \qquad\qquad T(H) = T_c \qquad (4)$$

Results for the hot spot length as a function of heating powers for a NbN HEB used as reference are plotted in Fig.1. A minimum heating power level is required to sustain a hot spot. These levels are different for RF and dc powers (in contrast to older models, see Ekström (1995)).

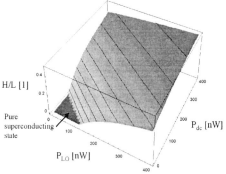

Fig. 1: Calculated normalized hot spot length as a function of RF and bias heating power. The values are obtained for a 2μm x 250nm NbN phonon-cooled HEB The hot spot length is normalized to the bridge length.

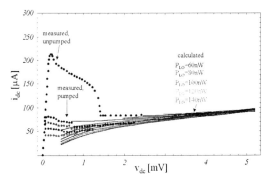

Fig. 2: Measured and calculated IV-curves for a 250nm ling 2micron wide NbN HEB

As the dc device resistance is at first order determined by the product of the film's normal conductivity and the hot spot length, current-voltage characteristics (iv curves) are easily obtained from hot spot length data (e.g. by cutting Fig 1 along a line of constant LO power). Results are shown in Fig.2 above. The resistance change of the bolometer with respect to small changes in heating powers is determined by the partial derivatives of the hot spot length with respect to the heating powers in the operating point. These slope factors (c.f. Arams (1966)) are thus given symbolically by:

$$C_{rf} = \frac{R_N}{L} \cdot \frac{\partial H}{\partial P_{LO}} \qquad\qquad C_{dc} = \frac{R_N}{L} \cdot \frac{\partial H}{\partial P_{DC}} \qquad (5)$$

respectively. They are used in the subsequent small signal expansion linking it to the large signal part.

3.   SMALL SIGNAL MODEL

Assume an ideally current biased HEB receiver with the topology from Arams (1966). A load resistance (in series with a dc block capacitor) is put parallel to the bolometer. Due to heating by RF and LO radiation the bolometer resistance varies with time. The RF and LO power's mean value causes a dc resistance shift and the power oscillation forces the bolometer to oscillate at the intermediate frequency. One obtains for the device resistance:

$$R(t) = R_0 + C_{dc} \cdot v(t) \cdot i(t) + \frac{1}{2} \cdot C_{RF} \cdot \left[P_{LO} \cdot P_{RF} + \sqrt{P_{LO} \cdot P_{RF}} \cdot \cos(\omega_{IF} \cdot t) \right] \qquad (6)$$

Here R_0 stands for the "current-free" resistance, $v(t)$ and $i(t)$ are the time varying device voltage and current with the (imposed) dc bias current I_0. The unknowns V_0, i_L denote dc device voltage and IF load current through the load resistance R_L The resistance in the operating point R_{b0} contains all dc resistance shifts and is given by:

$$R_{b0} = \frac{H}{L} R_N = R_0 + C_{dc} \cdot V_0 \cdot I_0 + \frac{1}{2} \cdot C_{RF} \cdot P_{LO} \cdot P_{RF} \qquad (7)$$

From (6) one obtains the conversion gain for the bolometer in the form:

$$G = \frac{2 R_L \cdot P_{LO} \cdot P_{dc}}{R_{b0} \cdot (R_L + R_{b0})^2} \cdot \left[\frac{C_{RF}}{1 - C_{dc} \cdot I_0^2 \cdot \frac{R_L - R_{b0}}{R_L + R_{b0}}} \right]^2 \qquad (8)$$

Fig. 3 summarizes curves for the conversion gain HEB for different heating powers.

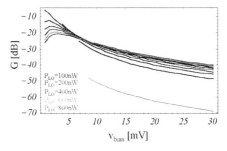

Fig. 3: Calculated small signal conversion gain for a 2μm x 250nm NbN phonon-cooled HEB. For this bolometer the optimal operating point at 0.8mV with a laser power of about 150nW at 1THz has been determined experimentally (Yagoubov 1998).

The dominant noise sources in a hot-electron bolometer is thermal fluctuation noise. Johnson noise being of the order of the critical temperature of the bolometer is not treated here. The fluctuation noise contribution is determined by the noise power produced by the statistical fluctuations of a piece of matter having the thermal capacitance c and being coupled to a substrate by a given thermal conductivity.This resistance fluctuation is assumed as "driving power" instead of a RF signal in (6). In a reasoning similar to the calculation of the conversion gain one obtains for the equivalent noise temperature due to thermal fluctuations:

$$kT_{FL} = \frac{I_0^2 \cdot R_L}{\left(R_L + R_{b0}\right)^2} \cdot \left[\frac{1}{1 - C_{dc} \cdot I_0^2 \cdot \frac{R_L - R_{b0}}{R_L + R_{b0}}}\right]^2 \cdot \left(2 \cdot \frac{R_N}{2L} \cdot \frac{1}{\left.\frac{dT_e}{dx}\right|_{Tc}}\right)^2 \cdot \frac{4kT_c^2}{c_e V} \cdot \tau_{e,relax}$$

(10)

Calculated device fluctuation noise temperatures are shown in Fig 4. There the noise temperature is evaluated along the IV curves shown in Fig 2.

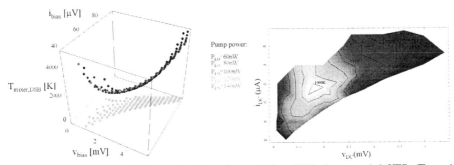

Fig. 4: Calculated and measured input noise for a 2μm x 250nm NbN phonon-cooled HEB. The noise temperature is calculated following the IV curves displayed in Figure 2. (The projections of the noise temperature on the IV-plane are indicated as gray dots) A wide valley around 1 mV for low LO power is clearly visible. The Johnson noise is assumed to be of the order of the critical temperature of the bolometer.

4.   CONCLUSION

A model describing phonon-cooled HEB mixers in terms of a hot spot mixer based on a nonlinear power balance equation system for electrons and phonons has been set up. There the generation of an IF signal in the load resistance is due to a parametric oscillation of the bolometer resistance together with a dc bias current through the device. The small signal model for parametric oscillation is connected to the hot spot model via hot spot change parameters. The model predicts measured IV curves, conversion gain and noise figures reasonably well. As a general trend, shorter bolometers tend to have worse gain and equal noise performance. Nevertheless short phonon cooled HEB offer an additional channel for electron cooling due to diffusion which may increase the mixer's IF frequency considerably.

5.   REFERENCES

H. Ekström, B. Karasik, E. Kollberg , IEEE Trans. MTT **45** (4), 1995, pp938-46
Yu.P. Gousev, et al., Appl.Phys, **75** (7) 1994, pp3695
E.M. Gershenzon, et al. , IEEE Trans. Magn. **27** (2),1991, pp1317-20
F. Moon, "Chaotic and Fractal dynamics ...", Wiley, New York 1992
H.F. Merkel, E. Kollberg, K.S. Yngvesson, Proc.9.th ISSTT, Pasadena, CA, 1998, pp81-97
H.F. Merkel, et al., IEEE Trans. ASC **9**(2) 1999, pp4201-4
C.P.Poole, H.A.Farach, R.J.Creswick, "Superconductivity" Acad. Press 1995
D.Wilms-Floet, et al., IEEE, Trans. ASC **9**(2), 1999, pp3749-52
F.Arams, C.Allen, B.Peyton, E.Sard, Proc. IEEE **54**(3) pp308-18 (1966)
P.Yagoubov et al., Appl. Phys. Lett. **73**(19), 1998

Inst. Phys. Conf. Ser. No 167
Paper presented at Applied Superconductivity, Spain, 14-17 September 1999
© 2000 IOP Publishing Ltd

Disordered suspensions of metastable superconducting microgranules: a simulation approach

A Peñaranda, C E Auguet and L Ramírez-Piscina

Departament de Física Aplicada, Universitat Politècnica de Catalunya, Av. Doctor Marañón 44, E-08028 Barcelona, SPAIN

ABSTRACT: We study disordered suspensions of superconducting superheated microgranules under an increasing external magnetic field by numerical simulations. The induced transitions produce an order in the system. We compare simulations with previous analytical perturbative results.

1. INTRODUCTION

Disordered suspensions of superconducting superheated microgranules (SSG) have been proposed as detectors of neutrinos and dark matter (Girard et al 1995). In these devices, a dispersion of a large number of superconducting micrometric spheres are maintained in a metastable state below T_c by an external magnetic field. When incident radiation with enough energy scatters over a granule, the normal phase can nucleate and the transition superconducting-to-normal of the granule can be detected. The loss of the Meissner effect produced by the transition induces a pulse on a surrounding wire loop to provide the interaction signal. The external magnetic field and the diamagnetic interactions generate a broad distribution of magnetic surface field values. This effect yields typically a 20% spreading of the transition fields of the ensemble, generating difficulties in interpreting the results of device response. On the other hand, the disorder of the suspension and the consequent effects change after each transition due to the long-range of the diamagnetic interaction (Peñaranda et al 1999a).

In our numerical simulations we study the successive transitions in systems of disordered dispersions of superconducting spheres when the external magnetic field is slowly increased from zero. We observe that the successive transitions induce a strong ordering mechanism on the system. They behave in such a way that both the spatial inhomogeneities and the surface magnetic field become more uniform in the system (Peñaranda et al 1999b). This has important practical consequences in SSG detection. Finally, we have compared our results with the perturbative theory proposed by Geigenmüller (1988), valid for diluted systems.

2. RESULTS

We have simulated dispersions of superconducting spheres of the same radius a sited at random in a thin sample. We consider that the transitions of each one to the normal phase is completed when the local magnetic field on any point of its surface reaches a threshold value B_{th}. This value for an isolated free of defects sphere could be equal to the superheating field value $B0_{SH} = 3/2 \, B_{ext}$. The effect of possible surface defects is introduced by an experimental distribution of values of B_{th} for tin microspheres diluted in parafine fitted by a parabolic distribution (Geigenmüller 1988, 1989). The computation of the magnetic fields on the spheres surface are performed by iteratively solving the Laplace equation, with the appropriate boundary conditions, in a multipole expansion scheme (Peñaranda et al 1999a). We monitor these local surface fields and the sequence of transitions when the external field B_{ext} increases slowly from zero (Peñaranda et al 1999b). We have employed configurations with a number of spheres up to 250 placed at random in a volume given by the desired filling factor ρ (fraction of volume occupied by the microgranules) from 0.001 up to 0.20.

We analyse the evolution of the system by increasing B_{ext} until all spheres have transited. The fraction f of still superconducting spheres versus B_{ext} for different values of ρ is represented by symbols in Fig. 1. The most dilute case follows the distribution of threshold values B_{th} very closely, as would be expected for isolated spheres. However, results of f show transitions for smaller external fields as concentrations of the system are larger. This is because the first ($i.e.$ at smaller B_{ext}) spheres that transite tend to be the closer ones, and the probability of finding this situation increases with ρ. Consequently the distance between the remaining superconducting spheres increases and the diamagnetic interactions decrease with the successive transitions. This produces a more homogeneous distribution of both positions and surface magnetic field. This is more patent in the evolution of system with higher concentration (Peñaranda et al 1999b).

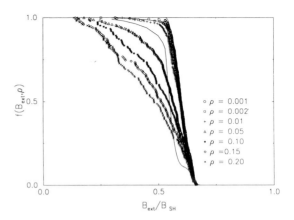

Fig. 1. f fraction of still superconducting spheres versus B_{ext}/B_{sh}, for different volume fractions ρ. Symbols represent simulation results. From right to left, continuous lines correspond to predictions of perturbative theory. ρ=0, 0.01 and 0.05.

We have compared configurations with different initial ρ when, after several transitions, they reach the same value ρ_{ef} of the remaining superconducting spheres. Previous simulations (Peñaranda et al 1999a) showed that, for the same external magnetic field, random configurations had surface field distributions whose width increased strongly with ρ. When these systems evolve towards the same ρ_{ef}, the order induced by the successive transitions produces a final situation with much smaller field spreading. This is shown in Fig. 2, where we can see the maximum surface field distribution for configurations that have reached the same $\rho_{ef} = 0.02$, compared to a configuration at random with ρ =0.02. Here, the field values occur within a narrow interval for the systems that have already suffered transitions. This

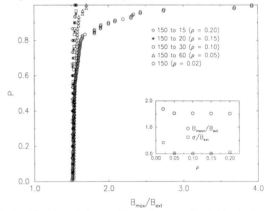

Fig. 2. P Fraction of spheres with maximum surface field value lower than the x-axe value (in units of B_{ext}) for configurations with different initial ρ and the same final value ρ_{ef}=0.02. In the inset the corresponding mean value (circles) and standard deviation (squares) as a function of ρ.

effect is stronger in configurations coming from higher initial ρ, that is, those having undergone a higher number of transitions. On the other hand, in the random configuration with ρ =0.02, the broadening of the field distribution is quite large, despite its low density. In the inset of the same figure, the mean value and standard deviation of the corresponding distributions are represented as a function of the initial ρ. The constancy of the mean value and the vanishing value of the standard deviation confirm the homogeneity of the transited systems compared with the random one. Similar behaviour has been obtained for higher values of ρ_{ef}. The spatial homogeneity is shown in Fig. 3, where the mean value of the distances between each sphere to the nearest one is represented as a function of the initial ρ. In this figure, all the configurations have the same effective filling factor, but the distances between superconducting spheres increase strongly with the initial ρ, and consequently with the number of transitions. We conclude that the successive transitions induced by external magnetic field is a strong ordering mechanism in disordered SSG configurations.

Previous theoretical works centered their attention on the study of effective magnetic properties of disordered systems. In the regime of dilute samples Geigenmüller (1988) constructed a perturbative theory not limited to effective properties, but which calculates statistics of local surface fields on the granules and of the transitions induced by the external field. This theory is formally based on a cluster expansion in powers of the volume fraction ρ. In practice, the expansion is performed up to first order; that is, considering only two-body and dipolar interactions.

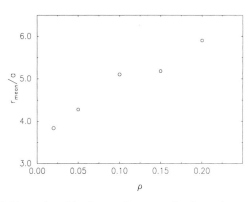

Fig. 3. Mean value of the distances between each sphere to its nearest one, in units of the radius of the spheres for systems evolved to the same final $\rho_{ef}=0.02$, versus the initial ρ.

We compare our results with those obtained from the perturbative theory for two quantities, the fraction of still superconducting spheres f, and the magnetic surface field distribution. The first is represented in Fig. 1, together with numerical results, versus the external magnetic field. We observe a reasonable agreement only for very diluted samples. The dependence of f on the density ρ for different values of B_{ext} is shown in Fig. 4 for both theoretical (continuous lines) and numerical (symbols) results. We can see that the perturbative theory provides a good qualitative prediction for behaviour of the systems, but quantitative agreement only for very dilute configurations, with concentrations smaller than 2% (Peñaranda et al 1999c).

For the distribution of maximum surface fields it is interesting to compare their evolution when the external magnetic field increases, and in consequence the system evolves to a more ordered configuration. Results for systems with different values of ρ present discrepancies between the perturbative theory and the simulations for the lowest values of B_{ext}, even for very dilute systems. The agreement between them improves as B_{ext} increases. Specially interesting are the results observed in Fig. 5 for an initial density of 0.20. The high discrepancies presented for the system at $B_{ext}=0.2\,B_{SH}$ evolve to a good agreement for $B_{ext}=0.6$

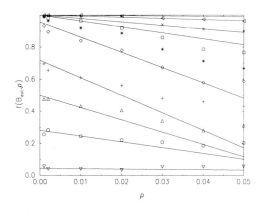

Fig. 4. f fraction of spheres still superconducting as function of ρ for several values of B_{ext}. From top to bottom: B_{ext} = 0.30, 0.40, 0.50, 0.53, 0.55, 0.58, 0.60, 0.62 and 0.65 in units of B_{SH}. Symbols represent simulation results, continuous lines correspond to predictions of perturbative theory.

B_{SH}, much better than for low densities. Then we can conclude that the perturbative theory implicitly includes the aforementioned ordering mechanism. To corroborate this, in the same figure is also represented the surface magnetic field distribution for a system with the same density of superconducting spheres than the coming from initial 0.20 at $B_{ext}=0.6\,B_{SH}$, but with the microgranules placed at random.

3. CONCLUSIONS

We have performed simulations of disordered suspensions of superconducting granules, transiting to normal when an external magnetic field is slowly increased from zero. Numerical simulations of these systems are made by solving the complete Laplace equation for the magnetic field. We find that the successive transitions induce a strong ordering mechanism on the systems, which is stronger for higher initial density configurations. This order behaves in such a way that both the spatial inhomogeneities and the surface magnetic field distributions tend to homogenize. As a consequence, the uncertainty in the energy threshold for transitions in SSG detectors can in principle be reduced.

We have compared results of both transitions and surface fields with the analytical cluster expansion of Geigenmüller (1988). We find that the theory qualitatively predicts the behaviour of the samples, but appears as quantitatively correct only for very diluted systems. Therefore such expansions, although providing a very useful framework for analyzing results from SSG experiments, should be used with caution for obtaining quantitative information.

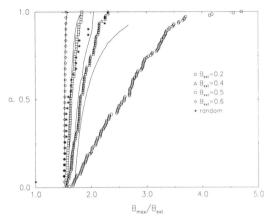

Fig. 5. P Fraction of spheres with maximum surface field lower than the x-axe value (in units of B_{ext}), corresponding to an initial $\rho=0.20$, for several values of external magnetic field. Continuous lines are the corresponding predictions of perturbative theory for the same values of B_{ext}. Corresponding to a system with the spheres placed at random is also represented.

ACKNOWLEDGMENTS

We acknowledge T. Girard for helpful discussions. We acknowledge financial support from Dirección General de Investigación Científica y Técnica (Spain) (Project PB96-0241-C02-02), Comissionat per a Universitats i Recerca (Spain) (Projects 1997SGR00439 and 1997XT00003), and European Commision (Project TMR-ERB4061PL951377). We also acknowledge computing support from Fundació Catalana per a la Recerca-Centre de Supercomputació de Catalunya (Spain).

REFERENCES

1995 Superconductivity and particle detection, eds T A Girard, A Morales and G Waysand (Singapore: World Scientific), and references therein.
Geigenmüller U 1988 J. Phys. (Paris) **49**, 405
Geigenmüller U 1989 Superconducting and Low-Temperature Particle Detectors, eds G Waysand and G Chardin (Amsterdam: Elsevier).
Peñaranda A, Auguet C E and Ramírez-Piscina L 1999a Nucl. Instrum. Methods A **424**, 512
Peñaranda A, Auguet C E and Ramírez-Piscina L 1999b Solid State Commun. **109**, 277

Inst. Phys. Conf. Ser. No 167
Paper presented at Applied Superconductivity, Spain, 14-17 September 1999
© 2000 IOP Publishing Ltd

Josephson arrays for DC and AC metrology

J Niemeyer

Physikalisch-Technische Bundesanstalt, Bundesallee 100, D 38116 Braunschweig

Abstract: Large series arrays of underdamped SIS tunnel junctions are the basic element of the primary DC voltage standards at present used. The development of versatile quantum voltmeters for DC and AC metrology requires the preparation of large and perfect series arrays of overdamped Josephson junctions. Overdamped junctions can be realised by externally shunted SIS junctions or by internally shunted SNS or SINIS junctions. Arrays of up to 8000 SINIS junctions or 30000 SNS junctions were successfully operated at the 1V DC level.

In addition to being used in large arrays for voltage metrology and oscillators, the described junction types may become very useful for the preparation of highly integrated single-flux quantum digital devices.

1. INTRODUCTION

If a Josephson junction is biased at a non-zero DC voltage V, the supercurrent across the junction oscillates with a frequency of $\omega = (2e/\hbar)V$. By phase-locking the Josephson oscillator to a frequency-stabilised microwave source, constant voltage steps at $V_n = n\Phi_0 f_e$, (n = 0,1,2,3, …) are generated over a certain current range ΔI in the DC characteristic. f_e denotes the frequency of the external oscillator and $\Phi_0 = h/2e$ the flux quantum. V_n can be used as the reference voltage for quantum-based voltage standards. As the output power and the impedance of a Josephson oscillator are very small and the DC reference voltage of a single Josephson junction is impractically low, it is obvious that arrays of junctions should be used together with a suitably designed microwave coupling circuit to realise oscillators and voltage standards with practical output data. For the purposes of DC voltage metrology the constant voltage steps generated by the array must be added and counted. An analysis of AC voltages with fundamental accuracy requires the DC Josephson voltage across the array to be switched between different values by rapid selection of the desired number of steps or by fast tuning of the RF drive frequency over a wide range.

Within the framework of the Stewart-McCumber model (1968), a Josephson junction is considered to be a damped LC resonator with non-linear Josephson inductance. The model describes a parallel connection of normal-state internal or external resistance R_n, junction capacitance C, and Josephson inductance L_J. Junction damping is characterised by the hysteresis parameter $\beta_c = (2e/\hbar)$ $I_c R_n^2 C$. β_c determines the quality factor $(\beta_c^{1/2})$ of the LC Josephson resonator. Another important junction parameter is the characteristic voltage $V_c = I_c R_n$. It determines the maximum operating frequency and the possible signal size of the circuit. The size of R_n and the impedances of the capacitive and inductive branches $Z_J = \omega L_J$, $Z_C = 1/\omega C$, and $Z_e = \omega L_e$ determine the flow of the RF current. The Josephson inductance can be approached with the aid of the Josephson equations $V = (\hbar/2e)d\varphi/dt$ and $I_J = I_c \sin\varphi$. For small values of the phase difference φ, $\sin\varphi$ may be replaced by φ. L_J $dI_J/dt = (\hbar/2eI_c) dI_J/dt$ is obtained in this case.

If the junctions are integrated into a microwave coupling circuit, the degree of matching between R_n and the impedance of the external circuit - for instance a micro-stripline in the case of a conventional voltage standard – together with the size of dissipative currents through R_n determines the amount of microwave power coupled from the external circuit to the junction or vice versa. Together with the attenuation of the coupling circuit this ratio governs the collective behavior of the array.

If all possible applications of arrays, including programmable voltage standards and circuits for RF voltage metrology are considered, a wide range of the junction hysteresis parameter β_c and the junction critical voltage V_c must be covered. More than just a single junction type is required to meet all requirements. As microbridges have not been practically used in arrays because sophisticated deep submicron lithography is required in order to pattern junctions with an $I_c R_n$ large enough, and a parameter spread small enough, for homogeneous large arrays, this article is restricted to the description of arrays of multilayer Josephson junctions which can be fabricated with a sufficiently small parameter spread like arrays of low critical current i.e. SIS (superconductor/-insulator/superconductor) junctions with $\beta_c \to \infty$, SNS (superconductor/normal metal/superconductor) junctions with $\beta_c \to 0$, and SINIS (superconductor/insulator/normal metal/insulator/superconductor) junctions with $\beta_c \approx 1$.

2. ARRAYS FOR VOLTAGE METROLOGY

2.1 SIS arrays

RF-driven series arrays of up to 20 000 completely underdamped SIS tunnel junctions form the essential part of the primary voltage standards at present used for reference voltages of up to 10 V. The junction material is usually Nb/Al$_2$O$_3$/Nb (Kohlmann et al 1997). The constant voltage steps are generated in the subgap part of the quasiparticle branch of the DC characteristic and, therefore, cross the voltage axis. This allows the standard to be operated at zero current, the consequence being that series resistances do not lead to reference voltage errors. As the maximum RF voltage amplitude of the microwave applied is not much larger than the highest constant voltage step, the normal state resistance R_n must be replaced by the subgap resistance with a value of about 10 R_n or even more, depending on the size of the subgap leakage current. The arrays are designed for typical drive frequencies of 70 GHz. A stable phase-lock requires critical currents I_c of about 100 µA at a junction area of about 1000 µm^2 and a junction capacitance C of 40 pF (for an overview cf. Kautz 1996). These values result in a large hysteresis parameter of $\beta_c \approx 1.3 \times 10^5$. At a typical drive frequency of $\omega_e = 2\pi f_e =$ (2π)70 GHz, this leads to an impedance for the capacitive branch of the junction of $Z_C \approx 6 \times 10^{-2}$. This value is much smaller than the subgap resistance and the Josephson impedance which becomes $Z_J \approx$ 1.3 Ω for small values of φ. The Rf currents are, therefore, mainly capacitive for the SIS junctions.

In the case of the conventional SIS array standard the SIS junctions are integrated into a microwave circuit which consists of 4 or 8 parallel superconducting periodic micro-striplines with relatively low impedances of between 2 and 5 Ω (Niemeyer et al 1984). The attenuation of this stripline configuration is very low: about 3×10^{-4} dB per stripline period or 2×10^{-3} dB/mm. The very small dissipative RF currents together with the large mismatch between R_n and the stripline impedance are the reason why only a very small amount of microwave power is transferred from the stripline to the individual junction. On the one hand, this guarantees a sufficiently homogeneous supply of all junctions. On the other hand, it prevents electromagnetic coupling between adjacent junctions strong enough for coherent operation of the array in the oscillator mode. Moreover, a relatively large amount of microwave power (about 10 mW at the circuit antenna) is needed to drive the array. Most of this power is fed to the terminating resistive loads at the end of the striplines.

The main advantages of the SIS arrays for voltage metrology are the high reference voltage per junction (up to 1 mV), the arrays' tolerance of series resistances and a relatively large parameter spread of the junctions. There is no undefined voltage state between adjacent steps, and switching between the steps should take place very rapidly.

However, as zero-current constant voltage steps completely overlap, the reference output voltage is multi-valued. This and the resulting relatively small current width of the step voltage lead to a difficult stability problem as regards the adjusted DC voltage. If, for instance, under the influence of an external noise signal, the phase-lock is lost for a short time, the system does not automatically return to the original voltage step. As the number of voltage steps which an individual junction contributes to the desired reference voltage is not exactly known, the total number of voltage steps cannot be determined by selecting the respective number of junctions. In practice, part of these problems are solved by biasing the array with a voltage source the load-line of which intersects only a small number of voltage steps so that the array cannot switch spontaneously to any voltage step. Although the stability of the output voltage was arbitrarily increased by improving the fabrication and, by this, the quality of the arrays (Müller et al 1997, 1999), the described difficulties prevent at present rapid switching between different voltage steps of conventional SIS arrays. It is not yet clear to what degree the use of perfectly fabricated arrays with no junctions shorted and with moderate junction damping (Kim and Niemeyer 1995) instead of strongly underdamped SIS arrays could help to solve the problems without the main advantages of the conventional SIS standard being lost. Recent experiments have shown that it is difficult to reduce the subgap resistance by the formation of a more transparent junction barrier without increasing the junction parameter spread (Schulze 1999, Kim and Niemeyer 1995).

In view of the above described problems it seems difficult at present to design a rapidly switchable and programmable voltage standard with conventional SIS arrays. It has, therefore, been suggested by Hamilton et al (1995) that series arrays of highly damped junctions be used in a suitable microwave coupling circuit to construct programmable voltage standards. In this case, an individual junction contributes an exactly defined number of constant voltage steps to the total reference voltage: in designs at present used the steps $\pm V_1$ besides zero voltage. Moreover, the output voltage is single-valued and the system automatically returns to the reference voltage chosen if the phase-lock was lost for a short time. Two different designs were suggested: the first version is an array divided into binary sections which allow the reference voltage NV_1 to be defined by combining the binary sections in such a way that the N junctions contribute to the reference voltage. The second version makes use of a non-sinusoidal RF pulse or bipolar drive which allows the reference voltage to be adjusted by tuning the repetition rate of the drive over a wide range (Benz and Hamilton 1996, Maggi 1996, Benz et al 1998 and 1999). No binary subsections are needed in these cases.

First experiments were performed with arrays of externally shunted SIS junctions (Hamilton et al 1995). In this case, the frequency of the external source can be kept at high values, e.g. 70 GHz. The higher the frequency is the less junctions are needed to reach a certain output voltage. At 70 GHz, a number of about 8 000 junctions is needed to reach a maximum output voltage of 1V, much more than for a conventional 1V standard. This number would be even increased if lower drive frequencies had be used because the I_cR_n product is low. It is of special importance to keep the junction number as low as possible bearing in mind that the junction parameter spread should be smaller than that for a conventional voltage standard and that programmable voltage standards do not tolerate series resistances. This demonstrates the need for well-developed technologies for the fabrication of programmable series array voltage standards. Unfortunately, the experiments with externally shunted SIS junctions have shown that it will be difficult to reach output voltages as high as 1V. The impedance of typical external shunts reaches the value of the DC resistance $R_n \approx 1\Omega$ of the shunts at drive frequencies of between 70 and 100 GHz. This already disturbs proper high-frequency operation. A recently published paper proposes to externally shunt blocks of junctions instead of shunting each individual junction (Hassel 1999). It is not clear yet if this design will overcome the difficulties.

In view of the fact that systems with external shunts are more difficult to fabricate with small parameter spread and that they are more space-consuming – a factor which becomes essential for large arrays driven at low frequencies - and that -, in this case, large rf filters are needed in the DC lines, it seems to be easier at present to use internally shunted junctions, such as SNS or SINIS junctions, instead of externally shunted SIS junctions.

2.2   SNS arrays

Binary divided SNS arrays of 32 768 Nb/PdAu/Nb junctions are successfully operated at 16 GHz and generate rf-induced reference voltages of 1V (Benz et al 1997). The minimum switching

710

time between different reference voltages is about 1 µs. However, while array sections are switched from the step voltage to zero voltage, undefined voltage states of the transient DC characteristic contribute to the reference voltage. This makes it difficult to achieve an accuracy sufficient for synthesising fast AC voltages. Typical junction parameters are $I_c \approx 10$ mA and $R_n \approx 2$mΩ. As the junction capacitance approaches zero, the RF currents are resistive. The array is part of a coplanar stripline configuration of 8 parallel lines, with a line impedance of about 50 Ω, so that a large impedance mismatch between junctions and stripline allows homogeneous microwave coupling to all junctions. First versions of pulse-driven and bipolar-driven arrays have reached output voltages in the range of a few mV.

The low product of I_cR_n of 10 to 40 µV is tolerable for voltage standards with relatively low drive frequencies. But in spite of this it would be advantageous to increase V_c to values which would make drive frequencies of 70 GHz, possible. The size of the circuits can be reduced arbitrarily. For the application of SNS junctions in RSFQ circuits, a higher value of V_c would be important, too. As the critical current depends exponentially on the thickness of the N barrier, it is possible to increase V_c by decreasing the junction area. Large series arrays of Nb/PdAu/Nb ramp-type junctions with a junction area A of about 0.25 x 1.3 µm² and a normal layer thickness of about 40 nm have, therefore, been prepared (Pöpel et al 1999). In spite of the relatively small junction area, the arrays show a small parameter spread (Fig. 1) and a magnetic field dependence which points a homogeneous current distribution (Fig. 2).

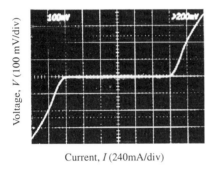

Current, I (240mA/div)

Fig. 1: DC characteristic of a series array with 10000 SNS ramp junctions.

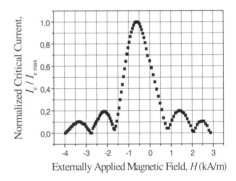

Externally Applied Magnetic Field, H (kA/m)

Fig. 2: Magnetic field dependence of an SNS ramp junction. $A = 0.25 \times 1.3$ µm².

Typical junction parameters are $I_c \approx 0.7$ mA, $R_n \approx 30$ mΩ, $I_cR_n \approx 20$ µV and $j_c \approx I_c/A = 200$ kA/cm². Arrays of 1000 ramp-type SNS junctions under 10 GHz microwave radiation provide constant voltage steps which show that no junction fails to contribute a fundamental step. The fact that, in this case, no integrated microwave coupling circuit is needed demonstrates the advantage of a small-sized circuit. For larger circuits a microwave coupling circuit will be needed. If the array forms a small stripline (width 1µm, thickness of the Si dielectric 365 µm) the attenuation results in 7x10⁻⁴ dB per stripline period or 0.25 dB/mm. This value is larger than that one for SIS arrays but the SNS junctions are much smaller than SIS junctions so that homogenous microwave coupling to a sufficiently large number of SNS junctions should be possible.

It should be possible to increase the critical current density j_c by about a factor of 10 without decreasing the thickness of the N-layer very much. I_cR_n would then reach values which allow RF operation of the circuit at frequencies of at least 70 GHz. With such values, SNS junctions may become interesting for highly integrated RSFQ circuits because of their large potential for size reduction.

It is possible to further increase the critical voltage by reducing the electronic mean free path of the N-barrier until it becomes semiconducting or by introducing disturbed S/N interfaces into the SNS structure. If the degree of disturbance is increased, this will finally lead to the SINIS structure (Kupriyanov and Lukichev 1988, Capogna and Blamire 1996, Maezawa and Shoji 1997, Sugiyama et al 1997).

2.3 SINIS arrays

Large series arrays of Nb /Al$_2$O$_3$/Al/Al$_2$O$_3$/Nb junctions with a small parameter spread were prepared for voltage metrology (Schulze et al 1998). Depending on the process parameters, the switching hysteresis of the DC characteristic can be adjusted to values down to less than 20% (Fig 3). The regular magnetic field dependence of the critical current demonstrates a very homogeneous current distribution over the junction area (Fig. 4).

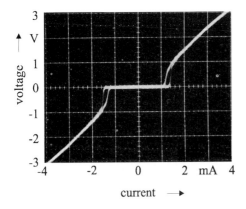

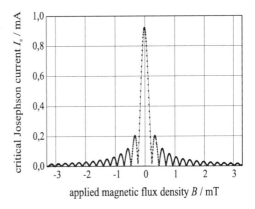

Fig. 3: DC characteristic of a series array with 8192 SINIS junctions.

Fig. 4: Magnetic field dependence of the critical current of a SINIS junction. $A = 25 \times 60 \; \mu m^2$.

The individual junction area is $20 \times 50 \; \mu m^2$ in size. The transparency of the barriers can be adjusted by the fabrication parameters over a wide range. This allows SINIS arrays to be operated over a frequency range of at least 10 to 100 GHz. Typical junction parameters are $I_c \approx 1$ mA , $j_c \approx 100$ A/cm^2, $R_n \approx 100$ mΩ and $I_cR_n \approx 100\mu$V. The relatively high value of V_c allows 70 GHz operation of the array. An 8192 junction array divided into binary subarrays is, therefore, large enough to reach a maximum output voltage of 1 V. Moreover, the microwave supply of the conventional voltage standard can be used for this type of programmable voltage standard. The arrays are integrated into a low-impedance micro-stripline comparable to that for the SIS arrays. However, in contrast to the SIS junctions, an arbitrary part of dissipative RF current here flows over the junction resistance because R_n is comparable to Z_c in the case of the typical SINIS junction. In addition, the mismatch between R_n and the stripline impedance is small enough to transfer RF power from the individual junction to the stripline to an amount large enough for phase-locking adjacent junctions. If the junction number is large enough, the array acts as a coherent microwave oscillator whose output power is added to the rf power of the external oscillator which is locked to the first junctions in the array (Schulze et al 1999a,

1999b, 1999c). It is, therefore, possible to generate constant voltage steps in very long arrays, although the attenuation of the stripline is rather high (about 0.05 dB per stripline period or 1 dB/mm). As a result of the microwave contribution from the active stripline, the microwave power required to generate a 1 V reference voltage is much lower than that needed to reach comparable output voltages by means of a conventional SIS array with much fewer junctions.

By chosing suitable fabrication process parameters SINIS junctions are useful for RSFQ logic circuits (Balashov et al 1998) and pulse driven arrays (Liefrink et al 1999), too.

References

Balashov D, Buchholz F-Im, Schulze H, Khabipov M I, Kessel W and Niemeyer J 1998 Supercond. Sci. Technol. **11** 1401

Benz S P, Burroughs C J, Harvey T E and Hamilton C A 1999 submitted to IEEE Trans. Appl. Supercond.

Benz S P and Hamilton C A 1996 Appl. Phys. Lett. **68** 3171

Benz S P, Hamilton C A, Burroughs C J, Christian L A and Harvey T E 1998 CPEM Conference Digest 437)

Benz S P, Hamilton C A, Burroughs C J and Harvey T E 1997 Appl. Phys. Lett. **71** 1866

Capogna L and Blamire M G 1996 Phys Rev. B **53** 5683

Hamilton C A, Burroughs C J and Kautz R L 1995 IEEE Trans. Instrum. Meas. **44** 223

Hassel J, Seppä H and Kiviranta M 1999 submitted to Physica B

Kautz R L Rep. Prog. Phys. **59** 935

Kim K T, Niemeyer J 1995 Appl. Phys. Lett. **66** 2567

Kohlmann J, Müller F, Gutmann P, Pöpel R, Dünnschede F-W, Meier W and Niemeyer J 1997 IEEE Trans. Appl. Supercond. **7** 3411

Kupriyanov M Yu and Lukichev V F 1988 Sov. Phys. JETP **67** 1163

Liefrink F, de Jong G, Teunissen P, Heimeriks J W, Royset A, Dyrseth A A, Schulze H, Behr R, Kohlmann J, Vollmer E and Niemeyer J 1999 BEMC Conference GB

Maezawa M and Shoji A 1997 Appl. Phys. Lett. **70** 3603

Maggi S J Appl. Phys. **79** 7860

McCumber D E 1968 J. Appl. Phys. **39** 2503 and 3223

Müller F, Kohlmann J, Hebrank F X, Weimann T, Wolf H, and Niemeyer J 1995 IEEE Appl. Supercond. **5** 2903

Müller F, Behr R, Kohlmann J, Pöpel R, Niemeyer J, Wende G, Fritsch L, Trum F, Meyer H G and Krasnopolin I Y 1997 ISEC'97 extended Abstracts **1** 95

Niemeyer J, Hinken J H and Kautz R L 1984 Appl. Phys. Lett. **45** 478

Pöpel R, Hagedorn D, Buchholz F-Im and Niemeyer J 1999 10-62 this issue, EUCAS '99

Schulze H 1999 Thesis: "Josephson-Kontakte mit intrinsischem Shunt für Josephson-Spannungs-normale", Eberhard-Karls-Universität Tübingen

Schulze H, Behr R, Kohlmann J, Müller F, Krasnopolin I Ya and Niemeyer J 1999c 16-114 this issue, EUCAS '99

Schulze H, Müller F, Behr R, Kohlmann J, Niemeyer J and Balashov D 1999a submitted to IEEE Trans. Appl. Supercond.

Schulze H, Müller F, Behr R, Kohlmann J and Niemeyer J 1999b 16-118 this issue, EUCAS '99

Schulze H, Müller F, Behr R and Niemeyer J 1998 Appl. Phys. Lett. **73** 996

Stewart W C 1968 Appl. Phys. Lett. **12** 277

Sugiyama H, Yanada A, Ota M, Fujimaki A and Hayakawa H 1997 Jpn. J. Appl. Phys. Part 2 **36** 1157

Acknowledgments

This work is supported by the European Communities (Project No. SMT4-CT98-2239) and by the BMBF (Projects 13N6835, 13N7259 and 13N7494/3). The author would like to thank R. Pöpel for helpful discussions.

Inst. Phys. Conf. Ser. No 167
Paper presented at Applied Superconductivity, Spain, 14-17 September 1999
© *2000 IOP Publishing Ltd*

Externally phase locked sub-mm flux flow oscillator for integrated receiver

V Koshelets[1], A Baryshev[1,2], J Mygind[3], V Vaks[4], S Shitov[1], L Filippenko[1], P Dmitriev[1], W Luinge[2], N Whyborn[2]

[1]Institute of Radio Engineering and Electronics RAS, Mokhovaya 11, Moscow, 103097, Russia
[2]SRON-Groningen, P.O. Box 800, 9747 AV, Groningen, the Netherlands
[3]Department of Physics, Technical University of Denmark, B309, DK-2800 Lyngby, Denmark
[4]Institute for Physics of Microstructure, RAS, GSP-105, Nizhny Novgorod, 603600, Russia

ABSTRACT: The feasibility of phase locking a Josephson Flux Flow Oscillator (FFO) to an external reference oscillator is demonstrated experimentally. A FFO linewidth as low as 1 Hz (determined by the resolution bandwidth of the spectrum analyzer) has been measured relatively to reference oscillator in the frequency range 270 - 440 GHz. This linewidth is far below the fundamental level given by shot and thermal noise of the free-running tunnel junction. The concept of the single-chip fully superconductive integrated receiver with phase-locked loop for spectral radio astronomy and aeronomy applications is discussed.

1. INTRODUCTION

Flux flow oscillator (Nagatsuma et al. 1984, 1985, 1987, 1988) has proven to be a reliable wideband and easy tunable local oscillator suitable for integration with a SIS-mixer in a single-chip sub-mm wave receiver (Koshelets et al, 1997a). A noise temperature (DSB) as low as 100 K has been achieved by Shitov et al (1998) for an integrated receiver with the FFO operating around 500 GHz. The antenna beam, approximately f/10 with sidelobes of about - 17 dB, makes the integrated receiver suitable for coupling to the real telescope. For spectral radio-astronomy applications besides a low noise temperature and a good antenna beam pattern a high frequency resolution of a receiver is required. It is determined by both the instant linewidth of the local oscillator and its long-time stability and should be much less than 1 ppm of the center frequency. Recently a reliable technique for linewidth measurements has been developed by Koshelets et al (1996) and an autonomous FFO linewidth of a few hundred kHz was measured (Koshelets et al, 1996, 1997b), this value should be decreased considerably to met the requirements of real applications. In this report a dependence of a microwave linewidth of Nb-AlO$_x$-Nb FFOs on the junction parameters has been measured in the frequency range 250 - 550 GHz. The linewidth were measured both for autonomous FFO and FFO locked to an external synthesizer via a wideband feedback loop.

2. EXPERIMENTAL DETAILS AND RESULTS

FFO is a long Josephson tunnel junction in which an applied dc magnetic field and a bias current I_B drive a unidirectional flow of fluxons. An integrated control line with current I_{CL} is used to generate the dc magnetic field applied to the FFO. The velocity and density of the fluxons and thus the power and frequency of the emitted mm-wave signal may be easily tuned by either of the two external parameters. According to Josephson relation the junction biased at voltage V_{FFO}

oscillates with a frequency f $=(2\pi/\Phi_0)$ V_{FFO}), where Φ_0 is the magnetic flux quantum = 2 10^{-15} Wb (that corresponds to 483.6 GHz/mV). Experimentally measured current-voltage characteristics (IVCs) of the Nb-AlO$_x$-Nb FFO are shown in Fig. 1. One can see that FFO IVCs are considerably modified at some threshold voltage of about 930 μV (which is 1/3 of the superconductor gap voltage for Nb-AlO$_x$-Nb tunnel junctions). A simple model based on Josephson radiation self-coupling (JSC) - Hasselberg et al (1974) - was introduced by Koshelets et al (1997b) to explain the experimentally measured FFO IVCs. The JSC caused by the absorption of the internal *ac* Josephson radiation photons by the quasiparticles results in current "bumps" at the voltage V_{JSC} = $V_g/(2n + 1)$, which gives $V_{JSC} = V_g/3$ for n = 1. The effect of self-pumping explains the abrupt vanishing of the Fiske steps (FS) at $V \approx V_g/3$ due to the strongly increased damping. This changing results in essential broadening of the FFO linewidth due to considerable increase of the FFO differential resistance (both for bias, $R_d^B = \partial V_{FFO}/\partial I_B$ and control line current, $R_d^{CL} = \partial V_{FFO}/\partial I_{CL}$).

A FFO linewidth is measured in the frequency range from 250 to 550 GHz by a new original experimental technique (Koshelets et al, 1996). The submm-wave signal coming from a FFO is mixed in a SIS mixer with n-th harmonic of an external reference synthesizer frequency f_{SYN} ($\approx$10 GHz, n = 26 - 55). A photo of the integrated circuit for linewidth measurements is shown in Fig. 2. In order to prevent the external reference oscillator from reaching the FFO a high-pass microstrip filter with a cut-off frequency of about 200 GHz is used. The intermediate frequency (IF) signal, $f_{IF} = \pm(f_{FFO} - n\, f_{SYN})$ is amplified in a cooled amplifier. After additional amplification the signal enters a PLL system. A small part of the signal is applied via a directional coupler to a spectrum analyzer, which is also phase locked to the synthesizer by a common 10 MHz reference signal. By using this technique a down-converted FFO spectrum is measured. The recorded spectrum is the difference between the FFO signal and the n-th harmonic of the synthesizer, and thus a FFO phase noise is measured relative to an appropriate synthesizer harmonic. In the PLL unit the signal is compared with a 100 MHz reference signal that is also phase locked to the main 10 GHz synthesizer. The output signal proportional to the phase difference is returned via a Loop Bandwidth Regulator (maximum bandwidth about 10 MHz) to the FFO current bias through a coaxial cable and a cold 50 Ω resistor mounted on a bias plate.

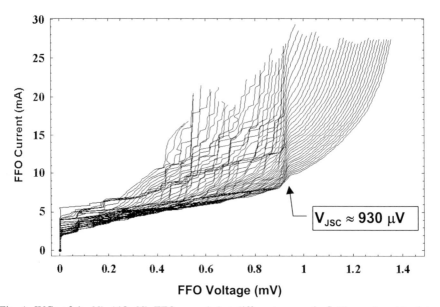

Fig. 1. IVCs of the Nb-AlO$_x$-Nb FFO recorded at different magnetic fields produced by integrated control line. Note an abrupt changing of the FFO behavior at boundary voltage $V_{JSC} \approx 930$ μV.

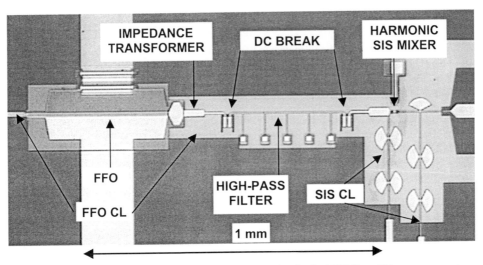

Fig. 2. Microphotograph of the central part of the microcircuits for FFO linewidth measurements.

The PLL system with a relatively low loop gain and a narrow bandwidth setting ($<$ 10 kHz) was used for *frequency locking* of the FFO to the synthesizer in order to measure an autonomous FFO linewidth, Δf_{AUT}. In this case a spectral shape of the measured linewidth is unchanged and is equal to linewidth of the free-running FFO Δf_{AUT} (see Fig. 3a, where a spectrum of frequency locked FFO is shown by dash-dotted line). It was experimentally found by Koshelets et al (1998) and Mygind et al (1999) that a PLL system can considerably narrow a FFO linewidth if Δf_{AUT} at the 3 dB level is smaller than the PLL regulation bandwidth B_{PLL}. In our experiment full phase locking takes place for $\Delta f_{AUT} <$ 2.5 MHz. Fig. 3 shows IF power spectra of the phase locked FFO measured at $f_{FFO} =$ 439 GHz for different settings of the spectrum analyzer. A FFO linewidth as low as 1 Hz (determined by the resolution bandwidth of the spectrum analyzer) is presented in Fig. 3b. It means that the FFO linewidth can be reduced below the value determined by the fundamental shot and thermal fluctuations of a free-running tunnel junction. A consequence of the phase locking is the appearance of a vertical step in the FFO IVC ($R_d^B = 0$). The position of this step is also insensitive to small changes in the control line current, and accordingly also $R_d^{CL} = 0$. The low linewidth $\Delta f_{AUT} \leq$ 2.5 MHz could be realized only in the resonant regime at $f_{FFO} <$ 450 GHz where values of both R_d^B and R_d^{CL} do not exceed 0.01 Ω

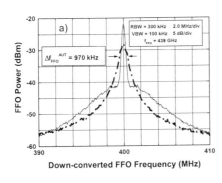
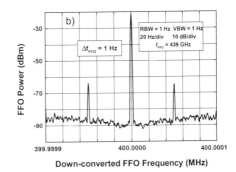

Fig. 3 IF power spectra of the FFO operating at 439 GHz for different settings of spectrum analyzer. Solid and dash-dotted lines in Fig. 3a show data for the optimal and the minimal B^{PLL}, respectively.

A Phase Locked Integrated all-superconducting receiver based on the developed technique for phase locking of the FFO has been proposed by Koshelets et al (1998). In this concept two separate SIS mixers are placed on one chip and connected to the same FFO. One SIS mixer serves as a heterodyne detector in the receiver while the other is used for phase locking of the FFO. Along with this concept an integrated superconducting heterodyne receiver for a frequency of about 350 GHz containing a phase-locked flux-flow oscillator has been designed and fabricated (Shitov et al, 1999). An alternative concept based on an already proven design of integrated receiver could be introduced. At this approach a submm signal from an external harmonic multiplier driven by 10-20 GHz synthesizer is applied to integrated receiver via beamsplitter. A small portion of the IF band (of about 100 MHz) will be used for monitoring of the mixing product between n-th harmonic of the synthesizer and FFO signal. This downconverted signal after narrow band filter will control the PLL system while the rest IF band will be used for analyzing an input signal.

3. CONCLUSION

The results given above demonstrate our ability to narrow the intrinsic linewidth of a Josephson oscillator using an external electronic PLL system, provided that the PLL bandwidth is larger than the linewidth of the autonomous oscillator. The combination of narrow linewidth and wide band tunability makes FFO a perfect on-chip local oscillator for integrated submm wave receiver intended for spectral radio astronomy aeronomy applications.

4. ACKNOWLEDGEMENTS

The work was supported in parts by the Russian SSP "Superconductivity", INTAS project 97-1712, the Danish Natural Science Foundation, the Netherlandse Organisatie voor Wetenshappelijk Onderzoek (NWO) grant and ESA TRP contract 11/653/95/NL/PB/SC.
Authors thank Th. de Graauw, H. Golstein, M. Samuelsen, and H. van de Stadt for fruitful and stimulating discussions as well as H. Smit and D. Van Nguyen for help in the experiment.

REFERENCES

Hasselberg L-E, Levinsen M T, Samuelsen M R 1974 Phys Rev B **9** 3757-65

Koshelets V P, Shitov S V, Shchukin A V, Filippenko L V, and Mygind J 1996 Appl Phys Lett **69**, 699-701

Koshelets V P, Shitov S V, Filippenko L V, Baryshev A M, Luinge W, Golstein H, van de Stadt H, Gao J-R, de Graauw T 1997a IEEE Trans on Appl Supercond **7** 3589-92.

Koshelets V P, Shitov S V, Shchukin A V, Filippenko L V, and Mygind J and Ustinov A V 1977b Phys Rev B **56** 5572- 7.

Koshelets V P, Shitov S V, Shchukin A V, Filippenko L V, Dmitriev P N, Vaks V L, Mygind J, Baryshev A M, Luinge W, Golstein H 1998 Presented at ASC-98 Palm Desert CA USA (Report EQB-04) to be published *IEEE Trans on Appl Supercond* (1999)

Mygind J, Koshelets V P, Shitov S V, Filippenko L V, Vaks V L, Baryshev A M, Luinge W, Whyborn N 1999 Presented at ISEC'99 Berkeley CA USA

Nagatsuma T, Enpuku K, Irie F, and Yoshida K 1983 J Appl Phys. **54** 3302-9; see also Pt. II 1984 J Appl Phys **56** 3284; Pt. III 1985 J Appl Phys **58** 441; Pt. IV 1988 J Appl Phys **63** 1130

Shitov S V, Ermakov A B, Filippenko L V, Koshelets V P, Baryshev A M, Luinge W, Gao J-R 1998 Presented at ASC-98 Palm Desert CA USA (Report EMA-09) to be published *IEEE Trans on Appl Supercond* (1999)

Shitov S V, Koshelets V P, Filippenko L V, Dmitriev P N, Vaks V L, Baryshev A M, Luinge W, Whyborn N, Gao J-R 1999 this conference

Inst. Phys. Conf. Ser. No 167
Paper presented at Applied Superconductivity, Spain, 14-17 September 1999
© 2000 IOP Publishing Ltd

MM wave Josephson radiation in High-Tc bicrystal junction arrays

K Y Constantinian*, A D Mashtakov*, G A Ovsyannikov*

V K Kornev[#], A V Arzumanov[#], N A Shcherbakov[#]

M Darula[$]
J Mygind[&] and N F Pedersen[&]

*Institute of Radio Engineering and Electronics RAS, 103907, Moscow, Russia

\# Moscow State University, Moscow, Russia

$Research Centre Juelish, Juelish, Germany

&Department of Physics, Technical University of Denmark, DK-2400, Lyngby, Denmark

ABSTRACT: High-T_c Josephson junction arrays with coupling by superconducting short-circuited (5 μm wide and 0.1 μm thick) coupled microstrip lines, provided dc bias in parallel, were fabricated on bicrystal sapphire substrates for experimental investigation of the phase-locking phenomenon. Josephson self radiation, emitted by the array was measured by an external 90 GHz receiver for different applied dc magnetic fields. The numerical simulation of array shows that strong interaction between Josephson junctions and the standing waves within structure require small enough damping $\alpha \leq 10^{-3}$ per coupling section for reduction of Josephson oscillation linewidth.

1. INTRODUCTION

The advantage of lock-in Josephson junction (JJ) arrays, consisted of lumped JJs, over a single JJ for applications in mm and submm wave oscillators and in low-noise front-ends of receivers are known (Han, et al 1994). Aiming at applications at frequencies laying above the energy gap of conventional superconductors, the high-T_c superconducting (HTS) JJs should have small enough spread of parameters. For present the minimal spread of 10-15% demonstrate HTS JJs on bicrystal SrTiO$_3$ substrates. SrTiO$_3$ has too large dielectric constant and thus is hardly applicable at submm waves. In this paper we report on fabrication, experimental and theoretical study of single layer JJ-arrays, based on YBaCuO thin film bicrystal JJs, deposited on promising for high frequency applications sapphire substrates. For lock-in operation all of JJs in array must oscillate at one desired frequency and have to deliver in-phase the emitted power to the common load. The first condition is reached by parallel voltage biasing, while the second one has been realized the JJ coupling by superconducting short-circuited coupled microstrip lines. Digital simulation of such circuit was carried out by PSCAN program (Kornev, et al 1997). The appropriate experimental multijunction Josephson structures were fabricated and measured by means of direct detection of emitted self-radiation. The obraned data are discussed and compared to digital simulations

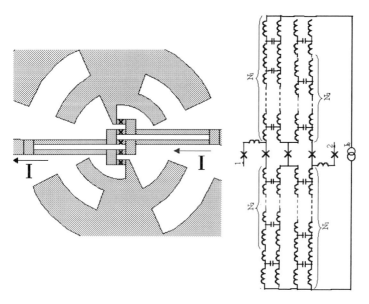

Fig.1. The layout and it's circuit model of the studied JJ array. Loop is 90 µm, strip width is 5µm. JJs are marked by crosses.

2.   EXPERIMENTAL

The design and topology of 5-junction array, fabricated using YBCO thin film deposition over sapphire bicrystal substrate is shown on Fig.1. The short-circuited superconducting (5 µm wide and 0.1 µm thick) coupled microstrip lines of dc bias in parallel were designed to provide mutual interaction of JJs in array. Josephson self radiation, emitted by the array was measured by an external 90 GHz receiver at different fixed levels of applied dc magnetic fields. The fabrication techique for array is analogous to that for single YBCO thin film JJs on sapphire bicrystal substrate (Mashtakov 1999). Fig.2 demonstrate a typical exeperimental dependence P(V) of emitted power from 3-junction array vs bias voltage. No any significant difference from the P(V), corrresponding to emission from single JJ was revealed

The behaviour of emission characteristics of array demonstrates very strong dependence from H. Within one modulation period by H (due to SQUID loops in array) the value of P changed up to 10 times and the Δf - up to 3 times (see Fig.3). These loops made emission process less stable, moreover, spontanious flux pinning could destroy emission mode dramatically. However, an average broadenning of linewidth for array was no larger than that of single junction, with some narrow intervals by H of essential Δf rise by additional 3-4 times. For single junction, varying the applied magnetic field H, the value of differential resistance $R_d(V)$ at bias voltage V was changed which gave monotonous change of power P and the oscillation linewidth Δf. Increasing the H, the P(V) dependense could be plotted with well accuracy and the estimated linewidth was of a order larger, than predicted by RSJ model. $\Delta f = 41TR_d^2/R_N$ MHz for T=10K and $R_N = 2.45 \ \Omega$.

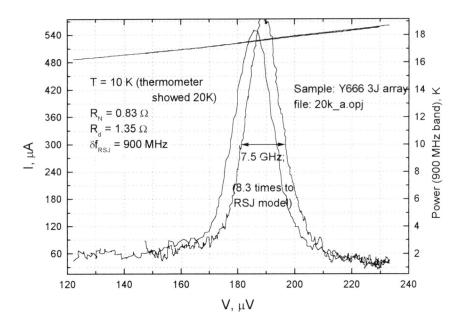

Fig. 2. Self-radiation power, emitted by 3-junction array, directly measured at 90 GHz

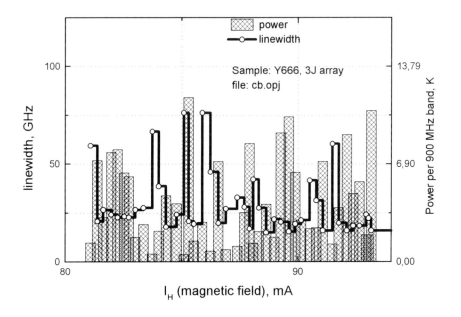

Fig. 3. Magnetic field dependencies for radiation power and linewidth for 3-junction array.

720

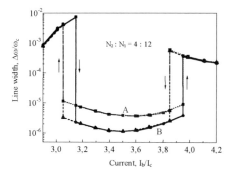

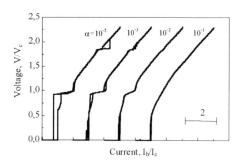

Fig.4. A set of I-V curves for Josephson junctions of the system at different values of high frequency loss coefficient α. McCumber parameter - $\beta = 1$, $\rho/R_n = 10$.

Fig.5 Oscillation linewidth dependence on bias current within resonance peak on I-V curve for the array at $\beta = 1$, $\rho/R_n = 10$, and $\alpha = 10^{-3}$. A – for Josephson junction oscillations, B – for output signal V_{out}.

3.   DIGITAL SIMULATION

We have performed numerical simulation of the experimentally studied system shown in Fig. 1. The long rectangular coupling loops between Josephson junctions in the real system imply shorted sections of a coupled microstrip line and hence can be modeled by the shorted sections of an equivalent two-conductor line, which in its turn, be modeled by the multi-element lumped LC-chain (Fig.1). This allowed us to perform digital simulation of the system by means of well known PSCAN program for the lumped Josephson junction circuits. According to the topology of the system shown in Fig.1 its equivalent scheme consists of identical LC-chains of number N_1 and N_2. An additional set of resistors connected to the capacitors are used for symmetric biasing. Fig. 4 shows a set of I-V curves for the JJ array with distributed coupling circuits at different values of the loss factor α. If α is 10^{-2} and less, one can see well pronounced resonance peculiarities on the curves that correspond to the standing wave excitation in the distributed coupling circuits. In this case the standing wave impact on JJ oscillation provide reasonable reduction in the oscillation linewidth. The linewidth reduction at loss factor $\alpha = 10^{-3}$ is presented in Fig.5. It should be emphasized that the standing waves in the distributed circuits are phase-locked just in-phase by their coupling via JJs. Therefore the amplitude of the output signal, applied to the antenna vibrators is about the sum of the amplitudes of the standing wave oscillations in the position where JJs are connected into the coupling lines. As it was found the optimal position providing maximum output signal corresponds to the ratio $N1/N2 \approx 0.25...0.4$. It can be easily explained if we take into account that the standing wave excitation is provided by Josephson junction interaction with the wave of the current, and the output signal is formed by the standing wave of the voltage.

The work was partially support by Russian Foundation of Fundamental Research, Russian State Program "Modern Problems of the Solid State Physics", "Superconductivity" division, and INTAS program of EU.

REFERENCES

Han S et al 1994 Applied Physics Letters **64**, 1424
Kornev V K, Arzumanov A V 1997 Inst. Phys. Conf. Ser. No158, pp 627-630.
Mashtakov A D et al 1999 Technical Physics Letter **25**, 249

Inst. Phys. Conf. Ser. No 167
Paper presented at Applied Superconductivity, Spain, 14-17 September 1999
© 2000 IOP Publishing Ltd

Basic RSFQ circuits and digital SQUIDs realised in standard Nb/Al-Al$_2$O$_3$/Nb technology.

M. Khabipov[1,2], L. Fritzsch[1], S. Lange[3], R. Stolz[1], F.H. Uhlmann[3], and H.-G. Meyer[1]

1. Institute for Physical High Technology, P.O. Box 100239, D-07702 Jena, Germany
2. Permanent address: IREE Russian Academy of Sciences, Mokhovyay St. 11, Moscow, Russia.
3. Technical University Ilmenau, P.O. Box 100565, D-98684 Ilmenau

ABSTRACT: Basic RSFQ circuits and digital SQUIDs with several designs were fabricated using a Nb/Al-Al$_2$O$_3$/Nb technology and were experimentally tested. The critical current density of the junctions is 500 A/cm^2.

The proper operation of the basic RSFQ circuits including dc/SFQ and SFQ/dc converters and switching elements is experimentally proved. Operation margins of the bias currents of ± 20% have been measured.

A digital SQUID circuit with the feedback loop integrated on the same chip has been realised. The proper response of the feedback circuit for clock frequencies up to 100 MHz was experimentally verified.

1. INTRODUCTION

Rapid Single Flux Quantum (RSFQ) logic circuits are characterised by high speed and large margins of operation (Likharev and Semenov 1991). The short switching time (a few ps) of Josephson junctions and low power consumption (a few µW per gate) together with low noise operation at 4.2 K enable them to be used in a large variety of applications in electronics for fast real-time digital signal processing. The realisation of SQUIDs and their feedback circuitry on the same chip increases the bandwidth of operation of the whole circuit in the real environment.

The Nb/Al-Al$_2$O$_3$/Nb technology has found widespread use in superconducting electronics (Likharev and Semenov 1991). Usually, the area of Josephson junctions is defined by a selective Nb etching process (SNEP) (Gurvitch et al.1983) or a selective Nb anodization process (SNAP) (Kroger et al. 1981) from a Nb/Al-Al$_2$O$_3$/Nb trilayer covering the whole substrate. Also variations and combinations of etching and anodization processes are used to fabricate high-quality tunnel junctions (Imamura and Hasuo 1988, Lee et al. 1990). We developed a process for the definition of Josephson junctions comprising lift-off and SNAP techniques combined with resistors of low parasitic inductance for shunting the junctions and biasing circuits.

2. FABRICATION TECHNOLOGY

A cross section of a shunted Josephson junction is shown in Fig. 1. The complete preparation process comprises nine photomask steps. Thermally oxidized silicon wafers of 3 inch diameter are used as substrates. At first a Nb groundplane with a thickness of 200 nm is deposited by dc magnetron sputtering and patterned by reactive ion etching (RIE) in a CF$_4$ plasma. The isolation of the groundplane to the following parts of the circuit is made by the standard anodization procedure (Kroger et al. 1981) up to a voltage of 25 V. This isolation is reinforced by an evaporated 200 nm thick SiO layer. In order to connect the groundplane to other metal layers, holes are opened by photoresist cover during anodization and by lift-off in the SiO isolation. The next fabrication step is

722

the deposition of the base-electrode consisting of 250 nm Nb and its structuring by CF_4-RIE. The Nb/Al-Al_2O_3/Nb (60nm/12nm/30nm) trilayer is deposited without vacuum interruption onto the Nb base-electrode and patterned by lift-off. The areas of the Josephson junctions are defined by the following complete anodization of the trilayer counter-electrode up to a voltage of 35 V. For the anodization all structures in the groundplane and in the Nb base-electrode have to be connected to the anodization terminal. For the not grounded structures of the circuitry this is done with additional wires, which are removed by a cut etch step using RIE in a CF_4 plasma. A second SiO layer with a thickness of 150 nm is evaporated onto the sample to strengthen the anodic oxide. Contact holes in this SiO layer to the junction surface and to the Nb base-electrode are opened by lift-off. Next, as resistive layer, a 75 nm thick Mo film is sputtered and patterned by lift-off to form the resistors. A third 150 nm thick SiO layer is evaporated onto the sample to strengthen the isolation of the different metal layers and to protect the Mo film against environmental influences. The circuits are completed by sputter deposition of a 12 nm Al etch stop and a 350 nm Nb wiring layer. The Nb is etched with a CF_4-RIE process. The Al etch stop layer prevents an overetch into the Mo resistive film and is removed by wet chemical etching after the Nb structuring process.

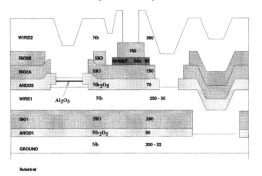

Fig. 1 Cross section of a shunted and grounded Josephson junction.

3.   RSFQ CIRCUITS

Basic RSFQ circuits have been realised in various designs and layouts comprising of dc/SFQ and SFQ/dc converters, Josephson transmission lines (JTLs), and on/off switching elements. The critical current density of the junctions has been set to a nominal value of $j_C = 500$ A/cm^2. The area of

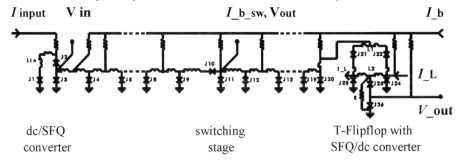

dc/SFQ
converter

switching
stage

T-Flipflop with
SFQ/dc converter

Fig. 2: Schematic diagram of the RSFQ circuit consisting of a dc/SFQ converter, a switching stage and a SFQ/dc converter.

the smallest junctions used in the circuits is $A = 24$ µm^2, and the nominal value of their critical current is $I_C = 125$ µA.

Fig. 2 shows the schematic circuit containing a dc/SFQ converter, a switching stage and a SFQ/dc converter. The design of the converter circuits is similar to the designs described by Kaplunenko et al. (1990), however, the parameter values are different. The JJs are biased at values of approximately 70% of their critical currents I_C using dc current and a resistive divider. A switching stage was included into the JTL of the circuit consisting of a dc/SFQ and a SFQ/dc converters. This switching circuit connects or disconnects two parts of RSFQ circuits depending on its bias current I_b_sw. It consists of two junctions, J_{10} and J_{11} (critical currents: $I_{C10} < I_{C11}$), included into the JTL (see

Fig. 2). It operates as the simplest type of a RSFQ buffer stage providing one-directional SFQ impulse transfer (Kaplunenko et al. 1989, Khabipov et al. 1999, Likharev and Semenov 1991). The switching circuit is connected to a resistive divider and biased by the common dc bias current of the whole circuit. The transfer of SFQ impulses is controlled by a separate bias current $I_{\_b\_sw}$ fed to the inductance between J_{10} and J_{11}. For values of $I_{\_b\_sw}$ close to zero, junction J_{11} switches at the arrival of a SFQ pulse, regenerating it afterwards and transferring it to the SFQ/dc converter. In the case of a definite negative value of $I_{\_b\_sw}$, junction J_{10} will switch upon arrival of a SFQ pulse which, in turn, stops the pulse propagation in the JTL.

Figs. 3 (a) and (b) illustrate the operation of the circuit of Fig. 2. Fig. 3a. shows that each period the input signal I_{in} fed to the dc/SFQ converter creates one SFQ pulse. For positive switching

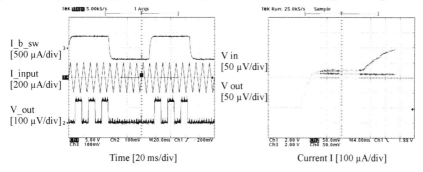

Fig. 3. Operation of the RSFQ circuit. The positive biased switching stage transferred the SFQ pulses created by dc/SFQ converter a) single SFQ pulses to the SFQ/dc converter and b) SFQ pulse train to the JTL.

stage bias current the pulses transferred to the output. Fig. 3b shows the current voltage characteristics (IVC) of the JTLs measured on both sides of the switching stage with the fed bias current on it. These experimental results prove the proper operation of the circuit up to 60 GHz. For a constant value of I_{b_sw} (about 300 µA), typical bias current margins have experimentally been determined to be larger than ± 20%.

4.   DIGITAL SQUID CIRCUITS

Digital SQUID circuits with several variations were fabricated using the Nb/Al-Al$_2$O$_3$/Nb Josephson junction technology described before. The process ensured the manufacturing of transformers with up to 25 turns over washer structures with line widths and spaces of 3 µm.

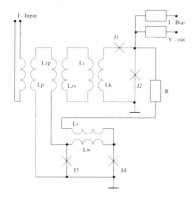

Fig. 4. Schematic diagram of the Digital SQUID circuit.

The digital SQUID structure (Fujimaki 1990, Radparvar and Rylov 1996, Fath et al. 1997) consists of a comparator circuit which is an unsymmetrical interferometer realised using unshunted Josephson junctions. The comparator circuit detects the flux in an input coil and generates voltage pulses of both polarity with an amplitude of about 2 mV. These pulses contain information about applied the flux and are used in the feedback circuit. The feedback loop contains a write gate which is converting voltage pulses to SFQ pulses of the same polarity. The flux created in the feedback loop is opposite to the flux created by the applied signal and stabilises the working point of the comparator circuit.

Fig.5 shows the response of the biased (clock frequency is 20 KHz in this experiment) comparator circuit to the applied signal I_in. The output signal has an amplitude of 2 mV and indicate

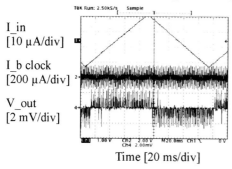

I_in
[10 µA/div]

I_b clock
[200 µA/div]

V_out
[2 mV/div]

Time [20 ms/div]

Fig. 5. Operation of the Digital SQUID circuit

the proper operation of the feedback circuit. The digital SQUID circuits were tested with clock frequencies of the bias current of the comparator circuit up to 100 MHz and proper response of the feedback circuit was experimentally verified. The tested bandwidth was limited by our experimental set up. The bandwidth of the digital SQUID circuits obtained in simulations is 5 GHz (cut frequency of the bias current). For characterisation of the digital SQUID circuits the measurements of the noise is important. These measurements are in progress.

5.   CONCLUSION

Basic RSFQ circuits and digital SQUIDs with several designs were fabricated using a Nb/Al-Al_2O_3/Nb technology and were experimentally tested. The critical current density of the junctions is 500 A/cm^2. The on chip homogeneity of the JJ current density has been measured on an array consisting of 1000 JJs with areas of 48 µm^2 and is about 6%.

The proper operation of the basic RSFQ circuits including dc/SFQ and SFQ/dc converters and switching elements is experimentally proved. Operation margins of bias currents of ± 20% have been measured. The proper operation of the switching circuit has been measured up to 60 GHz.

A digital SQUID circuit with the feedback loop integrated on the same chip has been realised. The proper response of the feedback circuit for clock frequencies up to 100 MHz was experimentally verified.

ACKNOWLEDGMENT

The authors wish to thank K. Pippardt and B. Steinbach for assistance in fabricating the devises and G. Hildebrand for his commitment in the simulations.

REFERENCES

Fath U, Hundhausen R, Fregin T, Gerigk P, Eschner W, Schindler A and Uhlmann F H *IEEE Trans. Appl. Superconduct 7* (2) pp 2747-2751
Fujimaki N *FUJITSU Sci. Tech. J. 27* (1) pp 59-83
Gurvitch M, Washington M A and Huggins H A 1983 *Appl. Phys. Lett.* **42**(5) 472-474
Imamura T and Hasuo S 1988 *J. Appl. Phys.* **64**(3), 1586-1588
Lee L P S, Arambula E R, Hanaya G, Dang C, Sandell R and Chan H 1990 *IEEE Trans. Magn.* **27** 3133-3136
Kaplunenko V K, Khabipov M I, Koshelets V P, Likharev K K, Mukhanov O A, Semenov V K, Serpuchenko I L, Vystavkin A N 1989 *IEEE Trans on Magn.*, **25** (2) pp. 861-864
Kaplunenko V K, Khabipov M I, Koshelets V P, Serpuchenko I L, Vystavkin A N 1989 *Extended Abstracts of ISEC-89* pp. 411-414
Khabipov M I, Balashov D, Buchholz F.-Im, Kessel W, Niemeyer J 1999 *IEEE Trans. Appl. Superconduct* in press
Kroger H, Smith L N and Jillie D W 1981 *Appl. Phys. Lett* **39**(3) 280-282
Likharev K K and Semenov V K 1991 *IEEE Trans. Appl. Superconduct.* **1**(1) 3-28
Radparvar M and Rylov S 1997 *IEEE Trans. Appl. Superconduct 7* (2) pp 3682-3685

Inst. Phys. Conf. Ser. No 167
Paper presented at Applied Superconductivity, Spain, 14-17 September 1999
© 2000 IOP Publishing Ltd

Ramp edge junctions on a ground plane for RSFQ applications

A H Sonnenberg[1], G J Gerritsma and H Rogalla

Low Temperature Division, Department of Applied Physics, University of Twente, P.O. box 217, 7500 AE Enschede, The Netherlands

ABSTRACT: Application of ramp edge junctions in RSFQ circuits requires junctions connected by low inductances. Fabricating the junctions on top of a buried ground plane has minimized stray inductances. Direct injection SQUIDs have been fabricated and inductance values have been measured both on and off the ground plane as a function of temperature and microstrip length. In order to test the junctions in an RSFQ environment a balanced comparator has been fabricated. Dc measurements of the switching properties show a minimum width of the gray zone of 8 μA at 4.2 K.

1. INTRODUCTION

RSFQ is mentioned as a very promising technology in order to improve both speed and power consumption of electronic circuits (Likharev et al. 1991). Up to now most RSFQ circuits have been fabricated in Nb-technology. With HTS-junctions higher I_cR_n-products and higher operation temperatures are achievable. Since RSFQ requires low inductances, we are fabricating the junctions on top of a ground plane. However, care must be taken that the junctions do not become hysteretic due to the increased shunt capacitance. Decreasing this shunt capacitance can be realized by choosing proper materials and/or increasing the thickness of the insulation layer. In this paper we report about junctions, about inductances and about a balanced comparator based on ramp edge junctions on a buried ground plane.

2. MULTI-LAYER PROCESS

The ground plane has been buried under the Josephson junctions in stead of placed on top of the junctions as proposed by Terai et al. (1997). In this way we avoid oxygenation problems in the junctions due to heating up to deposition temperatures twice after deposition of the barrier and counter electrode. However, we have to heat up the ground plane several times. We have minimized oxygenation problems in the ground plane by adding anneal steps during cooling down after each deposition.

In short, we start with a single $DyBa_2Cu_3O_{7-x}$ layer of 100 nm on a $SrTiO_3$ substrate. After patterning the ground plane we deposit an insulation layer. For the inductance measurements we used a $SrTiO_3$-layer of 120 nm thick. But in order to decrease the capacitive shunt we used a combination of $PrBa_2Cu_3O_{7-x}$ /$SrTiO_3$ with a thickness of 370 nm. The resistance values are of the order of 1 kΩ for a square of 200 by 200 μm². Next, vias are etched in order to make electrical contact to the ground plane. After a clean step by physical etching, we deposit the base electrode and the insulation layer in situ. This second insulation layer consists of a $SrTiO_3$/$PrBa_2Cu_3O_{7-x}$ combination of around 170 nm thick. Then, the ramp is formed by physical etching under an angle of 40°. Again the surface

---

[1] E-mail: a.h.sonnenberg@tn.utwente.nl

of the sample is cleaned by a short etch step and subsequently the $PrBa_2Cu_{2.9}Ga_{0.1}O_{7-x}$ barrier and the counter electrode are deposited. Typical thicknesses are 18 nm for the barrier and 150 nm for the counter electrode.

3.   MEASUREMENTS

Both single junctions as well as direct injection SQUIDs have been measured in order to check the junction properties and to measure the inductances of the strips lying above the ground plane. The SQUIDs have been fabricated with a relatively thin insulation layer of 120 nm of $SrTiO_3$. For the fabrication of the balanced comparator, we have increased the thickness of the insulation layers as described above.

3.1   Single junctions

Each chip contains 4 separate junctions in order to measure the IV-characteristics, see Fig. 1, and the magnetic field dependence. The IV-characteristics show almost no hysteresis. However, junctions with larger resistances do show hysteresis at low temperatures (4-10 K). Fig. 2 shows the Fraunhofer pattern for one of our junctions on top of a ground plane, measured at 16 K. The pattern has been made with a voltage criterion of 10 µV. It shows a good correspondence with the fit. The zero points are not reached due to the voltage criterion and possibly due to a small residual critical current.

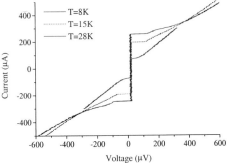

Fig. 1, Current voltage characteristics at different temperatures.

3.2   Direct injection SQUIDs

Direct injection SQUIDs have been fabricated in order to determine the inductance per micrometer of our strips on top of the ground plane. Fig. 3 shows the measurements of the inductance as a function of microstrip length at 10 K for strips with and without a ground plane. Good scaling has been achieved as could be expected. The ground plane works properly since it reduces the inductance with respect to ungrounded strips considerably, about a factor of three. Between 4.2 and 45 K we did not observe a temperature dependence of the inductance, which is explained by the small temperature dependence of the London penetration depth in this regime (Terai et al. 1997).

The inductances of the strips with ground plane have been fitted to the formula derived by Chang (1979):

$$L = \frac{l\mu_0}{w}\left(h + \lambda \coth\left(\frac{d_b}{\lambda}\right) + \lambda \coth\left(\frac{d_c}{\lambda}\right)\right),$$

where w is the width of the strip, l the length, h the thickness of the dielectric layer, λ the London penetration depth and d_b (d_c) the thickness of the bottom (counter) electrode. Given our

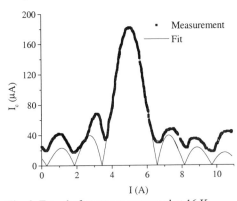

Fig. 2, Fraunhofer pattern measured at 16 K.

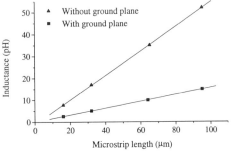

Fig. 3, Inductance reduction by addition of a buried ground plane.

layer thicknesses and assuming a London penetration depth of 160 nm this gives an inductance of 0.16 pH/μm in close agreement to the experimental fit.

3.3 Balanced comparator

A balanced comparator (Filippov et al. 1991, Oelze et al. 1996) has been fabricated and tested according to the diagram shown in Fig. 5. The SFQ pulses are generated by junction J_1 and transported by the Josephson transmission line J_2-J_4 to the comparator loop J_5-J_8. The operation frequency is determined from the voltage across J_1 according to $f=V/\Phi_0$. The junctions J_7 and J_8 are equally biased by the bias current and see the same sampling SFQ pulse. Therefore, for zero injected current I_x, J_7 and J_8 have an equal switching probability. In principle the threshold of the switching probability is sharply defined by the injected current. However, the threshold is smeared out by temperature fluctuations giving rise to a so-called gray zone. The width of this gray zone is defined as:

$$\Delta I_x = \frac{dI_x}{dP_{J8}}\bigg|_{P=1/2} ,$$

where P is the switching probability of junction J_8.

All the junctions are designed to have the same critical current and the inductance of the comparator loop has been minimized. The inductances, L, in the Josephson transmission line were designed to be 5 pH, but are somewhat larger because we have increased the thickness of the insulation layer in order to minimize the capacitive shunt of the junctions.

Fig. 4 shows the switching characteristic of the balanced comparator. For small positive signal currents junction J_8 switches (V_{J8} equals V_{J1}) and for small negative signal currents junction J_7 switches (V_{J7+J8} equals V_{J1}). The switching center is not located at 0 μA, because the critical currents of J_7 and J_8 are not exactly equal. For large positive signal currents junction J_8 reaches its critical current and the voltage across J_8 is no longer equal to the generator voltage. For negative voltages the same happens but now junction J_7 reaches its critical current.

Fig. 6 shows the temperature dependence of the gray zone width at a generator voltage of 80 μV. The line is calculated according a model developed by Filippov et al. (1995) for balanced comparators with a small loop inductance:

$$\Delta I_x = \sqrt{2\pi\mu I_T I_c} ,$$

where μ is a parameter close to 1, I_T the thermal current, 0.042 μA/K, and I_c the critical current. The critical current has been calculated by taking the average of four test junctions on the chip. As shown by the line the model agrees quite well with the measurement. Although three of the

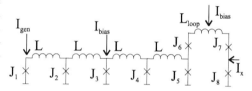

Fig. 5, Equivalent scheme of the balanced comparator.

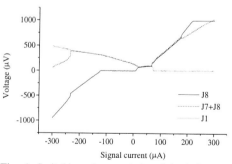

Fig. 4, Switching characteristic of the balanced comparator at 4.2K.

measurement points do not seem to agree well, they are only about 3-4 μA away from the prediction.

In Fig. 7 the width of the gray zone is shown as a function of operation frequency at T=19.5 K. The width of the gray zone increases rapidly with generation voltage. However, for the larger values the shape of the gray zone starts to deviate and determination of the gray zone become ambiguous as has been found by Semenov et al. (1997).

728

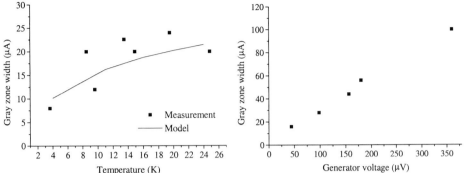

Fig. 6, Gray zone width as a function of temperature. The line shows the expected value calculated from junction I_c's

Fig. 7, Gray zone width as a function of operation frequency at T=19.5 K.

4. CONCLUSIONS

We have fabricated ramp edge junctions and direct injection SQUIDs on a buried ground plane. The ground plane functions properly and reduces the inductances of the strips roughly by a factor of three. We have measured an inductance of 0.16 pH/μm for strips above a ground plane using a 120 nm thick insulation layer made of $SrTiO_3$. The capacitive shunt of the junctions has minimized by decreasing the overlapping area between the ground plane and the counter electrode and increasing the thickness of the insulation layer by adding an additional layer of $PrBa_2Cu_3O_{7-x}$. In addition, a balanced comparator has been fabricated and tested. We have measured the gray zone width as a function of temperature and of operation frequency. The temperature dependence of the gray zone width agrees quite well with a model based on the thermal fluctuation limit.

ACKNOWLEDGEMENTS

This work has been supported by the European Community (EC ESPRIT program under contract No. 23429).

REFERENCES

Chang W H, 1979 J. Appl. Phys. **50(12)**, 8129
Filippov T V, Kornev V K, IEEE Trans. Mag. 1991 **27(2)**, 2452
Filippov T V, Polyakov Yu A, Semenov V K, Likharev K K, 1995 IEEE Trans. Appl. Superc. **5(2)**, 2240
Likharev K K, Semenov V K, 1991 IEEE Trans. Appl. Superc. **1(1)**, 3
Oelze B, Ruck B, Roth M, Dömel R, Siegel M, Kidiyarova-Shevchenko A Yu, Filippov T V, Kupriyanov M Yu, Hildebrandt G, Töpfer H, Uhlmann F H, Prusseit W, Appl. Phys. Lett. 1996 **68(19)**, 2732
Semenov V K, Filippov T V, Polyakov Y A, Likharev K K, IEEE Trans. Appl. Superc. 1997 **7(2)**, 3617
Terai H, Hidaka M, Satoh T, Tahara S 1997 Appl. Phys. Lett. **70(20)**, 2690

Inst. Phys. Conf. Ser. No 167
Paper presented at Applied Superconductivity, Spain, 14-17 September 1999
© *2000 IOP Publishing Ltd*

Fabrication and tests of RSFQ-logic-based D/A converter components

F Hirayama, H Sasaki, S Kiryu, M Maezawa, T Kikuchi and A Shoji

Electrotechnical Laboratory, 1-1-4 Umezono, Tsukuba, Ibaraki 305-8568, Japan

ABSTRACT: Main components of an RSFQ-logic-based D/A converter, a pulse-number multiplier (PNM), a pulse distributor (PD) and a voltage multiplier (VM), were designed and fabricated at Electrotechnical Laboratory using a Nb-based superconducting circuit technology. We confirmed correct operation of the PNM and the PD by low frequency (<1 kHz) functional tests. By average voltage measurement, operating margins for the dc bias current of the output stage of the VM were +/-15% at 10 GHz.

1. INTRODUCTION

Since Hamilton (1992), many efforts have been made (Bentz and Hamilton 1996, Semenov 1993, Smith *et al.* 1999) to develop D/A converters based on the Josephson frequency-voltage relationship. In order to synthesize a waveform with metrological accuracy by the converter, transient effects induced at the time when the output level changes should be adequately evaluated and suppressed. In the previous paper (Sasaki 1999), we proposed a D/A converter, which have advantages: 1) suppression of the transient error without severe requirement for room-temperature electronics, 2) no need of microwave devices and 3) isolation between the output of the converter and the other part. In order to realize the converter, we optimized schematic and layout design of main components for our own fabrication technology. Resulting circuits had operating margins as large as +/-30% by numerical simulations. In this paper, we describe fabrication and experimental results of main components of the D/A converter.

2. DESIGN

A block diagram of the *n*-bit D/A converter we proposed is shown in Fig. 1. By giving digital codes (*CODE0, CODE1*...) synchronized with reference-clock (*RCLK*) pulses, metrologically accurate voltage (*VOUT*) corresponding to the codes can be obtained between the output terminals of the converter. The details of the operation of the converter were described in the previous paper (Sasaki 1999).

Figure 2 (a) shows a block diagram of a binary counter with complementary inputs/outputs and shift-registers (DRBCSR), which is a main circuit in the pulse-number multiplier (PNM) using the latency-reduction technique described by Lin and Semenov (1995). We have fabricated 4-stage DRBCSR in the present study.

Figure 2 (b) is a block diagram of the pulse distributor (PD) circuit. The circuit consists of a pulse switch (PLSW) and a non-destructive read out cell (NDRO). The code signals from the waveform ROM are applied to the pair of the input terminals D+ and D- as a balanced current input to reduce cross talk of signals. At each trigger of *LATCH* pulse, the PLSW generates either set (S) or reset (R) pulse according to the polarity of the control signal D+/D-. The S (or R) signal makes the internal state of the NDRO to allow (or not to allow) the *PTRAIN* pulses to pass to the *OUTPUT*.

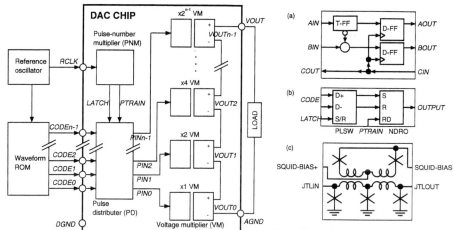

Fig. 1. Block diagram of an n-bit D/A converter.

Fig. 2. Block diagram of (a) binary counter with complementary input/output and shift-register (DRBCSR), (b) pulse distributor (PD) and schematic diagram of (c) voltage multiplier (VM).

Figure 2 (c) is a schematic diagram of one unit of the voltage multiplier (VM) circuit. The VM circuit consists of dc SQUIDs and JTLs coupled to the SQUIDs. In this design, two units of a JTL are coupled to the dc SQUID, resulting in wide operating margins in spite of relatively small coupling constant (k=0.4-0.5). Also, the optimization of the damping (McCumber) parameter of the junctions in the SQUID secures switching sequence and increases the operating margins.

For schematic and layout design of the circuits, we used the PSCAN and LMETER programs developed at SUNY at Stony Brook (SUNY) with CAD tools from Cadence Design Systems, Inc.

3.   FABRICATION

The PNM, PD and VM circuits were fabricated using a Nb/AlO$_x$/Nb junction integration process developed at Electrotechnical Laboratory. Hereafter, we call the process ETL standard process (ESP). The outline of ESP is as follows: (1) an MgO film is deposited on a 76 mm-diameter Si wafer as a stopping layer in following etching processes, (2) a Nb groundplane with moats is

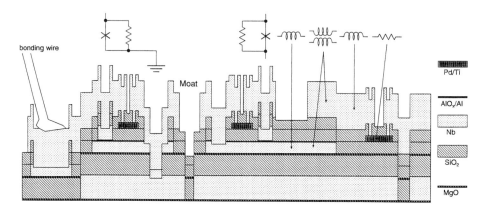

Fig. 3. Schematic cross section of a circuit fabricated by ETL standard process. Resistively shunted junctions (grounded and floating), self and mutual inductance and a resistor are indicated.

731

Table 1. Summary of parameters employed by ETL standard process.

Josephson critical current density	1.4 or 1.6 kA/cm²
Sheet resistance	1.2 Ω
Minimum junction size	3 x 3 μm²
Minimum width of the wiring	2 μm
Groundplane	Nb (200 nm, dc sputtering)
Groundplane insulation	MgO (20nm, EB evaporation) / SiO₂ (180nm, rf sputtering)
Base-electrode of the junction	Al (~8 nm, dc sputtering) / Nb (100nm, dc sputtering)
Counter-electrode of the jucntion	Nb (125 nm, dc sputtering)
Resistor	Pd (~55 nm, rf sputtering)
Junction insulation	SiO₂ (100 nm + 100 nm, rf sputtering)
Wiring	Nb (300 nm, dc sputtering)

fabricated by sputtering and etching, (3) an SiO₂ groundplane insulation with contact-holes is fabricated by sputtering and etching, (4) an MgO film with contact-holes is fabricated by evaporation and lift-off technique, (5) a Nb/AlOₓ/Nb trilayer is deposited, (6) base- and counter-electrodes of junctions are defined by etching, (7) an SiO₂ insulation layer is deposited, (8) Pd resistors with thin Ti adhesion layers are fabricated by sputtering and lift-off, (9) an SiO₂ insulation is deposited, (10) contact-holes are formed by etching and (11) Nb wiring are fabricated by sputtering and etching. We use an i-line stepper for all photolithography processes. A schematic cross section of a circuit fabricated by ESP is displayed in Fig. 3 and the process is summarized in Table 1.

4.   TESTING

All measurements presented here were done using a cryostat with double magnetic-shield placed in liquid helium.

Figure 4 shows the result of the low-speed test for a PNM circuit with four DRBCSR units. *AIN* and *CIN* are input signals to DC/SFQ circuits and current swing of those were both 0.7 mA. *AOUT* and *BOUT* are output signals measured through SFQ/DC read-out circuits. In this test, alternating input pulses of *AIN* and *CIN* were applied and the correct response, i.e., binary counting operation, were observed in waveforms *AOUT* and *BOUT*. However, the margins of global bias current for the circuit were less than +/-1%, which are much narrower than those (+/-25%) obtained from a computer simulation. Although the origin of the large reduction of the experimental margin is not clear now, some inadequate layout of the circuit patterns might be responsible for it.

Figure 5 shows results of a low-speed test for a PD circuit. The *CODE* is a current input

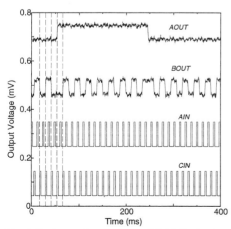

Fig. 4. Testing results of a 4-stage DRBCSR circuit. *AOUT* and *BOUT* show complementary output of a 4-bit (1/16) frequency divider.

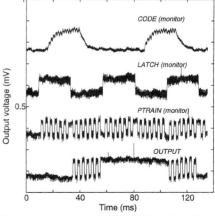

Fig. 5. Testing result of a PD. *CODE* and *LATCH* gate *PTRAIN* pulses to *OUTPUT*.

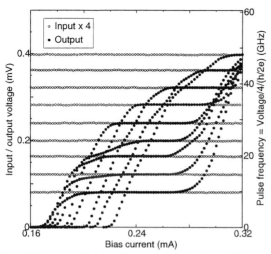

Fig. 6. Voltage multiplying characteristics of a 4-stage VM.

displayed with arbitrary unit. Other input signals of *LATCH* and *PTRAIN* were measured through SFQ/DC circuits. Correct operation expected from design was observed with the margins +/-7.5%, which were narrower than those by simulation (+/-26%) but practically large enough for prototype (4~5 bit) D/A converter.

Figure 6 shows the voltage multiplying characteristics of a 4-stage VM measured by average voltage measurement method. Y-axes show the input (x4) and the output voltages of the VM and the corresponding pulse frequencies. The dependence of the characteristics on the SQUID-bias current was measured changing the frequency of input pulses. From the results, the operating margins at 10GHz were +/-15%, about a half of that calculated by simulation (+/-30% at 10GHz). The degradation seems to be due to difficulties of extraction the mutual inductance accurately from the layout. So experimental extraction may improve the accuracy of the simulation, and then, may enhance the operating margins.

4. CONCLUSION

We have fabricated and tested three component circuits for an RSFQ-logic-based D/A converter: a pulse-number multiplier (PNM) circuit, a pulse distributor (PD) circuit and a voltage multiplier (VM) circuit. The results of low-speed tests for the PNM and PD circuits showed that both circuits operated correctly, but their bias margins were narrower than those expected from simulations. The bias margins for the 4-stage VM circuit were +/-15% at 10 GHz. To apply the component circuits to a D/A converter, further study is necessary to improve the bias margins for the PNM circuit.

REFERENCES

Benz S P and Hamilton C A 1996 Appl. Phys. Lett. **68**, 3171
Hamilton C A 1992 IEEE Trans. Appl. Superconductivity **2**, 139
Lin J and Semenov V K 1995 IEEE Trans. Appl. Superconductivity **5**, 3472
Sasaki H, Kiryu S, Hirayama F, Kikuchi T, Maezawa M and Shoji A 1999 IEEE Trans. Appl. Superconductivity **9**, 3561
Semenov V K 1993 IEEE Trans. Appl. Superconductivity **3**, 2637
Smith A D, Durand DJ and Dalymple BJ 1999 IEEE Trans. Appl. Superconductivity **9**, 63
SUNY (PSCAN, LMETER) http://pavel.physics.sunysb.edu/RSFQ/

Inst. Phys. Conf. Ser. No 167
Paper presented at Applied Superconductivity, Spain, 14-17 September 1999
© 2000 IOP Publishing Ltd

733

Measurements of a high-T_c RSFQ 4-stage shift register and the inductance of the shift register cell

J H Park, Y H Kim, J H Kang*, T S Hahn, C H Kim, J M Lee and S S Choi

P.O.Box 131, CheongRyang, Seoul, Korea 130-650, Superconductivity Research Laboratory, Korea Institute of Science and Technology
*University of Inchon, Inchon, Korea

ABSTRACT: A 4-bit rapid single flux quantum(RSFQ) shift register circuit using $YBa_2Cu_3O_x$ (YBCO) bicrystal junctions has been designed, fabricated and tested. The circuit consists of 4 shift register stages and a read SQUID placed next to each side of the stages to monitor the data shifts. Each SQUID was inductively coupled to the nearby shift register stage, respectively. We could observe that a data "1010" as well as a simpler data "0000" or "1000" was successfully shifted through the stages by controlled current pulses used as clock pulses.

1. INTRODUCTION

Over the past several years, there has been considerable development of superconducting digital electronics based on the RSFQ logic family. The main advantage of this new digital technology is its unparalleled speed and very low power consumption. Due to the continuous works on RSFQ circuits using low-T_c Nb multilayer technology, some devices reported by O A Mukhanov et al and J C Lin et al are on the stage of becoming practical system components. In contrast, circuits based on high-T_c technology to date are still quite simple because no mature fabrication technology has yet been found to implement more than 10^3 reproducible junctions on a chip with critical current spread of a few percent. And the multilayer process incorporating a superconducting ground plane with associated epitaxial insulators is also still immature. Nevertheless, continuous research on the simple high-T_c RSFQ circuits is very important in several aspects. First, possible higher operating temperature can reduce cooling costs. Second, the maximum operating speed of RSFQ devices, generally limited by the I_cR_n values of junctions, can be increased by using high-T_c superconductors instead of low-T_c superconductors. This is because high-T_c junctions have larger energy gaps than conventional superconductors and thus higher I_cR_n values. Finally, high-T_c digital circuits can have more advantage on a wide variety of applications where the less cooling load is required. Shift register is a basic component in a wide variety of RSFQ circuit and simple enough to be fabricated in a single YBCO layer. We have designed, fabricated and tested a 4-stage shift register consisting of 9 YBCO bicrystal Josephson junctions. Two read SQUID, in particular, were fabricated next to each side of the stages to correctly monitor the data shifts.

2. CIRCUIT DESIGN AND FABRICATION

Fig.1 shows the micrograph of the fabricated circuit (a) and its circuit schematic (b). The complete circuit consists of 4 shift register stages(five Josephson junctions), two read SQUID(4 Josephson junctions), each of them placed next to each side of the shift register stages, and two control lines tuning the two read SQUID to have the maximum transfer functions. All the Josephson junctions forming the shift register had a width of 8 μm and were separated by the holes with the dimension of 4×6 μm². The read SQUID junctions had a width of 4 μm and the SQUID loop had a hole with the dimension of 4×10 μm².

The correct operation of the circuit has been demonstrated in the following ways. In the first step, binary input data "1" was produced by the generation of a single SFQ pulse by pulsing the Data in(I_{in}) line(see Fig.1) with controlled current value and a flux quantum was stored in the first stage(leftmost) of the shift register. Then this loaded data was shifted to the second, to the third and to the final stage(rightmost) by the subsequent Shift 1(I_1)–Shift 2(I_2)–Shift 3(I_3) pulses, respectively. Then the shift register was reset by the Data out(I_{out}) pulse. Both of the read SQUID voltages were measured simultaneously after each pulse injection. We also generated a data sequence "1010" and stored in the shift register. Then we successfully shifted the data through the stages with well-determined values of the Shift pulses.

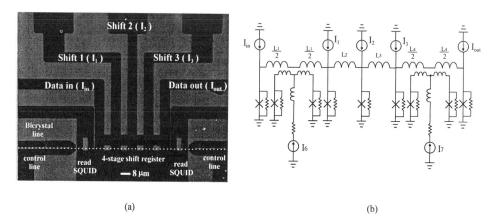

(a) (b)

Fig. 1. Microphotograph (a) and schematic (b) of the fabricated 4-stage shift register circuit.

The circuit was fabricated on an asymmetric 24° SrTiO₃ bicrystal substrate. An epitaxial 200 nm-thick YBCO film was deposited by a pulsed laser ablation technique and patterned by standard optical lithography and ion milling technique. A gold layer was evaporated and patterned by lift-off and annealed at 500 °C for 1 hour for good electrical contacts.

3.   DIGITAL MEASUREMENT SET-UP

The shift register circuit was tested in a liquid helium dewar with a single cryoperm magnetic shield. The temperature of the sample was controlled to 75±0.02 K. All the bias currents to the read SQUID and control lines were supplied by Keithley 220/224 current sources and the current pulses to the shift register bias lines were supplied by a set of D/A converters(HP E1328A), operating in voltage mode with 2 kΩ resistors in series. The voltages of two read SQUID were simultaneously measured using Keithley 181 nanovoltmeters and all the lines were filtered with RC filters with a cutoff frequency of 10 Hz.

4.   RESULTS AND DISCUSSIONS

The major obstacle in testing the circuit was the interference between the two read SQUID. There was a large difference in shape between the independently measured V-Φ modulation curves of the two read SQUID and the simultaneously measured curves obtained by applying the same control current. This could be explained by the fact that the read SQUID may feel the magnetic field caused by the control line at the opposite side to a certain extent. We solved the problem by finding the correct operation points of the SQUID from the simultaneously measured V-Φ modulations. To estimate the inductance values of the shift register SQUID loops, the periodic voltage modulation of the data SQUID in a flip-flop circuit was measured using a direct current injection method described elsewhere by M G Forrester et al. This flip-flop circuit was fabricated near the shift register on the same chip. The loop of this data SQUID had the same dimension as the shift register loop. As shown in Fig. 2, the measured period of 400 μA corresponds to the inductance value of 5 pH. Since this value was only about half of the full inductance value, we got the value of 10 pH for the full inductance of the data SQUID. The critical current of the data SQUID junction was measured to be 500 μA. With these values, the inductive parameter $\beta_L = 2\pi L I_c / \Phi_0$ could be calculated as 15.7, somewhat larger than the optimum value of 10 for a stable circuit operation.

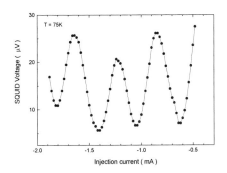

Fig. 2. The periodic voltage modulation of the data SQUID in a flip-flop circuit, which was fabricated near the shift register. The voltage modulated with a period of about 400 μA, which corresponds to an inductance of 5 pH for the half of the total inductance of the data SQUID.

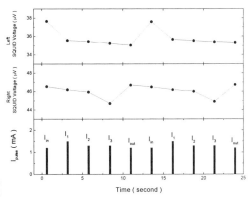

Fig. 3. Results of the operation of the shift register for data "1000".

In Fig. 3, the correct operation of the shift register is shown for data "1000". When the data "1" is loaded into the first stage(leftmost) by the Data in(I_{in}) pulse, the left SQUID voltage shifts to the high level corresponding to a change of one flux quantum. By the next shift 1(I_1) pulse, the data is moved into the second stage and the left SQUID voltage abruptly changes to the low level and subsequently decreases to the slightly lower values as the data is shifted away to the third and to the final stage. Finally, the SQUID voltage returns to its initial level as the data is shifted out from the shift register by the Data out(I_{out}) pulse. In the meantime, the right SQUID voltage decreases to the slightly lower value as the data is subsequently shifted to the third stage. When the data is moved into the final stage, the SQUID voltage shows abrupt change to the low level and returns to the initial level after the data is shifted out.

Fig. 4 shows the results of the same operations described in Fig.3, repeated 100 times with the changes in the amplitude the Data in pulse. In Fig. 4(a), we can see that if data "0" instead of "1" is loaded into the first stage by applying a sufficiently small Data in pulse, the voltages of both read SQUID remain at their initial levels. But the probability of generating the data "1" will rise as the amplitude is increased. Also, whenever a data "1" is generated it gets shifted rightward resulting in the change of the right read SQUID voltage. Fig. 4(b) shows the expected results that the right SQUID voltage changes only when the left SQUID voltage changes. The amplitude of the Data in pulse applied in Fig. 4(c) is only 25 μA larger than the one applied in Fig. 4(b). But the probability of a data "1" generation is strikingly different. This difference possibly results from a considerably narrow circuit margin of our shift register expected by the aforementioned larger

736

value 15.7 of β_L than the optimum one. Once the amplitudes of all the pulses were fixed to the appropriate values, the operation of the shift register was considerably stable. The circuit was correctly operated for 23 min. with only one time error and the result is shown in Fig. 4(d).

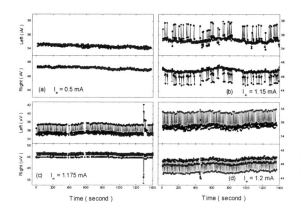

Fig. 4. The results of the same operations described in Fig.3, repeated 100 times with the changes in the amplitude of the Data in pulse(I_{in}). The current values used were $I_1 = 1.5$ mA, $I_2 = 1.3$ mA, $I_3 = 1.3$ mA, and $I_{out} = 1.2$ mA.

After confirming the correct operation of the shift register, we measured the operation of the circuit with a data sequence "1010". First, two read SQUID voltages corresponding to the "0000" data state were measured. And then, the data sequence "1010" was stored using subsequent I_{in}-I_1-I_2-I_{in} pulses. The I_3-I_1, I_{out}-I_2 and I_3 pulses moved the "1010" data to the right direction making the shift register data state to be "0101", "0010" and "0001", respectively. Finally, the shift register was reset by the I_{out} pulse, and returned the shift register to its initial "0000" state. This operation was repeated 100 times with about 90 % accuracy. The voltage responses of the two read SQUID corresponding to each data states were simultaneously measured. Fig. 5 shows the typical voltage responses of one cycle. In this figure, we can easily notice that the left SQUID shows its maximum response only when "1" is stored in the first stage and so the right SQUID does only when "1" is stored in the last stage.

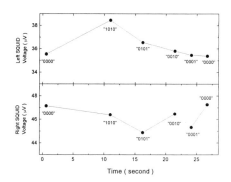

Fig. 5. The operation of the circuit for the data "1010".

4. SUMMARY

We have designed and fabricated a 4-stage shift register circuit using SrTiO$_3$ bicrystal and tested its correct operations by a computer-controlled digital measurement set-up. We got the value of 10 pH for the full inductance of the shift register data stage, somewhat larger than its optimum value. A data sequence "1010" as well as a simpler data "0000" or "1000" was successfully shifted through the shift register stage by controlled current pulses used as clock pulses.

REFERENCES

J C Lin, V K Semenov and K K Likharev 1995 IEEE Trans. Appl. Supercond. **5**, 2252
M G Forrester, J X Przybysz, J Talvacchio, J Kang, A Davidson and J R Gaveler 1995 IEEE Trans. Appl. Supercond. **5**, 3401
O A Mukhanov and A Kirichenko 1995 IEEE Trans. Appl. Supercond. **5**, 2461

Inst. Phys. Conf. Ser. No 167
Paper presented at Applied Superconductivity, Spain, 14-17 September 1999
© 2000 IOP Publishing Ltd

Σ-Δ modulator for HTS ADC based on the ramp-type Josephson junctions.

A Yu Kidiyarova-Shevchenko [1,2]**, M Huang**[1]**, P Komissinski** [1,3]**, Z Ivanov** [1,4]**, and T Claeson**[1]

[1] Chalmers University of Technology, Gothenburg, Sweden
[2] Nuclear Physics Institute, Moscow State University, Moscow, Russia
[3] Institute of Radio Engineering and Electronic RAS, Moscow, Russia
[4] SP, Swedish National Testing and Research Institute, Boras, Sweden

ABSTRACT: An all high-T_c modulator for Σ-Δ analog-to-digital converters (ADC) was optimized with parameters determined by a ramp type Josephson junction technology having a critical current spread less than 13% and using four superconducting layers. Equivalent circuit inductances were extracted from layouts with the help of a new multilayer 3D- MLSI software package. The resulting performance of the ADC is analyzed in detail and possible improvements are discussed.

1. INTRODUCTION

Numerous suggestions have been made to implement superconducting analog-to-digital converters (ADC). At the hi-end of that research are oversampling ADCs based on Σ-Δ and Δ modulation techniques (Przybysz et al. (1993) and Semenov et al. (1997)). The first fully operational Δ ADC with on chip RSFQ digital signal processing was successfully fabricated in niobium technology and the demonstrated performance was about 16 Spurious Free Dynamic Range (SFDR) at 5MHz signal bandwidth (Semenov et al (1997)).

High-T_c based ADCs may have advantage to their low-T_c analogs due to at least 10 times higher sampling frequency that is determined by the I_cR_n product of the Josephson junctions. An complicated technology that satisfies all requirements on HTS digital circuit still has to be developed to obtain such performance.

Among the different types of HTS Josephson junction technologies, the ramp type junction technology is one of the most promising due to its natural multilayer structure. It was recently reported by Sonnenberg (1999) and Huang (1999) and that this technology provides critical current spread of less than 13% and hence allows device fabrication using 30-50 Josephson junctions.

In this paper, we evaluate the performance of a Σ-Δ modulator that can be realized in a ramp type Josephson junction fabrication process with four superconducting layers.

2. BASIC CONCEPT

Following the approach of Przybysz et al. (1993), a first-order superconducting Σ-Δ modulator can be constructed as a combination of an R/L integrator, a quantizer formed by a balanced comparator *J0, J1* (Filippov et al. (1995)), sampling and output pulses conditioners *J3-J5* and *J6-J9* (see Fig. 1). The components are well known and will not be described in detail here.

The performance of the modulator is determined by the speed and accuracy of the quantizer. An independent optimization of the balanced comparator (the same schematic but without the R/L loop) shows that the maximum operational frequency of *0.24f$_c$* can be reached under the condition of 30% margins on the circuit parameters is equal. Under the same conditions, we kept the threshold

738

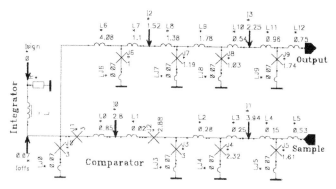

Fig. 1. Equivalent circuit of the Σ-Δ modulator. The parameters are given in units normalized to the minimal critical current (J6).

misplacement as small as possible. The optimization resulted in a minimum threshold misplacement of about $3\%I_c(J0)$. Corresponding optimal working parameters are indicated in the figure.

The R/L pole of the integrator is determined by the useful signal bandwidth and the value of the thermally induced threshold misplacement of the quantizer, ΔI_{thr}. It was shown by Przbysz et al. (1993), that at $\Delta I_{thr}=0.2\Delta$ ($\Delta=\Phi_0/L$) and input signal frequency f twice as large as the integrator pole $R/2\pi L$, the Σ-Δ modulator can have a regular increase in signal to noise ratio (SNR) by $9dB/octave$ of oversampling.

The output of the modulator represents the input signal together with its out-of-band components and has to be passed through a digital low-pass filter to attenuate it. Under the limiting condition on the number of Josephson junctions, the decimation filter can be replaced by a counter formed from a T-flip-flop with an SFQ/DC converter proposed by Likharev and Semenov (1991).

3. FABRICATION PROCESS

The HTS ramp junction technology that can be used to fabricate the Σ-Δ ADC fabrication has to satisfy minimal requirements on the circuit topology and inductive parameters. In order to achieve this, the structure has to be integrated over a ground plane; the Josephson junctions have to be formed with a rotating substrate during the etching to achieve their arbitrary orientation and an additional superconducting layer should provide a connection of the two electrodes of the Josephson junctions to the ground plane. A sketch of the process that employs all these features is shown in Fig. 2.

Since the quality of the ramp junctions is very sensitive to the roughness of the films, the four layer structure with ground plane M0, insulator I0, bottom electrode of the Josephson junction M1 and insulator I1 have to be prepared in situ in one technology step. Such a structure was deposited and he measured roughness was about 15nm. All films were grown by pulsed laser deposition at 800°C and 0.2mbar oxygen pressure.

The Josephson junction fabrication process was established independently. Smooth ramps with angles of 25-29° to the substrate were formed in M1/I1 layers using an Ar⁺ milling through a hard baked photoresist mask with an etch angle of 45° normal to the rotating substrate. The 25nm thick PBCGO barrier and the upper electrode M2 were then deposited and patterned. The junctions showed RSJ like behavior with $J_c(4.2K)\approx10^4A/cm^2$, $I_cR_n(4.2K)\approx4mV$ and 1σ spread measured over 27 junctions less than 13%. A detailed description of the technology process and experimental results can be found in [4].

An additional YBCO layer M3 is going to be used for "via" connection

M3, YBCO 130nm   **I1**, PBCO/STO/PBCO 100nm

M2, YBCO 130nm   **M1**, YBCO 185nm

Barr, PBCGO 25nm   **I0**, PBCO/STO/PBCO 100nm

M0, YBCO 130nm   **JJ**

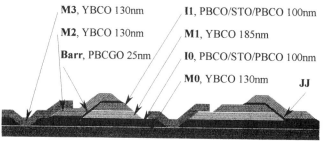

Fig. 2. A schematic view of the ramp type junction integrated circuit. Contacts to ground are made through the M3 electrode.

to the groundplane. The windows in insulators I0, I1 can be opened in one technology step. The etching conditions should be the same as in the ramp preparation to avoid the appearance of the step-edge Josephson junctions in M3 layer. Measurements of the parameters of the Josephson junctions, before and after heating them at the YBCO deposition temperature, indicated that the addition of the fourth superconducting layer to the process costs 20% decrease in the critical current density.

4. CIRCUIT DESIGN

Several circuits were designed for experimental testing: balanced comparator, T-flip-flop, DC/SFQ converter and SFQ/DC converter.

The minimum critical current of Josephson junctions, I_c^{min}, was chosen to give a sufficiently low bit error rate $P=10^{-16}$ $1/bit$. The incorrect switching, with probability P, in pairs of competitive Josephson junctions (balanced comparator) was considered as a major source of error in the circuits. To find the relation between the required value of I_c^{min} and the working temperature T, we can apply the equation for the probability P for a certain value of the measured current I near the threshold I_{thr} (Filippov (1995)):

$$P = 1 - erf(\sqrt{\pi}\frac{I_{thr}-I}{\Delta I_{thr}}), \quad \Delta I_{thr}[\mu A] = 10^3 \left(2\pi\frac{2\pi k_B T}{\Phi_0}\right)^{1/2}\left(I_c^{min}\right)^{1/2}$$

where ΔI_{thr} is the effective threshold uncertainty. When the current I is considerably far from the threshold I_{thr}-I=$30\%I_c^{min}$, the minimal critical current will be given by $I_c^{min}[\mu A]=34T[K]$.

The range of critical current is also limited by the criterion $w \leq 4\lambda_J$, where w is the width of the Josephson junction. If the normalized critical currents in the circuit range between I_c^{min} and $2I_c^{min}$, then the temperature dependence of I_c^{min} in ramp type junctions can be expressed by:

$$I_c^{min} \leq 2\lambda_J \frac{h}{\sin(\alpha)} J_c(T) = \left(\frac{\Phi_0 J_c(T)}{2\pi\lambda_L(T)\mu_0}\right)^{1/2}\frac{h}{\sin(\alpha)},$$

where h is the width of the bottom electrode film and α is the ramp angle.

Thus, for the given technological parameters of $J_c(4.2K)=8*10^3 A/cm^2$, $\lambda_L(4.2)=180nm$, $h=184nm$ and $\alpha=25°$, the available range of the critical currents vary from $I_c^{min}=203\mu A$ $(w=6.4\mu m)$ to $2I_c^{min}=410\mu A$ $(w=13\mu m)$. The estimated working temperature is about $T=7K$, which results in simulation units of voltage, resistance, inductance and frequency of $I_c R_n=3.1mV$, $R_n=15.3\Omega$, $\Phi_0/2\pi I_c=1.62pH$ and $2\pi I_c R_n/\Phi_0=9.47THz$ respectively.

Fig. 3 shows the schematic and the corresponding layout of the most complex part, an SFQ/DC converter. The equivalent scheme inductances were extracted with the help of the 3D-MLSI (3D MultiLayer Superconducting Inductances) program (Khapaev (1999)). This program takes into

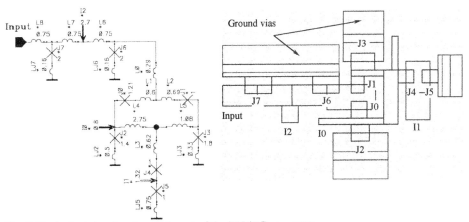

Fig. 3. Equivalent circuit and mask layout of the SFQ/DC converter.

account the 3D distribution of the magnetic field in the structure caused by high values of the London penetration depth compared to the thicknesses of the layers.

We have optimized again the parameters of the circuits using the extracted values of the inductances and the estimated critical currents of the Josephson junctions. As a result, the lowest margin is 19% on the critical current of the junction J4 in SFQ/DC converter with the optimal parameters shown in Figure 3. The variation of the parameters of the balanced comparator (see Table 1) did not affect the margins but it has reduced the sampling frequency to $0.1f_c$ and it has increased the threshold uncertainty to 7%(J0).

With the estimated working parameters the threshold uncertainty of the quantizer is $\Delta I_{thr}=17.8\mu A$ and it defines the minimum of integrator inductance $L=23pH$. It, in turn, limits the input resistor to $R=8m\Omega$ for $60MHz$ input signal bandwidth. A complete Σ-Δ modulator in our existing technology can not be fabricated as it is hard to achieve the required small value of the resistor in the integrator loop due to the high ($\approx 10^{-5}$-$10^{-6}\Omega cm^2$) contact resistance between YBCO and normal metal (Komissinski (1999)).

Table 1. Optimal parameters for balanced comparator in HTS ramp junction technology

J0=2	J1=2	J2=2	J3=2	J4=2
J5=1	J6=1	J7=1.63	J8=2	J9=2
I0=2.3	I1=2.7	I2=1.54	I3=2.7	L0=0.75
L1=0.42	L2=0.96	L3=1.08	L4=0.9	L5=0.75
L6=3.5	L7=1.08	L8=0.8	L9=1.5	L10=0.75
L11=0.75	L12=0.75	LJ0=0.18	LJ3=0.18	LJ4=0.18
LJ5=0.18	LJ6=0.18	LJ7=0.18	LJ8=0.18	LJ9=0.18

5.   DISCUSSION

The biggest difficulties in the application of HTS Josephson junctions to RSFQ digital circuits have been high inductances and topology restrictions caused by technology limitations. The four-layer ramp junction technology proposed here could overcome these difficulties.

A quantizer and one stage of the counter with SFQ/DC converter optimized for such a process are expected to operate with minimum 19% margins of the parameters and at a sampling frequency of $f_s=150GHz$. The resulting signal to noise ratio is equal to $SNR=f_s/2f=1250$ for a $60MHz$ input signal bandwidth and this provides 3 bits of extra resolution in comparison to low temperature oversampling ADCs. The working temperature is $7K$ and the level of thermal noise induced errors is $BER=10^{-16}$ 1/bit.

Improvements in the technology by increasing the critical current density up to $J_c=3*10^5 A/cm^2$ and correspondingly decreasing the junction linear size to $1.2\mu m$ should allow an operational temperature of about $30K$. To overcome the difficulties with a small resistor at the input of the Σ-Δ modulator, an alternative Δ modulator can be considered.

ACKNOWLEDGMENTS

The research presented in this paper was mainly funded by the ESPRIT grant HTS-RSFQ, the Material Consortium on Superconducting and in part by the grants INTAS-97-1712 and ISTC N11-99.

REFERENCES

Filippov T V et al 1995 IEEE Trans. Appl. Supercond 5 pp 2566-2569
Huang M Q et al. 1999 to be published in J. Low Temp. Phys.
Likharev K K and Semenov V K 1991 IEEE Trans. Appl. Supercond. 1 pp 3-28
Komissinski P. et al presenting at this conference
Khapaev M M 1999 to be published in IEEE Trans. Microwave Theory and Technique
Przybysz J X et al 1993 IEEE Trans. Appl. Supercond. 1 pp 2732-2735
Semenov V K, Polyakov Yu, Schneider D 1997 Ext. Abstract of ISEC'97 2 pp 344-46
Sonnenberg A H, Gerritsma G J, and Rogalla H 1999 to be published in Physica C

Inst. Phys. Conf. Ser. No 167
Paper presented at Applied Superconductivity, Spain, 14-17 September 1999
© 2000 IOP Publishing Ltd

Universal NAND gate based on single flux quantum logic

H Myoren†, S Ono and S Takada

Faculty of Engineering, Saitama University, 255 Shimo-Okubo, Urawa 338-8570, Japan

ABSTRACT: We propose a universal NAND logic gate based on single flux quantum (SFQ) logic. In the proposed gate, three superconducting loops share two Josephson junctions (JJs). The critical currents of the JJs were designed to allow each of any two loops to trap an SFQ at the same time. Using an SFQ pulse splitter, a clock SFQ pulse induces an SFQ pulse to set SFQ in one of the superconducting loop. Simulation results for dynamic operation of this NAND gate show that the NAND gate can operate with a delay time of 35 ps.

1. INTRODUCTION

Single flux quantum (SFQ) logic circuits have been reported to be able to operate at high frequency and at ultra-low power(Likharev and Seminov 1991, Nakajima *et al* 1991). The potential operation-speed performance of SFQ circuits has been realized using low (Bunyk *et al* 1995) and high (Kaplunenko *et al* 1995) temperature superconductor junctions.

To construct any logic circuits, AND gate, OR gate and NOT gates have been required. Universal gates such as NAND gate are attractive since all logical function can be constructed by combination of one kind of universal gate (Nakajima *et al* 1991). By constructing NAND gates based on the SFQ circuit, we could construct all logical functions using SFQ logic circuits. Therefore, we can adapt a standard-cell method for making layout of logic gates that is explored in MOS technology now used in semiconducting digital circuits. From this point of view, we have investigated the NAND gate so as to be able to integrate SFQ circuits.

In a previous study (Myoren *et al* in press), we described the principle of the NAND gate based on SFQ logic. The proposed NAND gate has a extra input terminal T that enable the NAND operation. To reduce complexity of clock timing, it is better to eliminate the extra terminal.

In this study, we proposed a practical NAND gate that can be used in the SFQ circuits, and confirm the correct NAND operation.

2. NAND GATE

2.1. Circuit design

We propose a NAND gate based on the SFQ circuit shown in Fig.1. This NAND gate has three superconducting loops $\ell_T(J_{T1}, J_{T2}, L_T, J_1$ and $J_2)$, $\ell_A(J_{A1}, J_{A2}, L_A, J_1$ and $J_2)$ and $\ell_B(J_{B1}, J_{B2}, L_B, J_1$ and $J_2)$ that share two Josephson junctions (JJs) of J_1 and J_2 in series. J_1 limits the number of SFQ trapped in the loops and J_2 outputs SFQ pulses. Each superconducting loop has its own input terminal and the three terminals are equivalent to each other. To get an output, there are a clock (**Clk**) and an output (**Out**) terminals. The output circuit contains the junctions of J_3 and J_2 in series. A clock SFQ pulse splits two SFQ pulses

---

† E-mail: myoren@super.ees.saitama-u.ac.jp

at an SFQ pulse splitter. One of the SFQ pulse propergate to the output circuits. The other pulse is trapped in the loop ℓ_T. This trapped SFQ in the loop ℓ_T enables the NAND operation. The loop ℓ_T is connected to the clock terminal via the SFQ pulse splitter. After applying an SFQ pulse to the loop ℓ_T, SFQ is trapped in the loop ℓ_T and the loop current i_ℓ flowing through J_1 increases. The critical current (I_{c1}) of J_1 is chosen to allow each of two loops to trap an SFQ at the same time. When input pulses are applied to both of terminals A and B, i.e. (A,B)=(1,1), J_1 switches because the loop current i_ℓ exceed the critical current I_{c1} and all loops are reset with no SFQ trapping. To avoid switching J_2, the critical current I_{c2} of J_2 should be larger than I_{c1}.

The output circuit senses i_ℓ flowing through J_2, corresponding to the number of SFQ trapped in the loops. Critical currents of J_2 and J_3 are chosen as follows: If one or two loops trap an SFQ, J_2 switches and all SFQ in the loops is reset when a clock SFQ pulse is applied. In this way we can get an output signal. Conversely if the loops trap no SFQ, J_3 switches and no SFQ pulse is output when the clock pulse is applied. After a clock SFQ pulse, the superconducting loops are reset to the no-SFQ trapping state. As an SFQ pulse is simultaneously applied to the loop ℓ_T, the superconducting loops become the one-SFQ trapping state. Operation of the proposed NAND gate is summarized in Table 1. It is clear that the proposed logic gate executes the NAND logic function.

The SFQ logic circuits reported so far are achieved to transfer a single flux quantum using a closed loop with a single Josephson junction and a single inductance element, and are not allowed to trap two or more magnetic quantum flux in the closed loop. The proposed NAND gate based on the SFQ logic consists of multi-closed superconducting loops with Josephson junctions. The structure of the NAND gate is similar to that of an exclusive OR gate reported by Likharev and Seminov (1991), but there is an additional superconducting loop of ℓ_T and the loops can trap two single flux quanta at the same time. Thus, the functional logic of the NAND gate is executed.

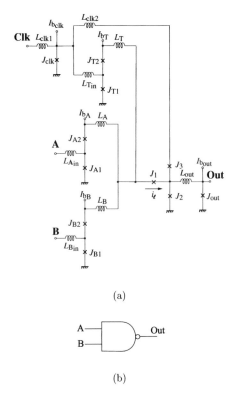

Fig. 1. The NAND gate. (a) Logic circuit. (b) Notation. The proposed NAND gate has three superconducting loops sharing J_1 and J_2. J_1 limits the number of SFQ trapped in the loops and J_2 outputs SFQ pulses.

Table 1. Operation of the proposed NAND gate.

A	B	J_1	J_2	J_3	Output
0	0	—	sw	—	1
0	1	—	sw	—	1
1	0	—	sw	—	1
1	1	sw	—	sw	0

sw=switches

2.2. Dynamic Simulation of the NAND gate

We simulated the dynamic behavior of the proposed NAND gate using the JSIM program (Fang and Duzer 1989). We assumed that the critical current density of Nb/AlO$_x$/Nb JJ is 0.8 kA/cm^2 and the McCumber parameter β_c is unity. Applying a 5 μm design rule, the critical current becomes 200 μA. Table 2 shows the parameters for the proposed NAND gate circuit. To get SFQ pulses, we used a dc/SFQ converter.

The superconducting loops have three stable states, that is, no-SFQ trapping state ("0"), one-SFQ trapping state ("1") and two-SFQ trapping state ("2"). From Fig. 3(b), it is shown that there is three stable current levels related to the stable states. When (A,B)=(0,0), the superconducting loops are in the state "1" before a clock pulse, as shown in Fig. 3(b). After a clock pulse, the current through J_2 exceeds I_{c2} and J_2 switches. Then an output SFQ pulse is obtained. When (A,B)=(1,0), the superconducting loops have two ϕ_0 (the state "2") before a clock pulse. With a clock pulse, the current flow through J_2 exceeds I_{c2} and J_2 switches. Then an output pulse is also obtained. When (A,B)=(1,1), the superconducting loops is reset to the state "0" by switching of J_1 after sequential SFQ pulses are applied from input terminals A and B. Then, the clock pulse switches J_3 and the induced current by the clock pulse is reset. Then, no SFQ pulse is output, as shown in Fig.3(a).

It should be noted that the junctions at input terminals named J_{A2} and J_{B2} play an important role in achieving NAND function. A current pulse propagates to reverse direction is also induced by switching of J_2. The current pulse propagates into each of the loops ℓ_A, ℓ_B and ℓ_T. The trapped magnetic flux in the loop is reset by the current pulse. If there is no trapped magnetic flux in the loop, the current pulse is reset by switching of junctions of J_{A2} and J_{B2}. Thus, the NAND gate is initialized for the next clock period without disturbing the input and output terminals.

We evaluated the delay time for the NAND gate. The delay time from input SFQ pulses to switching of J_1 is 7 ps. The delay time from clock pulses to initializing of the loops is 28 ps. Then, the minimum total delay time becomes 35 ps. However it should be evaluated at a lower speed due to timing skews between clock SFQ pulses and input SFQ pulses. On the other hand, it is expected in the future to be a shorter delay time since the current density of the

Table 2. Parameters for the NAND gate circuit.

junctions	$I_c(\mu A)$	$I_b(\mu A)$	$L(pH)$
J_{A1},J_{A2}	200	$I_{b_A}=170$	$L_A=12$
J_{B1},J_{B2}	200	$I_{b_B}=170$	$L_B=12$
J_{T0},J_{T1},J_{T2}	200	$I_{b_T}=170$	$L_T=12$
J_{clk}	600	$I_{b_{clk}}=500$	$L_{A_{in}}=5$
J_{out}	200	$I_{b_{out}}=160$	$L_{B_{in}}=5$
J_1	$I_{c1}=250$		$L_{T_{in}}=5$
J_2	$I_{c2}=280$		$L_{clk1}=2.5$
J_3	$I_{c3}=260$		$L_{clk2}=5$
			$L_{out}=5$
critical current density		$J_c=0.8kA/cm^2$	
McCumber parameter		$\beta_c=1$	
LI_c product for JTL		$LI_c=\Phi_0/2$	

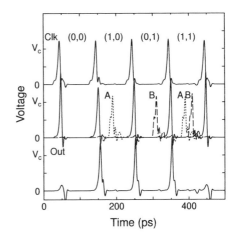

(a)

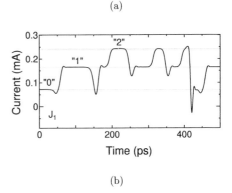

(b)

Fig. 3. Dynamics of the NAND gate with different inputs (A,B). (a) Relationship between input SFQ pulses and output SFQ pulses and (b) the loop current i_ℓ flowing through J_1. Note that V_c is the product of the critical current and the effective resistance of JJs where $V_c=0.234$mV.

744

JJs could improve the switching speed of the junctions. It is also estimated that power consumption is 0.06 μW/gate.

3. CONCLUSION

We have proposed a practical universal NAND gate based on the SFQ logic circuit. Simulations show correct NAND operation of the proposed NAND gate. The NAND gate can be expected to operate with a delay time of 35 ps. Power consumption was calculated to be about 0.06 μW/gate.

ACKNOWLEDGMENT

We would like to express our appreciation to Mr. Takahisa Kaneko for help with simulation and discussions. This work was supported by Grant-in-Aid for Scientific Research on Priority Area, "Vortex Electronics" given by the Ministry of Education, Science, Sports and Culture, Japan.

REFERENCES

Bunyk P I, Oliva A, Seminov V K, Bhushan M, Likharev K K, Lukens J E, Ketchen M B and Mallison W H 1995 *Appl. Phys. Lett.* **66** 646
Fang E S and Duzer T van 1989 *Extended Abstracts of 1989 International Superconductivity Electronics Conference* (ISEC '89), p. 407.
Kaplunenko V K, Ivanov Z G, Stepantsov E A, Cleason T and Wikborg E 1995 *Appl. Phys. Lett.* **67** 282
Likharev K K and Seminov V K 1991 *IEEE Trans. Appl. Supercond.* **1** 3
Myoren H, Ono S and Takada S *IEICE Trans. Eelectron.* **E83–c** in press.
Nakajima K, Mizusawa M, Sugahara H and Sawada Y 1991 *IEEE Trans. Appl. Supercond.* **1** 29

Inst. Phys. Conf. Ser. No 167
Paper presented at Applied Superconductivity, Spain, 14-17 September 1999
© 2000 IOP Publishing Ltd

Matching of Rapid Single Flux Quantum Digital Circuits and Superconductive Microstrip Lines.

N A Joukov [1], D E Kirichenko [1], A Yu Kidiyarova-Shevchenko [2,3], and M Yu Kupriyanov[3]

[1] Moscow State University, Physics Dept., Moscow, Russia,
[2] Chalmers University of Technology, Gothenburg, Sweden,
[3] Nuclear Physics Institute, Moscow State University, Moscow, Russia,

ABSTRACT: New solution for matching Superconducting Microstrip Lines (SMSL) with the RSFQ logic/memory family circuits is presented. The simulation shows that with new matching solution 20 Gbit/sec data transmission rate can be achieved with 2mm length SMSL and even higher for shorter SMSLs. Dependencies of the parameters margins vs. operational frequency are studied in detail and a set of circuits for experimental investigation of the proposed method is designed using niobium technology with bias margins as wide as 30%.

1. INTRODUCTION

Complex Rapid Single Flux Quantum (RSFQ) (Likharev, 1996) digital circuits use Superconductive Microstrip Lines (SMSL) for ballistic transfer of Single Flux Quantum (SFQ) pulses along the chip. The basic promise of SMSLs is in high speed, low power dissipation, high integration density and arbitrary low cost in design. The difficulties in dealing with them are proper matching of impedance between SMSL and RSFQ components and many degrees of freedom in simulations associated with different technologies and applications demands.

First solutions for matching between Josephson transmission line (JTL) and SMSL presented by Polonsky et al. (1993) used parallel scheme containing capacitors of large geometric size. The low frequency propagation of the SFQ pulse was simulated and experimentally tested with bias supply margins about 30%.

In this paper, we discuss a serial-matching scheme that makes possible to reduce significantly the geometrical area occupied on a chip by RSFQ drivers and receivers. Also we focus on the high data transmission rate and dependence between speed, parameters margins and SMSL length.

2. BASIC CONCEPT

We will consider an one directional pulse transfer along the SMSL with different matching to driver and receiver JTLs. Input matching between JTL and SMSL is not critical and can be implemented as a small inductance because of driver output junction is not very sensitive to the reflection after its phase has turned over π. The most difficult task here is matching between SMSL and receiver. The reflection from receiver runs back and forth along the SMSL with small dissipation. This signal causes harmful interference to the correct propagation of the information pattern.

Following the approach described by Polonsky et al (1993) we have used linearized representation of the JTL's impedance as a parallel connection of a capacitor $C=C_j/2\alpha^2$ and a resistor $Z=\alpha(L/C_j)^{1/2}$ (see inset (b) to the Fig. 1). Here C_j is the Josephson junction capacitance, L is the

746

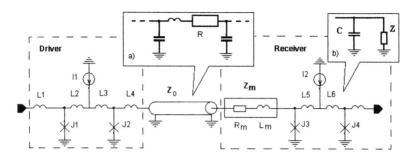

Fig. 1. The equivalent scheme of the simulated circuit with 2mm SMSL: $J1=J2=2$, $J3=J4=1$, $I1=1.88$, $I2=1.35$, $L1=1.5$, $L2=0.75$, $L3=0.45$, $L4=0.61$, $L5=2.75$, $L6=2.85$, $Lm=1.76$, $Rm=0$, $R=0.001$, $\beta_c(J2)=1.15$, $\beta_c(J3)=0.8$. The parameters are given in units normalized to the critical current $I_c=125\mu A$ and characteristic voltage $V_c=290\mu V$.

inductance of one stage of the JTL and $\alpha=(\beta_c\beta_l)^{1/2}(\tau_c/\tau)$ is the dimensionless parameter, which characterizes the time delay in JTL.

The proper matching conditions mean equivalence between the impedance of the SMSL, Z_0, and the impedance of the matching circuit, Z_m, that connected in series to JTL with impedance, Z_{jtl} (see Fig. 1):

$$Z_0 = Z_m + Z_{JTL} = Z_m - \frac{iZ}{\omega C(Z - \frac{i}{\omega C})}. \tag{1}$$

The solution for Z_m from equation (1) leads to a scheme with a resistor R_m connected in series with an inductance L_m given by:

$$Z_m = Z_0 - \frac{Z}{1+Z^2\omega^2C^2} + \frac{i\omega CZ^2}{1+Z^2\omega^2C^2} = R_m + i\omega L_m \tag{2}$$

$$R_m = Z_0 - \frac{Z}{1+Z^2\omega^2C^2}, \quad L_m = \frac{Z^2C}{1+Z^2\omega^2C^2}. \tag{3}$$

For the standard niobium process (see www.hypres.com) the impedance of SMSL formed in first superconducting layer M1 over a ground plane is given by $Z_0 = (L_0/C_0)^{1/2}1/w$, where w is the width of the microstrip, $L_0=0.43$ pH/sq. is the SMSL 2D inductance and $C_0=0.277$ fF/μm^2 is the capacitance of the film. For optimal parameters of JTL the parameter α is equal to 0.5 (see Polonsky et al. (1993)), the junction capacitance C_j is equal to $0.465pF$, JTL inductance L equal to 7.9 pH and resistance of JTL Z is equal to $2.05\ \Omega$.

The relations (3) define the optimal parameters of the matching circuit depending on values Z_0 and Z. However, it is desirable to avoid the resistor R_m. The resistor in the matching circuit contributes significant current noise at high frequencies and dissipates much energy. Thus we set an additional condition $R_m=0$, which combined with relation (3) relates values Z_0 and Z: $Z_0 = Z/(1 + Z^2\omega^2C^2)$.

The found optimal relations are frequency dependent. Thus they can be satisfied by constant values Z_0 and L_m only at some fixed frequency. So it is impossible to reach ideal matching without reflection of the part of the SFQ pulse power. We have chosen the frequency of the matching $\omega\approx0.3\omega_c$ ($\omega_c=150GHz$), that caries 80% power of SFQ pulse. The solution results in SMSL characteristic impedance $Z_0\approx Z$, SMSL width $w\approx19.3\mu m$ and $L_m\approx3.88pH\approx0.5L$.

3. NUMERICAL SIMULATIONS

During the simulation, a single stage of the SMSL was represented as the LC-chain with resistor R (see inset (a) to the Fig. 1). The resistor $R=0.001\Omega/sq.$ represents the surface resistance of

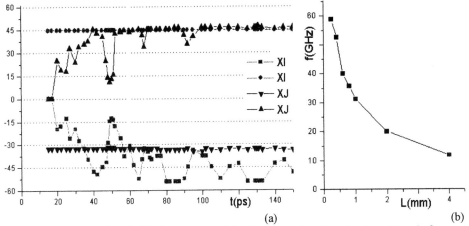

Fig. 2. The dependencies of critical margins vs. delay between pulses in test pattern (a) and of limiting operational frequency vs. SMSL length (b).

the microstrip at typical working frequencies $f > 1GHz$. We have investigated the SMSL with length scales from $0.25mm$ to $2mm$. The length of the SMSL stage in the equivalent scheme is determined by the SFQ pulse spectrum and was less than $50\mu m$.

To verify the high frequency operation of the circuit the test data pattern has been chosen as a sequence of 15 pulses with equal time delay T which determine the data frequency $f=1/T$. The optimization was done for each particular length of SMSL to reach maximum operational margins for circuit parameters and two global parameters XI and XJ (the last two represent the margins of the global bias supply and the critical current density of Josephson junctions).

A typical dependence of XI and XJ margins vs. pulse delay T is presented in Fig. 2a. There are a few distinguished peculiarities on the curves. The origin of these peculiarities is an interaction between the test pulses and their reflected part. The position of the most left peak in Fig. 2a corresponds to the sum of back and forth propagation time of the reflected part of a pulse. The maximum operational frequency of the circuit is limited by this time. The error in the circuit operation that bounds the working region is occurred when SFQ pulse completely reflects from the receiver. Numerical simulation has shown that for long SMSL we can either decrease the main peak depth and increase other peaks or decrease all of the peaks above 30% except the main peak. For SMSLs shorter than $\approx100\mu m$ all the peaks are disappeared and SMSL can be considered as a simple inductance. The optimal parameters of the circuit for 2mm length of SMSL are presented in the Figure 1. The resulting matching inductance $Lm=4.62pH$ is close to the value estimated theoretically. Having based on this approach we found the dependence of the maximum operational frequency vs. the length of the SMSL (see Fig. 2b).

4.   TEST CIRCUIT DESIGN

We have designed two test circuits for low frequency and high-frequency experimental investigation of matching between JTL and SMSL. For low frequency test the conventional procedure with the help of DC/SFQ and SFQ/DC converters is used. The looping testing scheme suggested by Goldobin et al. (1993) can be employed for high frequency experiment. The block diagram of the test circuit is presented in Fig. 3a. SFQ pulses are injected into the loop formed by driver, SMSL, and receiver from the DC/SFQ converter via confluence buffer. The averaged voltage $\overline{V}_{out}$ measured at any point of the loop corresponds to the rotation frequency f of the SFQ pulse $\overline{V}out=Nf2\pi/\Phi_0$, where N is the number of SFQ pulses in the loop. By changing the number of the SFQ pulses one can measure the dependence within a wide range of data frequencies. The limitation is the interaction between the pulses inside the JTL at very high frequencies.

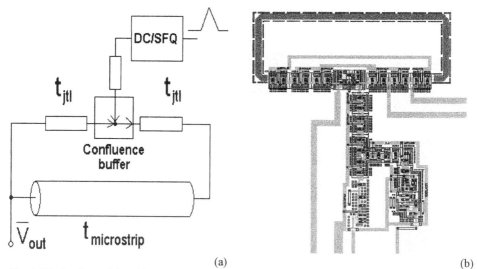

(a)
(b)

Fig. 4. Block scheme (a) and layout (b) of the high-frequency test circuit with 2mm SMSL.

The layout of the high frequency test circuit is presented in Fig. 3b. The layout was designed under the standard Nb/AlOx/Nb process design rules (www.hypres.com) with Josephson junction critical current density $J_c=1kA/cm^2$. The length of the SMSL is varied between $0.25mm$ and $2mm$. Driver, receiver, and delay lines contain 14 Josephson junctions that correspond to the frequency 8.9GHz (18.3μV) for $2mm$ SMSL.

5. DISCUSSION

We have introduced and simulated the serial matching scheme between RSFQ circuits and microstrip line. The simulations have shown that the data propagation rate is limited by the back and force propagation time of the reflection part of the pulses and therefore is dependent on the length of the microstrip.

The optimal parameters for driver and receiver are also dependent of the microstrip length that requires a reoptimization procedure each time when SMSLs are going to be used. Under the conditions of *30%* on the circuit parameter margins the maximum data rate about *20Gbit/sec* can be achieved for *2mm*-microstrip line. For a longer distance either the combination of the short lines with intermediate transmitters or application of the special designed high-voltage drivers and receivers can be used.

ACKNOWLEDGMENTS

Authors would like to thank K Yu Platov for fruitful discussion and numerous suggestions. The research presented in this paper was mainly funded by the grants INTAS-97-1712 and ISTC N11-99 and in part by the ESPIRIT grants HTS-RSFQ.

REFERENCES

Likharev K K 1996 Proceedings of the 21st International Conference on Low Temperature Physics. pp 3331-3338.

Polonsky S V, Semenov V K and Schneider D F 1993 IEEE Trans. on Appl. Supercond. **3** pp 2598-2600

Goldobin E B, Golomidov V M, Kaplunenko V K, et al 1993 IEEE Trans. on Appl. Supercond. **3** pp 2641-2644

Inst. Phys. Conf. Ser. No 167
Paper presented at Applied Superconductivity, Spain, 14-17 September 1999
© 2000 Marconi Electronic Systems Limited

High temperature superconducting low noise oscillator

T A Koetser and R B Greed

Marconi Research Centre, West Hanningfield Road, Great Baddow, Chelmsford. CM2 8HN, England

ABSTRACT: This paper details the development of a low noise oscillator at 1.8GHz. Low noise is achieved by the incorporation of a high temperature superconducting (HTS) planar disk resonator. The HTS resonator, which is etched YBCO on a $LaAlO_3$ substrate, is excited in the TM_{010} mode, in which the current density carries only a radial component, yielding a very high Q resonator. The planar resonator is incorporated into the microstrip feedback loop of a low-noise amplifier, together with phase and amplitude setting components. At 60 Kelvin, the resonator Q has been measured to be 80,000.

1. INTRODUCTION

Receiver modules for future communications CDMA system basestations require a low noise local oscillator. Important criteria are that it demonstrates high Q, low phase noise and is small in size. A HTS planar disk resonator implemented into a feed back loop lends itself well to this application. This paper discusses the design theory and realisation of the disk resonator, and its combination with other components to form the oscillator.

2. DISK RESONATOR THEORY AND DESIGN

Conventional planar resonators have a factor which limits the value of the unloaded quality factor, Q. Current flowing parallel to an edge is influenced by magnetic displacement, resulting in a high current density close to the edge. Above a certain power level, this will introduce an unwanted non-linear response, and consequently the unloaded Q is degraded. Planar disk resonators overcome this by utilising the TM_{010} mode. Not only is the current distributed over a larger conductor surface, but also this mode carries only radial components of current and therefore no current flows parallel to an edge. Therefore etching defects at the edges do not reduce the Q. The performance of the resonator is not degraded by non-linear effects caused by high current density close to the resonator edges. This ensures high values for unloaded Q are attained.

Resonant frequency for a circular disk operating in the TM_{010} mode is determined by the disk diameter and is given by:

$$D[cm] \approx \frac{36.6}{\sqrt{\varepsilon_r} f_0[GHz]}$$

Suppression of neighbouring modes which may produce spurious oscillations is important. Fig 1. illustrates the field patterns of three possible modes and Table 1. shows the six nearest modes and their centre frequency positions.

Acknowledgements: Support for this work is provided by Marconi Electronic Systems and the European Commission (ACTS Program).

TM$_{01}$ TM$_{11}$ TM$_{21}$

**Fig 1. Distribution of magnetic field (dashed lines)
and current (solid lines) for three modes.**

	m=0	m=1	m=2	m=3
n=1	1.8GHz	0.864GHz	1.534GHz	1.973GHz
n=2	3.294GHz	2.504GHz		

**Table 1. Resonant frequencies of neighbouring modes
to the TM$_{010}$ mode. Format: TM$_{m,n,0}$**

Neighbouring modes, especially the TM$_{310}$ mode, which lies within 10% of the TM$_{010}$ mode, must be suppressed. This is achieved by orthogonal positioning of the input and output ports. This does not affect the excitation of the symmetrical TM$_{010}$ mode. However excitation of other modes will be decreased as the orthogonal coupling port arrangement does not support these well.

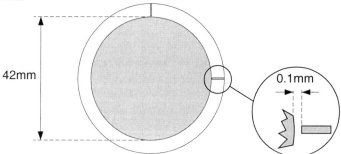

42mm   0.1mm

Fig 2. Disk Resonator Size and Coupling Gap

The resonator consists of YBa$_2$Cu$_3$O$_7$ sputtered on a 0.5mm thick LaAlO$_3$ substrate. This is electrically bonded to a carrier and mounted in a test box. Electromagnetic simulations were performed on "Ansoft Strata". This software package is a 2.5D planar EM solver which utilises adaptive meshing to analyse the field patterns. This simulation showed high Q$_U$ for the TM$_{010}$ mode. Q$_U$ of neighbouring modes, particularly the TM$_{310}$ mode was encouragingly low. The results are illustrated in Fig 4. These were conducted at 60K using network analyser HP8510C.

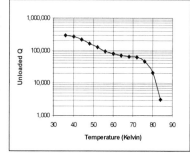

Fig 3.Unloaded Q Vs Temperature

For comparison with the simulation results, the Strata results have been overlaid. Refer to Table 1 for mode positions. Unloaded Q of the TM$_{010}$ mode was 80,000 at 60 Kelvin (Fig.3).

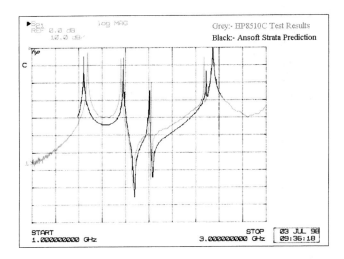

Fig 4. Measured Results vs. Simulation on Ansoft Strata.

3. OSCILLATOR

3.1 Oscillator Development

The disk resonator defines the frequency and provides a low noise, high Q output. For optimum noise performance the input and output coupling strengths to the resonator are made equal, and are set to realise a total coupling loss of 6dB.

The oscillator also requires three additional components to function: an amplifier and amplitude and phase setting components. At room temperature the amplifier showed a noise figure of 0.8dB. However at the specified working temperature of the oscillator of 60 Kelvin this figure decreased to <0.2dB. The amplifier has a typical gain of 18dB at 60K and a 1dB Gain Compression Point of −4.2dB. The amplifier package size of 16×9×4mm makes implementation straight forward.

The gain stage (amplifier) of the oscillator must provide a gain equal to the loss around the feedback loop. A closely matched gain and loss will result in lower noise output as the amplifier will not be driven hard into compression. Therefore a controlled degree of loss and phase must be introduced to match this level as closely as possible, while compensating for loss introduced by the inherent line loss. In this design power is coupled to the output by means of a linear quarter wavelength microstrip coupler. The fourth, redundant, port of this coupler is terminated with a 50Ω load.

For oscillation, the total phase around the feedback loop must be 0^0 (or multiples of 360^0). This is achieved by the use of one of a set of fixed microstrip phase lengths. The total phase then being set to 0^0. The amplitude is adjusted by a further set of fixed couplers. The phase and amplitude setting components were etched on alumina microstrip.

The above components were assembled in the oscillator housing. Carriers for the alumina circuits were fabricated from titanium which was then gold plated over a nickel flash. The circuit substrates were then bonded to the carriers with electrically conductive heat-cured epoxy film. Individual circuits were interconnected with gold tape bonded with silver loaded epoxy. Figs 5 and 6 show the prototype component layout.

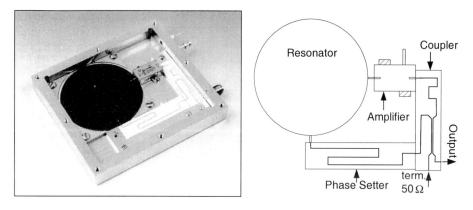

Fig 5. Prototype Oscillator   Fig 6. Oscillator Schematic.

3.2   Oscillator Results

Frequency domain measurements were carried out on the spectrum analyser HP8569B: Fig 7. shows a typical measurement taken at 60 Kelvin, and at a bias power of 384mW (6V, 64mA). Two spurious signals peaking at a power level of –40dBm were observed at 0.8GHz and 3.3GHz corresponding to the TM_{110} and TM_{020} modes respectively.

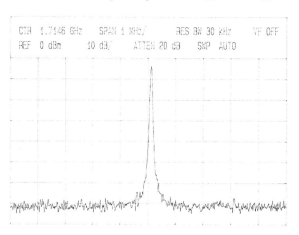

Fig 7. HP8569B Test Results

Phase Noise Measurements for the oscillator are awaiting completion.

4.   REFERENCES

Superconductive planar disk resonators and filters with high power handling capacity. Chaloupka H et al. Electronics Letters. **32**, 18. p1736.
Lower order modes of YBCO/STO/YBCO circular disk resonators, Gevorgian S et al. IEEE Transactions on Microwave Theory and Techniques. **44**, 10. p1739.

Inst. Phys. Conf. Ser. No 167
Paper presented at Applied Superconductivity, Spain, 14-17 September 1999
© *2000 IOP Publishing Ltd*

753

Josephson-junction arrays with lumped and distributed coupling circuits

V K Kornev, A V Arzumanov, and N A Shcherbakov

Moscow State University, Physics Department, 119899, Moscow

ABSTRACT: Josephson-junction arrays with lumped and distributed coupling circuits have been studied by means of numerical simulation technique. It has been shown that the finite coupling radius results in a "saturation" of the phase-locked oscillation linewidth reduction with number N of Josephson junctions in the case of lumped circuits. It has been found that an additional reduction in phase-locked oscillation linewidth can be provided by the impact of the standing electromagnetic waves excited in the distributed coupling circuits. Different ways of Josephson junction connection to the distributed circuits have been analyzed. The crucial role of high frequency losses has been also studied.

1. INTRODUCTION

Some promising phase-locked Josephson-junction arrays with lumped coupling circuits have been recently studied and the reduction in the oscillation linewidth with number N of Josephson junctions has been also analysed by Kornev et al (1999). It has been shown that the linewidth reduction takes place while N does not exceed a coupling radius (see, for example, Fig. 2), which has the finite value for all actual multi-junction systems. An additional linewidth reduction could be obtained by means of the distributed coupling circuits providing standing wave excitation.

The distributed circuits can constitute sections of either microstrip line, coplanar line or split line. All this circuits may be modeled by sections of an equivalent two-conductor line. In its turn the line can be modeled by the multi-element LC-chain. This allows us to perform numerical simulation of the systems as the structures with lumped elements by means of well-known PSCAN program (Polonsky et al 1991, Kornev et al 1997).

There are two basic ways of Josephson junction connection with the distributed coupling circuits. The first mode implies that the junctions should be connected between the conductors of the equivalent two-conductor line and hence should interact with the standing wave of the voltage. It may be a parallel biased array of tunnel Josephson junctions inserted into microstrip line as it is shown in Fig. 1a. The second mode implies that the junctions are connected just into the conductors of the line (Fig. 1b) and thus interact with the standing wave of the current. An example of the topology of this system based on high-Tc superconductor technology is presented by Konstantinian et al (1999).

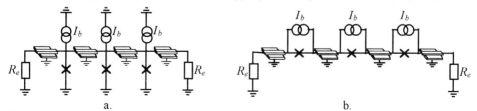

a.   b.

Fig. 1 Two basic ways of Josephson junction connection to the distributed coupling circuits

2. VOLTAGE WAVE INTERACTION SYSTEMS

One of the typical distributed systems shown in Fig. 1a is the parallel biased array of the tunnel Josephson junctions inserted into the superconducting microstrip line (Kaplunenko et al 1995). While the tunnel junction McCumber parameter $\beta = (2\pi / \Phi_0) I_C R_N^2 C >> 1$ (I_C – critical current, R_N – normal resistance, C – capacitance), the junction impedance Z_J at frequency $\omega \sim \omega_C$ (characteristic frequency) is usually much less then characteristic impedance of the microstrip line ρ, even though $R_N \geq \rho$. It leads to the high wave reflection from the junctions and hence the microstrip line presents a sequence of resonators, which are formed by the line sections shorted by Josephson junctions.

The oscillation linewidth reduction determined by dc coupling via superconducting loops takes place only within a few sections because the normalized inductance $l = (2\pi / \Phi_0) I_C L$ of the loop corresponding to one line section is usually far more then 1 (see Fig. 2). Quality factor of the resonators does not increase with number of the sections and hence the oscillation linewidth does not decrease with number of junctions in this multi-junction system. Moreover, the quality factor Q of the multi-resonator system is a bit less than Q of the single resonator because of their nonlinear interaction via Josephson junctions. The interaction results in nonlinear dependence of the boundary conditions and resonance frequency on the wave amplitude. The equivalent quality factor of the nonlinear system is always less than Q for the linear one. This feature is illustrated by oscillation linewidth behavior within resonance peak on I-V curve for the one- and two-resonator structures (two- and three-junction arrays correspondingly). Both structures provide about the same maximum linewidth reduction (Fig. 3) because dc coupling effect in the 3-junction array is balanced out by decreasing of the quality factor Q.

As for the opposite case, it is practically impossible to realize the situation when $Z_J >> \rho$ in the frame of the tunnel junction structure.

3. CURRENT WAVE INTERACTION SYSTEMS

Figure 4 presents the current standing wave structures and resonance peaks on I-V curves for two-junction system shown in Fig. 1b at $R_e = \infty$ (open terminals). If Josephson junction impedance Z_J is much less then ρ, the system constitutes one resonator, and I-V curve shows the sharp resonance peak. In the opposite case, when $Z_J >> \rho$ and the high wave reflection from the junctions takes place, the system should be considered as a three-resonator structure. The resonance peak on I-V curve is essentially flattened in this case. This can be explained by a strong nonlinear interaction between the resonators that results in a strong dependence of their resonance frequency on the bias current. In other words, this case is characterized by strong resonance mode interaction, which gives both smooth and abrupt changing in

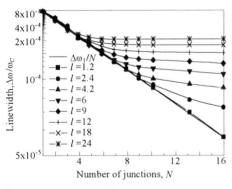

Fig. 2 Josephson oscillation linewidth dependence on number of junctions in the parallel lumped array for different values of inductance parameter l at $\omega / \omega_c = 1.8$, $\beta = 1$.

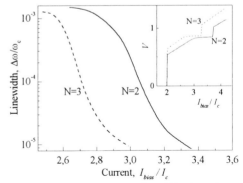

Fig. 3 Josephson oscillation linewidth dependence on bias current within resonance peak for 2- and 3-junction arrays inserted into matched superconducting microstrip line ($R_e = \rho$) at $\beta = 10$, $R_N / \rho = 0.8$. Inset shows I-V curves.

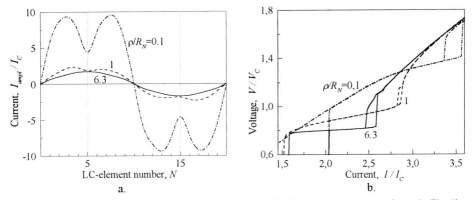

Fig. 4 Current standing wave structures (a) and I-V curves (b) for two-junction array shown in Fig. 1b at $R_e = \infty$ (open terminals), $\beta = 1$, loss factor $\alpha = 10^{-3}$ and different ratio ρ/R_N. Josephson junction positions correspond to N = 5 and N = 15.

resonance frequency with the bias current. Therefore just the case of $Z_J \ll \rho$, which fits the requirements of the oscillation linewidth reduction, is analyzed further.

The Josephson junction energy contribution into the excited standing wave is defined by amplitudes of both the current I through the junction and the voltage V across the one, and also by the cosine of the phase difference between I and V. Figure 5 shows the dependence of these components on the bias current within the resonance peak for both symmetric and asymmetric Josephson junction position in the resonator. One can see from Fig. 5a that the voltage amplitude increases with bias current whereas the current amplitude is almost constant. The voltage amplitude increase is balanced out by the decrease of cosine and therefore the energy contribution is about the same within the resonance peculiarity. Figure 5b shows more complex behavior of voltage amplitude and cosine for each junction in the asymmetric case. The fact that one of the cosines is negative means that the corresponding junction takes up the energy from the standing wave. Nevertheless, the energy contribution from the other junction dominates and the resulting energy contribution into the standing wave is also about the same within the resonance peculiarity.

The microwave loss impact has been studied by adding a set of resistors R_{loss} connected to the capacitors of LC-elements. It is known that the moving wave in LC-chain can be written as

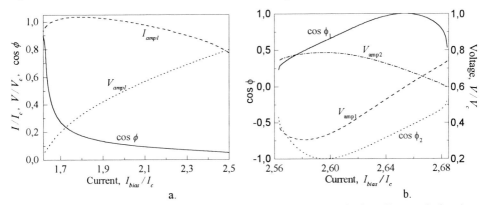

Fig. 5 Dependence of amplitude of current I through junctions, amplitude of voltage V across the junctions and cosine of the phase difference between I and V on bias current within resonance peculiarity for symmetric (a) and asymmetric (b) position of junctions in the resonator (Fig. 1b) at $R_e = \infty$, $\beta = 1$, $\alpha = 10^{-3}$, $\rho/R_N = 6,3$. The numbers of LC-elements between terminals and junctions are $N_1 = N_2 = 4$ at total number $N = 16$ (a), $N_1 = 4$, $N_2 = 11$ at total number $N = 23$ (b).

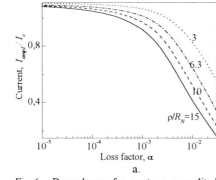
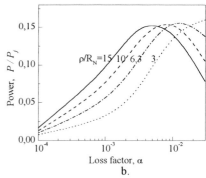

Fig. 6   Dependence of current wave amplitude (a) and Josephson junction power contribution (b) on loss factor for two-junction array shown in Fig. 1b at $R_e = \infty$, $\beta = 1$ and different ratio ρ/R_N. Junctions are placed symmetrically, total number $N = 16$.

$E_0 exp(-j\gamma x)exp(j\omega t)$, where $\gamma = [(R+j\omega L)(G+j\omega C)]^{1/2}$. If $R = 0$, $G = 1/R_{loss}$, then the loss factor per LC-element $\alpha = Re(\gamma) \approx 0.5\rho/R_{loss}$, where $\rho = (L/C)^{1/2}$.

Figure 6 presents dependence of the current wave amplitude and the energy contribution on the loss factor α. As the losses are reduced, the saturation in the current wave amplitude increase takes place. The saturation is accompanied by the decrease of Josephson junction energy contribution. While the energy contribution peaks at $\alpha = 10^{-2}$, the standing wave amplitude drastically falls and resonance peculiarity on I-V curve vanishes if α becomes greater then 10^{-2}. Therefore this value of loss factor is the threshold value for linewidth reduction effect.

3.   CONCLUSION

One can state two basic requirements for Josephson junction arrays with distributed coupling circuits to provide maximum oscillation linewidth reduction. First, the total microwave losses of the loaded system should be less then threshold value. Second, the relationship between Josephson junction impedance and characteristic impedance of the distributed circuits should correspond to the low wave reflection from the junctions.

The last requirement should provide high value of quality factor Q of the whole system and hence extremely sharp resonance peak on I-V curve, that gives maximum linewidth reduction. In this case the extra losses caused by an external load are spread on the whole system. It means that the losses per one Josephson junction section of the system will not essentially increase with the loading.

The system matched these requirements could be realised not only as the one-dimensional array shown in Fig. 1, but as quasi two-dimensional structure described by Konstantinian et al (1999).

This work was supported in part by INTAS foundation under grant № INTAS-OPEN-97-1940, Russian State Program Condensed Matter Physics under grant № 98051 and Ministry of Education under grant № 97-8.3-58.

REFERENCES

Kaplunenko V K,   Larsen B H,   Mygind J   and   Pedersen N F   1995,   IEEE   Trans.   on   Appl. Superconductivity **5 (2)** pp 2711-2714
Konstantinian K I, Mashtakov A D, Ovsyannikov G A, et. al. 1999 "MM wave Josephson radiation in High-T$_c$ bicrystal junction arrays" this conference.
Kornev V K and Arzumanov A V 1997 Inst. Phys. Conf. Ser. **158** (IOP Publishing Ltd)
Kornev V K and Arzumanov A V 1999 IEEE Trans. on Appl. Superconductivity **9 (2)**
Polonsky S V, Semenov V K and Shevchenko P N 1991 Proc. Int. Supercond. Electron. Conf. (ISEC'91) pp 160-163.

Inst. Phys. Conf. Ser. No 167
Paper presented at Applied Superconductivity, Spain, 14-17 September 1999

Linewidth Measurement of Josephson Array Oscillators with Microstrip Resonators

Akira Kawakami, Yoshinori Uzawa, and Zhen Wang

KARC, Communications Research Laboratory, 588-2 Iwaoka, Nishi-ku, Kobe 651-2401, Japan

ABSTRACT: The linewidth of a Josephson array oscillator was measured using an integrated receiver consisting of two Josephson array oscillators and an SIS mixer. The oscillator was formed with 30 Josephson junctions and Nb microstrip resonators. In the I-V characteristics of the oscillator, current steps resulting from the resonance of the microstrip resonators were observed around 1.17 mV. When the oscillators were biased at 0.96 mV and 1.17 mV, the composite linewidth of the Josephson array oscillators was measured to be about 7 MHz at 464 GHz and 8 MHz at 566 GHz.

1.   INTRODUCTION

For many sub-mm wave frequency applications, such as high-speed communication and radio astronomy, a low weight, low power dissipation, and long-lifetime wave source is required. In this frequency range, an SIS (superconductor-insulator-superconductor) mixer is currently the most sensitive detector. However, inefficient multiplication of millimeter wave source has been used as a local oscillator for the SIS mixer. At higher frequency, the lack of a compact source is a serious problem for sub-mm wave applications.

Josephson array oscillators have been studied as solid-state sub-mm wave sources. The output impedance of Josephson array oscillators is probably higher than other types of Josephson oscillators and can be tuned to match with a common load (~50 Ω) (Han *et. al.* 1994, Kawakami *et. al.* 1998, 1999). To improve the high-frequency performance of array oscillators, we have recently developed resistively shunted Nb/Al-AlOx/Nb tunnel junctions that have a small parasitic inductance of about 100 fH (Kawakami *et. al.* 1998). Using these junctions, we pushed the operating frequency of an array oscillator up to the Nb gap frequency. The estimated power of the Josephson oscillator was about 10 μW at 625 GHz and 50 nW at 1 THz (Kawakami *et. al.* 1998). Although enough power was generated to pump an SIS mixer in the wave region, the reported linewidth of the Josephson array oscillator was larger than that of FFO (Josephson flux-flow oscillator) (Bi *et. al.* 1993, Koshelets *et. al.* 1997, Zhang *et. al.* 1993).

A popular method of reducing the linewidth of array oscillators is to increase the number of junctions and decrease the resistance of each junction. However, this makes it difficult to fabricate a lot of Josephson junctions with good uniformity, and results in an increase in the surface loss of microstrip lines. We have been studying Josephson junctions with microstrip resonators and found that the interaction between Josephson junctions and microstrip resonators caused a strong phase-locking status and a decreased dynamic resistance. It is conceivable that the interaction may reduce the linewidth in the array oscillator without an increase in the number of junctions.

This paper reports the measurements of the linewidth of Josephson array oscillators with microstrip resonators. To measure the linewidth, we designed and fabricated an integrated receiver consisting of two Josephson array oscillators and an SIS mixer. The composite linewidth of the Josephson array oscillators was measured and discussed.

758

2. DESIGN AND FABRICATION

Figure 1 shows a photograph of the integrated receiver consisting of two Josephson array oscillators and an SIS mixer. The Josephson array oscillator was formed with 30 shunted tunnel junctions and Nb microstrip resonators. The junction area of the oscillators was set at 4 μm^2 and the capacitance C_j was estimated to be about 0.4 pF using a specific capacitance value of 100 fF/μm^2 (Booi 1995). By assuming a parasitic inductance of L_s=100 fH, we predicted the resonant frequency of the shunted tunnel junctions, which dominates the high-frequency performance of the oscillator was to be about 700 GHz.

The designed frequency of the oscillators was set at 550 GHz and the wavelength was 350 μm. The junctions were placed at intervals of half a wavelength along the microstrip line. The Nb ground-plane was made over the junctions and the microstrip lines, and a 1-μm-thick layer of SiO dielectric separated them from the Nb ground-plane. The width of the microstrip line was 20 μm and the characteristic impedance was estimated to be about 8.3 Ω. The dc bias lines used for the generation of Josephson oscillation were formed with 4-μm- and 20-μm-wide microstrip lines which served as $\lambda/4$ filters. Around the designed frequency, the impedance of the bias line seen from the oscillator junction is high enough to be ignored. The I-V characteristics of the array oscillators were observed through these lines in parallel.

Each oscillator was coupled to the SIS mixer by using a microstrip line and a capacitor coupler. The width of the line was 3 μm ($\cong$50 Ω) and the length was about 0.5 (oscillator (a)) and 3 mm (oscillator (b)). The junction for the SIS mixer had an area of 0.8 μm^2 and the junction capacitance was tuned out at about 550 GHz using an open stub tuner. Both the SIS mixer and the junctions of array oscillators were made from the same Nb/Al-AlOx/Nb trilayer.

A $\lambda/4$ low-pass filter was used as the IF output line and the SIS mixer was biased using this line. The IF output was transmitted through a coaxial cable from the mixer in liquid helium to room-temperature amplifiers and then detected by a spectrum analyzer. The total gain of the IF system was about 68 dB. Each oscillator was biased by using two battery-operated current sources.

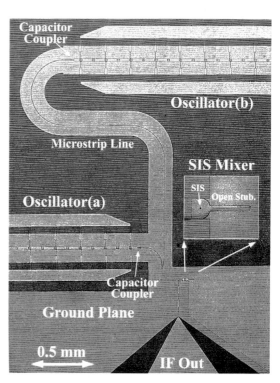

Fig. 1 Photograph of the integrated receiver for linewidth measurements. The receiver consists of two Josephson array oscillators and an SIS mixer.

3. LINEWIDTH MEASUREMENTS OF THE JOSEPHSON ARRAY OSCILLATORS.

Figure 2 shows the I-V characteristics of both Josephson array oscillators shown in Fig. 1. Here, the current density J_c of the junctions was about 25 kA/cm^2 and the shunt resistance R_s was about 1.0 Ω. In both characteristics, a current step resulting from the designed resonance of the microstrip resonators was observed at about 1.17 mV, which corresponds to 566 GHz. This step is a

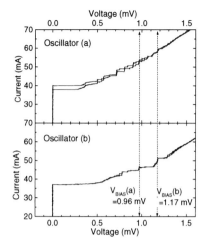

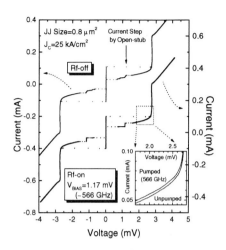

Fig. 2 I-V characteristics of Josephson array oscillators which comprise the integrated receiver for linewidth measurements. Around 1.17 mV, fundamental current steps were observed. The IF power spectrum was observed when both oscillators were biased at 0.96 mV (V_{BIAS}(a)) and 1.17 mV (V_{BIAS}(b)).

Fig. 3 I-V characteristics of the SIS junction. A current step caused by tuning out the junction capacitance was observed at about 1.2 mV corresponding to 570 GHz. The inset shows I-V characteristics under 566-GHz irradiation from oscillator (a).

fundamental step, because the microstrip resonator of the array oscillator served as a $30\lambda/2$ resonator at the designed frequency. Of course, the resonator is also able to work at $28\lambda/2$, $29\lambda/2$, and $31\lambda/2$ resonance modes. The small current steps, which we call resonator coupled steps, caused by these resonance modes were also observed around the fundamental step.

Figure 3 shows the I-V characteristics of the SIS mixer. In the I-V characteristics, a current step caused by tuning out the junction capacitance was observed at about 1.2 mV, which corresponds to 580 GHz. When oscillator (a) was biased at 1.17 mV, Shapiro steps and an increase in the dc current caused by photon-assisted tunneling were observed. The inserted I-V characteristics shows the pumped and unpumped I-V characteristics of the SIS mixer at 566 GHz.

The IF power spectrum was observed at 4.2 K when both oscillators were biased at 0.96 mV (464 GHz) and 1.17 mV (566 GHz). This is shown in Fig. 4. When the oscillators were biased at about 0.96 mV, the composite linewidth (-3 dB) of the two Josephson oscillators was about 7 MHz. Figure 5 shows the dynamic resistance of the oscillators at each bias point. At the biased point, the

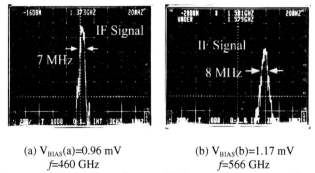

(a) V_{BIAS}(a)=0.96 mV
f=460 GHz

(b) V_{BIAS}(b)=1.17 mV
f=566 GHz

Fig. 4 The IF power spectrum of Josephson oscillation at 4.2 K. Both oscillators were biased around 0.96 mV, which corresponds to 460 GHz, and 1.17 mV, which corresponds to 566 GHz. The composite linewidth of the two Josephson array oscillators was about 7 MHz and 8 MHz.

760

dynamic resistance of the oscillators was decreased by the interaction between Josephson junctions and microstrip resonators and it was about 0.12 (oscillator (a)) and 0.10 Ω (oscillator (b)). By assuming that the linewidth was mainly dominated by the dynamic resistance 0.12 Ω, the linewidth was to be about 8 MHz. The decreased dynamic resistance acted the linewidth to be narrow.

When the oscillators were biased at about 1.17 mV, which was in the fundamental step, the composite linewidth was about 8 MHz. At the biased point, the dynamic resistance per shunted junction of the oscillators was about 0.39 (oscillator (a)) and 0.18 Ω (oscillator (b)). By assuming that the linewidth was mainly dominated by the dynamic resistance 0.39 Ω, we estimated the linewidth to be about 87 MHz by using Eq. (1) (Rogovin *et. al.* 1974). In this bias region of the fundamental current step, we think that the Josephson junctions of the array oscillator were in a phase-locking status caused by the interaction between the Josephson junctions and the microstrip resonators (Kawakami *et. al.* 1997). If the Josephson junction array acted as one Josephson junction, the linewidth would be estimated to be about 3 MHz. These results suggest that the phase-locking status caused the linewidth of the Josephson array oscillators to be narrow compared to the linewidth of the single junction.

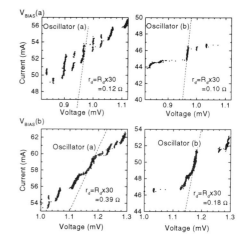

$$\Delta f = \frac{2\pi}{\Phi_0^2} R_{dyn.}^2 e \left[I_{qp}(V_{DC}) coth\left(\frac{eV_{DC}}{2k_B T}\right) + 2I_p \, coth\left(\frac{eV_{DC}}{k_B T}\right) \right] \quad (1)$$

Fig. 5 Dynamic resistance of the Josephson array oscillators at 0.96 mV (V_{BIAS}(a)) and 1.17 mV (V_{BIAS}(b)).

4. CONCLUSION

To measure the linewidth, we designed and fabricated an integrated receiver consisting of two Josephson array oscillators and an SIS mixer. When the oscillators were biased at 0.96 mV (464 GHz) and 1.17 mV (566 GHz), the composite linewidth of the oscillators was measured to be about 7 MHz and 8 MHz at 4.2 K. At the biased point of 0.96 mV, the decreased dynamic resistance resulting from the interaction between Josephson junctions and microstrip resonators acted the linewidth to be narrow. At 1.17 mV, the dynamic resistance per shunted junction of the oscillators was about 0.39 and 0.18 Ω. By assuming that the linewidth was mainly dominated by the dynamic resistance 0.39 Ω, we estimated the linewidth to be about 87 MHz. These results suggest that the array was in the phase-locking status in the fundamental step and this status caused the linewidth to be narrow compared to the linewidth of a single junction.

REFERENCES

Bi B, Han S, Lukens J E, and Wan K, 1993 *IEEE Trans. Appl. Supercond.*, **3**, 2303
Booi P A A, 1995 *Ph. D. thesis, University of Twente, Enschede,* The Netherlands
Han S, Bi B, Zhang W, and Lukens J E, 1994 *Appl. Phys. Lett* **64**, 1424
Kawakami A, Uzawa Y, and Wang Z, 1997 *IEEE Trans. Appl. Supercond.*, **7**, 3126
Kawakami A and Wang Z, 1998 *IEICE Trans. Electron.*, **E81-C**, 1595
Kawakami A, Uzawa Y, and Wang Z, 1999 *IEEE Trans. Appl. Supercond.*, **9**, 4554
Koshelets V P, Shitov S V, Shchukin A V, Filippenko L V, Mygind J, and Ustinov A V, 1997 Phys. Rev. **B 56**, 5572
Rogovin D and Scalapino D J, 1974 *Ann. Phys.,* **86**, 1
Zhang Y M, Winkler D, and Claeson T, 1993 *Appl. Phys. Lett.* **62**, 3195

Inst. Phys. Conf. Ser. No 167
Paper presented at Applied Superconductivity, Spain, 14-17 September 1999
© 2000 IOP Publishing Ltd

Microwave properties of SINIS Josephson junction series arrays

H Schulze, I Y Krasnopolin*, J Kohlmann, R Behr, F Müller, and J Niemeyer

Physikalisch-Technische Bundesanstalt, 38116 Braunschweig, Germany
* Permanent adress: Russian Research Institute for Metrological Service, Moscow, Russia

ABSTRACT: We have fabricated series arrays of several thousands of overdamped SINIS junctions integrated into a superconducting 5 Ω microstripline. In large series arrays, the junctions are mutually synchronized by the microwave power which they emit into the microstripline. Very low external microwave power is required in this case, as a major part of the microwave is generated by the Josephson array itself. Measurements performed to determine the operating conditions for this microwave coupling mechanism are described. A high-impedance stripline microwaveguide with silicon substrate acting as the dielectric layer is presented, wich allows the microwave attenuation to be reduced.

1. INTRODUCTION

Josephson junctions with a Superconductor-Insulator-Normal conductor-Insulator-Superconductor (SINIS) structure, fabricated in Nb-Al/AlO$_x$ technology, have been shown to be a suitable technology for electronic applications, where non-hysteretic current-voltage characteristics are required (Maezawa et al 1997, Sugiyama et al 1997, Schulze et al 1998). These are, for example, programmable Josephson voltage standards for dc and ac voltage metrology (Hamilton et al 1995 and Benz et al 1998,1999), Josephson oscillator arrays (for a review, see Darula et al 1999) and RSFQ logic circuits (Likharev et al 1987). SINIS junctions are intrinsically shunted and, therefore, overdamped without an external shunt resistor being used. They can be fabricated with low parameter spread and with good yield and reproducibility. The characteristic voltage $V_c = I_c R_n$, i. e. the product the junction's critical Josephson current I_c and the normal resistance R_n, can be tuned between 20 μV and 200 μV, which enables optimum operation of programmable voltage standard circuits at the most common microwave frequencies f of between 10 GHz and 100 GHz as well as in RSFQ logic circuits.

We have fabricated 1 V programmable Josephson voltage standard chips with 8192 SINIS junctions for 70 GHz operation. The junctions are integrated into a 5 Ω stripline microwaveguide with a width of 50 μm and a 2.2 μm sputtered SiO$_2$ dielectric layer. The microwave is splitted into 4 or 8 parallel branches. As shown by Behr et al (1999), under microwave irradiation, the circuits show a voltage step approximately 200 μA wide with a quantized voltage. This step can be used for calibration purposes.

2. MICROWAVE DISTRIBUTION IN SINIS JUNCTION SERIES ARRAYS

2.1 5 Ω stripline microwaveguide

According to the resistively capacitatively shunted junction (RCSJ) model, Josephson junctions can be modelled by a parallel connection of an ideal Josephson junction, a resistor and a

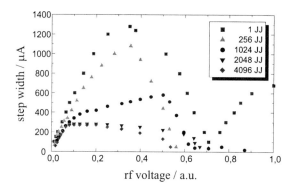

Fig. 1   Dependence of the first-order Shapiro step width on the microwave voltage applied, measured at SINIS junction series arrays of different length.

capacitor. The microwave current through it is distributed on these branches according to their microwave impedances. With $R_n = 100$ mΩ, the quasiparticle resistance of the SINIS junctions is by nearly 3 orders of magnitude lower than that of SIS junctions ($R_n = 50$ Ω) and equals the impedance of the capacitive branch at $f = 70$ GHz microwave frequency, $Z_d = (2\pi f C)^{-1} \approx 50$ mΩ. Here, the capacitance of the SINIS junctions, $C \approx 50$ pF, is estimated from the hysteresis parameter. The resistive fraction of the microwave current, leading to microwave dissipation, is therefore no longer negligible. According to Kautz (1992), the microwave attenuation of a superconducting microstripline with integrated Josephson junctions can be calculated on the basis of the RCSJ model. For a 5 Ω stripline microwaveguide, an attenuation of 50 dB per 1000 SINIS junctions is obtained.

Nevertheless, there is a flat common voltage step in series arrays of more than 4000 Josephson junctions arranged in a single branch of the stripline microwaveguide. This means that all Josephson junction are phase-locked to the external microwave frequency. As shown by Schulze et al (1999), in large series arrays the junctions are mutually synchronized by the microwave power which they emit into the microstripline.

Since the major part of the required microwave power is generated in the array itself, the arrays can be operated at a very low microwave power of about 100 μW, i. e. a power which is by a factor of 20 lower than that of conventional 1 V SIS Josephson voltage standards. This offers the great advantage that low-power microwave sources can be used, which could make on-chip integration possible.

The mechanism of microwave coupling has several consequences for the design (Kohlmann et al 1999) and operation of programmable Josephson voltage standard circuits. The maximum step width ΔI is limited by the microwave power emitted by the junctions and does not reach the Bessel limit $\Delta I \approx 1.2\ I_c$. Fig. 1 shows the width of the first-order Shapiro step of SINIS arrays of different length, integrated into a 5 Ω stripline microwaveguide. It is supplied at one end with microwave power and terminated on the other end with a resistive load to avoid reflections. For the single junctions and the short series array (256 JJ), the step width shows a Bessel-shaped behaviour. In the long arrays of 2048 and 4096 Josephson junctions, the externally applied microwave power is attenuated almost entirely in the array. Hence, the step width is limited by the mutual coupling of the junctions due to their self-generated microwave power. Nevertheless, with more than 200 μA the step is about five times wider than in conventional SIS 1-V Josephson voltage standards.

According to Kautz (1994), the optimum operating frequency is by a factor of 1 to 3 higher than the characteristic frequency $f_c = (2e/h)V_c$ of the junctions. In the 70 GHz voltage standard circuits, the junction's characteristic frequency is about 50 GHz; this condition is fulfilled. In addition, the microwave coupling mechanism of the long SINIS arrays leads to a second condition: If the microwave frequency exceeds the plasma frequency $f_p = (2eI_c/\pi hC)^{0.5} \approx 50$ GHz, the

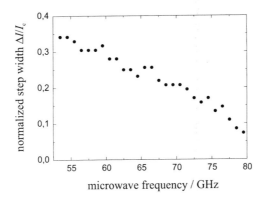

Fig. 2   Microwave frequency dependence of the first-order voltage step width of a 1 V SINIS voltage standard circuit. The maximum number of junctions per microwave branch is 2048 junctions.

microwave current generated by the junctions is increasingly shorted by the junctions' self-capacitance and does not contribute to mutual coupling. In Fig. 2, the microwave frequency dependence of the maximum step width of a 1 V SINIS array is plotted. This result agrees with measurements of the self-generated microwave of the SINIS array (Schulze et al 1999). Here, the width of the Shapiro step of the detector array becomes smaller with increasing oscillator bias current, i.e. with increasing microwave frequency. An additional contribution to the decreasing step width may arise from the junctions' parameter spread.

2.2   High-impedance microwaveguide

An increase in the width of the voltage step might be advantageous for some applications. This means that the whole array is supplied with externally applied microwave. Apart from distributing the microwave to a larger number of parallel branches, it is possible to reduce the waveguide attenuation. If the impedance mismatch between the junctions and the stripline waveguide is increased, a smaller part of the microwave power in the stripline couples into the junctions and can lead to microwave attenuation. At a fixed junction impedance of 100 mΩ, the microwave impedance of the microwave guide has to be increased.

For this purpose, we have aligned the Nb counter-electrode of the microstripline on the back-

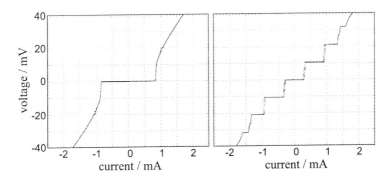

Fig. 3   Current-voltage characteristic of a series array of 512 Josephson junctions, operated at 10 GHz microwave frequency; (a) without microwave (b) with microwave, -3 dBm

side of the wafer. The Si substrate 400 µm in thickness then forms a waveguide dielectric of high quality. We obtain a stripline impedance of about 80 Ω with this geometry (Gutmann et al 1992). According to Kautz (1992), a microwave attenuation due to junction losses of about 3 dB per 1000 SINIS junctions is to be expected.

First tests of this stripline type were performed with binary divided series arrays of 512 Josephson junctions for operation at a microwave frequency of 10 GHz (Fig. 3). When the middle Al layer thickness of the junction is increased from 8 nm to 16 nm, the characteristic voltage of the junctions is reduced to 30 µV to match it to the microwave frequency. Special care was taken as regards the filter design at the leads to avoid standing waves due to reflections. Fig. 3 shows the current-voltage characteristic of the circuit. The width of the Shapiro step reaches the full width, as expected from the Bessel-like behaviour. Operation of the array at higher-order Shapiro steps is possible.

3. CONCLUSION AND OUTLOOK

As a result of the high microwave attenuation of the 5 Ω microstripline waveguide with integrated SINIS junctions, the contribution of the junctions to microwave generation is essential. This offers the advantage of operation at low externally applied microwave power levels. An important condition for an integration of the microwave supply into the Josephson voltage standard chip is, therefore, fulfilled. In view of the self-synchronisation of the SINIS arrays, it seems possible to operate a voltage standard array without any external microwave source. Only part of the microwave generated by the array has to be coupled out to exactly determine and stabilize the frequency with a phase-lock loop.

In addition to the stripline waveguide, we have fabricated and successfully operated first circuits with an 80 Ω high-impedance stripline waveguide for 10 GHz microwave frequency. Next, we will test the 80 Ω stripline microwaveguide for 1 V voltage standard circuits at 70 GHz.

The authors thank R. Harke, T. Weimann and P. Hinze for technical assistance and T. Funck and E. Vollmer for stimulating discussions. Financial support by BMBF (ref. No. 13N6835) and the EU (ref. No. SMT4-CT98-2239) is gratefully acknowledged.

REFERENCES

Behr R, Schulze H, Müller, F Kohlmann J and Niemeyer J 1999 IEEE Trans. Instrum. Meas. **48**, 270

Benz S P and Hamilton C A 1996 Appl. Phys. Lett. **68**, 3171

Benz S P, Hamilton C A, Burroughs C J, Harvey T E and Christian L A 1997 Appl. Phys. Lett. **71**, 1866

Benz S P, Hamilton C A, Burroughs C J and Harvey T E 1999 IEEE Trans. Instrum. Meas. **48**, 266

Darula M, Doderer T and Beuven S 1999 Supercond. Sci. Technol **12**, R1

Gutmann P, Vollmer E, Kohlmann J, Müller F, Weimann T, and Niemeyer J 1992 Electronics Letters **28**, 1422

Hamilton C A, Burroughs C J and Kautz R L 1995 IEEE Trans. Instrum. Meas. **44**, 223

Kautz R L 1992 Design and Operation of Series-Array Josephson Voltage Standards – Proc. of the International School of Physics 'Enrico Fermi', Course CX, North-Holland

Kautz R L 1994 J. Appl. Phys. **76**, 5538

Kohlmann J, Schulze H, Behr R, Krasnopolin I Yu, Müller F and Niemeyer J 1999 Programmable voltage standards using SINIS junctions, submitted to Appl. Supercond. (EUCAS'99)

Likharev K K, Mukhanov O A and Semenov V K 1987 IEEE Trans. Magn. **23**, 759

Maezawa M and Shoji A 1997 Appl. Phys. Lett. **70**, 3603

Schulze H, Behr R, Müller F and Niemeyer J 1998 Appl. Phys. Lett. **73**, 996

Schulze H, Müller F, Behr R, Kohlmann J, Niemeyer J and Balashov D 1999 IEEE Trans. Appl. Supercond. **9**, 4241

Sugiyama H, Yanada A, Ota M, Fujimaki A and Hayakawa H 1997 Jpn. J. Appl. Phys. **36**, 1157

Inst. Phys. Conf. Ser. No 167
Paper presented at Applied Superconductivity, Spain, 14-17 September 1999
© *2000 IOP Publishing Ltd*

Two-stacks of parallel arrays of long Josephson junctions

G Carapella, G Costabile, G Filatrella, and R Latempa

Unità INFM and Dipartimento di Fisica, Università di Salerno, I-84081 Baronissi, Italy

ABSTRACT: We propose and modelise a structure consisting of two parallel arrays of long Josephson junctions separated by an electrode that allows inductive coupling between the arrays. Voltage locked states inferred by the mathematical model are experimentally demonstrated in devices we fabricated. Appreciable microwave radiation from some cavity modes excited in the structure is also reported.

1.   INTRODUCTION

Performances of soliton oscillators can be improved using arrays of coherently operating devices. Stacks of magnetically coupled long Josephson junctions have been proven good candidates for this purpose. It has been shown theoretically [1] and experimentally [2-4] that linear as well as non linear (fluxons) excitations can be synchronised in these systems. Moreover, parallel arrays of long Josephson junctions (LJJ's) have been experimentally demonstrated [5] capable of mutual synchronisation. Here we propose an hybrid structure, consisting of a two-stack of parallel arrays of LJJ's. The mathematical model of the structure we propose indicates that synchronised states are possible; synchronous motion has been found in the fabricated devices.

2.   THE PHYSICAL SYSTEM AND ITS MODEL

To model the two-stack of parallel arrays of LJJ's in Fig. 1(b), we start from the description of the two-dimensional double overlap stack in Fig. 1(a) that is modelled as [6]

$$\varphi_{xx} + \varphi_{yy} - \varphi_{tt} = \sin\varphi + \alpha\varphi_t + \varepsilon\left(\psi_{xx} + \psi_{yy}\right)$$
$$\psi_{xx} + \psi_{yy} - \psi_{tt} = \sin\psi + \alpha\psi_t + \varepsilon\left(\varphi_{xx} + \varphi_{yy}\right)$$
$$\varphi_x(0) = \varphi_x(1) = (1+\varepsilon)\eta^e ; \; \psi_x(0) = \psi_x(1) = (1+\varepsilon)\eta^e \tag{1}$$
$$\varphi_y(0) = -(1+\varepsilon)\eta^T - \frac{(I_A - I_B)}{(1+\varepsilon)2\lambda_J J_0 L} ; \; \varphi_y(W) = -(1+\varepsilon)\eta^T + \frac{(I_A + I_B)}{(1+\varepsilon)2\lambda_J J_0 L}$$
$$\psi_y(0) = -(1+\varepsilon)\eta^T - \frac{(I_B - I_A)}{(1+\varepsilon)2\lambda_J J_0 L} ; \; \psi_y(W) = -(1+\varepsilon)\eta^T + \frac{(I_B + I_A)}{(1+\varepsilon)2\lambda_J J_0 L}$$

where we have accounted for an externally applied magnetic field $H=(H^T, H^e, 0)$, as well as for the field generated by the bias currents I_A and I_B. In (1) the length is normalised to the Josephson penetration length λ_J [6], and $-1 < \varepsilon < 0$ is the magnetic coupling coefficient [1]. The time is normalised to the inverse of the plasma frequency $\omega_J = \bar{c}/\lambda_J$, where $\bar{c}$ is the Swihart velocity [6].

The stack of parallel arrays of long Josephson junctions (that are assumed one-dimensional for the sake of simplicity) in Fig. 1(b) becomes the continuous system of Fig. 1(a) in the limit

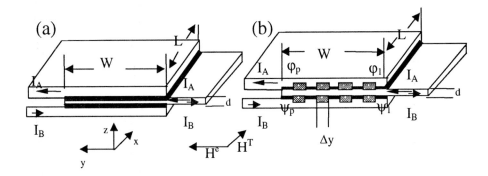

Fig. 1 (a) Sketch of a two-dimensional double overlap stack. (b) Two-stack of parallel arrays of p long Josephson junctions.

$p \to \infty$, $\Delta y \to 0$; $p\Delta y = W$. Thus, the model for our system can be obtained by discretising the model Eqs. (1) in the y-direction (physically, this accounts for the fluxoid quantization in the loops)

$$\phi[x, y, t] \to \phi\left[x, (n-1)\sqrt{\beta}, t\right] \equiv \phi_n(x, t); \psi[x, y, t] \to \psi\left[x, (n-1)\sqrt{\beta}, t\right] \equiv \psi_n(x, t),$$

where we have defined $\beta \equiv \Delta y^2 / \lambda_J^2$. Hence, our model is

$$\phi_{ntt} = \phi_{nxx} - \sin\phi_n - \alpha\phi_{nt} - \varepsilon\psi_{nxx} + \frac{1}{\beta}(\phi_{n+1} - 2\phi_n + \phi_{n-1}) - \frac{\varepsilon}{\beta}(\psi_{n+1} - 2\psi_n + \psi_{n-1})$$

$$\psi_{ntt} = \psi_{nxx} - \sin\psi_n - \alpha\psi_{nt} - \varepsilon\phi_{nxx} + \frac{1}{\beta}(\psi_{n+1} - 2\psi_n + \psi_{n-1}) - \frac{\varepsilon}{\beta}(\phi_{n+1} - 2\phi_n + \phi_{n-1})$$

$$\phi_{1tt} = \phi_{1xx} - \sin\phi_1 - \alpha\phi_{1t} - \varepsilon\psi_{1xx} + \frac{2}{\beta}(\phi_2 - \phi_1) + \frac{2}{\sqrt{\beta}}(1 - \varepsilon^2)\eta^T - \frac{2\varepsilon}{\beta}(\psi_2 - \psi_1) + p(\gamma_A - \gamma_B)$$

$$\psi_{1tt} = \psi_{1xx} - \sin\psi_1 - \alpha\psi_{1t} - \varepsilon\phi_{1xx} + \frac{2}{\beta}(\psi_2 - \psi_1) + \frac{2}{\sqrt{\beta}}(1 - \varepsilon^2)\eta^T - \frac{2\varepsilon}{\beta}(\phi_2 - \phi_1) + p(\gamma_B - \gamma_A)$$

$$\phi_{ptt} = \phi_{pxx} - \sin\phi_p - \alpha\phi_{pt} - \varepsilon\psi_{pxx} + \frac{2}{\beta}(\phi_{p-1} - \phi_p) - \frac{2}{\sqrt{\beta}}(1 - \varepsilon^2)\eta^T - \frac{2\varepsilon}{\beta}(\psi_{p-1} - \psi_p) + p(\gamma_A + \gamma_B)$$

$$\psi_{ptt} = \psi_{pxx} - \sin\psi_p - \alpha\psi_{pt} - \varepsilon\phi_{pxx} + \frac{2}{\beta}(\psi_{p-1} - \psi_p) - \frac{2}{\sqrt{\beta}}(1 - \varepsilon^2)\eta^T - \frac{2\varepsilon}{\beta}(\phi_{p-1} - \phi_p) + p(\gamma_A + \gamma_B)$$

$$\phi_{nx}(0) = \phi_{nx}(1) = (1 + \varepsilon)\eta^e; \psi_{nx}(0) = \psi_{nx}(1) = (1 + \varepsilon)\eta^e$$

(2)

The dispersion relation of the coupled structure is obtained substituting a plane wave solution

$$\phi_n(x, t) = A e^{i[k_x x + (n-1)\sqrt{\beta}k_y - \omega t]}; \psi_n(x, t) = B e^{i[k_x x + (n-1)\sqrt{\beta}k_y - \omega t]}$$

in the unperturbed $[\alpha = \sin(\phi_n) = \sin(\psi_n) = \eta^T = \eta^e = 0]$ model Eqs. (2) and cavity modes resonances are then found imposing open circuit boundary conditions, with the result

$$\omega^{\pm}_{j,m} = u^{\pm} \sqrt{\left(\frac{j\pi}{1}\right)^2 + \frac{4}{\beta}\sin^2\frac{m\pi}{2(p-1)}}$$

(3)

where m, j are integers, and $u^{\pm} = \sqrt{1 \pm |\varepsilon|}$ are the two characteristic velocities corresponding to the in-phase (A=B) mode and to the out-of-phase mode (A=-B) in the structure. These resonances are

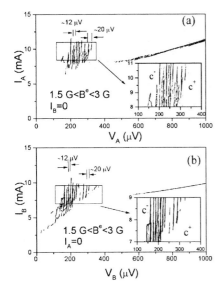

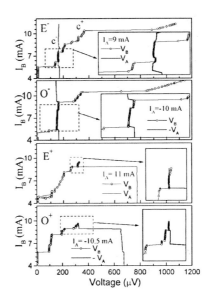

Fig. 2 Cavity mode resonances in the arrays of the stack excited by a magnetic field in the y-direction.

Fig. 3 Voltage locked states recorded in the two-stack. The magnetic field is B^e=2 G.

practically excited by a magnetic field and appear as current singularities in the I-V characteristic. So, if we apply only a magnetic field along the y-direction, the cavity modes will have voltages

$$V^\pm = j\Phi_0 \bar{c}^\pm / 2L = j\Phi_0 \bar{c}\sqrt{1 \pm |\epsilon|} / 2L \qquad (4)$$

where L is the physical length of the junctions, Φ_0 is the flux quantum and $\bar{c}$ is the Swihart velocity in the stack [1, 6]. In other words we expect two series of singularities, with characteristic voltage spacing $\Delta V^\pm$ corresponding to the two velocities present in the structure. If one applies a field only in the x-direction, the cavity modes will have voltages

$$V_m = \bar{c}^\pm \frac{\Phi_0}{\Delta y} \left| \sin \frac{m\pi}{2(p-1)} \right| \qquad (5)$$

i.e., two series of singularities not evenly spaced, and limited in voltage. So, due to the inductive coupling, the two stacked arrays exhibit a splitting in two branches of the dispersion relation of a parallel array of LJJ's [5]. Preliminary numerical simulations of model Eq. (2) confirm relations (4) and (5).

3. EXPERIMENTAL RESULTS AND DISCUSSION

We have fabricated and tested two-stacks of parallel arrays of long Nb/AlO$_x$/Nb Josephson junctions with the geometry shown in Fig. 1(b). The outer Nb electrodes are 300 nm thick, while the intermediate Nb electrode has a thickness d ≈ 100 nm. Due to the fabrication process, the estimated London penetration depth in our Nb films is λ_L ≈ 120 nm. In the two-stack on which we report here, the arrays have five (p=5) junctions (600×20) μm^2 separated by Δy=20 μm. The critical current per junction is I_{cB} = 4 mA for the bottom array and I_{cA} = 3.6 mA for the top array. We estimate Josephson

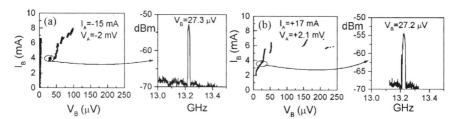

Fig.4 Microwave radiation received by cavity modes excited in the bottom array of the stack biasing the top array with a negative [in (a)] or positive [in (b)] current .

penetration lengths $\lambda_{JA} \approx \lambda_{JB} = 60$ μm, normalised lengths $l_A \approx l_B = 10$, and parameter $\beta = 0.1$. From the thickness of the intermediate electrode we expect a magnetic coupling $\varepsilon \approx -0.44$. Finally, the critical field is estimated to be $B_c \approx 0.5$ G.

Figure 2 shows the cavity modes resonances (or Fiske steps) excited in one array of the stack by a magnetic field in the y direction, while the other array is unbiased. The plots are obtained superimposing the I-V characteristics recorded for different values of the applied magnetic field. The cavity modes exhibit two characteristic voltage spacings $\Delta V^- \approx 12$ μV and $\Delta V^+ \approx 20$ μV. As noted above, this correspond to the splitting of the Swihart velocity $\bar{c}$ in two velocities $\bar{c}^{\pm} = \bar{c}\sqrt{1 \pm |\varepsilon|}$. From data in Fig. 2 we estimate $\varepsilon = -0.47$, that agrees reasonably with the expected value of $\varepsilon = -0.44$.

Biasing one array of the stack with a constant current and sweeping the other on the Fiske steps, it is possible to record up to four voltage locked states. In other words, it is possible to excite four different Fiske modes, with equal or opposite voltage polarities, both in the $\bar{c}^-$ and in the $\bar{c}^+$ family. This is shown in Fig. 3, where we labelled as $O^{\pm}$ the voltage locked states with opposite voltage polarities (with respect to the intermediate electrode) and as $E^{\pm}$ the voltage locked states with equal voltage polarities. In this structure a robust two-dimensional coherence is possible. For example, the locking in the $\bar{c}^+$ family means, from previous analysis, that adjacent junctions of the bottom array, the junctions in the top arrays, and also bottom and top array oscillate in phase. So, the structure exhibits coherence in both the y and z direction. From the applicative point of view, this should compel a great enhancement of the emitted radiation from the device. The frequency of the emitted signal falls, from the Josephson relation $v_{em} = \Phi_0^{-1} V$, in the range 70 GHz$< v_{em} <$200 GHz, that is, unfortunately, out of the band of our instrumentation. In order to fall in the band of our instrumentation, we can excite low voltage (low frequency) cavity modes steps in one array by passing a current through the other array of the stack (field generator), as shown in Fig. 4, where the top array is used as a magnetic field generator. The generated field is in the x direction, so these cavity modes should be of the family described by relation (5). Though these steps are not voltage locked, the radiation we coupled to a room temperature spectrum analyser when polarised on these steps was quite high: about 10 nW on a 50 Ω unmatched load. This preliminary result on the emitted radiation suggests that the structure is very promising as a microwave oscillator.

REFERENCES

[1] Sakai S, Bodin P, Pedersen N F 1993 J. Appl. Phys. **73**, 2411
[2] Ustinov A V, Kohlstedt H and Heiden C 1994 Appl. Phys. Lett. **65**, 1457
[3] Carapella G and Costabile G 1997 Appl. Phys. Lett. **71**, 3409
[4] Carapella G, Costabile G, Petraglia A, Pedersen N F, J. Mygind 1996 Appl. Phys. Lett. **69**, 1300
[5] Carapella G, Costabile G and Sabatino P 1998 Phys. Rev. B **58**, 15094
[6] Carapella G, Costabile G, Sakai S and Pedersen N F 1998 Phys. Rev. B **58**, 6497

Inst. Phys. Conf. Ser. No 167
Paper presented at Applied Superconductivity, Spain, 14-17 September 1999
© 2000 IOP Publishing Ltd

Programmable voltage standards using SINIS junctions

J Kohlmann, H Schulze, R Behr, I Y Krasnopolin*, F Müller, and J Niemeyer

Physikalisch-Technische Bundesanstalt, Bundesallee 100, D-38116 Braunschweig, Germany
* Permanent address: Russian Research Institute for Metrological Service, Moscow, Russia

ABSTRACT: A programmable voltage standard or a D/A converter with fundamental accuracy has been realized with a binary sequence of nonhysteretic junctions. The arrays of Superconductor - Insulator - Normal metal - Insulator Superconductor (SINIS) Josephson junctions have been fabricated using the reliable Nb-Al/AlO$_x$ technology. The arrays of up to 8192 junctions (14 bit) are operated at microwave frequencies of about 70 GHz.

1.   INTRODUCTION

Besides the conventional 1 V and 10 V dc Josephson voltage standard arrays consisting of about 2000 or 14000 Superconductor - Isolator - Superconductor (SIS) junctions, programmable voltage standard arrays as first proposed by Hamilton et al (1995) have increasingly gained in importance in recent years (e.g. Benz et al 1997, Schulze et al 1999a, and references cited therein). In contrast to the underdamped SIS arrays these new arrays show a nonhysteretic current voltage characteristic. The arrays are based on over-damped Josephson junctions which can be realized by different kinds of shunted junctions. Under microwave irradiation of frequency f, each junction generates steps of constant voltage at $V = n \cdot f/K_J$, where n = 0, +1 or -1 is the quantum step number, depending on the bias current, and $K_J = 483597.9$ GHz/V the Josephson constant. As the current voltage characteristic remains single-valued under microwave irradiation, the arrays ensure fast step selection and inherent stability. For use as programmable voltage standards and D/A converters, the arrays are divided in a binary sequence.

Various kinds of junctions have been proposed and realized to date. Hamilton et al (1995) fabricated arrays of resistively shunted SIS junctions, Benz et al (1997) used (intrinsically shunted) Superconductor - Normal metal - Superconductor (SNS) junctions. Superconductor - Insulator - Normal metal - Insulator Superconductor (SINIS) junctions were proposed by Kupriyanov and Lukichev (1988) and recently studied for electronic applications by Maezawa and Shoji (1997) and Sugiyama et al (1997). They are intrinsically shunted Josephson junctions with moderate current density and combine most of the advantages of SNS and SIS junctions. Due to their high characteristic voltage $V_c = I_c \cdot R_n$ of up to 200 μV (I_c being the critical current and R_n the normal state resistance), SINIS junctions can be operated at frequencies of more than 70 GHz. At PTB, the applications of SINIS junctions for programmable voltage standards have been studied recently (Schulze et al 1999a). Several

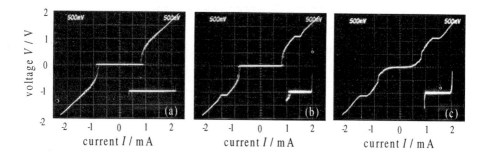

Fig 1: Current-voltage characteristic of a series array of 7552 SINIS junctions at different microwave power levels (f = 70 GHz): (a) without microwave irradiation, (b) 0.4 mW, (c) 3 mW.
The inset shows (a) the critical current (I: 500 µA/div; U: 1 mV/div), (b), (c) the step of constant voltage with high resolution (I: 100 µA/div; U: 100 µV/div).

improvements and new results of the development of programmable voltage standard arrays and D/A converters are described and discussed in this paper.

2. FABRICATION AND PROPERTIES OF SINIS JUNCTIONS

Arrays of SINIS junctions were fabricated at PTB using the reliable Nb-Al/AlO$_x$ technology. The fabrication process is more or less the same as for SIS junction arrays (Kohlmann et al 1997), an exception being the Nb/Al/AlO$_x$/Al/AlO$_x$/Al/Nb junction sandwich. The thickness of each Al-layers is 10 nm; for the two AlO$_x$ barriers the first two Al-layers are oxidized in 0.5 Pa O$_2$ at 20°C for 5 minutes. This process yields a current density of about 130 A/cm^2, a normal state resistance of about 100 mΩ, a characteristic voltage of about 130 µV and a McCumber parameter of about 1. A critical current of about 1.3 mA results from junction dimensions of 20 µm x 50 µm.

3. RESULTS AND DISCUSSION

SINIS junction series arrays of different designs have been fabricated. Fig. 1 shows the current-voltage characteristic of a series array with 7552 SINIS junctions. The parameter spread of the junctions is very small (Fig. 1a). Under 70 GHz microwave irradiation a step of constant voltage at the 1 V level is generated by a low microwave power of about 0.4 mW at the finline taper of the array (Fig. 1b). A representation of the 1 V step enlarged with high resolution is shown in the inset. The step of a SINIS array has been proved to be suitable for calibration purposes (Behr et al 1999). In a direct comparison with a conventional SIS Josephson voltage standard at 1.018 V, the voltages of both arrays agree to better than $1 \cdot 10^{-10}$. The step width can be increased only slightly by increasing the microwave power to 3 mW (Fig. 1c). This unusual behaviour results from the characteristics of a series array of SINIS junctions in a low-impedance stripline. According to Kautz (1992), the attenuation of the series array can be estimated at about 50 dB/1000 SINIS junctions (Schulze et al 1999a). The fact that, in spite of this high attenuation, steps are generated in series arrays of more than 1000 SINIS junctions can be explained by the active contribution of the SINIS junctions. The junctions operate as oscillators which compensate the attenuation in the stripline (Schulze et al 1999a).

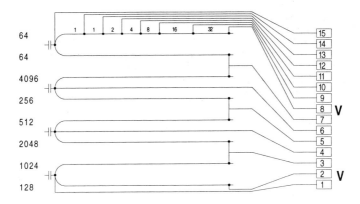

Fig. 2: Schematic layout of a 1 V series array divided into a binary sequence and consisting of 8192 junctions (14 bit).

As only dc-biased junctions can operate as oscillators, the design of a binary-divided array must take the attenuation of unbiased segments into account. It is not possible to operate two or more large segment (more than 250 junctions) in the same microwave path. If the first segment is not dc-biased, its attenuation is too high to supply sequencing segments with microwave power. The layout of a binary-divided array with 8192 junctions (14 bit) is schematically shown in Fig. 2. The microwave stripline is split into eight parallel paths. The six smallest bits are arranged in a single path, all other bits in separate paths.

The step width vs. microwave power for various bits is plotted in Fig. 3. For small bits the width is larger than the critical current of about 1 mA. Due to the increasing attenuation, the width decreases rapidly for arrays with more than 256 junctions. For the longest bits the width is nearly constant ($0.3\,I_c$) due to the active contribution of the SINIS junctions. However, the width of these steps of more than 200 μA is still sufficient for most of the possible applications.

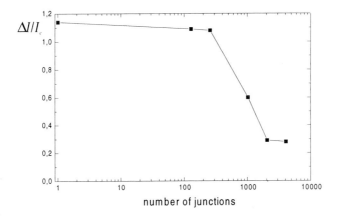

Fig. 3: Normalized step width $\Delta I/I_c$ vs. number of junctions of a binary-divided SINIS array. Due to the increasing attenuation the step width decreases with increasing number of junctions. The step width is nearly constant for the longest bits.

4.   CONCLUSIONS AND OUTLOOK

Series arrays of SINIS junctions have been fabricated for programmable voltage standards and D/A converters using the reliable Nb-AlO$_x$ technology. Steps of constant voltage of up to 1 V have been generated by arrays of up to 8192 junctions divided in a binary sequence (14 bit) and operated at microwave frequencies of about 70 GHz. The arrays are suitable for metrological purposes. The operation of the array using independent, computer-controlled bias sources is under development. Further possibilities of the arrays will then be tested and studied. The microwave behaviour of SINIS arrays will be further investigated (cf. Schulze et al 1999b).

ACKNOWLEDGEMENT

The authors thank R. Harke and Th. Weimann for technical assistance.

This work has been supported in part by the Bundesministerium für Bildung und Forschung (BMBF), Germany (ref. No. 13N6835 and 13N7259) and the European Community (ref. No. SMT4-CT98-2239)

REFERENCES

Behr R, Schulze H, Müller F, Kohlmann J and Niemeyer J 1999 IEEE Trans. Instrum. Meas. **48**, 270

Benz S P, Hamilton C A, Burroughs C J, Harvey T E and Christian L A 1997 Appl. Phys. Lett. **71**, 1866

Hamilton C A, Burroughs C J and Kautz R L 1995 IEEE Trans. Instrum. Meas. **44**, 223

Kautz R L 1992 Design and operation of series-array Josephson voltage standards, Proc. Int. School of Physics "Enrico Fermi", Course CX, North Holland

Kohlmann J, Müller F, Behr R, Krasnopolin I Y, Pöpel R and Niemeyer J 1997 Appl. Supercond. - Inst. Phys. Conf. Ser. **158**, 631

Kupriyanov M Yu, Lukichev V F 1988 Sov. Phys. JETP **67**, 1163

Maezawa M and Shoji A 1997 Appl. Phys. Lett. **70**, 3603

Schulze H, Müller F, Behr R, Kohlmann J, Niemeyer J and Balashov D 1999a IEEE Trans. Appl. Supercond. **9**, 4241

Schulze H, Krasnopolin I Y, Behr R, Kohlmann J, Müller F and Niemeyer J 1999b Microwave properties of SINIS Josepshon junction series arrays, submitted to Appl. Supercond. (EUCAS´99), this volume

Sugiyama H, Yanada A, Ota M, Fujimaki A and Hayakawa H 1997 Jpn. J. Appl. Phys. **36**, L1157

Inst. Phys. Conf. Ser. No 167
Paper presented at Applied Superconductivity, Spain, 14-17 September 1999

773

A 4-pixel YBCO mid-infrared bolometer array with an associated cryogenic CMOS circuit

F Voisin, G Klisnick, Y Hu, M Redon, J Delerue*, A Gaugue*, A Kreisler*

Laboratoire des Instruments et Systèmes - Université de Paris 6 - 4 Place Jussieu, 75252 Paris Cedex 05 - France
*Laboratoire de Génie Electrique de Paris - Universités de Paris 6 et Paris 11 - UMR 8507 CNRS, Supélec, Plateau de Moulon, 91192 Gif sur Yvette Cedex - France

ABSTRACT: The design of a high T_C superconducting bolometer geometry and its readout electronic for mid-infrared imaging array applications are presented. The sensor part consists of an YBCO thin film patterned into four meanders to obtain a 2×2 pixel matrix. In order to form a fully integrated system, we have realized a cryogenic CMOS ASIC which contains all the readout circuits and the programmable bias current sources that are needed to drive the bolometer array.

1. INTRODUCTION

Many experiments have been performed to study the photoresponse of superconducting detectors, but only a few realizations of multidetectors have been reported (Osterman et al 1991, Mai et al 1997, Li et al 1997).

Moreover, when using cooled sensors, room temperature low-level signal processing can lead to some difficulties due to microphonics, lead capacitance or noise pickup, for instance. In the case of a bolometer array, it is also of chief importance to integrate a set of preamplifiers (and preprocessing functions) in a same chip to obtain similar pixel characteristics. This can be achieved with standard integrated circuit technologies (such as CMOS). But in view of undegraded (if not improved) overall noise performance, the ability of such a circuit to operate correctly at low temperature must be evaluated.

In a former paper, we had described a CMOS cryogenic amplifier associated with an YBaCuO transition edge bolometer (Hu et al 1998). In order to integrate preprocessing electronics close to the sensor, an ASIC was designed that included a low noise preamplifier working from room temperature down to 77 K. It was successfully used with an YBaCuO bolometer elaborated on polycrystalline (yttria-stabilized) zirconia substrate (Ben Ayadi et al 1996), to form an integrated 1-pixel mid-infrared detector that was tested at 10 μm wavelength (CO_2 laser radiation). It was shown that cooling the preamplifier lowered the noise level by about 3 dB in the white noise region.

In this paper, we describe a compact mid-infrared imaging demonstrator. A new cryogenic CMOS ASIC was designed to work with a transition edge bolometric sensor array of 2x2 pixels, in view of CO_2 laser spot tracking. To further improve the noise performance, the design comprised bolometer bias circuits, thus achieving a whole cryogenic readout circuitry.

2. COMPACT 4-PIXEL MID-INFRARED DETECTOR TOPOLOGY

In the 1-pixel detector framework, described Fig. 1-a, the bolometer, which is driven by an external bias current, delivers a DC potential added to the AC signal. To amplify only the AC signal without excess noise, we had chosen to make a preamplifier with a single input coupled to the

774

bolometer by a capacitance. This capacitance has a large value, so that the contribution of the preamplifier equivalent input noise current had a negligible effect, even at low frequencies. To obtain the lowest possible equivalent input noise current, MOSFET was preferred to J-FET or to bipolar transistor.

Similar but optimized structures have been used to realize the 4-pixel detector (Fig. 1-b). This detector consists in two branches, each branch having two superconducting sensors in series. This topology allows to reduce by two the number of integrated current sources instead of four current sources used for four 1-pixel detectors, while reducing the global supply voltage compared to the case where the four superconducting sensors would have been connected in series.

To optimize the overall frequency response, it is convenient that the current sources be close to the sensor. So a programmable current source has been integrated in the ASIC, mirroring an external reference current.

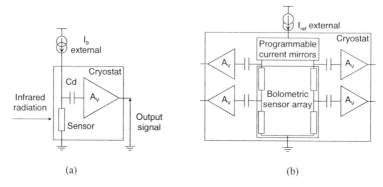

(a)   (b)

Fig. 1. 1-pixel mid-infrared detector principle (a) and 4-pixel imaging topology with readout electronics (b).

3. HIGH T$_C$ SUPERCONDUCTING BOLOMETER GEOMETRY

We have realized a 4-pixel detector structure forming a typical 2×2 bolometer array (with a total area of 2×2 mm^2), as shown in Fig. 2. The sensitive elements of the array are meanders. Each bolometer has connecting pads to measure the voltage across a single element.

YBCO film with a thickness of 200 nm was produced on MgO substrate of 500 μm thickness by *in situ* RF magnetron sputtering. The meanders were patterned by photolithography and ion milling. Two types of meanders have been etched. The first as shown in Fig. 3, had 100 μm line width (and line spacing), a total length of 5.5 mm and a normal state resistance of 300 - 500 Ω. The second had a line width of 40 μm (and line spacing), a total length of 12.5 mm and a normal state

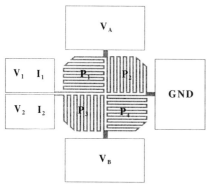

Fig. 2. Schematic of a 2×2 superconducting bolometer array for demonstration of spot tracking. Represented meander has 40 μm width and 12.5 mm total length.

Fig. 3. Photograph of a 2×2 YBCO bolometer array. The width and the length of the meander are 100 μm and 5.5 mm, respectively.

resistance of 1.5 - 3 kΩ. The leads to perform electrical measurements were formed from a 100 nm thick gold layer produced by RF sputtering and also patterned by photolithography and ion milling.

4.   LOW NOISE CRYOGENIC PREAMPLIFIER STRUCTURE

Each implemented preamplifier adopts a similar structure as was used in a former ASIC version (Klisnick et al 1998). It uses a single-ended folded-cascode topology (Fig. 4). The total noise of this amplifier is mainly determined by the input transistor contribution. To obtain better noise performances (for both thermal and flicker noise) than in our previous work, in particular, we have significantly increased the input transistor size to 36000 µm for gate width and 4.2 µm for gate length. The input transistor capacitance was therefore highly increased so imposing the use of an external input capacitance C_d of larger value than that had been integrated on the former ASIC version.

To achieve a low thermal noise, a high input transistor transconductance of 9 mA/V has been achieved by means of a high bias current of 450 µA. Due to its large size, this bias current then constrains this transistor to operate in moderate inversion regime. Furthermore, it is well-known (Sansen and Laker 1994), and experimentally verified by us down to 77 K, that a biasing point Vgs_0 exists for which a MOS transistor I_{DS} drain canal current is temperature independent. Here, the input transistor M1 was Vgs biased at a value lower than Vgs_0 whereas the load transistor M5 is Vgs biased at a value higher than Vgs_0. Thus, the preamplifier AC voltage gain magnitude within bandpass, which is given by the transconductance ratio of the input transistor M1 and the load transistor M5 ($|A_V| \approx g_{m1}/g_{m5}$), will decrease with decreasing temperature. However, the preamplifier gain value can be modified by adjusting the external DC bias current source, if necessary.

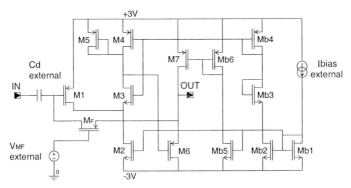

Fig. 4. Schematic of the preamplifier.

5.   EXPERIMENTAL RESULTS

The preamplifier simulated performances at 300 K are given in Fig. 5. A gain of 136 has been obtained with a -3 dB bandwidth from 8 Hz up to 4.6 MHz using an external 100 nF input capacitance C_d. An input equivalent noise voltage of 1.4 nV Hz$^{-1/2}$ at 10 kHz has been calculated with the flicker noise starting around 350 Hz. We do not present simulations results at 77 K due to the lack of accurate simulation models at low temperature.

The chip has been implemented in a 1.2 µm N-well CMOS process and has been tested. Measurement results (gain and noise) are also presented in Fig. 5. At 300 K, there is good agreement between simulations and measurements for both the AC voltage gain magnitude in the bandpass and the equivalent input voltage thermal noise. We can see that the low frequency noise appears below about 800 Hz which is close to the simulated frequency value in spite of the difficulty to characterize the low frequency noise in simulation models (Camin et al 1996).

Low temperature and room temperature measurements were performed in the same conditions without any change of the external DC bias current source value. At 77 K, the AC gain

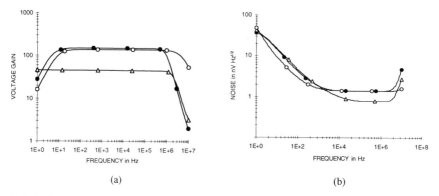

Fig. 5. AC gain magnitude (a) and input equivalent noise (b), simulation ("O" at 300 K) and measurement ("●" at 300 K and "△" at 77 K) using an spectrum analyzer with 3201 frequency points by decade.

magnitude is decreased to 45 and the thermal noise is reduced from $1.4\,nV\,Hz^{-1/2}$ at 300 K to $0.8\,nV\,Hz^{-1/2}$ at 77 K, as expected.

6. CONCLUSION

In view of improving the noise behaviour and the compactness of the overall infrared detector, we have designed a new cryogenic CMOS ASIC including a low noise preamplifier providing good performances at low temperature (77 K) and an integrated programmable current source to drive the bolometric sensor array.

The association of a bolometric matrix of 2×2 pixels shaped in an original design with preprocessing electronics working close to the sensor at cryogenic temperature, allowed to demonstrate the feasability of a fully compact mid-infrared imaging system using an ASIC designed in a standard low cost CMOS technology. As a further step, this detector unit will be tested using an infrared CO_2 laser radiation to achieve spot tracking.

REFERENCES

Ben Ayadi Z, Gaugue A, Degardin A, Caristan E, Kreisler A, Fourrier M and Redon M 1996 Proc. 2nd European Workshop on Low Temperature Electronics, Eds C Clayes and E Simoen (Editions de Physique, France) pp 289-294

Camin D V and Pessina G 1996 Proc. 2nd European Workshop on Low Temperature Electronics, Eds C Clayes and E Simoen (Editions de Physique, France) pp 225-230

Hu Y, Klisnick G, Voisin F, Redon M, Gaugue A and Kreisler A 1998 Proc. 3th European Workshop on Low Temperature Electronics, Eds L Brogiato D V Camin and G Pessina (Editions de Physique, France, 1998) pp 209-212

Klisnick G, Hu Y, Voisin F, Redon M, Gaugue A and Kreisler A 1998 11th Annual IEEE International ASIC Conference, Rochester NY, USA, Eds Schrader M E et al pp 405-408

Li H, Wang R, Wan F, Ping Y, He G and Yu M 1997 IEEE Trans. Appl. Supercond. AS-7(2) 2371

Mai Z, Zhao X, Zhou F and Song W 1997 Infrared Physics & Technology 38 13

Osterman D P, Marr P, Dang H, Yao C-T and Radparvar M 1991 IEEE Trans Mag. 27(2) 2681

Sansen W M C and Laker K R 1994 Design of Analog Integrated Circuits and Systems (New-York: McGraw-Hill) p 166

Inst. Phys. Conf. Ser. No 167
Paper presented at Applied Superconductivity, Spain, 14-17 September 1999
© 2000 IOP Publishing Ltd

RF magnetic shielding effects of an HTS cylinder

M Itoh, K Itoh, K Mori* and Y Hotta**

Dept. of Electronic Eng., Kinki Univ., Higashi-Osaka, Osaka 577-8502, Japan
*R&D Unit, Tokin Corp. Co. Ltd., Sendai, Miyagi 982-8510, Japan
**EMC Tech. Center, Tokin EMC Eng. Co. Ltd., Kawasaki, Kanagawa 213-0023, Japan

ABSTRACT: The detailed characteristics of electromagnetic shielding effects for HTS vessel are generally unknown. In the present research, the RF shielding of superconducting BPSCCO bulk cylinders employed as shielding vessels are measured. It was found that the RF magnetic shielding of the cylinder decreased as the applied frequency range increased from 1MHz to 1GHz. It was also found that the shielding displayed no evidence of dependence on the RF power (-17dBm ~ 43dBm).

1. INTRODUCTION

Electromagnetic shielding is required to improve electromagnetic environments. The shielding is used to control the propagation of an electromagnetic field from one region to an other. For example, shields are used to surround a noise source, and to also prevent noise infiltration. There has been an increasing need for a configuration of radio frequency (RF) shields capable of limiting regions to very low energy values of electromagnetic waves. Typical examples of such needs are medical instruments, industrial robots, vehicles that make use of superconductivity, and superconducting devices.

The ideal magnetic shielding vessel (Itoh 1992a, Itoh et al. 1992b) can be realized by use of a high-critical temperature superconductor (HTS), due to its property of perfect diamagnetism. The authors have evaluated the shielding effects of HTS vessels by measuring the value of the maximum shielded magnetic flux density B_s (Itoh et al. 1995). Little is known, however, of the characteristics and evaluation procedures for the electromagnetic shielding of an HTS vessel. In the present research, the RF shielding effect is measured, that is, the ratio of the output voltages of the receiving antenna with and without a superconducting Bi-Pb-Sr-Ca-Cu-O (BPSCCO) bulk cylinder employed as the RF shielding vessel. Use is also made of two BPSCCO cylinders having different l/r (10.3, 15.9). Here, l and r are the length and inner radius of the BPSCCO cylinder, respectively. In order to simplify the theoretical analysis, the authors have limited the evaluation of the RF shielding effects. The RF magnetic shielding effects are found to decrease with an increase in RF frequency (1MHz ~ 1GHz). The shielding also displays no evidence of dependence on the RF power (-17dBm ~ 43dBm). In the case of RF electric shielding, no remarkable characteristics were found.

The present paper examines the effects of RF electromagnetic shielding within the BPSCCO cylinder. This includes the RF magnetic and electric shielding effects as a function of frequency f, and the RF magnetic and electric shielding effects as a function of RF power.

2.   EXPERIMENTAL PROCEDURE

Fig. 1, is a schematic diagram illustrating the experimental configuration used to measure the electromagnetic shielding characteristics at the center of a superconducting BPSCCO cylinder, under conditions of the boiling point of liquid nitrogen (77.4K). The RF output of the tracking generator, including a spectrum analyser, in the frequency region of 1MHz to 1GHz, is amplified by 40dB by use of a broad band power amplifier, and then guided to a transmitting antenna. The output of the receiving antenna is amplified by 60dB, making use of a preamplifier, and guided to the spectrum analyser input terminal. The results from the spectrum analyser are then transferred through a GP-IB to a desk-top computer. The electromagnetic wave was always applied perpendicular to the axial direction of the BPSCCO cylinder. The coaxial cable used as the receiving line, as illustrated in Fig. 1, is threaded through ferrite rings in order to reduce any mutual interaction between the transmitting and receiving lines.

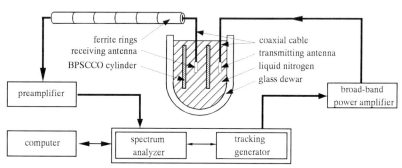

Fig. 1. Schematic diagram of the apparatus used to measure the electromagnetic shielding characteristics at the center of a superconducting BPSCCO cylinder.

The two BPSCCO cylinders were fabricated according to the methods discussed by Itoh et al. (1992b). The cylinders consist of cylinder A (12mm inner radius, 3mm thickness, 123mm length) and cylinder B (10mm inner radius, 2mm thickness, 159mm length). The outer surfaces of the BPSCCO cylinders were wrapped with several turns of fluoroplastic (PTFE) tape, making use of the troidal winding method, in order to avoid any sudden temperature change.

The shielding degree (SD) can be specified in terms of the reduction in magnetic and/or electric field strength due to the shielding vessel. In general, the SD is defined for magnetic fields as

$$SD_{H}=20\log\frac{H_0}{H_1} \text{ (dB)}, \tag{1}$$

and for electric fields as

$$SD_{E} = 20\log\frac{E_0}{E_1} \text{ (dB)}. \tag{2}$$

In the preceding equations, H_0 (E_0) is the incident field strength, and H_1 (E_1) the field strength of the transmitted wave as it emerges from the shielding vessel. The input power of the transmitting antenna is held constant. Loop and probe antennas are used for measuring the degree of the magnetic and electric shielding respectively.

3. RESULTS AND DISCUSSION

3.1 RF Magnetic Shielding Effect

Fig. 2, shows the typical magnetic shielding characteristics of two BPSCCO cylinders as a function of radio frequency over the RF range 1MHz ~ 1GHz at a temperature of 77.4K, under a constant RF output power P_H (-7dBm) of the transmission antenna. In this figure, the open and solid circles represent cylinder A (l/r=10.3) and cylinder B (l/r=15.9) respectively. The magnetic shielding degree SD_H of the two cylinders decreases with an increase in RF fre-quency. Furthermore, it can be seen that the characteristics of cylinder A are almost the same as those for cylinder B. The authors are, therefore, now attempting to improve the characteristics of magnetic shielding by use of a radio wave absorber, such as wood ceramics (Hotta et al. 1998).

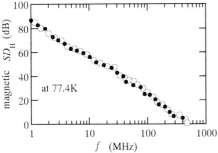

Fig. 2. Typical magnetic shielding character-istics of two BPSCCO cylinders at 77.4K. The open and solid circles represent the change in RF frequency for cylinder A (l/r=10.3) and cylinder B (l/r=15.9) re-spectively.

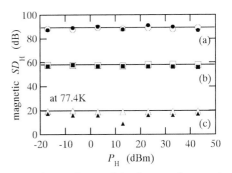

Fig. 3. Dependence of the degree of magnetic shielding SD_H on the RF output power P_H of the transmission antenna for the two BPSCCO cylinders at 77.4K. Curves (a), (b), and (c) represent the results for RF values of 1MHz, 10MHz, and 100MHz respectively.

The dependence of the degree of SD_H on the value of RF output power P_H for cylinders A and B are displayed in Fig. 3. The curves (a), (b), and (c) represent the results for RF values of 1MHz, 10MHz, and 100MHz respectively. In this figure, the open circles, open squares, and open triangles represent results for cylinder A. The solid circles, solid squares, and solid triangles denote results for cylinder B. It can be seen that the characteristics of SD_H for the two BPSCCO cylinders display no evidence of dependence on the values of P_H in the region between -17dBm and 43dBm.

3.2 RF Electric Shielding Effect

Fig. 4, shows the typical electric shielding SD_E characteristics for cylinders A and B as a function of radio frequency over the frequency range of 1MHz ~ 1GHz at 77.4K, under constant transmission antenna RF output power P_E (-7dBm). In this figure, the open and solid squares represent the results for cylinder A and cylinder B, respectively. The degree of electric shielding SD_E of the two BPSCCO cylinders exhibits no remarkable characteristics, averaging about 14dB in the frequency region from 20MHz to 500MHz. Furthermore, in Fig. 5, it can be seen that the characteristics of SD_E for the two BPSCCO cylinders (cylinder A, open squares and cylinder B, solid squares) display no evidence of dependence on the values of P_E in the region between -17dBm and 43dBm under conditions of 200MHz of RF frequency. The authors are now attempting to improve the characteristics of electric shielding by the use of low resistance metals, such as copper and aluminum cylinders.

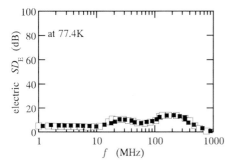

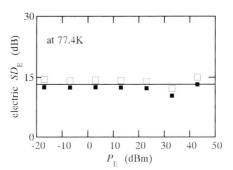

Fig. 4. Typical electric shielding characteristics of the two BPSCCO cylinders at 77.4 K. The open and solid squares represent values of the shielding effect versus radio frequency for cylinders A and B, respectively.

Fig. 5. Dependence of the degree of electric shielding SD_E on the RF output power P_E of the transmission antenna for the two BPSCCO cylinders under constant RF of 200MHz, at 77.4K. The open and solid squares represent the results for cylinders A and B, respectively.

4.   CONCLUSIONS

As one of the basic areas of research for the development of a highly effective HTS shielding vessel, the present paper has examined the electromagnetic shielding effects of a fabricated BPSCCO cylinder. The degree of magnetic shielding SD_H of the two cylinders decreased with an increase in the applied radio frequency. Furthermore, it was seen that the characteristics of cylinder A were almost identical to those of cylinder B. It was found that the degree of shielding SD_H for the two BPSCCO cylinders displayed no evidence of dependence on the values of RF output power P_H in the range between -17dBm and 43dBm.

The degree of electric shielding SD_E of the two BPSCCO cylinders exhibited no remarkable characteristics, and averaged about 14dB in the frequency range from 20MHz to 500MHz. Furthermore, it was found that the characteristics of SD_E for the two BPSCCO cylinders displayed no evidence of dependence on the values of RF output power P_E in the region between -17dBm and 43dBm.

The authors are now investigating methods for the improvement of the magnetic and electric shielding effects in the RF region, making use of radio wave absorbers and low resistance metals respectively.

REFERENCES

Hotta Y, Mori K and Itoh M 1998 Proc. ICEC 17 (London: IOP) pp 507-10

Itoh M 1992a Advances in High Temperature Superconductivity eds D Andreone R S Connelli and E Mezzetti (Singapore: World Sci. Pub.) pp 232-43

Itoh M, Ohyama T, Minemoto T, Numata K and Hoshino K 1992b J. Physics D: Appl. Phys. **25** 1630

Itoh M, Ohyama T, Mori K and Minemoto T 1995 Trans. IEE in Japan **115-C** 696

Inst. Phys. Conf. Ser. No 167
Paper presented at Applied Superconductivity, Spain, 14-17 September 1999
© 2000 IOP Publishing Ltd

Estimation of the magnetic field within a highly effective HTS magnetic shielding vessel

M Itoh, Y Horikawa, K Mori* and T Minemoto**

Dept. of Electronic Eng., Kinki Univ., Higashi-Osaka, Osaka 577-8502, Japan
*R&D Unit, Tokin Corp. Co. Ltd., Sendai, Miyagi 982-8510, Japan
**Div. of System Science Kobe, Hyogo 657-8501, Japan

ABSTRACT: The behavior of the magnetic field within a shielding vessel B_{in} for an excitation magnetic field B_{ex} is unknown. The present research examines and clarifies the magnetic behavior of a highly effective magnetic shielding vessel, measured with the use of the HTS dc-SQUID. A long time is required, however, to measure the flux density within the shielding vessel. Therefore, a method is developed for estimating the value of B_{in} from the B_{in}-B_{ex} characteristics of the BPSCCO cylinder, and single- or double-layered soft-iron cylinders by use of a gaussmeter.

1. INTRODUCTION

The value of the maximum shielded magnetic flux density B_s of a high-critical temperature superconductor (HTS) vessel, in general, does not satisfy the criteria required for practical use. The present authors have improved the magnetic shielding effects associated with the superposition of single- or double-layered soft-iron cylinders over a Bi-Pb-Sr-Ca-Cu-O (BPSCCO) bulk cylinder; termed the single-superimposed and double-superimposed cylinders, respectively. Little is known, however, of the characteristics and evaluation procedures used to determine the behavior of the magnetic flux density B_{in} within shielded cylinders when exposed to a magnetic flux density B_{mag} of less than the value of B_s. The value of B_{in} for an applied B_{ex} can be measured by an HTS dc-SQUID magnetometer, although, a long period of time is required to measure the flux density within the BPSCCO cylinder. Therefore, it would be helpful if the value of B_{in} could be estimated for the single- and double-superimposed cylinders from the B_{in}-B_{ex} characteristics of the single-BPSCCO, the single-layered, and double-layered soft-iron cylinders, where the magnetic characteristics of the soft-iron cylinders are measured by use of a gaussmeter. It was found that the estimated values of B_{in} agreed well with the experimental values. The present paper examines the method of estimation and optimum shielding conditions for constructing an ideal magnetic shielding vessel. The results of a magnetic step response of B_{in} to B_{ex} are also examined.

2. EXPERIMENTAL PROCEDURE

The BPSCCO cylinders were fabricated according to the methods reported in Itoh et al. (1992). The outer surfaces of the cylinders were wrapped with several turns of fluoroplastic

(PTFE) tape, making use of the troidal winding method, in order to avoid sudden temperature changes (Itoh et al. 1995). After being exposed to 100 cycles of temperatures between room temperature (300K) and the boiling point of liquid nitrogen (77.4K), the characteristics of the BPSCCO cylinders, such as shown in Figs. 1 and 2, exhibited no significant changes in the degree of magnetic shielding. In addition, all characteristics of the BPSCCO cylinders in the present experiment were obtained by zero field cooling.

The soft-iron cylinders, used as the single-superimposed and double-superimposed cylinders, were constructed of commercial soft-iron, and degaussed by an ac magnetic field (60Hz) at room temperature, prior to conducting the present experiment. Table 1 lists the inner radius r_1, outer radius r_2, and length l of the BPSCCO cylinder, soft-iron cylinder A, and soft-iron cylinder B. Also listed is the value of B_s for the BPSCCO cylinder.

Table 1. Dimensions of the cylinders and the value of B_s for the BPSCCO cylinder.

Cylinder	r_1 (mm)	r_2 (mm)	l (mm)	$B_s (\times 10^{-4}\text{T})$
BPSCCO	12.0	15.0	123	90.0
soft-iron cylinder A	17.5	19.1	120	—
soft-iron cylinder B	23.8	25.4	120	—

The magnetic shielding effects for the single- and double-superimposed cylinders were then evaluated. The B_{ex} was applied parallel to the axial direction of the cylinders. The effects were measured with the use of an HTS dc-SQUID magnetometer (Conductus, iMC-303) such as discussed by Mori et al. (1998). The results were processed through a GP-IB to a desk-top computer. The magnetic shielding effects of the cylinders were measured by use of a gaussmeter (LakeShore, 450) for values of B_{ex} greater than B_s.

3.   RESULTS AND DISCUSSION

3.1   Magnetic Shielding Effect of the Single BPSCCO Cylinder

Figure 1 displays the temporal magnetic step responses of B_{in} to applied values of B_{ex} for the single BPSCCO cylinder at 77.4K. The values of B_{in} were measured by an HTS dc-SQUID magnetometer. In this figure, the curves labelled (a), (b), (c), (d), and (e) are the measured results of B_{in} at the center of the cylinder for the application of B_{ex} values of 3×10^{-4}T, 10×10^{-4}T, 30×10^{-4}T, 50×10^{-4}T, and 70×10^{-4}T, respectively. The values of B_{in} within the single BPSCCO cylinder become rapidly saturated, and maintain a constant state, as shown in Fig. 1. Prior to this experiment, this response was not generally evident, but has now been clarified by this experiment.

Figure 2 shows the characteristics of the magnetic shielding at the center of the single BPSCCO cylinder at 77.4K. The values of B_{in} for applied values of B_{ex} greater than that of B_s (open circles) were measured by a gaussmeter. The values of B_{in} for applied values of B_{ex} less than that of B_s (solid circles) were measured by the HTS dc-SQUID magnetometer. The plotted values of the constant state are the same as those shown in Fig. 1. Prior to this experiment, little was known of the dependence of B_{in} on values of B_{ex} less than the value of B_s. It is clear from these measurements that B_{in} is significantly influenced by B_{ex} for values of B_{ex} less than the value of B_s.

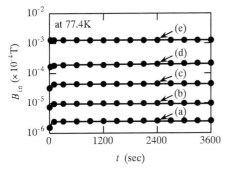

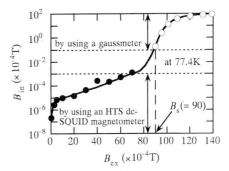

Fig. 1. Temporal magnetic step responses of B_{in} to applied values of B_{ex} for the single BP-SCCO cylinder at 77.4K. Curves (a), (b), (c), (d), and (e) are the measured results of B_{in} for applications of B_{ex} values of 3×10^{-4} T, 10×10^{-4}T, 30×10^{-4}T, 50×10^{-4}T, and 70×10^{-4}T, respectively.

Fig. 2. Typical characteristics of the magnetic shielding within the single BP-SCCO cylinder at 77.4K. Solid and open circles are the measured values of B_{in} by using an HTS dc-SQUID magnetometer and those by a gaussmeter, respectively.

3.2   Magnetic Shielding Effect of the Soft-Iron Cylinder

The values of the magnetic field B_{mag} and relative permeability μ_s at the center of the soft-iron cylinder as a function of B_{ex} are displayed in Figs. 3 and 4, respectively. In both figures, the curves (a) and (b) represent the results for the soft-iron cylinder A, and the superposition of the soft-iron cylinder B over cylinder A at a temperature of 77.4K, respectively. Similar results such as those shown in Fig. 3 (a) are obtained for cylinder B (not shown). The values of μ_s were measured according to the method discussed by Mori et al. (1997). The magnetic characteristics of the single- and double-layered soft-iron cylinders were measured by use of a gaussmeter.

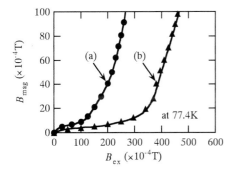

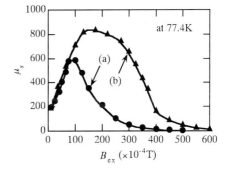

Fig. 3. Values of B_{mag} as a function of B_{ex} for (a) the single-layered soft-iron cylinder (solid circles), and (b) the double-layered soft-iron cylinder (solid triangles), at 77.4K.

Fig. 4. Values of μ_s as a function of B_{ex} for (a) the single-layered soft-iron cylinder (solid circles), and (b) the double-layered soft-iron cylinder (solid triangles), at 77.4K.

3.3　Estimation of the Magnetic Field Within the Superimposed Cylinder

From the results of Figs. 2 and 3, the values of B_{in} were estimated for the single- and double-superimposed cylinders, such as shown in Fig. 5. In this figure, curves (a) and (b) are the shielding characteristics for the single-superimposed and double-superimposed cylinders, respectively, under the condition that the values of B_{mag} are less than that of B_s. The solid squares and solid triangles represent experimental values, while the open squares and triangles are estimated values. As an example, the value of B_{mag} to the value of B_{ex} of 150×10^{-4}T is 21×10^{-4}T, as shown in Fig. 3 for the single-superimposed cylinder (a). Therefore, the value of B_{in} (= 2.0 $\times 10^{-9}$T) can be simply estimated for the single-superimposed cylinder from the curve (a) plotted in Fig. 3. It is found that estimated values agree well with experimental values, under the condition that the values of B_{in} are for applied values of B_{ex} less than that of B_s.

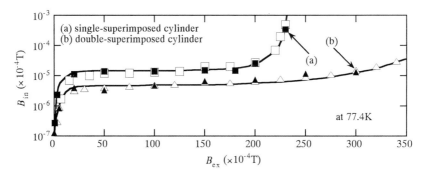

Fig. 5. Magnetic shielding characteristics of (a) the single-superimposed cylinder (squares) and (b) the double-superimposed cylinder (triangles), at 77.4K. The solid squares and solid triangles are experimental values, the open squares and triangles the estimated values.

From the results shown in Fig. 2, it can be seen that the values of B_{in} increase as the values of B_{ex} increase. From Fig. 5 curves (a) and (b), however, it is found that the region of constant B_{in} for the single-superimposed and double-superimposed cylinders holds over a wide range of values of B_{ex}. The region of constant B_{in} demonstrates the stability of the magnetic shielding effect. It is found, furthermore, that the value of B_{in} in the region of constant values for the double-superimposed cylinder is approximately 65 percent less than that of the single-superimposed cylinder.

The method used, which simplifies the estimation of the value of B_{in} within the single-superimposed and double-superimposed cylinders, is important to determine criteria fundamental in the design of an effective and reliable magnetic shielding vessel.

REFERENCES

Itoh M, Ohyama T, Minemoto T, Numata K and Hoshino K 1992 J. Physics D: Appl. Phys., **25**, 1630

Itoh M, Ohyama T, Mori K and Minemoto T 1995 Trans. IEE in Japan **115-C**, 696

Mori K, Minemoto T and Itoh M 1997 IEEE Trans. Appl. Superconductivity **7**, 378

Mori K, Itoh M and Minemoto T 1998 Proc. 17th Int. Cryogenic Engineer. Conf. eds D Dew-Hughes R G Scurlock and J H P Watson (Bristol: IOP)

Inst. Phys. Conf. Ser. No 167
Paper presented at Applied Superconductivity, Spain, 14-17 September 1999
© 2000 IOP Publishing Ltd

Low noise hybrid TlBaCaCuO / GaAs 5.1 GHz transponders for satellite communications

K-C Huang, A Jenkins, D Edwards, and D Dew-Hughes

Department of Engineering Science, University of Oxford, Parks Road, Oxford, UK

ABSTRACT: This paper examines the combination of high temperature superconducting (HTS) passive circuits with GaAs FETs to construct a low noise and small size RF repeater at 77K. Two novel HTS interdigital filters, a two-stage low-noise amplifier, a mixer, and a self-feedback local oscillator with a novel two-dimensional HTS resonator were designed, patterned and interconnected to form a low-noise microwave transponder. The advantages of small size and low power consumption are observed because of the utilization of thin film HTS materials on high dielectric constant substrates.

1. INTRODUCTION

Since the discovery of high-temperature superconductor (HTS) Tl-Ba-Ca-Cu-O 2212 material, there has been a growing interest in applying its low microwave loss and finite penetration depth characteristics in the manufacture of new and radical types of satellite communication components (Barner et al 1995). In space communications, one of the critical subsystems is a transponder. Normally, a number of transponders are employed in a satellite to provide an essential function in communication link between earth stations.

Transponders usually connect the transmitter and receiver antennas and may provide functions such as frequency translation and channel rearrangement. Figure 1 shows the standard configuration. It consists of an HTS input filter, a cryogenic low-noise amplifier (LNA) using GaAs field effect transistor (FET), a cryogenic mixer, and a cryogenic local oscillator with an HTS resonator. The frequency translation effect is implemented by feeding a microwave mixer with the uplink frequency and an on-board generated CW signal (at 2.73 GHz) produced by a local-oscillator unit. In this transponder, we proposed is designed to fit into a Leybold-Heraeus cyrocooler (Diameter = 10mm, height = 17mm).

2. HTS FILTERS

To meet the limited space inside the cryocooler, miniaturised narrow-band bandpass filters are necessary. These superconducting filters are microstrip circuits on a 20mm long, 20mm wide and 0.5 mm thick single crystal MgO (100) substrate. The patterned microstrip conductor on the top surface is a thin film (~0.6 μm) of Tl-Ba-Ca-Cu-O 2212 (Fig. 2).

This filter combines a dual-mode band-pass filter and a interdigital capacitor. A small triangular patch is attached to an inner corner of a square loop. A pair of degenerate modes are excited and coupled to each other by the patch so as to form a dual-mode filter. The size of this

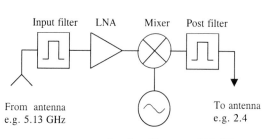

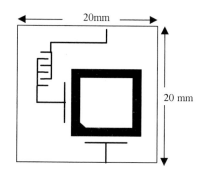

Fig. 1. The block diagram of the hybrid High Temperature Superconducting / Semi-conducting 5.13 GHz transponder.

Fig. 2. A circuit layout for the TBCCO input filter on MgO.

triangle patch determines the mode splitting and the degree of coupling. The inter-digital capacitor is a counterpart of a series-connected parallel L-G-C circuit and works effectively as a one-pole notch filter. Thus, the skirt of the band-pass signal will be improved and will become sharper. Figure 3 shows a typical response of this 5.13 GHz TBCCO filter, together with the result of the dual-mode filter without a inter-digital capacitor. This filter can reject the unwanted image frequency and protect the input amplifier. The rolloff at upper frequency for this filter is 217.8 dB/GHz while that for the dual-mode filter alone, without inter-digital capacitor, is 80.0 dB/GHz. The initial design of this filter was simulated by an electromagnetic analysis simulation package (HP EEsof).

The post filter is a combination of interdigital capacitors (Huang et al 1999a); its frequency response is shown in Fig. 4. This filter rejects the undesired harmonics, and also unwanted mixing products. It is possible to advance the performance of the filters further by optimisation design.

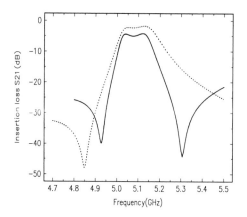

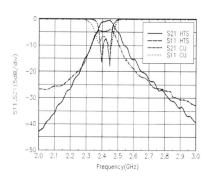

Fig. 3. Simulated frequency response of two 5.13 GHz input filters. (dotted line) the dual-mode filter without an inter-digital capacitor. (solid line) the dual-mode filter with an inter-digital capacitor.

Fig. 4. Measured frequency response of the 2.4GHz post-mixer filter.

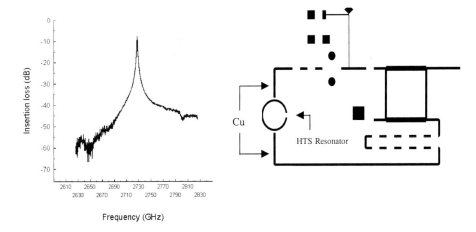

Fig. 5. Frequency response of the resonator.

Fig. 6. Layout of a HTS resonator-based local oscillator.

3. RESONATOR-BASED LOCAL OSCILLATOR

A gapped-ring resonator was chosen to stabilise the oscillator. This resonator was made from a single-sided TBCCO(2212) thin film, deposited onto a 10mm long × 10mm wide × 0.5mm thick LaAlO3 (110) substrate. The resonator layout was patterned on the film using standard photolithographic techniques and a wet etch of citric acid. The fundamental mode resonates while the total length of the ring circumference equals half-wavelength. This resonator was designed for 2.73 GHz operation and a nominal 50 Ω impedance. The frequency response is shown in Fig. 5. Its unloaded Q value is 3367 at 77 K, -10dBm applied power.

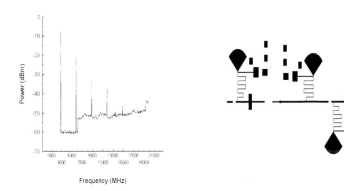

Fig. 7. Frequency response of the local oscillator.

Fig. 8. A circuit layout for the low-noise amplifier.

The layout of the oscillator is shown in Fig. 6. Amplification was provided by one GaAs monolithic microwave integrated circuits (MMIC) amplifier (HP MGA 86576) with sufficient gain at 2.73 GHz and 77 K for the operation of the oscillator. The amplified signal is divided by a branch-line coupler with equal power splits (Huang et al 1999b). The resonator is employed in the transmission mode in order to reduce the influence of bias drift and load pulling. A trade-off was made between circuit performance and the limited space inside the cryocooler. This microstrip circuit was patterned on a copper-coated Duroid substrate and connected to the HTS resonator by silver electrodag. The output of this oscillator is about 2.7GHz with output power levels of –7 dBm (see Fig. 7).

4.   AMPLIFIER AND MIXER

A two-stage low-noise amplifier was designed to amplify the filtered signal. The layout is shown in Fig. 8. Each stage uses a HP ATF-10135 GaAs FET device. Because the noise introduced by the first stage has the greatest impact on the sensitivity of the entire system, the first FET was matched for minimum noise figure while the second one was design for maximum gain (Gonzalez 1997). Bias is supplied both at the gate and at the drain. This bias circuit provides a path for the DC voltage supply and avoids the unwanted ac coupling. In addition, it is also designed as a bandpass filter for stopping unwanted amplified signal. Chip capacitors are placed at the input and the output of each stage to prevent DC voltages connecting to the input / output port. The gain was measured as 17 dB. In practice, As a fully optimised HTS amplifier and mixer combination was not available, the amplifier was fabricated on copper-clad Duroid while the mixer was a commercial semiconductor device.

5.   INTEGRATION

All the above devices were housed in aluminium boxes. SMA connectors were used for the input and output of the boxes. The effects of losses of the connectors was considered as well as that of the coaxial cables inside and outside the cryocooler. The preliminary results for the integrated devices were obtained at 77K by tuning of the bias voltage to optimise the centre frequency and gain such as shown in Fig. 9. The output power of -14 ± 3 dBm was obtained in a 300 MHz bandwidth around an centre frequency of 2.4 GHz while the input power is -19 dBm.

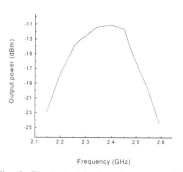

Fig. 9. Final electrical performance of the HTS transponder as a function of RF.

6.   CONCLUSION

We have incorporated TBCCO(2212) devices with semiconductor devices for a 5.13 GHz transponder. These components were tested separately and were integrated into a cryocooler and shown to operate at 77K. This shows that TBCCO material can be introduced into portions of a transponder subsystem and demonstrates substantial advantages to satellite communication in terms of volume, mass, power consumption, and performance.

REFERENCES

Barner J B et al, 1995 IEEE Transaction on Applied Superconductivity, **5,** pp 2075-2078
Gonzalez G 1997 Microwave transistor amplifiers analysis and design (New Jersey Prentice Hall)
Huang K-C et al 1999a IEEE Transaction on Applied. Superconductivity,. **9**, pp 3889-92
Huang K-C et al 1999b Superconductor Science and Technology,**12**, pp. 717-719.

Inst. Phys. Conf. Ser. No 167
Paper presented at Applied Superconductivity, Spain, 14-17 September 1999
© *2000 Marconi Electronic Systems Limited*

HTS filters and cooled electronics for communications - system performance and cooling needs.

R B Greed[1] and J Tilsley[2]

[1] Marconi Research Centre, Great Baddow, Chelmsford, Essex, CM2 8HN, England.
[2] Marconi Infra-red, Southampton, Hampshire, SO15 0EG, England.

ABSTRACT: Future Third-Generation mobile communication systems, to support the growth in multi-media services, will demand improved sensitivity and selectivity. Increased sensitivity provides extended coverage, higher capacity CDMA systems, longer 'hand-set' talk-time, while better selectivity reduces interference and improves bandwidth utilisation.
A base-station transceiver to meet the increased performance needs, in which key components are fabricated using thin-film High Temperature Superconductor (HTS) technology, is described.
For HTS techniques to be commercially viable a high performance cryo-package is essential. The cryo-packaging described is compact, maintenance free, and includes an integrated cooling engine designed to provide >5watt heat lift at 60K.

1. INTRODUCTION

Microwave hardware design based on high temperature superconductor (HTS) technology is a complex mix of electronic circuit and mechanical design, thermal management and vacuum and cryo-cooler technologies, which must be jointly and severally optimised. This can lead to design conflicts. The system and HTS component designs will only briefly be described, being are fully described in [1,2,3]. This paper concentrates on the design philosophy and implementation of a microwave sub-system comprising a vacuum encapsulation (dewar), and integrated closed cycle cryo-cooler. A thorough understanding of these design aspects and interrelationships is crucial for the future exploitation of HTS technology. The system considered here is for the application to the next generation, mobile communications base transceiver station (BTS). In addition to the r.f. requirements, the specification places stringent demands on the total input power, the vacuum life and cooling system. The design must require less than 200watts of input power, be compact and low mass, have a 5 year maintenance free life, be tower top mountable and employ a low vibration, non-water cooled, cooling engine.

2. TRANSCEIVER SYSTEM

The transceiver follows conventional microwave front end design practices, and is shown schematically in Fig. 1. In operation the HTS components must be cooled. However, the global noise figure of the receiver is significantly improved by also cooling the low noise amplifier. For the LNA used the measured noise figure was reduced from 0.8dB at 293K to <0.2dB at 77K.
Two planar filter topologies have been implemented. A pseudo-lumped element [1] design on a high dielectric constant substrate provides a compact solution. An alternative distributed design [2] on a lower dielectric material is larger. However, by meandering the resonators the size is reduced. Ideally the filters are designed to eliminate the need for mechanical tuning screws. Unfortunately the accuracy of current design software and the dimensional control of the substrate preclude this ideal.

Acknowledgement: This work was part-funded by the European Commission under the ACTS programme.

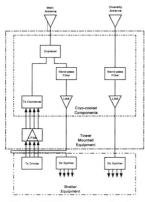

Fig. 1, BTS Transceiver showing tower mounted components

The high power duplexer [3] comprises a pair of identical HTS stop-band filters, tuned to the receive channel, and coupled using a pair of broadband hybrid couplers. Owing to the size constraints of the available lanthanum aluminate substrates, and deposition process, the four circuits are fabricated individually. The four substrates are mounted using conductive epoxy to a single expansion matched titanium carrier

Two designs of the high power transmitter combiner are currently being evaluated. Both are designed to meet the common modular approach of the cryo-package.

To allow for the different operators' requirements, the various functional elements are assembled into separate 'connectorised' modules. Thus system functionality can easily be interchanged to meet each operator's requirement.

3. FILTER TUNING

To facilitate 'bench top' tuning, a mimic of the encapsulation has been implemented on a laboratory Gifford-McMahon cryo-cooler. Though a convenient test vehicle, the GM cooler has a long 'cold finger' which introduces significant differential contraction between the r.f. module and the outer encapsulation wall. This dictates that the r.f. interconnect must be capable of compensating for 0.6mm differential movement.

The r.f. components are mounted in the final assembly (Fig. 2) which is then mounted to the mimic encapsulation, sealed using O-rings and evacuated by continuous pumping. A special 'screwdriver' tuning plate allows the three filters on one side of the module to be tuned, after which the module is inverted and the remaining filters are tuned.

4. ENCAPSULATION DESIGN

The design of the functional circuits, the cryo-cooler and the encapsulation are inseparably linked. Circuit dissipation and size of the r.f. module impinge on the cryo-cooler and the encapsulation designs. Likewise, the encapsulation, including the input and output microwave feeds, dictates the cryo-cooler capacity; while the available cryo-cooler capacity influences the r.f. module and encapsulation designs. The design is therefore an iterative process of compromise and refinement as shown in Fig. 3. In practice the process requires several iterations to achieve the best electrical and thermal design.

A general design maxim for microwave circuits for these applications is that the smaller the components are the better. This leads to a reduction in performance

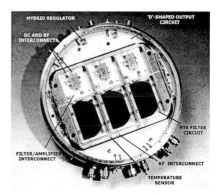

Fig. 2, Filter/LNA module

if conventional materials are used. HTS planar circuits enable a high performance to be restored.

The challenge is then to design the circuit housing and the encapsulation to maintain electrical and mechanical integrity over a large temperature range (+75°C to 60K), to have low mass and thermal capacity (essential in rapid cool down applications), and high thermal resistance to the ambient. For normal materials these requirements are in conflict. A further challenge to the designer is the lack of detailed design data for many engineering materials at such low temperatures.

In this design, titanium is used for the magnesium oxide substrate carrier. Referring to Fig. 4, the expansion of magnesium oxide and titanium are ideally matched over the whole operating range. Titanium, although not quite as well matched to lanthanum aluminate, has also been successfully used as a carrier for these substrates. In both cases, on cooling, the substrates are put into compression, which is the preferred stress mode. The residual differential expansion between the substrate and the carrier is taken up in the epoxy bonding film. However, the other thermal properties, such as the thermal conductivity and specific heat capacity of titanium are not optimum.

Fig. 3, RF - Dewar - Cryo-cooler design cycle

In practice little is known of the thermal characteristics of conductive epoxies at low temperatures, particularly in the longer term. Practical experience, over a number of years, has identified two epoxy systems that appear to survive repeated thermal cycling to 60K, one is a two part system in a paste form, the other is an unsupported single part film epoxy. In both cases the bond layer is thin enough not dominate the expansion characteristics of the assembly. Following established practice, the initial housing was designed in titanium. The cold head interface was also machined from titanium. In this application the low thermal conductivity of titanium severely limited the ability of the miniature cooler to achieve the operating temperature. Using titanium for the cold head in place of copper reduced the heat lift by a factor of three at 60K. The housing and cold head had to be redesigned in high conductivity copper. The high expansion mismatch between the copper housing and the substrates is overcome by the use of intermediate titanium carriers.

Initially, the large microstrip input feed circuits (Fig. 2) were designed to be bonded directly to the 'warm side' of the encapsulation. However, reference to Fig. 4 shows a significant expansion mismatch between alumina and stainless steel which is sufficient to cause the circuits to fracture on cooling to −50°C. To match the expansions the alumina substrates were bonded to gold plated kovar carriers. The film epoxy was used, as this proved to be more 'resilient' over larger areas at low temperature.

By contrast the design of the r.f. housing incorporates small alumina circuits adjacent to the HTS substrates. These circuits provide a robust interconnection point for the wire bond interconnects. A simple epoxied tape bond is made between the gold and the HTS circuits. Although there is a large differential expansion between alumina circuit and copper housing, due to the small

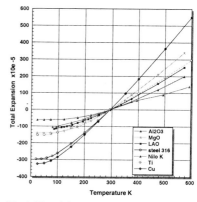

Fig. 4, Materials expansion

circuit size and the ability of the epoxy to absorb the induced stresses, the assemblies have proved reliable over many thermal cycles. As with many of the 'engineering' practices used in the design, the reliability of the assembly over an extended life cycle is yet to be proved.

A design aim of the encapsulation was to incorporate a fully welded construction to ensure that a high vacuum could be achieved and maintained over a five-year period. Furthermore, as the design incorporated many new and unproven features the design also included a rework capability. Each section of the encapsulation included flanges (Fig. 5), which after welding could be machined away and subsequently re-welded. This approach follows techniques well established in the infra-red industry. However, it has yet to be proven for the very much larger packages used in this application. External r.f interfaces are made via standard K-type connectors. These connectors include a separate solder-in, matched type glass-to-metal seal, which provides the vacuum interface. Initially it was

792

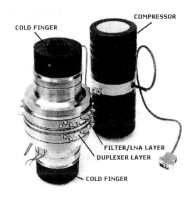

COMPRESSOR
COLD FINGER
FILTER/LNA LAYER
DUPLEXER LAYER
COLD FINGER

Fig. 5, Encapsulation and compressor

difficult to guarantee a good vacuum seal owing to the large number of seals that had to be soldered simultaneously. A significant number of failures were traced to leaks along the centre conductor of the proprietary seals. For future applications an improved method of soldering multiple glass-to-metal seals into the encapsulation is required. A vacuum tight, single packaged multiple connector would provide a better solution.

The r.f. and thermal interconnects across the vacuum space, are made using microstrip lines and wire bonds (Fig. 2). The alumina circuits also include d.c. bias lines and links to the temperature sensors. The LNA regulators are also mounted on these 'warm side' circuits. Thus this major source of dissipation does not present a direct heat load to the cold finger. The r.f interconnect to the components mounted on the cold finger is formed from small wires to minimise the thermal conductivity and hence heat leak to the cold finger (estimated to be 8mW for each link). The wires link both the signal line and ground plane and are looped to cater for the differential movement of the cold finger and the outer encapsulation wall. Each link is electrically matched using a LC π-network. The r.f loss per link was measured to be <0.3dB.

All components within the vacuum space are gold plated to provide a reliable grounding, and also to provide a low emissivity surface to reduce heat transfer via radiation.

5. INTEGRATED CRYO-COOLER.

For the emerging microwave applications a cryo-cooler with a low mechanical vibration, to eliminate induced electrical noise, is perceived as prime requirement. The compressor and cold-finger use dual opposed pistons to reduce vibration. As the compressor is the major source of vibration it is decoupled from the cold-fingers by a long (300mm maximum) small-bore pressure line. (see Fig. 5.)

To ensure a low thermal heat leak via the cold-fingers these are fabricated from very thin material and are as a result very fragile. In a static application such as a BTS the cooler can be preferentially mounted to minimise mechanical loading on the cold finger. This allows the r.f. modules to be rigidly mounted to the cold-fingers to maximise cooling efficiency. For future vehicular applications the r.f. modules may be mechanically decoupled from the cold-fingers and independently supported.

6. CONCLUSION

A cryogenically packaged HTS sub-system has been designed considering the electrical, thermal, mechanical and vacuum engineering requirements concurrently. Trade-offs have to be made to achieve an optimum design. The work has emphasised the importance of a collaborative design process between constituent disciplines and has established design guidelines. Changes to materials and refinements to assembly processes are expected to evolve. The processes and assembly techniques described have been successfully demonstrated but their long-term reliability remains to be proven.

7. REFERENCES

1. Reppel M and Chaloupka H J IEEE MMT-S Digest 1998
2. Hong J S, Lancaster M J, Jedamzik D and Greed R B IEEE MTT-S International Microwave Symposium, Baltimore June 1998.
3. Hong J S, Lancaster M J, Greed R B, Jedamzik D, Mage J C and. Chaloupka. H. J. IEEE MTT 1998

Author Index